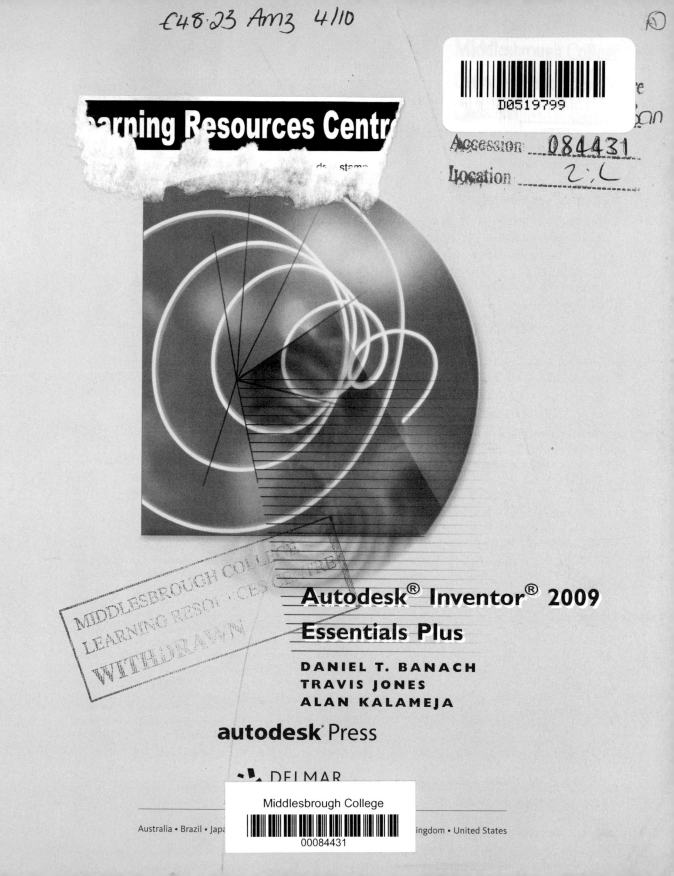

Autodesk® Inventor® 2009 Essentials Plus

DANIEL T. BANACH
TRAVIS JONES
ALAN KALAMEJA

autodesk Press

DELMAR

Australia • Brazil • Japa… …ingdom • United States

Autodesk Inventor 2009 Essentials Plus, 1e
Daniel T. Banach, Travis Jones, and Alan Kalameja

Vice President, Career and Professional Editorial: David Garza

Director of Learning Solutions: Sandy Clark

Acquisitions Editor: John Fedor

Managing Editor: Larry Main

Senior Product Manager: John Fisher

Senior Editorial Assistant: Dawn Daugherty

Vice President, Career and Professional Marketing: Jennifer McAvey

Marketing Director: Deborah S. Yarnell

Marketing Manager: Jimmy Stephens

Marketing Specialist: Mark Pierro

Production Director: Wendy Troeger

Production Manager: Stacy Masucci

Content Project Manager: Angela Iula

Art Director: David Arsenault

Technology Project Manager: Christopher Catalina

Production Technology Analyst: Thomas Stover

Library of Congress Control Number: 2008923422

ISBN-13: 978-1-4354-0255-3

ISBN-10: 1-4354-0255-3

Delmar
5 Maxwell Drive
Clifton Park, NY 12065-2919
USA

Cengage Learning products are represented in Canada by Nelson Education, Ltd.

For your lifelong learning solutions, visit **delmar.cengage.com**

Visit our corporate website at **cengage.com**.

NOTICE TO THE READER

Publisher does not warrant or guarantee any of the products described herein or perform any independent analysis in connection with any of the product information contained herein. Publisher does not assume, and expressly disclaims, any obligation to obtain and include information other than that provided to it by the manufacturer. The reader is expressly warned to consider and adopt all safety precautions that might be indicated by the activities herein and to avoid all potential hazards. By following the instructions contain herein, the reader willingly assumes all risks in connection with such instructions. The publisher makes no representation or warranties of any kind, including but not limited to, the warranties of fitness for particular purpose or merchantability, nor are any such representations implied with respect to the material set forth herein, and the publisher takes no responsibility with respect to such material. The publisher shall not be liable for any special, consequential, or exemplary damages resulting, in whole or part, from the readers' use of, or reliance upon, this material.

Printed in the United States of America
1 2 3 4 5 XXX 11 10 09 08

CONTENTS

INTRODUCTION

INTRODUCTION

Welcome to the *Autodesk Inventor 2009 Essentials Plus* manual. This manual provides a thorough coverage of the features and functionalities offered in Autodesk Inventor.

Each chapter in this manual is organized with the following elements:

Objectives Describes the content and learning objectives.

Topic Coverage Presents a concise, thorough review of the topic.

Exercises Presents the workflow for a specific tool or process through illustrated, step-by-step instructions.

Applying Your Skills Checks your skills and understanding of the material covered in the chapter using challenge exercises. These exercises describe a design challenge but may not provide step-by-step instructions.

Checking Your Skills Tests your understanding of the material using True/False and multiple-choice questions.

NOTE TO THE LEARNER

Autodesk Inventor is designed for easy learning. Autodesk Inventor's help system provides you with ongoing support as well as access to online documentation.

As described above, each chapter in this manual has the same instructional design, making it easy to follow and understand. Each exercise is task-oriented and based on real-world mechanical engineering examples.

WHO SHOULD USE THIS MANUAL?

The manual is designed to be used in instructor-led courses, although you may also find it helpful as a self-paced learning tool.

RECOMMENDED COURSE DURATION

Four days (32 hours) to seven days (56 hours) are recommended, although you may use the manual for specific Autodesk Inventor topics that may last only a few hours.

USER PREREQUISITES

It is recommended that you have a working knowledge of Microsoft® Windows XP Professional® or Windows Vista™ as well as a working knowledge of mechanical design principles.

MANUAL OBJECTIVES

The primary objective of this manual is to provide instruction on how to create part and assembly models, document those designs with drawing views, and automate the design process.

Upon completion of all chapters in this manual, you will be proficient in the following tasks:

- Basic and advanced part modeling techniques
- Drawing view creation techniques
- Assembly modeling techniques
- Sheet metal design

While working through these materials, we encourage you to make use of the Autodesk Inventor help system, where you may find solutions to additional design problems that are not addressed specifically in this manual.

MANUAL DESCRIPTION

This manual provides the foundation for a hands-on course that covers basic and advanced Autodesk Inventor features used to create, edit, document, and print parts and assemblies. You learn about the part and assembly modeling tools through online and print documentation and through the real-world exercises in this manual.

PROJECT EXERCISES

The self-paced, step-by-step project exercises found in Appendix A provide opportunities for you to work through real-world modeling, assembly, and documentation tasks. The geometry used in the project exercises will flow from chapter to chapter, utilizing the functionality that you learned in that chapter.

ESSENTIALS EXERCISE FILES

The files for the exercises can be installed from the CD-ROM attached to the back cover of your book.

INSTALLING THE EXERCISE FILES

To install the exercise data files:

1. Browse to the *Autodesk Inventor 2009 Essentials Plus Exercises* CD-ROM, and copy the class files to C:\.
2. The files will be placed in the folder C:\INV 2009 Ess Plus.
3. To be able to save the exercise files, remove the Read Only attribute for all exercise files.

PROJECTS

Most engineers work on several projects at a time, with each project consisting of a number of files. To accommodate this, Autodesk Inventor uses projects to help organize related files and maintain links between files.

Each project has a *project file* that stores the paths to all files related to the project. When you attempt to open a file, Autodesk Inventor uses the paths in the current project file to locate other necessary files.

For convenience, a project file is provided with the *exercises*.

USING THE PROJECT FILE

Before starting the exercise, you must complete the following steps:

1. Start Autodesk Inventor.
2. In the Getting Started dialog box, click Projects in the What to Do section.
3. In the Projects window, right-click the background, and select Browse. Navigate to the folder where you installed the Essentials Exercises, and double-click the *C:\INV 2009 Ess Plus\INV 2009 Ess Plus.ipj* file.
4. Double-click the *INV 2009 Ess Plus* project in the Projects window to make it the active project.
5. You can now start doing the exercises.

Note: Projects are reviewed in more detail in the Getting Started section in Chapter 1.

TERMS AND PHRASES

To help you to understand Autodesk Inventor better, the following section explains a few of the terms and phrases that are used in this book.

Parametric Modeling *Parametric modeling* is the ability to drive the size of the geometry by dimensions, and it is sometimes referred to as "dimension-driven design." For example, if you want to increase the length of a plate from 5 to 6, change the 5= dimension to 6=, and the geometry will be updated. Think of it as the geometry being "along for a ride" and driven by the dimensions. Associative dimensions work in the opposite fashion. As lines, arcs, and circles are drawn, they are created to the exact length or size; when they are dimensioned, the dimension reflects the exact value of the geometry. If you need to change the size of the geometry, you stretch the geometry, and the dimension is automatically updated. Think of it as the dimension being "along for a ride" and driven by the geometry.

Feature-Based Parametric Modeling *Feature-based* means that as you create your model, each hole, fillet, chamfer, extrusion, and so on is an independent feature whose dimensional values can be changed without having to re-create the feature.

Adaptivity *Adaptivity* allows parts to have a physical relationship between each other. An example might be when one plate is created with a dimensioned cutout, and another plate is created with the same cutout, but no dimensions are created. You can constrain the second cutout to match the size of the dimensioned cutout. When the dimensions change on the first cutout, the second cutout will be updated or adapt to show the change.

Bidirectional Associativity *Bidirectional associativity* means that the model and the drawing views are linked. If the model changes, the drawing views will be updated automatically. Additionally, if the dimensions in a drawing view change, the model is updated, and the drawing views are updated based on the updated part.

Bottom-Up Assembly *Bottom-up assembly* refers to an assembly whose parts are created in individual part files outside of the assembly and then referenced into the assembly.

Top-Down Assembly *Top-down assembly* refers to an assembly modeling technique where parts are created while working in the assembly file.

Skeletal Modeling *Skeletal modeling* involves the creation of numerous parts and features based on information from a single part file.

ACKNOWLEDGMENTS

TECHNICAL EDIT

The authors would like to thank Kevin Robinson for his in-depth and thoughtful technical edit. His knowledge and attention to detail have added a great deal to this book.

The authors would also like to thank Lori Stephens, Verbatim Editorial, for her detailed copyediting.

Getting Started

INTRODUCTION

This chapter provides a look at the user interface, projects, application options, starting tools (commands), and instructions on how to view parts in Autodesk Inventor.

OBJECTIVES

In this chapter, you will gain an understanding of the following:

- How to open files
- How to start new files
- Reasons for which a project file is used
- How to create a project file for a single user
- Autodesk Vault
- Different file types used in Autodesk Inventor
- Application options
- The Help system
- The user interface
- How to issue commands
- Different viewing tools

GETTING STARTED WITH AUTODESK INVENTOR

The default Autodesk Inventor Open dialog box looks similar to the following image and allows you to open Autodesk Inventor files. In the dialog box, you can select or create a new project file that helps Autodesk Inventor to look for files, or you can start a new Autodesk Inventor file. The default screen can be changed to the New dialog box or start a specified file, which can be set via the Application Options in the General tab under Start-up action.

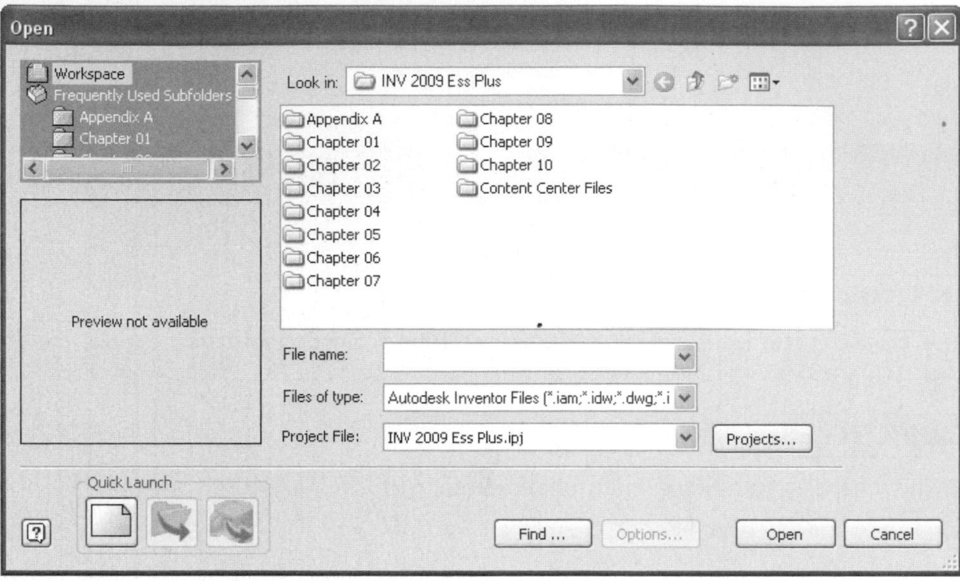

Figure 1.1

OPEN

In the Open dialog box, the directory that opens by default is set in the current project file. You can open files from other directories that are not defined in the current project file, but this is not recommended. Part, drawing, and assembly relationships may not be resolved when you reopen an assembly containing components outside the locations defined in the current project file. To open an Autodesk Inventor file, you can also click Open from the File menu, click the Open icon from the standard menu, as shown in the following image, or press CTRL + O.

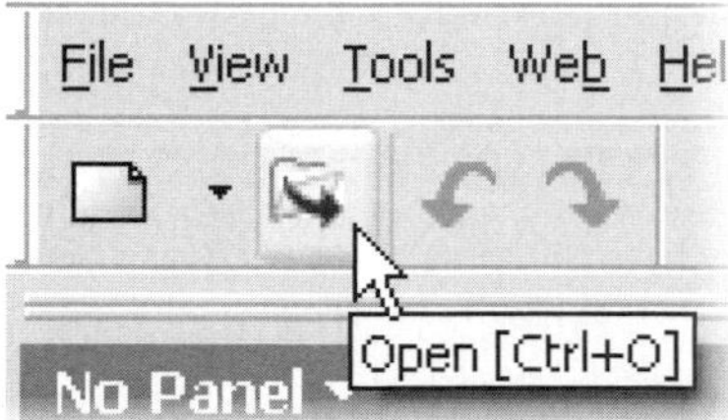

Figure 1.2

NEW FILES: QUICK LAUNCH

Click the New icon in the lower left corner of the dialog box to start a new file while in the Open dialog box. The New dialog box will appear as shown in the following image. Begin by selecting the type of file to create or one of the drafting standards named on the tabs, and then select a template for a new part, assembly, presentation file, sheet metal part, or drawing. If Autodesk Inventor Professional is installed, a Professional tab will exist.

Create a subdirectory in the *Autodesk\Inventor(version number)\templates* directory and add a file to it. A new template tab with the same name as the subdirectory is created automatically.

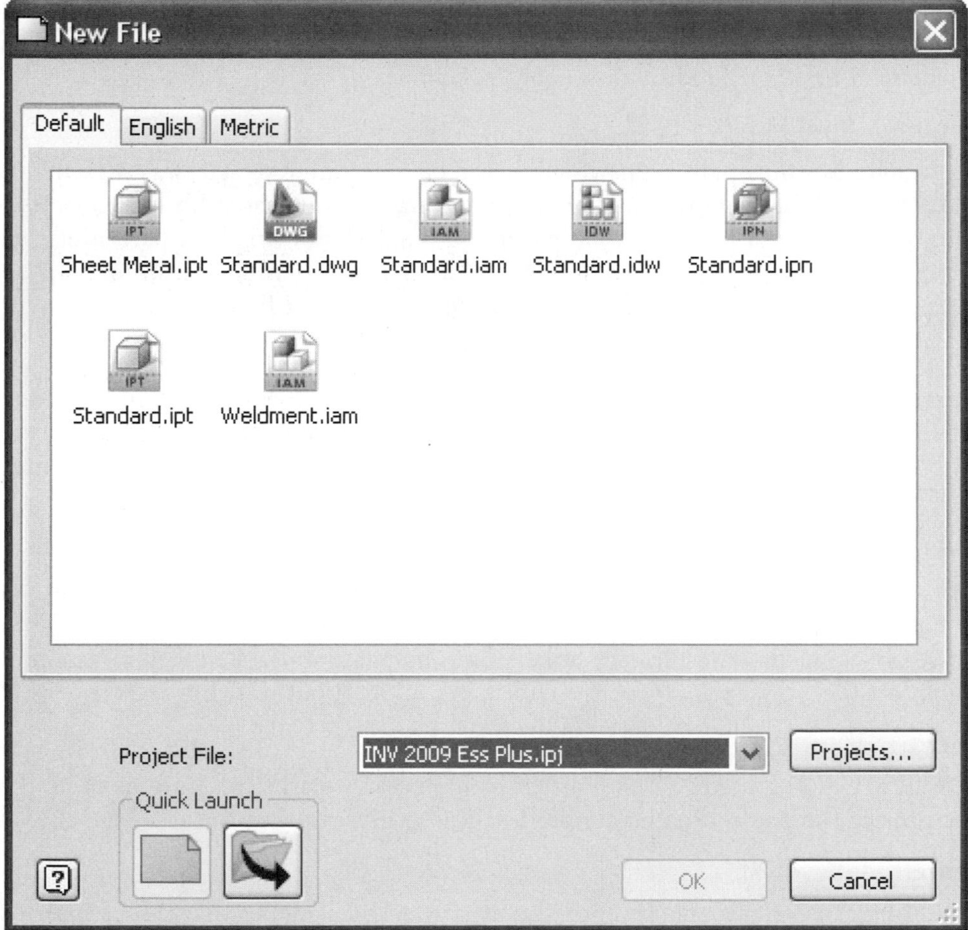

Figure 1.3

FILE INFORMATION

While creating parts, assemblies, presentation files, and drawing views, data is stored in separate files with different file extensions. This section describes the different file types and the options for creating them.

FILE TYPES

The following section describes the main file types that you can create in Autodesk Inventor, their file extensions, and descriptions of their uses.

Part (.ipt)

Part files contain only one part, which can be either 2D or 3D.

Assembly (.iam)

Assembly files can consist of a single part, multiple parts, or subassemblies. The parts themselves are saved to their own part file and are referenced (linked) in the assembly file. See chapter 6 for more information about assemblies.

Presentation (.ipn)

Presentation files show parts of an assembly exploded in different states. A presentation file is associated with an assembly, and any changes made to the assembly will be updated in the presentation file. A presentation file can be animated, showing how parts are assembled or disassembled. The presentation file extension is *ipn*, but you save animations as AVI files. See chapter 6 for more information about presentation files.

Sheet Metal (.ipt)

Sheet metal files are part files that have the sheet metal environment loaded. In the sheet metal environment, you can create sheet metal parts and flat patterns. You can create a sheet metal part while in a regular part. This requires that you load the sheet metal environment manually. See chapter 10 for more information about creating sheet metal parts.

Drawing (.idw)

Drawing files can contain 2D projected drawing views of parts, assemblies, and/or presentation files. You can add dimensions and annotations to drawing views. The parts and assemblies in drawing files are linked, like the parts and assemblies in assembly and presentation files. See chapter 5 for more information about drawing views.

Project (.ipj)

Project files are structured XML files that contain search paths to locations of all the files in the project. The search paths are used to find the files in a project.

iFeature (.ide)

iFeature files can contain one or more 3D features or 2D sketches that can be inserted into a part file. You can place size limits and ranges on iFeatures to enhance their functionality. See chapter 9 for more information about creating iFeatures.

Design Views (.idv)

Private Design Views are configuration files in which the following information about an assembly file is saved: component visibility, component selection status, color settings, zoom magnification, and viewing angle. See chapter 9 for more information about creating design views.

OPENING MULTIPLE DOCUMENTS

You can open multiple Autodesk Inventor files at the same time by holding down the CTRL key and selecting the files to open as shown in the following image. Each file will be opened in its own window in a single Autodesk Inventor session. To switch between the open documents, click the file on the Windows menu. The files can also be arranged to fit the screen or to appear cascaded. If the files are arranged or cascaded, click a file to activate it. Only one file can be active at a time.

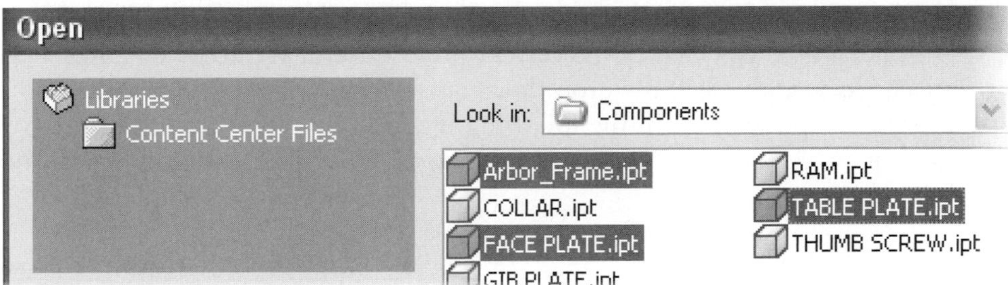

Figure 1.4

SAVE OPTIONS

There are three options on the File menu for saving your files: Save, Save Copy As, and Save All, as shown in the following image.

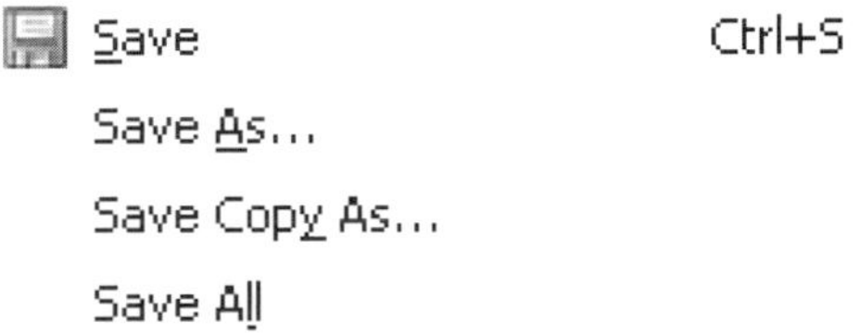

Figure 1.5

Save

The Save command saves the current document with the same name and to the location where you created it. If this is the first time that a new file is saved, you are prompted for a file name and file location.

Tip: To run the Save command, click the Save icon on the standard toolbar, use the shortcut keys CTRL-S, or click Save on the File menu.

Save As

Use the Save As command to save the active document with a new name and location, if required. A new file is created and is made active.

Save Copy As

Use the Save Copy As command to save the active document with a new name and location, if required. A new file is created but is not made active.

Save All

Use the Save All command to save all open documents and all of their dependents. The files are saved with the same name to the location where you created them. The first time that a new file is saved, you will be prompted for a file name and file location.

PROJECTS

In the Open dialog box, you can set the current project file or create a new project file. To make a project file current, click the drop-down arrow next to Project File entry and select an existing project file as shown in the following image. Click the Project button to create a new project or modify the existing project file options. Project concepts are introduced in the next section.

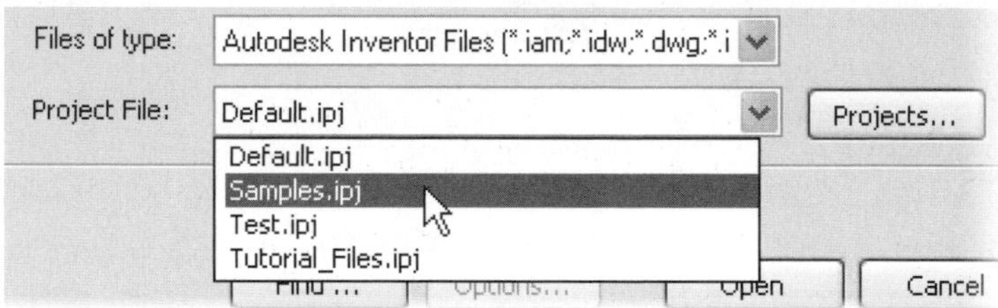

Figure 1.6

PROJECTS IN AUTODESK INVENTOR

Almost every design that you create in Autodesk Inventor involves more than a single file. Each part, assembly, presentation, and drawing created is stored in a separate file. Each of these files has its own unique file extension. There are many times when a design will reference other files. An assembly file, for example, will reference a number of individual part files and/or additional subassemblies. When you open the parent or top-level assembly, it must contain information that allows Autodesk Inventor to locate each of the referenced files. Autodesk Inventor uses a project file to organize and manage these file-location relationships. There is no limit to the number of projects you can create, but only one project can be active at any given time

You can structure the file locations for a design project in many ways. A single-person design shop has different needs than a large manufacturing company or a design team with multiple designers working on the same project. In addition to project files, Autodesk Inventor includes a program called Autodesk Vault on the DVD that controls basic check-out and check-in file-reservation mechanisms; these control file access for multiuser design teams. Autodesk Inventor always has a project named Default. Specifically, if all the files defining a design are located in a single folder, or in a folder tree where each referenced part is located with its parent or in a subfolder underneath the parent, the Default project may be all that is required.

Note: It is recommended that files in different folders never have the same name to avoid the possibility of Autodesk Inventor resolving a reference to a file of the same name but in a different folder.

PROJECT SETUP

To reduce the possibility of file resolution problems later in the design process, always plan your project folder structure before you start a design. A typical project might consist of parts and assemblies unique to the project, standard components that are unique to your company, and off-the-shelf components such as fasteners, fittings, or electrical components.

PROJECT FILE SEARCH OPTIONS

Before you create a project, you need to understand how Autodesk Inventor stores cross-file reference information and how it resolves that information to find the referenced file. Autodesk Inventor stores the file name, a subfolder path to the file, and a library name (optionally) as the three fundamental pieces of information about the referenced file.

When you use the Default project file, the subfolder path is located relative to the folder containing the referencing file. It may be empty, or may go deeper in the subfolder hierarchy, but it can never be located at a level above the parent folder.

If you create your own project file and specify project search locations, the subfolder path is initially applied to your specified project locations. If Autodesk Inventor does not find the file there, it will continue to look relative to the folder containing the source file.

A project file can have zero or more file search path locations defined. These paths are searched in a specific order when a document, such as an assembly, needs to find referenced components, such as the part files in the assembly. Autodesk Inventor searches the locations in the following order: library paths, workspace, workgroups, and the folder containing the parent file.

CREATING PROJECTS

To create a new project or edit an existing project, use the Autodesk Inventor Project File Editor. The Project File Editor displays a list of shortcuts to previously active projects. A project file has an *.ipj* file extension and typically is stored in the home folder for the design-specific documents, while a shortcut to the project file is stored in the Projects Folder. The Projects Folder is specified on the Files tab of the Options dialog box, as shown in the following image. All projects with a shortcut in the Projects Folder are listed in the top pane of the Project File Editor.

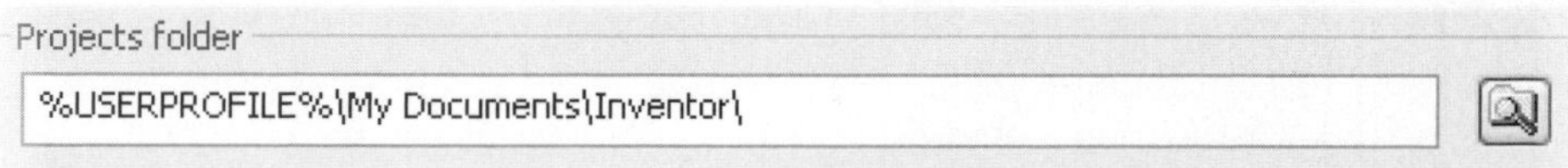

Figure 1.7

You create or edit a project file by clicking the Project button in the Open dialog box as shown in the following image on the left or by clicking Projects from the File menu as shown in the following image on the right.

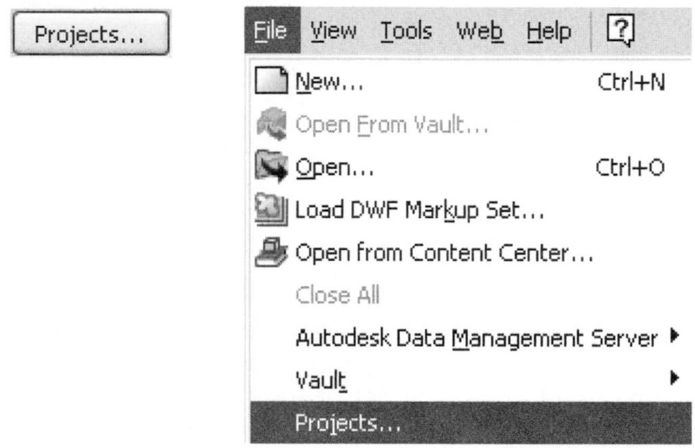

Figure 1.8

The Projects dialog box will appear as shown in the following image. The Projects dialog box is divided into two panes. The top pane lists shortcuts to the project files that have been active previously. Double-click on a project's name to make it the active project. All Inventor files must be closed before making a project current. Only one project file can be active in Autodesk Inventor at a time. The bottom portion reflects information about the project selected in the top pane. If a project file already exists, click on the Browse button on the bottom of the dialog box, then navigate to and select the project file. The bottom pane of the dialog box lists information about the highlighted project. To see more information about the project, click the More button (>>) on the bottom-right side of the dialog box.

Note: When defining a path to a folder on a network, it is better to define a Universal Naming Convention (UNC) path starting with the server name (*Server*\...) and not to use shared (mapped) network drives.

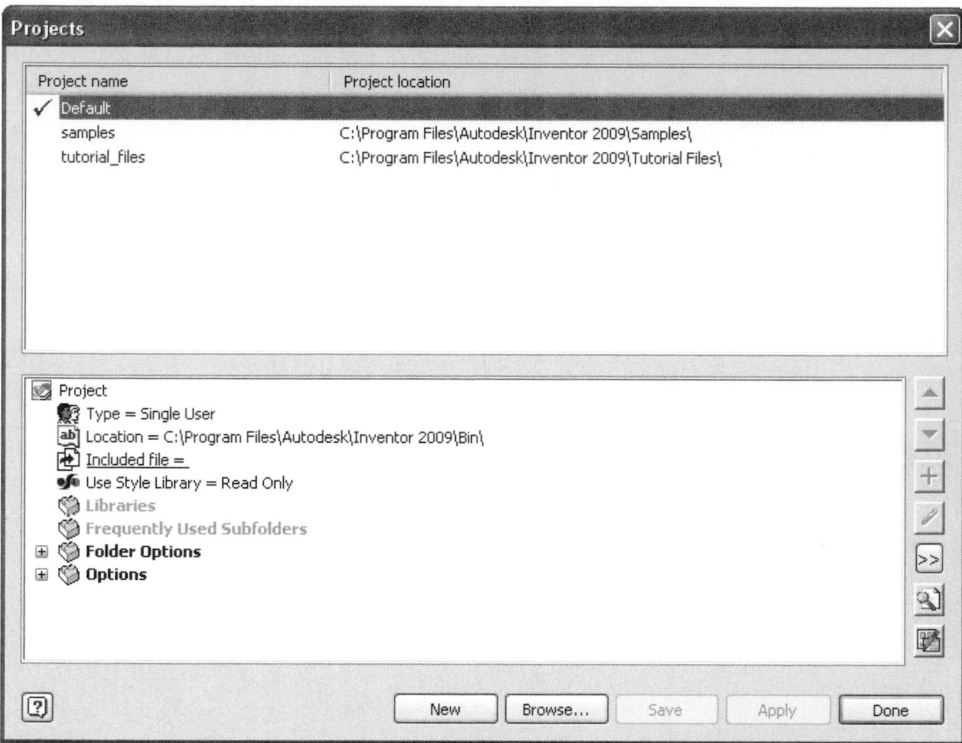

Figure 1.9

To create a new project, follow these steps:

1. Click the New icon in the Projects dialog box.

Tip: You can also create a new project when the Autodesk Inventor application is not running. Click on the Microsoft Windows Start menu and then click Programs > Autodesk > Inventor (version number) > Tools > Project Editor.

2. In the Projects dialog box, click the New button at the bottom to initiate the Inventor project wizard.

3. In the Inventor project wizard, follow the prompts to the following questions.

WHAT TYPE OF PROJECT ARE YOU CREATING?

If Autodesk Vault is installed, you will be prompted to create a New Vault project or a New Single-User project. If Autodesk Vault is not installed, only a New Single-User project type will appear in the list.

New Vault Project

This project type is used with Autodesk Vault and is not available until you install Autodesk Vault. It creates a project with one workspace and any needed library location(s), and it sets the multiuser mode to Vault. More information about Autodesk Vault appears later in this section.

New Single-User Project

This is the default project type, which is used when only one user will reference Autodesk Inventor files. It creates one workspace where Autodesk Inventor files are stored and any needed library location(s), and it sets the Project Type to Single User. No workgroup is defined but can be defined later.

The next section covers the steps for creating a new single-user project. For more information on projects, consult the online Help system.

Creating a New Single-User Project

Click on New Single User Project as shown in the following image. If Autodesk Vault is not installed, New Single User Project will be selected for you.

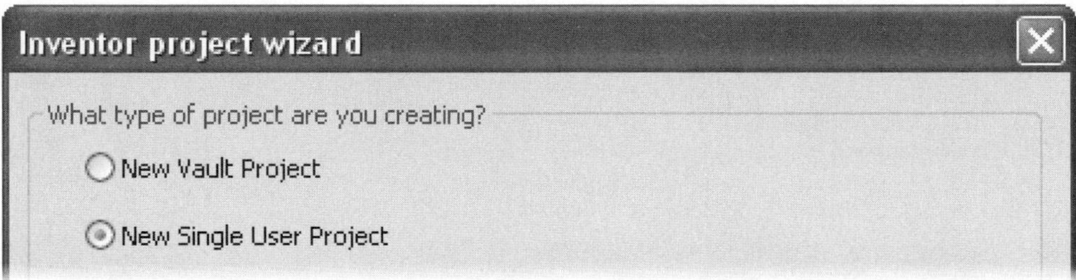

Figure 1.10

Click the Next button, and specify the project file name and location on the second page of the Inventor project wizard as shown in the following image.

Name

Enter a descriptive project file name in the Name field. The project file will use this name with an *.ipj* file extension.

Inventor project wizard
Project File
Name
Essentials Book
Project (Workspace) Folder
C:\Essentials Book\
Project File to be created
C:\Essentials Book\Essentials Book.ipj

Figure 1.11

Project (Workspace) Folder

This specifies the path to the home or top-level folder for the project. You can accept the suggested path, enter a path, or click the Browse button (...) to manually locate the path. The default home folder is a subfolder under My Documents, or wherever you last browsed, that is named to match the project file name.

Project File to be created

The full path name of the project file is displayed below the Location field.

Note: You can specify the home or top-level folder as the location of your workspace, but it is preferable for the project file (*.ipj*) to be the only Autodesk Inventor file stored in the home folder. This makes it easier to create other project locations, such as library and workspace folders, as subfolders without creating nesting situations that can lead to confusion.

Click the Next button at the bottom of the Inventor project wizard, and specify the project library search paths.

You can add library search paths from existing project files to this new project file. The library search paths from every project with a shortcut in your Projects Folder are listed on the left in the Inventor project wizard dialog box, as shown in the following image. You can add and remove libraries from the New Project area by clicking on their names in either the All Projects area or the New Project area and then clicking the arrows in the middle of the dialog box. The libraries listed by default in the All Projects list will match those in the project file that you selected prior to starting the New Project process.

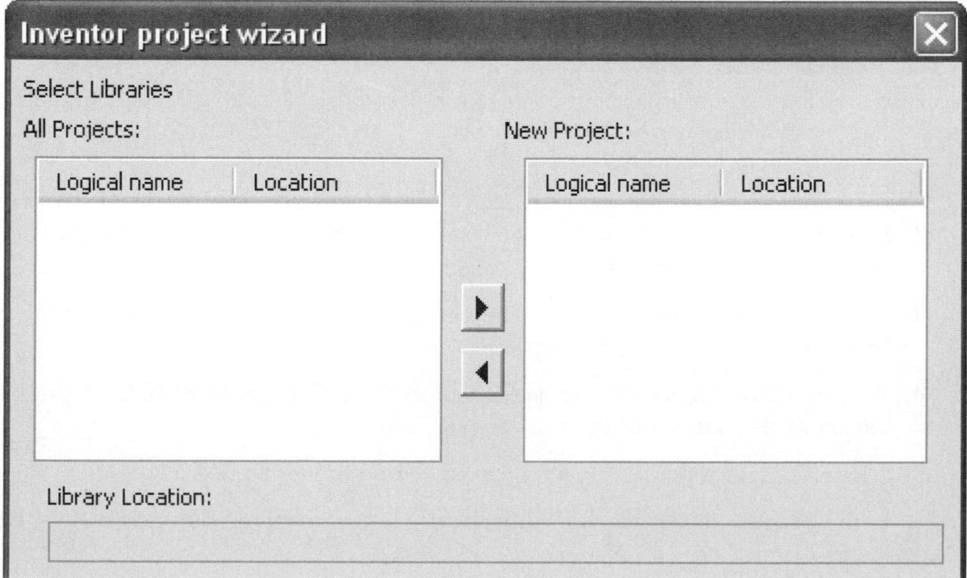

Figure 1.12

Click the Finish button to create the project. If a new directory will be created, click OK in the Inventor Project Editor dialog box. The new project will appear in the Open dialog box. Double-click on a project's name in the Open dialog box to make it the active project. A checkmark will appear to the left of the active project, as shown in the following image.

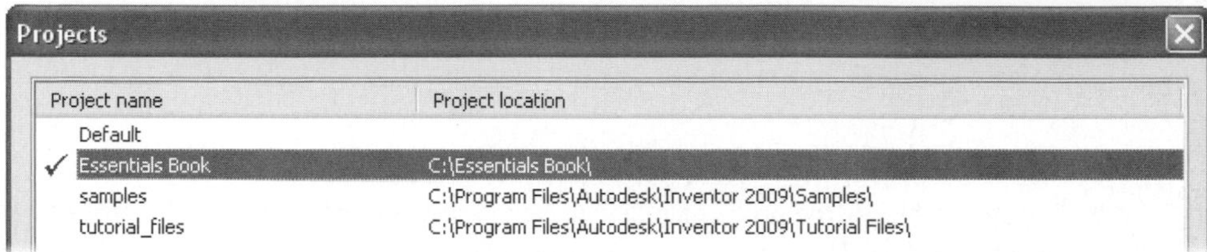

Figure 1.13

AUTODESK VAULT

Autodesk Vault is available on the Autodesk Inventor DVD. The Autodesk Vault enhances the data-management process by managing more than just Autodesk Inventor files and by tracking file versions as well as team member access. Controlling access to data, tracking modifications, and communicating the design history are important aspects of managing collaborative data. When working with a vault project, your data files are stored in a central repository that records the entire development history of the design. The vault manages Inventor and non-Inventor files alike. In order to modify a file, it first must be checked out of the vault. When the file is checked back into the vault, the modifications are stored as the most recent version for the project, and the previous version is sequentially indexed as part of the living history of the design.

EXERCISE 1-1: PROJECTS

In this exercise, you create a project file for a single-user project, open existing files, delete the project file, and make an existing project file current.

1. Prior to creating a new project file, close all Inventor files.
2. From the main menu, select File > Projects. You can also access the Project File Editor from the Open dialog box.
3. Double-click the Default project. The default project contains no search paths. You base the new project on the active project.
4. You now create a new project file based on the default project file. Click the New button at the bottom of the Project File Editor.
5. If it is not selected, click New Single User Project, and click Next.
6. In the next dialog box of the Inventor project wizard, enter **Essentials Plus Book** in the Name field.

7. Click the Browse button (...) to the right of the Project (Workspace) Folder field.
8. Browse to and open the *C:\INV 2009 Ess Plus* folder as shown in the following image. The project file (.ipj) will be placed in this folder. The selected folder is the top-level folder for the project. It is good practice to place project component files such as parts, assemblies, and drawings in folders below the top-level folder, but not in the top-level folder.

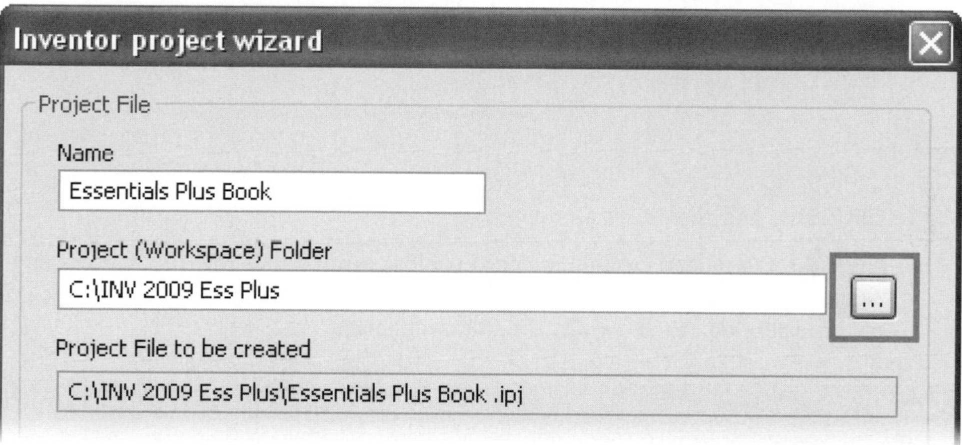

Figure 1.14

9. Click Next to add a library search path. The New Project list on the right side should be blank.
10. Click Finish. The new project file is highlighted in the upper pane of the Project File Editor. The search paths for the new project are listed in the lower pane.
11. Activate the new project by double-clicking on **Essentials Plus Book** in the upper pane of the Project File Editor.
12. Click the More button (>>) on the bottom-right side of the dialog box.
13. Expand Workspace and notice that the Workspace search path is listed as a period "." as shown in the following image. The "." denotes that the workspace location is relative to the location where the project file is saved.

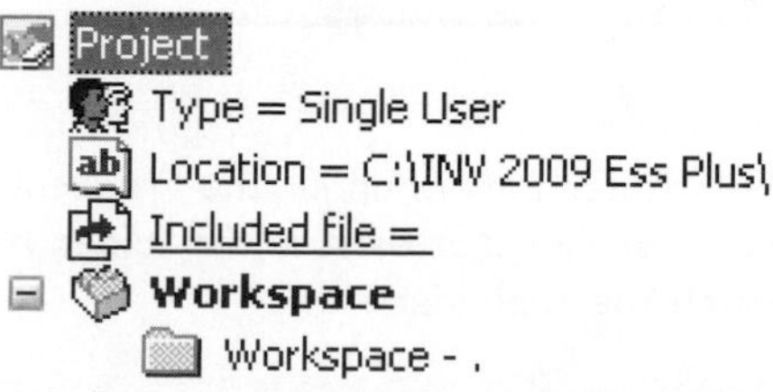

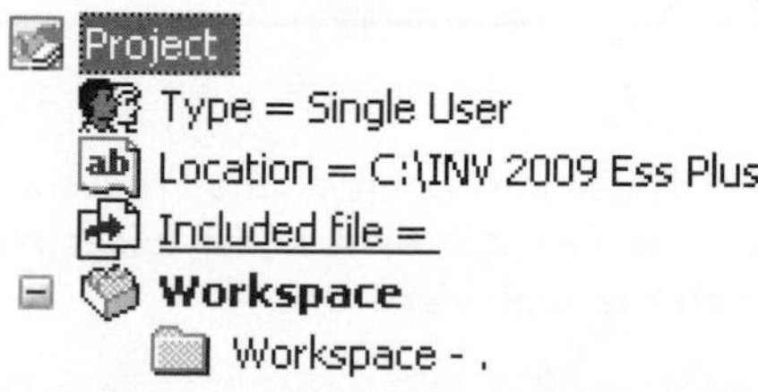

Figure 1.15

14. To make it easier to use subfolders below the workspace, you can add them to the Frequently Used Subfolders. Right-click on the Frequently Used Subfolders option, and click Add Paths from Directory as shown in the following image. Then navigate to and select the folder C:*INV 2009 Ess Plus* folder. Click OK to add the subfolders.

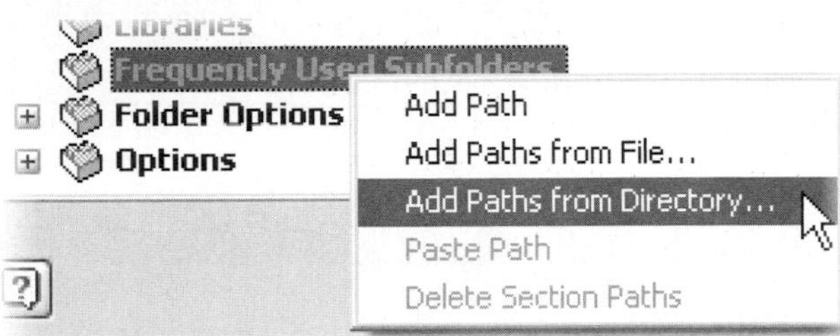

Figure 1.16

15. Click Save, and then click Done.

16. Click the Open icon on the Standard toolbar. Notice that the 10 chapters and Appendix A appear in the Frequently Used Subfolders area in the upper-left corner of the dialog box as shown in the following image.

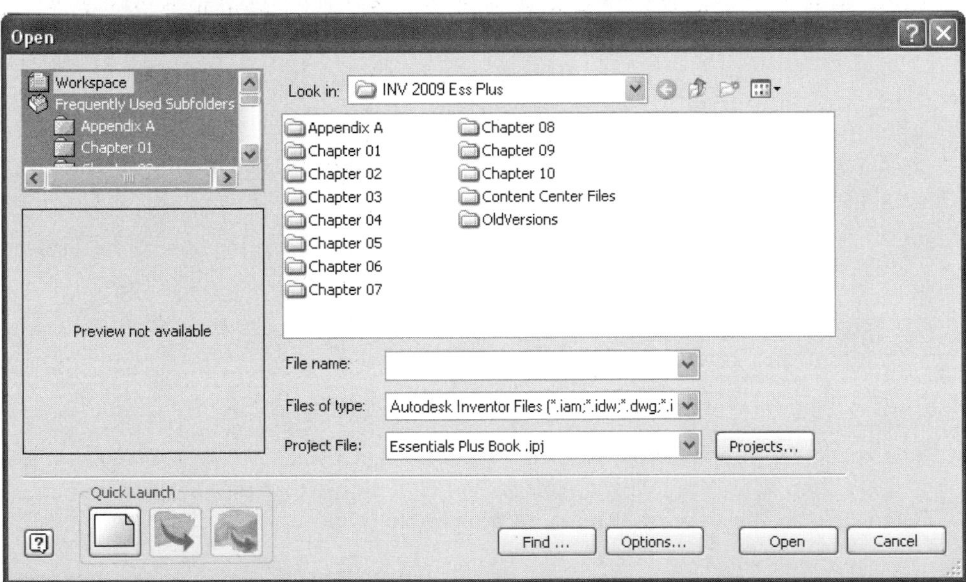

Figure 1.17

17. Click on any of the chapter folders under the Frequently Used Subfolders, and then open any of the Inventor files that exist in the folder. Notice that the Look in: location changes to match the subfolder name.

18. You must change the active project file to successfully complete the remaining exercises. Close all files currently open in Autodesk Inventor.

19. Select File > Projects from the main menu.

20. To add an existing project file to the current list, click the Browse button (...) at the bottom of the dialog box. Navigate to and select *C:\INV 2009 Ess Plus\INV 2009 Ess Plus.ipj*. This project should now be the current project. This project will be used for the remaining exercises. Expand the Frequently Used Subfolders section to verify that each chapter has its own subfolder.

21. Right-click on the project **Essentials Plus Book** in the upper pane and select delete as shown in the following image.

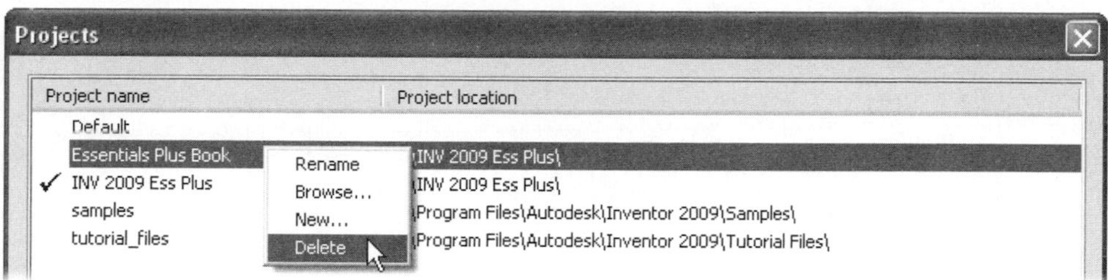

Figure 1.18

22. End of exercise.

APPLICATION OPTIONS

Autodesk Inventor can be customized to your preferences. On the Tools menu, click Application Options to open the Options dialog box as shown in the following. You set options on each of the tabs to control specific actions in the Autodesk Inventor software. Each section is covered in more detail in the pertinent sections throughout this book. For more information about application options, see the online Help system.

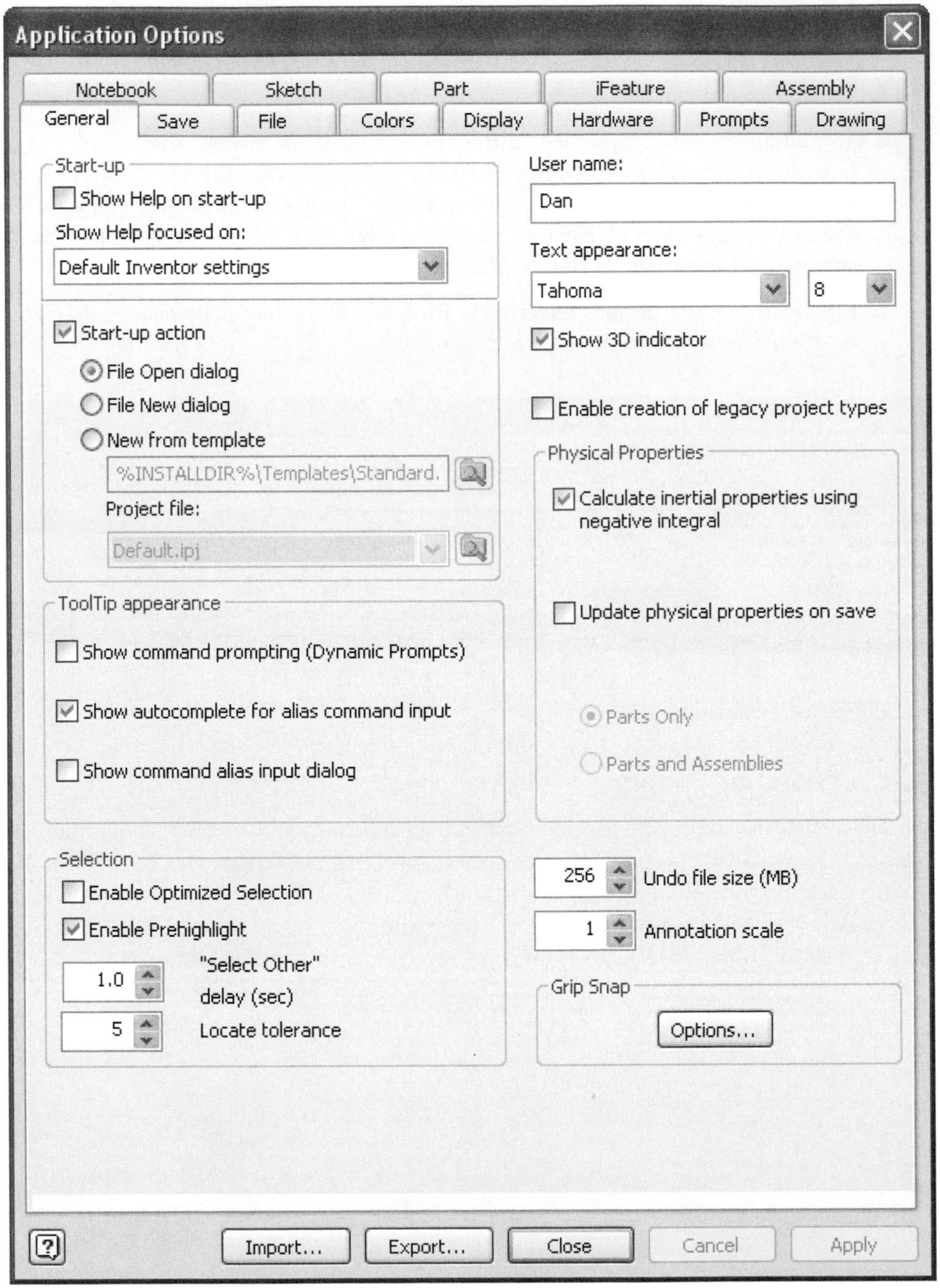

Application Options

Notebook | Sketch | Part | iFeature | Assembly

General | Save | File | Colors | Display | Hardware | Prompts | Drawing

Start-up

Show Help on start-up

Show Help focused on:

Default Inventor settings

Start-up action

File Open dialog

File New dialog

New from template

%INSTALLDIR%\Templates\Standard.

Project file:

Default.ipj

ToolTip appearance

Show command prompting (Dynamic Prompts)

Show autocomplete for alias command input

Show command alias input dialog

Selection

Enable Optimized Selection

Enable Prehighlight

1.0 "Select Other" delay (sec)

5 Locate tolerance

User name:

Dan

Text appearance:

Tahoma 8

Show 3D indicator

Enable creation of legacy project types

Physical Properties

Calculate inertial properties using negative integral

Update physical properties on save

Parts Only

Parts and Assemblies

256 Undo file size (MB)

1 Annotation scale

Grip Snap

Options...

Import... Export... Close Cancel Apply

Figure 1.19

General

Set general options for how Autodesk Inventor operates.

Save

Set how files are saved.

File

Set where files are located.

Colors

Change the color scheme and color of the background on your screen. Determine if reflections and textures will be displayed.

Display

Adjust how parts look. Your video card and your requirements affect the appearance of parts on your screen. Experiment with different settings to achieve maximum video performance.

Hardware

Adjust the interaction between your video card and the Autodesk Inventor software. The software is dependent upon your video card. Take time to make sure that you are running a supported video card and the recommended video drivers. If you experience video-related issues, experiment with the options on the Hardware tab. For more information about video drivers, click Graphics Drivers on the Help menu.

Prompts

Modify the response given to messages that are displayed.

Drawing

Specify the way that drawings are created and displayed.

Notebook

Specify how the Engineer's Notebook is displayed.

Sketch

Modify how sketch data are created and displayed.

Part

Change how parts are created.

iFeature

Adjust where iFeatures data is stored.

Assembly

Specify how assemblies are controlled and behave.

HELP SYSTEM

The online Help system in Autodesk Inventor goes beyond basic command definition by offering assistance while you design. The following help mechanisms make up the Help System:

- Help Topics
- Visual Syllabus
- New Features Workshop
- Tutorials
- Learning Paths
- Design Doctor
- Sketch Doctor
- Autodesk Online (Skill Builders)

To access the online Help system, use one of these methods:

- Press the F1 key, and the Help system helps you with the active operation.
- Click an option on the Help menu.
- Click a Help option on the right side of the standard toolbar.
- In any dialog box, click the "?" icon.
- Click on How To on a shortcut menu initiated within an active command.

Sketch Doctor or Design Doctor appears on your screen when a problem exists with the current operation. The following image shows the Design Doctor examining a problem extrusion.

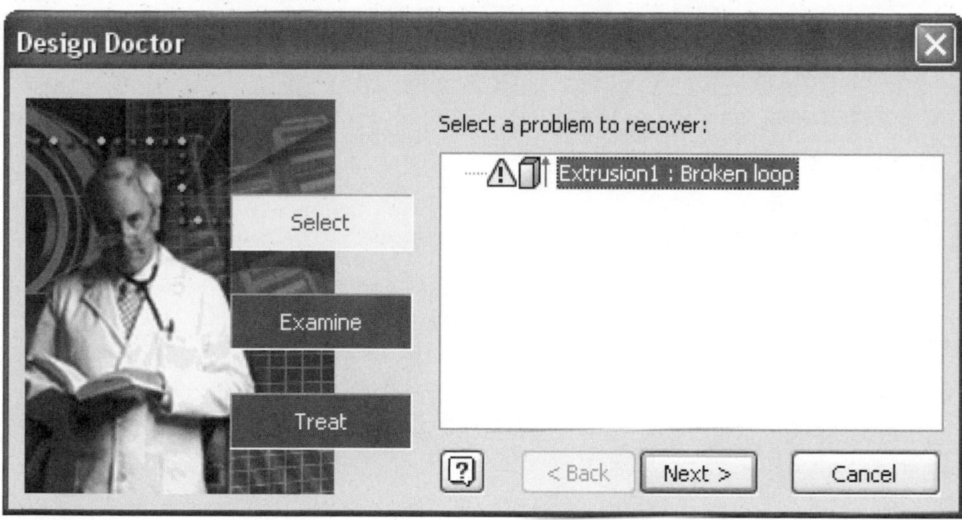

Figure 1.20

Another way to access Design Doctor and Sketch Doctor is to right-click on a sketch in the browser. Then click Sketch Doctor from the menu, or click on the cross (+) on the right side of the standard toolbar when it appears red, or when a red cross appears in a dialog box. This alerts you to a problem. Design Doctor or Sketch Doctor opens and provides options for resolving the problem.

USER INTERFACE

The default sketch environment of a part (*.ipt*) in the Autodesk Inventor application window is shown in the following image.

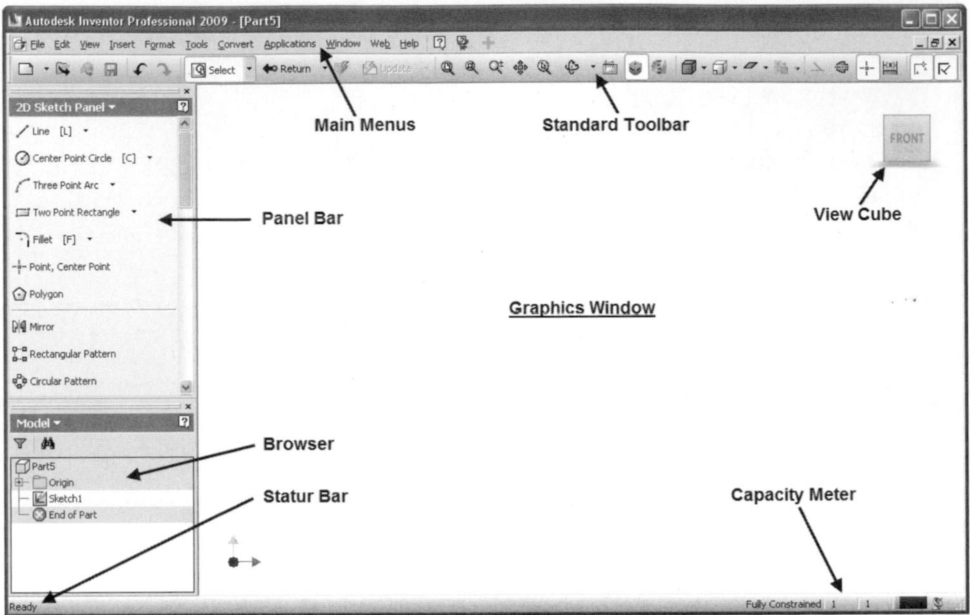

Figure 1.21

The screen is divided into the following areas:

Main Menus Access tools via text menus.

Standard Toolbar Access basic Windows and Autodesk Inventor tools.

Panel Bar Activate most of the Autodesk Inventor tools. The set of tools in the panel bar changes to reflect the environment in which you are working.

Browser Shows the history of how the contents in the file were created. The browser can also be used to edit objects.

Graphics Window Displays the graphics of the current file.

Status Bar View text messages about where you are in a process.

View Cube Displays current viewpoint and allows you to change the orientation of the view.

Capacity meter Displays how many occurrences (parts) are in the active document, the number of open documents in the current session, and how much memory is being used.

TOOLBARS, PANEL BAR, AND COMMAND ENTRY

Toolbars, the panel bar, and the browser can be moved to another docked or undocked location. To move a toolbar, click the two horizontal lines on the top of the toolbar, keep the mouse button depressed, and drag the toolbar to a new location. You get a preview of what the toolbar looks like in the new location. To turn the toolbar visibility on and off, click Toolbar on the View menu.

The tools in the panel bar adjust to the environment in which you are working, whether it is a part, an assembly, or sheet metal. You can change the status of the panel bar manually. Click the title line of the panel bar that has a down arrow, or right-click in the panel bar and select another environment from the menu.

The tool icons in the panel bar are displayed with text on by default as shown in the following image on the left, with a line of text describing the icon. To turn the text off, either click the title area of the panel, or right-click in the panel bar; then select Display Text with Icons. The following image on the right shows the icons without text.

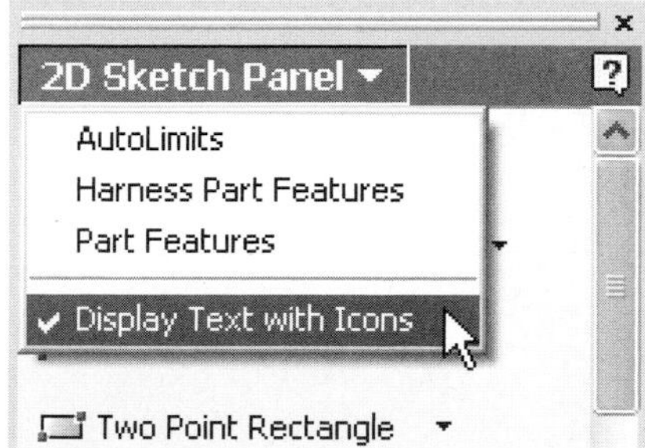

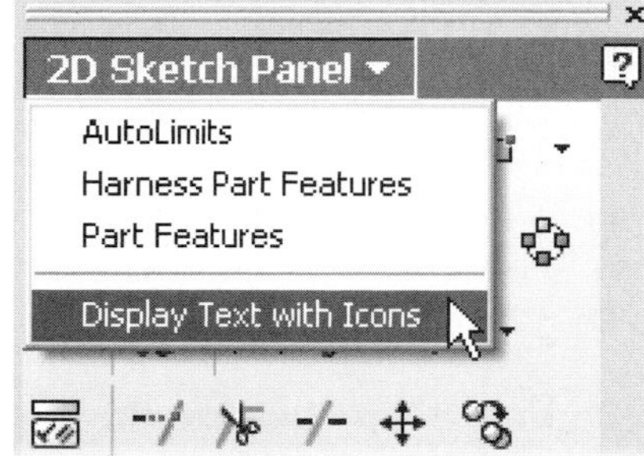

Figure 1.22

COMMAND ENTRY

There are several methods to issue commands in Autodesk Inventor. In the following sections you will learn how to start a command. There are no right or wrong methods for starting a command, and with experience, you will develop your own preference.

TOOLBARS AND THE PANEL BAR

In the last section, you learned how to control toolbars and the panel bar. To use a tool from a toolbar, move the cursor over the desired tool, and a tool tip appears with the name of the tool. The following image shows the Line tool from the Sketch toolbar and the tool tip. Some of the icons in the Panel Bar have a small down arrow in the lower-right corner. Select the arrow to see additional tools. To activate a tool, move the cursor over a tool icon and click.

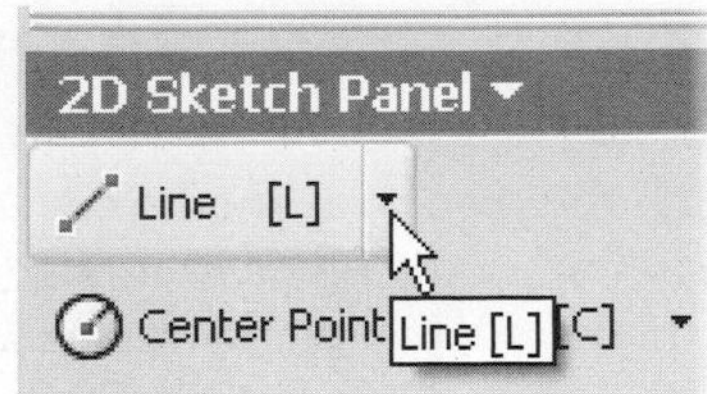

Figure 1.23

SHORTCUT MENUS

Autodesk Inventor also uses shortcut menus. These are text menus that pop up when you press the right mouse button. The shortcut menus are context sensitive; you open a menu with tools or options that are relevant to the current task.

WINDOWS SHORTCUTS

Another way to activate tools is to use Windows shortcut keys. Windows shortcuts use a two-key combination. Press both keys at the same time to activate the following operations:

CTRL-C Copy

CTRL-N Create a new document

CTRL-O Open a document

CTRL-P Print the active document

CTRL-S Save the current document

CTRL-V Paste

CTRL-Y Redo

CTRL-Z Undo

AUTODESK INVENTOR SHORTCUT/ALIAS KEYS

Autodesk Inventor has keystrokes called shortcuts or aliases that are preprogrammed. To start a command via a shortcut key, press the desired preprogrammed key(s). The following image shows a few of the predefined shortcut keys A list of shortcut commands can be found by clicking Shortcut Quick Reference from the Help menu as shown in the following image on the left. Either select the correct command from the menu, or continue typing until the alias is the only possible solution. If there is only one possible command, it will start without a dialog box.

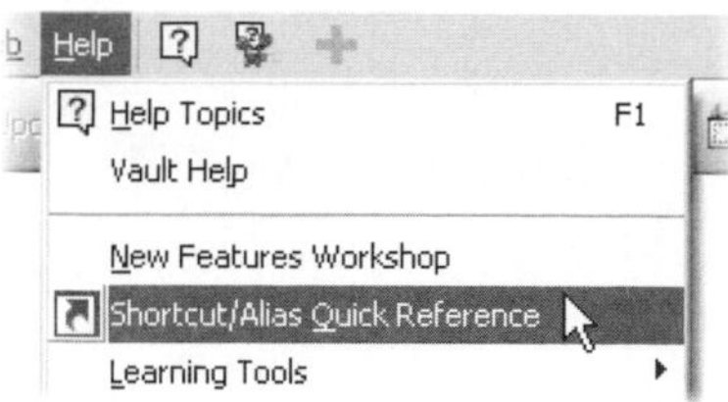

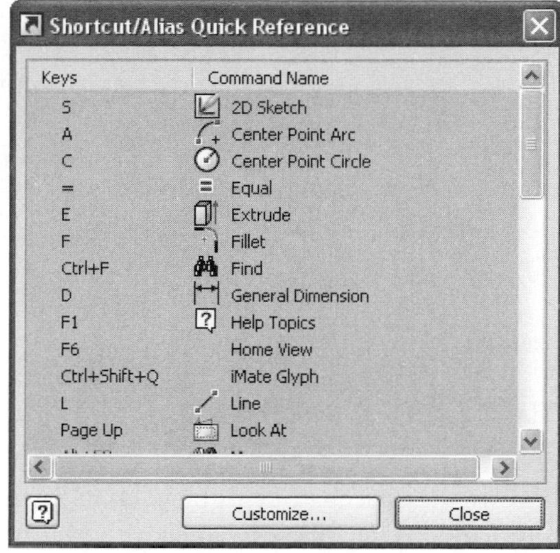

Figure 1.24

The Command Alias Input dialog box display can be turned off by clicking Tools > Application Options. On the General tab of the Options dialog box, uncheck Show command alias input dialog as shown on the left in the following image. If the dialog box is turned off, the keystroke that is typed will appear on the lower-left corner of the status bar. When the up and down arrows appear, like what is shown on the right in the following image, you can type the entire command alias and then press ENTER, or use the up and down arrows on the keyboard to cycle through the possible command. In the next section, you will learn how to modify shortcut keys.

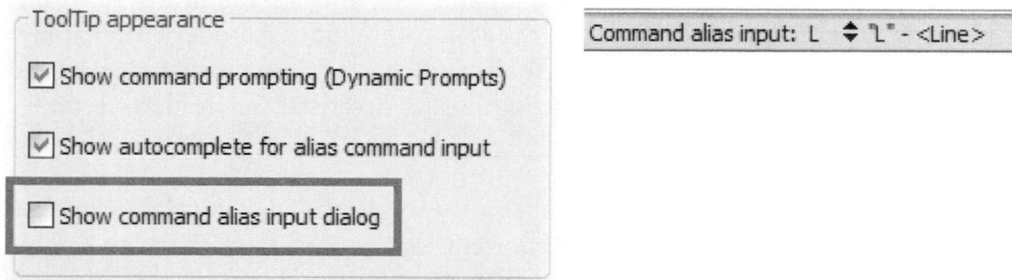

Figure 1.25

COMMAND TOOLTIPS

When a command is started, a command prompt will appear by default at the location of the cursor as shown in the following image on the left. The prompt will guide you for the next step. This prompt can be turned off by clicking Tools > Application Options. On the General tab of the Options dialog box, uncheck Show command prompting (Dynamic Prompts). The following image on the right shows the prompt checked.

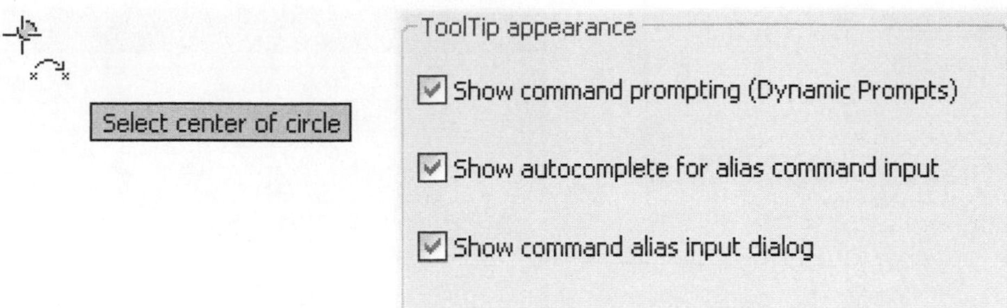

Figure 1.26

REPEAT LAST COMMAND

To restart a command and instead of reselecting the tool in a panel bar, either press ENTER or the spacebar, or right-click and click the top entry in the menu Repeat "the last command." The following image shows the Line command being restarted.

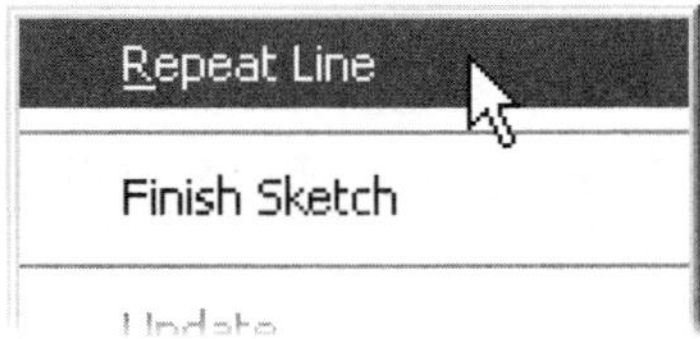

Figure 1.27

CUSTOMIZING SHORTCUTS AND ALIASES

You can customize many of the predefined command shortcuts and aliases. A shortcut is used for transparent commands such as Zoom and Pan, and for file utility functions such as printing. An alias is used for general Inventor commands. Autodesk Inventor can be customized by clicking Customize on the Tools menu or by clicking the Customize button in the Shortcut/Alias Quick Reference dialog box as described in the last section.

KEYBOARD TAB

The Keyboard tab gives you access to all of the tools that are available in Autodesk Inventor, and it allows you to modify the shortcut that is used to start the tool. The same key combinations can be used as long as they are used in different environments. For example, "C" is used to create a circle in the Sketch environment, and to create assembly constraints in the Assembly environment. The following image shows the Keyboard tab of the Customize dialog box.

Check the Use default multi-character Command Aliases option if you prefer to use multiple characters for aliases.

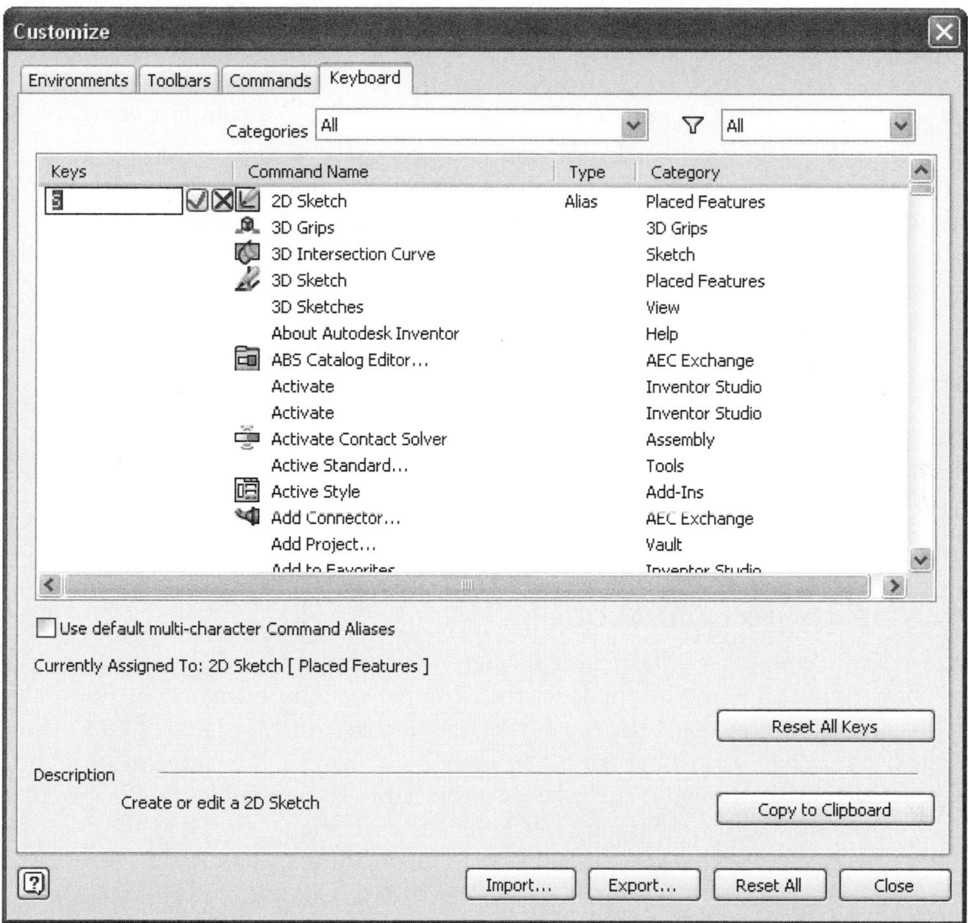

Figure 1.28

When you select an item from the Categories list, the tools that are available in that category are displayed in the dialog box.

To add a new shortcut or alias or to change an existing shortcut, follow these steps:

1. Click in the Keys area of the tool that you want to change. Information about that command will appear on the lower portion of the menu. The previous image shows the information that is displayed after selecting 2D Sketch.
2. On the keyboard, press the combination of keys that will affect the shortcut. A single letter or number can be used. You can also use the SHIFT, CTRL, and ALT keys, and a letter or number. The ALT key can be combined with the SHIFT and/or CTRL key(s) and a letter or number. The ALT key cannot be used with a single letter.
3. Press ENTER on the keyboard or click the green checkmark to create the shortcut or alias as shown in the following image.

To delete a shortcut, click on the shortcut and then press DELETE or BACKSPACE on your keyboard.

Note: To switch an existing shortcut to a different tool, you must first delete the shortcut from the existing tool.

General Options The Export, Import, Reset All, and Close buttons reside at the bottom of the dialog box and are available regardless of which tab is active. These buttons function as follows:

Export Exports an XML file containing customized settings to a selected name and location.

Import Imports an XML file containing customized settings.

Reset All Resets all customized settings to their original installation setting.

Close Closes the dialog box.

AUTO-HIDE COMMAND DIALOGS

You can control whether an individual dialog box appears as normal or auto-hide (rolled up) to show only the name of the box when the cursor is not located in the dialog box itself. To auto-hide a dialog box, move the cursor over the title area of the dialog box, right-click, and click Auto-hide from the menu. The dialog box then displays only the horizontal title bar, as shown in the image on the right. To maximize the dialog box, move the cursor over the title bar. This option is set for each dialog box.

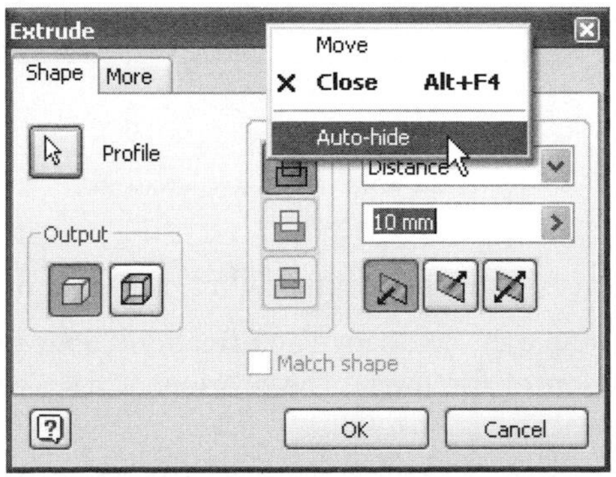

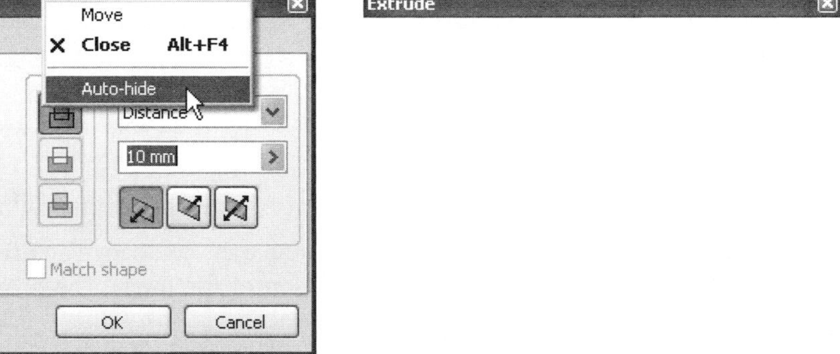

Figure 1.29

UNDO AND REDO

You may want to undo an action that you just performed, or undo an undo. The Undo tool backs up Autodesk Inventor one function at a time. If you undo too far, you can use the Redo tool to move forward one step at a time. The Zoom, Rotate, and Pan tools do not affect the Undo and Redo tool. To start the tools, either use the shortcut keys, using CTRL-Y for Redo or CTRL-Z for Undo, or select the tool from the standard toolbar, as shown in the following image. The Undo tool is to the left, and the Redo tool is to the right.

Note: To set the Undo file size allocation, click Tools > Application Options. On the General tab of the Options dialog box, change the Maximum size of Undo file (MB).

Figure 1.30

VIEWPOINT OPTIONS

When you work on a 2D sketch, the default view is looking straight down at the XY plane, or the plane view. When you work in 3D, it is helpful to view objects from a different viewpoint, and to zoom in and out or pan the objects on the screen. The next section guides you through the most common methods for viewing objects from different perspectives and viewpoints. As you use these tools, the physical objects remain unmoved. Your perspective or viewpoint of the objects is what creates the perceived movement of the part. If you are performing an operation while a viewing command is issued, the operation resumes after the transition to the new view is completed.

HOME (ISOMETRIC) VIEW

Change to an isometric viewpoint by pressing the F6 key, or by right-clicking in the graphics window and then selecting Home View from the menu, as shown in the following image. The view on the screen transitions to a predetermined home view. You can redefine the home view with the View Cube option. View Cube is explained later in this section.

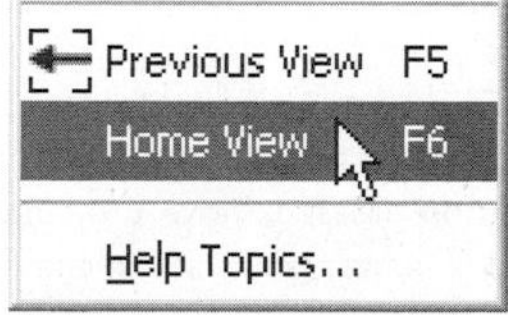

Figure 1.31

VIEWTOOLS

To zoom, pan, and rotate the geometry on the screen, use the View tools on the standard toolbar. Descriptions of the View tools follow.

Viewing Tools

Button	Tool	Function
	Zoom All	Maximizes the screen with all parts that are in the current file. The screen transitions to the new view. Issue the Zoom All tool, or press the Home key.
	Zoom Window	Zooms in on an area that is designated by two points. Issue the Zoom Window tool, or press SHIFT + F3 and select the first point. With the mouse button depressed, move the cursor to the second point. A rectangle representing the window appears. When the correct window is displayed on the screen, release the mouse button, and the view transitions to it.
	Zoom In-Out	Zooms in or out from the parts. Issue the Zoom In-Out tool, or press the F3 key. Then, in the graphics window, press and hold the left mouse key. Move the mouse toward you to make the parts appear larger, and away from you to make the parts appear smaller. If you have a mouse with a wheel, roll the wheel toward you, and the parts appear larger; roll the wheel away from you, and the parts appear smaller.
	Pan View	Moves the view to a new location. Issue the Pan View tool, or select the F2 key. Press and hold the left mouse button, and the screen moves in the same direction that the cursor moves. If you have a mouse with a wheel, hold down the wheel, and the screen moves in the same direction that the cursor moves.
	Zoom Selected	Fills the screen with the maximum size of a selected face, faces, or a part. Either select the face or faces and then issue the Zoom Selected tool, or press the END key, or launch the Zoom Selected tool, or press the END key. Then select the face or faces to which you wish to zoom. Use the Select Other tool to select a part to zoom to.
Free Orbit Constrained	Free Orbit and Constrained	Rotates your viewpoint dynamically. Issue the Free Orbit tool; a circular image with lines at the quadrants and center appears. To rotate your viewpoint, click a point inside the circle, and keep the mouse button pressed as you move the cursor. Your viewpoint rotates in the direction of the cursor movement. When you release the mouse button, the viewpoint stops rotating. To accept the view orientation, either press the ESC key or right-click and select Done from the menu. Click the outside of the circle to rotate the viewpoint about the center of the circle. To rotate the viewpoint about the vertical axis, click one of the horizontal lines on the circle and, with the mouse button pressed, move the cursor sideways. To rotate the viewpoint about the horizontal axis, click one of the vertical lines on the circle and, with the mouse button pressed, move the cursor upward or downward. Use the Constrained Orbit tool to rotate the model about the horizontal screen axes like a turn table. Click one of the horizontal lines on the circle and, with the mouse button pressed, move the cursor sideways. The model is rotated about the model space center point set in the SteeringWheels CENTER command.
	Look At	Changes your viewpoint so that you are looking parallel to a plane, or rotates the screen viewpoint to be horizontal to an edge. Issue the Look At tool or press the PAGE UP key, then select a plane or edge. The Look At tool can also be issued by selecting a plane or edge, right-clicking while the cursor is in the graphics window, and selecting Look At from the standard toolbar.
	View Cube	Turn the display of the view cube on/off. The view cube is covered in the next section.

Viewing Tools

Icon	Tool	Description
	Steering Wheel	Click the Steering Wheel icon to turn on the steering wheel tool. Click one of the eight options; that option will be active. Orbit – Rotate the viewpoint. Zoom – Zoom in and out. Rewind – Change to a pevious view. A series of images appears representing the previous views. Move the cursor over the image that represents the view to restore. Pan – Move the view to a new location. Center – Center the view based on the position of the cursor over the wheel. Walk – Swivel the viewpoint. Up/Down – Change the viewpoint vertically. Look – Move the view without rotating the viewpoint.
F4	F4 Shortcut to Dynamic Rotation	Rotates viewpoint dynamically. While working, press and hold down the F4 key. The circular image appears with lines at the quadrants and center. With the F4 key depressed, rotate the viewpoint. When you finish rotating the viewpoint, release the F4 key. If you are performing an operation while the F4 key is pressed, that operation will resume after you release the F4 key.
F5	Previous View	Returns to the view that was previously on the screen. Press the F5 key or right-click in the graphics window, and select Previous View from the menu. Continue this process until the view to which you want to return appears on the screen.
	Display Options	Accesses the three options from the standard toolbar to display 3-D parts: Shaded Display, Hidden Edge Display, and Wireframe Display. You can choose which mode works best for you, and switch between the modes as you see fit. Each display mode is described below.
	Shaded Display	Shades objects in the color or material that was assigned to them. Parts or faces that are behind other parts or faces are not displayed.
	Hidden Edge Display	Shades objects and displays the edges that are behind other parts or faces. In complex parts and assemblies, this display can be confusing.
	Wireframe Display	Displays only the outline of objects. In wireframe display, objects are not shaded.
	Camera Views	Set the camera viewpoint to be parallel, meaning that lines are projected perpendicular to the plane of projection, or perspective. This means that the geometry on the screen converges to a vanishing point similar to the way the human eye sees. The parallel viewpoint is set by default. To change the camera viewpoint, click either the Parallel Camera icon from the Standard toolbar or the Perspective Camera icon, which is the option just below Parallel Camera as shown in the following image. For clarity, this book shows all the images in the parallel view.

Viewing Tools		
	Shadow	To give your model a realistic look, you can choose to have shadows displayed on the ground. From the Standard toolbar, click Ground Shadow (as shown in the following image), No Ground Shadow, or X-Ray Ground Shadow; the top icon is the active shadow type. The default is No Ground Shadow. With shadows on, only model features cast shadows. Work geometry, sketches, origin indicators, engineer's notes, and trails in presentation documents do not cast shadows. X-Ray Ground Shadows are the same as Ground Shadows, but they additionally show detail information about hidden features.

View Cube

The view cube allows you to quickly change the viewpoint of the screen. With the view cube turned on, move the cursor over the view cube and the opacity becomes 100%. Move the cursor over the home in the view cube to return to the default isometric view as shown in the following image.

Figure 1.32

Change the viewpoint by using one of the following techniques.

Isometric Change to a different isometric view by clicking a corner on the view cube as shown in the following image on the left.

Face Change the viewpoint so it looks directly at a plane by clicking a plane on the View Cube as shown in the following left-middle image.

Rotate in 90 Degree Increments When looking plan at a view you can rotate the view by 90 degrees by clicking one of the arcs with arrows or one of the four inside facing triangles as shown in the following middle-right image.

Edge You can also orientate the viewpoint to an edge. Click an edge on the view cube as shown in the following image on the right.

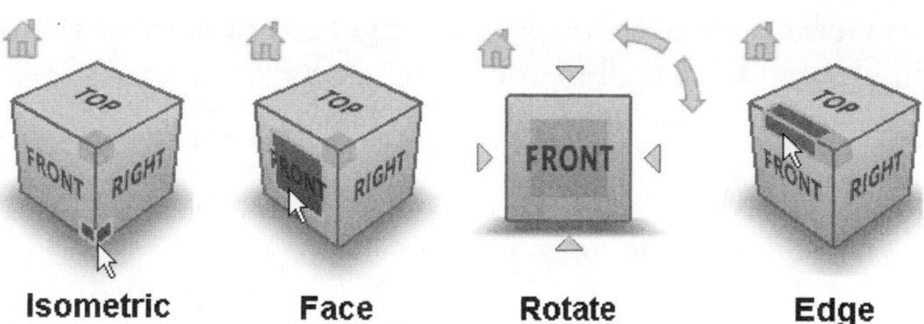

Figure 1.33

Dynamic Rotate To dynamically rotate your viewpoint, click anywhere on the view cube and keep the mouse button pressed as you move the cursor. Your viewpoint rotates in the direction of the cursor movement. When you release the mouse button, the viewpoint stops rotating. When the viewpoint does not match a defined viewpoint, the view cube edges appear in dashed lines from one of the corners as shown in the following image.

Figure 1.34

View Cube Options

To change the View Cube's options move the cursor over the View Cube and right-click. From the menu you can change the following.

Go Home Change the viewpoint to the default Home View.

Orthographic Set the display viewpoint to be parallel; the lines are projected perpendicular to the plane of projection.

Perspective Set the display viewpoint to be perspective; the geometry on the screen converges to a vanishing point similar to the way the human eye sees.

Perspective with Ortho Faces Set the display to perspective except when looking directly at a plane of the view cube.

Lock to current selection Select a part(s) and click this option; the view cube options will center and display only the selected part(s).

Set current view as Home - Fixed Distance Defines the current view as the default home view. The view will return to the same size and orientation as when it was set.

Set current view as Home - Fit to View Defines the current view as the default home view. The view size will always display all of the parts but have the same orientation as when it was set.

Set current view as Front Defines the current view as the front view. The view does not need to be plan to a plane on the view cube.

Reset Front Resets the front view to the default setting.

Options Opens the ViewCube Options dialog box.

Help Topics Launches the online Help system and displays the topic on the ViewCube.

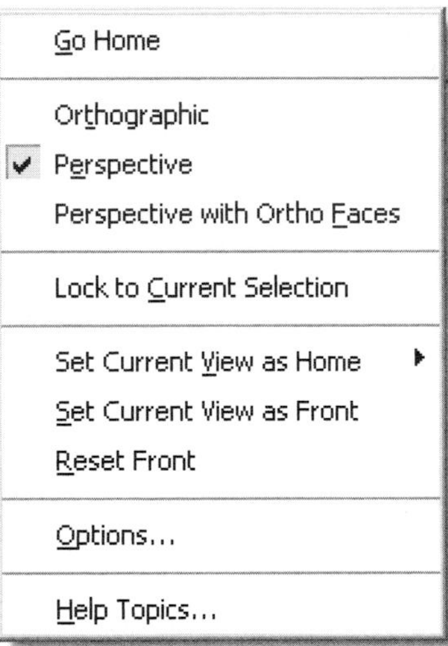

Figure 1.35

EXERCISE 1-2: VIEWING A MODEL

In this exercise, you use View Manipulation tools that make it easier to work on your designs.

1. Open C:\INV 2009 Ess Plus\Chapter 01\ESS_E01_02.iam, the file contains a model of a clamp.

Tip: Click the Chapter 1 subfolder from the Frequently Used Subfolders area, and then click the file in the file area as shown in the following image.

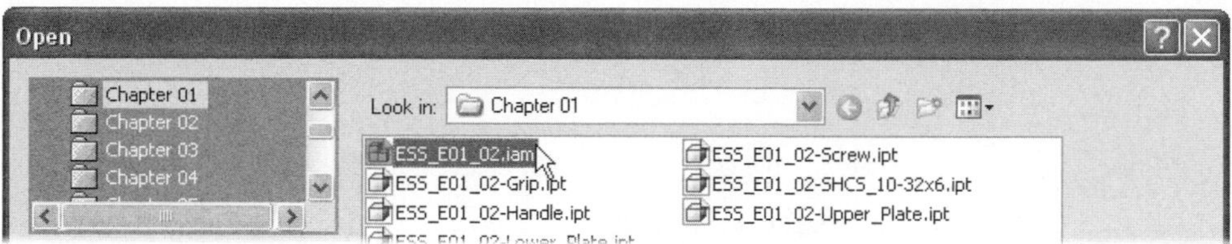

Figure 1.36

2. Click the Zoom Window tool.
3. Click a point below one of the clamp grips, move the cursor to expand the preview window, and click again to define the zoom window as shown in the following image.

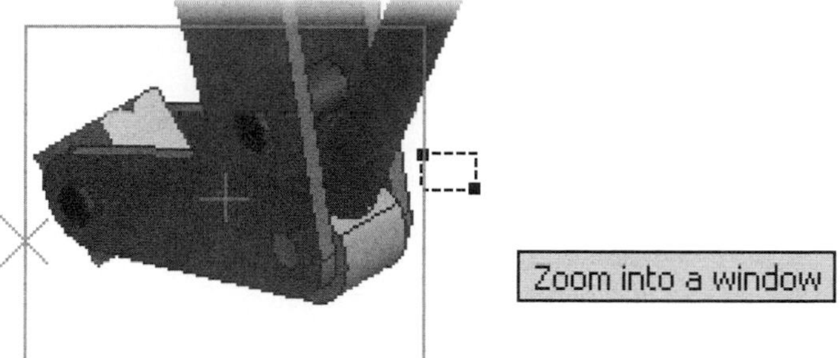

Figure 1.37

4. Press F5 to return to the previous view.
5. Use the Orbit tool to rotate the viewpoint. Click the Free Orbit tool. The 3D Rotate symbol appears in the window as shown in the following image.

Figure 1.38

6. Move the cursor inside the rim of the 3D Rotate symbol, noting the cursor display.
7. Click and drag the cursor to rotate the model.
8. To return to the Home View view, press F6.
9. Place the cursor on one of the horizontal handles of the 3D Rotate symbol, noting the cursor display.
10. Drag the cursor right or left to rotate the model about the Y axis. As you rotate the viewpoint, you will see the bottom of the assembly.
11. To again rotate the viewpoint about the vertical axis like a turntable, click the Constrained Rotate tool and click one of the horizontal lines on the circle. With the mouse button pressed, move the cursor sideways. The view will rotate like a turntable; notice that you do not see the bottom of the assembly.
12. Return to the default view and click the Home symbol near the view cube.
13. If you have a wheel mouse, spin the wheel toward you to make the model appear larger, or spin the wheel away from you to make the model appear smaller.
14. Click the Zoom tool, click and hold the mouse button down, and drag the cursor toward the bottom edge of the screen. Then drag the cursor toward the top of the screen.
15. Return to the default view and click the Home symbol near the view cube.

Tip: While you are performing another process, you can click on the viewing tools, spin the wheel to zoom, press and hold the wheel to pan, or press and hold F4 to rotate the view.

16. To change the viewpoint to predetermined isometric views, or directly at a plane, use the view cube. If the View Cube is not visible, click the View Cube icon on the Standard toolbar.

17. Click the Top-Front left corner of the view cube as shown in the following image on the left. Continue clicking the other corners of the view cube as shown in the following middle and right images.

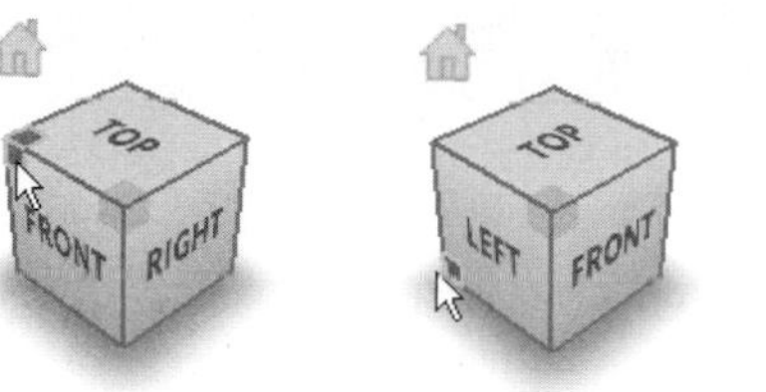

Figure 1.39

18. Change to the front view by clicking the Front plane on the view cube as shown in the following image.

Figure 1.40

19. Rotate the view 90 degrees by clicking the arc with arrow in the view cube as shown in the following image on the left.

20. To look at the Top view that is also rotated by 90 degrees click the left arrow as shown in the following image on the right.

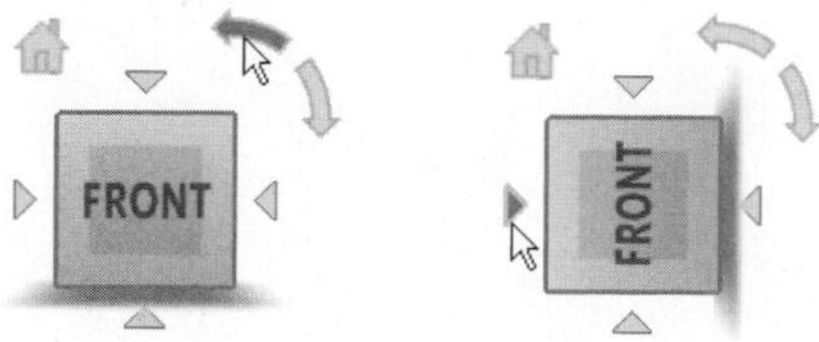

Figure 1.41

21. Practice changing views by clicking the corners, faces, and edges of the view cube.
22. Click the SteeringWheels icon on the Standard toolbar.
23. Click the different options on the Steering Wheel and experiment with them. The first time the Steering Wheels tool is started, additional information is displayed to help you use the tool.

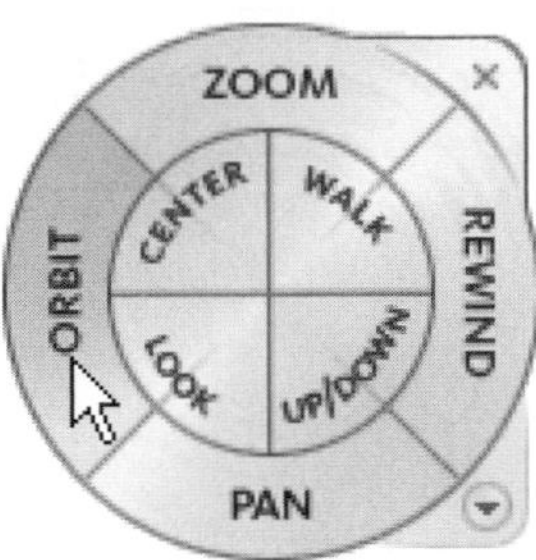

Figure 1.42

24. Change the view back to the default view by clicking the home image in the view cube.
25. View the model shaded or with hidden edges visible or wireframe. Click the Hidden Edge Display tool, and the view will change as shown in the following image.

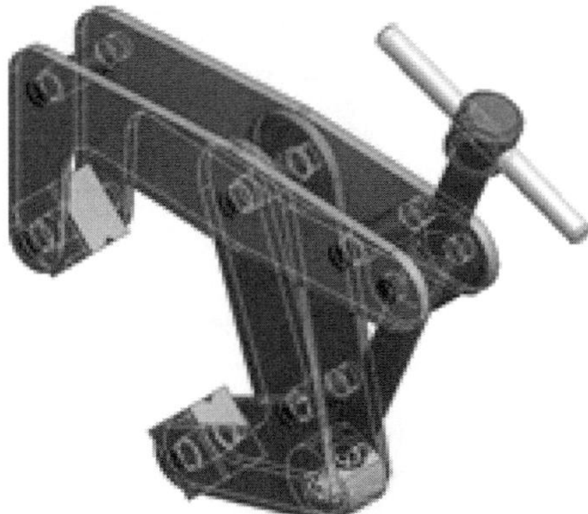

Figure 1.43

26. Click the arrow beside Shaded Display. Click the Wireframe Display tool, and the view will change as shown in the following image.

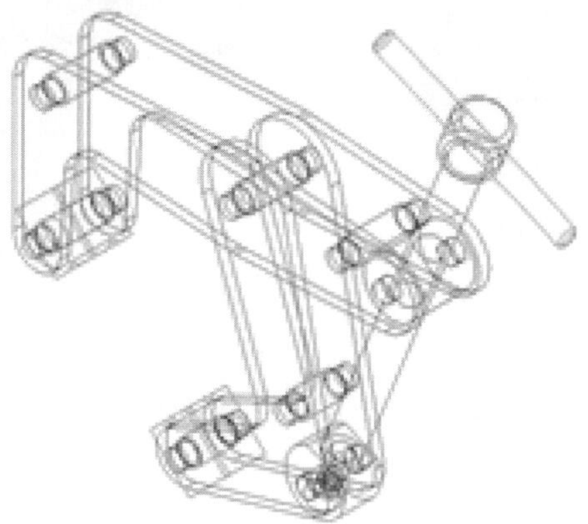

Figure 1.44

27. Click the Shaded Display tool.

28. Set the view mode to Perspective Camera mode and display shadows. Click the arrow beside Parallel Camera, then click Perspective Camera. To change the perspective, press CTRL + Shift, and spin the wheel on the mouse. Change the camera view back to parallel.

29. Click the arrow beside No Ground Shadow, then click Ground Shadow. Shadows are projected onto a floor plane below the model as shown in the following image on the left.

30. Click the arrow beside No Ground Shadow, then click X-Ray Ground Shadow. Individual components are distinguishable (as shown in the following image on the right).

Figure 1.45

31. Rotate the model and review the shadow effects.

32. Set the view mode back to Parallel. Click the arrow beside Perspective, and then click Parallel.

33. Click the arrow beside X-Ray Ground Shadow, and then click No Ground Shadow.

34. Close the file. Do not save changes. End of exercise.

Project Exercise: Chapter 1

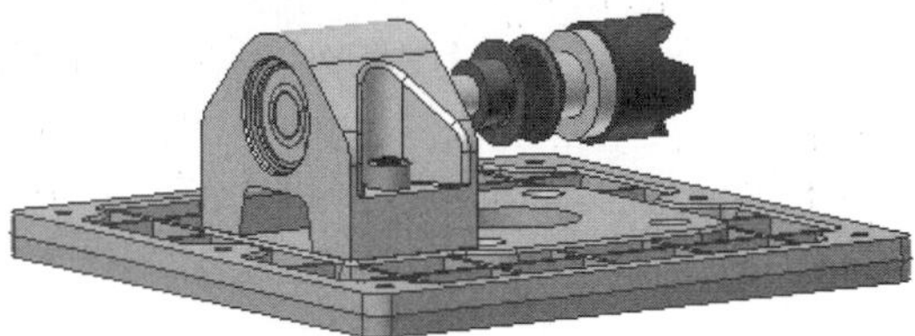

Figure 1.46

Self-paced step-by-step project exercises are found in Appendix A. They provide opportunities for you to work through real-world modeling, assembly, and documentation tasks. The geometry used in the project exercises will flow from chapter to chapter utilizing the functionality that you learned in that chapter. The project exercises are based on a coating system gripper station design shown in the previous image. The specifications for the design define a modular gripper station that can be cycled through several cleaning, coating spin, and curing stations. All stations need to work within a particulate-free, laminar air-flow environment.

CHECKING YOUR SKILLS

Use these questions to test your knowledge of the material covered in this chapter.

1. Explain the reasons why a project file is used.

   ____

   ____

2. **True** _ **False** _ Only one project can be active at any time.

3. List the sequence of locations searched when a file is opened in Autodesk Inventor.

   ____

   ____

4. **True** _ **False** _ Autodesk Inventor stores the part, assembly information, and related drawing views in the same file.

5. **True** _ **False** _ Press and hold down the F4 key to dynamically rotate a part.

6. **True** _ **False** _ The Save Copy As command saves the active document with a new name, and then makes it current.

7. List four ways to access the Help system.

   ____

   ____

8. Explain how to create a shortcut or an alias.

   ____

   ____

9. **True** _ **False** _ The Look At tool changes the viewpoint to an isometric view.

10. **True** _ **False** _ You can only edit a part while it is in Shaded Display.

CHAPTER 2

Sketching, Constraining, and Dimensioning

INTRODUCTION

Most 3D parts in Autodesk Inventor start from a 2D sketch. This chapter first provides a look at the application options for sketching and part creation. It then covers the three steps in creating a 2D parametric sketch: sketching a rough 2D outline of a part, applying geometric constraints, and then adding parametric dimensions.

OBJECTIVES

After completing this chapter, you will be able to do the following:

- Change the sketch and part options as needed
- Sketch an outline of a part
- Create geometric constraints
- Use construction geometry to help constrain sketches
- Dimension a sketch
- Create dimensions using the automatic dimensioning tool
- Change a dimension's value in a sketch
- Import AutoCAD DWG data

SKETCHING AND PART APPLICATION OPTIONS

Before you create a sketch, examine the sketch and part options in Autodesk Inventor that will affect sketching and part modeling. While learning Autodesk Inventor, refer back to these option settings to determine which ones work best for you—there are no right or wrong settings.

SKETCH OPTIONS

Autodesk Inventor sketching options can be customized to your preferences. Click Tools > Application Options, and then click on the Sketch tab, as displayed in the following image. Descriptions of the Sketch options follow. These settings are global, and they affect all currently active Autodesk documents and Autodesk Inventor documents you open in the future.

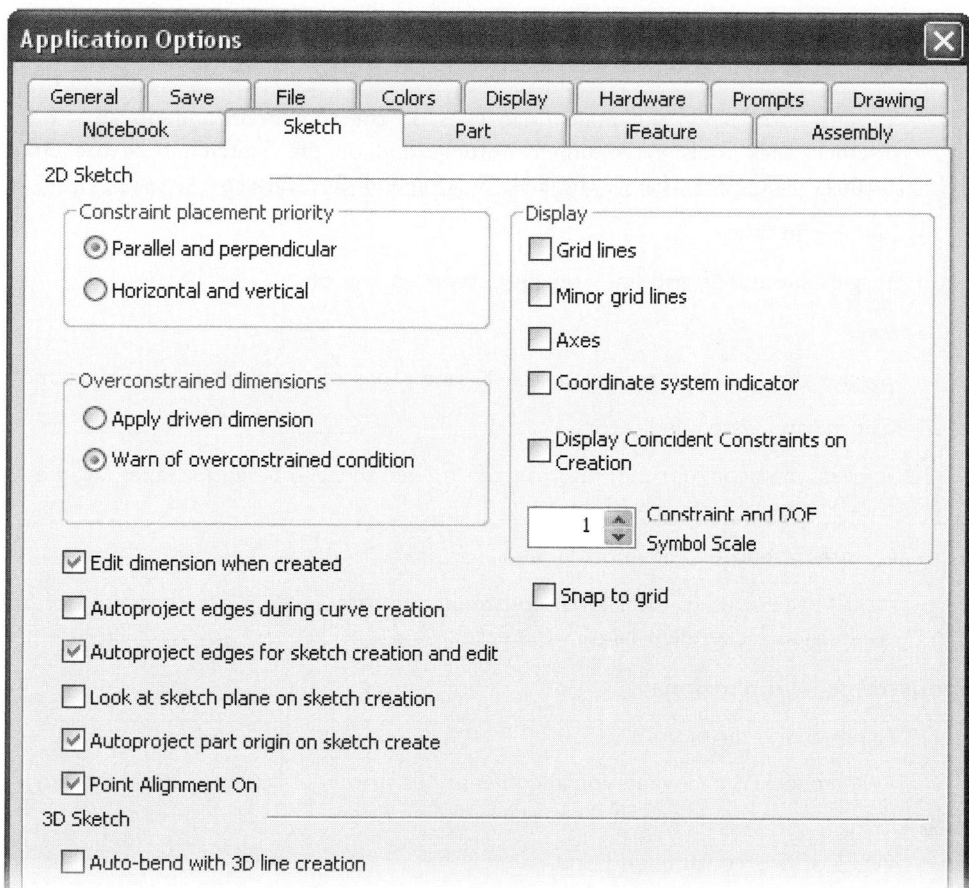

Figure 2.1

Constraint Placement Priority

- Parallel and perpendicular

  When checked, and when a parallel or perpendicular condition exists while sketching, applies a parallel or perpendicular constraint before any other possible constraints that affect the geometry being created.

- Horizontal and vertical

  When checked, and when a horizontal or vertical condition exists while sketching, applies a horizontal or vertical constraint before any other possible constraints that affect the geometry being created.

Display

- Grid lines

  Toggles both minor and major grid lines on the screen on and off. To set the grid distance, click Tools > Document Settings, and on the Sketch tab of the Document Settings dialog box, change the Snap Spacing and Grid Display.

- Minor grid lines

  Toggles the minor grid lines on the screen on and off.

- Axes

  Toggles the lines that represent the X- and Y-axis of the current sketch on and off.

- Coordinate system indicator

  Toggles the icon on and off that represents the X-, Y-, and Z-axes at the 0, 0, 0 coordinates of the current sketch.

- Constraint and DOF Symbol Scale

  Adjusts the scale of the sketch constraint toolbars and the sketch degree of freedom symbols to make them larger or smaller.

Overconstrained Dimensions

- Apply driven dimensions

  When checked, and when you add dimensions that would overconstrain the sketch, adds the dimension as a driven (reference) dimension.

- Warn of overconstrained condition

  When checked, and when you add dimensions that would overconstrain the sketch, a dialog box appears warning of the condition. This allows you to accept the placement of a driven dimension or cancel the dimension placement.

Snap to grid

When checked, endpoints of sketched objects snap to the intersections of the grid as the cursor moves over them.

Edit dimension when created

When checked, edits the values of a dimension in the Edit Dimension dialog box immediately after you position the dimension.

Autoproject edges during curve creation

When checked, and while sketching place the cursor over an object and it will be projected onto the current sketch. You can also toggle Autoproject on and off while sketching by right-clicking and selecting Autoproject from the menu.

Autoproject edges for sketch creation and edit

When checked, automatically projects all of the edges that define that plane onto the sketch plane as reference geometry when you create a new sketch.

Look at sketch plane on sketch creation

When checked, automatically changes the view orientation to look directly at the new or active sketch.

Autoproject part origin on sketch create

When checked, the part's origin point will automatically be projected when a new sketch is created.

Point Alignment On

When checked, automatically infers alignment (horizontal and vertical) between endpoints of newly created geometry. No sketch constraint is applied. If this option is not checked, points can still be inferred; this technique is covered later in this chapter in the Inferred Points section.

3D Sketch

- Auto-bend with 3D line creation

  When checked, automatically places a tangent arc between two 3D lines as they are sketched. To set the radius of the arc, click Tools > Document Settings, and change the Auto-Bend Radius on the Sketch tab of the Document Settings dialog box.

PART OPTIONS

You can customize Autodesk Inventor Part options to your preferences. Click Tools > Application Options, and click on the Part tab, as shown in the following image. Descriptions of the Part options follow. These settings are global—they will affect all active and new Autodesk Inventor documents.

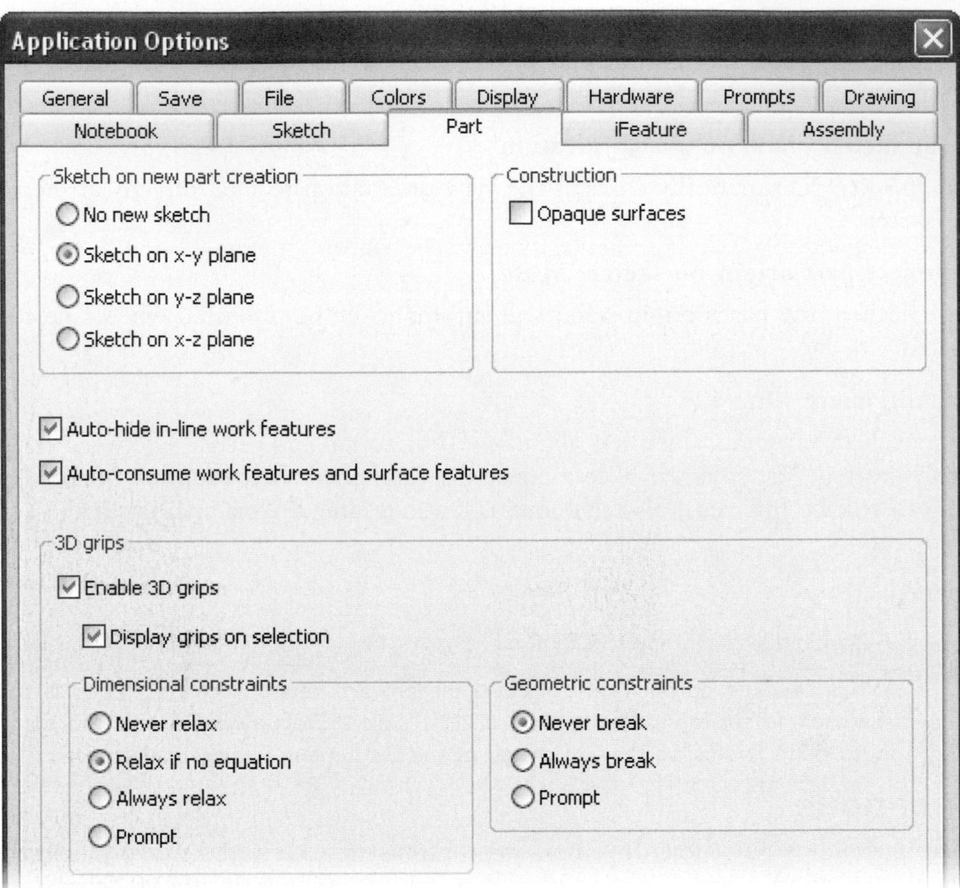

Figure 2.2

Sketch on New Part Creation

- No new sketch

  When checked, does not set sketch plane when you create a new part.

- Sketch on x-y plane

  When checked, sets the x-y plane as the current sketch plane when you create a new part.

- Sketch on y-z plane

  When checked, sets the y-z plane as the current sketch plane when you create a new part.

- Sketch on x-z plane

  When checked, sets the x-z plane as the current sketch plane when you create a new part.

Construction

- Opaque surfaces

  When checked, surfaces you create are displayed as opaque; otherwise, they will be translucent. After you create a surface, its translucency can be controlled from a menu.

Auto-hide in-line work features

When checked, automatically hides work features that are consumed by another work feature.

Auto-consume work features and surface features

When checked, work features and surfaces will become children of the feature that were used to create them.

3D grips

Sets the preferences for using 3D grips.

Enable 3D grips When checked, enables 3D grips. When unchecked, 3D grips will not be utilized.

Display grips on selection When checked, displays the grip when selecting a face or an edge in a part (.ipt) or an assembly (.iam) file. The grip displays when the selection priority is set to faces and edges, and the face can be edited with 3D Grips. Clicking on a grip launches the 3D Grips command. Clear the check box to turn off the display of the grip.

Dimensional constraints

- Never relax

  When checked, a feature cannot be grip edited in a direction that has a linear or angular dimension.

- Relax if no equation

  When checked, a feature cannot be grip edited in a direction that is defined by an equation that is based on linear or angular dimension. Dimensions without equations are not affected.

- Always relax

  When checked, a feature can be grip edited, whether or not a linear, angular, or equation-based dimension is applied. Equation-based dimensions are converted to numeric values.

- Prompt

  When checked, similar to Always relax, a dialog box is displayed warning if a grip edit will affect either dimensions or equation-driven dimensions. When accepted, dimensions and equations are relaxed, and both are updated as numerical values after grip editing.

Geometric constraints

- Never break

  When checked, a feature cannot be grip edited if a constraint exists.

- Always break

  When checked, a feature can be grip edited, even if a constraint exists.

- Prompt

  When checked, similar to Always break, a dialog box is displayed warning if a grip edit will break one or more constraints.

UNITS

Autodesk Inventor uses a default unit of measurement for every part and assembly file. The default unit is set from the template file from which you created the part or assembly file. When specifying numbers in dialog boxes with no unit, the default unit will be used. You can change the default unit in the active part or assembly document by clicking Tools > Document Settings and selecting the Units tab as shown in the following image. The unit system values change for all of the existing values in that file.

Note: In a drawing file, the appearance of dimensions is controlled by dimension styles. Drawing settings will be covered in chapter 5.

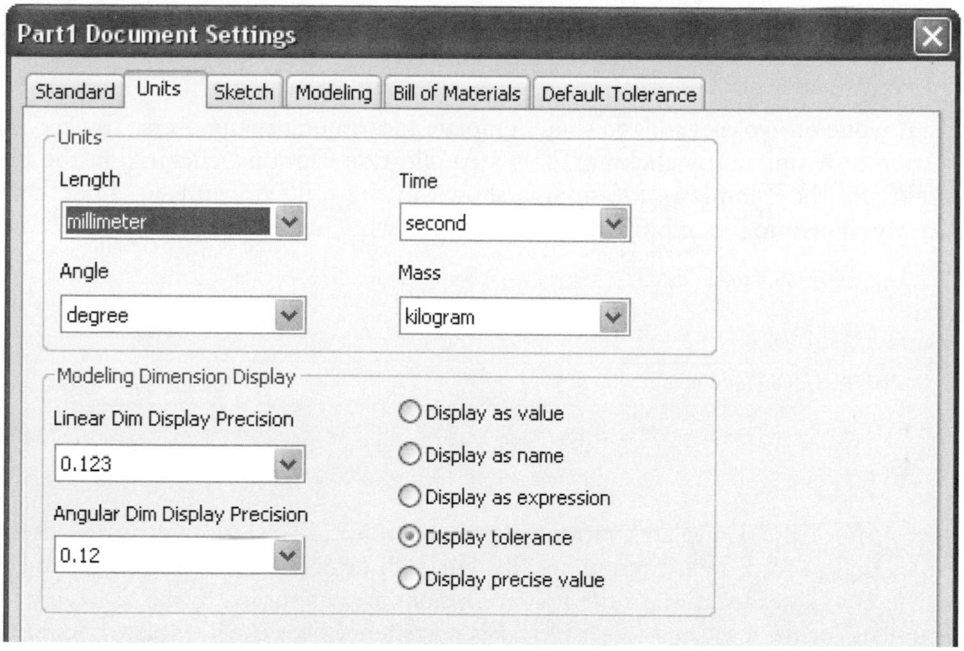

Figure 2.3

You can override the default unit for any value by entering the desired unit. If you were working in an mm file, for example, and you placed a horizontal dimension whose default value was 50 mm, you could enter 2 in. Dimensions appear on the screen in the default units.

For the previous example, 50.8 mm would appear on the screen. When you edit a dimension, the overridden unit appears in the Edit Dimension dialog box as shown in the following image.

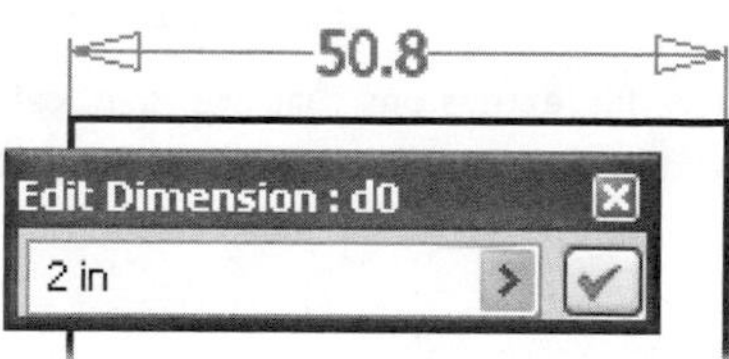

Figure 2.4

TEMPLATES

Each new file is created from a template. You can modify existing templates or add your own templates. As you work, make note of the changes that you make to each file. You then create a new template file or modify an existing file that contains all of the changes and save that file to your template directory, which by default is located at C:\Program Files*Autodesk\Inventor(version number)\templates* directory. You can also create a new subdirectory under the templates folder, and place the file there. After you create the new template subdirectory and add a template file to it, a new tab will appear with the directory name. You can place any Autodesk Inventor file in this new directory, and it will be available as a template.

You can use one of two methods to share template files among many users. You can modify the location of templates by clicking Tools > Application Options, clicking on the File tab and modifying the Templates location as shown in the following image. The Templates location will need to be modified for each user who needs access to these common templates.

Default templates

%INSTALLDIR%\Templates\

Figure 2.5

You can also set the Templates location in each project file. This method is useful for companies that need different templates files for each project. While editing a project file, change the Templates location in the Folder Options area as shown in the following image. The Template location in the project file takes precedence over the Templates option in the Application Options, File tab.

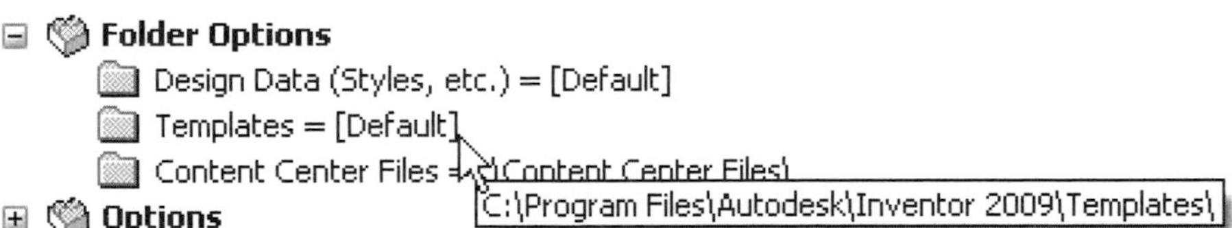

Figure 2.6

Note: Template files have file extensions that are identical to other files of the same type, but they are located in the template directory. Template files should not be used as production files.

CREATING A PART

The first step in creating a part is to start a new part file or create a new part file in an assembly. The first four chapters in this book deal with creating parts in a new part file, and chapter 6 covers creating and documenting assemblies. You can use the following methods to create a new part file:

- Click New on the File menu, and then click the *Standard.ipt* icon on the Default tab, as shown in the following image on the left. You can also click *Standard (unit).ipt* on one of the other tabs.

- Click the down arrow of the New icon, and select Part from the left side of the standard toolbar as shown in the following image on the right. This creates a new part file based on the units that were selected when Autodesk Inventor was installed or on the *Standard.ipt* that exists in the Templates folder.
- Click the New icon in the Quick Launch area of the Open dialog box, and then click the *Standard.ipt* icon on the Default tab, as shown in the following image on the left, or click *Standard (unit).ipt* on one of the other tabs.
- Use the shortcut key, CTRL-N, and then click the *Standard.ipt* icon on the Default tab, as shown in the following image. You can also click *Standard (unit).ipt* on one of the other tabs.

After starting a new part file using one of the previous methods, Autodesk Inventor's screen will change to reflect the part environment.

Note: The units for the files located in the Default tab are based upon the units you selected when you installed Autodesk Inventor.

Figure 2.7

SKETCHES AND DEFAULT PLANES

Before you start sketching, you must have an active sketch on which to draw. A sketch is a plane on which 2D objects are sketched. You can use any planar part face or work plane to create a sketch. A sketch is automatically set by default when you create a new part file. You can change the default plane on which you will create the sketch by selecting Tools > Application Options and clicking on the Part tab. Choose the sketch plane to which new parts should default.

Each time you create a new Autodesk Inventor, there are three planes (XY, YZ, and XZ), three axes (X, Y, and Z), and the center (origin) point at the intersection of the three planes. You can use these default planes to create an active sketch. By default, visibility is turned off to the planes, axes, and center point. To see the planes, axes, or center point, expand the Origin entry in the browser by clicking on the + to the left side of the text. You can then move the cursor over the names, and they will appear in the graphics area. The following image illustrates the default planes, axes, and center point in the graphics area shown in an isometric view with their visibility on. The browser shows the Origin menu expanded. To leave the visibility of the planes or axes on, right-click in the browser while the cursor is over the name, and select Visibility from the menu.

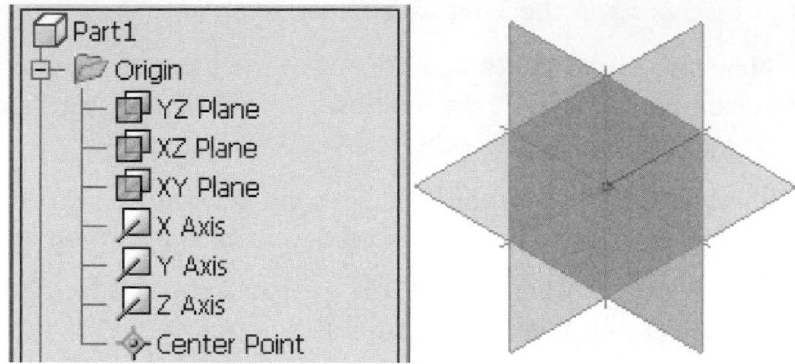

Figure 2.8

AUTOPROJECT CENTER (ORIGIN) POINT

To access another option that will automatically project the origin point for each new sketch of a part, click Tools > Application Options, click on the Sketch tab, and then check Autoproject part origin on sketch create as displayed in the following image. The origin point can be used to constrain a sketch to the 0, 0 point of the sketch.

☑ Autoproject part origin on sketch create

Figure 2.9

NEW SKETCH

Issue the Sketch tool to create a new sketch on a planar part face, a work plane, or to activate a nonactive sketch in the active part. To create a new sketch or make an existing sketch active, use one of these methods:

- Click the Sketch tool on the standard toolbar as shown in the following image. Then click a face, a work plane, or an existing sketch in the browser.
- Click a face, a work plane, or an existing sketch in the browser. Then click the Sketch tool on the standard toolbar.
- Press the hot key (a keyboard shortcut) S and click a face of a part, a work plane, or an existing sketch in the browser.
- Click a face of a part, a work plane, or an existing sketch in the browser, and then press the hot key S.
- While not in the middle of an operation, right-click in the graphics area, and select New Sketch from the menu. Then click a face, a work plane, or an existing sketch in the browser.

- While not in the middle of an operation, click a face of a part, a work plane, or an existing sketch in the browser. Then right-click in the graphics area, and select New Sketch from the menu.

Figure 2.10

After you have activated the sketch, the X- and Y-axes will align automatically to this plane, and you can begin to sketch.

Note: This book assumes that when you installed Autodesk Inventor, you selected mm as the default unit. If you selected inch as the default unit, then select the *Standard (mm).ipt* template file from the Metric tab. This book will use the XY plane as the default sketch plane.

STEP 1—SKETCH THE OUTLINE OF THE PART

As stated at the beginning of this chapter, 3D parts usually start with a 2D sketch of the outline shape of the part. You can create a sketch with lines, arcs, circles, splines, or any combination of these elements. The next section will cover sketching strategies, tools, and techniques.

SKETCHING OVERVIEW

When deciding what outline to start with, analyze how the finished shape will look. Look for the shape that best describes the part. When looking for this outline, try to look for a flat face. It is usually easier to work on a flat face than on a curved edge. As you gain modeling experience, you can reflect on how you created the model and think about other ways that you could have built it. There is usually more than one way to generate a given part.

When sketching, draw the geometry so that it is close to the desired shape and size—you do not need to be concerned about exact dimensional values. Autodesk Inventor allows islands in the sketch (closed objects that lie within another closed object). An example would be a circle that is drawn inside of a rectangle. When you extrude the sketch, the island may become a void in the solid. A sketch can consist of multiple closed objects that are coincident.

The following guidelines will help you to successfully generate sketches:

- Select an outline that best represents the part. It is usually easier to work from a flat face.
- Draw the geometry close to the finished size. If you want a 20 mm square, for example, do not draw a 200 mm square.
- Create the sketch proportional in size to the finished shape. When drawing the first object verify, its size in the lower-right corner of status bar. Use this information as a guide.

- Draw the sketch so that it does not overlap. The geometry should start and end at a single point, just as the start and end points of a rectangle share the same point.
- Do not allow the sketch to have a gap; all of the connecting endpoints should be coincident.
- Keep the sketches simple. Leave out fillets and chamfers when possible. You can easily place them as features after the sketch turns into a solid. The simpler the sketch, the fewer the number of constraints and dimensions that will be required to constrain the model.

If you want to create a solid, the sketch must form a closed shape. If it is open, it can only be turned into a surface.

SKETCHING TOOLS

Before you start sketching the part, examine the 2D sketching tools that are available. By default, the 2D sketch tools appear on the Panel Bar with Display Text with Icons turned on (text descriptions shown). As you become more proficient with Autodesk Inventor, you can turn off the text, as shown in the following image. Do this by clicking on the title area of the menu or by right-clicking on the Panel Bar and selecting Display Text with Icons from the menu. You can also use the 2D Sketch Panel to access the 2D sketch tools. The most frequently used tools will be explained throughout this chapter.

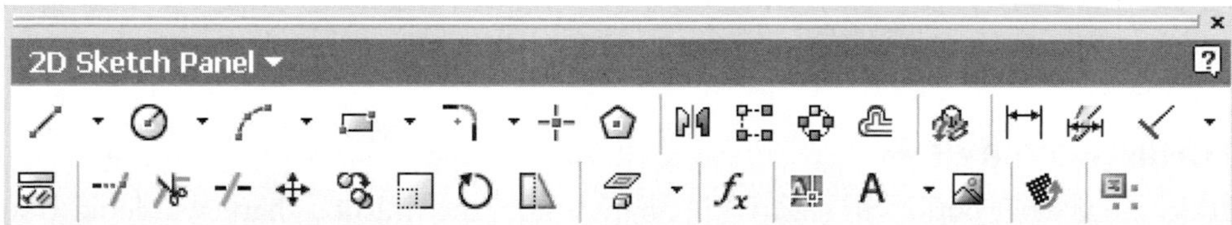

Figure 2.11

Using the Sketch Tools

After starting a new part, use the sketch tools to draw the shape of the part. As you are sketching, Autodesk Inventor can give you visual feedback about what is happening on the screen with text messages near the cursor. The following image on the left shows the message displayed when using the Line tool. To turn on the text message click Tools > Application Options, click on the General tab and check Show command prompting (Dynamic Prompts). If this option is unchecked, the text message will appear in the lower-left area of the Status Bar. To start sketching, issue the sketch tool that you need, click a point in the graphics area, and follow the prompt on the status bar. The sections that follow will introduce techniques that you can use to create a sketch.

Figure 2.12

Line Tool

The Line tool is one of the most powerful tools that you will use to sketch. Not only can you draw lines with it, but you can also draw an arc from the endpoint of a line segment. After issuing the Line tool as shown in the following image on the left, you will be prompted to click a first point, select a point in the graphics window and then click a second point. You can continue drawing line segments, or you can move the cursor over the endpoint of a line segment or arc, and a small gray circle will appear at that endpoint. The following image on the right shows how the endpoint of a line segment looks when the cursor moves over it.

Figure 2.13

Click on the small circle, and with the left mouse button pressed down, move the cursor in the direction that you want the arc to go. Up to eight different arcs can be drawn, depending upon how you move the cursor. The arc will be tangent to the horizontal or vertical edges that are displayed from the selected endpoint. The following image shows an arc that is normal to the sketched line being drawn.

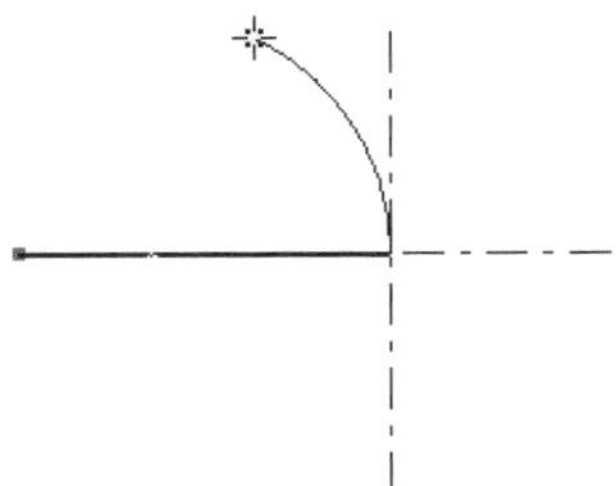

Figure 2.14

Tip: When sketching, look at the bottom-right corner of the status bar (bottom of the screen) to see the coordinates, length, and angle of the objects that you are drawing. The following image shows the status bar when a line is being drawn.

13.158 mm, 7.785 mm | Length=8.118 mm | Angle=106.48 deg

Figure 2.15

Object Tracking – Inferred Points

If the Point Alignment On option is checked from the Sketch tab of the Application Options, as you sketch dashed lines will appear on the screen. These dashed lines represent the endpoints, midpoints, and theoretical intersections of lines, arcs, and center points of arcs and circles that represent their horizontal, vertical, or perpendicular positions. As the cursor gets close to these inferred points, it will snap to that location. If that is the point that you want, click that point; otherwise, continue to move the cursor until it reaches the desired location.

When you select inferred points, no constraints (geometric rules such as horizontal, vertical, colinear, etc.) are applied from them. If the Point Alignment On option is unchecked, you can still infer points, move the cursor over a point, and move the cursor. Using inferred points helps create more accurate sketches. The following image shows the inferred points from two endpoints that represent their horizontal and vertical position.

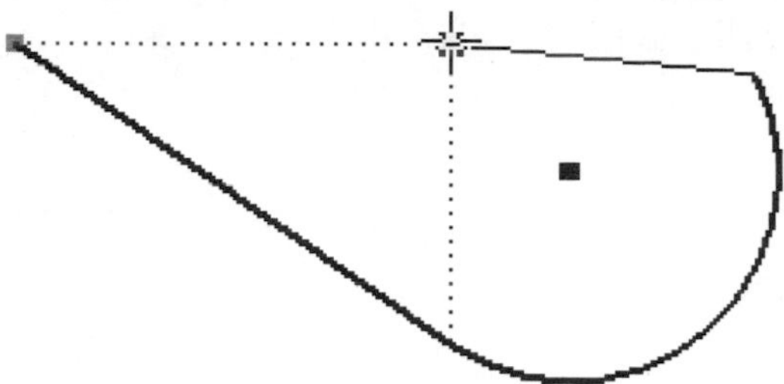

Figure 2.16

Automatic Constraints

As you sketch, small constraint symbols appear that represent geometric constraint(s) that will be applied to the object. If you do not want a constraint to be applied, uncheck the Constraint Persistent option on the Standard toolbar or hold down the CTRL key when you select the point. If you want only coincident geometric constraints applied to the sketch, click Tools > Application Options, click on the Sketch tab, and check None from the Constraint Placement Priority before sketching. The following image shows a line being drawn from the arc, tangent to the arc and parallel to the angled line. The symbol appears near the object from which the constraint is coming. Constraints will be covered in the next section.

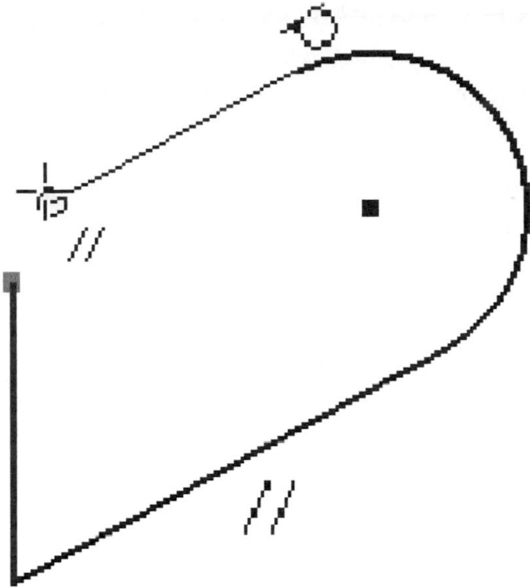

Figure 2.17

Scrubbing

As you sketch, you may prefer to apply a constraint different from the one that automatically appears on the screen. You may want a line to be perpendicular to a given line, for example, instead of being parallel to a different line. The technique to change the constraint is called *scrubbing*. To place a different constraint while sketching, move the cursor so it touches (scrubs) the other object to which the constraint should be related. Move the cursor back to its original location, and the constraint symbol changes to reflect the new constraint. The same constraint symbol will also appear near the scrubbed object, representing that it is the object to which the constraint is matched. Continue sketching as normal. The following image shows the top horizontal line being drawn with a perpendicular constraint that was scrubbed from the left vertical line. Without scrubbing the left vertical line, the applied constraint would have been parallel to the bottom line.

Figure 2.18

CONTROLING SKETCH CONSTRAINTS

While sketching you can control whether or not the sketch constraints are applied. From the Standard toolbar select or unselect the following options.

Constraint Persistence

Constraint Persistence turns sketch constraints on/off. When on, sketch constraints are applied while sketching. When off, the constraint icon will appear on the screen, but it will not be applied to the sketch. Coincident constraints are applied regardless of this setting. The default setting is on.

Constraint Inference

Constraint Inference turns the preview of the constraint on/off. When off, Constraint Persistence is also turned off and sketch constraints are not applied. The default setting is on.

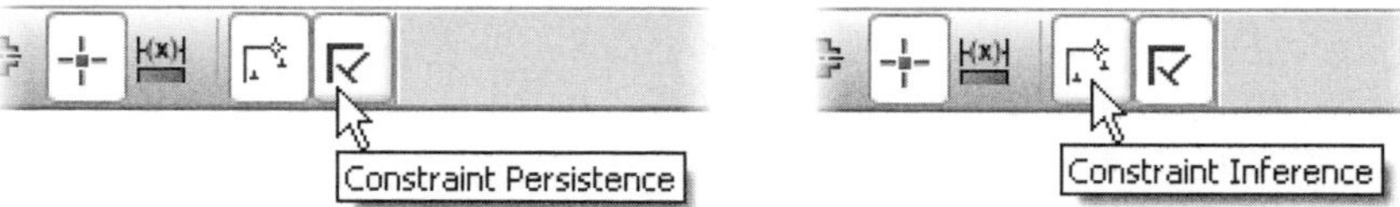

Figure 2.19

With Constraint Inference on you can control which sketch constraints can be inferred and which geometry they should be inferred from. While in a sketch, right-click and choose the Constraint Options from the menu as shown in the following image on the left. The Constraint Options dialog box will appear as shown in the following image on the right. In the Selection for Constraint Inference section, uncheck constraints that you do want to be inferred when sketching. In the Scope of Constraint Inference section you can uncheck the All geometry option and then select the geometry that constraints can be inferred from. The constraint types are covered in the next section of this chapter.

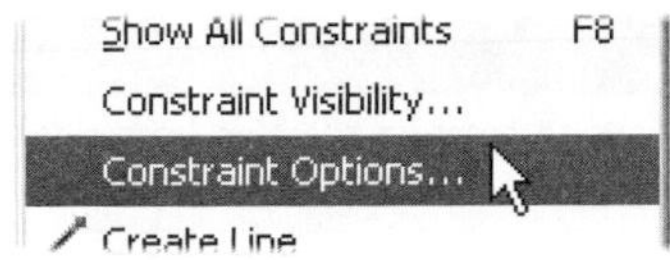

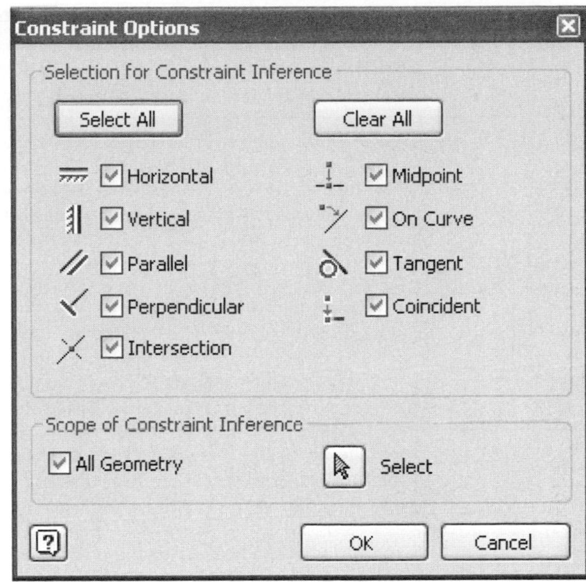

Figure 2.20

COMMON SKETCH TOOLS

The following chart lists common 2D sketch tools that are not covered elsewhere in this chapter. Some tools are available by clicking the down arrow in the lower-right corner of the top tool.

Tool	Function
Center-Point Circle	Creates a circle by clicking a center point for the circle and then a point on the circumference of the circle.
Tangent Circle	Creates a circle that will be tangent to three lines or edges by clicking the lines or edges.
Three Point Arc	Creates an arc by clicking a start and endpoint and then a point that will lie on the arc.
Tangent Arc	Creates an arc that is tangent to an existing line or arc by clicking the endpoint of a line or arc and then clicking a point for the other endpoint of the arc.
Center-Point Arc	Creates an arc by clicking a center point for the arc and then clicking a start and endpoint.
Two Point Rectangle	Creates a rectangle by clicking a point and then clicking another point to define the opposite side of the rectangle. The edges of the rectangle will be horizontal and vertical.
Three Point Rectangle	Creates a rectangle by clicking two points that will define an edge and then clicking a point to define the third corner.
Fillet	Creates a fillet between two nonparallel lines, two arcs, or a line and an arc at a specified radius. If you select two parallel lines, a fillet is created between them without specifying a radius. When the first fillet is created, a dimension will be created. If many fillets are placed in the same operation, you choose to either apply or not apply an equal constraint.

Tool	Function
Chamfer	Creates a chamfer between lines. There are three options to create a chamfer: both sides equal distances, two defined distances, or a distance and an angle.
Polygon	Creates an inscribed or a circumscribed polygon with the number of faces that you specify. The polygon's shape is maintained as dimensions are added.
Mirror	Mirrors the selected objects about a centerline. A symmetry constraint will be applied to the mirrored objects.
Rectangular Pattern	Creates a rectangular array of a sketch with a number of rows and columns that you specify.
Circular Pattern	Creates a circular array of a sketch with a number of copies and spacing that you specify.
Offset	Creates a duplicate of the selected objects that are a given distance away. By default, an equal-distance constraint is applied to the offset objects.
Trim	Trims the selected object to the next object it finds. Click near the end of the object that you want trimmed. While using the Trim tool, hold down the SHIFT key to extend objects. If desired, hold down the CTRL key to select boundary objects.
Extend	Extends the selected object to the next object it finds. Click near the end of the object that you want extended. While using the Extend tool, hold down the SHIFT key to trim objects. If desired, hold down the CTRL key to select boundary objects.

SELECTING OBJECTS

After sketching objects, you may need to move, rotate, or delete some or all of the objects. To edit an object, it must be part of a selection set. There are two methods that you can use to place objects into a selection set.

- You can select objects individually by clicking on them. To select multiple individual objects, hold down the CTRL or SHIFT key while clicking the objects. You can remove selected objects from a selection set by holding down the CTRL or SHIFT key and reselecting them. As you select objects, their color will change to show that they have been selected.
- You can select multiple objects by defining a selection window. To define the window, click a starting point. With the left mouse button depressed, move the cursor to define the box. If you draw the window from left to right, only the objects that are fully enclosed in the window will be selected. If you draw the window from right to left, as shown in the following image, all of the objects that are fully enclosed in the window *and* the objects that are touched by the window will be selected.
- You can use a combination of the methods to create a selection set.

When you select an object, its color will change according to the color style that you are using. To remove all of the objects from the selection set, click in a blank section of the graphics area.

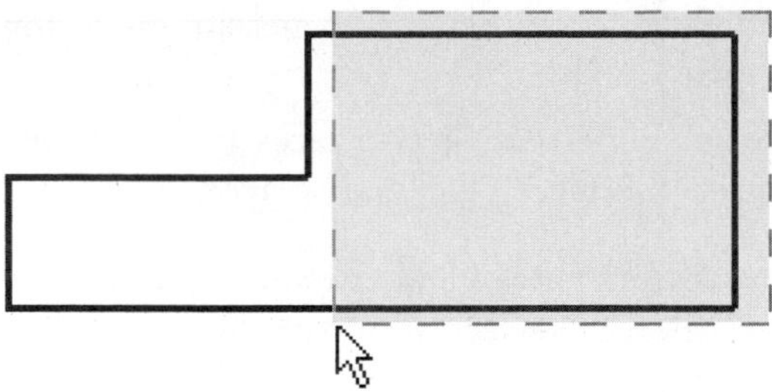

Figure 2.21

DELETING OBJECTS

To delete objects, select them, and then either press the DELETE key, or right-click and select Delete from the menu as shown in the following image.

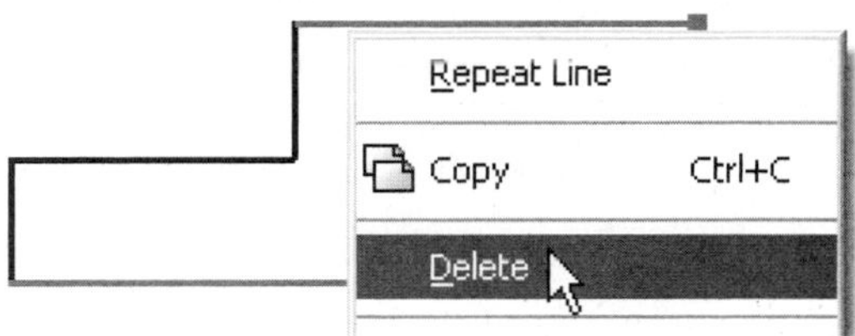

Figure 2.22

MEASURE TOOLS

Measure tools that assist in analyzing sketch, part, and assembly models are available. You can measure distances, angles, and loops, and you can perform area calculations.

The Measure tools are on the Tools menu, as shown in the following image. The following sections discuss these tools in greater detail.

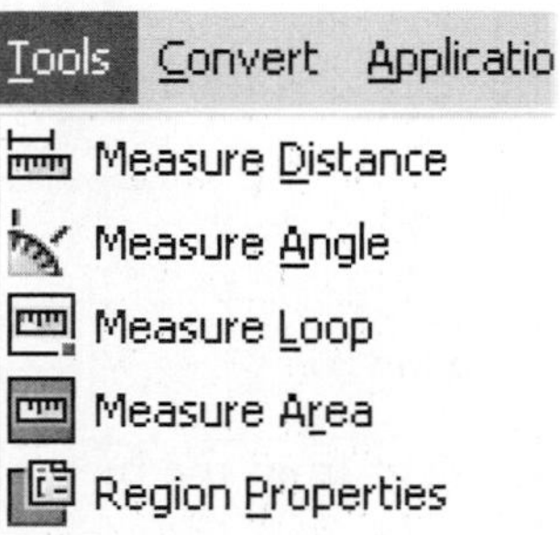

Figure 2.23

Measure Distance Click to measure the length of a line or edge, length of an arc, distance between points, radius and diameter of a circle or the position of elements relative to the active coordinate system. A temporary line designating the measured distance appears, and the Measure Distance dialog box displays the measurement for the selected length, as shown in the following image.

If two disjointed faces of a single part or two faces from different parts are selected, the minimum distance between the faces will be displayed.

Measure Angle Click to measure the angle between two lines or edges. The measurement box displays the angle based on the selection of two lines or edges, or two lines defined by the selection of three points.

Measure Loop Click to measure the length of closed or open loops defined by face boundaries or other geometry. When moving your cursor over a part face, all edges of the face will become highlighted. Clicking on this face will calculate the closed loop or perimeter of the shape.

Measure Area Click to measure the area of enclosed regions or faces, as shown in the following image. Moving your cursor inside the closed outer shape will cause the outer shape and all holes (referred to as "islands") to also become highlighted. Clicking inside this shape will calculate the area of the shape.

When you click the arrow beside the measurement box, a menu similar to the following image will appear.

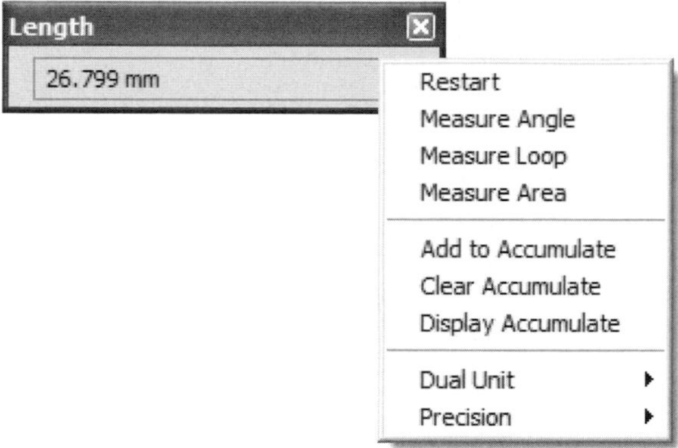

Figure 2.24

Brief explanations of each option follow:

Restart Click to clear the measurement from the measurement box so that you can make another measurement.

Measure Angle Click to change the measurement mode to Measure Angle.

Measure Loop Click to change the measurement mode to Measure Loop.

Measure Area Click to change the measurement mode to Measure Area.

Add to Accumulate Click to add the measurement in the measurement box to accumulate a total measurement.

Clear Accumulate Click to clear all measurements from the accumulated sum, resetting the sum to zero.

Display Accumulate Click to display the sum of all measurements you have added to the accumulated sum.

Dual Unit Click to display the measurement in another unit. The measurement in the second unit will be displayed below the first unit measurement.

Precision Click to change the decimal display between showing all decimal places and showing the number of decimal places specified in the document settings for the active part or assembly.

Region Properties While in a sketch you can determine the properties such as the area, perimeter, and Moment of Inertia properties of a closed 2D sketch. All measurements are taken from the sketch coordinate system. The properties can be displayed in dual units, the default unit of the document, or a unit of your choice. The following image shows the region properties of two circles.

Figure 2.25

EXERCISE 2-1: CREATING A SKETCH WITH LINES

In this exercise, you create a new part file, and then create sketch geometry using basic construction techniques. You also learn how you can use the Autodesk Inventor Help System to assist in the design process.

1. To automatically have the origin point projected for the part, click Tools > Application Options, click the Sketch tab, and then check Autoproject part origin on sketch create.
2. Click OK to close the dialog box.
3. Click the New tool, click the Metric tab and then double-click *Standard (mm).ipt.*
4. Click the Line tool in the Panel Bar.
5. Click near the origin point in the graphics window, move the cursor to the right approximately 100 units, and, when the horizontal constraint symbol displays, click to specify a second point as shown in the following image.

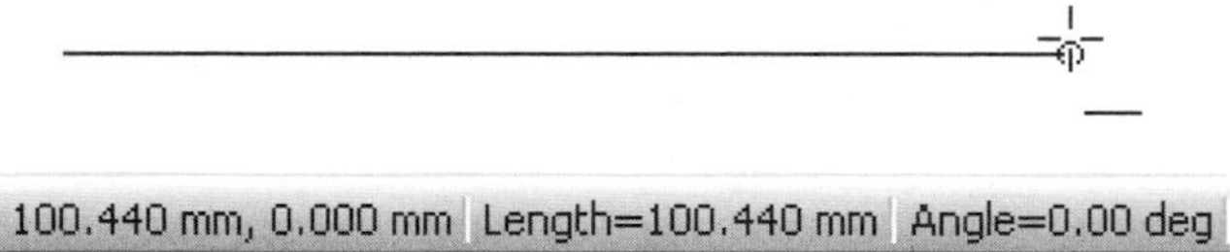

Figure 2.26

Note: Symbols indicate the geometric constraint. In the figure above, the symbol indicates that the line is horizontal. When you create the first entity in a sketch, make it close to final size. The length and angle of the line are displayed in the lower-right corner of the window to assist you.

6. To learn how to create lines, use the Autodesk Inventor Help System. With the Line tool still active, right-click in the graphics window and click How To as shown in the following image.

Figure 2.27

7. Click the three tabs and click a few topics. Close the Autodesk Inventor Help dialog box when done.
8. Move the cursor up until the perpendicular constraint symbol displays beside the first line and then click to create a perpendicular line that is approximately 100 mm as shown in the following image.

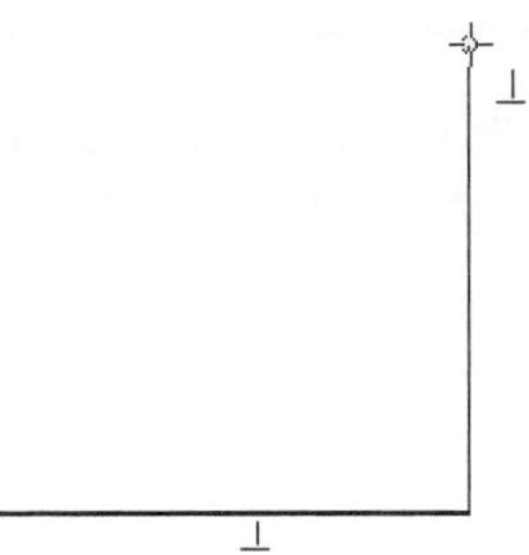

Figure 2.28

9. Move the cursor to the left and create a horizontal line approximately 30 mm that is parallel to the first horizontal line. The parallel constraint symbol is displayed.

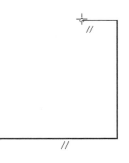

Figure 2.29

10. Move the cursor down and create a vertical line of approximately 60 mm that is perpendicular to the first vertical line.
11. Move the cursor left to create a horizontal line of approximately 20 mm.
12. Move the cursor up until the parallel constraint symbol is displayed, and a dotted alignment line appears as shown in the following image. If the parallel constraint does not appear, move (or scrub) the cursor over the inside vertical line to create a relationship to it.

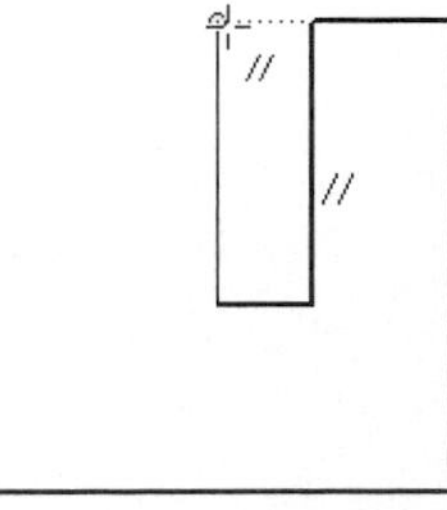

Figure 2.30

13. Click to specify a point.
14. Move the cursor left until the parallel constraint symbol is displayed, and a dotted alignment line appears as shown in the following image. Then click to specify a point.

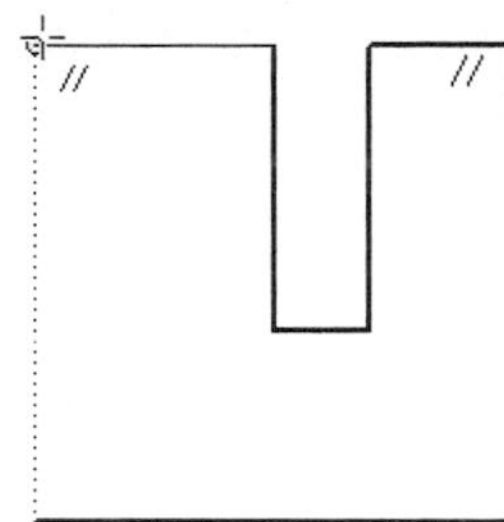

Figure 2.31

15. Move the cursor down until it touches the first point of the sketch. When the green circle, the coincident constraint symbol, is displayed, click to place the line. Your screen should resemble the following image.
16. Right-click in the graphics screen, and click Done to end the Line tool.
17. Right-click in the graphics screen again, and click Finish Sketch.
18. Close the file. Do not save changes. End of exercise.

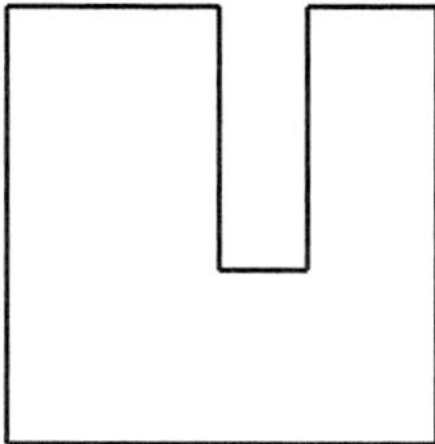

Figure 2.32

EXERCISE 2-2: CREATING A SKETCH WITH TANGENCIES

In this exercise, you create a new part file, and then you create a profile consisting of lines and tangent arcs.

1. If needed, turn on the origin point projected for the part, click Tools > Application Options, click on the Sketch tab, and then check Autoproject part origin on sketch create.
2. Click the New tool, click the Metric tab, and then double-click *Standard (mm).ipt.*
3. Click the Line tool in the Panel Bar.
4. Click the projected origin point in the graphics window, and create a horizontal line approximately 125 mm to the right of the origin point.

Note: If the second point of the line lies off the screen, roll the mouse wheel away from you to zoom out, hold down the mouse wheel, and drag to pan the view.

5. Move the cursor up and create a vertical line of approximately 90 mm.
6. Move the cursor left to create a horizontal line of approximately 40 mm as shown in the following image.

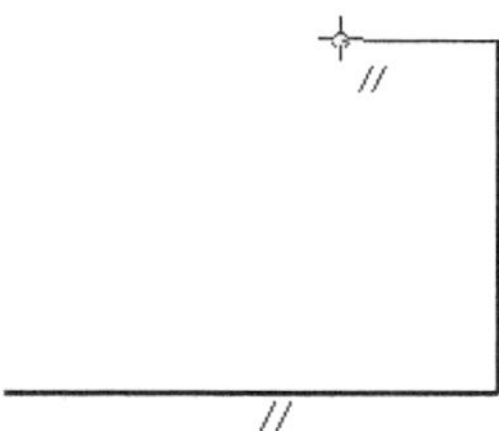

Figure 2.33

7. In this step, you infer points, meaning that no sketch constraint is applied. Move the cursor to the intersection of the midpoints of the right-vertical line and bottom horizontal line. Dashed lines (inferred points) appear as shown in the following image, and then click to create the line.

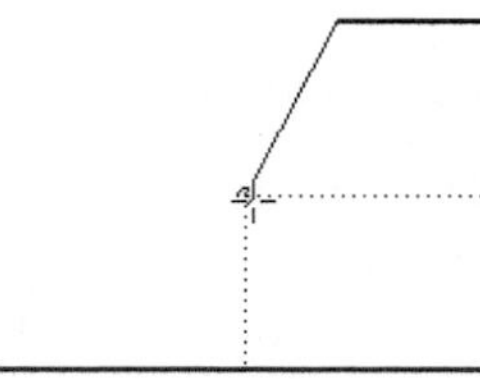

Figure 2.34

8. Move the cursor to the left until the vertical alignment line and the parallel constraints displays as shown in the following image, and then click to place the line.

EXERCISES

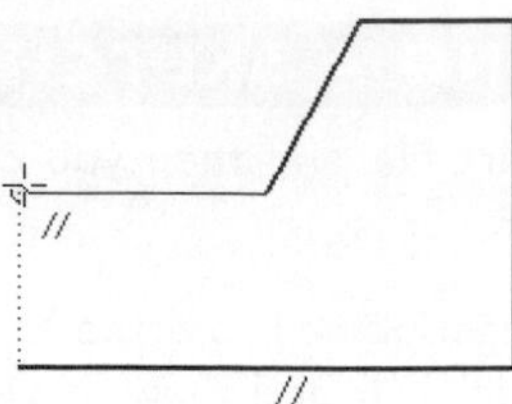

Figure 2.35

9. Right-click in the graphics window, and then select How To.
10. From the Procedure tab click Create an arc and click To create an arc tangent to a curve. To watch the animation, click Show Me how to create a tangent arc. When done, close the Animation and Help dialog boxes.
11. Click the gray dot at the end of the line and hold and drag the endpoint to create a tangent arc. **Do not release the mouse button.**
12. Move the cursor over the endpoint of the first line segment until a coincident constraint (green circle) and the two tangency constraints at start and end points of the arc are displayed as shown in the following image.

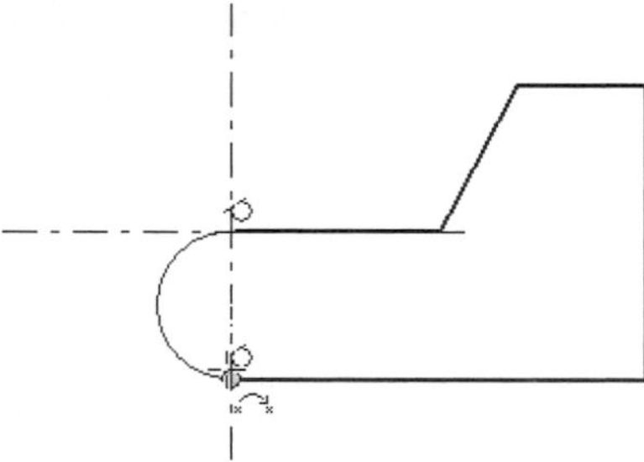

Figure 2.36

13. Release the mouse button to place the arc.
14. Right-click in the graphics window, and then click Done.
15. Close the file. Do not save changes. End of exercise.

STEP 2—CONSTRAINING THE SKETCH

After you draw the sketch, you may want to add geometric constraints to it. Geometric constraints apply behavior to a specific object or create a relationship between two objects. An example of using a constraint is applying a vertical constraint to a line so that it will always be vertical. You could apply a parallel constraint between two lines to make them parallel to one another; then, as one of the line's angles changes, so will the other's. You can apply a tangent constraint to a line and an arc or to two arcs.

When you add a constraint, the number of constraints or dimensions that are required to fully constrain the sketch will decrease. On the bottom-right corner of Autodesk Inventor, the number of constraints or dimensions will be displayed similar to what is shown in the following image. A fully constrained sketch is a sketch whose objects cannot move or stretch.

Figure 2.37

To help you see which objects are constrained, Autodesk Inventor will change the color of constrained objects if a point on the sketch has a coincident constraint applied to the origin point or another point of an existing sketch. If you have not sketched a point coincident to the projected origin point or another point of an existing sketch, the color of the sketch will not change. You could apply a fix constraint instead of using the origin point, but it is not recommended. If you have not sketched a point coincident to the origin point or applied a fix constraint to the sketch, objects are free to move in their sketch plane.

Note: Autodesk Inventor does not force you to fully constrain a sketch. However, it is recommended that you fully constrain a sketch, as this will allow you to better predict how a part will react when you change dimensions values.

CONSTRAINT TYPES

Autodesk Inventor has 11 geometric constraints that you can apply to a sketch. The following image shows the constraint types and the symbols that represent them. Descriptions of the constraints follow.

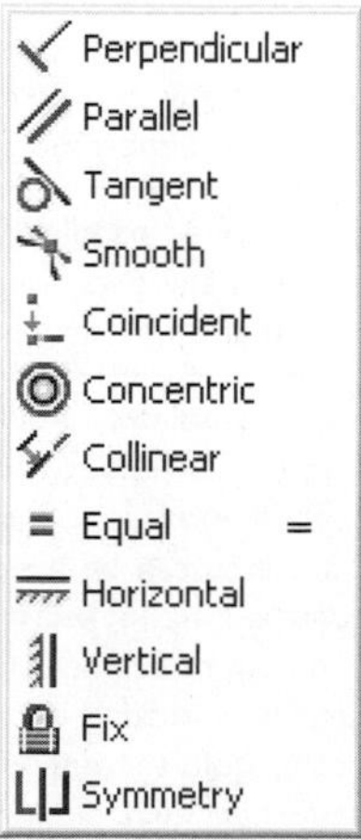

Figure 2.38

Constraint Tools

Button	Tool	Function
	Perpendicular	Lines will be repositioned at 90° angles to one another. The first line sketched will stay in its position, and the second will rotate until the angle between them is 90°.
	Parallel	Lines will be repositioned so that they are parallel to one another. The first line sketched will stay in its position, and the second will move to become parallel to the first.
	Tangent	An arc, circle, or line will become tangent to another arc or circle.
	Smooth (G2)	A spline and another spline, line, or arc that connect at an endpoint with a coincident constraint will represent a smooth G2 (continuous curvature) condition.
	Coincident	A point is constrained to lie on another point or curve (line, arc, etc.).
	Concentric	Arcs and/or circles will share the same center point.
	Colinear	Two selected lines will line up along a single line; if the first line moves, so will the second. The two lines do not have to be touching.
	Equal	If two arcs or circles are selected, they will have the same radius or diameter. If two lines are selected, they will become the same length. If one of the objects changes, so will the other object to which the Equal constraint tool has been applied. If the Equal constraint tool is applied after one of the arcs, circles, or lines has been dimensioned, the second arc, circle, or line will take on the size of the first one. If you select multiple similar objects (lines, arcs, etc.) before selecting this tool, the constraint is applied to all of them. This rule applies to some of the other sketch constraints as well.
	Horizontal	Lines are positioned parallel to the X-axis, or a horizontal constraint can be applied between any two points in the sketch. The selected points will be aligned such that a line drawn between them will be parallel to the X-axis.
	Vertical	Lines are positioned parallel to the Y-axis, or a vertical constraint can be applied between any two points in the sketch. The selected points will be aligned such that a line drawn between them will be parallel to the Y-axis.
	Fix	Applying a fixed point or points will prevent the selected sketch entities from moving. The Fix constraint tool overrides any other constraint. Any endpoint or segment of a line, arc, circle, spline segment, or ellipse can be fixed. Multiple points in a sketch can be fixed. If you select near the endpoint of an object, the endpoint will be locked from moving. If you select near the midpoint of a segment, the entire segment will be locked from moving. If applying constraints and the profile is moving in directions that are undesirable, you can apply fix constraints to hold the endpoints of the objects in place. You can remove a fix constraint as needed. Deleting constraints will be covered later in this chapter.
	Symmetry	Selected points defining the selected geometry are made symmetric about the selected line.

Note: To fully constrain a base sketch, constrain or dimension a point in the sketch to the center point or apply a fix constraint.

ADDING CONSTRAINTS

As stated previously in this chapter, you can apply constraints while you sketch objects. You can also apply additional constraints after the sketch is drawn. However, Autodesk Inventor will not allow you to overconstrain the sketch or add duplicate constraints. If you add a constraint that would conflict with another, you will be warned with the message, "Adding this constraint will overconstrain the sketch." For example, if you try to add a vertical constraint to a line that already has a horizontal constraint, you will be alerted. To apply a constraint, follow these steps:

1. Click a constraint from the Constraint menu in the Sketch Panel Bar or Sketch toolbar, or right-click in the graphics window. Select Create Constraint from the menu, and click the specific constraint from the menu as shown in the previous image before the chart.
2. Click the object or objects to apply the constraints.

SHOWING AND DELETING CONSTRAINTS

To see the constraints that are applied to an object, use the Show Constraints tool from the Sketch Panel Bar, as shown on the left side of the following image. After issuing the Show Constraints tool, select an object, and a constraint icon and yellow squares that represent coincident constraints will appear with the constraints that are applied to the selected object—similar to what is shown in the middle of the following image. You can modify the size of the constraint toolbar by clicking Tools > Application Options, clicking on the Sketch tab, and modifying the size of the Constraint and DOF Symbol Scale. The following image on the right shows the scale increased from 1.0 to 1.25.

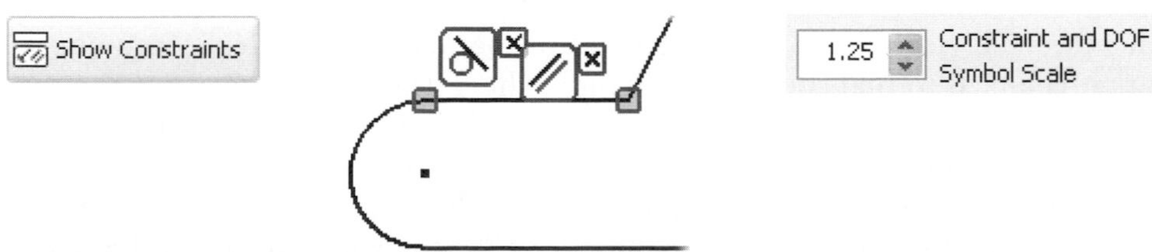

Figure 2.39

To show all the constraints for all of the objects in the sketch, do the following:

- While not in an operation, right-click in the graphics window, and from the menu, click Show All Constraints, or press the F8 key. To hide all the constraints, right-click in the graphics window, and click Hide All Constraints from the menu or press the F9 key as shown in the following image on the left.
- When the constraints are shown, all the constraints in the sketch will appear. You can control which constraints are visible by right-clicking in the graphics window while not in a command and then clicking Constraint Visibility from the menu as shown in the following image on the left. The Constraint Visibility dialog will appear; uncheck the constraint type that you do not want to see. These setting apply only to the current sketch.

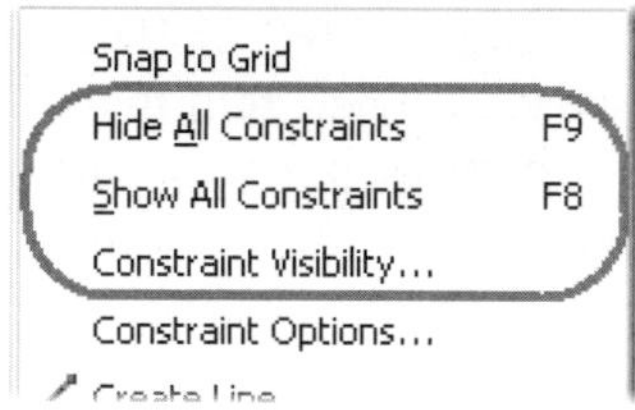

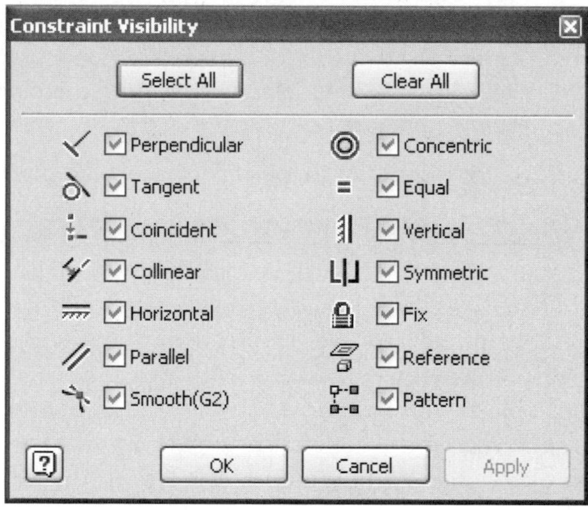

Figure 2.40

As you move the cursor over a constraint icon, the matching sketch constraint and the object that is linked to that constraint will highlight. The coincident constraint will appear as a yellow square at the point that the constraint exists. The perpendicular constraint will appear once at the location the constraint is applied. The following image on the left shows the cursor over the perpendicular constraint on the right vertical edge; the bottom horizontal and right vertical lines are highlighted. The following image on the right shows the cursor over the bottom horizontal line, and the constraints that are related to the line are highlighted.

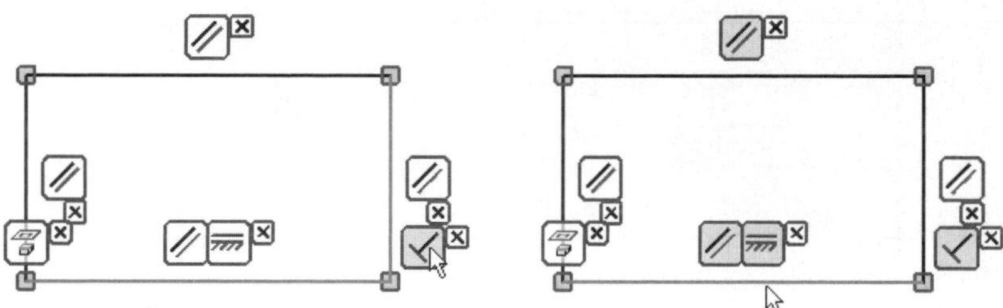

Figure 2.41

To delete the constraint, either click on it and then right-click, or right-click while the cursor is over the constraint and select Delete from the menu. You can also click on it and press the DELETE key. The following image shows the parallel constraint being deleted.

To close constraint icons, click the × on the right side of the constraint symbols toolbar.

Figure 2.42

CONSTRUCTION GEOMETRY

Construction geometry can help you create sketches that would be otherwise difficult to constrain. You can constrain and dimension construction geometry like normal geometry, but the construction geometry will not be recognized as a profile edge in the part when you turn the sketch into a feature. When you sketch, the sketches by default have a normal geometry style, meaning that the sketch geometry is visible in the feature. Construction geometric can reduce the number of constraints and dimensions required to constrain a sketch fully, and it can help to define the sketch. For example, a construction circle inside a hexagon can drive the size of the hexagon. Without construction geometry, the hexagon would require six constraints and dimensions. With construction geometry, it would require only three constraints and dimensions; the circle would have tangent or coincident constraints applied to it and the hexagon. You create construction geometry by changing the line style before or after you sketch geometry in one of the following two ways:

- Before sketching, click the Construction icon on the standard toolbar, as shown in the following image on the left. While creating geometry you can also right-click and select Construction from the menu, as shown in the following image on the right.
- After creating the sketch, select the geometry that you want to change and click the Construction icon on the standard toolbar.

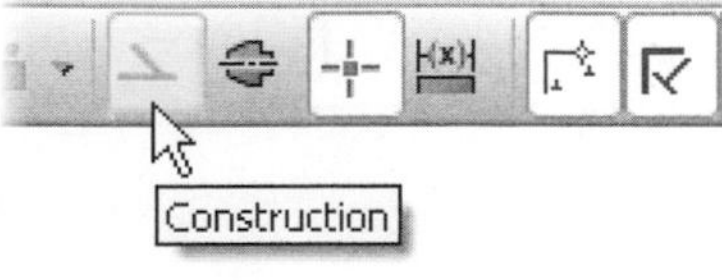

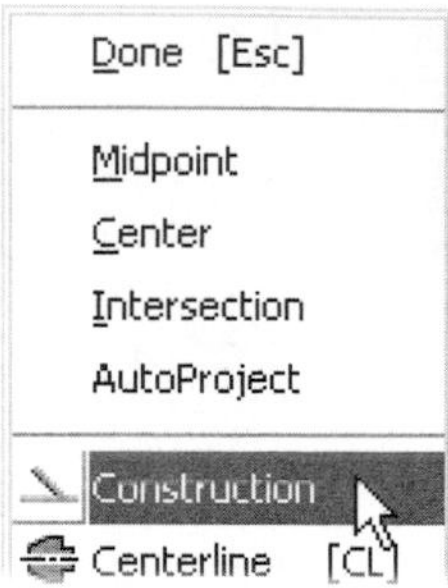

Figure 2.43

After turning the sketch into a feature, the construction geometry will disappear or be consumed. When you edit a feature's sketch that you created with construction geometry, the construction geometry will reappear during editing and disappear when the part is updated. You can add or delete construction geometry to or from a sketch just like any geometry that has a normal style. In the graphics window, construction geometry will be displayed as a dashed line, lighter in color and thinner in width than normal geometry. The following image on the left shows a sketch with a construction line for the angled line. The angled line has a coincident constraint applied to it at every point that touches it. The image on the right shows the sketch after it has been extruded. Note that the construction line was not extruded.

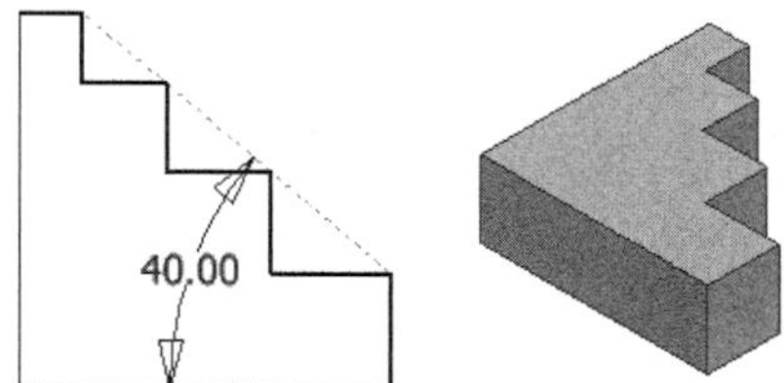

Figure 2.44

SNAPS

Another method used to place geometry with a coincident constraint is snaps: midpoint, center, and intersection. After using a snap, a coincident constraint will be applied, and it will maintain the relationship that you define. For example, if you use a midpoint snap, the sketched point will always be in the middle of the selected object, even if the selected object's length changes.

To use snaps, follow these steps:

1. Select a sketching tool, and right-click in the graphics window.
2. From the menu, select the desired snap.
3. Click on the object to which the sketched object will be constrained. For the intersection snap, select two objects.

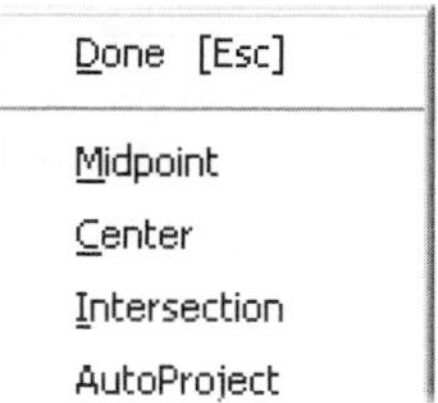

Figure 2.45

SKETCH DEGREES OF FREEDOM

While constraining and dimensioning a sketch there are multiple ways to determine the open degrees of freedom. When you add a constraint or dimension the number of constraints or dimensions that are required to fully constrain, the sketch will decrease.

Status Bar

On the bottom-right corner of Autodesk Inventor, the number of constraints or dimensions will be displayed similar to what is shown in the following image. A fully constrained sketch is a sketch whose objects cannot move or stretch.

Figure 2.46

Degrees of Freedom

To see the areas in the sketch that are NOT constrained, you can display the degrees of freedom. While in a sketch right-click and click Show Degree of Freedom as shown in the following image in the middle. Lines and arcs with arrows will appear as shown in the following image on the right. As constraints and dimensions are added to the sketch, degrees of freedom will disappear. To remove the degree of freedom symbols from the screen, right-click and click Hide All Degrees of Freedom from the menu as shown in the following image on the right.

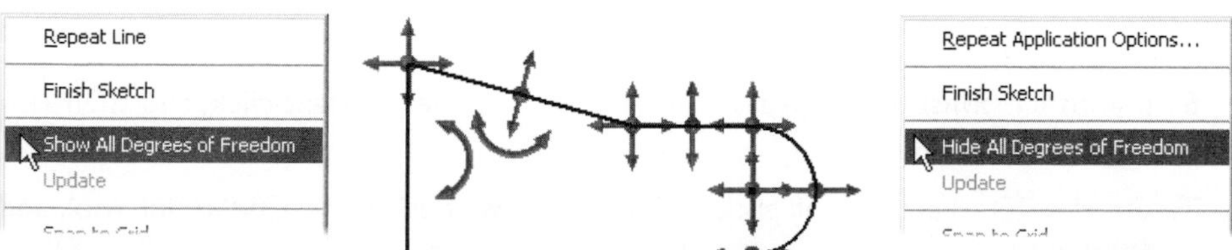

Figure 2.47

Dragging a Sketch

Another method to determine whether or not an object is constrained is to try to drag it to a new location. While not in a command, click a point or an edge, or select multiple objects on the sketch. With the left mouse button depressed, drag it to a new location. If the geometry stretches, it is underconstrained. For example, if you draw a rectangle that has two horizontal and two vertical constraints applied to it and you drag a point on one of the corners, the size of the rectangle will change, but the lines will maintain their horizontal and vertical behavior. If dimensions are set on the object, they too will prevent the object from stretching.

EXERCISES

EXERCISE 2-3: ADDING AND DISPLAYING CONSTRAINTS

In this exercise, you add geometric constraints to sketch geometry to control the shape of the sketch.

1. If not already done, click Tools > Application Options, and then click on the Sketch tab. Check Autoproject part origin on sketch create to automatically have the origin point projected for the part.
2. Click OK to close the dialog box.
3. Click the New tool, click the Metric tab and double-click *Standard (mm).ipt.*
4. Sketch the geometry as shown in the following image, with an approximate size of 35 mm in the X (horizontal) and 20 mm in the Y (vertical). Place the upper-left corner of the sketch at the projected origin point. Right-click in the graphics window, and then click Done. By starting the line at the origin point, that point is constrained to the origin.
5. Right-click in the graphics window, and click Show All Constraints or press the F8 key. Your screen should resemble the following image.

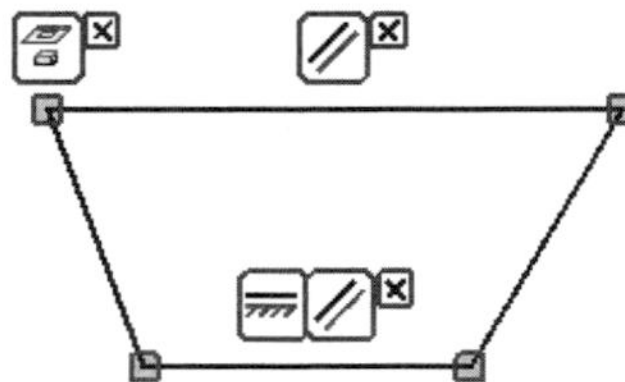

Figure 2.48

6. If another constraint appears, place the cursor over it, right-click, and then click Delete from the menu.
7. On the 2D Sketch panel, click the down arrow beside the Constraint tool, and click Parallel.
8. Click the two angled lines. Depending upon the order in which you sketched the lines, the angles may be opposite of the following image on the left.

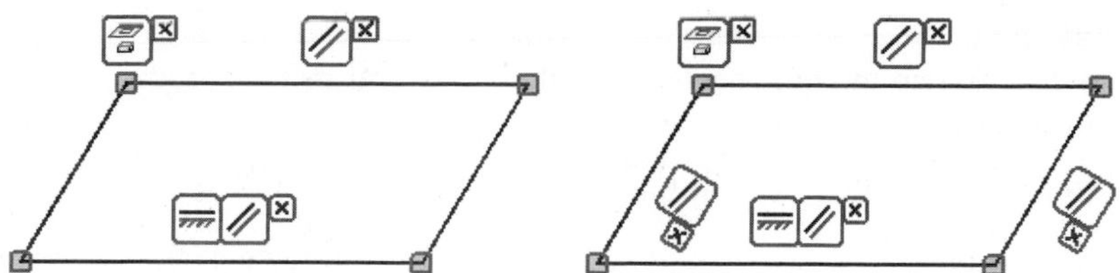

Figure 2.49

9. Press the ESC key to stop adding constraints.
10. Press the F8 key to refresh the visible constraints. Your screen should resemble the following image on the right.
11. Click on the bottom horizontal line in the sketch and drag the line. Notice how the sketch changes its size, but not its general shape. Try to drag the top horizontal line. The line cannot be dragged as it is constrained.
12. Click on an endpoint on the bottom horizontal line, and drag the endpoint. The lines remain parallel due to the parallel constraints.
13. Place the cursor over an icon for the parallel constraint for the angled lines, right-click, and click Delete from the menu as shown in the following image on the left. The parallel constraint that applies to both angled lines is deleted.
14. On the 2D Sketch panel, click the down arrow beside the Constraint tool, and click Perpendicular.
15. Click the bottom horizontal line and the angled line on the right side. Even though it may appear that the rectangle is fully constrained, the left vertical line is still unconstrained and can move.
16. Click the bottom horizontal line and the left vertical line, and your screen.
17. Press the F8 key to refresh the visible constraints. Your screen should resemble the following image on the right.

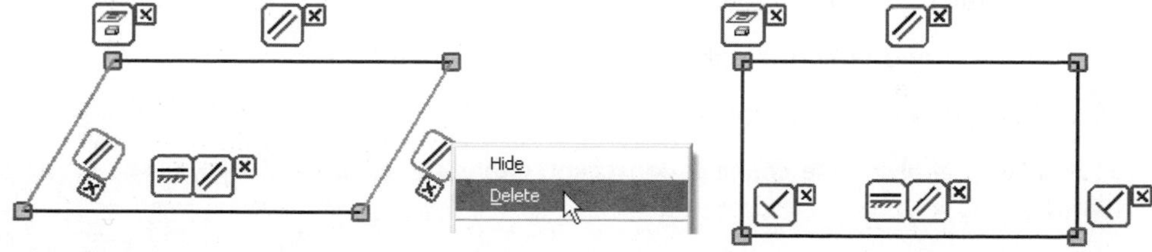

Figure 2.50

18. Right-click in the graphics window, and click Hide All Constraints or press the F9 key.

19. Drag the point at the lower-right corner of the sketch to verify that the rectangle can change size in both the horizontal and vertical directions, but its shape is still maintained.
20. Close the file without saving the changes.
21. Click the New tool, click the Metric tab, and then double-click *Standard (mm).ipt.*
22. Sketch the geometry as shown with an approximate size of **50 mm** in the X and **35 mm** in the Y. Place the lower-left corner of the sketch coincident to the projected origin point. Right-click in the graphics window, and then click Done.

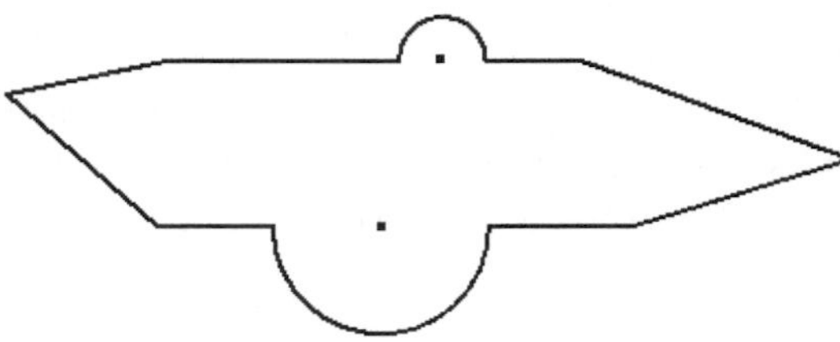

Figure 2.51

23. On the 2D Sketch panel, click the down arrow beside the Constraint tool, and click Equal or press the = key on the keyboard.
24. Click the two left angled lines.
25. Click the two right angled lines.
26. Click an angle line on the left and right side
27. Click the two arcs.
28. On the 2D Sketch panel, click the down arrow beside the Constraint tool, click Colinear or right-click in the graphics window, click Create Constraint, and then click Colinear. Note: In the following steps, if the constraint cannot be placed, you have a constraint that prevents it from being placed. Delete the constraint that is preventing the colinear constraint from being placed.
29. Click the two bottom horizontal lines.
30. Click the two top horizontal lines.
31. To stop applying the colinear constraint, either right-click and click Done or press the ESC key.
32. On the 2D Sketch panel, click the down arrow beside the Constraint tool, and click Vertical.
33. Click the center point of the bottom arc, and then click the center point of the top arc.
34. On the 2D Sketch panel, click the down arrow beside the Constraint tool, and click Horizontal.

35. Click the vertex of the left angled lines, and then click the vertex of the right angled lines.
36. Right-click in the graphics window, and then click Done.
37. Display all of the constraints by pressing the F8 key. Your screen should resemble the following image.

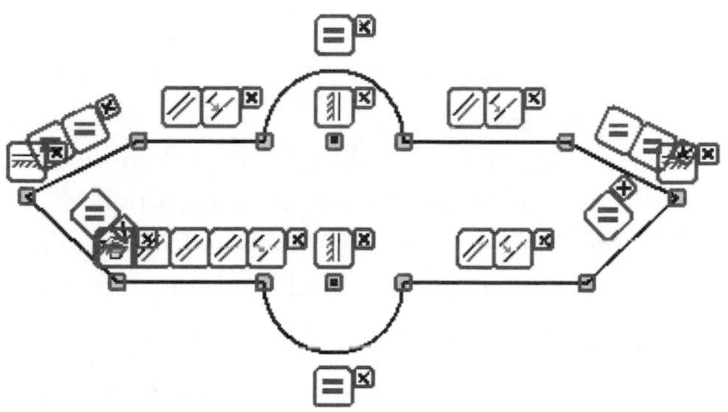

Figure 2.52

38. Click on a line in the sketch and drag the line. Try dragging different lines, and notice how the sketch changes shape.
39. Click on an endpoint in the sketch and drag the endpoint. Try dragging different points, and notice how the sketch changes shape.
40. Close the file. Do not save changes. End of exercise. Note that dimensions would be added to further define the sketch. Dimensions are covered in the next section.

STEP 3—ADDING DIMENSIONS

The last step to constraining a sketch is to add dimensions. The dimensions you place will control the size of the sketch and can also appear in the part drawing views when they are generated. When placing dimensions, try to avoid having extension lines go through the sketch, as this will require more cleanup when drawing views are generated. Click near the side from which you anticipate the dimensions will originate in the drawing views.

All dimensions that you create are parametric, which means that they will change the size of the geometry. All parametric dimensions are created with either the General Dimension or Auto Dimension tools.

GENERAL DIMENSIONING

The General Dimension tool can create linear, angle, radial, or diameter dimensions one at a time. To start the General Dimension tool, follow one of these techniques:

- Click the Create Dimension tool from the Sketch toolbar as shown in the following image.

- Right-click in the graphics area and select Create Dimension from the menu.
- Press the hot key D; you may need to press ENTER if command aliases have been added that use D and another key.

Figure 2.53

When you place a linear dimension, the extension line of the dimension will snap automatically to the nearest endpoint of a selected line; when an arc or circle is selected, it will snap to its center point. To dimension to a quadrant of an arc or circle, see "Dimensioning to a Tangent of an Arc or Circle" later in this chapter.

After you select the General Dimension tool, follow these steps to place a dimension:

1. Click a point or points to locate where the dimension is to start and end.
2. After selecting the point(s) to dimension, a preview image will appear attached to your cursor showing the type of dimension. If the dimension type is not what you want, right-click, and then select the correct style from the menu. After changing the dimension type, the dimension preview will change to reflect the new style.
3. Click a point on the screen to place the dimension.

The next sections cover how to dimension specific objects and how to create specific types of dimensioning with the General Dimension tool.

Dimensioning Lines

There are two techniques for dimensioning a line. Issue the General Dimension tool and do one of the following:

- Click near two endpoints, move the cursor until the dimension is in the correct location and click.
- To dimension the length of a line, click anywhere on the line; the two endpoints will be selected automatically. Move the cursor until the dimension is in the correct location and click.
- To dimension between two parallel lines, click one line and then the next, and then click a point to locate the dimension.
- To create a dimension whose extension lines are perpendicular to the line being dimensioned, click the line and then right-click. Click Aligned from the menu, and then click a point to place the dimension.

Dimensioning Angles

To create an angular dimension, issue the General Dimension tool, click near the midpoint of two lines between which you want the angle dimension, move the cursor until the dimension is in the correct location and click.

Dimensioning Arcs and Circles

To dimension an arc or circle, issue the General Dimension tool, click on the circle's circumference, move the cursor until the dimension is in the correct location and click. To dimension the angle of an arc, click on the arc, click the arc center point, and then locate the dimension. By default, when you dimension an arc, the result is a radius dimension. When you dimension a circle, the default is a diameter dimension. To change the radial dimension to diameter or a diameter to radial, right-click before you place the dimension, and select the other style from the menu.

You can dimension the angle of the arc. Starting the General Dimension tool, click on the arc's circumference, click the center point of the arc, and then place the dimension or click the center point and then click the circumference of the arc.

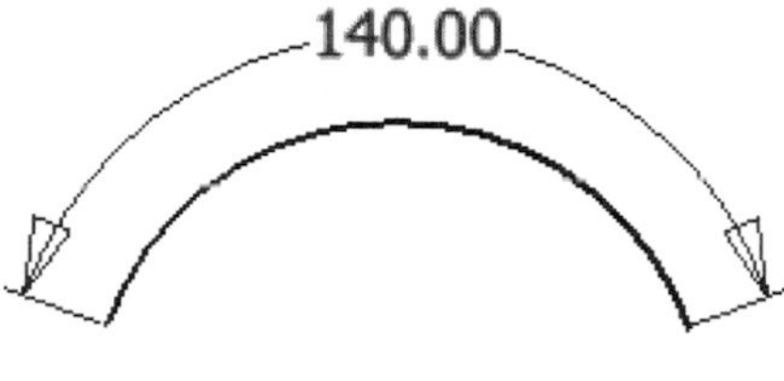

Figure 2.54

Dimensioning to a Tangent of an Arc or Circle

To dimension to a tangent of an arc or circle, follow these steps:

1. Issue the General Dimension tool.
2. Click a line that is parallel to the tangent arc or circle that will be dimensioned.
3. Place the cursor over the tangent arc or circle that should be dimensioned.
4. Move the cursor over the tangent until the constraint symbol changes to reflect a tangent, as shown on the left side of the following image.
5. Click to accept the dimension type, and then move the cursor until the dimension is in the correct location. Click as shown on the right side of the following image.

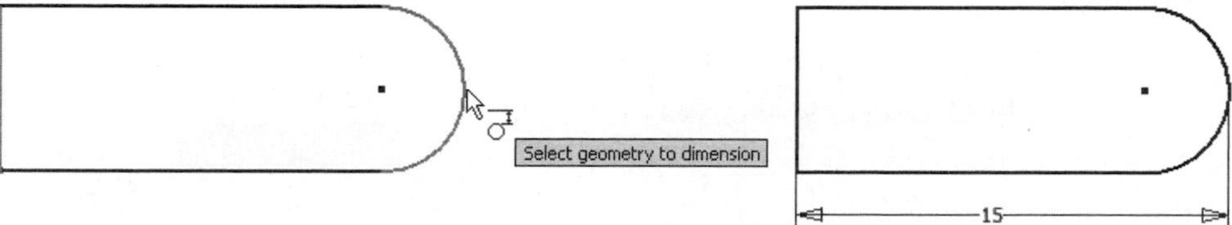

Figure 2.55

To dimension to two tangents, follow these steps:

1. Issue the General Dimension tool.
2. Click an arc or circle that includes one of the tangents to which it will be dimensioned.
3. Place the cursor over the tangent edge of the second arc or circle to which it will be dimensioned.
4. Move the cursor over the tangent until the constraint symbol changes to *tangent*, as shown on the left in the following image.
5. Click and then move the cursor until the dimension is in the correct location, and then click as shown in the following image on the right.

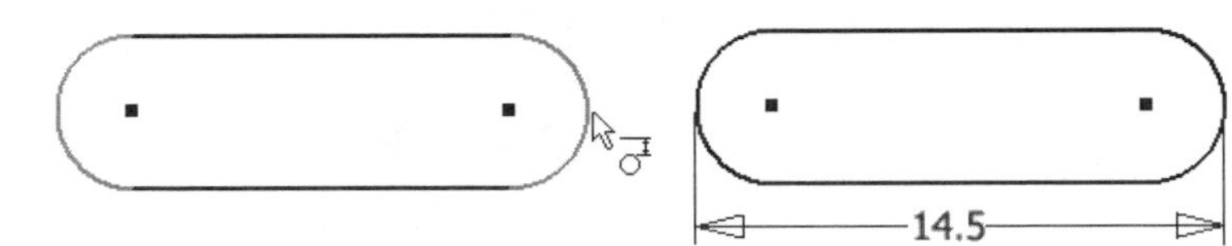

Figure 2.56

ENTERING AND EDITING A DIMENSION VALUE

After placing the dimension, you can change its value. Depending on your setting for editing dimensions when you created them, the Edit Dimension dialog box may or may not appear automatically after you place the dimension. To set the Edit Dimension option, do the following:

1. Click Tools > Application Options.
2. On the Sketch tab of the Options dialog box, select or deselect Edit dimension when created as shown on the left side of the following image.

If the Edit dimension when created option is checked, the Edit Dimension dialog box will appear automatically after you place the dimension. Otherwise, the dimension will be placed with the default value, and you will not be prompted for a different value. You can also set this option by right-clicking in the graphics area while placing a dimension and selecting or deselecting Edit Dimension from the menu, as shown on the right side of the following image.

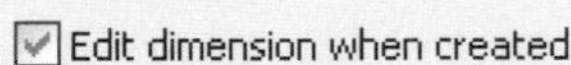

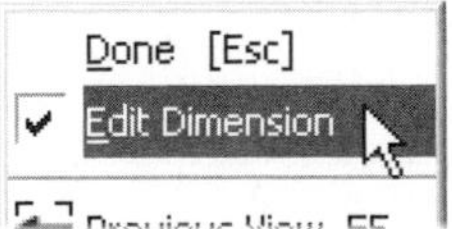

Figure 2.57

To edit a dimension that has already been created, double-click on the value of the dimension, and the Edit Dimension dialog box will appear, as shown in the following image. Enter the new value and unit for the dimension; then either press ENTER or click the checkmark in the Edit Dimension dialog box. If no unit is entered, the units that the file was created with will be used. When inputting values, enter the exact value—do not round up or down. The accuracy of the dimension is from the current dimension style. Autodesk Inventor parts are accurate to six decimal places; for example, 1.0625 is more accurate than 1.06.

Note: When placing dimensions, it is recommended that you place the smallest dimensions first. This will help prevent the geometry from flipping in the opposite direction.

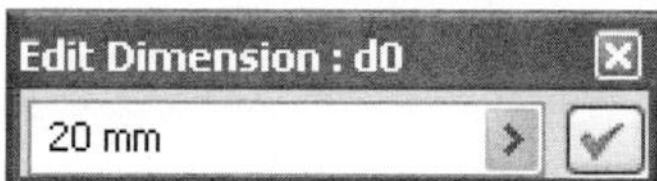

Figure 2.58

Repositioning a Dimension

Once you place a dimension, you can reposition it, but the origin points cannot be moved. Follow these steps to reposition a dimension:

1. Exit the current operation either by pressing ESC twice or right-clicking and then selecting Done from the menu.
2. Move the cursor over the dimension until the move symbol appears as shown in the following image.
3. With the left mouse button depressed, move the dimension or value to a new location and release the button.

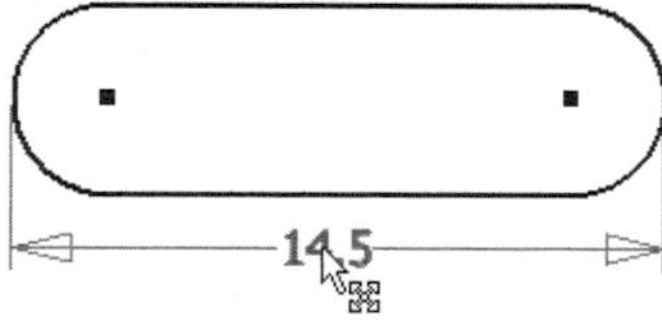

Figure 2.59

Overconstrained Sketches

As explained in the "Adding Constraints" section, Autodesk Inventor will not allow you to overconstrain a sketch or add duplicate constraints. The same is true when adding dimensions. If you add a dimension that will conflict with another constraint or dimension, you will be warned that this dimension will overconstrain the sketch or that it already exists. You will then have the option to either not place the dimension or to place it as a driven dimension.

A driven dimension is a reference dimension. It is not a parametric dimension—it just reflects the size of the points to which it is dimensioned. A driven dimension will appear with parentheses around the dimensions value—for example, (30). When you place a dimension that will overconstrain a sketch, a dialog box will appear similar to the one in the following image. You can either cancel the operation and no dimension will be placed, or accept the warning and a driven dimension will be created.

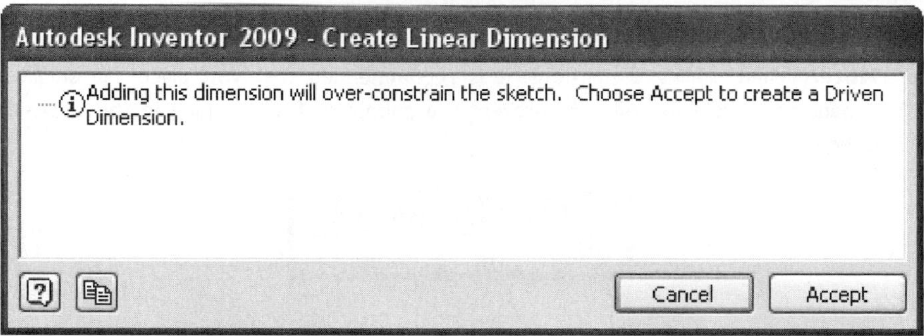

Figure 2.60

Autodesk Inventor gives you an option for handling overconstrained dimensions. To set the overconstrained dimensions option, click Tools > Application Options, and on the Sketch tab of the Options dialog box, change the Overconstrained dimensions option as shown in the following image. You have the following two options:

- Apply driven dimension

  When checked, this option automatically creates a driven dimension without warning you of the condition.

- Warn of overconstrained condition

  When checked, this option causes a dialog box to appear stating that the dimension will overconstrain the sketch. Click Cancel to not place a driven dimension, or click Accept to place a driven dimension.

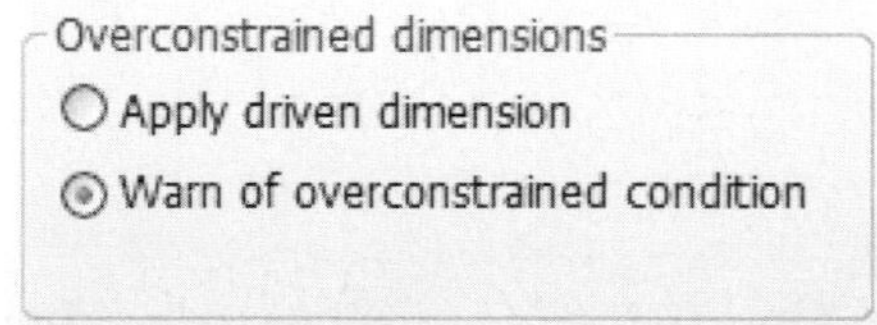

Figure 2.61

Another option for controlling the type of dimension that you create is to use the Driven Dimension tool on the standard toolbar. If you issue the Driven Dimension tool, any dimension you create will be a driven dimension. If you do not issue the tool, you create a regular dimension. The following image shows the Driven Dimension tool on the standard toolbar in its normal condition. The same Driven Dimension tool can be used to change an existing dimension to either a normal or driven dimension by selecting the dimension and clicking the Driven Dimension tool.

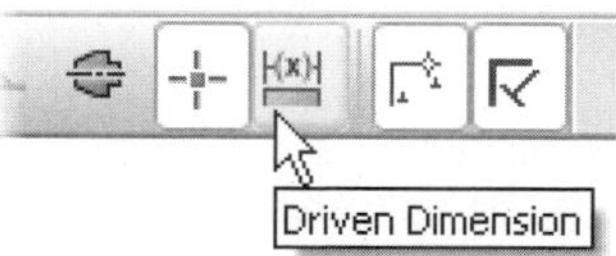

Figure 2.62

AUTO DIMENSION

Adding constraints and dimensions to a sketch or removing dimensions from a sketch, can be a time-consuming task. To automate this process, you can use the Auto Dimension tool to create or remove dimensions or to add constraints to selected geometry automatically. Before using the Auto Dimension tool, you should apply critical constraints and dimensions using the appropriate sketch constraint or Dimension command. The Auto Dimension tool will not override or replace any existing constraints or dimensions. Click the Auto Dimension tool on the Sketch Panel Bar, as shown in the following image on the left. The Auto Dimension dialog box will appear as shown on the right.

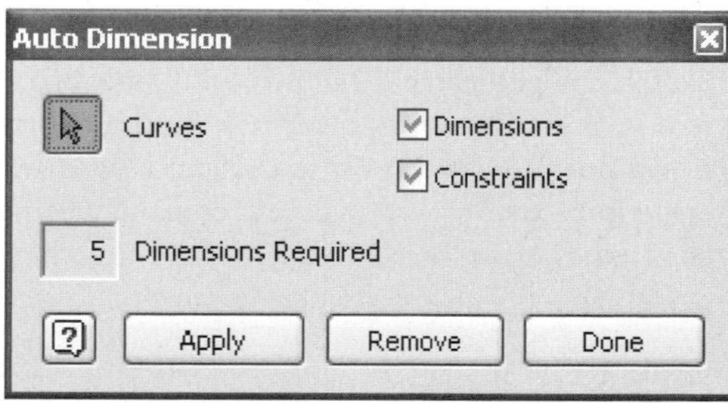

Figure 2.63

To use the Auto Dimension tool, follow these steps:

1. Click the Auto Dimension tool on the Sketch Panel Bar.
2. The number of constraints and dimensions required to fully constrain the sketch appear in the lower-left corner of the dialog box.
3. Determine if you want to create dimensions or constraints or remove the dimensions or constraints that you previously added using the Auto Dimension tool.
4. Click the Dimensions box and/or the Constraints box.
5. Click the Curves option, and then select the objects with which to work in the graphics window.

6. Click the Apply button to create the dimensions and/or apply constraints to the selected curves, or click the Remove button to delete the selected dimensions and/or constraints.
7. If you clicked Apply, you can change the values of the dimensions that you placed by double-clicking on the dimension text and entering in a new value in the Edit Dimension dialog box.
8. After the dimensions are placed, you can change their values by double-clicking on the numbers and entering in a new value.

Note: If you use the Auto Dimension tool on the first sketch, but do not connect the sketch to the projected origin point, two additional dimensions or constraints will be required to fully constrain the sketch. Drag a point on the sketch to the projected origin, or apply a coincident constraint between the origin point and a point on the sketch to fully constrain the sketch.

EXERCISE 2–4: DIMENSIONING A SKETCH

In this exercise, you add dimensional constraints to a sketch. Note: this exercise assumes the "Edit dimension when created" and "Autoproject part origin on sketch create" options are checked in the Options dialog box under the Sketch tab. Experiment with Autodesk Inventor's color schemes to see how the sketch objects changes color to represent if they are constrained.

1. Click the New tool and click the Metric tab, and then double-click *Standard (mm).ipt.*
2. Sketch the geometry as shown with an approximate size of 90 mm in the X and 40 mm in the Y. Place the lower-right corner of the sketch coincident to the projected origin point. Right-click in the graphics window, and then click Done. When sketching, ensure that a perpendicular constraint is not applied between the two angled lines. Hold down the CTRL key while sketching to prevent sketch constraints from being applied. The arc should be tangent to both adjacent lines.

Figure 2.64

3. Add a horizontal constraint between the midpoint of the right vertical line and the center of the arc, as shown in the following image on the left.
4. Add a vertical constraint between the endpoints of the angled lines nearest the center of the sketch, as shown, in the following image on the right.

EXERCISES

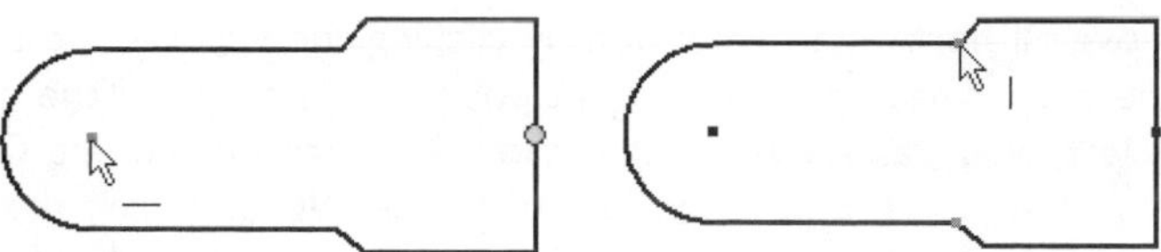

Figure 2.65

5. Add an equal constraint between the two angled lines by pressing the = key and then select the two angled lines.
6. Click the General Dimension tool in the 2D Sketch Panel Bar.
7. Click the top horizontal line, drag the dimension up, click a point to locate it, enter **30** (if the Edit Dimension dialog box did not appear, double-click on the dimension), and click the checkmark as shown in the following image.

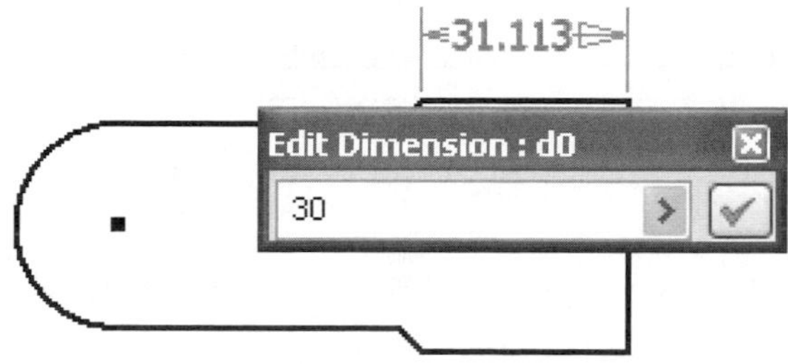

Figure 2.66

8. Add an angle dimension by clicking the bottom horizontal line and the lower angled line. Drag the dimension to the left, click a point above the top horizontal line to locate it, enter **30**, and click the checkmark. If you are unable to add the angle dimension, you may have a perpendicular constraint that needs to be deleted.
9. Add a radial dimension by clicking the arc. Drag the dimension to the left, click a point to locate it, enter **15**, and click the checkmark.
10. Add a vertical dimension by clicking the vertical line. Drag the dimension to the right, click a point to locate it, enter **40**, and click the checkmark. When complete, your sketch should resemble the following image on the left. Notice on the bottom right of the status bar that the sketch requires 1 dimension as shown in the following image on the right.

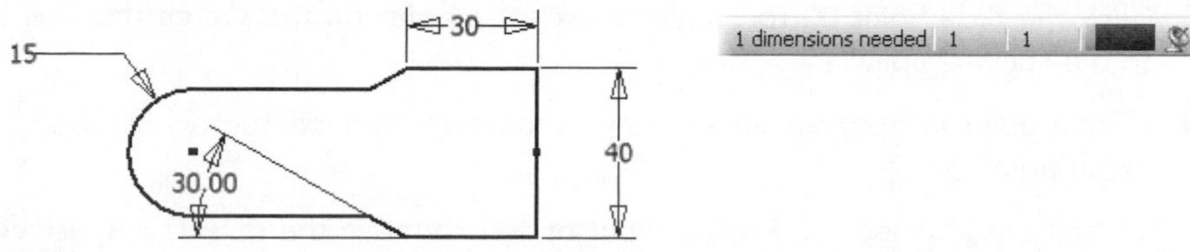

Figure 2.67

11. Add an overall horizontal dimension by clicking the vertical line (not an endpoint). Move the cursor near the left tangent point of the arc until the glyph of dimension to a circle appears, as shown in the following image on the left. Click, drag the dimension down, click a point to locate it, enter **90**, and click the checkmark. When complete, your sketch should resemble the following image on the right.

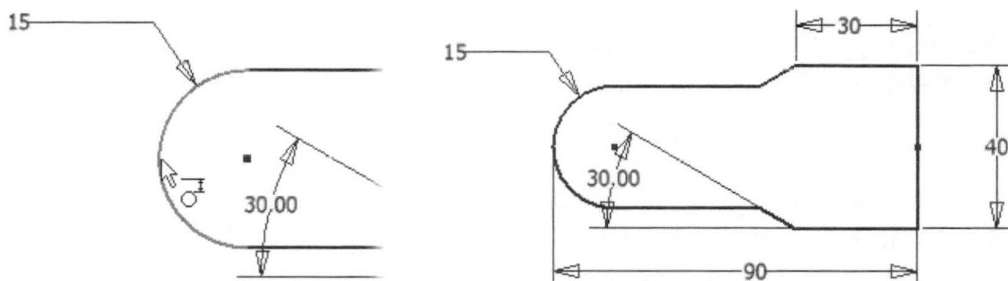

Figure 2.68

12. Edit the value of the dimensions, and examine how the sketch changes. The arc should always be in the middle of the vertical line.
13. Delete the horizontal constraint between the center of the arc and the midpoint of the vertical line.
14. Delete the vertical constraint between the left endpoints of the angled lines.
15. Practice adding other types of constraints and dimensions.
16. Close the file. Do not save changes. End of exercise.

EXERCISES

MOVE AND SCALE TOOLS

After importing data into Autodesk Inventor, you may need to move the object to another location. For example, you may wish to reposition it so that a point on the sketch is coincident to the origin point. You may need to scale the geometry if it was imported with incorrect units.

MOVE

To use the move tool, follow these steps.

1. Click the Move tool from the Sketch Panel.
2. In the graphics window, select the objects to move.
3. Click the Base Point button, and then select a point on the sketch that will be used as the starting point.
4. Click a point in the graphics window to move the objects to, such as the origin point.
   a. You can also use the Precise Input option to move the objects a specified distance.

b. Click the Copy option to make a copy of the selected objects.

c. If the move would not be possible because a dimension or constraint is not allowing it, you will be prompted to relax these dimensions or constraints. To set how dimensions or constraints will be handled, click the >> button, and select an option to relax the dimensions and/or constraints.

d. Check the option Optimize for Single Selection if you want to use only one selection and then automatically be prompted to select the base point.

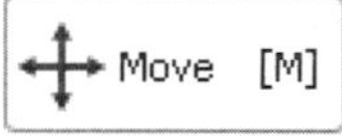

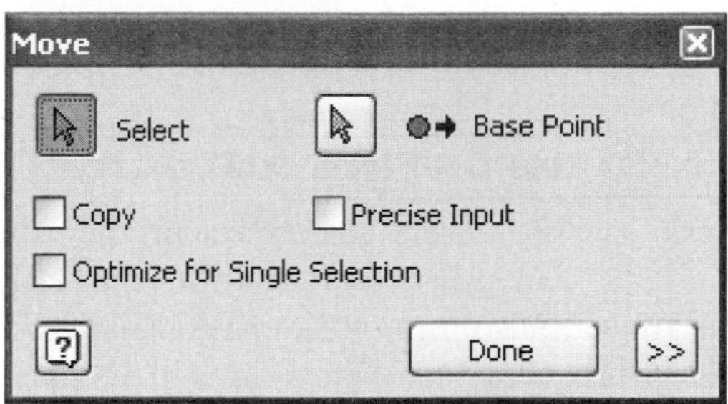

Figure 2.69

SCALE

To use the scale tool, follow these steps.

1. Click the Scale tool from the Sketch Panel.
2. In the graphics window, select the objects to scale.
3. Click the Base Point button, and then select a point on the sketch from which the scale will be based.

   a. Type a Scale Factor or move the cursor and click a second point.

   b. You can also use the Precise Input option to scale the objects a specified distance by using the Precise Input toolbar; the distance between the two points determines the scale factor.

   c. If the scale would not be possible because a dimension or constraint is not allowing it, you will be prompted to relax these dimensions or constraints. To set how dimensions or constraints will be handled, click the >> button and select an option to relax the dimensions and/or constraints.

   d. Check the option Optimize for Single Selection if you want to use only one selection and then automatically be prompted to select the base point.

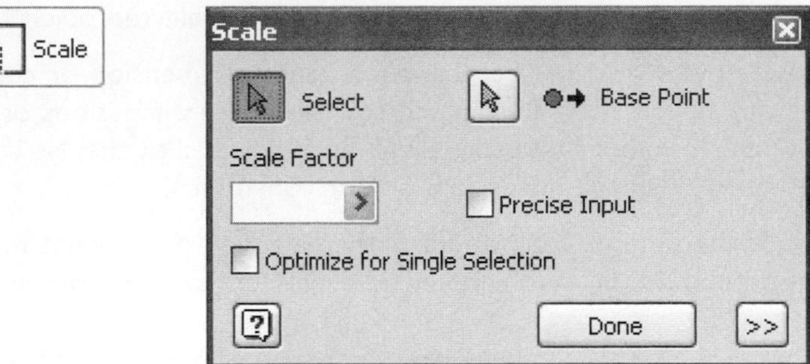

Figure 2.70

OPENING AND IMPORTING AUTOCAD FILES

Many Autodesk Inventor users store data in the AutoCAD DWG format. Instead of redrawing this data, you can import it into Autodesk Inventor drawings or into a part feature sketch. You can also import AutoCAD files containing 3D solids using the wizard; the AutoCAD solids will be opened in a new part or assembly file depending upon whether or not there are multiple AutoCAD solids. When importing a 2D DWG file into Autodesk Inventor, you can either copy the contents from the DWG file to the clipboard via Autodesk Inventor or AutoCAD and paste into Autodesk Inventor, or use an import wizard that guides you through the process. The following sections will introduce you to the options available when importing 2D AutoCAD data into Autodesk Inventor.

COPY AND PASTE A 2D DWG FILE FROM AUTODESK INVENTOR OR AUTOCAD

The first method for opening a DWG or DXF data is to open a drawing file from Autodesk Inventor. Copy it to the clipboard and paste it into Autodesk Inventor by following these steps. This same procedure can be done by opening the file in AutoCAD and copying it to the clipboard.

1. From in Autodesk Inventor, click File > Open on the main menu, or click Open on the standard toolbar.
2. In the Open dialog box, navigate to and select the DWG file, as shown in the following image.

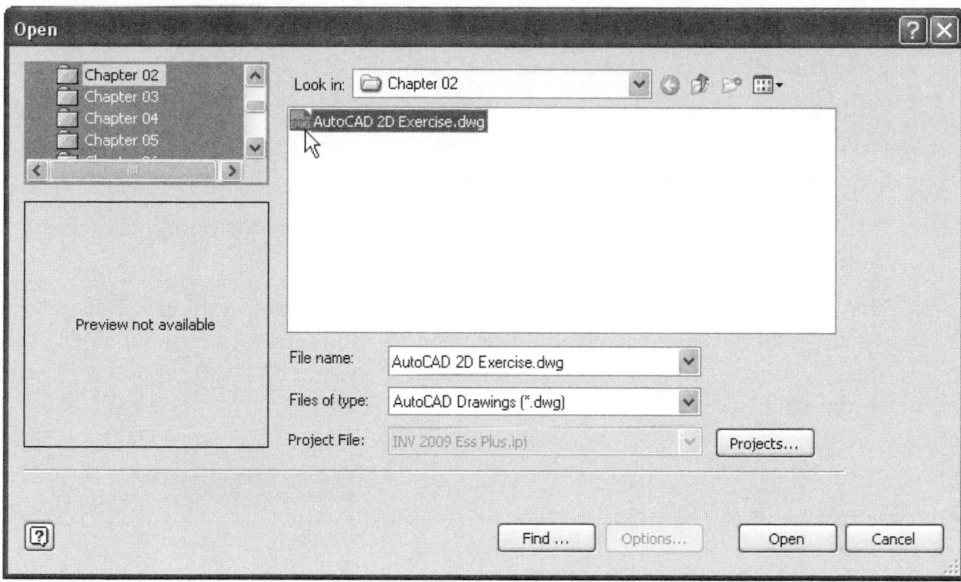

Figure 2.71

3. Click Open from the dialog box, and the DWG file will open inside of Autodesk Inventor, as shown in the following image.

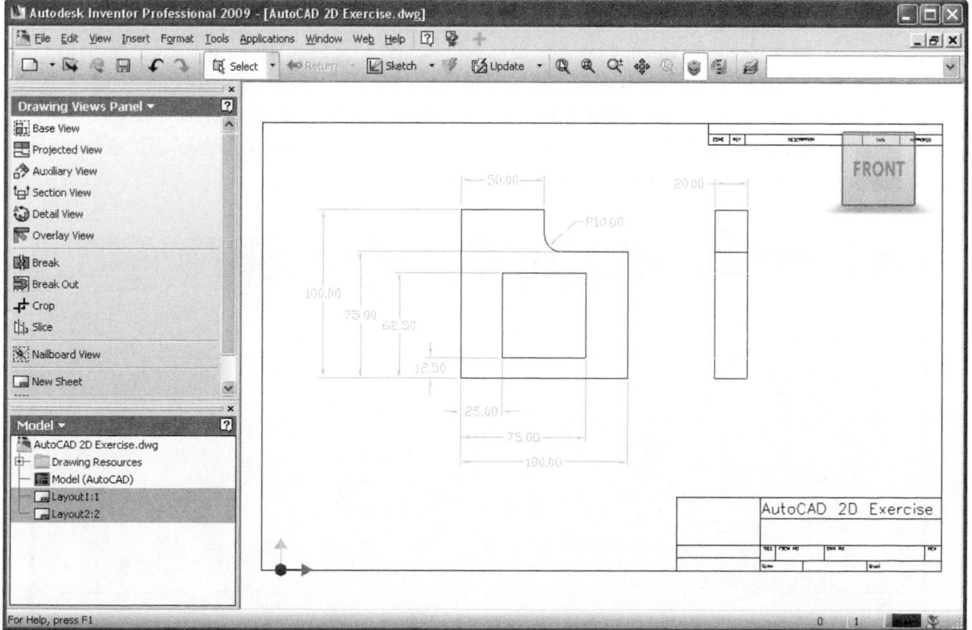

Figure 2.72

4. Select the geometry and dimensions you want to copy into Autodesk Inventor.
5. Copy the data to the clipboard by using the shortcut key CTRL-C, or right-click and click Copy from the menu.
6. Make the Autodesk Inventor part file active, or start a new Autodesk Inventor part file in which the DWG data will be used. A sketch must be active.
7. Paste the data by using the shortcut key CTRL-V, or right-click and click Paste from the menu.
8. The bounding box appears as shown in the following image on the left. Depending upon the size of the copied objects, you may need to zoom out to see the bounding box. Then right-click, and click Paste Options from the menu, as shown in the following image on the right.

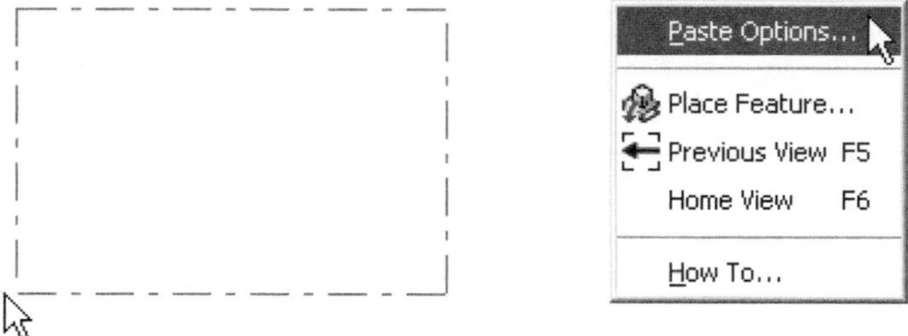

Figure 2.73

9. The Paste Options dialog box appears as shown in the following image on the right. Select the unit in which the AutoCAD geometry was created, and check Constrain End Points if you want a coincident constraint to be applied to geometry where two endpoints touch. Select the Apply geometric constraints option to have 2D sketch constraints automatically applied to the sketch.

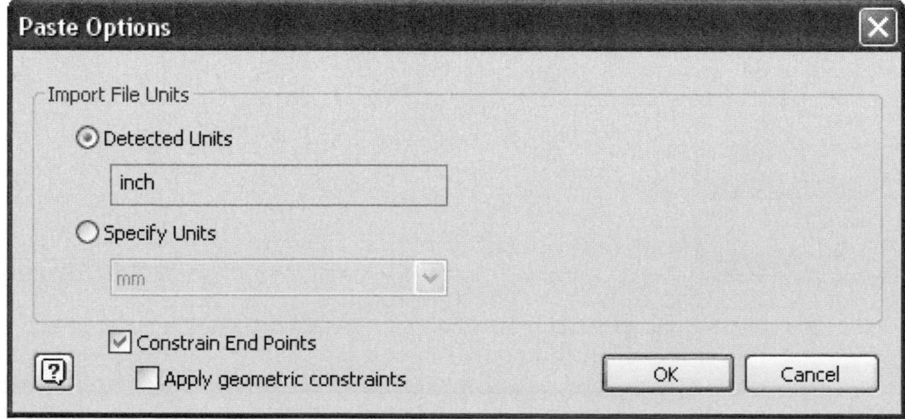

Figure 2.74

10. Click the OK button.
11. Click a point in the graphics window where you want the copied geometry to be placed. The DWG data is now in the active sketch.

INSERTING 2D AUTOCAD DATA INTO A SKETCH

Another method that utilizes existing AutoCAD 2D data inserts the data into the active sketch in a part or drawing. To insert AutoCAD data into the active sketch, follow these steps:

1. Make a sketch active in a part file or a draft view active in a drawing file.
2. Click the Insert AutoCAD file icon on the 2D Sketch Panel Bar, as shown in the following image. Alternatively, in a drawing, click the Insert AutoCAD file tool on the Drawing Sketch Panel Bar.

Figure 2.75

3. Browse to and select the desired DWG file. The Open dialog box will appear.

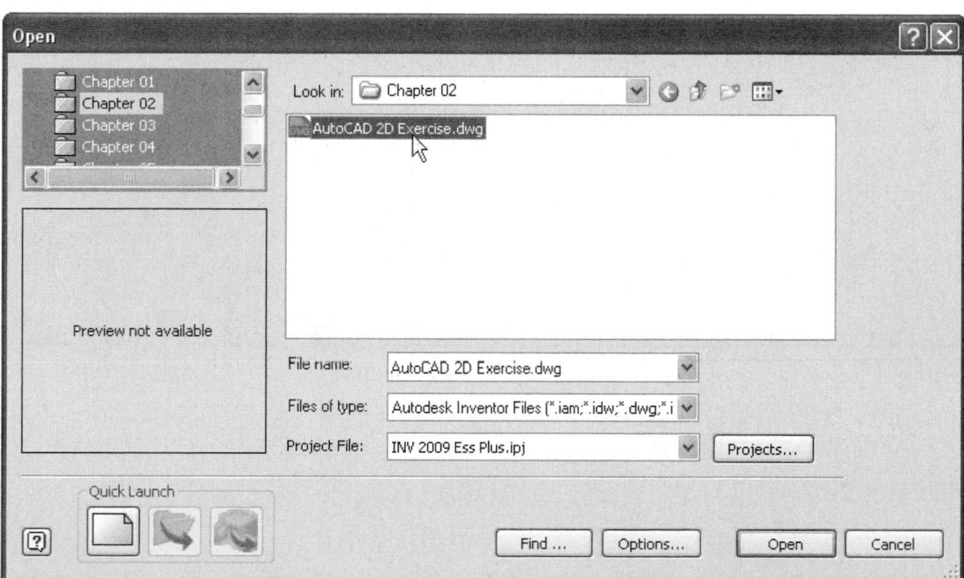

Figure 2.76

4. Click the Open button. To select specific objects, uncheck the All option, and then select the desired data in the preview window.

5. You can change the background color of the preview image by clicking the black or white icon at the top of the dialog box. Import objects from Model Space or from a layout within the DWG file by clicking the different tabs at the bottom of the screen. The names of the tabs are identical to the tab names in the AutoCAD file.

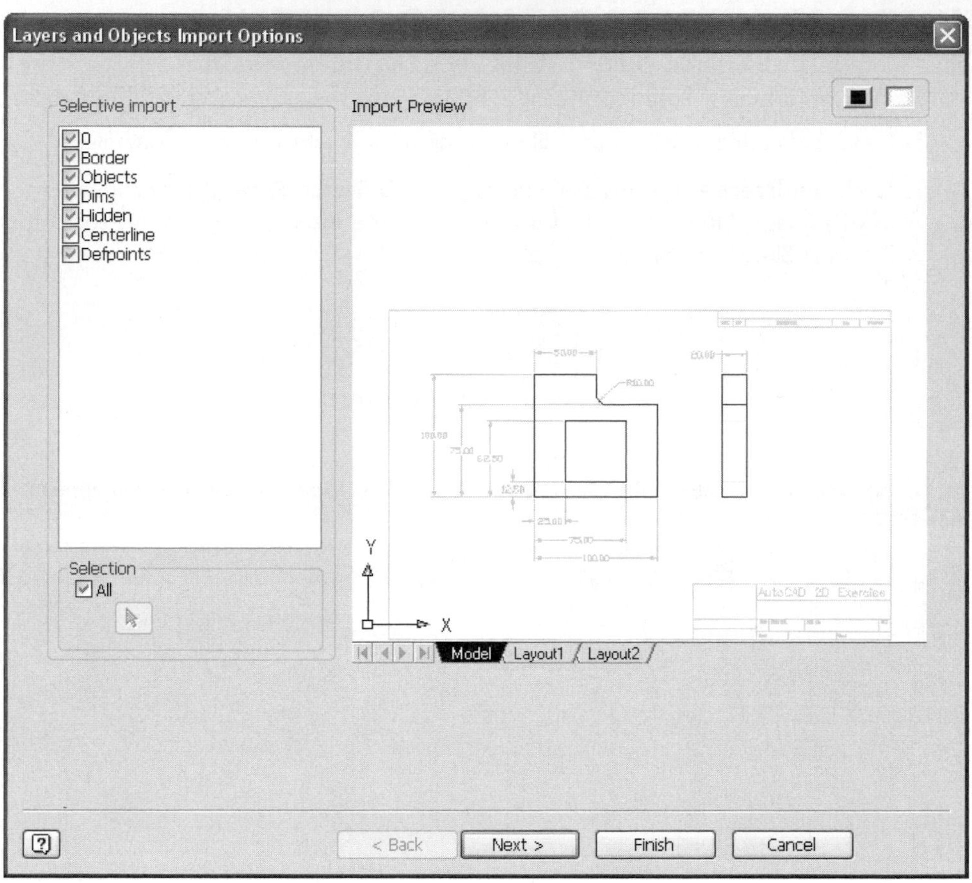

Figure 2.77

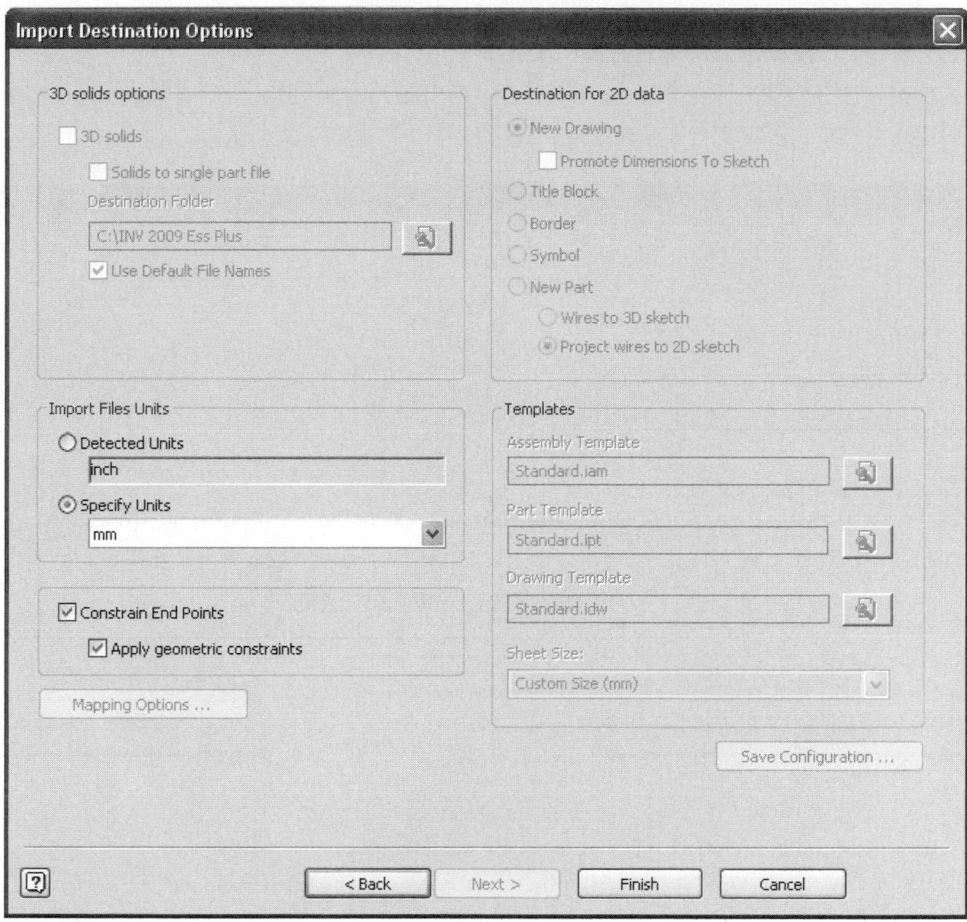

Figure 2.78

6. Specify the units in which the data was created.
7. Check or uncheck the options to Constrain End Points and Apply geometric constraints to the sketch.
8. To complete the operation, click Finish.

IMPORT OTHER FILE TYPES

Autodesk Inventor can also import parts and assemblies exported from other CAD systems. Autodesk Inventor models created from these formats are base solids or surface models, and no feature histories or assembly constraints are generated when you import a file in any of these formats. You can add features to imported parts, edit the base solids using Autodesk Inventor's solids editing tools, and add assembly constraints to the imported components. To open file types such as SAT, STEP, PRO/E, Parasolids,

SolidWorks, Unigraphics, DXF, IDF Board File, and IGES, click File > Open on the main menu or click Open on the standard toolbar.

In the Open dialog box, click the desired file format in the Files of type list. See the help system for more information about the different file types.

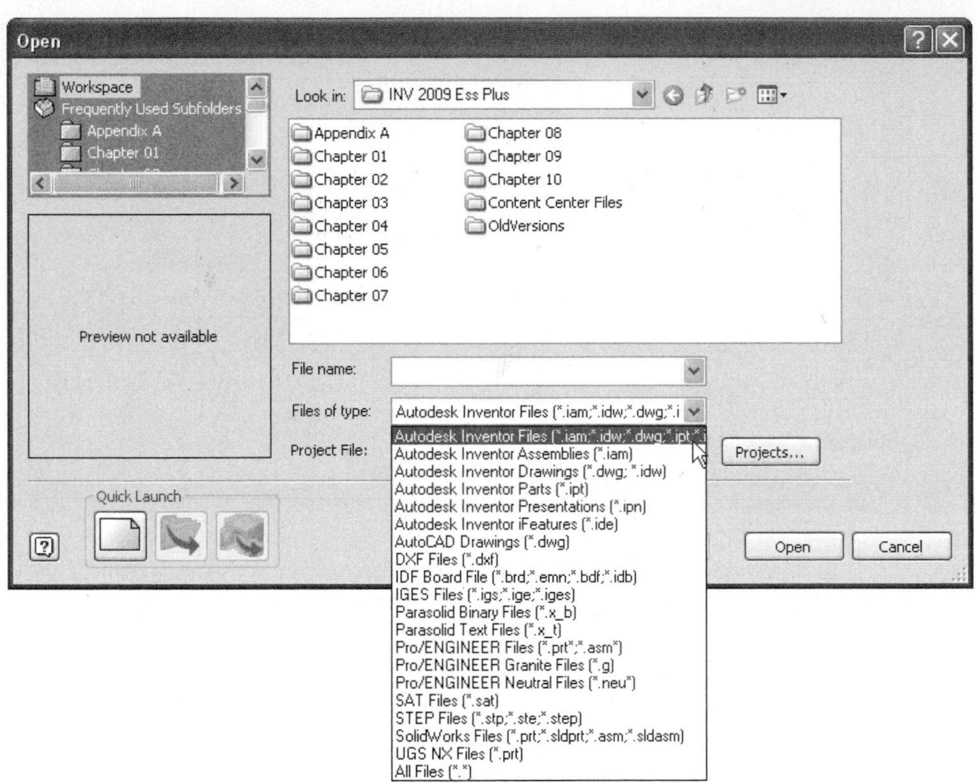

Figure 2.79

EXERCISE 2-5: INSERTING AN AUTOCAD FILE

In this exercise, you open an AutoCAD drawing, copy data to the clipboard and then paste it into a sketch.

1. From in Autodesk Inventor, click File > Open on the main menu or click Open on the standard toolbar.
2. Open C:\INV 2009 Ess Plus\Chapter 02\ AutoCAD 2D Exercise.dwg.

Tip: Click the Chapter 02 subfolder from the Frequently Used Subfolder area and then click the file in the file area.

3. The DWG file will open in a new window; window select the geometry and dimensions as shown in the following image on the left.

4. Right-click, and click Copy from the menu as shown in the following image on the right.

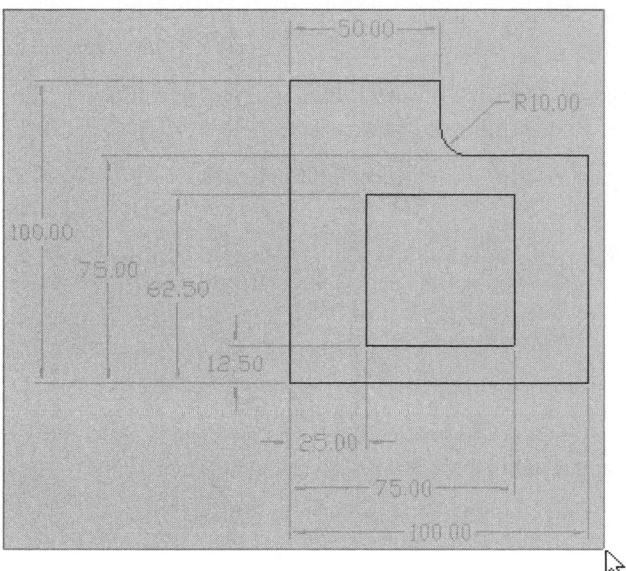

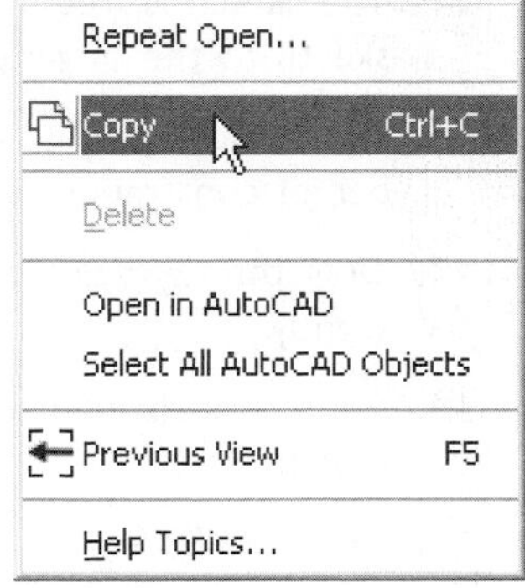

Figure 2.80

5. Click the New tool, click the Metric tab, and then double-click *Standard (mm).ipt*.
6. Move the cursor into the graphics window, right-click, and click Paste from the menu.
7. Right-click, and click Paste Options as shown in the following image on the left.
8. In the Paste Options dialog box, click Specify Units and verify that mm is selected. If needed, check both Constrain End Points and Apply geometric constraints, and then click OK.

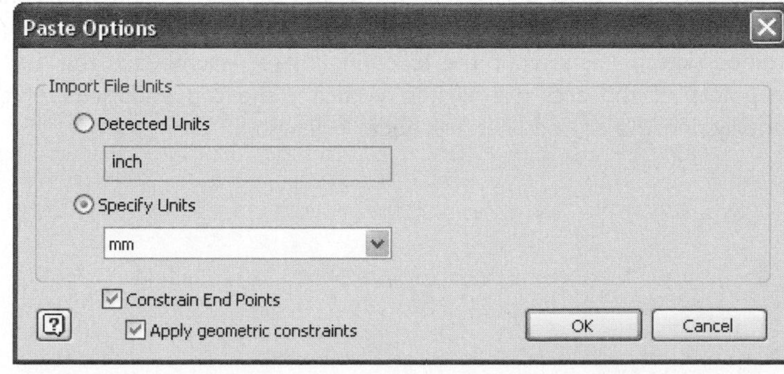

Figure 2.81

EXERCISES

9. To paste the data, click a point in the graphics window
10. If needed, zoom out to see the data.
11. The sketch is free to move. To constrain the sketch drag the lower-left corner of the sketch to the origin point
12. On the lower-right corner of Autodesk Inventor, the text should state that the number of constraints needed to constrain the sketch is 1.
13. Drag the right-side endpoint on the lower horizontal line; the sketch will be rotated slightly.
14. Apply a horizontal constraint to the lower line, and this will fully constrain the sketch.
15. Click the F8 key to see all the constraints.
16. Click the F9 key to hide all the constraints.
17. The dimensions on the sketch are now parametric and can be edited as any other parametric dimension. Practice editing the values of the dimensions.
18. Close the file. Do not save changes. End of exercise.

Project Exercise: Chapter 2

The self-paced, step-by-step project exercises found in Appendix A provide opportunities for you to work through real-world modeling, assembly, and documentation tasks. The geometry used in the project exercises will flow from chapter to chapter, utilizing the functionality that you learned in that chapter.

Applying Your Skills

SKILL EXERCISE 2–1

In this exercise, you create a sketch and then add geometric and dimensional constraints to control the size and shape of the sketch. Start a new part file based on the *Standard (mm).ipt,* and create the fully constrained sketch as shown in the following image. Assume that the horizontal lines are colinear, the center points of the arcs are aligned vertically, the top endpoints of the bottom arc are aligned horizontally, and the angled lines are equal in length.

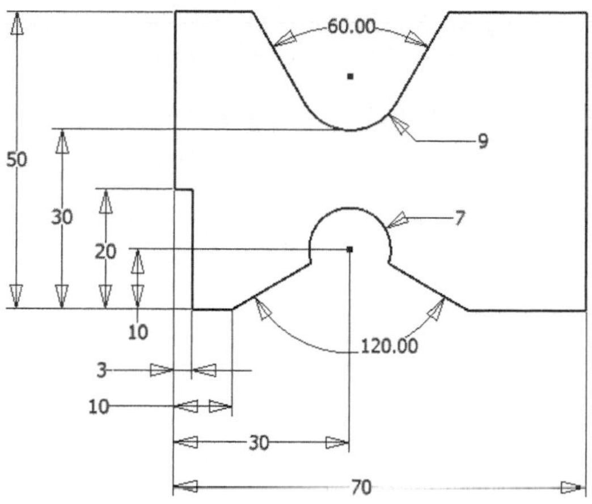

Figure 2.82

SKILL EXERCISE 2–2

In this exercise, you create a sketch with linear and arc shapes, and then add geometric and dimensional constraints to fully constrain the sketch. Start a new part file based on the *Standard (mm).ipt*, and create the fully constrained sketch as shown in the following image. First create the two outer circles and align them horizontally. Create the two lines, trim the two circles, and place a vertical constraint between the line endpoints on both ends.

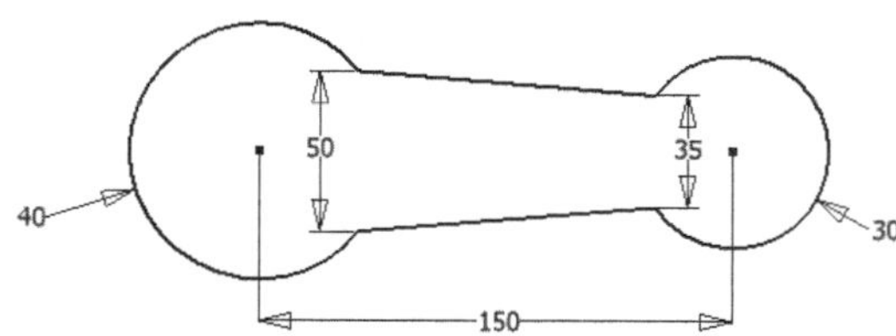

Figure 2.83

CHECKING YOUR SKILLS

Use these questions to test your knowledge of the material in this chapter.

1. **True _ False _** When you sketch, constraints are not applied to the sketch by default.
2. **True _ False _** When you sketch and a point is inferred, a constraint is applied to represent that relationship.
3. **True _ False _** A sketch does not need to be fully constrained.
4. **True _ False _** When working on an mm part, you cannot use English units.
5. **True _ False _** After a sketch is constrained fully, you cannot change a dimension's value.
6. **True _ False _** A driven dimension is another name for a parametric dimension.

EXERCISES

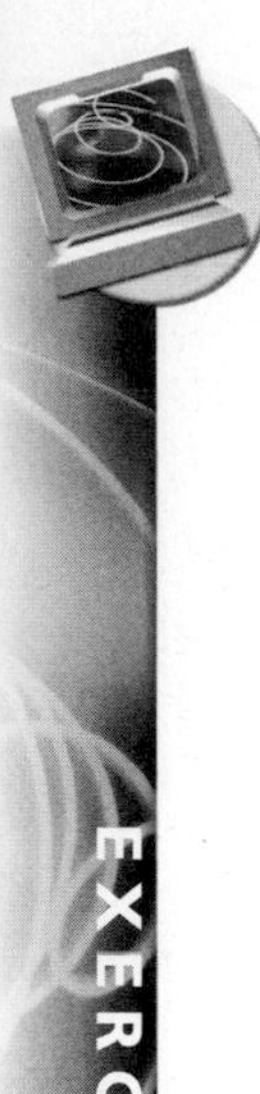

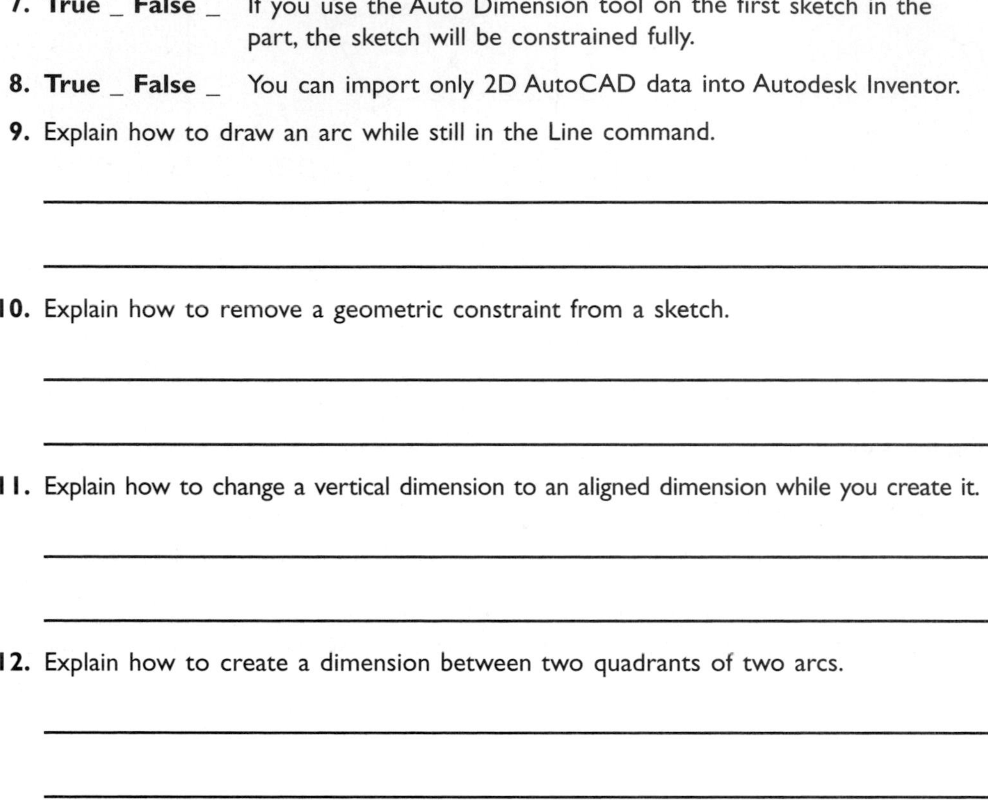

7. **True** _ **False** _ If you use the Auto Dimension tool on the first sketch in the part, the sketch will be constrained fully.

8. **True** _ **False** _ You can import only 2D AutoCAD data into Autodesk Inventor.

9. Explain how to draw an arc while still in the Line command.

____

____

10. Explain how to remove a geometric constraint from a sketch.

____

____

11. Explain how to change a vertical dimension to an aligned dimension while you create it.

____

____

12. Explain how to create a dimension between two quadrants of two arcs.

____

____

CHAPTER 3

Creating and Editing Sketched Features

INTRODUCTION

After you have drawn, constrained, and dimensioned a sketch, your next step is to turn the sketch into a 3D part. This chapter takes you through the process to create and edit sketched features.

OBJECTIVES

After completing this chapter, you will be able to perform the following:

- Understand what a feature is
- Use the Autodesk Inventor browser to edit parts
- Extrude a sketch into a part
- Revolve a sketch into a part
- Edit features of a part
- Edit the sketch of a feature
- Make an active sketch on a plane
- Create sketched features using one of three operations: cut, join, or intersect
- Project edges of a part

UNDERSTANDING FEATURES

After creating, constraining, and dimensioning a sketch, the next step in creating a model is to turn the sketch into a 3D feature. The first sketch of a part that is used to create a 3D feature is referred to as the base feature. In addition to the base feature, you can create sketched features in which you draw a sketch on a planar face or work plane, and you can either add or subtract material to or from existing features in a part. Use the Extrude, Revolve, Sweep, or Loft tools to create sketched features in a part. You can also create placed features such as fillets, chamfers, and holes by applying them to features that have been created. Placed features will be covered in chapter 4. Features are the building blocks in creating parts.

A plate with a hole in it, for example, would have a base feature representing the plate and a hole feature representing the hole. As features are added to the part, they appear in the browser, and the history of the part or assembly, that is, the order in which the features are created or the parts are assembled, is shown. Features can be edited, deleted from, or reordered in the part as required.

CONSUMED AND UNCONSUMED SKETCHES

You can use any sketch as a profile in feature creation. A sketch that has not yet been used in a feature is called an unconsumed sketch. When you turn a 2D sketched profile into a 3D feature, the feature consumes the sketch. The following image shows an unconsumed sketch in the browser on the left and a consumed sketch in the browser on the right.

Figure 3.1

Although a consumed sketch is not visible as you view the 3D feature, you may need to access sketches and change their geometric or dimensional constraints in order to modify their associated features. A consumed sketch can be accessed from the browser by right-clicking and selecting Edit Sketch from the menu. You may also access the sketch by starting the Sketch command and selecting the sketch from the browser. The following image on the left shows the unconsumed sketch, and the image on the right shows the extruded solid that consumes the sketch.

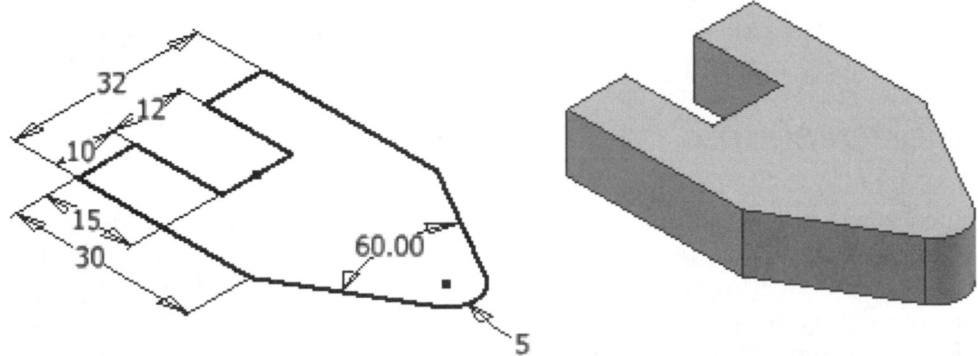

Figure 3.2

USING THE BROWSER FOR CREATING AND EDITING

The Autodesk Inventor browser, by default, is docked along the left side of the screen and displays the history of the file. In the browser you can create, edit, rename, copy, delete, and reorder features or parts. You can expand or collapse the browser to display the history of the features (the order in which the features were created) by clicking the + and - on the left side of the part or feature name in the browser. An alternate method of expanding the browser is to place your cursor on top of a feature icon but not click. The item in the browser automatically expands. To expand all the features, move the cursor into a blank area in the browser, right-click, and click Expand All from the menu.

The following image shows a browser with features of the part expanded.

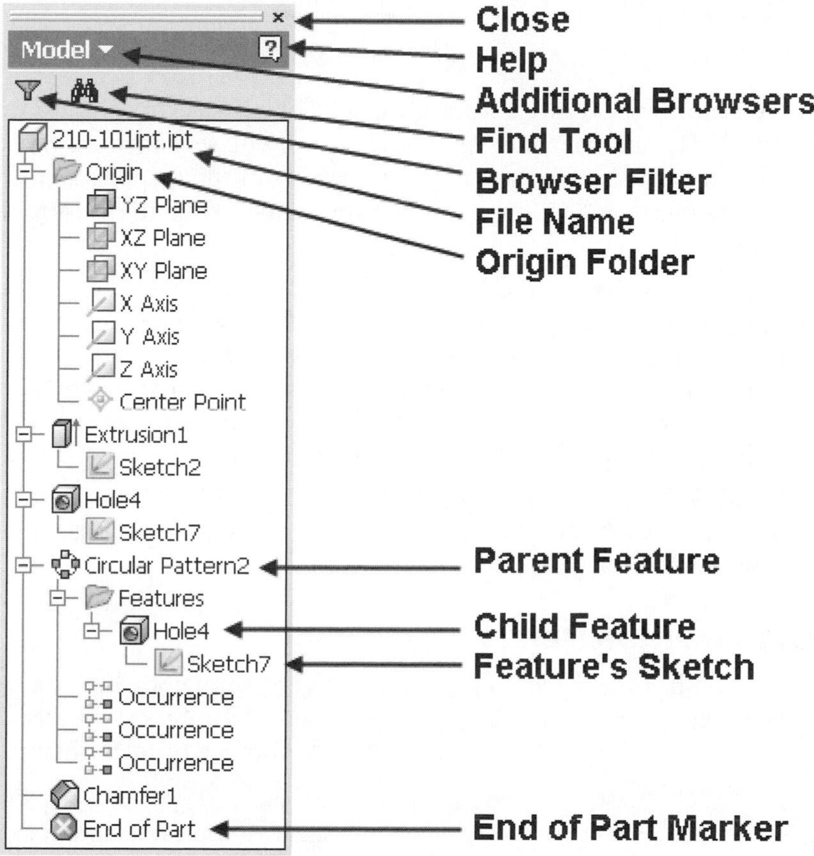

Figure 3.3

As parts grow in complexity, so will the information found in the browser. Dependent features are indented to show that they relate to the item listed above it. This is referred to as a *parent-child relationship*. If a hole is created in an extruded rectangle, for example, and the extrusion is deleted, the hole will also be deleted. To help filter out some of the object types that appear in the browser, you can click the icon that looks like a funnel from the top of the browser. After clicking the funnel, you can select object type(s) to hide in the browser, as shown in the following image.

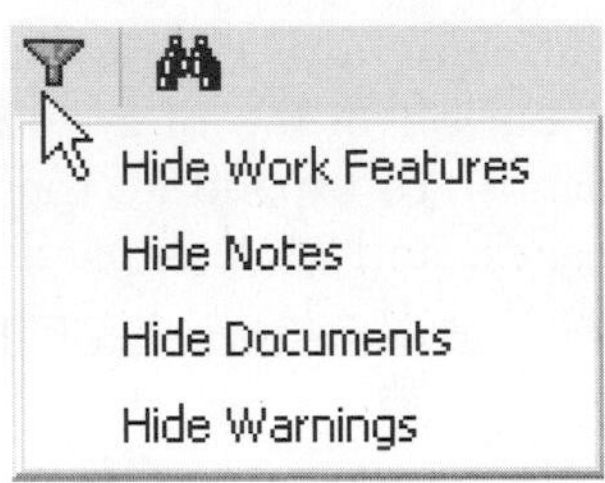

Figure 3.4

Each feature in the browser is given a default name. The first extrusion, for example, will be named Extrusion1, and the number in the name will sequence as you add similar features. The browser can also help you to locate features in the graphics area. To highlight a feature in the graphics window, simply move your cursor over the feature name in the browser.

To zoom in on a selected feature, right-click on the feature's name in the browser and select Find in Window on the menu or press the END key on your keyboard. The browser itself functions similarly to a toolbar except that you can resize it while it is docked. To close the browser, click the X in its upper-right corner. If the browser is not visible on the screen, you can display it by clicking the browser bar on the Toolbar pop-up menu on the View menu. You can also right-click while over a toolbar and selecting the browser bar from the list.

Specific functionality of the browser will be covered throughout this book in the pertinent sections. A basic rule, however, is to either right-click or double-click the feature's name or icon to edit or perform a function on the feature.

SWITCHING ENVIRONMENTS

Up to this point, you have been working in the sketch environment where the work is done in 2D. The next step is to turn the sketch into a feature. To do so, you need to exit the *sketch environment* and enter the *part environment*. A number of methods can be used to accomplish this transition:

- Click the Return tool on the left side of the standard toolbar as shown in the following image. From the drop-down list you can access the Parent and Top option. The Parent option returns you to the top level of the part file you are working on, and the Top option returns you to the top level of the assembly file. The Top option is available while you are working in an assembly file.

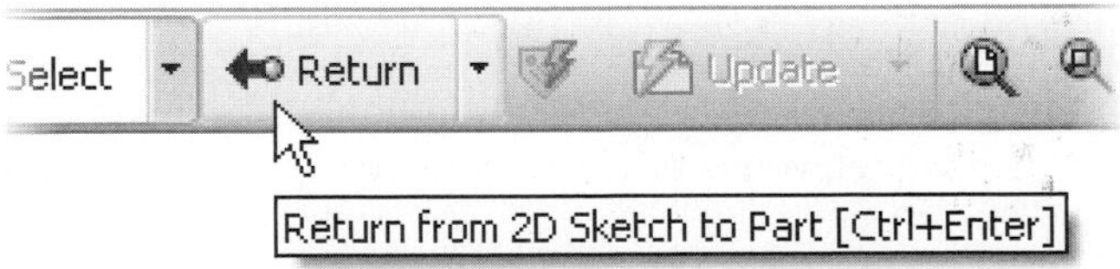

Figure 3.5

- Click the arrow on the Panel Bar near 2D Sketch Panel, and click Part Features from the drop-down list as shown in the following image.

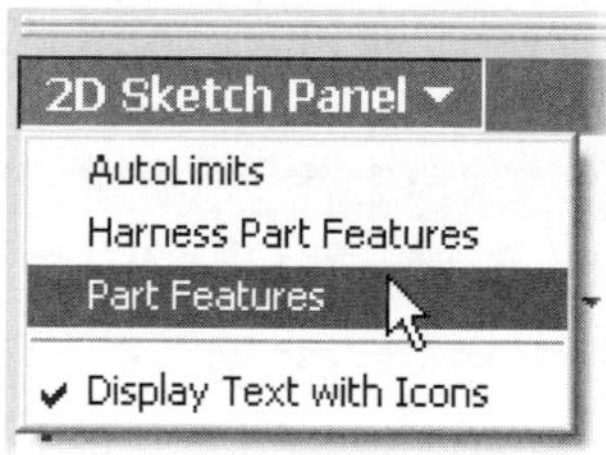

Figure 3.6

Note: If you exit the sketch environment with the Return tool, this is done automatically.

- Right-click in the graphics area and select Finish Sketch from the menu. You can also right-click in the graphics area, select Create Feature from the menu as shown in the following image, and then click the tool you need to create the feature. Only the tools that are applicable to the current situation will be available.
- Enter a shortcut key to initiate one of the feature tools.

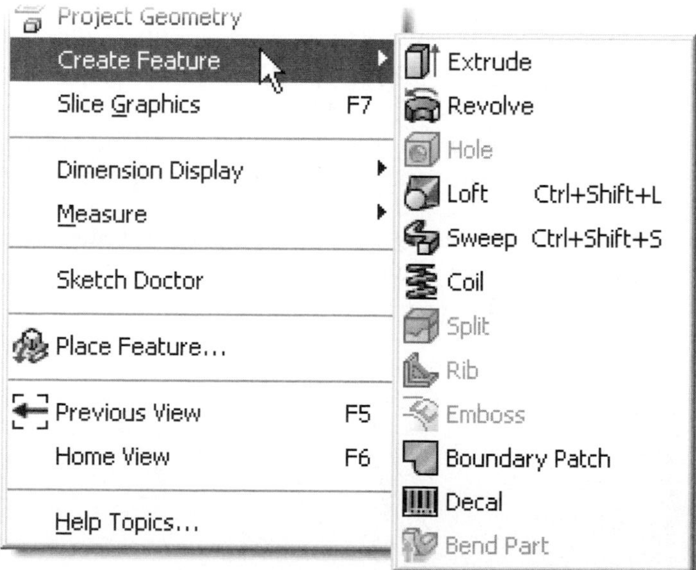

Figure 3.7

FEATURE TOOLS

Once you are in the part environment, the Panel Bar icons will change to the part feature tools. The following image shows the Part Features panel bar with the Display Text with Icons feature turned off. Many of these features will be covered throughout this book.

Figure 3.8

EXTRUDING A SKETCH

The most common method for creating a feature is to extrude a sketch and give it depth along the Z-axis. Before extruding, it is helpful to view the part in an isometric view. Autodesk Inventor previews the extrusion depth and direction in the graphics window.

To extrude a sketch, click the Extrude tool on the Part Features panel bar as shown in the following image on the left, and press the hot key E. Alternately, you can right-click in the graphics screen and select Create Feature > Extrude on the menu. After you issue the tool, the Extrude dialog box appears as shown in the following image on the right.

Note: When an arrow in a dialog box is red, the color indicates that Autodesk Inventor needs input from you for that specific function.

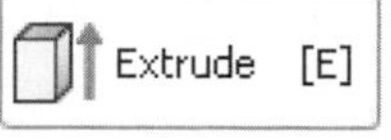

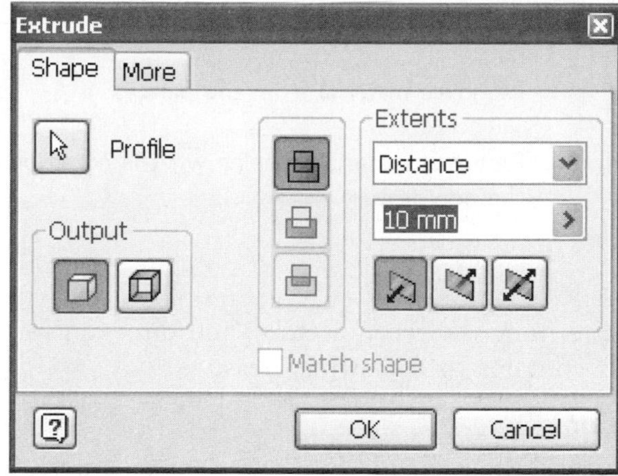

Figure 3.9

The Extrude dialog box has two tabs: Shape and More. When you make changes in the dialog box, the shape of the sketch will change in the graphics area to represent these values and options. When you have entered the values and the options you need, click the OK button to create the extruded feature.

SHAPE

The Shape tab gives you options to specify the profile to use, operation, extents, and output type. The options are described below.

 Profile — Click this button to choose the sketch that you wish to extrude. If there are multiple closed profiles, you will need to select which sketch area you want to extrude. If there is only one possible profile, Autodesk Inventor will select it for you, and you can skip this step. If you select the wrong profile or sketch area, click the Profile button again and choose the desired profile or sketch area. To remove a selected profile, hold the CTRL key and click the area you wish to remove.

Operation

This is the unlabeled middle column of buttons. If this is the first sketch that you create a solid from, it is referred to as a base feature, and only the top button is available. The operation defaults to Join, which is the top button. Once the base feature has been established, you can extrude a sketch, adding or removing material from the part by using the Join or Cut options, or you can retain the common volume between the existing part and the newly defined extrude operation using the Intersect option.

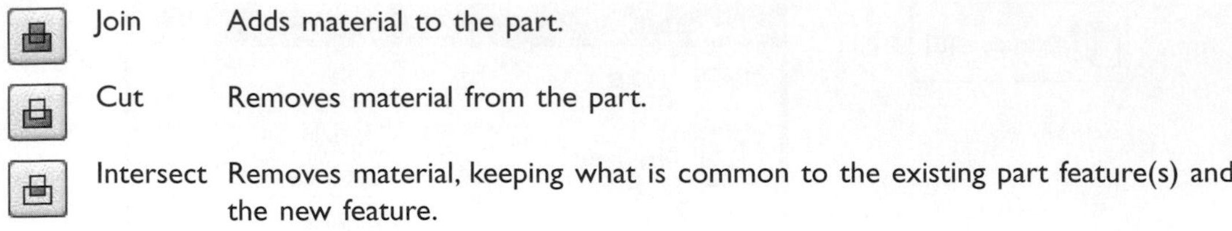

	Join	Adds material to the part.
	Cut	Removes material from the part.
	Intersect	Removes material, keeping what is common to the existing part feature(s) and the new feature.

Extents

Extents determine the type, distance, and direction for an extrusion. This section of the dialog box contains the following areas:

Termination

The termination determines how the extruded sketch will be terminated. There are five options from which to choose, but like the operation section, some options may not be available until a base feature exists.

Distance This option determines that the sketch will be extruded a specified distance.

To Next This option determines that the sketch will be extruded until it reaches a plane or face. The sketch must be fully enclosed in the area to which it is projecting; if it is not fully enclosed, use the To termination with the Extend to Surface option. Click the Direction button to determine the extrusion direction.

To This option determines that the sketch will be extruded until it reaches a selected face or plane. To select a point (midpoint or endpoint), plane or face to end the extrusion, click the Select Surface to end the feature creation button, as shown in the following image, and then click a face or plane at which the extrusion should terminate.

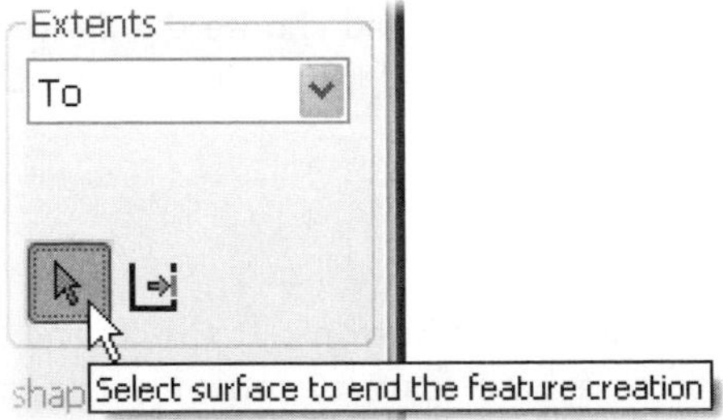

Figure 3.10

From To This option determines that the extrusion will start at a selected plane or face and stop at another plane or face. Click the Select surface to start feature creation button as shown in the following image on the left, and then click the face or plane where the extrusion will start. Then click the Select surface to end the feature creation button, as shown in the following image on the right and then click the face or plane where the extrusion will terminate.

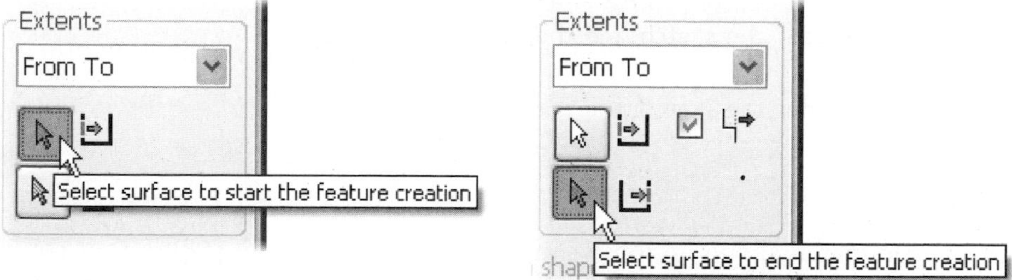

Figure 3.11

All This option determines that the sketch will be extruded all the way through the part in one or both directions.

Distance

If Distance is selected as the termination, enter a value at which the sketch will be extruded, click the arrow to the right, measure two points to determine a value, display the dimensions of previously created features to select from, or select from the list of the most recent values used. After you enter a value, a preview image appears in the graphics area to show how the extrusion will look. Another method is to click the edge of the extrusion preview shown in the graphics area and drag it.

A preview image appears in the graphics area, and the corresponding value appears in the distance area.

If values and units appear in red when you enter them, the defined distance is incorrect and should be corrected. For example, if you entered too many decimal places (e.g., 2.12.5) or an incorrect unit for the dimension value, the value will appear in red. You will need to correct the error before the extrusion can be created.

Direction

There are three buttons from which to choose for determining the direction. Choose from the first two to flip the extrusion direction, or click the last button (midplane) to have the extrusion go equal distances in the negative and positive directions. If the extrusion distance is 2 mm, for example, the extrusion will go 1 mm in both the negative and positive Z directions when using the midplane option.

Output

Two options are available to define the type of output that the Extrude tool will generate:

	Solid	Extrudes the sketch, and the result is a solid body.
	Surface	Extrudes the sketch, and the result is a surface.

MATCH SHAPE

You can use the match shape option when working with an open profile that you want to extrude. The edges of the open profile are extended until they intersect geometry of the model. This provides a "flood-fill"-type effect for the extrude feature. Open profiles are covered in chapter 7.

When you convert a sketch into a part or feature, the dimensions on the sketch are consumed (they disappear). When you edit the feature, the dimensions reappear. The dimensions can also be displayed when drawing views are made. For more information on editing parts or features, see the "Editing 3D Parts" section later in this chapter.

MORE

The More tab, as shown in the following image, contains additional options to refine the feature being created:

Alternate Solution

Alternate Solution terminates the feature on the most distant solution for the selected surface. An example is shown in the following image.

Figure 3.12

Minimum Solution

Minimum Solution terminates the feature on the first possible solution for the selected surface. An example is shown in the following image.

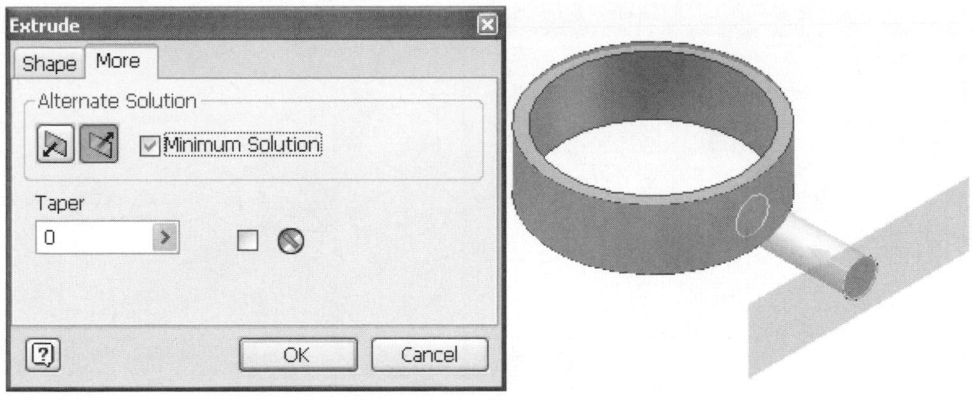

Figure 3.13

Taper

Taper extrudes the sketch and applies a taper angle to the feature. To extend the taper angle out from the part, give the taper angle a positive number. This increases the volume of the resulting extruded feature. This step is also known as "reverse draft."

Infer iMates

Check this box to automatically create an iMate on a full circular edge. Autodesk Inventor attempts to place the iMate on the closed loop most likely to be used.

EXERCISE 3-1: EXTRUDING A SKETCH

In this exercise, you create a base feature by extruding an existing profile. You will examine the direction options available in the Extrude dialog box.

1. Open *ESS_E03_01.ipt* from the Chapter 03 subfolder.
2. Click the Extrude tool. Because there is only one possible sketch, the profile is automatically selected.
3. In the Extrude dialog box, set the Distance to **15 mm**.
4. Select the back edge of the extrusion in the graphics window.
5. Drag the edge until a distance of **25 mm** is displayed in the Distance field of the Extrude dialog box as shown in the following image, and then release the mouse button.

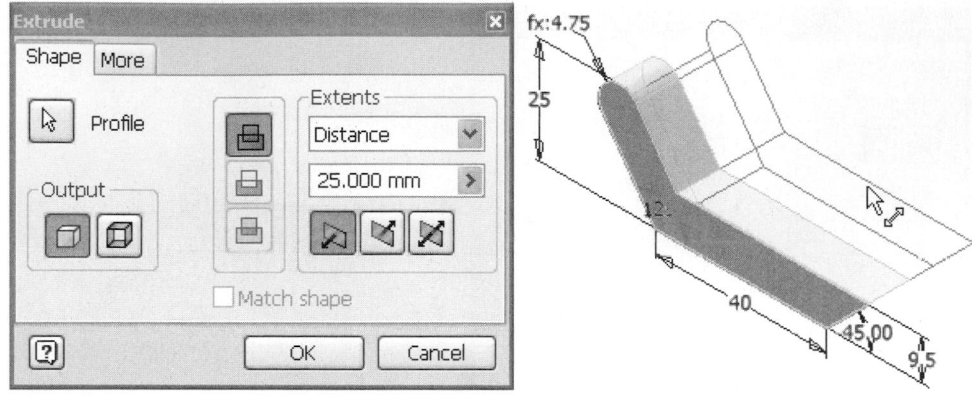

Figure 3.14

6. Click the More tab in the Extrude dialog box.
7. Adjust the value of the Taper to **10**.
8. Press and hold the F4 key to rotate the part until you can see the arrow previewing that the taper will add mass to the part as shown in the following image.
9. Release the F4 key.

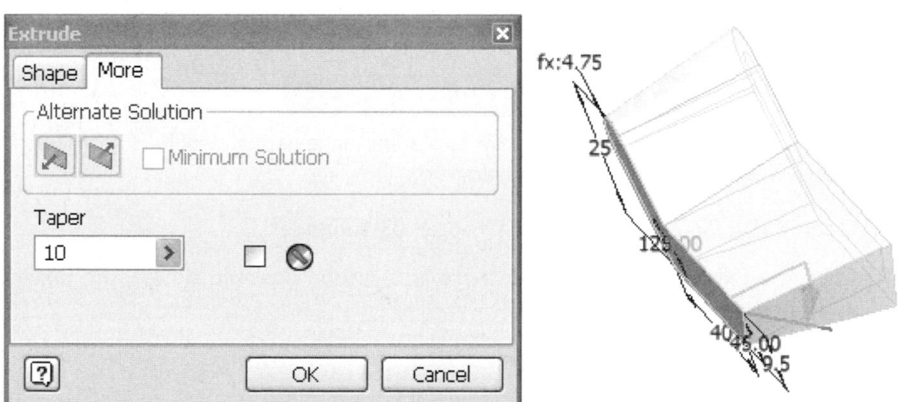

Figure 3.15

10. Press F6 to return to a Home View of the model.
11. Click the Shape tab in the Extrude dialog box.
12. Change the direction of the extrusion by selecting the middle button on the direction area. This will flip the direction of the extrusion.
13. Change the direction to go evenly in both directions by selecting the rightmost (midplane) button on the direction area as shown in the following image.

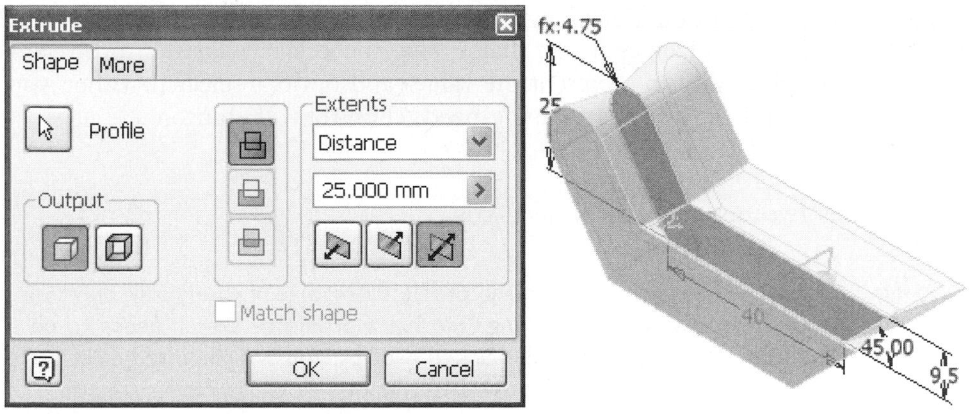

Figure 3.16

14. Practice changing the values and directions to see the results. When done, click OK, and the extrusion is created. Later in this chapter you will learn how to edit features.

15. Close the file. Do not save changes. End of exercise.

REVOLVING A SKETCH

Another method for creating a part is to revolve a sketch around a straight edge or axis (centerline). You can use revolved sketches to create cylindrical parts or features. To revolve a sketch, you follow the same steps that you did to extrude a sketch. Create the sketch and add constraints and dimensions, and then click the Revolve tool on the Part Features panel bar as shown in the following image on the left; press R or right-click in the graphics window and select Create Feature > Revolve from the menu. The Revolve dialog box appears, as shown in the following image on the right.

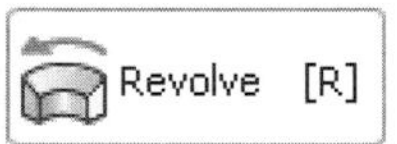

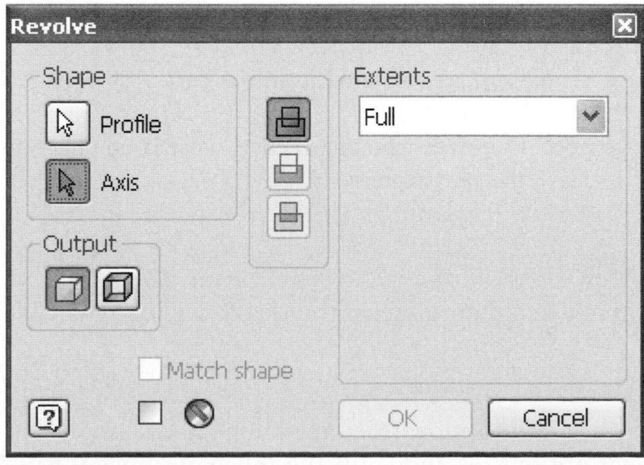

Figure 3.17

The Revolve dialog box has five sections: Shape, Operation, Extents, Output, and Match Shape. When you make changes in the dialog box, the preview for the revolved feature changes in the graphics area to represent the values and options selected. When you have entered the values and the options that you need, click the OK button.

SHAPE

This section has two options: Profile and Axis.

Profile Click this button to choose the profile to revolve. If the Profile button is shown depressed, this is telling you that a profile or sketch needs to be selected. If there are multiple closed profiles, you will need to select the profile that you want to revolve. If there is only one possible profile, Autodesk Inventor will select it for you, and you can skip this step. If the wrong profile or sketch area is selected, click the Profile button, and choose the new profile or sketch area. To remove a selected profile, hold the CTRL key and click the profile to remove.

Axis Click a straight edge or centerline in the same sketch about which the profile(s) should be revolved. See the section below on how to create a centerline and create diametric dimensions.

OPERATION

This is the unlabeled middle column of buttons. If this is the first sketch that you create a solid from, it is referred to as a base feature, and only the top button is available. The operation defaults to Join (the top button). Once the base feature has been established, you can then revolve a sketch, adding or removing material from the part using the Join or Cut options, or you can keep what is common between the existing part and the completed revolve operation using the Intersect option.

Join Adds material to the part.

Cut Removes material from the part.

Intersect Removes material, keeping what is common to the existing part feature(s) and the new feature.

EXTENTS

The Extents area determines if the sketch will be revolved 360° or another specified angle.

Full Full is the default option; it will revolve the sketch 360° about a specified edge or axis.

Angle Click this option from the drop-down list, and the Revolve dialog box displays additional options, as shown in the following image. Enter an angle for the sketch to be revolved. The three buttons below the degree area will determine the direction of the revolution. Choose the first two to flip the revolution direction, or click the right-side button to have the revolution go an equal distance in the negative and positive directions. If the angle was set to 90°, for example, the revolution will go 45° in both the negative and positive directions.

To Next This option determines that the sketch will be revolved until it reaches a plane or face. The sketch must be fully enclosed in the area to which it is projecting; if it is not fully enclosed, use the To termination with the Extend to Surface option. Click the Direction button to determine the revolve direction.

To This option determines that the sketch will be revolved until it reaches a selected face or plane. To select a plane or face to end the revolve, click the Select Surface button, as shown in the following image, and then click a face or plane at which the extrusion or revolve should terminate.

From To This option determines that the revolve will start at a selected plane or face and stop at another plane or face. Click the Select surface to start feature creation button, and then click the face or plane where the revolve will start. Click the Select surface to end the feature creation button, and then click the face or plane where the extrusion or revolve will terminate.

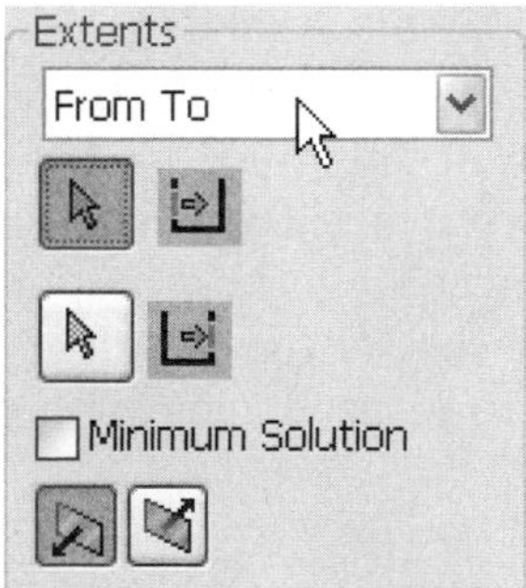

Figure 3.18

OUTPUT

Two options are available to select the type of output that the Revolve tool will generate.

	Solid	Revolves the sketch, and the resulting feature is a solid body.
	Surface	Revolves the sketch, and the resulting feature is a surface.

MATCH SHAPE

You can use the Match shape option when working with an open profile. The edges of the open profile are extended until they intersect the geometry of the model. This provides a "flood-fill"-type effect for the revolve feature.

INFER IMATES

Check this box to automatically create an iMate on a full circular edge. Autodesk Inventor attempts to place the iMate on the closed loop most likely to be used.

CENTERLINES AND DIAMETRIC DIMENSIONS

When revolving a sketch, you may want to specify diametric linear dimensions instead of radial dimensions. The sketches you revolve are usually a half section of the completed part. The following image on the left shows a sketch that represents a quarter section of the completed part with a centerline and the diametric linear dimensions.

The dimensions placed on a sketch can be used for drawing views. If you want to place a diametric linear dimension on the sketch, you should select a centerline. To create a centerline, activate the Centerline tool on the standard toolbar, as shown in the following image on the right. You can activate the Centerline tool before sketching the line to be displayed as a centerline, or you can select an existing sketched entity and then select the Centerline tool.

To create a diametric linear dimension, use the General Dimension tool and select either the centerline and the other point or line to be dimensioned, or click a point or edge and then the centerline to place the diametric dimension. When selecting the centerline, be certain to select the entire centerline, not just an endpoint of the centerline. If you do not use a centerline as part of the dimension, you can right-click after selecting the geometry to be dimensioned and then select Linear Diameter from the menu to place a diametric dimension.

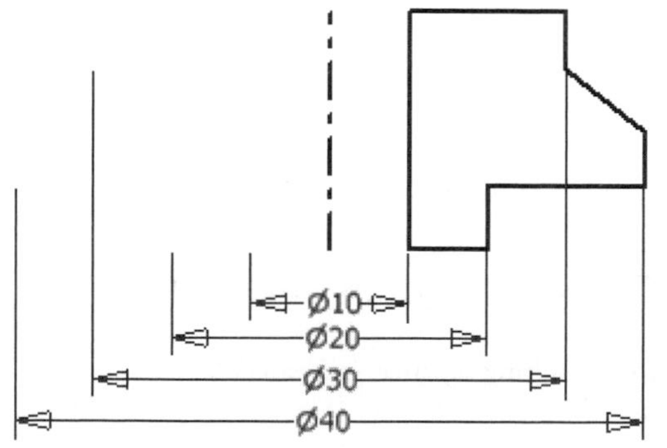

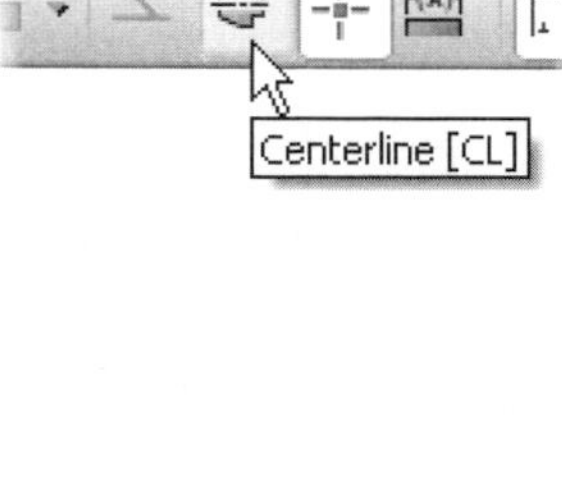

Figure 3.19

LINEAR DIAMETRIC DIMENSIONS

When a centerline will not be used to revolve the sketch around, you can create diametric dimensions for sketches that represent a half outline of a revolved part. To create a diametric dimension, follow these steps:

1. Draw a sketch that represents a quarter section of the finished part.
2. Draw a line, if needed, around which the sketch will be revolved. This line can be on the closed profile of the sketch.
3. Issue the General Dimension tool.
4. Click the line (not an endpoint) that will be the axis of rotation.
5. Click the other point to be dimensioned.
6. Right-click, and select Linear Diameter from the menu.
7. Move the cursor until the diameter dimension is in the correct location and click.

The following image on the left shows a sketch with the menu for changing the dimension to Linear Diameter. The image on the right shows the placed diametric dimension—the left vertical line will be the axis of rotation.

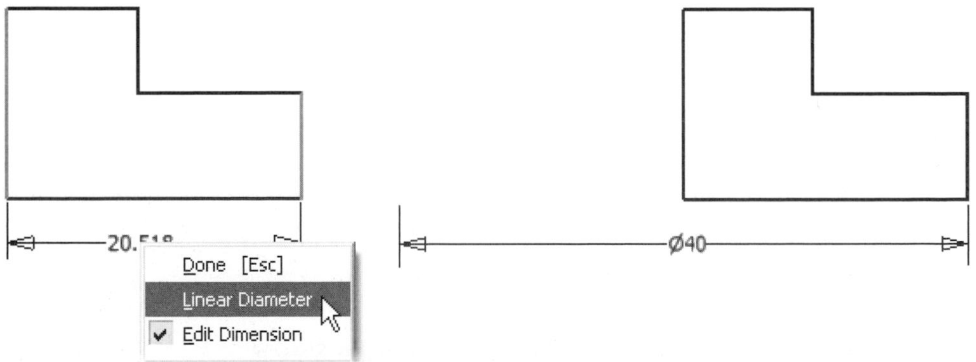

Figure 3.20

EXERCISE 3-2: REVOLVING A SKETCH

In this exercise, you create a sketch and then create a revolved feature to complete a part. This exercise demonstrates how to revolve sketched geometry about an axis to create a revolved feature.

1. Click the New tool.
2. Select the Metric tab, and then double-click *Standard (mm).ipt*.
3. Create the sketch geometry as shown. Place the lower endpoint of the centerline at the projected origin point. Note: Use the Centerline tool on the Standard toolbar to change the left vertical line to a centerline as shown in Figure 3.19.

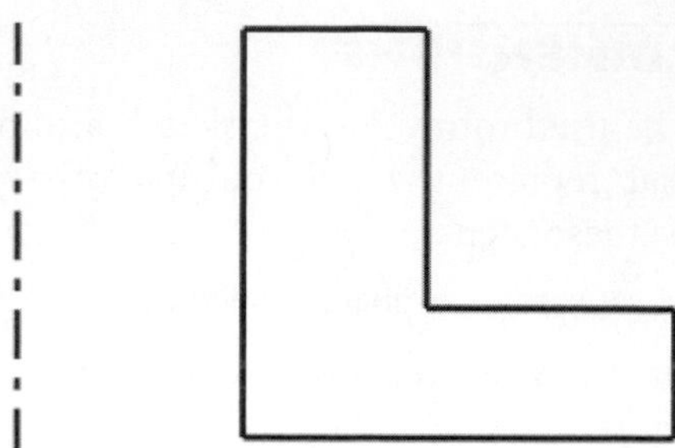

Figure 3.21

4. Click the General Dimension tool.
5. Add the linear diameter dimensions shown by selecting the centerline and selecting an endpoint; then click a point to place the dimension.

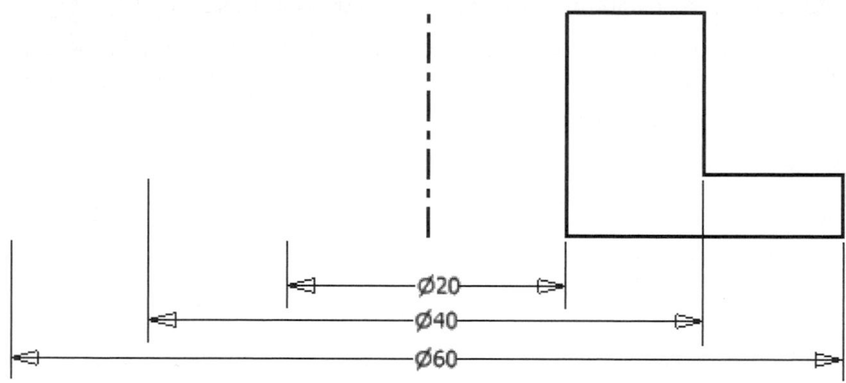

Figure 3.22

6. Finish the dimension tool, right-click, and click Done.
7. Right-click in the graphics window, and click Finish Sketch.
8. Change to an isometric view, on the View Cube click the corner of the Top, Front, and Right.
9. From the Part Features panel bar, click the Revolve tool. Since there is only one possible profile, the profile is selected for you. The centerline is also automatically selected as the axis.
10. Change the extents to Angle.
11. Enter **45°**. The preview of the model updates to reflect the change as shown in the following image.

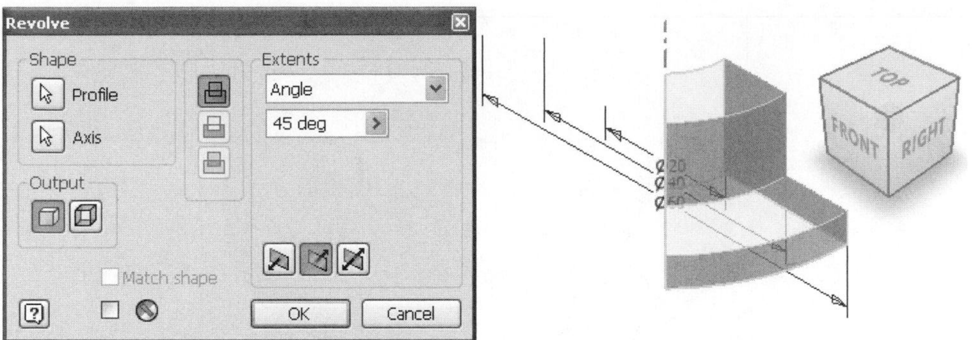

Figure 3.23

12. Change the direction of the revolve by selecting the middle flip button. The preview image will reverse the direction.
13. Set the revolve direction to midplane.
14. Enter **90 deg** for the angle. The preview should resemble the following image.

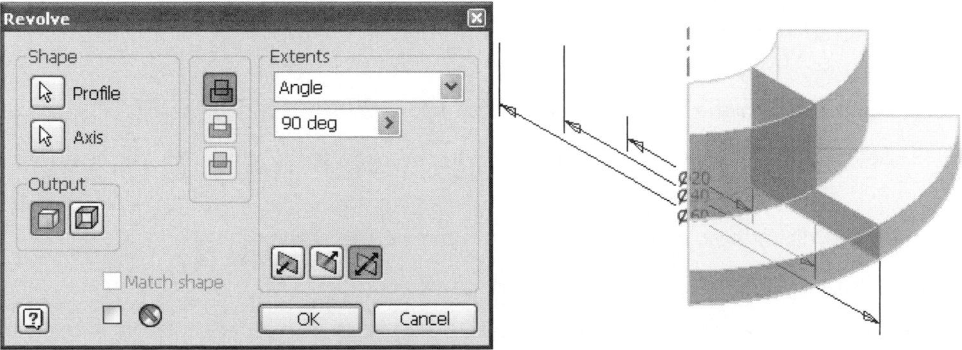

Figure 3.24

15. Change the extent type to Full.
16. Click the OK button to create the feature.
17. Your part should resemble the following image.

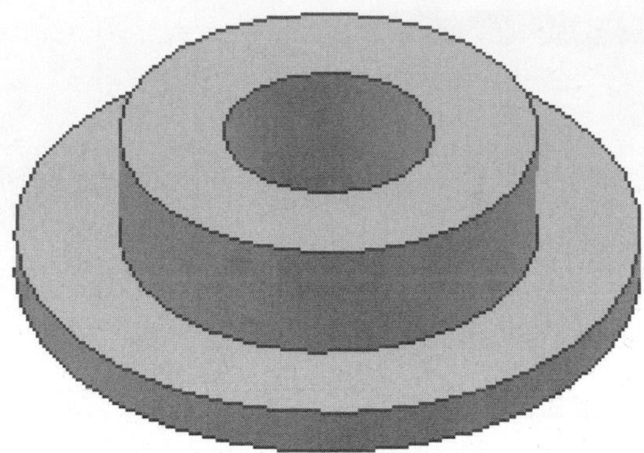

Figure 3.25

18. Close the file. Do not save changes. End of exercise.

EDITING A FEATURE

After you create a feature, the feature consumes all of the dimensions that were visible in the sketch. If you need to change the dimensions' values, taper, operation, or termination that were entered in the dialog box during the feature creation, you will need to edit the feature. To do so, follow these steps:

1. Right-click on the feature's name in the browser.
2. Select Edit Feature from the menu or, in the browser, double-click the feature's name or icon.
3. In the dialog box, enter new values or change the settings.
4. Click the OK button to complete the edit.

Everything can be changed in the dialog box except the join operation on a base feature and the output. The following image on the left shows the menu that appears after right-clicking on Extrusion1.

EDITING FEATURE SIZE

To change the dimensional values of a feature, you need to edit the feature so the dimensions are visible. There are multiple methods that you can use to edit the dimensions. There is no preferred method, so use the method that works best for your workflow.

- Double-click the features name in the browser; the dialog box that was used to create the feature will appear.
- In the browser, right-click the feature's name, and select Edit Feature from the menu as shown in the following image on the left.
- In the browser, expand the children of the feature, right-click the name of the sketch, and choose Edit Sketch from the menu as shown in the following image on the right.

- In the browser, right-click the feature's name, and select Show Dimensions from the menu.

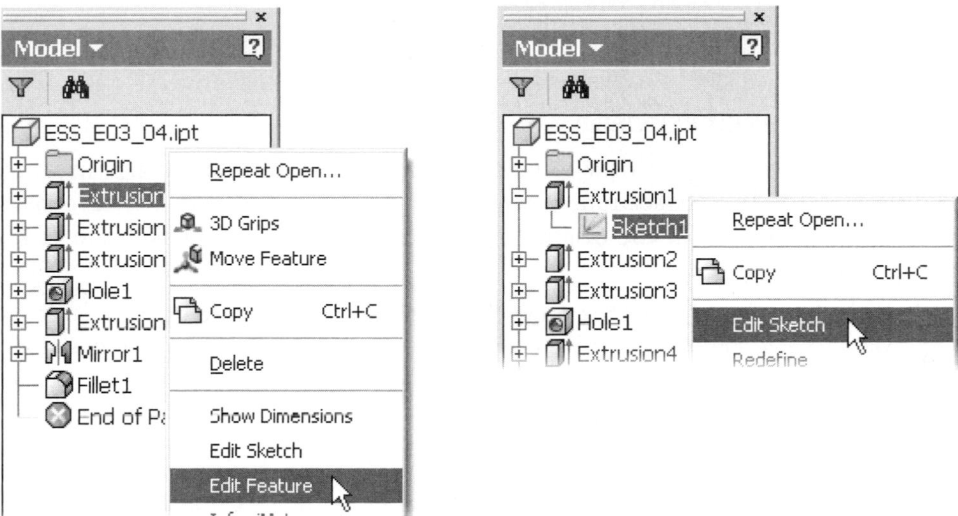

Figure 3.26

To edit the features by selecting them in the graphics window, click the down arrow next to the Select option on the Standard toolbar, and click Feature Priority from the menu as shown in the following image. Double-click the feature that you want to edit, in either the browser or graphics window, and the dimensions will appear on the part. You can also access the Select tools by holding down the SHIFT key and right-clicking in the graphics window.

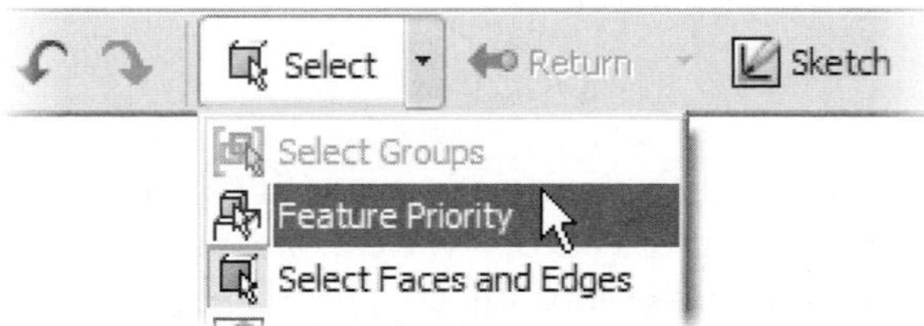

Figure 3.27

When the dimensions are visible on the screen, double-click the dimension text that you want to edit. The Edit Dimension dialog box appears. Enter a new value, and then click the checkmark in the dialog box or press the ENTER key. Continue to edit the dimensions and, when finished, click the Update button on the command bar as shown in the following image. The dimensions will disappear, and the new values will be used to regenerate the part.

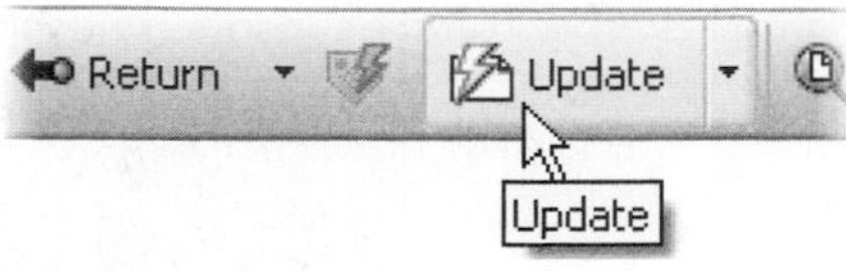

Figure 3.28

EDITING A FEATURE SKETCH

In the last section, you learned how to edit the dimensions and the settings in which the feature was created. In this section, you will learn how to add and delete constraints, dimensions, and geometry in the original 2D sketch. To edit the 2D sketch of a feature, do the following:

1. In the browser, right-click the name or icon of the sketch feature or sketch that you want to edit, and select Edit Sketch from the menu, as shown in the previous section.
2. Add or remove geometry from the sketch, and add or delete constraints and dimensions, as described in chapter 02.

While editing the sketch, you can both add and remove objects. You can add geometry lines, arcs, circles, and splines to the sketch. To delete an object, right-click it and select Delete from the menu, or click it and press the Delete key. If you delete an object from the sketch that has dimensions associated with it, the dimensions are no longer valid for the sketch, and they will also be deleted. You can also delete the entire sketch and replace it with an entirely new sketch. When replacing entire sketches, you should first delete other features that would be consumed by the new objects and re-create them. Once you modify the sketch, update the part by clicking the Return or the Update button on the Standard toolbar.

Note: If you receive an error after updating the part, make sure that the sketch forms a closed profile. If the appended or edited sketch forms multiple closed profiles, you will need to reselect the profile area.

3D GRIPS

An alternate method for editing a feature size is to use 3D Grips. The 3D Grips tool can be used to push or pull the faces of an extruded, revolved, or a sweep feature. Edits done with 3D Grips may alter the parametric dimensions values. The 3D Grips tool is accessed by selecting a face from the shortcut menu after you have selected a feature in the browser as shown in the following image on the left. You can also click a face in the graphics window; after the face is highlighted, right-click and click 3D Grips from the menu as shown in the following image on the right.

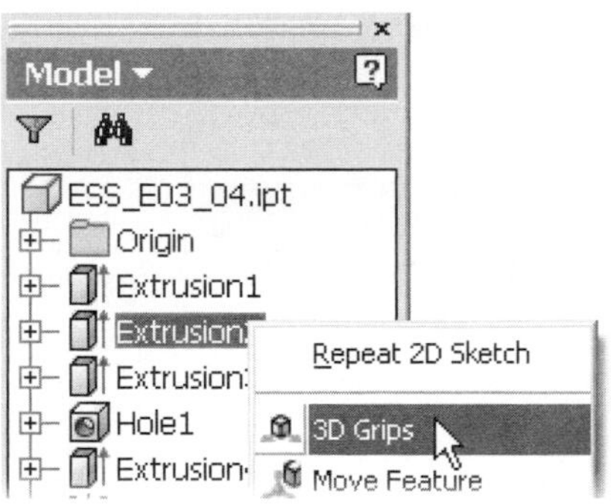

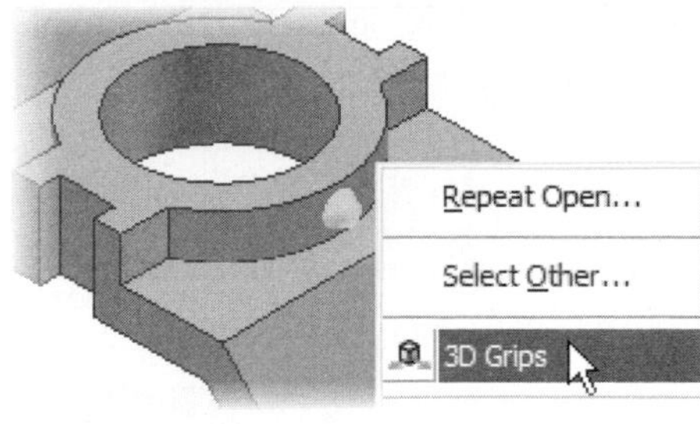

Figure 3.29

After the 3D Grips are displayed, the outline of the feature will be represented by a color correlated to the operation that was used to create it.

Red = Cut operation

Green = Joined operation

Blue = Intersect operation

Arrows are displayed as you move the cursor over existing faces of the feature that are being modified as shown in the following image. To modify the size of a feature, click and drag the grip arrow that is associated with the geometry that you want to modify as shown in the following image on the left. Right-click, and click Done from the menu as shown in the following image on the right. You can also click a point or face for the selected geometry to align to.

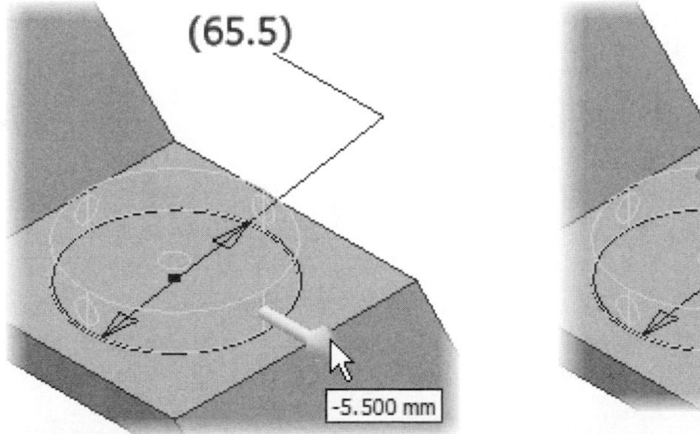

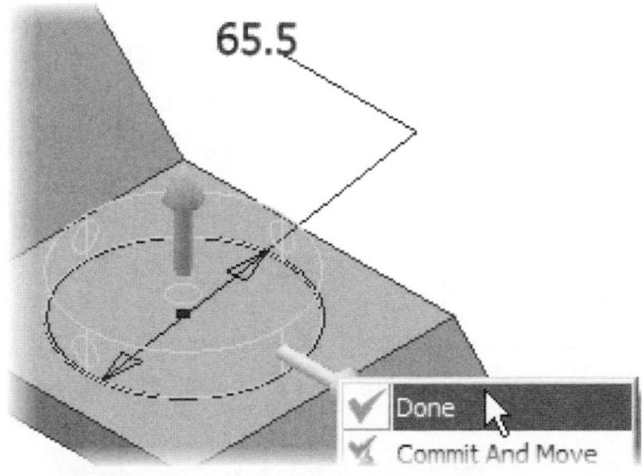

Figure 3.30

You can also use the grip edit capabilities to modify the feature by a specific distance or angle. This is done by right-clicking the arrow and choosing the appropriate option from the menu as shown in the following image on the left. Depending on the available geometry of the feature, the menu will contain different options such as Edit Angle, Edit Offset, Edit Radius, Edit Extent, and so on. After selecting a menu option, enter the new value in the edit dialog box, and click Done from the menu as shown in the following image on the right.

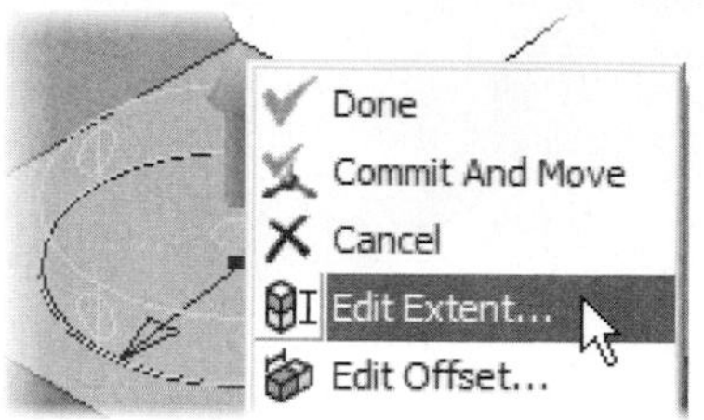

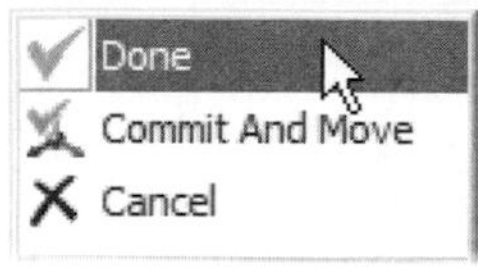

Figure 3.31

When using the grip edit functionality, parametric dimensions are ignored by default and are modified as if they were reference dimensions. You can customize how 3D Grips interact with existing dimensions and sketch constraints by clicking Tools > Application Options and then clicking on the Part tab. When the grip edit is complete, the dimension values are updated to reflect the new values.

MOVE FEATURE

The Move Feature tool can also be used to edit a feature. This tool allows you to click and drag a feature to a new location or move it from one part face to another. Similar to the 3D Grips tool, the Move Feature tool is accessed by right-clicking after selecting a feature in the browser as shown in the following image. You can also change to the Feature Priority option from the Select area of the standard toolbar, move the cursor over a desired feature, right-click, and then click Move Feature from the menu.

Figure 3.32

After dragging the feature to the new location or face, right-click and select Done from the menu. If you choose Commit and 3D Grip Edit, the feature is relocated and is ready for modification using grips as described above and as shown in the following image.

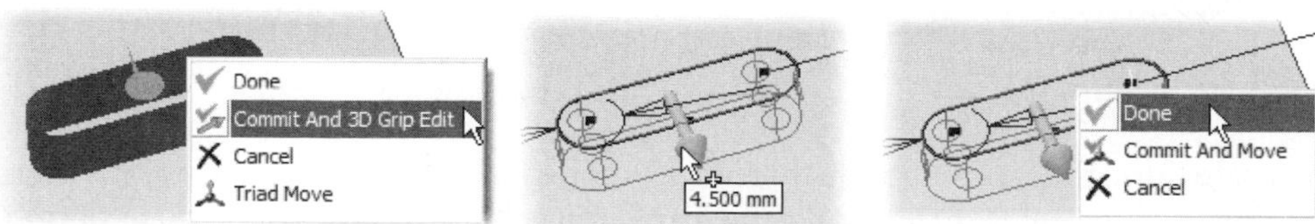

Figure 3.33

You can also use the 3D Move/Rotate tool to move the feature if the Triad Move option is selected from the menu. The 3D Move/Rotate tool is discussed in chapter 04..

RENAMING FEATURES AND SKETCHES

By default, each feature is given a name. These feature names may not help you when trying to locate a specific feature of a complex part, as they will not be descriptive to your design intent. The first extrusion, for example, is given the name Extrusion1 by default, whereas the design intent may be that the extrusion is the thickness of a plate. To rename a feature, slowly double-click the feature name and enter a new name. Spaces are allowed.

FEATURE AND FACE COLOR

When parts become complex—and, for example, if you want to differentiate between a cast and a machined surface—you may want to change the colors of specific features. To change a feature color, right-click the feature name in the browser and select Properties from the menu as shown in the following image on the left. In the Feature Color Style dialog box, select a new color from the drop-down menu, as shown in the following image on the right, and then click OK. You can also change a feature's name in the top area of the Properties dialog box.

You can change the color of model faces using a similar method. Select one or more faces on the part. Right-click a selected face in the graphics area, and select Properties from the menu. You can select a new face color from the drop-down menu in the Feature Properties dialog box.

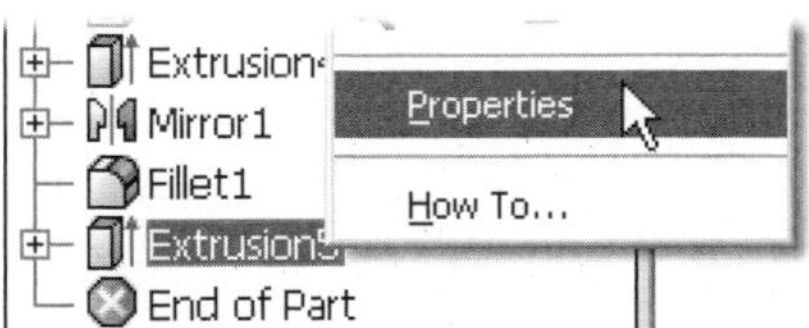

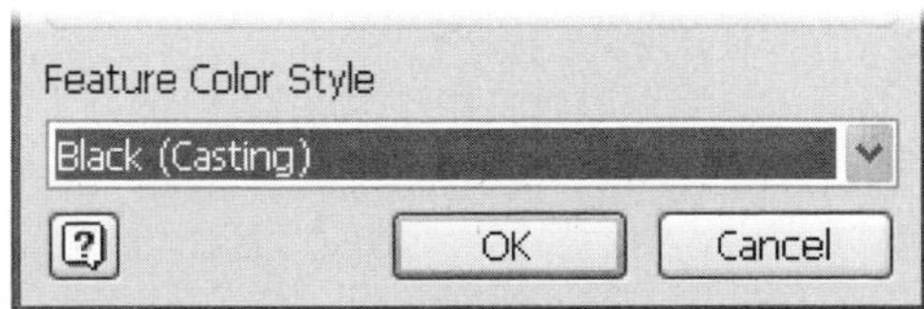

Figure 3.34

DELETING A FEATURE

You may choose to delete a feature after it has been placed. To delete a feature, right-click the feature name in the browser, and select Delete from the menu, as shown in the following image. The Delete Features dialog box will then appear, and you should choose what you want to delete from the list. You can delete multiple features by holding down the CTRL or SHIFT key, clicking their names in the browser, right-clicking one of the names, and then selecting Delete from the menu. The Delete Features dialog box appears; it allows you to delete consumed or dependent sketches and features.

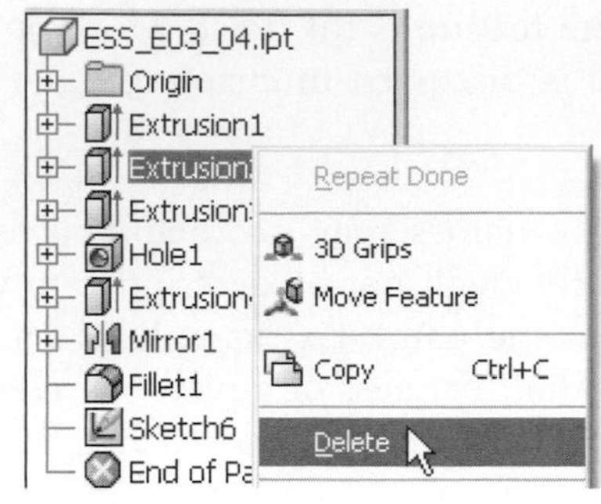

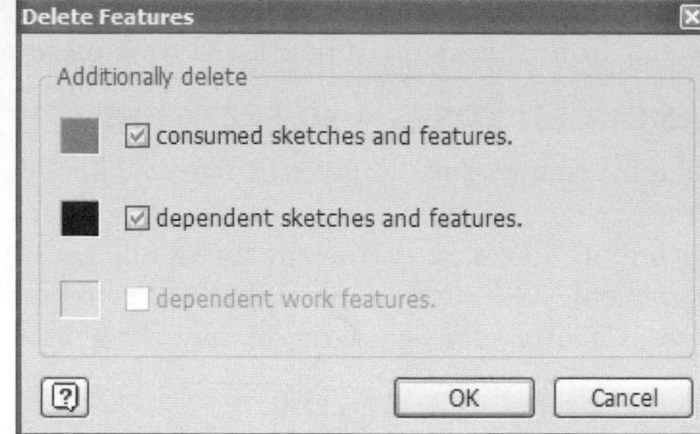

Figure 3.35

FAILED FEATURES

If the feature in the browser turns red after updating the part, this is an alert that the new values or settings were not regenerated successfully. You can then edit the values, enter new values, or select different settings to define a valid solution. Once you define a valid solution, the feature should regenerate without error.

EXERCISE 3-3: EDITING FEATURES AND SKETCHES

In this exercise, you edit a consumed sketch in an extrusion and update the part.

1. Open *ESS_E03_03.ipt* from the Chapter 03 folder.

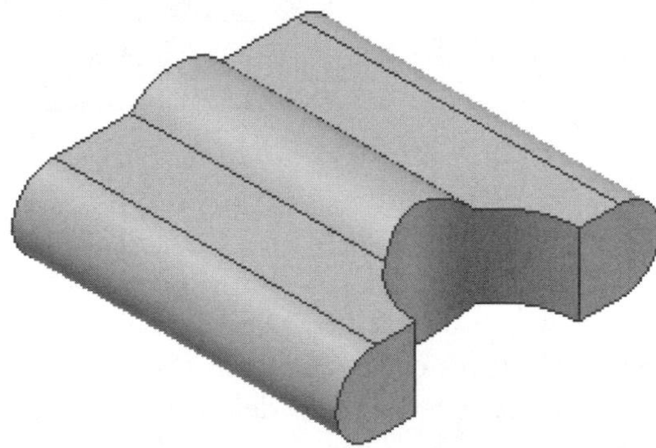

Figure 3.36

2. In the browser, right-click Extrusion1, and Click Edit Sketch. The consumed feature sketch is displayed.
3. Double-click the 10 radius dimension. In the field, enter **8**, and then press Enter or click the checkmark in the dialog box.
4. Double-click the 60 dimension, and then edit the value to **100** as shown in the following image on the left.

5. Click Return or Update tool on the Standard toolbar. The feature is updated with the new values as shown in the following image on the right.

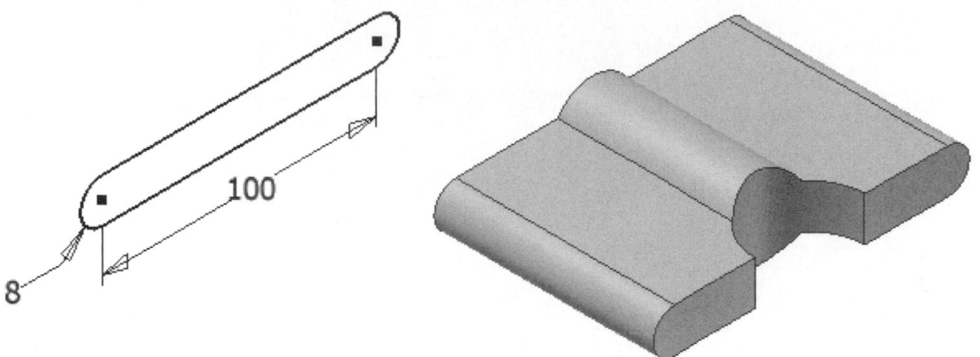

Figure 3.37

6. You now edit the termination method for an extrusion. In the browser, double-click on Extrusion2. The Extrude dialog box is displayed.
7. Change the Operation to Join as shown in the following image.
8. Change the Extents option to To and select the bottom face, which will stop the extrusion at this face no matter what dimensional changes occur to the model.

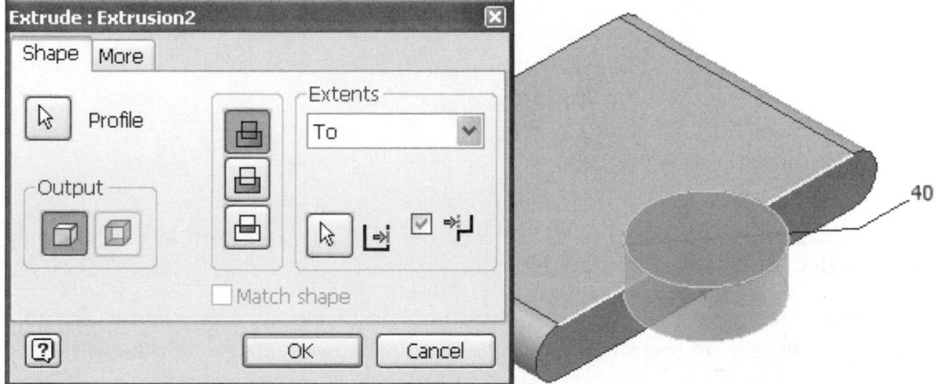

Figure 3.38

9. Click OK. The feature is updated, as shown in the following image.

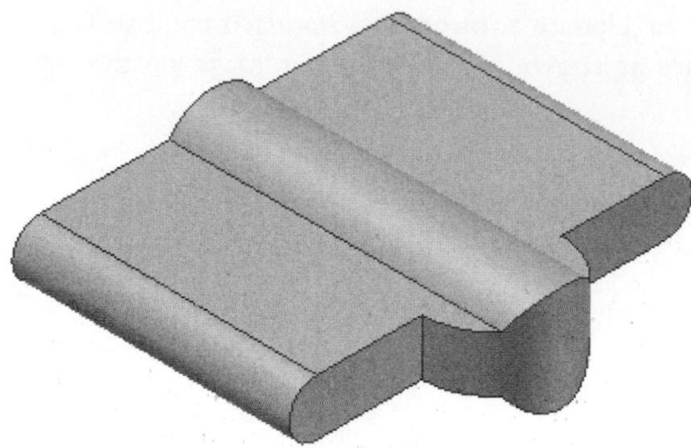

Figure 3.39

10. In the browser, right-click Extrusion3. Click Delete and then click OK to delete consumed sketches and features. When done, your part should resemble the following image.

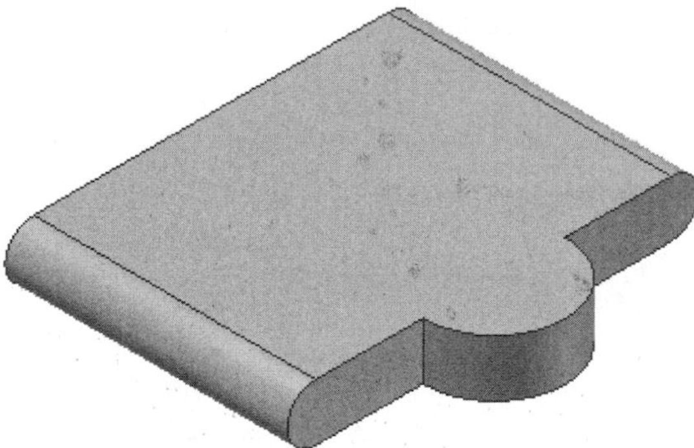

Figure 3.40

11. Click the top-horizontal face, and click the 3D Grip as shown in the following image on the left.
12. Move the cursor over the back circle until an arrow appears. Right-click and select Edit Offset from the menu as shown in the following image on the right.

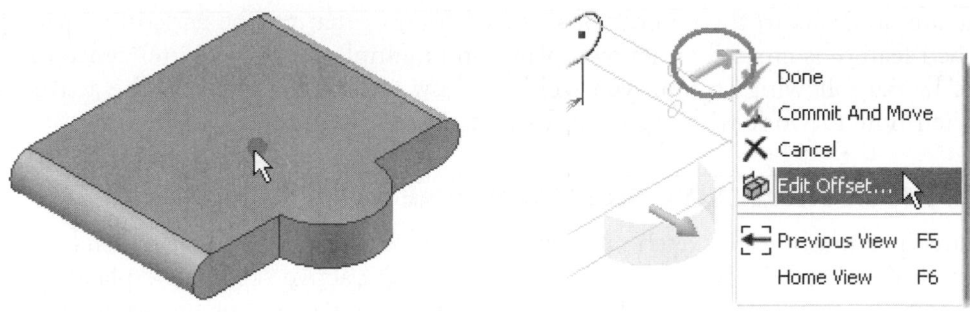

Figure 3.41

13. Enter **-40** in the Edit Offset dialog box, and click OK.
14. Right-click, and click Done from the menu.
15. In the browser, expand Extrusion1, and move the cursor over Sketch1 to verify that the length of the slot has been changed from 100 to 60 as shown in the following image. To clear the dimensions, click twice in a blank area of the graphics window.

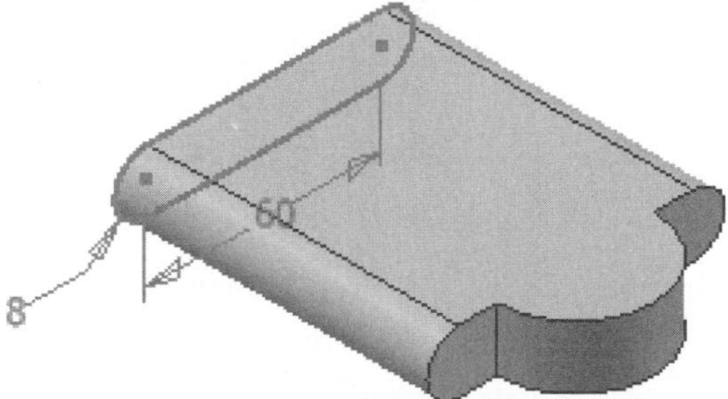

Figure 3.42

16. Practice editing the sketch dimensions and features.
17. Close the file. Do not save changes. End of exercise.

SKETCHED FEATURES

A sketched feature is created from a sketch that you draw on a plane or face. The basic steps to create a sketched feature are as follows:

1. Create or make an existing sketch active.
2. Draw the geometry that defines the sketch.
3. Add constraints and dimensions.
4. Perform a Boolean operation that will either add material to or remove material from the part or that will keep whatever is common between the part and the completed feature.

There are no limits to the number of sketched features that can be added to a part. Each sketched feature is created on its own plane, and multiple features can reference the same plane. In the following section, you will learn how to assign a plane to the active sketch and then how to work with sketched features.

DEFINING THE ACTIVE SKETCH PLANE

As stated previously, each sketch must exist on its own plane. The active sketch has a plane on which the sketch is drawn. To assign a plane to the active sketch, the plane on which the sketch will be created must be a planar face, a work plane, or an origin plane. The planar face does not need to have a straight edge. A cylinder has two faces, one on the top and the other on the bottom of the part. Neither has a straight edge, but a sketch can be placed on either face.

To make a sketch active, use one of the following methods:

- Issue the Sketch tool from the command bar, as shown in the following image, and then click the plane where you want to place the sketch.

Figure 3.43

- Press the hot key S, and then click the plane where you want to place the sketch.
- Click a plane that will contain the active sketch, and issue the Sketch tool from the command bar.
- Click a plane that will contain the active sketch, and press the hot key S.
- Click a plane that will contain the active sketch, right-click in the graphics area, and select New Sketch from the menu.
- Issue the Sketch tool from the command bar, expand the Origin folder in the browser, and click one of the default work planes.
- Expand the Origin folder in the browser, right-click on one of the default work planes, and select New Sketch on the menu.
- To make a previously created sketch active, issue the Sketch tool and click the sketch name in the browser.

Once you have created a sketch, it appears in the browser with the name Sketch#, and sketch tools appear in the 2D Sketch panel bar. The number will sequence for each new sketch that is created. To rename a sketch, go to the browser, slowly double-click the existing name, and enter the new name. When you have created a new sketch on a plane, it is sometimes easier to work in a plan view, that is, looking straight at, or normal to, the sketch plane. This can be done by using the Look At tool and by clicking the plane or the sketch in the browser. You can place sketch curves, apply constraints, and apply dimensions exactly as you did with the first sketch. In addition to constraining and dimensioning the new sketch, you can also constrain the new sketch to the existing part. You can place dimensions to geometry that does not lie on the current plane; the dimensions, however, will be placed on the current plane. When you look at a part from different viewpoints, you may see arcs and circular edges appearing as lines. After the sketch has been constrained and dimensioned, it can be extruded, revolved, swept, or lofted. Exit the sketch environment by right-clicking in the graphics area and selecting Finish Sketch or by clicking the Return tool in the Standard toolbar. Only one sketch can be active at a time.

FACE CYCLING

Autodesk Inventor has dynamic face highlighting that helps you to select the correct face to activate and to select objects. As you move the cursor over a given face, the edges of the face are highlighted. If you continue to move the cursor, different faces are highlighted as the cursor passes over them.

To cycle to a face that is behind another one, move the cursor over the face that is in front of one that you want to select and hold the cursor still. The Select Other tool appears, as shown in the following image on the left. Select the left or right arrow to cycle through the faces until the correct face is highlighted, and then press the left mouse button or click the green rectangle in the middle of the tool. You can also access the Select Other tool by right-clicking the desired location in the graphics area and clicking Select Other from the menu.

You can specify the amount of time before the Select Other tool will appear automatically. Click Application Options on the Tools menu and click the General tab. As shown in the following image, the "Select Other" delay (sec) feature can be specified in tenths of a second. If you do not want the Select Other tool to open automatically, specify OFF in the field. The default value is 1.0 second.

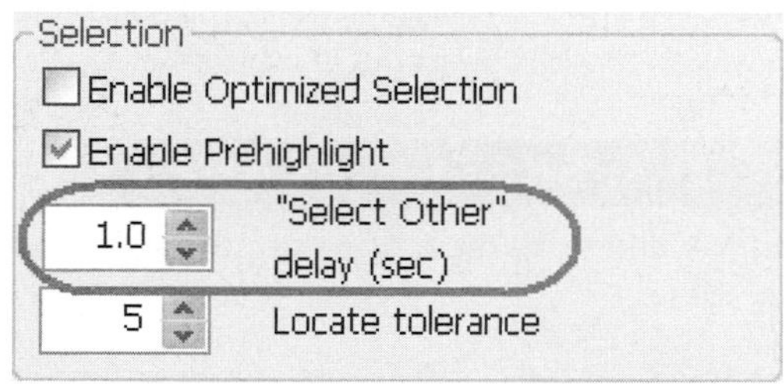

Figure 3.44

SLICE GRAPHICS

While creating parts, you may need to sketch on a plane that is difficult to see because features are obscuring the view. The Slice Graphics option will temporarily slice away the portion of the model that obscures the active sketch plane on which you want to sketch. The following image on the left shows a revolved part with an origin plane visible and the

Slice Graphics menu. The image on the right shows the graphic sliced and the origin plane's visibility turned off. To temporarily slice the graphics screen, follow these steps:

1. Make a plane that the graphics of the active sketch will be sliced through.
2. Rotate the model so the correct side will be sliced, i.e., the side of the model that faces you will be sliced away.
3. While editing the sketch, right-click and select Slice Graphics from the menu as shown in the following image, and then press the F7 key or click Slice Graphics on the View menu. The model will be sliced on the active sketch plane.
4. Use sketch tools from the Sketch toolbar to create geometry on the active sketch.
5. To restore the sliced graphics, right-click and select Slice Graphics, click Slice Graphics on the View menu, and then press the F7 key or click the Sketch or Return button on the Command bar to leave sketch mode.

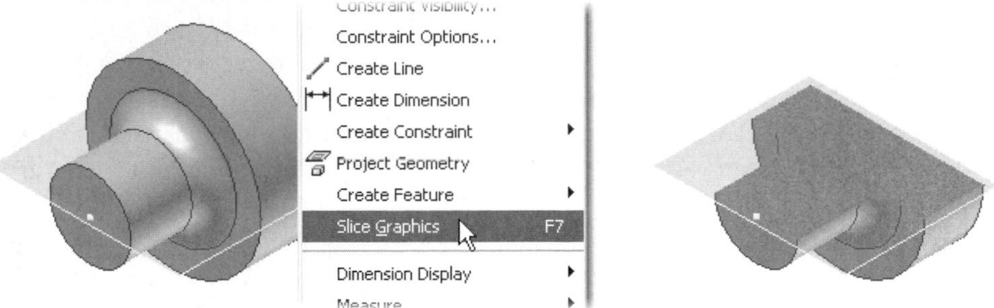

Figure 3.45

Note: When working in an assembly, additional slice graphics tools (Assembly Section Views) are available from the Assembly toolbar.

EXERCISE 3-4: SKETCH PLANES

In this exercise, you create a sketch plane on the angled face of a part and then create a slot using the new sketch plane.

1. Open *ESS_E03_04.ipt* from the Chapter 03 folder.
2. Click the Sketch tool on the Standard toolbar, and then click the top-inside angled face as shown in the following image.

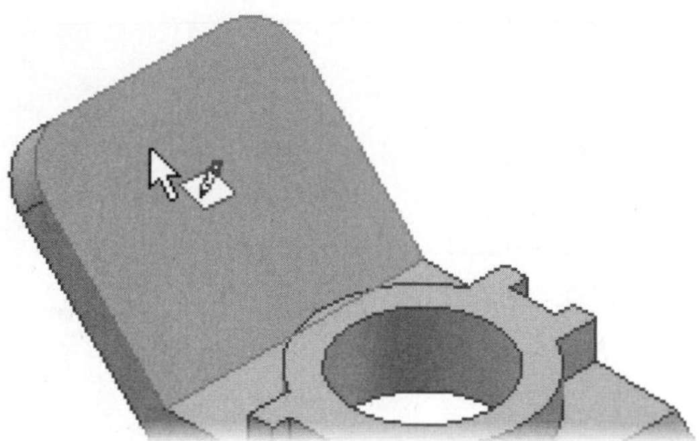

Figure 3.46

3. Click the Look At tool from the Standard toolbar, and then select Sketch6 in the browser.

4. Next you create a sketch for a slot in the part and center the slot vertically. Sketch a slot as shown in the following image. Both arcs should be tangent to the adjacent lines.

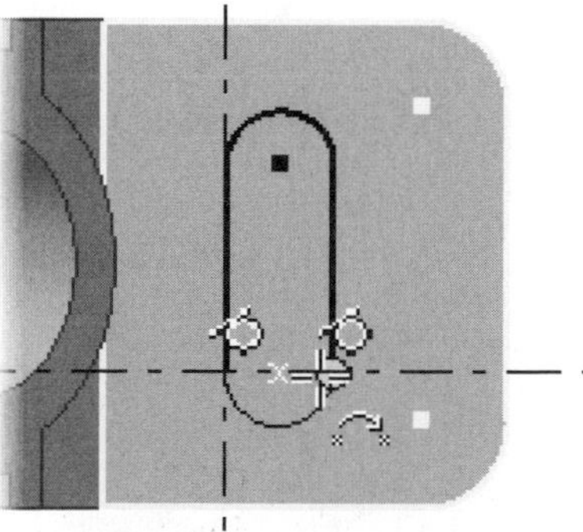

Figure 3.47

5. Click the Horizontal constraint tool from the 2D Sketch Panel bar, click the drop-down arrow to the right of Perpendicular or the last constraint that was applied, and select Horizontal as shown in the following image on the left.

6. Select the midpoint of the right vertical edge of the rectangle and the midpoint of the right vertical edge of the part as shown in the following image on the right. The rectangle is centered vertically using just geometric constraints.

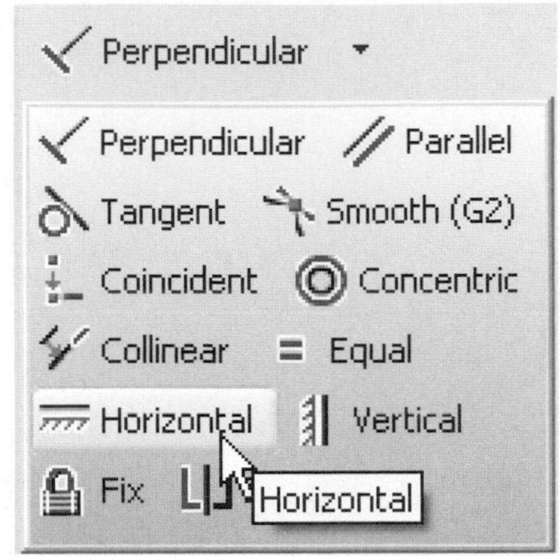

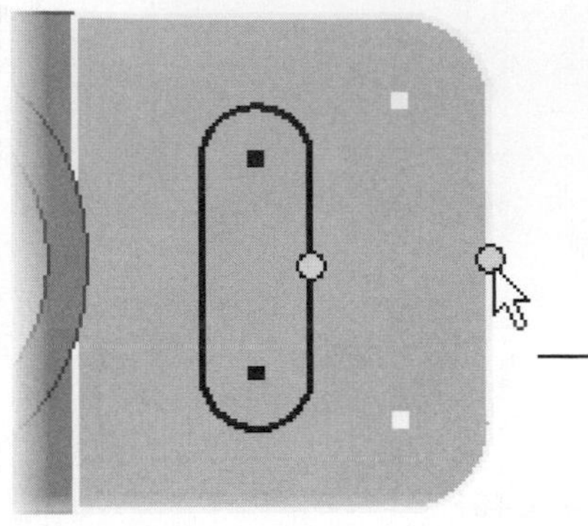

Figure 3.48

7. Click the General Dimension tool. Place three dimensions as shown in the following image.

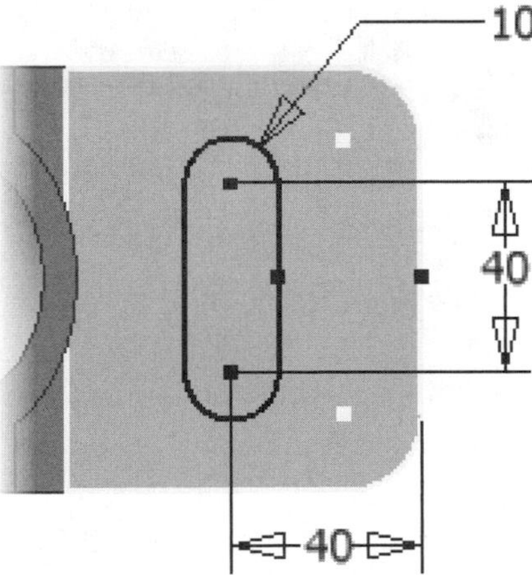

Figure 3.49

8. Press the F6 key to change to the Home View.
9. Press E to start the Extrude tool.
10. For the profile, click inside the slot.
11. Click the Cut operation, and change the Extents to select All. Ensure that the direction is into the part as shown in the following image on the left.
12. Click OK. The completed part is shown in the following image on the right.

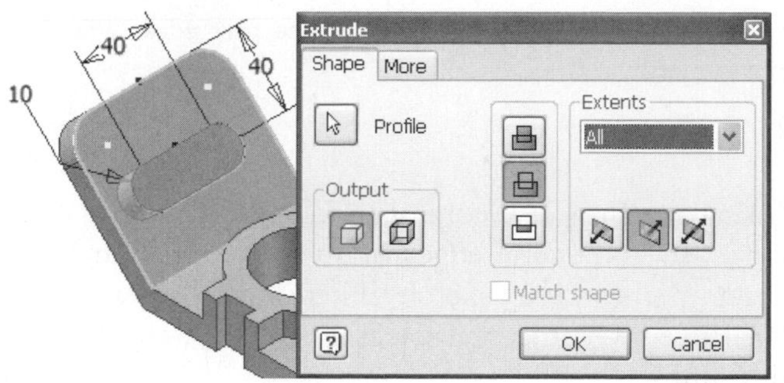

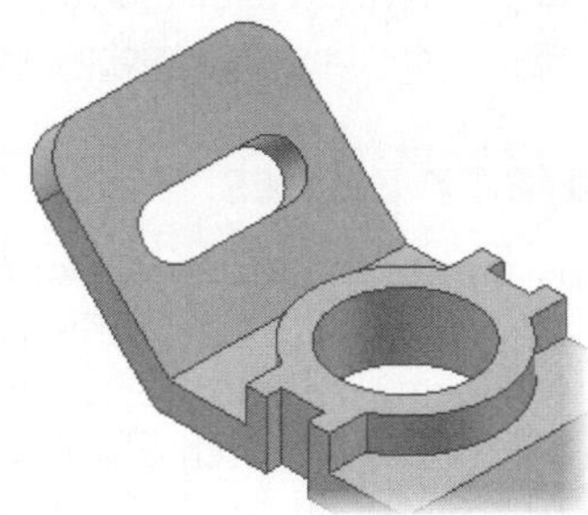

Figure 3.50

13. Close the file. Do not save changes. End of exercise.

PROJECTING PART EDGES

Building parts based partially on existing geometry is done often, and you will frequently need to reference faces, edges, or loops from features that have been created.

While in a part file, you can project an edge, face, point, or loop onto a sketch. Projected geometry can maintain an associative link to the original geometry that is projected. If you project the face of a feature onto another sketch, for example, and the parent sketch is modified, the projected geometry will update to reflect the changes.

DIRECT MODEL EDGE REFERENCING

While you sketch, you can use direct model edge referencing to

- automatically project edges of the part to the sketch plane as you sketch a curve;
- create dimensions and constraints to edges of the part that do not lie on the sketch plane;
- control the automatic projection of part edges to the sketch plane.

Creating Reference Geometry

There are two ways to automatically project part edges to the sketch plane:

- Move the cursor on an edge of the part while sketching a curve. Note: This process requires the Application Option "Autoproject edges during curve creation and edit" to be on.
- Click an edge of the part while creating a dimension or constraint.

  On the Sketch tab in the Application Options dialog box, you can

- Autoproject edges during curve creation, which controls the ability to rub and project edges while sketching a curve;
- Autoproject edges for sketch creation and edit, which controls the automatic projection of the edges of the selected face when you start a sketch on a planar face of the part.

Note: Neither of these options disables the ability to reference part edges when creating dimensions and constraints.

PROJECT EDGES

In this section, you learn about using the Project Geometry tool that can project selected edges, vertices, work features, curves, the silhouette edges of another part in an assembly, or other features in the same part to the active sketch. There are three project tools available on the Sketch Panel Bar: Project Geometry, Project Cut Edges, and Project Flat Pattern, as shown in the following image.

Project Geometry Use to project geometry from a sketch or feature onto the active sketch.

Project Cut Edges Use to project part edges that touch the active sketch. The geometry is only projected if the uncut part would intersect the sketch plane. For example, if a sphere has a sketch plane in the center of the part and the Project Cut Edges tool is initialized, a circle will be projected onto the active sketch.

Project Flat Pattern Use to project a selected face or faces of a sheet metal part flat pattern onto the active sheet metal part sketch plane. Creation of sheet metal parts is covered in chapter 10.

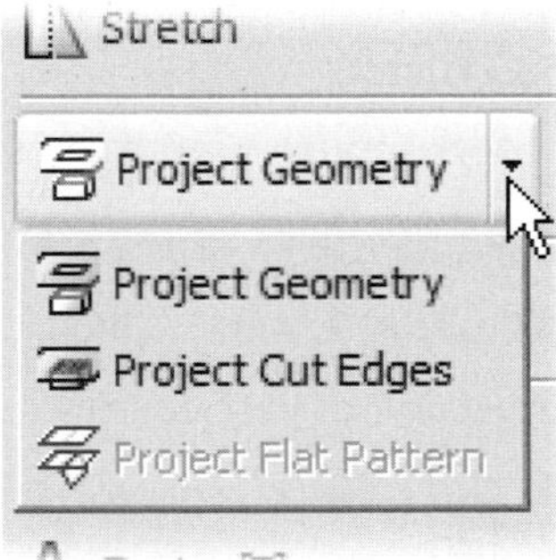

Figure 3.51

To project geometry, follow these steps:

1. Make a plane the active sketch to which the geometry will be projected.
2. Click the Project Geometry tool on the 2D Sketch Panel Bar.
3. Select the geometry to be projected onto the active sketch. Click a point in the middle of a face, and all edges of the face will be projected onto the active sketch plane. If you want to project all edges that are tangent to an edge (a loop), use the Select Other tool to cycle through until they all appear highlighted, as shown in the following image on the left. The following image on the right shows the projected geometry.

4. To exit the operation, press the ESC key or click another tool.

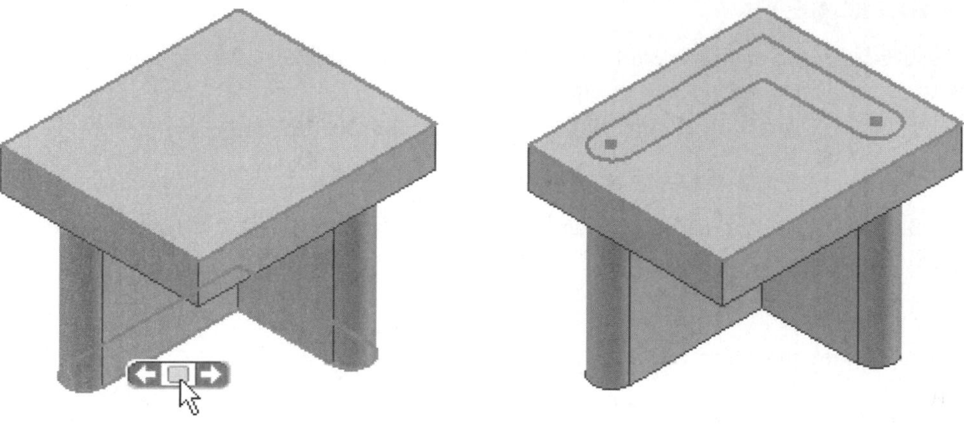

Figure 3.52

If you clicked a loop for the projection, the sketch is updated to reflect the modification when any part of the profile changes. If a face is projected, the internal islands that are defined on the face are also projected and will update accordingly. For example, if the face of the following image on the left is projected, the outer loop of the face and all of the circles that define the hole pattern on the face will be projected and updated if they are modified. If a loop is projected, you can break the association to the projected loop by making the sketch active. Then right-click in the browser on the Projected Loop, and click Break Link from the menu as shown in the following image on the right.

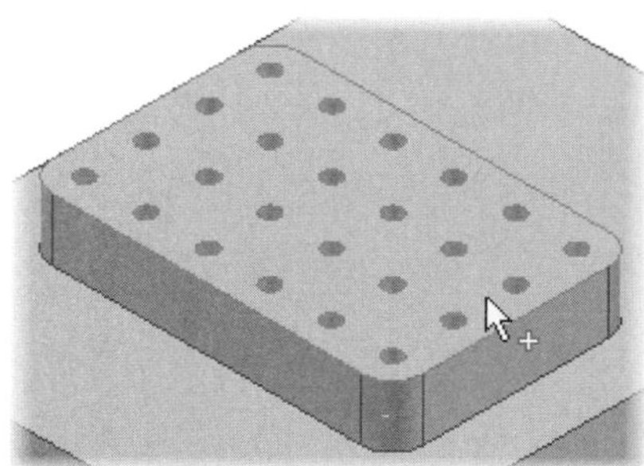

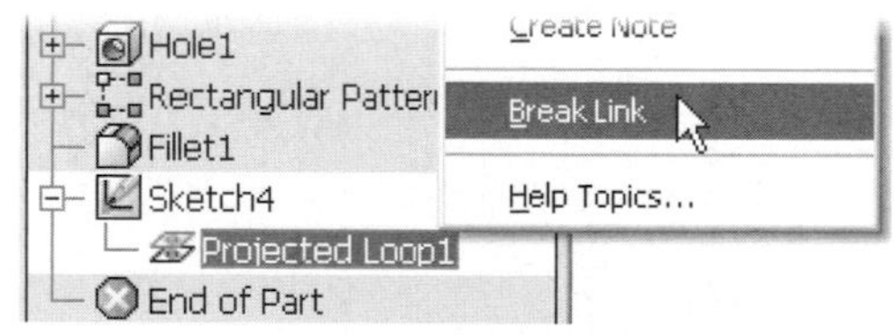

Figure 3.53

Project Exercise: Chapter 3

The self-paced, step-by-step project exercises found in Appendix A provide opportunities for you to work through real-world modeling, assembly, and documentation tasks. The geometry used in the project exercises will flow from chapter to chapter, utilizing the functionality that you learned in that chapter.

Applying Your Skills

SKILL EXERCISE 3–1

In this exercise, you create a bracket from a number of extruded features. Assume that the part is symmetrical about the center of the horizontal slot.

1. Start a new part based on the metric *Standard (mm).ipt* template.
2. Create the sketch geometry for the base feature.
3. Add geometric constraints and dimensions. Make sure that the sketch is fully constrained.
4. Extrude the base feature.
5. Create the three remaining features to complete the part.

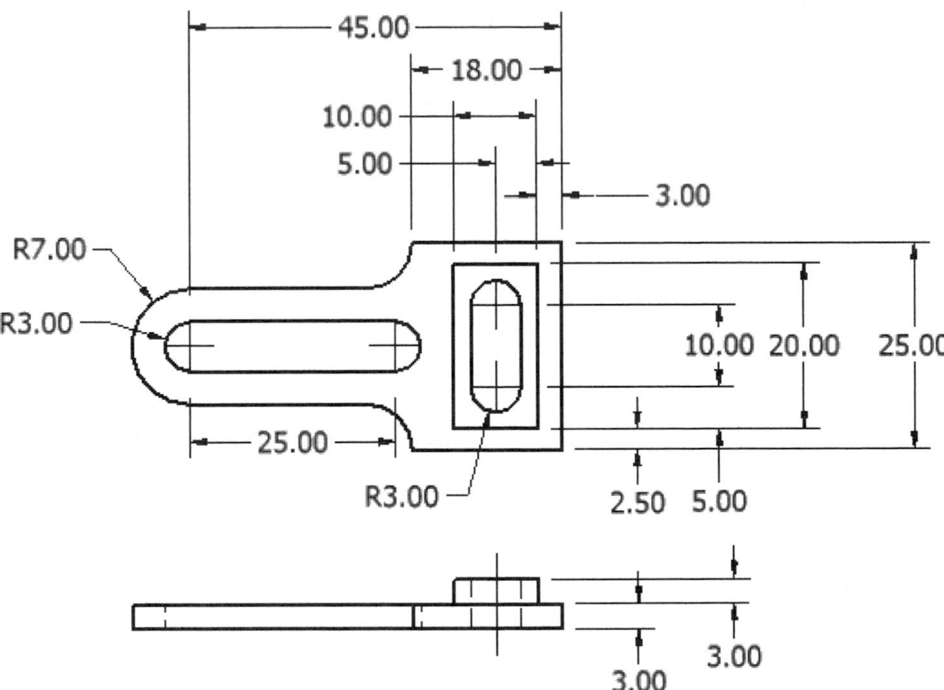

Figure 3.54

The completed part should resemble the following image.

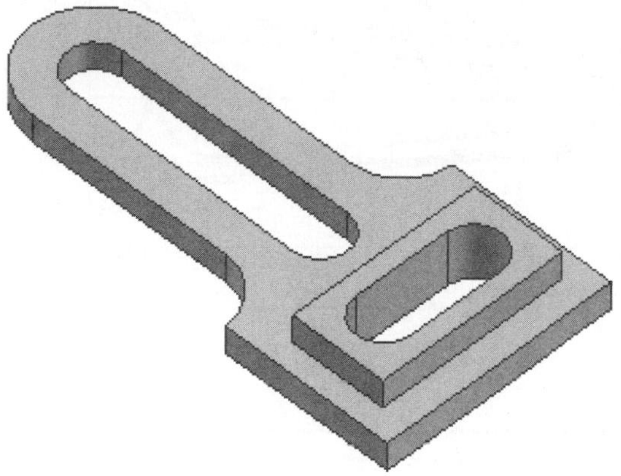

Figure 3.55

SKILL EXERCISE 3–2

In this exercise, you create a connecting rod and add draft to extrusions during feature creation.

1. Start a new part based on the metric *Standard (mm).ipt* template.
2. Create the sketch geometry for the outside of the connecting rod.
3. Add geometric constraints and dimensions to fully constrain the sketch.
4. Extrude the base feature using the midplane option, adding a **-5** degree taper.
5. Create a sketch for the two holes and extrude (cut) them.

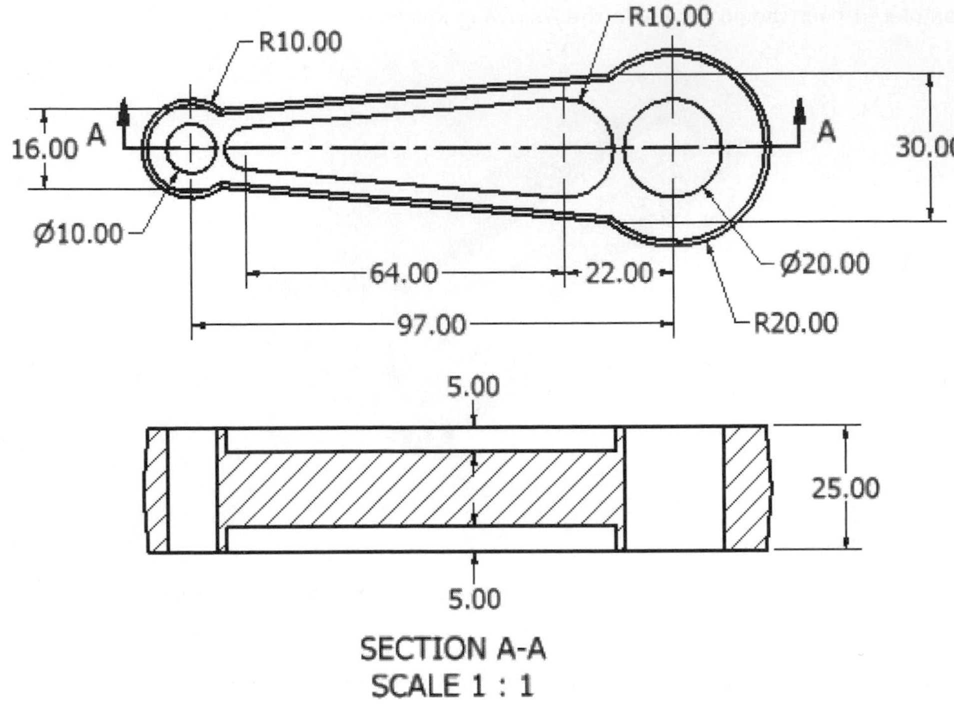

Figure 3.56

6. Create separate sketches for each pocket; then constrain the sketch geometry and add dimensions. The sides of the pocket are parallel to the sides of the connecting rod.
7. Extrude each pocket using a draft of -3 degrees.

The completed part should resemble the following image.

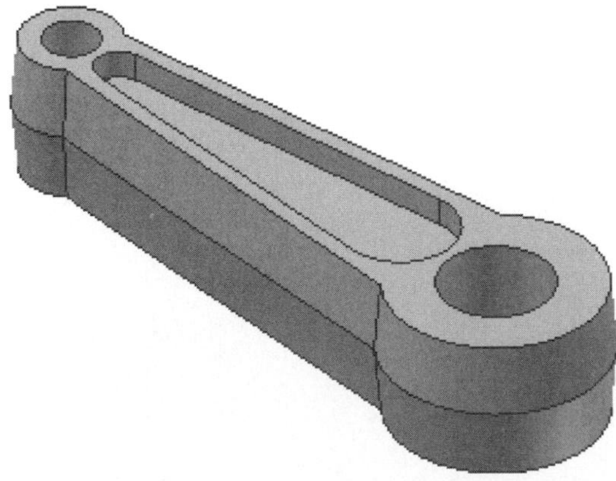

Figure 3.57

SKILL EXERCISE 3-3

In this exercise, you create a pulley using a revolved feature. Assume that the part is symmetrical.

1. Start a new part based on the metric *Standard (mm).ipt* template.
2. Create the sketch geometry for the cross-section of the pulley.

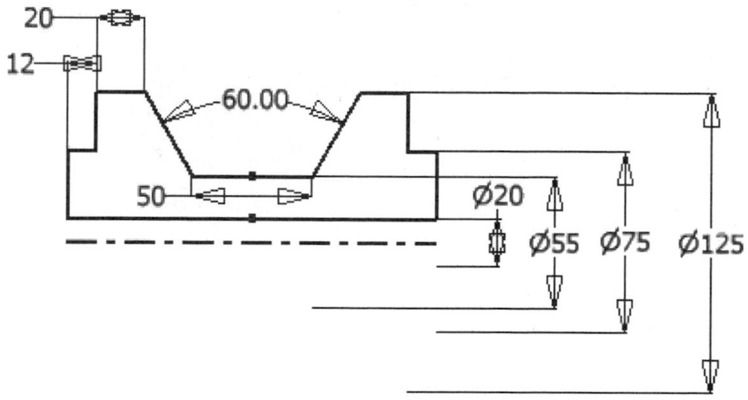

Figure 3.58

Tip: To create a centerline, draw a line, select it, and then select the Centerline tool on the Standard toolbar.

3. Apply appropriate geometric constraints.
4. Add dimensions. To create a linear diameter dimension, select the centerline and then the line to which you want to apply the dimension.
5. Revolve the sketch.
6. Close the file. Do not save changes.

The completed part should resemble the following image.

EXERCISES

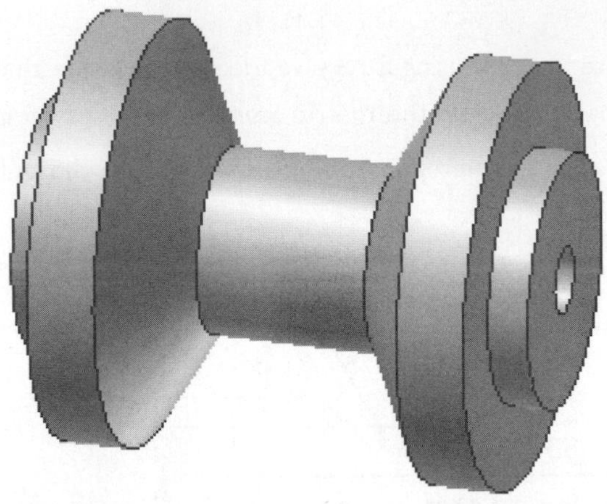

Figure 3.59

CHECKING YOUR SKILLS

Use these questions to test your knowledge of the material covered in this chapter.

1. What is a base feature?

2. **True _ False _** When creating a feature with the Extrude or Revolve tool, you can drag the sketch to define the distance or angle.

3. Which objects can be used as an axis of revolution?

4. Explain how to create a diametric dimension on a sketch.

5. Name two ways to edit an existing feature.

6. **True _ False _** Once a sketch becomes a base feature, you cannot delete or add constraints, dimensions, or objects to the sketch.

7. Name three operation types used to create sketched features.

____

____

8. True _ False _ A cut operation cannot be performed before a base feature is created.

9. True _ False _ Once a sketched feature exists, its termination cannot be changed.

10. True _ False _ Geometry that is projected from one feature to a sketch that defines another feature will update automatically based on changes to the original projected geometry.

EXERCISES

Creating Placed Features

INTRODUCTION

In chapter 3, you learned how to create and edit base and sketched features. In this chapter, you will learn how to create *placed* features. These are features that are predefined except for specific values and only need to be located. You can edit placed features in the browser like sketched features. When you edit a placed feature, either the dialog box that you used to create it will open or feature values will appear on the part.

When creating a part, it is usually better to use placed features instead of sketched features wherever possible. To make a through hole as a sketched feature, for example, you can draw a circle profile, dimension it, and then extrude it with the cut operation, using the All extension. You can also create a hole as a placed feature—you can select the type of hole, size it, and then place it using a dialog box. When drawing views are generated, the type and size of the hole are easy to annotate, and they automatically update if the hole type or values change.

OBJECTIVES

After completing this chapter, you will be able to perform the following:

- Create fillets
- Create chamfers
- Create holes
- Create internal and external threads
- Shell a part
- Add face draft to a part
- Create work axes
- Create work points
- Create work planes
- Pattern features

FILLETS

Fillet features consist of fillets and rounds. Fillets add material to interior edges to create a smooth transition from one face to another. Rounds remove material from exterior edges. The following image shows a part without fillets on the left and the part with fillets on the right.

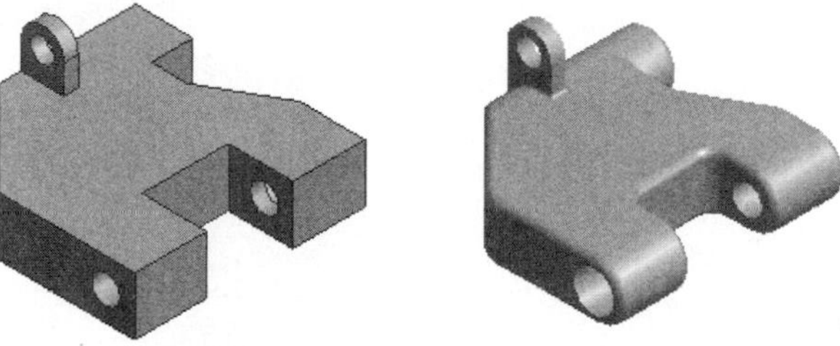

Figure 4.1

To create fillets in 3D, you select the edge that needs to be filleted; the fillet is created between the two faces that share the edge, or you can select two faces that a fillet will go between. When placing a fillet between two faces, the faces do not need to share a common edge. This is different from placing a fillet in 2D. In the 2D environment, you click two objects, and a fillet is created between them. When creating a part, it is good practice to create fillets and chamfers as some of the last features in the part. Fillets add complexity to the part, which in turn adds to the size of the file. They also remove edges that you may need to place other features.

To create a fillet feature, click the Fillet tool from the Part Features panel bar, as shown in the following image on the left, or press the hot key F. After you click the tool, the Fillet dialog box appears. The following image on the right shows the Fillet dialog box with all options displayed.

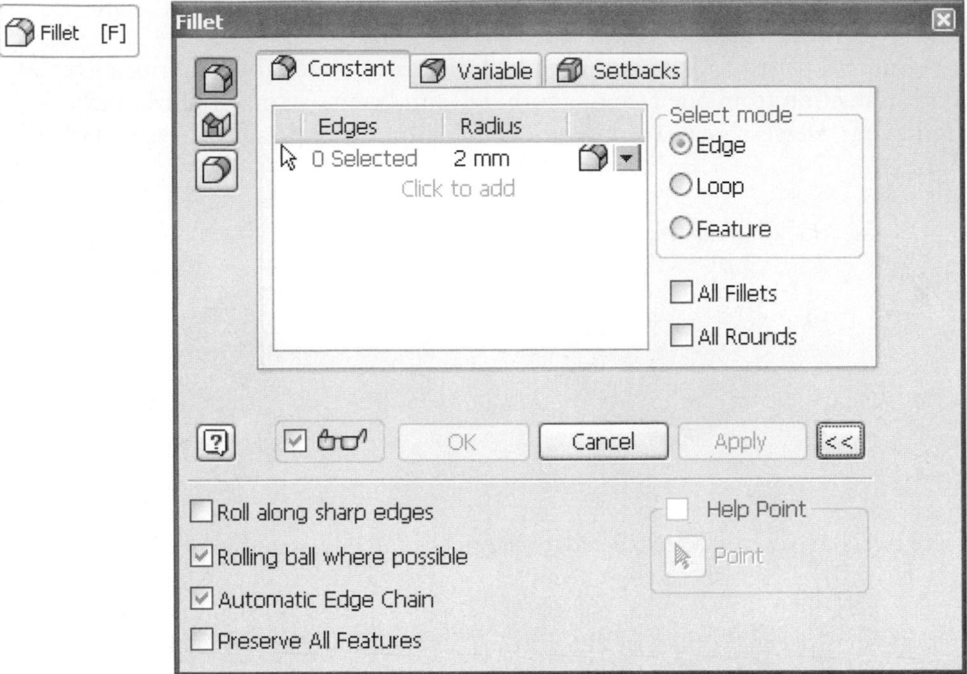

Figure 4.2

Along the left side of the Fillet dialog box, there are three types of fillets: Edge, Face, and Full Round. When the Edge option is selected, you see three tabs: Constant, Variable, and Setbacks. Each tab creates a fillet along an edge(s) with different options. The options for each of the tabs and fillet types are described in the following sections. A preview option appears on the bottom of the dialog box. When checked, and a valid fillet can be created from the input data, the fillet will be previewed.

Before we look at the tabs, let's discuss the methodology that you will use to create fillets. You can click an edge or edges, or faces to fillet, or select the type of fillet to create. If you click an edge or face before you issue the Fillet feature tool, it is placed in the first selection set. Selection sets contain the edges or faces that will be filleted when the OK button is clicked.

Each fillet feature can contain multiple selection sets, each having its own unique fillet value. There is no limit to the number of selection sets that can exist in a single instance of the feature. An edge, however, can only exist in one selection set. All of the selection sets included in an individual fillet command appear as a single fillet feature in the browser. Click the type of fillet that you want to create. To add edges to the first selection set, select the edges that you want to have the same radius. To create another selection set, select Click to add, and then click the edges that will be part of the next selection set. To remove an edge or face that you have selected, click the selection set that includes the edge or face, and the edges or faces will be highlighted. Hold down the CTRL key, and click the edge(s) or face(s) to be removed from the selection set. Enter the desired values for the fillet. As changes are made in the dialog box, a representation of the fillet is previewed in the graphics area. When the fillet type and value are correct, click the OK button to create the fillet.

To edit a fillet's type and radius, follow these steps:

1. Issue the Edit Feature tool by right-clicking on the fillet's name in the browser and selecting Edit Feature from the menu. You can also double-click on the feature's name in the browser or change the select priority to Feature Priority and double-click on the fillet that is on the part. The Fillet dialog box appears with all of the settings that you used to create the fillet.
2. Change the fillet settings as needed.
3. Click the Update tool to update the part.

EDGE FILLET

With the Edge fillet type selected, the Constant tab will be activated. With the options on the Constant tab, as shown in the previous image, you can create fillets that have the same radius from beginning to end. There is no limit to the number of part edges that you can fillet with a constant fillet. You can select the edges as a single set or as multiple sets, and each set can have its own radius value. The order in which you select the edges of a selection set is not important. You do need to select the edges that are to be filleted individually, and the use of the window or crossing selection method is not allowed. If you change the value of a selection set, all of the fillets in that group will change. To remove an edge from a group, choose the group in the select area, and then hold down the CTRL key and click the edge to be removed.

Constant Tab

The following section describes the options that are available on the Constant tab.

Select Edge, Radius, and Continuity

Edge By default, after issuing the Fillet tool, you can click edges, and they appear in the first selection set. You can continue to select multiple edges. To remove an edge from the set, hold down the CTRL key and select the edge.

Radius Enter a size for the fillet. The size of the fillet will be previewed on the selected edges.

Continuity To adjust the continuity of the fillet, select either a tangent or smooth (G2) condition (continuous curvature), as shown in the following image.

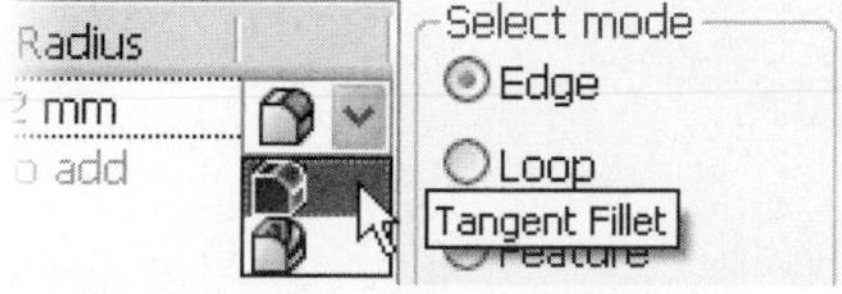

Figure 4.3

To create another selection set, select Click to add, and then click the edges that will be part of the next selection set. After clicking an edge, a preview image of the fillet appears on the edge that reflects the current values, as shown in the following image.

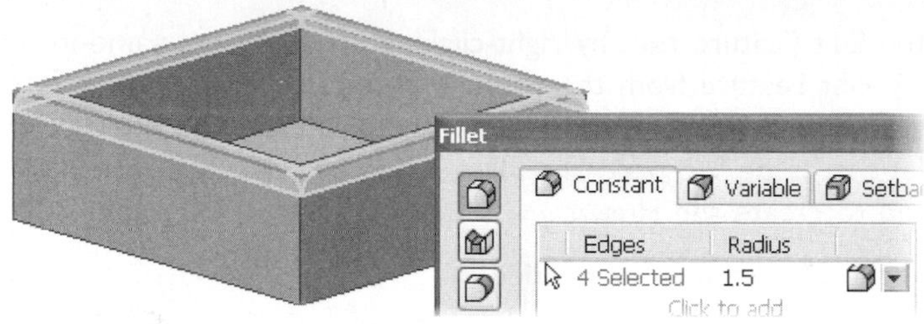

Figure 4.4

Select Mode

Edge Click the Edge mode to select individual edges to fillet. By default, any edge that is tangent to the clicked edges is also selected. If you do not want to have tangent edges automatically selected, uncheck the Automatic Edge Chain option in the More (>>) section.

Loop Click the Loop mode to have all of the edges that form a closed loop with the selected edge filleted.

Feature Click the Feature mode to select all of the edges of a selected feature.

All Fillets Click the All Fillets option to select all concave edges of a part that you have not filleted already—see the following image on the left. The All Fillets option adds material to the part, and it requires a separate edge selection set to remove material from the remaining edges using the All Rounds option.

All Rounds Click the All Rounds option to select all convex edges of a part that you have not filleted already, as seen in the following image on the right. The All Rounds option removes material from the part, and it requires a separate edge selection set from All Fillets.

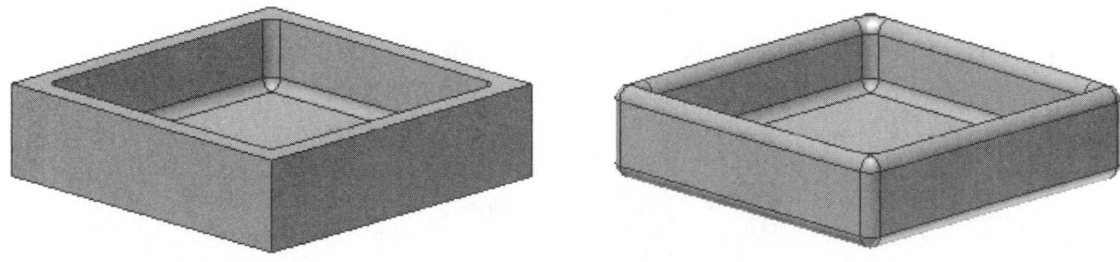

Figure 4.5

More (>>) Options Click the More (>>) button to access other options, as shown in the following image.

Roll along sharp edges Click this option to adjust the specified radius when necessary to preserve the edges of adjacent faces.

Rolling ball where possible Click this option to create a fillet around a corner that looks like a ball has been rolled along the edges that define the corners, as shown in the following (middle) image. When the rolling ball solution is possible but you have not selected it, a blended solution is used, as shown in the following image on the right.

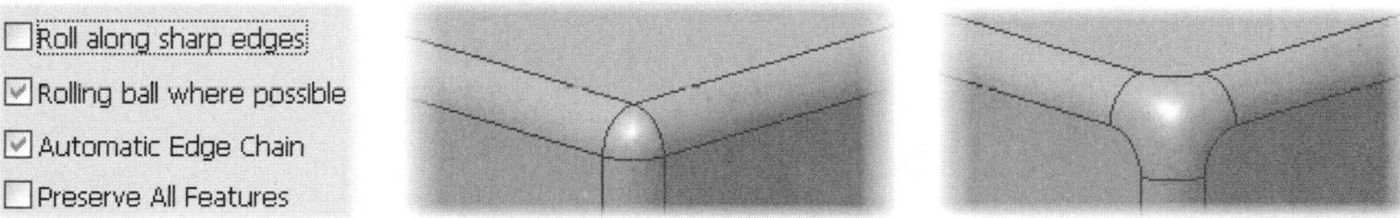

Figure 4.6

Automatic Edge Chain Click this option to select tangent edges automatically when you click an edge.

Preserve All Features Click to check all features that intersect with the fillet and to calculate their intersections during the fillet operation. If the option's checkbox is clear, only the edges that are part of the fillet operation are calculated during the operation.

For example, if Preserve All Features is checked, a fillet is placed on the outside edge of a shelled box, and the fillet is larger than the shell thickness, a gap will exist in the fillet, as shown in the image on the left. If the Preserve All Features is unchecked, the inside edges for the shell will appear in the fillet, as shown in the following image on the right.

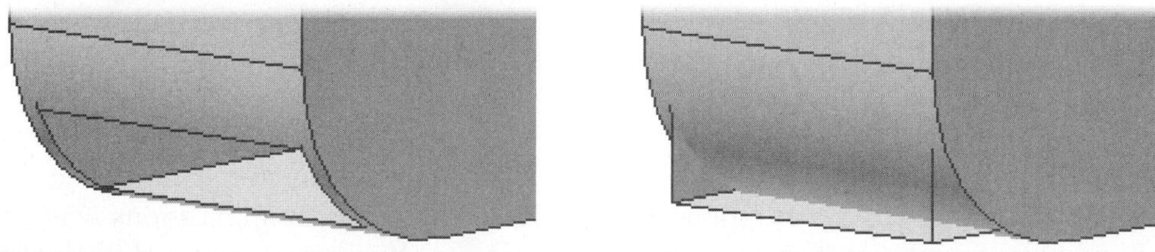

Figure 4.7

Variable Tab

You can also create a variable radius fillet that has a different starting and ending radius and/or a different radius between the starting and ending radius. To create a variable radius fillet, click the Edge Fillet option in the upper-left corner of the dialog box, and click the Variable tab, as shown in the following image.

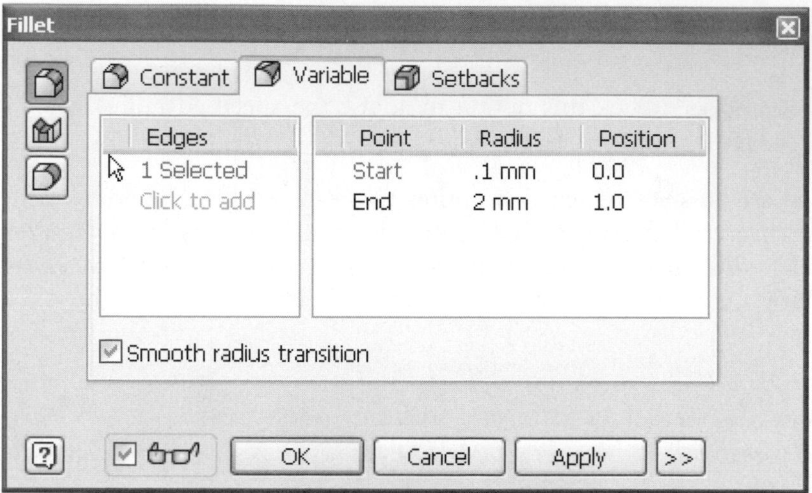

Figure 4.8

The following image on the left shows a variable fillet with the smooth option, and the image on the right shows a variable fillet blending in a straight line.

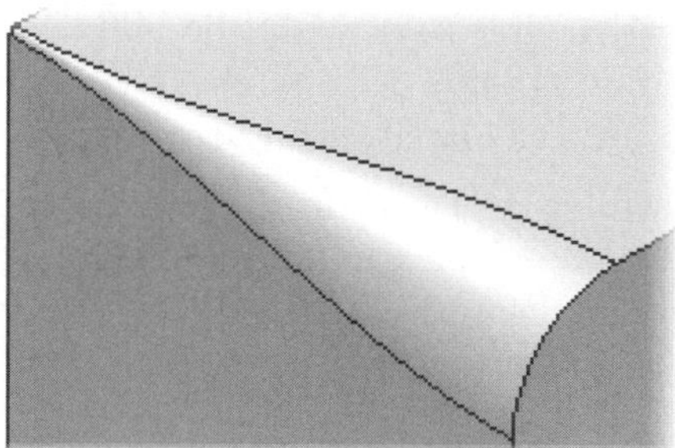

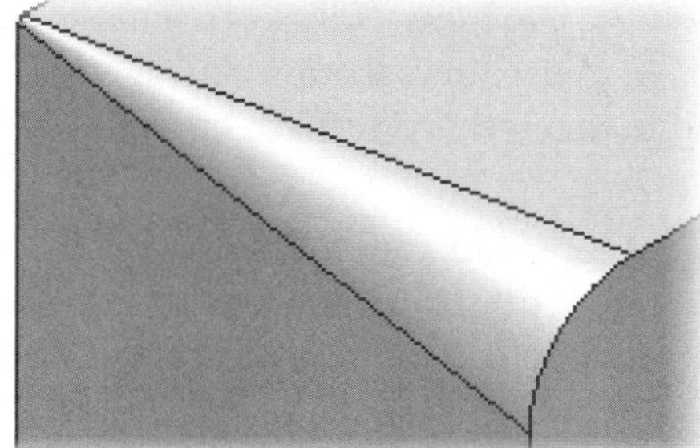

Figure 4.9

Setbacks Tab

You can specify the distance at which a fillet starts its transition from a vertex with the options on the Setbacks tab. Using these options, you can model special fillet applications where three or more edges converge, as shown in the following image. You can choose a different radius for each converging edge if needed. Click the minimal option to create a setback with the smallest possible fillet. You can only use setbacks where three or more filleted edges form a vertex.

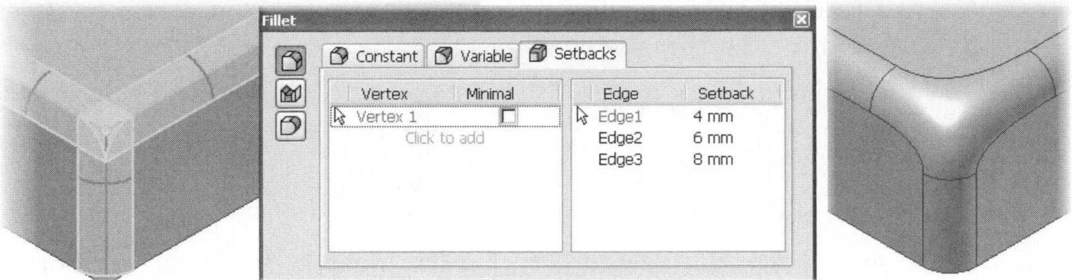

Figure 4.10

FACE FILLET

With the Face fillet type selected, the dialog box will change, as shown in the following image. Create a face fillet by selecting two or more faces; the faces do not need to be adjacent. If a feature exists that will be consumed by the fillet, the volume of the feature will be filled in by the fillet.

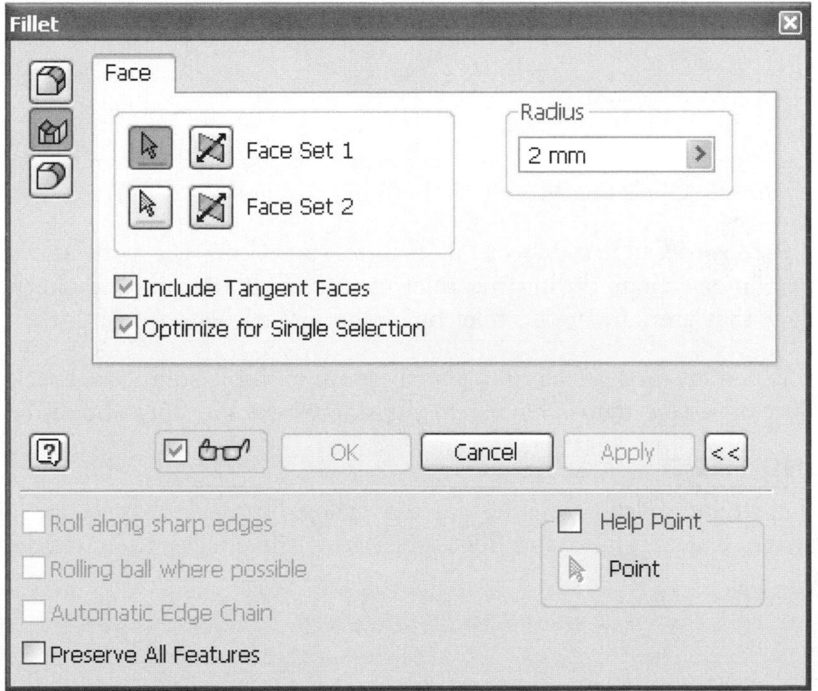

Figure 4.11

To create a face fillet, follow these steps.

1. With the Face Set 1 button active, select one or more tangent contiguous faces on the part to which the fillet will be tangent.

2. With the Face Set 2 button active, select one or more tangent contiguous faces on the part to which the fillet will be tangent.

Check the Include Tangent Faces option to automatically chain all faces that are tangent to faces in the selection set.

Check the Optimize for Single Selection to automatically make the next selection set button active after selecting a face.

The following image on the left shows a part that has a gap between the bottom and top extrusion, the middle image shows the preview of the face fillet, and the image on the right shows the completed face fillet. Notice the rectangular extrusion on the top face is consumed by the fillet.

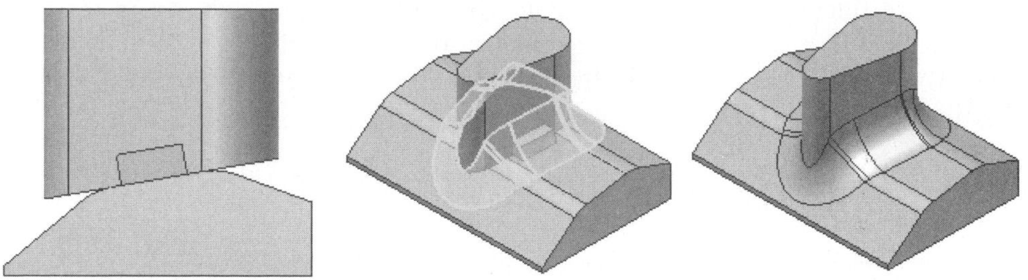

Figure 4.12

More (>>) Options Click the More (>>) button to access other options.

Preserve All Features Click to check all features that intersect with the fillet and to calculate their intersections during the fillet operation. If the option's checkbox is clear, only the edges that are part of the fillet operation are calculated during the operation.

Help Point When creating a face fillet and there are multiple solutions, check this option, and then click on a face that is closest to the side where the fillet should be created.

FULLROUND FILLET

With the FullRound fillet type selected, the dialog box will change, as shown in the following image. Create a fullround fillet by selecting three faces; the faces do not need to be adjacent.

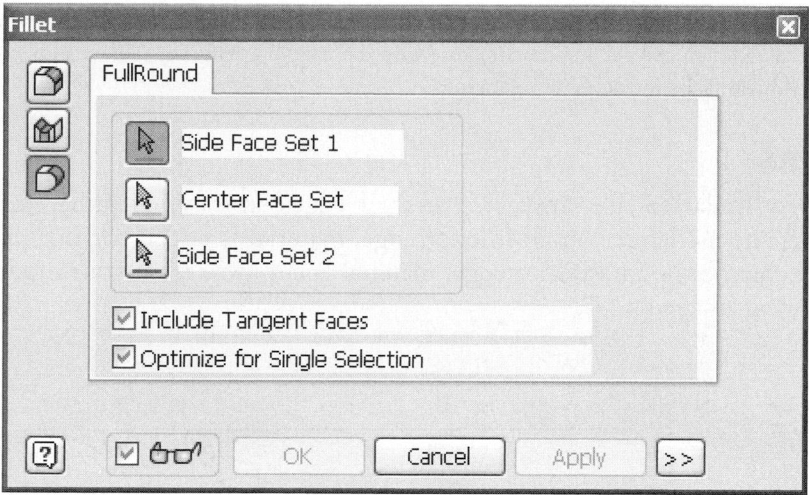

Figure 4.13

To create a FullRound fillet, follow these steps:

1. With the Side Face Set 1 button active, select a face on the part that the fillet will start at and to which it will be tangent.
2. With the Center Face Set button active, select a face on the part that the middle of the fillet will be tangent.
3. With the Side Face Set 2 button active, select a face on the part that the fillet will end at and to which it will be tangent.

Check the Include Tangent Faces option to automatically chain all faces that are tangent to faces in the selection set.

Check the Optimize for Single Selection to automatically make the next selection set button active after selecting a face.

The following image shows three faces selected on the left and the resulting fullround fillet on the right.

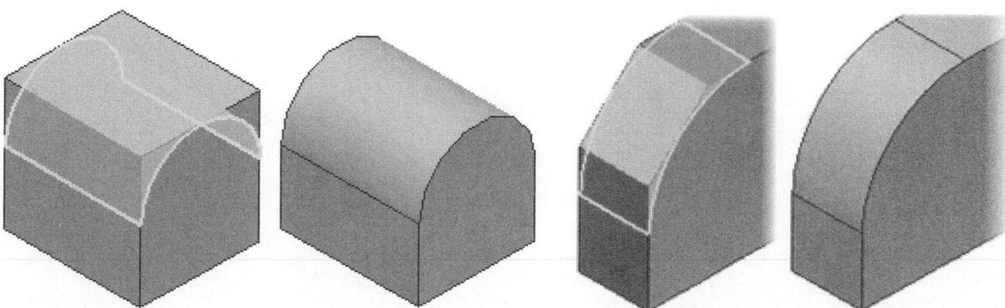

Figure 4.14

Tip: If you get an error when creating or editing a fillet, try to create it with a smaller radius. If you still get an error, try to create the fillet in a different sequence or create multiple fillets in the same operation.

CHAMFERS

Chamfers are similar to fillets except that their edges are beveled rather than rounded. When you create a chamfer on an interior edge, material is added to your model. When you create a chamfer on an exterior edge, material is cut away from your model, as shown in the following image.

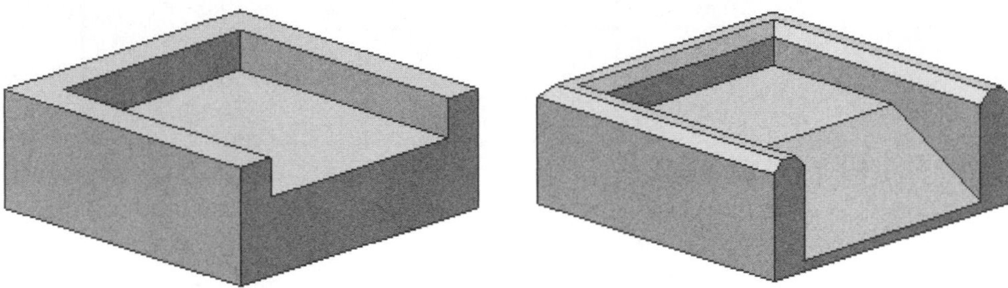

Figure 4.15

To create a chamfer feature, follow the same steps that you used to create the fillet features. Click the common edge, and the chamfer is created between the two faces sharing the edge. To create a chamfer feature, click on the Chamfer tool on the Part Features panel bar, as shown in the following image on the left, or press the hot key CTRL + SHIFT + K. After you click the tool, the Chamfer dialog box appears, as shown in the following image on the right. As with fillet features, you can select multiple edges to be included in a single chamfer feature. From the dialog box, click a method, click the edge or edges to chamfer, enter a distance and/or angle, and then click OK.

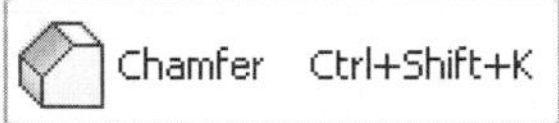

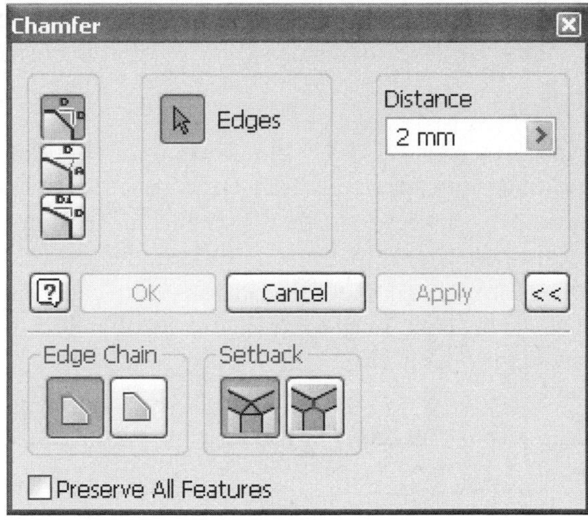

Figure 4.16

To edit the type of chamfer feature or distances, use one of the following methods:

- Double-click the feature's name or icon in the browser to edit a distance.
- Right-click the chamfer's name in the browser, and select Edit Feature from the menu. The Chamfer dialog box appears with all the settings that you used to create the feature. Adjust the settings as desired.
- Change the select priority to Feature Priority, and then double-click the chamfer on the part. The Chamfer dialog box appears with all the settings that you used to create the feature. Adjust the settings as desired.

METHOD

Distance

Click the Distance option to create a 45° chamfer on the selected edge. You determine the size of the chamfer by typing a distance in the dialog box. The value is the offset from the common edge of the two adjacent faces. You can select a single edge, multiple edges, or a chain of edges. A preview image of the chamfer appears on the part. If you select the wrong edge, hold down the CTRL key and select the edge to remove. The following image illustrates the use of the Distance option on the left.

Distance and Angle

Click the Distance and Angle option to create a chamfer offset from a selected edge on a specified face, at the defined angle. In the dialog box, enter an angle and distance for the chamfer, then click the face to which the angle is applied and specify an edge to be chamfered. You can select one edge or multiple edges. The edges must lie on the selected face. A preview image of the chamfer appears on the part. If you selected the wrong face or edge, click on the Edge or Face button, and choose a new face or edge. The following image in the middle illustrates the use of the Distance and Angle option.

Two Distances

Click the Two Distances option to create a chamfer offset from two faces, each being the amount that you specify. Click an edge first, and then enter values for Distance1 and Distance2. A preview image of the chamfer appears. To reverse the direction of the distances, click the Flip button. When the correct information about the chamfer is in the dialog box, click the OK button. You can only use a single edge or chained edges with the Two Distances option. The following image illustrates the use of the Two Distances option on the right.

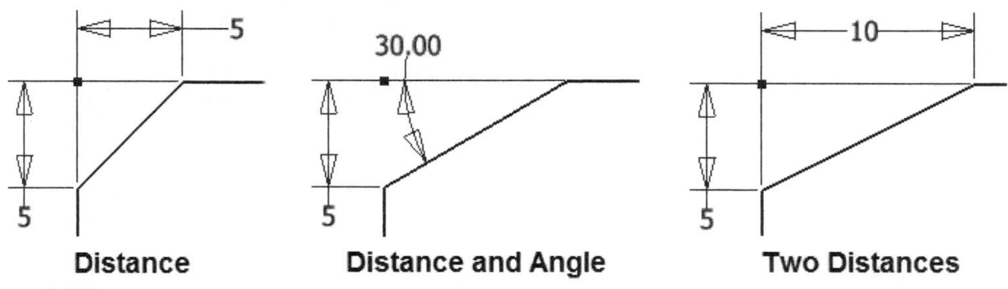

Figure 4.17

EDGE AND FACE

Edges

Click an edge or edges to be chamfered.

Face

Click a face on which the chamfer will be based.

Flip

Click the button to reverse the direction of the distances for a Two Distances chamfer.

DISTANCE AND ANGLE

Distance

Enter a distance to be used for the offset.

Angle

Enter a value that will be used for the angle if creating the Distance and Angle chamfer type.

EDGE CHAIN AND SETBACK

Edge Chain Click this option to include tangent edges in the selection set automatically, as shown in the following image.

Setback When the Distance method is used and three chamfers meet at a vertex, click this option to have the intersection of the three chamfers form a flat edge (left button) or to have the intersection meet at a point as though the edges were milled (right button), as shown in the following image.

Preserve All Features Click this option to check all features that intersect with the chamfer and to calculate their intersections during the chamfer operation, as shown in the following image. If the option's checkbox is clear, only the edges that are part of the fillet operation are calculated during the operation.

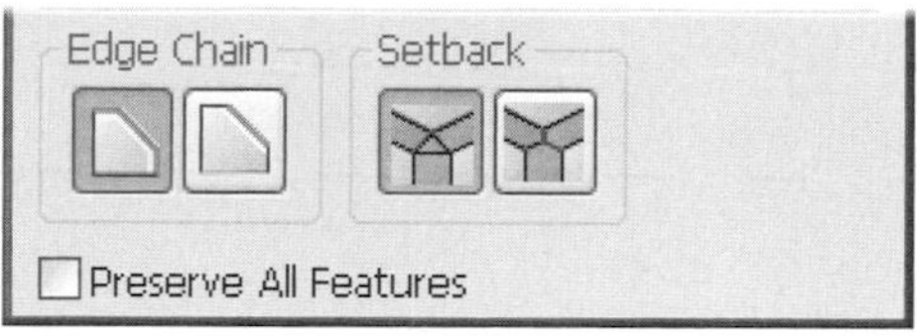

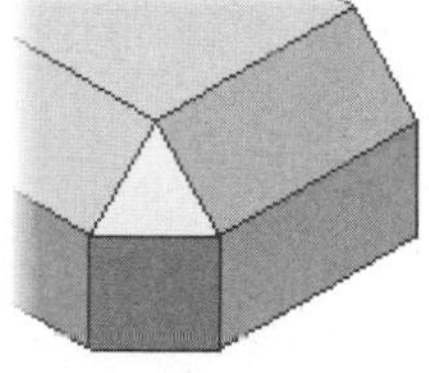

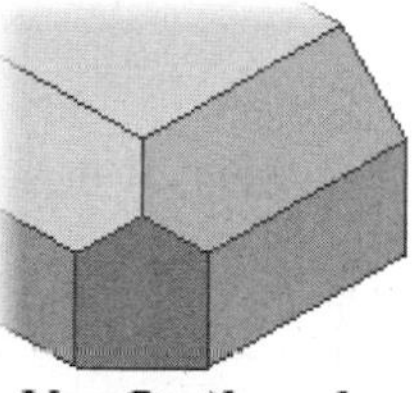

Setback No Setback

Figure 4.18

EXERCISE 4-1: CREATING FILLETS AND CHAMFERS

In this exercise, you create constant radius fillets, variable radius fillets, and chamfers.

1. Open *ESS_E04_01.ipt* in the Chapter 04 folder.
2. Click the Fillet tool in the Part Features panel bar.
3. Click the inside edge of the slot, as shown in the following image on the left.
4. In the Fillet dialog box, click on the first entry in the Radius column, and then type **4**, as shown in the following image on the right.

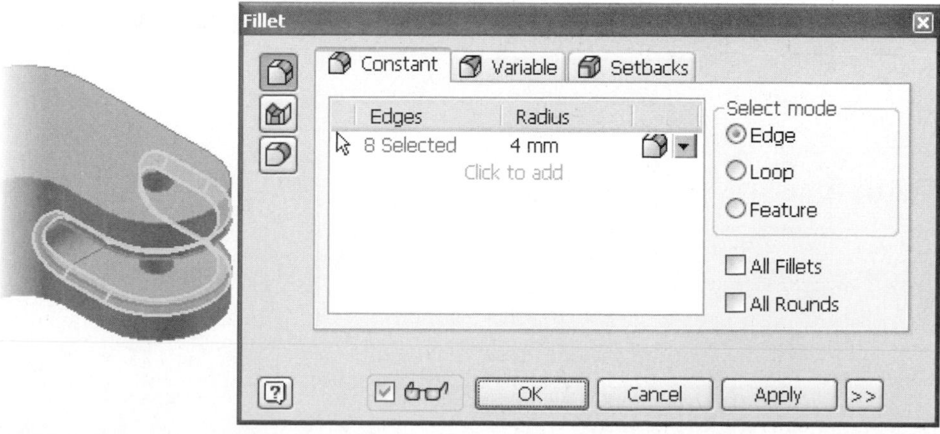

Figure 4.19

5. Click Apply to create the fillet.
6. Next, create a Full Round Fillet. In the Fillet dialog box, click the Full Round Fillet option in the left column.
7. Click the front-inside face, then the left-front face, and then the back-vertical face of the model, as shown in the following image.

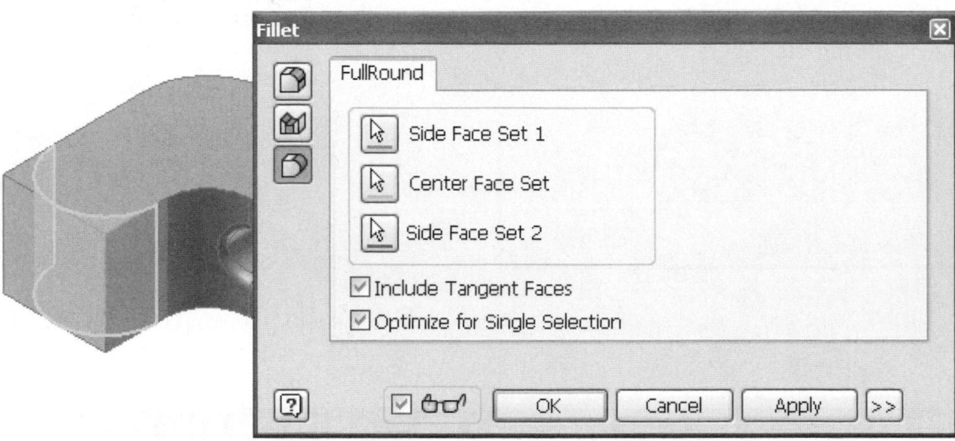

Figure 4.20

8. Click OK to create the fillet.
9. In the browser, right-click on Extrusion4, and click Unsuppress Features.
10. Next, create a Face Fillet. Click the Fillet tool in the panel bar. In the Fillet dialog box, click the Face Fillet option in the left column.
11. Click the cylindrical face of Extrusion4 and then the full round fillet in the model you created in the last step, as shown in the following image on the left.
12. In the Fillet dialog box, type **5 mm** in the Radius field, as shown in the following image on the right.

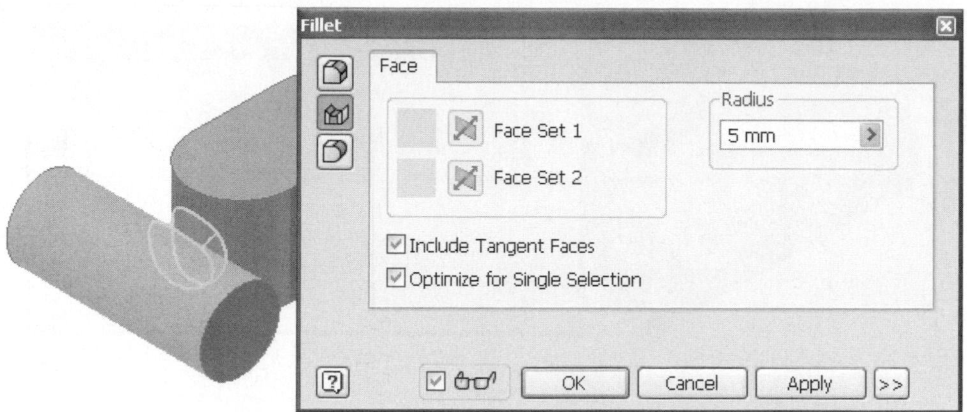

Figure 4.21

13. Click OK to create the fillet.
14. Move the cursor into a blank area in the graphics window, right-click, and click Repeat Fillet from the menu.
15. Create two edge fillets by clicking the top and bottom edges of the model, as shown in the following image.

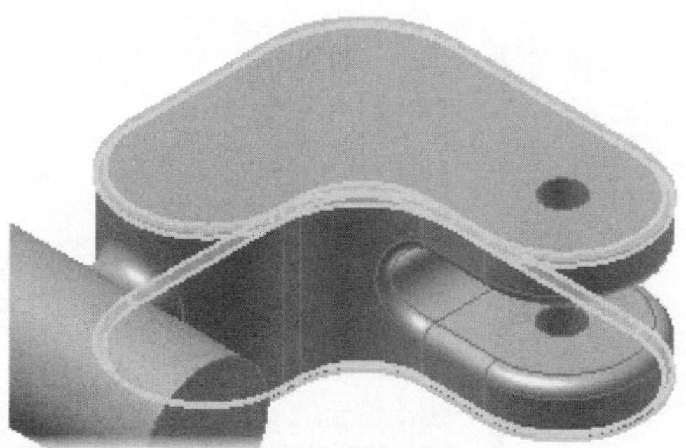

Figure 4.22

16. In the Fillet dialog box, type **2 mm** in the Radius field.
17. Click OK to create the fillet.
18. Click the Chamfer tool in the Part Features panel bar.
19. Click the back cylindrical edge on Extrusion4, as shown in the following image on the left.
20. In the Chamfer dialog box, type **5 mm** in the Distance field, as shown in the following image on the right.

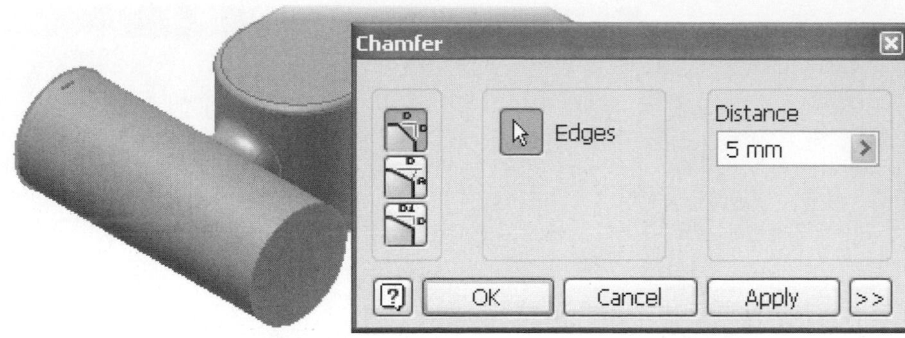

Figure 4.23

21. Click Apply to create the chamfer.
22. In the Chamfer dialog box, click the Distance and Angle option.
23. Click the front-circular face on Extrusion4.
24. Click the front-circular edge of the same face.
25. In the Chamfer dialog box, enter **6 mm** in the Distance field and **60 deg** in the Angle field, as shown in the following image.

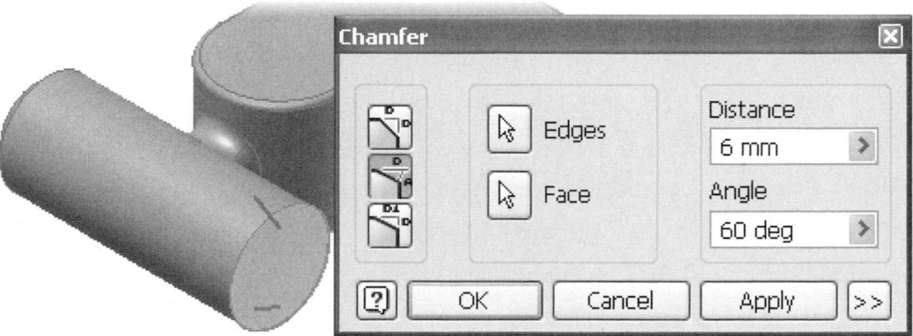

Figure 4.24

26. Click OK to create the chamfer. When done, your model should resemble the following image.

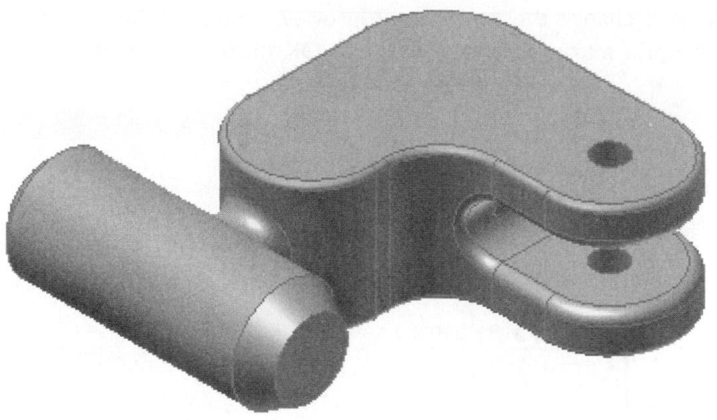

Figure 4.25

27. Close the file. Do not save changes. End of exercise.

HOLES

The Hole tool lets you create drilled, counterbored, spotface, countersunk, clearance, tapped, and taper tapped holes, as shown in the following image. You can place holes using sketch geometry or existing planes, points, or edges of a part. You can also specify the type of drill point and thread parameters.

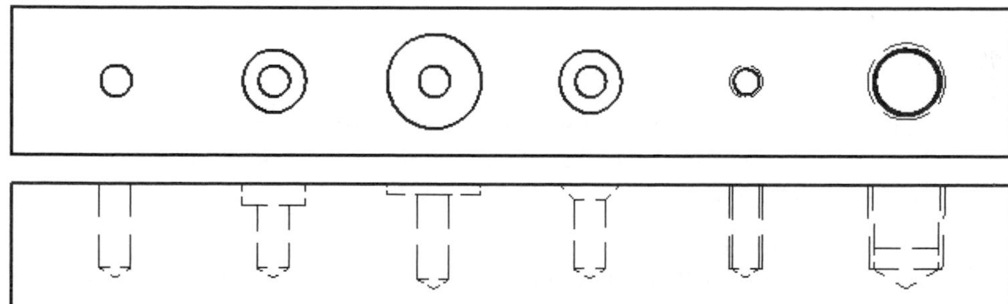

Figure 4.26

To create a hole feature, follow these steps:

1. Create a part that contains a face or plane where you want to place a hole.
2. Click the Hole tool from the Part Features panel bar, as shown in the following image on the left, or press the hot key H.
3. The Holes dialog box appears, as shown in the following image on the right. Four placement options are available: From Sketch, Linear, Concentric, and On Point. The Placement options are covered in the next section. After you have chosen the hole placement options, select the desired hole style options from the Holes dialog

box. As you change the options, the preview image of the hole(s) is updated. When you are done making changes, click the OK button to create the hole(s).

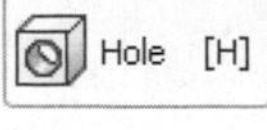

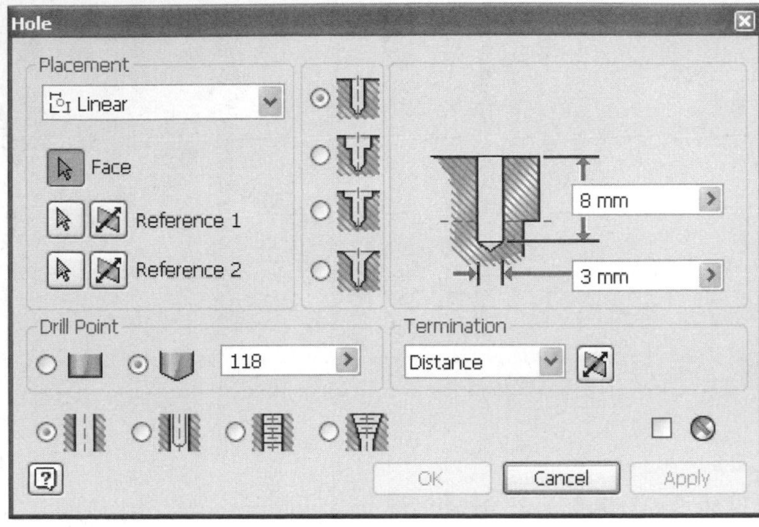

Figure 4.27

EDITING HOLE FEATURES

To edit the type of hole feature or distances, use one of the following methods:

- Double-click the feature's name or icon in the browser to display the Holes dialog box with the dimensions and option you used to create the hole feature.
- Change the select priority to Feature Priority, and then double-click the hole feature on the part in the graphics area. This will display the Holes dialog box with the dimensions and option you used to create the hole feature.
- Right-click the hole's name or icon in the browser, and select Edit Feature from the menu to display the Holes dialog box with the dimensions and option you used to create the hole feature.

HOLES DIALOG BOX

In the Holes dialog box, you establish the placement method, type of hole, its termination, and additional options such as type of drill point, angle, and tapped properties.

Placement

Select the appropriate placement method. If you select From Sketch, a sketch containing hole centers or any point, such as an endpoint of a line, must exist on the part. Hole centers are described in the next section. The Linear, Concentric, and On Point options do not require an unconsumed or shared sketch to exist in the model and are based on previously created features. Depending on the placement option you select, the input parameters will change, as shown in the following images and as described below.

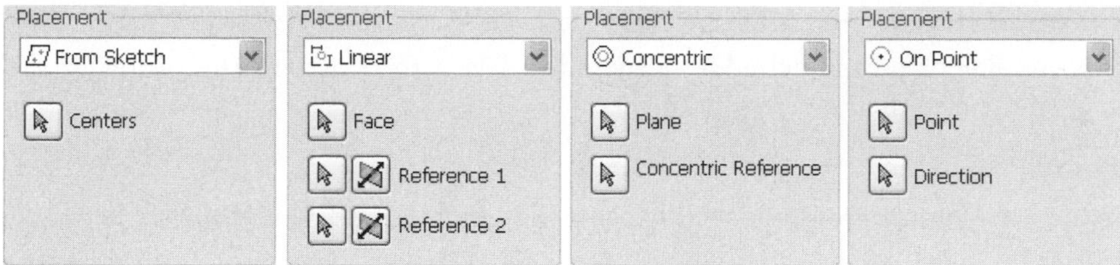

Figure 4.28

From Sketch

Select the From Sketch option to create holes that are based on a location defined within an unconsumed or shared sketch. You can base the center of the hole on a point/hole center or endpoints of sketched geometry like endpoints, centers of arcs and circles, and spline points. You can also use points from projected geometry that resides in the unconsumed or shared sketch.

Centers

Select the hole center point or sketch points where you want to create a hole.

Linear

Select the Linear option to place the hole relative to two selected face edges.

Face

Select the face on the part where the hole will be created.

Reference 1

Select a face edge as a positional reference for the center of the hole. When you select the edge, a dimension appears that can be edited to constrain the center of the hole dimensionally.

Reference 2

Select a face edge as a positional reference for the center of the hole. When you select the edge, a dimension appears that can be edited to constrain the center of the hole dimensionally.

Flip Side

Click this button to position the hole on the opposite side of the selected edge.

Concentric

Select the Concentric option to place the hole on a planar face and concentric to a circular or arc edge or a cylindrical face.

Plane

Select a planar face or plane where you want to create the hole.

Concentric Reference

Select a circular or arc model edge or cylindrical face to constrain the center of the hole to be concentric with the selected entity.

On Point

Select the On Point option to place the center of the hole on a work point. The work point must exist on the model prior to selecting this option.

Point

Select a work point to position the center of the hole.

Direction

Select a plane, face, work axis, or model edge to specify the direction of the hole. When selecting a plane or face, the hole direction will be normal to the face or plane.

Hole Option

Click the type of hole that you want to create: drilled, counterbore, spotface, or countersink, and enter the appropriate dimensions. The following image shows the counterbore option.

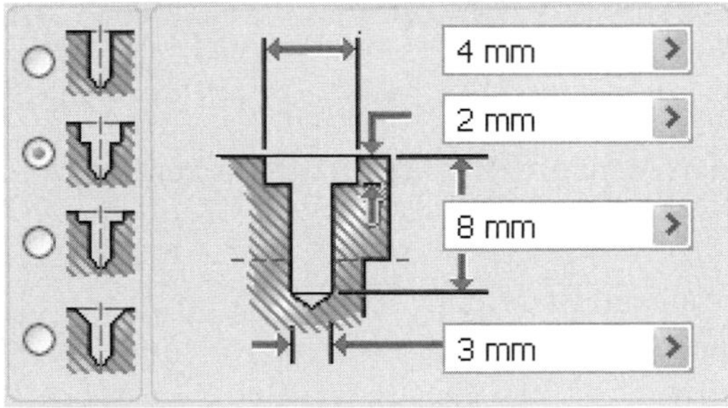

Figure 4.29

Termination Select how the hole will terminate.

- *Distance:* Specify a distance for the depth of the hole.
- *Through All:* Choose to extend the hole through the entire part in one direction.
- *To:* Select a plane at which the hole will terminate.
- *Flip:* Reverse the direction in which the hole will travel.

Drill Point Select either a flat or angle drill point. If you select an angle drill point, you can specify the angle of the drill point.

Infer iMates Check this box to automatically create an iMate on a full circular edge. Autodesk Inventor attempts to place the iMate on the closed loop most likely to be used.

Dimensions To change the diameter, depth, countersink, counterbore diameter, countersink angle, or counterbore depth of the hole, click the dimension in the dialog box, and enter a desired value.

Hole Type

Click the type of hole you want to create. There are four options: Simple Hole, Clearance Hole, Tapped Hole, and Taper Tapped Hole. The following image shows the tapped hole selected. After selecting the hole type, fill in the dialog box with the specific data for the hole you need to create.

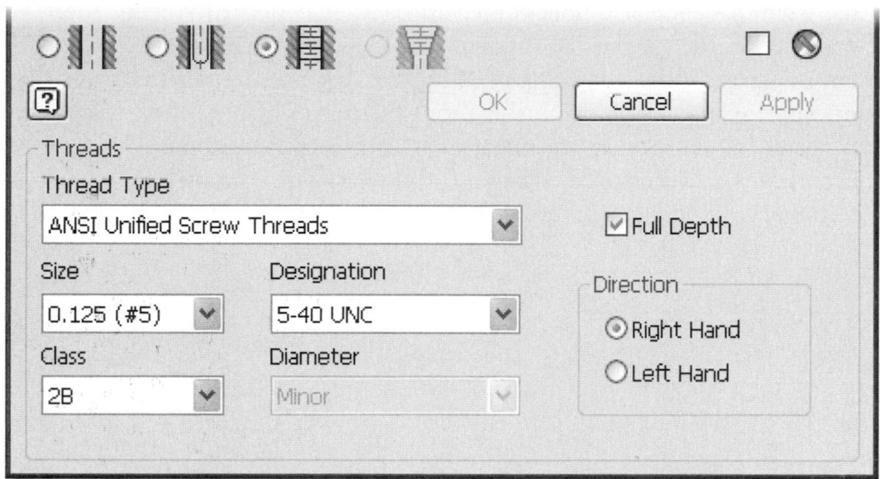

Figure 4.30

Simple Hole Click the Simple Hole option to create a (drilled) hole feature with no thread features or properties.

Clearance Hole Click the Clearance Hole option to create a (drilled) hole feature with no thread features or properties to match a specified fastener.

Tapped Hole Click the Tapped Hole option if the hole is threaded. Thread information appears in the dialog box area so that you can specify the thread properties, as shown in the above image.

Taper Tapped Hole Click the Taper Tapped Hole option if the hole is tapered thread. The taper tapped hole information appears in the dialog box area so you can specify the thread properties.

CENTER POINTS

Center points are sketched entities that can be used to locate hole features. To create a hole center, follow these steps:

1. Make a sketch active.
2. Click the Point, Center Point tool in the 2D Sketch panel bar, as shown in the following image.
3. Click a point to locate the hole center where you want to place the hole, and constrain it as desired.

You can also switch between a hole center and a sketch point by using the Center Point tool on the Standard toolbar, as shown in the following image, when the sketch environment is active. When you press the button, you create a center point. If you use a center point style when you activate the From Sketch option of the Hole tool, all center points that reside in the sketch are selected as centers for the hole feature automatically. This can expedite the process of creating multiple holes in a single hole feature. Points can be deselected by holding down the CTRL or Shift key and then selecting the points.

Figure 4.31

EXERCISE 4–2: CREATING HOLES

In this exercise, you add drilled, tapped, and counterbored holes to a cylinder head.

1. Open *ESS_E04_02.ipt* in the Chapter 04 folder.
2. The first hole you place is a linear hole. Click the Hole tool in the panel bar.

- For the Face, click the top face of the part.
- For Reference 1 and 2, select the bottom left and right edge, and place the point **10 mm** in from each edge.
- Verify that the Drilled hole option is selected.
- Change the Termination to Through All.
- Change the hole's diameter to **5 mm**, as shown in the following image.
- Click OK to create the hole.

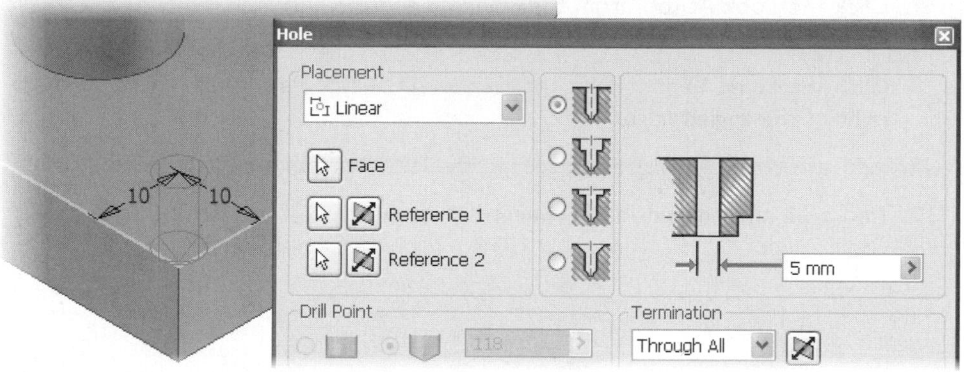

Figure 4.32

3. Next, you create a hole based on the Sketch option. Click the Sketch tool on the Standard toolbar, and then click the top face. The center point of the arc is projected onto the sketch.
4. Right-click the graphics window, and then click Finish Sketch.
5. Click the Hole tool in the panel bar or press the H key.

- Click the projected center point.
- Change the hole option to counterbore.
- Change the Termination to Through All.
- Change the hole's values as shown in the following image.

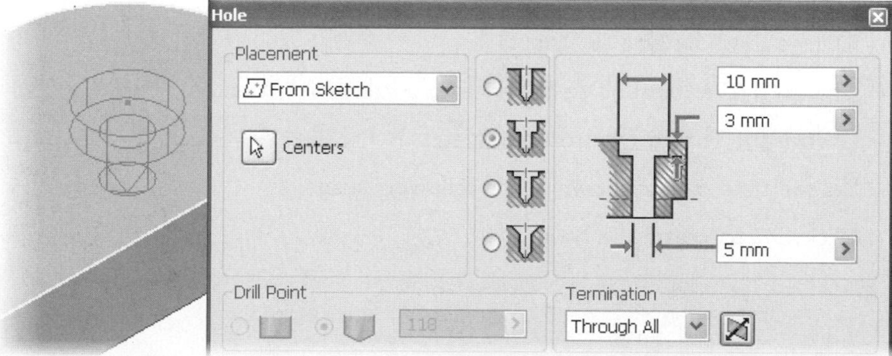

Figure 4.33

- Click OK to create the hole.

6. Next, you create a tapped hole based on the Sketch option but place it by a center point. Click the Sketch tool on the Standard toolbar, and then click the top face.

7. Click the Look At tool from the Standard toolbar, and click the top face or click the Front face in the View Cube.
8. Click the Point, Center Point tool in the 2D Sketch Panel, and place a point to the right of the angled edges.
9. Add a horizontal constraint between the left-most point and the center point.
10. Add a **15 mm** dimension between the left-most point and the center point.

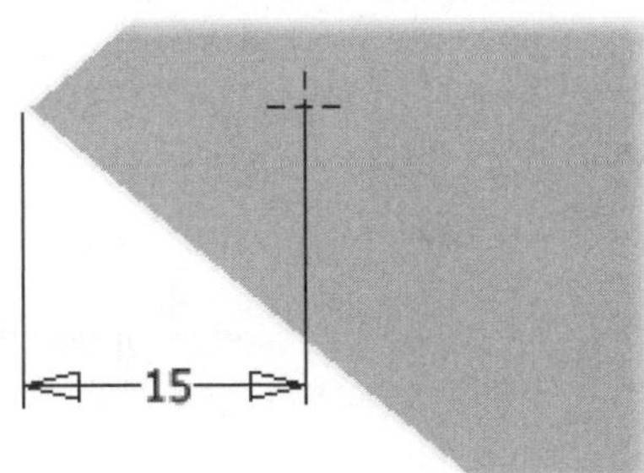

Figure 4.34

11. Finish the dimension tool, right-click and click Done.
12. Right-click in the graphics window, click Finish Sketch, and then press the F6 key to change to the Home View.
13. Click the Hole tool in the panel bar or press the H key.

- The Center Point is automatically selected as the Center.
- Change the hole option to Drilled, if needed.
- Change the hole type to Tapped.
- Change the Termination to Distance to a value of **7 mm.**
- Change the Thread Type to ANSI Metric M Profile.
- Set the size to **5**, as shown in the following image.
- Click OK to create the hole.

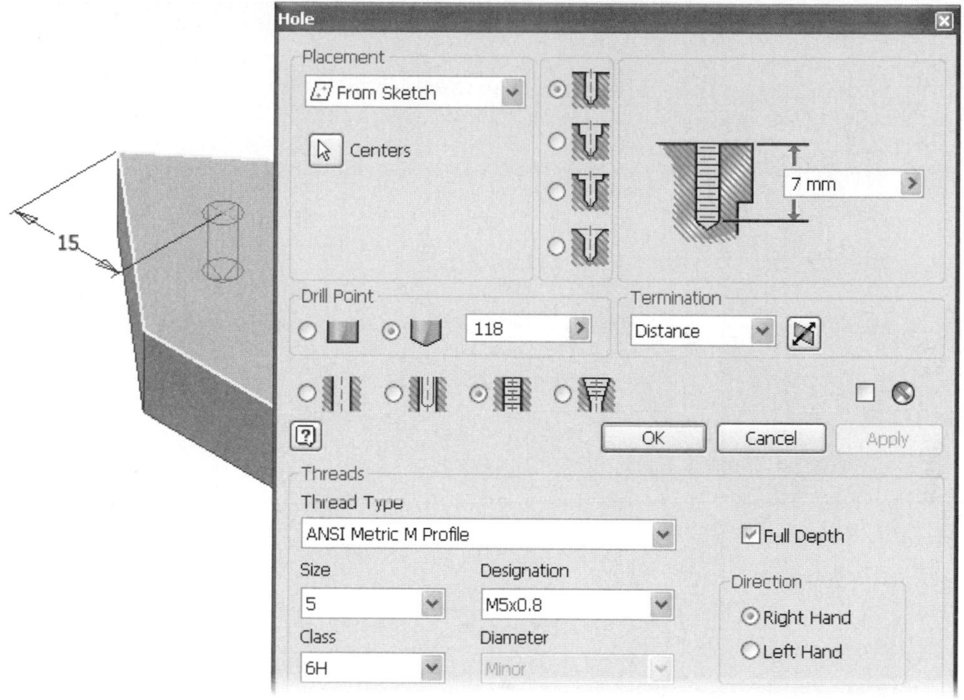

Figure 4.35

14. Next, you create a Taper Tapped hole that is concentric to the top of the cylinder.

- Press the ENTER key to start the Hole tool.
- For the Plane, select the top face of the cylinder.
- For the Concentric Reference, select the top circular edge of the cylinder.
- Change the hole option to Drilled, if needed.
- Click the Taper Tapped Hole type.
- Change the Thread Type to NPT.
- Change the size to **1/8**.
- Change the Termination to Through All, as shown in the following image.
- Click OK to create the hole.

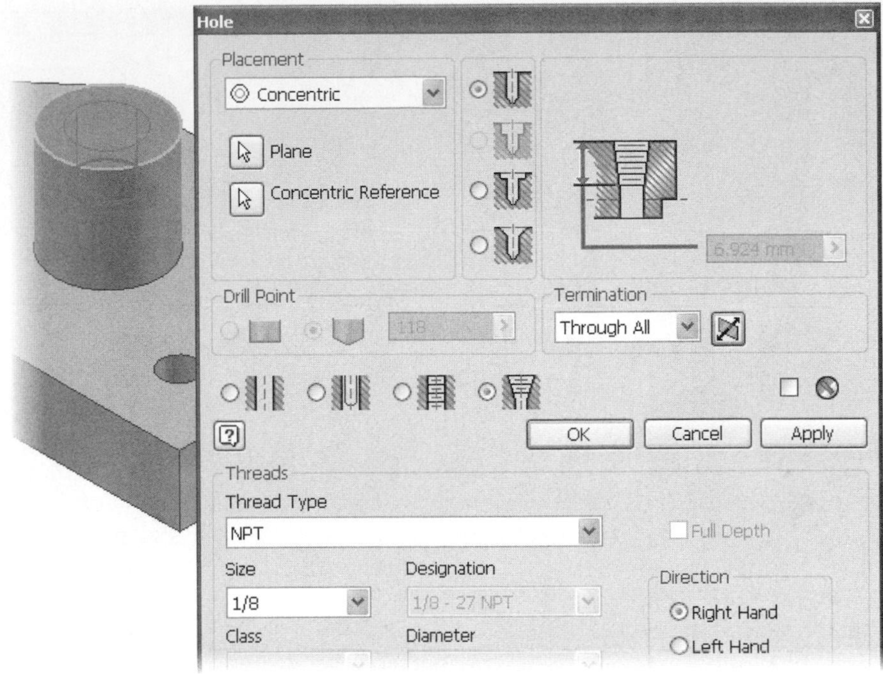

Figure 4.36

15. The completed part is shown in the following image. Rotate the part and examine the holes.

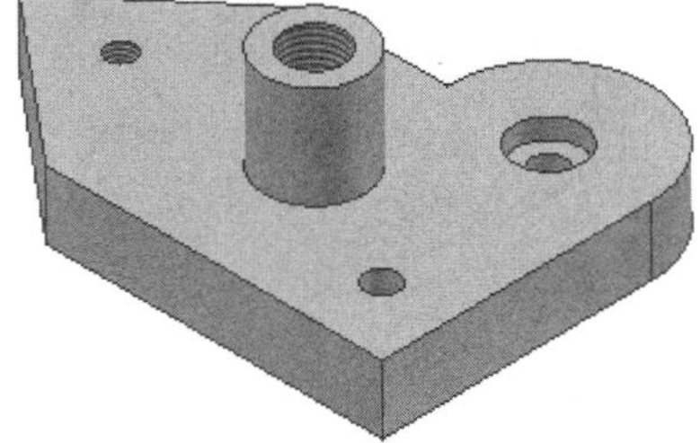

Figure 4.37

16. Practice editing the holes and placing new holes.

17. Close the file. Do not save changes. End of exercise.

THREADS

You use thread features to create both internal and external threads. The threads appear on the parts with a graphical representation, and when you create a drawing view, you can call out the thread per drafting standards. Since the threads are graphical representations, they do not physically exist on the part—if a model were to be cast directly from the part, no threads would exist on the finished model. You can add thread features to any cylindrical or conical face. If you are adding threads to a hole, you can do so using either the Hole or Thread tool. If you add it using the Thread tool, it is a separate feature in the browser. For this reason, it is recommended that you use the Hole tool with the tapped option when creating a threaded hole.

To create external threads, follow these steps:

1. Create a cylinder, and dimension it to the size that will represent the major diameter of the thread. This value must be between the maximum and minimum major diameter.
2. To create the thread feature, issue the Thread tool from the Part Features panel bar, as shown in the following image on the left.
3. Enter the data into the Thread Feature dialog box as needed.

There are two tabs in the Thread Feature dialog box: Location and Specification. The thread data come from an Excel spreadsheet named *Thread.xls*. By default, the spreadsheet is in the *Program Files\ Autodesk\ Inventor 2009\ Design Data* directory. You can modify this spreadsheet to match your company's standards. The thread data are used when the thread appears on the part and when the thread is annotated in a drawing view. The thread data are not associative to the threads on existing parts. When you make changes to the spreadsheet, the new values are only used when new threads are created. If you make a dimensional change to the diameter of the cylinders where you placed a thread, a warning dialog box appears when you update the part. The dialog box notifies you that you used an inappropriate thread size. Accept the warning message, and then modify the thread feature to a size that fits the corresponding diameter.

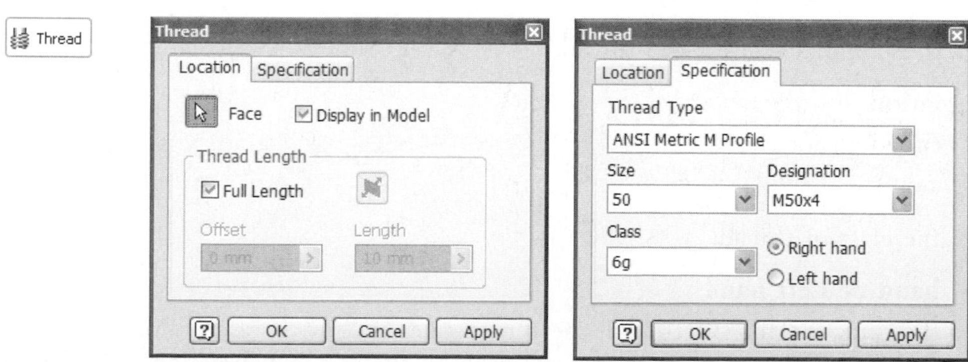

Figure 4.38

LOCATION TAB

The previous image shows the options available on the Location tab on the left, and the following sections describe them.

Face

Click this button, and then select a cylindrical or conical face on or in which to place a thread.

Display in Model Click this option to display a visual representation of the thread on the part. If the option's box is clear, the thread only appears when you create a drawing view.

Thread Length

Full Length Click this option so that the thread continues the entire length of the selected face.

Flip Click this button to reverse the direction of the thread.

Length When you do not click Full Length (the box is clear), enter a value for the length of the thread.

Offset When you do not click Full Length, enter a value to which the thread will be offset. The offset distance is from the closest end plane relative to the selected cylindrical or conical face.

SPECIFICATION TAB

The previous image on the right shows the options available on the Specification tab, and the following sections describe them.

Thread Type

Select the thread type; the types are defined by the tabs in the *Thread.xls* file.

Size

Select the nominal size for the thread.

Designation

Select the pitch for the thread.

Class

Select the class of thread.

Right hand or Left hand

Select the direction for the thread, the thread size, and its representation. The graphical representation of the thread will change based on the selected direction.

EXERCISE 4-3: CREATING THREADS

In this exercise, you use the Thread tool to create thread features on a cylinder.

1. Open *ESS_E04_03.ipt* in the Chapter 04 folder.
2. Click the Thread tool in the panel bar.
3. Select the inside face of the cylinder. On the Specification tab, verify that the nominal Size is **25** and change the Designation to **M25x1**, as shown in the following image.

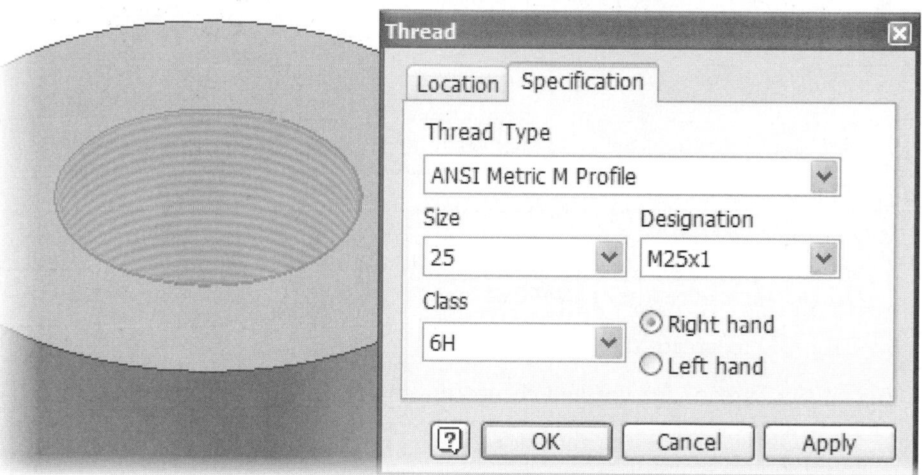

Figure 4.39

4. Click OK to create the thread.
5. Press the ENTER key to restart the Thread tool.
6. Select the outside face near the bottom of the cylinder, and uncheck Full Length. Set the offset to **10 mm** and the length to **20 mm**, as shown in the following image. The offset is based from the closest edge of the selected face.

Figure 4.40

7. On the Specification tab, verify that ANSI Metric M Profile is current, Size is **50** and change the Designation to **M50x2**.
8. Click OK to create the thread.
9. Edit the Thread2 feature and change the thread length to Full Length.
10. Click OK to update the thread feature.
11. Edit the sketch for the first extrusion, change the diameter of the outside circle to **75 mm**, and then update the part.
12. A warning dialog box is displayed, alerting you to a design problem, as shown in the following image.

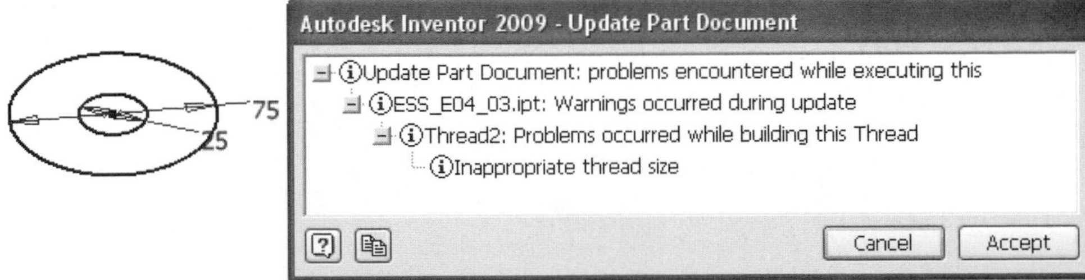

Figure 4.41

13. Accept the warning, and notice the warning symbol in the browser next to the second thread feature.
14. Edit the Thread2 feature; from the Specification tab, change its nominal size to **75** and the Designation to **M75x4**.

15. Click OK to complete the edit.
16. Close the file without saving changes. End of exercise.

SHELLING

As you design parts, you may need to create a model that is made of thin walls. The easiest way to create a thin-walled part is to create the main shape and then use the Shell tool to remove material.

The term *shell* refers to giving a wall thickness to the outside shape of a part and removing the remaining material. Essentially, you are scooping out the inside of a part and leaving the walls a specified thickness, as shown in the following image. You can offset the wall thickness in, out, or evenly in both directions. If the part you shell contains a void, such as a hole, the feature will have the thickness built around it.

A part may contain more than one shell feature, and individual faces of the part can have different thicknesses. If a wall has a different thickness than the shell thickness, it is referred to as a unique face thickness. If a face that you select for a unique face thickness has faces that are tangent to it, those faces will also have the same thickness. You can remove faces from being shelled, and these faces remain open. If no face is removed, the part is hollow on the inside.

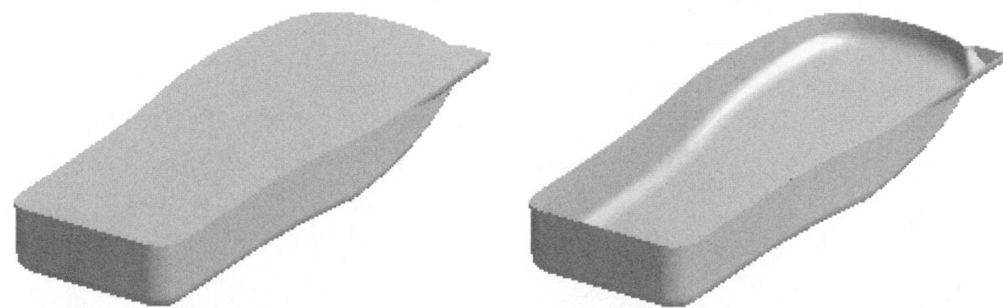

Figure 4.42

To create a shell feature, follow these steps:

1. Create a part that will be shelled.
2. Issue the Shell tool from the Part Features panel bar, as shown in the following image on the left.
3. The Shell dialog box appears, as shown in the following image on the right. Enter the data as needed.

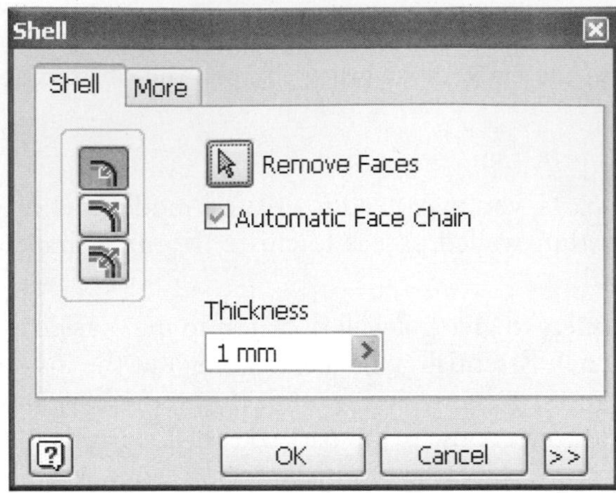

Figure 4.43

4. After filling in the information in the dialog box, click OK, and the part is shelled.

To edit a shell feature, use one of the following methods:

- Right-click the name of the shell feature in the browser, and select Edit Feature from the menu. Alternately, you can double-click the feature's name or icon in the browser, and the Shell dialog box appears with all of the settings you used to create the feature. Change the settings as needed.
- Change the select priority to Feature Priority, double-click the shell feature, and the Shell dialog box appears with all of the settings you used to create the feature.

The following sections explain the options that are available for the Shell tool.

DIRECTION

Inside

Click this button to offset the wall thickness into the part by the given value.

Outside

Click this button to offset the wall thickness out of the part by the given value.

Both

Click this button to offset the wall thickness evenly into and out of the part by the given value.

REMOVE FACES

Click the Remove Faces button, and then click the face or faces to be left open. To deselect a face, click the Remove Faces button, and hold down the CTRL key while you click the face.

AUTOMATIC FACE CHAIN

When you are removing faces and this option is checked, faces that are tangent to the selected face are automatically selected. Uncheck this option to select only the selected face.

THICKNESS

Enter a value or select a previously used value from the drop-down list to be used for the shell thickness.

UNIQUE FACE THICKNESS

Unique face thickness is available by clicking the More (>>) button that is located on the lower-right corner of the dialog box, as shown in the following image on the left.

To give a specific face a thickness, select Click to add, click the face, and enter a value. A part may contain multiple faces that have a unique thickness, as shown in the following image on the right.

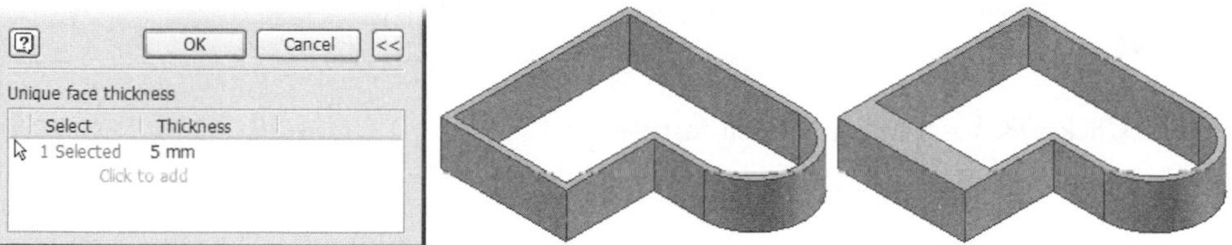

Figure 4.44

EXERCISE 4-4: SHELLING A PART

In this exercise, you use the Shell tool to create a shell on a part.

1. Open *ESS_E04_04.ipt* in the Chapter 04 folder.
2. Click the Shell tool in the panel bar.
3. Remove the top face by selecting the top face of the part.
4. Type a thickness of **3 mm**, as shown in the following image.

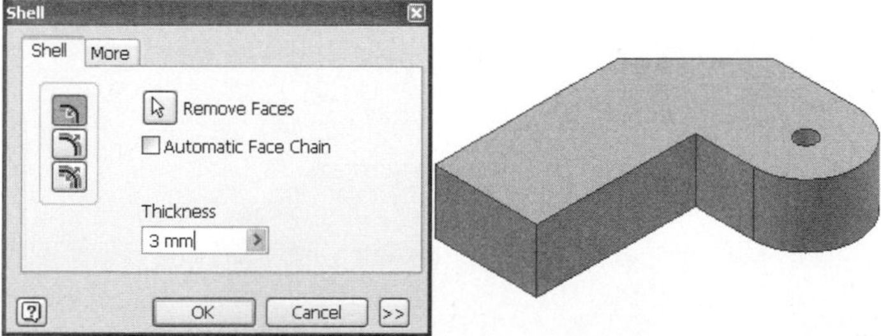

Figure 4.45

5. Click OK to create the shell feature.
6. Rotate the part and examine the shell.
7. Click the Home View.

8. Edit the Shell feature that you just created.
9. Select the More button (>>), click in the "Click to add" area, and select the left-vertical face.
10. Enter a value of **10 mm**, as shown in the following image.

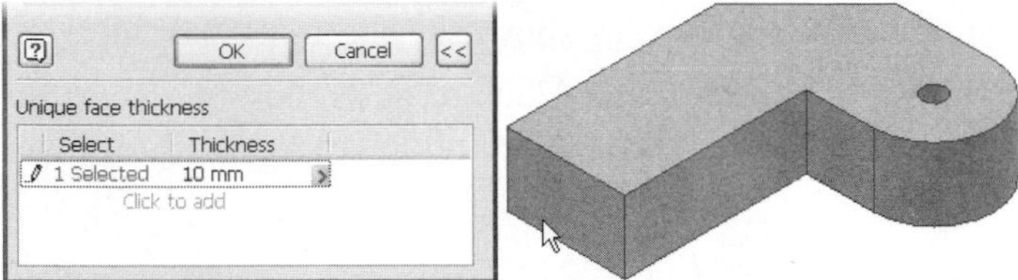

Figure 4.46

11. Click OK to create the shell feature.
12. Click the Shell tool in the panel bar.
13. If the Remove Faces button is not active, click the Remove Face button. Then select the same face, the one to which you applied a unique thickness, and type a thickness of **3 mm**, as shown in the following image.

Figure 4.47

14. Click OK to create the shell.

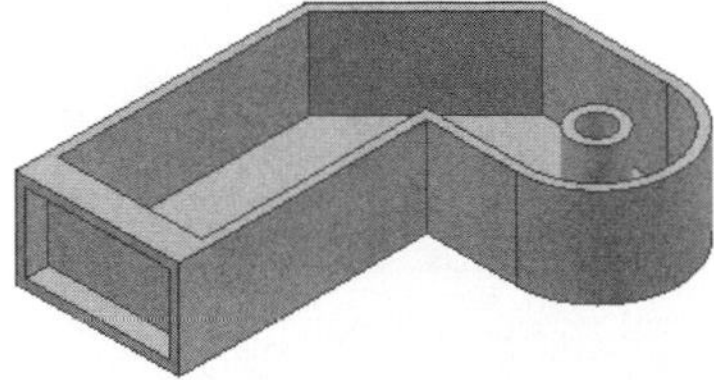

Figure 4.48

15. Close the file without saving changes. End of exercise.

FACE DRAFT

The Face Draft feature applies an angle to a face. You can apply face draft to any specified internal or external face, including shelled parts. When you apply face draft, any tangent face also has the face draft applied to it.

To create a face draft feature, follow these steps:

1. Issue the Face Draft tool from the Part Features panel bar, as shown in the following image on the left.
2. The Face Draft dialog box appears, as shown in the following image on the right. The following section describes the options in the Face Draft dialog box.
3. Click the Pull Direction that shows how the mold will be pulled from the part.
4. Click a face or faces to which to apply the face draft. If you select an incorrect face, you can deselect it by holding down the CTRL key and clicking the face.

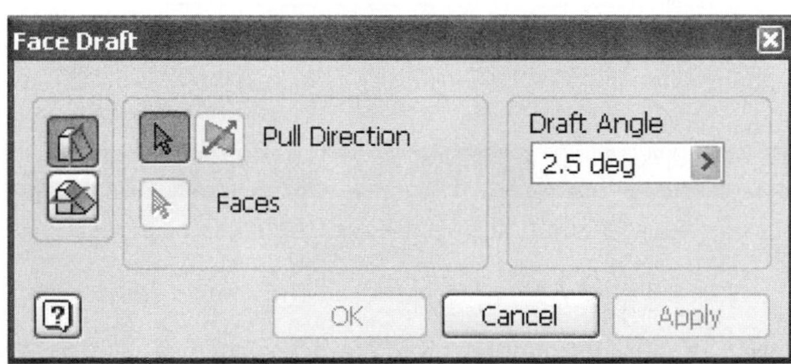

Figure 4.49

DRAFT TYPE

Click the type of draft that you want to create. Two types of drafts are available.

Fixed Edge

Click the Fixed Edge button to allow the draft to be created from an edge or series of contiguous edges.

Fixed Plane

Click the Fixed Plane button to allow the draft to be created from a selected plane. The plane you select is used to specify both the pull direction and the fixed plane.

PULL DIRECTION AND FIXED PLANE

Click the direction that the mold will be pulled from the part. When you select the Fixed Plane draft type, click a planar face or work plane from which to draft the faces.

Direction

Click the arrow, and move the cursor around the part. A dashed line appears—this line points 90° from the highlighted face or along the highlighted linear edge, and it shows the direction that the mold will be pulled. When the correct direction appears on the screen, left-click.

Flip

Click the Flip button to reverse the pull direction 180°.

FACES

Click the face or faces to which to apply the face draft. As you move the cursor over the face, a symbol with an arrow appears that shows how the draft will be applied. As you move the cursor to different edges and faces, the arrow direction preview shows how the draft will be applied to that specific face.

DRAFT ANGLE

Enter a value for the draft angle, or select a previously used draft angle from the drop-down list.

EXERCISE 4-5: CREATING FACE DRAFTS

In this exercise, you apply draft angles to the faces of a part.

1. Start a new part file based on the metric *standard(mm).ipt* template file.
2. Sketch and dimension a rectangle that measures **50 mm** horizontally by **75 mm** vertically.
3. Change to the default home view.
4. Extrude the rectangle **25 mm**.
5. Shell the part **3 mm**, and remove the top face.
6. Click the Face Draft tool in the Part Features panel bar.
7. Select near the middle of the inside-horizontal face, and a dashed line will appear, as shown in the following image. When it is displayed, press the left mouse button, and an arrow that points up will appear.

Note: If the arrow is pointing in the wrong direction, pick the Flip button in the Face Draft dialog box.

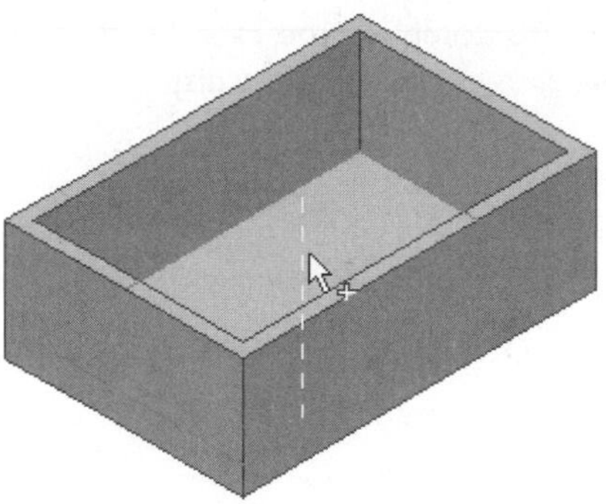

Figure 4.50

8. Change the draft angle to **5**.

9. In the Face Draft dialog box, select the Faces button, and move the cursor near the inside top right edge of the box, as shown in the following image, and click. This is the edge that will be fixed.

Figure 4.51

10. While still in the same operation, select near the inside back bottom edge of the box, as shown in the following image, and click. This is the edge that will be fixed.

Figure 4.52

11. Click the OK button to complete the Face Draft feature. Your part should resemble the following image, shown in wireframe display.

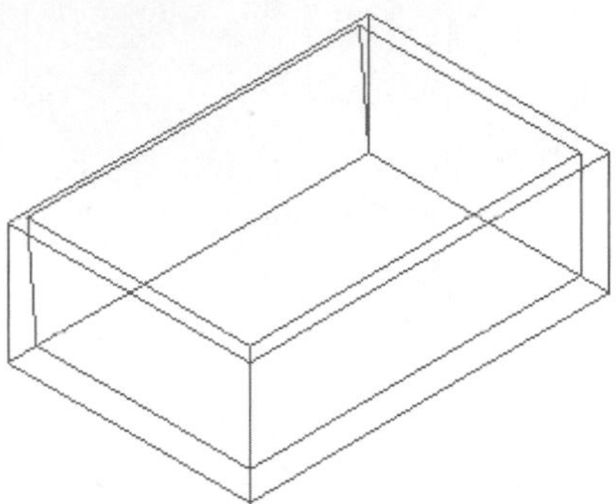

Figure 4.53

12. Delete the Face Draft feature that you just created. In the browser, right-click the feature FaceDraft1and select Delete from the menu.
13. Place a 5 mm fillet in each of the four inside vertical edges of the box.
14. Click the Face Draft tool in the Part Features panel bar.
15. Define the pull direction by clicking near the middle of the inner-bottom face of the shell.
16. Change the draft angle to **5.**
17. In the Face Draft dialog box, click the Faces button, and move the cursor near the inside top back edge of the box, as shown in the following image. Since all the faces are tangent, all the top edges that are inside the part will be highlighted. These edges will be fixed.

Figure 4.54

18. Click OK to complete the operation. The part should resemble the following image, shown in wireframe display.

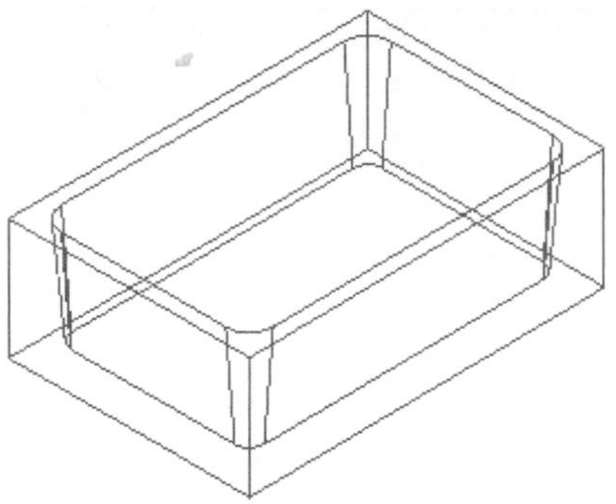

Figure 4.55

19. Close the file without saving changes. End of exercise.

WORK FEATURES

When you create a parametric part, you define how the features of the part relate to one another; a change in one feature results in appropriate changes in all related features. Work features are special construction features that are attached parametrically to part geometry or other work features. You typically use work features to help you position and define new features in your model. There are three types of work features: work planes, work axes, and work points.

Use work features in the following situations:

- To position a sketch for new features when a planar part face is not available.
- To establish an intermediate position that is required to define other work features. You can create a work plane at an angle to an existing face, for example, and then create another work plane at an offset value from that plane.
- To establish a plane or edge from which you can place parametric dimensions and constraints.
- To provide an axis or point of rotation for revolved features and patterns.
- To provide an external feature termination plane off the part, such as a beveled extrusion edge, or an internal feature termination plane in cases where there are no existing surfaces.

CREATING A WORK AXIS

A work axis is a feature that acts like a construction line. In the database, it is infinite in length but displayed a little larger than the model, and you can use it to help create work planes, work points, and subsequent part features. You can also use work axes as axes of rotation for polar arrays or to constrain parts in an assembly using assembly constraints. Their length always extends beyond the part—as the part changes size, the work axis also changes size. A work axis is tied parametrically to the part. As changes occur to the part, the work axis will maintain its relationship to the points, edge, or cylindrical face from which you created it. To create a work axis, use the Work Axis tool on the Part Features panel bar, as shown in the following image, or press the hot key / (forward slash).

Figure 4.56

Use one of the following methods to create a work axis:

- Click a cylindrical face to create a work axis along the axis of the face.
- Click two points on a part to create a work axis through both points.
- Click a linear edge on a part to create a work axis on the part edge.
- Click a work point or sketch point and a plane or face to create a work axis that is normal to the selected plane or face and that passes through the point.
- Click two nonparallel planes or faces to create a work axis at their intersection.
- Click a work point and a surface to create a work axis that is normal to the surface and passes through the work point. The work point does not need to be on the surface.
- Click a point or work point and a linear edge to create a work axis that goes through the point and that is parallel to the edge. The point does not need to lie on one of the planes bounded by the linear edge.

CREATING WORK POINTS

A work point is a feature that you can create on the active part or in 3D space. You can create a work point any time a point is required. To create a work point, use the Work Point tool on the Part Features panel bar, as shown in the following image, or press the hot key (.) (period). You can also create a point when using the Work Plane or Work Axis tools; right-click and select Create Point when one of these work feature tools is active. Then, after the work point is created, the tool you were running will be active. The created point will be indented as a child of the work axis or the work plane in the browser.

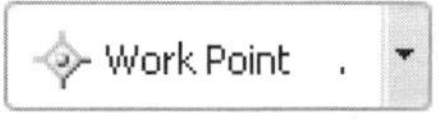

Figure 4.57

Use one of the following methods to create a work point:

- Click an endpoint or the midpoint of an edge.
- Click on two intersecting edges or axes to create a work point at the intersection, or theoretical intersection, of the two.
- Click an edge and plane to create a work point at the intersection, or theoretical intersection, of the two.

- Click a spline that intersects a face or plane to create a work point at the intersection of the spline and face or plane.
- Click three nonparallel faces or planes to create a work point at their intersection or theoretical intersection.

Grounded Work Points

You can create grounded work points that are positioned in 3D space. Grounded work points are not associated with the part or any other work features, including the original locating geometry. When you modify surrounding geometry, the grounded work point remains in the specified location. To create a grounded work point, use the Grounded Work Point tool on the Part Features panel bar, located by clicking the arrow next to the Work Point tool, as shown in the following image, or press the hot key (;) (semicolon).

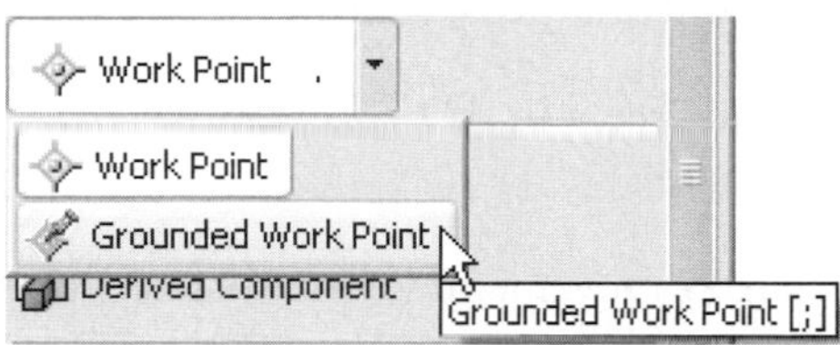

Figure 4.58

After clicking the Grounded Work Point tool, select a vertex, midpoint, sketch point, or work point on the model. When you have selected the vertex or point, the 3D Move/Rotate dialog box and a triad appear, as shown in the following image. The initial orientation of the triad matches the principle axes of the part. These colors represent the three axes: red = X, green = Y, and blue = Z.

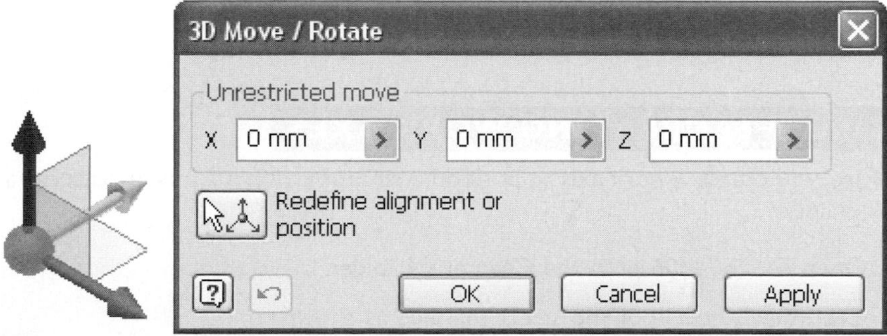

Figure 4.59

Enter values in the 3D Move/Rotate dialog box to precisely position the grounded work point relative to the selected point. You can also select areas of the triad to move the triad and locate the grounded work point in the desired direction, as shown in the following image and as described in the following sections.

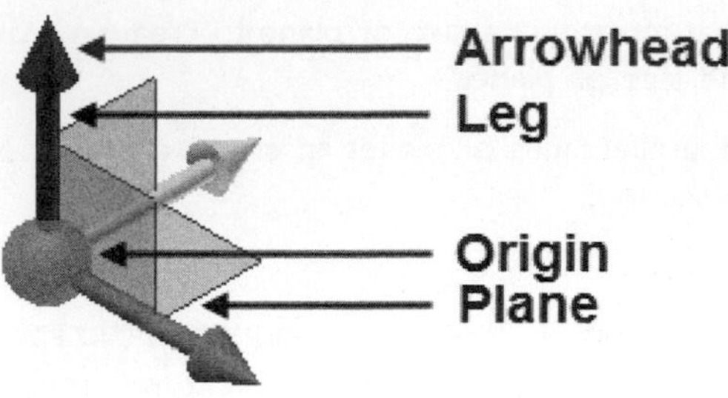

Figure 4.60

Arrowheads Select an arrowhead to specify a position along a particular axis, and move the cursor or enter a value.

Legs Select a leg to rotate about that axis, and move the cursor or enter an angle.

Origin Click to move the triad freely in 3D space, move it to a selected vertex or point, or to enter X, Y, or Z coordinates.

Planes Select a plane to restrict movement to the selected plane.

Once you position the triad properly, click Apply or OK in the 3D Move/Rotate dialog box to create the grounded work point. You can identify a grounded work point in the browser by the thumbtack icon that is placed on the work point. The following image shows a regular work point (Work Point1) and a grounded Work Point (Work Point2) represented in the browser.

Figure 4.61

EXERCISE 4-6: CREATING WORK AXES

In this exercise, you create a work axis to position a circular pattern. Circular patterns are covered later in this chapter

1. Open *ESS_E04_06.ipt* in the Chapter 04 folder.
2. Click the Sketch tool and select the face as shown.

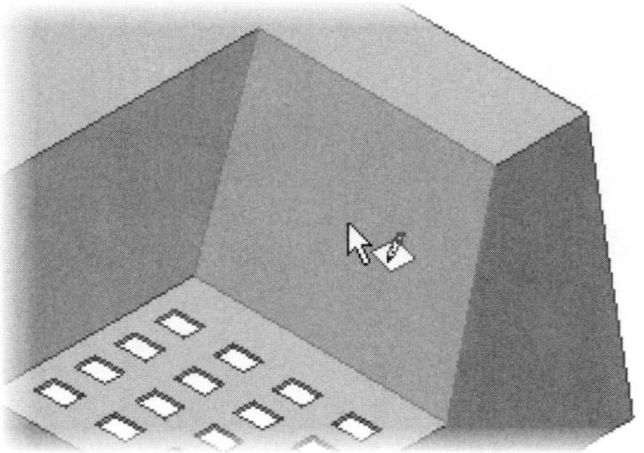

Figure 4.62

3. Click the Look At tool, and then select the new sketch.
4. Click the Point, Center Point tool in the panel bar, and place a point near the middle of the sketch.
5. Place a vertical constraint between the center point you just created and the midpoint on the top line. Apply a horizontal constraint between the center point and the midpoint of the line on the right, as shown in the following image.

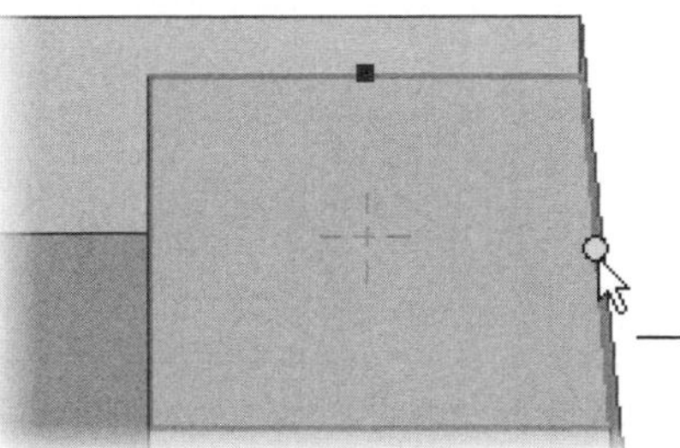

Figure 4.63

6. Right-click in the graphics window, and click Done to finish the constraint tool.
7. Right-click in the graphics window, and then select Finish Sketch.
8. Right-click in the graphics window, and then select Home View.
9. Click the Work Axis tool from the Part Features panel bar.
10. Select the angled plane, and then select the center point. The work axis is created through this point and normal to the plane, as shown in the following image on the left. For clarity, it is shown in wireframe.
11. One use of a work axis is to use it as a Rotation Axis when creating a circular pattern. The following image on the right shows a hole that was patterned around the work axis. Patterns are covered later in this chapter.

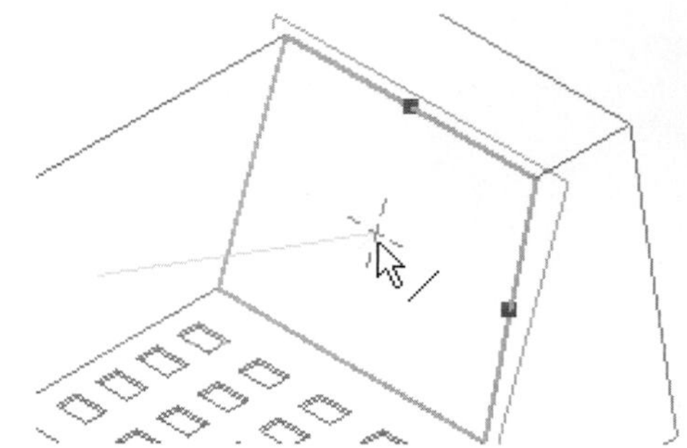

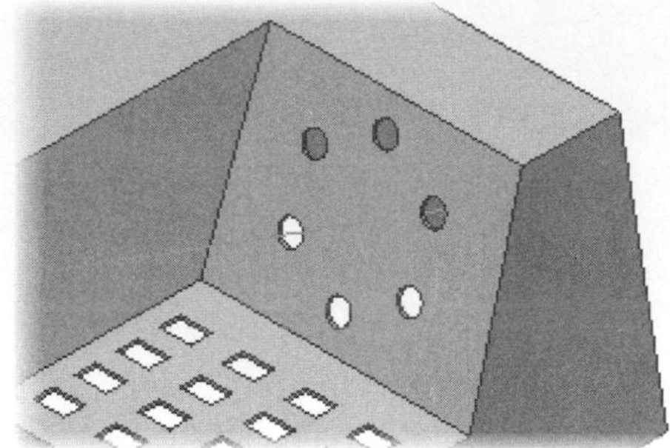

Figure 4.64

12. Practice creating work axes by trying these three methods; selecting two points, a circular face, and a plane and a point.

13. Close the file. Do not save changes. End of exercise.

CREATING WORK PLANES

Before introducing work planes, it is important that you understand when you need to create a work plane. You can use a work plane when you need to create a sketch and no planar face exists at the desired location, or if you want a feature to terminate at a plane and no face exists to select. If you want to apply an assembly constraint to a plane on a part and no part face exists, you will need to create a work plane. If a face exists in any of these scenarios, you should use it and not create a work plane. A new sketch can be created on a work plane.

A work plane looks like a rectangular plane. It is tied parametrically to the part. Though extents of the plane will always appear slightly larger than the part, the plane is in fact infinite. If the part or related feature moves or resizes, the work plane will also move or resize. For example, if a work plane is tangent to the outside face of a 1" diameter cylinder and the cylinder diameter changes to 2", the work plane moves with the outside face of the cylinder. You can create as many work planes on a part as needed, and you can use any work plane to create a new sketch. A work plane is a feature and is modified like any other feature.

Before creating a work plane, ask yourself where this work plane needs to exist and what you know about its location. You might want a plane to be tangent to a given face and parallel to another plane, for example, or to go through the center of two arcs. Once you know what you want, select the appropriate options and create a work plane. There are times when you may need to create an intermediate work plane before creating the final work plane. You may need to create a work plane, for example, that is at 30° and tangent to a cylindrical face. You should first create a work plane that is at a 30° angle and located at the center of the cylinder; then create a work plane parallel to the angled work plane that is also tangent to the cylinder.

To create a work plane, click the Work Plane tool on the Part Features panel bar, as shown in the following image, or press the hot key (]) (end bracket). A work plane is created depending on what, where, and how you select options. Work planes can also be associated to the default reference (origin) planes that exist in every part. These origin planes initially have their visibility turned off, but you can make them visible by expanding the Origin folder in the browser, right-clicking on a plane or planes, and selecting Visibility from the menu. You can use these default planes to create a new sketch or to create other work planes.

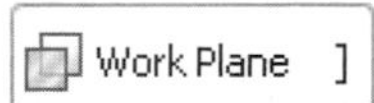

Figure 4.65

Use one of the following methods to create a work plane:

- Click three points.
- Click a plane and a point.
- Click a plane and an edge or axis.
- Click two edges, two axes, or an edge and an axis.
- To create a work plane that is tangent to a face, click an axis, plane, or face and a cylindrical face. The resulting work plane is created parallel or coincident with the selected axis or plane or parallel to the face and tangent to the selected cylindrical face.
- To create an angled work plane, click a plane or face and an edge; a dialog box appears for you to enter the angle.
- To create an offset work plane from an existing face, click a plane and then drag the new work plane to a selected location. While dragging the work plane, an Offset dialog box appears that displays the offset distance. Click a point, enter a precise value for the offset distance, and click the checkmark in the dialog box or press the ENTER key.
- To create a work plane at the midplane defined by the two selected planes or faces, click the two parallel planes or faces.

When you are creating a work plane, and more than one solution is possible, the Select Other tool appears. Click the forward or reverse arrows from the Select Other tool until you see the desired solution displayed. Click the checkmark in the selection box. If you clicked a midpoint on an edge, the resulting work plane links to the midpoint. If the selected edge's length changes, the location of the work plane will adjust to the new midpoint.

Note: The order in which points or planes are selected is irrelevant.

Use the Show Me animations located in the Visual Syllabus to view animations that display how to create certain types of work planes.

TYPES OF WORK PLANES

You can use the following work plane construction methods in the modeling process:

- Angled
- Edge and face normal
- Edge and tangent
- Offset
- Point and face normal
- Point and face parallel
- Tangent and face parallel
- Tangent and edge or axis
- 3-point
- 2-edge or 2-axis
- Through line endpoint, perpendicular to line
- Normal to arc at a point on the arc
- Normal to a spline or work curve at a point on the spline or curve
- Midway between two parallel planes

FEATURE VISIBILITY

You can control the visibility of the origin planes, origin axes, origin point, user work planes, user work axes, user work points, and sketches by either right-clicking on them in the graphics area or on their name in the browser and selecting Object Visibility from the menu. You can also control the visibility for all origin planes, origin axes, origin point, user work planes, user work axes, user work points, and sketches from the Object Visibility option in the View menu. Visibility can be checked to turn visibility on or cleared to turn it off, as shown in the following image.

Figure 4.66

EXERCISE 4–7: CREATING WORK PLANES

In this exercise, you create work planes in order to create a boss on a cylinder head and a slot in a shaft.

1. Open *ESS_E04_07.ipt* in the Chapter 04 folder.
2. Create an angled work plane. Click the Work Plane tool in the Part Features panel bar.
3. Click the top rectangular face and the top-back left edge, and then enter a value of **-30** as shown in the following image. Click the checkmark to create the work plane.

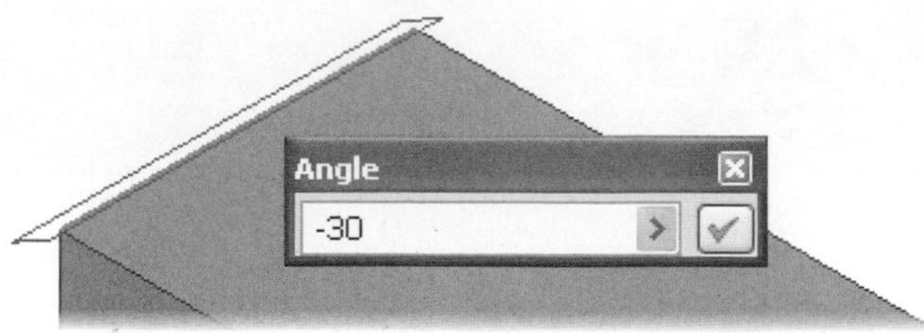

Figure 4.67

4. Make the top rectangular face the active sketch. Click the Sketch tool in the Standard toolbar and click the top rectangular face of the extrusion.
5. Create and dimension a circle as shown in the following image.

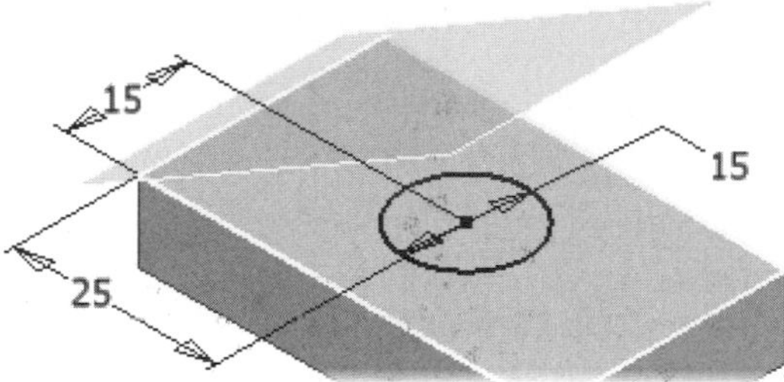

Figure 4.68

6. Extrude the circle with the Extents set to To, and select the work plane, as shown in the following image.

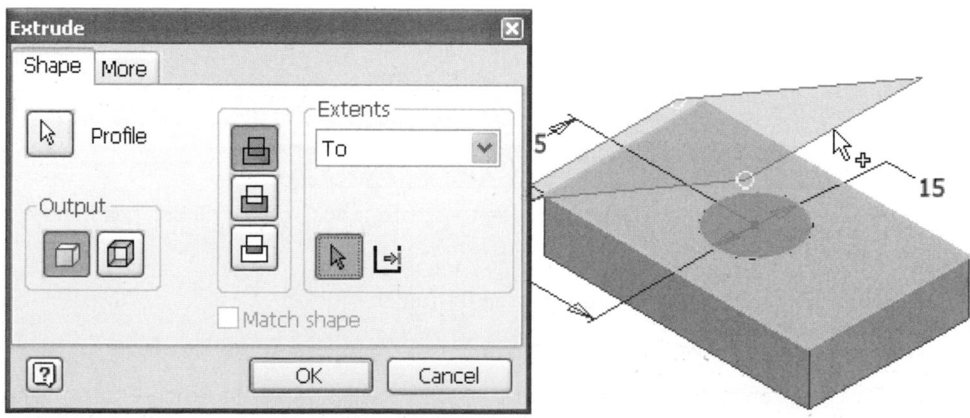

Figure 4.69

7. The following image shows the completed extrusion.

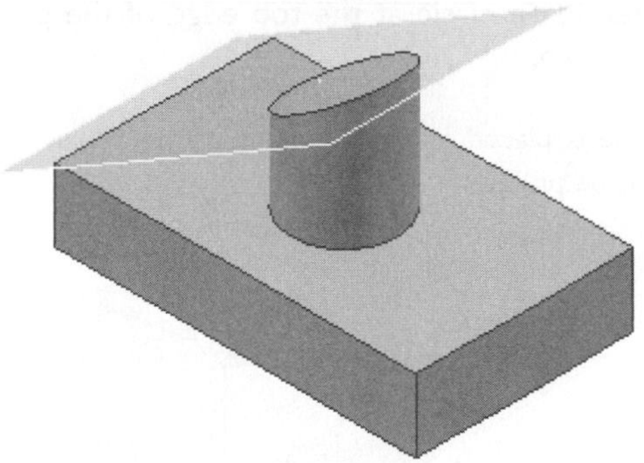

Figure 4.70

8. In the browser, double-click on the angled Work Plane you create in step 3 and enter different values for the angle; click the Update tool on the Standard toolbar to update the model.
9. Turn off the visibility of the work plane by moving the cursor over an edge of the work plane in the graphics window, right-click, and click Visibility from the menu.
10. Create a work plane that is centered between two parallel planes. Click the Work Plane tool in the Part Features panel bar.
11. Click the front-left face and the back-right face, as shown in the following image.

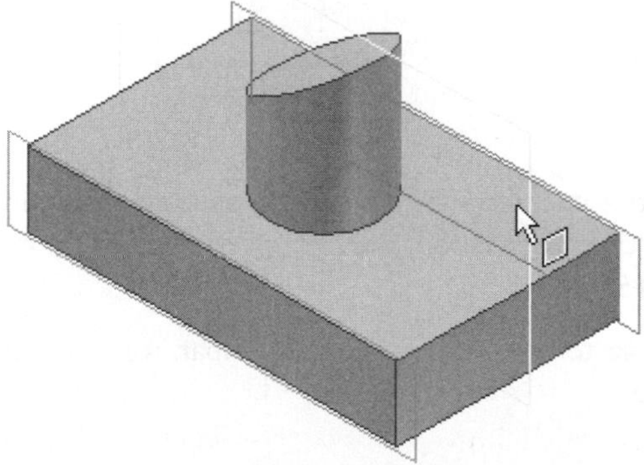

Figure 4.71

12. Click the Sketch tool on the Standard toolbar, and then click the work plane you just created.
13. To project the edges of the part, click the Project Cut Edges tool from the 2D Sketch panel bar.

14. Create and dimension a circle at the top edge of the projected geometry, as shown in the following image.

Note: If the circle is placed at the midpoint of the projected edge, the horizontal dimension cannot be placed.

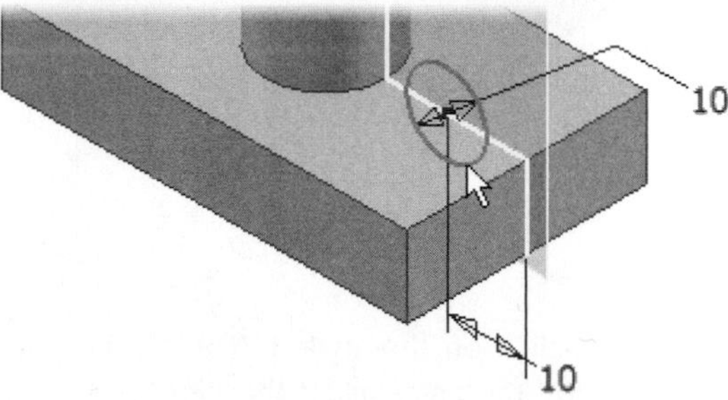

Figure 4.72

15. Extrude the circle **10 mm** with the midplane option, as shown in the following image.

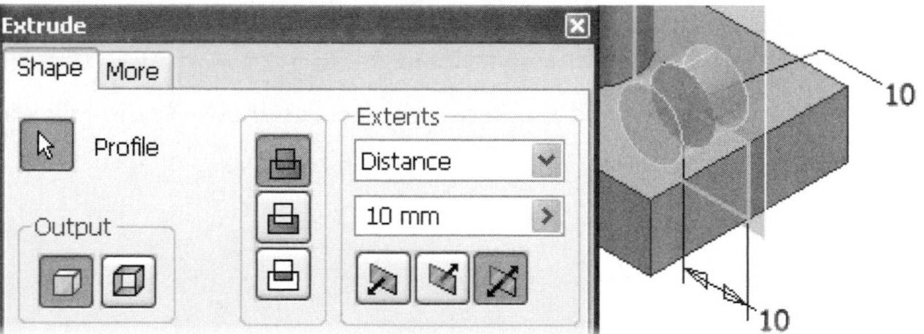

Figure 4.73

16. To verify that the midplane extrusion will be maintained in the center of the part even if the width changes, edit Sketch1 of Extrusion1, and change the **30 mm** dimension to **45 mm**.

17. Click the Update tool on the Standard toolbar. Your screen should resemble the following image.

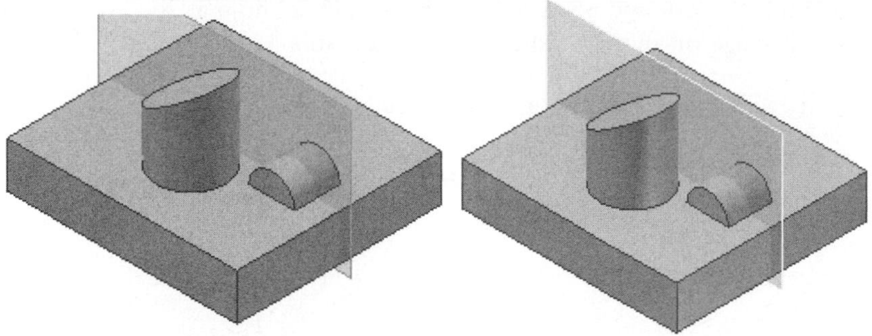

Figure 4.74

18. Close the file. Do not save changes.

19. In this portion of the exercise, you will place two holes on a cylinder. Open *ESS_E04_07_2.ipt* from the Chapter 04 folder.

20. Create a work plane that is parallel to an origin plane and tangent to the cylinder.

- In the browser, expand the Origin folder.
- Click the Work Plane tool from the Part Features panel bar.
- Click the YZ plane under the Origin folder.
- Click a point to the front right of the cylinder, as shown in the following image.

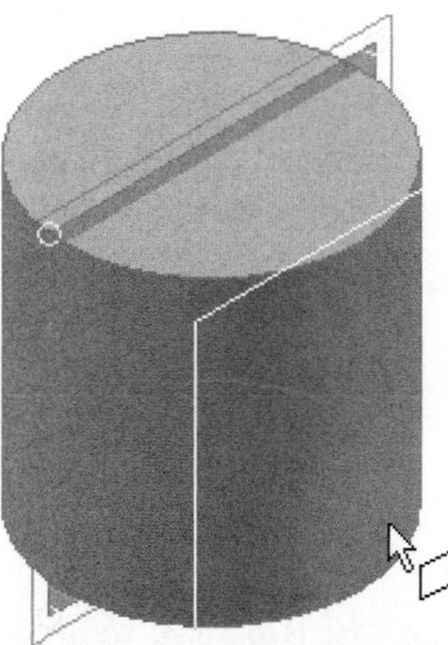

Figure 4.75

21. Make the new work plane the active sketch.

22. Project the Z axis of the origin folder onto the sketch.

23. Place a Point, Center Point on the projected axis—this will center the point in the center of the cylinder—and dimension it, as shown in the following image on the left.

24. Click the Hole tool and place a **10 mm** through all hole at the center point. The following image on the right shows the placed hole.

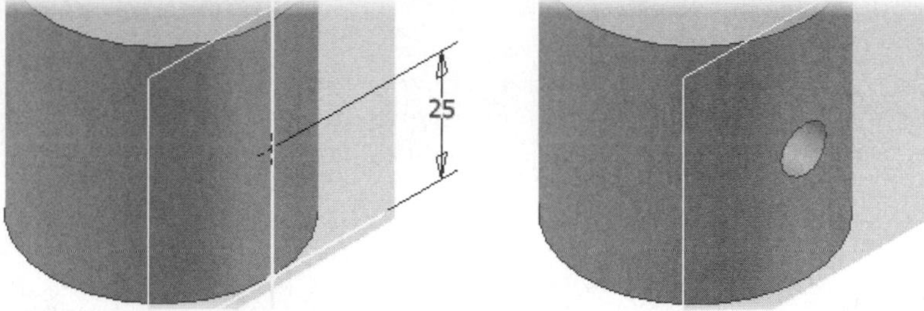

Figure 4.76

25. Next, place a hole on the cylinder at an angle. Create an angled work plane.

- Click the Work Plane tool from the Part Features panel bar.
- Click the YZ plane under the Origin folder.
- Click the Z Axis under the Origin folder.
- In the Angle dialog box, enter **-45.**

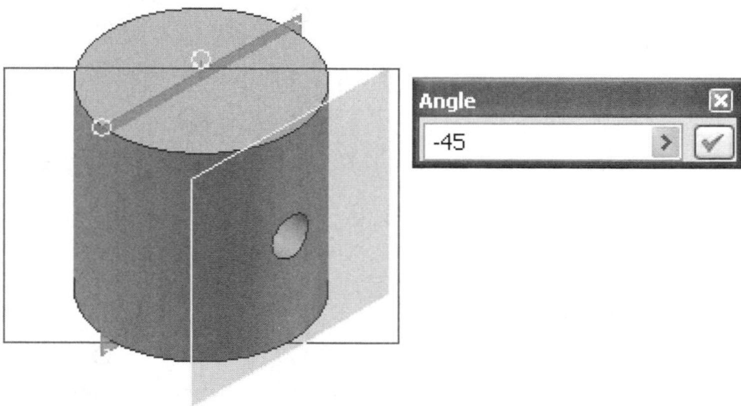

Figure 4.77

26. Next, create a work plane that is parallel to an angled plane and tangent to the cylinder.

- Click the Work Plane tool from the Part Features panel bar.
- Click the angle work plane that you just created.
- Click a point to the front of the cylinder, as shown in the following image.

EXERCISES

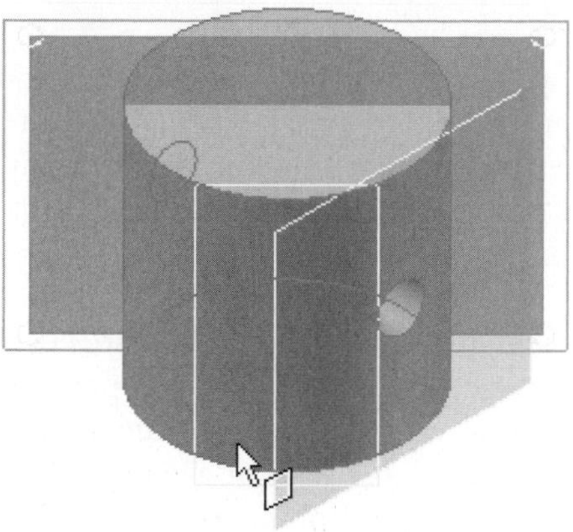

Figure 4.78

27. Turn off the visibility of the first two work planes you created. Move the cursor over the work plane. Right-click, and click Visibility from the menu.
28. Create a new sketch on the new work plane.
29. Use the Project Geometry tool to project the Z axis of the origin folder onto the sketch.
30. Place a Point, Center Point on the projected axis, and dimension it, as shown in the following image on the left.
31. Click the Hole tool, and place a hole of your choice at the center point. The following image shows a counterbore hole on the right.

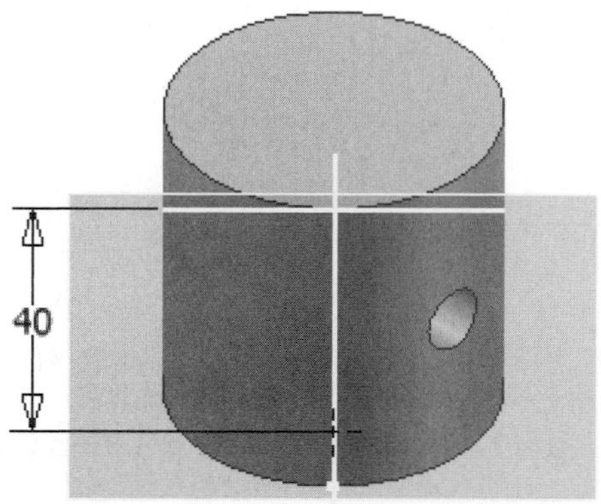

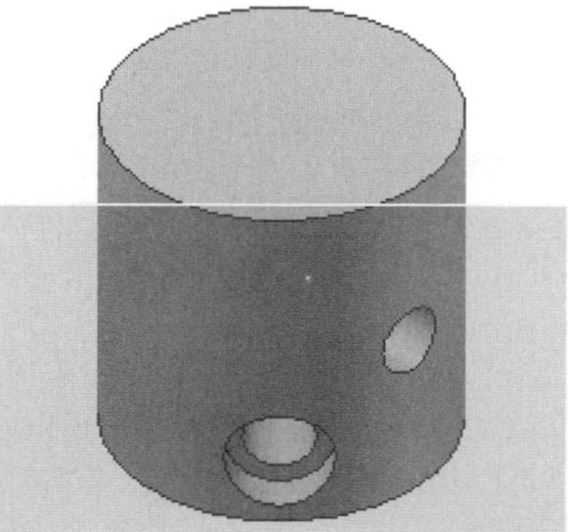

Figure 4.79

32. Expand Work Plane3 in the browser, double-click on Work Plane2, and enter a new value. Update the model to see the change.
33. Close the file. Do not save changes. End of exercise.

EXERCISES

PATTERNS

There are two types of pattern methods: rectangular and circular. The pattern is represented as a single feature in the browser, but the original feature and individual feature occurrences are listed under the pattern feature. You can suppress the entire pattern or individual occurrences except for the first occurrence. Both rectangular and circular patterns have a child relationship to the parent feature(s) that you patterned. If the size of the parent feature changes, all of the child features will also change. If you patterned a hole, and the parent hole type changes, the child holes also change. Because a pattern is a feature, you can edit it like any other feature. You can pattern the base part or feature, as well as patterns. A rectangular pattern repeats the selected feature(s) along the direction set by two edges on the part or edges that reside in a sketch. These edges do not need to be horizontal or vertical, as shown in the following image. A circular pattern repeats the feature(s) around an axis, a cylindrical or conical face, or an edge.

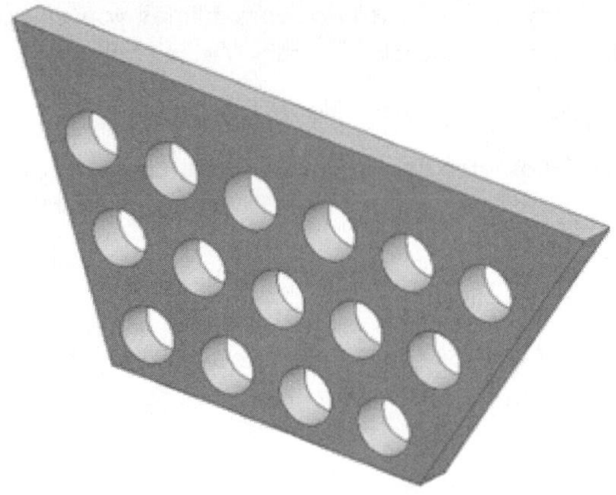

Figure 4.80

RECTANGULAR PATTERNS

When creating a rectangular pattern, you define two directions by clicking an edge or line segment in a sketch to define alignment. After you click the Rectangular Pattern on the Part Features panel bar, as shown in the following image on the left, or press the hot key CTRL + SHIFT + R (rectangular pattern), the Rectangular Pattern dialog box appears, as shown in the following image on the right. Click one or more features to pattern, enter the values, and click the edges or axes as needed. A preview image of the pattern appears in the graphics area.

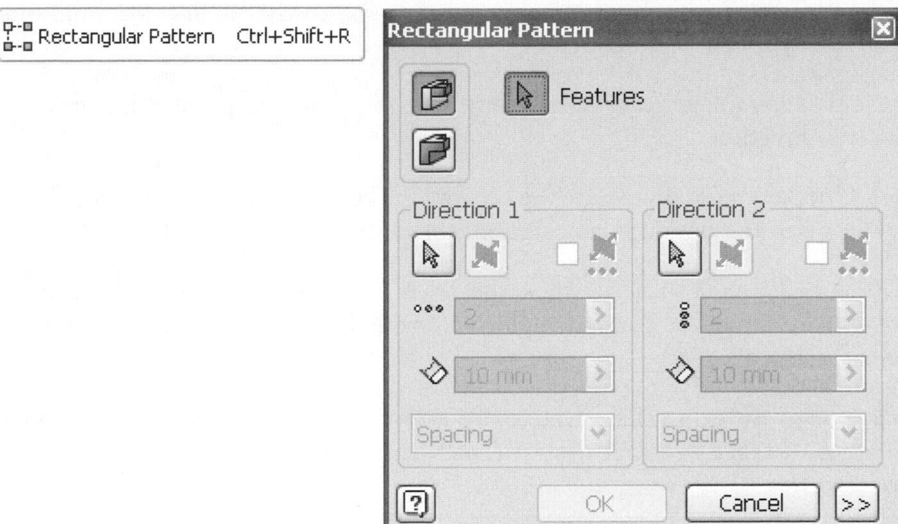

Figure 4.81

The following options are available for rectangular patterns.

Pattern Individual Features

Click this button to pattern a feature or features. When you select this option, you activate a features button, as described below.

Pattern the Entire Solid

Click this button to pattern a solid body. When you select this option, you select the entire part as the item to pattern. You also have the Include Work Features option when patterning an entire solid.

Features

Click this button, and then click a feature or features to be patterned from either the graphics window or the browser. You can add or remove features to or from the selection set by holding down the CTRL key and clicking them.

Include Work Features

Click this button, and then click a work feature or work features to include in the pattern. You can add or remove work features to or from the selection set by holding down the CTRL key and clicking them.

Direction 1

In Direction 1, you define the first direction for the alignment of the pattern. It can be an edge, an axis, or a path.

Path Click this button, and then click an edge or sketch that defines the alignment along which you will pattern the feature.

Flip If the preview image shows the pattern going in the wrong direction, click this button to reverse its direction.

Midplane Check this option to have the occurrences be patterned on both sides of the selected feature. The midplane option is independent for both Direction 1 and Direction 2.

Column Count Enter a value or click the arrow to choose a previously used value that represents the number of feature(s) you will include in the pattern along the selected direction or path.

Column Spacing Enter a value or click the arrow to choose a previously used value that represents the distance between the patterned features or the total overall distance for the patterned features.

Distance Define the occurrences of the pattern using the provided dimension as the spacing between the occurrences or the total overall distance for the patterned features.

Curve Length Create the occurrences of the pattern at equal spacing along the length of the selected curve.

Direction 2

In Direction 2, you can define a second direction for the alignment of the pattern. It can be an edge, an axis, or a path, but it cannot be parallel to Direction 1.

Path Click this button, and then click an edge or sketch that defines the alignment along which you will pattern the feature.

Flip If the preview image shows the pattern going in the wrong direction, click this button to reverse its direction.

Midplane Check this option to have the occurrences be patterned on both sides of the selected feature. The midplane option is independent for both Direction 1 and Direction 2.

Row Count Enter a value or click the arrow to choose a previously used value that represents the number of feature(s) that you will include in the pattern in the second direction.

Row Spacing Enter a value or click the arrow to choose a previously used value that represents the distance between the patterned features or the total overall distance for the patterned features.

Distance Define the occurrences of the pattern using the provided dimension as the spacing between the occurrences or the total overall distance for the patterned features.

Curve Length Create the occurrences of the pattern at equal spacing along the length of the selected curve.

Options for the start point of the direction, compute type, and orientation method, as shown in the following image, are available by clicking the More (>>) button located in the bottom-right corner of the Rectangular Pattern dialog box.

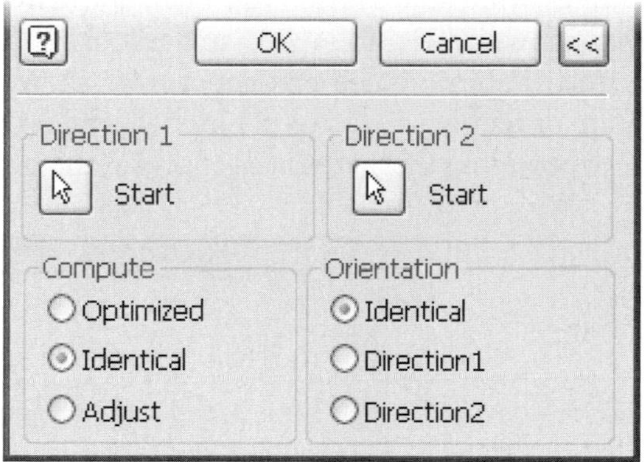

Figure 4.82

Start

Click the Start button to specify where the start point for the first occurrence of the pattern will be placed. The pattern can begin at any selectable point on the part. You can select the start points for both Direction 1 and Direction 2.

Compute

Optimized Click this option to use faces instead of features to calculate all of the occurrences in the pattern. This option is ideal when the occurrences you are creating do not intersect and are all identical. It can improve the performance of pattern creation.

Identical Click this option to use the same termination as that of the parent feature(s) for all of the occurrences in the pattern. This is the default option.

Adjust Click this option to calculate the termination of each occurrence individually. Since each occurrence is calculated separately, the processing time can increase. You must use this option if a parent feature terminates to a face or plane.

Orientation

Identical Click this option to orient all of the occurrences in the pattern the same as the parent feature(s). This is the default option.

Direction1 Click this option to control the position of the patterned features by the selected direction. Each occurrence of the pattern is rotated to maintain proper orientation with the 2D tangent vector of the path.

Direction2 Click this option to control the position of the patterned features by the selected direction. Each occurrence of the pattern is rotated to maintain proper orientation with the 2D tangent vector of the path.

CIRCULAR PATTERNS

When creating a circular pattern, you must have a work axis, a part edge, or a cylindrical face about which the features will rotate. After you click the Circular Pattern tool on the Part Features panel bar, as shown in the following image on the left, or press the hot key CTRL + Shift + O a Circular Pattern dialog box appears, as shown in the following image on the right. Click a feature or features to pattern, enter the values, and click the edges and axis as needed. A preview image of the pattern appears in the graphics area.

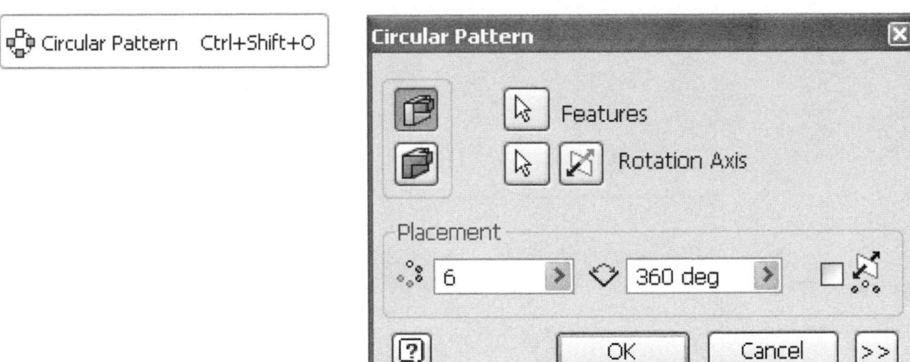

Figure 4.83

The following options are available for circular patterns.

Pattern Individual Features

Click this button to pattern a feature or features. When you select this option, the features button is available and operates as described below.

Pattern the Entire Solid

Click this button to pattern a solid body. When you select this option, you select the entire part as the item to pattern. You also have the Include Work Features option when patterning an entire solid.

Features

Click this button, and then click a feature or features to be patterned. You can add or remove features to or from the selection set by holding down the CTRL key and clicking them.

Include Work Features

Click this button, and then click a work feature or work features to include in the pattern. You can add or remove work features to or from the selection set by holding down the CTRL key and clicking them.

Rotation Axis

Click the button and then click an edge, axis, or cylindrical face (center) that defines the axis about which the feature(s) will rotate.

Flip If the preview image shows the pattern going in the wrong direction, click this button to reverse its direction.

Placement

Occurrence Count Enter a value or click the arrow to choose a previously used value that represents the number of feature(s) that you will include in the pattern. A positive number will pattern the feature(s) in the clockwise direction; a negative number will pattern the feature in the counterclockwise direction.

Occurrence Angle Enter a value or click the arrow to choose a previously used value that represents the angle that you will use to calculate the spacing of the patterned features.

Midplane Check this option to have the occurrences be patterned evenly on both sides of the selected feature.

By clicking the More (>>) button, located in the bottom-right corner of the Circular Pattern dialog box, you can access options for the creation method and positioning method of the feature, as shown in the following image.

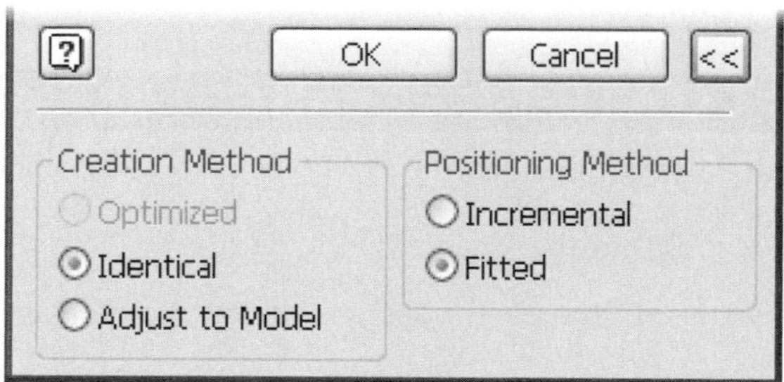

Figure 4.84

Creation Method

Optimized Click this option to use faces instead of features to calculate all of the occurrences in the pattern. This option is ideal when the occurrences you are creating do not intersect and are all identical. It can improve the performance of pattern creation.

Identical Click this option to use the same termination as that of the parent feature(s) for all of the occurrences in the pattern. This is the default option.

Adjust to Model Click this option to calculate each occurrence termination individually. Because each occurrence is calculated separately, the processing time can increase. You must use this option if a parent feature terminates to a face or plane.

Positioning Method

Incremental Click this option to separate each occurrence by the number of degrees specified in Angle in the dialog box.

Fitted Click this option to space each occurrence evenly within the angle specified in Angle in the dialog box.

Tip: A work axis, an edge, or a cylindrical face about which the feature will rotate must exist before you create a circular pattern.

LINEAR PATTERNS—PATTERN ALONG A PATH

There are many modeling cases in which you need to create a pattern that follows a path. You can define a path by a complete or partial ellipse, an open or closed spline, or a series of curves (lines, arcs, splines, etc.).

To pattern along a path, click the Path button, and use the options described above for rectangular patterns. The path you use can be either 2D or 3D.

To create a pattern along 3D paths, follow these steps:

1. Create a 3D sketch.
2. Create a path; include model edges and work curves.
3. Create the pattern.

EXERCISES

EXERCISE 4-8: CREATING RECTANGULAR PATTERNS

In this exercise, you create a rectangular pattern of holes to a cover plate.

1. Open *ESS_E04_08.ipt* in the Chapter 04 folder.
2. Rotate your view until the part looks like the following image.

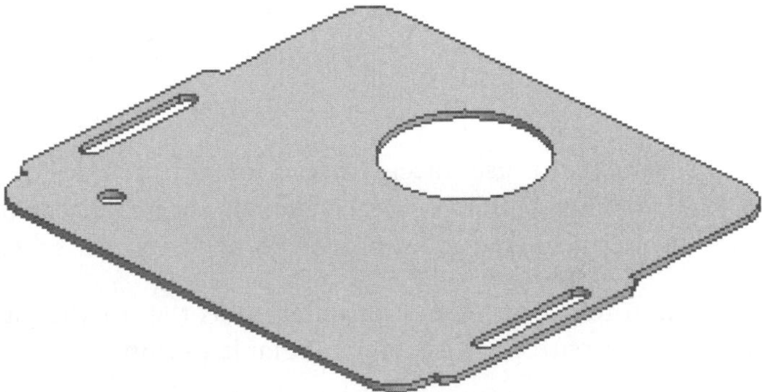

Figure 4.85

3. You now add the hole pattern. Click the Rectangular Pattern tool in the Part Features panel bar.
4. Click the small hole feature as the feature to be patterned.
5. In the Direction 1 area of the dialog box, click the Path button, and then click the bottom horizontal edge of the part. A preview of the pattern is displayed.
6. Click the Flip button.
7. Enter **5** Count and **17.5 mm** Spacing.
8. Click the Direction 2 Path button, and then click the vertical edge on the left side of the part.
9. Enter **4** Count and **17.5 mm** Spacing.

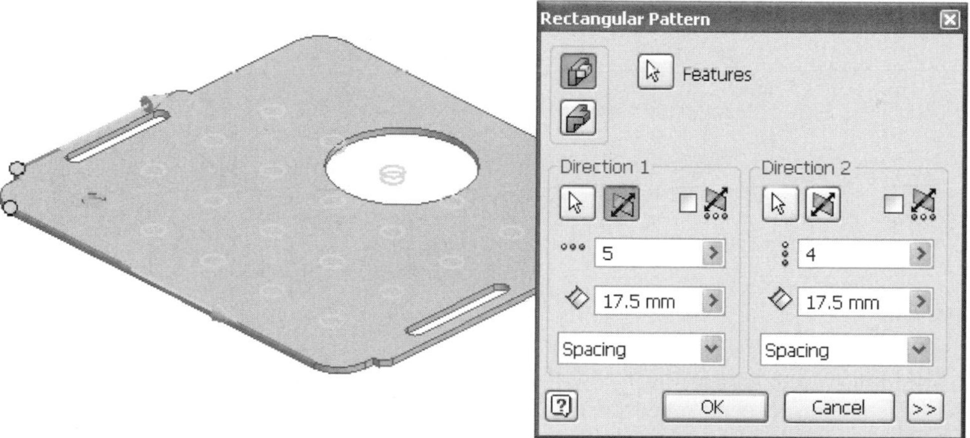

Figure 4.86

10. Click OK to create the pattern.

You now suppress three of the holes that are not required in the design.

11. Expand the rectangular pattern feature entry in the browser to display the occurrences.
12. In the browser, move the cursor over the occurrences. Each occurrence highlights in the graphics window as you point to it in the browser.
13. Hold the CTRL key down, and click the three occurrences the holes will highlight on the model, as shown in the following image on the left.
14. Right-click on any one of the highlighted occurrences in the browser, and click Suppress, as shown in the following image on the right.

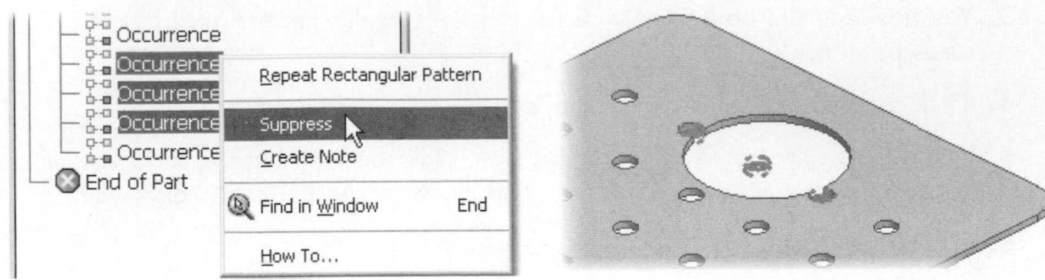

Figure 4.87

15. The holes are suppressed in the model, as shown in the following image.

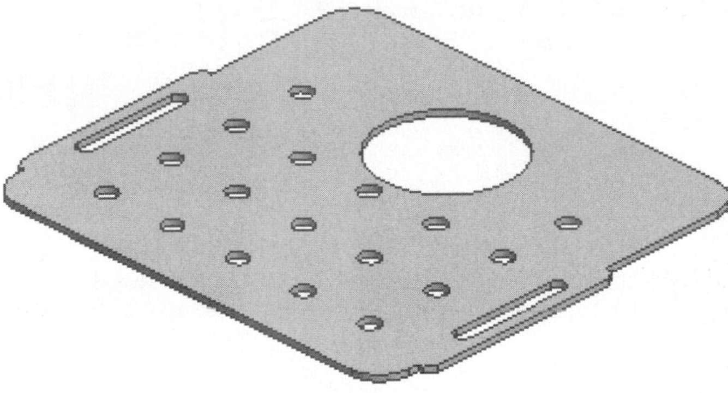

Figure 4.88

16. Edit the feature pattern, and change the count and spacing for each direction. The suppressed occurrences are still suppressed

17. Close the file. Do not save changes. End of exercise.

EXERCISE 4–9: CREATING CIRCULAR PATTERNS

In this exercise, you create a circular pattern of eight counterbore holes on a flange.

1. Open *ESS_E04_09.ipt* in the Chapter 04 folder.
2. Edit the Hole1 feature to verify that the termination for the hole is Through All.
3. Click the Cancel button to close the dialog box.
4. Click the Circular Pattern tool in the Part Features panel bar.
5. Click the hole feature.
6. Click the Rotation Axis button in the Circular Pattern dialog box.
7. Click the work axis to specify the rotation axis. A preview of the pattern is displayed.
8. Type **8** in the Count field.

9. Click the More button (>>).

10. Under Creation Method, verify that Identical is selected. Under the Positioning Method, verify that Fitted is selected, as shown in the following image.

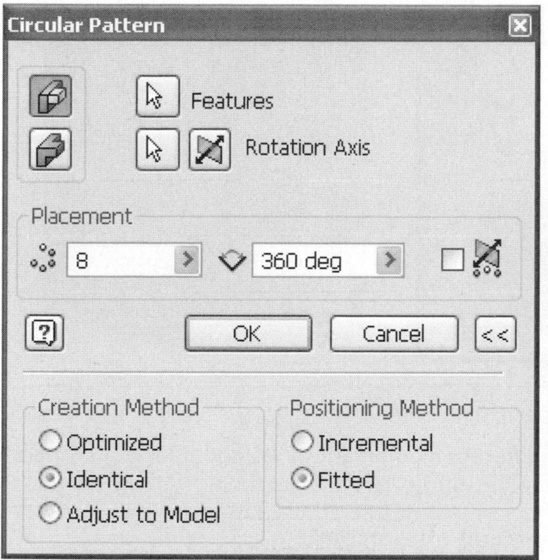

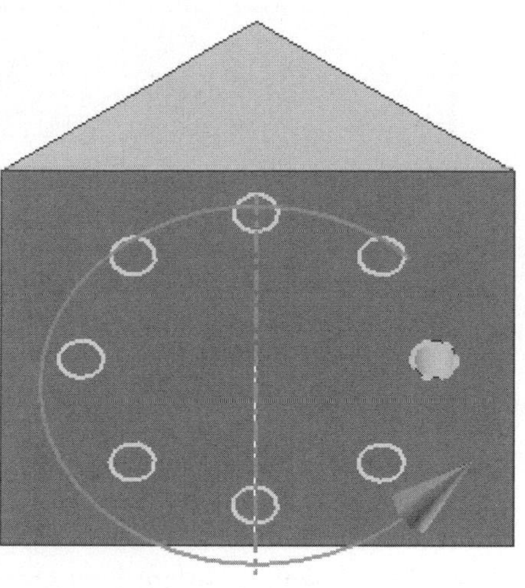

Figure 4.89

11. Click the OK button to create the pattern.

12. Rotate the model to see the back side, as shown in the following image. The other six holes do not go through because the Identical Creation Method was selected; that is, the hole that is patterned is identical to the original hole. If desired, change to wireframe display to verify that all the holes are exactly the same.

Figure 4.90

13. Move the cursor over the Circular Pattern feature in the browser, right-click, and click Edit Feature from the menu.

14. Click the More button (>>).

15. Under Creation Method, select the Adjust to Model option, as shown in the following image on the left.
16. Click OK to create the circular pattern. When done, your model should resemble the following image on the right.

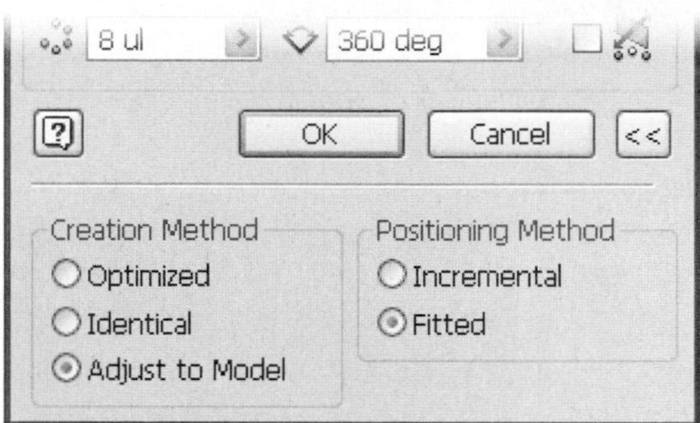

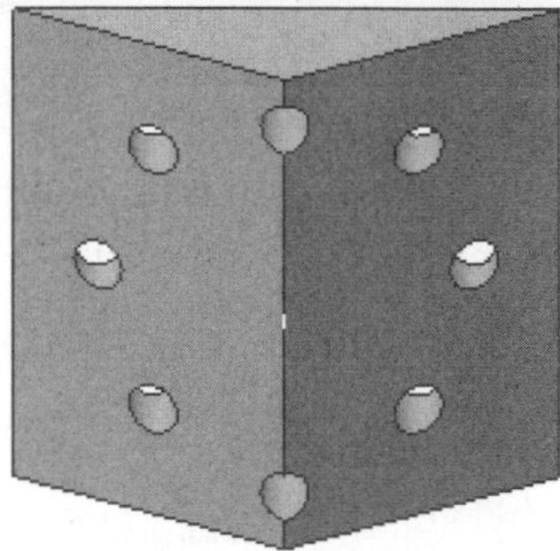

Figure 4.91

17. Edit the circular pattern, and try different combinations of count, angle, creation method, and positioning method.
18. Close the file. Do not save changes. End of exercise.

EXERCISE 4–10: CREATING PATH PATTERNS

In this exercise, you pattern a boss and hole along a nonlinear path, and in the second portion of the exercise, you pattern a hole around a 3D helix.

1. Open *ESS_E04_10.ipt* in the Chapter 04 folder. Note that the visibility of the sketch that the pattern will follow is on.
2. Click the Rectangular Pattern tool in the panel bar.
3. In the browser, click both Extrusion2 and Hole2 feature.
4. In the Direction 1 area in the dialog box, click the Path arrow button. In the graphics window, click the sketch line near the extrusion feature to select the entire path. A preview of the pattern is displayed, as shown in the following image.

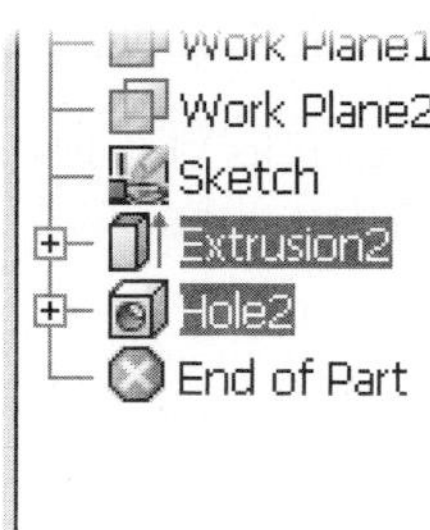

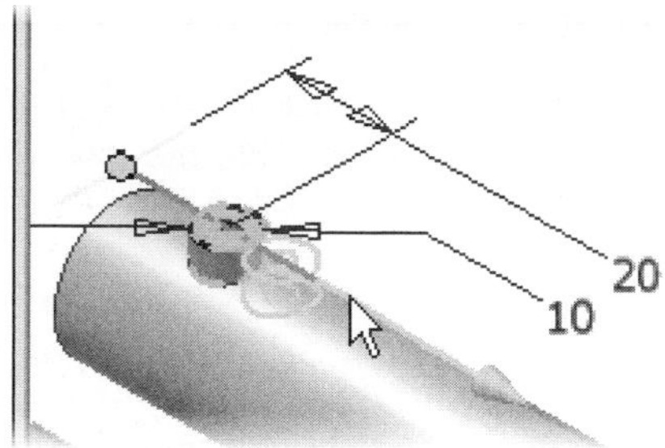

Figure 4.92

5. In Count, enter **40**, and in Spacing, enter **36**. The preview shows 36 mm between each occurrence.

6. From the Spacing drop-down list, select Distance. The preview shows 40 occurrences fit within 36 mm.

7. From the Distance drop-down list, select Curve Length. The preview shows the 40 occurrences fitting within the entire length of the path.

8. Rotate the model to verify that the occurrences on the right side are hanging over the model; this is because the occurrence are spaced 10 mm away from the start of the path.

9. To solve the starting point issue, click the More (>>) button.

10. In the Direction 1 area in the dialog box, click the Start button, and then click the center point of the first hole, as shown in the following image on the left. The preview updates to show that all occurrences are now located on the part, as shown in the following image on the right. However, the last occurrence is on the edge of the part.

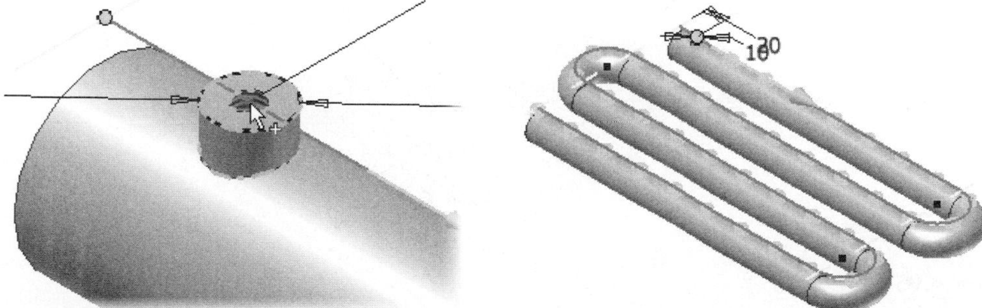

Figure 4.93

11. From the Curve Length drop-down list, select Distance. The curve length is left in the distance area but can now be edited.

12. Click in the distance area and arrow to the right of the value. Subtract **–20 mm** from the distance, as shown in the following image on the left.

13. Click OK to create the pattern.
14. In the browser, right-click on the entry Sketch, and turn its Visibility off. When done, your model should resemble the following image on the right.

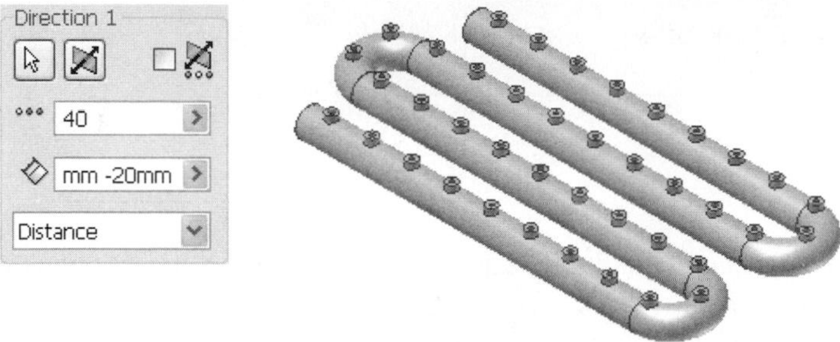

Figure 4.94

15. Edit the pattern trying different combinations.
16. Close the file. Do not save changes.
17. In this portion of the exercise you pattern a hole around a 3D helical path. Open *ESS_E04_10_2.ipt* in the Chapter 04 folder. In this part, a 3D helix has been created, and a hole has been placed on the helix.
18. Click the Rectangular Pattern tool in the Part Features panel bar.
19. In the browser or in the graphics window, click the Hole1 feature.
20. In the Direction 1 area in the dialog box, click the Path arrow button. In the graphics window, click the 3D helix, and a preview of the pattern is displayed, as shown in the following image.

Figure 4.95

21. In Count, enter **20** and in Spacing, enter **25**. The preview shows 25 mm between each occurrence, but the pattern is off the part, as shown in the following image.

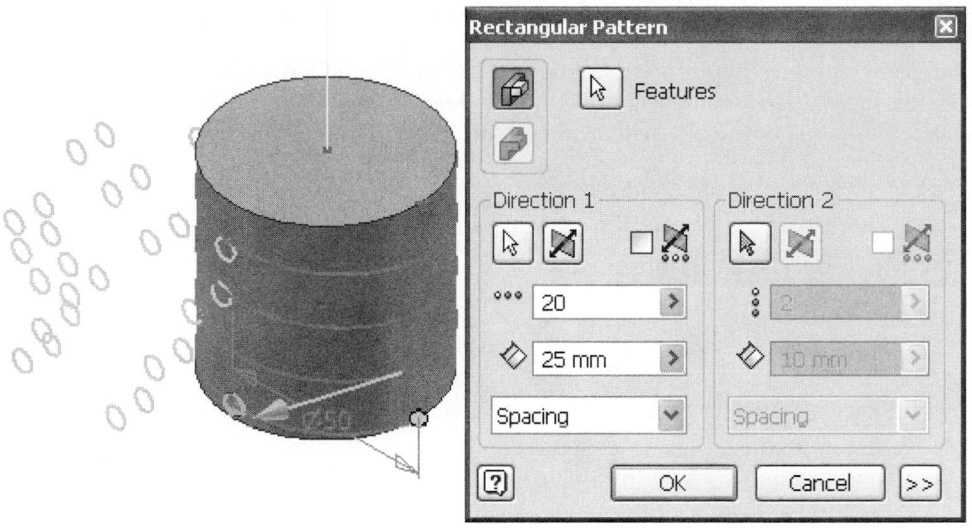

Figure 4.96

22. To solve the starting point issue, click the More (>>) button.
23. In the Direction 2 area in the dialog box, click the Direction1 button, as shown in the following image on the left. The preview updates to show that all occurrences are now located on the part.
24. Click OK to create the pattern, and your part should resemble the following image on the right.

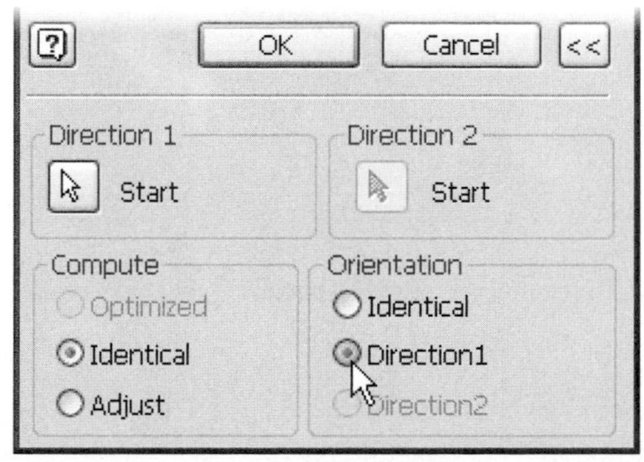

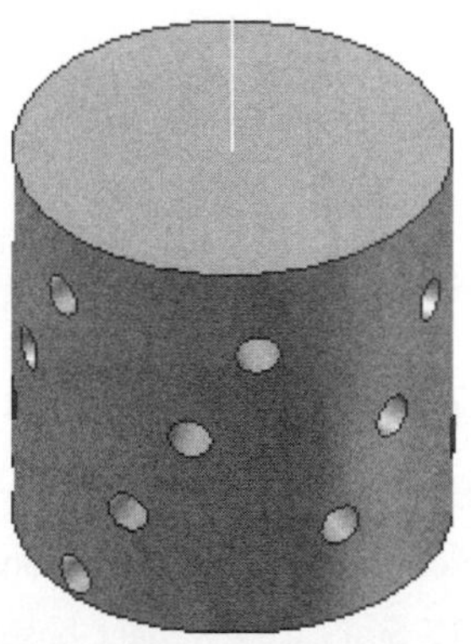

Figure 4.97

25. Edit the pattern by trying different combination of settings.
26. Close the file. Do not save changes. End of exercise.

Project Exercise: Chapter 4

The self-paced, step-by-step project exercises found in Appendix A provide opportunities for you to work through real-world modeling, assembly, and documentation tasks. The geometry used in the project exercises will flow from chapter to chapter, utilizing the functionality that you learned in that chapter.

Applying Your Skills

SKILL EXERCISE 4–1

In this exercise, you create a drain plate cover.

1. Start a new part based on the metric *Standard (mm).ipt* template.

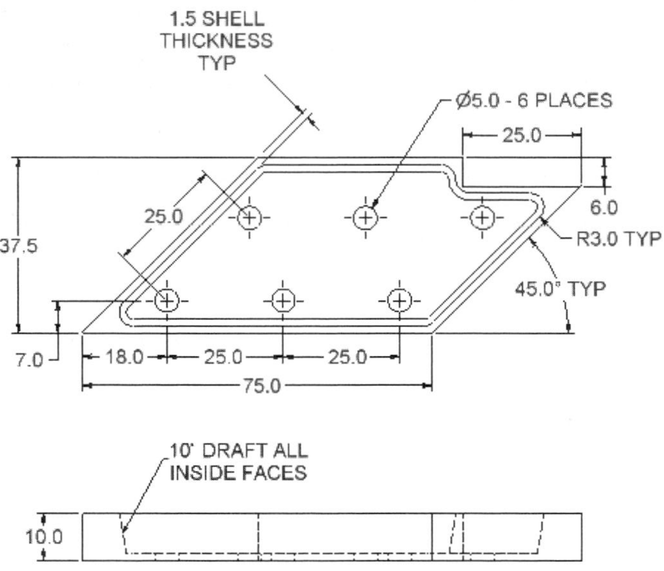

Figure 4.98

2. Use the extrude, shell, face draft, hole, and rectangular pattern tools to create the part.

SKILL EXERCISE 4–2

In this exercise, you create a connector part.

1. Start a new part based on the metric *Standard(mm).ipt* template.
2. Use the extrude, work plane, hole, thread, chamfer, fillet, and circular pattern tools to complete the part.

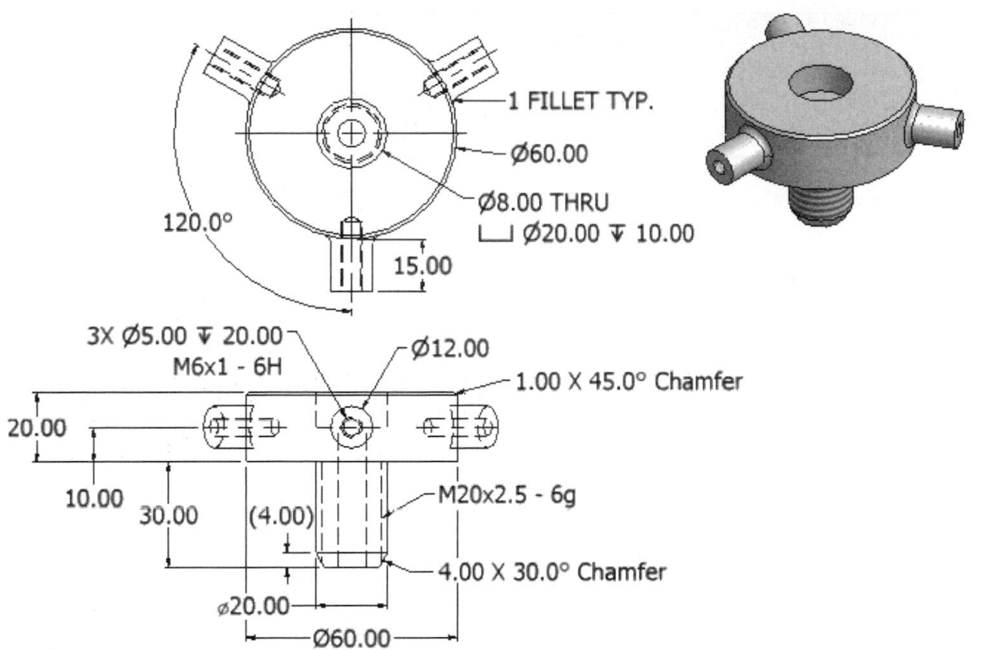

Figure 4.99

CHECKING YOUR SKILLS

Use these questions to test your knowledge of the material covered in this chapter.

1. **True _ False _** When creating a fillet feature that has more than one selection set, each selection set appears as an individual feature in the browser.

2. In regard to creating a fillet feature, what is a smooth radius transition?

____

____

3. **True _ False _** When you are creating a fillet feature with the All Fillets option, material is removed from all concave edges.

4. **True _ False _** When you are creating a chamfer feature with the Distance and Angle option, you can only chamfer one edge at a time.

5. **True _ False _** When you are creating a hole feature, you do not need to have an active sketch.

6. What is a Point, Center Point used for?

7. **True** _ **False** _ Thread features are represented graphically on the part and will be annotated correctly when you generate drawing views.

8. **True** _ **False** _ A part may contain only one shell feature.

9. When you are creating a face draft feature, what is the definition of pull direction?

10. **True** _ **False** _ The only method to create a work axis is by clicking a cylindrical face.

11. **True** _ **False** _ You need to derive every new sketch from a work plane feature.

12. Explain the steps to create an offset work plane.

13. **True** _ **False** _ You cannot create work planes from the default work planes.

14. **True** _ **False** _ When you are creating a rectangular pattern, the directions along which the features are duplicated must be horizontal or vertical.

15. **True** _ **False** _ When you are creating a circular pattern, you can only use a work axis as the axis of rotation.

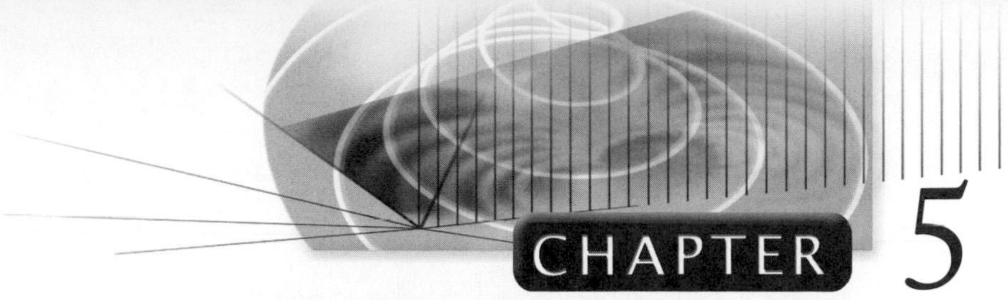

Creating and Editing Drawing Views

INTRODUCTION

After creating a part or assembly, the next step is to create 2D drawing views that represent that part or assembly. To create drawing views, start a new drawing file, select a 3D part or assembly on which to base the drawing views, insert or create a drawing sheet with a border and title block, project orthographic views from the part or assembly, and then add annotations to the views. You can create drawing views at any point after a part or assembly exists. The part or assembly does not need to be complete, because the part and drawing views are associative in both directions (bidirectional). This means that if the part or assembly changes, the drawing views will automatically be updated. If a parametric dimension changes in a drawing view, the part will be updated before the drawing views get updated. This chapter will guide you through the steps for setting up styles, creating drawing views of a single part, editing dimensions, and adding annotations.

OBJECTIVES

After completing this chapter, you will be able to perform the following:

- Understand drawing options
- Create and edit drawing borders and title blocks
- Create base and projected drawing views from a part
- Create auxiliary, section, detail, broken, breakout, cropped, and perspective views
- Edit the properties and location of drawing views
- Retrieve model dimensions to use in drawing views
- Edit, move, and hide dimensions
- Select drawing objects using a window or a crossing window
- Add automated centerlines
- Add general dimensions
- Add annotations such as text, leaders, Geometric Dimensioning & Tolerancing (GD&T), surface finish symbols, weld symbols, and datum identifiers
- Create hole notes

- Create chamfer notes
- Open a model from a drawing to edit it
- Manage drawing sheets
- Create baseline dimensions
- Create ordinate dimensions
- Create hole tables
- Create general tables
- Create revision tables
- Plot multiple drawing sheets

DRAWING OPTIONS, CREATING A DRAWING, AND DRAWING TOOLS

In this section, you will learn about the available drawing options, how to create a new drawing, and the tools that are used to create drawing views.

DRAWING OPTIONS

Before drawing views are created, the drawing options should be set to your preferences. To set the drawing options, click Application Options on the Tools menu. The Options dialog box will appear. Click the Drawing tab; your screen should resemble the following image. Make any changes to the options before creating the drawing views, or the changes may not affect drawing views that you have already created. The following sections describe the drawing options.

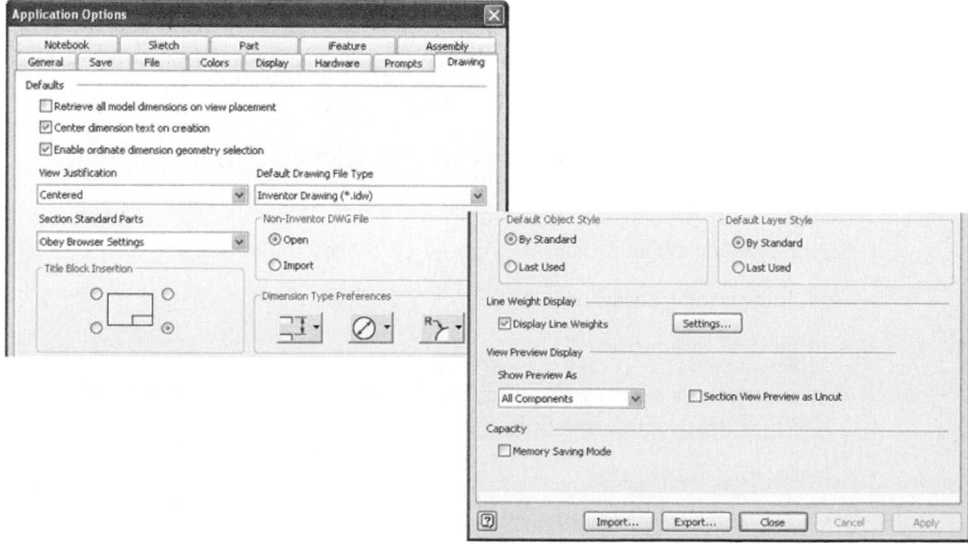

Figure 5.1

Retrieve All Model Dimensions on View Placement

Click this box to add applicable model dimensions to drawing views when they are placed. If the box is clear, no model dimensions will be placed automatically. You can override this setting by manually selecting All Model Dimensions in the Drawing View dialog box when creating base views.

Center Dimension Text on Creation

Click this option to have dimension text centered as you create the dimension.

Dimension Type Preferences

Use this area to set the preferred type of dimensions, referring to the following image as a guide.

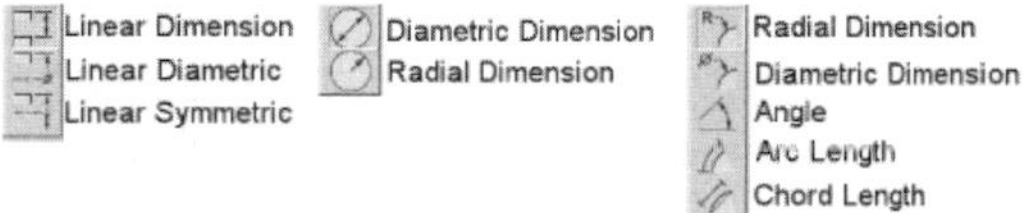

Figure 5.2

View Justification

Use this option to set the default justification for drawing views. Two modes are available: Centered and Fixed.

Section Standard Parts

Use this option to control whether standard parts placed into an assembly from the supplied Inventor parts library, such as nuts, bolts, and washers, are sectioned. Three options are available in this area: Always, Never, and Obey browser settings.

Line Weight Display Options

Use this option to control the display of line weights in a drawing.

Display Line Weights

Click this option to allow line weights to appear in drawings. Visible lines will appear in drawings with the line weights defined in the active drafting standard. Clear the box to display all lines without weights.

Display Line Weights TRUE When clicked, line weights appear on the screen as they would appear when plotted on paper.

Display Line Weights by Range (Millimeter) When selected, line weights appear according to the values entered by the user. These values range from the smallest value on the left to the largest value on the right.

Note: This setting does not affect line weights when you print the views.

Title Block Insertion

Click to select the title block insertion point for the first and subsequent drawing sheets.

CREATING A DRAWING

The first step in creating a drawing from an existing part or assembly is to create a new drawing IDW or DWG file by one of the following methods. Click the New icon in the What To Do section of the Getting Started page. You could also click New on the File menu, and then click the desired drawing template from one of the template tabs, as shown in the following image. As a third option, you could click the down arrow on the New icon on the left side of the Standard toolbar and then click Drawing.

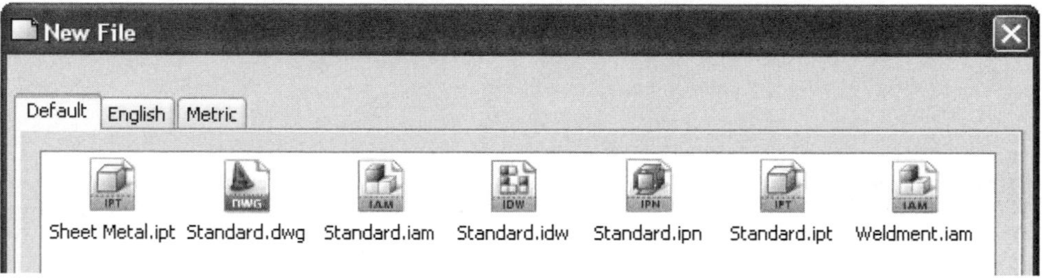

Figure 5.3

Six drafting templates, included with Autodesk Inventor, are presented under the Metric tab, as shown in the following image. They include the following:

- ANSI (American National Standards Institute)
- BSI (British Standards Institute)
- DIN (The German Institute for Standardization)
- GB (The Chinese National Standard)
- GOST (The Russian Standard)
- ISO (International Organization for Standardization)
- JIS (Japan Industrial Standard)

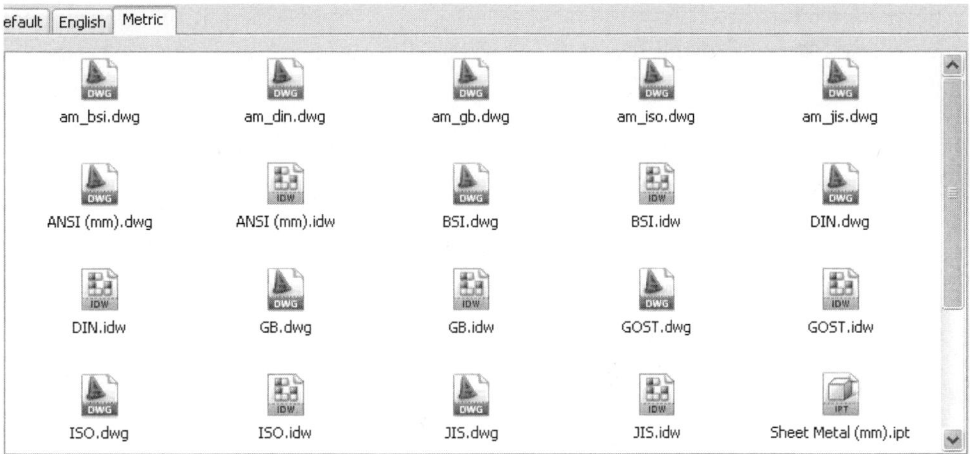

Figure 5.4

DWG TRUECONNECT

DWG TrueConnect allows you to work directly with AutoCAD DWG files without the need for translations. This feature allows you to view, plot, and measure AutoCAD drawing data while using Autodesk Inventor and perform the same actions to Autodesk Inventor drawing data while using AutoCAD. In this way, you can supply AutoCAD customers with DWG drawings that contain familiar features such as layers and object types such as blocks and title blocks.

DRAWING TOOLS

After you start a new drawing, Autodesk Inventor's screen will change to reflect the new drawing environment. There are three toolbars available in the panel bar that you can use to create drawings: Drawing Views, Sketch, and Drawing Annotation. By default, when a new drawing is created, the panel bar will show the Drawing Views tools. These tools allow you to create drawing views, sheets, and draft views. The Sketch tools enable you to create custom symbols and add 2D geometry and dimensions to draft views. The Drawing Annotation tools enable you to add annotations to drawing sheets and existing drawing views. These tools will be explained throughout this chapter.

DRAWING SHEET PREPARATION

When you create a new drawing file using one of the provided template files, the program displays a default drawing sheet with a default title block and border. The template DWG or IDW file that is selected determines the default drawing sheet, title block, and border. The drawing sheet represents a blank piece of paper on which you can alter the border, title block, and drawing views. There is no limit to the number of sheets that can exist in the same drawing, but you must have at least one drawing sheet. To create a new sheet, click the New Sheet tool on the Drawing Views panel bar as shown on the left in the following image.

Alternately, you can right-click in the browser or on the current sheet in the graphics window and select New Sheet from the menu as shown on the right in the following image.

Figure 5.5

A new sheet will appear in the browser, and the new sheet will appear in the graphics window with a default border and title block. You can rename the sheet by slowly double-clicking on its name in the browser and then entering a new name or right-clicking on the sheet in the browser and selecting Edit Sheet from the menu. You can then enter a new name in the Edit Sheet dialog box. To create a sheet of a different size, you can double-click on one of the Sheet Formats in Drawing Resources in the browser, as shown in the following image, or right-click in the graphics window and select New Sheet from the menu.

Figure 5.6

If a sheet is selected from the Sheet Formats list, predetermined drawing views will be created. If the necessary sheet size is not on the list, right-click on the sheet name in the browser, and select Edit Sheet from the menu, as shown in the following image on the left. Then select a size from the list, as shown in the following image on the right. To use your own values, select Custom Size from the list, and enter values for height and width.

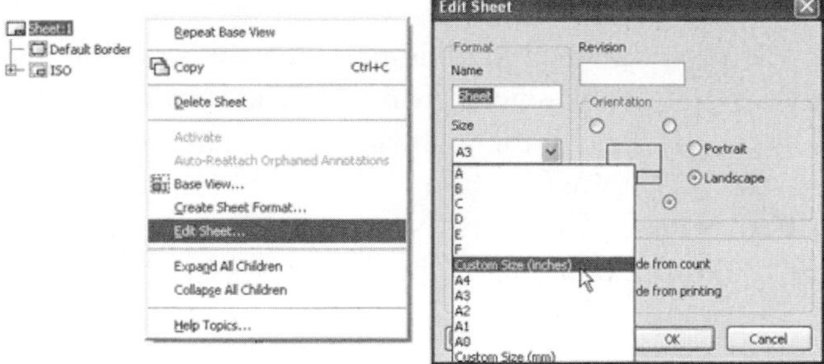

Figure 5.7

Note: The sheet size is inserted full scale (1:1) and should be plotted at 1:1. The drawing views will be scaled to fit the sheet size.

TITLE BLOCKS

To add a title block to the drawing sheet, you can either insert a default title block or construct a customized title block and insert it into the drawing sheet.

INSERTING A DEFAULT TITLE BLOCK

To insert a default title block, follow these steps:

1. Make the sheet active, and then place the title block by double-clicking on its name in the browser.
2. Insert the title block by expanding Drawing Resources > Title Blocks in the browser, as shown in the following image. Either double-click on the title block's name, as shown in the following image, or right-click on the title block's name and select Insert from the menu.

Note: If a title block already exists in a drawing, it must be deleted before a new title block can be inserted.

Figure 5.8

EDIT PROPERTY FIELDS DIALOG BOX

The time will come when you will need to fill in title block information. Expanding the default title block in the browser and double-clicking on the Field Text category will display the Edit Property Fields dialog box. By default, the following information will already be filled in: Sheet Number, Number of Sheets, Author, Creation Date, and Sheet Size. To fill in other title block information such as Part Number, Company Name, Checked By, and so on, select Properties-Drawing from the drop-down list, and click the iProperties button, as shown in the following image on the left.

The Properties dialog box will appear, as shown in the following image on the right, and you can fill in the information as needed. You can find most title block information under the Summary, Project, and Status tabs. The following image shows the Properties dialog box and the Status tab with the Checked By and Eng. Approved By categories filled in.

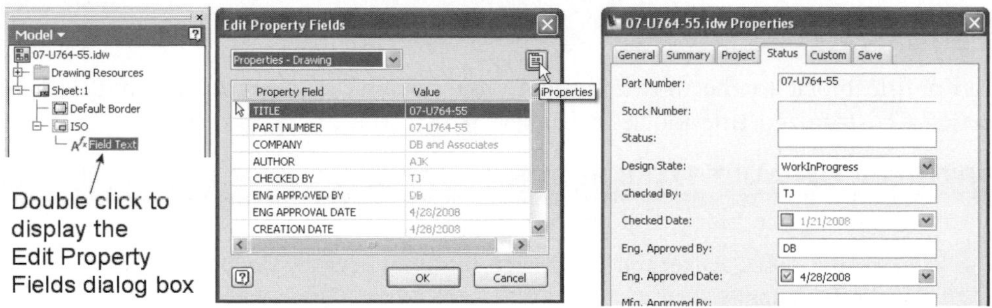

Figure 5.9

STYLES

Autodesk Inventor uses styles to control how objects appear. Styles are saved within an Autodesk Inventor file or to a project library location, and they can be saved to a network location so that many users can access the same styles. This section introduces you to styles. Autodesk Inventor uses Extensible Markup Language (XML) files for storing style information externally from Autodesk Inventor documents. Autodesk Inventor does not support the editing or use of these XML files with anything other than the tools provided inside Autodesk Inventor and the Style Library Manager. Once a style is used in a document, it is stored in the document.

STYLE NAME/VALUE

Autodesk Inventor uses the style's name as the unique identifier of that style: only one name for the same style type can exist. For example, in a drawing, only one dimension style with a specific name "Default ANSI" can exist. However, a text style in the same file with the name "Default ANSI" could exist.

CREATING A NEW STYLE

To create a new style, follow these steps:

1. Click the Style and Standard Editor tool on the Format menu.
2. Click and expand the style section for which you want to create a new style.
3. Right-click the style on which the new style will be based, and select New Style, as shown in the following image on the left. This image shows a new style being created from the Default (ANSI) dimension style.
4. The New Style Name dialog box will appear, as shown in the following image on the right. Enter a style name, and if you do not want the style to be used in the standard, uncheck Add to standard.
5. Make changes to the style, and save the changes.

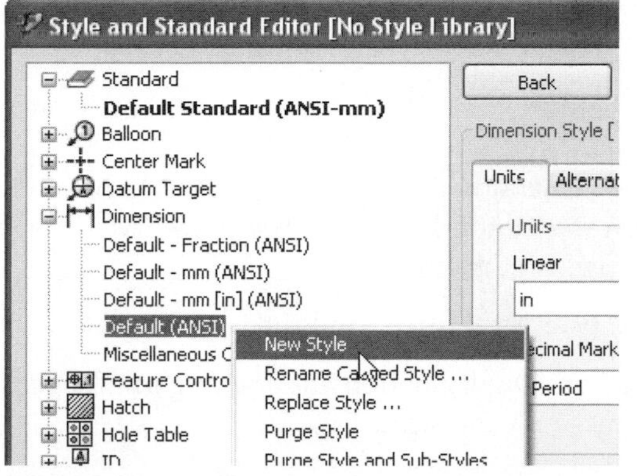

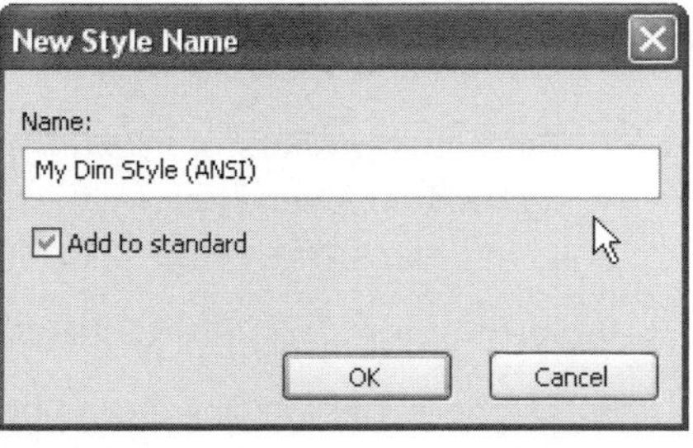

Figure 5.10

EXERCISES

EXERCISE 5-1: CREATING TEXT AND DIMENSION STYLES

In this exercise, you create new text and dimension styles.

1. Open the file *ESS_E05_01.idw*. The drawing consists of three orthographic views in addition to an isometric view.
2. Begin the process of creating a new text style. From the Format menu, click Style and Standard Editor, as shown in the following image on the left.
3. When the Style and Standard Editor dialog box appears, expand the Text style in the left pane. Then right-click on Current-ANSI and click New Style, as shown in the following image on the right.

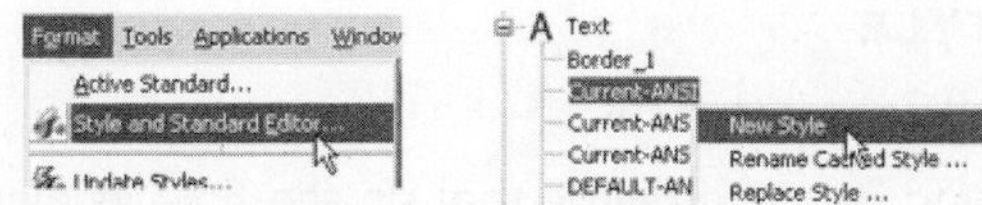

Figure 5.11

4. In the New Style Name dialog box, type **My ANSI Text**, and then click OK.
5. In the Font list, select Arial, and in the Size list, select **3.50 mm**, as shown in the following image.
6. When you have finished modifying this new text style, click the Save button.

Figure 5.12

7. Begin the process of creating a new dimension style. In the left pane, expand the Dimension style. Then right-click on the Metric style, and click New Style from the menu, as shown in the following image on the left.
8. In the New Style Name dialog box, type **My Metric**, and then click OK.
9. Uncheck the Leading Zeros box for the Display option and the Angular Display option, as shown in the following image on the right.

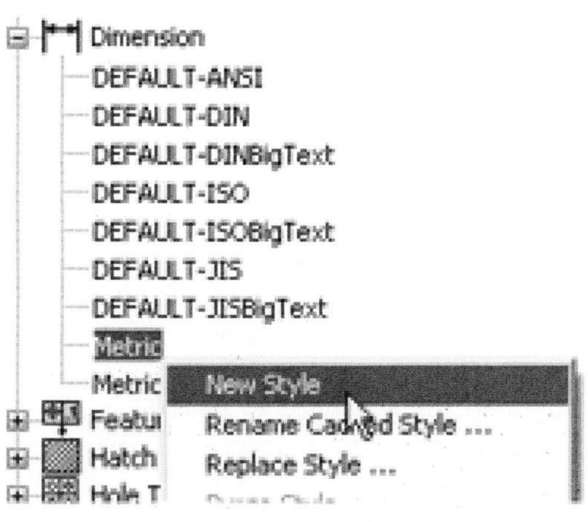

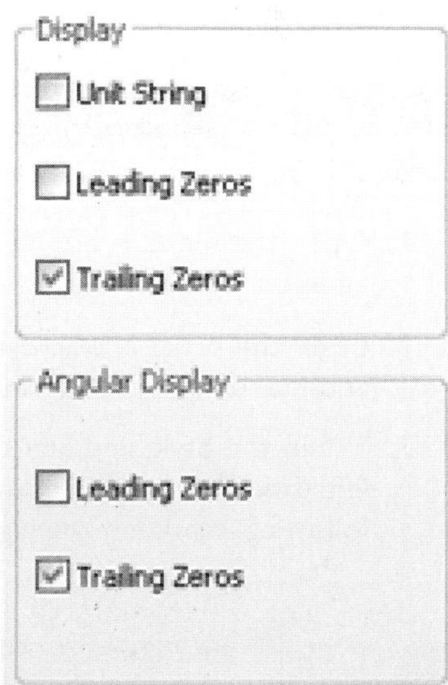

Figure 5.13

10. Next click the Display tab. Type **3.00 mm** for A: Extension and B: Origin Offset, as shown in the following image on the left.

11. Next click the Text tab. In the Primary Text Style list, select My ANSI Text, as shown in the following image on the right.

12. Click the Save button and then the Done Button.

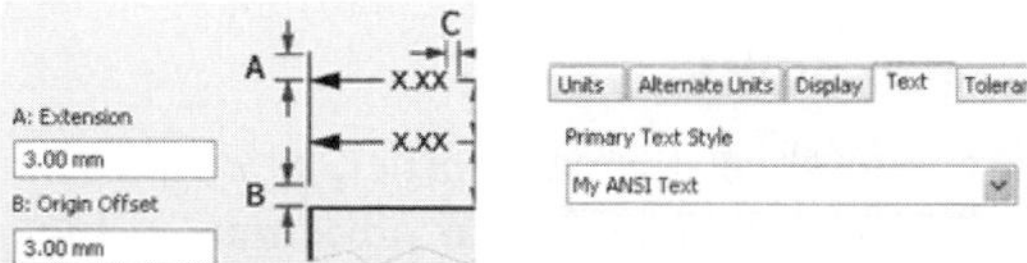

Figure 5.14

13. You will now change some dimensions from one style to another. Zoom in on the top view, and examine a few dimensions. Notice the text style and leading zeros.

14. Select all the dimensions in the top view. To do this, hold down the CTRL key while picking the dimensions, but do not select any of the edges.

15. From the dimension style drop-down list, click My Metric, as shown in the following image on the left.

16. Zoom in and examine the updated dimensions. Your display should appear similar to the following image on the right.

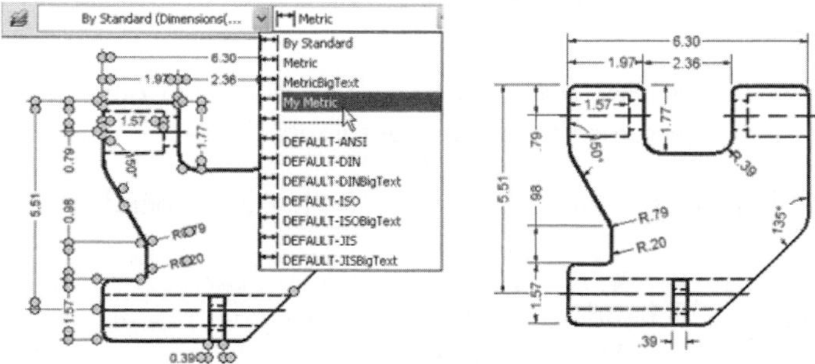

Figure 5.15

Note: The new dimension style will pertain only to this drawing. To add a new style to the style library, the Autodesk Inventor Project file Use Styles Library option must be set to Yes. Then from the Format menu, click Save Styles to Style Library.

17. Close all open files. Do not save changes. End of exercise.

TEMPLATES

After you have created a drawing sheet, border, and title block and have set various styles, you can save the file as a template and use this template as a basis for new drawing files. To save a file as a template, click Save Copy As on the File menu, enter a new template name, and save the file to the appropriate Autodesk Inventor template folder. This new file is now available as a template when creating new drawings.

CREATING DRAWING VIEWS

After you have set the drawing sheet format, border, title block, and styles, you can create drawing views from an existing part, assembly, or presentation file. The file from which you will create the views does not need to be open when a drawing view is created. It is suggested, however, that both the file and the associated drawing file be stored in the same directory and that the directory be referenced in the project file. When creating drawing views, you will find that there are many different types of views you can create. The following sections describe these view types.

Base View This is the first drawing view of an existing part, assembly, or presentation file. It is typically used as a basis for generating the following dependent view types. You can create many base views in a given drawing file.

Projected View This is a dependent orthographic or isometric view that is generated from an existing drawing view.

- Orthographic (Ortho View): A drawing view that is projected horizontally or vertically from another view.
- Iso View: A drawing view that is projected at a 30° angle from a given view. An isometric view can be projected to any of four quadrants.

Auxiliary View This is a dependent drawing view that is perpendicular to a selected edge of another view.

Section View This is a dependent drawing view that represents the area defined by a slicing plane or planes through a part or assembly.

Detail View This is a dependent drawing view in which a selected area of an existing view will be generated at a specified scale.

Broken View This is a dependent drawing view that shows a section of the part removed while the ends remain. Any dimension that spans over the break will reflect the actual object length.

Break Out View This is a drawing view that has a defined area of material removed in order to expose internal parts or features.

Crop View This drawing view allows a view to be clipped based on a defined boundary.

Overlay View This drawing view uses positional representations to show an assembly in multiple positions in a single view.

USING THE DRAWING VIEW DIALOG BOX

You use the Drawing View dialog box to create the various drawing views described above. Activate the dialog box by clicking the Base View tool on the Drawing Views panel bar, as shown in the following image, or by right-clicking in the graphics window and selecting Base View from the menu.

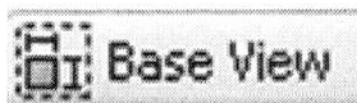

Figure 5.16

As shown in the following image, the Drawing View dialog box has three tabs: Component, Model State, and Display Options. The following sections describe these tabs.

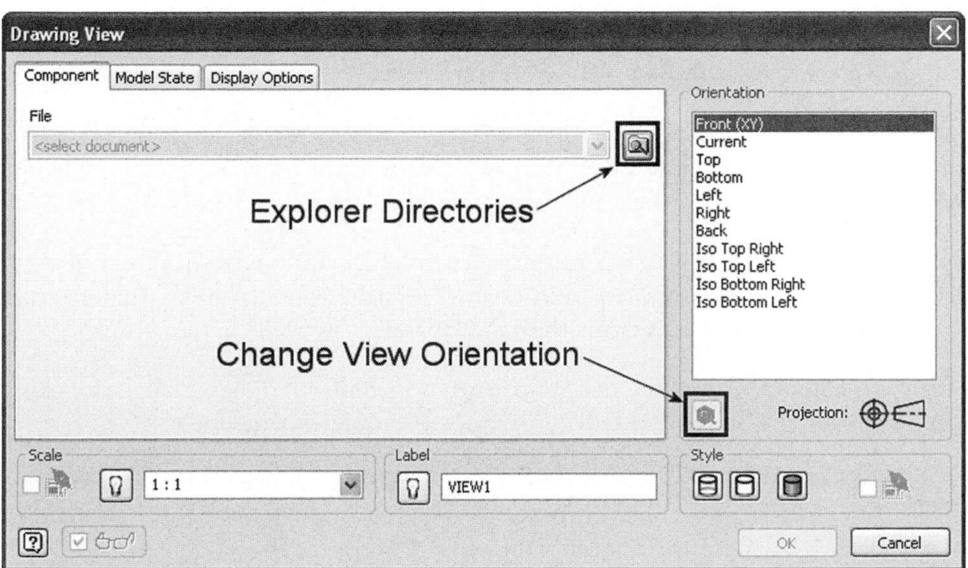

Figure 5.17

THE COMPONENT TAB

The following categories are available under the Drawing View Component tab.

File Any open part, assembly, or presentation files will appear in the drop-down list. You can also click the Explore directories icon and navigate to and select a part, assembly, or presentation file.

Orientation After selecting the file, choose the orientation in which to create the view. After selecting an orientation, a preview image will appear in the graphics window attached to your cursor. If the preview image does not show the view orientation that you want, select a different orientation view by selecting another option. A symbol that identifies the type of drawing projection is also displayed under the Orientation area and is a symbol that identifies the type of drawing projection, such as First Angle or Third Angle Projection. The following view orientations are available:

- Front, Current (current orientation of the part, assembly, or presentation in its file), Top, Bottom, Left, Right, Back, Iso Top Right, Iso Top Left, Iso Bottom Right, and Iso Bottom Left. A Change view orientation button is also present that allows you to create a custom view.

Scale Enter a number for the scale in which to create the view. Note that the drawing sheet will be plotted at full scale (1:1), and the drawing views are scaled as needed to fit the sheet. You can edit the scale of the views after the views have been generated.

Scale from Base Click to set the scale of a dependent view to match that of its base view, as shown in the following image on the left. To change the scale of a dependent view, clear the option's box, and modify the scale value for that view. This feature activates when you edit existing views.

Scale Label Visibility Light Bulb Click to display or hide the view scale label. Clicking the checkbox displays the scale label, and leaving the box unchecked hides the scale label.

Label Use to include and/or change the label for the selected view. When you create a view, a default label is determined by the active drafting standard. To change the label, select the label in the box and enter the new label.

View Label Visibility Light Bulb Click to display or hide the view label. Clicking the checkbox displays the label, and leaving the box clear hides the label.

Style Choose how the view will appear. There are three choices: Hidden Line, Hidden Line Removed, and Shaded. The preview image will not update to reflect the style choice. When the view is created, the chosen style will be applied. The style can be edited after the view has been generated.

Style from Base Click to set the display style of a dependent view to be the same as that of its base view, as shown in the following image on the right. To change the display style of a dependent view, clear the checkbox and modify the style for that view. This feature activates when you edit existing views.

Figure 5.18

Assembly Drawing Views

When generating drawing views from an assembly model, a Representation area is displayed on the Component tab of the Drawing View dialog box as shown in the following image. The elements of the Representation area are explained below.

View After selecting an assembly file that has design view representations, the available representation names will be listed in the View area. If you selected a presentation file that has presentation views, the names of the presentation views will be listed in the View area.

Position Used to create a drawing view based on an assembly file's positional representation. This feature is covered in greater detail in chapter 09.

Level of Detail Used to create a drawing view based on an assembly file's level of detail representation. This feature is covered in greater detail in chapter 09.

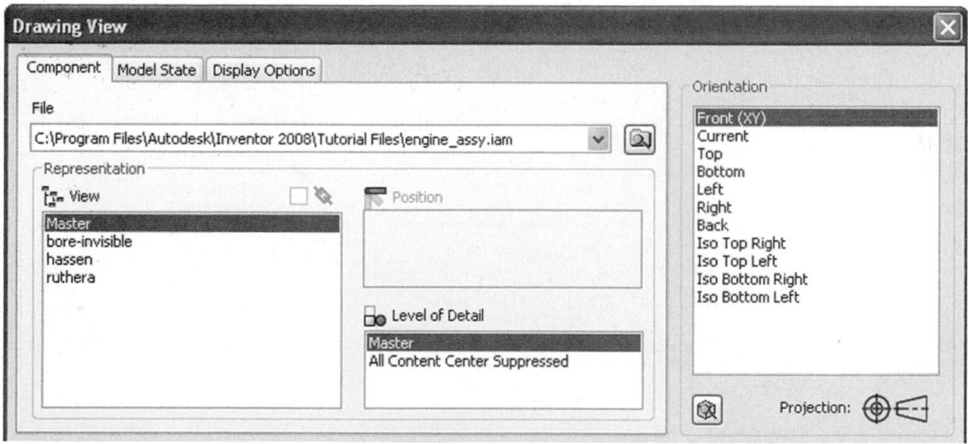

Figure 5.19

THE MODEL STATE TAB

Use this tab to specify the weldment state and member states of an iAssembly or iPart to use in a drawing view, as shown in the image below. Other items are used to control line style and hidden line calculations.

Weldment Enable this area when you select a document that is a weldment. Weldments have four states: Assembly, Preparations, Welds, and Machining.

Member Allows you to select the member of an iAssembly or iPart to represent in a drawing view.

Line Style Three options are presented: As Reference Parts, As Parts, and Off. When you select As Reference Parts for shaded view types, the reference data will appear transparently shaded with tangent edges turned off and all part edges as phantom lines. The reference data will appear on top of the product data that is shaded. When you select As Parts, reference data will have no special display characteristics. The reference data will appear on top of product data, and the reference data will appear with tangent edges turned off and all other edges as the phantom line type. When you select Off, reference data will not appear.

Hidden Line Calculation Specifies if hidden lines are calculated for Reference Data Separately or for All Bodies.

Margin To see more reference data, set the value to expand the view boundaries by a specified value on all sides.

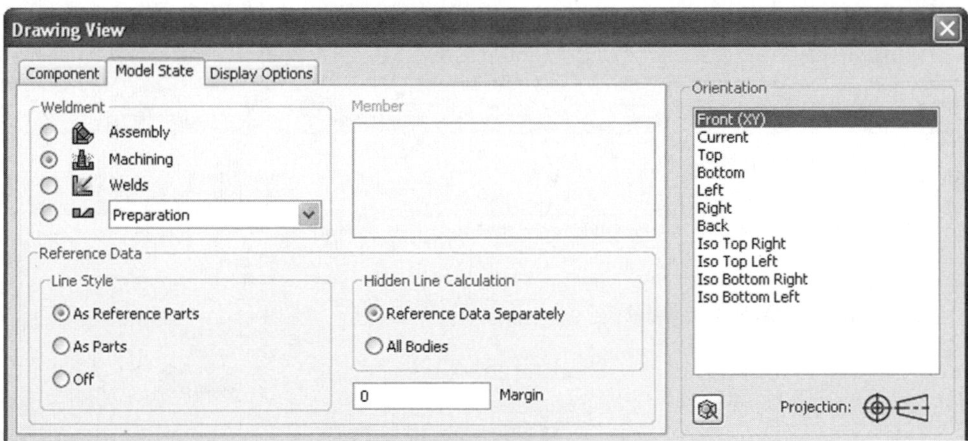

Figure 5.20

THE DISPLAY OPTIONS TAB

The following image shows the Display Options tab of the Drawing View dialog box. The following sections describe the tab's options.

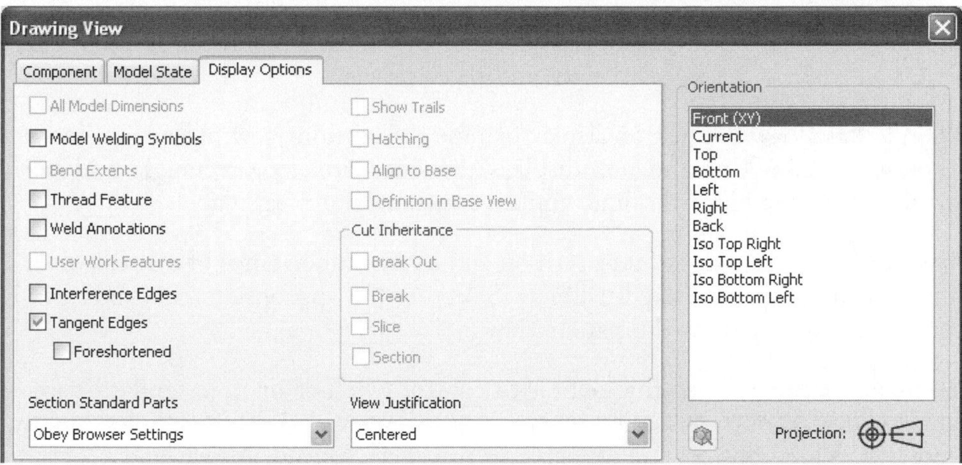

Figure 5.21

All Model Dimensions Click to see model dimensions in the view, which is only active upon base view creation. If the option's box is clear, model dimensions will not be placed automatically upon view creation. When checked, only the dimensions that are parallel to the view and have not been retrieved in existing views on the sheet will appear.

Model Welding Symbols This box is active only if you are creating a drawing view of a weldment. Click to retrieve welding symbols placed in the model in the drawing.

Bend Extents This box is active only if you are creating a drawing view of a sheet metal part flat pattern. Click to control the visibility of bend extent lines or edges.

Thread Feature This box is active only if you are creating a view of an assembly model. Click to set the visibility of thread features in the view.

Weld Annotations This box is active only if creating a view of a weldment. Click to control the display of weld annotations.

User Work Features Click to display work features in the drawing view.

Interference Edges When selected, a drawing view will display both hidden and visible edges due to an interference condition such as a press or interference fit.

Tangent Edges Click to set the visibility of tangent edges in a selected view. Checking the box displays tangent edges, and leaving the box clear hides them. If enabled, tangent edges can also be adjusted to display Foreshortened.

Show Trails Click to control the display of trails in drawing views based on presentation files.

Hatching Click to set the visibility of the hatch lines in the selected section view. Checking the box displays hatch lines, and leaving the box clear hides them.

Align to Base Click to remove the alignment constraint of a selected view to its base view. When the box is checked, alignment of views exists. Leaving the box clear breaks the alignment and labels the selected view and the base view.

Definition in Base View Click to display or hide the section view projection line or detail view boundary. Checking the box displays the line, circle, or rectangle and text label; leaving the box clear hides the line, circle, or rectangle and text label.

Cut Inheritance Use this option to turn on and off the inheritance of a Break Out, Break, Section, and Slice cut for the edited view. Selecting the appropriate checkbox will inherit the corresponding cut from the parent view.

Section Standard Parts Use this option to control whether or not standard parts placed into an assembly from the supplied Inventor parts library, such as nuts, bolts, and washers, are sectioned. Three options are available in this area: Always, Never, and Obey Browser Settings.

View Justification Use to control the position of drawing views when the size or position of the model changes. This area is especially helpful with creating drawing views from assembly models. This area contains two modes: Center and Fixed. The Center mode keeps the model image centered in the drawing view. If, however, the model's view increases or decreases in size, the drawing view will shift on the drawing sheet. In some cases, this shift could overlap the drawing border or title block. The Fixed mode keeps the drawing view anchored on the drawing sheet. In the event that the model image increases, an edge of the drawing view will remain fixed.

BASE VIEWS

A base view is the first view that you create from the selected part, assembly, or presentation file. When you create a base view, the scale is set in the dialog box, and from this view, you can project other drawing views. There is no limit to the number of base views you can create in a drawing based on different parts, assemblies, or presentation files. As you create a base view, you can select the orientation of that view from the Orientation list found on the Component tab of the Drawing View dialog box.

To create a base view, follow these steps:

1. Click the Base View tool on the Drawing Views panel bar, or right-click in the graphics window and select Base View.
2. The Drawing View dialog box will appear. On the Component tab, click the Explore directories icon to navigate to and select the part, assembly, or presentation file from which to create the base drawing view. After making the selection, a preview image will appear attached to your cursor in the graphics window. Do not place the view until the desired view options have been set.
3. Select the type of view to generate from the Orientation list.
4. Select the scale for the view.

5. Select the style for the view.
6. Locate the view by selecting a point in the graphics window.

PROJECTED VIEWS

A projected view can be an orthographic or isometric view that you project from a base view or any other existing view. When you create a projected view, a preview image will appear, showing the orientation of the view you will create as the cursor moves to a location on the drawing. There is no limit to the number of projected views you can create. To create a projected drawing view, follow these steps:

1. Click the Projected View tool on the Drawing Views panel bar, as shown in the following image. Select Projected View in Model Views on the Insert menu, or right-click inside the bounding area of an existing view box, shown as dashed lines when the cursor moves into the view. Select Create View > Projected from the menu.

Figure 5.22

2. If you selected the Projected View tool, click inside the desired view to start the projection.
3. Move the cursor horizontally, vertically, or at an angle to get a preview image of the view you will generate. Keep moving the cursor until the preview matches the view that you want to create, and then press the left mouse button. Continue placing projected views.
4. When finished, right-click, and select Create from the menu.

EXERCISES

EXERCISE 5-2: CREATING A MULTIVIEW DRAWING

In this exercise, you will create an independent view to serve as the base view, and then you will add projected views to create a multiview orthographic drawing. Finally, you will add an isometric view to the drawing.

1. Open the file *ESS_E05_02.idw*. This drawing file contains a single sheet with a border and title block as shown in the following image.

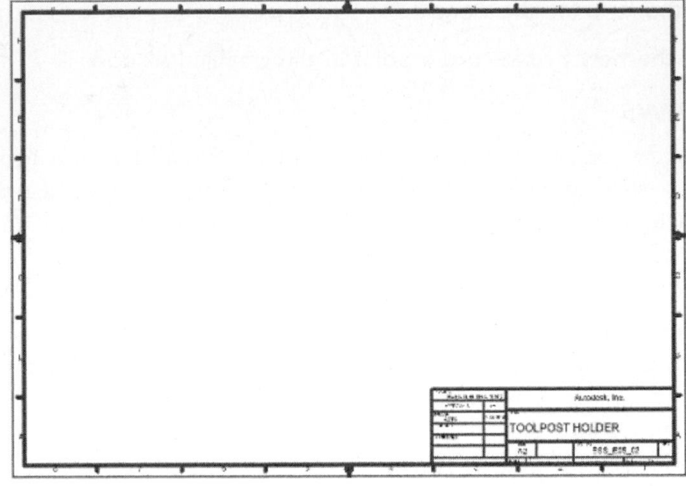

Figure 5.23

2. Create a base view by clicking on the Base View tool; this will display the Drawing View dialog box.
3. Under the File area of the Component tab, click the Explore directories button, and double-click the file *ESS_E05_02.ipt* in the Chapter 5 folder to use it as the view source.
4. In the Orientation area, verify that Front is selected.
5. In the Scale list, select 1:1.
6. In Style, click the Hidden Line button.
7. Click the Display Options tab, and ensure that All Model Dimensions is not checked.
8. Position the view preview in the lower-left corner of the sheet (in Zone C6), and then click to place the view as shown in the following image.

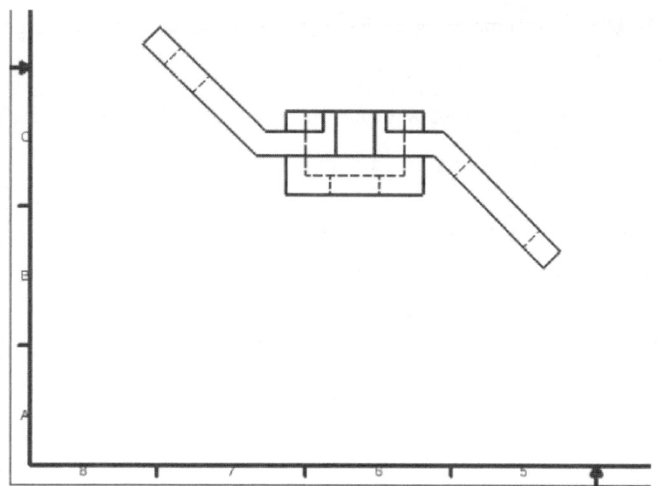

Figure 5.24

9. Click the Projected View tool in the panel bar. This tool will be used to create the top and right-side views from the front view.
10. Click the base view, and move the cursor vertically to a point above the base view. Click in Zone E6 to place the top view.
11. Move the cursor to the right of the base view. Click in Zone C2 to place the right-side view.
12. Right-click, and choose Create from the menu to create the new views, as shown in the following image.

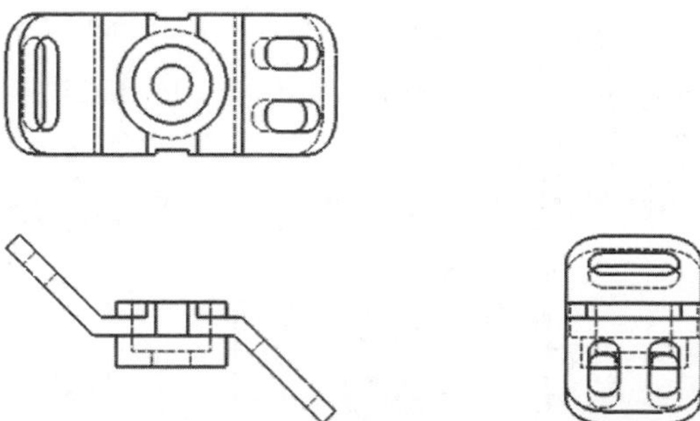

Figure 5.25

13. The views are crowded with a scale of 1:1. In the steps that follow, you will reduce the size of all drawing views by setting the base view scale to 1:2. The dependent views will update automatically. Right-click the base view or front view, and select Edit View.

Note: To activate this menu, right-click on the view border or inside the view. Do not right-click on the geometry.

14. When the Drawing View dialog box displays, select 1/2 from the Scale list, and click OK.

15. The scale of all views updates, as shown in the following image.

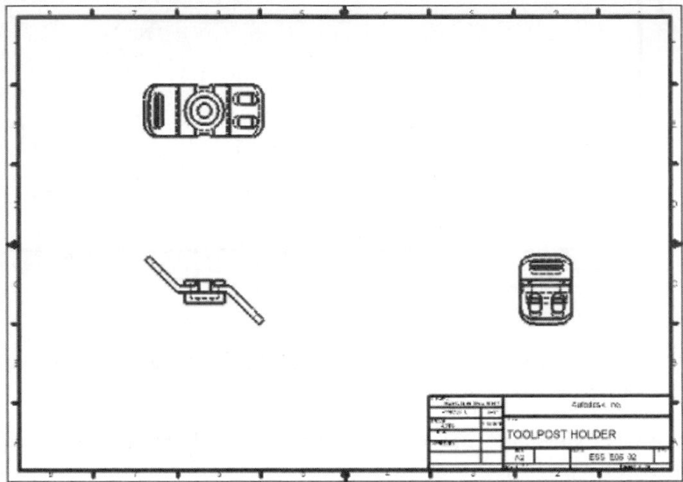

Figure 5.26

16. Now change the drawing scale in the title block. Begin this process by expanding Sheet:1 in the browser.

17. While still in the browser, expand ANSI-Large under Sheet:1.

18. Right-click on the Field Text icon, and choose Edit Field Text from the menu.

19. The Edit Property Fields dialog box will display. In the SCALE cell, click on 1:1 and enter a new scale of **1:2**.

20. Click OK. The title block updates to the new scale, as shown in the following image.

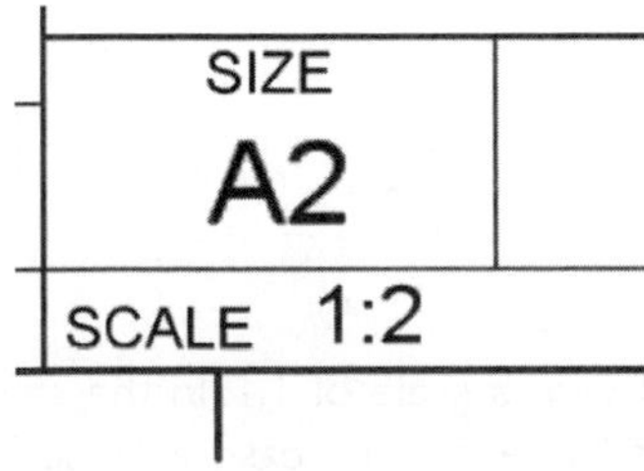

Figure 5.27

21. To complete the multiview layout, an isometric view will be created. Begin by clicking the Projected View tool in the panel bar.

EXERCISES

22. Select the base view or front view, and move your cursor to a point above and to the right of the base view.

23. Click in Zone E3 to place the isometric view. Right-click the sheet and choose Create. Your drawing should appear similar to the following image.

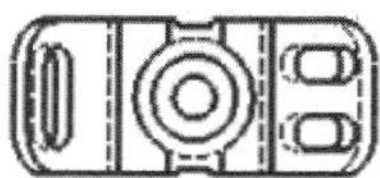

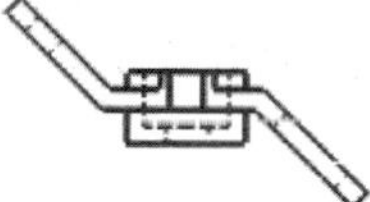

Figure 5.28

24. Double-click the isometric view and click the Display Options tab.

25. In the Display area, clear the checkbox for Tangent Edges, and then click OK.

26. Complete this exercise by moving drawing views to better locations. Select the right-side view, and drag it to Zone C4. Select the isometric view, and drag it to Zone E4. Your drawing should appear similar to the following image.

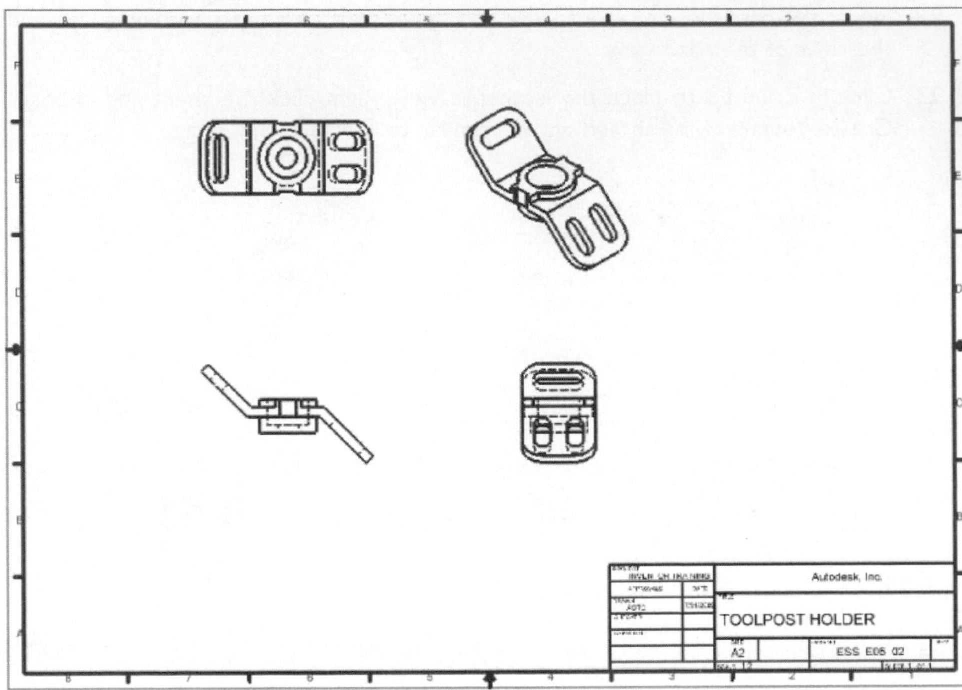

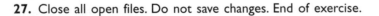

Figure 5.29

27. Close all open files. Do not save changes. End of exercise.

AUXILIARY VIEWS

An auxiliary view is a view that is projected perpendicular to a selected edge or line in a base view. It is designed primarily to view the true size and shape of a surface that appears foreshortened in other views.

To create an auxiliary drawing view, follow these steps:

1. Click the Auxiliary View tool on the Drawing Views panel bar, as shown in the following image. You can also right-click inside the bounding area of an existing view box, shown as a dotted box when the cursor moves into the view, and then select Create View > Auxiliary from the menu.

Figure 5.30

2. If you selected the Auxiliary View tool, click inside the view from which the auxiliary view will be projected. The Auxiliary View dialog box will appear, as shown in the following image on the left. Type in a name for Label and a value for Scale, and then select one of the Style options: Hidden, Hidden Line Removed, or Shaded.

3. In the selected drawing view, select a linear edge or line from which the auxiliary view will be perpendicularly projected, as shown in the following image on the right.

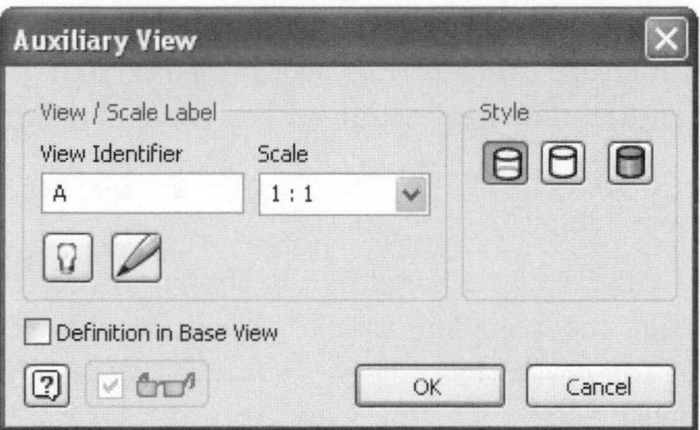

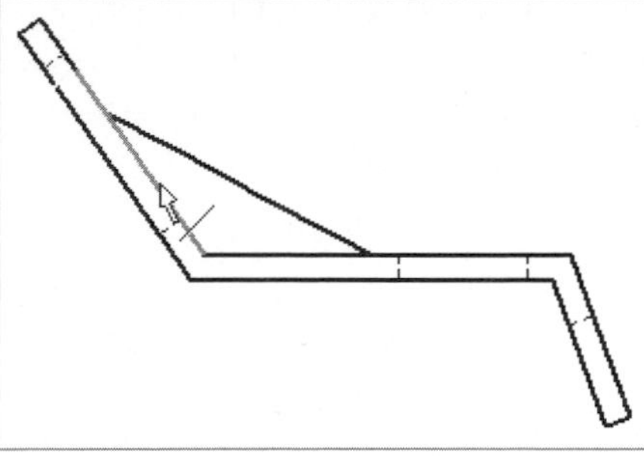

Figure 5.31

4. Move the cursor to position the auxiliary view, as shown in the following image on the left. Notice that the view takes on a shaded appearance as you position it.
5. Click a point on the drawing sheet to create the auxiliary view. The completed auxiliary view layout is shown in the following image on the right.

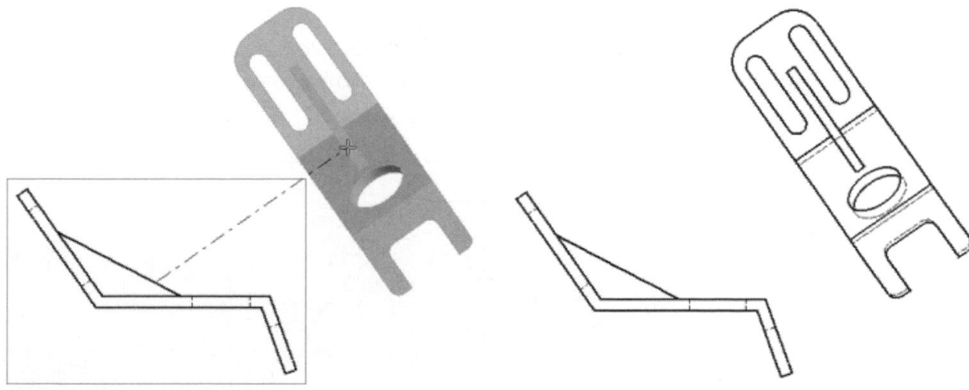

Figure 5.32

SECTION VIEWS

A section view is a view you create by sketching a line or multiple lines that will define the plane(s) that will be cut through a part or assembly. The view will represent the surface of the cut area and any geometry shown behind the cut face from the direction being viewed. When defining a section, you sketch line segments that are horizontal, vertical, or at an angle. You cannot use arcs, splines, or circles to define section lines. When you sketch the section line(s), geometric constraints will automatically be applied between the line being sketched and the geometry in the drawing view. You can also infer points by moving the cursor over (or scrubbing) certain geometry locations, such as centers of arcs, endpoints of lines, and so on, and then moving the cursor away to display a dotted line showing that you are inferring or tracking that point. To place a geometric constraint between the drawing view geometry and the section line, click in the drawing when a green circle appears; the glyph for the constraint will appear. If you do not want the section lines to have constraint(s) applied to them automatically when they are created, hold down the CTRL key when sketching the line(s). Because the area in the section view that is solid material appears with a hatch pattern by default, you may want to set the hatching style before creating a section view.

To create a section drawing view, follow these steps:

1. Click the Section View tool on the Drawing Views panel bar, as shown in the following image. You can also right-click inside the bounding area of an existing view box, shown as a dotted box when the cursor moves into the view. You can then select Create View > Section from the menu or select Section View from the Insert > Drawing Views menu.

Figure 5.33

2. If you selected the Section View tool, click inside the view from which to create the section view.

3. Sketch the line or lines that define where and how you want the view to be cut. In the following image, a vertical line is sketched through the center of the object.

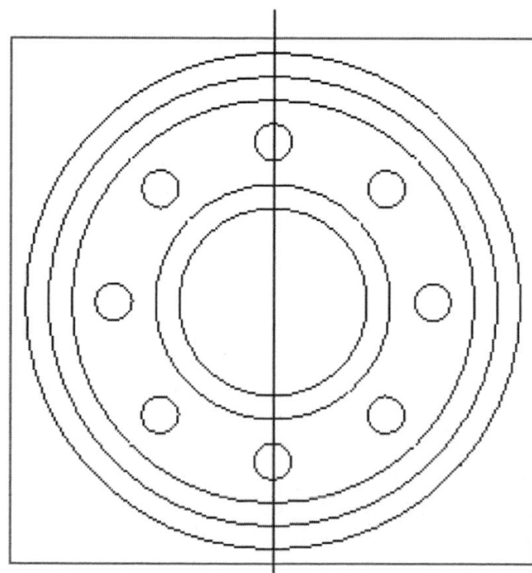

Figure 5.34

4. When you finish sketching the section line(s), right-click and select Continue from the menu.
5. The Section View dialog box will appear, as shown in the following image. Fill in the information for how you want the label, scale, and style to appear in the drawing view. When the Include Slice option is checked, a section view will be created with some components sliced and some components sectioned, depending on their browser attribute settings. Placing a check in the box next to Slice All parts will override any browser component settings and will slice all parts in the view according to the Section line geometry. Components that are not crossed by the Section Line will not be included in the slicing operation.

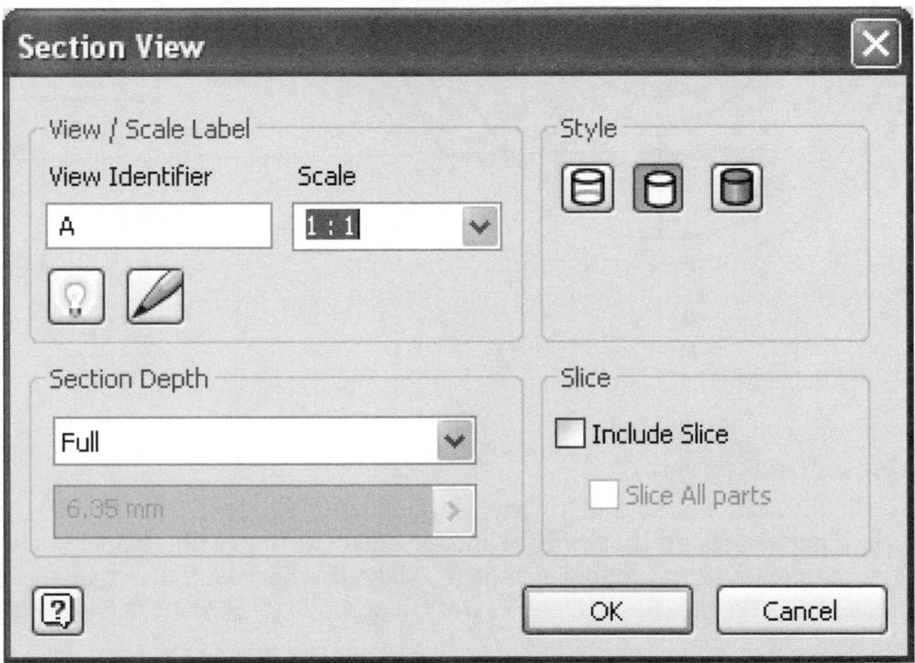

Figure 5.35

6. Move the cursor to position the section view, as shown in the following image, and select a point to place the view.

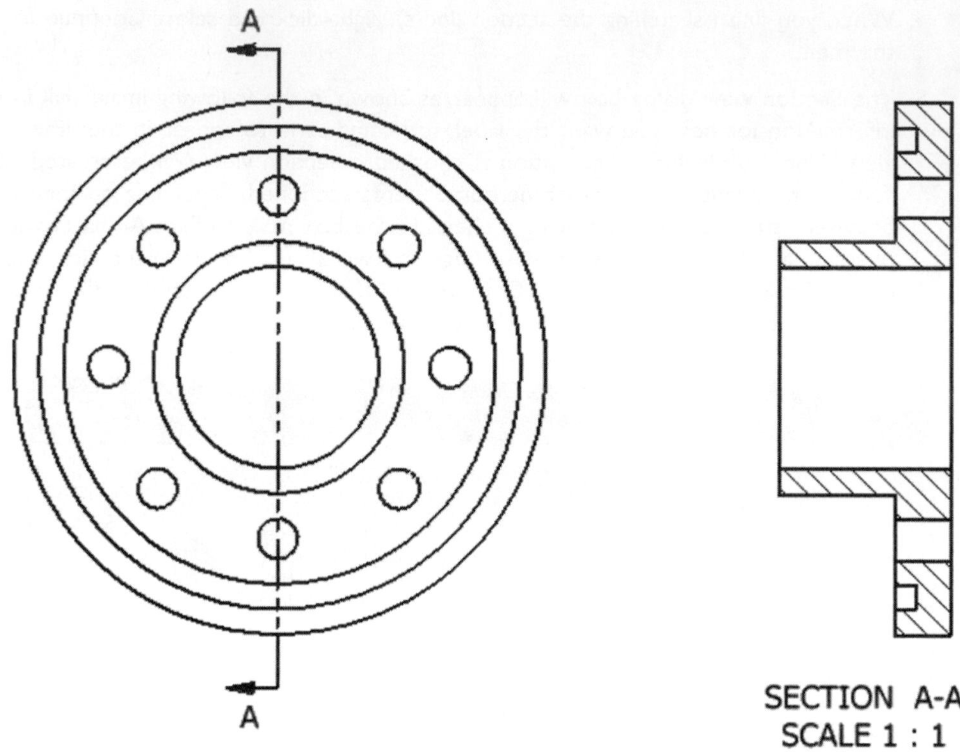

Figure 5.36

Modifying a Hatch Pattern

You can edit the section view hatch pattern by right-clicking on the hatch pattern in the section view and selecting Edit from the menu. This will launch the Edit Hatch Pattern dialog box, as shown in the following image. Make the desired changes in the Pattern, Angle, Scale, Line Weight, Shift, Double, and Color areas. Click the OK button to reflect the changes in the drawing.

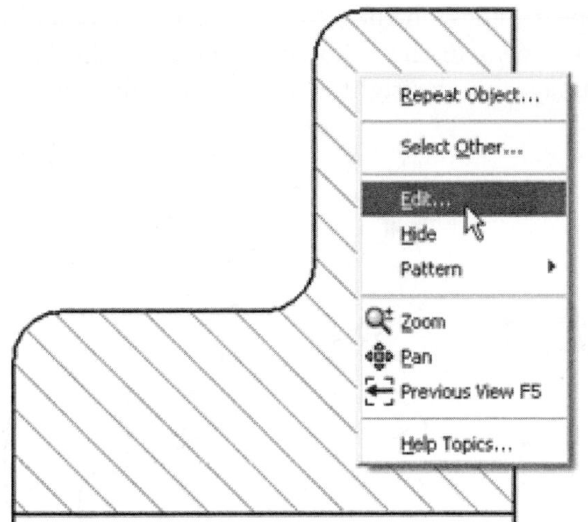

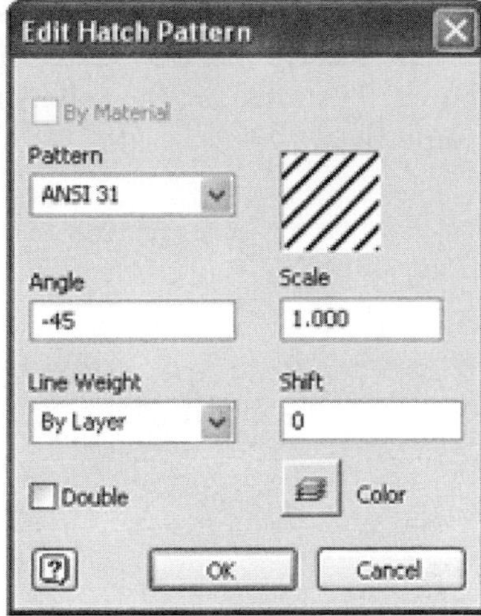

Figure 5.37

Note: The Shift mode of the Edit Hatch Pattern dialog box shifts the hatch pattern to offset it slightly from the hatch pattern on a different part. This option would be ideal for creating sections that involve assembly models. Enter the distance for the shift.

HATCHING ISOMETRIC VIEWS

In addition to applying crosshatching to orthographic view, as shown in the following image on the left, you can also place crosshatching into an isometric view. To perform this operation, right-click inside of the bounding box that holds the isometric view, and pick Edit View from the menu, as shown in the following image on the right.

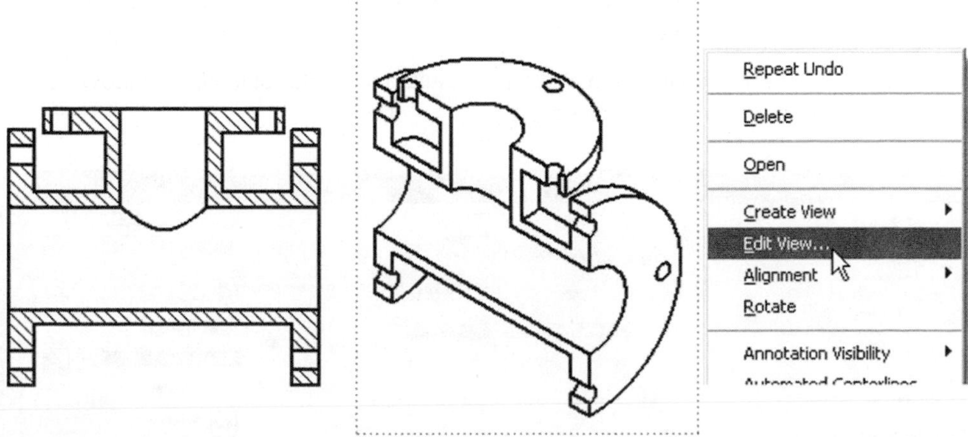

Figure 5.38

When the Drawing View dialog box appears, click on the Display Options tab, and place a check in the box next to Hatching, as shown in the following image on the left. The results are illustrated in the following image on the right with the hatching applied to the isometric view.

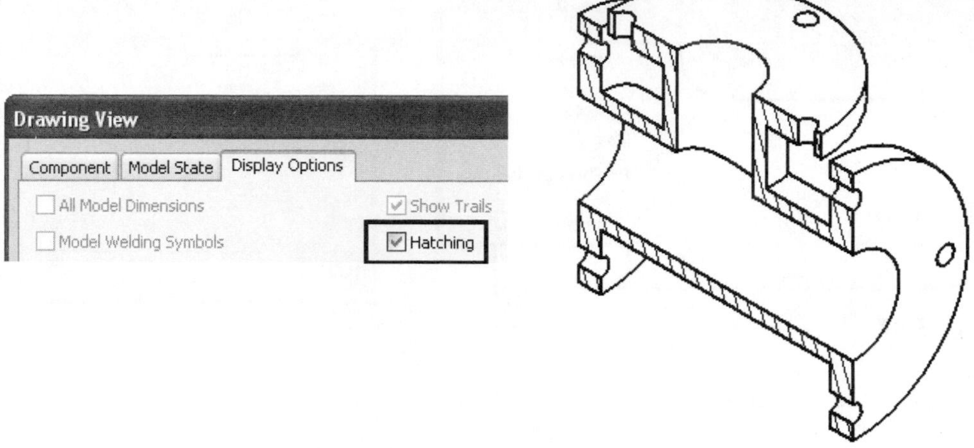

Figure 5.39

CROSSHATCHING BY MATERIAL TYPE

Hatch patterns can be associated with component materials. In this way, the hatching generated by a section, breakout, or slice view will correspond with a hatch pattern that is defined for the sectioned material. Follow the next series of steps for performing this operation.

1. Select Format from the drop-down menu, followed by Style and Standard Editor to open the Styles and Standard Editor dialog box, as shown in the following image. In the styles list, expand Standard, and then click one of the listed standards as shown in the following image on the left.
2. Click on the Material Hatch Pattern Defaults tab in the Standard window, as shown in the following image on the right.

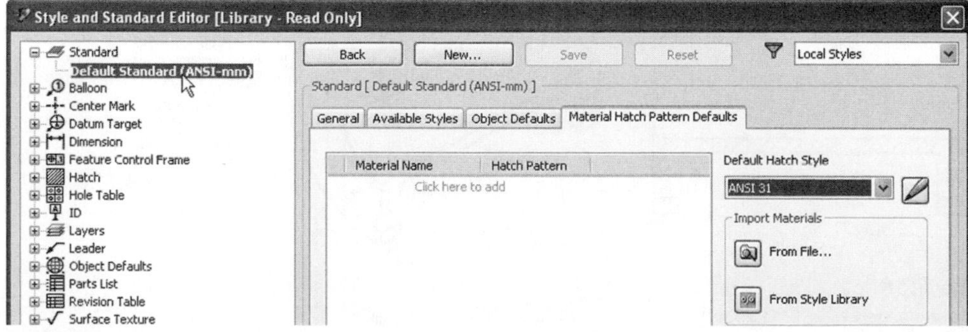

Figure 5.40

3. You will now create a list of materials in the mapping table. The material Titanium, as shown in the following image on the left, was created by first clicking on the From File button found under the Imported Materials area, as shown in the following image on the right, and then selecting the part file that contains this material.
4. With the material Titanium listed, map the desired material, as shown in the following image.
5. Click on the Save button to save these changes, and then click Done to close the Style and the Standard Editor dialog box.

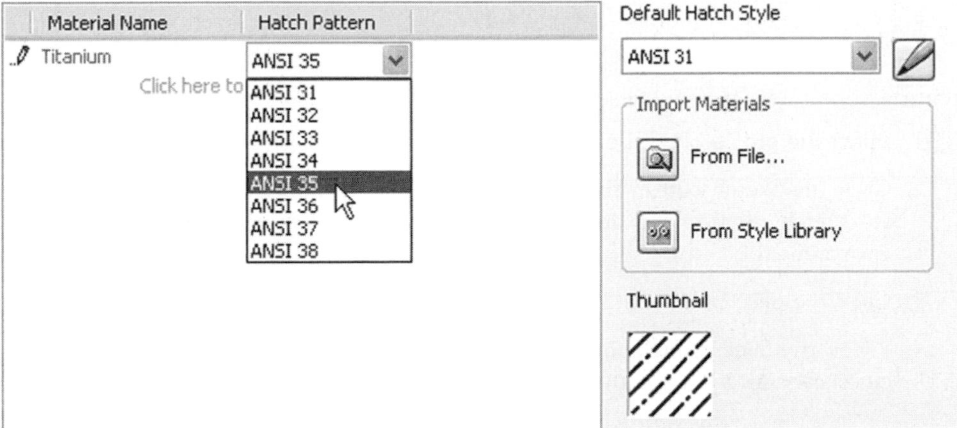

Figure 5.41

6. Once back in the drawing editor, right-click on the hatch pattern, and choose Edit from the menu, as shown in the following image on the left. When the Edit Hatch Pattern dialog box displays, place a check next to By Material to map the new material to the object, as shown in the following image on the right.

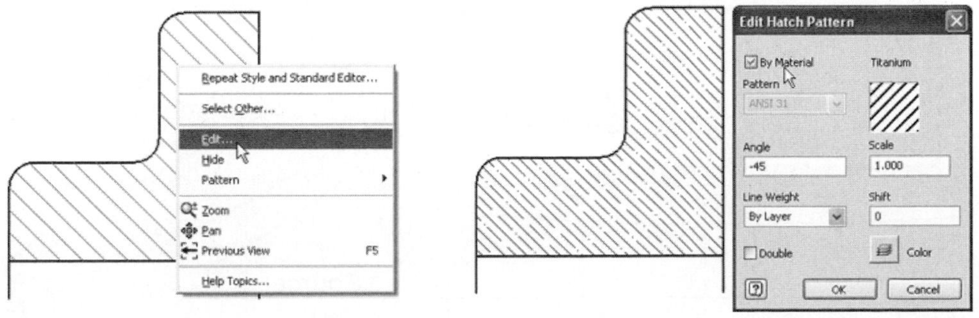

Figure 5.42

SLICE VIEWS

A sliced view is a different type of section cut that allows you to create slices at key locations based on a sketch located in the source view. The target view is the view displaying the various slices. The Slice tool can be activated from the Drawing Views panel bar, as shown in the following image. You can also right-click inside the bounding area of an existing view box, shown as dashed lines when the cursor moves into the view, and select Create View > Slice from the menu.

Figure 5.43

To create a slice view, follow these steps:

1. Select the profile of the boat hull as the Source view, as shown in the following image.
2. Click the Sketch button to open a drawing sketch associated with the view, and create sketch geometry to define an open slice sketch. When finished, exit the sketch environment.
3. Click the Slice button, and select the isometric view as the Target view.
4. When the Slice dialog appears, select the previously defined sketch geometry as the cut sketch, as shown in the following image.

Note: When dealing with multiple parts, the Slice All Parts checkbox will activate. Use this feature to override any component settings made in the browser and slice all parts in the view, depending on the sketch geometry. Components that are not crossed by this geometry will not be sliced.

5. Click OK to perform the Slice operation.

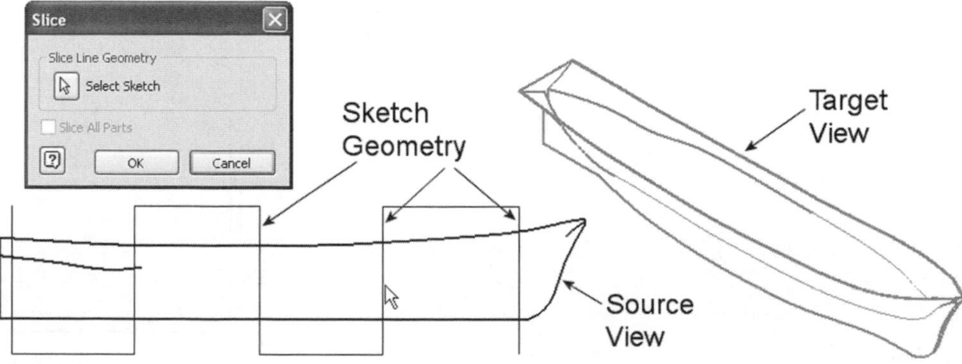

Figure 5.44

The results of the slice operation are illustrated in the following image on the right. A cross-section was created from every location of the boat hull where vertical sketch geometry was present.

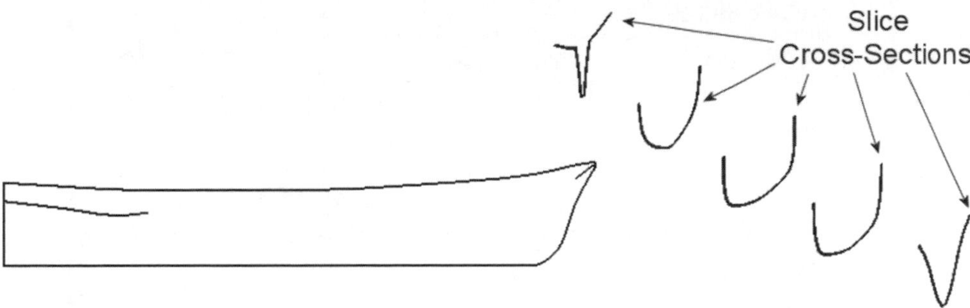

Figure 5.45

DETAIL VIEWS

A detail view is a drawing view that isolates an area of an existing drawing view and can reflect a specified scale. You define a detailed area by a circle or rectangle and can place it anywhere on the sheet. To create a detail drawing view, follow these steps:

1. Click the Detail View tool on the Drawing Views panel bar, as shown in the following image. You can also right-click inside the bounding area of an existing view box, shown as dashed lines when the cursor moves into the view, and select Create View > Detail from the menu.

Figure 5.46

2. If you selected the Detail View tool, click inside the view from which you will create the detail view.
3. The Detail View dialog box will appear, as shown in the following image. Fill in information according to how you want the label, scale, and style to appear in the drawing view, and pick the desired view fence shape to define the view boundary. Do not click the OK button at this time, as doing so will generate an error message.

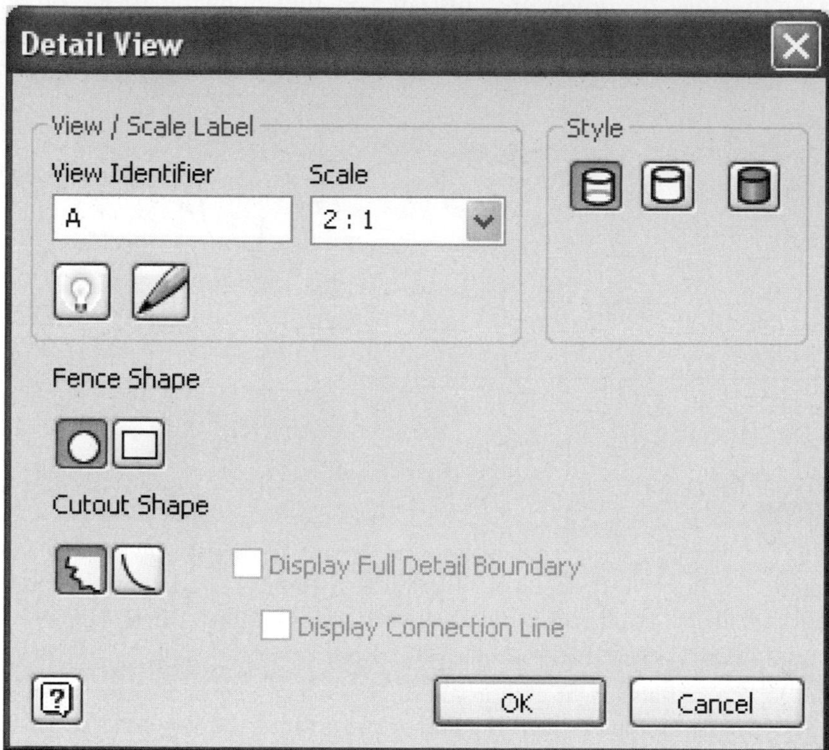

Figure 5.47

4. In the selected view, select a point to use as the center of the fence shape that will describe the detail area. The following image on the left shows the icon that appears when creating a circular fence. The image on the right shows the icon used for creating a rectangular fence.

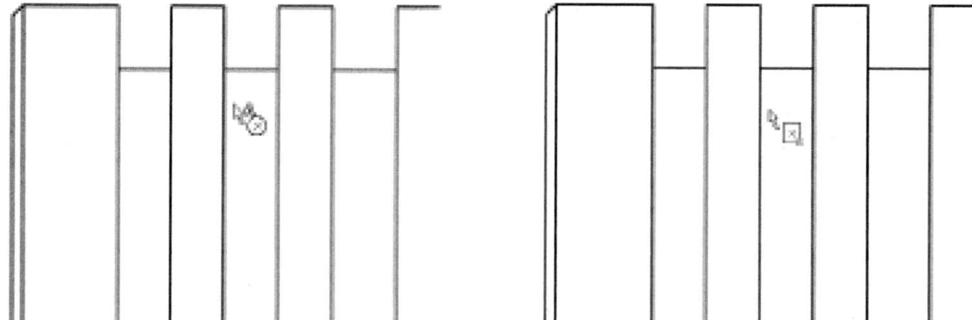

Figure 5.48

Note: You can right-click at this point and select Rectangular Fence to change the bounding area of the detail area definition.

5. Select another point that will define the radius of the detail circle, as shown in the following image on the left, or corner of the rectangle, as shown in the following image on the right. As you move the cursor, a preview of the boundary will appear.

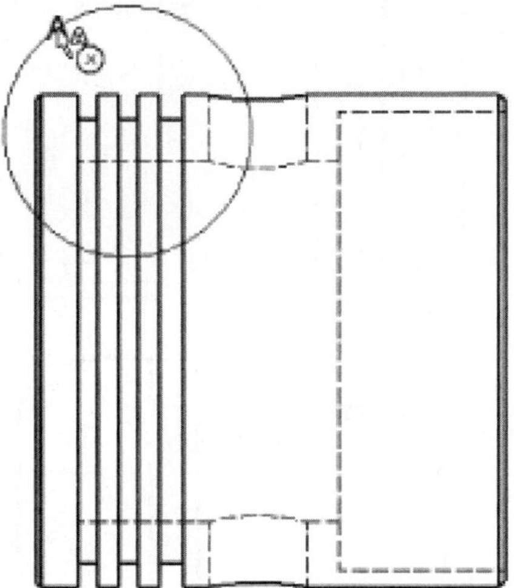

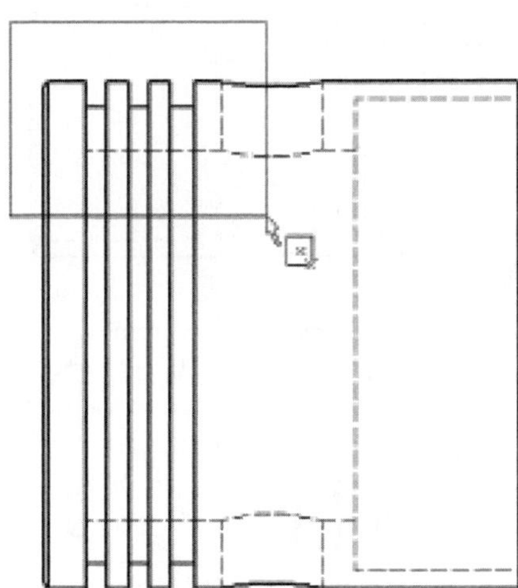

Figure 5.49

6. Select a point on the sheet where you want to place the view. There are no restrictions on where you can place the view. The following image on the left shows the completed detail view based on a circular fence; a similar detail based on a rectangular fence is shown on the right.

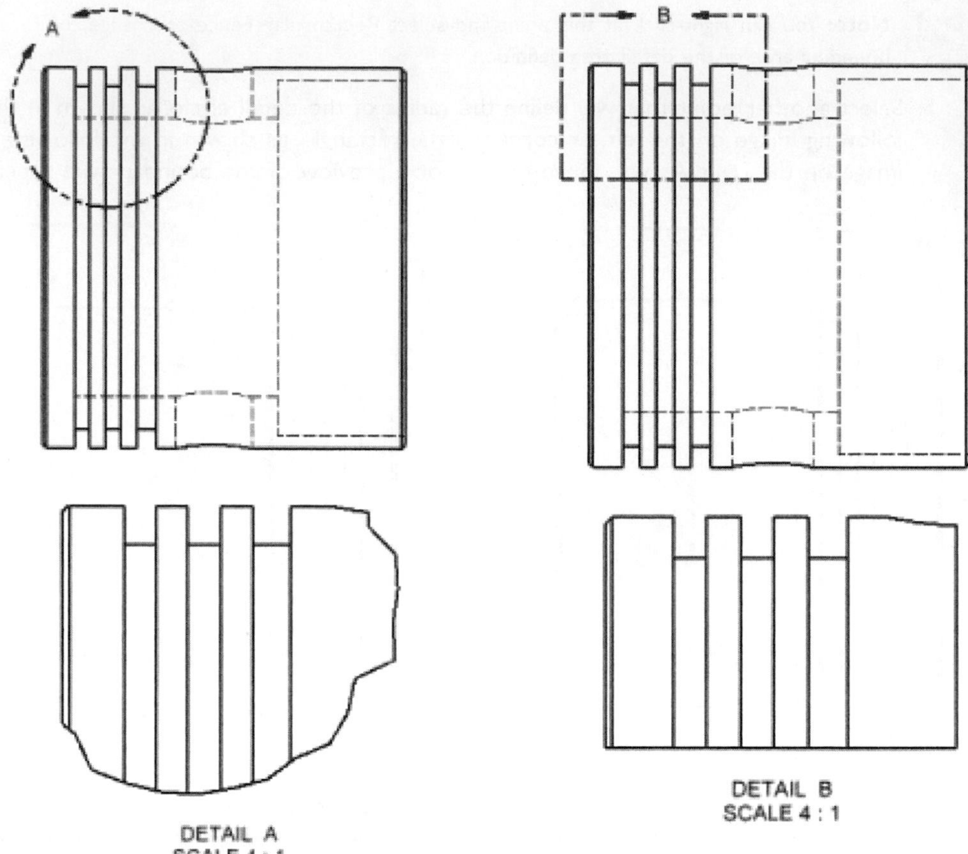

Figure 5.50

Detail View Options

After a detail view is created, numerous options are available to fine-tune the detail. To access these controls, right-click on the edge of the detail circle and choose options, as shown in the following image on the right. The three detail view options are explained below.

Smooth Cutout Shape: Placing a check in this box will affect actual detail view. In the following image, instead of the cutout being ragged, a smooth cutout shape is present.

Full Detail Boundary: This option displays a full detail boundary in the detail view when checked, as shown in the following image.

Connection Line: This option will create a connection line between the main drawing and the detail view, as shown in the following image.

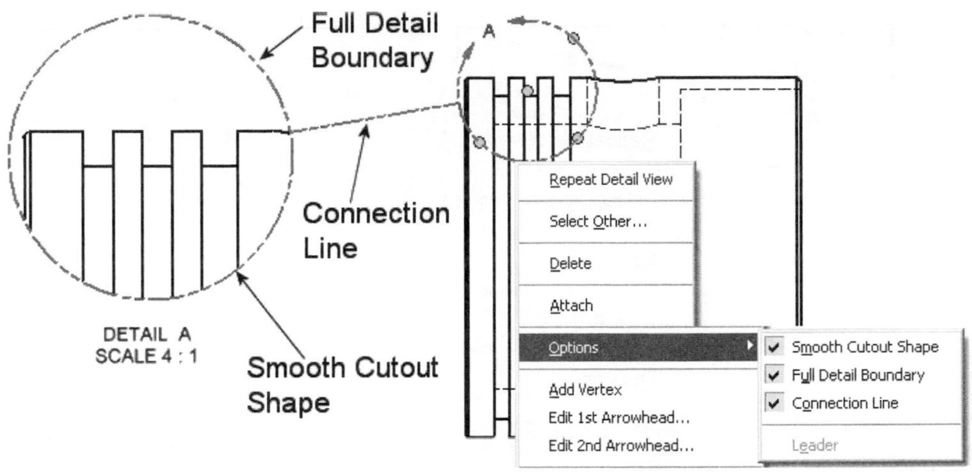

Figure 5.51

Note: Once a connection line has been created in a detail view, you can add a new vertex by right-clicking the detail boundary or connection line and clicking Add Vertex from the menu. Picking a point on the connection line will allow you to add the vertex and then move the new vertex to the new position. You can even remove a vertex by right-clicking a vertex and choosing Delete Vertex from the menu.

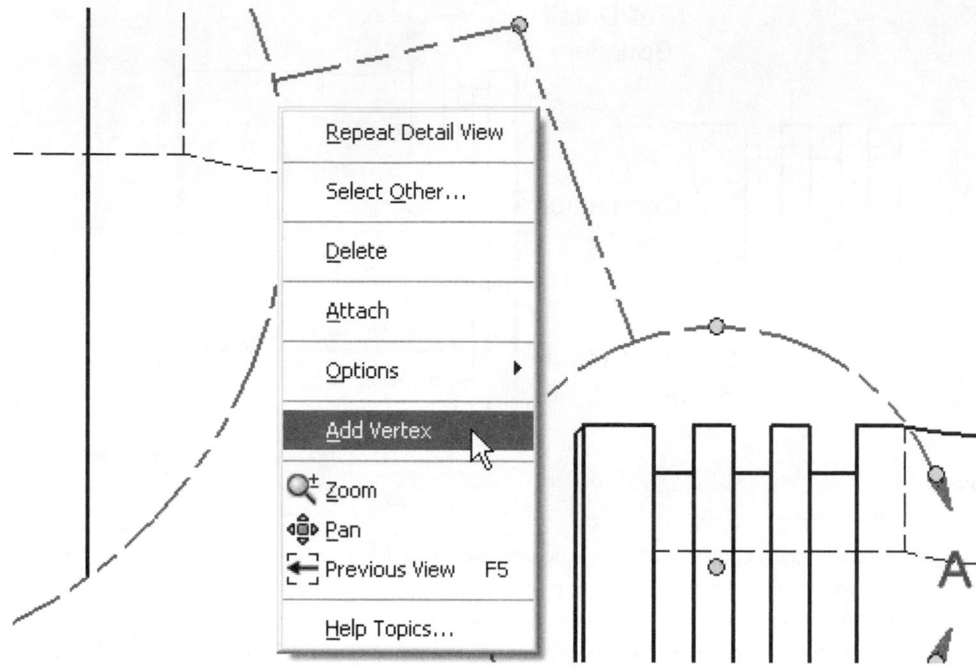

Figure 5.52

BREAK VIEWS

When creating drawing views of long parts, you may want to remove a section or multiple sections from the middle of the part and show just the ends. This type of view is referred to as a "break view." You may, for example, want to create a drawing view of a 50×50×6 angle, as shown in the following image, which is 1500 mm long and has only the ends chamfered. When you create a drawing view, the detail of the ends is small and difficult to see because the part is so long.

Figure 5.53

In this case, you can create a break view that removes a 1000 mm section from the middle of the angle and leaves 250 mm on each end. When you place an overall length dimension that spans the break, it appears as 1500 mm, and the dimension line shows a break symbol to note that it was derived from a break view, as shown in the following image.

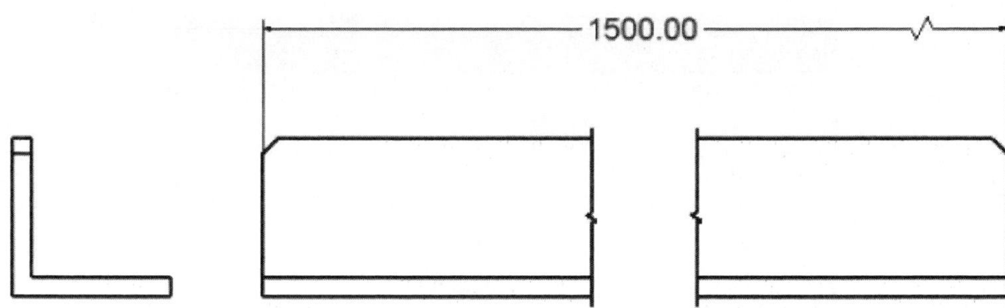

Figure 5.54

You create a break view by adding breaks to an existing drawing view. The view types that can be changed into break views are as follows: part views, assembly views, projected views, isometric views, auxiliary views, section views, breakout views, and detail views. After creating a break view, you can move the breaks dynamically to change what you see in the broken view.

To create a break view, follow these steps:

1. Create a base or projected view or one that will eventually be shown as broken.
2. Click the Break tool on the Drawing Views panel bar, as shown in the following image. You can also right-click inside the bounding area of an existing view box, shown as a dotted rectangle when the cursor is moved into the view. Select Create View > Break from the menu.

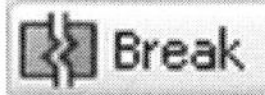

Figure 5.55

3. If you selected the Break tool from the Drawing Views panel bar, click inside the view from which to create the break view.
4. The Break dialog box will appear, as shown in the following image. Do not click the OK button at this time, as this will end the operation.

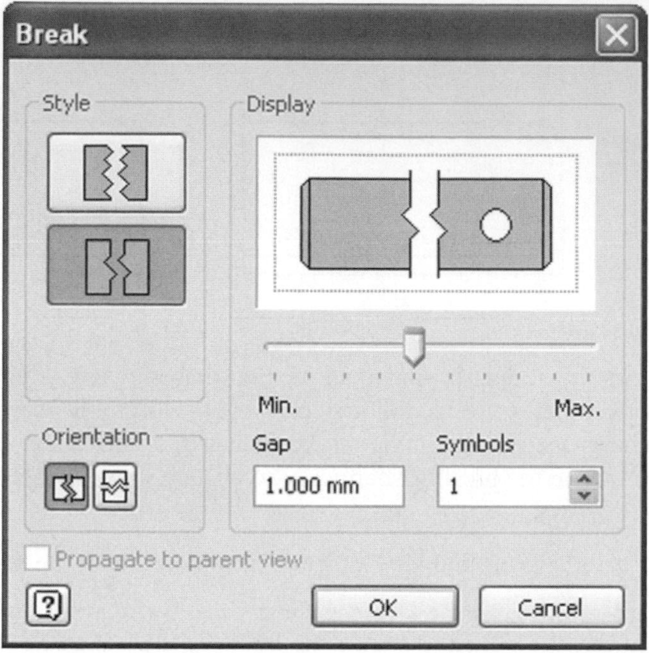

Figure 5.56

5. In the drawing view that will be broken, select a point where the break will begin, as shown in the following image.

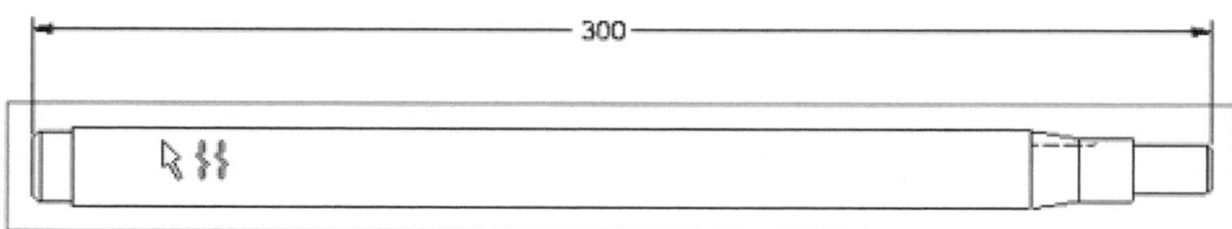

Figure 5.57

6. Select a second point to locate the second break, as shown in the following image. As you move the cursor, a preview image will appear to show the placement of the second break line.
7. The following image illustrates the results of creating a broken view. Notice that the dimension line appears with a break symbol to signify that the view is broken.

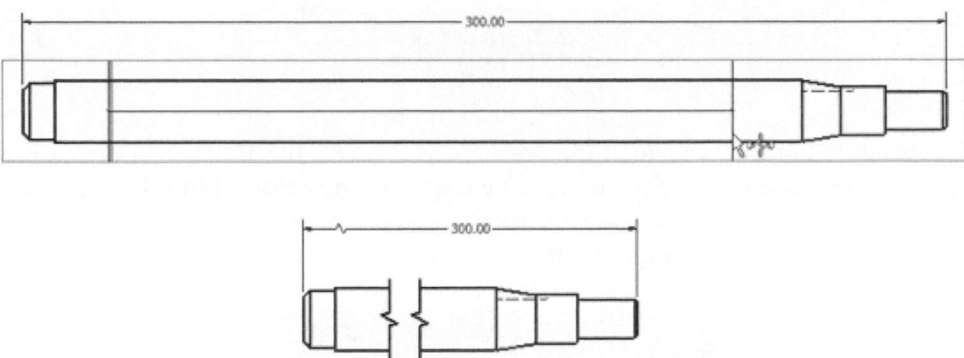

Figure 5.58

8. To edit the properties of the break view, move the cursor over the break lines, and a green circle will appear in the middle of the break. Right-click and select Edit Break from the menu. The same Break dialog box will appear. Edit the data as needed, and click the OK button to complete the edit.
9. To move the break lines, click on one of the break lines and, with the left mouse button pressed down, drag the break line to a new location, as shown in the following image. The other break line will follow to maintain the gap size.

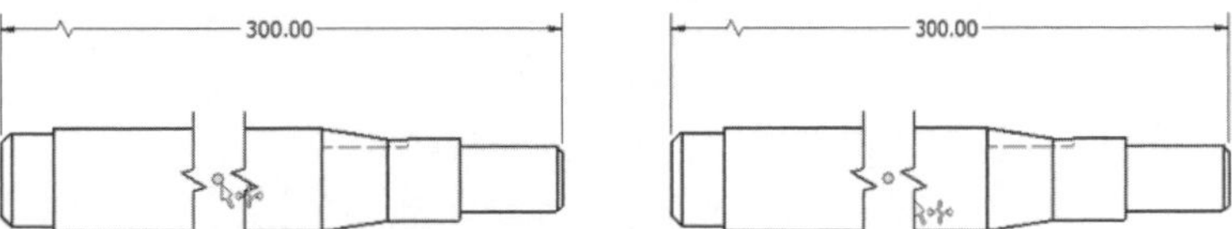

Figure 5.59

BREAKOUT VIEWS

When you want to expose internal components or features, Autodesk Inventor allows you to cut or peel away a body and expose those internal parts; this is called a "breakout view." It is not unusual for assemblies to have housings or covers that hide internal components. Breakout views make it possible to expose these components. You can create breakout views on assemblies as well as on part files.

To use breakout views, you will need to be able to define two items: a closed profile boundary over the area to break and the depth of material to remove or the portion of a component to remove in order to see other components. As with broken views, the breakout view is defined and displayed on the same view.

To create a breakout view, click the Break Out tool on the Drawing Views panel bar, as shown in the following image.

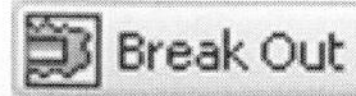

Figure 5.60

After you select the view in which the breakout will occur, the Break Out dialog box will appear, as shown in the following image. The termination options, From Point, To Sketch, To Hole, and Through Part, are available to give you control over how the breakout section is created. The following sections explain these options.

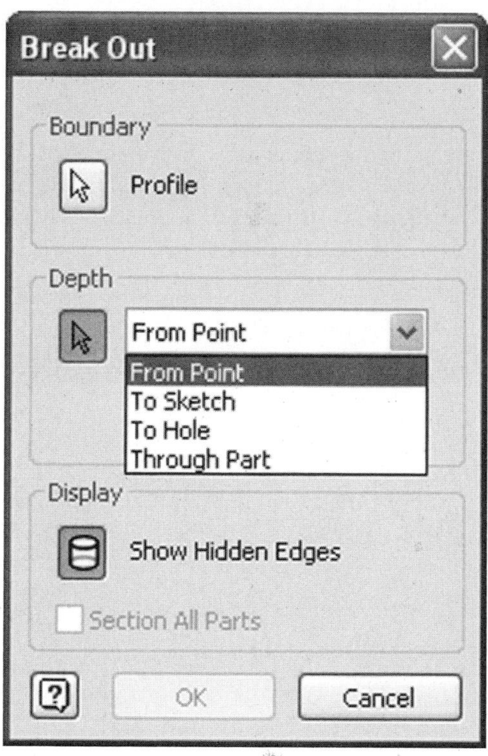

Figure 5.61

From Point Select this to make the breakout occur at a specified distance from a point located in a view. The point could be located in the base view or in a projected view.

To Sketch Select this to use sketched geometry associated with another view to define the depth of the breakout.

To Hole Select this to base the breakout depth on a hole feature whose axis determines the termination of the break.

Through Part Select this to remove drawing content from inside a closed profile through selected components located in the browser. This termination option is especially useful when you want to reveal internal components or features.

CREATING A BREAKOUT VIEW USING THE FROM POINT OPTION

To create a breakout view using the From Point option, follow these steps:

1. Click on the view in which to create the breakout section.
2. Click on the Sketch button on the Standard toolbar, and sketch a closed profile over the area you want broken out, as shown in the following image. When done, right-click, and select Finish Sketch from the menu.

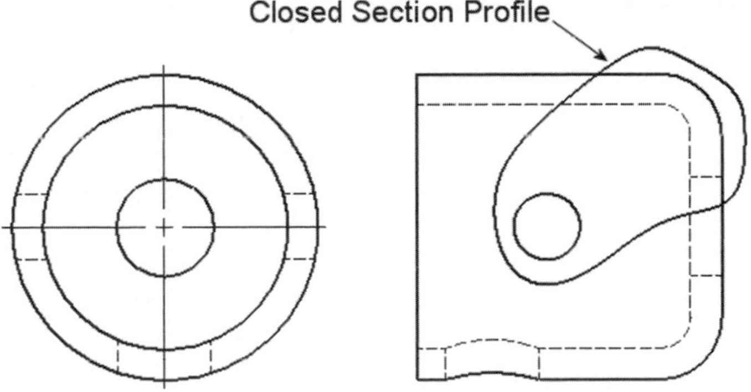

Figure 5.62

3. Click the Break Out button; this will activate the Break Out dialog box. Select the view in which the breakout will occur, and select the defined boundary. Select the sketch you just created as the boundary in the view that will contain the breakout.
4. Select the From Point option in the dialog box, click on the Depth arrow, and select a point in the adjacent projected view. This point will be used to calculate the depth of cut based on the distance, as shown in the following image.

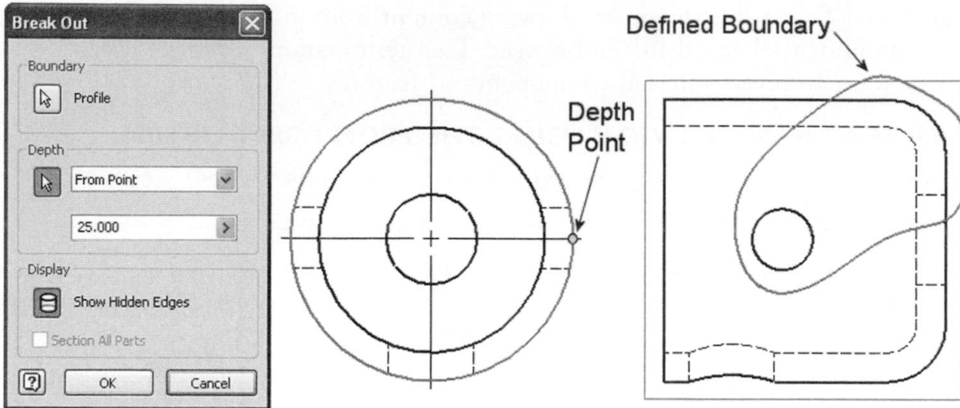

Figure 5.63

5. The following image shows the results of using the From Point option for creating a breakout. The depth of 25 units from the selected point cuts the view at the middle of the circular features, thus displaying the wall thickness of the part.

Note: You can also select a point at the center of the circle for the depth of the cut. In this case, the distance value would change to 0, because the depth of the cut was specified by a point.

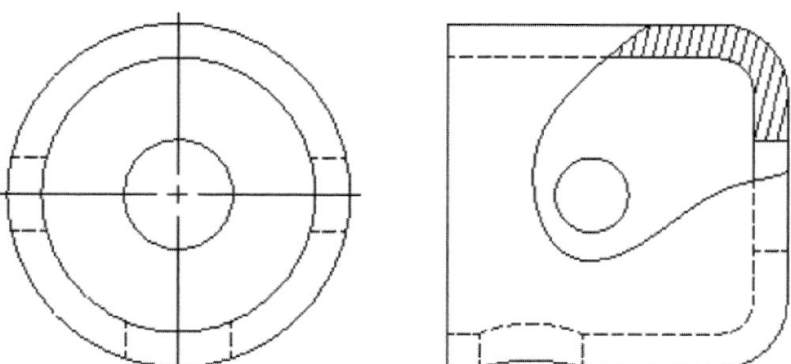

Figure 5.64

Creating a Breakout View Using the To Sketch Option

To create a breakout view using the To Sketch option, follow these steps:

1. Activate the view in which the breakout will occur, and sketch a closed profile over the area you want broken out. In the following image, two closed profiles will be used to create the breakout view. When finished with this operation, right-click, and select Finish Sketch.
2. Activate the adjacent projected view and create another sketch. This sketch will be used to determine the depth of the cut for the breakout, as shown in the following image. When finished with this operation, right-click, and select Finish Sketch.

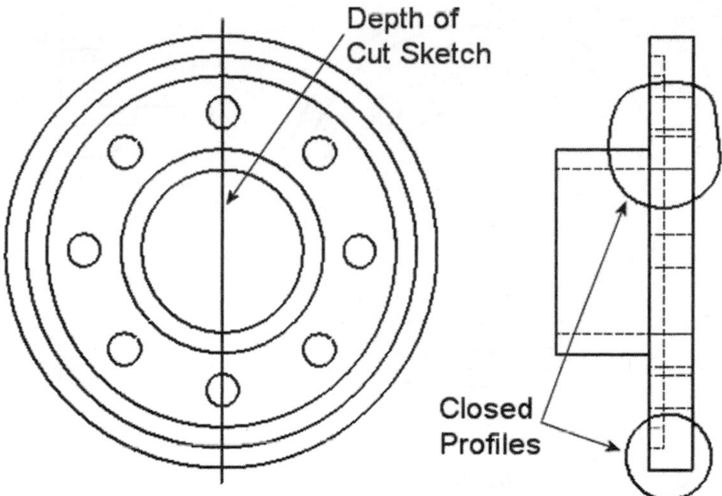

Figure 5.65

3. Click the Break Out tool, and select a view as the base for the breakout view.
4. When the Break Out dialog box appears, select the initial sketch or sketches to use as the boundary profile. In the Depth area of the dialog box, change the option to Sketch. In the adjacent projected view, select the sketch that will determine the depth of the cut for creating the breakout view, as shown in the following image.

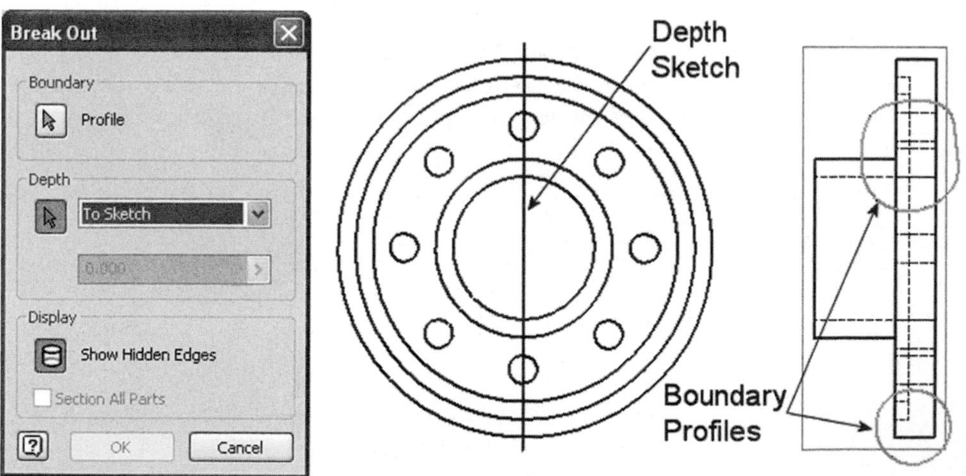

Figure 5.66

5. The following image shows the breakout view created based on the first group of sketched boundaries as well as the second sketch, which determined the depth of the cut.

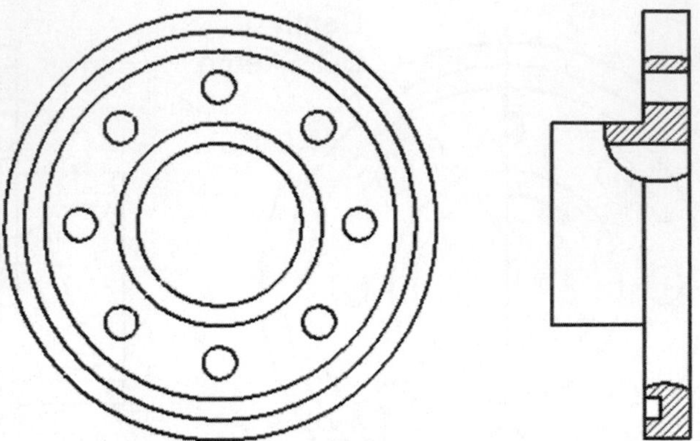

Figure 5.67

Creating a Breakout View Using the To Hole Option

The To Hole option for creating breakout views is similar to the To Sketch option. Instead of basing the breakout on a sketch, it will base it on a hole feature. The axis of the hole feature determines the termination depth.

To create a breakout view using the To Hole option, follow these steps:

1. Click the Break Out button on the Drawing Views panel bar.
2. Activate the view in which the breakout will occur, and sketch a closed profile around the hole you want broken out.
3. In the graphics area, click to select the view. When the Break Out dialog box appears, click to select the defined boundary. The boundary profile must be on a sketch associated to the selected view. In the Break Out dialog box, click the arrow next to the Depth type box, and select the To Hole option, as shown in the following image.

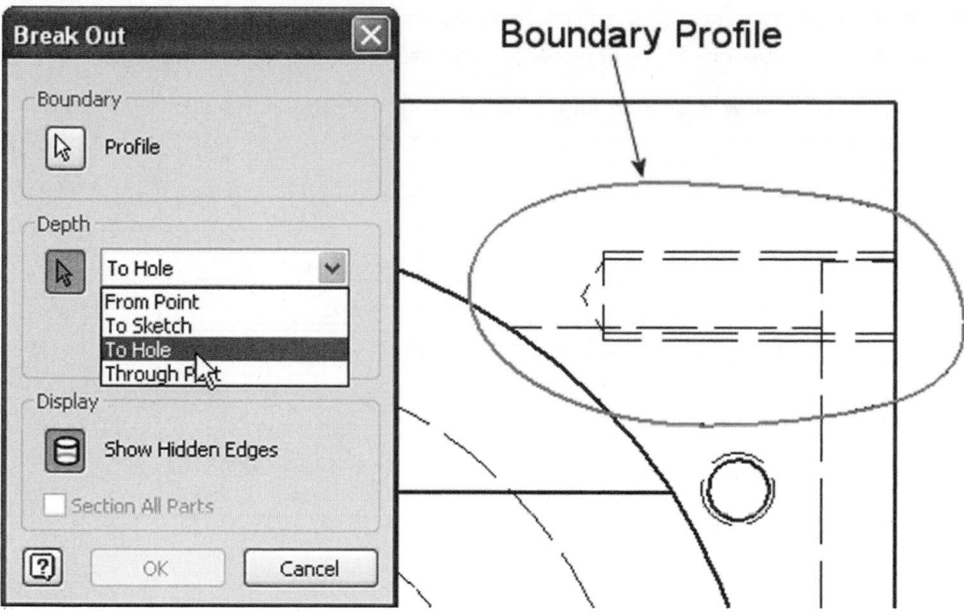

Figure 5.68

4. Click the select arrow, and then click to select the hole feature in the graphic window, as shown in the following image. The depth is defined by the axis of the hole.

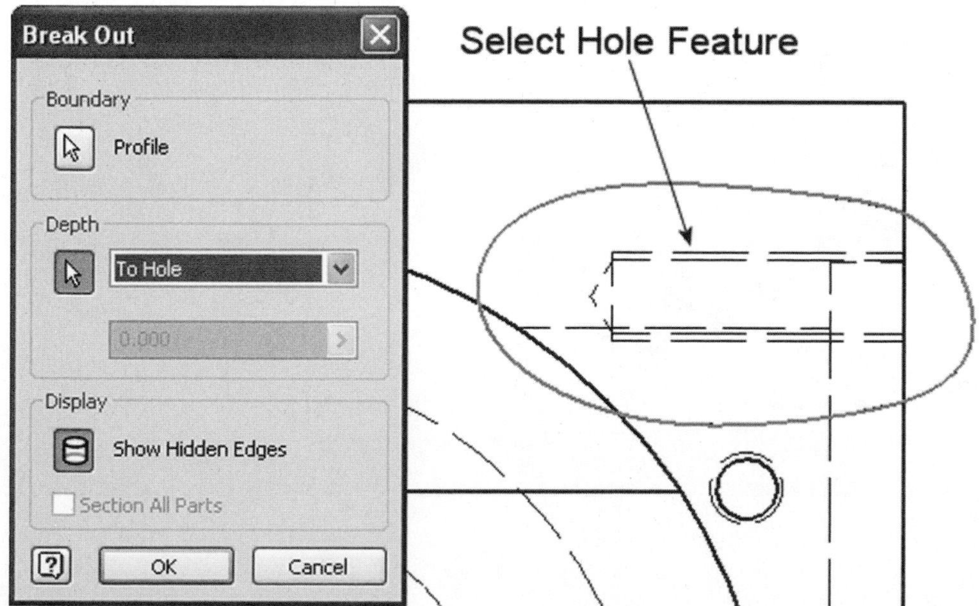

Figure 5.69

Note: If the hole feature is hidden, edit the view, and click the Show Hidden Edges button to temporarily display hidden lines. You can also select the hole in another view.

5. When the view is defined fully, click OK to create the view, as shown in the following image.

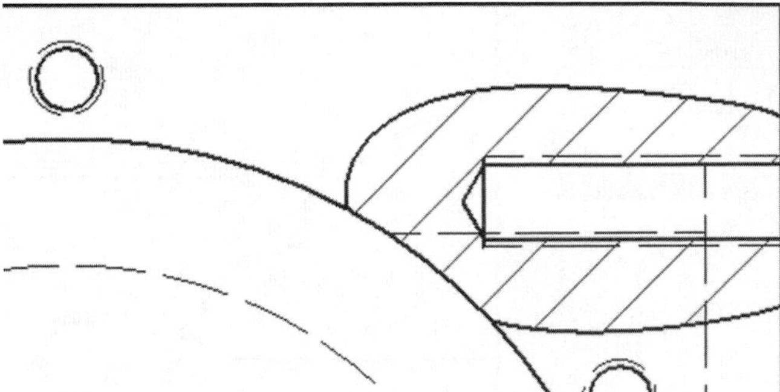

Figure 5.70

Creating a Breakout View Using the Through Part Option

In the following image, a housing hides internal details of an assembly. You can cut away a segment of the housing to expose obscured parts or features using the Through Part option of the Break Out tool.

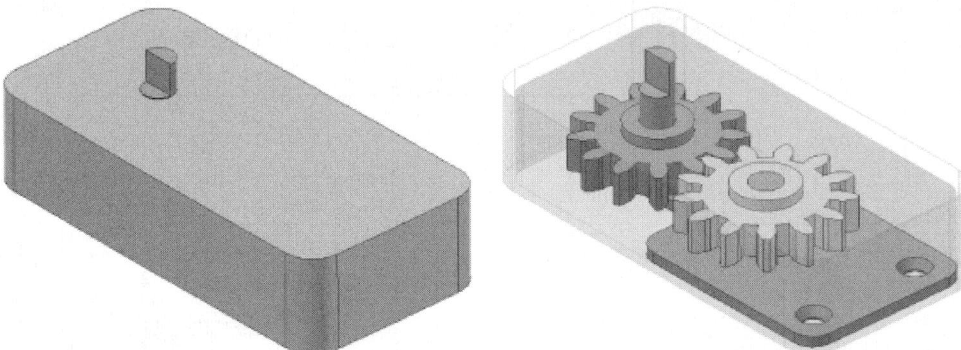

Figure 5.71

To create a breakout view using the Through Part option, follow these steps:

1. Activate the view through which you wish to cut, and click on the Sketch button in the Standard toolbar.
2. Next sketch a closed profile, as shown in the following image, as the area through which you wish to cut. When finished, right-click, and select Finish Sketch.

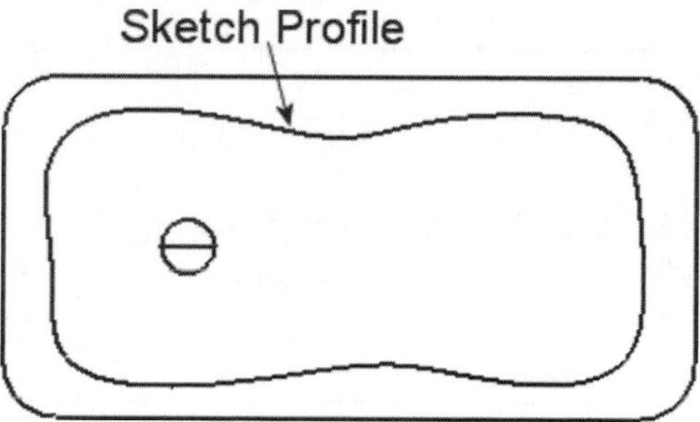

Figure 5.72

3. Click on the Break Out tool. Selecting the view in which to perform the operation will activate the Break Out dialog box. The boundary profile consisting of the previous sketch will be selected automatically. In the Depth area of the dialog box, click on the Through Part option, as shown in the following image.

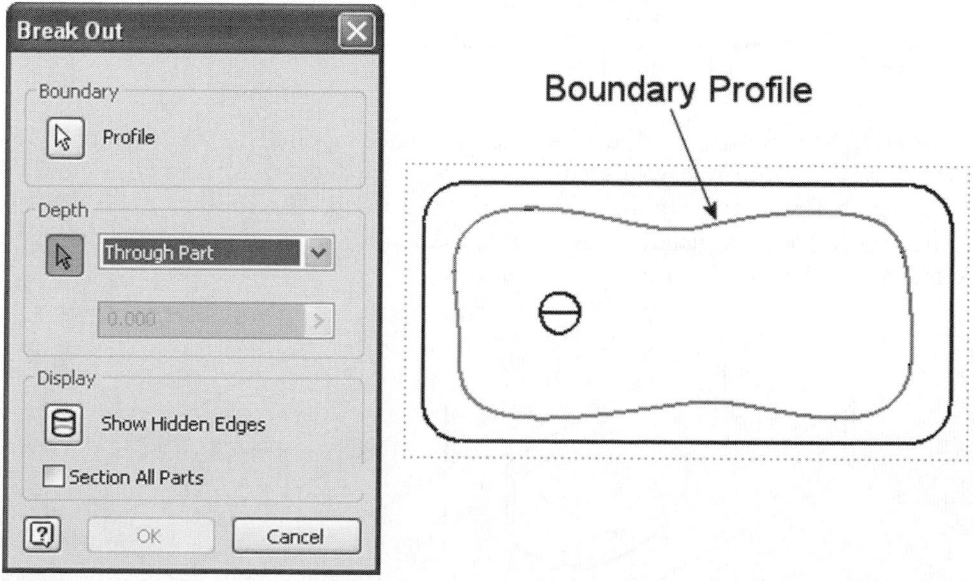

Figure 5.73

4. Move your cursor over the edge of the part, and select this edge, as shown in the following image.

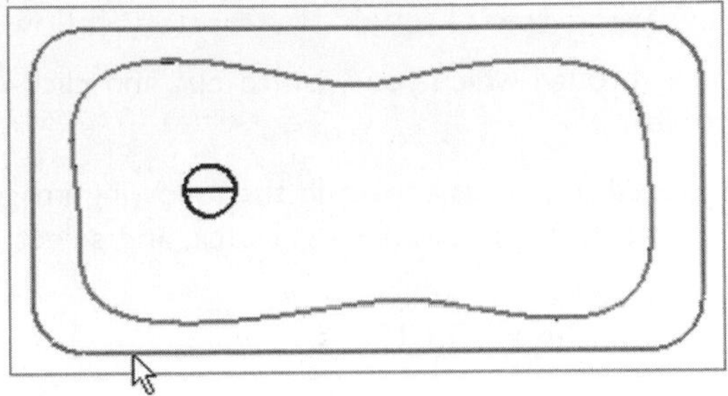

Figure 5.74

5. When finished, click the OK button. The following image illustrates the results of using the Through Part option. The gear cover has been sliced away, based on the sketched boundary, to expose the internal workings of the gear mechanism.

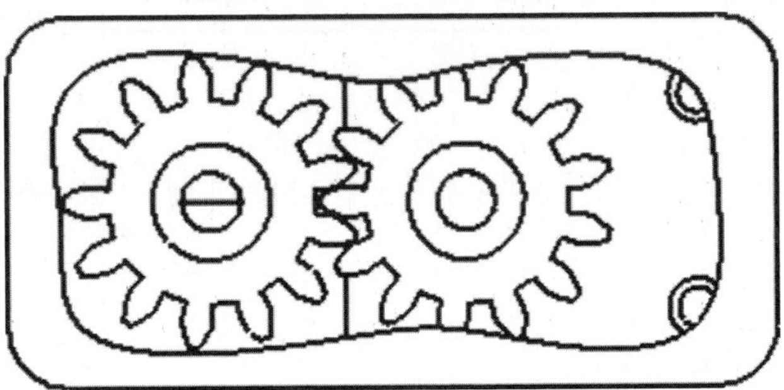

Figure 5.75

The Through Part option can be very effective when applied to an isometric view. As shown in the following image, a spline sketch was created on an isometric view. Notice how the sketch covers the top and sides of the isometric. The image on the right in the figure illustrates the completed breakout view. All sides surrounded by the spline sketch break out to display the internal components.

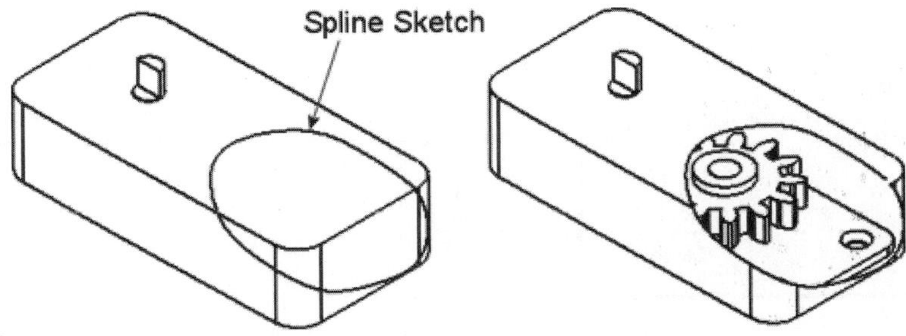

Figure 5.76

CREATING CROPPED VIEWS

Cropped views are used to show a portion of a part compared to a larger portion. The following image on the left illustrates a complete view showing object and hidden lines. After cropping this view, only a portion of the view is shown. This is different from a detail view, which was explained previously. A detail view is generated by a parent view and references this parent, complete with callout information. The cropped view will not create a callout and is meant for stand-alone use.

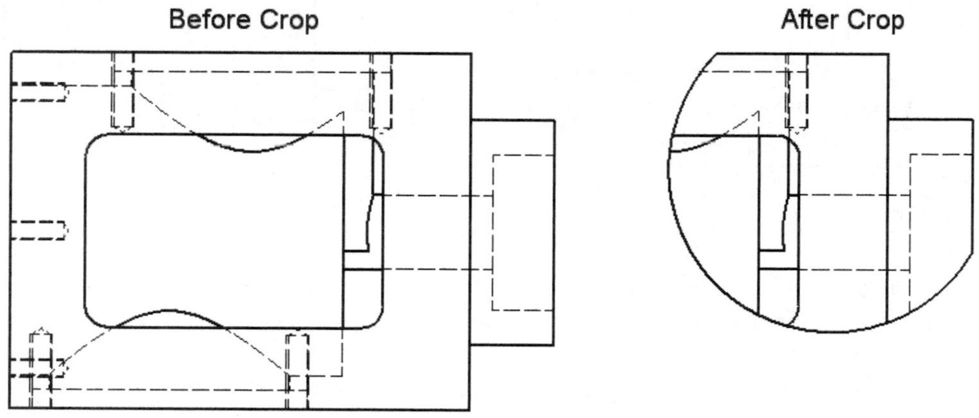

Figure 5.77

The steps used for creating a cropped view are as follows.

1. On the Drawing Views panel bar, click the Crop tool as shown in the following image.

Figure 5.78

2. Select the view to crop by clicking the edge of the view.
3. Right-click the menu to select the boundary type that can be either circular or rectangular. When specifying the circular boundary, a center point acts as the boundary center; for rectangular boundaries, two diagonal points are used to create the rectangle as shown in the following image.

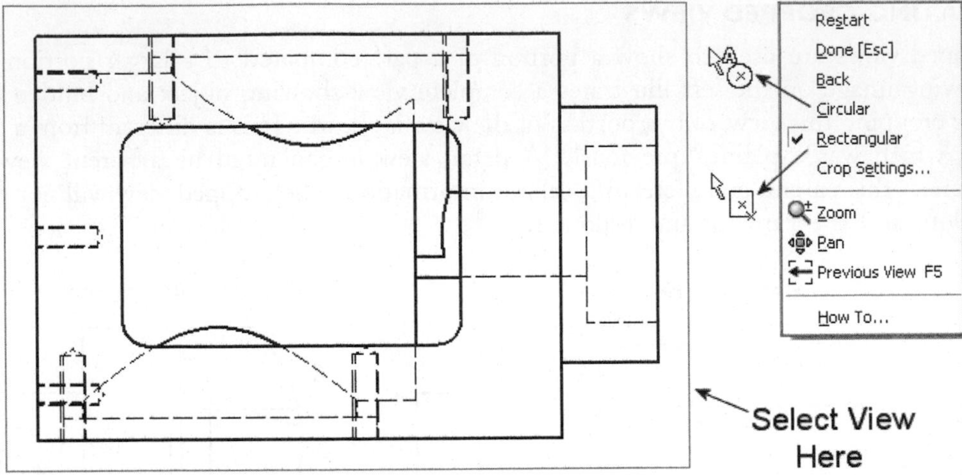

Figure 5.79

You could also control the display of the crop cut lines in the cropped view. Click Crop Settings from the menu to display the Crop Settings dialog box as shown on the left in the following image. In addition to the circular and rectangular boundary type already mentioned, you can turn on or off the display of crop cut lines as shown in the following image.

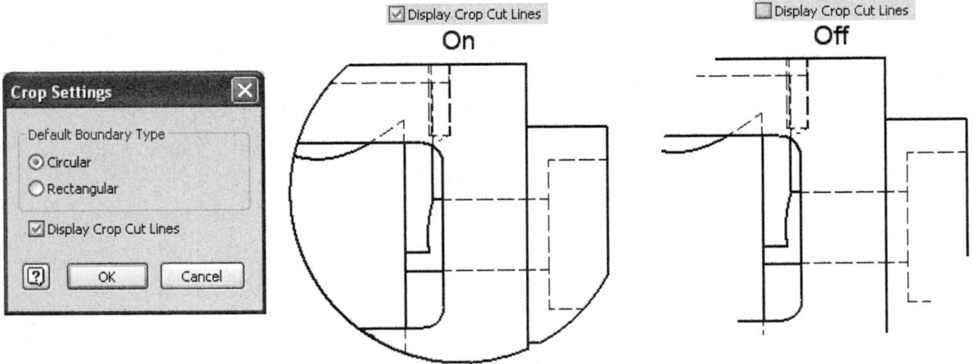

Figure 5.80

You can also create a cropped view that is based on an existing sketch. A few rules to follow are be sure that the sketch does not contain any non-self-intersecting loops and be sure the sketch is associated with the view that is currently being cropped. The illustration on the left in the following image shows a spline that was sketched. As long as this sketch is associated with this view, you will be able to create a cropped view by clicking the sketch as shown on the right in the following image.

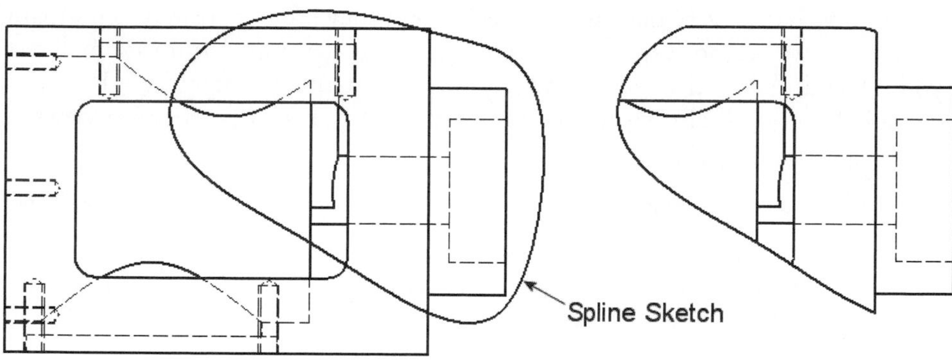

Figure 5.81

Note: To edit a cropped sketch, open the cropped view in the browser, select the sketch, right-click, and select edit. Make the necessary changes to the sketch and click the Return button. This will leave the sketch environment and update the cropped view.

When working with cropped views, you must be aware that dimensions and other annotations may affect how cropped views are displayed. For example, illustrated on the left in the following image is an engine block that has a few dimensions and a centerline through the view. However, once a cropped view is created, the anchors used as endpoints for the dimensions will no longer use these anchors, resulting in an incomplete dimension as shown on the right in the following image. Either the anchor would need to be re-established by editing the cropped view sketch or the entire dimension would have to be deleted.

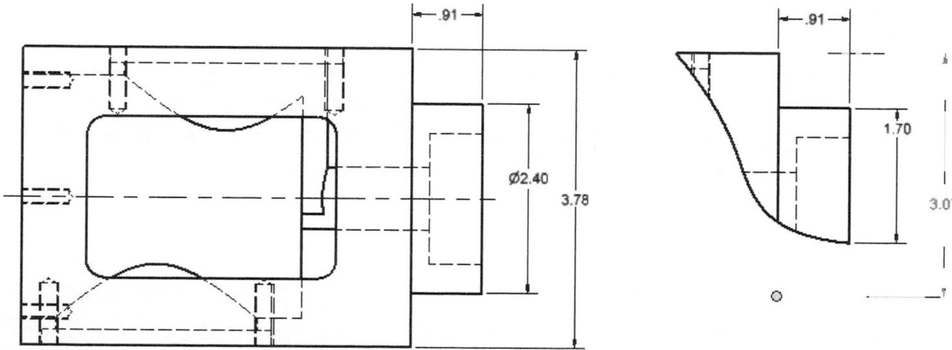

Figure 5.82

CREATING PERSPECTIVE VIEWS

In addition to creating isometric drawing views, you can create perspective drawing views based on part, assembly, presentation, and custom view information. Perspective views provide a more natural or realistic view of an assembly or component.

Perspective views have a very narrow and specific use; they are not treated as normal views, because other views are not expected to be projected or developed from them. Therefore, when you select a perspective view, view creation commands that require a base view will not be enabled. You will also be unable to apply dimensions to perspective views.

To create a perspective view, follow these steps:

1. While in drawing mode, click on the Base View button on the Drawing Views panel bar.
2. While in the Drawing View dialog box, select the model file to use for the perspective view. Then click on the Custom View button, as shown in the following image. This button is labeled Change view orientation.

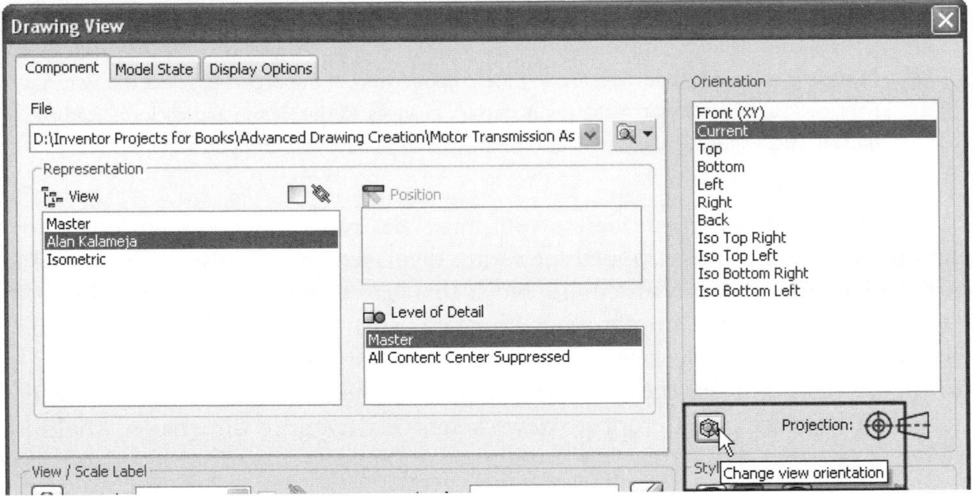

Figure 5.83

3. The Custom View window containing the current model image will appear. Use Pan, Zoom, Rotate, or other viewing tools on the Standard toolbar to position the view, as shown in the following image. The figure also illustrates the Incremental View Rotate dialog box, which is used for fine-tuning the position of the view.

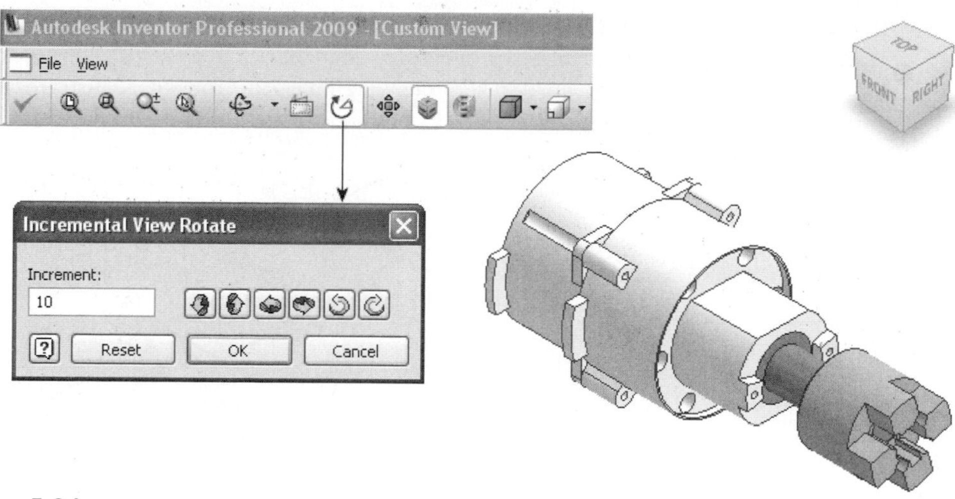

Figure 5.84

4. Use the Perspective Camera tool, as shown in the following image, to switch the view to perspective.

Note: You can resize the Custom View panel by double-clicking on its title bar.

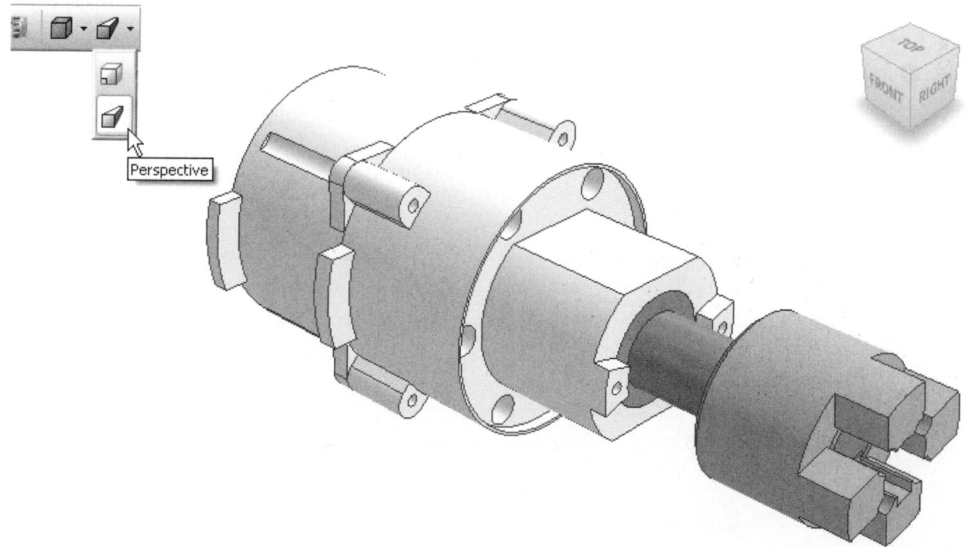

Figure 5.85

5. When you are satisfied with the view displayed in the Custom View window, click on the green checkmark located on the Standard toolbar to accept this view and close

the window, as shown in the following image. You are returned to the Drawing View dialog box, where you can make any additional changes, such as scale, before placing the view inside the drawing border. The following image on the right illustrates the created perspective drawing view.

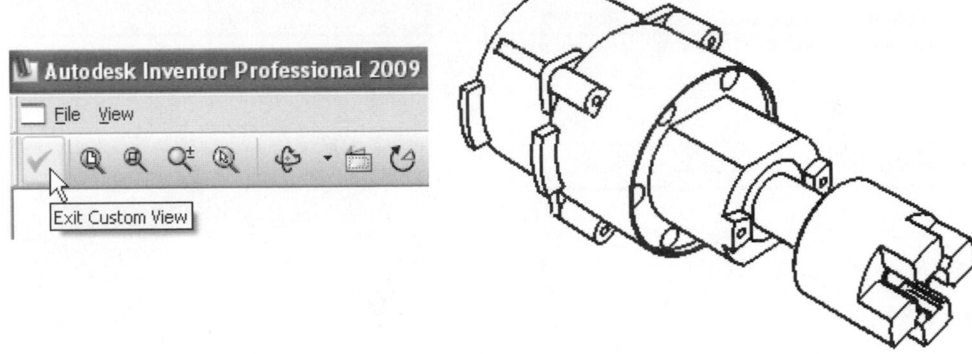

Figure 5.86

Note: If the model file is open and the current display is set to Camera Perspective, you can set up the perspective view in the drawing by selecting Current from the Orientation list of the Drawing View dialog box and then placing the view in the drawing.

EXERCISE 5-3: CREATING BREAK, SECTION, AUXILIARY, AND DETAIL VIEWS

In this exercise, you create a variety of complex drawing views from a model of a cover.

1. Open the file *ESS_E05_03.idw*. Use the Zoom tool to zoom in on the base view, as shown in the following image.

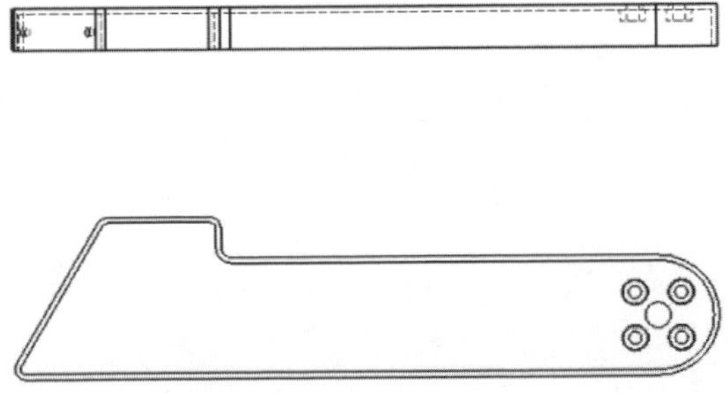

Figure 5.87

2. You will now create a break view. First click the Break tool on the Drawing Views panel bar.

3. In the graphics window, select the lower-left view to break.
4. Select the start and end points for the material to be removed, as shown in the following image. Change the break gap in the view to **10 mm.**

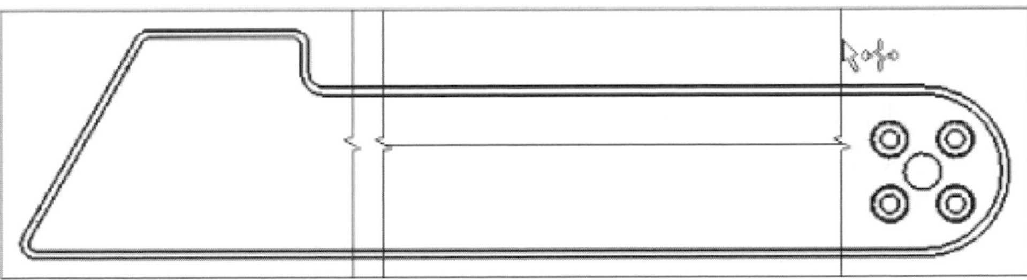

Figure 5.88

5. The results are displayed in the following image. The base and projected views are broken.

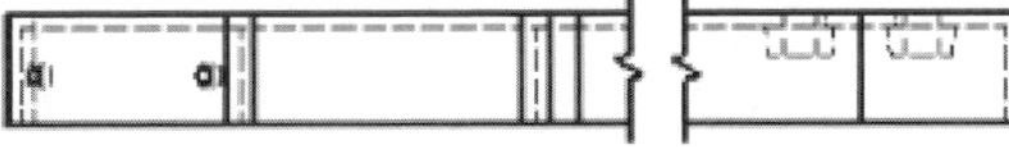

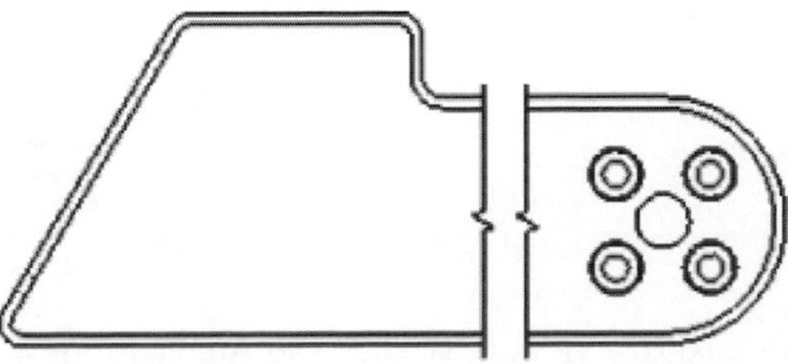

Figure 5.89

6. Next create a section view. First click the Section View tool.

7. In the graphics window, select the lower-right portion of the break view.
8. To define the first point of the section line, hover the cursor over the center hole, and then click a point directly above the hole, as shown in the following image.

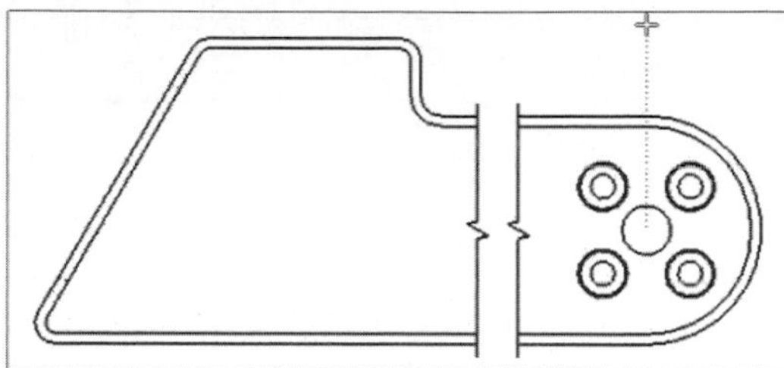

Figure 5.90

9. Complete the three-segment section line, as shown in the following image on the left. Right-click and select Continue.
10. Place the section view with the default settings to the right of the break view, as shown in the following image on the right.

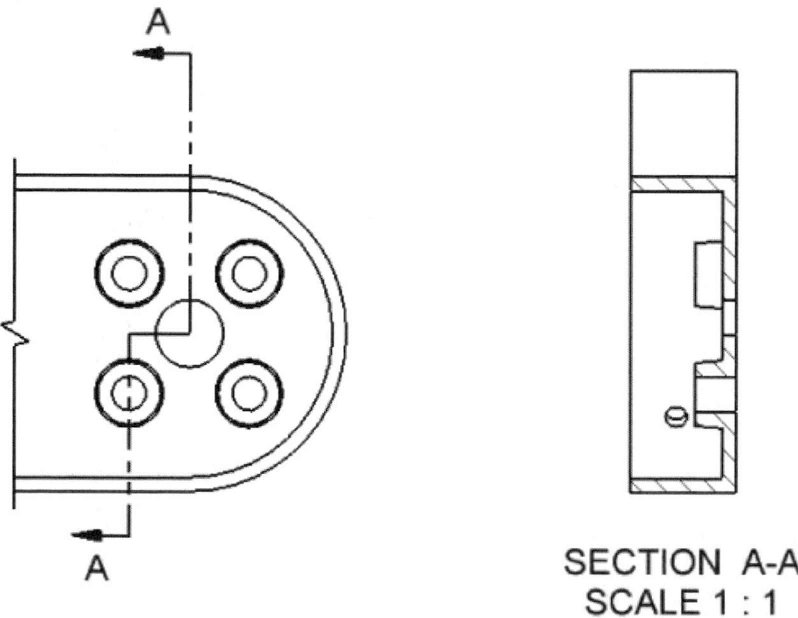

Figure 5.91

11. Next create an auxiliary view. First click the Auxiliary View tool.
12. In the graphics window, select the break view.

13. Select the outside angled line, as shown in the following image, to define the projection direction.

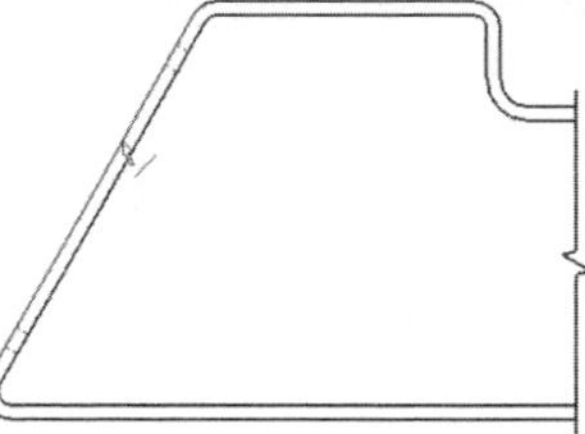

Figure 5.92

14. Place the view to the left of the break view, as shown in the following image.

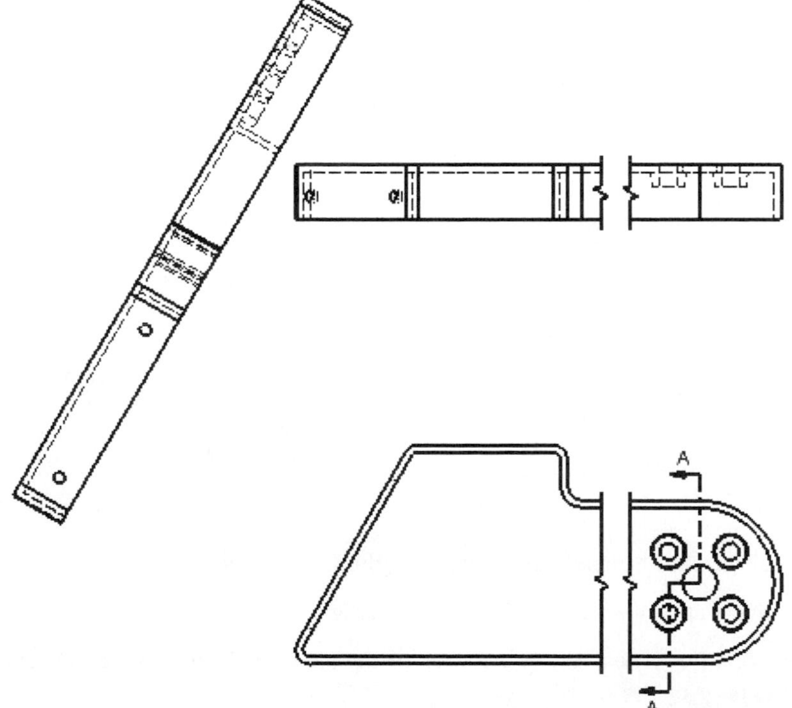

Figure 5.93

15. Next break the alignment of the auxiliary view to the parent view, and reposition the auxiliary view. In the graphics window, double-click in the auxiliary view geometry to edit the auxiliary view.

16. In the Drawing View dialog box, click the Display Options tab, clear the Align to Base option. This will also check the box next to Definition in Base View. Click OK to continue.

17. In the graphics window, select and drag the auxiliary view to the right side of the drawing sheet, as shown in the following image.

Note: You must click and drag the view border.

18. Adjust the length of the Auxiliary View cutting plane line by clicking on and dragging the green handles.

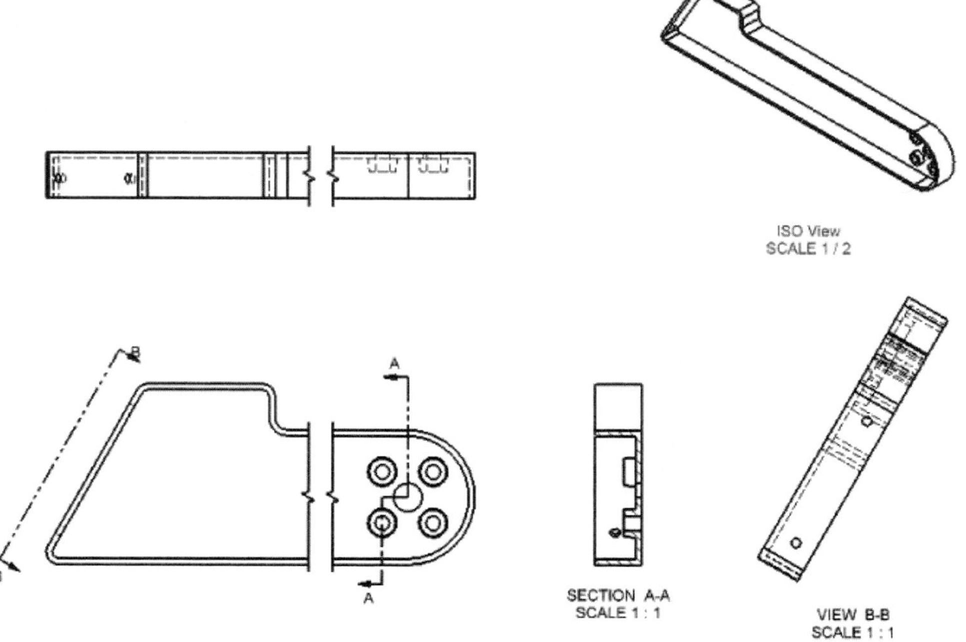

Figure 5.94

19. Finally, create a detail view. First click the Detail View tool.

20. In the graphics window, select the lower-left break view.

21. Select a point near the left corner of the view to define the center of the circular boundary of the detail view.

22. Drag the circular boundary to include the entire lower-left corner of the view geometry, as shown in the following image.

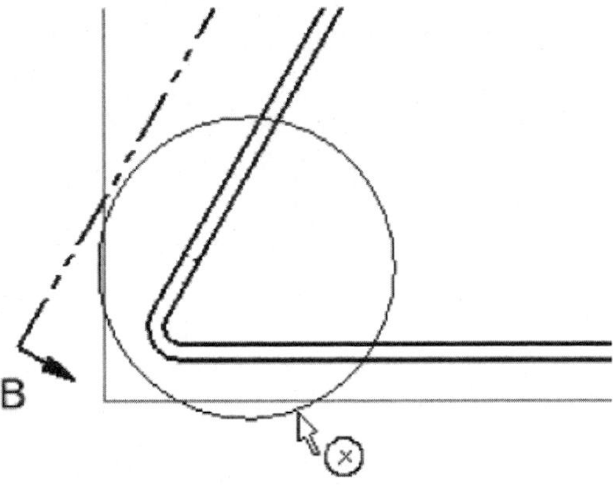

Figure 5.95

23. Click to position the detail view below the break view, as shown in the following image.

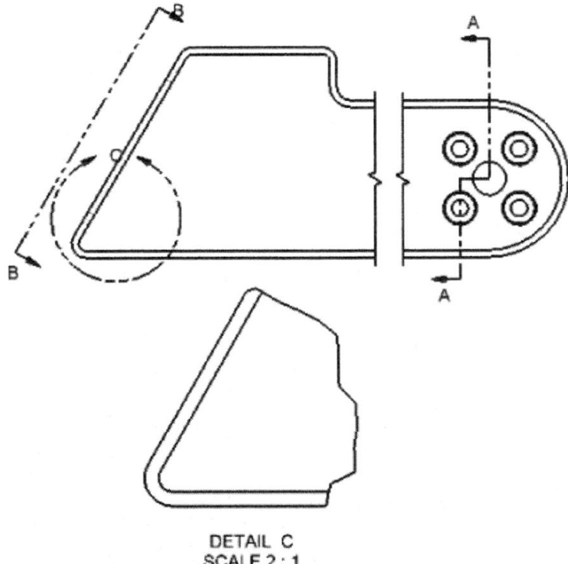

Figure 5.96

24. To finish this exercise, reposition the drawing views as required. The finished drawing is displayed, as shown in the following image.

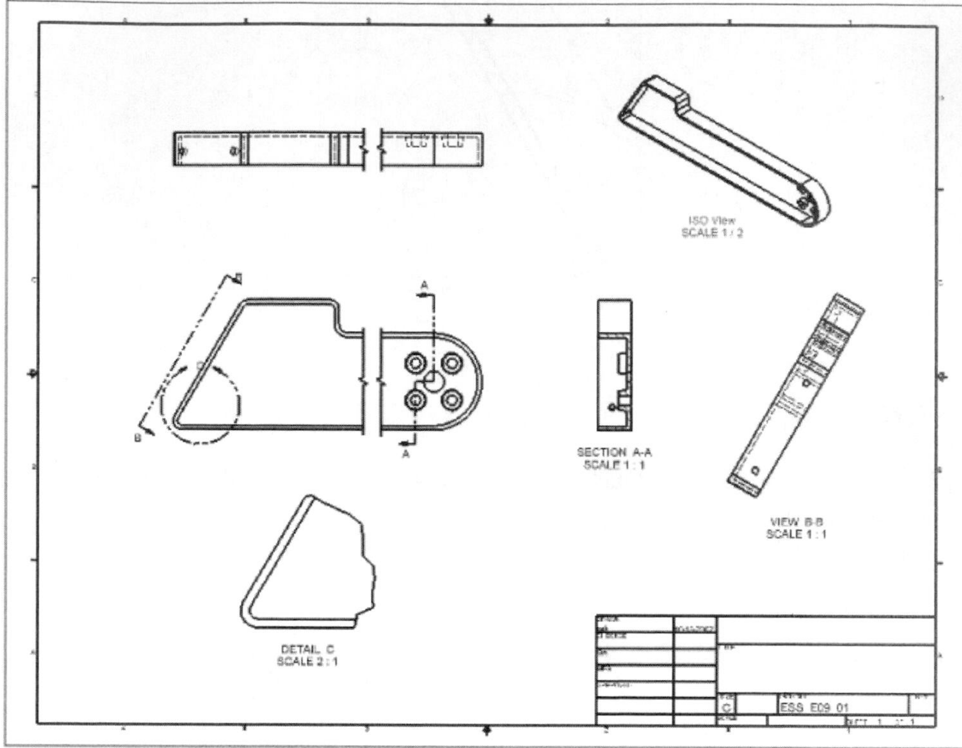

Figure 5.97

25. Close all open files. Do not save changes. End of exercise.

EDITING DRAWING VIEWS

After creating the drawing views, you may need to move, edit the properties of, or delete a drawing view. The following sections discuss these options.

MOVING DRAWING VIEWS

To move a drawing view, click inside the view until a bounding box consisting of red dotted lines appears, as shown in the following image. Move your cursor to the red dotted boundary, press and hold down the left mouse button, and move the view to its new location. Release the mouse button when you are finished. As you move the view, a rectangle will appear that represents the bounding box of the drawing view. If you move a base view, any projected children or dependent views will also move with it as required to maintain view alignment. If you move an orthographic or auxiliary view, you will only be able move it along the axis in which it was projected from the part edge or face. You can move detail and isometric views anywhere in the drawing sheet.

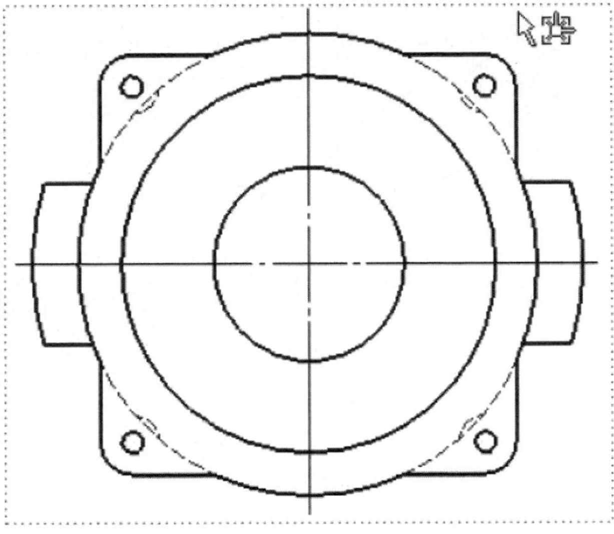

Figure 5.98

Note: You can also move a view by picking the view boundary and dragging it directly without first activating the view with a pick.

EDITING DRAWING VIEW PROPERTIES

After creating a drawing view, you may need to change the label, scale, style, or hatching visibility, break the alignment constraint to its base view, or control the visibility of the view projection lines for a section or auxiliary view. To edit a drawing view, follow one of these steps:

- Double-click in the bounding area of the view.
- Double-click on the icon of the view you want to edit in the browser.
- Right-click in the drawing view's bounding area or on its name, and select Edit View from the menu, as shown in the following image.

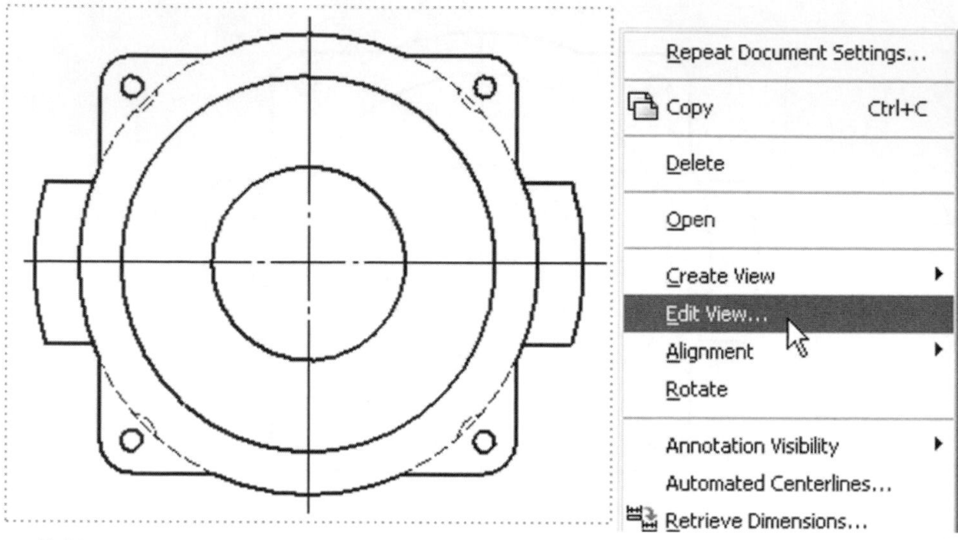

Figure 5.99

When you perform the operations listed above, the Drawing View dialog box will appear, as shown in the following image. This is thc samc dialog box uscd to create drawing views. Depending on the view that you selected, certain options may be grayed out from the dialog box. Make the necessary changes, and click the OK button to complete the edit. When you change the scale in the base view, all the dependent views will be scaled as well.

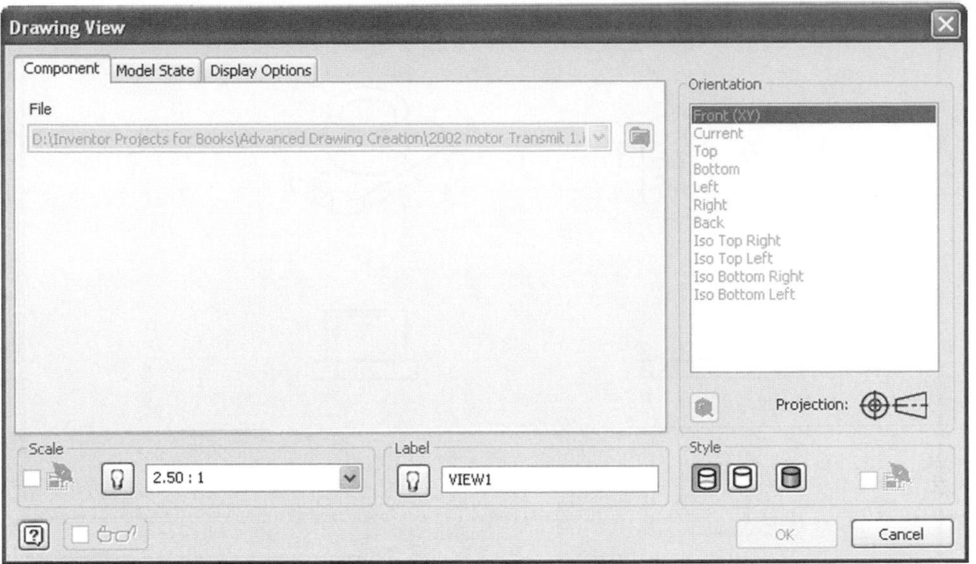

Figure 5.100

See the Using the Drawing View Dialog Box section in this chapter for detailed descriptions of Component and Options tab elements.

DELETING DRAWING VIEWS

To delete a drawing view, either right-click in the bounding area of the drawing view or on its name in the browser, and select Delete from the menu. You can also click in the bounding area of the drawing view, and press the DELETE key on the keyboard. A dialog box will appear, asking you to confirm the deletion of the view. If the selected view has a view that is dependent on it, you will be asked if the dependent views should also be deleted. By default, the dependent view will be deleted. To exclude a dependent view from the delete operation, expand the dialog box, and click on the Delete cell to change the selection to No.

EXERCISE 5-4: EDITING DRAWING VIEWS

In this exercise, you delete the base view while retaining its dependent views. The base view is not required to document the part, but its dependent views are. Next you align the section view with the right-side view to maintain the proper orthographic relationship between the views. You also modify the hatch pattern of the section to better represent the material.

1. Open the drawing *file ESS_E05_04.idw*. This drawing contains three orthographic views, an isometric view, and a section view, as shown in the following image.

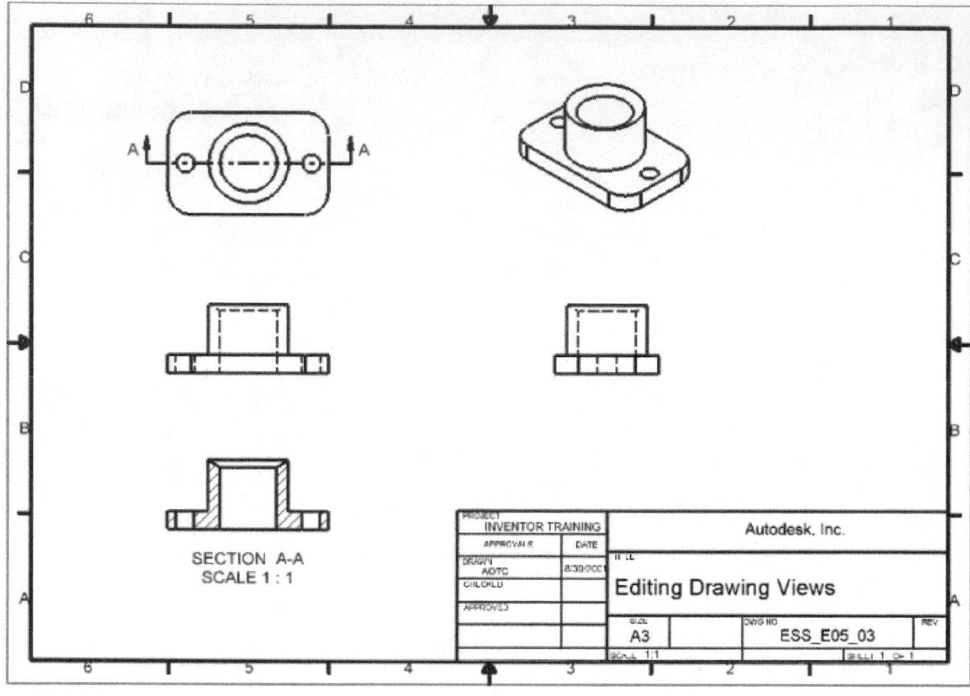

Figure 5.101

2. Right-click on the base view, as shown in the following image, and choose Delete from the menu.

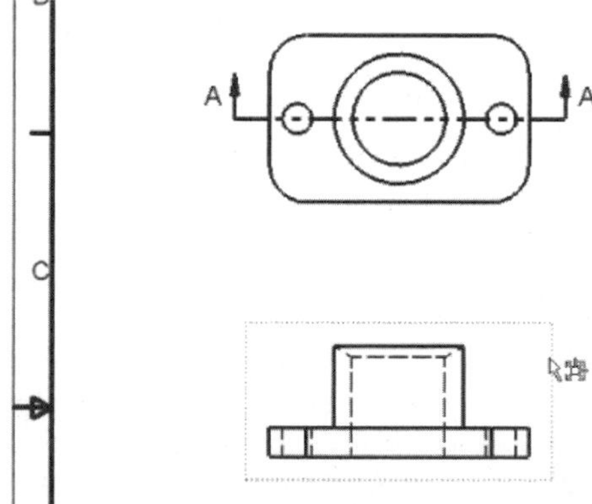

Figure 5.102

3. In the Delete View dialog box, notice that the projected views are highlighted on the drawing sheet. Select the More (<<) button in the dialog box.

4. Click on the word Yes in the Delete column for each dependent view to toggle each to No. This will delete the base view but leave the dependent views present on the drawing sheet.
5. Click OK to delete the base view and retain the two dependent views. Your drawing should appear similar to the following image.

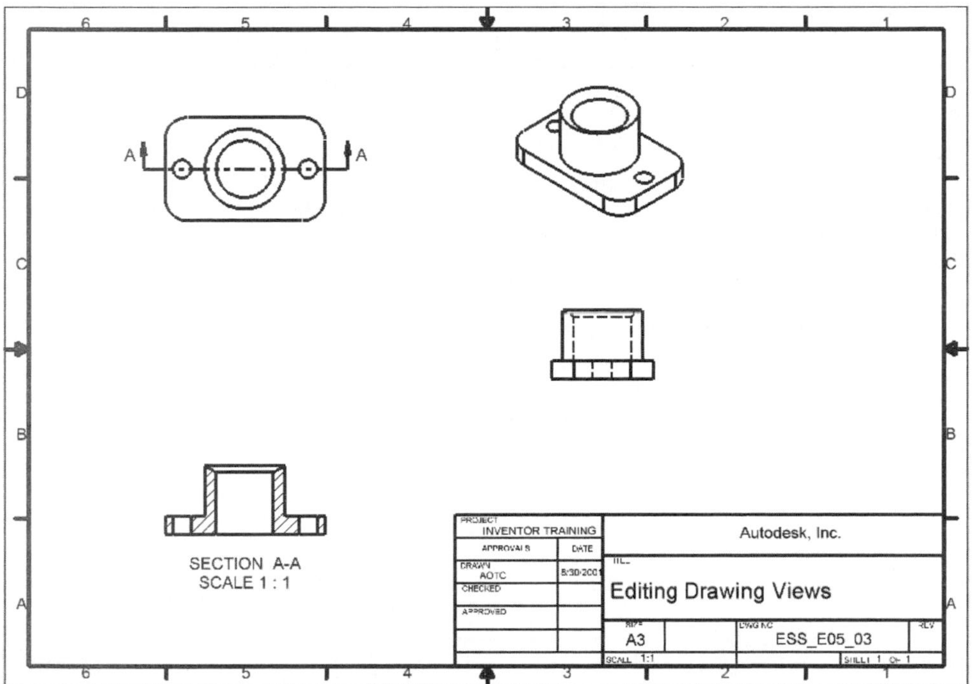

Figure 5.103

With the front view deleted, the section view needs to be identified as the new base view and aligned with the right-side view.

6. Right-click the right-side view, and choose Alignment > Horizontal from the menu.
7. Select the section view as the base view.
8. Select the section view, and drag the view vertically to the area previously occupied by the front view. Notice how the right-side view remains aligned to the section view.
9. Right-click the isometric view, and select Alignment > In Position.
10. Select the section view as the base view. Move the section view, and notice that the isometric view now moves with the section view. Your drawing should appear similar to the following image.

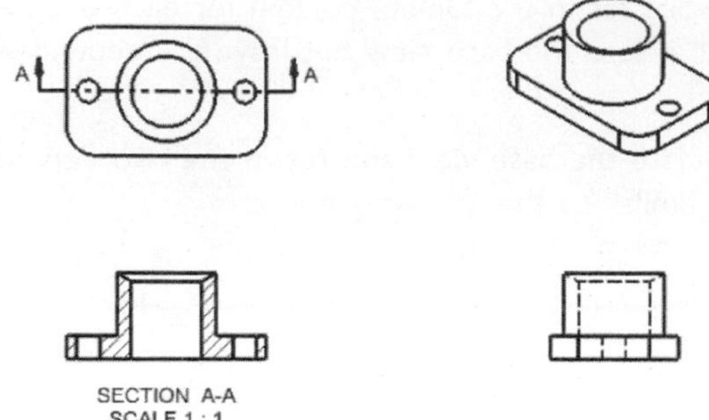

Figure 5.104

You now edit the section view hatch pattern to represent the material as bronze using the ANSI 33 hatch pattern.

11. Right-click the hatch pattern in the section view, and choose Edit from the menu.
12. The Edit Hatch Pattern dialog box is displayed. Select ANSI 33 from the Pattern list, and click OK. The hatch pattern changes, as shown in the following image.

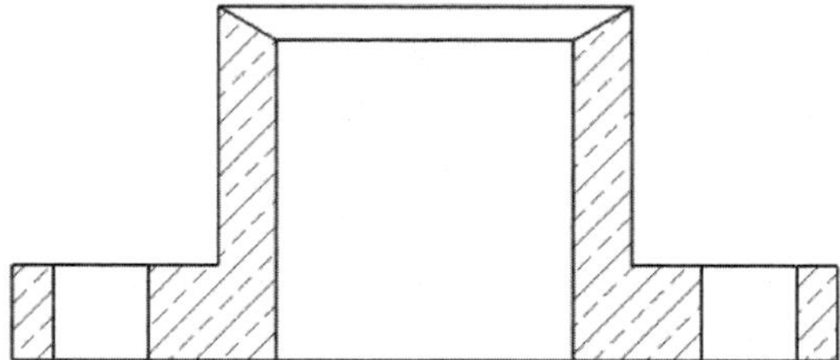

Figure 5.105

13. Close all open files. Do not save changes. End of exercise.

DIMENSIONS

Once you have created the drawing view(s), you may need to change the value of a model dimension. If the dimensions did not appear when you created the view, you may want to alter the model dimensions so they appear in a view, hide certain dimensions, add drawing (general) dimensions, or move dimensions to a new location. The following sections describe these operations.

RETRIEVING MODEL DIMENSIONS

Model dimensions may not appear automatically when you create a drawing view. You can use the Retrieve Dimensions tool to select valid model dimensions for display in a drawing view. Only those dimensions that you placed in the model on a plane parallel to the view will appear.

To activate this command, use one of the following three methods:

- Right-click in the bounding area of a drawing view, and select Retrieve Dimensions.
- Right-click on the name of the view in the browser, and select Retrieve Dimension from the menu.
- Click on the Retrieve Dimension button located at the bottom of the Drawing Annotations panel bar.

The following image shows all three methods.

The appropriate dimensions for the view that were used to create the part will appear.

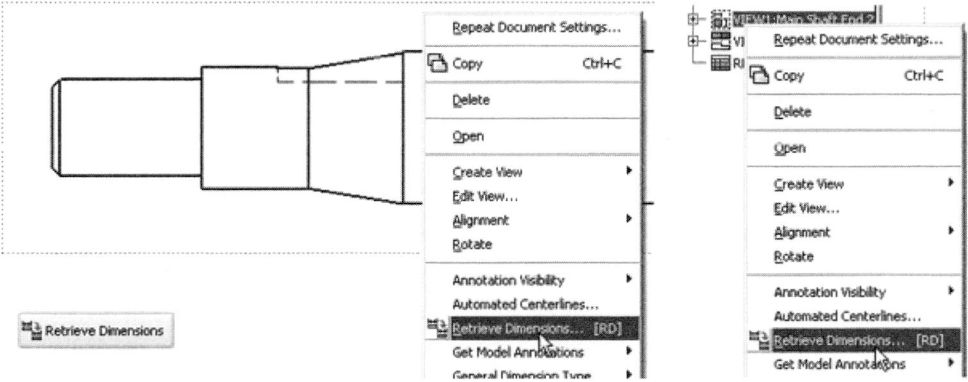

Figure 5.106

Selecting Retrieve Dimensions will display the Retrieve Dimensions dialog box, as shown in the following image. In this dialog box, you select a drawing view in which to retrieve the model dimensions. The Select Source area of the dialog box allows you to retrieve model dimensions based on an entire part or upon part features. The Select Parts option allows you to retrieve all model dimensions based on a part. When you use either of these methods, only valid model dimensions appear in preview mode. Once you have previewed the model dimensions, click the Select Dimensions button to pick the dimensions you want to retrieve.

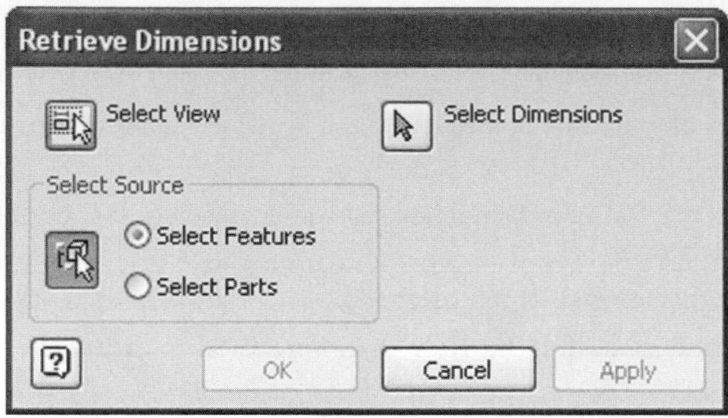

Figure 5.107

To retrieve model dimensions, follow these steps:

1. Click Retrieve Dimensions on the Drawing Annotations panel bar.
2. Select the drawing view in which to retrieve model dimensions.
3. To retrieve model dimensions based on one or more features, click the Select Features radio button. Then select the desired features.
4. If you want to retrieve model dimensions based on the entire part in a drawing view, click the Select Parts radio button. Then select the desired part or parts. You can select multiple objects by holding down the SHIFT or CTRL key. You can also select objects with a selection window or crossing selection box.
5. When the model dimensions appear in the drawing view in preview mode, click the Select Dimensions button, and select the desired dimensions. The selected dimensions will be highlighted.
6. Click the Apply button to retrieve the model dimensions, and leave the dialog box active for further retrieval operations.
7. Click the OK button to retrieve the selected dimensions, and dismiss the Retrieve Dimensions dialog box.

The following image shows the effects of retrieving model dimensions using the Select Features mode. In this example, the large rectangle of the engine block was selected. This resulted in the horizontal and vertical model dimensions being retrieved.

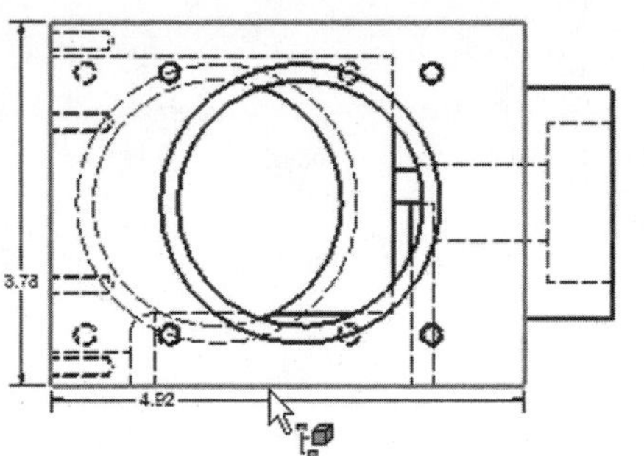

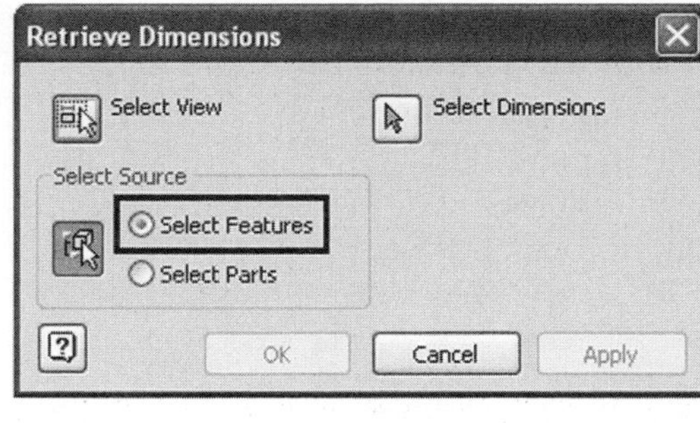

Figure 5.108

The following image shows the effects of retrieving model dimensions using the Select Parts mode. In this example, the entire engine block was selected. This resulted in all model dimensions parallel to this drawing view being retrieved.

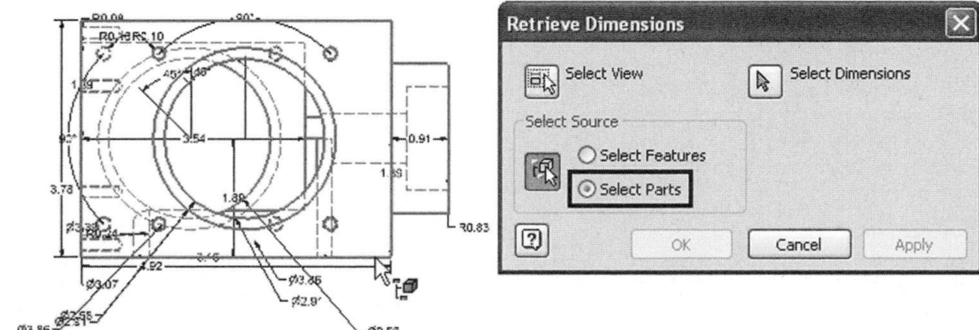

Figure 5.109

You can also retrieve model dimensions automatically during the drawing view creation process. To perform this task, follow these steps:

1. Launch the Options dialog box by clicking Tools > Application Options.
2. Click on the Drawing tab, and locate the option Retrieve all model dimensions on view placement, as shown in the following image.
3. Click in the box to automatically retrieve model dimensions in all placed views. Clear the box if you want to create drawing views without retrieving model dimensions automatically.
4. When you want to retrieve model dimensions in a base drawing view, click the All Model Dimensions option on the Options tab of the Drawing View dialog box.

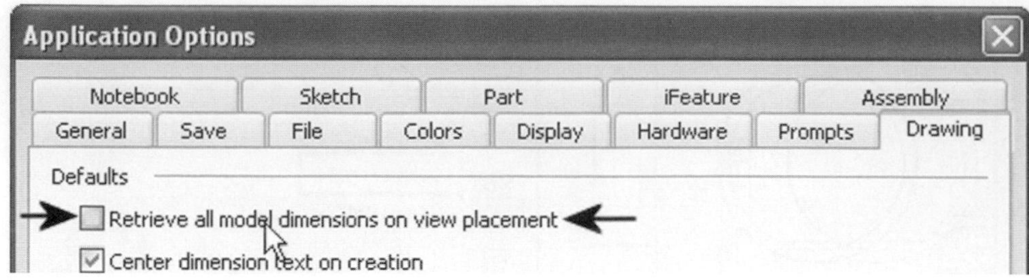

Figure 5.110

DIMENSION VISIBILITY

To hide all of the model or drawing dimensions for a drawing view, right-click in the bounding area of a drawing view, and select Annotation Visibility from the menu, as shown in the following image. The appearance of a check means the specific feature is turned on. Clicking the check adjacent to Model Dimensions will remove the check and turn off the model dimensions in a drawing. Use the same operation to turn off drawing dimensions.

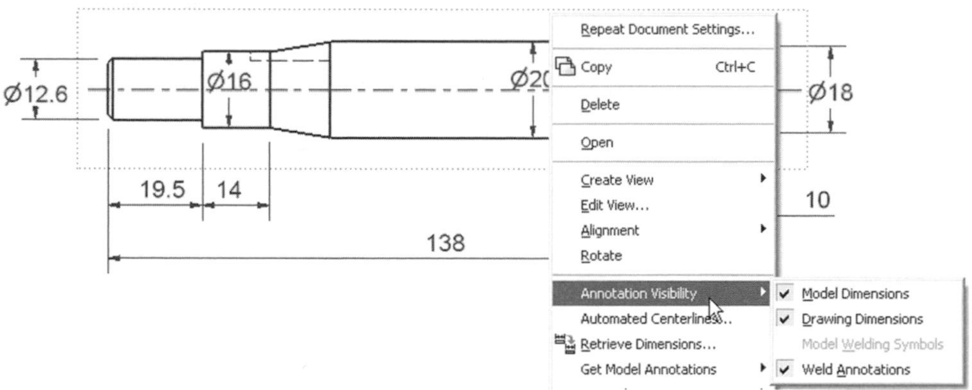

Figure 5.111

CHANGING MODEL DIMENSION VALUES

While working in a drawing view, you may find it necessary to change the model (parametric) dimensions of a part. You can open the part file, change a dimension's value, and save the part, and the change will then be reflected in the drawing views. You can also change a dimension's value in the drawing view by right-clicking on the dimension and selecting Edit Model Dimension from the menu, as shown in the following image.

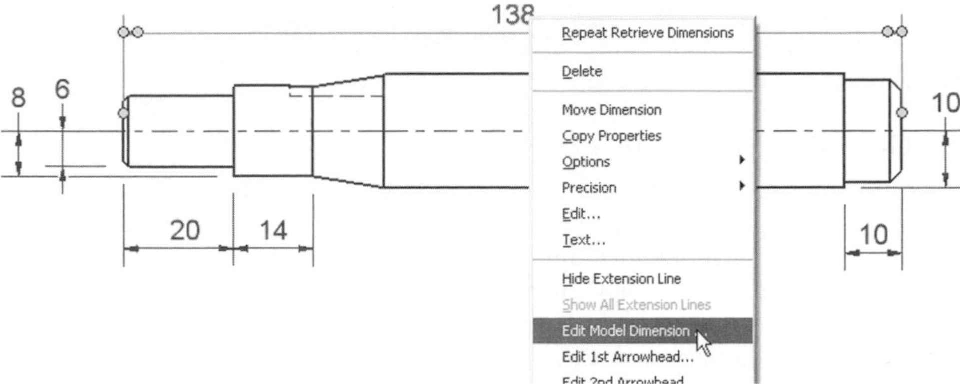

Figure 5.112

The Edit Dimension text box will appear. Enter a new value and click the checkmark, as shown in the following image, or press ENTER. The associated part will be updated and saved, and the associated drawing view(s) will be updated automatically to reflect the new value.

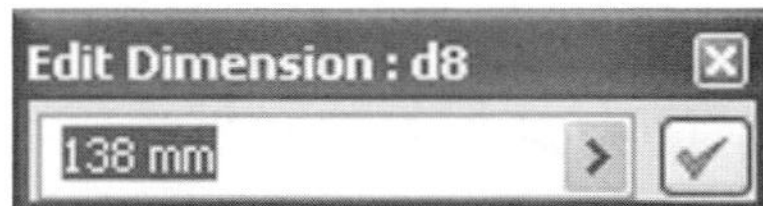

Figure 5.113

Note: It is possible to edit model dimensions found in drawing views. This feature is possible when you install Autodesk Inventor and enable the option to edit model dimensions from the drawing. However, it is considered poor practice to make major dimension changes from the drawing. It is highly recommended that you make all dimension changes through the part model.

GENERAL DIMENSIONS

After laying out the drawing views, you may find that a dimension other than an existing model dimension or an additional dimension is required to better define the part. You can go back and add a parametric dimension to the sketch if the part was underconstrained, add a driven dimension if the sketch was fully constrained, or add a drawing (general) dimension to the drawing view. A general dimension is not a parametric dimension; it is associative to the geometry to which it is referenced. The general dimension reflects the size of the geometry being dimensioned. After you create a general dimension, and the value of the geometry that you dimensioned changes, the general dimension will be updated to reflect the change. You add a general dimension by using the General Dimension tool on the Drawing Annotation panel bar, as shown in the following image. Create a general dimension in the same way you would create a parametric dimension.

General Dimension [D]

Figure 5.114

Adding General Dimensions to a Drawing View

The following image illustrates a simple object with various dimension types. General dimensions can take the form of linear dimensions as shown at (A) and (B), diameter dimensions as shown at (C), radius dimensions as shown at (D), and angular dimensions as shown at (E). When using the General Dimension tool, Autodesk Inventor automatically chooses the dimension type depending on the object you chose. In this example, the counterbored hole is dimensioned using a special tool called a hole note. This tool will be described later in this chapter.

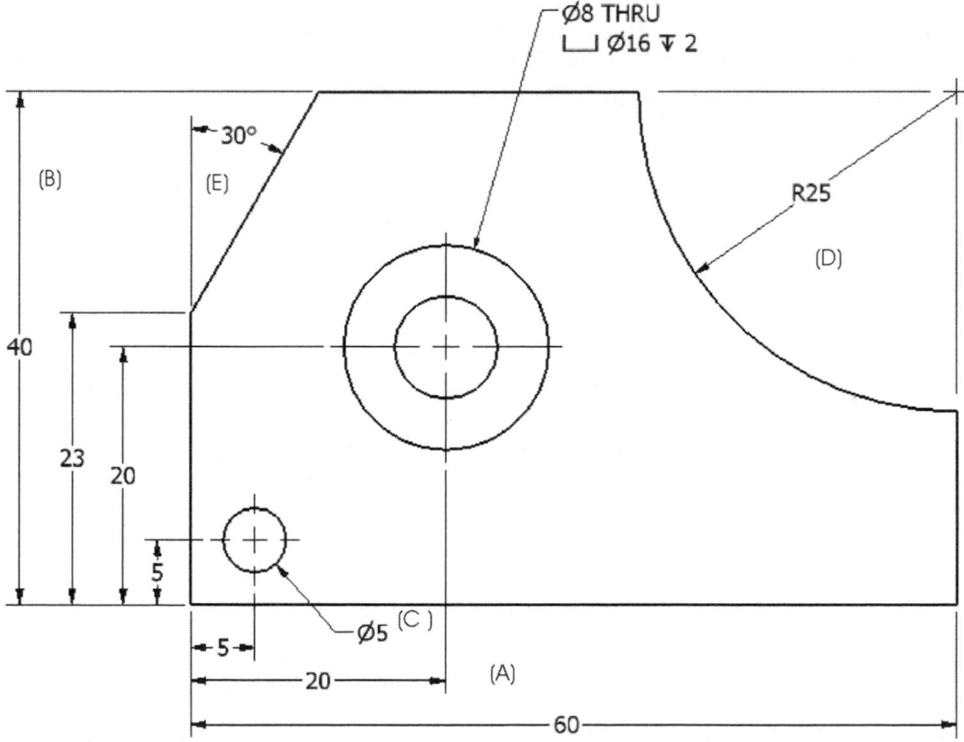

Figure 5.115

MOVING AND CENTERING DIMENSION TEXT

To move a dimension by either lengthening or shortening the extension lines or moving the text, position the cursor over the dimension until it becomes highlighted, as shown in the following image. An icon consisting of four diagonal arrows will appear attached to your cursor. Notice the appearance of a centerline. This represents the center of the dimension text. You can drag the dimension text up, down, right, or left in order to better position it.

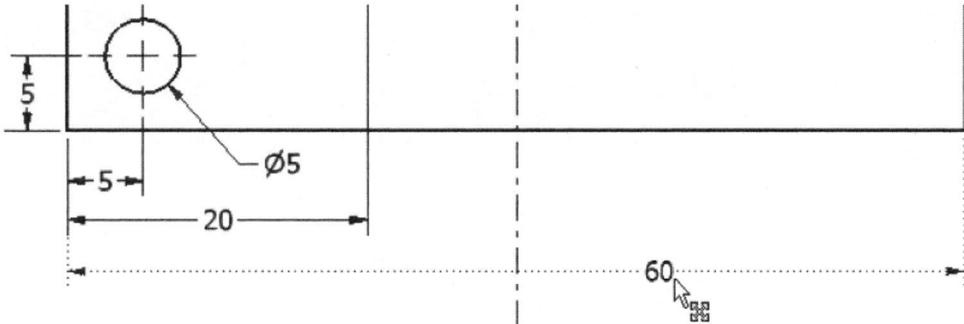

Figure 5.116

If you want to recenter the dimension text, slide the text toward the centerline. The dimension will snap to the centerline. In the following image, the centerline changes to a dotted line to signify that the dimension text is centered. This centerline action will work on any type of linear dimension text including horizontal, vertical, and aligned. This feature is not active when repositioning the text of radius, diameter, or hole note dimensions.

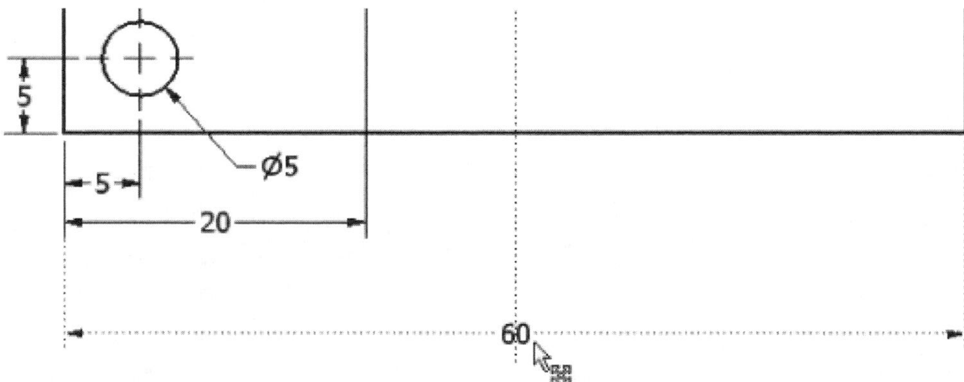

Figure 5.117

ADDING DIMENSIONS TO ISOMETRIC VIEWS

The use of the General Dimension tool can be extended to adding dimensions to isometric drawing views as well as to standard orthographic views. When placing a dimension on an isometric view, the dimension text, dimension lines, extension lines, and arrows are all displayed as oblique and are aligned to the geometry being dimensioned. The following image shows a typical isometric view complete with oblique dimensions.

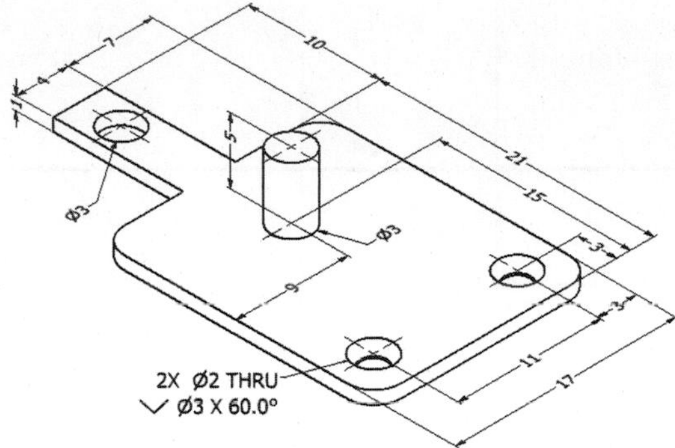

Figure 5.118

Depending on the object being dimensioned, additional controls are available to manipulate the isometric dimension being created. Once edges or points in a drawing are selected, the dimension displays based on an annotation plane, as shown in the following image on the left. If more than one annotation plane exists, you can toggle between these planes by pressing the spacebar. The results are shown in the following image on the right.

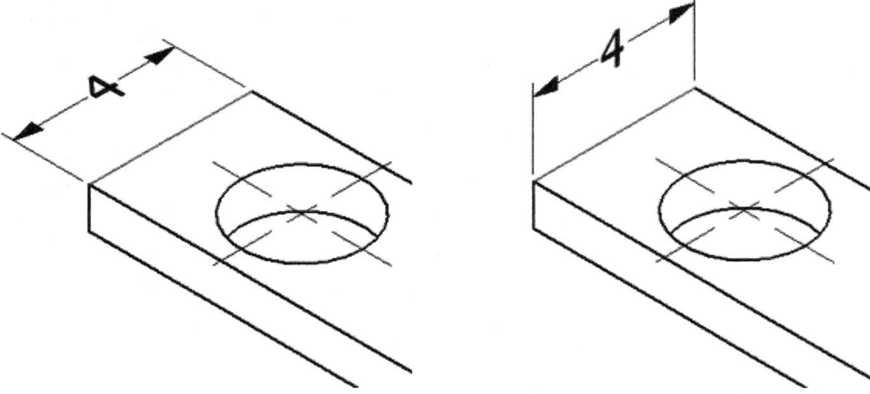

Figure 5.119

Note: You can select other annotation planes by right-clicking when placing the dimension, selecting Annotation Plane, as shown in the following image, and selecting one of the following options:

- Show All Part Work Planes
- Displays all default and user-defined work planes. These planes can then be selected as new annotation planes.
- Show All Visible Work Planes
- Displays only those work planes that are currently visible in the model. These planes can then be selected as new annotation planes.
- Use Sheet Plane
- Uses the current drawing sheet as a plane for placing the isometric dimension. Linear dimensions placed on the sheet plane are not true values.

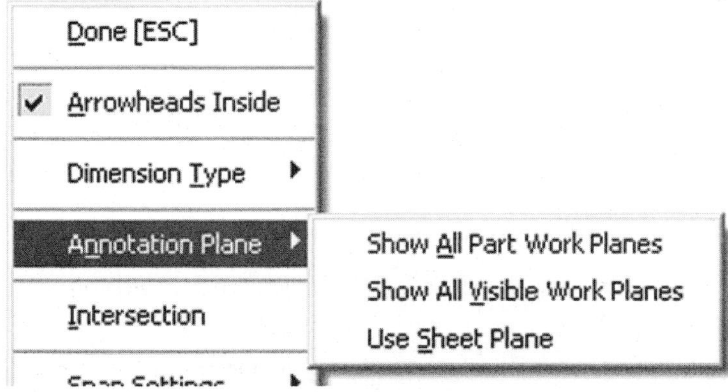

Figure 5.120

ANNOTATIONS

To complete an engineering drawing, you must add annotations such as centerlines, surface texture symbols, welding symbols, geometric tolerance symbols, text, bills of materials, and balloons. Before adding annotations to a drawing, make the desired drawing style active and add annotations as needed.

CENTER MARKS AND CENTERLINES

When you need to annotate the centers of holes, circular edges, or the middle (center axis) of two lines, four methods allow you to construct the needed centerlines. Use the Center Mark, Centerline, Centerline Bisector, and Centered Pattern tools on the Drawing Annotation panel bar, as shown in the following image. The centerlines are associated to the geometry that you select when you create them. If the geometry changes or moves, the centerlines update automatically to reflect the change. This section outlines the steps for creating the different types of centerlines.

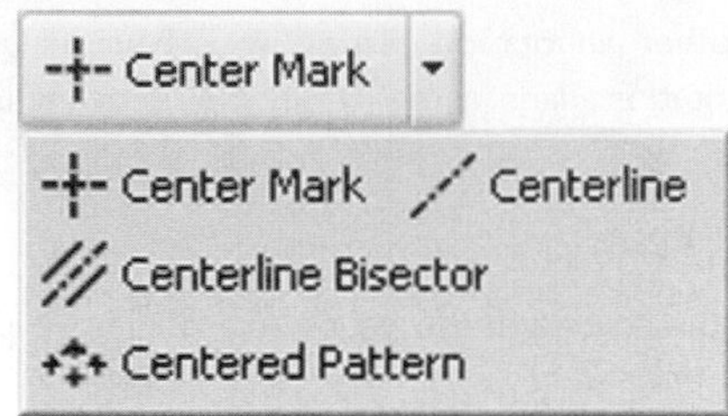

Figure 5.121

Center Mark

To add a center mark, follow the steps and refer to the following image:

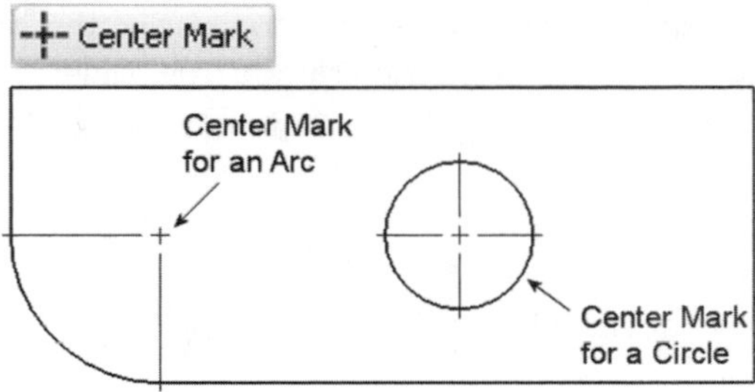

Figure 5.122

1. Click the Center Mark tool on the Drawing Annotation panel bar.
2. In the graphics window, select the circle and/or arc geometry in which you want to place a center mark.
3. Continue placing center marks by selecting geometry.
4. To complete the operation, right-click, and select Done from the menu.

Centerline Bisector

To add a centerline bisector, follow the steps and refer to the following image:

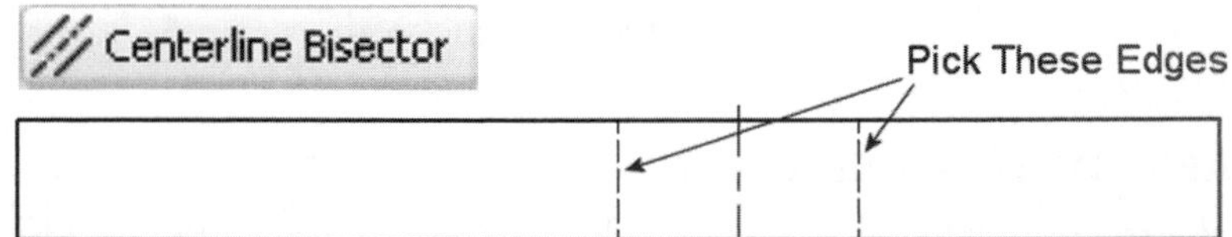

Figure 5.123

5. Click the Centerline Bisector tool on the Drawing Annotation panel bar.
6. In the graphics window, select two lines between which you want to place the centerline bisector.

7. Right-click, and select Create from the menu to place the centerline bisector.
8. Continue placing centerline bisectors by selecting geometry.
9. To complete the operation, right-click, and select Done from the menu.

Centerline

To add a centerline, follow these steps and refer to the following image:

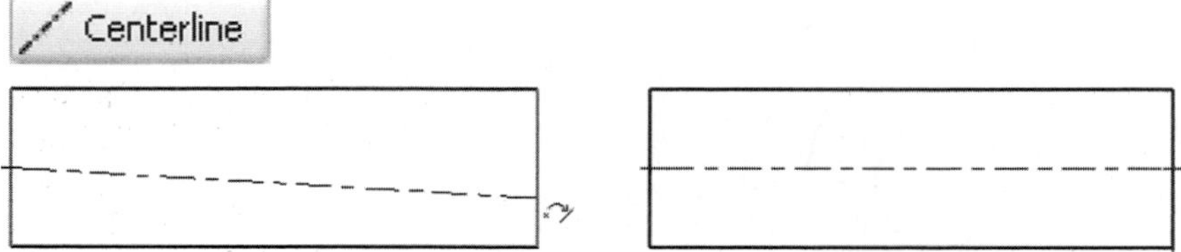

Figure 5.124

10. Click the Centerline tool on the Drawing Annotation panel bar.
11. In the graphics window, select a piece of geometry for the start of the centerline.
12. Click a second piece of geometry for the ending location.
13. Right-click, and select Create from the menu to create the centerline. The centerline will be attached to the midpoints of the selected geometry.
14. Continue placing centerlines by selecting geometry.
15. To complete the operation, right-click, and select Done from the menu.

Centered Pattern

To add a centered pattern, follow the steps and refer to the following image:

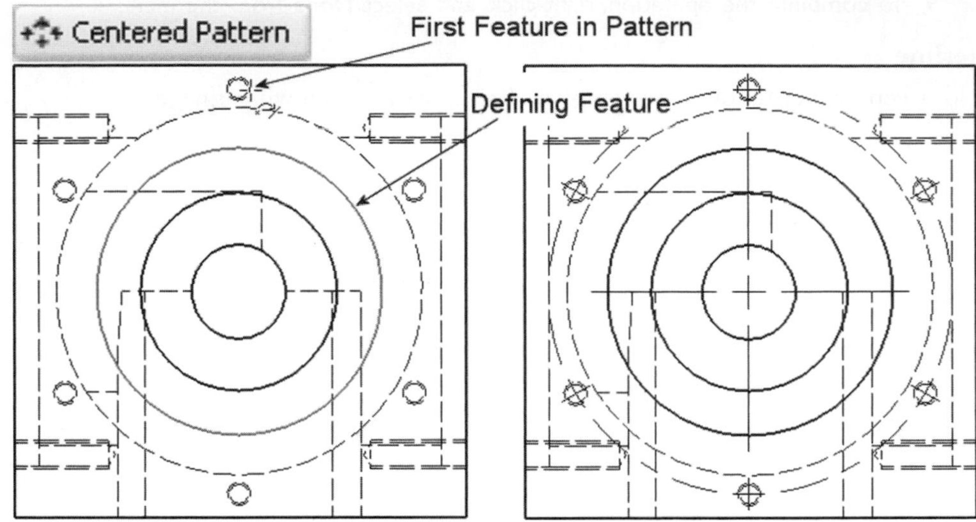

Figure 5.125

16. Click the Centered Pattern tool on the Drawing Annotation panel bar.
17. In the graphics window, select the defining feature for the pattern to place its center mark.
18. Click the first feature of the pattern.
19. Continue selecting features in a clockwise direction until all of the features are added to the selection set.
20. Right-click, and select Create from the menu to create the centered pattern.
21. To complete the operation, right-click, and select Done from the menu.

AUTOMATED CENTERLINES FROM MODELS

Creating centerlines automatically can eliminate a considerable amount of work. You can control which features automatically get centerlines and marks and in which views these occur.

You can apply automated centerlines and marks as a drawing template default by clicking Active Standard on the Format menu, as shown in the following image. This will activate the Document Settings dialog box for the active document. Click the Automated Centerline Settings button on the Drawing tab, as shown in the following image. You can also set automated centerlines for drawing views by right-clicking on the view boundary or hold down the Ctrl key and select multiple view boundaries and right-click to display the menu and selecting Automated Centerlines.

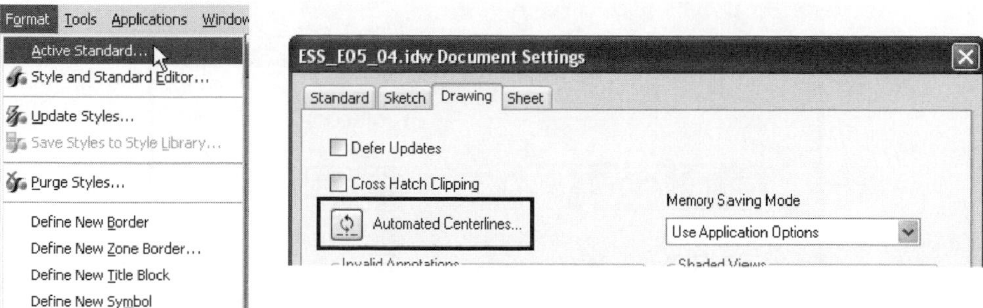

Figure 5.126

Clicking the Automated Centerline button will launch the Centerline Settings dialog box, as shown in the following image. Use this dialog box to set the type of feature(s) to which you will apply automated centerlines, such as holes, fillets, cylindrical features, etc., and to choose the projection type (plan or profile). The following sections discuss these action in greater detail.

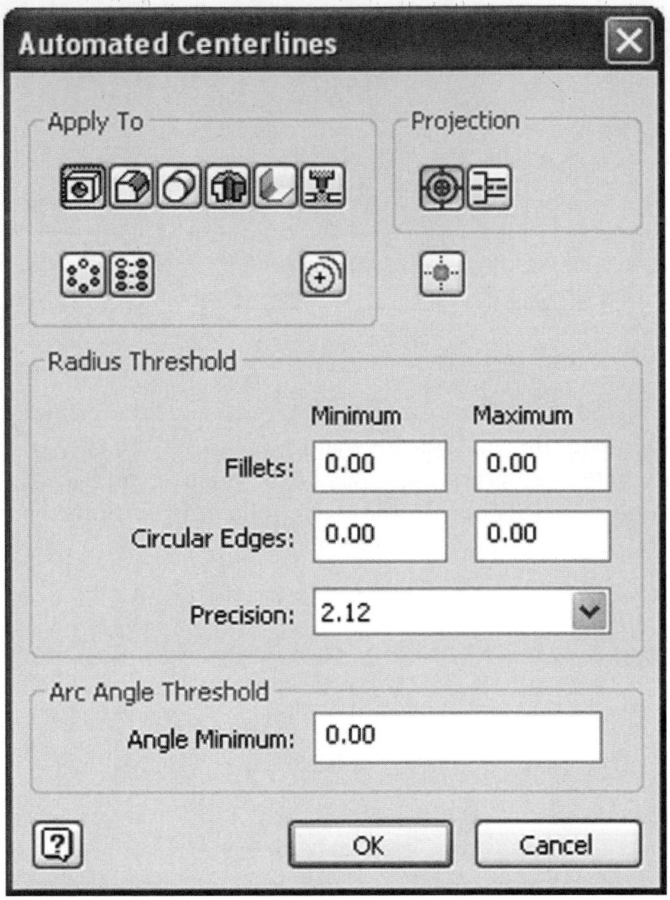

Figure 5.127

Apply To

This area controls the feature type to which you want to apply automated centerlines. Feature types include holes, fillets, cylindrical features, revolved features, circular patterns, rectangular patterns, sheet metal bends, punches, and circular sketched geometry. Click on the appropriate button to activate it, and automated centerlines will be applied to all features of that type in the drawing. You can click on multiple buttons to apply automated centerlines to multiple features. To disable centerlines in a feature, click on the feature button a second time.

Projection

Click on the projection buttons to apply automated centerlines to plan (axis normal) and/or profile (axis parallel) views.

Radius Threshold

Thresholds are minimum and maximum value settings and are provided for fillet features, arcs, and circles. Any object residing within a range should get the appropriate center mark. The values are based upon the model values, not the drawing values. This allows you to know what will or will not receive a centerline regardless of the view scale in the document. For example, if you set a minimum value of 0.50 for the fillet feature, a fillet that has a radius of 0.495 will not receive a center mark. A zero value on both threshold settings (min/max) denotes no restriction. This means that center marks will be placed on all fillets regardless of size.

Arc Angle Threshold

This option sets the minimum angle value for creating a center mark or centerline on circles, arcs, or ellipses.

Note: Automated centerlines can easily be created by preselecting the edges of multiple views and right-clicking the menu to display Automated Centerlines as shown in the following image.

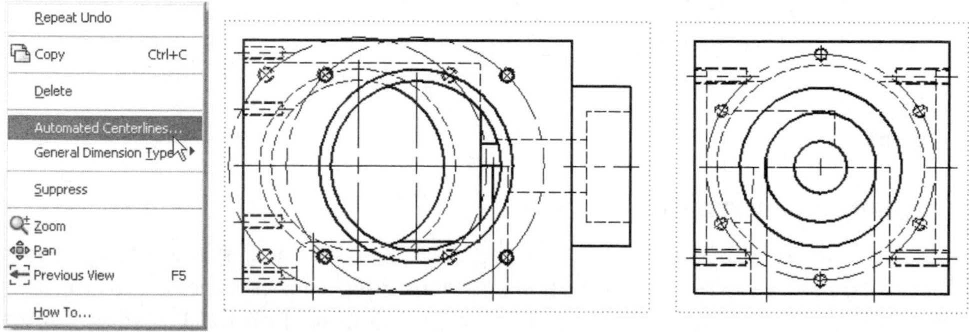

Figure 5.128

TEXT AND LEADERS

To add text to the drawing, click either the Text or the Leader Text tool on the Drawing Annotation panel bar, as shown in the following image.

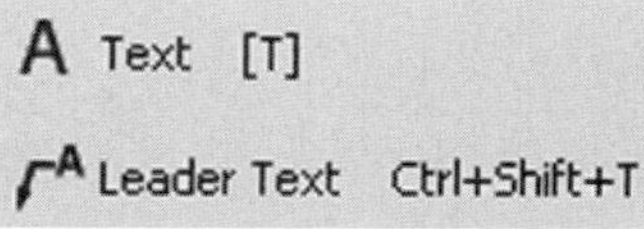

Figure 5.129

The Text tool will add text while the Leader Text tool will add a leader and text. Select the desired text tool, and define the leader points and/or text location. Once you have chosen the location in the graphics window, the Format Text dialog box will appear. When placing text through the Format Text dialog box, select the orientation and text style as needed and type in the text, as shown in the following image. Click the OK button to place the text in the drawing. To edit the text or text leader position, move your cursor over it, click one of the green circles that appear, and drag to the desired location. To edit the text or text leader content, right-click on the text, and select Edit Leader Text or Edit Text from the menu.

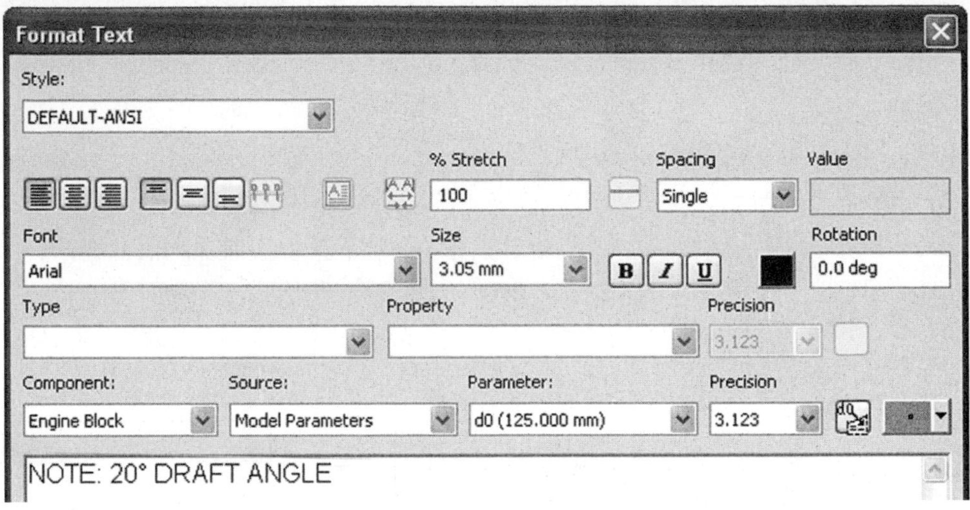

Figure 5.130

ADDITIONAL ANNOTATION TOOLS

To add more detail annotations to your drawing, you can add surface texture symbols, welding symbols, feature control frames, feature identifier symbols, datum identifier symbols, and datum targets by clicking the corresponding tool on the Drawing Annotation panel bar, as shown in the following image.

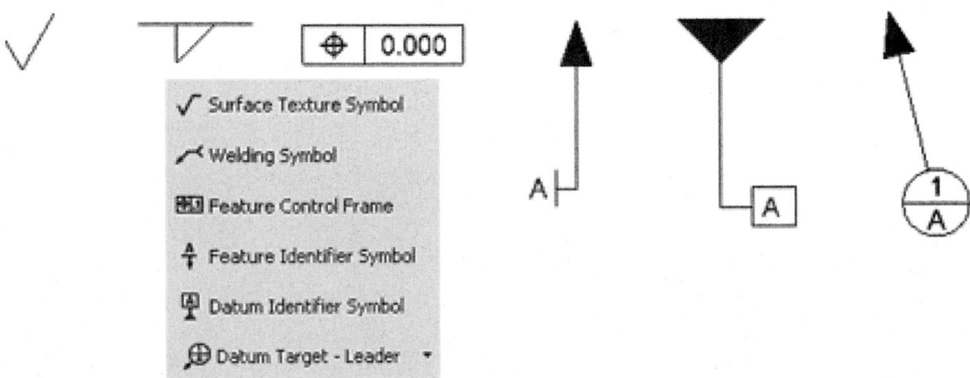

Figure 5.131

Follow these steps for placing these symbols:

1. Click the tool on the Drawing Annotation panel bar.
2. Select a point at which the leader will start.
3. Continue selecting points to position the leader lines.
4. Right-click, and select Continue from the menu.
5. Fill in the information as needed in the dialog box.
6. When done, click the OK button in the dialog box.
7. To complete the operation, right-click, and select Done from the menu.
8. To edit a symbol, position the cursor over it. When the green circles appear, right-click, and select the corresponding Edit option from the menu.

HOLE AND THREAD NOTES

Another annotation you can add is a hole or thread note. Before you can place a hole or thread note in a drawing, a hole or thread feature must exist. You can also annotate extruded circles using the Hole or Thread Note tool. If the hole or thread feature changes, the note will be updated automatically to reflect the change. The following image shows examples of counterbore, countersink, and threaded/tapped hole notes.

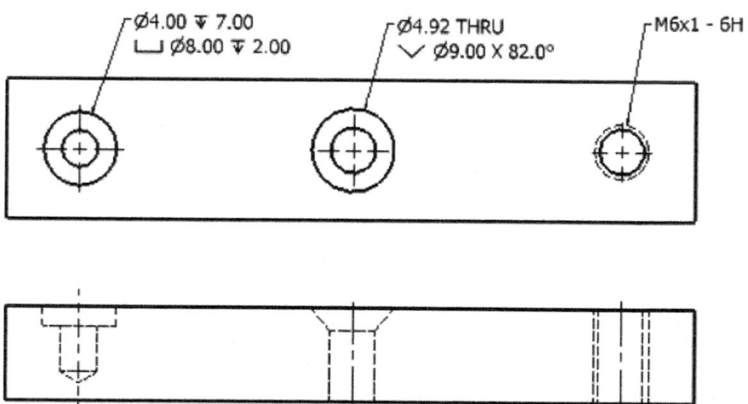

Figure 5.132

To create a hole or thread note, follow these steps:

1. Click the Hole/Thread Notes tool on the Drawing Annotation panel bar, as shown in the following image.

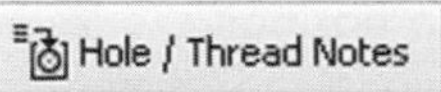

Figure 5.133

2. Select the hole or thread feature to annotate.
3. Select a second point to locate the leader and the note.
4. To complete the operation, right-click, and select Done from the menu.
5. To edit a note, position the cursor over it. When the green circles appear, right-click, and select the corresponding Edit option from the menu.

Hole Note Styles

Hole note styles enable you to define global formatting of a hole note as well as edit existing hole notes and apply specific formatting on an as-needed basis. This gives you the flexibility to conform to a standard or create your own standard for hole notes.

To control a hole note as part of an existing dimension style, follow these steps:

1. In the Style and Standard Editor, select an existing dimension style, and click on the Notes and Leaders tab, as shown in the following image.

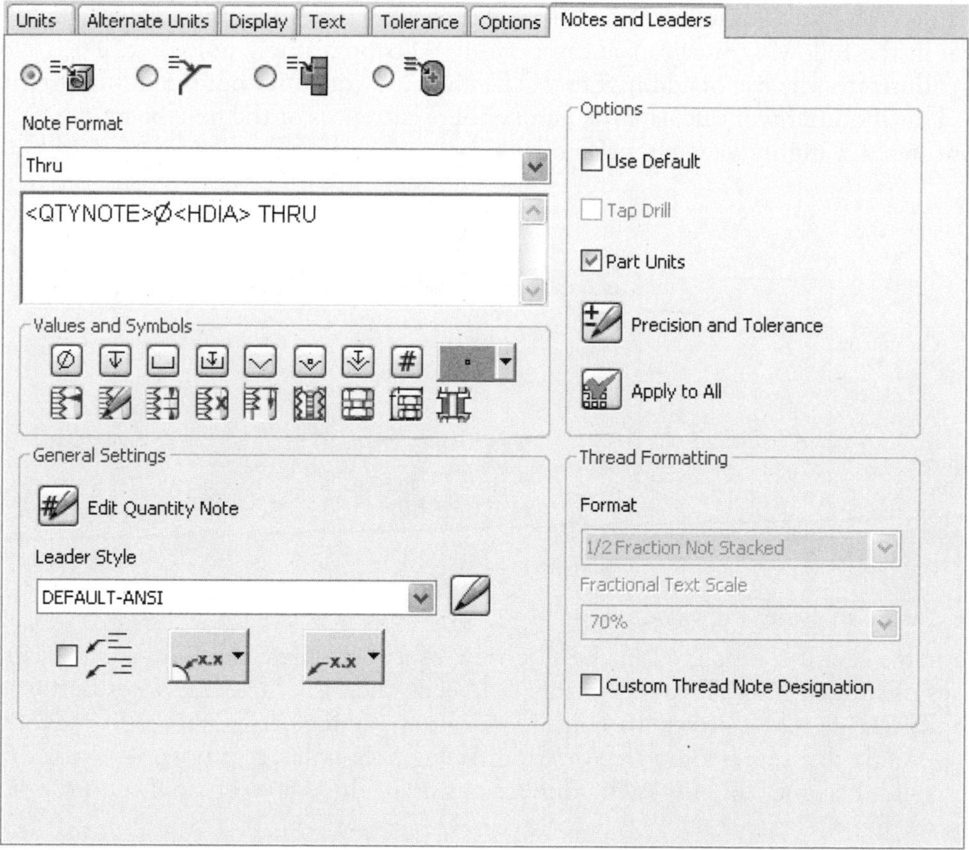

Figure 5.134

2. Either modify the hole note as part of an existing dimension style or click the new button to create a new dimension style.
3. Select the hole type to format from the Hole Type drop-down list.
4. Change the Value or Symbol to reflect the desired hole format.
5. Click the Save button to save any changes to the new or existing dimension style.

Editing Hole Notes

A hole note edit is invoked by right-clicking on an existing hole note to display the menu, as shown in the following image.

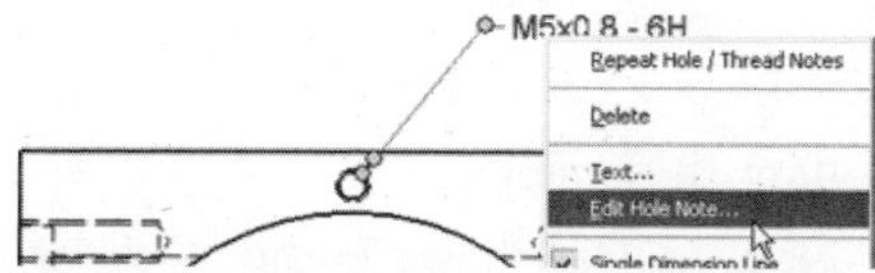

Figure 5.135

Selecting Edit Hole Note from this menu will display the Edit Hole Note dialog box, as shown in the following image. You can perform edits on a single hole note. The following image illustrates the use of adding the <QTYNOTE> (quantity note) modifier to the hole note. This modifier will calculate the number of occurrences of the hole being noted, which is considered a highly valuable productivity tool.

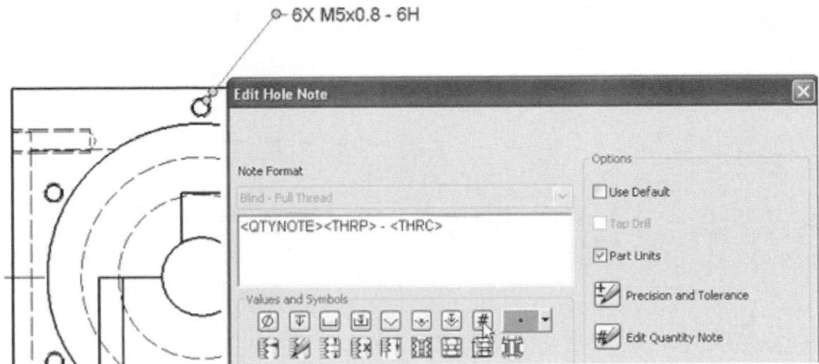

Figure 5.136

Hole Notes in Side Views

Hole notes can also be placed in the side view of a part where the hole is represented by a series of hidden lines. To perform this task, click the Hole/Thread Notes button, select the hidden edge that represents a hole, and locate the hole note, as shown in the following image. While this represents a rare case for placing hole notes, it is possible that a drawing view could become so busy with dimensions that this type of application would be necessary for clarity.

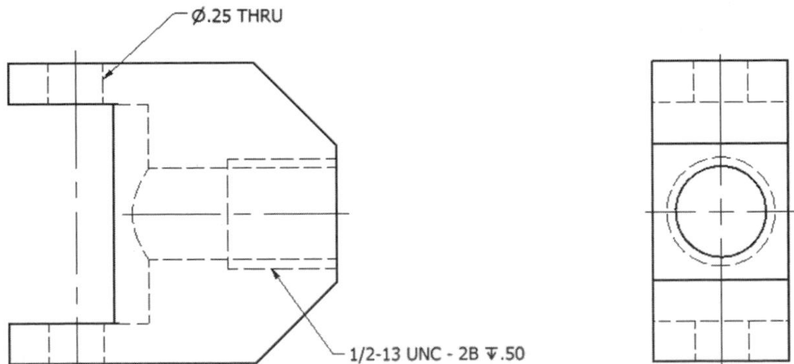

Figure 5.137

THE CHAMFER NOTE TOOL

To place a chamfer note, follow these steps:

1. Click the Chamfer Note tool, as shown in the following image, located in the Drawing Annotations panel bar.

Figure 5.138

2. In the drawing view, select the chamfered line or edge, as shown in the following image on the left.
3. In the same drawing view, select a reference line or edge that shares end points or intersects with the chamfer, as shown in the following image on the left. Parallel or perpendicular lines cannot be selected.
4. Click to place the chamfer note. The default attachment point is the midpoint of the chamfer, as shown in the following image on the right. After the chamfer note is created, you can click the attachment point, and drag the point to a new location on the same edge.

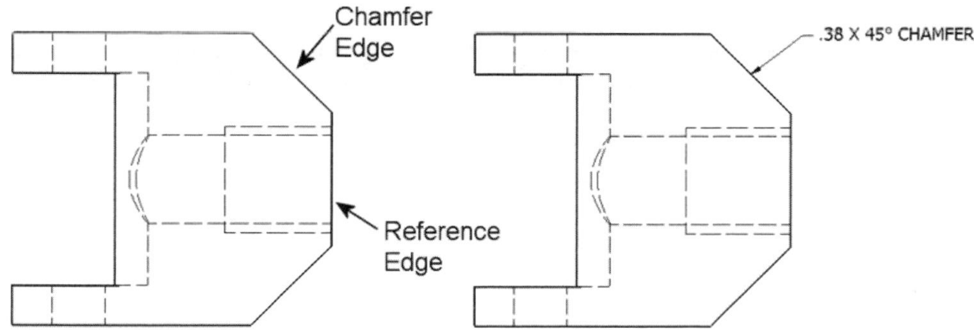

Figure 5.139

Once a chamfer note is placed, it can be easily edited. Double-clicking on the chamfer note will launch the Edit Chamfer Note dialog box, as shown in the following image. In this example, the text (TYP) has been added to the end of the note, signifying a typical chamfer distance. You can also add extra information to this note by clicking on the buttons under Values and symbols. The three buttons will add Distance 1, Distance 2, and Angle information to the chamfer note.

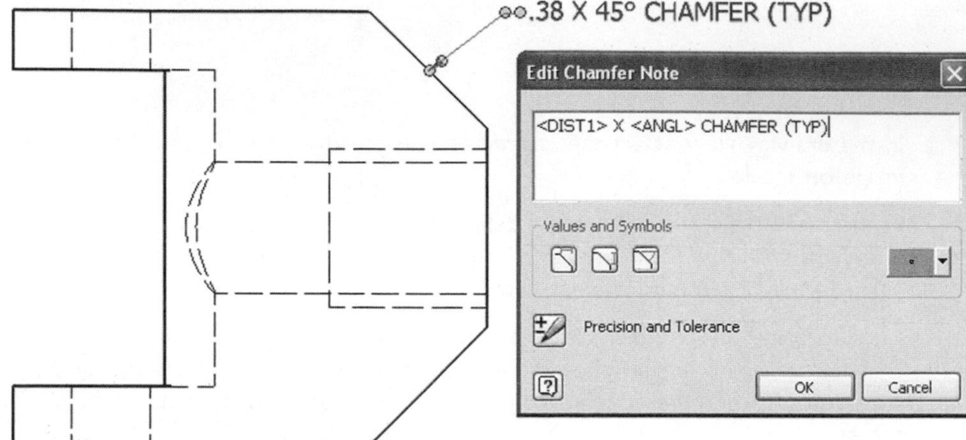

Figure 5.140

EXERCISE 5-5: ADDING DIMENSIONS AND ANNOTATIONS

In this exercise, you add dimensions and annotations to a drawing of a clamp that is used to hold a work piece in position during machining operations.

Both model dimensions and drawing dimensions are used to document feature size.

1. Open the drawing *ESS_E05_05.idw.* The drawing file contains a single sheet with 4 drawing views.

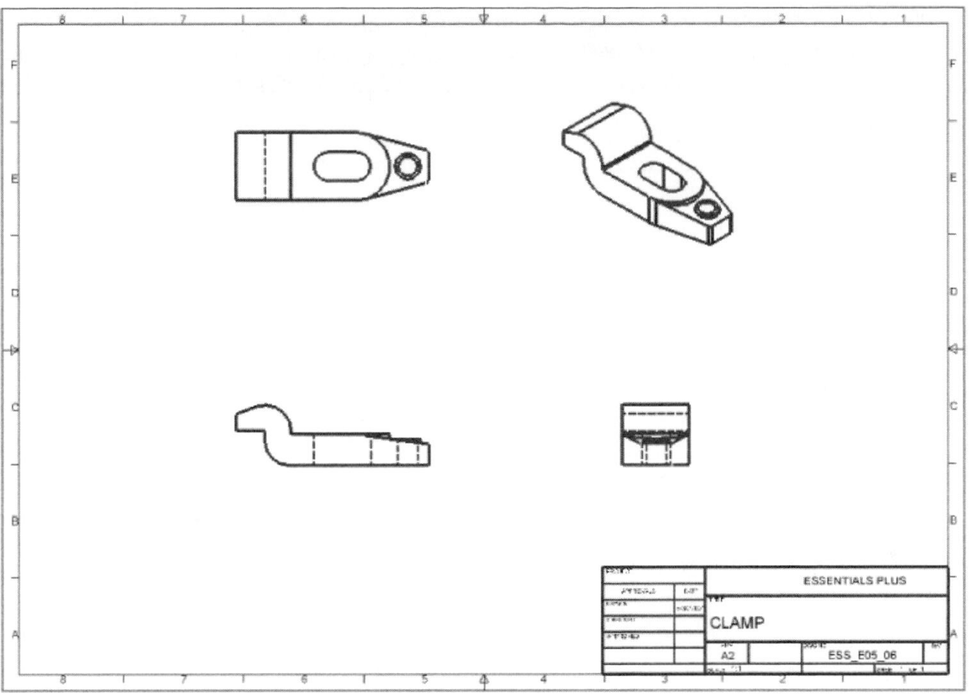

Figure 5.141

2. In this step, you retrieve model dimensions for use in the drawing view. Begin by zooming in on the front view.
3. Right-click the front view, and choose Retrieve Dimensions from the menu. This will launch the Retrieve Dimensions dialog box.
4. In this dialog box, click the Select Dimensions button. The model dimensions will appear on the view.

EXERCISES

5. Drag a selection box around all dimensions. This will select the dimensions to retrieve. When finished, click the OK button. The model dimensions that are planar to the view are displayed, as shown in the following image:

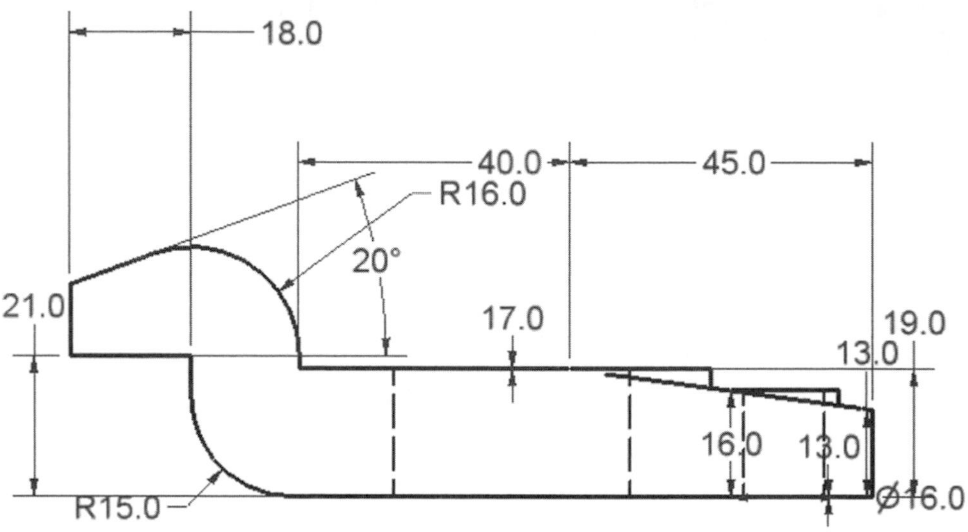

Figure 5.142

6. You will now delete certain model dimensions that do not accurately describe the feature being dimensioned. First hold down the CTRL key, and select the **17.0**, **13.0**, and **16.0** diameter dimensions, as shown in the following image on the left. Right-click, and choose Delete from the menu. Your display should appear similar to the following image on the right.

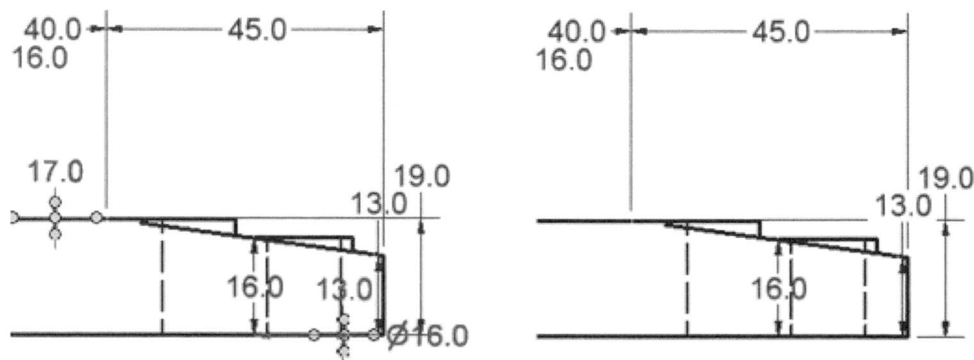

Figure 5.143

7. Reposition the model dimensions to better locations by dragging the dimensions until they appear as shown in the following image.

EXERCISES

Note: To reposition dimension text, click a dimension text object, and drag it into position. The dimension will be highlighted when it is a preset distance from the model.

Radial dimensions can be repositioned by selecting the handle at the annotation end of the leader.

To drag dimensions, make sure that no command is active.

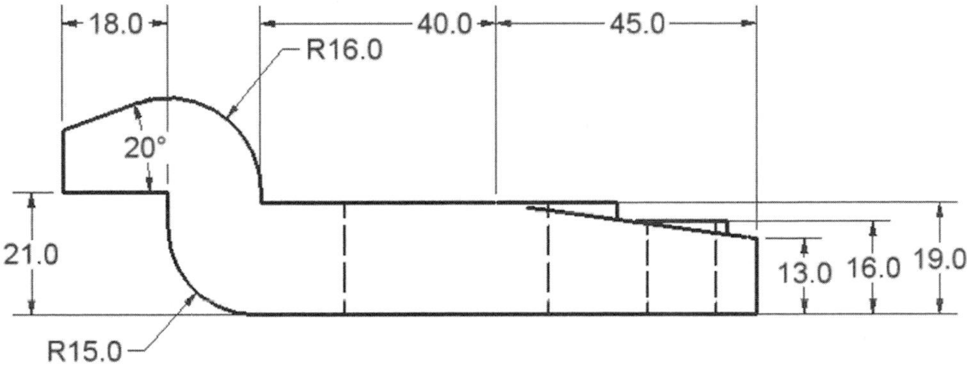

Figure 5.144

8. Centerlines and center marks need to be added to aid in the placement of drawing dimensions. To display the center mark annotation tools, click the title area of the panel bar, and select Drawing Annotation panel.

9. Click the Centerline Bisector tool in the panel bar (it may be under the Center Mark tool), and select the two hidden lines that represent the drilled hole through the boss. Your display should be similar to the following image.

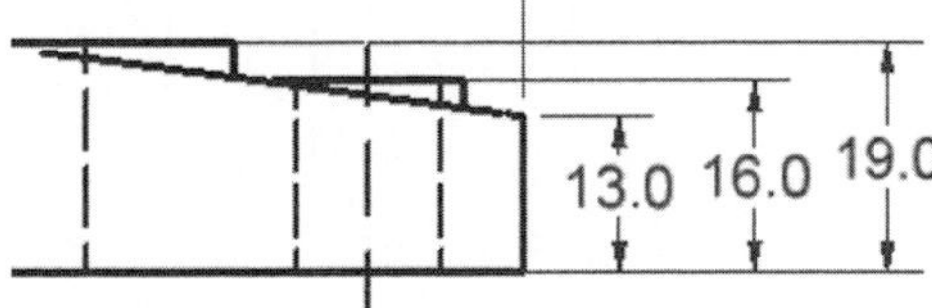

Figure 5.145

10. Pan to display the top view. Then click the Center Mark tool in the panel bar.

11. Click the outer circle of the boss and the two arcs of the slot to place center marks on these features, as shown in the following image.

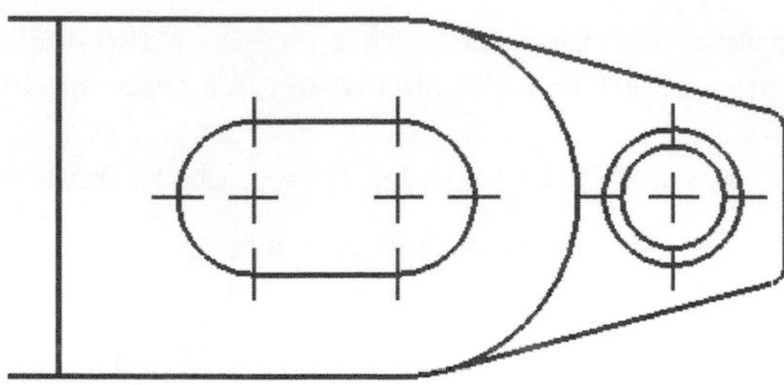

Figure 5.146

12. Next manually add dimensions to the top view. Pan to display the top view.

13. Click the General Dimension tool in the panel bar.

14. Click the endpoints of the centerlines in the slot, and drag above the model to place the **17.0** dimension.

15. Click the arc on the slot, and place the **9.0** radial dimension, as shown in the following image.

16. Place the **16.0** diameter dimension on the boss.

17. Click the top horizontal line, then click the sloped line and place the **15°** angular dimension. Your display should appear similar to the following image.

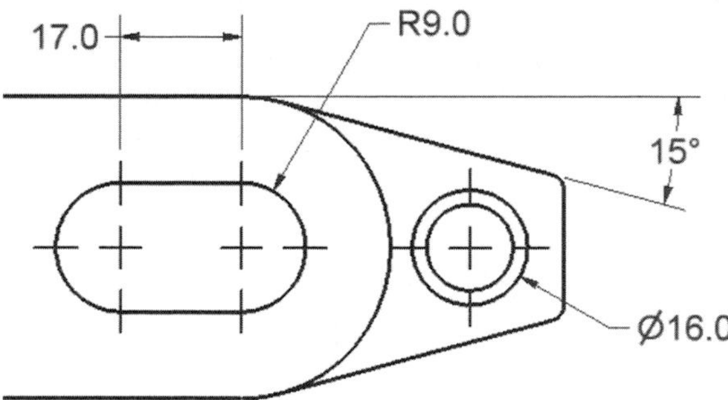

Figure 5.147

18. Add the **13.0** horizontal dimension, as shown in the following image.

Tip: To align a dimension when dragging it, move the cursor over an existing dimension line, and acquire an alignment point. Move the cursor back to the dimension being placed. The dotted line indicates an alignment inference. Click to place the dimension.

19. Add the **21.0** radial dimension. When finished, right-click and choose Done.

20. Next, you will move two dimensions from the front view to the top view. Zoom out until you are able to see both the front and top views. Right-click on the **40.0** horizontal dimension in the front view, and pick Move Dimension from the menu. Then pick anywhere inside of the top view. After the **40.0** dimension appears in the top view, it will need to be moved to a new location. First drag this dimension to the bottom of the top view. Then drag the ends of the extension line to intersect with the outside of the centerline and the part. Perform these same steps on the **45.0** horizontal dimension located in the front view. Your display should appear similar to the following image.

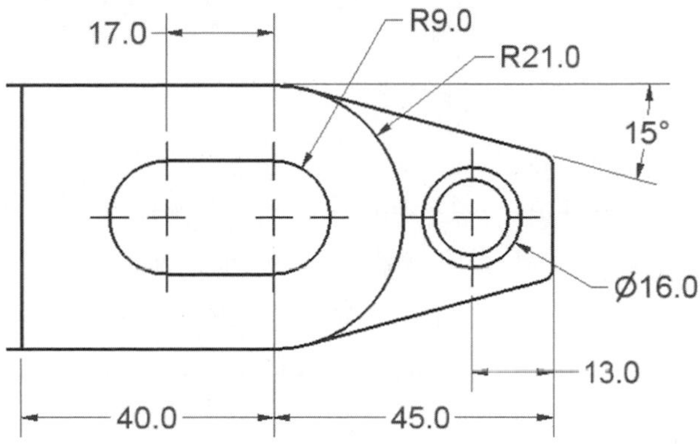

Figure 5.148

21. You need to add a note to the end of the angle dimension. Right-click the 15° dimension, and choose Text.
22. At the insertion point, press the space bar then type **TYP**, as shown in the following image on the left, and click OK. Your display should appear similar to the following image on the right.

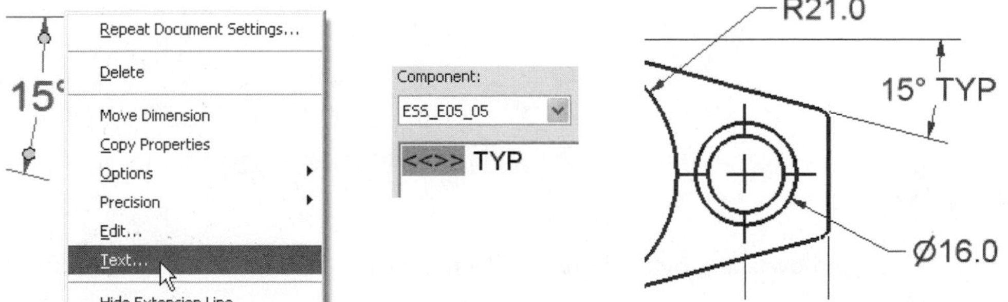

Figure 5.149

23. Additional notes need to be added to other dimensions. Right-click the 16.0 diameter dimension, and choose Text.
24. At the insertion point, press the space bar, enter **BOSS**, and press ENTER.
25. Select the diameter symbol from the symbol list in the dialog box, as shown in the following image on the left, and type **12.0 THRU**. When finished, click OK. Your display should appear similar to the following image on the right.

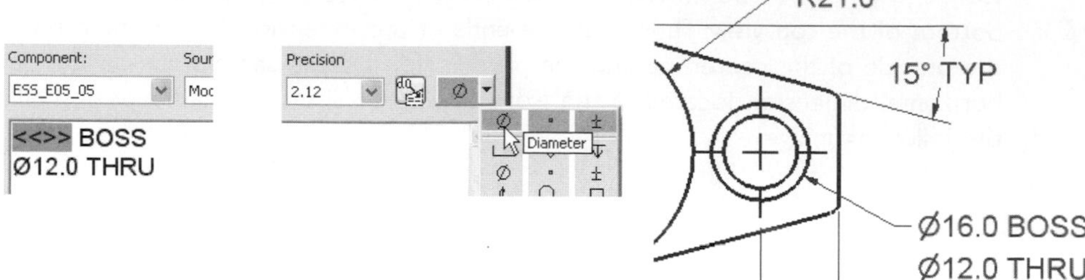

Figure 5.150

26. In the following steps, you add a general note. First click the Text tool in the panel bar.
27. Click a point below and to the right of the top view.
28. In the text entry area, enter **TOLERANCE FOR**, and press ENTER.
29. On the next line, enter **ALL DIMENSIONS**, press the space bar, and select ± from the symbol list.
30. Type **0.5,** and click OK. Right-click, and choose Done. Your display should appear similar to the following image.

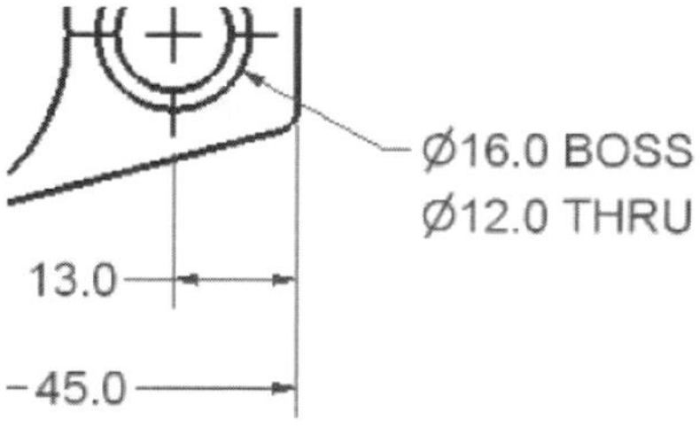

TOLERANCE FOR
ALL DIMENSIONS ± 0.5

Figure 5.151

31. You will now add a radial dimension. First click the General Dimension tool in the panel bar.
32. Select the bottom arc on the right end to define the leader start point. You may have to zoom into this area to assist in selecting the arc.
33. Click a point below and to the right to define the end of the leader, and place the dimension.
34. Double-click on this dimension, enter **ROUNDS** in front of the <<>> in the Edit Dimension dialog box, and click OK. Your display should appear similar to the following image. You may have to move the 13.0 dimension text to a more convenient location.

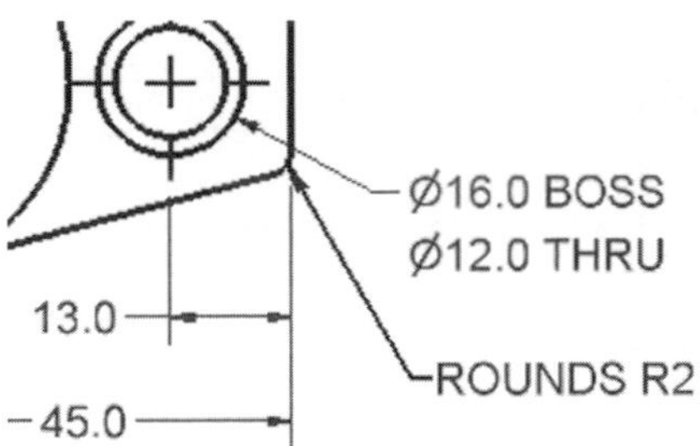

TOLERANCE FOR
ALL DIMENSIONS ± 0.5

Figure 5.152

35. You will now finish documenting this drawing by using drawing properties to complete the title block information.

36. From the File menu, select iProperties. This will display the Drawing Properties dialog box.

37. On the Summary tab, enter your name in the Author field.

38. Click on the Status tab, and select the current date from the Checked Date list.

39. Enter your initials in the Checked By field, and click OK to update the title block. Zoom in to the title block to examine the changes.

40. Close all open files. Do not save changes. End of exercise.

OPENING A MODEL FROM A DRAWING TO EDIT

While working inside of a complex drawing view, you may need to make changes to a part or assembly file. Rather than close down the drawing and open the individual part file, you can open a part or assembly file directly from the drawing manager. Use the following steps to accomplish this task:

One technique is to right-click a drawing view to display the menu in the following image, and then click Open to open the model file. In this example, notice that an assembly file was opened from the drawing view.

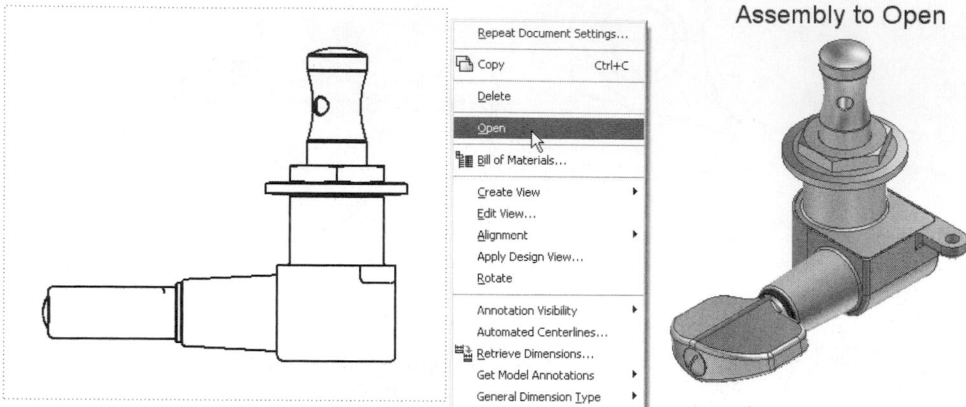

Figure 5.153

Another technique involves expanding the drawing view in the browser to expose the assembly. Then continue expanding the browser until the part you want to open is visible in the list. Right-click the desired part file in the browser and select Open from the menu to display the part model as shown in the following image.

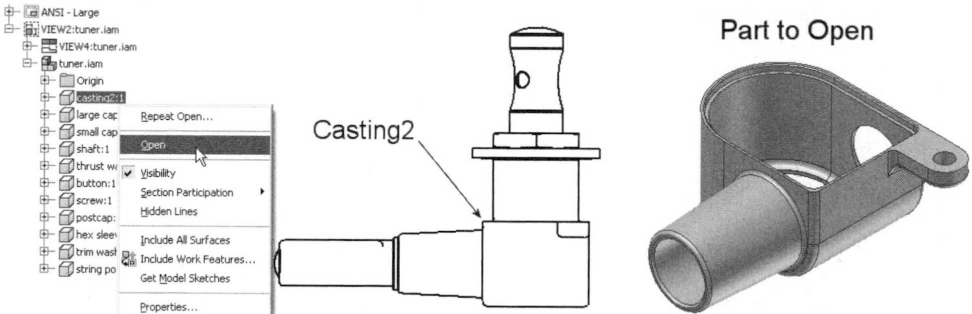

Figure 5.154

A third technique used to open a part file from a drawing is to first change the selection filter to Part Priority. After doing so, right-click on the specific part in the drawing view and select Open from the menu, as shown in the following image. This will open the source part file.

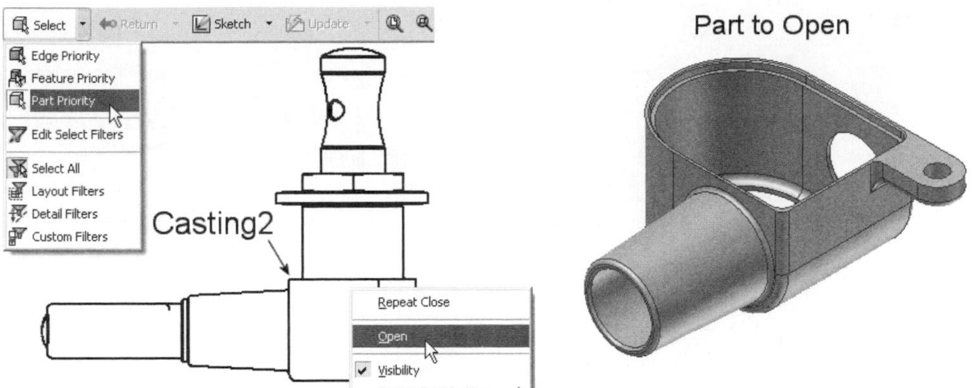

Figure 5.155

MANAGING DRAWING SHEETS

An Autodesk Inventor drawing file can have more than one sheet. A multiple-sheet drawing would typically be used when the model cannot be represented on a single sheet. Each sheet is numbered, and a count of the total number of sheets is maintained. This information can be displayed in the title block; for example, the title block can indicate that the current sheet is Sheet:2 of a five-sheet drawing.

CREATING MULTIPLE SHEETS

You can add sheets to a drawing by activating commands located in the panel bar, from dedicated menus, or through tools accessed in the browser.

- On the drawing Views panel bar, click the New Sheet button, as shown in the following image at "A."
- From the Insert menu, select Sheet, as shown in the following image at "B." Notice that you can also create a new sheet by holding down the SHIFT key and pressing N.
- Right-click the sheet background, and select New Sheet from the menu, as shown in the following image at "C."
- Right-click in the browser background and select New Sheet from the menu, as shown in the following image at "D."

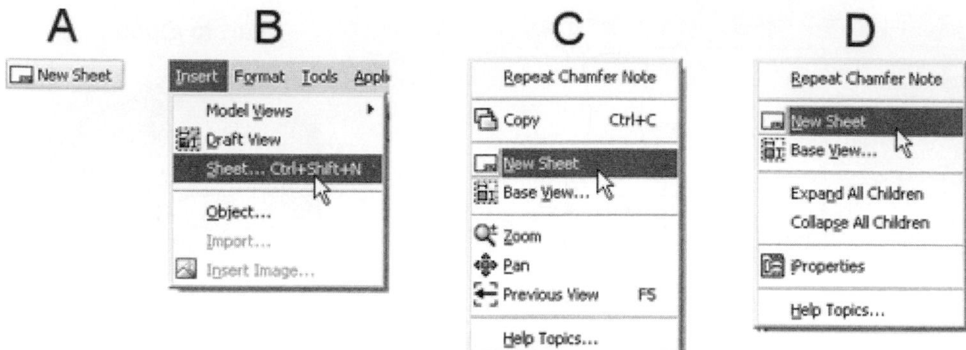

Figure 5.156

You can also create a new sheet with a predefined layout. In the browser, expand Drawing Resources > Sheet Formats to list the defined formats. Pick a format, right-click, and select New Sheet, as shown in the following image. Next browse to select the model to document on the new sheet in the resulting Select Component dialog box.

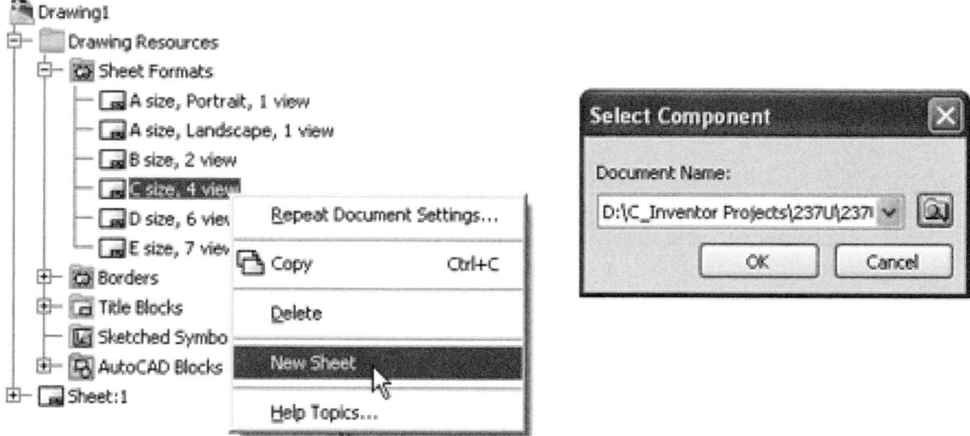

Figure 5.157

You can create a new sheet with an existing layout from another drawing. Right-click a sheet border in the graphics window or the browser, and select Copy, as shown in the following image.

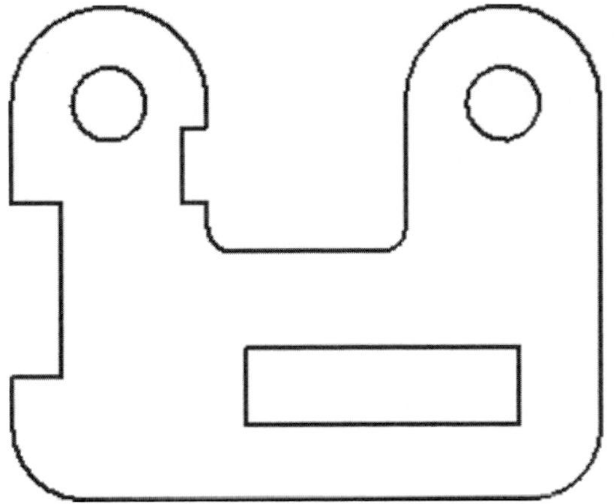

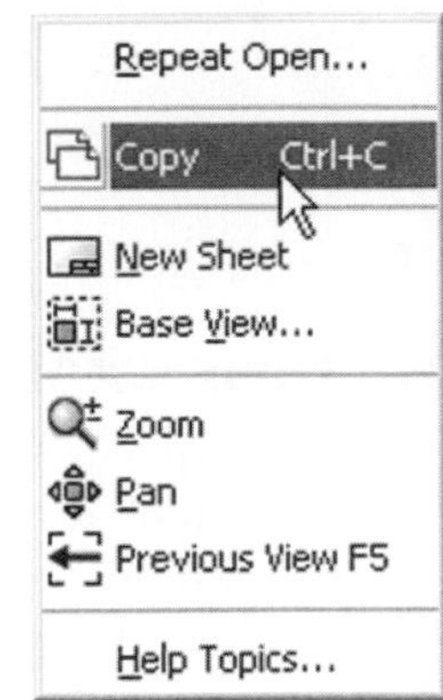

Figure 5.158

Switch to another open drawing, right-click the drawing name in the browser, and select Paste, as shown in the following image. A copied sheet contains the same views as the source.

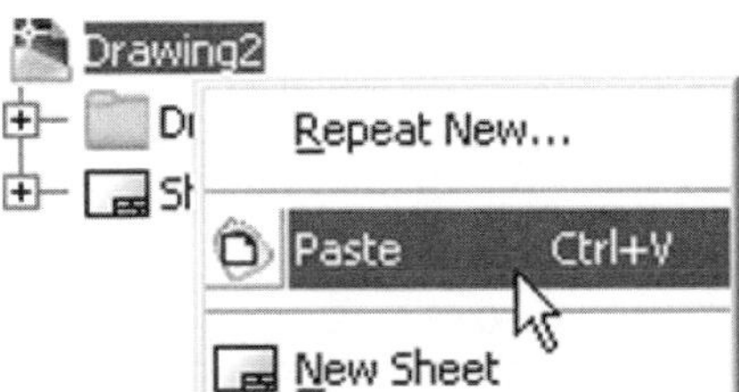

Figure 5.159

Note: You cannot copy and paste a sheet within the same drawing.

While numerous drawing sheets can be created, only one sheet can be active at a time. In the following image, Sheet:2 is currently the active sheet while Sheet:1 and Sheet:3 are inactive, as identified by the gray background. To make another sheet active, double-click on the sheet in the browser, or right-click on the sheet in the browser, and select Activate.

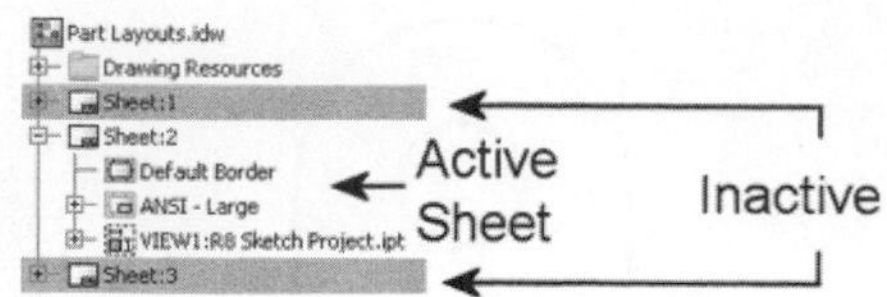

Figure 5.160

BASELINE DIMENSIONS

To add multiple drawing general dimensions to a drawing view in a single operation, use the Baseline Dimension tool, which will place a baseline dimension about the selected geometry. The dimensions can be either horizontal or vertical. Two tools are available to assist with the creation of these dimensions: Baseline Dimension and Baseline Dimension Set.

The Baseline Dimension tool allows you to create a group of horizontal or vertical dimensions from a common origin. However, the group of dimensions is not considered one object; rather, each dimension that makes up a Baseline Dimension is considered a single object.

To create baseline dimensions, follow these steps:

1. Select the Baseline Dimension tool from the Drawing Annotation panel bar, as shown in the following image.

Figure 5.161

2. Define the origin line by selecting an edge of the object to be dimensioned. This edge will act as the baseline for other dimensions.
3. Individually select or drag a selection window around the geometry that you want to dimension. To window select, move the cursor into position where the first point of the box will be. Press and hold down the left mouse button, and move the cursor so the preview box encompasses the geometry that you want to dimension, as shown in the following image, and then release the mouse button.

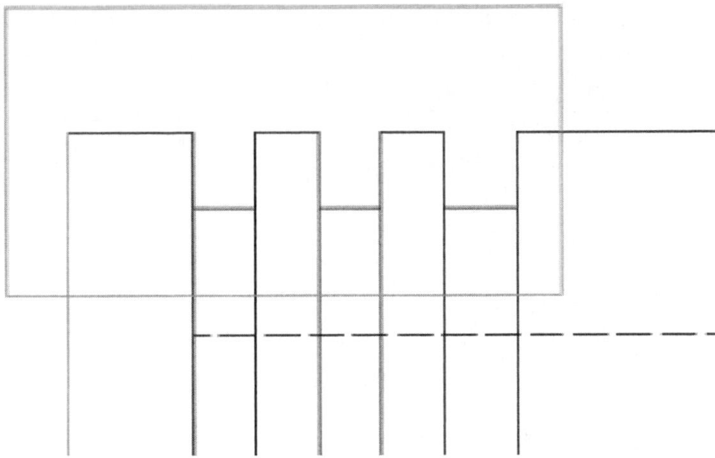

Figure 5.162

4. When you are finished selecting geometry, right-click, and select Continue from the menu.
5. Move the cursor to a position where the dimensions will be placed, as shown in the following image. When the dimensions are in the correct place, click the point with the left mouse button to anchor the dimensions.

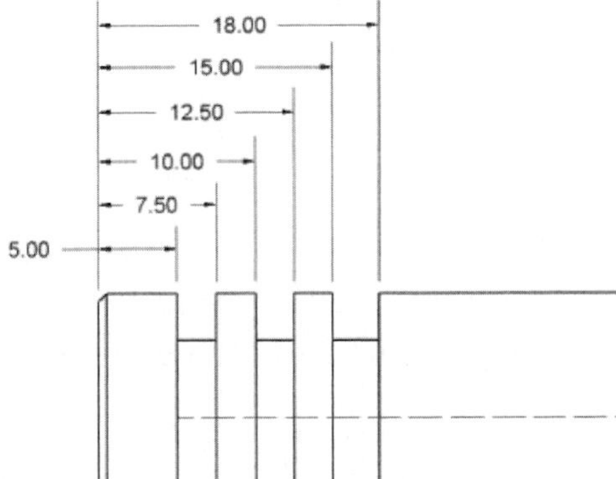

Figure 5.163

6. To complete the operation, either press the ESC key or right-click and select Done from the menu.

After creating dimensions using the Baseline Dimension tool, various edit operations can be performed beyond the basic dimension options. To perform these edits, move the cursor over the baseline dimension, and small green circles will appear on the dimension. Right-clicking will display the menu, as shown in the following image.

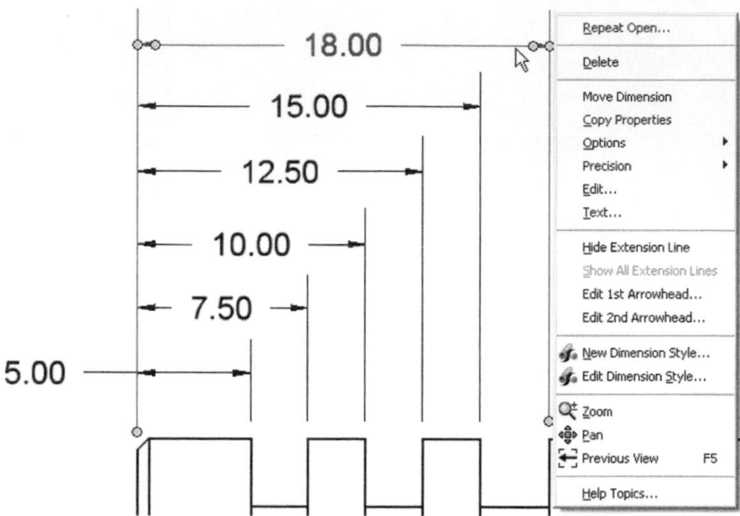

Figure 5.164

BASELINE DIMENSION SET

Creating a Baseline Dimension Set is similar to the Baseline Dimension as explained in the previous example. With the Baseline Dimension Set tool, however, dimensions are grouped together into a set. Select the Baseline Dimension tool from the Drawing Annotation panel bar, as shown in the following image.

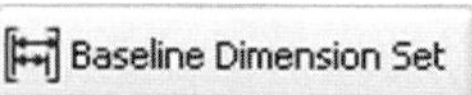

Figure 5.165

Members can be added and deleted from the set, a new origin can be placed, and the dimensions can be automatically rearranged to reflect the change. Right-clicking on any dimension in the set will highlight all dimensions and display the menu, as shown in the following image.

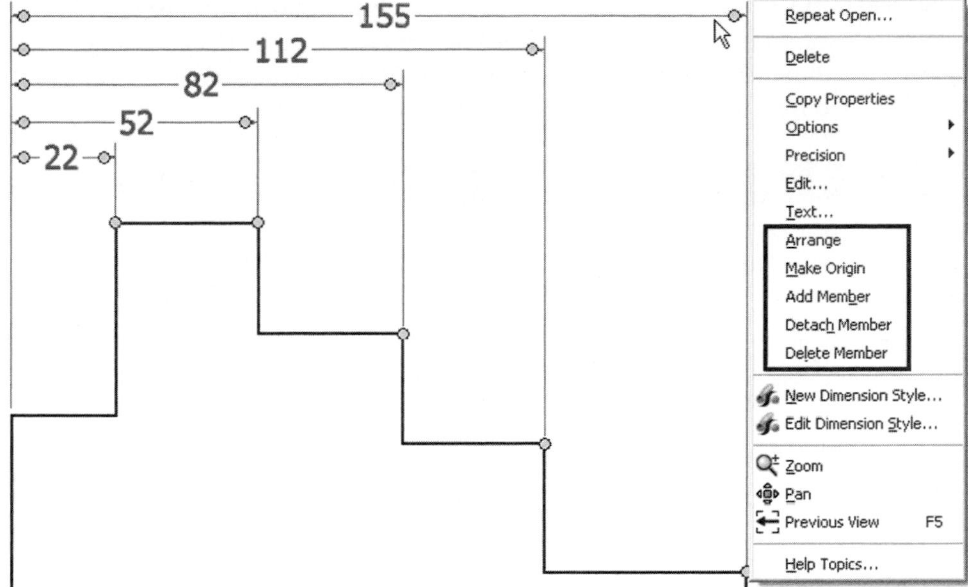

Figure 5.166

Options for editing a baseline dimension set are described as follows:

Delete This option will delete all baseline dimensions in the set. Move your cursor over the top of the baseline dimension set. When the green circles appear on each dimension, right-click, and select Delete from the menu.

Arrange This tool will rearrange the dimensions to the spacing defined in the Dimension Style.

Make Origin This option will change the origin of the baseline dimensions. Move the cursor over the top of the baseline dimension set. When the green circles appear on each dimension, select an extension line that will define the new baseline, right-click, and select Make Origin from the menu.

Add Member This option will add a drawing dimension to the baseline dimensions. After creating a drawing dimension, move the mouse over the baseline dimensions, and when the green circles appear on each dimension, right-click, and select Add Member from the menu. Now select the drawing dimension. The drawing dimension will be added to the set of baseline dimensions and rearranged in the proper order.

Detach Member This option will remove a dimension from the set of baseline dimensions. Move your cursor over the top of the baseline dimension set. When the green circles appear on each dimension, right-click on the green circle on the dimension that you want to detach, right-click, and select Detach Member from the menu. The detached dimension can be moved and edited like a normal drawing dimension.

Delete Member This option will erase a dimension from the set of baseline dimensions. Move your cursor over the top of the baseline dimension set. When the green circles appear on each dimension, right-click on the dimension that you want to erase, and then select Delete Member from the menu. The dimension will be deleted.

EXERCISE 5–6: CREATING BASELINE DIMENSIONS

In this exercise, you create a drawing for the shim plate shown and then add annotations using baseline dimensions. You create and apply dimension styles to alter the appearance of the dimensions.

1. Open ESS_E05_06.idw.
2. Create a new dimension style based on the existing Default-ISO style.

- Choose Format > Style and Standard Editor.
- When the Style and Standard Editor dialog box appears, expand the Dimension category.
- Select DEFAULT-ISO from the list.
- Click the New button. Change the newly created style name from Copy of DEFAULT-ISO to Baseline-ISO.
- Click the OK button.
- Change the characteristics of the new dimension style.
- Under the Units tab, change the Decimal Marker to . (period).
- ISO dimension standards require that dimension text appear aligned with and above dimension lines. You can easily change the text orientation to that used by another standard, as in this case, ANSI.
- Click the Text Tab.
- Under Orientation, click Horizontal Dimensions.
- Click the button to place text inside the dimension lines, as shown in the following image on the left.
- Click Vertical Dimensions.
- Click the button to rotate the text horizontally and force it inside the dimension lines, as shown in the following image on the right.
- Click the Save button to save the changes to the new style.
- Click Done to exit the Style and the Standard Editor dialog box.

Note: You can easily compare settings between existing styles by clicking different styles names while viewing the tab settings.

3. You will now add the first baseline dimension set. Baseline dimensions are used to

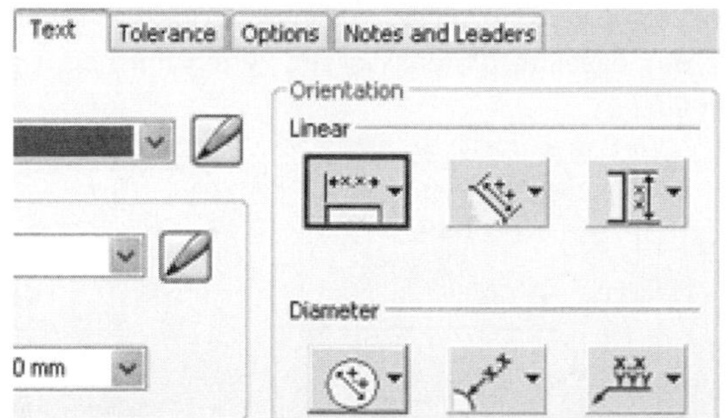

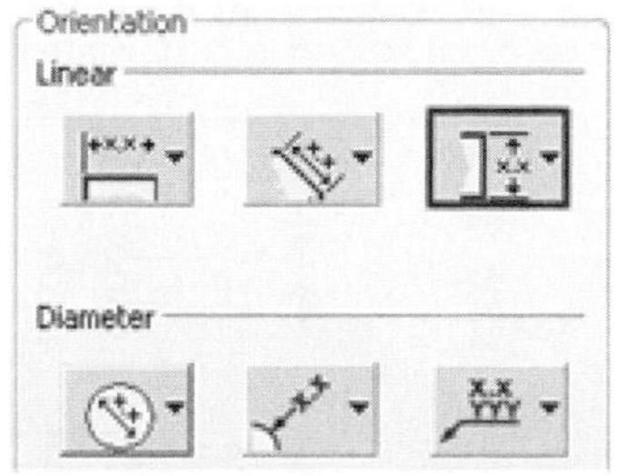

Figure 5.167

quickly annotate the drawing. First select the Drawing Annotation panel bar, and then click the Baseline Dimension Set tool. Before picking edges for the baseline dimension set, you must make the Baseline-ISO dimension style current. Perform this operation by clicking the arrow in the Dimension Style edit box on the Standard toolbar. Change By Standard (DEFAULT-ISO) to Baseline-ISO, as shown in the following image.

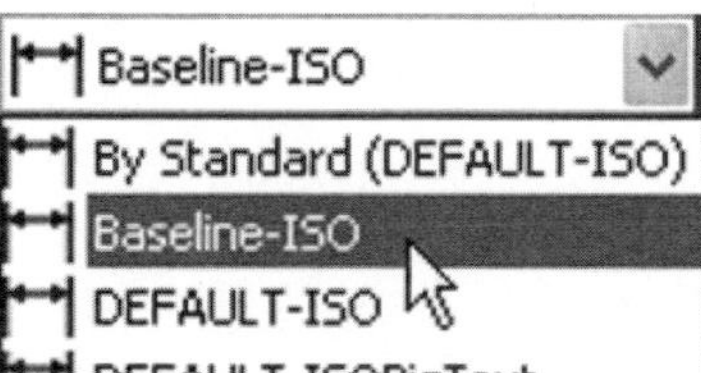

Figure 5.168

4. Continue with the Baseline Dimension Set tool, and select the part edges, as shown in the following image on the left. Right-click, and choose Continue. Click to set the position of the dimension set. Right-click, and pick Create from the menu. Notice the creation of the first baseline dimension set. However, all baseline dimensions need to reference the opposite edge of the object. To perform this task, move the cursor over the extension line, as shown in the following image in the middle. Right-click, and choose Make Origin from the cursor menu. Notice how the dimensions are regenerated from the new origin, as shown in the following image on the right.

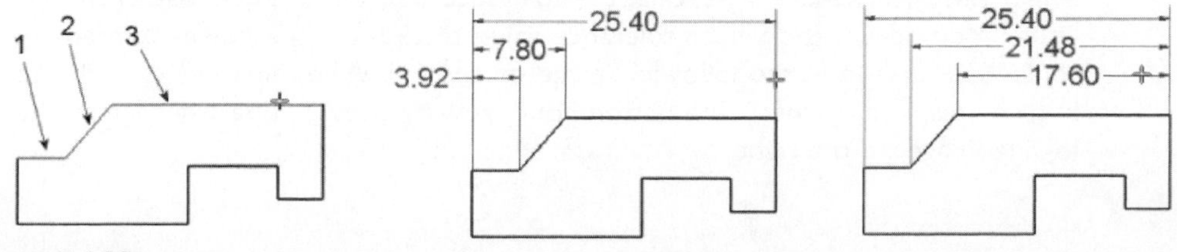

Figure 5.169

5. With the first baseline dimension set created, click on the Baseline Dimension Set tool again, make the Baseline-ISO dimension style current, and select the part edges, as shown in the following image on the left. Right-click, and pick Continue from the menu. Click to set the position of this baseline dimension set. When finished, right-click, and pick Create from the menu.

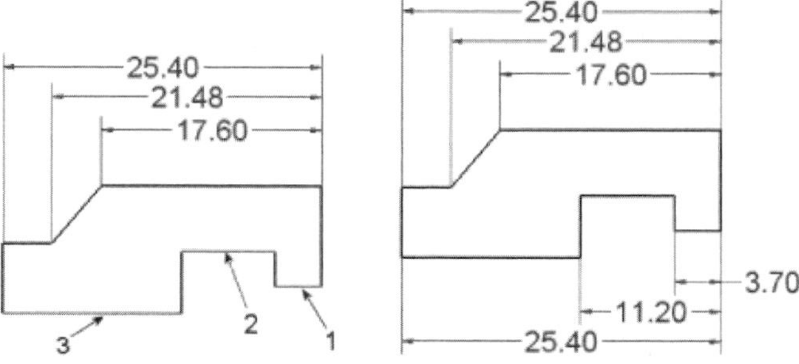

Figure 5.170

6. There are now duplicate dimensions at the top and bottom of the part. The next step describes how to remove one of them even though they are both members of baseline sets. Right-click the **25.40** dimension at the bottom of the part, as shown in the following image on the left. Then click Delete Member from the menu. Your drawing should appear similar to the following image on the right.

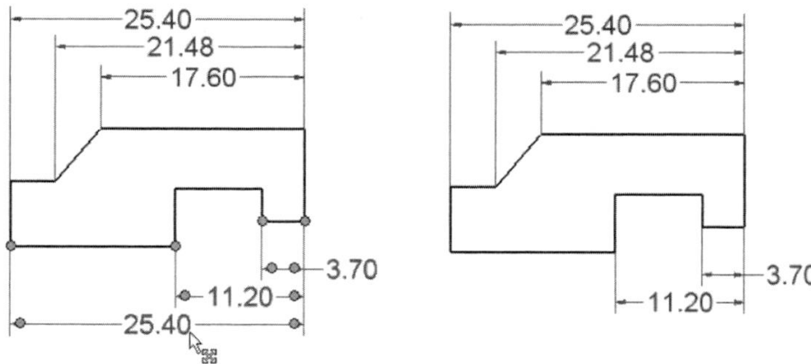

Figure 5.171

7. You will now modify all baseline sets by adding tolerances to each dimension. First choose Format > Style and Standard Editor and expand the Dimension category. Select the Baseline-ISO style, click the Tolerance tab, and then choose Deviation under Method. Set the Upper tolerance value to **.50** and the Lower tolerance value to **1.00**, as shown in the following image on the left. When finished, click the Save button, followed by the Done button. Your drawing should appear similar to the following image on the right.

EXERCISES

Note: You could also right-click on one of the dimensions and select Edit Dimension Style to make these changes.

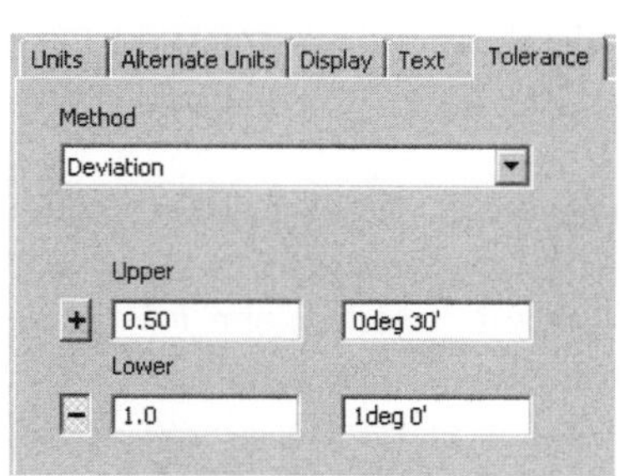

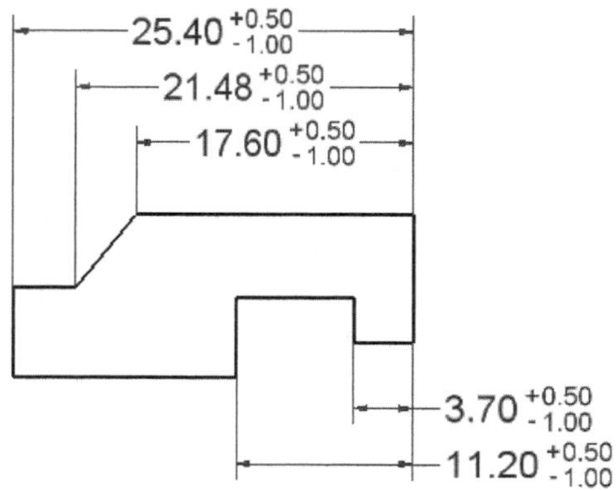

Figure 5.172

8. The overall length of the shim plate need not be manufactured to tolerances. Double-click the **25.40** dimension at the top of the part, switch to the Precision and Tolerance tab of the Edit Dimension dialog box, and click Basic from the Tolerance Method list, as shown in the following image on the left. Your drawing should appear similar to the following image on the right.

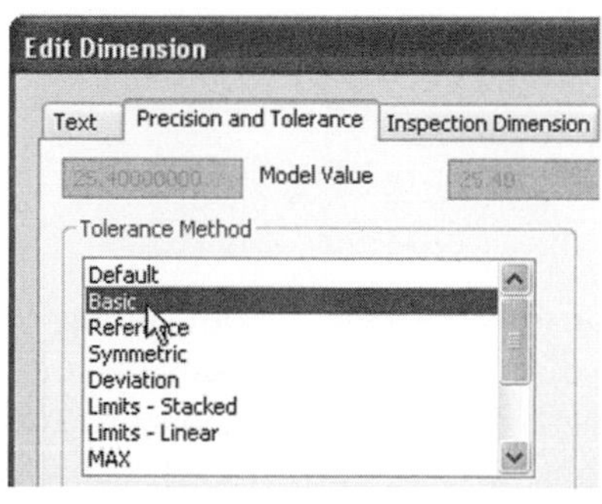

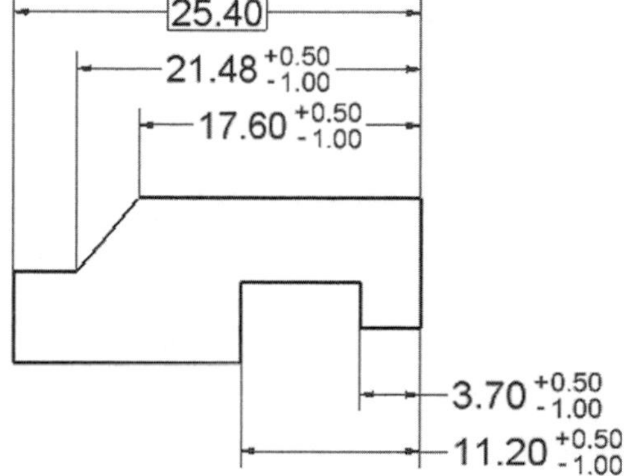

Figure 5.173

9. Two vertical dimensions need to be added to the part. Click the General Dimension tool, change the dimension style to Baseline-ISO, and add the vertical dimensions, as shown in the following image.

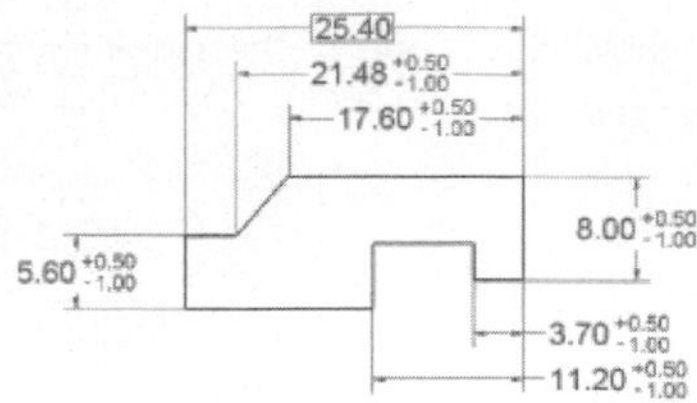

Figure 5.174

10. Complete the drawing by adding a vertical dimension to the side view, as shown in the following image on the left. To remove the tolerance from the dimension, double-click the dimension, and then choose Default from the Precision and Tolerance tab of the Edit Dimension dialog box. Your dimension should appear similar to the following image in the middle. Add the final horizontal dimension, as shown in the following image on the right.

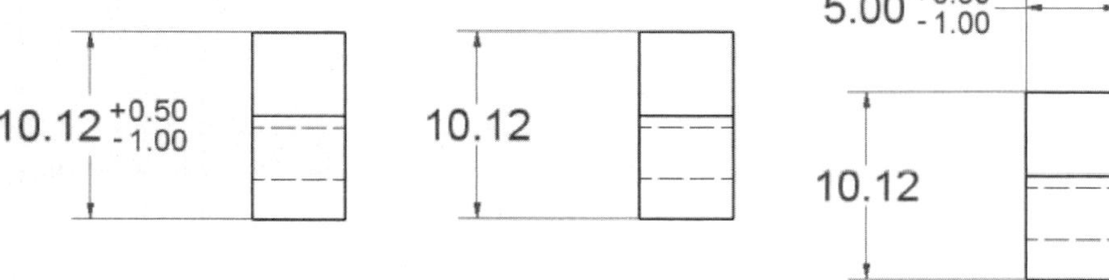

Figure 5.175

11. Close all open files. Do not save changes. End of exercise.

ORDINATE DIMENSIONS

Ordinate dimensions are used to indicate the location of a particular point along the X- or Y-axis from a common origin point. This type of dimensioning is especially suited for describing part geometry for numerical control tooling operations. Two tools are available for creating ordinate dimensions: Ordinate Dimension Set and Ordinate Dimension.

In an ordinate dimension set, all the ordinate dimensions that are created in a single operation will be grouped together and can be edited individually or as a set. When creating an ordinate dimension set, the first dimension created will be used as the origin. The origin dimension needs to be a member of the set and can later be changed to a different dimension. If the location of the origin or the origin member changes, the other members will be updated to reflect the new location.

Ordinate dimensions created with the Ordinate Dimension tool are recognized as individual objects, and an origin indicator will be created as part of the operation. If the origin indicator location is moved, the other ordinate dimensions will be updated to reflect the change.

Both methods will create drawing dimensions that reference the geometry and will be updated to reflect any changes in the geometry to which they are dimensioned. Ordinate dimensions can be placed on any point, center point, or straight edge. Before adding ordinate dimensions to a drawing view, create or edit a dimension style to reflect your standards. When you place ordinate dimensions, they will automatically be aligned to avoid interfering with other ordinate dimensions.

To create an ordinate dimension set, follow these steps:

1. Create a drawing view.
2. Select the Ordinate Dimension Set tool from the Drawing Annotation panel bar, as shown in the following image.

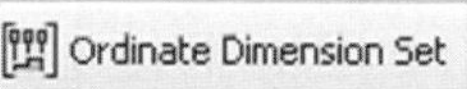

Figure 5.176

3. Select the desired dimension style from the Style list in the Standard toolbar.
4. Select a point to set the origin for the dimensions.
5. Select a point on the drawing view to locate the dimension, and then move the mouse to place the dimension in a horizontal or vertical orientation. Click to place the dimension.
6. Continue selecting points and placing dimensions that will make up a dimension set, as shown in the following image.

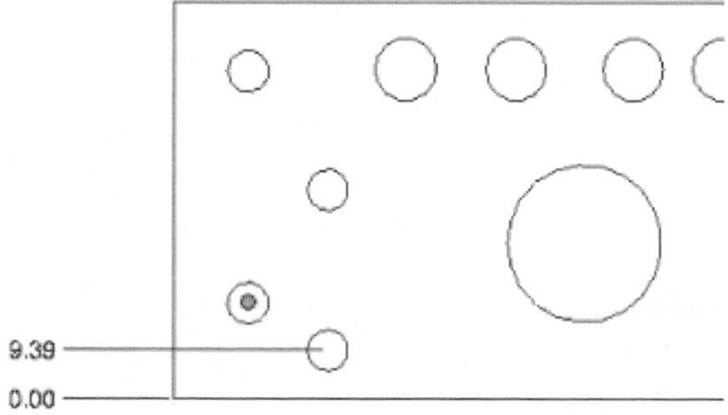

Figure 5.177

7. To change options for the dimension set, right-click at any time, and then select Options, as shown in the following image.

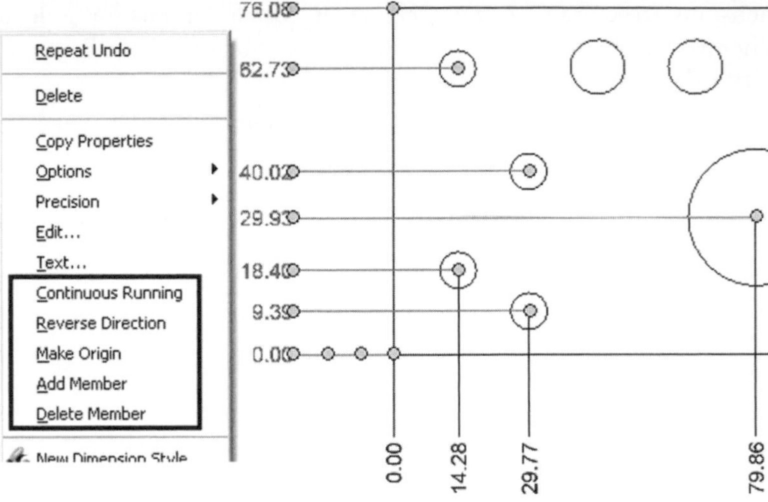

Figure 5.178

8. To create the dimensions, right-click and select Create from the menu.
9. To edit the origin, right-click on the dimension that will be the origin and click Make Origin from the menu (see following image). The other dimension values will be updated to reflect the new origin.
10. To edit a dimension set, right-click on a dimension in the set and select the desired option from the menu.

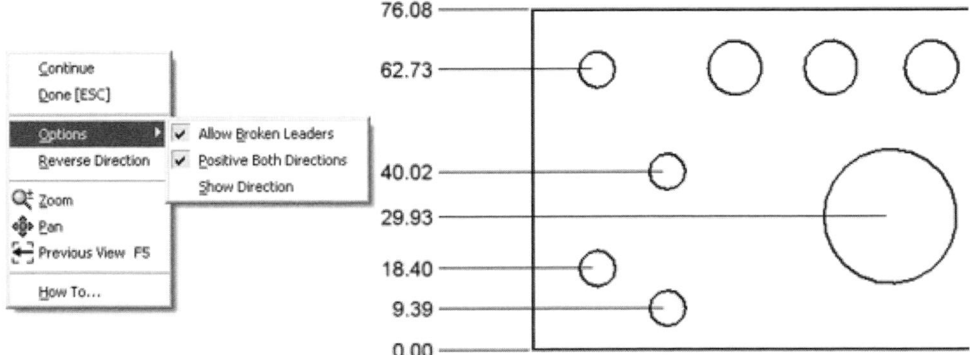

Figure 5.179

To create an ordinate dimension using the Ordinate Dimension tool, follow these steps:

11. Create a drawing view.

12. Click the Ordinate Dimension tool from the Drawing Annotation panel bar, as shown in the following image.

Figure 5.180

13. Select the desired dimension style from the Style list in the Standard toolbar.
14. In the graphics window, select the view to dimension.
15. Select a point to set the origin indicator for the dimensions, as shown in the following image.

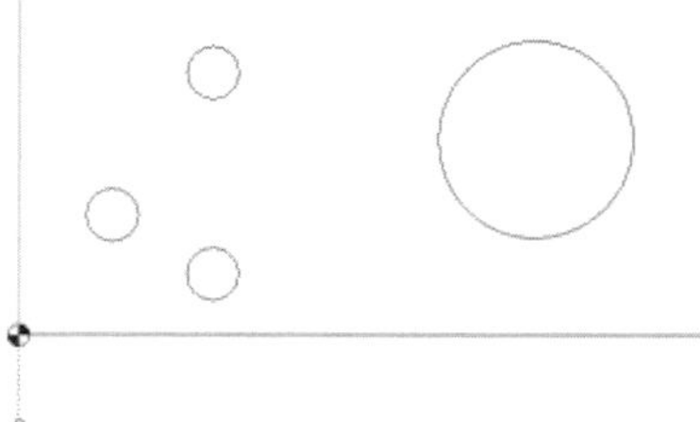

Figure 5.181

16. Select a point to locate the dimension and then move the mouse to place the dimension in a horizontal or vertical orientation. Click to place the dimension.
17. Continue selecting points and placing dimensions.
18. To create the dimensions and end the operation, right-click, and select Done from the menu.
19. To edit a dimension, right-click on a dimension and select the desired option from the menu, as shown in the following image.

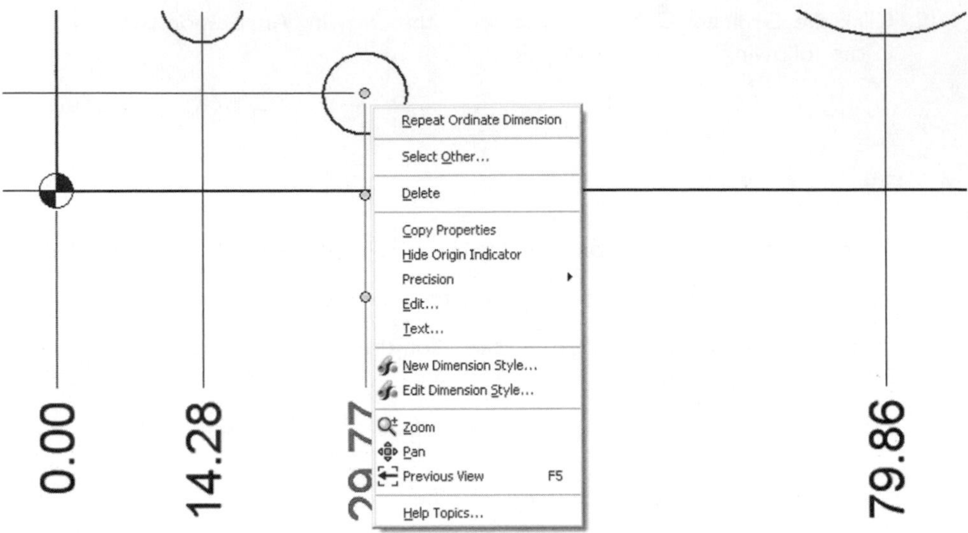

Figure 5.182

20. To edit an ordinate dimension leader, grab an anchor point (a green circle), and drag it to the desired location.
21. To edit the origin indicator, do one of the following:
    - Grab the origin indicator, and drag it to the desired location.
    - Double-click the origin indicator, and enter the precise location in the Origin Indicator dialog box.

HOLE TABLES

If the drawing view that you are dimensioning contains holes, you can locate them by placing individual dimensions or by creating a hole table that will list the location and size of all the holes, or just the selected holes, in a view. The hole locations will be listed in both X- and Y-axis coordinates with respect to a hole datum that will be placed before creating the hole table.

After placing a hole table, if you add, delete, or move a hole, the hole table will automatically be updated to reflect the change. Each hole in the table is automatically given an alphanumeric tag as its name. It can be edited by double-clicking on the tag in the drawing view or hole table. Alternately, you can right-click on the tag in the drawing view or hole table, and select Edit Tag from the menu. Type in a new tag name, and the change will appear in the drawing view and hole table. There are three tools for creating a hole table:

- Hole Table – Selection
- Hole Table – View
- Hole Table – Selected Feature

The tools are located in the Drawing Annotation panel bar, as shown in the following image.

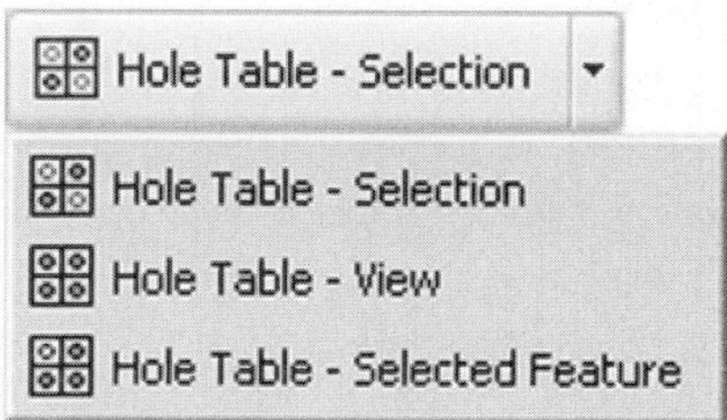

Figure 5.183

When using the Hole Table – Selection mode, a hole table will be created from only the selected holes, as shown in the following image. You control the holes that are selected.

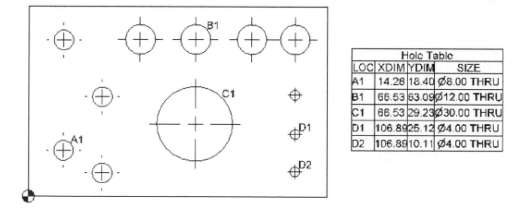

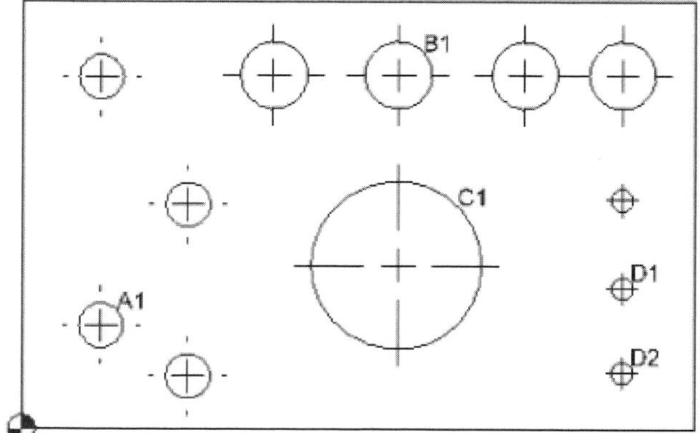

Hole Table			
LOC	XDIM	YDIM	SIZE
A1	14.28	18.40	Ø8.00 THRU
B1	66.53	63.09	Ø12.00 THRU
C1	66.53	29.23	Ø30.00 THRU
D1	106.89	25.12	Ø4.00 THRU
D2	106.89	10.11	Ø4.00 THRU

Figure 5.184

When using the Hole Table – View mode, a hole table will be created based on all the holes in a selected view, as shown in the following image. Notice that all holes are given an alphanumeric identifier, which locates each hole by X- and Y-axis coordinates.

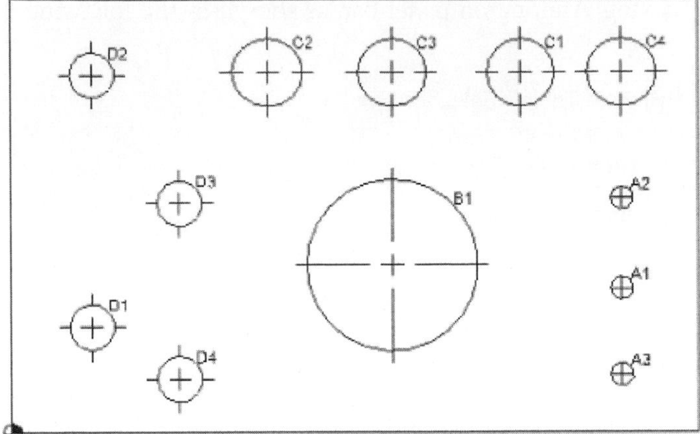

Hole Table			
LOC	XDIM	YDIM	SIZE
A1	106.86	25.12	Ø4.00 THRU
A2	106.85	40.99	Ø4.00 THRU
A3	106.83	10.11	Ø4.00 THRU
B1	66.53	29.23	Ø30.00 THRU
C1	89.23	63.39	Ø12.00 THRU
C2	44.91	63.39	Ø12.00 THRU
C3	66.53	63.39	Ø12.00 THRU
C4	106.85	63.39	Ø12.00 THRU
D1	14.28	18.40	Ø8.00 THRU
D2	14.28	62.73	Ø8.00 THRU
D3	29.77	40.32	Ø8.00 THRU
D4	29.77	9.39	Ø8.00 THRU

Figure 5.185

When using the Hole Table – Selected Feature mode, the hole table will be created based on a single grouping of holes. In the following image, one of the larger holes located in the upper part of the object is selected using the Hole Table – Selected Feature tool.

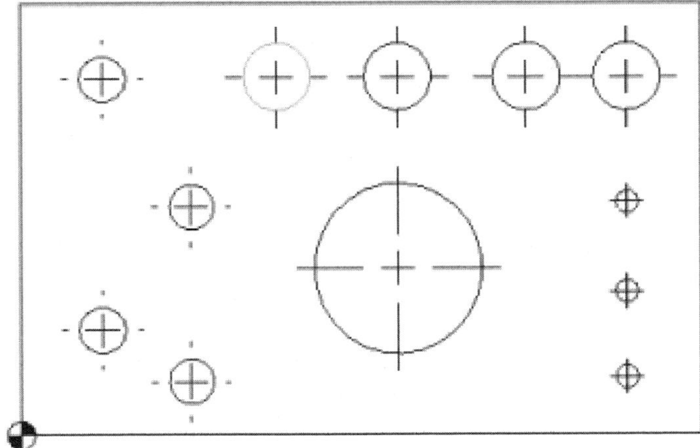

Figure 5.186

When the hole table is created, all holes that share the same type and size will be added to the hole table, as shown in the following image.

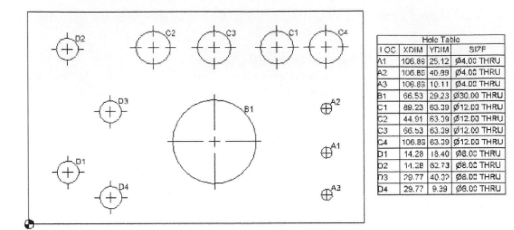

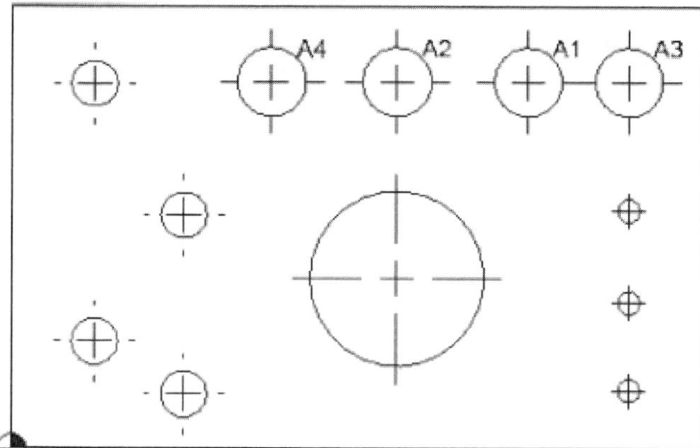

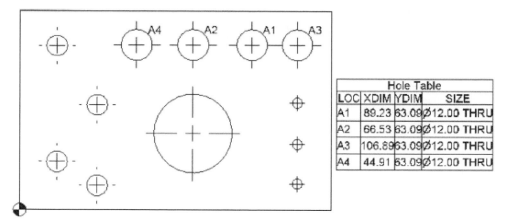

Hole Table			
LOC	XDIM	YDIM	SIZE
A1	89.23	63.09	Ø12.00 THRU
A2	66.53	63.09	Ø12.00 THRU
A3	106.89	63.09	Ø12.00 THRU
A4	44.91	63.09	Ø12.00 THRU

Figure 5.187

To create a hole table based on an existing drawing view that shows holes in a plan view, follow these steps:

1. Select the desired Hole Table tool from the Drawing Annotation panel bar.
2. Select the view on which the hole table will be based.
3. Select a point to locate the origin. Typical origins include the corners of rectangular objects and the centers of drill holes used for data.
4. If the Hole Table - Selection tool is used, select the holes to include in the hole table. The holes can be selected individually, or you can drag a selection window around the holes. If the Hole Table - Selected Feature is used, select the individual hole or holes that will act as references for all other holes of the same type.
5. Select a point to locate the table.
6. The contents of the hole table can be edited by either right-clicking on the hole table, on the tag name in the hole table, or on the tag name in the drawing view. Next choose the desired option from the menu, as shown in the following image.

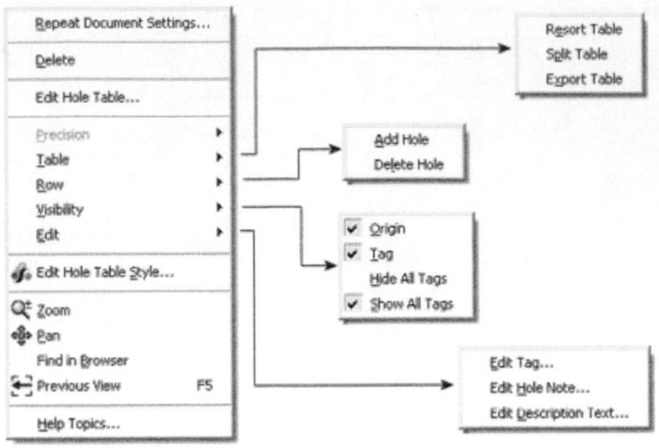

Figure 5.188

EDITING THE HOLE TABLE

Right-clicking on a hole table and then picking Edit Hole Table displays the Edit Hole Table dialog box, as shown in the following image. Use the Formatting tab of this dialog box to change items such as the following:

- Location of the Hole Table Heading, either above or below the coordinate information.
- The types of hole properties used in the table through the Column Chooser. You can add or remove properties from a hole table, and you can create new fields to add to the hole table.

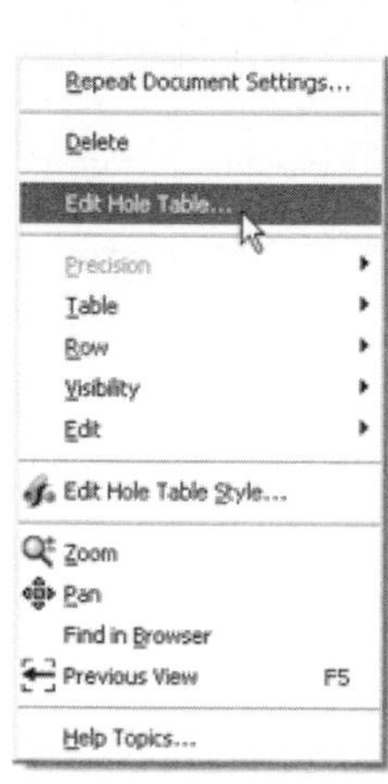

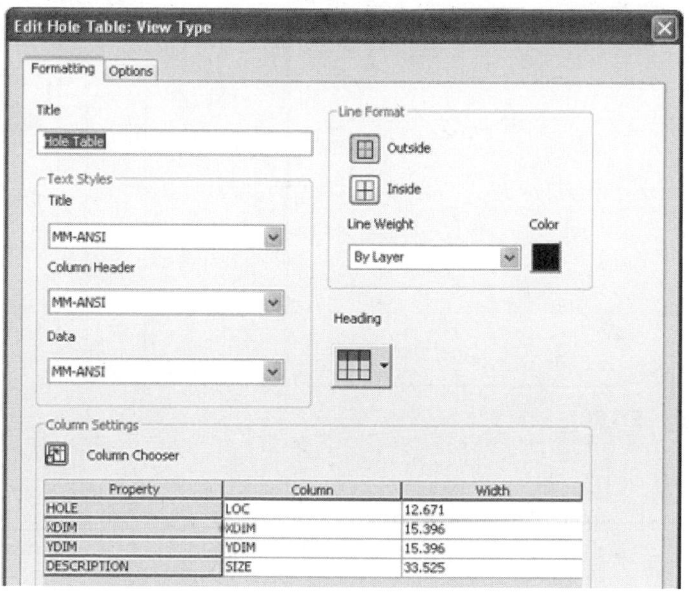

Figure 5.189

EXERCISE 5-7: CREATING HOLE TABLES

This exercise walks you through the creation of a hole chart for a shim plate. After creating the hole table, you modify the plate model to add new holes and update the hole table accordingly. After applying new text styles, the drawing will appear similar to the one below.

1. In preparation for creating the hole table, first open the existing drawing, *ESS_E05_07.idw*.
2. To create the hole table, first activate the Drawing Annotation panel bar, and click the Hole Table – View tool. Select the front view of the plate as the view to which the hole table will be associated. To place the hole table datum, select the center of the tooling hole, as shown in the following image on the left. Then place the hole table near the upper-right corner of the sheet, as shown in the following image on the right.

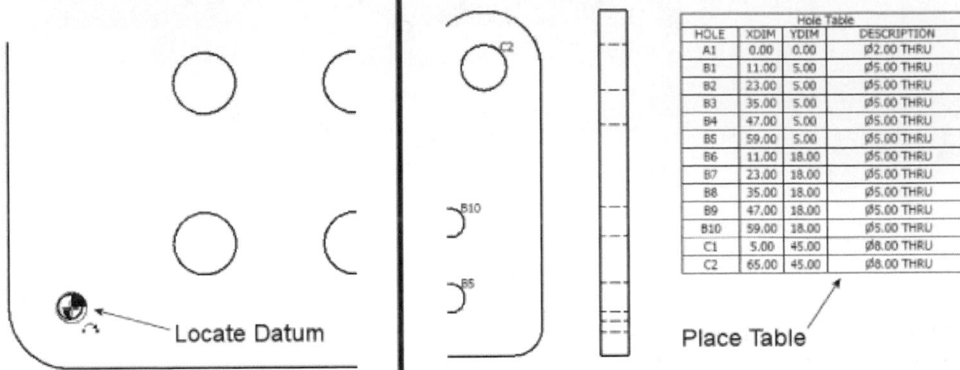

HOLE	XDIM	YDIM	DESCRIPTION
A1	0.00	0.00	Ø2.00 THRU
B1	11.00	5.00	Ø5.00 THRU
B2	23.00	5.00	Ø5.00 THRU
B3	35.00	5.00	Ø5.00 THRU
B4	47.00	5.00	Ø5.00 THRU
B5	59.00	5.00	Ø5.00 THRU
B6	11.00	18.00	Ø5.00 THRU
B7	23.00	18.00	Ø5.00 THRU
B8	35.00	18.00	Ø5.00 THRU
B9	47.00	18.00	Ø5.00 THRU
B10	59.00	18.00	Ø5.00 THRU
C1	5.00	45.00	Ø8.00 THRU
C2	65.00	45.00	Ø8.00 THRU

Figure 5.190

3. The plate lug holes need to be changed to counterbore holes. You will now switch from the drawing to the part file by using the Window drop-down menu located at the top of the Inventor display screen. From the Drawing browser, expand View3. When the part file icon appears for ESS_E05_07.ipt, right-click this icon and click Open from the menu. This will open the part file associated with the drawing.

4. In the browser right-click the Hole1 feature, and choose Edit Feature, as shown in the following image on the left. While in the Hole dialog box, change the hole type by clicking the Counterbore button and changing the hole specifications, as shown in the following image in the middle. Click the OK button to update the holes. Switch back to the drawing, and note the updated hole table, as shown in the following image on the right.

Note: Holding down the CTRL key and pressing the TAB key will switch you back and forth from the drawing to the part file.

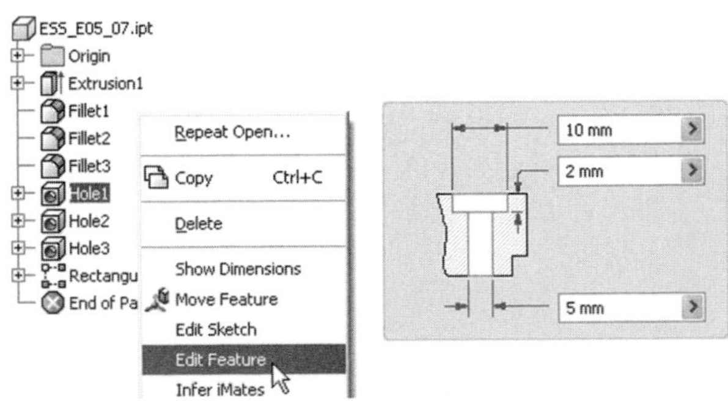

Hole Table			
HOLE	XDIM	YDIM	DESCRIPTION
A1	0.00	0.00	Ø2.00 THRU
B1	11.00	5.00	Ø5.00 THRU
B2	23.00	5.00	Ø5.00 THRU
B3	35.00	5.00	Ø5.00 THRU
B4	47.00	5.00	Ø5.00 THRU
B5	59.00	5.00	Ø5.00 THRU
B6	11.00	18.00	Ø5.00 THRU
B7	23.00	18.00	Ø5.00 THRU
B8	35.00	18.00	Ø5.00 THRU
B9	47.00	18.00	Ø5.00 THRU
B10	59.00	18.00	Ø5.00 THRU
C1	5.00	45.00	Ø5.00 THRU ⌴ Ø10.00 ↧ 2.00
C2	65.00	45.00	Ø5.00 THRU ⌴ Ø10.00 ↧ 2.00

Figure 5.191

5. A series of mounting holes needs to be created in the model consisting of countersink holes. From the Window drop-down menu, click *ESS_E05_07.ipt* to view the part. Add two countersink mounting holes using the following specifications, as shown in the following image. Note that both countersink holes are set to the Through All termination option.

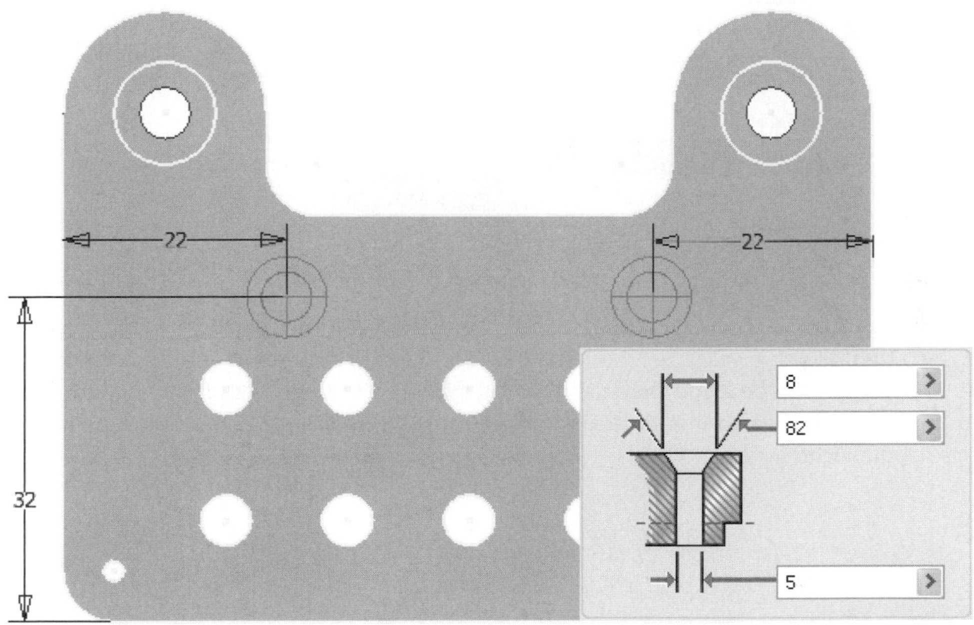

Figure 5.192

6. From the Window drop-down menu, click *ESS_E05_07.idw* to view the drawing. Note in the following image that the new holes have been automatically added to the hole chart.

C1	[illegible]	[illegible]	⌴ Ø10.00 ↧ 2.00
C2	65.00	45.00	Ø5.00 THRU ⌴ Ø10.00 ↧ 2.00
D1	53.00	26.34	Ø5.00 THRU ⌵ Ø8.00 X 82.0°
D2	17.00	27.00	Ø5.00 THRU ⌵ Ø8.00 X 82.0°

Figure 5.193

7. The hole needs to be modified. Right-click on the D1 hole in the table, and choose Row, followed by Delete Hole, from the menu to remove it from the table, as shown in the following image. Perform the same operation on the D2 hole in the table.

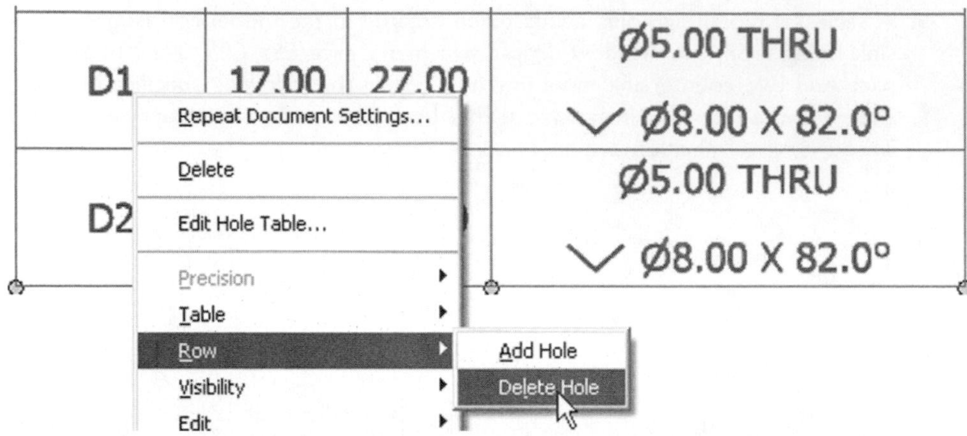

Figure 5.194

8. The hole tags will now be modified. Drag and position the A1 hole tag, as shown in the following image on the left. Next double-click on the A1 tag to open the Format Text dialog box, add the text **Tooling Hole** to the hole tag, and click OK. The hole table updates to include the Tooling Hole label, as shown in the following image on the right.

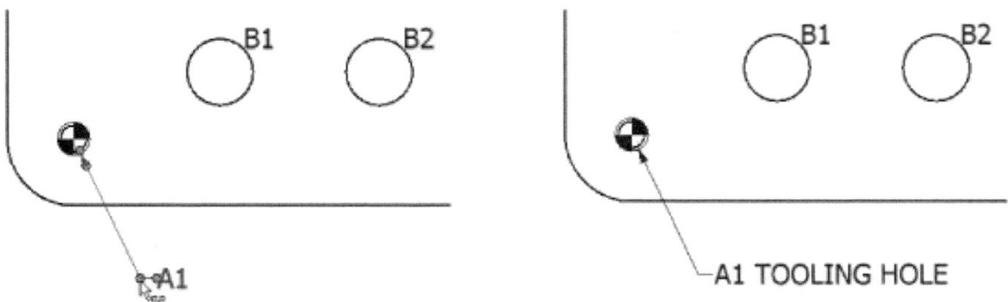

Figure 5.195

9. Pan the drawing over to the hole table, and notice that the information is updated to include the Tooling Hole label, as shown in the following image.

Hole Table			
HOLE	XDIM	YDIM	DESCRIPTION
A1 TOOLING HOLE	0.00	0.00	Ø2.00 THRU
B1	11.00	5.00	Ø5.00 THRU

Figure 5.196

10. Hole table information can be exported into a *.txt* file. To perform this task, right-click on the hole table, and choose Table, followed by Export Table, as shown in the following image on the left. When prompted for a file name, enter **Shim Plate** in the File name text box. The contents of the text file should appear similar to the following image on the right.

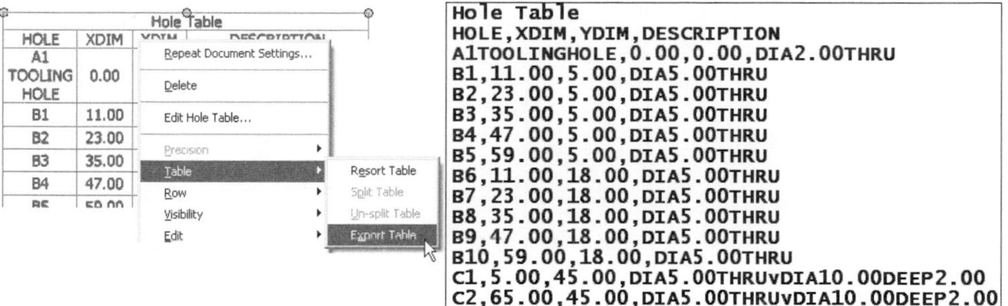


```
Hole Table
HOLE,XDIM,YDIM,DESCRIPTION
A1TOOLINGHOLE,0.00,0.00,DIA2.00THRU
B1,11.00,5.00,DIA5.00THRU
B2,23.00,5.00,DIA5.00THRU
B3,35.00,5.00,DIA5.00THRU
B4,47.00,5.00,DIA5.00THRU
B5,59.00,5.00,DIA5.00THRU
B6,11.00,18.00,DIA5.00THRU
B7,23.00,18.00,DIA5.00THRU
B8,35.00,18.00,DIA5.00THRU
B9,47.00,18.00,DIA5.00THRU
B10,59.00,18.00,DIA5.00THRU
C1,5.00,45.00,DIA5.00THRUvDIA10.00DEEP2.00
C2,65.00,45.00,DIA5.00THRUvDIA10.00DEEP2.00
```

Figure 5.197

11. Close all open files. Do not save changes. End of exercise.

THE TABLE COMMAND

In addition to organizing hole patterns, tables can also contain data from an iPart or iAssembly. They can be generated manually or created from an Excel spreadsheet.

CREATING A GENERAL TABLE

To create a general table, use the following steps:

1. On the Drawing Annotation panel bar, click the Table button, as shown in the following image.

Figure 5.198

2. When the Table dialog box appears, verify that <Empty table> is selected from the source list. Then specify the number of rows and columns, as shown in the following image on the left. On the Table dialog box, select <Empty table> from the source list. Clicking the OK button will create the table, as shown in the following image on the right. Notice that the title of the table is blank. Notice also the column headers labeled Column 1 and so on. Both of these items are derived from the current table style.

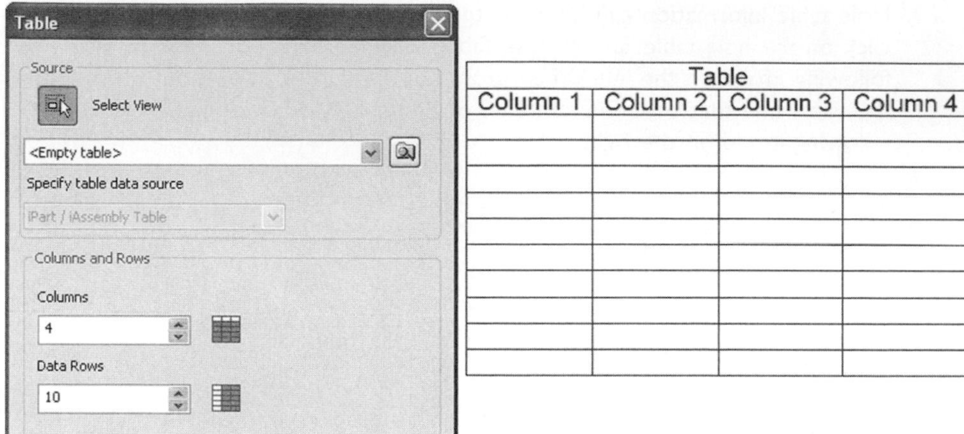

Figure 5.199

3. To add a title to the table, select the table, right-click, and pick Edit Table Style from the menu, as shown in the following image.

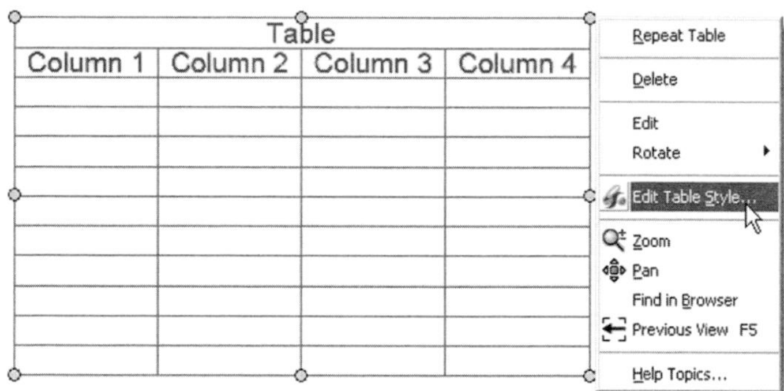

Figure 5.200

4. When the Style and Standard Editor dialog box appears, expand the Table category on the left side of the dialog box, and pick the Table standard. Enter the title of the table, as shown in the following image. Click the Save button, followed by the Done button, to exit this dialog box and create the table title.

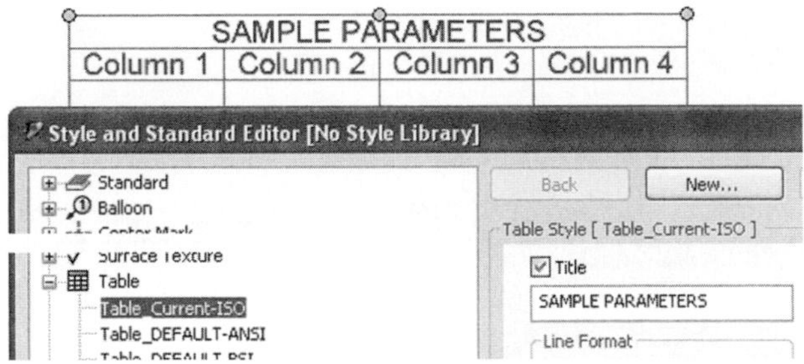

Figure 5.201

5. To change the column headers and add information to the individual table cells, select the table, right-click, and select Edit from the menu, as shown in the following image.

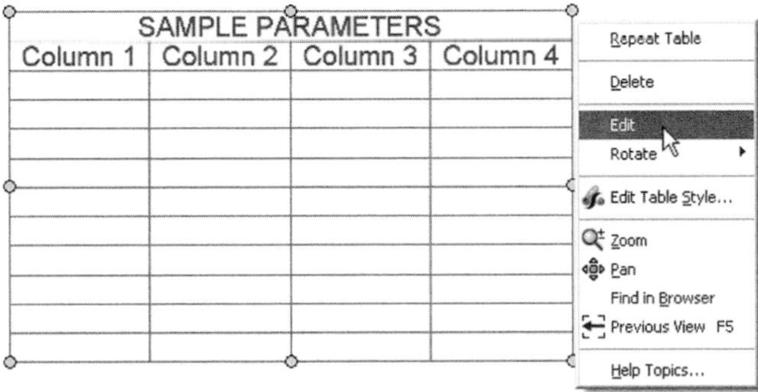

Figure 5.202

6. When the Table dialog box appears, move your cursor over the column heading, right-click, and select Format Column from the menu. This will launch the Format Column dialog box where you can change the name of the heading. In the following image, the Column 1 heading was changed to a new heading called Parameter by deselecting the Name From Data Source checkbox and typing in the desired header name, as shown in the following image.

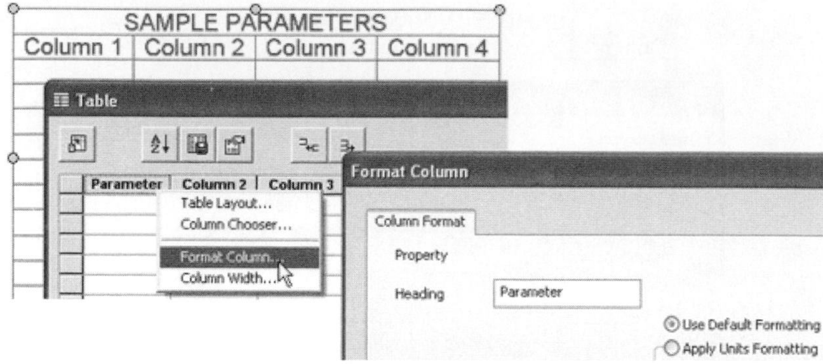

Figure 5.203

7. While still in the Table dialog box, click in an individual cell and begin adding information, as shown in the following image. After information is entered and as you move on to another cell, clicking on the Apply button will update the table in the drawing while still keeping you in the Table dialog box. When finished, click the OK button.

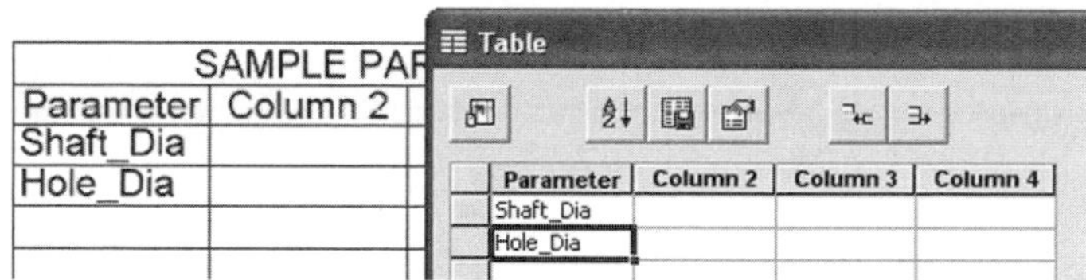

Figure 5.204

CREATING A TABLE FROM AN EXCEL SPREADSHEET

Tables can also be created from an external Excel spreadsheet using the following steps:

1. On the Drawing Annotation panel bar, click the Table tool.
2. When the Table dialog box appears, click on the Browse for file button to locate the Excel file as the source.
3. In the Table dialog box, specify the start cell and column header row, as shown in the following image on the left.
4. Click OK to place the table on the drawing sheet, as shown in the following image on the right.

Note: Make sure that the Excel sheet is located in the current project path(s).

FLANGE COUPLING		
quation	Units	Description
).9375	in	Shaft Diameter
00-.1875	in	Key Length
5	in	Outer Diameter
1.5	in	Flange Height
2	in	Inner Diameter
1.1875	in	Rim Height
).5625	in	Base Thickness
3.25	in	Bolt Circle Diameter
2.375	in	Base Relief Diameter
4.5	in	Male Connect Diameter
4.501	in	Female Connect Diameter
).1875	in	Wall Thickness

Figure 5.205

CREATING A REVISION TABLE

During the course of documenting mechanical designs, the drawing document typically undergoes numerous changes. For legal, technical, and other reasons, companies have adopted the practice of maintaining records of these changes, also known as revisions. Revision records come in the form of Engineering Change Orders (ECO) and Engineering Change Notices (ECN). To formally display and keep track of these changes on the drawing, you should insert a revision table. Two tools are available for tracking revisions in a drawing: the Revision Table tool and the Revision Tag tool. You can find the Revision Table tool in the Drawing Annotation panel bar, as shown in the following image.

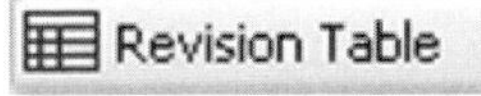

Figure 5.206

Clicking on the Revision Table tool will launch the Revision Table dialog box, as shown in the following image. Use this dialog box to define the scope and revision indexing methods. These items are described as follows:

Two settings are contained under the Table Scope heading. The Entire Drawing setting will create a revision table for the entire drawing no matter how many sheets are created. The Active Sheet setting will create a revision table for the active sheet. In this manner, different revision tables can be applied to the different drawing sheets.

Under the Revision Index heading, Auto-Index is checked by default. This setting automatically indexes the revisions. If this feature is not checked, the revision cell for any new revision rows will remain empty or blank. The Alpha setting uses alphabetic indexing, and the Numeric setting uses numeric indexing. The Start Value sets the initial revision number or letter.

When the Update property on revision number edit box is selected, the revision number in the active row of the revision table is connected with the revision number property saved in drawing iProperties or sheet properties.

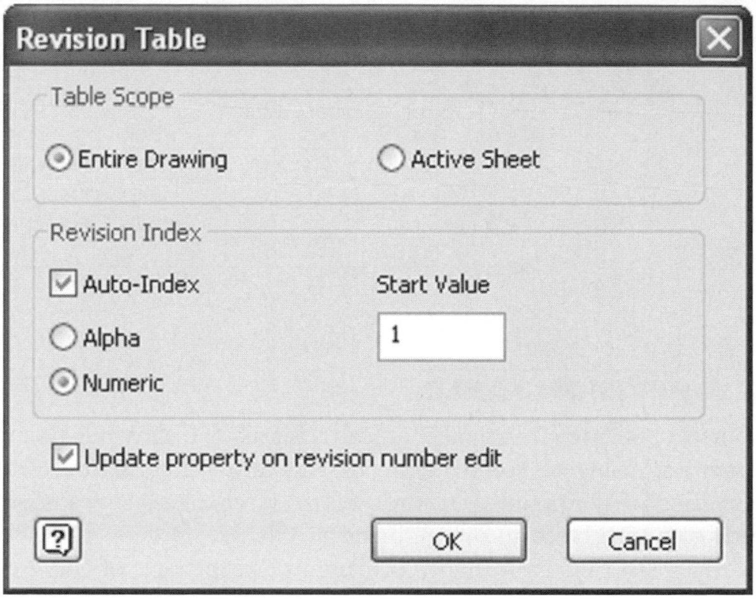

Figure 5.207

After making changes in the Revision Table dialog box, the revision table attaches to your cursor for placement in a drawing. The following image shows the default revision table. As with a parts list, the revision table displays a series of green dots at its corners. These dots allow you to place the revision table accurately in any corner of the border. All examples of revision blocks in this chapter will be placed at the upper-right corner of the border.

REVISION HISTORY				
ZONE	REV	DESCRIPTION	DATE	APPROVED
	1			AJK

Figure 5.208

When the revision table is placed in the desired location, you edit the contents of the various fields in order to reflect the most current changes in the drawing. To do this, double-click on the text in the revision table. This will launch the Revision Table: Drawing Scope dialog box, as shown in the following image. Add the appropriate text to the dialog box, and click the OK button when finished. The specific field of the revision table should reflect these changes.

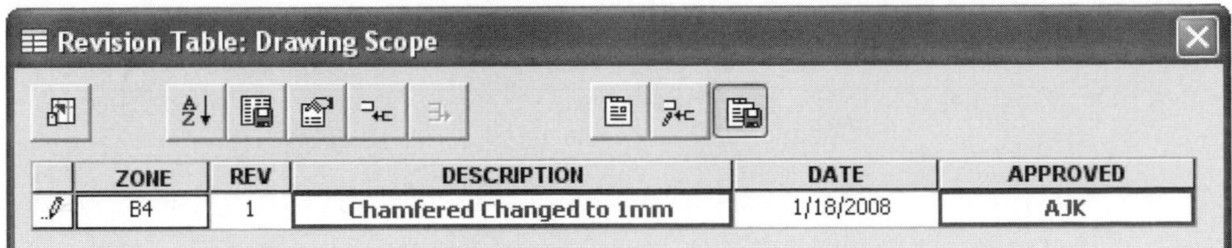

ZONE	REV	DESCRIPTION	DATE	APPROVED
B4	1	Chamfered Changed to 1mm	1/18/2008	AJK

Figure 5.209

Extra rows can be added to the revision table by right-clicking on the revision table and picking Add Revision Row from the menu, as shown in the following image.

REVISION HISTORY			
ZONE	REV	DESCRIPTION	DAT
B4	1	Chamfered Changed	/18/2

Repeat Document Settings...
Delete
Edit
Add Revision Row
Delete Row

Figure 5.210

To lend further support for the revision table, a revision tag is also available. You can access the Revision Tag tool, as shown in the following image, through the Drawing Annotation panel bar. This tool allows you to tag an item on the drawing. The tag information is formatted according to the options selected in the Format text and edit Revision Table dialog box. Revision numbers or letters will increment automatically.

Figure 5.211

The following image illustrates a revision table and tag applied to an object in a drawing file.

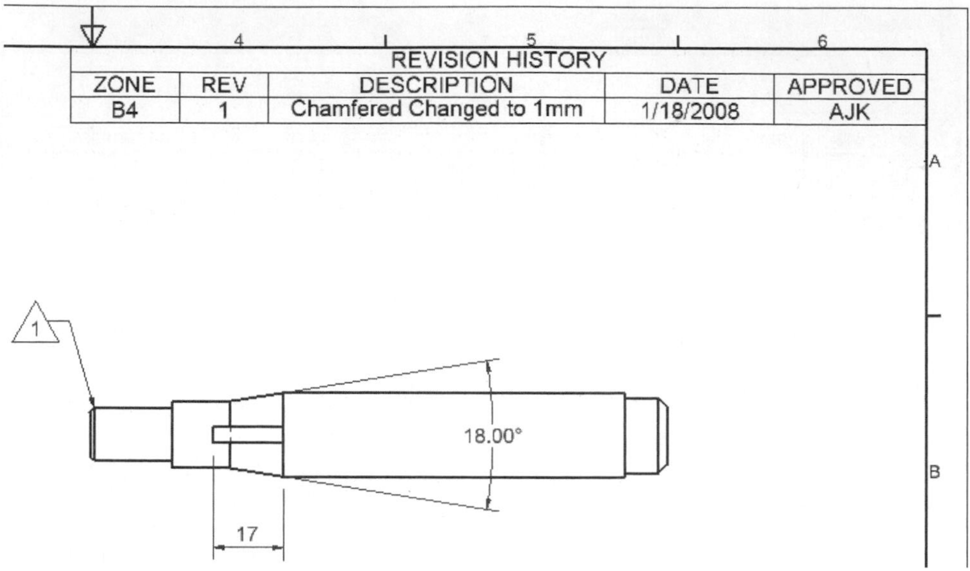

Figure 5.212

Note: Revision tables can also be rotated at 90° intervals.

MULTI-SHEET PLOTTING

Use the Multi-Sheet Plot wizard to plot multiple drawing sheets. Choose this wizard by clicking on Start > All Programs > Autodesk > Autodesk Inventor 2008 > Autodesk Multi-Sheet Plot, as shown in the following image.

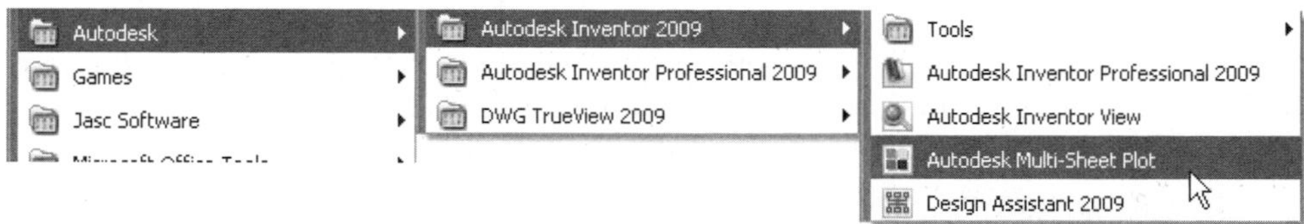

Figure 5.213

When launching the wizard, the initial dialog box deals with the current print/plot device. The second dialog box, Select Drawing Files, is shown in the following image on the left. Use this area to specify the project that contains the drawings to plot. Once the drawing names appear in the Source files column, select a drawing and click on the >> key to move the drawing to the Selected files column, as shown in this image.

The next dialog box in the Multi-Sheet Plot wizard, Preview, is shown in the following image on the right. Use this dialog box to view the multiple pages. When you are satisfied with the results, click the Submit button to continue.

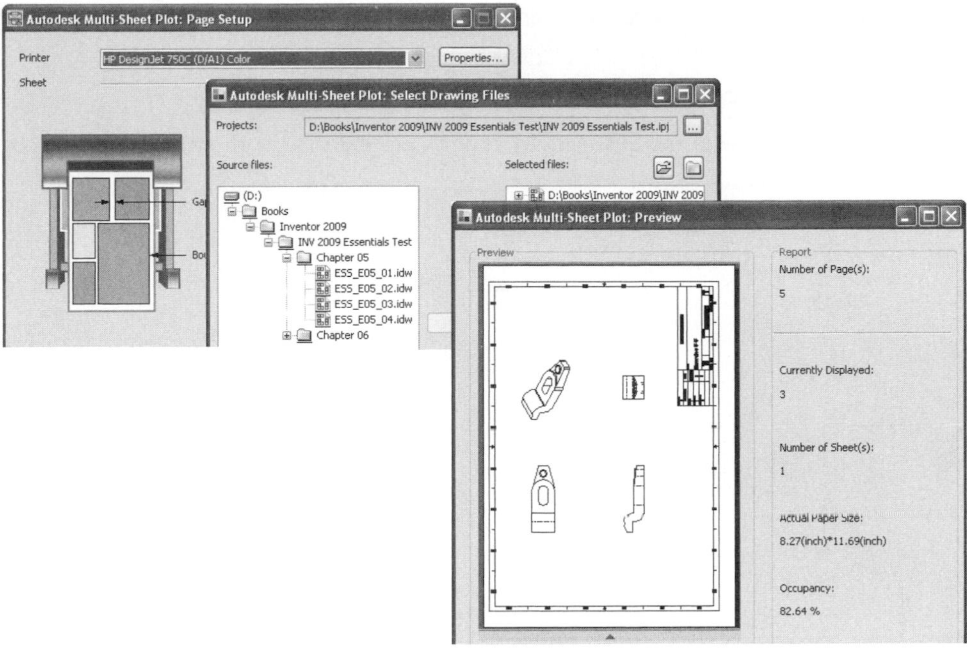

Figure 5.214

Project Exercise: Chapter 5

The self-paced, step-by-step project exercises found in Appendix A provide opportunities for you to work through real-world modeling, assembly, and documentation tasks. The geometry used in the project exercises will flow from chapter to chapter, utilizing the functionality that you learned in that chapter.

EXERCISES

Applying Your Skills

SKILL EXERCISE 5-1

In this exercise, you create a working drawing of a part.

1. Begin by starting a new drawing based on the *metric ANSI (mm).idw* template.
2. Use Edit Sheet to select an A2-sized sheet.
3. Use the Base View and Projected View tools to create the required views of *ESS_E05_08.ipt.*
4. Use a scale of 2 for all views.
5. Add a center mark and centerline bisector to the drawing views.
6. Use the Retrieve Dimensions tool to display the model dimensions.

7. Delete dimensions as required, and then use the General Dimension tool to add the dimensions.
8. Use the Leader Text tool to place the thread note **M8x1.25 – 6H.**
9. Continue on with the Leader Text tool by placing the depth symbol followed by **20.00 TYP.**

Your display should appear similar to the following image.

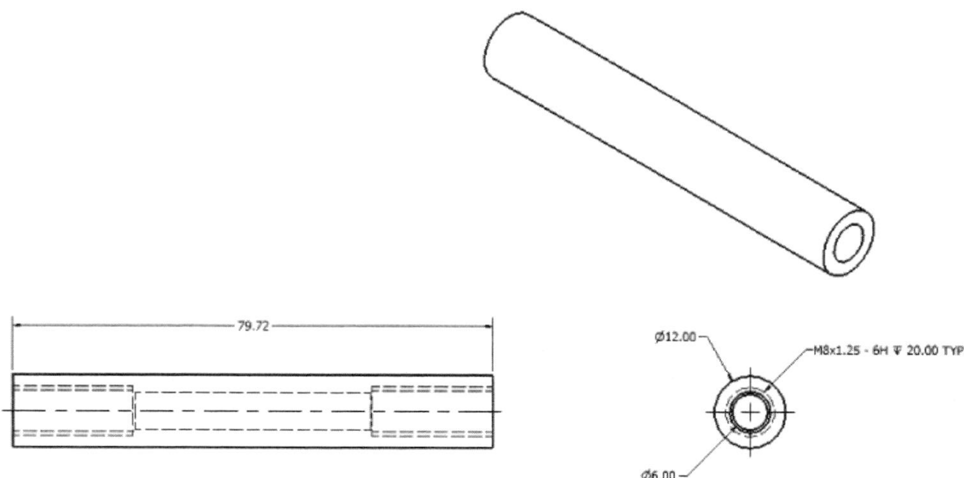

Figure 5.215

In this exercise, you create a drawing for a drain plate cover.

1. Begin by starting a new drawing using a *Metric ANSI (mm).idw* template.
2. Create three views of the part *ESS_E05_09.ipt* on an A2-sized sheet.
3. Use a scale of 2 for all views.
4. Add center marks to the views.
5. Create a new dimension style. Set all text to align horizontally. Set all trailing zeros to display. Set all text height to **3.50 mm**.
6. Add dimensions and annotations. Your display should appear similar to the following image.

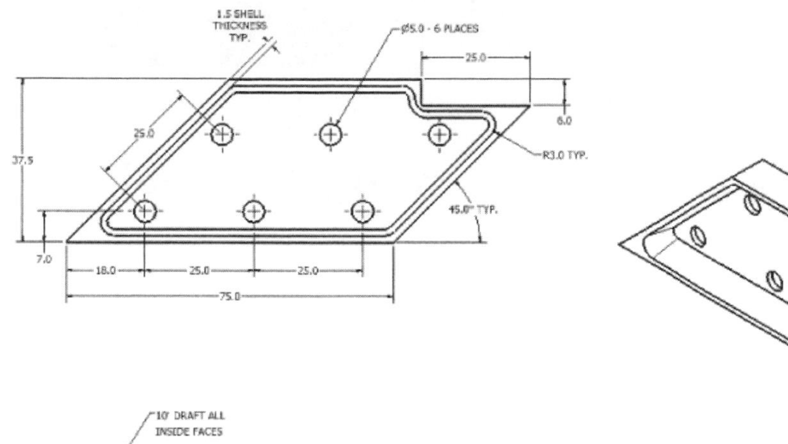

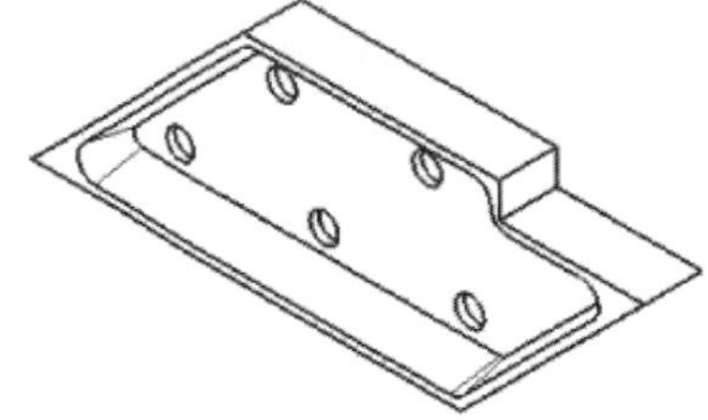

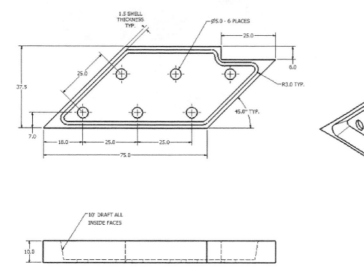

Figure 5.216

CHECKING YOUR SKILLS

Use these questions to test your knowledge of the material covered in this chapter.

1. **True** _ **False** _ A drawing can have an unlimited number of sheets.
2. **True** _ **False** _ A drawing's sheet size normally is scaled to fit the size of the drawing views.
3. **True** _ **False** _ There can only be one base view per sheet.
4. **True** _ **False** _ An isometric view can only be projected from a base view.
5. **True** _ **False** _ Drawing dimensions can drive dimensional changes parametrically back to the part.
6. **True** _ **False** _ Explain how to shade an isometric drawing view.
7. **True** _ **False** _ When creating a hole note using the Hole/Thread Notes tool, circles that are extruded to create a hole can be annotated.

CHAPTER 6

Creating and Documenting Assemblies

INTRODUCTION

In the first three chapters, you learned how to create a component in its own file. In this chapter, you will learn how to place individual component files into an assembly file. This process is referred to as *bottom-up* assembly modeling. You will also learn to create components in the context of the assembly file, which is referred to as *top-down* assembly modeling. After creating components, you will learn how to constrain the components to one another using assembly constraints, edit the assembly constraints, check for interference, and create presentation files that show how the components are assembled or disassembled.

OBJECTIVES

After completing this chapter, you will be able to perform the following:

- Understand the assembly options
- Create bottom-up assemblies
- Create top-down assemblies
- Create subassemblies
- Constrain components together using assembly constraints
- Edit assembly constraints
- Create adaptive parts
- Pattern components in an assembly
- Check parts in an assembly for interference
- Drive constraints
- Create a presentation file
- Manipulate and edit the Bill of Materials (BOM)
- Create individual and automatic balloons
- Create and perform edits on a Parts List of an assembly

CREATING ASSEMBLIES

As you have already learned, component files have the *.ipt* extension, and they can only have one component each. In this chapter, you will learn how to create assembly files (*.iam* file extension). An assembly file holds the information needed to assemble the components together. All of the components in an assembly are *referenced in*, meaning that each component exists in its own component IPT file, and its definition is linked into the assembly. You can edit the components while in the assembly, or you can open the component file and edit it. When you have made changes to a component and saved the component, the changes will be reflected in the assembly after you open or update it. There are three methods for creating assemblies: bottom-up, in-place, and a combination of both. *Bottom-up* refers to an assembly in which all of the components were created in individual component files and are referenced into the assembly. The *in-place* approach refers to an assembly in which the components are created from within the context of the assembly. In other words, the user creates each component from within the top-level assembly. Each component in the assembly is saved to its own IPT file. The following sections describe the bottom-up and in-place assembly techniques.

Note that assemblies are not made up only of individual parts. Complex assemblies are typically made up of subassemblies, and therefore you can better manage the large amount of data that is created when building assemblies.

To create a new assembly, click New on the File menu, and then click the Standard.iam icon, as shown in the following image. Alternately, you can click the down arrow of the New icon on the left side of the Standard toolbar and select Assembly. After issuing the new assembly operation, Autodesk Inventor's tools will change to reflect the new assembly environment. The assembly tools will appear on the panel bar, and these tools will be covered throughout this chapter.

Note: There are a number of valid approaches to creating an assembly. You will determine which method works best for the assembly that you are creating based on experience. You can create an assembly using the bottom-up technique, the in-place technique, or a combination of both. Whether you place or create the components in the assembly, all of the components will be saved to their own individual IPT files, and the assembly will be saved as an IAM file.

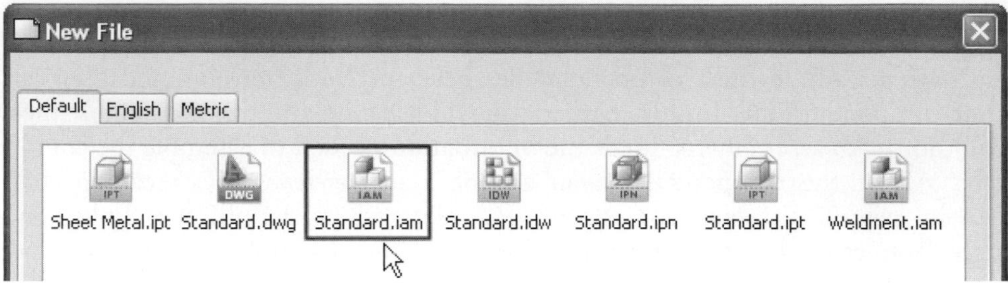

Figure 6.1

ASSEMBLY OPTIONS

Before creating an assembly, review the assembly option settings. On the Tools menu, click Application Options, and the Options dialog box will appear. Click on the Assembly tab, as shown in the following image. The following sections describe the various assembly settings. These settings are global and will affect how new components are created, referenced, analyzed, or placed in the assembly.

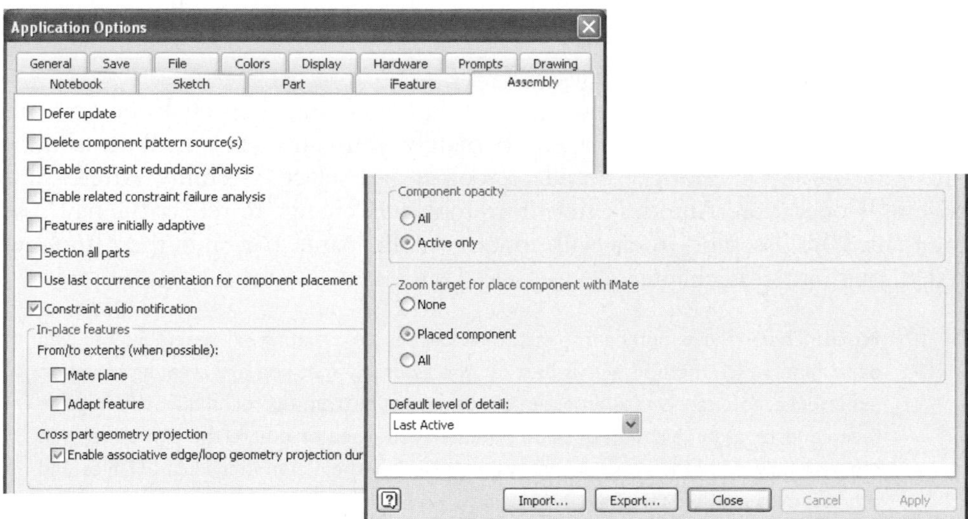

Figure 6.2

Defer Update Click this option so that when you change a component that will affect the assembly, you will have to click the Update button manually to update the assembly. If you leave the box clear, the assembly will update automatically to reflect the change to a component.

Delete Component Pattern Source(s) Click to delete the source component of a component pattern when you delete the pattern. If you leave the box clear, the source component will not be deleted when you delete the pattern.

Enable Constraint Redundancy Analysis Click to perform a secondary analysis of all assembled component constraints. You are notified when redundant constraints exist. The component degrees of freedom (DOF) are also updated but not displayed.

Enable Related Constraint Failure Analysis This setting enables you to perform an analysis to identify the constraints and components if constraints fail. By default, this setting is turned off. Placing a check in the box will turn the setting on and activate an option in the Constraint Doctor dialog box to enable the analysis.

Section All Parts Click to section standard parts placed in an assembly from the Content Center libraries when you create a section drawing view through them. To prevent the sectioning of standard parts, leave this box unchecked.

Use Last Orientation Position For Component Placement This setting controls the orientation of multiple instances of the same component placed into an assembly. When checked, the orientation of the last occurrence in the browser is used to determine the orientation of new instances.

Constraint Audio Notification Click to turn off the sound that is made once a constraint is created in an assembly. By default, this option is turned on.

Note: Features are Initially Adaptive, In-Place Features, and Cross Part Geometry Projection will be covered later in this chapter under the section on Adaptivity.

Component Opacity When a component is edited inside an existing assembly, the remainder of the assembly takes on a faded appearance; this action will discussed later in this chapter. In this section, you determine if all components or only the active component will be opaque when it is edited in place in the assembly. You can also toggle this setting from the Standard toolbar.

All Click to make all components in the assembly opaque when the active component is edited in place in the assembly.

Active Only Click to make only the active component in the assembly opaque when the active component is edited in place in the assembly.

Zoom Target for Place Component with iMate Set the default zoom behavior for the graphics window when placing components with iMates.

None Click to perform no zooming and leave the graphics display as is.

Placed Component Click to zoom in on the placed part so that it fills the graphics window.

All Click to zoom in on the assembly so that all elements in the model fill the graphics window.

THE ASSEMBLY CAPACITY METER

To provide feedback on the resources used by an assembly model, use the Assembly Capacity Meter. This meter is located in the lower-right corner of the display screen and is present only in the assembly environment. Three different types of information are displayed in the status bar of this meter. The first numbered block deals with the number of occurrences or instances in the active assembly. The second numbered block displays the total number of files open in order to display the assembly. The third block is a graphical display of the total amount of memory being used to open and work in this assembly.

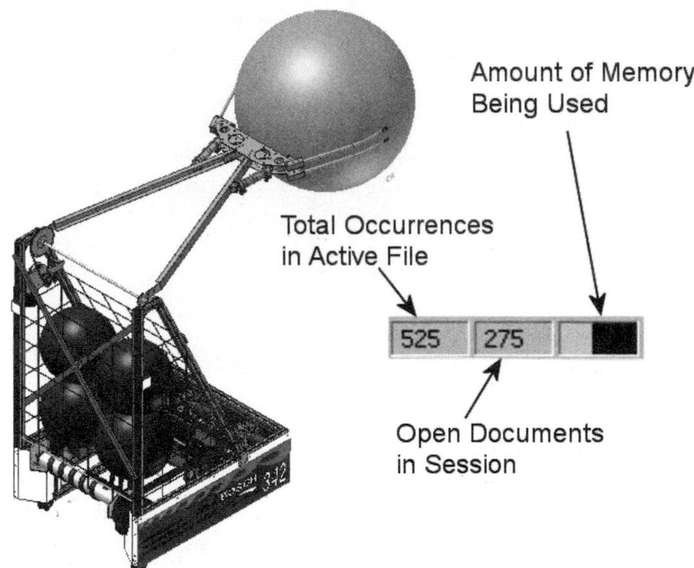

Figure 6.3 *(Courtesy US FIRST Team 342—Robert Bosch Corp., Trident Technical College, Fort Dorchester High School, Summerville High School; Charleston, South Carolina)*

THE ASSEMBLY BROWSER

The Assembly browser displays the hierarchy of all part and subassembly occurrences and assembly constraints in the assembly, as shown the following image. Each occurrence of a component is represented by a unique name. In the browser, you can select a component for editing, move components between assembly levels, reorder assembly components, control component status, rename components, edit assembly constraints, and manage design views and representations.

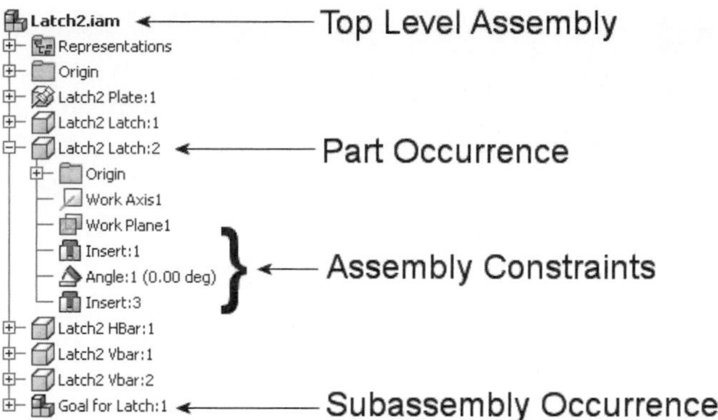

Figure 6.4

BOTTOM-UP APPROACH

The bottom-up assembly approach uses existing files that are referenced to an assembly file. To create a bottom-up assembly, create the components in their individual files. If you place an assembly file into another assembly, it will be brought in as a subassembly. Any drawing views created for these files will not be brought into the assembly file. Be sure to include the path(s) for the file location(s) of the placed components in the project file; otherwise, Autodesk Inventor may not be able to locate the referenced component when you reopen the assembly.

To insert a component into the current assembly, click the Place Component tool on the Assembly panel bar, as shown in the following image, press the hot key P, or right-click in the graphics window, and select Place Component from the menu.

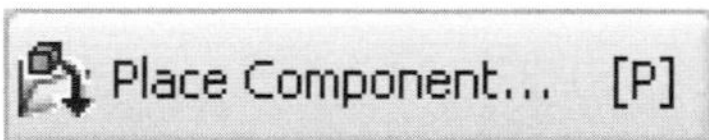

Figure 6.5

The Place Component dialog box will appear, as shown in the following image. If the component you are placing is the first in the assembly, an instance of it will be placed into the assembly automatically. If the component is not the first, you pick a point in the assembly where you want to place the component. If you need multiple occurrences of the component in the assembly, continue selecting placement points. When done, press the ESC key, or right-click and select Done from the menu.

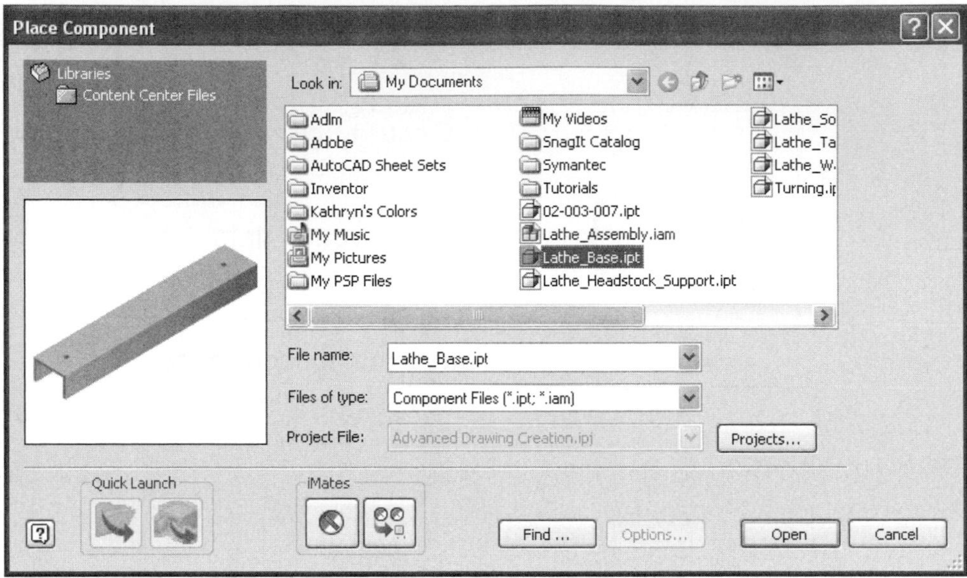

Figure 6.6

CREATE COMPONENTS IN PLACE

You can create new components while in an assembly. This method is referred to as the in-place approach. To create a new component in the current assembly, click the Create Component tool on the Assembly panel bar, as shown in the following image, press the hot key N, or right-click and select Create Component from the menu.

Figure 6.7

The Create In-Place Component dialog box will appear, as shown in the following image. Enter a new file name, a file type, a location where you will save it, and the template file on which to base it. Also determine if the component will be constrained to a planar face on another component in the assembly. If you click the Constrain sketch plane to selected face or plane box option, a flush constraint will be applied between the selected face and the new component's initial sketch plane. If you leave the box clear, no constraint will be applied. However, you will still select a face on which to start sketching the new component.

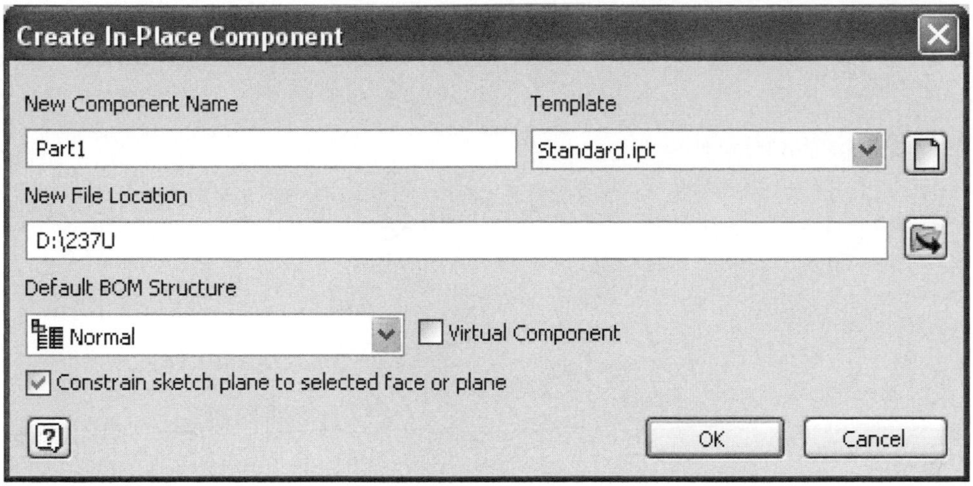

Figure 6.8

Occurrences

An occurrence, or instance, is a copy of an existing component and has the same name as the original component with a colon and a sequenced number. If the original component is named Bracket, for example, the next occurrence in the assembly will be Bracket:1 and a subsequent occurrence will be Bracket:2, as shown in the following image. If the original component changes, all of the component instances will reflect the change.

To create an occurrence, place the component. If the component already exists in the assembly, you can click the component's icon in the browser, and drag an additional occurrence into the assembly. You can also use the copy-and-paste method to place additional components by right-clicking on the component name in the browser, or on the component itself in the graphics window, and then right-clicking and selecting Copy from the menu. Then right-click and select Paste from the menu. You can also use the Windows shortcuts CTRL-C and CTRL-V to copy and paste the selected component. If you want an occurrence of the original component to have no relationship with its source component, make the original component active, click the Save Copy As tool on the File menu, and enter a new name. The new component will have no relationship to the original, and you can place it in the assembly using the Place Component command.

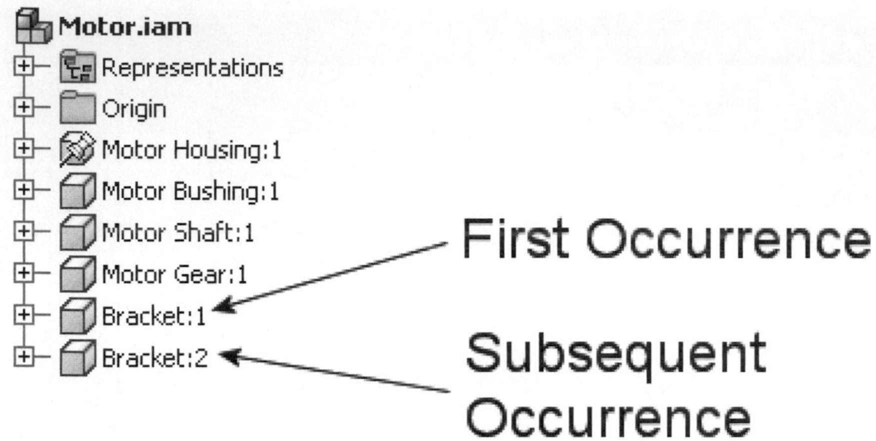

Figure 6.9

Active Component

To edit a component while in an assembly, activate the component. Only one component in the assembly can be active at a time. To make a component active, double-click on the component in the graphics window, and then double-click on the file name or an icon in the browser. Alternately, right-click on the component name in the browser or graphics window, and select Edit from the menu. Once the component is active, the other component names in the browser will appear shaded, as shown in the following image. If Component Opacity, in Application Options on the Assembly tab, is set to Active Only, the other components in the assembly will take on a faded appearance in the graphics window.

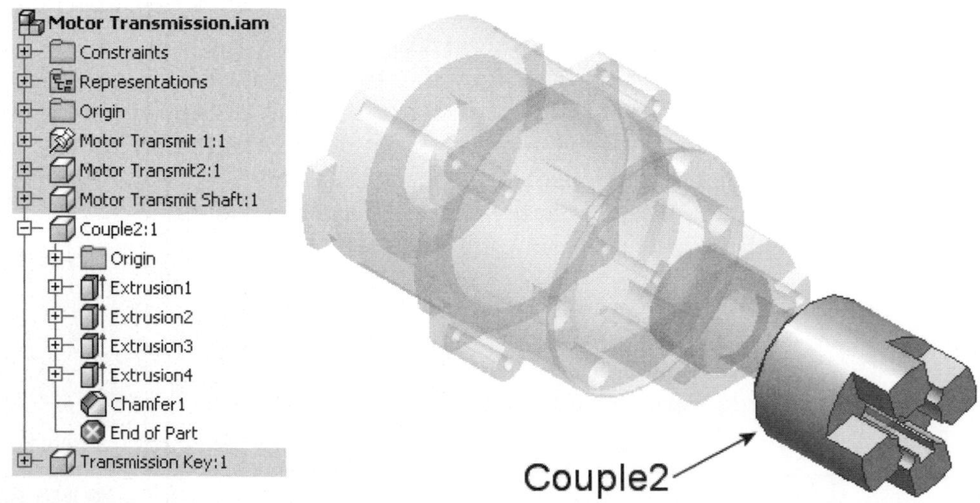

Figure 6.10

You can edit the component and save the changes by using the Save tool. Only the active component will be saved. To make the assembly active, click the Return button on the command bar, as shown in the following image on the left, and double-click on the assembly name in the browser, or right-click in the graphics window and select Finish Edit from the menu, as shown in the following image on the right.

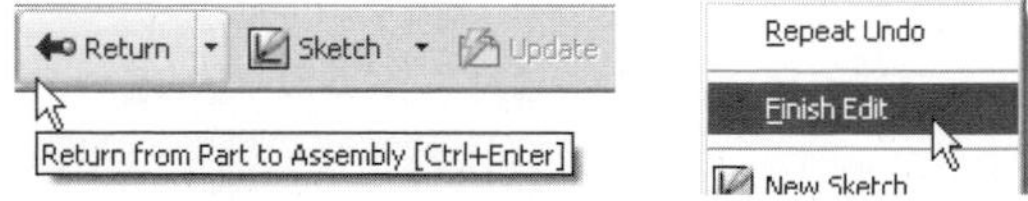

Figure 6.11

Open and Edit

Another way to edit a component in the assembly is to open the component in another window. Click the Open tool on the Standard toolbar, click Open on the File menu, or right-click on the component's name in the browser or on the component in the graphics window. Select Open from the menu, as shown in the following image. The component will appear in a new window. Edit the component as needed, save the changes, activate the assembly file, and the changes will appear in the assembly.

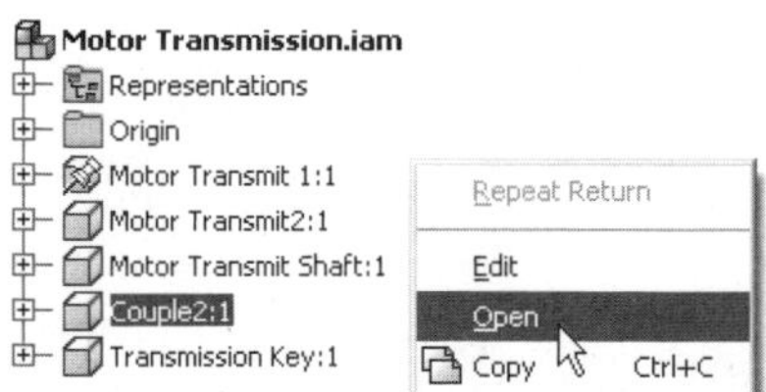

Figure 6.12

GROUNDED COMPONENTS

When assembling components, you may want to make a component or multiple components grounded or stationary, meaning that they will not move. When applying assembly constraints, the unconstrained components will be moved to the grounded component(s). By default, the first component placed in an assembly is grounded. There is no limit to how many components can be grounded. It is strongly recommended that at least one component in the assembly be grounded; otherwise, the assembly can move. A grounded component is represented with a pushpin superimposed on its icon in the browser, as shown in the following image on the left. To ground or unground a component, right-click on the component's name in the browser, and select or deselect Grounded from the menu, as shown in the following image on the right.

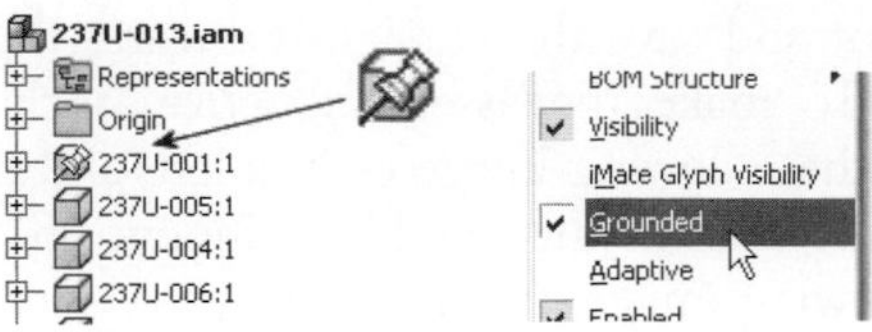

Figure 6.13

INSERTING MULTIPLE COMPONENTS INTO AN ASSEMBLY

Multiple components can also be placed into an assembly. In the following image, a number of files are selected for placement. When the files are in sequence, they can be easily selected as a group by holding down the SHIFT key. If numerous files need to be placed but are not in sequence, use the CTRL key to select the individual components.

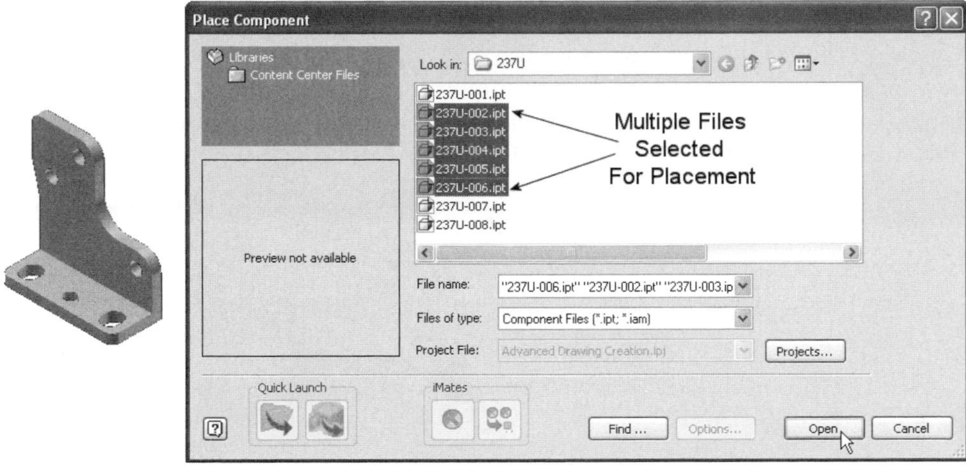

Figure 6.14

Clicking the Open button will place all selected files into the assembly, as shown in the following image.

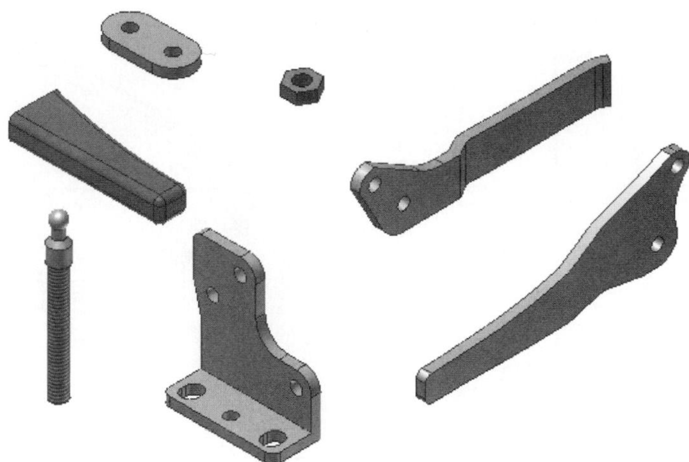

Figure 6.15

ASSEMBLY CONSTRAINTS

You have now learned how to create assembly files, but the components had no relationship to one another except for the relationship that was defined when you created a component from the context of an assembly and in reference to a face on another component. For example, if you placed a bolt in a hole and the hole moved, the bolt would not move to the new hole position. Use assembly constraints to create relationships between components. With the correct constraint(s) applied, if a hole moves, the bolt will move to the new hole location.

In a previous chapter, you learned about geometric constraints. When you apply geometric constraints to sketches, they reduce the number of dimensions or constraints required to constrain a profile fully. When you apply assembly constraints, they reduce the degrees of freedom (DOF) that allow the components to move freely in space.

There are six degrees of freedom: three are translational and three are rotational. Translational means that a component can move along an axis X, Y, or Z. Rotational means that a component can rotate about an axis X, Y, or Z. As you apply assembly constraints, the number of the DOF decreases.

Autodesk Inventor does not require components to be fully constrained. By default, the first component created or added to the assembly will be grounded and will have zero DOF. As discussed earlier, more than one component can be grounded. Other components will move in relation to the grounded component(s). To see a graphical display of the DOF remaining on all of the components in an assembly, select Degrees of Freedom on the View menu, as shown in the following image on the left.

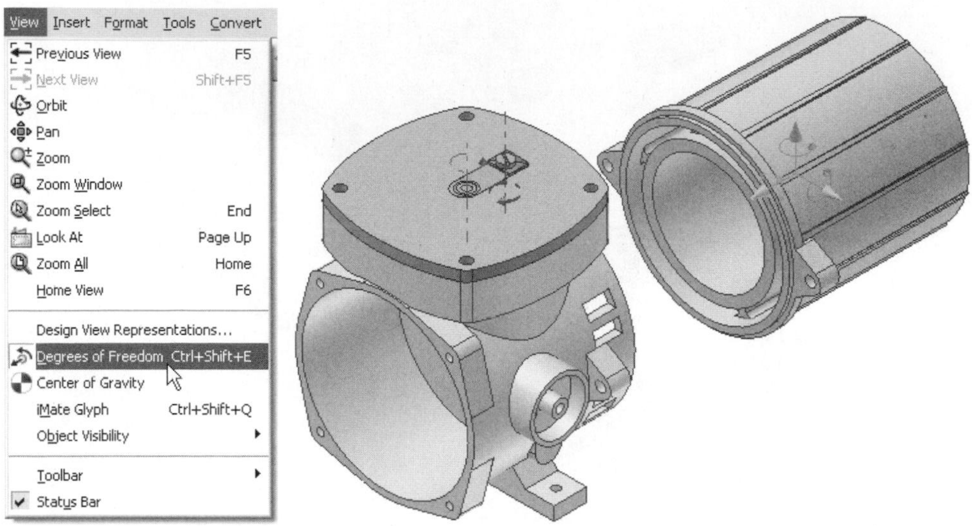

Figure 6.16

An icon will appear in the center of the component that shows the DOF remaining on the component. The line and arrows represent translational freedom, and the arc and arrows represent rotational freedom. To turn off the DOF icons, again click Degrees of Freedom on the View menu.

Tip: You can turn on the DOF symbols for single or multiple component(s) by right-clicking on the component's name in the browser, selecting Properties from the menu, and clicking Degrees of Freedom on the Occurrence tab. If the symbols are turned on, following the same steps will toggle them off.

When placing or creating components in an assembly, it is recommended to list them in the order in which they will be assembled. The order will be important when placing assembly constraints and creating presentation views. When constraining components to one another, you will need to understand the terminology. The following terminology is used with assembly constraints:

Line This can be the centerline of an arc, a circular edge, a cylindrical face, a selected edge, a work axis, or a sketched line.

Normal This is a vector that is perpendicular to a planar face.

Plane This can be defined by the selection of a plane or face to include the following: two noncolinear but coplanar lines or axes, three points, or one line or axis and a point that does not lie on the line or axis. When you use edges and points to select a plane, this creates a work plane, and it is referred to as a construction plane.

Point This can be an endpoint or midpoint of a line, the center or end of an arc or circular edge, or a vertex created by the intersection of an axis and a plane or face.

Offset This is the distance between two selected lines, planes, or points, or any combination of the three.

TYPES OF CONSTRAINTS

Autodesk Inventor uses four types of assembly constraints (mate, angle, tangent, and insert), two types of motion constraints (rotation and rotation-translation), and a transitional constraint. You can access the constraints through the Constraint tool found on the Assembly panel bar, as shown in the following image, by right-clicking and selecting Constraint from the menu, or by using the hot key C.

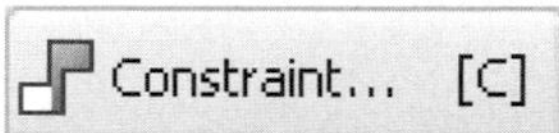

Figure 6.17

The Place Constraint dialog box appears, as shown in the following image. The dialog box is divided into four areas, which are described in the following sections. Depending upon the constraint type, the option titles may change.

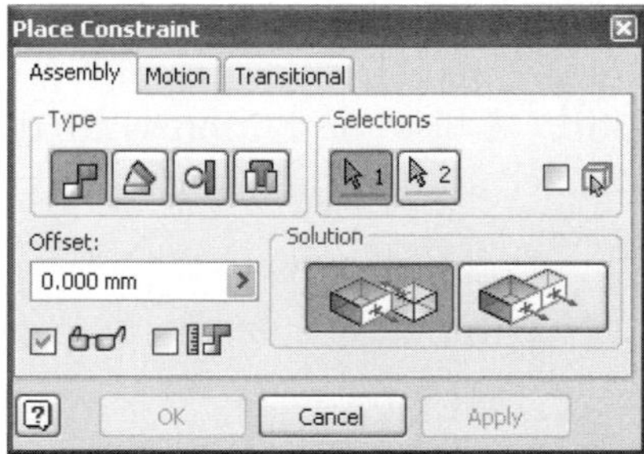

Figure 6.18

The Assembly Tab

Type Select the type of assembly constraint to apply: mate, angle, tangent, or insert.

Selections Click the button with the number 1, and select a component's edge, face, point, and so on, on which to base the constraint type. Then click the button with the number 2, and select a component's edge, face, point, and so on, on which to base the constraint type. By default, the second arrow will become active after you have selected the first input.

Color coding is also available to assist with the assembly process. For example, when picking a face with the number 1 button, the color blue is associated with this selection. In the same way, the color green is associated with the number 2 button selection. This schema allows you to better recognize the selections, especially if they need to be edited.

You can edit an edge, face, point, and so on, of an assembly constraint that has already been applied by clicking the number button that corresponds to the constraint and then selecting a new edge, face, point, and so on. While working on complex assemblies, you can click the box on the right side of the Selections section called Pick part first. If the box has a check, select the component before selecting a component's edge, face, point, and so on.

Offset/Angle Enter or select a value for the offset or angle from the drop-down list.

Solution Select how the constraint will be applied; the normals will be pointing in the same or opposite directions.

Show Preview Click, and when constraints are applied to two components, you will see the underconstrained components previewed in their constrained positions. If you leave the box clear, you will not see the components assembled until you click the Apply button.

Predict Offset and Orientation Click to display the existing offset distance between two components. This allows you to accept this offset distance or enter a new offset distance in the edit box.

The Motion Tab

Type Select the type of assembly constraint to apply: rotation or rotation-translation.

Selections Click the button with the number 1 and select a component's face or axis on which to base the constraint type. You will see a glyph in the graphics window previewing the direction of rotation motion. Click the button with the number 2, and select the component's axis or face on which to base the constraint type. A second glyph appears showing the direction of rotation.

Ratio Enter or select a value for the ratio from the drop-down list.

Solution Select how the constraint type will be applied. The components will rotate in the same or opposite directions as previewed by the graphics window glyphs.

The Transitional Tab

A transitional constraint will maintain contact between the two selected faces. You can use a transitional constraint between a cylindrical face and a set of tangent faces on another part.

Type Select transitional as the type of assembly constraint to apply.

Selections Click the First Selection button, and select the first face on the part that will be moving. Click the Second Selection button, and select a face around which the first part will be moving. If there are tangent faces, they will become chained automatically as part of the selected face.

ASSEMBLY CONSTRAINT TYPES

This section explains each of the assembly constraint types.

Mate

There are three types of mate constraints: plane, line, and point.

Mate Plane The mate plane constraint assembles two components so that the faces on the selected planes will be planar to and opposing one another. In the following image, the mate condition is being applied; it is selected in the Solution area of the Place Constraint dialog box.

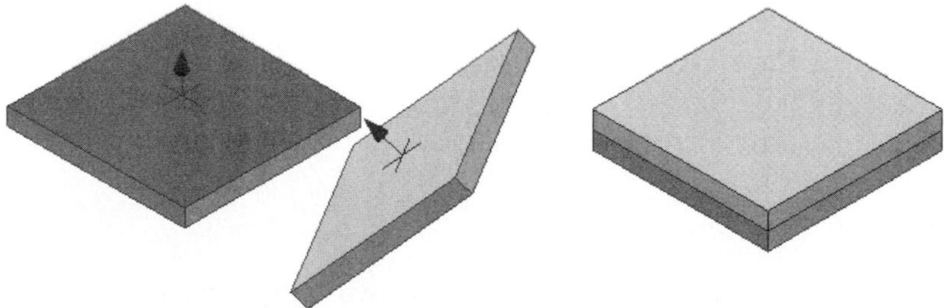

Figure 6.19

Mate Line The mate line constraint assembles the edges of lines to be collinear, as shown in the following image.

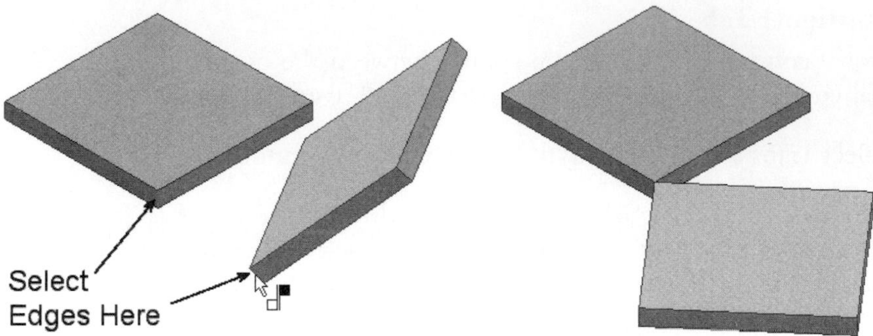

Figure 6.20

You can also use the mate line constraint to assemble the center axis of a cylinder with a matching hole feature or axis, as shown in the following image.

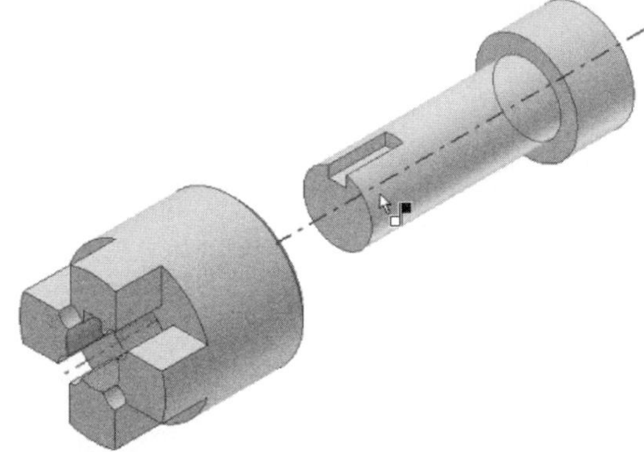

Figure 6.21

Mate Point The mate point constraint assembles two points, such as centers of arcs and circular edges, endpoints, and midpoints, to be coincident, as shown in the following image.

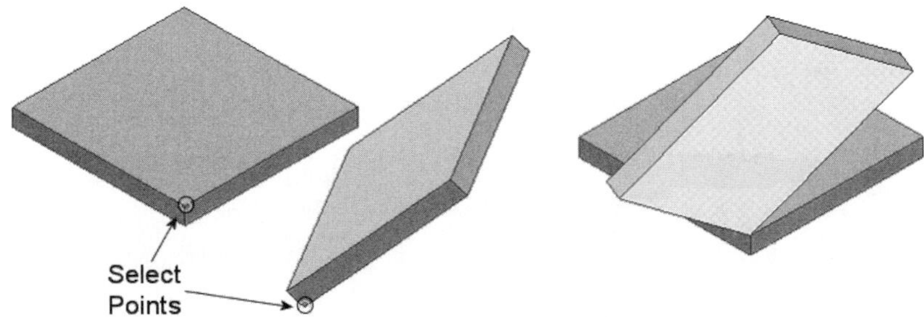

Figure 6.22

Mate Flush Solution

The mate flush solution constraint aligns two components so that the selected planar faces or work planes face the same direction or have their surface normals pointing in the same direction, as shown in the following image. Planar faces are the only geometry that can be selected for this constraint.

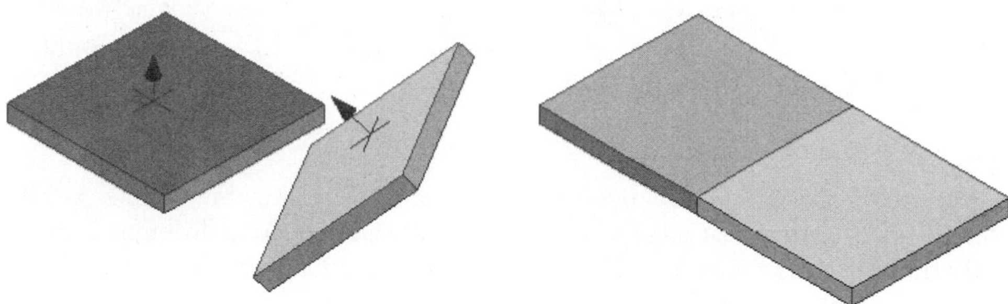

Figure 6.23

Angle

The angle constraint specifies the degrees between selected planes or faces or axes. The following image shows the angle constraint with two planes selected and a 30° angle applied.

Three solutions are available when placing an angle constraint: directed angle; undirected angle and Explicit Reference Vector. The Explicit Reference Vector option requires a third selection that defines the Z axis. You can experiment with both solutions, especially when driving the angle constraint and observing the behavior of the assembly.

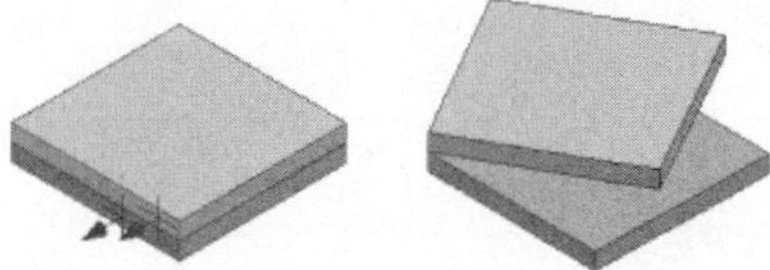

Figure 6.24

Tangent

The tangent constraint defines a tangent relationship between planes, cylinders, spheres, cones, and ruled splines. At least one of the faces selected needs to be a curve, and you can apply the tangency to the inside or outside of the curve.

The following image shows the tangent constraint applied to one outside curved face and a selected planar face, as well as the piece with the outer and inner solutions applied.

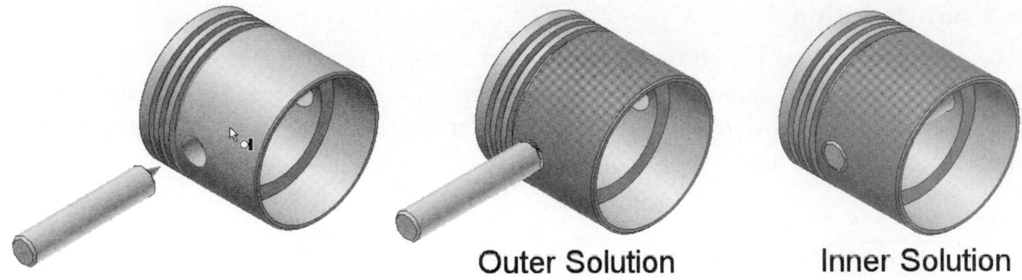

Figure 6.25

Insert

The insert constraint takes away five DOF with one constraint, but it only works with components that have circular edges. Select the circular edges of two different components. The centerlines of the selected circles or arcs will be aligned, and a mate constraint will be applied to the planes defined by the circular edges. Circular edges define a centerline/axis and a plane. The following image shows the insert constraint with two circular edges selected and the opposed solution applied.

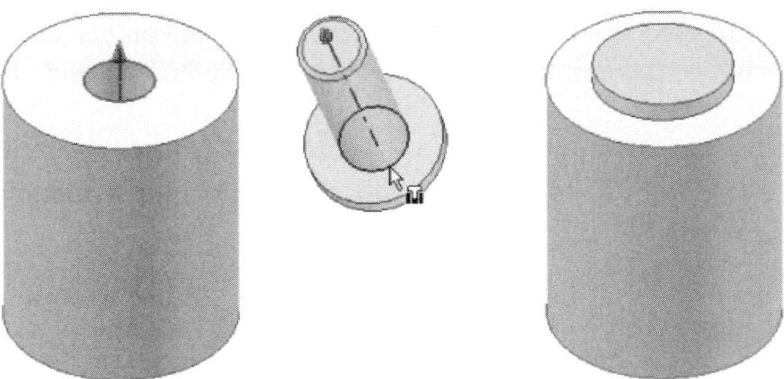

Figure 6.26

MOTION CONSTRAINT TYPES

There are two types of motion constraints: rotation and rotation-translation, as shown in the Type section of the following image. Motion constraints allow you to simulate the motion relationships of gears, pulleys, rack and pinions, and other devices. By applying motion constraints between two or more components, you can drive one component and cause the others to move accordingly.

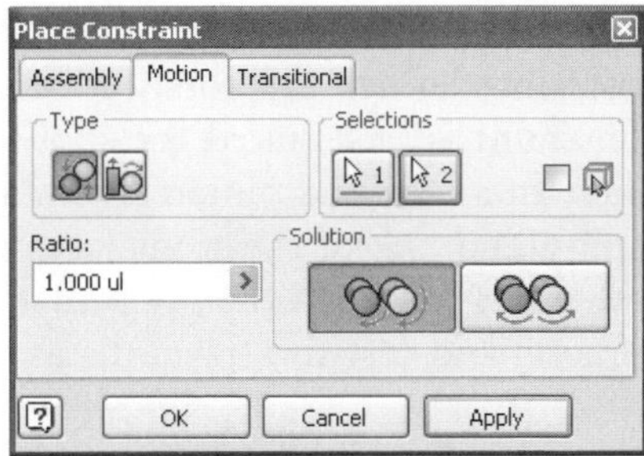

Figure 6.27

Both types of motion constraints are secondary constraints, which means that they define motion but do not maintain positional relationships between components. Constrain your components fully before you apply motion constraints. You can then suppress constraints that restrict the motion of the components you want to animate.

Rotation

The rotation constraint defines a component that will rotate in relation to another component by specifying a ratio for the rotation between the two components. Use this constraint for showing the relationship between gears and pulleys. Selecting the tops of the gear faces displays the rotation glyph, as shown in the following image. You may also have to change the solution type from Forward to Backward, depending on the desired results.

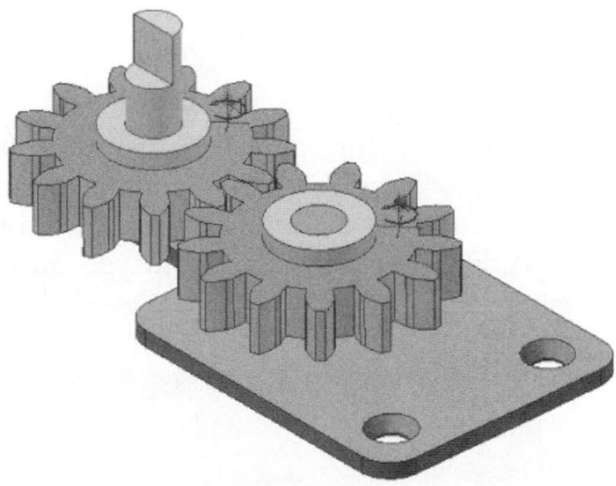

Figure 6.28

Rotation-Translation

The rotation-translation constraint defines the rotation relative to translation between components. This type of constraint is well suited for showing the relationship between rack and pinion gear assemblies. In a rack and pinion assembly, as shown in the following image, the top face of the pinion and one of the front faces of the rack are selected. You supply a distance the rack will travel based on the pitch diameter of the pinion gear, and then you can drive the constraints and test the travel distance of the mechanism.

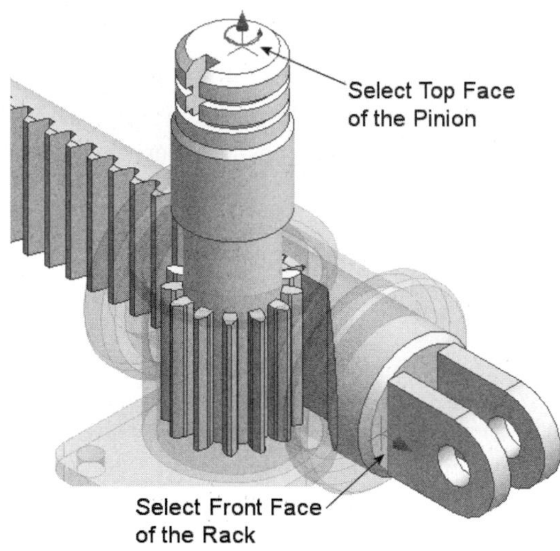

Figure 6.29

TRANSITIONAL CONSTRAINT

The transitional constraint specifies the intended relationship between, typically, a cylindrical part face and a contiguous set of faces on another part, such as a cam follower in a cam slot. The transitional constraint maintains contact between the faces as you slide the component along open DOF. Access this constraint type through the Transitional tab of the Place Constraint dialog box, as shown in the following image.

Figure 6.30

Select the moving face first, followed by the transitional face, as shown on the cam and follower in the following image.

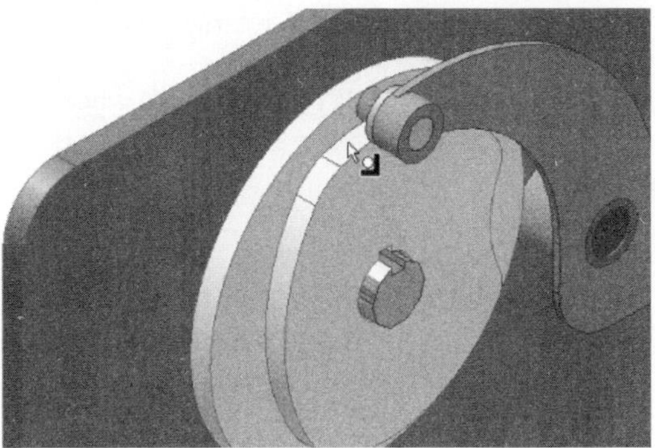

Figure 6.31

APPLYING ASSEMBLY CONSTRAINTS

After selecting the type of assembly constraint that you want to apply, the Selections button with the number 1 will become active; if it does not automatically become active, click the button. Position the cursor over the face, edge, point, and so on, to apply the first assembly constraint.

You may need to cycle through the selection set using the Select Other tool, as shown in the following image, until the correct location is highlighted. Cycle by clicking on the left or right arrows of the tool until you see the desired constraint condition, and then press the left mouse button or the green rectangle in the Select Other tool to place the constraint.

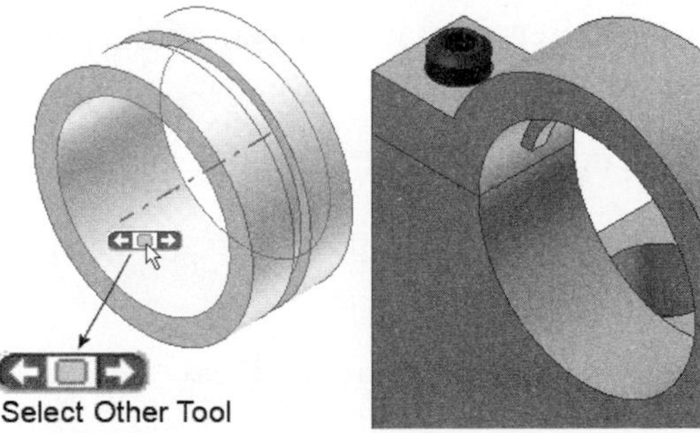

Figure 6.32

The next step is to position the cursor over the face, edge, point, and so on, and select the second geometry input for the assembly constraint. Again, you may need to cycle through the selection set until the correct location is highlighted. If the Show Preview option is selected in the dialog box, the components will move to show how the assembly constraint will affect the components, and you will hear a snapping sound when you preview the constraint. To change either selection, click on the button with the number 1 or 2, and select the new input. Enter a value as needed for the offset or angle, and select the correct Solution option until the desired outcome appears. Click the Apply button to complete the operation. Leave the Constraint dialog box active to define subsequent constraint relationships.

Note: If your mouse is equipped with a rotating middle wheel, you can roll the wheel when the Select Other tool is active to more efficiently cycle through the selection set of faces, edges, or points.

ALT + Drag Constraining

Another way to apply an assembly constraint is to hold down the ALT key while dragging a part edge or face to another part edge or face; no dialog box will appear. The key to dragging and applying a constraint is to select the correct area on the part. Selecting an edge will create a different type of constraint than if a face is selected. If you select a circular edge, for example, an insert constraint will be applied. To apply a constraint while dragging a part, you cannot have another tool active.

To apply an assembly constraint, follow these steps:

1. While holding down the ALT key, select the face, edge, or other part on the part that will be constrained.
2. Select a planar face, linear edge, or axis to place a mate or flush constraint. Select a cylindrical face to place a tangent constraint. Select a circular edge to place an insert constraint.
3. Drag the part into position. As you drag the part over features on other parts, you will preview the constraint type. If the face you need to constrain to is behind another face, pause until the Select Other tool appears. Cycle through the possible selection options, and then click the center dot to accept the selection.

To change the constraint type previewed while you drag the part, release the ALT key and press one of the following shortcut keys:

M or 1 Use to change to a mate constraint. Press the space bar to flip to a flush solution.

A or 2 Use to change to an angle constraint. Press the space bar to flip the angle direction on the selected component.

T or 3 Use to change to a tangent constraint. Press the space bar to flip between an inside and outside tangent solution.

I or 4 Use to change to an insert constraint. Press the space bar to flip the insert direction.

R or 5 Use to change to a rotation motion constraint. Press the space bar to flip the rotation direction.

S or 6 Use to change to a rotation-translation constraint. Press the space bar to flip the translation direction.

X or 8 Use to change to a transitional constraint.

Note: A work plane can also be used as a plane with assembly constraints, a work axis can be used to define a line, and a work point can be used to define a point.

MOVING AND ROTATING COMPONENTS

Use the Move Component tool, as shown in the following image, to drag individual components in any direction in the viewing plane.

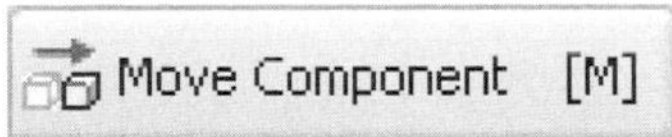

Figure 6.33

To perform a move operation on a component, activate the Move Component command. Click and hold the left mouse button on the component to drag it to a new location. Drop the component at its new location by releasing the button.

Moved components will follow these guidelines:

- An unconstrained component remains in the new location when moved until you constrain it to another component.
- A partially constrained component initially remains in the new location. When the assembly is updated, the component adjusts its location to comply with the constraints that you have already applied.

From the Assembly panel bar, use the Rotate Component tool, as shown in the following image, to rotate an individual component. This tool is very useful when constraining faces that are hidden from your view.

Figure 6.34

Follow these steps for rotating a component in an assembly:

1. Activate the Rotate Component command, and select the component to rotate. Notice the appearance of the 3D rotate symbol on the selected component in the following image.

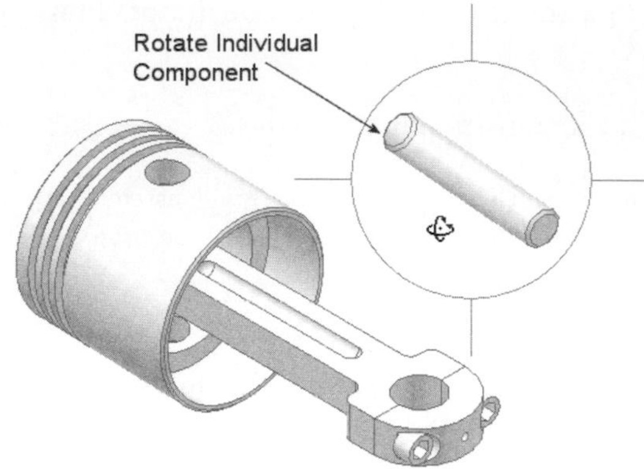

Figure 6.35

2. Drag your cursor until you see the desired view of the component.

- For free rotation, click inside the Dynamic Rotate tool, and drag in the desired location.
- To rotate about the horizontal axis, click the top or bottom handle of the Dynamic Rotate tool, and drag your cursor vertically.
- To rotate about the vertical axis, click the left or right handle of the Dynamic Rotate tool, and drag your cursor horizontally.
- To rotate planar to the screen, hover over the rim until the symbol changes to a circle, click the rim, and drag in a circular direction.
- To change the center of rotation, click inside or outside the rim to set the new center.

Release the mouse button to drop the component into the rotated position.

Note: If you click the Update button after moving or rotating components in an assembly, any components constrained to a grounded component will snap to their constrained positions in the new location. A fully constrained component can be moved or rotated temporarily. Once the assembly constraints are updated, the component will resolve the constraints and return to a fully constrained location/orientation.

EDITING ASSEMBLY CONSTRAINTS

After you have placed an assembly constraint, you may want to edit, suppress, or delete it to reposition the components. There are two ways to edit assembly constraints.

Both methods are executed through the browser. In the browser, activate the assembly or subassembly that contains the component that you want to edit. Expand the component name, and you will see the assembly constraints, as shown in the following image on the left. Double-click on the constraint name, and an Edit Dimension dialog box will appear, allowing you to edit the constraint offset value. You can also right-click on the assembly constraint's name in the browser and select Delete, Edit, or Suppress from the menu, as shown in the following image on the right.

Note: When editing offset dimension values, click once with the left mouse button on the constraint with the offset value. Then change the offset value from the edit box that will appear at the bottom of the Assembly browser.

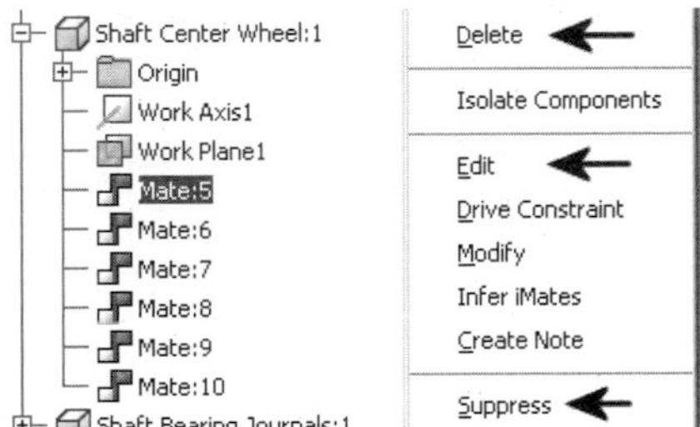

Figure 6.36

If you select Edit, the Edit Constraint dialog box will appear. If you select Suppress, the assembly constraint will not be applied. Select Drive Constraint to drive a constraint through a sequence of steps, simulating mechanical motion. Select Delete, and the assembly constraint will be deleted from the component. If you try to place or edit an assembly constraint and it cannot be applied, an alert box that explains the problem will appear. You will have to either select new options for the operation or suppress or delete another assembly constraint that conflicts with it.

If an assembly constraint is conflicting with another, a triangular, yellow icon with an exclamation point will appear in the browser, as shown in the following image on the left. To edit a conflicting constraint in the browser, either double-click on its name or right-click on its name and select Recover from the menu, as shown in the following image on the right. The Design Doctor will appear and walk you through the steps to fix the problem.

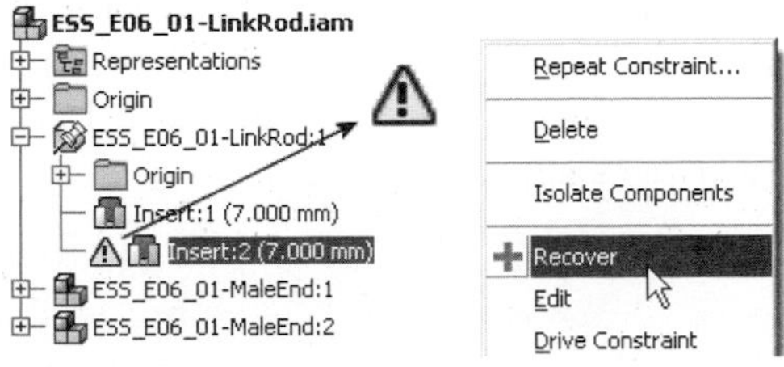

Figure 6.37

ADDITIONAL CONSTRAINT TOOLS

Additional tools available to manipulate, navigate, and edit assembly constraints include browser views, Other Half, Constraint Offset Value Modification, Isolating Assembly Components, and Isolating Constraint Errors. The following sections describe these tools.

BROWSER VIEWS

Two modes for viewing assembly information are located in the top of the browser toolbar: Assembly View and Modeling View. Assembly View, the default mode, displays assembly constraint symbols nested below both constrained components, as shown in the following image on the left. In this mode, the features used to create the part are hidden.

When Modeling View is set, assembly constraints are located in a Constraints folder at the top of the assembly tree, as shown in the following image on the right. In this mode, the features used to create the part are displayed just as they are in the original part file.

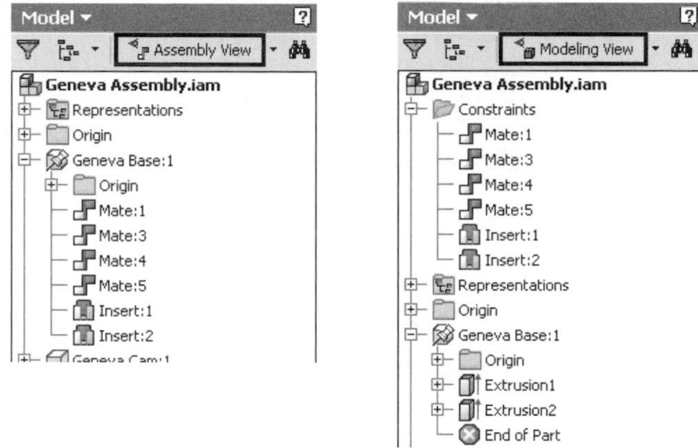

Figure 6.38

OTHER HALF

You can use the Other Half tool to find the matching part that participates in a constraint placed in an assembly. As you add parts over time, you may wish to highlight an assembly constraint and find the part(s) to which it is constrained. As shown in the following image on the left, a mate constraint has been highlighted. Half of this constraint has been applied to a part called Engine Block:1. To view the part sharing a common constraint, right-click on the constraint in the browser, and select Other Half from the menu, as shown in the following image in the middle. The browser will expand and highlight the second half of the constraint, as shown in the following image on the right. In this example, the other half of the mate constraint is a part called Cylinder:1.

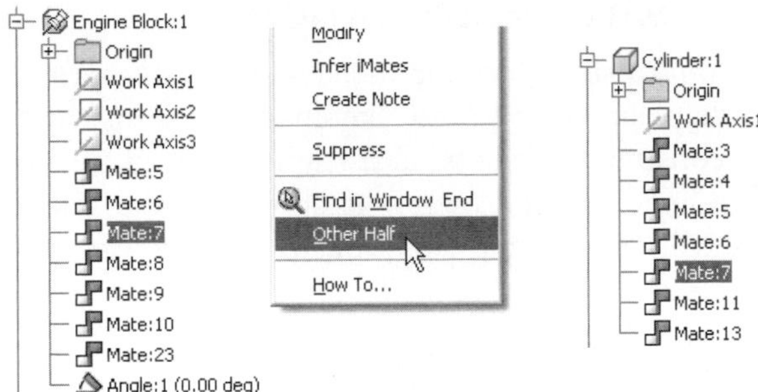

Figure 6.39

CONSTRAINT TOOL TIP

To display all property information for a specific constraint, move your cursor over the constraint icon, and a tooltip will appear, as shown in the following image. Although the constraint name is highlighted in the image, you must hover your cursor over the constraint icon to view the tooltip information.

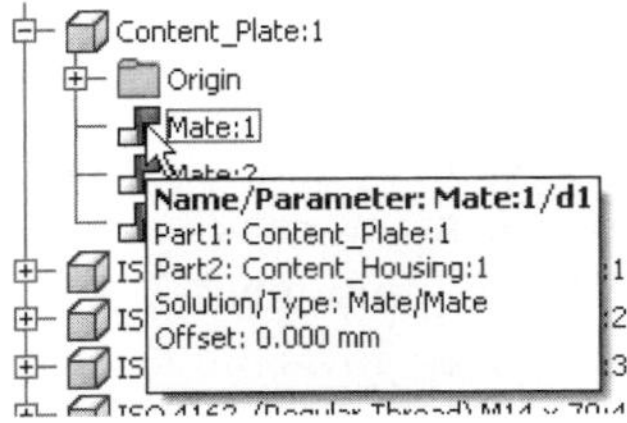

Figure 6.40

The following information is displayed in the tool tip:

- Constraint name and parameter name, applicable to offset and angle parameters
- Constrained components, that is, the two part names from the Assembly browser
- Constraint solution and type
- Constraint offset or angle value

CONSTRAINT OFFSET VALUE MODIFICATION

When editing a constraint offset value, use the standard value edit control. This process is similar to editing work plane offsets and sketch dimensions, and it will allow you to measure while editing constraint offset values. Right-click on the constraint to edit in the browser, and select Modify from the menu, as shown in the following image. The Edit Dimension dialog box will appear, allowing you to edit the offset value.

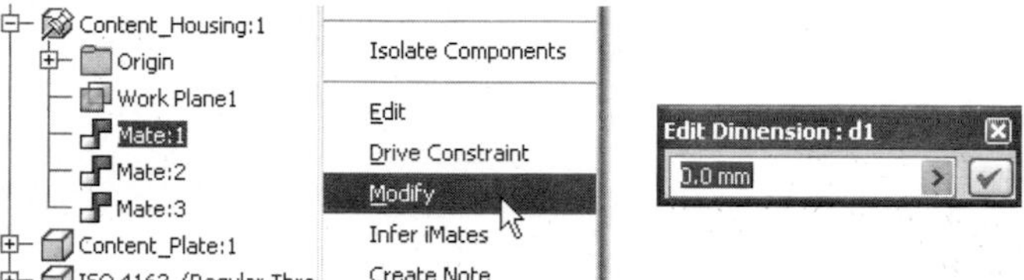

Figure 6.41

Note: When modifying an offset value, the Offset Edit box is also present at the bottom of the Assembly browser, enabling you to make changes to the constraint.

ISOLATING ASSEMBLY COMPONENTS

Components can be isolated as a means of viewing smaller sets of components, especially in a large assembly. In the assembly browser, click on the components to isolate, right-click to display a menu, and select Isolate Components, as shown in the following image. All unselected components will be set to an invisible status.

Components can be isolated based on an assembly constraint. As shown in the following image on the left, right-clicking on the Mate:7 constraint and selecting Isolate Components from the menu will display the engine assembly that is shown on the right.

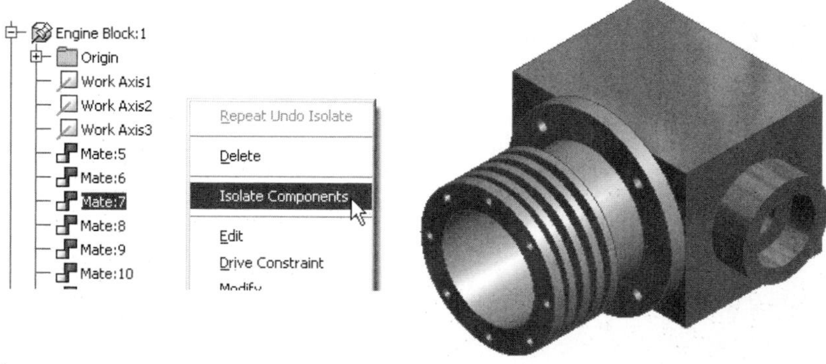

Figure 6.42

To return the assembly to its previous assembled state, right-click inside the browser, and select Undo Isolate from the menu, as shown in the following image.

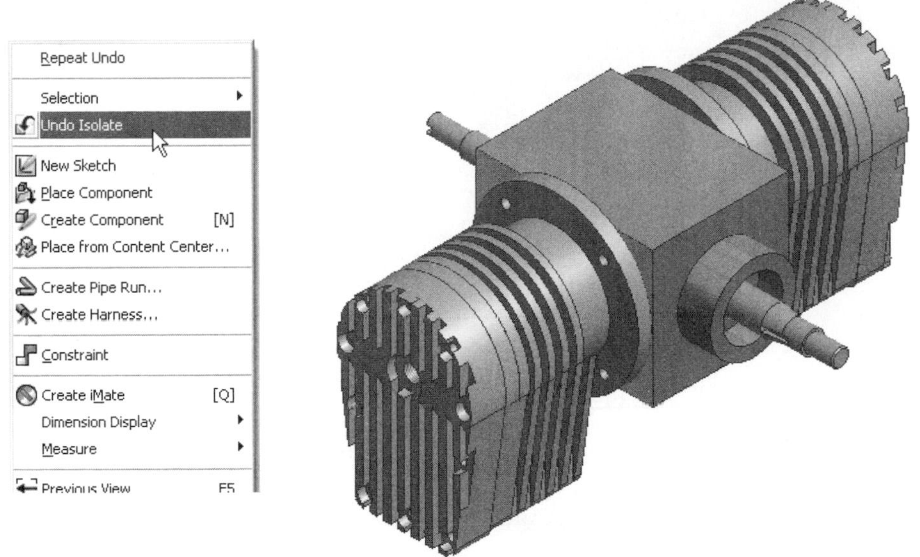

Figure 6.43

ISOLATING CONSTRAINT ERRORS

Assembly constraints can be isolated when errors in their placement occur. When editing a constraint through the Design Doctor, use the Isolate and Edit Constraint tool, as shown in the following image in the middle. This action will turn off all components except those that participate in the common constraint, and it will display the Constraint dialog box, allowing you to edit the constraint with the errors.

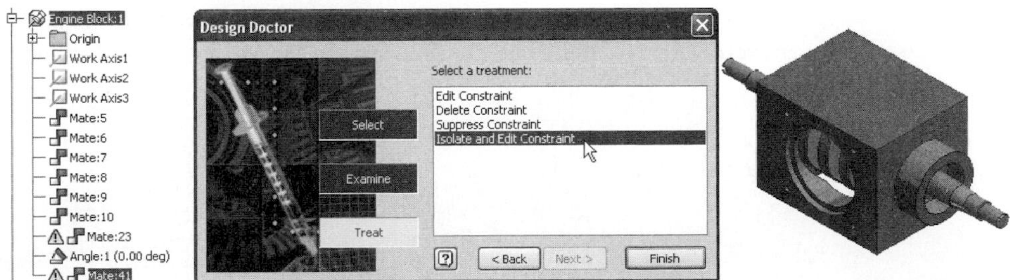

Figure 6.44

EXERCISE 6-1: ASSEMBLING PARTS

In this exercise, you assemble a lift mechanism.

1. Open the assembly file *ESS_E06_01.iam* in the chapter 06 folder, as shown in the following image.

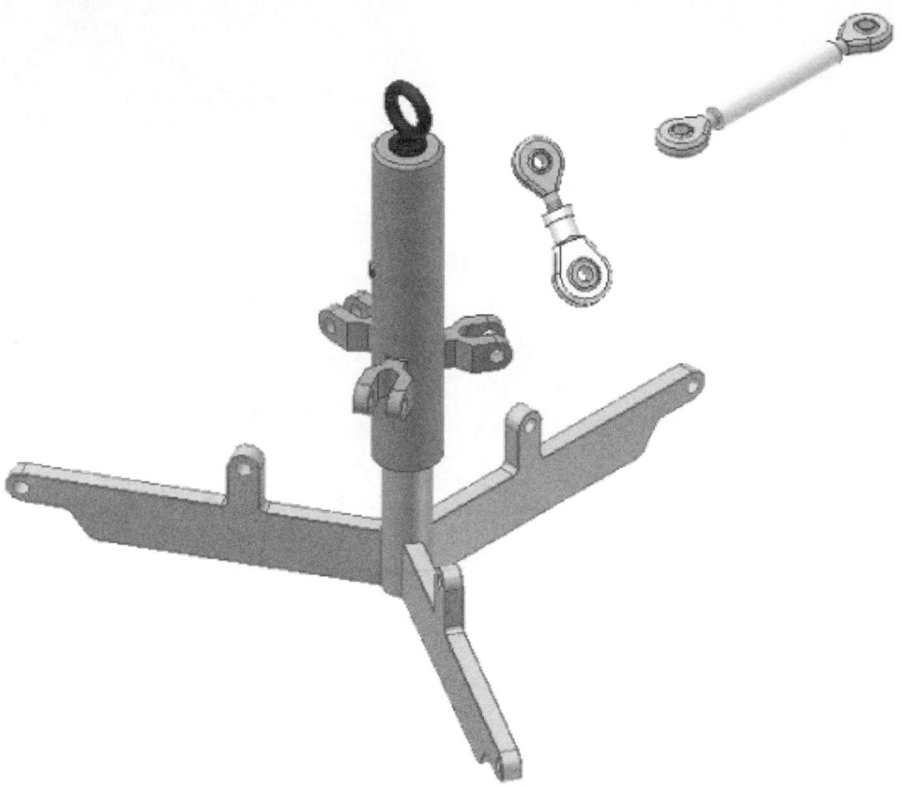

Figure 6.45

2. Begin assembling the connector and sleeve. Right-click in the graphics window, select Home View (F6 function key), and then zoom in on the small connector and sleeve. Drag the connector so that the small end is near the sleeve, as shown in the following image.

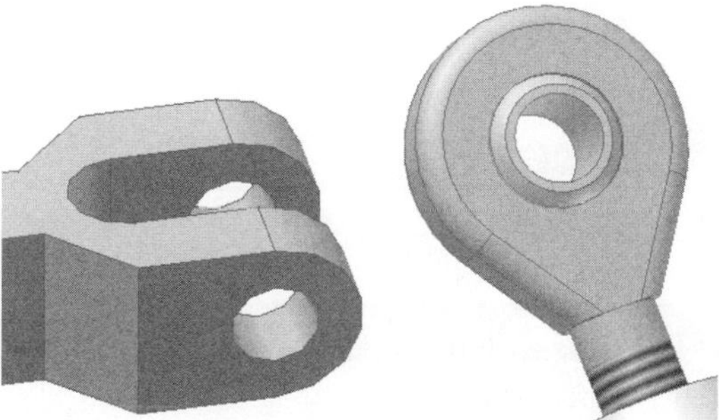

Figure 6.46

3. Next add a mate between the centerlines of both components. Click the Constraint tool in the panel bar. Mate is the default constraint.
4. Move the cursor over the hole in the arm on the sleeve. Click when the centerline displays, as shown in the following image.

Note: If the green dot displays, move the cursor until the centerline displays, or use Select Other to cycle through the available choices.

5. Move the cursor over the hole in the link. Click when the centerline displays, as shown in the following image.
6. Click Apply to accept this constraint.

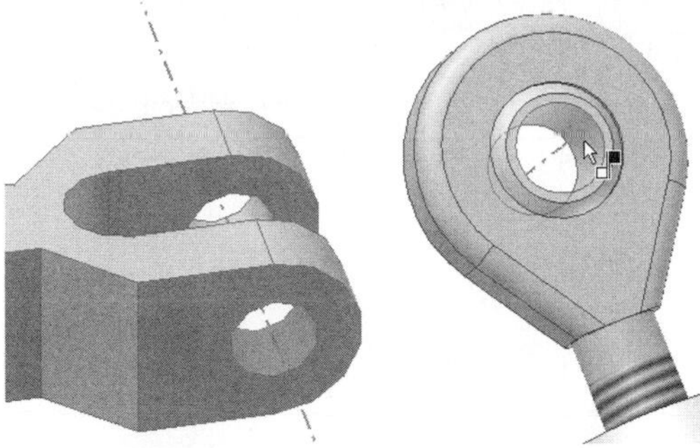

Figure 6.47

7. Now add a mate between the faces of both parts. Click the small flat face on the ball in the link end, as shown in the following image.
8. Click the inner face of the slot on the sleeve, as shown in the following image.
9. Click Apply to accept this constraint.

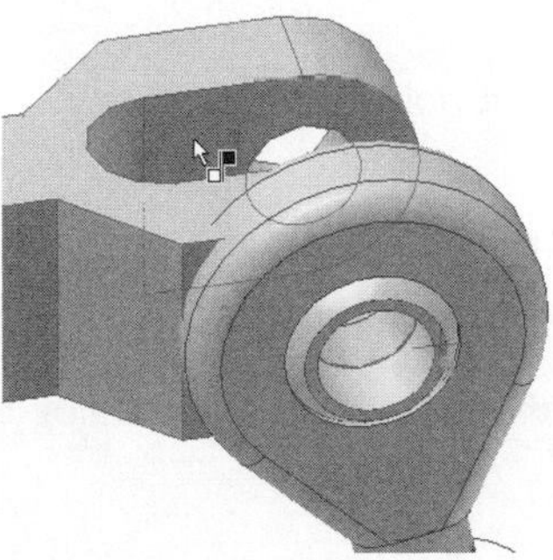

Figure 6.48

10. Click Cancel to close the Place Constraint dialog box.

11. Click on the small link arm, and drag it to view the effect of the two constraints, as shown in the following image.

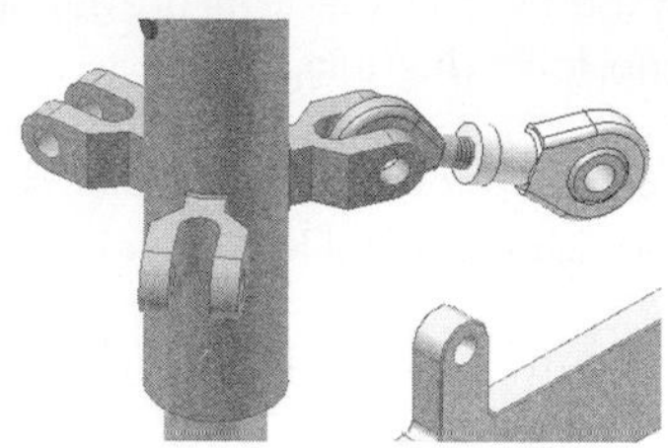

Figure 6.49

12. Place the crank in the assembly by clicking the Place Component tool in the panel bar and opening the file *ESS_E06_01-Crank.ipt.*
13. Move the component near the spyder arm, and click to place the part, as shown in the following image. Right-click, and select Done.

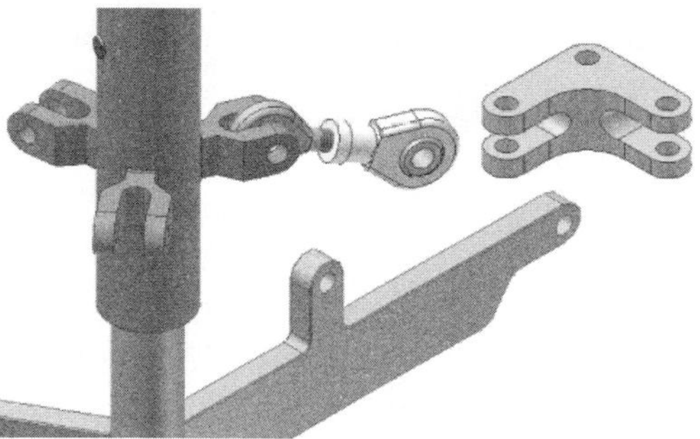

Figure 6.50

14. Begin the process of constraining the crank arm by clicking the Constraint tool in the panel bar.
15. Move the cursor over the hole in the arm on the crank, and click when the centerline displays.
16. Move the cursor over the hole in the spyder, and click when the centerline displays.
17. Click Apply to accept this constraint.
18. Click Cancel to close the Place Constraint dialog box.

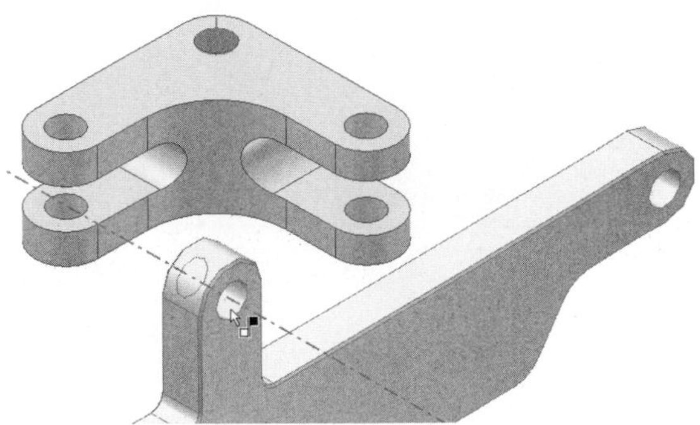

Figure 6.51

19. Drag the crank away from the spyder arm to make it easier to apply the next constraint.

20. Next constrain the faces of the crank arm and the spyder.

21. Click the Constraint tool in the panel bar.

22. Click the inner face of slot on the crank, as shown in the following image on the left.

23. Rotate the model, click the face of the spyder arm, as shown in the following image on the right, and choose the face.

24. Click Apply to accept this constraint. Click Cancel to close the Place Constraint dialog box.

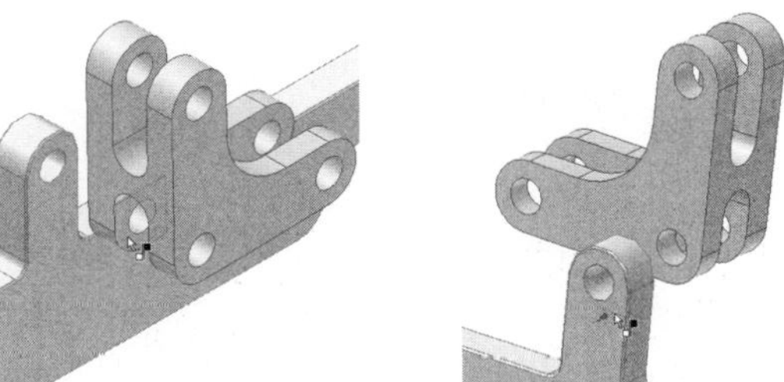

Figure 6.52

25. You will now assemble the crank and link. Return to the Home View.

26. Drag the end of the small link arm close to the crank, as shown in the following image.

27. Zoom in to the crank and link.

28. Place a mate constraint between the centerlines of the two holes, as shown in the following image. Click OK in the Place Constraint dialog box.

Figure 6.53

29. You will now place a claw in the assembly. Click the Place Component tool in the panel bar, and open the file *ESS_E06_01-Claw.ipt*.

30. Move the component near the end of the spyder arm, and click to place the part. Right-click, and select Done. Your display should appear similar to the following image.

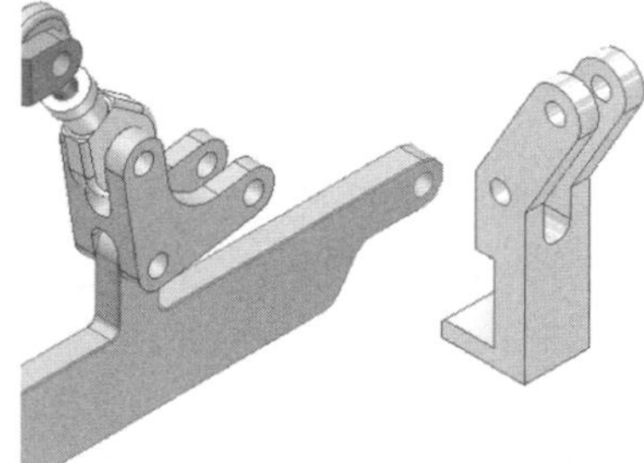

Figure 6.54

31. Constrain the claw to the spyder by first placing a mate constraint between the centerlines of the holes, as shown in the following image on the left.

32. Place another mate constraint between the two faces, as shown in the following image on the right.

Note: When the centerline mate constraint is applied, the claw may be oriented incorrectly. The orientation will correct when the face-to-face mate is applied.

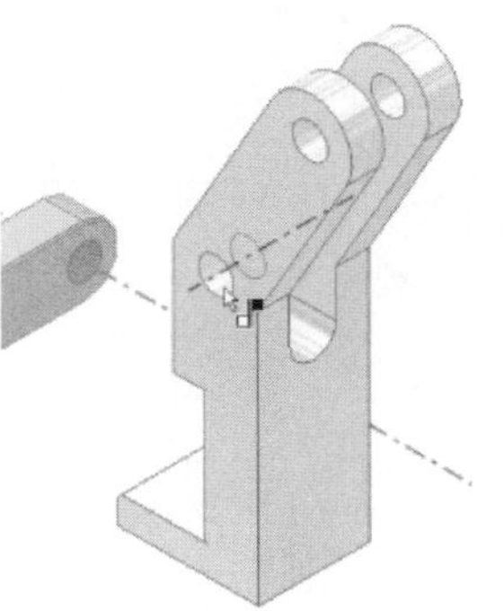
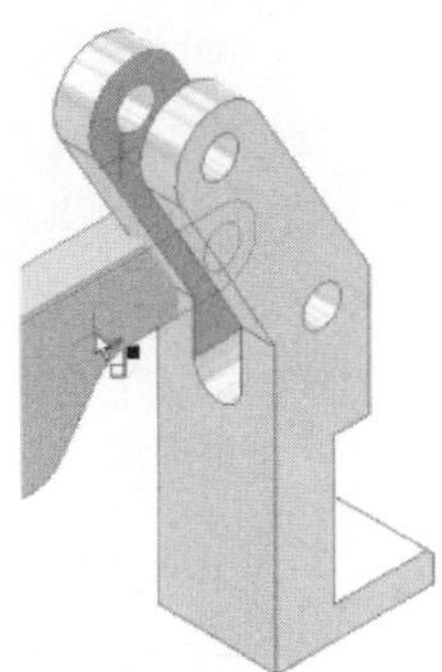

Figure 6.55

33. Assemble the link rod to the crank and claw. First drag the claw and link rod into the position, as shown in the following image.

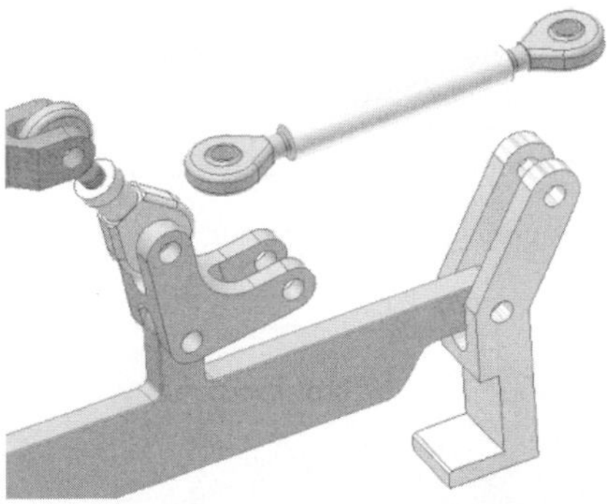

Figure 6.56

34. Assemble the link arm to the claw and crank using mate constraints between centerlines at each end and a mate between faces at one end. Close the Place Constraint dialog box, and drag the claw to view the affect of the assembly constraints, as shown in the following image.

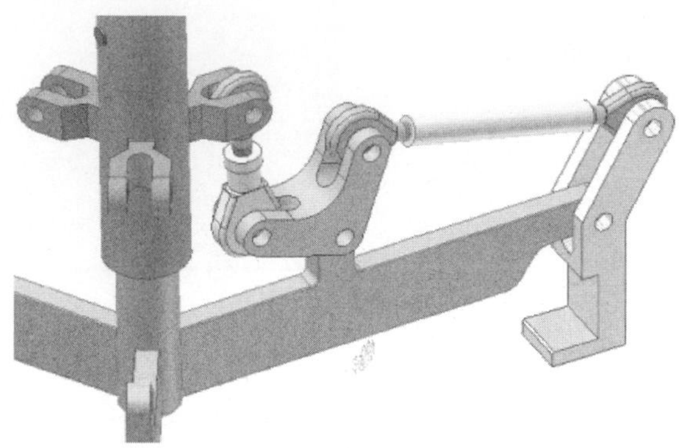

Figure 6.57

35. You will now add bolts and nuts to the spyder assembly. Begin by clicking the Place Component tool in the panel bar and opening the file *ESS_E06_01-Bolt.ipt*.

36. Place 6 bolts near their final position in the assembly, as shown in the following image. Right-click and select Done.

37. Click the Place Component tool in the panel bar, and open the file *ESS_E06_01-Nut.ipt*.

38. Place 6 nuts in the assembly, as shown in the following image. Right-click and select Done.

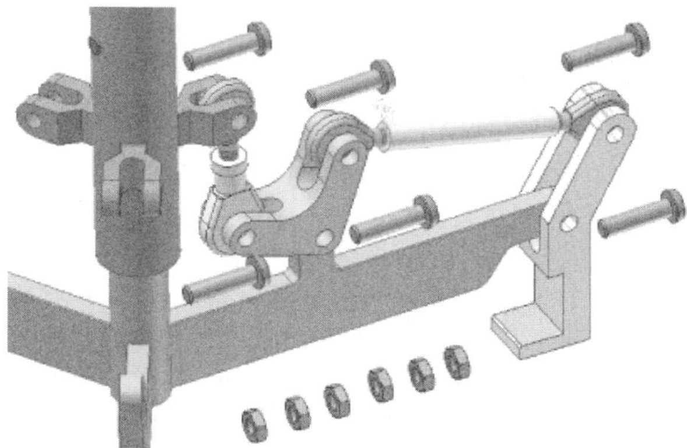

Figure 6.58

39. Insert the bolts into the assembly by clicking the Constraint tool in the panel bar.

40. In the Type area, click Insert to select an insert constraint.

41. Insert each bolt by selecting the top of the bolt's shank and then the edge of the hole, as shown in the following image on the left.

42. Close the Place Constraint dialog box.

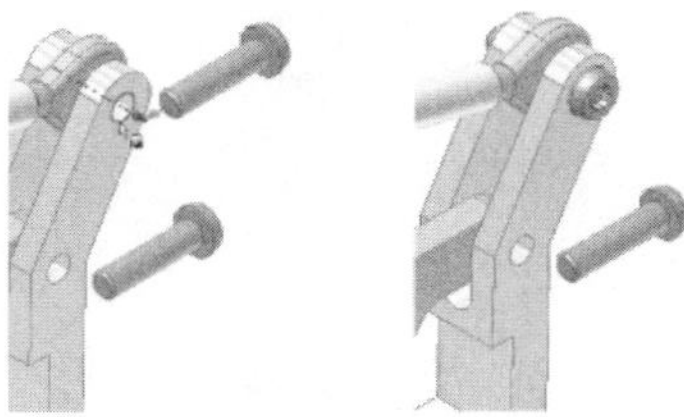

Figure 6.59

43. Now add the nuts to the end of the bolts.
44. Rotate the model so that you can see the back of the claw. Drag 2 nuts near their final location, as shown in the following image.
45. Click the Constraint tool in the panel bar. Click Insert to select an insert constraint.
46. Insert each nut by selecting the edge of the hole in the nut and the edge of the hole on the claw, as shown in the following image on the left.

Note: You must select an edge on the face of the nut. Do not select the inner chamfered edge.

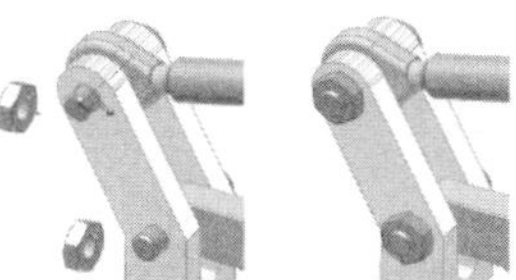

Figure 6.60

47. Assemble the remaining bolts and nuts using the same procedure. The completed assembly is displayed in the following image.

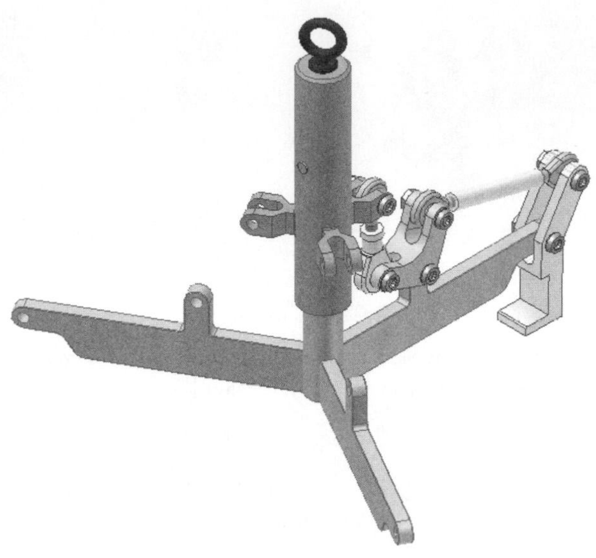

Figure 6.61

48. Close all open files. Do not save changes. End of exercise.

DESIGNING PARTS IN PLACE

Most components in the assembly environment are created in relation to existing components in the assembly. When creating an in-place component, you can sketch on the face of an existing assembly component or a work plane. You can also click the graphics window background to define the current view orientation as the XY plane of the assembly in which you are presently working. If the YZ or XZ plane is the default sketch plane, you must reorient the view to see the sketch geometry. Click Application Options on the Tools menu, and then click on the Part tab to set the default sketch plane.

When you create a new component, you can select an option in the Create In-Place Component dialog box to constrain the sketch plane to the selected face or work plane automatically, as shown in the following image. After you specify the location for the sketch, the new part immediately becomes active, and the browser, panel bar, and toolbars switch to the part environment.

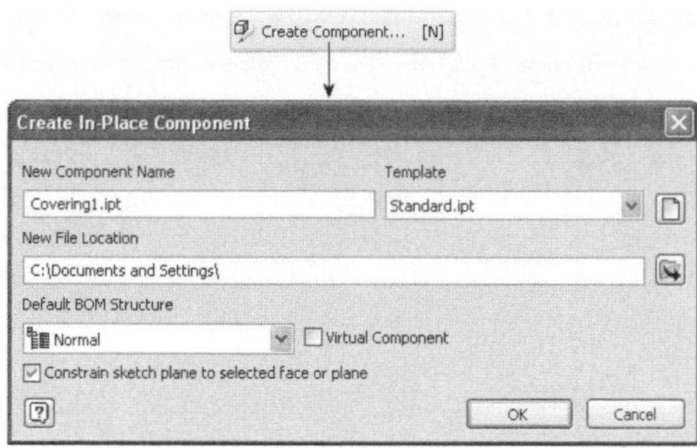

Figure 6.62

Notice also that the 2D Sketch Panel, as shown in the following image, is available to create sketch geometry in the first sketch of your new part.

Figure 6.63

After you create the base feature of your new part, you can define additional sketches based on the active part or other parts in the assembly. When defining a new sketch, you can click a planar face of the active part or another part to define the sketch plane on that face. You can also click a planar face and drag the sketch away from the face to create the sketch plane automatically on the resulting offset work plane. When you create a sketch plane based on a face of another component, Autodesk Inventor automatically generates an adaptive work plane and places the active sketch plane on it. The adaptive work plane moves as necessary to reflect any changes in the component on which it is based. When the work plane adapts, your sketch moves with it. Features based on the sketch then adapt to match its new position.

After you finish creating a new part, you can return to assembly mode by double-clicking the assembly name in the browser. In assembly mode, assembly constraints become visible in the browser. If you selected the Constrain Sketch Plane to Selected Face option when you created your new part, a flush constraint will appear in the Assembly browser. As with all constraints, you can delete or edit this constraint at any time. No flush constraint is generated if you create a sketch by clicking in the graphics window or if the box is clear when selecting an existing part face.

EXERCISE 6–2: DESIGNING PARTS IN THE ASSEMBLY CONTEXT

In this exercise, you create a lid for a container based on the geometry of the container. This cross part sketch geometry is adaptive, and it automatically updates to reflect design changes in the container.

1. Start with an existing assembly. You will then create a new part based on cross part sketch geometry. Begin by opening the file *ESS_E06_02.iam,* as shown in the following image.

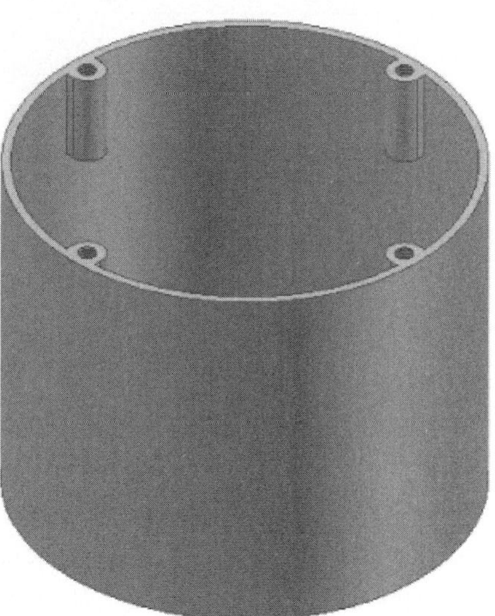

Figure 6.64

2. Begin the process of creating a new component in the context of an assembly by clicking the Create Component tool.
3. Under New File Name, enter ESS_E06_02-LID.
4. Click the Browse Templates button, and in the Metric tab, click *Standard (mm).ipt.* Click OK. There should be a checkmark beside Constrain Sketch Plane to Selected Face or Plane at the bottom of the Create In-Place Component dialog box.
5. Click the OK button to exit the dialog box, and then select the top face of the container, as shown in the following image. This face becomes the new sketch plane for the new part.

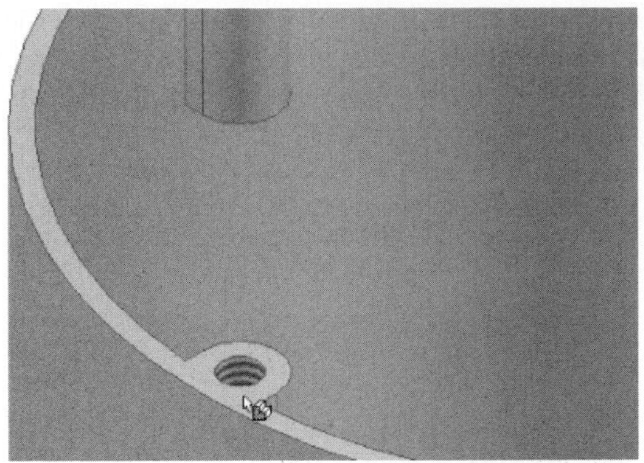

Figure 6.65

6. You will now project all geometry contained in this face. First click the Project Geometry tool.
7. Move the cursor onto the face of the base part until the profile of the entire face is highlighted, as shown in the following image.

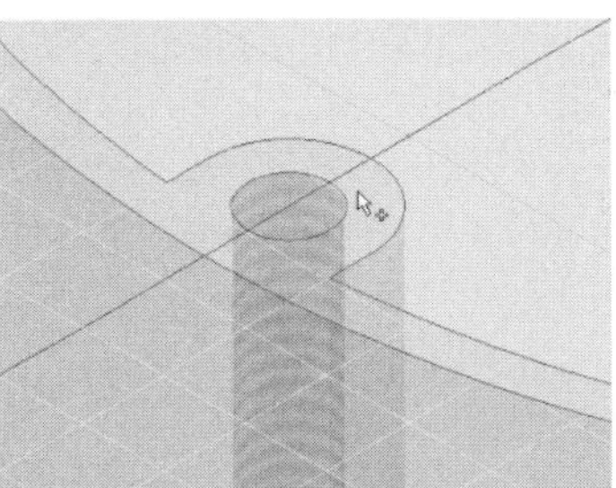

Figure 6.66

8. Click to project the edges. Your display should appear similar to the following image. (The default yellow color of the projected edges has been changed for clarity.)

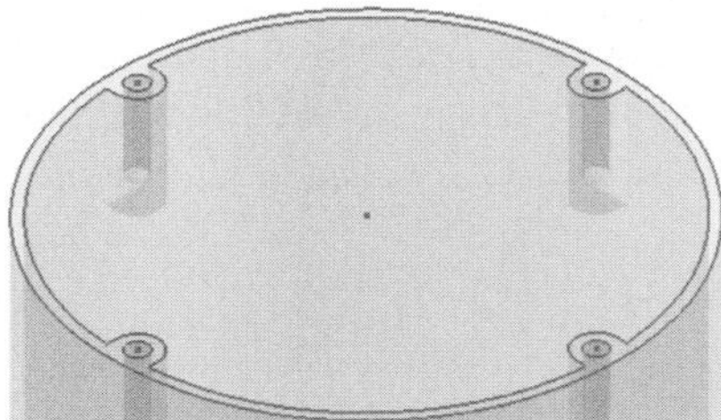

Figure 6.67

9. You will now create a series of clearance holes. First zoom into a tapped hole.
10. Click the Center Point Circle tool, and click the projected circles center point.

11. Move the cursor, and click to create a circle larger than the projected circle, as shown in the following image.

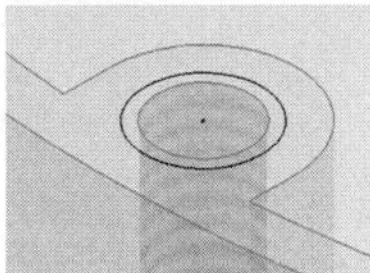

Figure 6.68

12. Create three additional circles centered on the remaining projected holes, as shown in the following image.

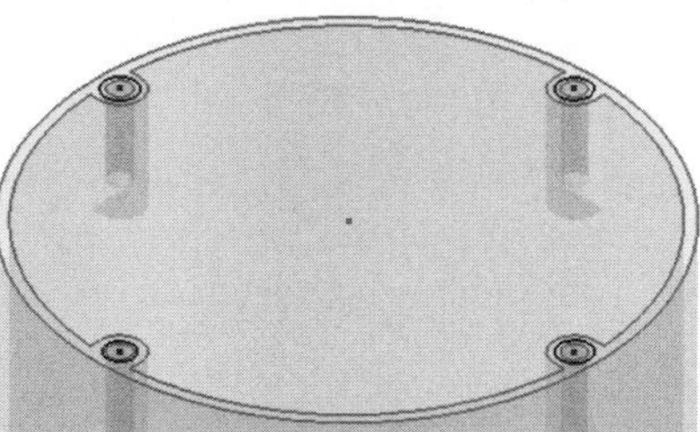

Figure 6.69

13. All circle diameters need to be made equal. Perform this operation by clicking the Equal constraint tool.
14. Click one circle, and then click one of the other circles. Click the first circle again, and click a third circle. Click the first circle again, and then click the fourth circle. These steps will make all circles equal to each other in diameter.
15. Now add a dimension to one of the circles to fully constrain the sketch. First click the General Dimension tool.
16. Click the edge of one of the circles, and click outside the circle to place the dimension.
17. Enter **2.5 mm** in the dimension edit box, and press ENTER to apply the dimension. Your display should appear similar to the following image.

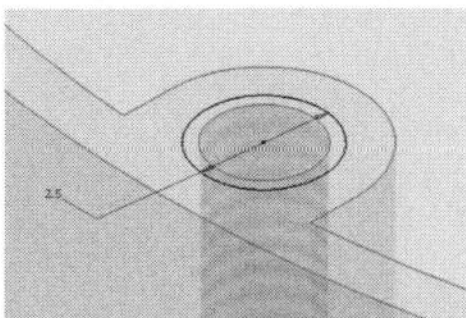

Figure 6.70

18. Next click the Return button to exit Sketch mode. Click the Extrude tool, and select the profile, as shown in the following image.

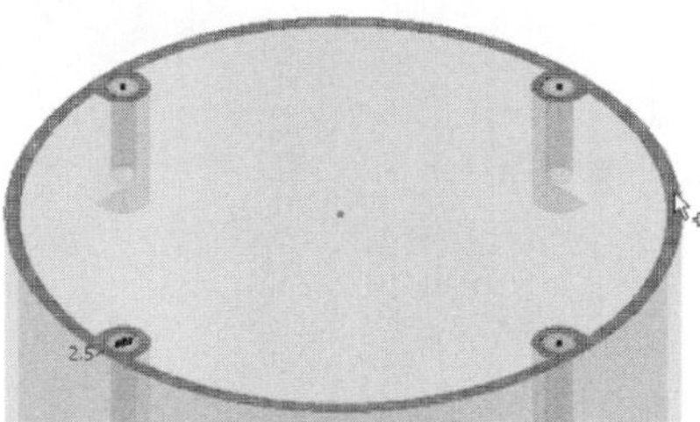

Figure 6.71

19. In the Extrude dialog box, enter a distance of **5,** and then click OK. Your display should appear similar to the following image.

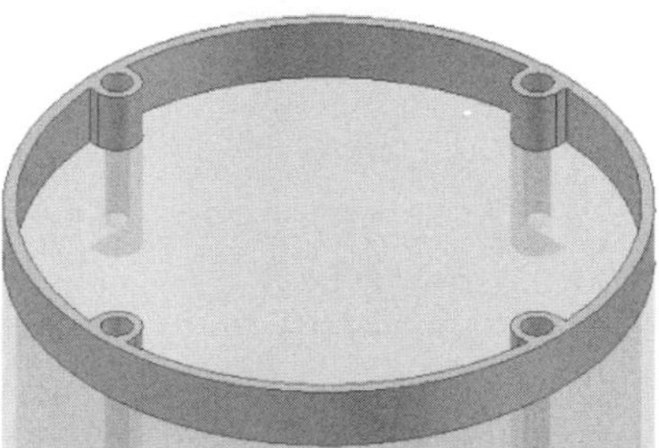

Figure 6.72

20. You will now use the Extrude tool to create the top of the lid. Click the Sketch tool, and select the top face of the lid, as shown in the following image.

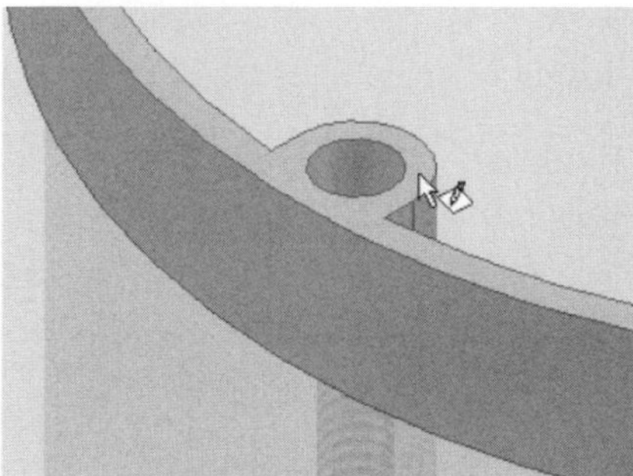

Figure 6.73

21. Press E to start the Extrude tool, and select the two profiles, as shown in the following images.

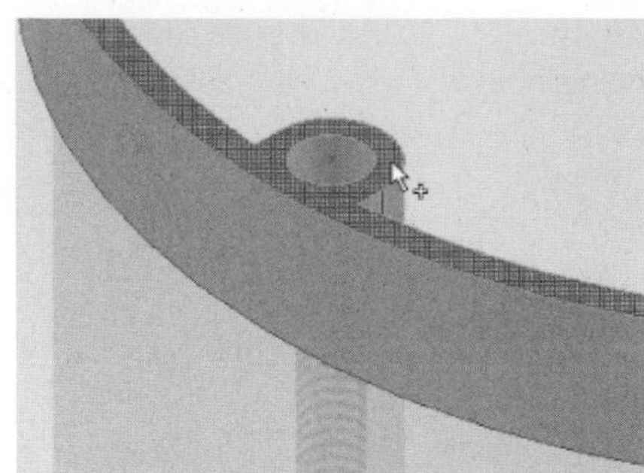
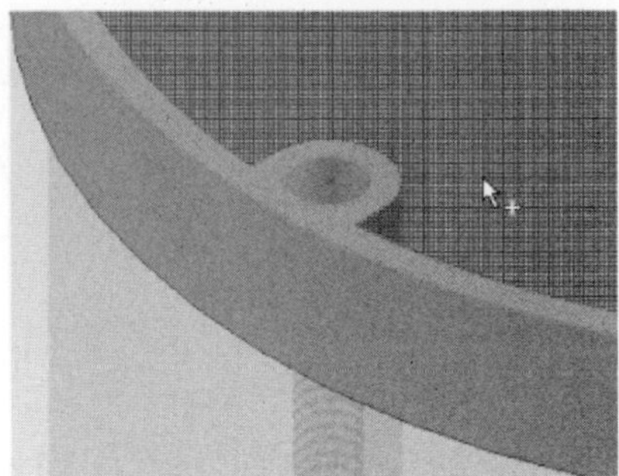

Figure 6.74

22. While inside the Extrude dialog box, enter a distance of **3**. You may also have to change the direction of the extrusion to point to the inside of the lid. Click OK to create the top of the lid, as shown in the following image.

Figure 6.75

23. Click the Return tool to activate the assembly. Your display should appear similar to the following image.

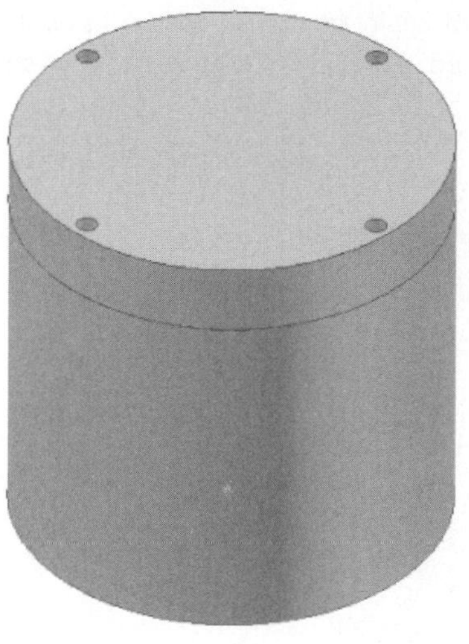

Figure 6.76

24. You will now modify the container base and observe how this affects the lid. First view the model as Hidden Edge Display, as shown in the following image on the left.
25. In the browser, right-click *ESS_E06_02-Container:1*, and then select Edit. Your display should appear similar to the following image on the right.

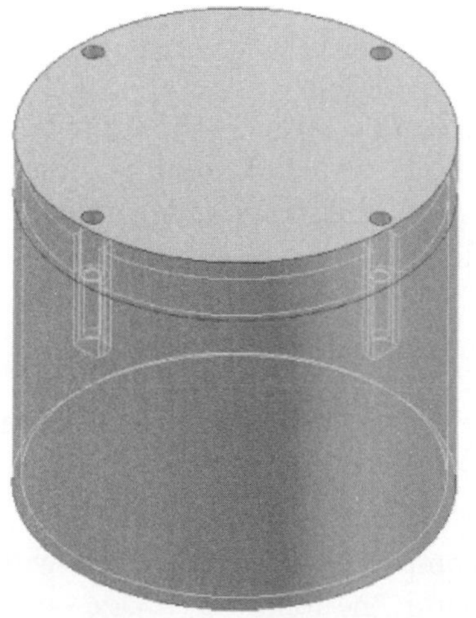

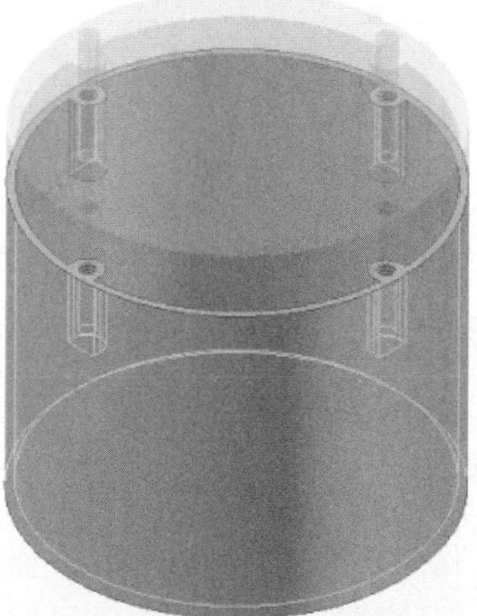

Figure 6.77

26. In the browser, right-click Extrusion1, and then select Edit Sketch.
27. Double-click the 50 dimension, and change the value to **60**.
28. Click the Return tool. Switch back to Shaded Display, as shown in the following image on the left.

29. Click the Return tool again to return to the assembly context. Notice that the lid adapts to the modified dimensions of the base, as shown in the following image on the right.

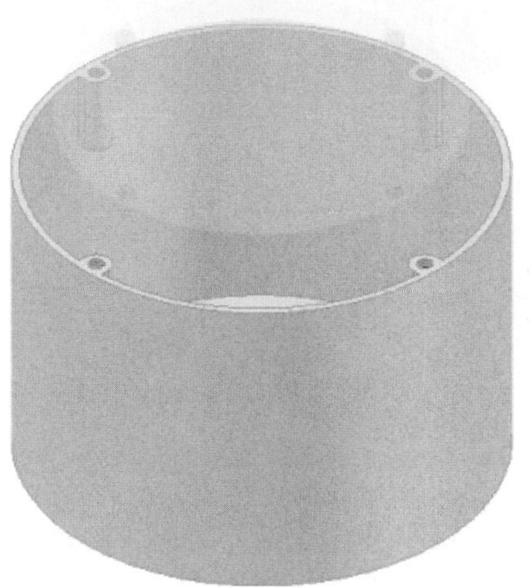

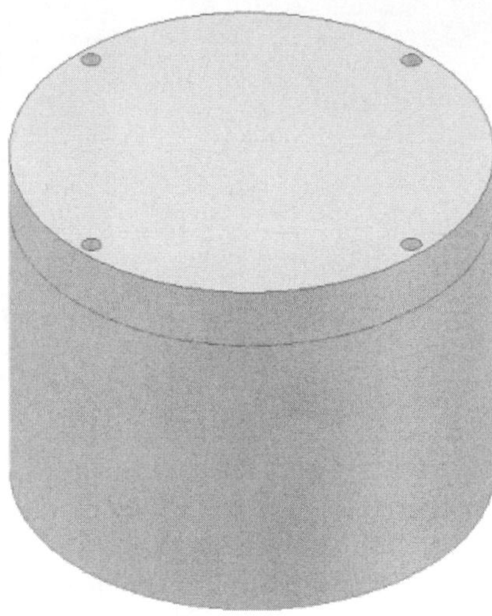

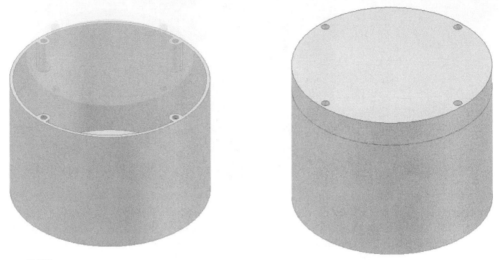

Figure 6.78

30. Close all open files. Do not save changes. End of exercise.

ASSEMBLY BROWSER TOOLS

Additional tools are available through the Assembly browser as a means of better controlling and managing data in an assembly file. These tools include In-Place Activation, Visibility Control, Assembly Reorder, Restructuring an Assembly, demoting and promoting assembly components, and using browser filters.

IN-PLACE ACTIVATION

The level of the assembly that is currently active determines whether or not you can edit components or features. You can take some actions only in the active assembly and its first-level children, while other operations are valid at all levels of the active assembly.

Double-click any subassembly or part occurrence in the browser to activate it, or right-click the occurrence in the browser and select Edit. All components not associated with the active component appear shaded in the browser, as shown in the following image on the left.

Note: Double-clicking directly on a component in the graphics window will also activate it for editing.

EXERCISES

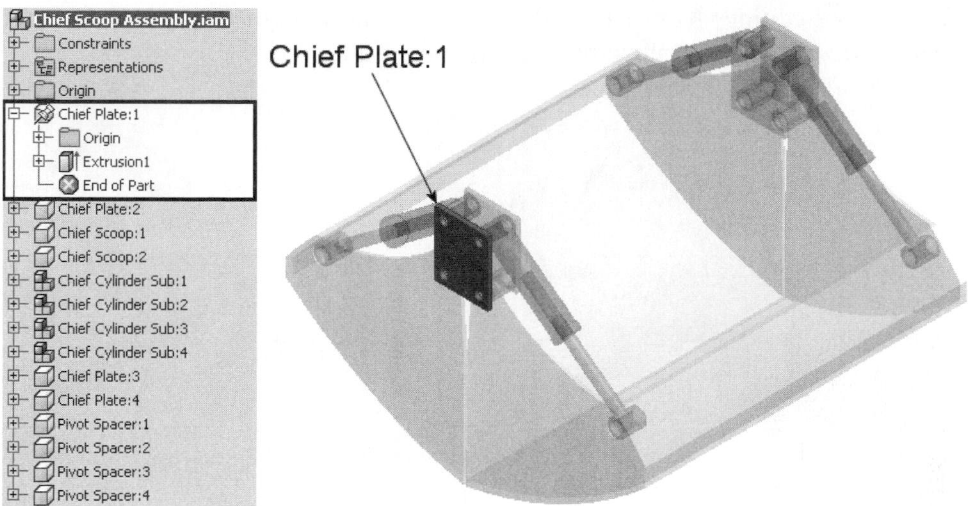

Figure 6.79

If you are working with a shaded display, the active component appears shaded in the graphics window, and all other components appear translucent. If you are working with a wireframe display, the active component appears in a contrasting color.

You can perform the following actions on the first-level children of the active assembly:

- Delete a component
- Display the DOF of a component
- Designate a component as adaptive
- Designate a component as grounded
- Edit or delete the assembly constraints between first-level components

You can edit the features of an activated part in the assembly environment. The panel bar and toolbars change to reflect the part environment when a part is activated.

Note: Double-click a parent or top-level assembly in the browser to reactivate it.

THE ASSEMBLY RETURN TOOL

After an assembly has been activated for in-place editing, a number of tools are available to quickly return to the desired environment. The Return tool is used to return to the previous editing state.

Suppose that you are editing an assembly in place. The following image on the left shows a partial engine assembly consisting of a number of subassemblies: a cylinder head, a crankshaft, and a piston. Double-clicking on the crankshaft activates the crankshaft subassembly for in-place editing. Similarly, double-clicking on the drive shaft activates this part for in-place editing. Notice how the entire assembly fades out except for the drive shaft part, as shown in the following image on the right.

Figure 6.80

With the drive shaft active, other Return tools are available to exit in-place editing and return you to the desired environment. The following image on the left shows the result of clicking the Return > Parent mode. Since the parent of the drive shaft is the crankshaft subassembly, this item returns to its original state. The following image on the right shows the result of clicking the Return > Top mode. No matter how deep you are inside of an assembly, this mode will return you to the top model in the browser.

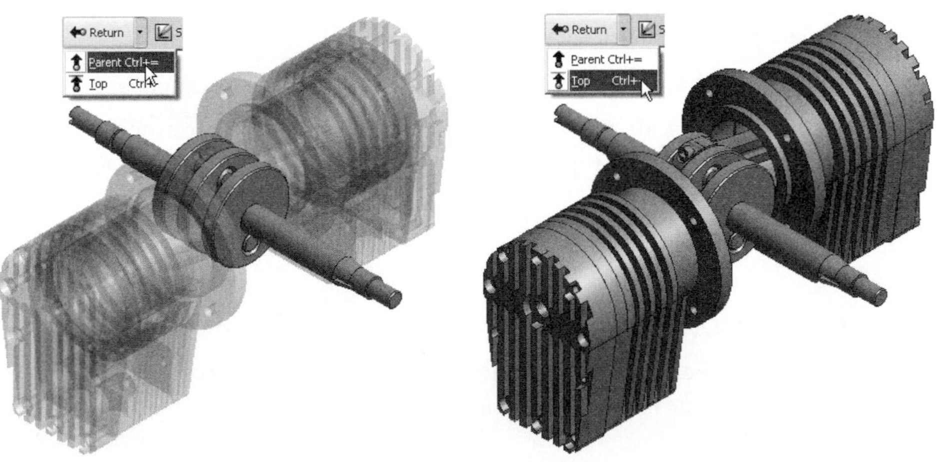

Figure 6.81

VISIBILITY CONTROL

Controlling the visibility of components is critical to managing large assemblies. You may need some components only for context, or the part you need may be obscured by other components. Assembly files open and update faster when the visibility of nonessential components is turned off.

You can change the visibility of any component in the active assembly even if the component is nested many layers deep in the assembly hierarchy. To change the visibility of a component, expand the browser until the component occurrence is visible, right-click the occurrence, and select Visibility, as shown in the following image.

Note: You can also right-click on a component in the graphics window and select Visibility.

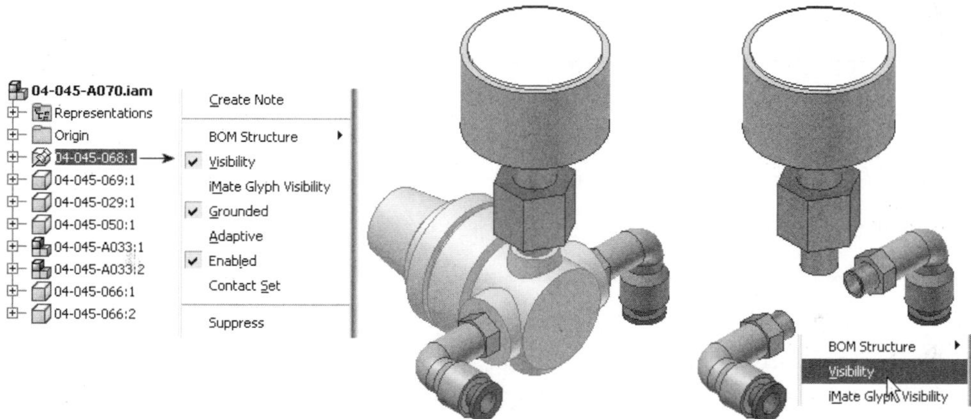

Figure 6.82

ADAPTIVITY

Adaptivity is the Autodesk Inventor function that allows the size of a part to be determined by setting up a relationship between the part and another part in the assembly. Adaptivity allows underconstrained sketches—features that have undefined angles or extents, hole features, and subassemblies, which contain parts that have adaptive sketches or features—to adapt to changes. The adaptivity relationship is defined by applying assembly constraints between an adaptive sketch or feature and another part. If a sketch is fully constrained, it cannot be made adaptive. However, the extruded length or revolved angle of the part can be. A part can only be adaptive in one assembly at a time. In an assembly that has multiple placements of the same part, only one occurrence can be adaptive. The other occurrences will reflect the size of the adaptive part.

An example of adaptivity would be determining the diameter of a pin from the size of a hole. You could determine the diameter of the hole from the size of the pin, and you can turn adaptivity on and off as needed. Once a part's size is determined through adaptivity, and adaptivity is no longer useful, you may want to turn its adaptivity off. If you want to create adaptive features, Autodesk Inventor includes features that will speed up the process of creating them. On the Tools menu, click Application Options. On the Assembly tab, three areas relate to adaptivity, as shown in the following image.

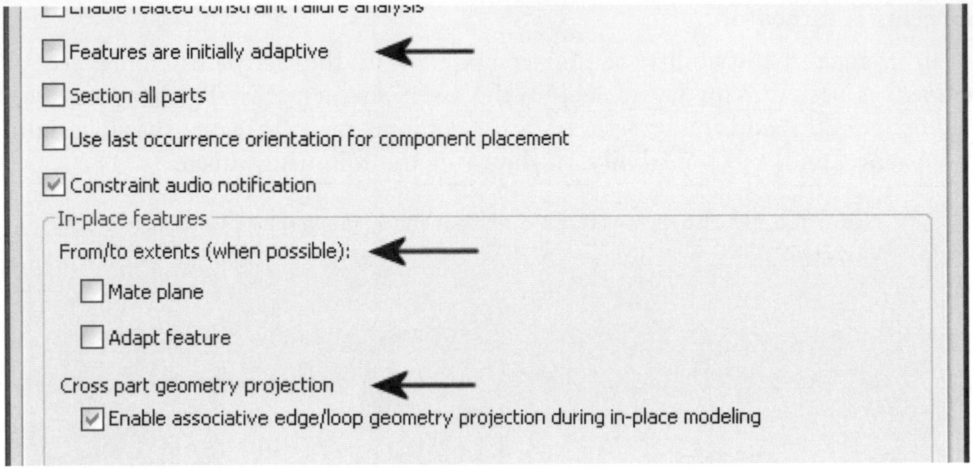

Figure 6.83

ASSEMBLY TAB OPTIONS

The following sections describe the adaptivity options found on the Assembly tab of the Options dialog box.

Features Are Initially Adaptive

By default, this option is unchecked. Click inside the box to make features adaptive when they are created. This step is useful when you are creating a large number of adaptive features. However, having too many adaptive features may result in an unstable assembly, and for this reason, beginning users leave this mode turned off or unchecked.

In-Place Features

From/To Extents (when possible) This tool determines whether or not a feature will be adaptive when the To or From/To option is selected for the extrusion extent. If both options are selected, Autodesk Inventor will try to make the feature adaptive. If it cannot, it will terminate at the selected face.

Mate Plane Click when you create a new component to have a mate constraint applied to the plane on which it was constructed. It will not be adaptive.

Adapt Feature Click when you want to create a new component to have it adapt to the plane on which it was constructed.

Cross Part Geometry Projection

Enable Associative Edge/Loop Geometry Projection during In-Place Modeling Click when geometry is projected from another part onto the active sketch to make the projected geometry associative, meaning that it has sketch associativity, and to update it when changes are made to the parent part. You can use projected geometry to create a sketched feature.

UNDERCONSTRAINED ADAPTIVE FEATURES

The next series of images show how parts adapt when you apply assembly constraints. In the following image, a rectangular sketch for a small plate is not dimensioned (unconstrained) along its length.

Figure 6.84

The extruded feature is then defined as adaptive by right-clicking on the extrusion in the browser to display the appropriate menu, as shown in the following image. Selecting Adaptive from the menu applies the adaptive property to the sketch and the extrusion.

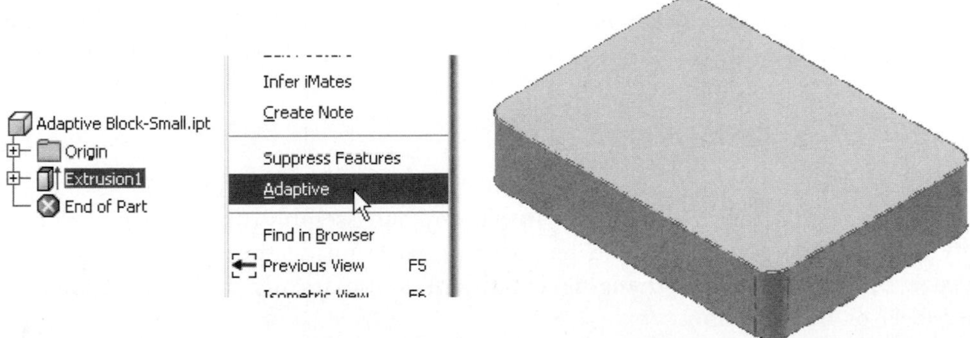

Figure 6.85

Once you have placed the parts in an assembly, right-clicking on the small box in the browser activates a menu. Selecting Adaptive from the menu adds an icon consisting of two arcs with arrows next to the small box part, as shown in the following image on the left. The presence of this icon in the browser means the small box part is now adaptive.

As flush constraints are placed along the edge faces of the plates, the small plate will adapt its length to meet the length of the large plate.

The results are shown in the following image on the right.

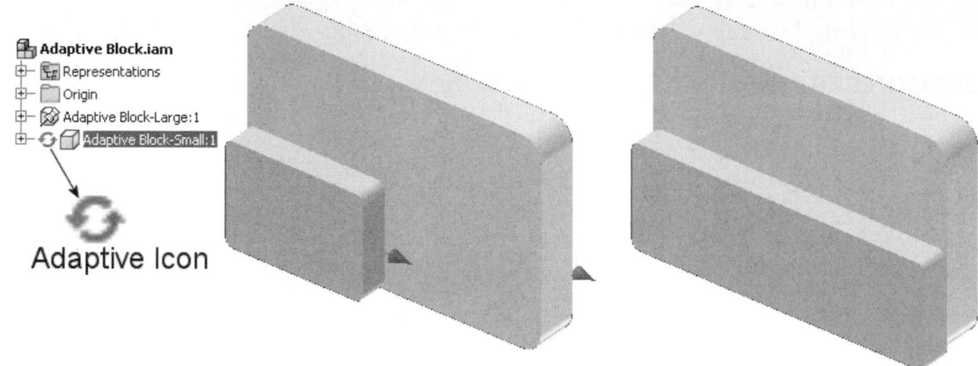

Figure 6.86

When you place or create a part containing adaptive features in an assembly, it is not initially adaptive. To specify the part as adaptive in the assembly context, right-click the component in the Assembly browser or graphics window and select Adaptive from the menu. You can also use the Assembly browser to specify a feature of the part as adaptive, and the part will become adaptive automatically.

When you constrain an adaptive part to fixed features on other components, underconstrained features on the adaptive part resize when the assembly is updated.

Only one occurrence of a part can define its adaptive features. If you use multiple placements of the same part in an assembly, all occurrences are defined by the one adaptive occurrence, including placements in other assemblies. If you want to adapt the same part to different assemblies, save the part file with a unique name using Save Copy As before defining any occurrences as adaptive.

ADAPTIVE SUBASSEMBLIES

In Autodesk Inventor, you can use adaptive subassemblies in your models to control assembly constraints for moving parts inside any subassembly nesting level. When you specify a subassembly occurrence to be adaptive, parts inside the subassembly can adjust their size or position to fit changing conditions automatically and independently in a higher level of the assembly.

Subassemblies, when merged into assembly files, are typically defined as rigid bodies. Drag constraints are used to work on underconstrained subassembly components. A typical example of this concept in action is an air cylinder. All parts of the assembly are fully dimensioned; however, the rod can translate along the axis of the cylinder.

In the following image, the air cylinders are constrained to the industrial scoop. Unfortunately, the air cylinder motion is restricted due to the rigid nature of the subassemblies.

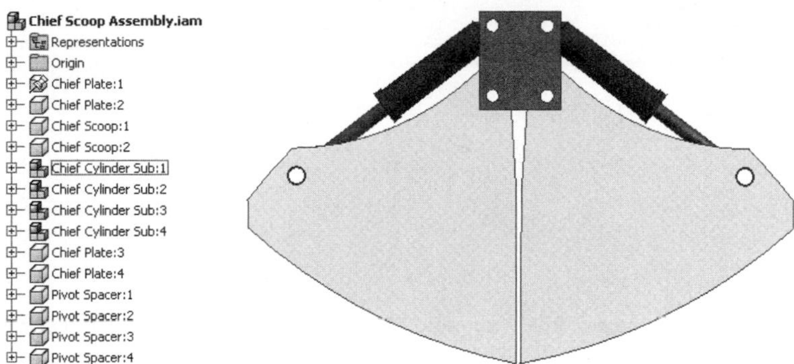

Figure 6.87

In the following image, one of the air cylinders is toggled to adaptive. Notice the appearance of the adaptive icon next to the subassembly in the browser. Multiple occurrences of the same subassembly are now controlled by the initial subassembly that has been made adaptive.

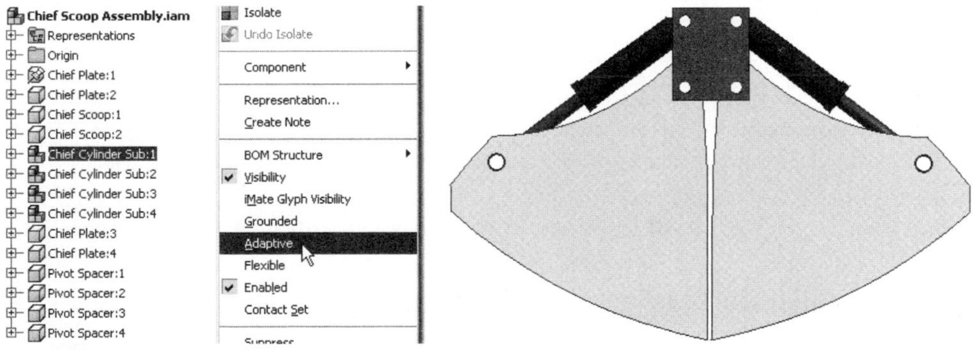

Figure 6.88

With the air cylinder toggled to adaptive, the underconstrained rod of the air cylinder subassembly can now move along the cylinder axis and affect the other shovel components of the assembly, as shown in the following image.

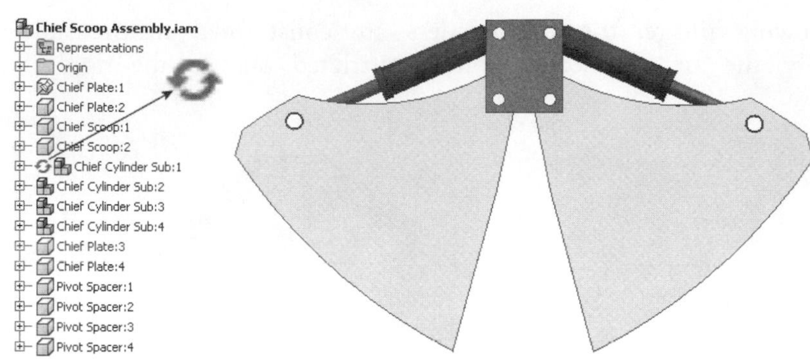

Figure 6.89

Tip: To turn off the adaptivity for sketches, features, and subassemblies, right-click on the sketch, feature, or subassembly, and deselect Adaptivity on the menu or in the Feature Properties dialog box.

ADAPTING THE SKETCH OR FEATURE

After making a sketch or feature adaptive, you must make the part itself adaptive at the assembly level. To make a part adaptive, make the assembly where the part exists active. Right-click on the part's name, and select Adaptive from the menu. Apply assembly constraints that will define the relationship for the adaptive sketch or feature. As you do so, degrees of freedom are being removed.

Note: You cannot make parts that are imported from an SAT or STEP file format adaptive, because they are static and do not have underconstrained sketches and features. However, you can make assemblies created from these parts adaptive.

In assemblies that have multiple adaptive parts, two updates may be required to solve correctly.

For revolved features, use only one tangency constraint.

Avoid offsets when applying constraints between two points, two lines, or a point and a line. Incorrect results may occur.

EXERCISE 6-3: CREATING ADAPTIVE PARTS

In this exercise, you create a link arm that adapts to fit between existing components in an assembly.

1. This exercise begins by creating a link arm. The length of the link and the diameters of the circles are not dimensioned in the sketch so they can adapt to fit components in the assembly.
2. Click the New tool, select the Metric tab, and double-click *Standard(mm).ipt*. A new part is created.

3. Click the Line tool, and click in the graphics screen to start the line.
4. Move the cursor to the right, and then click when a horizontal symbol is shown.
5. Click and drag off the point to create a 180° arc.
6. Move the cursor to the left. Click to create a line the same length and parallel to the first line.
7. Drag off the point to create an arc, and close the profile. Your link arm should appear similar to the following image.

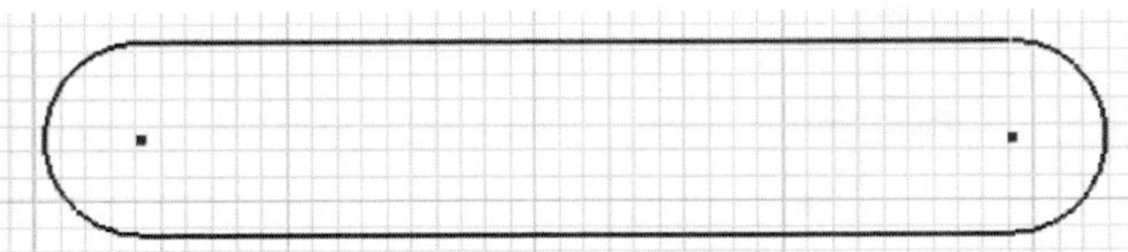

Figure 6.90

8. Continue with the creation of the link arm by clicking the down arrow beside the Constraint tool, and then click the Tangent constraint tool.
9. Click the last arc and first line.
10. Next click the Center Point Circle tool. On the left side of the sketch, select the center point, and then create a circle. On the right side of the sketch, select the center point, and then create a circle. Your sketch should appear similar to the following image.

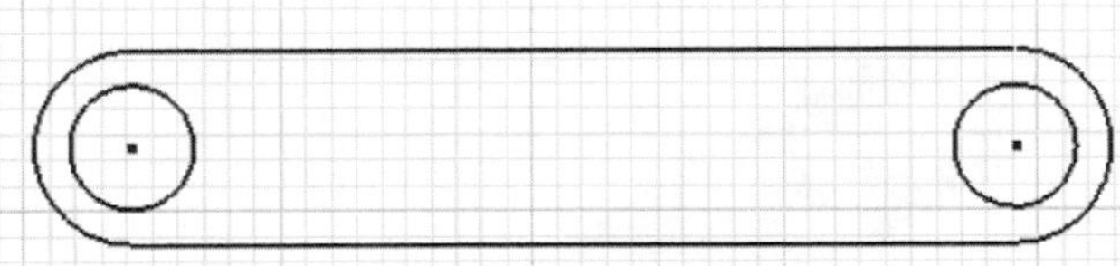

Figure 6.91

11. You will now dimension the arc. Click the General Dimension tool, and add a **6 mm** dimension to the left arc, as shown in the following image. This is the only dimension needed for this new part file. The length of the link arm and the diameter of the holes will be determined when the arm is assembled.

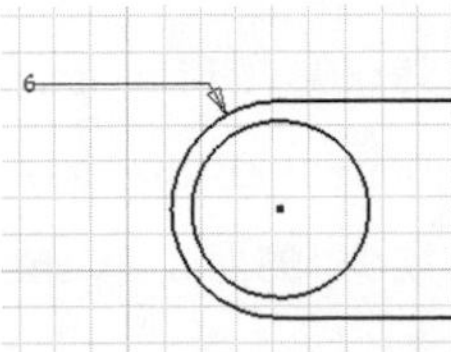

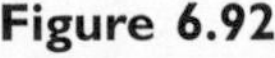

Figure 6.92

12. With the sketch completed but underconstrained, continue by right-clicking in the graphics window and selecting Home View.
13. Create an extruded feature by pressing E to start the Extrude tool.
14. Select the profile to extrude. The inside of the holes should not be selected.
15. Enter a value of **3** for the depth of the extrusion, and click the OK button. The extrusion is created, as shown in the following image, and an entry is added to the browser.

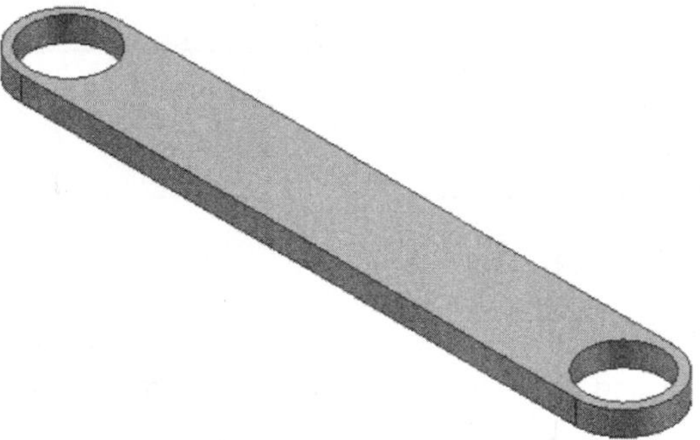

Figure 6.93

16. Change the color of the new part by selecting Style and Standard Editor from the Format drop-down menu and double-clicking Metal-Titanium from the list. When finished, click Done. Your part should now appear similar to the following image. It is acceptable if your part looks different when compared to the one in the illustration.

Figure 6.94

17. Save this new part file as *ESS_E06_03-Link.ipt.*
18. Close the part file, and continue with the assembly phase of this exercise.
19. Open the assembly file *ESS_E06_03.iam,* as shown in the following image.

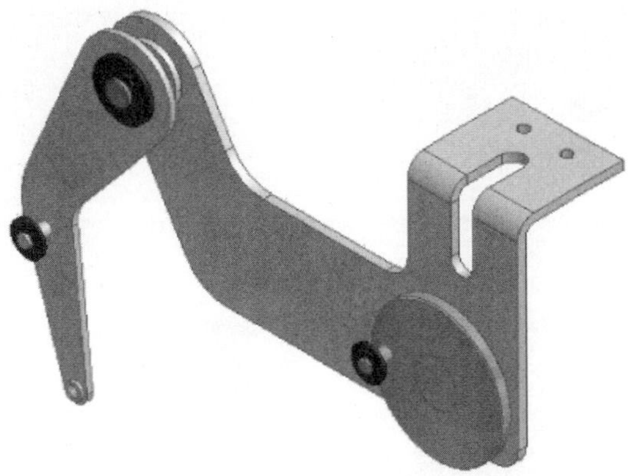

Figure 6.95

20. Place the link arm in the assembly by clicking the Place Component tool and opening the file *ESS_E06_03-Link.ipt*.

21. Click to place an occurrence of the link arm in the assembly, as shown in the following image. Right-click in the graphics window, and select Done.

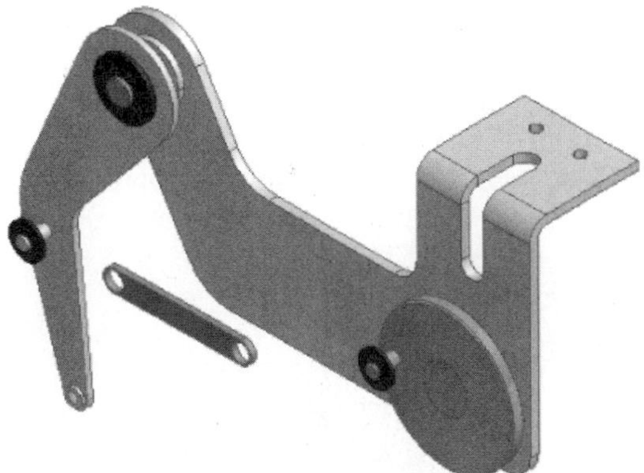

Figure 6.96

22. You will now make the link arm adaptive inside of the assembly. Click the Zoom Window tool, and define a window to zoom around the link arm and lower-left pin, as shown in the following image.

23. In the browser, right-click *ESS_E06_03-Link.ipt,* and select Edit from the menu.

24. Right-click Extrusion1, and select Adaptive from the menu. The adaptive symbol is added to the extrusion and the part in the browser.

25. Click the Return tool.

Figure 6.97

26. Now use assembly constraints to constrain the link arm to the main assembly. First click the Constraint tool.

27. Select the inside cylindrical surface of the hole in the link arm, as shown in the following image.

Note: Be sure not to select the centerline axis by mistake. Use the Select Other tool to scroll through the different solutions until the inner cylindrical surface highlights without the presence of the axis.

28. Select the outside surface of the pin, and click Apply.

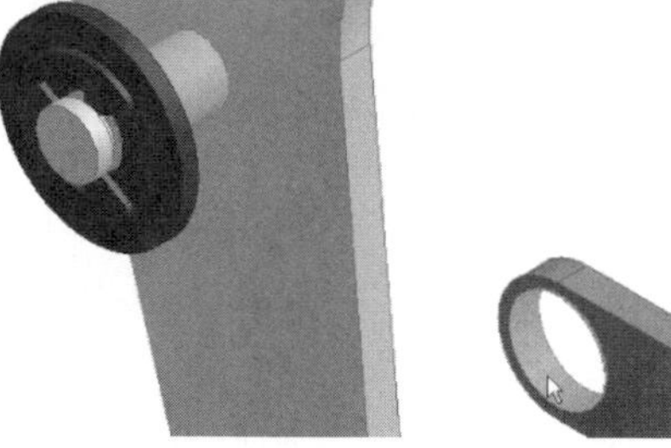

Figure 6.98

29. The link arm assembles to the pin, and the diameter of the hole adapts to fit the pin, as shown in the following image.

Figure 6.99

30. Continue assembling the link arm by adding a mate constraint between the face of the link arm and the main assembly arm. Do this by selecting the front face of the main assembly arm and the front face of the link arm with the mate constraint command active, as shown in the following image on the left.

31. Click Apply to accept the constraint, and then click Cancel to exit the Constraints dialog box. Your display should appear similar to the following image on the right.

Figure 6.100

32. Click the Zoom All tool on the Standard toolbar to view the entire assembly. Click the Zoom Window tool on the Standard toolbar, and then define a window to zoom around the right end of the link arm and the pin on the eccentric, as shown in the following image.

Figure 6.101

33. Constrain this end of the link arm to the pin on the eccentric. Complete this task by clicking the Constraint tool in the panel bar or from the Assembly toolbar.
34. Use the same technique to constrain the link arm and pin. Select the inside cylindrical surface of the link arm, then the outside surface of the pin, and then click Apply. Notice how the link arm adapts to fit between the two pins, as shown in the following image. Click Cancel to exit the Constraints dialog box.

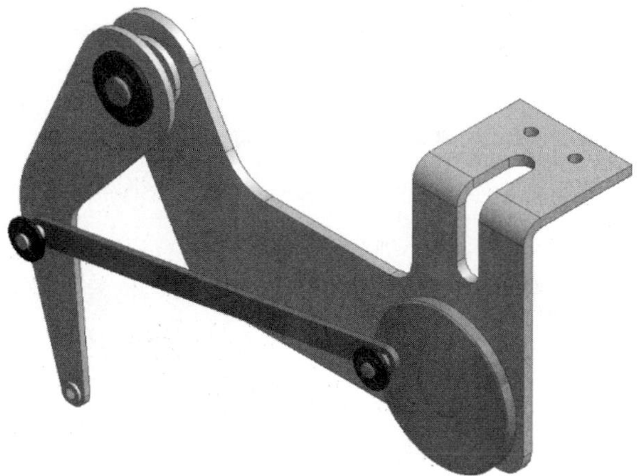

Figure 6.102

35. Use the workflow you learned in this exercise to place an assembly constraint between the front face of the link arm and the back face of the washer, using adaptivity to thicken the link arm.
36. Close all open files. Do not save changes. End of exercise.

2D DESIGN LAYOUT

When you create a new design, you often begin with design criteria and create components that meet those criteria. From a list of known parameters, you may create an engineering layout or a 2D design that evolves throughout the design process, and then build components that reference this layout.

The following image illustrates an assembly that consists of an arm designed to move a slide. The link between the arm and the slide has not yet been defined. Instead of creating the link as a solid model part, a 2D sketch can be used as a layout to test for motion.

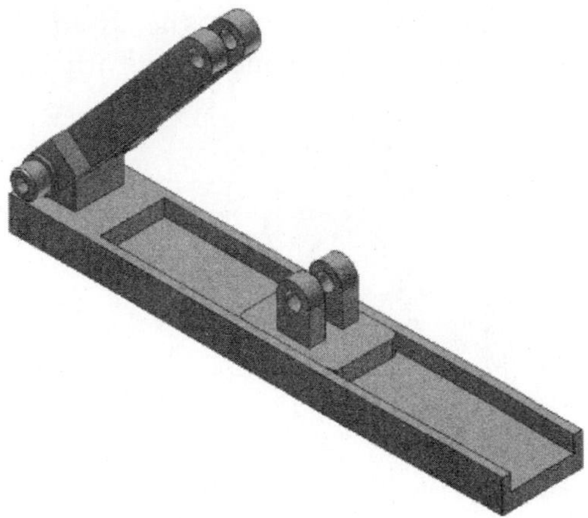

Figure 6.103

Before constructing the 2D design layout of the link, a work plane is added to the slide part and a new part is created that uses this plane as its base sketch plane, as shown in the following image.

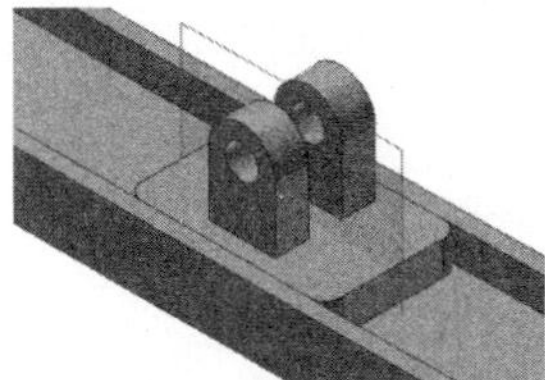

Figure 6.104

With the new part created relative to the work plane, the link geometry is sketched using a combination of sketch, constraint, and dimension tools, as shown in the following image.

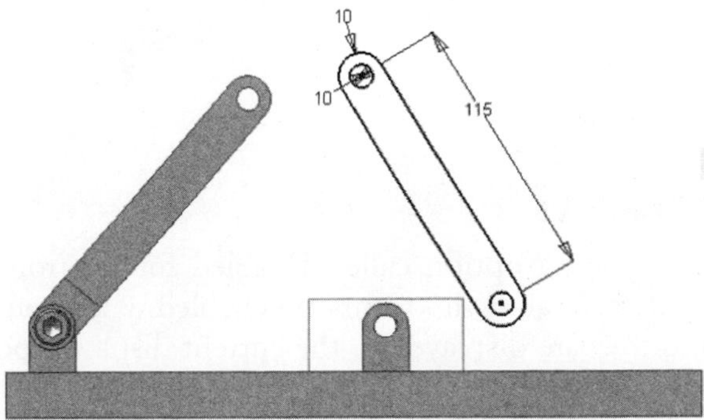

Figure 6.105

After the sketch is finished and you return to the assembly environment, assembly constraints are added between the sketch geometry and other parts.

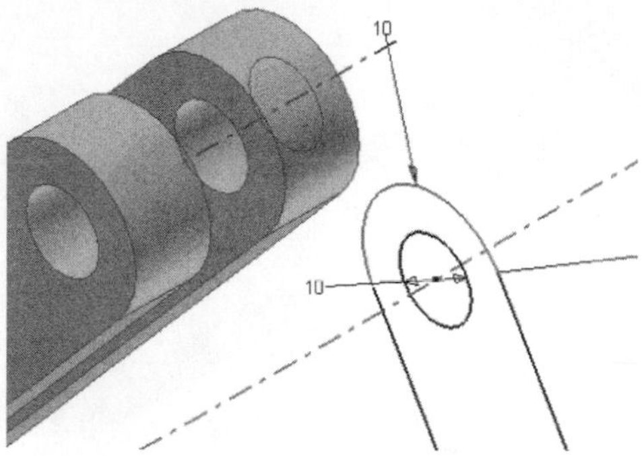

Figure 6.106

The completed sketch and assembly layout are shown in the following image. By using a top-down layout and constraining the components to the assembly, multiple design configurations can be evaluated by making simple changes to the layout parameters. You can quickly examine minimum and maximum values for component parameters, enabling you to determine if the working envelope of the mechanism meets the design criteria.

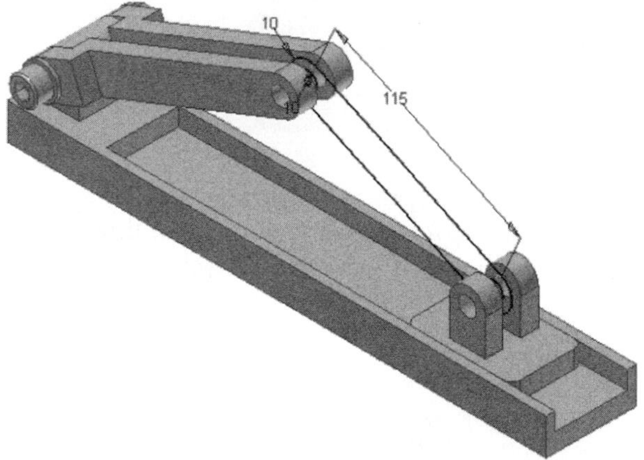

Figure 6.107

ENABLED COMPONENTS

Autodesk Inventor gives you an option called Enabled for controlling how components look in an assembly. By default, all components are enabled when you place or create them; their visibility is on, and they are displayed in the current display mode. When you disable a component, it appears transparent and cannot be selected in the graphics window. However, a disabled component can have geometry projected from it. To disable a component, select its name in the browser, or click on the component itself, right-click, and select Enabled from the menu, as shown in the following image.

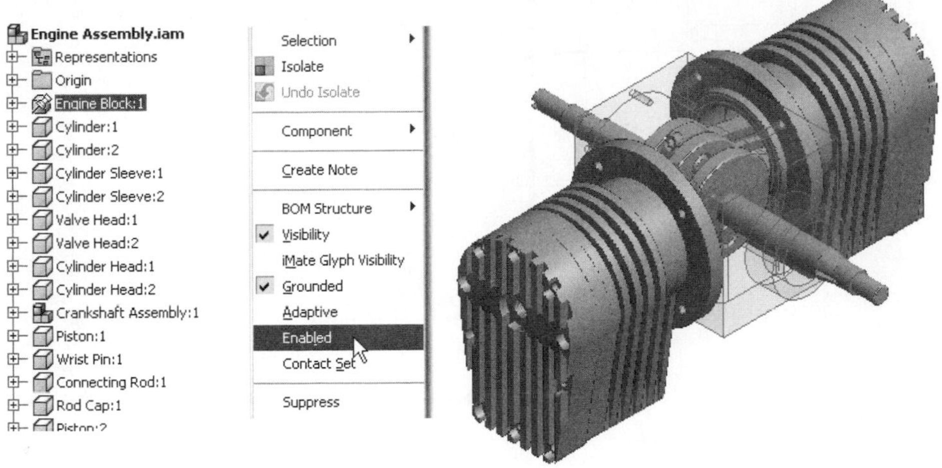

Figure 6.108

PATTERNING COMPONENTS

You can use the Pattern Component tool on the Assembly panel bar, as shown in the following image, when you place multiple occurrences of selected parts and subassemblies that match a feature pattern on another part (a component pattern) or that have a set of circular or rectangular part patterns in an assembly (an assembly pattern).

Three tabs are available in the Pattern Component dialog box: Associative, Rectangular, and Circular, as shown in the following image. The Associative tab is the default.

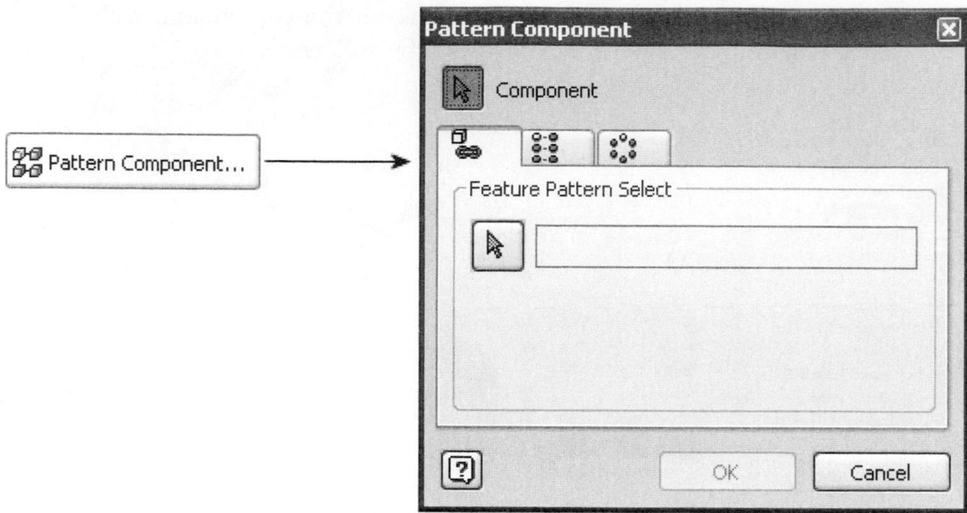

Figure 6.109

ASSOCIATIVE PATTERNS

An associative component pattern will maintain a relationship to the feature pattern that you select. For example, a bolt is component-patterned to a part bolt-hole circular pattern that consists of four holes. If the feature pattern, the bolt-hole, changes to six holes, the bolts will move to the new locations, and two new bolts will be added for the two new holes. To create an associative component pattern, there must be a feature-based rectangular or circular pattern, and the part that will be patterned should be constrained to the parent feature, (that is, the original feature that was patterned in the component. Issue the Pattern Component tool from the Assembly panel bar, and the Pattern Component dialog box will appear, as shown in the following image on the left.

By default, the Component selection option is active. Select the component, such as the cap screw, or components to pattern. Next click the Feature Pattern Select button in the dialog box, and select a feature, such as a hole, that is part of the feature pattern. Do not select the parent feature. After selecting the pattern, it will highlight on the part, and the pattern name will appear in the dialog box. When done, click OK to create the component pattern, as shown in the following image on the right.

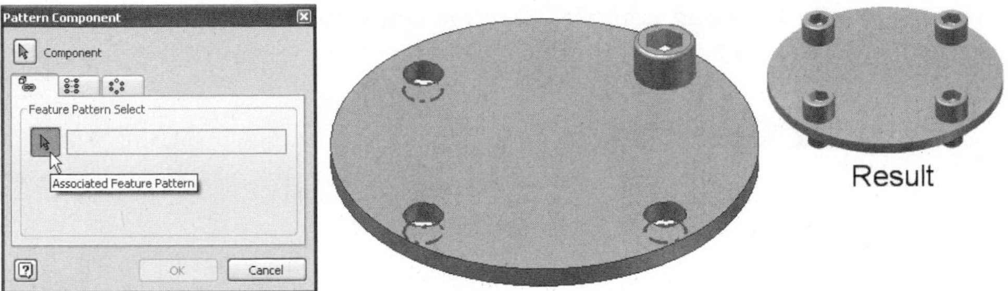

Figure 6.110

RECTANGULAR PATTERNS

A rectangular pattern operation is illustrated in the following image. When performing this operation, two edges are chosen that define the direction of the columns and rows of the pattern. Enter the number of columns and rows and then the spacing or distance between these rows and columns. The resulting pattern acts like a feature pattern. After creating it, you can edit it to change its numbers, spacing, and so on.

In the following image, because one of the part edges is inclined, work axes are turned on and used for defining the X and Y directions of the pattern.

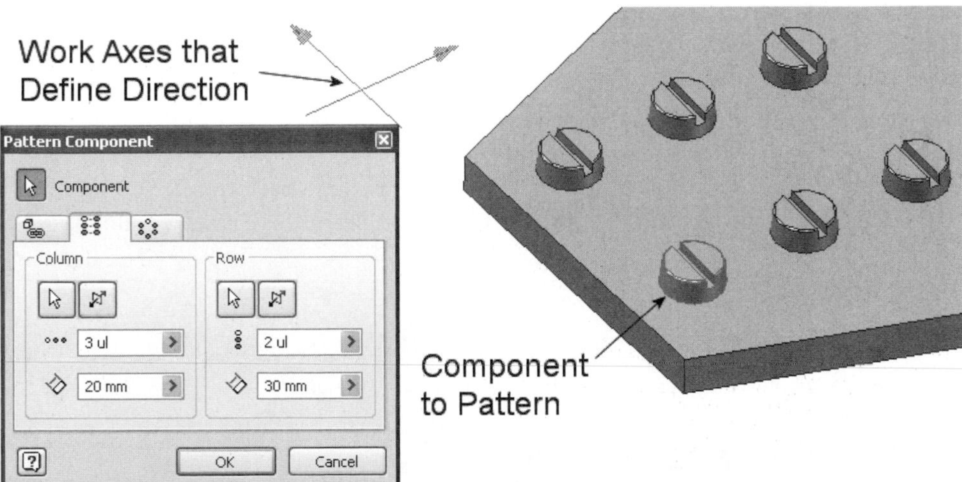

Figure 6.111

CIRCULAR PATTERNS

Circular patterns will copy selected components in a circular direction. After selecting the component or components to pattern, select an axis direction. In the following example, this element takes the form of the centerline that acts as a pivot point for the pattern. You then enter the number of occurrences or items that will make up the pattern and the circular angle. In the following example, since the bolt component is being patterned in a full circle, enter 360° as the value for the circular angle, as shown in the following image.

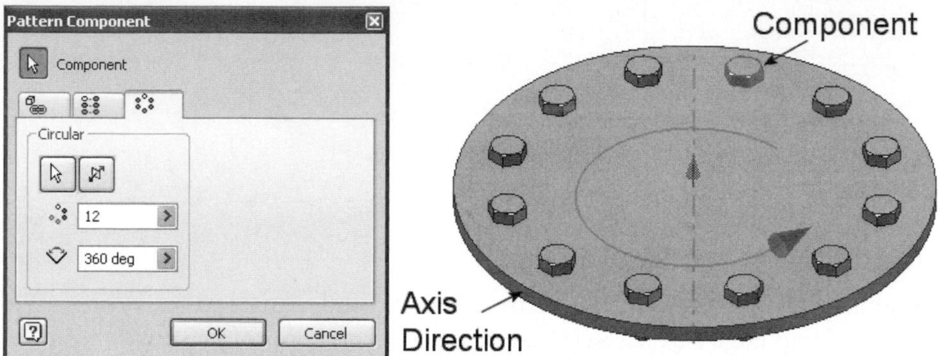

Figure 6.112

The completed pattern then acts as a single part. If one part moves, all parts move. The patterned component will be consumed into a component pattern in the browser, and each of the part occurrences will also appear as an element that you can expand.

PATTERN EDITING

Edit the component pattern by selecting the pattern in the graphics window and right-clicking, or by right-clicking on the pattern's name in the browser and selecting Edit from the menu. In the browser, the component that you patterned will be consumed into a component pattern, and the part occurrences will appear as elements, which can be expanded to see the part. You can suppress an individual pattern component by right-clicking on it and selecting Suppress from the menu, as shown in the following image.

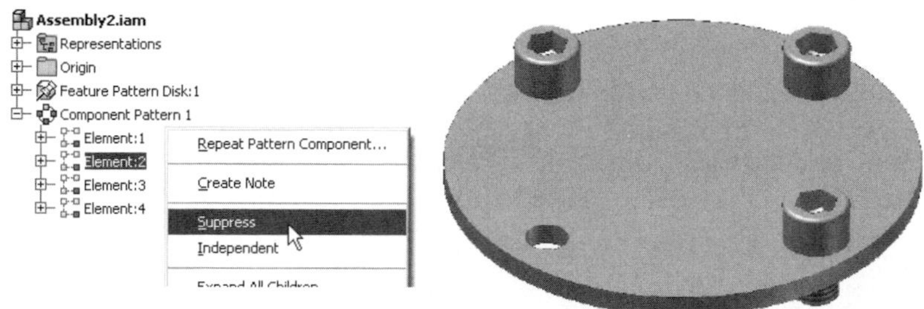

Figure 6.113

You can break an individual part out of the pattern by right-clicking on it and selecting Independent from the menu, as shown in the following image. Once a part is independent, it no longer has a relationship with the pattern.

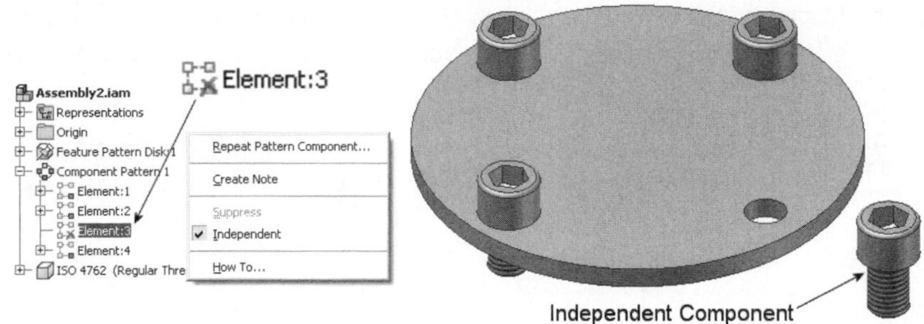

Figure 6.114

Deleting a Pattern Component

To delete a pattern, select the pattern in the graphics window or in the browser. Right-click and select Delete from the menu, or press the DELETE key on the keyboard. If the Delete component pattern source(s) option (under the Assembly tab, from the Applications Options dialog box), is selected, the source component will be deleted automatically. If the source component should not be deleted when the pattern elements are deleted, you should leave this option deselected, as shown in the following image.

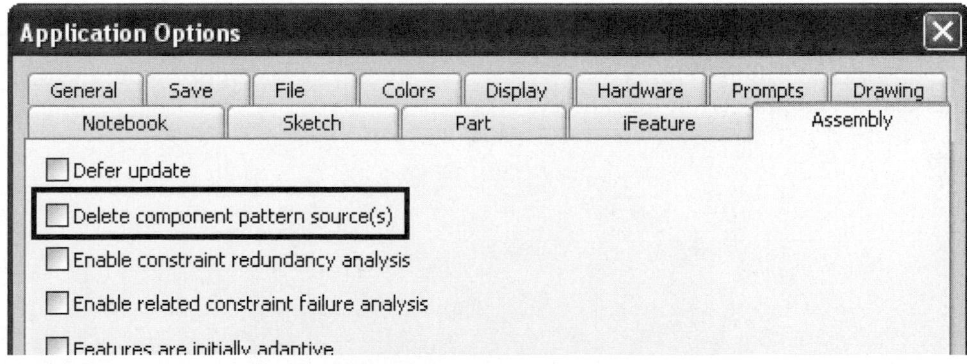

Figure 6.115

ADDITIONAL COMPONENT PATTERN OPTIONS

Additional options available when editing and manipulating component patterns include replacing all occurrences in an assembly pattern in a single step, restructuring a component pattern, and controlling the component pattern visibility. The following sections describe these features.

Component Pattern Replace

When replacing a component in a component pattern, you replace all occurrences in the selected component pattern with a newly selected component. Expand one of the elements in the browser, and select the component to replace. With this item highlighted in the browser, right-click and select Replace from the menu, as shown in the following image. The Open dialog box will appear and will enable you to select the replacement component.

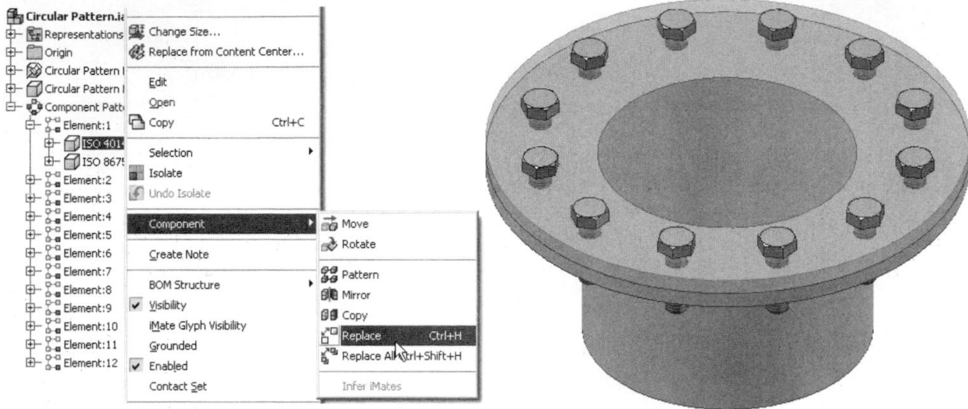

Figure 6.116

The following image shows the result of performing the operation to replace a pattern component. Since the use of component patterns allows for better capture of design intent, replacement of all occurrences maintains this design intent without manually replacing each component in the pattern. Overall ease of assembly use is improved, and component patterns can be completed more quickly.

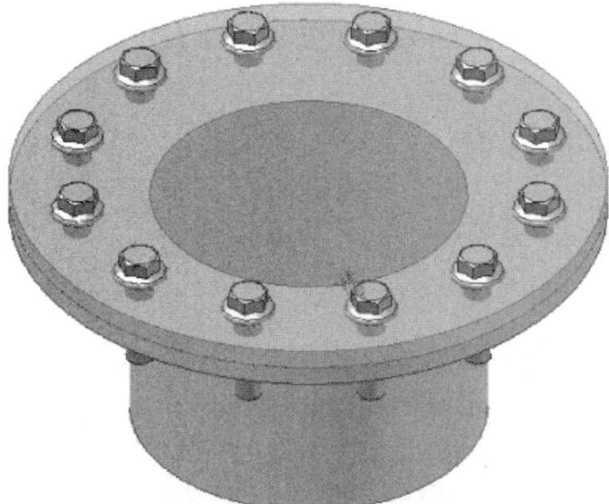

Figure 6.117

EXERCISE 6-4: PATTERNING COMPONENTS

In this exercise, you create a fastener component pattern to match an existing hole pattern and then replace the bolt in the pattern. You complete the exercise by demoting components to create a subassembly with the cap plate and patterned fasteners.

1. Open an existing assembly called *ESS_E06_04.iam,* as shown in the following image on the left. The file contains a T pipe assembly.
2. In the browser, expand the parts *ESS_E06_04-Cap_Bolt.ipt:1* and *ESS_E06_04-Cap_Nut.ipt:1*. Notice that the parts already have insert constraints applied to them, as shown in the following image on the right.

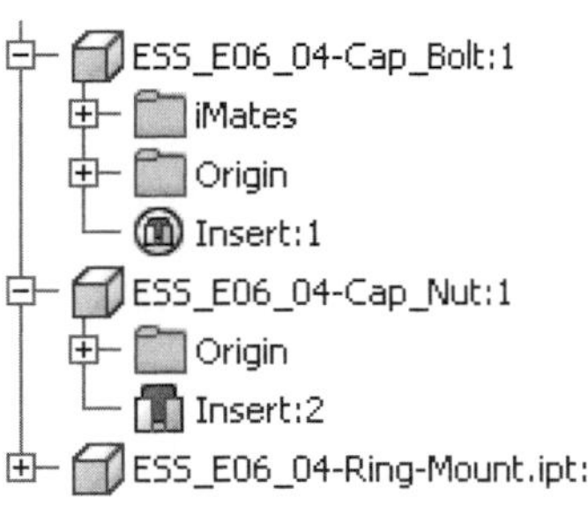

Figure 6.118

3. Now create a pattern consisting of a collection of nuts and bolts in a circular pattern. Click the Pattern Component tool to launch the Pattern Component dialog box, as shown in the following image on the left.
4. Press and hold down CTRL and select the *ESS_E06_04-Cap_Bolt.ipt:1* and *ESS_E06_04-Cap_Nut.ipt:1* parts in the browser, as shown in the following image on the right.

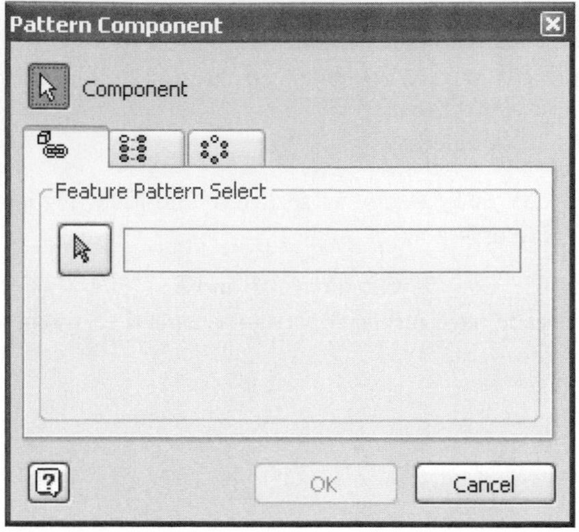

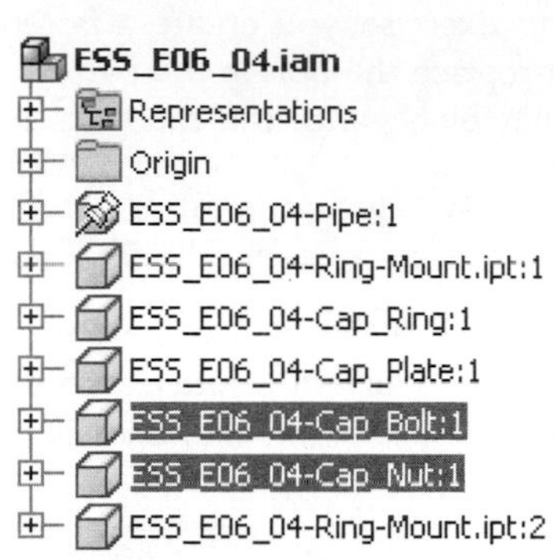

Figure 6.119

5. Click the Select button found under the Feature Pattern Select area of the Pattern Component dialog box. In the graphics window, point to any hole in the Cap_Plate part. When the circular pattern of holes in the cap plate is highlighted, pick the edge of one of the holes, as shown in the following image on the left.
6. You should see a preview of the component pattern, as shown in the following image on the right.

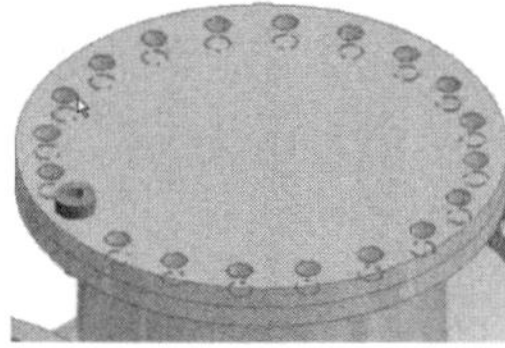

Figure 6.120

7. Click the OK button to close the Pattern Component dialog box and create the component pattern. Expand the display of the Component Pattern 1 feature in the browser, as shown in the following image on the left.
8. In the browser, expand Element:1 and the parts beneath it. Notice that the constraints on the original components were retained, as shown in the following image on the right.

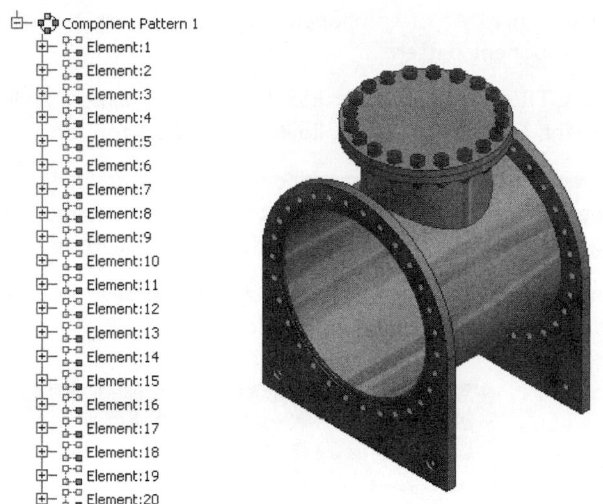

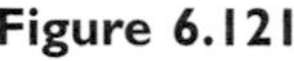
Figure 6.121

9. Now replace the bolt in the pattern with a similar but different fastener. Under Element:1 in the browser, right-click the *ESS_E06_04-Cap_Bolt,* select Component, and click Replace from the menu.
10. In the Open dialog box, select *ESS_E06_04-Cap_Hex_Bolt.ipt,* and click Open. A warning message box will appear to notify you that some assembly constraints may be lost.
11. Click the OK button. The pattern is updated to reflect the new bolt, as shown in the following image.

Note: The fastener in this exercise is constrained to the plate with an Insert iMate constraint. Since the replacement fastener has a similar Insert iMate, the constraint to the plate is retained.

Figure 6.122

12. Now demote a number of components; this will create a subassembly with the cap plate and component pattern.

13. Hold down CTRL while you click *ESS_E06_04-Cap_Plate:1* and Component Pattern 1 in the browser, as shown in the following image on the left.

14. Right-click, and select Component followed by Demote from the menu. In the Create In-Place Component dialog box, go to the New Component Name field. Enter Cap_Kit_Assembly.iam, as shown in the following image on the right.

15. When finished, click the OK button. A warning message box appears to notify you that some assembly constraints may be lost. Click Yes.

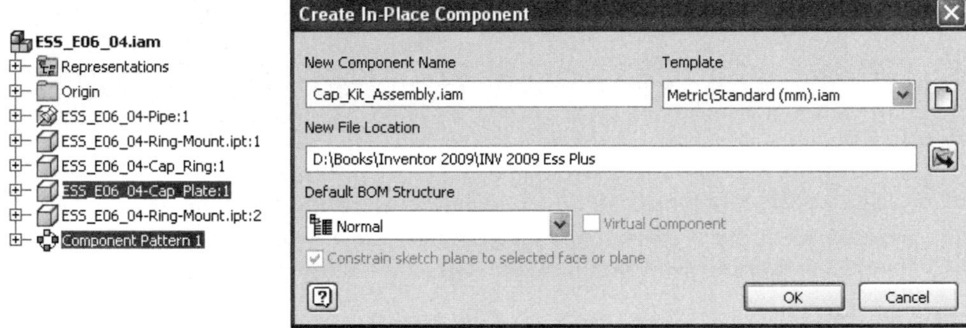

Figure 6.123

16. In the browser, expand Cap_Kit_Assembly:1, as shown in the following image on the left. Notice that the cap plate and component pattern are now part of the new subassembly.

17. In the graphics window, drag any bolt. Notice how the subassembly moves because constraints between the cap plate and cap ring are broken. Your display should appear similar to the following image on the right.

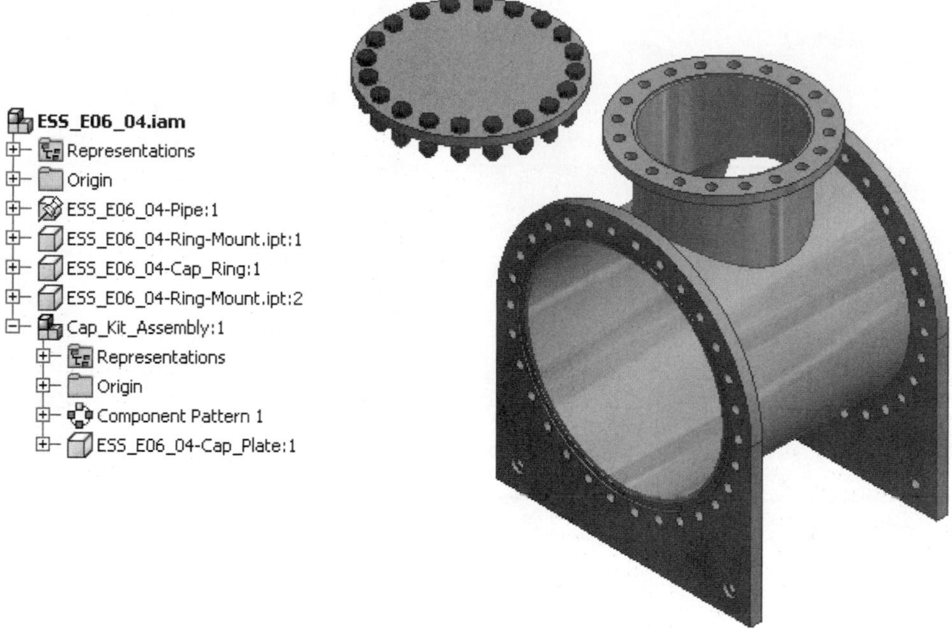

Figure 6.124

18. Place an Insert constraint between the bottom outside edge cap plate and the top outside edge of the cap ring.

19. Rotate the underside of the cap plate. Then apply a Mate constraint from the centerline of one hole to the centerline of one of the bolt threads as shown in the following image on the left. This will complete the assembly phase of the Cap_Kit_Assembly. Your display should appear similar to the following image on the right.

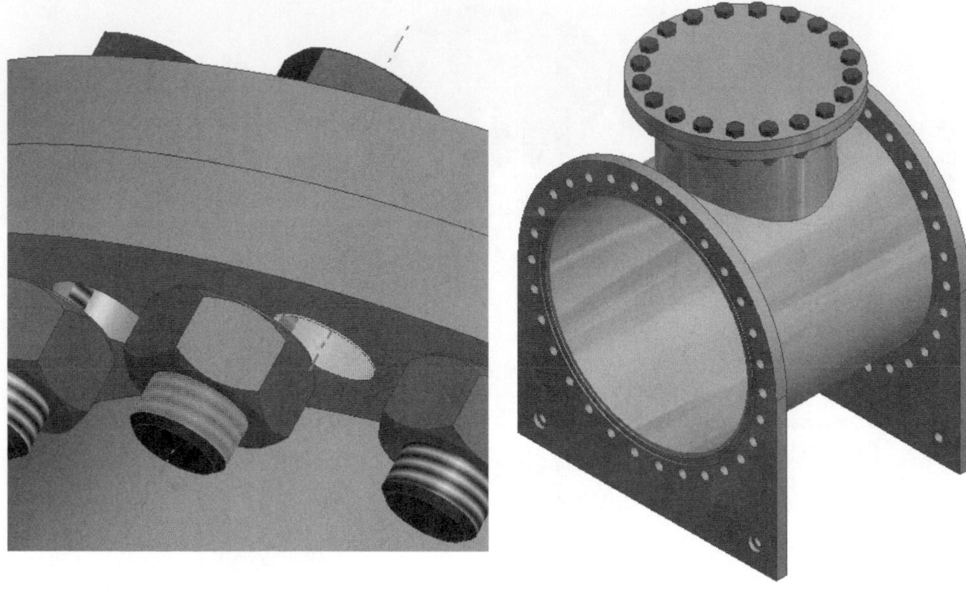

Figure 6.125

20. Close all open files. Do not save changes. End of exercise.

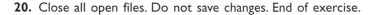

ANALYSIS TOOLS

Various tools are available that assist in analyzing sketch, part, and assembly models. You can calculate minimum distance between components, calculate the center of gravity of parts and assemblies, and perform interference detection.

MINIMUM DISTANCE TOOL

You can easily obtain the minimum distance between any components, parts, or faces in an assembly. While in the assembly, select the Measure Distance command and change the selection priority, depending on what you are trying to measure. The three available selection modes are shown in the following image.

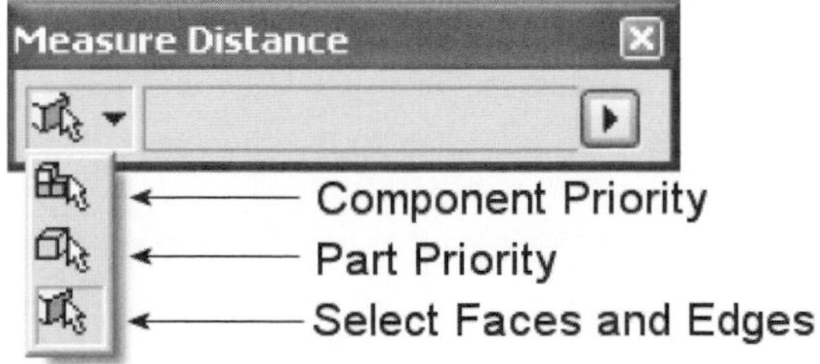

Figure 6.126

EXERCISES

CENTER OF GRAVITY

A center of gravity placed in an assembly could be critical to the overall design process of that assembly, whether it is used in the next assembly or in the main assembly. From the View menu, click the Center of Gravity tool as shown in the following image on the left. The image on the right shows the center of gravity icon applied to an assembly model. This icon actually consists of a triad displaying the X, Y, and Z directions. Three selectable work planes and a selectable work point area are also available for the purpose of measuring distances and angles.

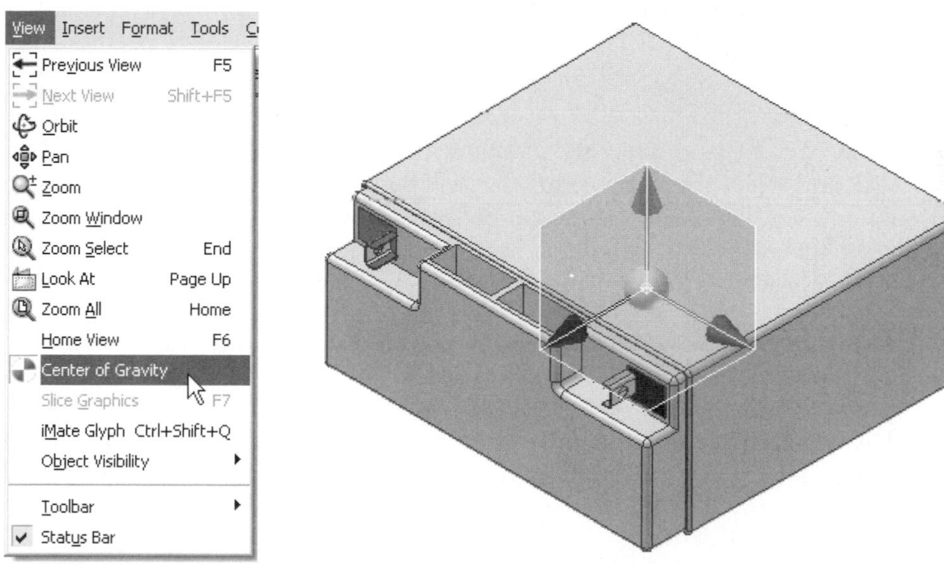

Figure 6.127

INTERFERENCE CHECKING

You can check for interference in an assembly using one or two sets of objects. To check the interference among sets of stationary components, make the assembly or subassembly in question active. Then click the Analyze Interference tool on the Tools menu, as shown in the following image on the left. The Interference Analysis dialog box will appear, as shown in the following image on the right.

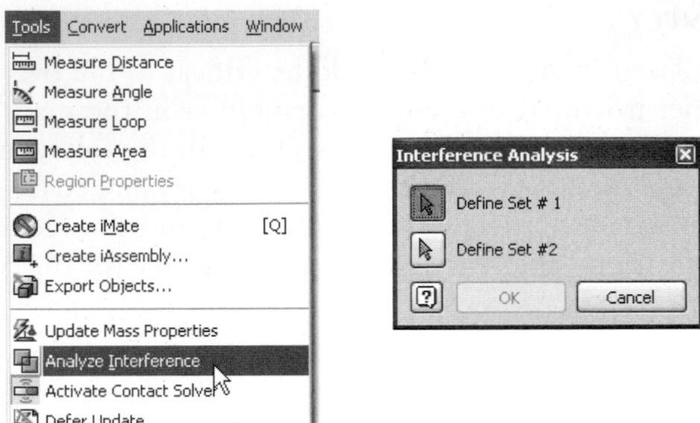

Figure 6.128

Click on Define Set #1, and select the components that will define the first set. Click on Define Set #2, and select the components that will define the second set. A component can exist in only one set. To add or delete components from either set, select the button that defines the set that you want to edit. Click components to add to the set, or press the CTRL key while selecting components to remove from the set.

Note: Use only Define Set #1 if you want to check for interference in a single group of objects.

Once you have defined the sets, click the OK button. The order in which you selected the components has no significance. If interference is found, the Interference Detected dialog box will appear, as shown in the following image.

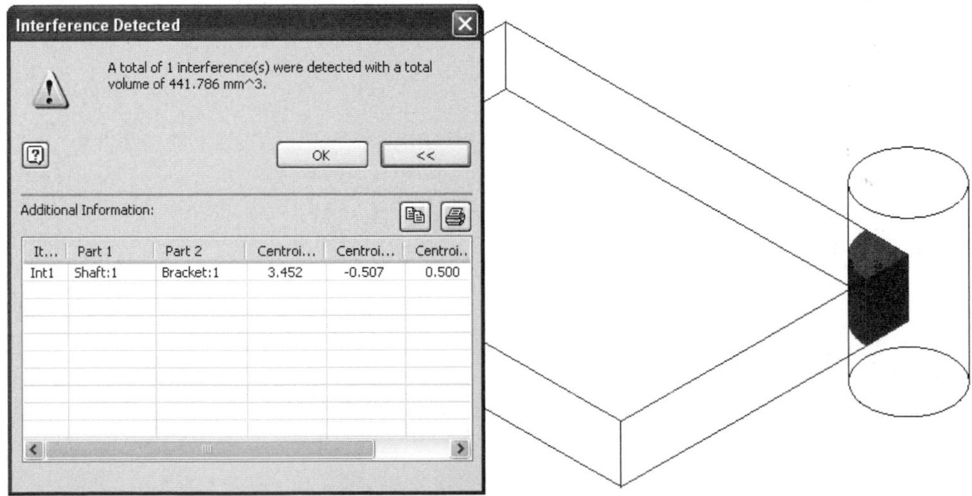

Figure 6.129

The information in the dialog box defines the X, Y, and Z coordinates of the centroid of the interfering volume. It also lists the volume of the interference and the components that interfere with one another. A temporary solid will also be created in the graphics window that represents the interference. You can copy the interference report to the clipboard or print it from the tools in the Interference Detected dialog box. When the operation is complete, click the OK button, and the interfering solid will be removed from the screen.

Note: Performing an interference detection does not fix the interfering problem; it only presents a graphical representation of the problem. After analyzing and finding an interference, edit the assembly or components to remove the interference. You can also detect interference when driving constraints.

EXERCISE 6–5: ANALYZING AN ASSEMBLY

In this exercise, you analyze a partially completed assembly for interference between parts and then check the physical properties to verify design intent.

1. Open the assembly *ESS_E06_05.iam*. The linkage is displayed, as shown in the following image.

Figure 6.130

2. Begin the process of checking for interferences in the assembly. First zoom into the linkages.
3. From the Tools menu, click Analyze Interference.
4. When the Analyze Interference dialog box displays, select the two components for Set #1, as shown in the following image.

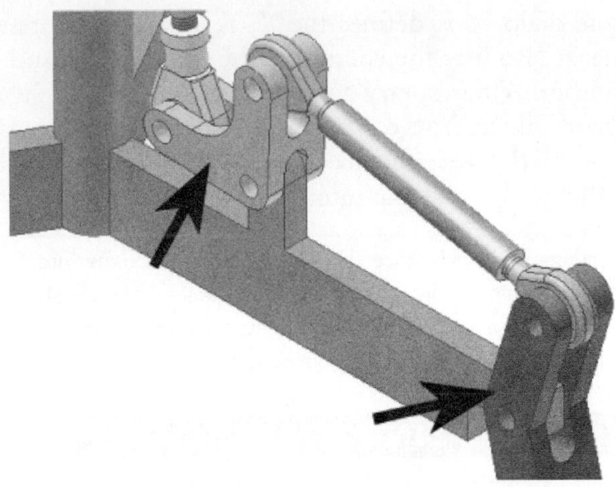

Figure 6.131

5. For Set #2, select the spyder, as shown in the following image.

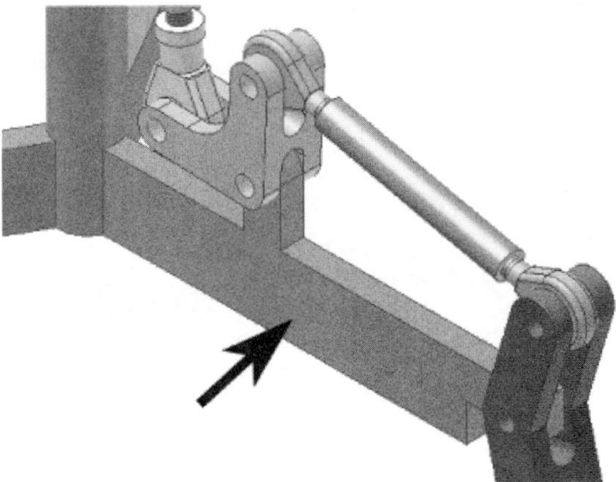

Figure 6.132

6. Click the OK button. Notice that the Interference Detected dialog box is displayed, and the amount of interference displays on the parts of the assembly, as shown in the following image.

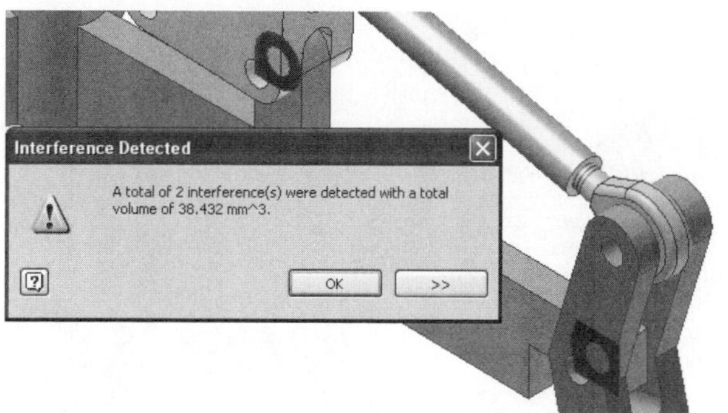

Figure 6.133

7. Continue using more features of the Interference Detected dialog box. Click the More (>>) button to expand the dialog box and display additional information.

8. Click the Copy to Clipboard button, as shown in the following image. This is a quick way of sharing the interference detection information with other applications such as Microsoft Word or Notepad.

9. Open a text editor, and paste the information into a document.

10. Exit the text editor. You do not have to save the text file.

11. Click the OK button to close the Interference Detected dialog box. Keep this assembly file open.

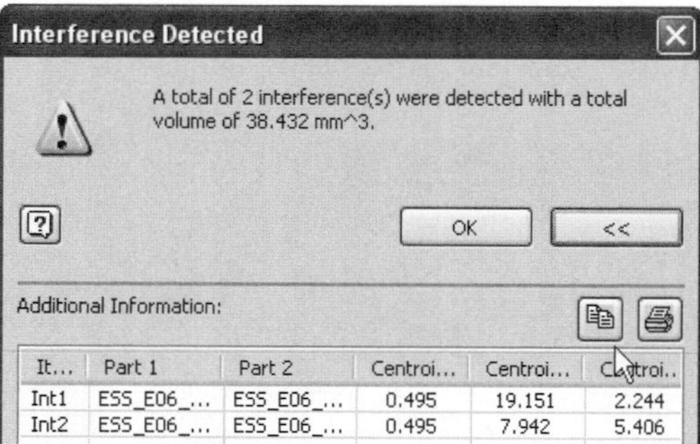

Figure 6.134

12. The design intent of the lifting mechanism relies on the correct selection of materials. This impacts the strength and mass of the assembly. You will now display the physical properties of the entire assembly.

13. View the assembly with the Home View. In the browser, right-click *ESS_E06_05.iam,* and then select iProperties, as shown following image.

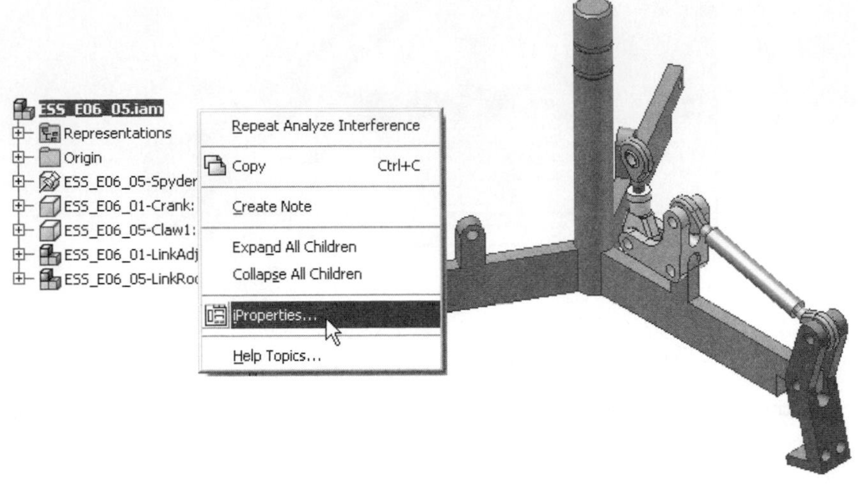

Figure 6.135

14. When the Properties dialog box displays, click the Physical tab.

15. Click the Update button. Notice that the physical properties of the assembly are displayed, as shown in the following image.

Note: To get a more accurate account of the assembly properties, click on Low under the Requested Accuracy area, and change this setting to Very High. The updating action may take some time, depending on the size of the assembly being analyzed.

16. Click the OK button to dismiss this dialog box.

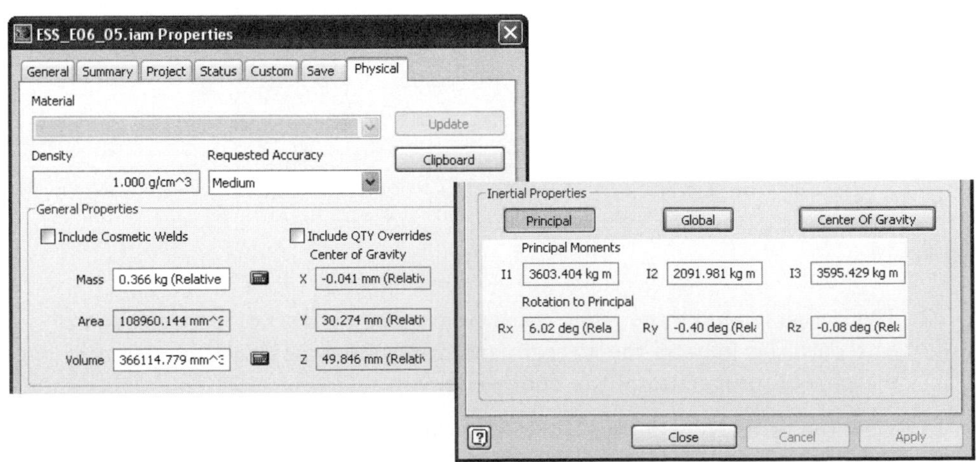

Figure 6.136

17. Close all open files. Do not save changes. End of exercise.

DRIVING CONSTRAINTS

You can simulate or drive mechanical motion using the Drive Constraint tool. To simulate motion, an angle, mate, tangent, or insert assembly constraint must exist. You can only drive one assembly constraint at a time, but you can use equations to create relationships to drive multiple assembly constraints simultaneously. To drive a constraint, right-click on the desired constraint in the browser, and select Drive Constraint from the menu, as shown in the following image.

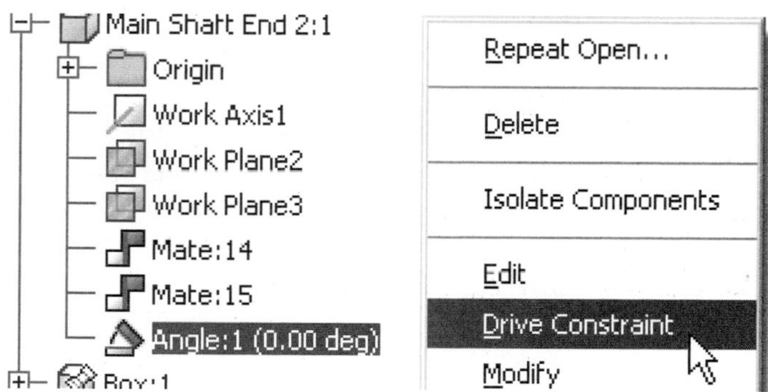

Figure 6.137

The Drive Constraint dialog box will appear, as shown in the following image. Depending on the constraint that you are driving, the units may be different. Enter a Start value; the default value is the angle or offset for the constraint. Enter a value for End and a value for Pause Delay if you want a dwell time between the steps. In the dialog box, you can also choose to create an animation (AVI) file that will record the assembly motion. You can replay the AVI file without having Autodesk Inventor installed.

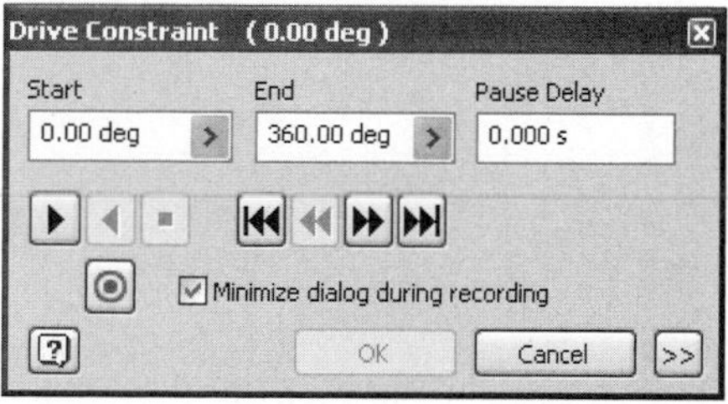

Figure 6.138

The following sections describe the Start, End, and Pause Delay controls.

Start Sets the start position of the offset or angle.

End Sets the end position of the offset or angle.

Pause Delay Sets the delay between steps. The default delay units are seconds.

Use the following Motion and AVI control buttons to control the motion and to create an AVI file.

Forward	Drives the constraint forward, from the start position to the end position.
Reverse	Drives the constraint in reverse, from the end position to the start position.
Pause	Temporarily stops the playback of the constraint drive sequence.
Minimum	Returns the constraint to the starting value and resets the constraint driver. This button is not available unless the constraint driver has been run.
Reverse Step	Reverses the constraint driver one step in the sequence. This button is not available unless the drive sequence has been paused.
Forward Step	Advances the constraint driver one step in the sequence. This button is not available unless the drive sequence has been paused.
Maximum	Advances the constraint sequence to the end value.
Record	Begins capturing frames at the specified rate for inclusion in an animation file named and stored to a location that you define.

To set more conditions on how the motion will behave, click the More (<<) button. This action will display the expanded Drive Constraint dialog box, as shown in the following image. The following sections describe the options in the dialog box.

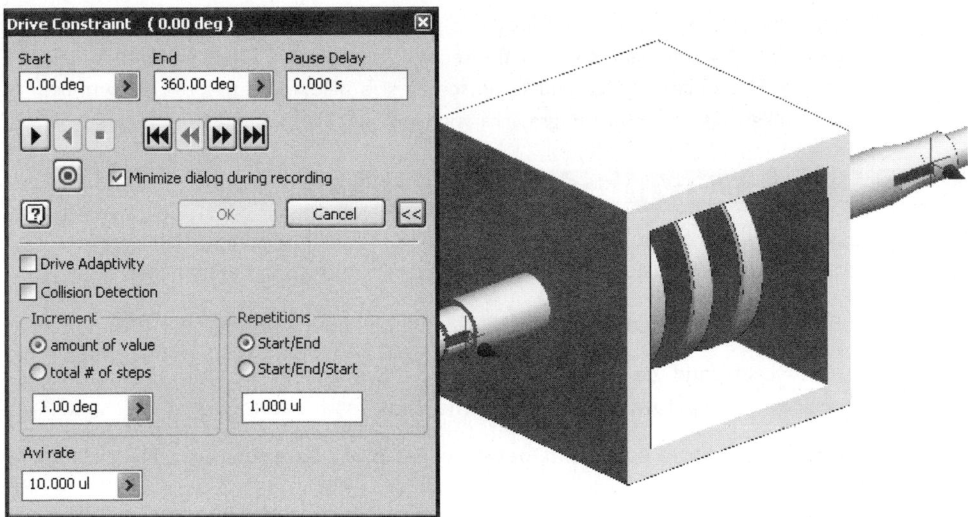

Figure 6.139

Drive Adaptivity Click this option to adapt the component while the constraint is driven. It only applies to assembly components for which adaptivity has been defined and enabled.

Collision Detection Click this option to drive the constraint until a collision is detected. When interference is detected, the drive constraint will stop, and the components where the collision occurs will be highlighted. It also shows the constraint value at which the collision occurred.

Increment Determines the value that the constraint will be incremented during the animation.

Amount of Value Increments the offset or angle value by this value for each step.

Total # of Steps The constraint offset or angle is incremented by the same value per step, based on the number of steps entered and the difference between the start and stop values.

Repetitions Sets how the driven constraint will act when it completes a cycle and how many cycles will occur.

Start/End Drives the constraint from the start value to the end value and resets at the start value.

Start/End/Start Drives the constraint from the start value to the end value, and then drives it in reverse from the end value to the start value.

AVI Rate Specifies how many frames are skipped before a screen capture is taken of the motion that will become a frame in the completed AVI file.

Note: If you try to drive a constraint and it fails, you may need to suppress or delete another assembly constraint to allow the required component degrees of freedom. To reduce the size of an AVI file, reduce the screen size before creating the file, and use a solid background color in the graphics window.

EXERCISE 6-6: DRIVING CONSTRAINTS

In this exercise, you drive an angle constraint to simulate assembly component motion. Next you drive the constraint and use collision detection to determine if components interfere.

1. Open the existing assembly file *ESS_E06_06.iam* from the chapter 6 folder.
2. Use the Rotate and Zoom tools to examine the assembly. When finished, right-click in the graphics window, and select Home View from the menu.
3. Now create an angle constraint between the pivot base and arm. First click the Constraint tool.
4. In the Place Constraint dialog box, click the Angle button.
5. Select the Base face and then arm face, as shown in the following image.
6. Enter an angle of **45°**. Click the OK button.

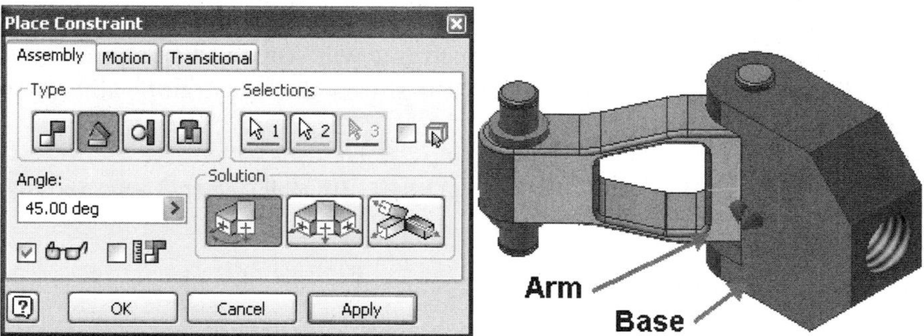

Figure 6.140

7. Now drive the angle constraint you just placed. In the browser, expand *ESS_E06_06-Pivot_Base:1*.
8. Right-click the Angle constraint, and then select Drive Constraint from the menu.
9. In the Drive Constraint dialog box, enter **45.00°** as the Start value and **120.00°** as the End value.
10. Click the More (<<) button.
11. In the Increment section, enter **2.00 Deg**, as shown in the following image.
12. Click the Forward button. Notice that the arm assembly interferes with the pivot base between 45.00° and 120.00° but Inventor continues to drive the constraint.

13. Click the Reverse button.

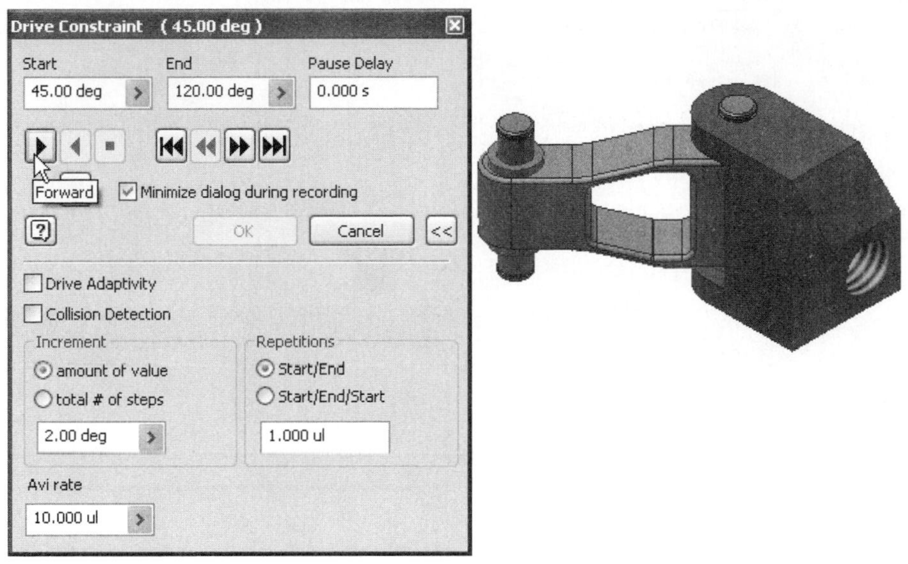

Figure 6.141

14. Now perform a collision detection operation. In the Drive Constraint dialog box place a check in the Collision Detection box.

15. Change the Increment value to **.1**.

16. Click the Forward button. Notice that a collision is detected at 80.7° and that the interfering parts are highlighted, as shown in the following image. Click OK to return to the Drive Constraint dialog box.

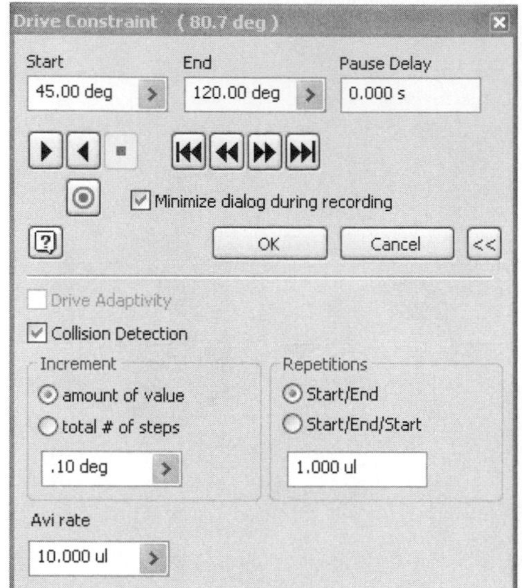

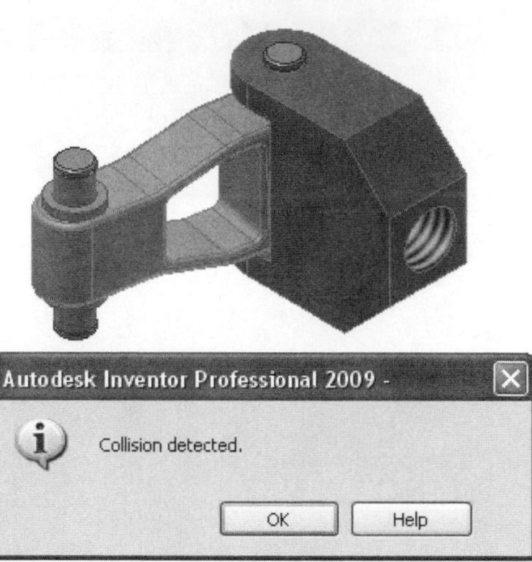

Figure 6.142

17. Now perform a check on the full range of motion for the arm. In the Drive Constraint dialog box, enter **80.00** for the Start value and **-80.00** for the End value.

18. Ensure that the Collision Detection option is checked.

19. Change the Increment value to **2.00**.

20. Click the Forward button to test the full range of motion for interference. No collision is detected in this range of motion.

21. In the Drive Constraint dialog box, click the OK button. The 80.00°, which represents the maximum angle before collision, is applied to the Angle constraint, as shown in the following image.

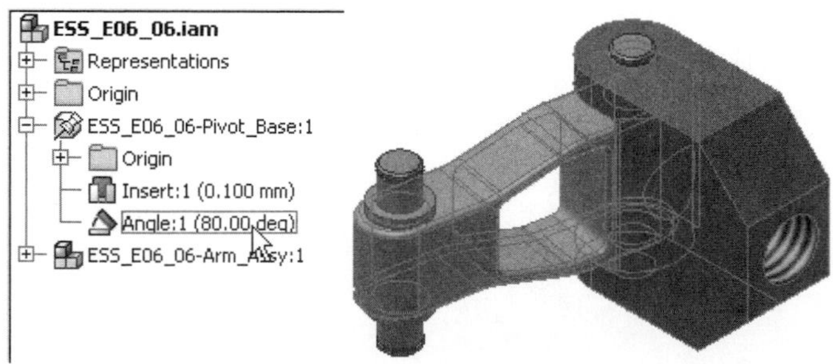

Figure 6.143

22. Close all open files. Do not save changes. End of exercise.

CREATING PRESENTATION FILES

After creating an assembly, you can create drawing views based on how the parts are assembled. If you need to show the components in different positions, similar to an exploded view, hide specific components, or create an animation that shows how to assemble and disassemble the components, you need to create a presentation file. Components can also be hidden and later retrieved through design view representations. This topic is covered in a later chapter. A presentation file is separate from an assembly and has a file extension of *.ipn*. The presentation file is associated with the assembly file, and changes made to the assembly file will be reflected in the presentation file. You cannot create components in a presentation file. To create a new presentation file, click New on the File menu the New icon in What To Do. When the New File dialog box appears, select the Standard.ipn icon, as shown in the following image.

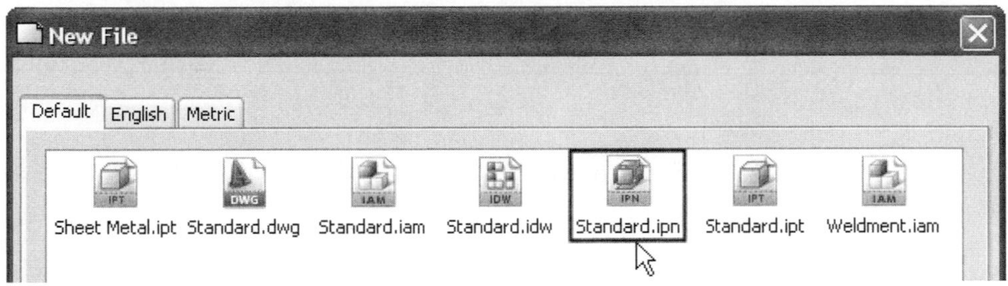

Figure 6.144

By default, the panel bar will show the presentation tools available to you, as shown in the following image on the left. The following sections explain these tools.

Create View Click to create views of the assembly. Created views are based on the same assembly model. Once created, a view will be listed as an explosion in the browser.

Tweak Components Click to move and/or rotate parts in the view.

Precise View Rotation Click to rotate the view by a specified angle and direction using a dialog box.

Animate Click to create an animation of the assembly tweaks and saved camera views; an AVI file can be output.

The basic steps for creating a presentation view are to reposition (tweak) the parts in specific directions and then create an animation, if needed. The following sections discuss these steps.

CREATING PRESENTATION VIEWS

The first step in creating a presentation is to create a presentation view. Issue the Create View tool on the Presentation panel bar, or right-click and select Create View from the menu. The Select Assembly dialog box appears, as shown in the following image on the right. The dialog box is divided into two sections: Assembly and Explosion Method.

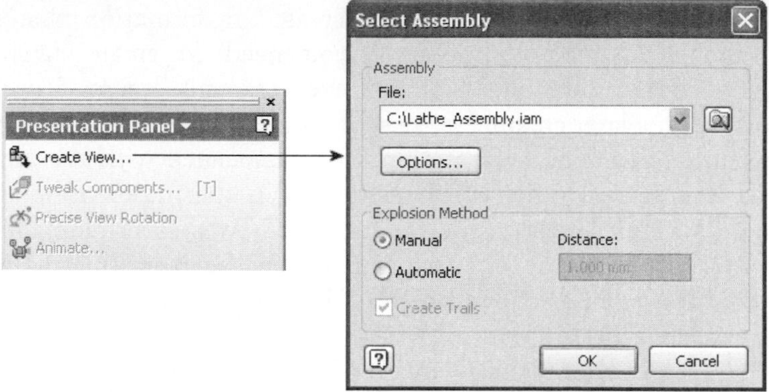

Figure 6.145

Assembly

In this section, you determine on which assembly and design view to base the presentation view.

File Enter or select an assembly file on which to base the presentation view.

Options Click to display the File Open Options dialog box. Use this dialog box to select design view representations, which are covered in chapter 9, and positional representations, which are covered in chapter 9.

Explosion Method

In this section, you can choose to manually or automatically explode the parts, meaning that you separate the parts a given distance.

Manual Click so that the components will not be exploded automatically. After the presentation view is created, you can add tweaks that will move or rotate the parts.

Automatic Click so that the part will be exploded automatically with a given value.

Create Trails If you clicked Automatic, click to have visible trails, or lines that show how the parts are being exploded.

Distance If Automatic is selected, enter a value that the parts will be exploded.

After making your selection, click the OK button to create a presentation view. The components will appear in the graphics window, and a presentation view will appear in the browser. If you clicked the Automatic option, the parts will be exploded automatically. To determine how the parts will be exploded, Autodesk Inventor analyzes the assembly constraints. If you constrained parts using the mate plane option and the arrows perpendicular to the plane (normals) for both parts were pointing outward, they will be exploded away from each other the defined distance. Think of the mating parts as two magnets that want to push themselves in opposite directions. The grounded component in the browser will be the component that stays stationary, and the other components will move away from it. If you are creating trails automatically, they will generally come from the center of the part, not necessarily from the center of the holes. After expanding all of the children in the browser, you can see numerically how the parts exploded. Once the parts are exploded, the distance is referred to as a tweak. The number by each tweak reflects the distance that the component is moved from the base component.

TWEAKING COMPONENTS

After creating the presentation view, you may choose to edit the tweak of a component or move or rotate the component an additional distance. To edit an individual tweak, click on its value in the browser, as shown in the following image. Enter a new value in the text box in the lower-left corner of the browser, and press ENTER to use the new value.

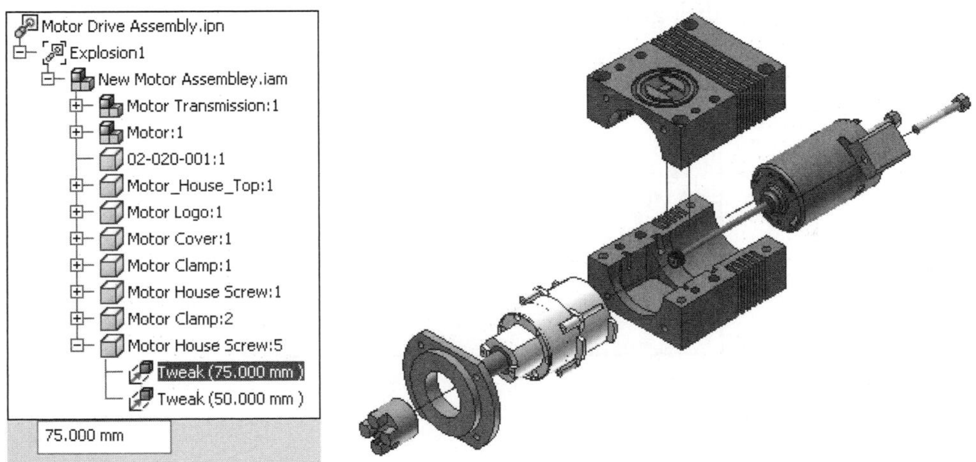

Figure 6.146

To extend all of the tweaks the same distance, right-click on the assembly's name in the browser, and select Auto Explode from the menu, as shown in the following image on the left. Enter a value, and click the OK button, as shown in the following image on the right, and all of the tweaks will be extended the same distance. You cannot use negative values with the Auto Explode method.

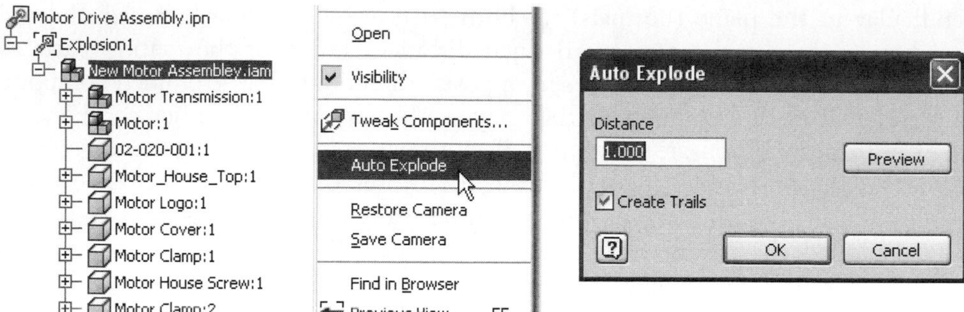

Figure 6.147

Another way to reposition the components manually is to tweak them using the Tweak tool. To manually tweak a component, click the Tweak Component tool on the Presentation panel bar, press the hot key T, or right-click and select Tweak Component from the menu. The Tweak Component dialog box will appear, as shown in the following image. The Tweak Component dialog box has two sections: Create Tweak and Transformations.

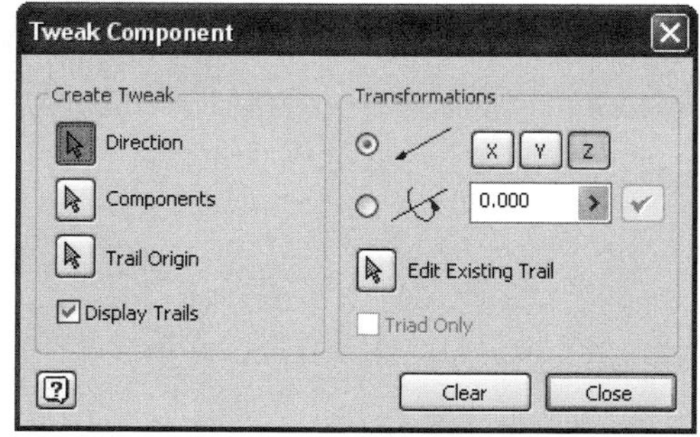

Figure 6.148

Create Tweak

In this section, you select the components to tweak, set the direction and origin of the tweak, and control trail visibility.

Direction Determine the direction or axis of rotation for the tweak. After clicking the Direction button, select an edge, face, or feature of any component in the graphics window to set the direction triad (X, Y, and Z) for the tweak. The edge, face, or feature that you selected does not need to be on the components you are tweaking.

Components Select the components to tweak. Click the Components button, and then click the components in the graphics window or browser to tweak. If you selected a component when you started the Create Tweak operation, it will be included in the components automatically. To remove a component from the group, press and hold the CTRL key and click the component.

Trail Origin Set the origin for the trail. Click the Trail Origin button, and then click in the graphics window to set the origin point. If you do not specify the trail origin, it will be placed at the center of mass for the part.

Display Trails Click if you want to see the tweak trails for the selected components. Clear the checkbox to hide the trails.

Transformations

In this section, you set the type and value of a tweak.

Linear Click the button next to the arrow and line to move the selected components in a linear fashion.

Rotation Click the button next to the arrow and arc to rotate the selected components.

X, Y, Z Click the X, Y, or Z coordinate button to determine the direction for a linear tweak or the axis for a rotational tweak. Alternately, you can select the arrow on the triad that represents the X, Y, or Z direction.

Text Box Enter a positive or negative value for the tweak distance or rotation angle, or click a point in the graphics window and move the cursor with the mouse button depressed to set the distance.

Apply After making all selections, click the Apply button to complete the tweak.

Edit Existing Trail To edit an existing tweak, click the Edit Existing Trail button. Select the tweak in the graphics window, and change the desired settings.

Triad Only Click the Triad Only option to rotate the direction triad without rotating selected components. Enter the angle of rotation, and then click the Apply button. After you rotate the triad direction, you can use it to define tweaks.

Clear Click the Clear button to remove all of the settings and set up for another tweak.

To tweak a component, follow these steps:

1. Issue the Tweak Component tool.
2. Determine the direction or axis of rotation for the tweak by clicking the Direction button and selecting an edge, face, or feature.
3. Select the components to tweak by clicking the Components button and clicking the components in the graphics window or browser that will be tweaked.
4. Select any additional trail origin points.
5. Determine whether or not you want trails to be visible.
6. Set the type of tweak to linear or rotation.

7. Click the X, Y, or Z coordinate button to determine the direction for a linear tweak or the axis for a rotational tweak.
8. Enter a value for the tweak in the text box, or select a point on the screen and drag the part into its new position.
9. Click the Apply button in the Tweak Component dialog box.

ANIMATION

After you have repositioned the components, you can animate the components to show how they assemble or disassemble. To create an animation, click the Animate tool on the Presentation panel bar, or right-click and select Animate from the menu. The Animation dialog box will appear, as shown in the following image. The Animation dialog box has three sections: Parameters, Motion, and Animation Sequence, which appears under the More (<<) button.

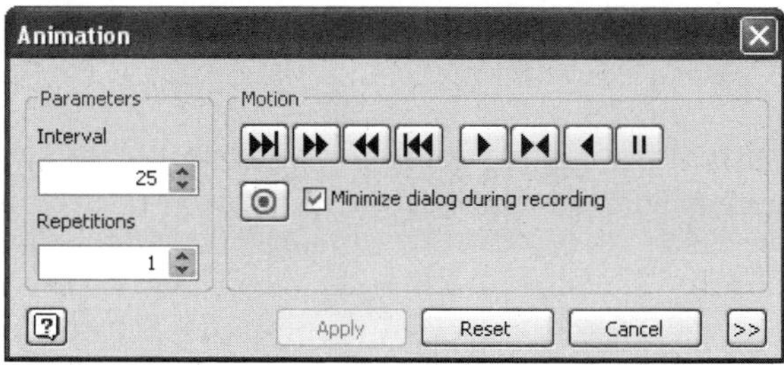

Figure 6.149

Parameters

In this section, you specify the playback speed and the number of repetitions for the animation.

Interval Set this value for the playback speed of the animation in frames. The higher the number, the greater the number of steps in the tweak and the slower the animation. A smaller number will speed up the animation.

Repetitions Set the number of times to repeat the playback. Enter the desired number of repetitions, or use the up or down arrow to select the number.

Note: To change the number of repetitions, click the Reset button on the dialog box, and then enter a new value.

Click to record the specified animation to a file so that you can replay it later.

To animate components, follow these steps:

1. Issue the Animation tool.
2. Set the number of repetitions.

3. Adjust the tweaks as needed.
4. Click one of the play buttons to view the animation in the graphics window.
5. To record the animation to a file, click the Record button, and then click one of the Play buttons to start recording.

Changing the Animation Sequence

In the Animation Sequence section, you can change the sequence in which the tweaks happen, select the tweak, and then select the needed operation. Expanding the Animation dialog box displays the following image.

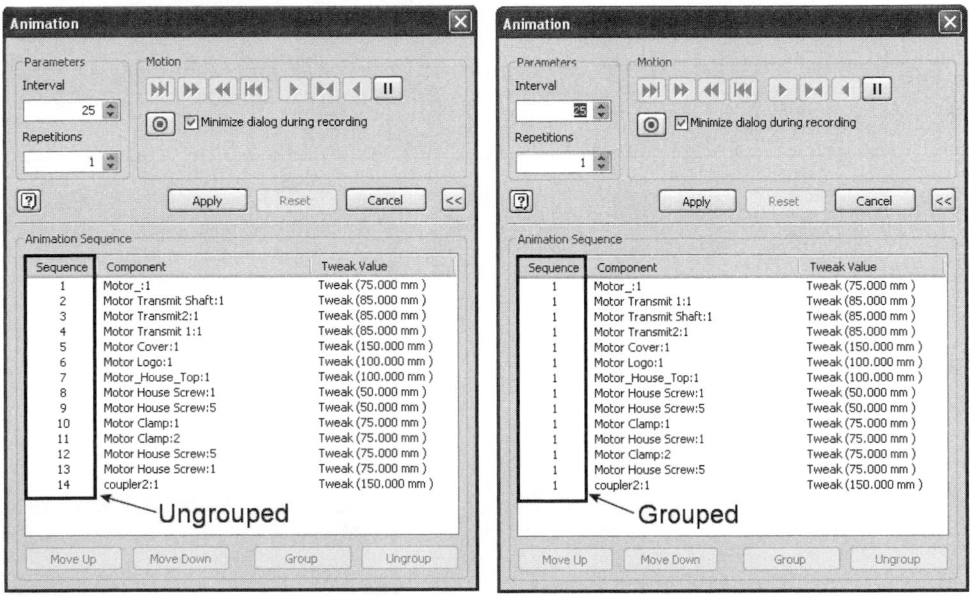

Figure 6.150

Move Up Click to move the selected tweak up one place in the list.

Move Down Click to move the selected tweak down one place in the list.

Group Select a number of tweaks, and then click the Group button. When tweaks are grouped, all of the tweaks in the group will move together as you change the sequence. The group assumes the sequence order of the lowest tweak number.

Ungroup After selecting a tweak that belongs to a group, you can click the Ungroup button, and the tweak can then be moved individually in the list. The first tweak in the group assumes a number that is one higher than the group. The remaining tweaks are numbered sequentially following the first.

EXERCISE 6-7: CREATING PRESENTATION VIEWS

In this exercise, you create a presentation view of an existing assembly, and you add tweaks to the assembly components to create an exploded view.

1. Begin creating a presentation by clicking the New tool.
2. Select the Metric tab, and then double-click *Standard(mm).ipn*.
3. While in the presentation graphics screen, click the Create View tool in the panel bar.
4. In the Select Assembly dialog box, verify that the Explosion Method is set to Manual. Then click the Explore button next to the File field. Open *ESS_E06_07.iam* from the chapter 06 folder, and click the Options button, as shown in the following image on the left.
5. In the Design View Representation field, select Internal Components, as shown in the following image on the right.

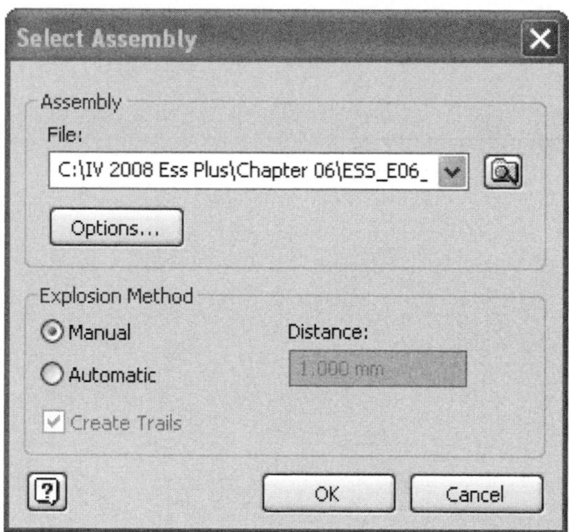

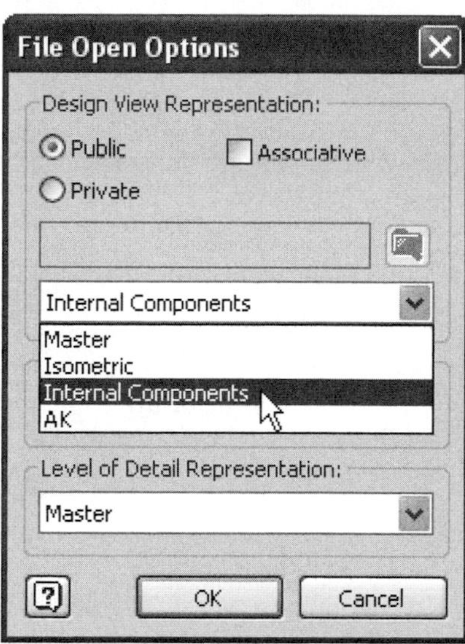

Figure 6.151

6. Click OK to create the view. Your display should appear similar to the following image.

Figure 6.152

7. Now set the tweak direction used to create an exploded presentation. In the browser, click the plus sign in front of Explosion1 and the plus sign in front of *ESS_E06_07.iam* to expand their displays. The assembly components are displayed so you can select components in the graphics screen or in the browser.
8. Click the Tweak Components tool in the panel bar.
9. In the Tweak Component dialog box, verify that Direction is selected and Display Trails is checked.
10. Position your cursor over a cylindrical face until the temporary Z axis aligns with the axis of the assembly, as shown in the following image.
11. Click to accept the axis orientation.

Figure 6.153

12. Click on the nut at the base of the assembly to select it. In the browser, observe that *ESS_E06-NutB* is highlighted.
13. In the Tweak Components dialog box, verify that the Z button is selected, and then enter **-40** in the Transformations field.
14. Click Apply, with a green checkmark, to create the tweak. Your display should appear similar to the following image.

Figure 6.154

Tip: After you define the tweak axes, you can select a component or group of components and then drag them to create a tweak. This way, you can quickly arrange your components visually. You can later edit the values of the tweaks from the browser to define their final positions.

15. Next click the valve plate to select it. Hold down the CTRL key, and click the nut to unselect it. Observe that the valve plate is the only entry highlighted in the browser.
16. Enter **-20** in the Transformations field of the dialog box.
17. Click Apply to create the tweak. Your display should appear similar to the following image.

Figure 6.155

Tip: You can also select components by clicking on them directly in the browser, which is useful when trying to select hidden components.

18. Hold down the CTRL key to unselect the valve plate. The next component to tweak is the top nut. In the browser, click *ESS_E06-NutA* to select it. Notice that the valve plate is no longer selected.
19. Enter **115** in the Transformations field, and click Apply to create the tweak. Your display should appear similar to the following image.

Figure 6.156

Tip: Examine the browser. Expand the components that have a plus sign in front of them to view the tweaks you have applied.

20. Complete this phase of the exercise by adding the following tweaks:

Part Name	Tweak
ESS_E06-WasherB.ipt:1	105
ESS_E06-Pin.ipt:1	85
ESS_E06-Spring.ipt:1	45
ESS_E06-WasherA.ipt:1	30
ESS_E06-Diaphragm.ipt:1	15

21. Close the Tweak Component dialog box when you are finished. Your display should appear similar to the following image.

Figure 6.157

22. Now adjust an existing tweak. In the browser, expand *ESS_E06-Diaphragm.ipt:1* and *ESS_E06-WasherA.ipt:1*. The hierarchy displays tweaks nested below each component definition.

23. Right-click each of the tweaks, and select Visibility to turn off the trail display for the tweak.

24. Click Tweak (15.000 mm) under *ESS_E06-Diaphragm.ipt:1*.

25. Below the browser, enter a new value of **10** in the Offset field, and then press ENTER. Observe that the name of the tweak reflects the change, as shown in the following image on the left, and the diaphragm moves to its new position, as shown in the following image on the right.

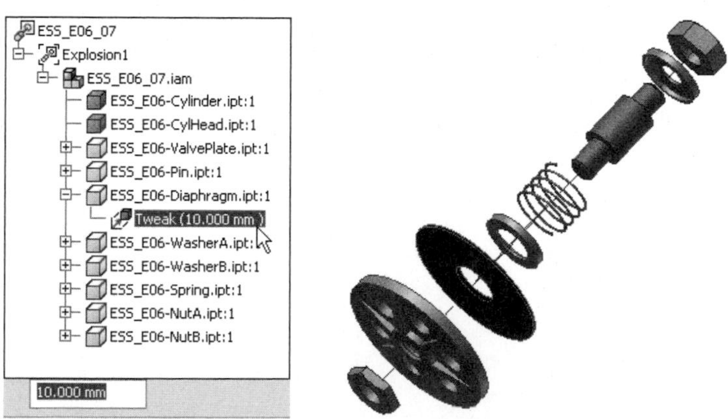

Figure 6.158

26. Close all open files. Do not save changes. End of exercise.

CREATING DRAWING VIEWS FROM ASSEMBLIES AND PRESENTATION FILES

After creating an assembly or presentation, you may want to create drawings that use the data. You can create drawing views from assemblies using information from an assembly or a presentation file. You can select from them a specific design view or presentation view. To create drawing views based on assembly data, start a new drawing file or open an existing drawing file. Use the Base View tool and select the IAM or IPN file from which to create the drawing. If needed, specify the design view or presentation view in the dialog box. The following image shows a presentation view being selected after a presentation file was selected.

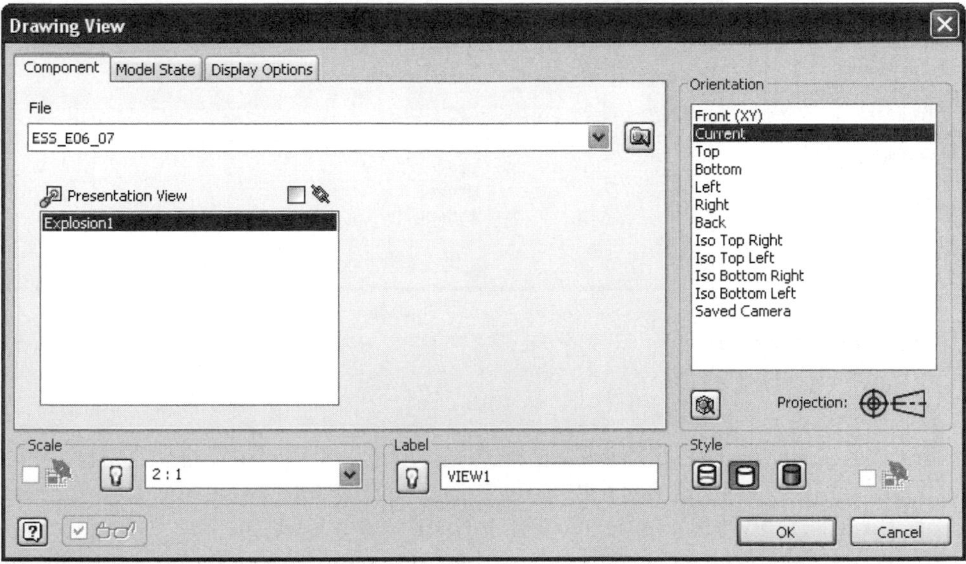

Figure 6.159

THE BOM EDITOR

A Bill of Materials, usually referred to as a BOM, is a table that contains information about the components inside an assembly. It can include item number, quantity, part number, description, vendor, and other information needed to describe the assembly. The BOM is also considered associative; when a component is added or removed from the assembly, the quantity field in the BOM will update to reflect this change in the assembly.

To facilitate managing item numbers, every item is automatically assigned a number, and the item number can be changed or reordered if necessary. Whenever these types of changes are made in the BOM, the same changes are updated automatically in Parts List and balloons.

Values in the BOM can be changed several ways. The first way is to add information while inside each individual part through the Properties dialog box. Another way is to add the part information directly into the BOM with the BOM Editor; then, when the BOM is saved, the information supplied inside of the BOM is also saved to each individual part. This is an efficient way to edit the properties of multiple parts at the same time.

In order to activate the BOM Editor, you must be inside an assembly model. Under the Tools menu, click Bill of Materials, as shown in the following image on the left, to launch the Bill of Materials dialog box, as shown in the following image on the right.

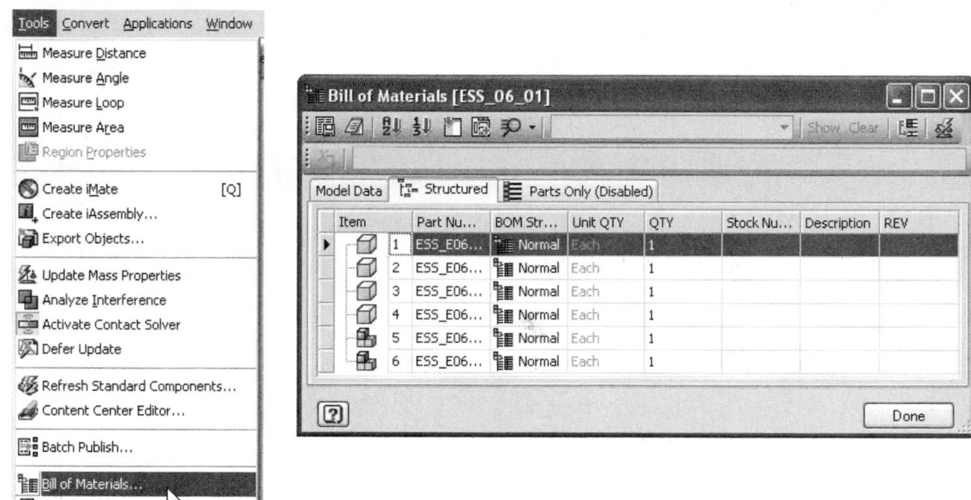

Figure 6.160

There are three tabs in the BOM Editor: Model Data, Structured, and Parts Only. The Model Data tab is shown in the following image on the left. Here the components that make up the assembly are arranged in a format that resembles the assembly browser. Crankshaft Assembly can be expanded to display its individual parts. The information under this tab is not meant to be reported to a parts list.

The Structured tab for this example is shown in the following image in the middle. This is the actual BOM data that is reported to the parts list. When arranging an assembly in Structured mode, Crankshaft Assembly is considered an individual item along with the other parts listed.

When the Parts Only tab is selected, Crankshaft Assembly is not listed. However, the individual parts that make up Crankshaft Assembly are listed along with the other parts, as shown in the following image on the right. This arrangement of parts is referred to as a flat list.

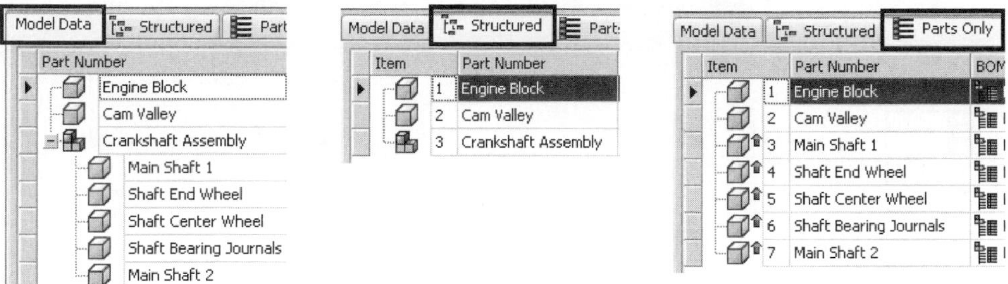

Figure 6.161

The BOM Editor also contains several buttons that provide further control over the information supplied in the BOM. These buttons are described as follows:

Button	Tool	Function
	Export Bill of Materials	Bill of material information can be exported out to an external file. The following file formats are supported: mdb, xls, dbf, txt, and csv.
	Engineer's Notebook	The BOM can be added to the Engineer's Notebook in the form of a table as a note that is linked to an Excel spreadsheet.
	Sort Items	Used for sorting the values in one or more BOM columns in ascending or descending order.
	Renumber Items	Allows items to be sorted based on the current sort order of the BOM.
	Choose Columns	Allows extra informational columns to be added or removed from the BOM.
	Add Custom iProperties Column	Adds the new values to the Custom tab in the corresponding iProperties dialog box.
	View Options	Enables or disables the BOM information from being displayed in the Model Data, Structured, and Parts Only tabs.
	Part Number Merge Settings	Allows different components with the same part number to be treated as a single component.
	Update	Calculates the mass and volume of all items in the BOM.

SORTING ITEMS IN THE BOM

Clicking on the Sort button located in the BOM Editor will display the Sort dialog box, as shown in the following image. Sorting can be performed only when in the Structured and Parts Only tabs. In the Sort by area of the dialog box, select the first column to sort by followed by selecting ascending or descending order. Columns can also be arranged by a secondary sort and, if desired, a tertiary sort.

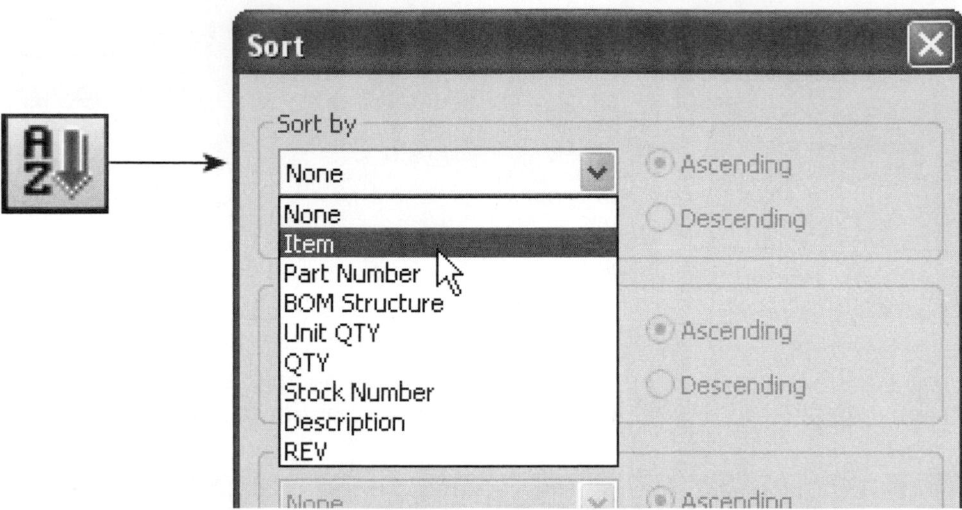

Figure 6.162

RENUMBERING ITEMS IN THE BOM

Clicking on the Renumber Items button will display the Item Renumber dialog box, as shown in the following image. This operation is usually performed directly after sorting information in the BOM Editor.

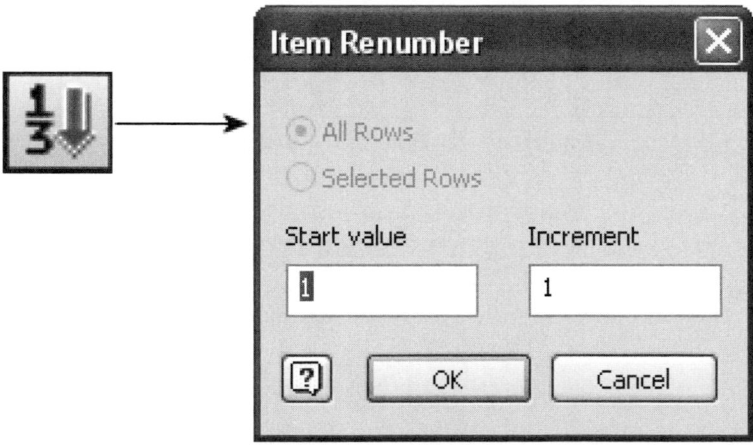

Figure 6.163

CHOOSING EXTRA COLUMNS FOR THE BOM

Clicking on the Choose Columns button will display a Customization dialog box in the lower-right corner of the BOM Editor, as shown in the following image. Choose the desired column from the list, and then drag and drop this listing onto an existing column in the BOM Editor. This action will add the custom column to the BOM. You can drag existing column headers to remove them from the BOM Editor; drop the column when a large X appears at your cursor position. Removed headers are added to the Customization dialog so that you can restore them to the BOM Editor.

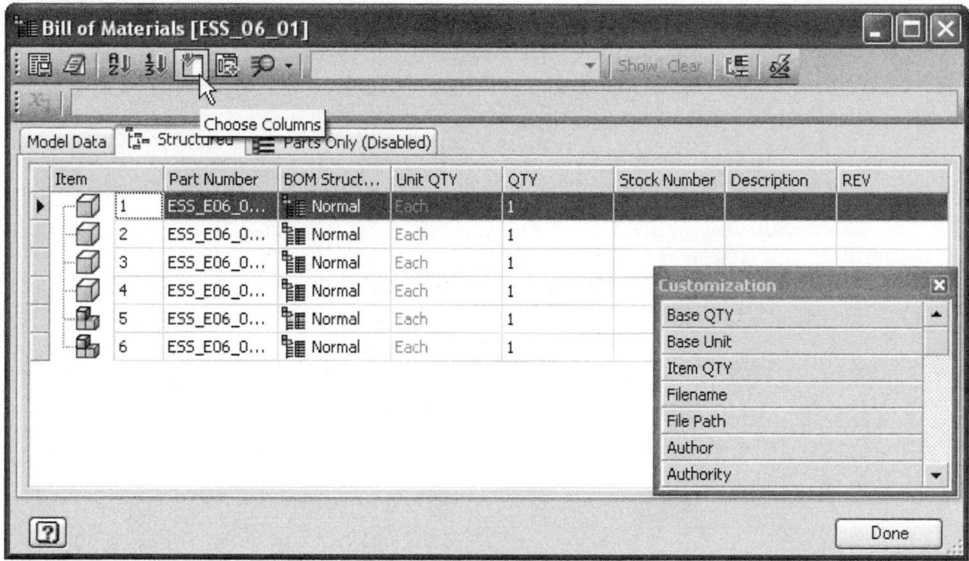

Figure 6.164

CHANGING THE VIEW OPTIONS OF THE BOM

Clicking on the View Options button activates a pull-down menu, as shown in the following image. You can enable or disable the BOM view based on the current tab. In this example, the current tab is Structured. Clicking on View Properties from the menu will activate the Structured Properties dialog box, as shown in the following image. Use this dialog to switch the BOM view from First Level to All Levels.

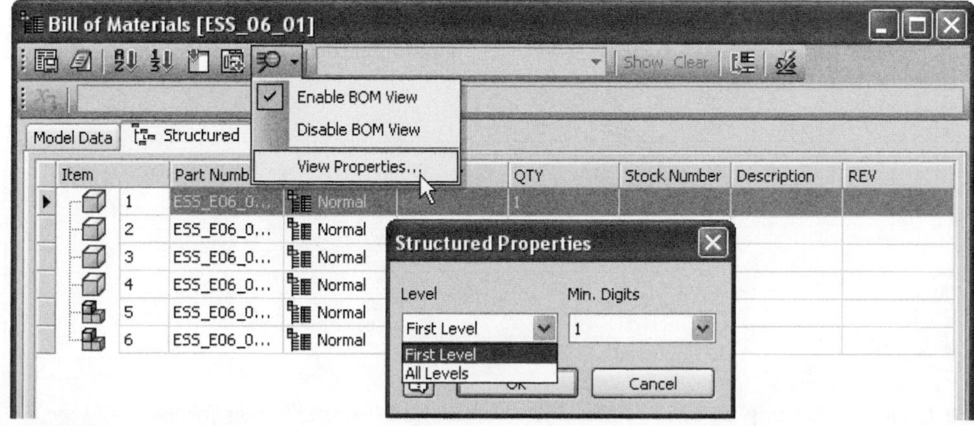

Figure 6.165

EDITING A BOM COLUMN

Right-clicking on a column in the BOM Editor will display a menu, as shown in the following image. Use this menu to sort the information in the column by ascending or descending order. Best Fit will resize the column width to fit the information that occupies each cell; you could also double-click on the divider between the column headers. In this case, the column to the left of the divider is resized to fit the contents of the cells in that column. Clicking on Best Fit (all columns) will resize the information in all columns.

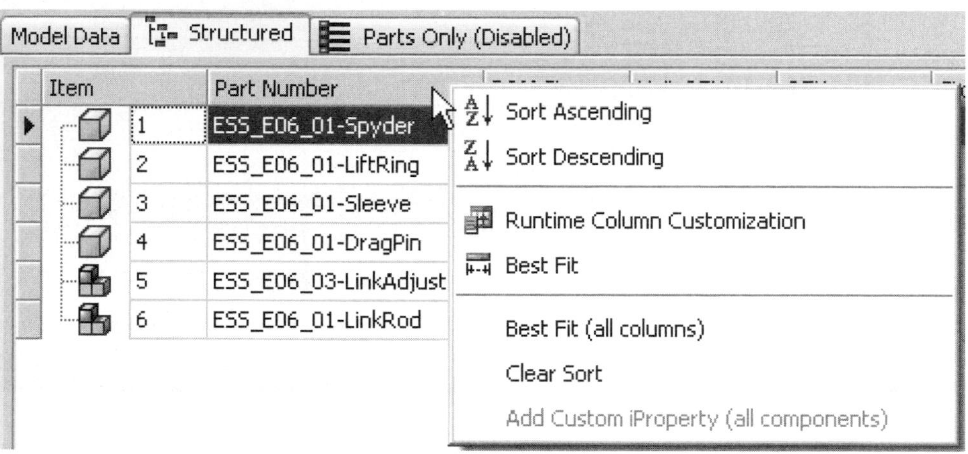

Figure 6.166

EXERCISE 6-8: EDITING THE BOM

In this exercise, you manipulate a number of items in the BOM Editor and see these changes reflected in the current drawing parts list.

1. Open the drawing file *ESS_ E06_08.idw*.
2. Examine the existing Parts List as shown in the following image. Notice that the Description column is blank. The fields are populated from the iProperties (meta data) of the components.

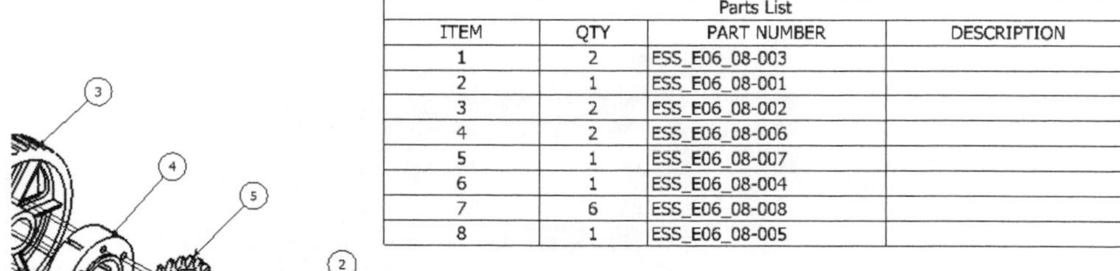

Parts List			
ITEM	QTY	PART NUMBER	DESCRIPTION
1	2	ESS_E06_08-003	
2	1	ESS_E06_08-001	
3	2	ESS_E06_08-002	
4	2	ESS_E06_08-006	
5	1	ESS_E06_08-007	
6	1	ESS_E06_08-004	
7	6	ESS_E06_08-008	
8	1	ESS_E06_08-005	

Figure 6.167 *(Courtesy US FIRST Team 342—Robert Bosch Corp., Trident Technical College, Fort Dorchester High School, Summerville High School; Charleston, South Carolina)*

3. The individual descriptions of each part can be easily added through the BOM Editor while working in the *.idw* file. Using this method, you will not have to open each individual part file to add a description. To activate this dialog box, right-click on the Parts List in the browser or in the graphics window, as shown in the following image on the left, and click Bill of Materials from the menu, as shown in the following image on the right.

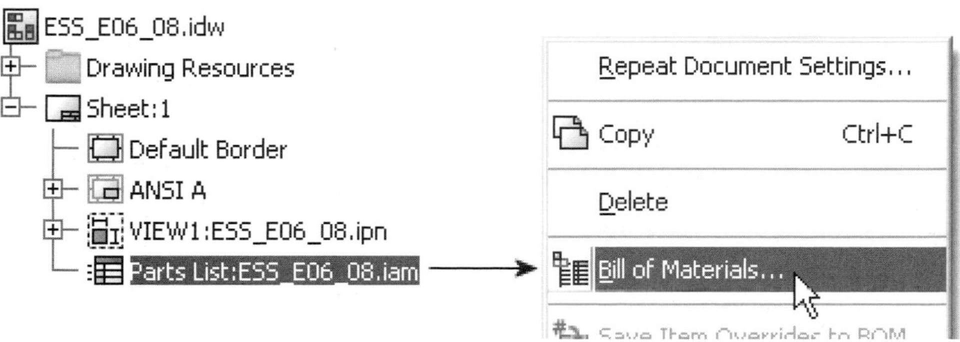

Figure 6.168

4. This action will launch the BOM Editor, as shown in the following image. Of the three tabs available, (Model Data, Structured, and Parts Only), the Parts Only tab will be made active for the remainder of this exercise. Notice the appearance of the Description column.

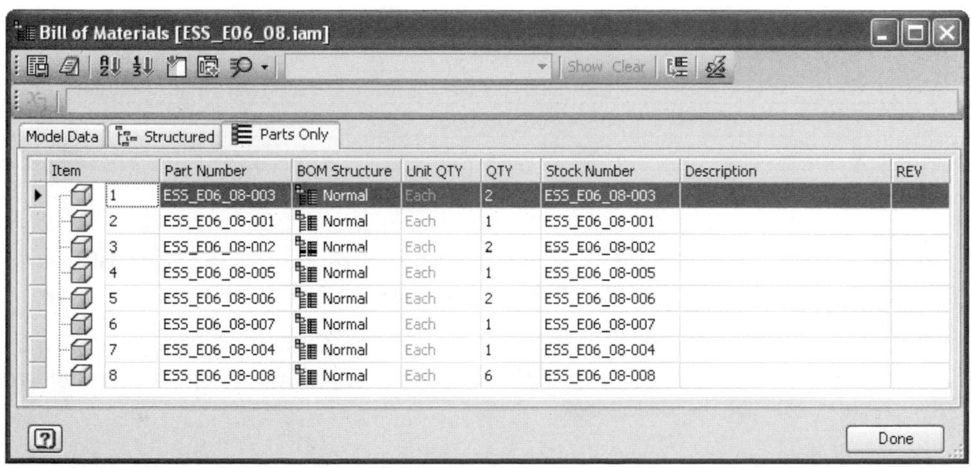

Item	Part Number	BOM Structure	Unit QTY	QTY	Stock Number	Description	REV
1	ESS_E06_08-003	Normal	Each	2	ESS_E06_08-003		
2	ESS_E06_08-001	Normal	Each	1	ESS_E06_08-001		
3	ESS_E06_08-002	Normal	Each	2	ESS_E06_08-002		
4	ESS_E06_08-005	Normal	Each	1	ESS_E06_08-005		
5	ESS_E06_08-006	Normal	Each	2	ESS_E06_08-006		
6	ESS_E06_08-007	Normal	Each	1	ESS_E06_08-007		
7	ESS_E06_08-004	Normal	Each	1	ESS_E06_08-004		
8	ESS_E06_08-008	Normal	Each	6	ESS_E06_08-008		

Figure 6.169

5. Complete the Description column by clicking in each cell and filling in all needed information, as shown in the following image. When finished, click the Done button to close the BOM Editor.

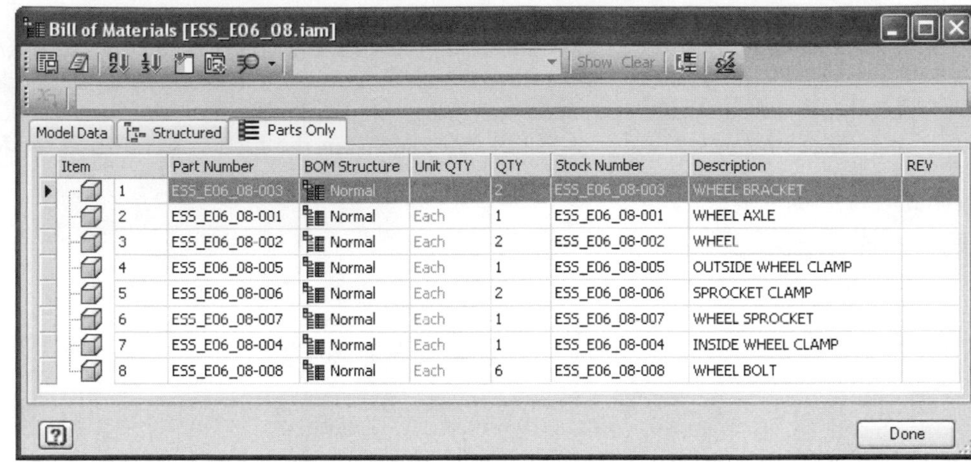

Figure 6.170

6. When you return to the drawing file, notice that the Description column of the Parts List updates to reflect the new information, as shown in the following image.

Parts List			
ITEM	QTY	PART NUMBER	DESCRIPTION
1	2	ESS_E06_08-003	WHEEL BRACKET
2	1	ESS_E06_08-001	WHEEL AXLE
3	2	ESS_E06_08-002	WHEEL
4	2	ESS_E06_08-006	SPROCKET CLAMP
5	1	ESS_E06_08-007	WHEEL SPROCKET
6	1	ESS_E06_08-004	INSIDE WHEEL CLAMP
7	6	ESS_E06_08-008	WHEEL BOLT
8	1	ESS_E06_08-005	OUTSIDE WHEEL CLAMP

Figure 6.171

7. Next add a Material column to the BOM Editor to capture the material type for each part. Activate the BOM Editor again from the Parts List in the browser.

8. To add a column called Material, click the Choose Columns button as shown in the following image. This will display the Customization box of properties in the lower-right corner of the dialog box.

9. Scroll down in the Customization box and locate Material as shown in the following image.

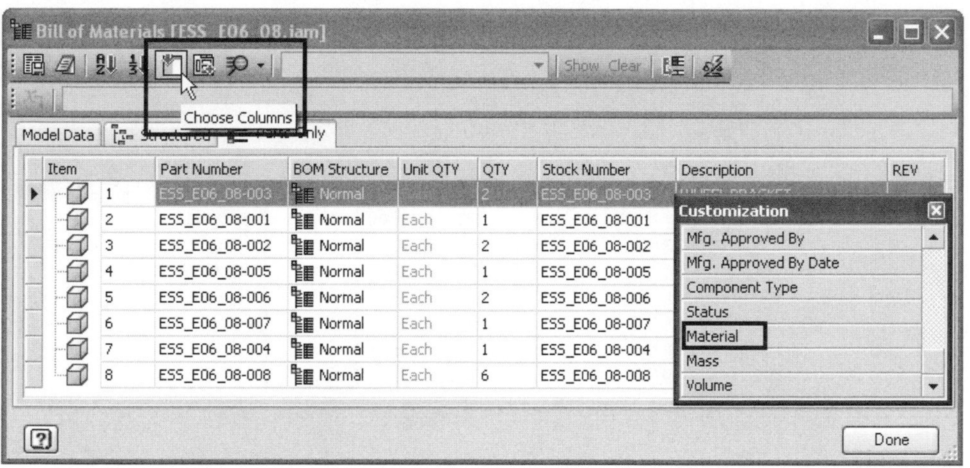

Figure 6.172

10. Click and drag the Material listing from the Customization box, and drop it into the Description Column heading as shown in the following image.

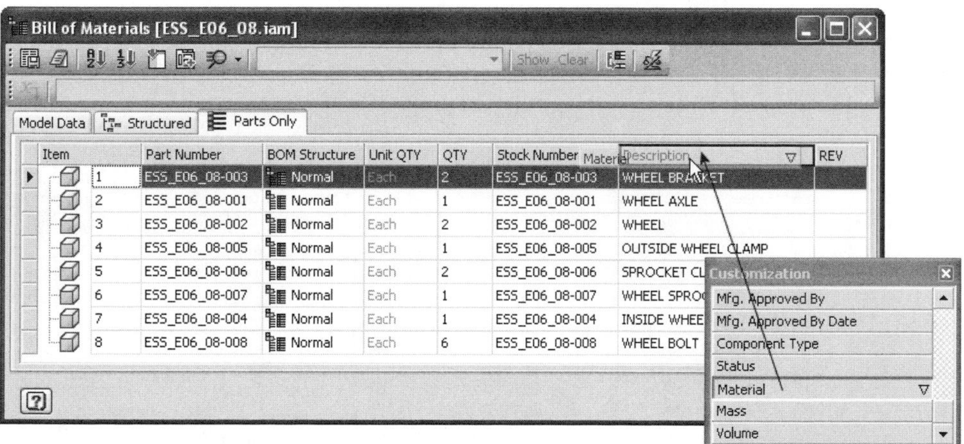

Figure 6.173

11. When finished, dismiss the Customization box by clicking on the Close button (X).

12. With the addition of the Material column, your display should appear similar to the following image. Notice that all Materials are listed as default. You will now add a material to each part.

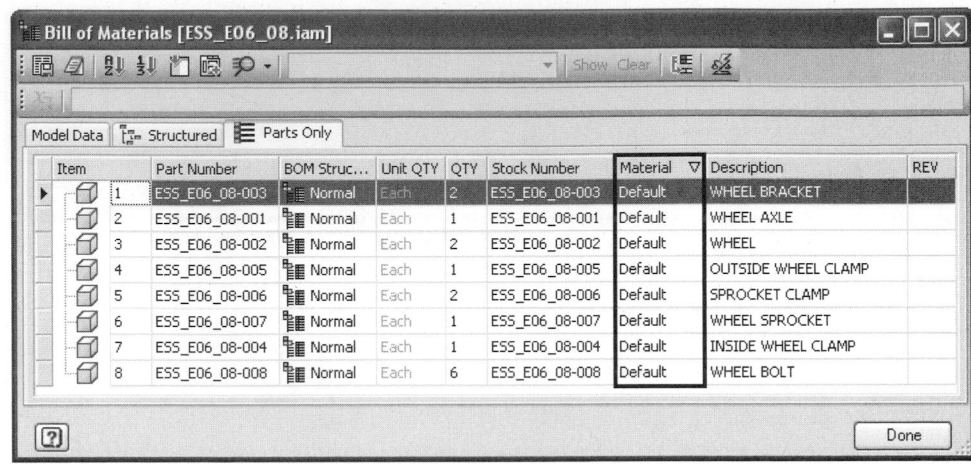

Bill of Materials [ESS_E06_08.iam]

Model Data | Structured | Parts Only

Item	Part Number	BOM Struc…	Unit QTY	QTY	Stock Number	Material	Description	REV
1	ESS_E06_08-003	Normal	Each	2	ESS_E06_08-003	Default	WHEEL BRACKET	
2	ESS_E06_08-001	Normal	Each	1	ESS_E06_08-001	Default	WHEEL AXLE	
3	ESS_E06_08-002	Normal	Each	2	ESS_E06_08-002	Default	WHEEL	
4	ESS_E06_08-005	Normal	Each	1	ESS_E06_08-005	Default	OUTSIDE WHEEL CLAMP	
5	ESS_E06_08-006	Normal	Each	2	ESS_E06_08-006	Default	SPROCKET CLAMP	
6	ESS_E06_08-007	Normal	Each	1	ESS_E06_08-007	Default	WHEEL SPROCKET	
7	ESS_E06_08-004	Normal	Each	1	ESS_E06_08-004	Default	INSIDE WHEEL CLAMP	
8	ESS_E06_08-008	Normal	Each	6	ESS_E06_08-008	Default	WHEEL BOLT	

Show Clear Done

Figure 6.174

13. Begin filling in the material type by clicking on Default for Item 1. When the drop-down list appears, select the proper material from the list supplied. Make material assignments that correspond to the part numbers, as shown in the following image.

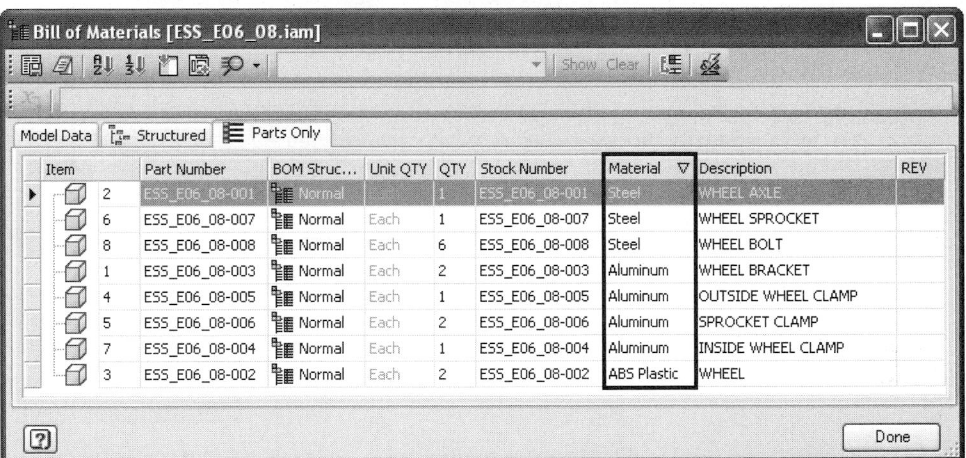

Bill of Materials [ESS_E06_08.iam]

Model Data | Structured | Parts Only

Item	Part Number	BOM Struc…	Unit QTY	QTY	Stock Number	Material	Description	REV
2	ESS_E06_08-001	Normal	Each	1	ESS_E06_08-001	Steel	WHEEL AXLE	
6	ESS_E06_08-007	Normal	Each	1	ESS_E06_08-007	Steel	WHEEL SPROCKET	
8	ESS_E06_08-008	Normal	Each	6	ESS_E06_08-008	Steel	WHEEL BOLT	
1	ESS_E06_08-003	Normal	Each	2	ESS_E06_08-003	Aluminum	WHEEL BRACKET	
4	ESS_E06_08-005	Normal	Each	1	ESS_E06_08-005	Aluminum	OUTSIDE WHEEL CLAMP	
5	ESS_E06_08-006	Normal	Each	2	ESS_E06_08-006	Aluminum	SPROCKET CLAMP	
7	ESS_E06_08-004	Normal	Each	1	ESS_E06_08-004	Aluminum	INSIDE WHEEL CLAMP	
3	ESS_E06_08-002	Normal	Each	2	ESS_E06_08-002	ABS Plastic	WHEEL	

Show Clear Done

Figure 6.175

14. When finished, click the Done button to return to the drawing file. Note: Upon leaving the BOM Editor, the iProperty data is written to the individual files.

15. When applying material types to the part numbers, the item numbers may change order. To correct this, click on the Sort button.

16. When the Sort dialog box appears, click the down arrow and pick Item from the list, as shown in the following image.

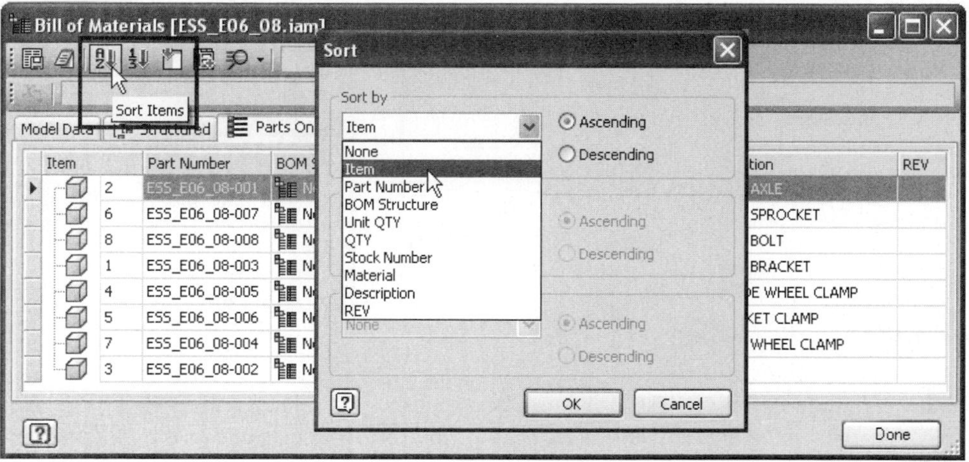

Figure 6.176

17. The results of the sort operation on the item number are shown in the following image.

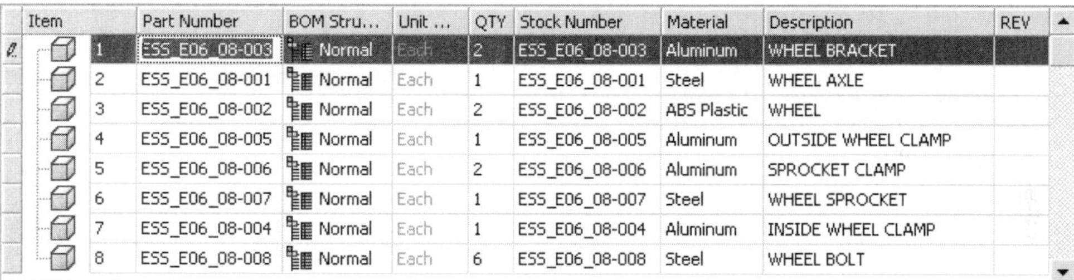

Item	Part Number	BOM Stru...	Unit ...	QTY	Stock Number	Material	Description	REV
1	ESS_E06_08-003	Normal	Each	2	ESS_E06_08-003	Aluminum	WHEEL BRACKET	
2	ESS_E06_08-001	Normal	Each	1	ESS_E06_08-001	Steel	WHEEL AXLE	
3	ESS_E06_08-002	Normal	Each	2	ESS_E06_08-002	ABS Plastic	WHEEL	
4	ESS_E06_08-005	Normal	Each	1	ESS_E06_08-005	Aluminum	OUTSIDE WHEEL CLAMP	
5	ESS_E06_08-006	Normal	Each	2	ESS_E06_08-006	Aluminum	SPROCKET CLAMP	
6	ESS_E06_08-007	Normal	Each	1	ESS_E06_08-007	Steel	WHEEL SPROCKET	
7	ESS_E06_08-004	Normal	Each	1	ESS_E06_08-004	Aluminum	INSIDE WHEEL CLAMP	
8	ESS_E06_08-008	Normal	Each	6	ESS_E06_08-008	Steel	WHEEL BOLT	

Figure 6.177

18. When you return to the drawing, the Material column will not automatically display in the Parts List. To display this column of information, right-click on the Parts List in the browser or in the graphics window, and click Edit Parts List from the menu, as shown in the following image on the left.

19. When the Parts List displays, click the Column Chooser button, as shown in the following image on the right.

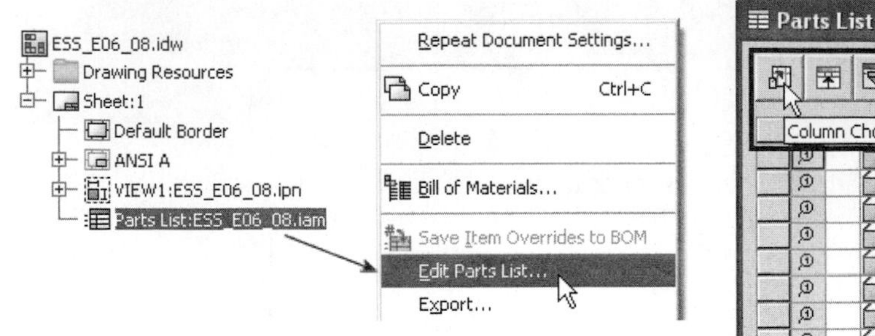

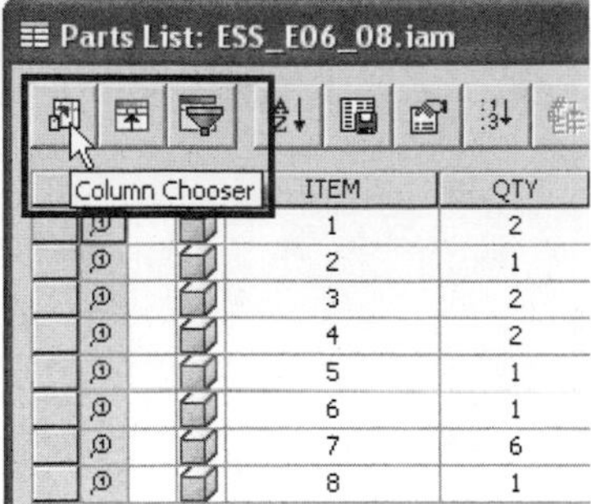

Figure 6.178

20. While in the Parts List Column Chooser dialog box, locate the MATERIAL property in the list on the left, and click the Add button to add it under the Selected Properties list, as shown in the following image on the left.

21. Select the MATERIAL property in the Selected Properties window, and click the Move Up button to place it above the Description property, as shown in the following image on the right. Click the OK button to return to the drawing.

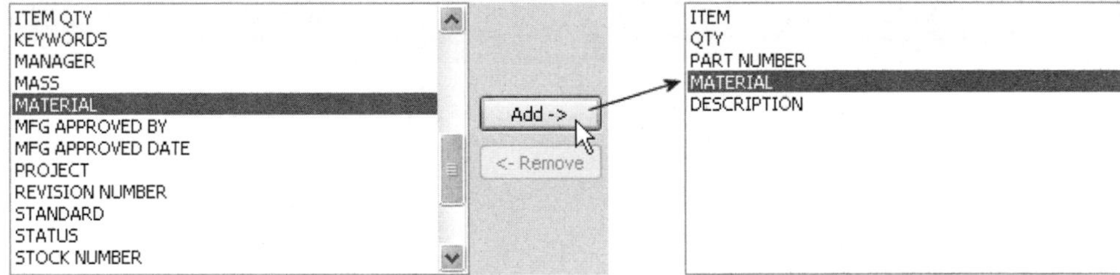

Figure 6.179

22. Notice that the Parts List now displays the MATERIAL column and that all information is filled in, as shown in the following image.

Parts List				
ITEM	QTY	PART NUMBER	MATERIAL	DESCRIPTION
1	2	ESS_E06_08-003	Aluminum	WHEEL BRACKET
2	1	ESS_E06_08-001	Steel	WHEEL AXLE
3	2	ESS_E06_08-002	ABS Plastic	WHEEL
4	2	ESS_E06_08-006	Aluminum	SPROCKET CLAMP
5	1	ESS_E06_08-007	Steel	WHEEL SPROCKET
6	1	ESS_E06_08-004	Aluminum	INSIDE WHEEL CLAMP
7	6	ESS_E06_08-008	Steel	WHEEL BOLT
8	1	ESS_E06_08-005	Aluminum	OUTSIDE WHEEL CLAMP

Figure 6.180

23. Next adjust the width of all columns by the amount of information contained under the heading. To perform this task, right-click the Parts List located in the browser and click Edit Parts List.

24. When the Parts List dialog box displays, right-click the column and pick Column Width from the menu, as shown in the following image on the left. When the Column Width dialog box appears, enter a value. Change the column headings to the following widths:

- ITEM = **25**, as shown in the following image on the right
- QTY= **20**
- PART NUMBER= **40**
- MATERIAL= **30**
- DESCRIPTION= **55**

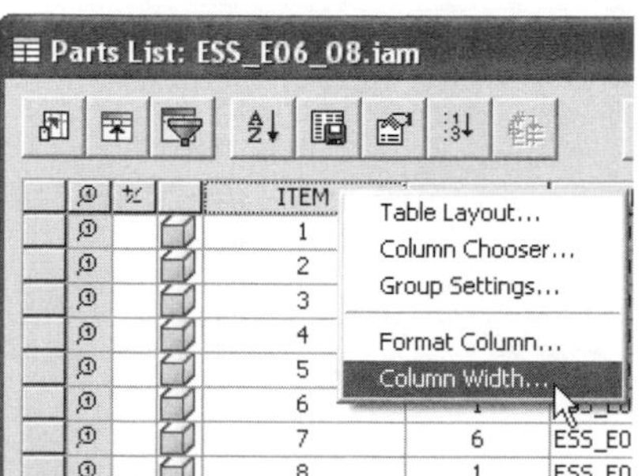

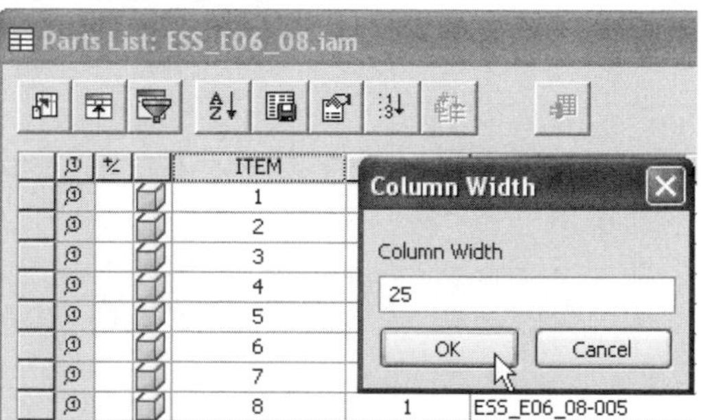

Figure 6.181

25. The Parts List now reflects the new column widths, as shown in the following image.

Parts List				
ITEM	QTY	PART NUMBER	MATERIAL	DESCRIPTION
1	2	ESS_E06_08-003	Aluminum	WHEEL BRACKET
2	1	ESS_E06_08-001	Steel	WHEEL AXLE
3	2	ESS_E06_08-002	ABS Plastic	WHEEL
4	2	ESS_E06_08-006	Aluminum	SPROCKET CLAMP
5	1	ESS_E06_08-007	Steel	WHEEL SPROCKET
6	1	ESS_E06_08-004	Aluminum	INSIDE WHEEL CLAMP
7	6	ESS_E06_08-008	Steel	WHEEL BOLT
8	1	ESS_E06_08-005	Aluminum	OUTSIDE WHEEL CLAMP

Figure 6.182

26. Close the file. Do not save changes. End of exercise.

CREATING BALLOONS

After you have created a drawing view of an assembly, you can add balloons to the parts and/or subassemblies. You can add balloons individually to components in a drawing, or you can use the Auto Balloon tool to automatically add balloons to all components.

To add individual balloons to a drawing, follow these steps:

1. Activate the Drawing Annotation panel bar by clicking Drawing Views on the panel bar and selecting Drawing Annotation from the menu.
2. On the Drawing Annotation panel bar, click the arrow next to the Balloon tool, as shown in the following image. Two icons will appear: the first is used for ballooning components one at a time, and the second is used for automatic ballooning components in a view in a single operation.

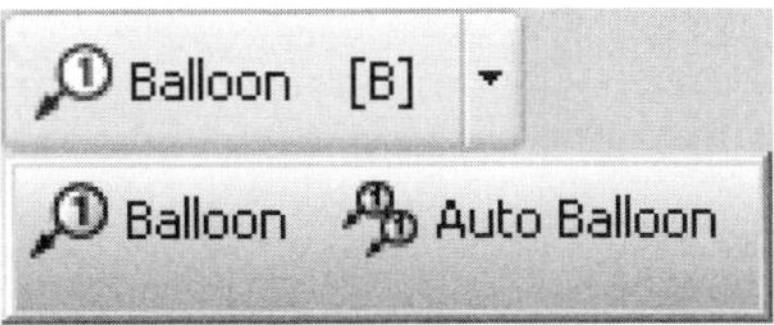

Figure 6.183

3. To balloon a single component, click the Balloon tool or use the hot key B. A preview image of the balloon will appear attached to the cursor.
4. Select a component to balloon.
5. The BOM Properties dialog box will appear, as shown in the following image.

Note: This dialog box will not appear if a Parts List or balloons already exist in the drawing.

6. Select which Source and BOM Settings to use, as shown in the following image, and select OK.

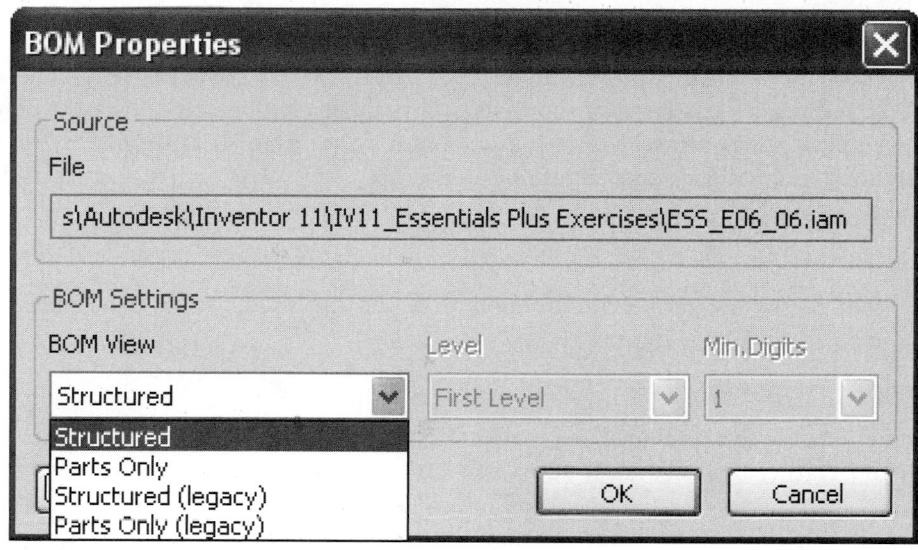

Figure 6.184

7. Position the cursor to place the second point for the leader, and then press the left mouse button.
8. Continue to select points to add segments to the balloon's leader, if desired.
9. When finished adding segments to the leader, right-click, and select Continue from the menu to create the balloon.
10. Select Done from the menu to cancel the operation without creating an additional balloon, or select the Back option to undo the last step that was created for the balloon.

BOM PROPERTIES DIALOG BOX OPTIONS

Source

Specify the source file on which the BOM will be based.

BOM Settings and BOM View

Structured A Structured list refers to the top-level components of an assembly in the selected view. Subassemblies and parts that belong to the subassembly will be ballooned, but parts that are in a subassembly will not be.

Parts Only View Click to balloon only the parts of the assembly in the selected view. Subassemblies will not be ballooned, but the parts in the subassemblies will be.

Note: In a Parts-Only list, components that are assemblies are not presented in the list unless they are considered inseparable or purchased.

Level

First Level or All Levels determines the level of detail that the Parts List and balloons will display. For example, if an assembly consists of a number of subassemblies and you choose First Level, each subassembly will be considered a single item. If you choose All Levels, the individual parts that make up the subassembly will also be treated as individual components when ballooned.

Min.Digits

Use this tool to set the minimum amount of digits displayed for item numbering. The range is fixed from 1 to 6 digits.

AUTO BALLOONING

In complex assembly drawings, it will become necessary to balloon a number of components. Rather than manually add a balloon to each component, you can use the Auto Balloon tool to perform this operation on several components in a single operation automatically. Clicking on the Auto Balloon button on the Drawing Annotation panel bar will display the Auto Balloon dialog box, as shown in the following image. The areas in this dialog box allow you to control how you place balloons.

Selection

This area requires you to select where to apply the balloons. With the view selected, you then add or remove components to the balloon. The Ignore Multiple Instances option appears in this area, and when the box is checked, multiple instances of the same component will not be ballooned. This action greatly reduces the number of balloon callouts in the drawing. If your application requires multiple instances to each have a balloon, remove the check from this box.

Placement

This area allows you to display the balloons along a horizontal axis, along a vertical axis, or around the view being ballooned.

BOM Settings

This area allows you to control if balloons are applied to structured, or first-level, components or to parts only.

Style Overrides

The Style overrides area allows you to change the shape of the balloon and assign a user-defined balloon shape. Click Balloon Shape to enable balloon shape style overrides.

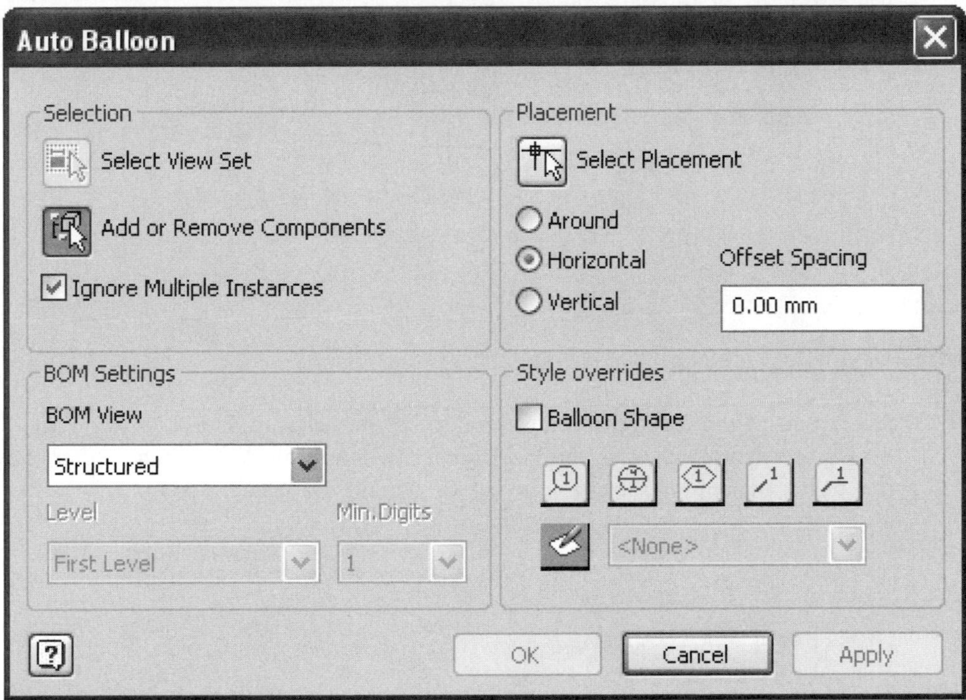

Figure 6.185

To create an auto balloon, follow these steps:

11. Select the view to which to add the balloons, as shown in the following image.

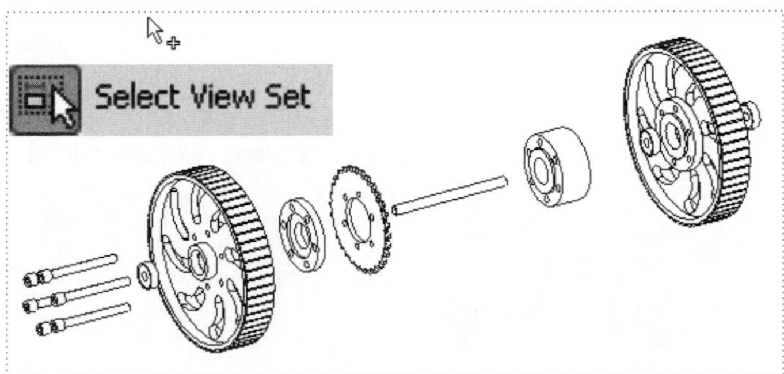

Figure 6.186

12. Select the components to which to add balloons. In the following image, all parts in the isometric drawing were selected using a window box. You will notice the color

of all selected objects change. Balloons will be applied to these selected objects. To remove components, hold down the SHIFT or CTRL key, and click a highlighted component. This action will deselect the component. You can also select components using window or crossing window selections.

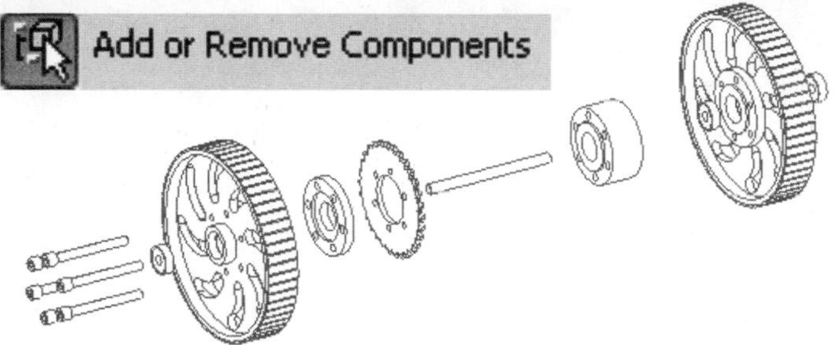

Figure 6.187

13. Select one of the three placement modes for the balloons. The following image shows an example of using the Horizontal Placement mode. You can further locate this balloon group by moving your cursor around the display screen. This balloon arrangement will retain its horizontal order.

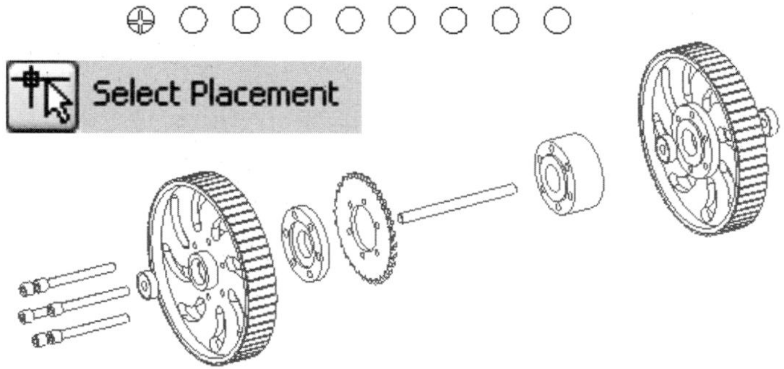

Figure 6.188

The following image shows an example of using the Vertical Placement mode. You can further locate this balloon group by moving your cursor around the display screen. This balloon arrangement will retain its vertical order.

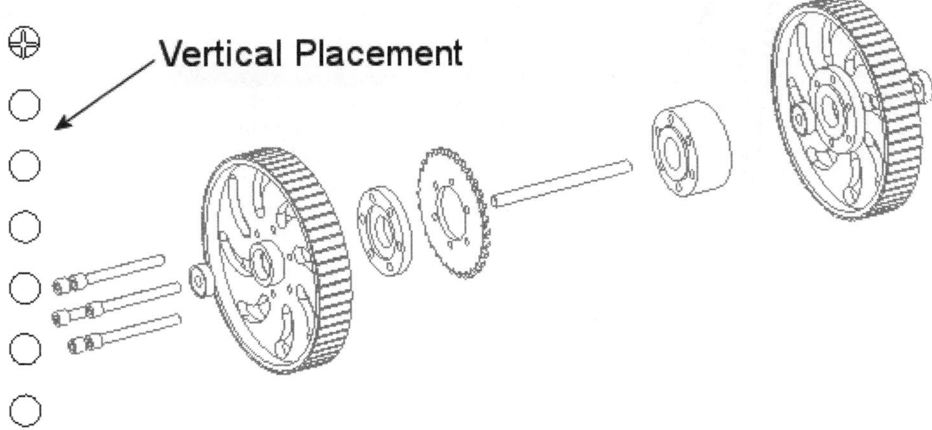

Figure 6.189

The following image shows an example of using the Around Placement mode. Moving your cursor around the display screen will readjust the balloons to different locations. In this example, the balloons are located in various positions outside of the view box. After you have placed these types of balloons, you may want to drag the balloons and arrow terminators to fine-tune individual balloon locations.

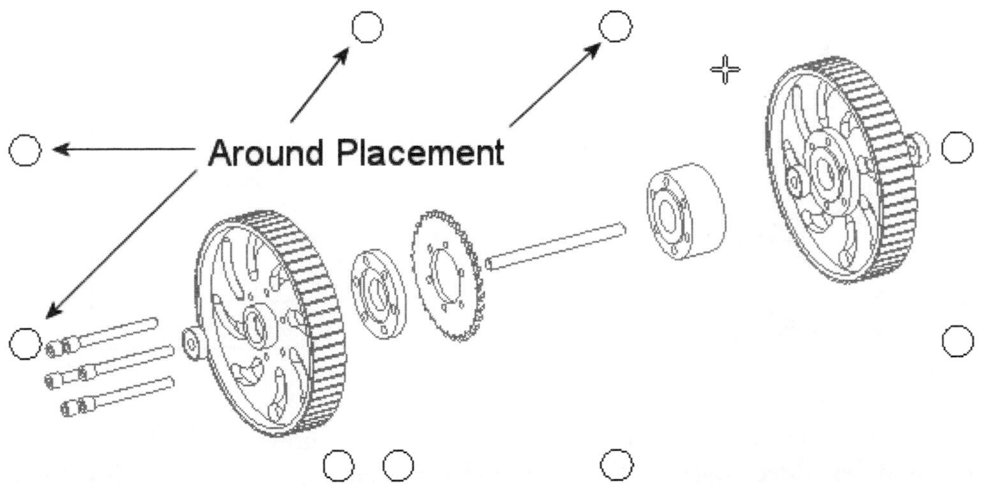

Figure 6.190

Editing Balloons

Edit a balloon by right-clicking on it. This action will activate a menu, as shown in the following image.

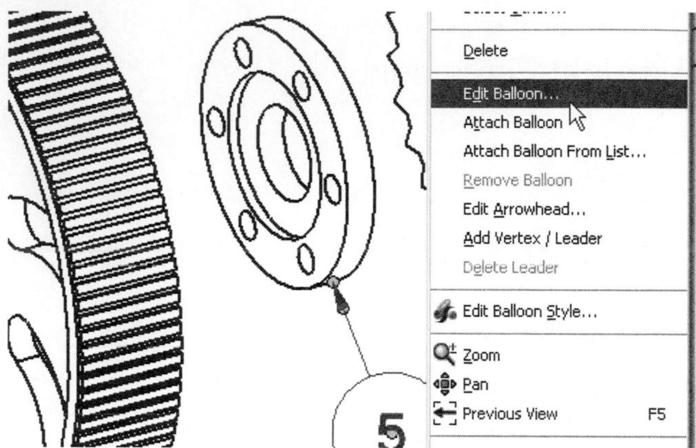

Figure 6.191

Selecting Edit Balloon from the menu will activate the Edit Balloon dialog box, as shown in the following image.

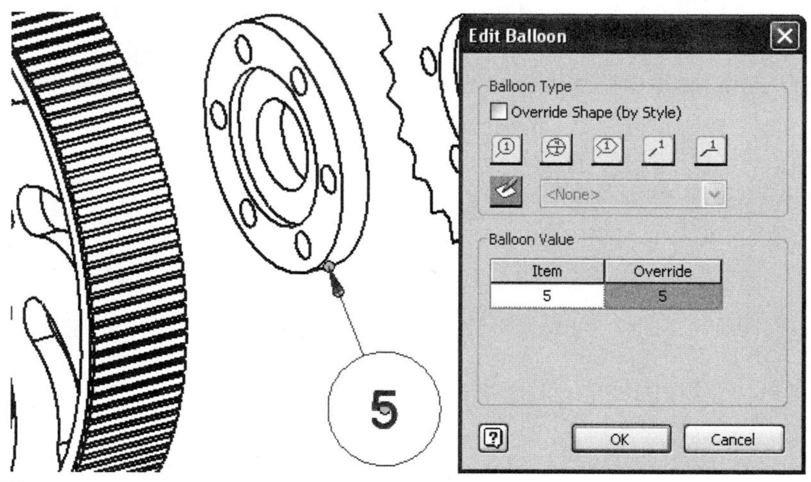

Figure 6.192

Use this dialog box to make changes to a balloon on an individual basis. The following features may be edited: Balloon Type and Balloon Value.

Editing the Balloon Value

The columns in the Balloon Value area of the Edit Balloon dialog box in the previous illustration specify Item and Override. When you edit the Item value, changes will be made to the balloon and will be reflected in the Parts List. In the following image, an Item was changed to 2A. Notice in the following image, on the left, that both the balloon and the Parts List have updated to the new value.

When you make a change in the Override column, the balloon will update to reflect this change, but the Parts List remains unchanged. The following image, on the right, shows this. The balloon was overridden with a value of 2A, but the Parts List has the item Rod Cap listed as 2.

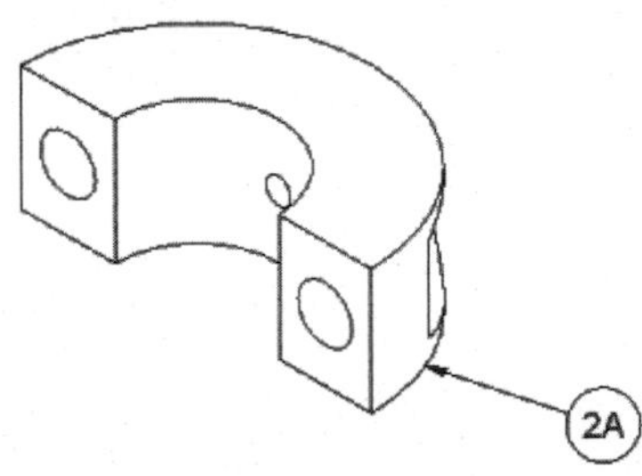

		Parts	
ITEM	QTY	PART NUMBER	
1	1	Connecting Rod	
2A	1	Rod Cap	
3	2	Rod Cap Screw	
4	1	Piston	
5	1	Wrist Pin	

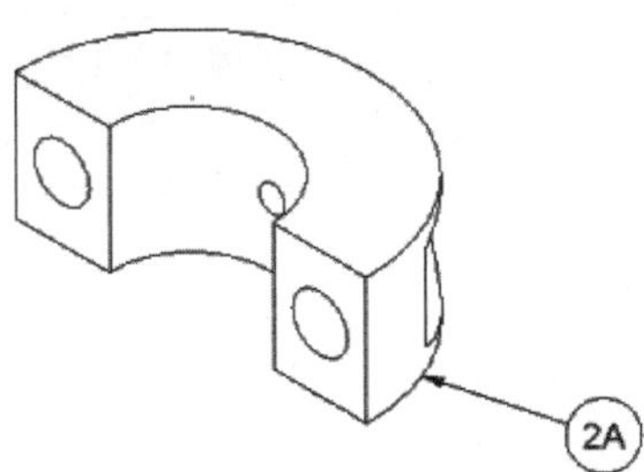

		Parts	
ITEM	QTY	PART NUMBER	
1	1	Connecting Rod	
2	1	Rod Cap	
3	2	Rod Cap Screw	
4	1	Piston	
5	1	Wrist Pin	

Figure 6.193

Note: Any changes performed with the Balloon Edit dialog box will be reflected in all parts lists associated with that view.

Attaching a Custom Balloon

If you have predefined a custom part, you can select it from a list of custom parts for custom balloon content. To attach a custom balloon, right-click on the balloon, and select Attach Balloon From List, as shown in the following image on the left. If no custom part is available, this menu item will be grayed out.

The Attach Balloon dialog box, as shown in the following image on the right, will prompt you for your selection of custom balloons. You can select only one custom item at a time.

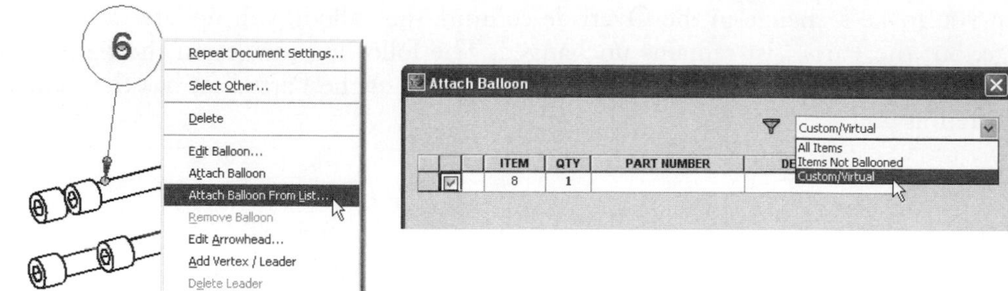

Figure 6.194

After selecting the custom part, click the OK button to return to the drawing. Drag the cursor around for placement of the attached balloon. Click to place the balloon in the desired location, as shown in the following image.

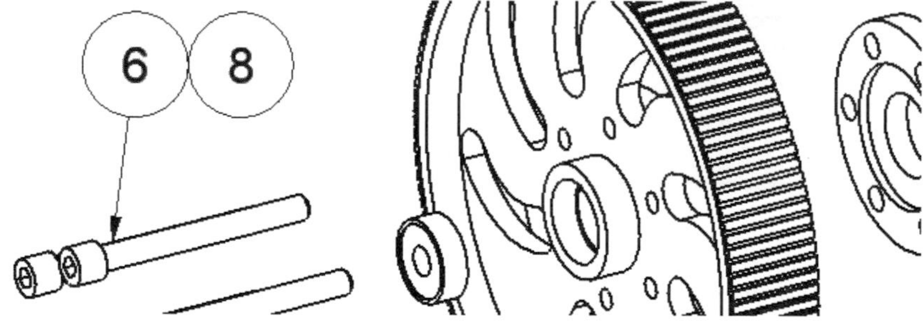

Figure 6.195

PARTS LISTS

After you have created the drawing views and placed balloons, you can also create a Parts List, as shown in the following image. As noted earlier in this section, components do not need to be ballooned before creating a Parts List.

Parts List			
ITEM	QTY	PART NUMBER	DESCRIPTION
1	1	Arbor Press	ARBOR PRESS
2	1	FACE PLATE	FACE PLATE
3	1	PINION SHAFT	PINION SHAFT
4	1	LEVER ARM	LEVER ARM
5	1	THUMB SCREW	THUMB SCREW
6	1	TABLE PLATE	TABLE PLATE
7	1	RAM	RAM
8	2	HANDLE CAP	HANDLE CAP
9	1	COLLAR	COLLAR
10	1	GIB PLATE	GIB PLATE
11	1	GROOVE PIN	GROOVE PIN
14	4	ANSI B18.3 - 1/4 - 20 UNC - 7/8	Hex Cap Screw
15	4	BS 4168 - M5 x 16	Hex Cap Screw
16	1	ISO 4766 - M5 x 5	Hex Cap Screw

Figure 6.196

To create a Parts List, select the Drawing Annotation panel bar, and click the Parts List tool, as shown in the following image.

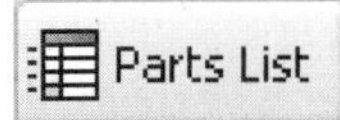

Figure 6.197

The Parts List dialog box will appear, as shown in the following image. Select a view on which to base the Parts List. After specifying your options, click the OK button, and a Parts List frame preview will appear attached to your cursor. If the components were already ballooned, the list preview will appear without displaying the Parts List dialog box. Click a point in the graphics window to place the Parts List. The information in the list is extracted from the properties of each component.

You can split or wrap a long Parts List into multiple columns through the Parts list dialog box. Notice the Table Wrapping area, as shown in the following image on the right. Place a check in the Enable Automatic Wrap box to activate this feature, and then enter the number of sections to break the Parts List into. In this example, 2 is used.

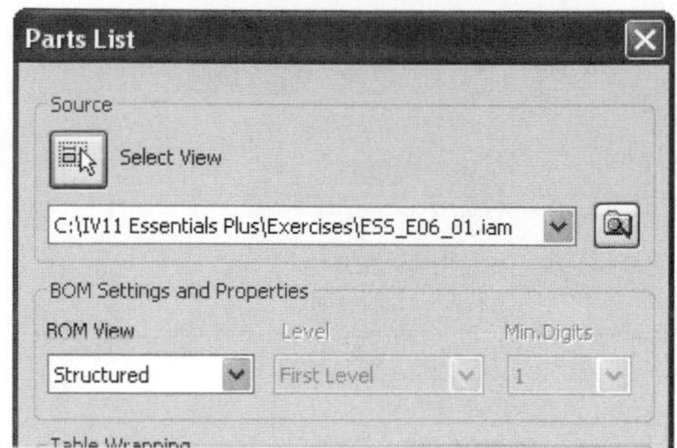

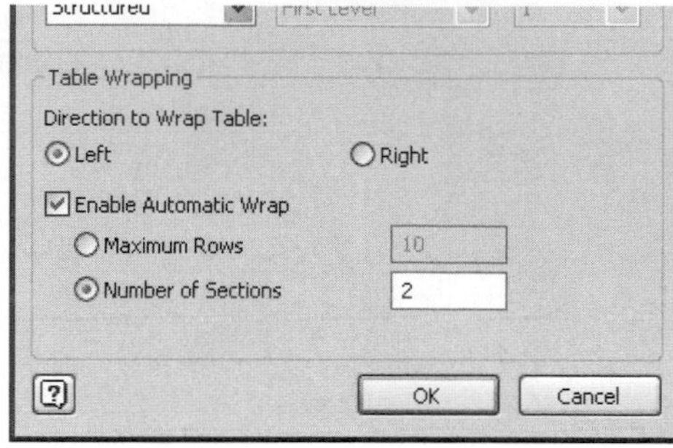

Figure 6.198

The following image shows the results of performing the wrapping operation on a Parts List. Notice the two distinct sections that have been created.

Parts List				Parts List			
ITEM	QTY	PART NUMBER	DESCRIPTION	ITEM	QTY	PART NUMBER	DESCRIPTION
8	2	HANDLE CAP	HANDLE CAP	1	1	Arbor Press	ARBOR PRESS
9	1	COLLAR	COLLAR	2	1	FACE PLATE	FACE PLATE
10	1	GIB PLATE	GIB PLATE	3	1	PINION SHAFT	PINION SHAFT
11	1	GROOVE PIN	GROOVE PIN	4	1	LEVER ARM	LEVER ARM
14	4	ANSI B18.3 - 1/4 - 20 UNC - 7/8	Hex Cap Screw	5	1	THUMB SCREW	THUMB SCREW
15	4	BS 4168 - M5 x 16	Hex Cap Screws	6	1	TABLE PLATE	TABLE PLATE
16	1	ISO 4766 - M5 x 5	Hex Cap Screws	7	1	RAM	RAM

Figure 6.199

EDITING PARTS LIST BOM DATA

Once a Parts List is created, you can open the BOM Editor directly from the drawing environment and edit the assembly BOM. All changes are saved in the assembly and the corresponding component files. Right-click on the Parts List in the Drawing browser, and then select Bill of Materials from the menu, as shown in the following image.

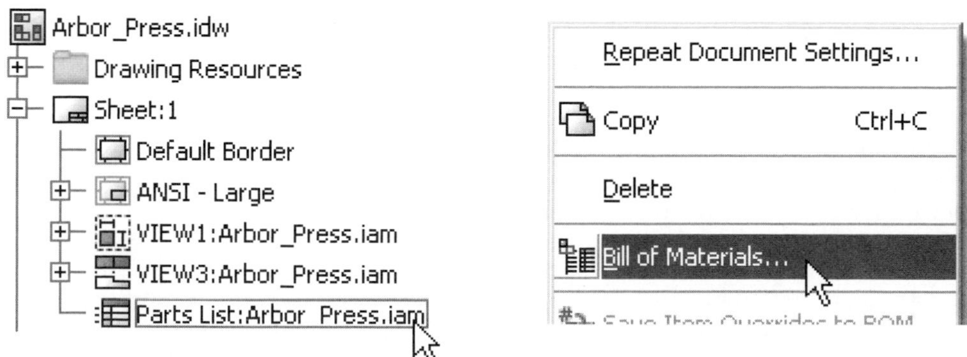

Figure 6.200

This action will launch the BOM Editor based on a specific assembly file. The dialog box provides a convenient location to edit the iProperties and Bill of Material properties for all components in the assembly. You can even perform edits on multiple components at once. Notice in the following image that the BOM Editor displays the BOM Structure. In the BOM Editor, perform the desired edits. When finished, click Done to close the dialog box. All changes will be updated in the Parts List and balloons.

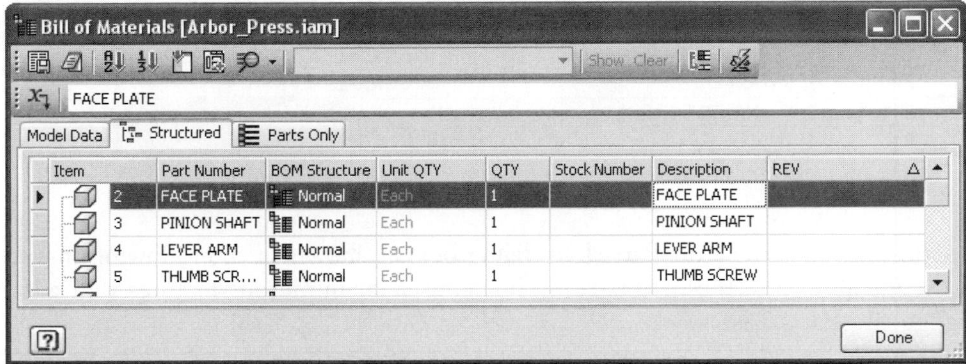

Figure 6.201

PARTS LIST TOOLS

The information supplied earlier in this chapter covered the basics of inserting and manipulating a Parts List. This section will expand on the capabilities of the Parts List and will cover the following topics:

- A detailed account of the Parts List Operations tools available at the top of the Edit Parts List dialog box
- The function of the spreadsheet view
- Additional sections of the Edit Parts List dialog box
- Expanded capabilities of the Edit Parts List dialog box
- Nested parts lists

PARTS LIST OPERATIONS (TOP ICONS)

Once you have placed a Parts List in the drawing, controls are available to help you modify the Parts List. Right-click on the Parts List in a drawing, and select Edit Parts List from the menu, as shown in the following image, or double-click on the Parts List in the drawing.

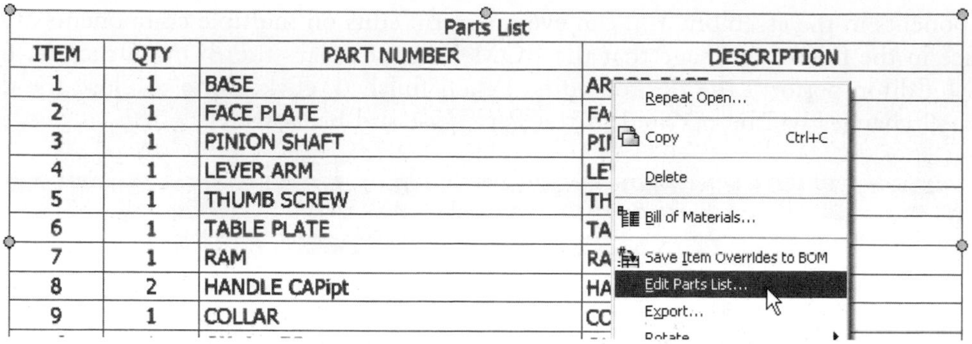

Parts List			
ITEM	QTY	PART NUMBER	DESCRIPTION
1	1	BASE	AR
2	1	FACE PLATE	FA
3	1	PINION SHAFT	PI
4	1	LEVER ARM	LE
5	1	THUMB SCREW	TH
6	1	TABLE PLATE	TA
7	1	RAM	RA
8	2	HANDLE CAPipt	HA
9	1	COLLAR	CC

Repeat Open...
Copy Ctrl+C
Delete
Bill of Materials...
Save Item Overrides to BOM
Edit Parts List...
Export...

Figure 6.202

This action displays the Edit Parts List dialog box, as shown in the following image. The row of icons along the top of the dialog box consist of numerous features that allow you to change the Parts List.

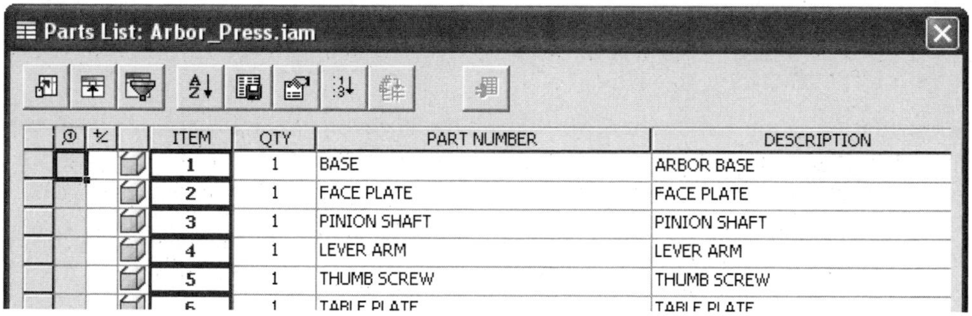

Parts List: Arbor_Press.iam

ITEM	QTY	PART NUMBER	DESCRIPTION
1	1	BASE	ARBOR BASE
2	1	FACE PLATE	FACE PLATE
3	1	PINION SHAFT	PINION SHAFT
4	1	LEVER ARM	LEVER ARM
5	1	THUMB SCREW	THUMB SCREW

Figure 6.203

The Parts List table gives detailed explanations of all Parts List icons:

Parts List Button	Function	Explanation
	Column Chooser	Opens the Parts List Column Chooser dialog box, where you can add, remove, or change the order of the columns for the selected Parts List. Data for these columns is populated from the properties of the component files.
	Group Settings	Opens the Group Settings dialog box, where you select Parts List columns to be used as a grouping key and group different components into one Parts List row. This button is active only when generating a Parts List using the Structured mode.
	Filter Settings	Opens the Filter Settings dialog box, where you can define such filter settings as Assembly View Representations, Ballooned Items Only, Item Number Range, Purchased Items, and Standard Content from which to filter the information. Once a filter is selected, you then filter out rows without changing the data in the parts list. You can also add parts list filters to a parts list style.
	Sort	Opens the Sort Parts List dialog box, where you can change the sort order for items in the selected Parts List.
	Export	Opens the Export Parts List dialog box so that you can save the selected Parts List to an external file of the file type you choose.
	Table Layout	Opens the Parts List Table Layout dialog box where you can change the title text, spacing, or heading location for the selected Parts List.
	Renumber	Renumbers item numbers of parts in the Parts List consecutively.
	Save Item Overrides to BOM	Saves item overrides back the assembly Bill of Materials.
	Member Selection	Used with iAssemblies. Clicking this button opens a dialog box where you can select which members of the iAssembly to include in the Parts List.

CREATING CUSTOM PARTS

Custom parts are useful for displaying parts in the Parts List that are not components or graphical data, such as paint or a finishing process. To add a custom part, right-click on an existing part in the Edit Parts List dialog box, as shown in the following image, and click Insert Custom Part. When a custom part no longer needs to be documented in the Parts List, you can right-click in the row of the custom part, and click Remove Custom Part from the menu, as shown in the following image.

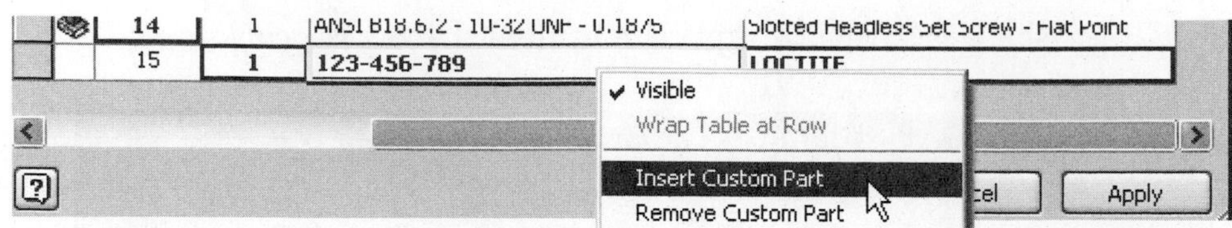

Figure 6.204

EXERCISE 6-9: CREATING ASSEMBLY DRAWINGS

In this exercise, you create drawing views of an assembly and then add balloons and a Parts List.

1. This exercise begins with formatting a drawing sheet. Open the drawing file *ESS_E06_09.idw* from the chapter 06 folder. This drawing consists of one blank A1 sheet with a border and title block.
2. In the browser, right-click Sheet1, and select Edit Sheet.
3. In the Edit Sheet dialog box, rename the sheet by typing **Assembly** in the Name field.
4. In the Size list, select A2. When finished, click the OK button.
5. The next phase of this exercise involves updating the file properties in order to update the title block fields. On the File menu, click iProperties.
6. Click on the Summary tab, and make the following changes:

- In Title, enter **Pump Assembly**.
- In Author, enter your name or initials.

7. Next click on the Project tab, and make the following changes:

- In Part number, enter **123-456-789**.
- In Creation Date, click the down arrow, and then select today's date.

8. When finished, click the OK button. Zoom in to the title block, and notice that the entries have updated, as shown in the following image.

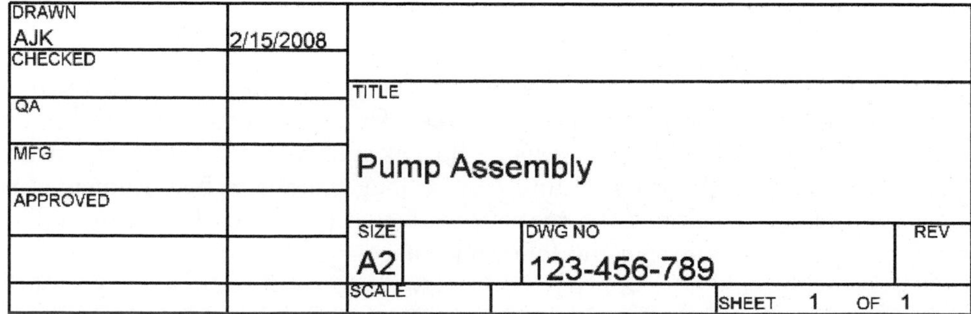

DRAWN
AJK
2/15/2008
CHECKED
QA
MFG
APPROVED
TITLE
Pump Assembly
SIZE
A2
DWG NO
123-456-789
REV
SCALE
SHEET 1 OF 1

Figure 6.205

9. Now create drawing views. Zoom out to view the entire sheet.
10. Click the Base View tool in the panel bar to display the Drawing View dialog box.
11. Click the Explore Directories button next to the File list, select Assembly Files from the Files of Type drop-down list, and double-click *ESS_E06_09.iam* from the Chapter 06 folder as the view source.
12. From the Orientation list, select Top.
13. Set the Style to Hidden Line. Do not select Hidden Line Removed.
14. Set the Scale to 1.
15. Move the view to the upper-left corner of the sheet, and then click to place the view. The drawing view should appear similar to the following image.

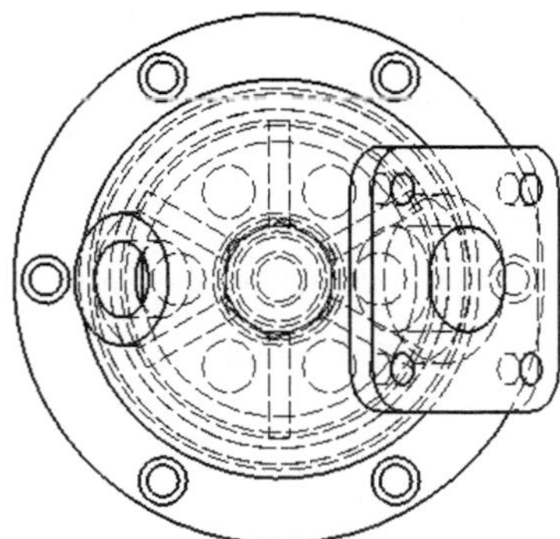

Figure 6.206

16. Now begin to generate a section view from the existing top view.
17. Before you create the section view, you will need to turn off the Section property for some components so that they are not sectioned by following the next series of directions:

- In the browser, expand the Assembly sheet.
- In the browser, expand View1:*ESS_E06_09.iam*. Your view number may be different.
- Expand *ESS_E06_09.iam*.
- Select *ESS_E06-Pin.ipt*, *ESS_E06-Spring.ipt*, *ESS_E06-NutA.ipt*, and *ESS_E06-NutB.ipt*.

Tip: To select multiple items, hold down the CTRL key while clicking the items from the browser.

Right-click any of the highlighted parts in the browser, select Section Participation from the menu, and check the box next to None, as shown in the following image.

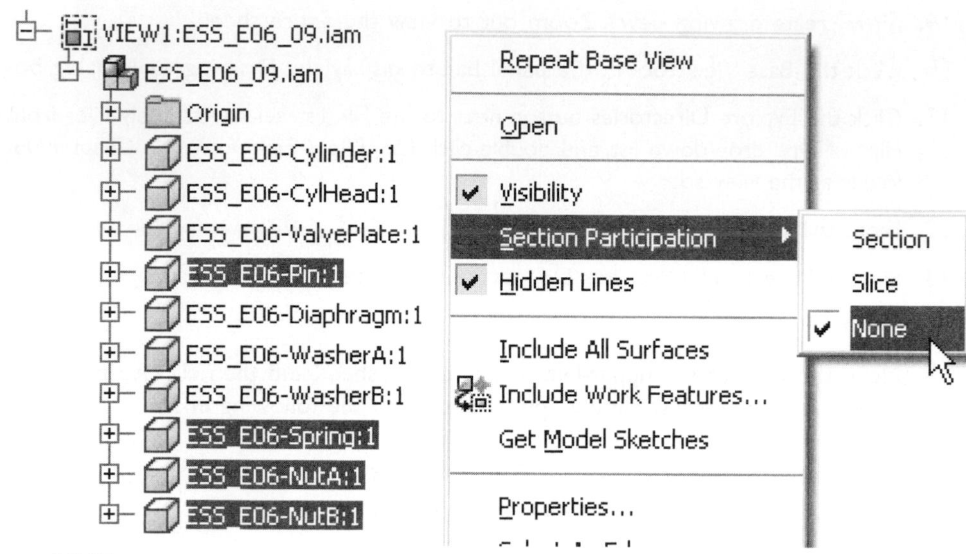

Figure 6.207

18. Now create the section view by clicking the Section View tool in the panel bar.
19. Select the top view of the assembly, and draw a horizontal section line through the center of the top view, as shown in the following image.

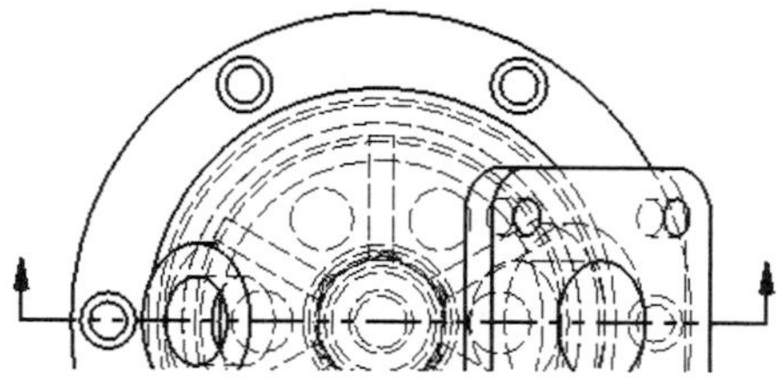

Figure 6.208

20. Right-click, and select Continue.
21. Move the preview below the top view, and click to place the view.
22. Notice that the nuts, spring, and pin are visible in the view but not sectioned, as shown in the following image on the left. Notice also the presence of the section label and drawing view scale.
23. Now edit the section view, and turn off the label and scale. On the sheet, select the sectioned assembly view, right-click, and select Edit View from the menu to display the Drawing View dialog box.

24. Click both light bulb buttons (Toggle Scale Visibility and Toggle Label Visibility) to turn off the scale and label in the section view. Click the OK button. The label and scale no longer display at the bottom of the view, as shown in the following image on the right.

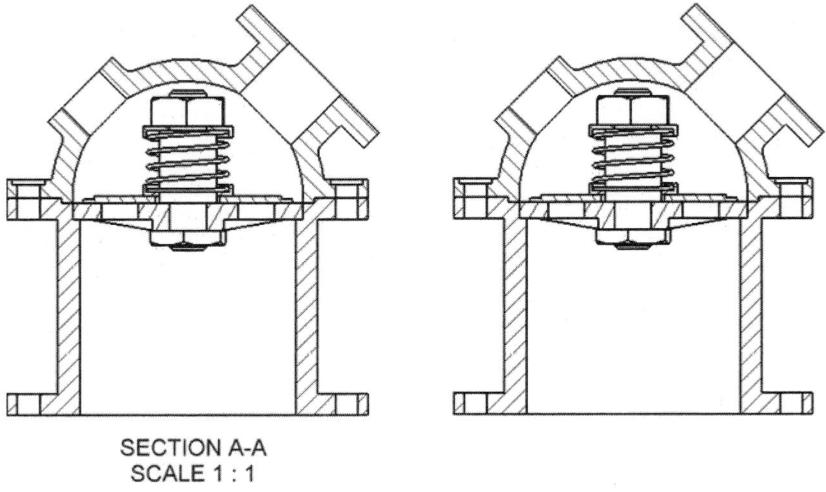

Figure 6.209

25. Now hide certain components from displaying in the top view. In the browser, under the expanded component listing for *ESS_E06_09.iam*, select both *ESS_E06-CylHead:1* and *ESS_E06-Spring:1*, as shown in the following image on the left.

26. Right-click, and clear the checkmark from Visibility. The cylinder head and the spring do not display in the top view, as shown in the following image on the right.

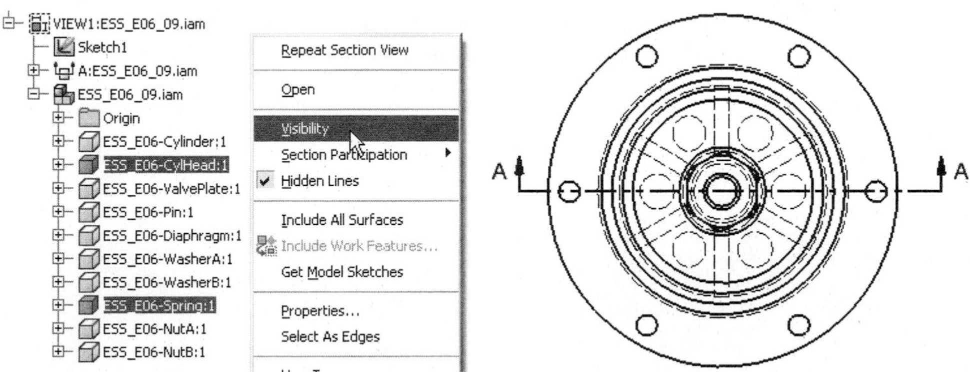

Figure 6.210

27. Next add a text note to the top view stating that the cylinder head and spring are not visible. Click the Text tool in the Drawing Annotation panel bar.

Note: To change panel bars, click on the title in the panel bar, and then select Drawing Annotation Panel from the context menu.

28. Select a point to the left of the top assembly view.

29. In the text entry area, type **CYLINDER HEAD AND**, press ENTER, and type **SPRING REMOVED**. Click OK. The text will be added to the top view, as shown in the following image. If necessary, select and drag the text to a better location.

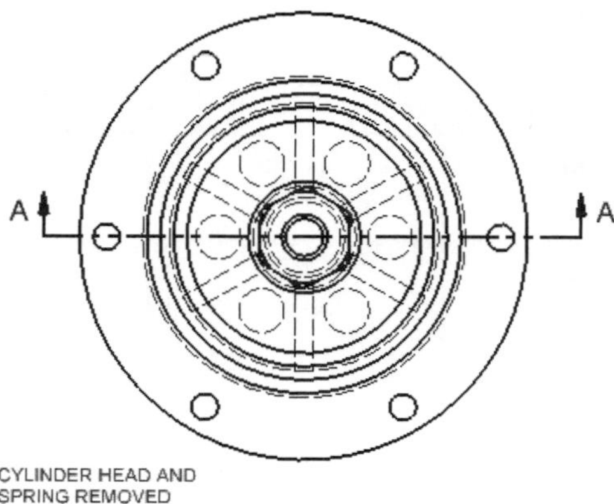

Figure 6.211

30. Now create an isometric view that will be based on the section view. Click the Projected View tool in the Drawing Views panel bar.

31. Select the sectioned assembly view as the base view.

32. Move the preview up and to the right, and then click the sheet.

33. When the view is positioned, right-click, and select Create from the menu. The isometric pictorial view is created, as shown in the following image.

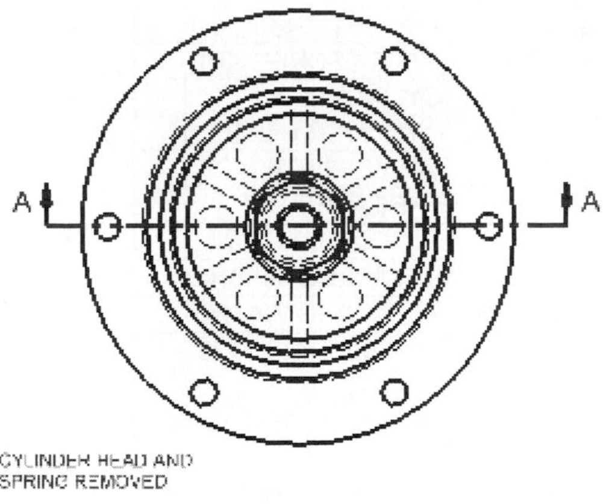

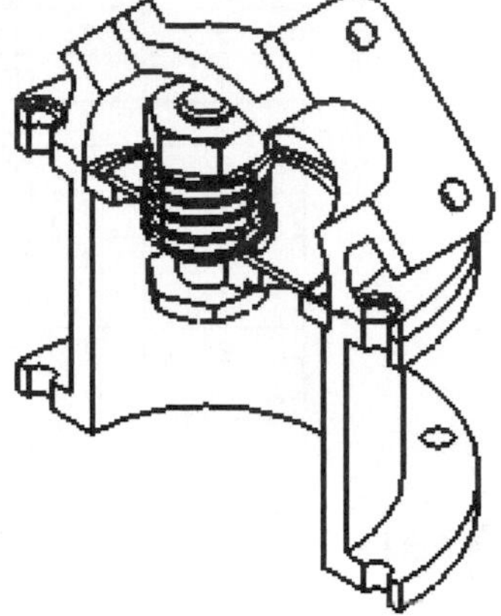

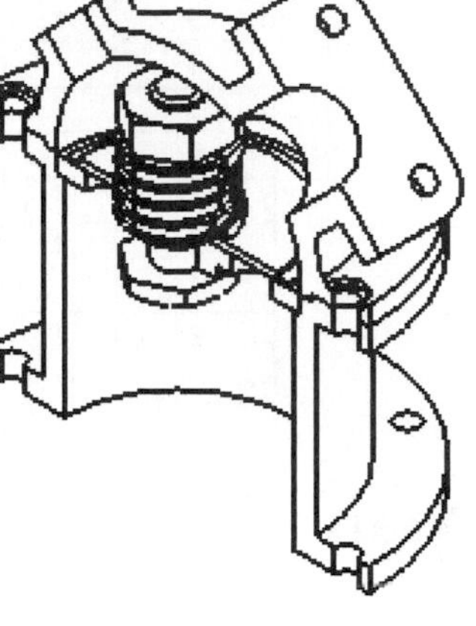

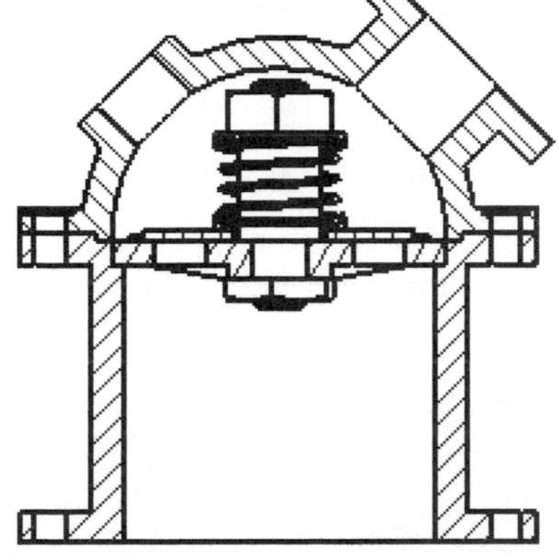

Figure 6.212

34. Now create a Parts List to identify the item number, quantity, part number, and description. Begin by clicking the Parts List tool in the Drawing Annotation panel bar.
35. The Parts List dialog box will display. Select the sectioned assembly view.
36. Verify that the BOM Settings area is set to Structured components, and click the OK button.
37. Move the Parts List until it joins the title block and border, and then click to place it. Zoom in to display the Parts List, as shown in the following image.

Parts List			
ITEM	QTY	PART NUMBER	DESCRIPTION
1	1	Cylinder	CYLINDER
2	1	Cyl Head	CYLINDER HEAD
3	1	Valve plate	VALVE PLATE
4	1	Pin	THREADED PIN
5	1	Diaphragm	DIAPHRAGM
6	1	Washer A	RETAINING WASHER
7	1	ESS_E06_05-WasherB	RETAINING WASHER
8	1	Spring	SPRING - Ø1 X Ø22 X 25
9	1	Nut A	FLAT NUT, REG - M16 X 1.5
10	1	Nut B	FLAT NUT, THIN - M16 X 1.5

Figure 6.213

38. In the following steps, you will edit the Parts List to reorder the columns and adjust the column widths. Begin by selecting the Parts List in the graphics window or from the browser.

39. Right-click, and select Edit Parts List from the menu to display the Edit Parts List dialog box.

40. Click the Column Chooser button to display the Parts List Column Chooser dialog box and make the following changes:

- In the Available Properties list, select MATERIAL.
- Click Add.
- In the Selected Properties list, select DESCRIPTION.
- Click Move Up two times.

41. Now edit the column widths. While still in the Edit Parts List dialog box, right-click on the DESCRIPTION header, and choose Column Width from the menu. Make the following column width changes:

- Enter **65** in the Column Width dialog box.
- Change the QTY field's width to **20**.
- Change the MATERIAL field's width to **42**.

42. Click the OK button to close the dialog box.

43. Move the Parts List so that it is flush with the border and the title block. Your display should appear similar to the following image.

Parts List				
ITEM	DESCRIPTION	QTY	PART NUMBER	MATERIAL
1	CYLINDER	1	Cylinder	Cast Iron
2	CYLINDER HEAD	1	Cyl Head	Cast Iron
3	VALVE PLATE	1	Valve plate	Steel, Mild
4	THREADED PIN	1	Pin	Steel, Mild
5	DIAPHRAGM	1	Diaphragm	Stainless Steel
6	RETAINING WASHER	1	Washer A	Steel, Mild
7	RETAINING WASHER	1	ESS_E06_05-WasherB	Steel, Mild
8	SPRING - Ø1 X Ø22 X 25	1	Spring	Stainless Steel
9	FLAT NUT, REG - M16 X 1.5	1	Nut A	Steel, Mild
10	FLAT NUT, THIN - M16 X 1.5	1	Nut B	Steel, Mild

Figure 6.214

44. Now place balloons in the assembly drawing. The item number in the balloon corresponds to the item number in the Parts List.
45. Pan and zoom to display the sectioned assembly view.
46. Change to the Drawing Annotation panel bar, and click the Balloon tool.
47. Select the edge of the component, as shown in the following image, as the start of the leader.
48. Click a point on the sheet to define the end of the first leader segment.
49. Right-click, and select Continue from the menu to place the balloon, as shown in the following image on the left. Report the previous steps to place the remaining balloons, as shown in the following image on the right.

Tip: To add a balloon for the spring, place the balloon for the washer (6), and then turn off the Balloon tool. Right-click balloon (6), and select Attach Balloon. Click the spring, position the new balloon, and then click to place the balloon.

50. Close all open files. Do not save changes. End of exercise.

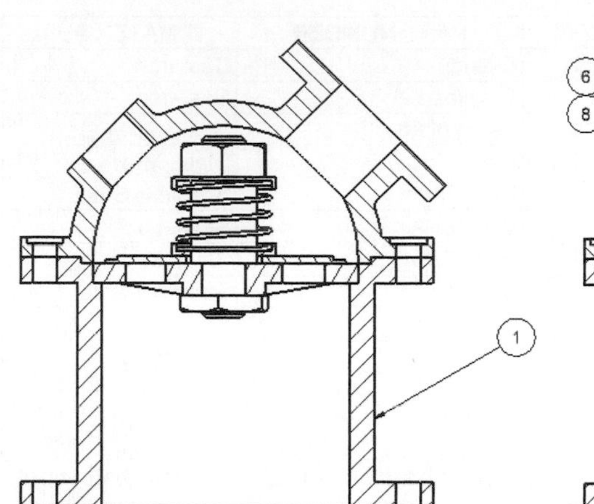

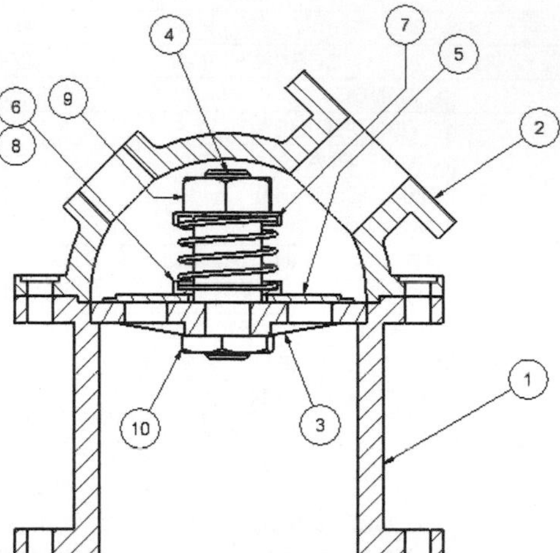

Figure 6.215

Project Exercise: Chapter 6

The self-paced, step-by-step project exercises found in Appendix A provide opportunities for you to work through real-world modeling, assembly, and documentation tasks. The geometry used in the project exercises will flow from chapter to chapter, utilizing the functionality that you learned in that chapter.

Applying Your Skills

SKILL EXERCISE 6–1

In this exercise, you create a new component for a charge pump and then assemble the pump.

1. This exercise uses the skills you have learned in previous exercises to assemble a pump and create a new part in place. Open the assembly file *ESS_E06_10.iam* from the chapter 06 folder, as shown in the following image.

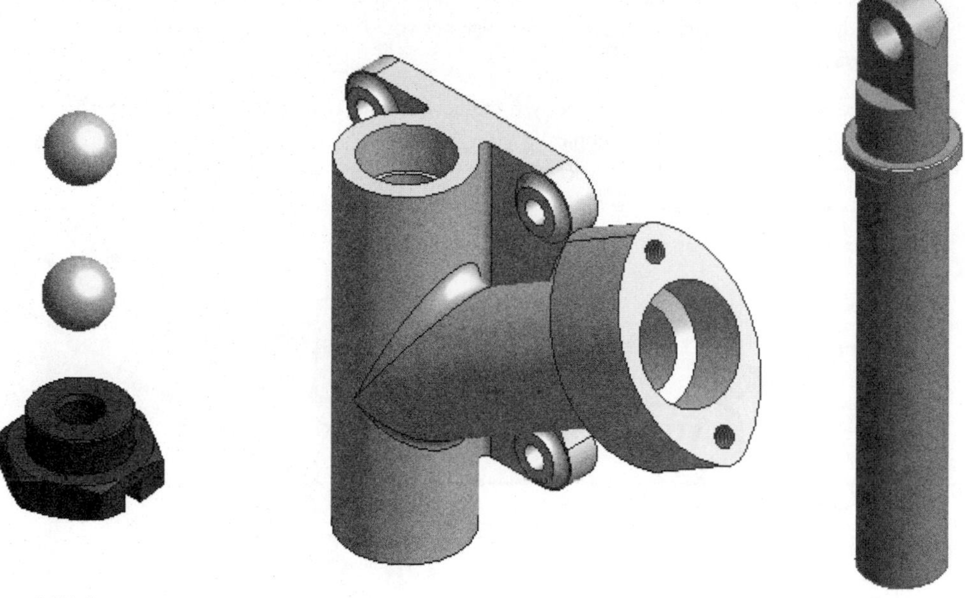

Figure 6.216

2. Place the following predefined components into the assembly:

- 1 occurrence of *ESS_E06_10-Union.ipt*
- 1 occurrence of *ESS_E06_10-Seal.ipt*
- 2 occurrences of *ESS_E06_10-M8x30.ipt*

3. Next create the gland in place. Project edges from the pump body to define the flange. See the following image for the dimensions needed to complete the gland.

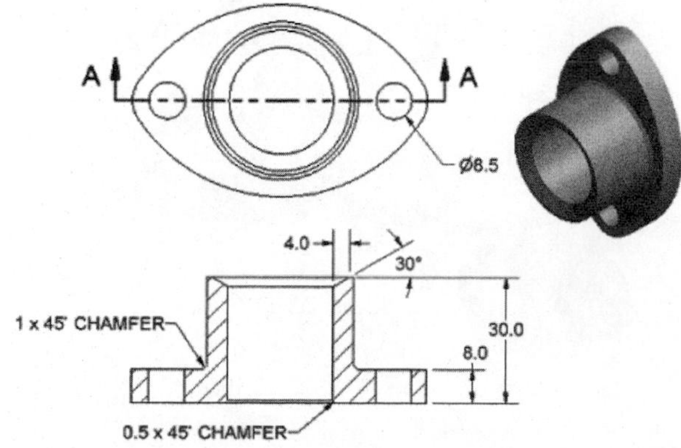

Figure 6.217

4. Use assembly constraints to build the model, as shown in the following image.

Tip: To assemble the balls, mate the center point of the ball with the centerline of the body, and then place a tangent constraint between the ball's surface and the sloped surface of the seat.

To assemble the seal into the body, first apply a mate constraint between the conical surface of the seal and the conical surface of the seal's seat in the body. You must use the Select Other tool to select the first conical surface. Apply a similar constraint between the gland and the seal.

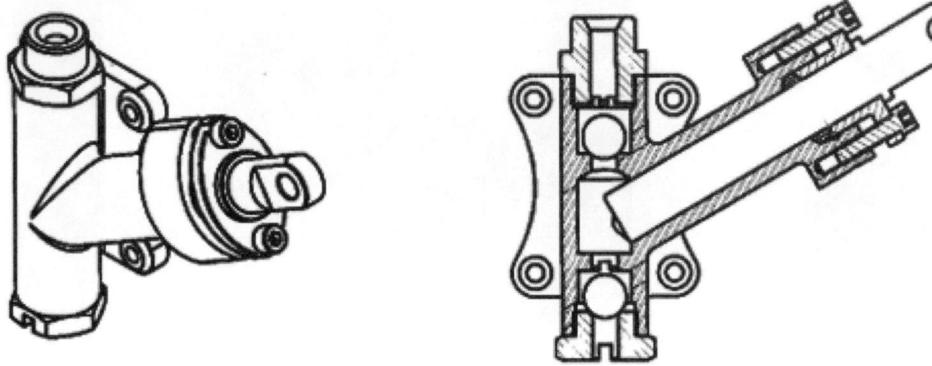

Figure 6.218

5. The completed assembly model should appear similar to the following image.

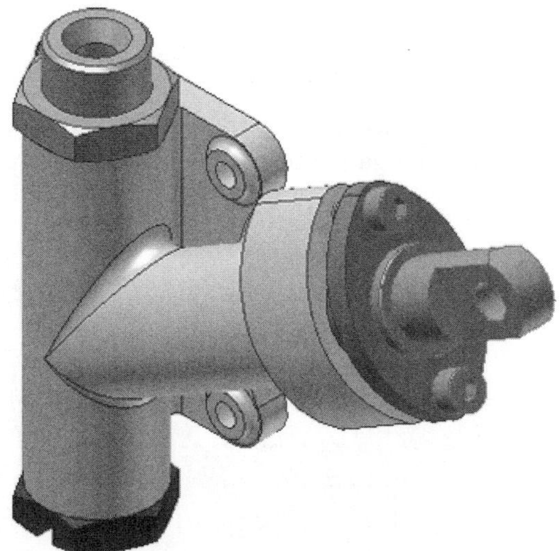

Figure 6.219

6. Close all open files. Do not save changes. End of exercise.

SKILL EXERCISE 6–2

In this exercise, you create an assembly drawing, a parts list, and balloons for a charge pump. The charge pump assembly is shown in the following image.

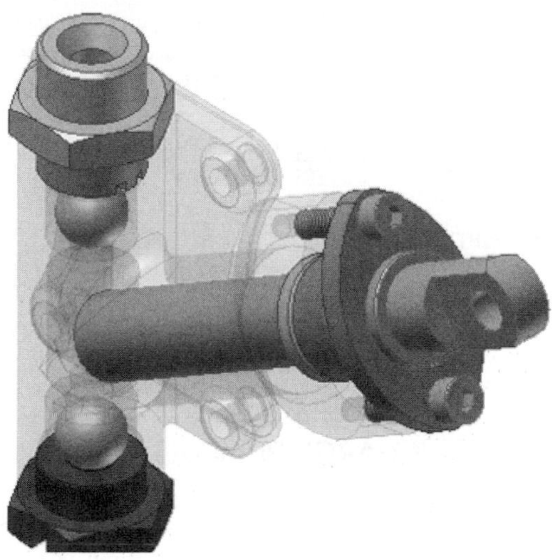

Figure 6.220

1. Create a new drawing using the ANSI (mm) template with a single A2 sheet named Assembly.
2. Insert a top view of *ESS_E06_11.iam* with a scale of 1:1.
3. Using the top view as the base view, create a sectioned front view. Exclude the ram, valve balls, and machine screws from sectioning.
4. Use the Create View tool to create an independent top-right isometric view of the pump assembly.
5. Insert the Parts List, add a Material column and then modify its format according to the column widths shown in the following table.

Column	Width
ITEM	15
QTY	15
DESCRIPTION	57
PART NUMBER	50
MATERIAL	60

6. The results should appear similar to the following image. Note how the item numbers are arranged in ascending order.

8	1	GLAND	ESS_E06_11-Gland	Brass, Soft Yellow Brass
7	2	PURCHASED	ESS_E06_10-M8x30	Default
6	1	SEAL	ESS_E06_10-Seal	Rubiconium
5	1	UNION	ESS_E06_10-Union	Rubiconium
4	1	SEAT	ESS_E06_10-Seat	Rubiconium
3	1	RAM	ESS_E06_10-Ram	Rubiconium
2	2	BALL	ESS_E06_10-Ball	Rubiconium
1	1	BODY	ESS_E06_10-Body	Rubiconium
ITEM	QTY	DESCRIPTION	PART NUMBER	MATERIAL
Parts List				

DRAWN AJK	1/24/2008	ESSENTIALS PLUS			
CHECKED		TITLE CHARGE PUMP ASSEMBLY			
QA					
MFG					
APPROVED					
		SIZE A2		DWG NO ESS_E06_11	REV
		SCALE		SHEET 1 OF 1	

Figure 6.221

7. Add balloons to identify the components.

8. Complete the title block to include the title and author (your name), as shown in the following image.

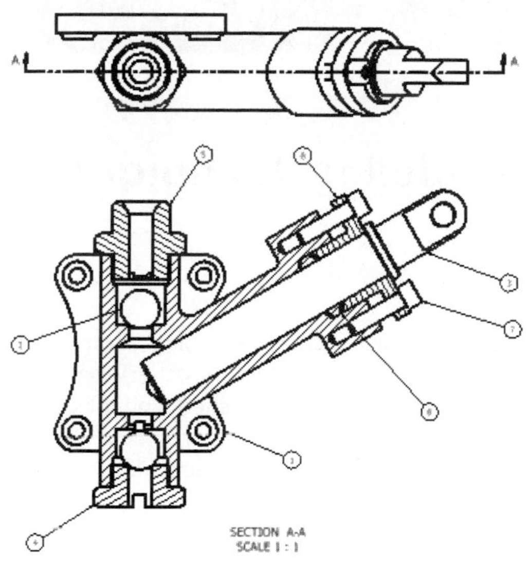

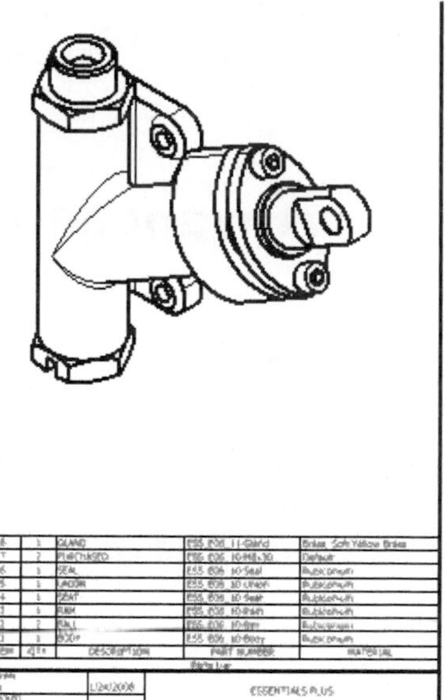

Figure 6.222

9. Close all open files. Do not save changes. End of exercise.

CHECKING YOUR SKILLS

Use these questions to test your knowledge of the material covered in this chapter.

1. **True** _ **False** _ The only way to create an assembly is by placing existing parts into it.
2. **True** _ **False** _ Explain top-down and bottom-up assembly techniques.
3. **True** _ **False** _ An occurrence is a copy of an existing component.
4. **True** _ **False** _ Only one component can be grounded in an assembly.
5. **True** _ **False** _ Autodesk Inventor does not require components in an assembly to be fully constrained.
6. **True** _ **False** _ A sketch must be fully constrained to adapt.
7. **True** _ **False** _ What is the purpose of creating a presentation file?
8. **True** _ **False** _ A presentation file is associated with the assembly file on which it is based.
9. **True** _ **False** _ When creating drawing views from an assembly, you can create views from multiple presentation views or design views.
10. **True** _ **False** _ A 2D design layout consists of a series of 2D sketches that are constrained to 3D parts in an assembly file.

Advanced Part Modeling Techniques

INTRODUCTION

In this chapter, you will learn how to use advanced modeling techniques. Using advanced modeling techniques, you can create transitions between parts that would otherwise be difficult to create. You can edit advanced features such as other sketched and placed features, but typically their creation requires more than one unconsumed sketch. This chapter also introduces you to techniques that will help you be more productive in your modeling.

OBJECTIVES

After completing this chapter, you will be able to perform the following:

- Extrude an open profile
- Create ribs, webs, and rib networks
- Emboss text and profiles
- Create sweep features
- Create coil features
- Create loft features
- Split a part or split faces of a part
- Bend a part
- Copy features within a part
- Reorder part features
- Mirror model features
- Suppress features of a part
- Create a derived part based on another part or an assembly
- Use Auto Limits to alert you when a part or assembly exceeds a target range for a specific property

USING OPEN PROFILES

In chapter 3, you learned how to extrude a closed profile. In this section, you will learn how to extrude an open profile: a sketch that does not form a closed area or boundary. You extrude an open profile bidirectionally in a positive or negative direction normal to the profile plane and in a fill direction. A fill direction will extend the profile until it touches a termination face in all directions, creates a closed area, and encloses any existing feature. The following image shows the original part on the left and with a line, based on a work plane, in the middle of the sketch. The part in the middle of the image shows the open profile (the line) extruded to the left: the Match shape option was used and the cylindrical boss was enclosed, respecting the hole through the boss. The part on the right of the image shows the open profile extruded to the right.

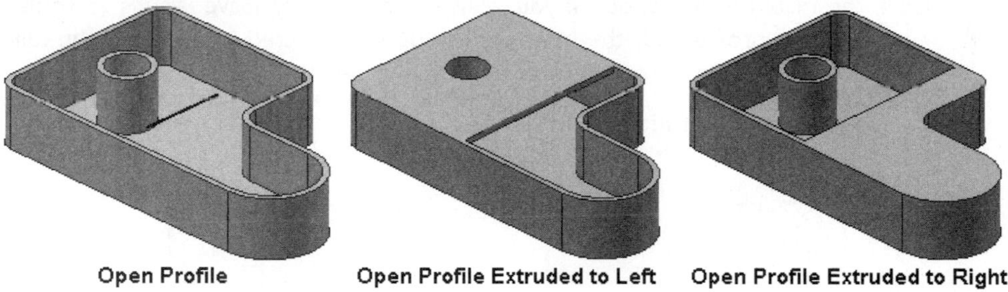

Figure 7.1

When working with an open profile, you can only use the Extrude or Revolve tool. The Sweep tool can use an open profile, but the result will be a surface. The open profile can be consumed inside or extend beyond the part or a combination of the two. When you extrude or revolve the open profile, it will either be extended or trimmed back to be enclosed with the part. You can constrain and dimension an open profile like any other profile.

To extrude an open profile, follow these steps:

1. Make a sketch active.
2. Draw an open profile.
3. Add geometric constraints and dimensions as needed.
4. Issue the Extrude tool.
5. Click the open profile on the part.
6. Click on the side of the part that will be filled in. The following image shows the left side of the part selected as the fill side, as shown in the following image on the left, and the right side of the part selected as the fill side, as shown in the following image on the right.

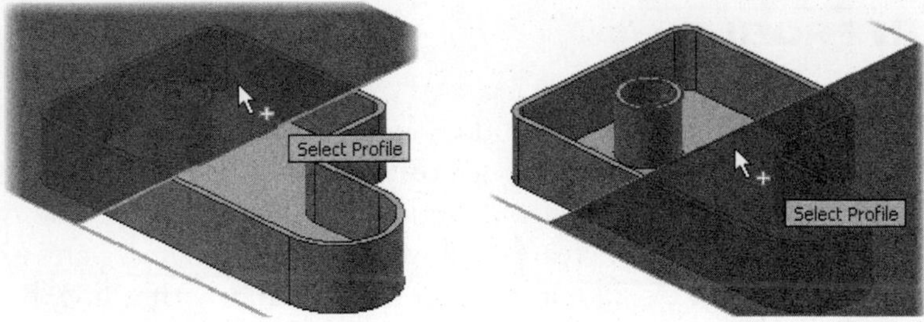

Figure 7.2

7. Change the Extents as needed: the options are Distance, To Next, To, From To, or All. Then specify the distance and the extrusion direction, if required.
8. Click the Match Shape option if you want the extrude to leave islands as voids. The extrude open profile with the Match Shape option selected is shown in the following image on the left. The image on the right side shows it without the option selected.
9. To complete the operation, click OK.

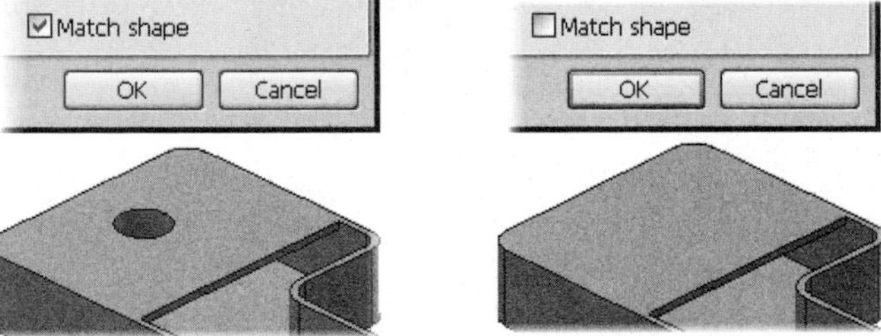

Figure 7.3 *(Left) Extrude open profile with Match shape checked. (Right) Extrude open profile with Match shape not checked.*

RIB AND WEB FEATURES

Ribs and webs are used primarily to reinforce or strengthen features in mold and cast parts, but you can also use them in machined parts and in other cases where additional support and minimal weight are required. The following image shows a part with a sketch and with the sketch used to create a rib and a web.

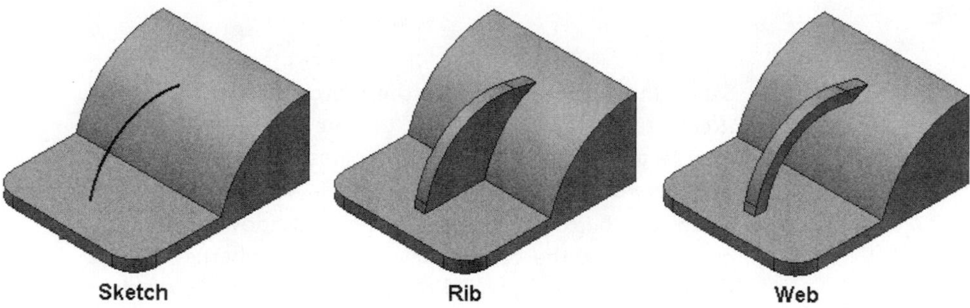

Figure 7.4

Using the Rib tool located on the Part Features panel bar, as shown in the following image on the left, you can create ribs, webs, and rib networks.

- A rib is a thin-walled feature that is typically closed.
- A web is similar in width but usually has an open shape.
- A rib network consists of a series of thin-walled support features.

A rib or web feature is defined by a single, open, unconsumed profile that is refined using the options in the Rib dialog box. If no unconsumed sketch exists in the part file, Autodesk Inventor will warn you with the message: "No unconsumed visible sketches." After starting the Rib tool, the Rib dialog box will appear, as shown in the following image on the right. The following sections describe the options in the dialog box.

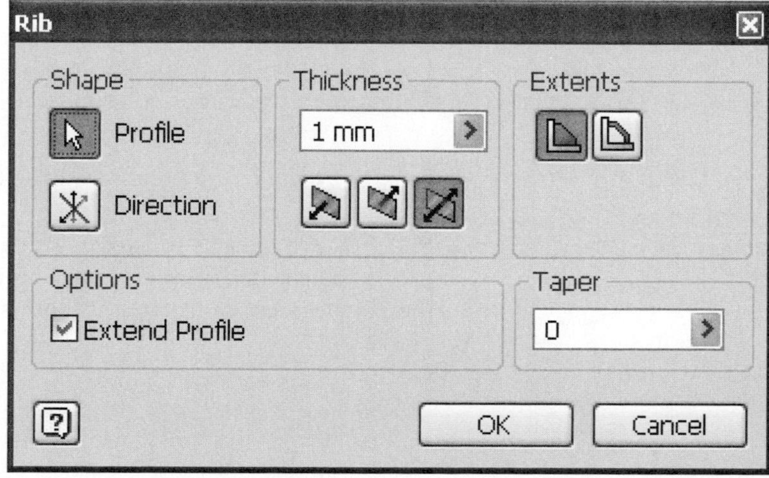

Figure 7.5

SHAPE

Profile Select this button to choose the sketch to extrude. The sketch can be an open profile, or you can select multiple-intersecting or nonintersecting profiles to define a rib or web network.

Direction Select this button, and in the graphics window, position the cursor around the open profile to specify whether or not the rib will extend parallel or perpendicular and in which direction relative to the sketch geometry. Once the correct direction is displayed, click in the graphics window. The following image shows a rib in three different directions.

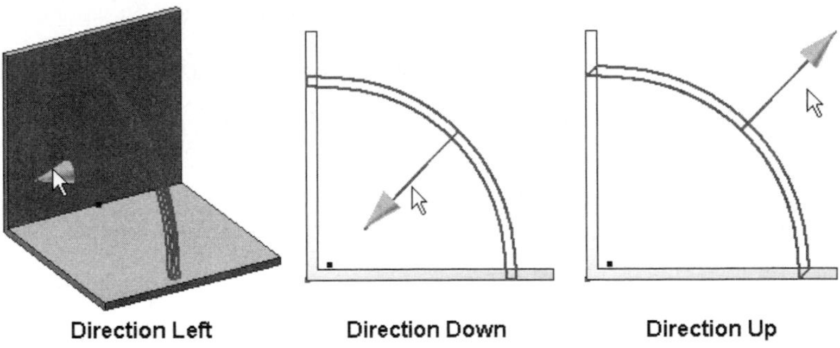

Figure 7.6

THICKNESS

In this section, you specify the thickness and thickness direction of the rib.

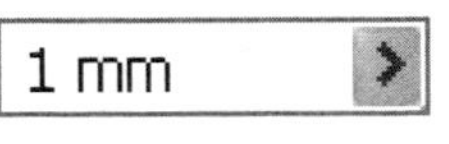

Value Enter the width of the rib feature using this edit box.

Flip Buttons Select the flip buttons to specify which side of the profile to apply the thickness value to or to add the same amount of material to both sides of the profile.

EXTENTS

In this section, you specify a rib (To Next) or a web (Finite).

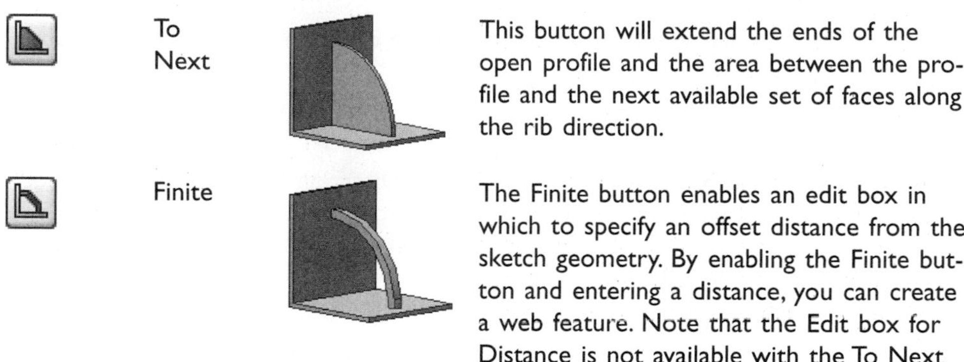

To Next		This button will extend the ends of the open profile and the area between the profile and the next available set of faces along the rib direction.
Finite		The Finite button enables an edit box in which to specify an offset distance from the sketch geometry. By enabling the Finite button and entering a distance, you can create a web feature. Note that the Edit box for Distance is not available with the To Next option.

OPTIONS

Extend Profile

The Extend Profile checkbox specifies whether to extend the endpoints of the sketch to the next available face or to leave the ends of the open profile as determined by the end of the rib feature. If you click the Extend Profile option, the ends are extended; if you leave the box clear, the ends cap at the end of the sketched profile. The Extend Profile option is always available for a finite web, but it is only active for the To Next option when the direction and face(s) on the model meet appropriate conditions.

TAPER

Taper applies a draft (taper) angle to the profile if it is normal to the sketch plane. The taper is created evenly in both directions. The following image shows the preview of the rib with taper.

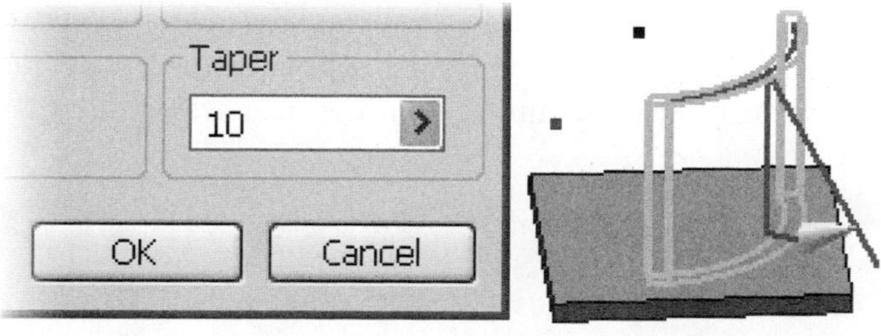

Figure 7.7

CREATING RIBS AND WEBS

To create a rib or web, follow these steps:

1. Create an active sketch in the location where you will place the rib or web.
2. Sketch an open profile that defines the basic shape of one edge of the rib or web.
3. Add constraints and dimensions to the sketch as needed.
4. Click the Rib tool located on the Part Features panel bar.
5. Specify the direction of the rib or web. Click the Direction button in the Rib dialog box. In the graphics window, position the cursor around the open profile to specify whether or not the rib or web will extend parallel or perpendicular and in which direction relative to the sketch geometry. Once the correct direction is previewed, click in the graphics window.
6. Enter a value for the rib or web in the Thickness edit box. This will specify the width of the rib or web.
7. Also in the Thickness section, use the Flip buttons to choose which side of the profile to apply the thickness value or to add the same amount of material to both sides of the profile.
8. Specify the depth of the profile by clicking either the To Next or the Finite buttons in the Extents area of the Rib dialog box.
9. If you use the Finite option, enter an offset distance, and click Extend Profile if the endpoints are to be extended to the next available face.
10. To complete the operation, click the OK button.

RIB NETWORKS

You can also create rib networks using the Rib tool. You can use multiple intersecting or nonintersecting sketch objects within a single profile to create a rib network. The creation process is the same for creating a single rib, except that the thickness is applied to all objects within the profile. When you select the profile objects, you will need to select them individually. If the rib network is to have equal spacing between the objects, use the 2D Rectangular Pattern or the 2D Circular Pattern tool to create the profile. The following image shows a part with multiple intersecting lines defined in a single profile, the same profile used to create a rib network, and the profile as a web network.

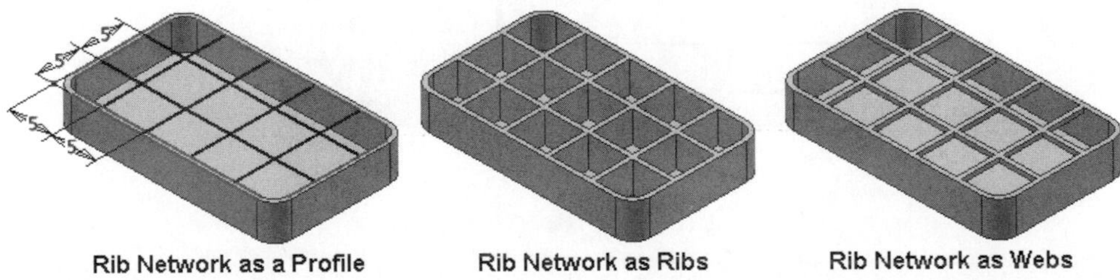

Figure 7.8

EXERCISE 7-1: CREATING RIBS AND WEBS

In this exercise, you sketch an open profile, use the Rib tool to create a rib, and then edit the rib feature to change the rib to a web. You complete the exercise by sketching overlapping lines and creating a rib network.

1. Open *ESS_E07_01.ipt* from the Chapter 07 folder.
2. Use the Zoom and Rotate tools to examine the part.
3. Create a new sketch on WorkPlane1.
4. Click the Look At tool, and then click the work plane.
5. Turn off the visibility of the work plane.
6. Use the Line and 3-Point arc tool to sketch the line and arc, as shown in the following image. Make sure that the arc you sketch, if extended to the right, would intersect the top of the part. Dimensions could be added as required.

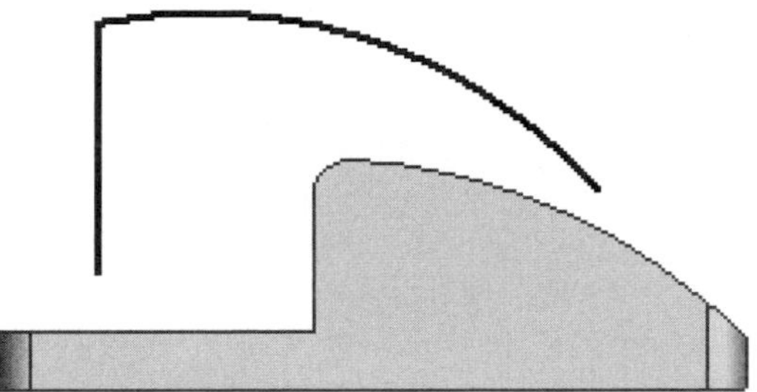

Figure 7.9

7. Click the Return tool from the panel bar to finish the sketch.
8. Change to the Home View.
9. Click the Rib tool on the Part Features panel bar. The Direction button is active. Move the cursor down, as shown in the following image on the left, and then click a point to get the preview.
10. In the Rib dialog box, enter **8 mm** in the Thickness edit box. Click OK. When complete, the rib should resemble the following image on the right.

Figure 7.10

11. To edit the rib, in the browser double-click the Rib feature.

12. In the Rib dialog box, click the Finite button, enter **25 mm** in the Extents edit box, as shown in the following image. Click OK.

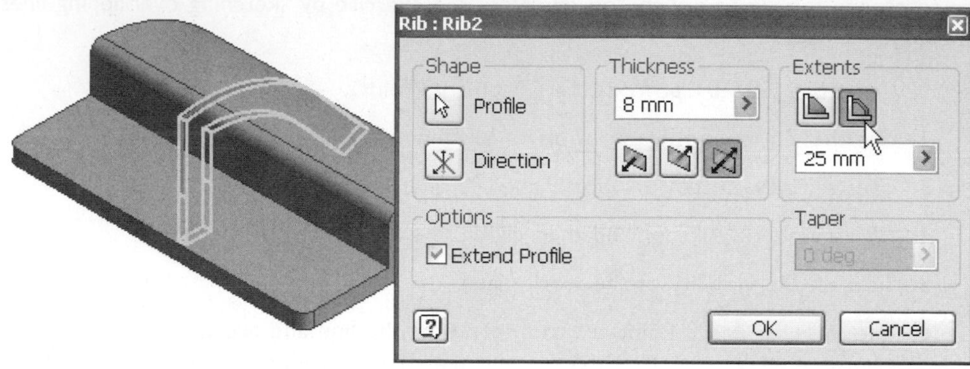

Figure 7.11

13. When complete, the rib should resemble the following image.

Figure 7.12

14. Use the Free Orbit tool to view the bottom of the part.
15. Create a new sketch on the inside flat face, as shown in the following image on the left.
16. Sketch three, overlapping line segments, as shown in the following image on the right. The line should be parallel and perpendicular to the existing lines and each other. Dimensions could be added as required and the Rectangular Pattern tool from the Sketch panel could be used to create more complex webs.

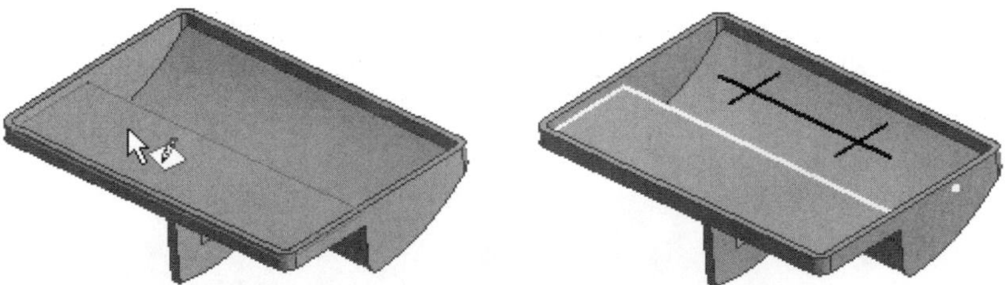

Figure 7.13

17. Click the Return tool.
18. Click the Rib tool.

19. Select the three line segments for the profile.

20. In the Rib dialog box:

- Verify that the Thickness field still shows a value of **8 mm**.
- Click the Finite button.
- Enter **6 mm** in the Extents edit box.

21. Click the Direction button and move the cursor around your profile until the preview shows the rib direction going down into the pocket, as shown in the following image, then click to accept that direction.

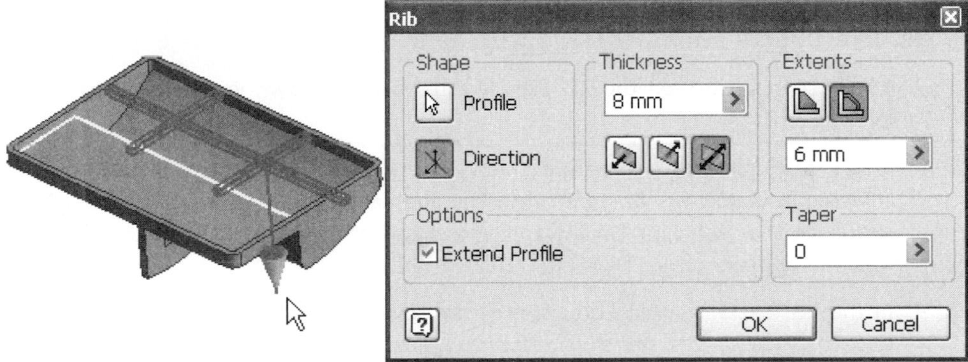

Figure 7.14

22. Click OK in the Rib dialog box, and your rib network should resemble the following image.

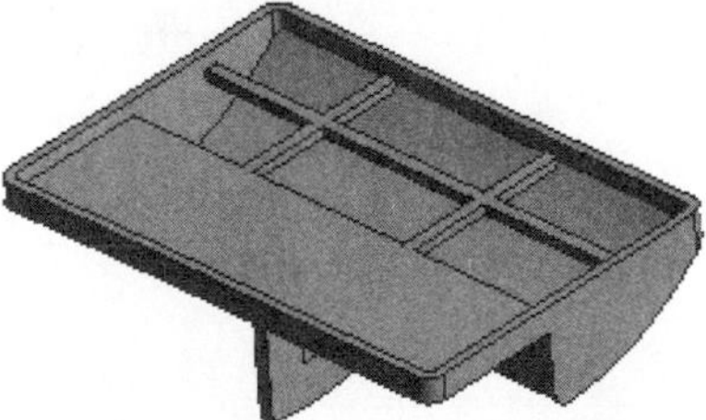

Figure 7.15

23. Use the Zoom and Rotate tools to examine the rib network.

24. Close the file. Do not save changes. End of exercise.

EMBOSSED TEXT AND CLOSED PROFILES

To better define a part, you may need to have a shape or text either embossed (raised) or engraved (cut) into a model. In this section you will learn how you can emboss or engrave a closed shape or text onto a planar or curved face. You can define a shape using the sketch tools on the 2D Sketch panel bar. The shape needs to be closed. There are two steps to emboss or engrave: first create the shape or text. Second, you emboss the shape or text onto the part. The text can be orientated in a rectangle or about an arc, circle, or line. The following sections describe the steps.

STEP 1: CREATING RECTANGULAR TEXT

To place rectangular text onto a sketch, follow these steps:

1. Make a sketch active.
2. Click the Text tool on the 2D Sketch panel bar, as shown in the following image on the left.
3. Pick a point or drag a rectangle where you will place the text. If a single point is selected, the text will fit on a single line and can be grip edited. If a rectangle is used, it defines the width for text wrapping and can be modified. The Format Text dialog box will appear.
4. In the Format Text dialog box, specify the text font and format style, and enter the text to place on the sketch. The following image shows the dialog box with text entered in the bottom pane.

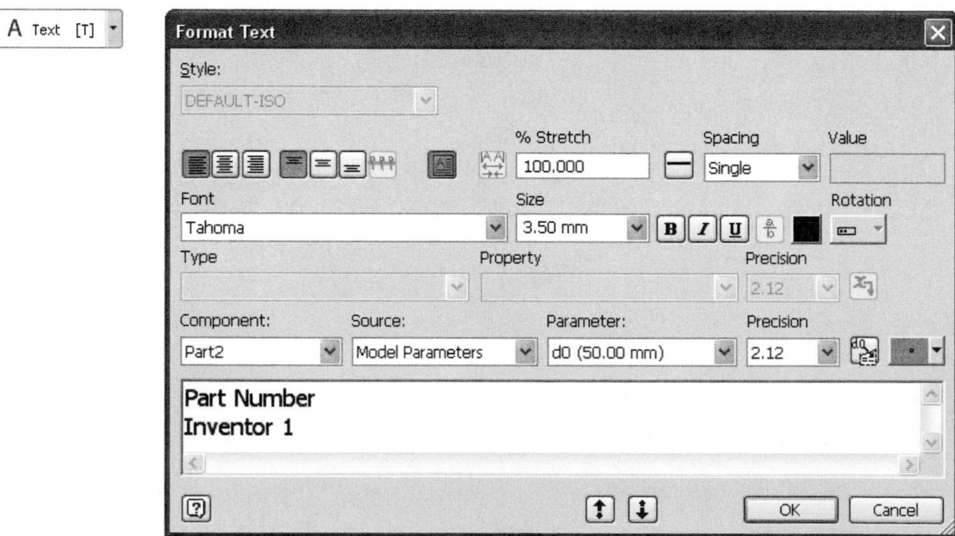

Figure 7.16

5. When done entering text, click OK.

6. When the text is placed, a rectangular set of construction lines defines the perimeter of the text. You can click and drag a point on the rectangle and change the size of the rectangle. Dimensions or constraints can be added to these construction lines to refine the text's location and orientation, as shown in the following image.
7. To edit the text, move the cursor over the text in the graphics window, right-click, and select Edit Text from the menu. The same Format Text dialog box that was used to create the text will appear.

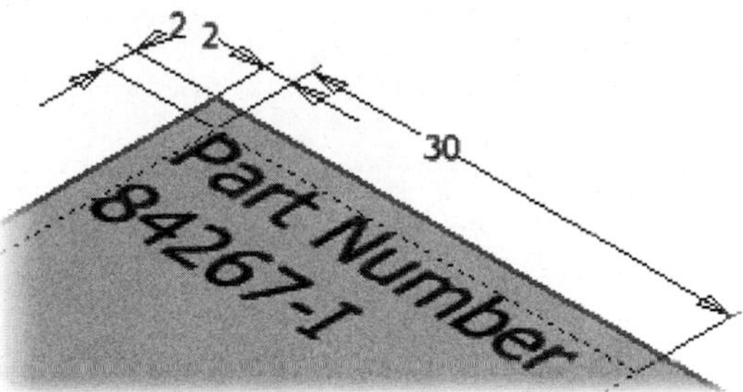

Figure 7.17

STEP 1: CREATING TEXT ABOUT GEOMETRY

To place text about an arc, circle, or line in a sketch, follow these steps:

1. Make a sketch active.
2. Sketch and constrain an arc, circle, or line that the text will follow.
3. Click the Geometry Text tool on the 2D Sketch panel bar, as shown in the following image on the left. The tool may be under the text tool.
4. In the sketch select the arc, circle, or line and the Geometry Text dialog box will appear.
5. In the Geometry Text dialog box, specify the text font and format style, direction, position, and start angle and enter the text to place on the sketch. The start angle is relative to the left quadrant point of a circle or the start point of the arc. The following image shows the dialog box with text entered in the bottom pane.

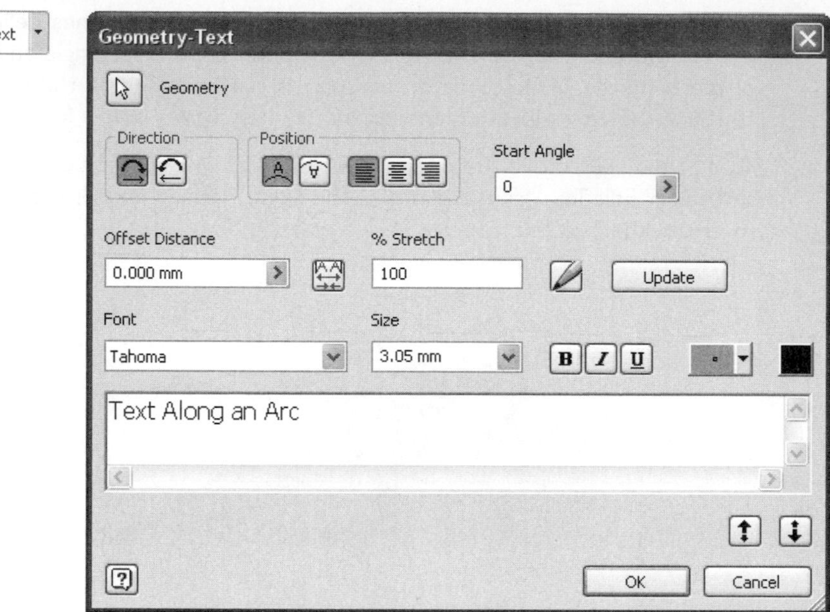

Figure 7.18

6. When done entering text, click the Update button in the dialog box to see the preview in the graphics window.
7. Click OK to complete the operation. The following image shows text placed along an arc.
8. To edit the text, move the cursor over the text in the graphics window, right-click, and click Edit Geometry Text from the menu. The same Format Text dialog box that was used to create the text will appear.

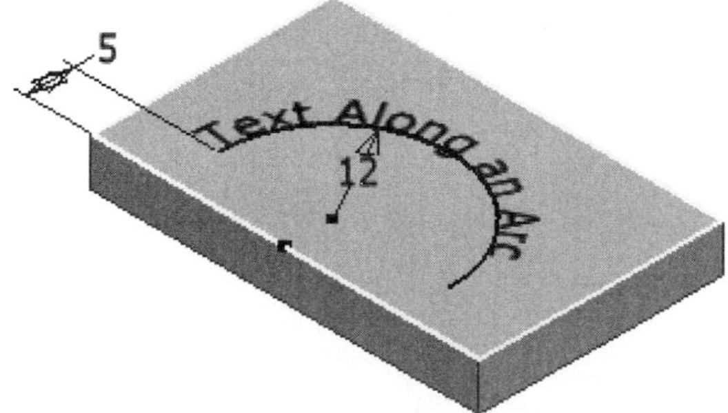

Figure 7.19

STEP 2: EMBOSSING TEXT

To emboss text or a closed profile, click the Emboss tool on the Part Features panel bar, as shown in the following image on the left. The Emboss dialog box will appear, as shown in the following image on the right.

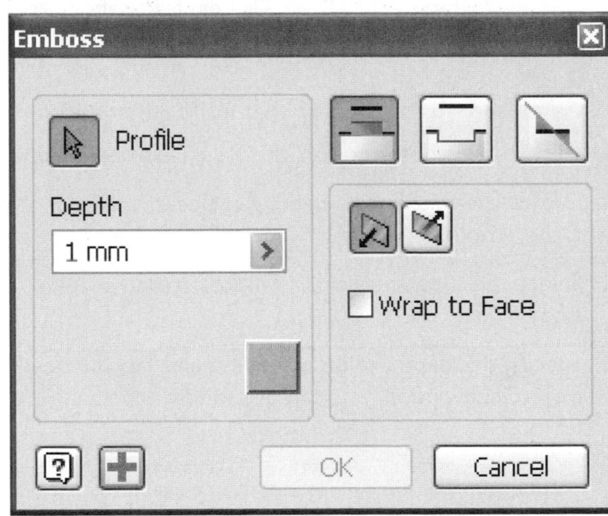

Figure 7.20

The Emboss dialog has the following options:

	Profile	Select a profile, meaning closed shape or text, to emboss. You may need to use the Select Other tool to select the text.
1 mm	Depth	Enter an offset depth to emboss or engrave the profile if you selected the Emboss from Face or Engrave from Face types.
	Top Face Color	Select a color from the drop-down list to define the color of the top face of the embossed area, not its lateral sides.
	Emboss from Face	Select this option to add material to the part.
	Engrave from Face	Select this option to remove material from the part.
	Emboss/ Engrave from Plane	Select this option to add and remove material from the part by extruding both directions from the sketch plane. Direction changes at the tangent point of the profile to a curved face.
	Flip Direction	Select either of these buttons to define the direction of the feature.

Wrap to Face	Wrap to Face	Check this box for Emboss from Face or Engrave from Face types to wrap the profile onto a curved face. Only a single cylindrical or conical face can be selected. The profile will be slightly distorted as it is projected onto the face. The wrap stops if a perpendicular face is encountered.
	Face	Click this button after checking the Wrap option, and then select the face around which the profile will be wrapped.

To emboss a closed shape or text, follow these steps:

1. Click the Emboss tool on the Part Features panel bar.
2. Define the profile by selecting a closed shape or text. If needed, use the Select Other tool.
3. Select the type of emboss: Emboss from Face, Engrave from Face, or Emboss/Engrave from Plane.
4. Specify the depth, color, direction, and face as needed. If you selected Emboss/Engrave from plane option, you can also add a taper angle to the created emboss/engrave feature.

Note: You edit the embossed feature like any other feature.

EXERCISE 7-2: CREATING TEXT AND EMBOSS FEATURES

In this exercise, you emboss and engrave sketched profile objects on faces of a razor handle model. You then create a sketch text object and engrave it on the handle.

1. Open *ESS_E07_02.ipt* from the chapter 07 folder.

2. Use the Free Orbit and Zoom tools to examine the part.

3. Turn on visibility of Sketch11 and rotate the view to display the top triangular face, as shown in the following image.

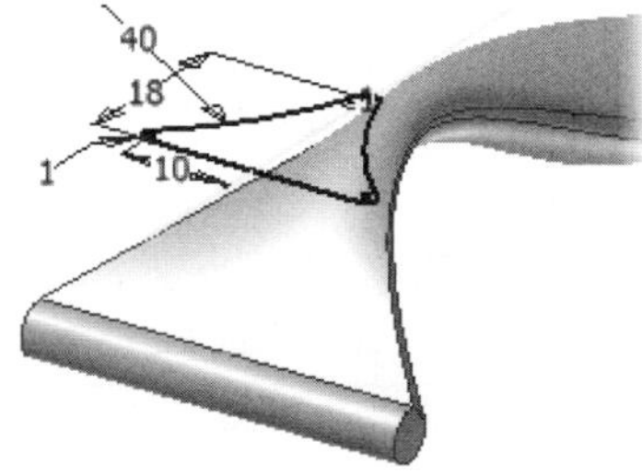

Figure 7.21

4. Click the Emboss tool to engrave a sketched profile. Notice that the only visible profile is automatically selected.

5. In the Emboss dialog box, click the Engrave from Face option, and change the Depth to **.9 mm**.
6. Click the Top Face Color button, and choose Aluminum (Flat) from the Color dialog box drop-down list, as shown in the following image on the left.
7. Click OK twice to close both dialog boxes and create the engraved feature, as shown in the following image on the right.

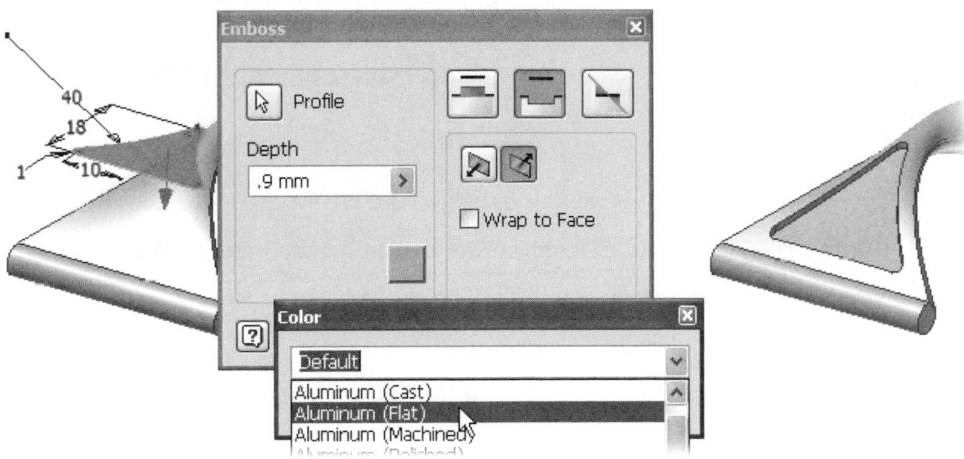

Figure 7.22

8. Change to the Home View, and turn on the visibility of Sketch10.
9. Click the Emboss tool.
10. For the profile, click the oblong and the six closed profiles. Be sure to select the left and right halves of the herringbone profiles, as shown in the following image on the left.
11. In the Emboss dialog box, click the Emboss from Face option, and change the Depth to **.2 mm**.
12. Click the Top Face Color button, and choose Black from the Color dialog box drop-down list, as shown in the following image on the right.

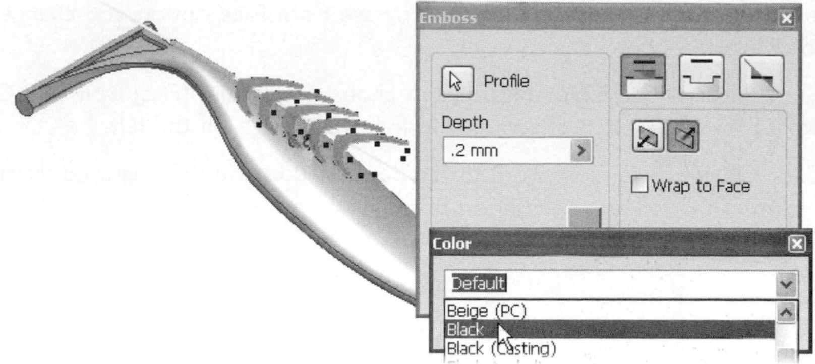

Figure 7.23

13. Click OK twice to close both dialog boxes and create the embossed feature.
14. Use the Free Orbit tool to examine the engraved and embossed features.
15. Next create a text sketch object. Turn on the visibility of Sketch9.
16. In the browser, double-click Sketch9 to edit it.
17. Use the Look At tool to set up a view normal to the sketch.
18. Click the Text tool.
19. Click in an open area below the part and under the left ledge of the construction rectangle to specify the insertion point and display the Format Text dialog box.
20. Click the Center and Middle Justification buttons and click the Italic option. Change the Stretch value to **120**.
21. Click in the text field, and type The SHARP EDGE.
22. In the text field, double-click on the word "The," and change the text size from 3.50 mm to **2.50 mm**, as shown in the following image.

EXERCISES

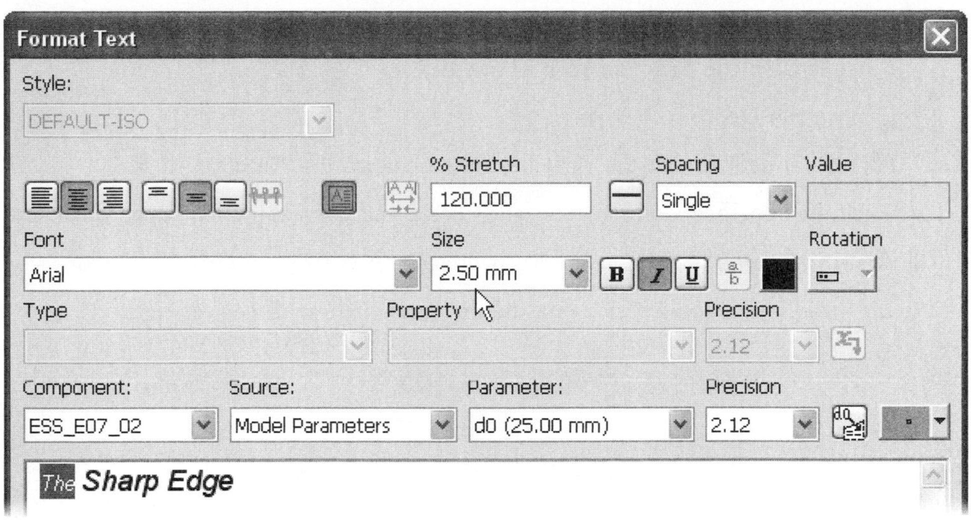

Figure 7.24

23. Click OK to place the text.
24. Draw a diagonal construction line coincident with the text bounding box corners, as shown in the following image.

Figure 7.25

25. Place a coincident constraint between the midpoint of the diagonal construction line of the text and Sketch9, as shown in the following image on the left. The completed operation is shown in the following image on the right.

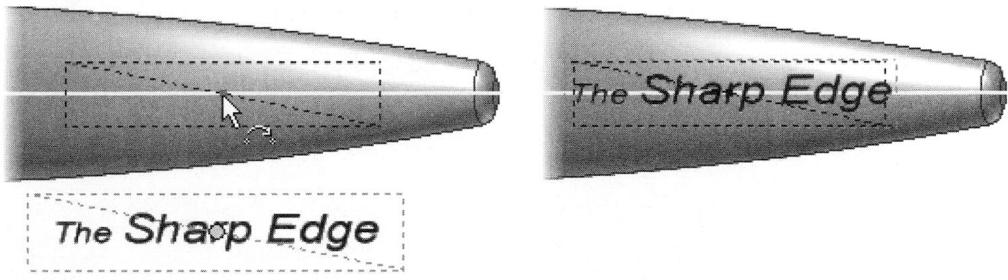

Figure 7.26

26. Click the Return button to finish editing the sketch, and then change to the Home View.
27. Click the Emboss tool.
   - For the profile, select the text object.

- In the Emboss dialog box, click the Engrave from Face option.
- If needed, flip the direction so it points down.
- Change the Depth to **0.5 mm**.
- For the direction, click the right button to change the direction down.
- Click the Top Face Color button, and click Black from the Color drop-down list.

28. Click OK twice to close both dialog boxes and create the engraved feature.

29. Use the Rotate tool to examine the engraved and embossed features.

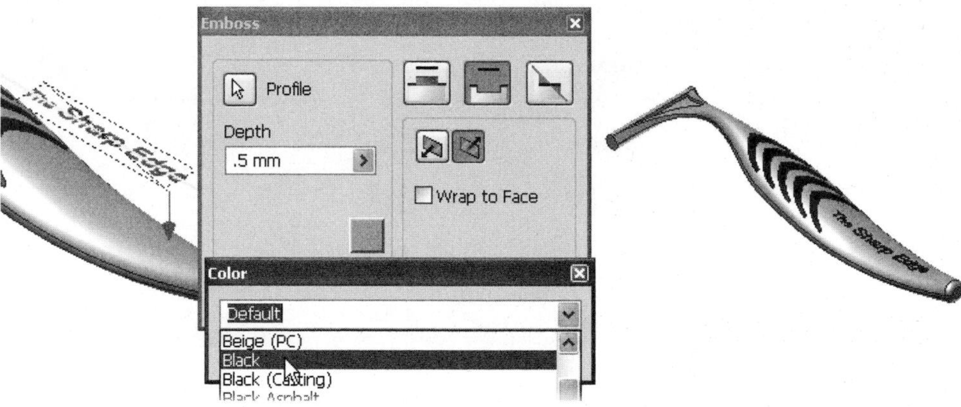

Figure 7.27

30. Delete the Emboss3 feature.

31. Edit Sketch9 by double-clicking it in the browser.

32. Delete the existing text, rectangles and angled lines; do NOT delete the yellow line (your color may be different depending upon your color scheme).

33. Use the Look At tool to set up a view normal to Sketch9.

34. Sketch and dimension a circle as shown. If needed project the vertical edge for the **100 mm** dimension.

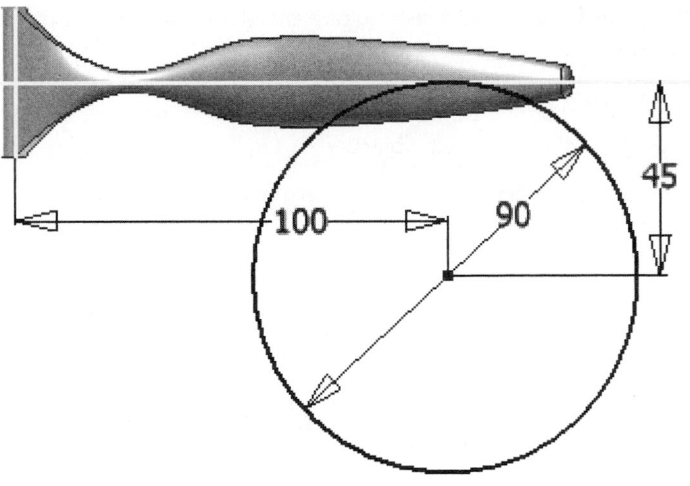

Figure 7.28

35. Click the Geometry Text tool and select the arc.

- Click the Center Justification button.
- Change the Start Angle to **90.00 deg**.
- Click in the text field, and enter **The Sharp Edge** as shown in the following image.

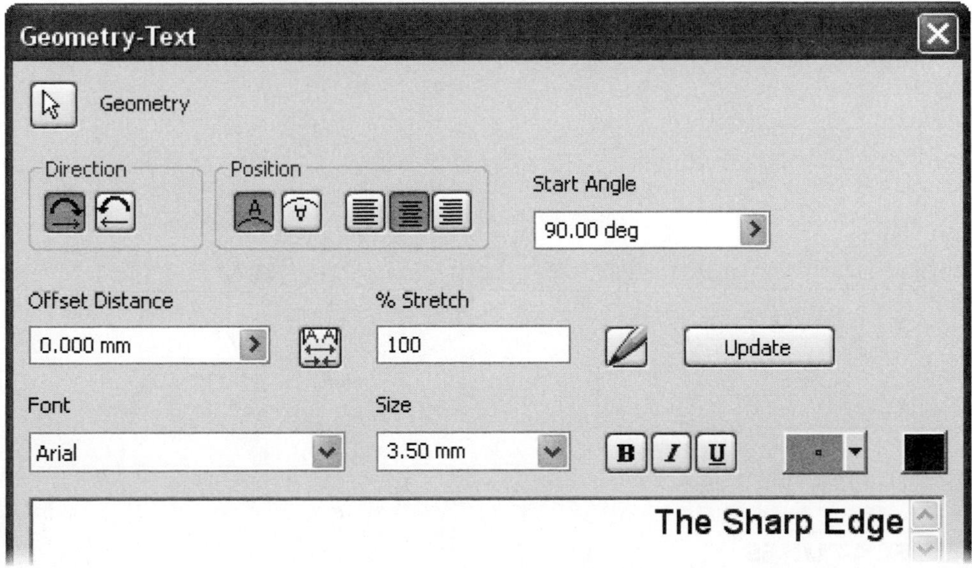

Figure 7.29

- Click the Update button in the Geometry dialog box; the text will appear in the sketch.

36. Click the Return button to finish editing the sketch, and then change to the Home View.

37. Click the Emboss tool.
   - For the profile, select the text object.
   - In the Emboss dialog box, click the Engrave from Face option.
   - Flip the direction so it points down.
   - Change the Depth to **0.5 mm**.

38. Click OK twice to close both dialog boxes and create the engraved feature.

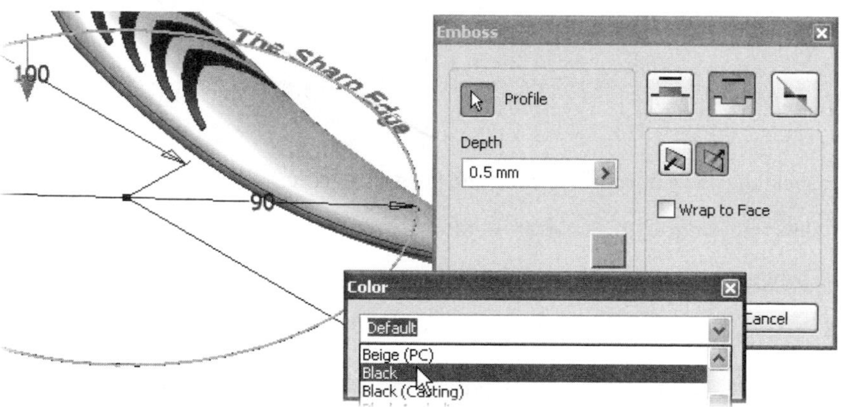

Figure 7.30

39. Use the Rotate tool to examine the engraved and embossed features as shown in the following image.

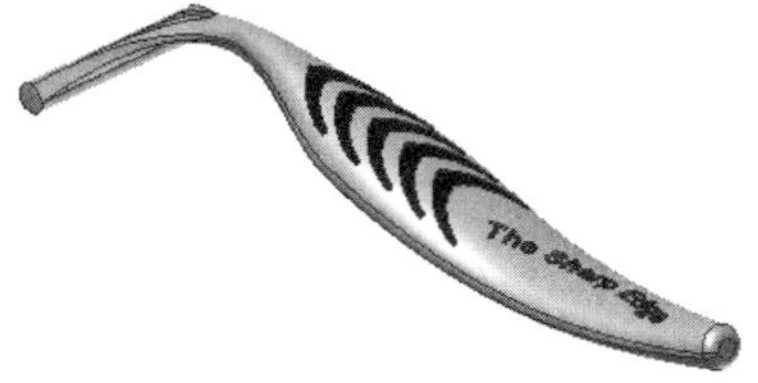

Figure 7.31

40. Close the file. Do not save changes. End of exercise.

SWEEP FEATURES

Unlike other sketched features, a sweep feature requires two unconsumed sketches: a profile to be swept and a path that the profile will follow. Additional profiles can be used as a guide rails or a surface can also be used to help shape the feature. The profile sketch and the path sketch cannot lie on the same or parallel planes. The path can be an irregular shape or can be based on a part edge by projecting and including the edges onto the active sketch. The path can be either an open or closed profile and can lie in a plane or lie in multiple planes (3D Sketch). Handles, cabling, and piping are examples of sweep features. A sweep feature can be a base or a secondary feature. To create a sweep feature, use the Sweep tool on the Part Features panel bar, as shown in the following image on the left. The Sweep dialog will appear, as shown in the following image on the right. The following list includes descriptions of the options in the Sweep dialog box.

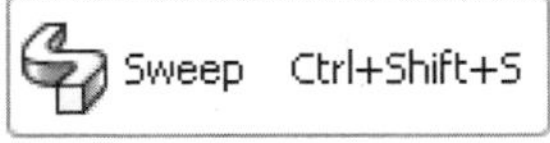

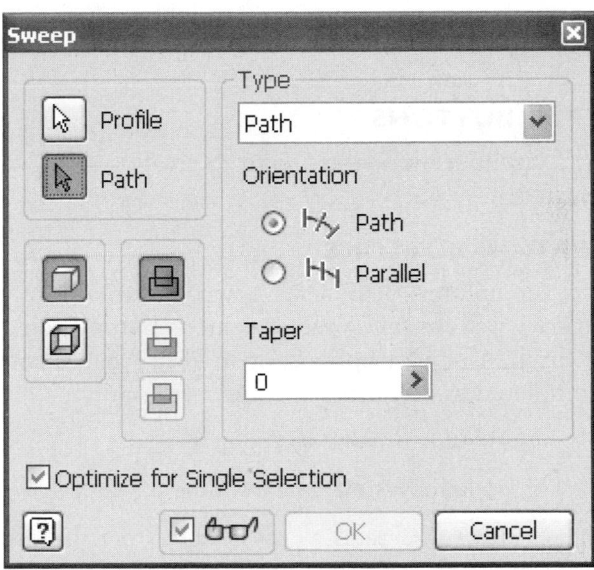

Figure 7.32

PROFILE

Click to choose the sketch profile to sweep. If the Profile button is depressed and red, it means that you need to select a sketch or sketch area. If there are multiple closed profiles, you will need to select the profile that you want to sweep. If there is only one possible profile, Autodesk Inventor will select it for you and you can skip this step. If you selected the wrong profile or sketch area, depress the Profile button, and deselect the incorrect sketch by clicking it while holding down the CTRL key. Release the CTRL key, and select the desired sketch profile.

PATH

Click this button to choose the path along which to sweep the profile. The path can be an open or a closed profile. The profile is typically perpendicular to and intersects with the start point of the path. The start point of the path is often projected into the profile sketch to provide a reference point.

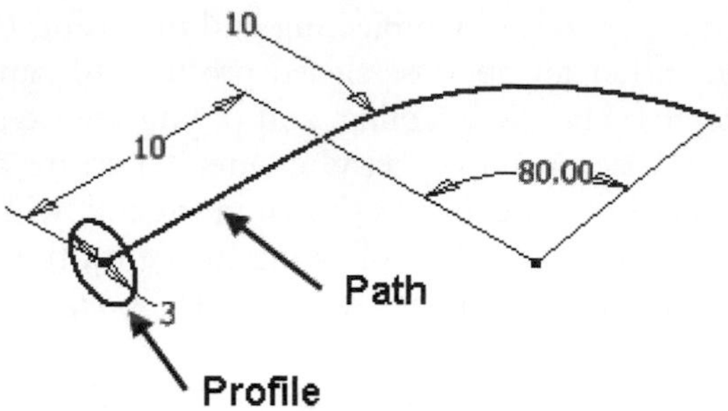

Figure 7.33

OUTPUT BUTTONS

In the Output section, click Solid to create a solid feature or Surface to create the feature as a surface.

OPERATION BUTTONS

This is the column of buttons down the middle of the dialog box. By default, the Join operation is selected. Use the operations buttons to add or remove material from the part, using the Join or Cut options, or to keep what is common between the existing part and the completed sweep using the Intersect option.

- *Join:* Adds material to the part.
- *Cut:* Removes material from the part.
- *Intersect:* Creates a new feature from the shared volume of the sweep feature and existing part volume. Material not included in the shared volume is deleted.

TYPE BOX

Path

Creates a sweep feature by sweeping a profile along a path.

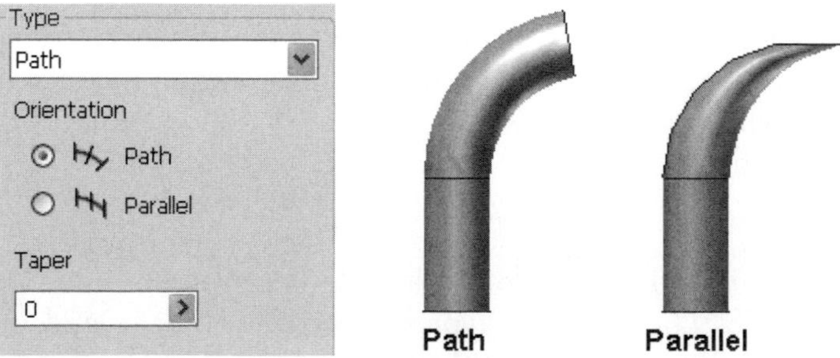

Figure 7.34

Orientation

Path Holds the swept profile constant to the sweep path. All sweep sections maintain the original profile relationship to the path, as shown in the previous image.

Parallel Holds the swept profile parallel to the original profile as shown in the previous image.

Taper Enter a value for the angle you want the profile to be drafted. By default, the taper angle is 0, as shown in the following image.

Path & Guide Rail

Creates a sweep feature by sweeping a profile along a path and uses a guide rail to control scale and twist of the swept profile. The following image shows the options for the path and guide rail type.

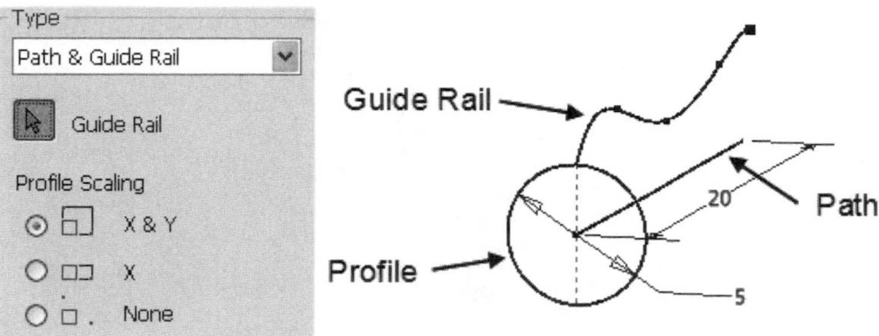

Figure 7.35

Guide Rail Select a guide curve or rail that controls the scaling and twist of the swept profile. The guide rail must touch the profile plane. If you project an edge to position the guide rail and the projected edge will not be the path, change the projected edge to a construction line.

Profile Scaling Specify how the swept section scales to meet the guide rail. The following image shows the path and the profile and guide rail from the previous image with the three different profile scaling options.

X and Y Scales the profile in both the X and Y directions as the sweep progresses.

X Scales the profile in the X direction as the sweep progresses. The profile is not scaled in the Y direction.

None Keeps the profile at a constant shape and size as the sweep progresses. Using this option, the rail controls only the profile twist and is not scaled in the X or Y direction.

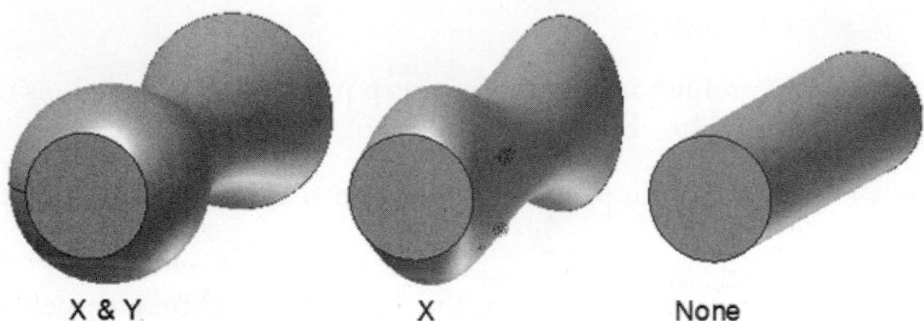

Figure 7.36

Path & Guide Surface

Creates a sweep feature by sweeping a profile along a path and a guide surface. The guide surface controls the twist of the swept profile. For best results, the path should touch or be near the guide surface.

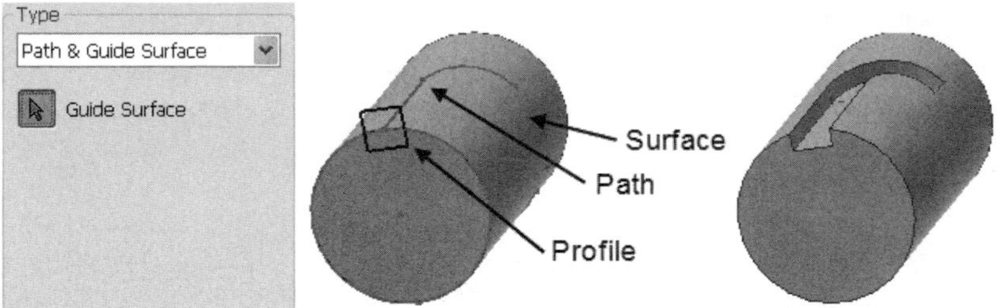

Figure 7.37

The following image on the left shows a sweep with just the path. Notice that the back inside edge of the cut is angled from the surface. The following image on the right shows the same sweep with a guide surface. Notice that the back inside edge is parallel to the top of the surface.

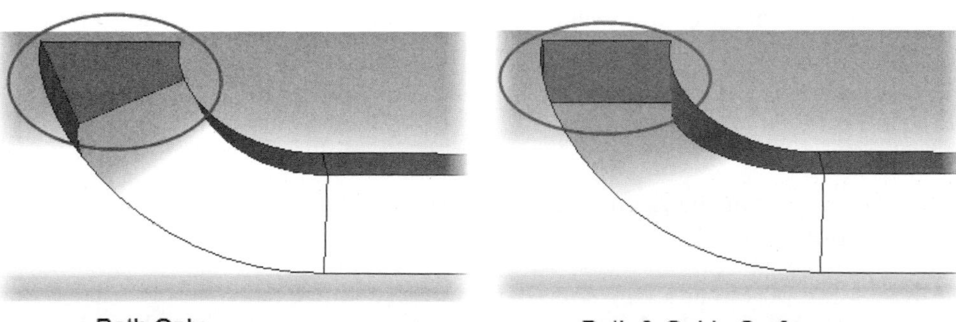

Figure 7.38

OPTIMIZE FOR SINGLE SELECTION

When this option is checked, Autodesk Inventor automatically advances to the next selection option once a single selection is made. Clear the checkbox when multiple selections are required.

PREVIEW

The button with the eyeglasses provides a solid preview of the sweep based on the current selections. If Preview is enabled and no preview appears in the graphics window, that the sweep feature will not be created.

CREATING A SWEEP FEATURE

In this section, you will learn how to create sweep features. You first need to have two unconsumed sketches. One sketch will be swept along the second sketch that represents the path. To create a sweep feature, follow these steps:

1. Create two unconsumed sketches: one for the profile and the other for the path. The profile and path must lie on separate nonparallel planes. It is recommended that the profile intersect the path. Use work planes to place the location of the sketches, if required. The sketch that you use for the path can be open or closed. Add dimensions and constraints to both sketches as needed. If needed, create a sketch that will be used as a rail, or create a surface to be used as a guide surface.
2. Click the Sweep tool on the Part Features panel bar.
3. The Sweep dialog box will appear. If two unconsumed sketches do not exist, Autodesk Inventor will notify you that two unconsumed sketches are required.
4. Click the Profile button, and then select the sketch that will be swept in the graphics window. If only one closed profile exists, this step is automated for you.
5. If it is not already depressed, click the Path button, and then select the sketch to be used as the path in the graphics window.
6. Select the type of sweep you are creating; Path, Path & Guide Rail, or Path & Guide Surface.
7. Define whether or not the resulting sweep will create a solid or a surface by clicking either the Solid or the Surface button in the Output area. Select the required options, as outlined in the previous descriptions.
8. If this is a secondary feature, click the operation that will specify whether material will be added or removed or if what is common between the existing part and the new sweep feature will be kept.
9. If you want the sweep feature to have a taper, click on the More tab, and enter a value for the taper angle.
10. Click the OK button to complete the operation.

EXERCISE 7-3: CREATING SWEEP FEATURES

In this exercise, you create a component with the sweep tool. Three sketches have been created, one each for the profile, the path, and the guide rail.

1. Open *ESS_E07_03.ipt* from the Chapter 07 folder.
2. Click the Sweep tool from the Part Features panel bar. The profile and path are automatically selected since only one closed profile (ellipse) and one open sketch (arc) are visible. Note that the profile or path can be open or closed, but if it is closed, you will need to manually select it; if the profile is open, a surface will be created.
3. Change the Orientation from Path to Parallel, as shown in the following image. Notice how the profile changes orientation.

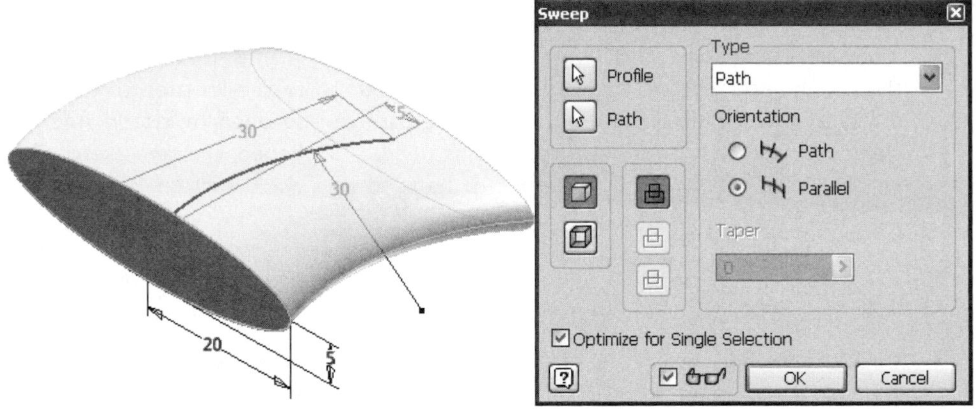

Figure 7.39

4. Click OK to create the sweep.
5. Turn on the visibility of the guide rail by moving the cursor over the entry Sketch Guide Rail in the browser. Right-click, and click Visibility from the menu.
6. Rotate the viewpoint, and notice that one end of the guide rail is coincident to the profile. To better see the geometry, you can change the display to Hidden Edge Display or Wireframe Display from the Standard toolbar.
7. Right-click in the graphics window, and click Home View from the menu.
8. Edit the sweep feature by moving the cursor over the "Sweep1" entry in the browser and double-click.
9. Change the type to Path & Guide Rail, and in the graphics window, select the spline, as shown in the following image.

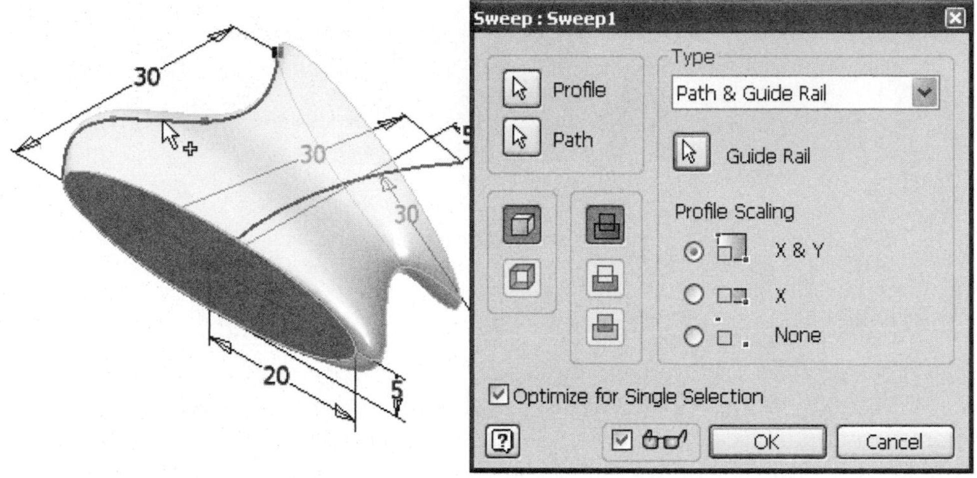

Figure 7.40

10. Notice that the Profile Scaling is set to X & Y. Click OK to create the part, as shown in the following image. Note that the sweep could have been completed as a Path & Guide Rail immediately.

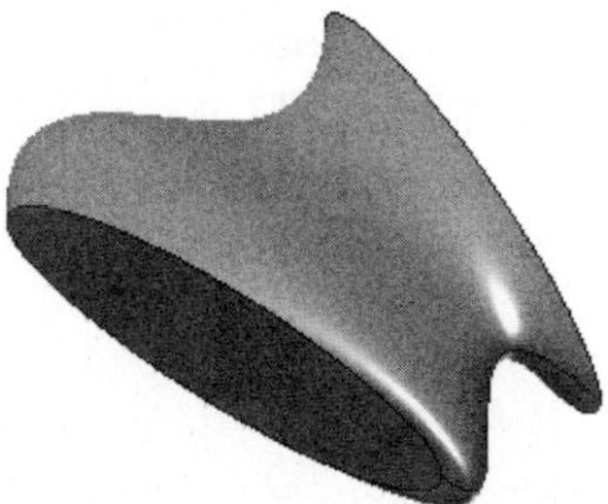

Figure 7.41

11. Rotate the model, and notice that the profile stretches in both X and Y directions.
12. Right-click in the graphics window, and click Isometric View from the menu.
13. Edit the sweep feature by moving the cursor over the "Sweep1" entry in the browser. Right-click, and click Edit Feature from the menu.
14. Change the Profile Scaling to X, as shown in the following image on the left, and click OK to update the sweep feature. The following image on the right shows the sweep feature looking at it head on.
15. Rotate the viewpoint, and notice that the profile stretches only in the X direction.

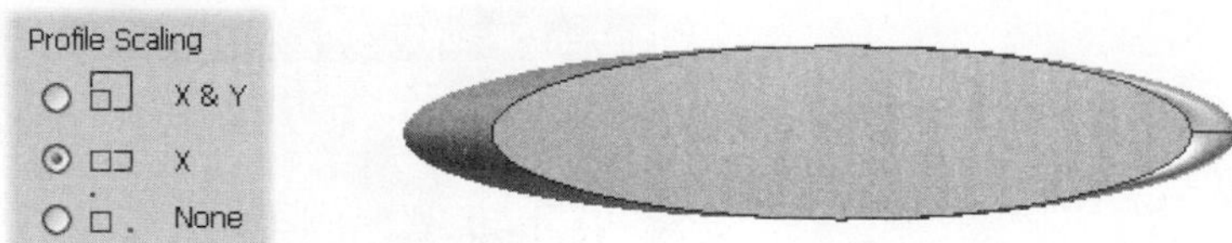

Figure 7.42

16. Close the file. Do not save changes. End of exercise.

3D SKETCHING

To create a sweep feature whose path does not lie on a single plane, you need to create a 3D sketch that will be used for the path. You can use a 3D sketch to define the path for a lip or to define the routing path for an assembly component, such as a pipe or duct work that crosses multiple faces on different planes. You need to define a 3D sketch in the part environment, and you can do this within an assembly or in its own part file. You can use the Autodesk Inventor adaptive technology during 3D sketch creation to create a path that updates automatically to reflect changes to referenced assembly components. In this section, you will learn strategies for creating 3D sketches.

3D Sketch Overview

When creating a 3D sketch, you use many of the sketching techniques that you have already learned with the addition of a few tools. 3D sketches use work points and model edges or vertices to define the shape of the 3D sketch by creating line or spline segments between them. You can also create bends between line segments. When creating a 3D sketch, you use a combination of lines, splines, fillet features, work features, constraints, and existing edges and vertices.

3D Sketch Environment

The 3D sketch environment is used to create 3D or a combination of both 2D and 3D curves. Before creating a 3D sketch, change the environment to the 3D sketch environment by clicking the 3D Sketch tool on the Standard toolbar beneath the 2D Sketch tool, as shown in the following image on the left. The panel bar's tools will change to the 3D Sketch tools, as shown in the center of the following figure. The tools that are unique to the 3D Sketch panel bar are explained throughout this section. While in the 3D Sketch environment, all features appear in the browser with a 3D sketch name. The following image on the right shows an example of a 3D Sketch in the browser. Once a feature uses the 3D sketch, it will be consumed under the new feature in the browser.

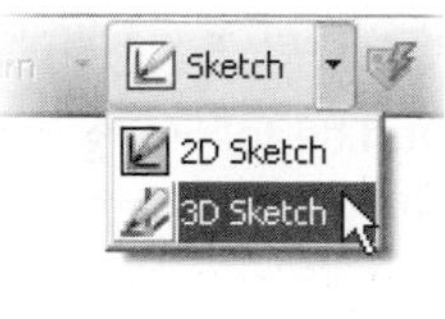

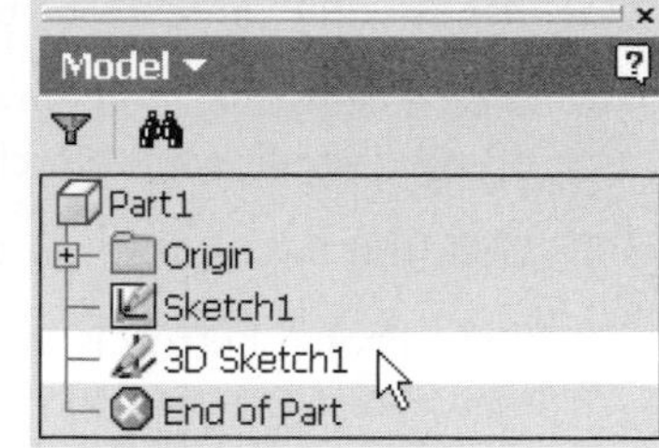

Figure 7.43

3D Path from Existing Geometry

One way to create a 3D path is to use existing geometry. If you wanted to create a lip on an existing part, for example, you would use the existing edges to define the path. You can include existing geometry by projecting part edges, vertices, and geometry from visible sketches into a 3D sketch. To use existing geometry to create a 3D path, follow these steps:

1. Activate the 3D Sketch environment by clicking the 3D Sketch tool on the Standard toolbar under the 2D Sketch tool.
2. Click the Include Geometry tool on the 3D Sketch panel bar, as shown in the following image on the left. This tool will add existing sketch geometry and existing part edges to a 3D sketch. The projected geometry is updated to reflect changes to the original geometry.
3. Click each of the model edges that you want to use for the 3D sketch. When finished, right-click and select Done from the menu, or press the ESC key on the keyboard. If you click an incorrect edge, you can delete it from the sketch manually. The following image on the right shows an example of including outside edges of a part in a 3D sketch.

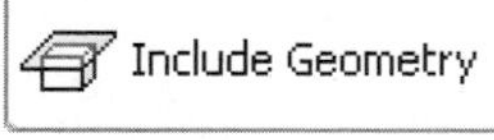

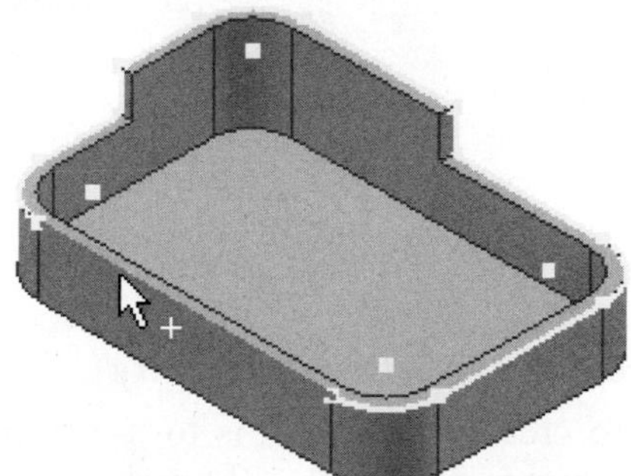

Figure 7.44

4. If a plane does not exist where you want to place the profile to be swept, create a work plane that defines the plane.

5. Use the 2D Sketch tool to create a 2D sketch on an existing plane or work plane.
6. Sketch, constrain, and dimension the profile that will be swept. The following image on the left shows a sketch created, constrained, and dimensioned on the work plane.
7. Click the Sweep tool on the Part Features panel bar.
8. If the Profile button is not already depressed in the Sweep dialog box, click it, and then select the sketch that will be swept in the graphics window.
9. Click the Path button, and then select the 3D sketch to use as the path in the graphics window.
10. Select the operation that will specify whether or not material will be added or removed or if what is common between the existing part and the new sweep feature will be kept.
11. If you want the sweep feature to have a taper, click on the More tab, and enter a value for the taper.
12. Click the OK button to complete the operation. The following image on the right shows the completed part.
13. To delete an edge that was included, exit the command, select the edge(s), and click the Delete key.

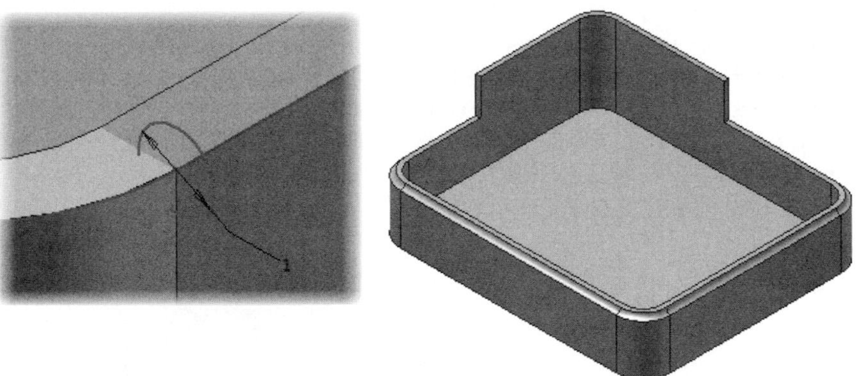

Figure 7.45

3D Sketch from Intersection Geometry

Another way to create a 3D path is to use geometry that intersects with the part. If the intersecting geometry defines the 3D path, you can use it. The intersection can be defined by a combination of any of the following: planar or nonplanar part faces, surface faces, a quilt, or work planes. To create a 3D path from an intersection, follow these steps:

1. Create the intersecting features that describe the desired path.
2. Change to the 3D Sketch environment by clicking the 3D Sketch tool on the Standard toolbar under the 2D Sketch tool.

3. Click the 3D Intersection Curve tool, shown in the following image on the left, on the 3D Sketch panel bar.
4. The 3D Intersection Curve dialog box appears, as shown in the following image in the middle.
5. Select the two intersecting features.
6. To complete the operation, click OK.

The following image on the right shows a 3D path being created on a cylindrical face at the edge defined by a work plane that intersects it at an angle.

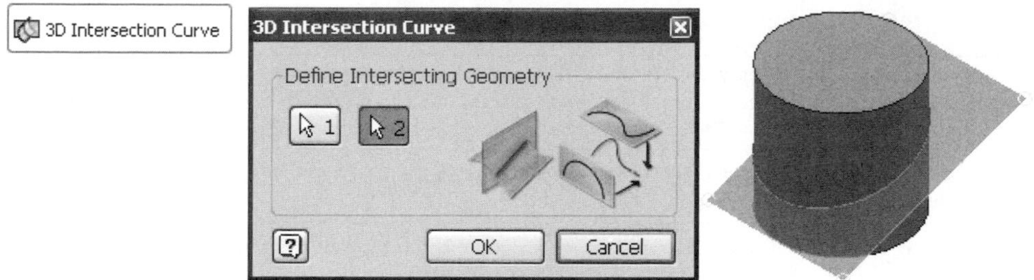

Figure 7.46

Project to Surface

While in a 3D sketch, you can project curves, 2D or 3D geometry, part edges, and points onto a face or onto selected faces of a surface or solid. To project curves onto a face, follow these steps:

1. Create the solid or surface onto which the curves will be projected.
2. Create the curves that will be projected onto the face(s).
3. Click the 3D Sketch tool from the Standard toolbar.
4. Click the Project Curve to Surface tool from the panel bar, as shown in the following image on the left.
5. The Project Curve to Surface dialog box will appear, as shown in the in the following image in the middle. The Faces button will be active. In the graphics window, select the face(s) onto which the curves will be projected.
6. Click the Curves Button, and then in the graphics window, select the individual objects to project. The following image on the right shows the face and curves selected.

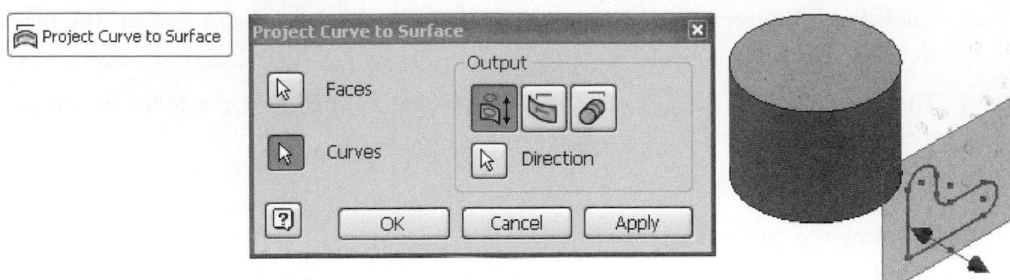

Figure 7.47

7. In the Output area, select one of the following options.

	Project along Vector	Specify the vector by clicking the Direction button and selecting a plane, edge, or axis. If a plane is selected, the vector will be normal (90°) from the plane. The curves will be projected in the direction of the vector.
	Project to Closest Point	Projects the curves onto the surface normal to the closest point.
	Wrap to Surface	The curves are wrapped around the curvature of the selected face or faces.

8. Click OK.

By default, the projected curves are linked to the original curve. If the original curves change size, they will be updated. To break the link, move the cursor into the browser over the name of the Projected to Surface entry, right-click, and select Break Link in the menu, as shown in the following image. You can also change the way the curves were projected by following the same steps as those used to break the link, but click Edit Projection Curve from the menu, and the Project Curve to Surface dialog box will appear. Make the changes as required.

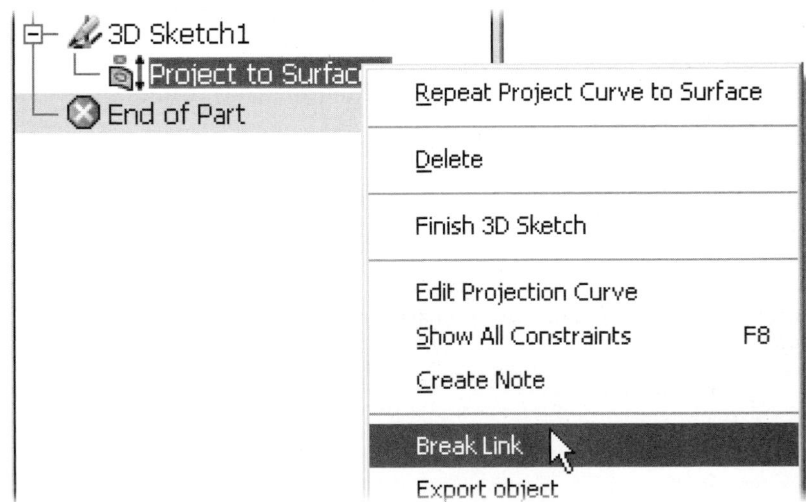

Figure 7.48

Constructed Paths

Another option for creating 3D paths is to define a path by creating work points at locations where the 3D path will intersect and then connecting the points using a 3D line or spline. Because this method of constructing a 3D path depends on work points, you should review the Creating Work Points section in chapter 4 to become comfortable creating and editing work points. The following is a brief review of the methods used to create work points:

- Click an endpoint or midpoint of an edge or sketch line. A work point is also generated automatically if you select a vertex while creating a 3D line.
- Click two edges or sketch lines that lie on the same plane. A work point is created at the intersection, or theoretical intersection, of the two.
- Click an edge or sketch line and plane. A work point is created at the intersection, or theoretical intersection, of the two.
- Click three nonparallel faces or planes, and a work point is created at their intersection or theoretical intersection.

You can also use grounded work points, but they are not associated dynamically with the part or any other work features including the original locating geometry. When you modify surrounding geometry, the grounded work point remains in the specified location.

There are two ways to create a 3D path. The first method is to use the line tool with the Inventor precise input dialog box. Follow these steps:

1. While in a 3D Sketch, click the line tool, and enter data into the Inventor Precise Input dialog box, as shown in the following image.

Figure 7.49

2. Parametric dimensions can be applied to the lines.
3. By default, a bend is not applied between 3D line segments, but this option can be toggled on and off by right-clicking while in the 3D line command and selecting or deselecting Auto-Bend on the menu, as shown in the following image on the left.
4. To set the default radius of the bend, Click Tools > Document Settings, and change the 3D Sketch Auto-Bend Radius setting on the Sketch tab, as shown in the following image in the middle.
5. To manually add a bend between two 3D lines, use the Bend tool on the 3D Sketch panel bar, shown in the following image on the right. In the 3D Sketch Bend dialog box, enter a value for the bend, and then select two 3D lines or the endpoint where they meet. Once the bend is placed, you can edit it by double-clicking on the dimension and entering a new bend radius value.

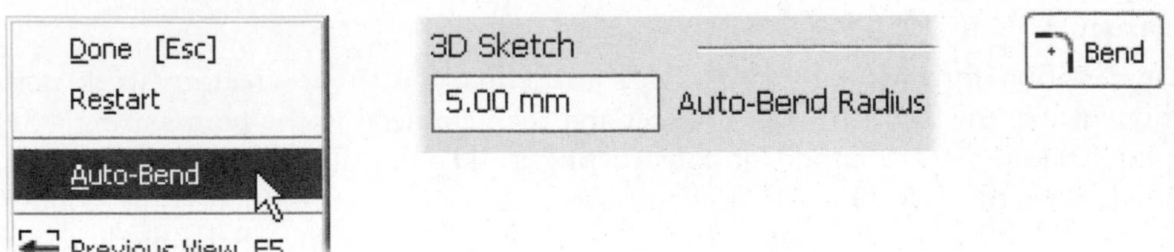

Figure 7.50

The second method to create a 3D path is to use work points and the line tool. The lines will be linked to the work points, and if a location changes, so will the line. Follow these steps:

1. Create work points and grounded work points, as needed, to define the 3D path.
2. Change to the 3D Sketch environment by clicking the 3D Sketch tool on the Standard toolbar under the 2D Sketch tool.
3. If you want the 3D path to place a bend automatically, you can set the size of the bend as described in Step 4 in the last procedure.
4. Click the Line tool on the 3D Sketch panel bar.
5. Select the work points in the order that the path will follow. By default, there is no bend between the line segments. To create a bend between line segments as they are created, right-click while in the 3D line command, and select Auto-Bend from the menu.
6. To manually add a bend between two 3D lines, use the Bend tool, as described in Step 5 of the previous procedure,
7. When you are done selecting points, right-click, and select Done from the menu.
8. Next create a new sketch that defines the desired profile that will be swept along the 3D path.
9. Click the Sweep tool, and create the 3D sweep by selecting the profile and path sketches.

The following image on the left shows a part with 3D lines and dimensions. The image on the right shows the completed sweep.

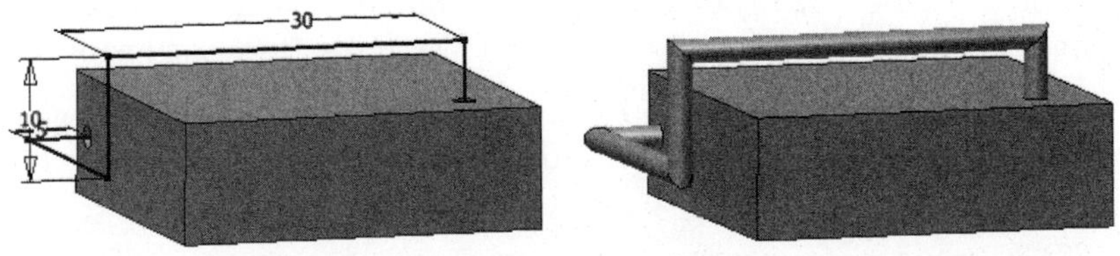

Figure 7.51

Splines

You can also create a spline between work points or vertices on a part, or you can use the precise input method, as outline above.

To create a 3D path using work points and a 3D spline, follow these steps:

1. Create work points and/or grounded work points, as needed, to define the 3D path.
2. Change to the 3D Sketch environment by clicking the 3D Sketch tool on the Standard toolbar under the 2D Sketch tool.
3. Click the Spline tool on the 3D Sketch panel bar, as shown in the following image on the left.
4. Select the work points in the order that the path will follow, as shown in the following image in the middle.
5. To exit the spline command, right-click, and select Continue from the menu. Then either press the ESC key or right-click and select Done from the menu.
6. You can add constraints between the spline and a part edge by clicking the desired constraint tool on the 3D Sketch panel bar. The following image on the right shows the available 3D sketch constraints.

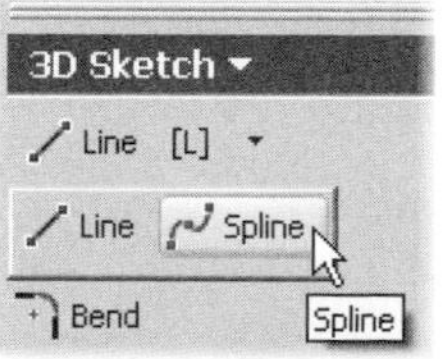

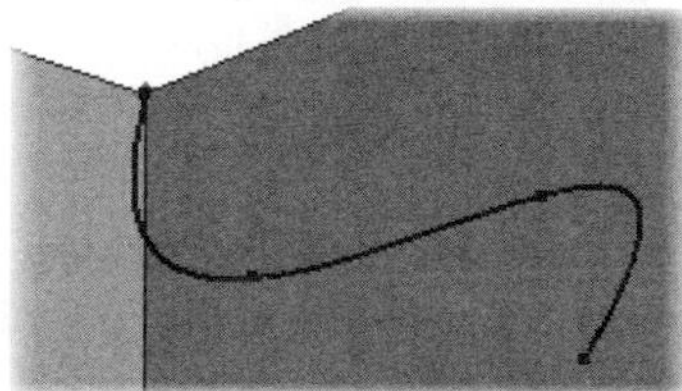

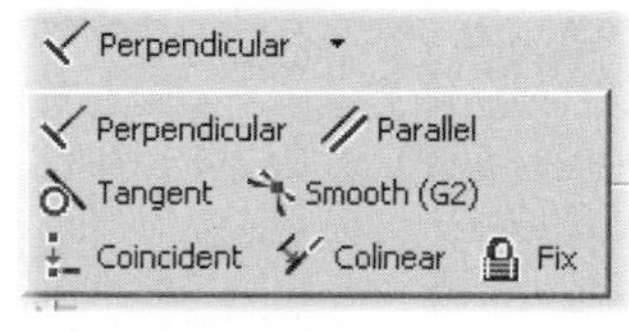

Figure 7.52

EXERCISE 7-4: 3D SKETCH—SWEEP FEATURES

In this exercise, you construct a 3D path entirely from existing part edges and use the Sweep tool to create a lip.

1. Open *ESS_E07_04.ipt* from the Chapter 07 folder.
2. From the Tools menu, click Application Options. In the Options dialog box: Click the Sketch tab.

- Uncheck the option Autoproject edges for sketch creation and edit.
- Click the OK button.

3. Right-click in the graphics window, and then click New 3D Sketch. You can also click the arrow next to Sketch on the Standard toolbar. Then click 3D Sketch to enter 3D Sketch mode.
4. Click the Include Geometry tool from the 3D Sketch panel bar.

5. Click the 13 edges that define the top outside edges, as shown in the following image. It is shown in wireframe display for clarity.

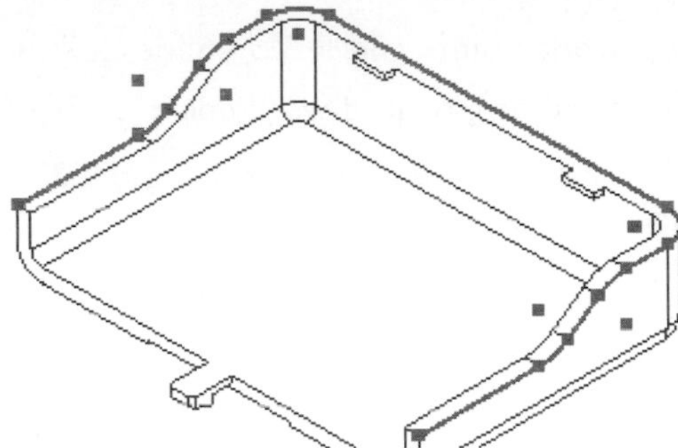

Figure 7.53

6. Right-click, and click Done.
7. Click the Return tool from the Standard toolbar.
8. Zoom in on the lower right-side of the part, as shown in the following image.
9. Right-click in the graphics window, and then click New Sketch from the menu. Click the front face of the part, as shown in the following image.

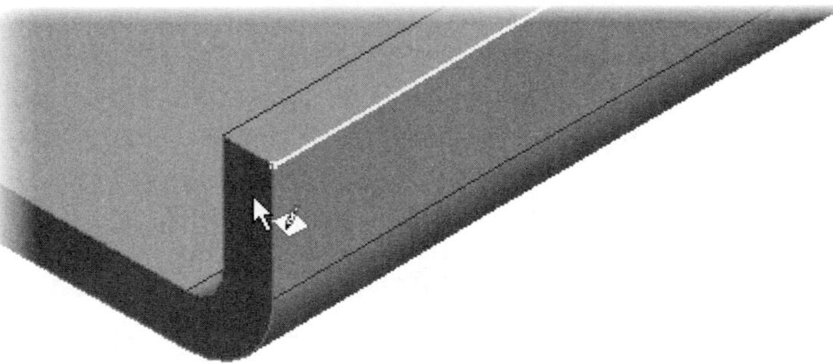

Figure 7.54

10. Start creating the 2D sketch by projecting the start point of the 3D path to the sketch plane. Click the Project Geometry tool.
    - Click the endpoint of the right 3D line.
    - Right-click, and then click Done. This point is used to orient a rectangular sketch that you create next.
11. Click the 2-Point Rectangle tool, and create a two point rectangle, as shown in the following image, starting at the projected point.
12. Add 1 mm horizontal and 2 mm vertical dimensions, as shown in the following image.

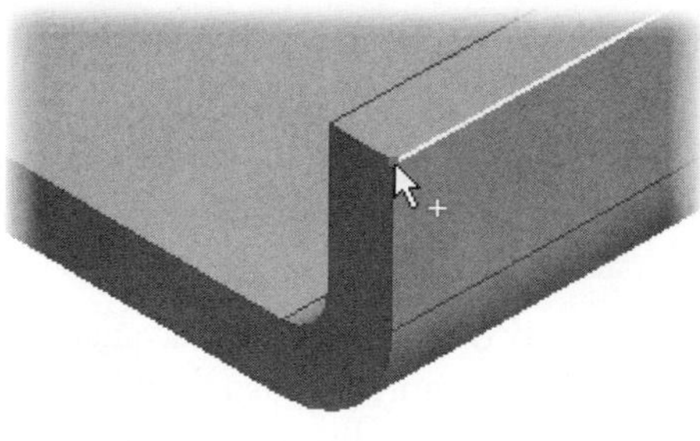

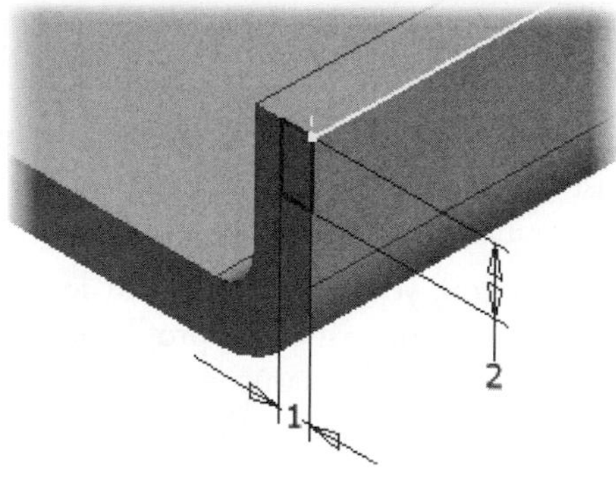

Figure 7.55

13. Click the Return tool.
14. Sweep the 2D profile along the 3D path using the Sweep tool. Click the Sweep tool from the panel bar.
15. The rectangular profile is selected for you, and the Path button is active. Click the 3D sketch geometry and click the Cut operation to remove material.
16. Click the OK button in the Sweep dialog box. The following image shows the completed part.
17. Use the Zoom and Rotate tools to examine the Sweep feature.

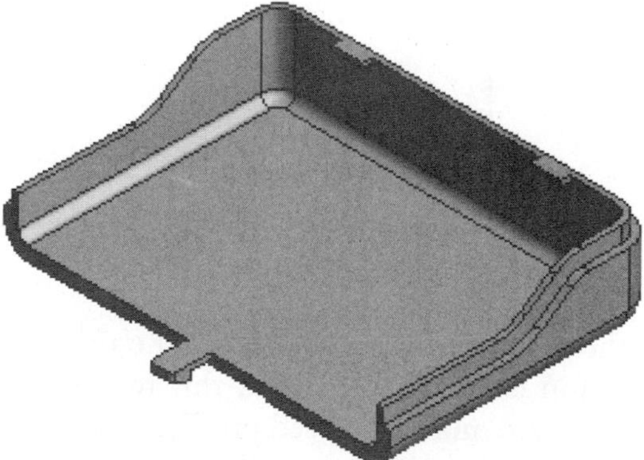

Figure 7.56

18. From the Tools menu, click Application Options. In the Options dialog box:
    - Click the Sketch tab.
    - Check the option Autoproject edges for sketch creation.
    - Click the OK button.
19. Close the file. Do not save changes. End of exercise.

COIL FEATURES

Using the Coil feature, you can easily create many types of helical, coil, or spiral geometry. You can create various types of springs by selecting different settings in the Coil dialog box. You can also use the Coil feature to remove or add a helical shape around the outside of a cylindrical part to represent a thread profile.

To create a coil, you need to have at least one unconsumed or shared sketch available in the part. This sketch describes the profile or shape of the coil feature and can also describe the coil's axis of revolution. If no unconsumed sketch is available, Autodesk Inventor will prompt you with an error message stating, "No unconsumed visible sketches on the part." After an unconsumed sketch is available, you can click the Coil tool from the Part Features panel bar, as shown in the following image on the left. The Coil dialog box appears. The following sections explain the tabs.

COIL SHAPE TAB

The Coil Shape tab allows you to specify the geometry and orientation of the coil, as shown in the following image.

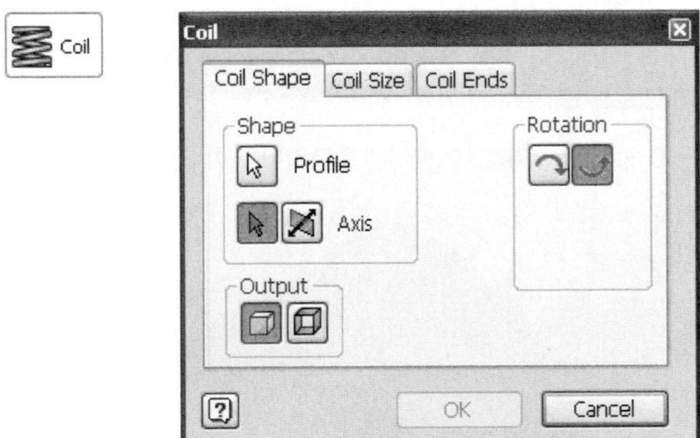

Figure 7.57

Profile Click to select the sketch that you will use as the profile shape of the coil feature. By default, the Profile button is shown depressed; this tells you that you need to select a sketch or sketch area. If there are multiple closed profiles, you will need to select the profile that you want to revolve. If there is only one possible profile, Autodesk Inventor will select it for you, and you can skip this step. If you select the wrong profile or sketch area, click the Profile button and select a new profile or sketch area. You can only use one closed profile to create the coil feature.

Axis Click to select a sketched line or centerline, a projected straight edge, or an axis about which to revolve the profile sketch. If selecting an edge or sketched centerline, it must be part of the sketch. If selecting a work axis, it cannot intersect the profile.

Flip Click to change the direction in which the coil will be created along the axis. The direction will be changed on either the positive or negative X or Y axis, depending upon the edge or axis that you selected. You will see a preview of the direction in which the coil will be created.

Rotation Click to specify the direction in which the coil will rotate, either clockwise or counterclockwise.

Operation The operation buttons are the column of buttons, along the center of the dialog box, that will appear if a base feature exists, as shown in the following image. The operation buttons are only available if the Coil feature is not the first feature in the part. By default, the Join operation is selected. You can select the other operations to either add or remove material from the part using the Join or Cut options or to keep what is common between the existing part volume and the completed coil feature using the Intersect option.

- *Join:* Adds material to the part.
- *Cut:* Removes material from the part.
- *Intersect:* Keeps what is common to the part and the coil feature.

Output Click to create a solid or a surface.

COIL SIZE TAB

The Coil Size tab, as shown in the following image on the left, allows you to specify how the coil will be created. You have various options for the type of coil that you can create. Based on the type of coil that you select, the other parameters for Pitch, Height, Revolution, and Taper will become active or inactive. Specify two of the three available parameters, and Autodesk Inventor will calculate the last field for you.

Type Select the parameters that you want to specify: Pitch and Revolution, Revolution and Height, Pitch and Height, or Spiral. If you select Spiral as the Coil Type, only the Pitch and Revolution values are required.

Pitch Type in the value for the height to which you want the helix to elevate with each revolution.

Revolution Specify the number of revolutions for the coil. A coil cannot have zero revolutions, and fractions can be used in this field. For example, you can create a coil that contains 2.5 turns. If end conditions are specified, as mentioned in the Coil Ends tab section, the end conditions are included in the number of revolutions.

Height Specify the height of the coil. This is the total coil height as measured from the center of the profile at the start to the center of the profile at the end.

Taper Type an angle at which you want the coil to be tapered along its axis.

Note: A spiral coil type cannot be tapered.

COIL ENDS TAB

The Coil Ends tab, as shown in the following image on the right, lets you specify the end conditions for the start and end of the coil. When selecting the Flat option, the helix, not the profile that you selected for the coil, is flattened. The ends of a coil feature can have unique end conditions that are not consistent between the start and the end of the coil.

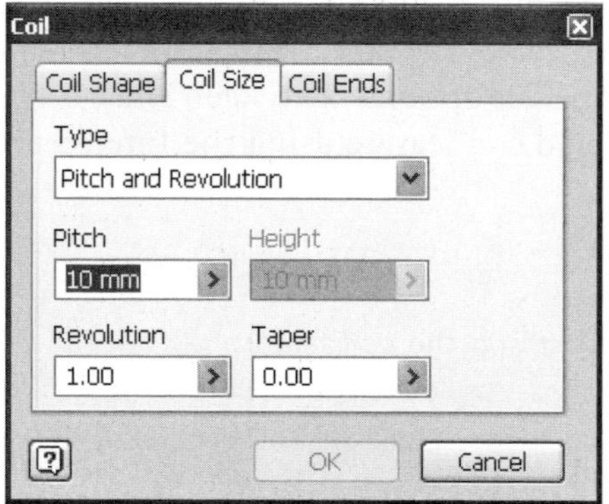

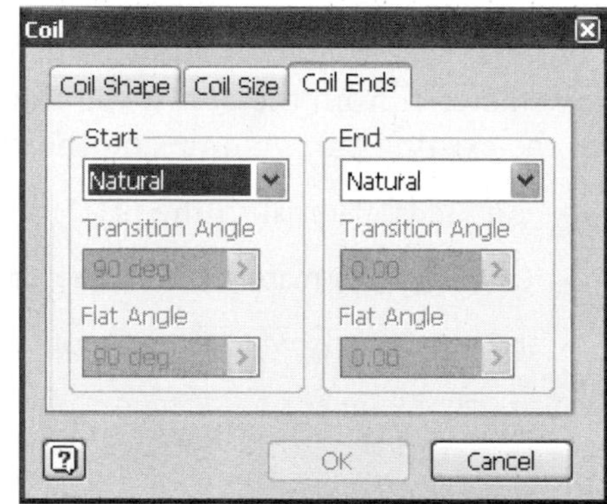

Figure 7.58

Start Select either Natural or Flat for the start of the helix. Click the down arrow to change between the two options.

End Select either Natural or Flat for the end of the helix. Click the down arrow to change between the two options.

Transition Angle This is the rotational angle, specified in degrees, in which the coil achieves the coil start or end transition. It normally occurs in less than one revolution.

Flat Angle This is the rotational angle, specified in degrees, that describes the amount of flat coil that extends after the transition. It specifies the transition from the end of the revolved profile into a flattened end.

The following image shows a coil created as the base feature. The image on the left shows the coil in its sketch stage; the rectangle will be used as the profile, and the centerline will be used as the axis of rotation. The finished part, as shown on the right, shows the coil with flat ends.

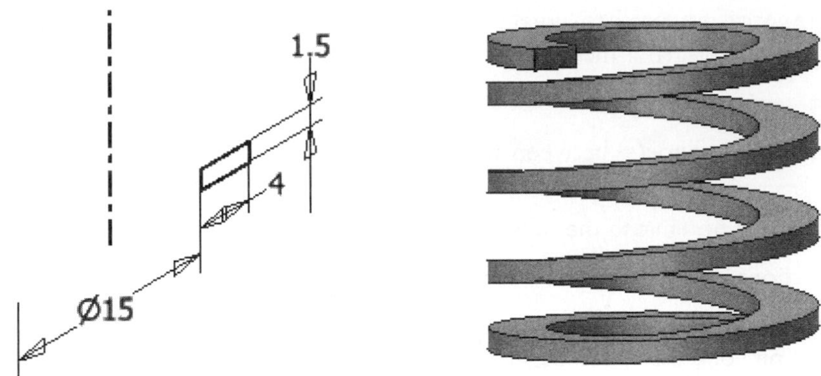

Figure 7.59

The following image shows a coil created as a secondary feature. The image on the left shows the coil in its sketch stage with the Coil dialog box displayed. The sketch, a rectangle, is drawn tangent to the cylinder. The rectangle will be used as the profile, and the work axis will be used as the axis of rotation. The finished part, as shown on the right, shows the coil with the flat end on the top.

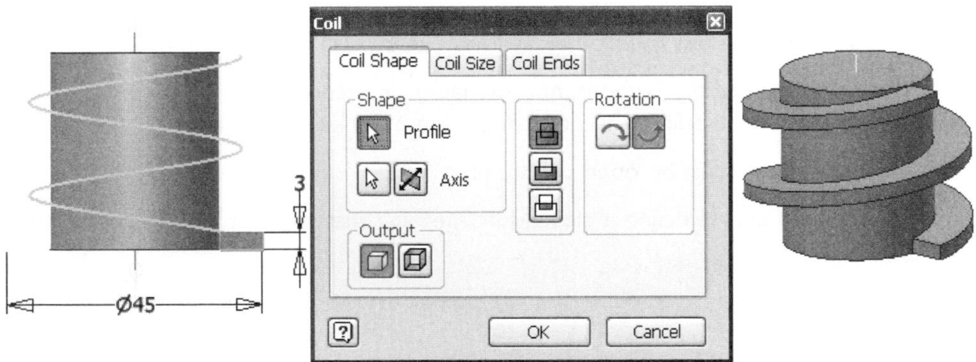

Figure 7.60

LOFT FEATURES

The Loft tool creates a feature that blends a shape between two or more different sections or profiles. A point can also be used to define the beginning and ending section of the loft. Loft features are used frequently when creating plastic or molded parts. A loft is similar to a sweep, but it can have multiple sections and rails. Many of these types of parts have complex shapes that would be difficult to create using standard modeling techniques. You can create loft features that blend between two or more cross-section profiles that reside on different planes. You can also control the area of a specific section in the loft. You can use a rail, multiple rails, or a centerline to define a path(s) that the loft will follow. There is no limit to the number of sections or rails that you can include in a loft feature. Four types of geometry are used to create a loft: sections, rails, centerlines, and points. The following sections describe these types of geometry.

SECTIONS

You can define the shape(s) between which the loft will blend. The following rules apply to sections:

1. There is no limit to the number of sections that you can include in the loft feature.
2. Sections do not have to be sketched on parallel planes.
3. You can define sections with 2D sketches (planar), 3D sketches (nonplanar), and planar or nonplanar faces, edges on a part, or points.
4. All sections must be either open or closed. You cannot mix open and closed profile types within the same loft operation. Open profiles result in a lofted surface.

RAILS

You can define rails by the following elements: 2D sketches (planar), 3D sketches (nonplanar), or part faces and edges. The following rules apply to rails:

1. There is no limit to the number of rails that you can create or include in the loft feature.
2. Rails must not cross each other and must not cross mapping curves.
3. Rails affect all of the sections, not just faces or sections that they intersect. Section vertices without defined rails are influenced by neighboring rails.
4. All rail curves must be open or closed.
5. Closed rail curves define a closed loft, meaning that the first section is also the last section.
6. No two rails can have identical guide points, even though the curves themselves may be different.
7. Rails can extend beyond the first and last sections. Any part of a rail that comes before the first section or after the last is ignored.
8. You can apply a 2D or 3D sketch tangency or smooth constraint between the rail and the existing geometry on the model.

CENTERLINE

A centerline is treated like a rail, and the loft sections are held normal to the centerline. When a centerline is used, it acts like a path used in the sweep feature, and it maintains a consistent transition between sections. The same rules apply to centerlines as to rails except that the centerline does not need to intersect sections and only one centerline can be used.

POINTS

A point can be used to help define the loft. The following rules apply to points used for a loft profile:

1. A point can be used to define the start or end of the loft.
2. An origin point, sketch point, center point, edge point or work point can be used.

MAP POINTS

You can map points to help define how the sections will blend into each other. The following rules apply to mapping points:

1. Two mapping points constitute a mapping curve between loft profiles.
2. A mapping point can lie anywhere on the loft profile.
3. The points in a set must either follow the order of the sections, with the first point on the first section, and so on, or follow the reverse of this order, with the first point on the last section, and so on.
4. Mapping curves must not cross other mapping curves or rail curves.
5. Mapping curves can share their first and/or last points with adjacent mapping curves.

CREATE A LOFT

To create a loft feature, follow these steps:

1. Create the profiles or points that will be used as the sections of the loft. Use work features, sketches, or projected geometry to position the profiles, if required.
2. Create rails or a centerline that will be used to define the direction or control the shape between sections.
3. Click the Loft tool on the Part Features panel bar, as shown in the following image on the left.
4. On the Curves tab of the Loft dialog box, the Sections option will be the default. In the graphics window, click the sketches, face loops, or points in the order in which the loft sections will blend.
5. If rails are to be used in the loft, click Click to add in the Rails section of the Curves tab, and then click the rail or rails.
6. If needed, change the options for the loft on the Conditions and Transition tabs.

The following image shows a loft created from two sections and a centerline rail. The image on the left shows two sections and a centerline in their sketch stages, shown in the top view. The image on the right shows the completed loft.

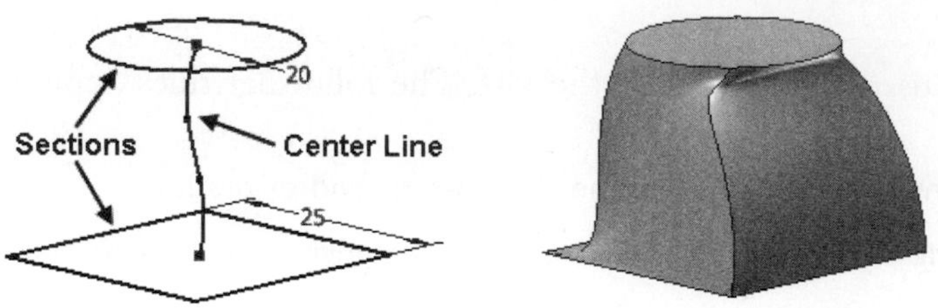

Figure 7.61

The following sections explain the options in the Loft dialog box and on its tabs.

CURVES TAB

The Curves tab, as shown in the following image on the right, allows you to select which sketches, part edges, part faces, or point will be used as sections, to select whether or not a rail or centerline will be used, and to determine the output condition.

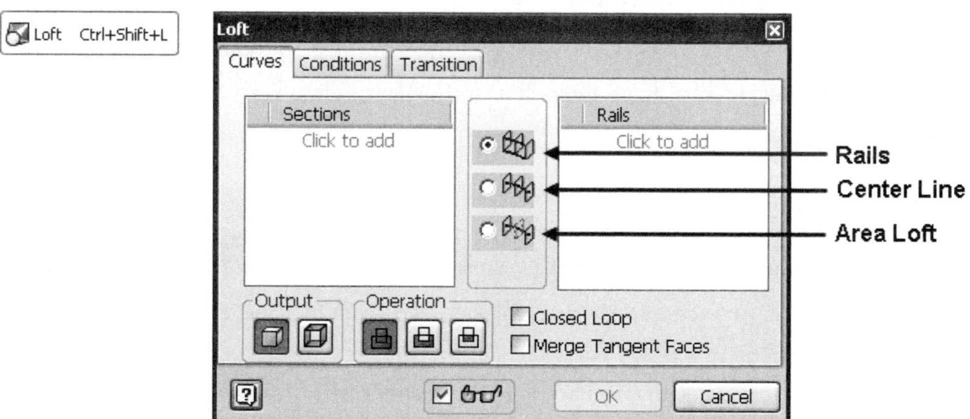

Figure 7.62

Sections Select two or more 2D or 3D sketches or part faces that make up the loft feature. The sketches and part faces can be on any plane or face.

Rails Select a sketch or sketches to be used as rails.

Center Line Select a sketch to be used as the centerline.

Area Loft Select a sketch to be used as the centerline, and then click a point on the centerline to define the area of the profile at the selected point. After picking a point, the Section Dimensions dialog box appears, as shown in the following image. You define its position either as a proportional or absolute distance, and you can define the section's size by area or scale factor related to the area of the original profile.

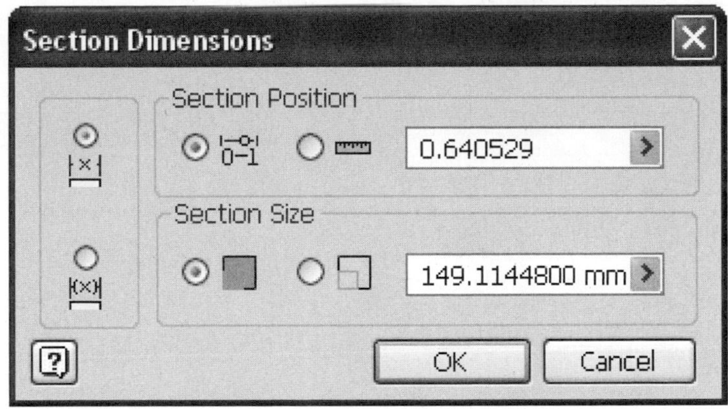

Figure 7.63

Output Select whether the loft feature will be a solid or a surface. Open sketch profiles selected as loft sections will define a lofted surface.

Operation Select an Operations button to add or remove material from the part, using the Join or Cut options, or to keep what is common between the existing part and the completed loft using the Intersect option. By default, the Join operation is selected.

Closed Loop Click to join the first and last sections of the loft feature to create a closed loop.

Merge Tangent Faces Click to join tangent faces on the completed loft into a single face.

CONDITIONS TAB

The Conditions tab, as shown in the following image, allows you to control the tangency condition, boundary angle, and weight condition of the loft feature. These settings affect how the faces on the loft feature relate to geometry at the start and end profiles of the loft. This may be existing part geometry or the plane or work plane containing the loft section sketch.

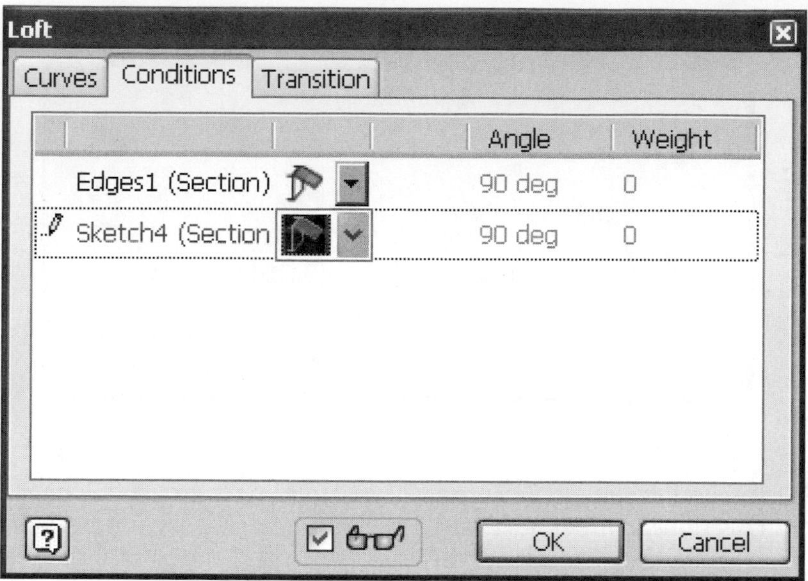

Figure 7.64

Conditions The column on the left lists the sketches and points specified for the sections. To change a sketch's condition, click on its name, and then select a condition option.

Condition Boundaries Sketches Two boundary conditions are available when the first or last section is a point, as shown in the following image.

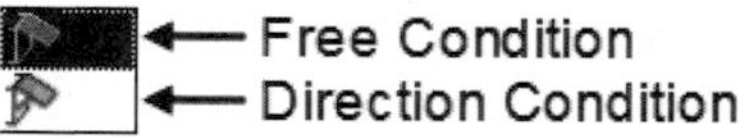

Figure 7.65

Free (top button): With this option, there is no boundary condition, and the loft will blend between the sections in the most direct fashion.

- *Direction* (bottom button): This option is only available when the curve is a 2D sketch. When selected, you specify the angle at which the loft will intersect or transition from the section.

Condition Boundaries Face Loop When a face loop or edges from a part are used to form a section, as shown in the following image on the left, three conditions are available, as shown in the following image on the right.

Figure 7.66

Free (top button): With this option, there is no boundary condition, and the loft will blend between the sections in the most direct fashion.

- *Tangent Condition* (middle button): With this option, the loft will be tangent to the adjacent section of face.
- *Smooth (G2) Condition* (bottom button): With this option, the loft will have a curvature continuity to the adjacent section of face.

Condition Boundaries Point Three conditions are available when a point is selected for the loft profile, as shown in the following image.

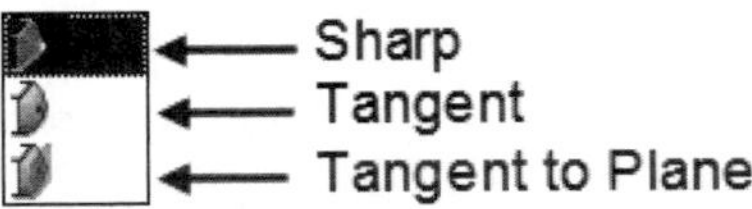

Figure 7.67

Sharp (top button): With this option, there is no boundary condition, and the loft will blend from the previous section to the point in the most direct fashion.

- *Tangent* (middle button): When selected, the loft transitions to a rounded or domed shape at the point.
- *Tangent to plane* (bottom button): When selected, the loft transitions to a rounded dome shape. You select a planar face or work plane to be tangent to. This option is not available when a centerline loft is used.

Angle This option is enabled for a section only when the boundary condition is Tangent or Direction. The default is set to 90° and is measured relative to the profile plane. The option sets the value for an angle formed between the plane that the profile is on and the direction to the next cross-section of the loft feature. Valid entries range from 0.0000001° to 179.99999°.

Weight The default is set to 0. The weight value controls how much the angle influences the tangency of the loft shape to the normal of the starting and ending profile. A small value will create an abrupt transition, and a large value creates a more gradual transition. High weight values could result in twisting the loft and may cause a self-intersecting shape.

TRANSITION TAB

The Transition tab, as shown in the following image with the Automatic mapping box unchecked, allows you to specify point sets. A point set is used to define section point relationships and control how segments blend from one section to the segments of the adjacent sections. Points are reoriented or added on two adjacent sections.

Point Set The name of the point set appears here.

Map Point The corresponding sketch for the selected point set appears here.

Position The location of the selected map point appears here. You can modify the position by entering a new value or by dragging the point to a new location in the graphics window.

To modify the default point sets or to add a point set, follow these steps:

1. Click on the Transition tab, and clear the Automatic Mapping box. The dialog box will populate the automatic point data for each section. The list is sorted in the order in which the sections were specified on the Curve tab.
2. To modify a point's position, select the point set in which the point is specified. When you select its name, it will become highlighted in the graphics window.
3. Click in the Position section of the map point that you want to modify, and enter a new value or drag the point to a new location in the graphics window.
4. To add a point set, click Click to add in the point set area.
5. Click a point on the profile of two adjacent sections. As you move the cursor over a valid region of the active section, a green point appears. As the points are placed, they are previewed in the graphics window, as shown in the following image on the right.
6. You can modify the new point set in a similar way to the default point set.

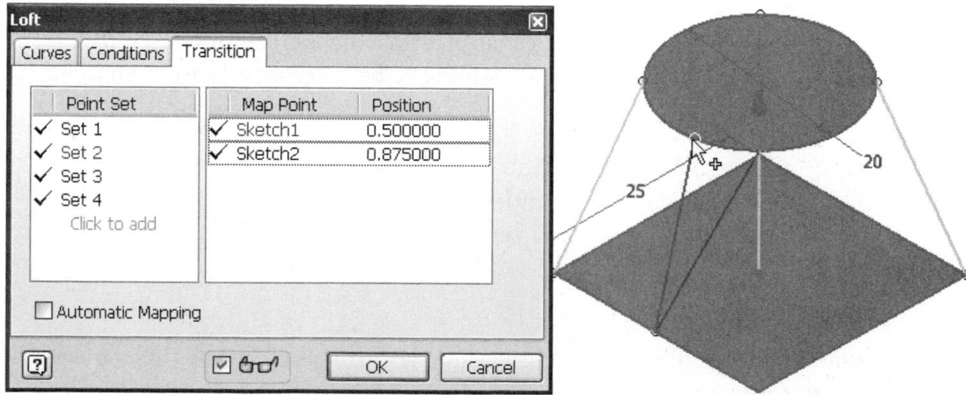

Figure 7.68

EXERCISE 7-5: CREATING LOFT FEATURES

In this exercise, you use loft options and controls to define the shape of a razor handle.

1. Open *ESS_E07_05.ipt* from the Chapter 07 folder.
2. In the browser, double-click Sketch1 to edit the sketch, and then use the Look At tool to change the view normal to the sketch, which will rotate the view to the back of the current view.
3. Click the Point, Center Point tool in the 2D Sketch panel bar.
4. Click a point coincident with the spline near the bottom of the curve, making sure that the coincident glyph is displayed as you place the point.

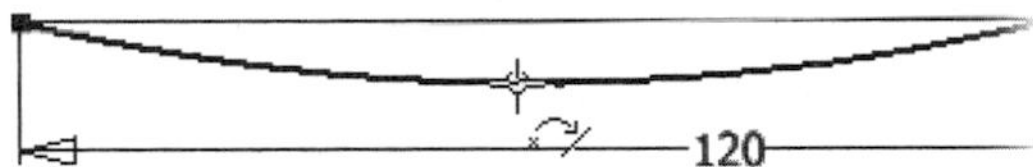

Figure 7.69

5. Draw two construction lines coincident with the sketched point and the nearest spline points on both sides of the Point that you just created.

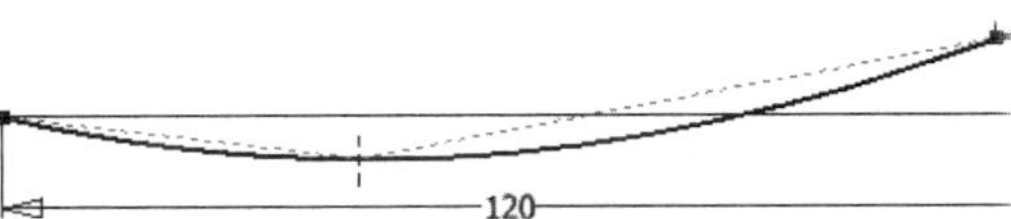

Figure 7.70

6. To parametrically position the sketched point midway between the spline points, place an equal constraint between the two construction lines.
7. Change the 120 mm dimension to **130 mm**, and verify that the sketched point moves along the spline to maintain its position midway between the spline points.
8. Click Return to finish editing the sketch.
9. Change to the Home View.
10. Click the Work Plane tool. To create a work plane perpendicular to the spline, click the Center Point you just created, and then click the spline, not the construction lines.

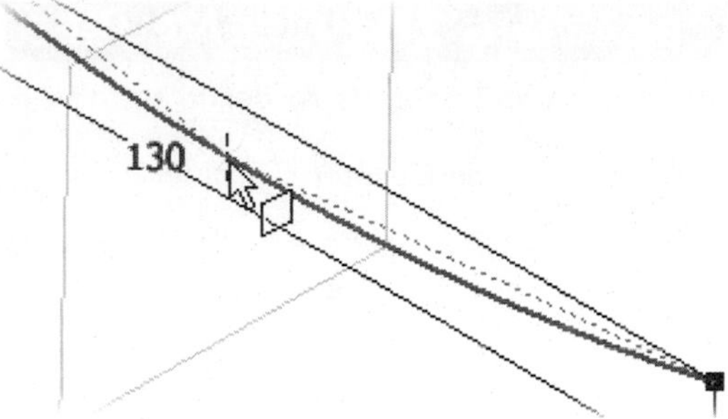

Figure 7.71

11. Next create a loft profile section on the new work plane. Create a new sketch on the new work plane.

12. Project the Center Point that you created in an earlier step onto the sketch.

13. Draw an ellipse with its center coincident with the projected point and the second point so the ellipse's axis is horizontally constrained to the Point. Click a third point, as shown in the following image.

14. Place **4 mm** and **9 mm** dimensions to control the ellipse size, as shown in the following image.

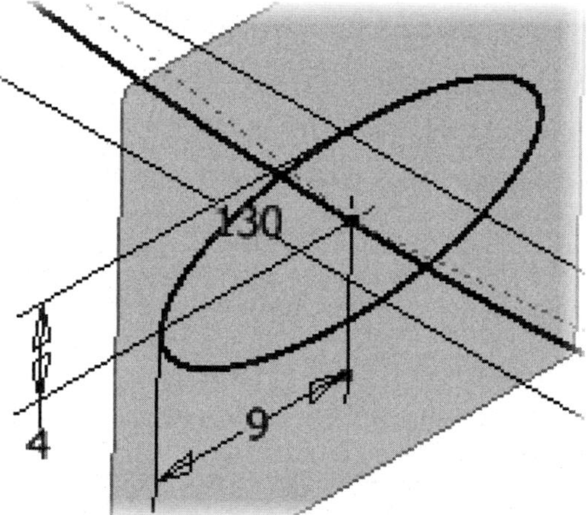

Figure 7.72

15. Click Return to finish the sketch.

16. Click the Loft tool from the panel bar.

For the first section, click the concave 3D face, as shown in the following image.

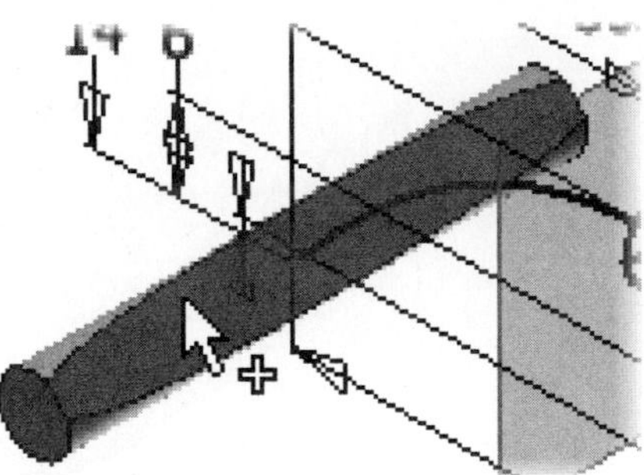

Figure 7.73

17. Click the other three profile sections in left to right order, and then click the point on the right end of the spline. Use the Select other tool to select only the point and not the spline.
18. Click the Center Line option in the dialog box, and then select the spline. When finished, the preview looks like the following image on the left.

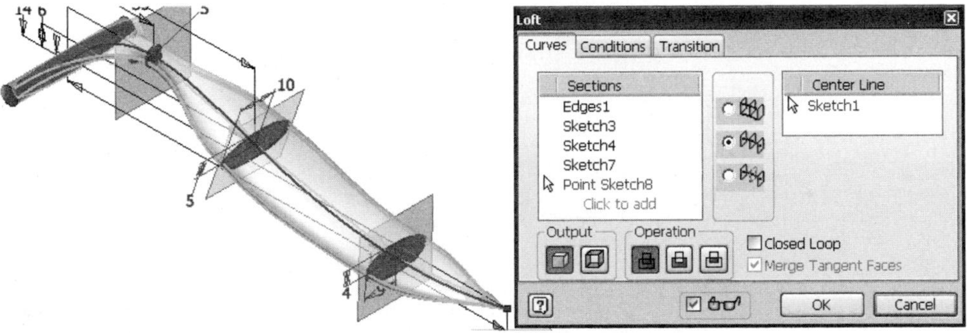

Figure 7.74

19. Click OK to create the loft.
20. Turn off the visibility of all four work planes by pressing the ALT and] keys, and then use the Rotate and other view tools to examine the loft. Notice that the end is sharp at the point.

EXERCISES

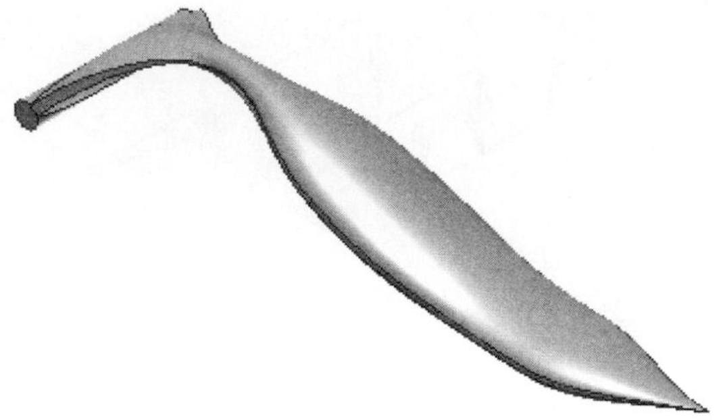

Figure 7.75

21. To edit the loft move the cursor over the entry Loft1 in the browser, and double-click. Click the Conditions tab, and change the Point entry to a Tangent condition, as shown in the following image on the left.

22. Click OK to update the loft. The right side should resemble the following image on the right.

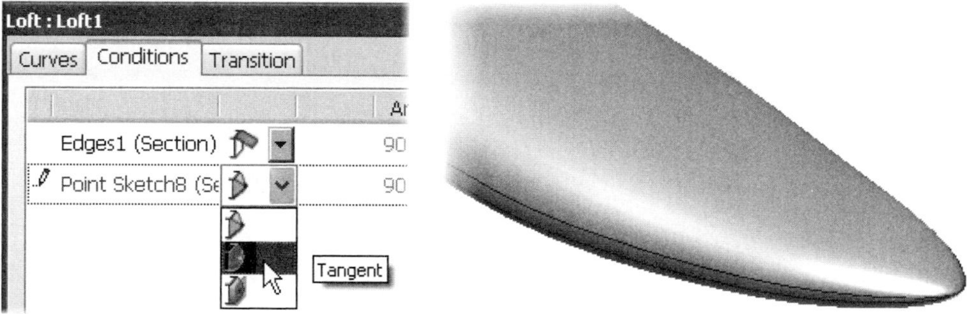

Figure 7.76

23. Next add another section that defines the section's area. In the browser, move the cursor over the entry Loft1, right-click, and click Edit Feature.

- From the Curves tab, click Area Loft option.
- In the Placed Sections area, click Click to add.
- In the graphics window, click a point on the spline or the centerline, as shown in the following image.

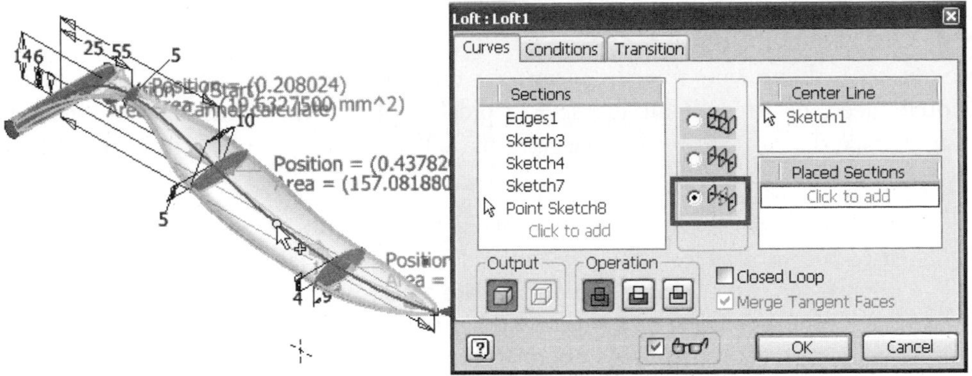

Figure 7.77

24. In the Section Dimensions dialog box:

- Change the Section Position proportional distance to **.5**.
- Change the Section Size area to **250 mm^2**.
- Click OK.
- In the graphics window, notice the values of each section.

25. In the Loft dialog box, click OK. The following image on the right shows the updated loft.

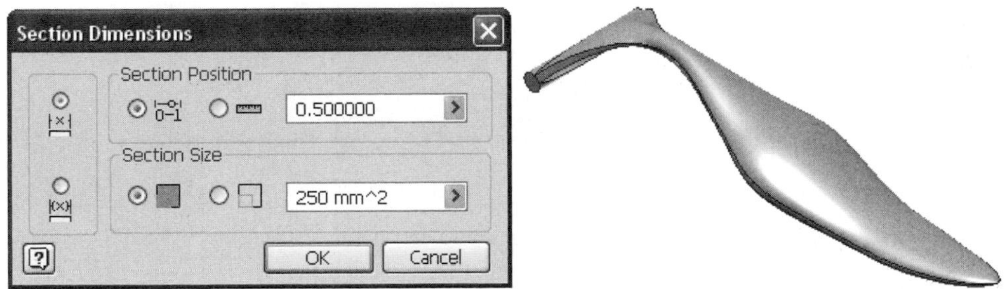

Figure 7.78

26. Close the file. Do not save changes. End of exercise.

SPLIT PART AND SPLIT FACE

The Split tool allows you to split a part by removing one portion of the part, to split individual faces, or to split all faces. A typical application of this tool is to allow the creation of face drafts to the split faces of a part. You can use the Split tool to perform the following:

- Split an entire part by sketching a parting line or by placing a work plane and then using it to cut material from the part in the direction you specify. The side that is removed is suppressed rather than deleted. To create a part with the other side removed, edit the split feature, redefine it to keep the other side, save the other half of the part to its own file using the Save Copy As option, or create a derived part.
- Split individual faces by using a surface, sketching a parting line, or placing a work plane, and then selecting the faces to split. You can edit the split feature and modify it to add or remove the desired part faces to be split.

The Split tool is located on the Part Features panel bar. Once selected, the Split dialog box will appear, as shown in the following image.

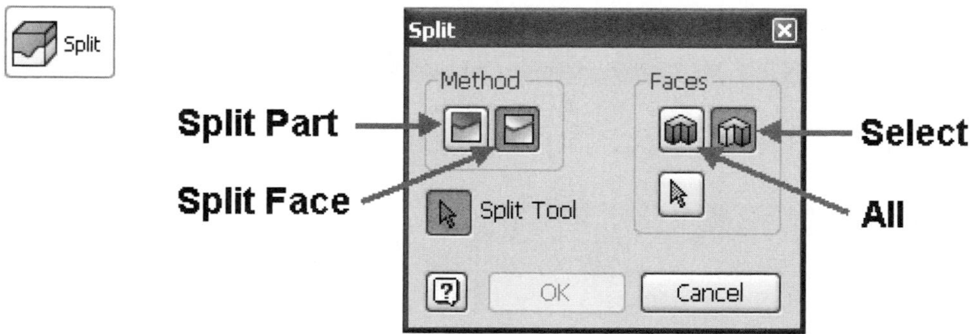

Figure 7.79

The Split dialog box contains the following sections:

METHOD

Split Part Click the left button in the Method section to split an entire part using a selected work plane, surface, or sketched geometry to cut or remove material. If you select this option, you are prompted to choose the direction of the material that you want to remove.

Split Face Click the right button in the Method section to split individual faces of a part by selecting a work plane, surface, or sketched geometry, and then selecting the faces to split. The split face method can split individual faces or all the faces on the part. When you select the split face method, the Remove area in the dialog box will be replaced with the Faces area, which allows you to select all or individual faces.

Remove The option to remove material is only available when you use the split part method. After splitting a part, you can retrieve the cut material by editing the split feature and clicking to remove the opposite side or by deleting the split feature.

FACES

The Faces option is available only when you select the split face method.

All Click this button to split all faces of the part that intersect the Split tool.

Select Click this button to enable the selection of specific faces that you want to split. After clicking the Select button, the Faces to Split tool becomes active.

Faces to Split Click this button, and select the faces of the part that you want to split. Any faces selected will be split where they intersect the Split tool.

Split Tool Click this button, and select a surface, work plane, or sketch that you want to use to split the part.

EXERCISE 7-6: SPLITTING A PART

In this exercise, you use split features to complete a model of a wireless phone handset. Three mating parts of the phone will be created from a single model.

1. Open *ESS_E07_06.ipt* from the Chapter 07 folder.
2. Next use a work plane to split the handset into upper and lower halves. Click the Split tool in the Part Features panel bar.

- Click Work Plane1 in the graphics window or browser.
- In the Split dialog box, click the split part method.
- Click the left Side to Remove button so the dialog box and preview resemble the following image.

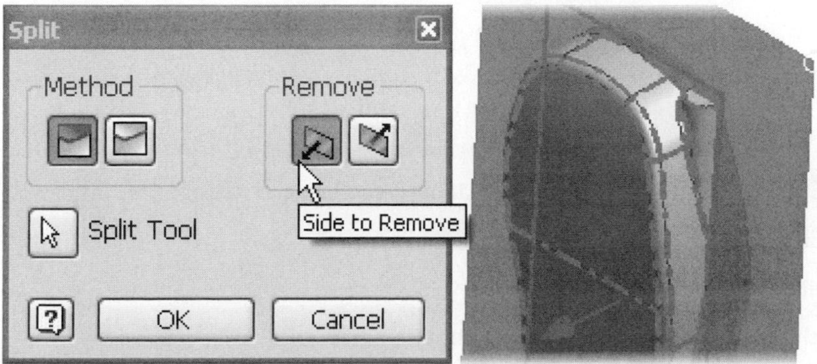

Figure 7.80

3. Click OK to remove the top half.
4. Right-click Work Plane1 in the graphics window or browser, and click Visibility from the menu to turn off its visibility.
5. Next use a sketch to split the part again to create the battery cover. From the browser, turn on the visibility of the YZ plane from the Origin folder.
6. Create a new sketch on the YZ plane.
7. Click the Look At tool, and click the YZ plane.
8. Click the Right face in the Home View and rotate the view 90 degrees as shown in the following image.

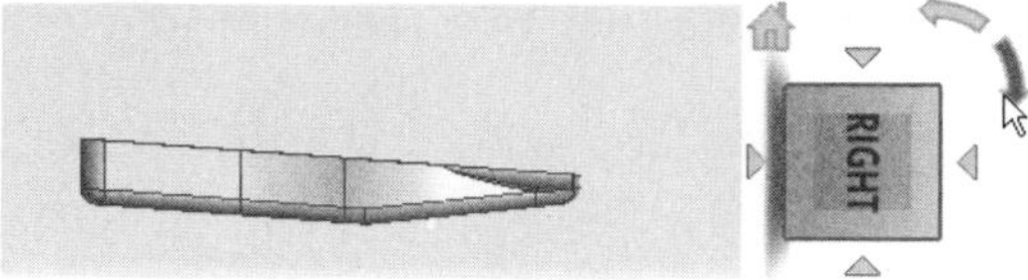

Figure 7.81

9. In the graphics window, right-click, and select Slice Graphics.
10. Click the Project Cut Edges tool from the 2D Sketch panel bar. Notice that the edges of the sliced graphic are automatically projected and displayed in a different color.
11. Press the L key to start the Line tool, and draw two lines, as shown in the following image. The sketch lines should be parallel to both the projected left vertical edge and the lower-left angled line.
12. Click the General Dimension tool. Place **8 mm** and **90 mm** dimensions, as shown in the following figure. Select the lines to automatically align them. Edit or apply constraints as needed.

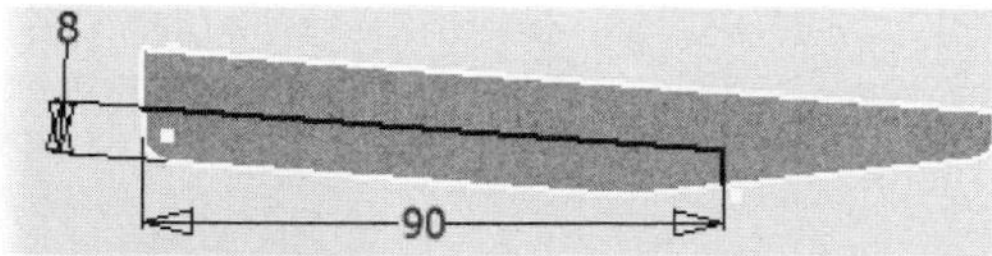

Figure 7.82

13. Click the Return tool on the Standard toolbar, and change the display to Wireframe.
14. Click the Split tool.
  - Select the split part method.
  - Select the split tool arrow if it's not selected.
  - Select the profile you drew in the previously.

- In the Remove area of the dialog box, click the right Side to Remove button to retain the battery cover, and remove the rest of the part.
- When your preview resembles the following figure, click OK.

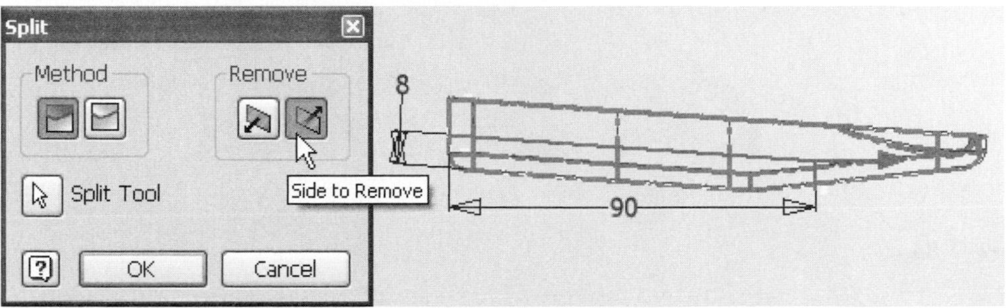

Figure 7.83

15. The part is split to create the battery cover. Press F4 and rotate the part to examine it. Note: The file now contains model data for all three parts: upper cover, lower cover, and battery cover. In the following steps, you edit features and save copies of this file to create separate files for each part.

16. In the browser or graphics window, turn off the visibility of the YZ plane.

17. Next create mating part files from the original file. From the File menu, select Save Copy As. Save the file under the name *ESS_E07_06-BatCover.ipt* in the same directory as the original file. The battery cover part is now complete.

18. In the browser, double-click the Split3 feature to edit it.

- Click the left Side to Remove button to reverse the remove direction, as shown in the following image.
- Click OK to finish editing the split feature and create the mating part.

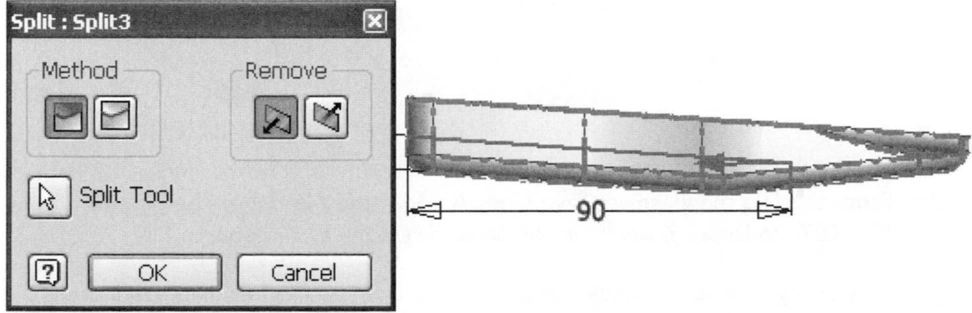

Figure 7.84

19. Use the Orbit tool to examine the part as shown in the following image.

EXERCISES

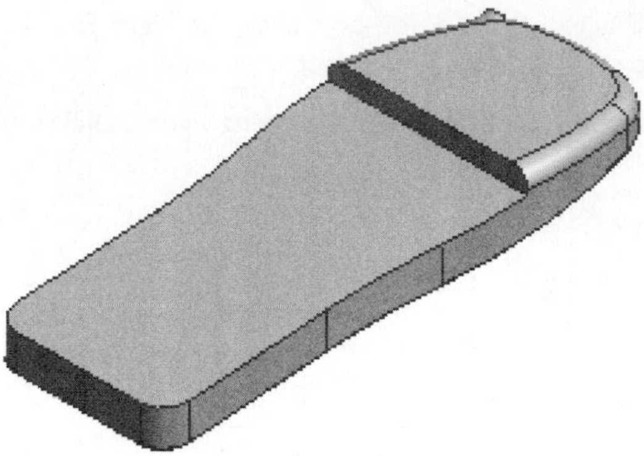

Figure 7.85

20. From the File menu, select Save Copy As. Save the file under the name *ESS_E07_06-Lower_Cover.ipt* in the same directory as the source file.
21. In the browser, right-click Split3, and click Delete.
22. When the Delete Features dialog box is displayed, click OK to delete the related sketch data.
23. In the browser, double-click the Split2 feature.
    - In the Remove area of the dialog box, select the right Side to Remove button to reverse the remove direction as shown in the following image on the left.
    - Click OK to complete the edit. The image on the right shows the complete part.

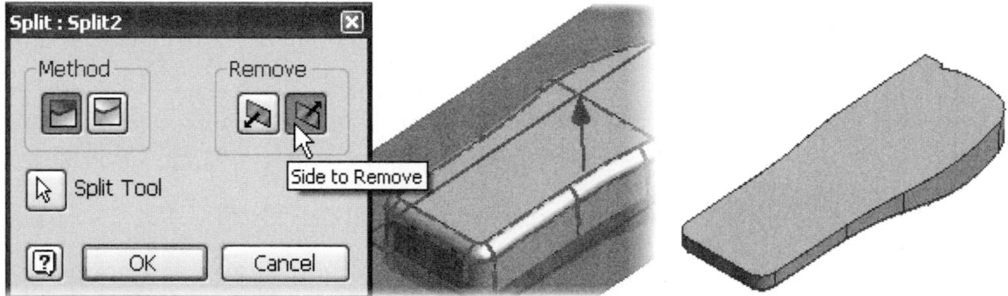

Figure 7.86

24. From the File menu, select Save Copy As. Save the file under the name *ESS_E07_06-Upper_Cover.ipt* in the same directory as the source file.

Note: In your files, you would complete the parts by opening the individual files and changing the file properties. The individual parts could also be created using the derived parts. Derived parts are covered later in this chapter.

25. Close the file. Do not save changes. End of exercise.

BEND PART

While designing parts that are bent, it may be easier to create the part in a flattened state and then bend it. The Bend Part tool will bend a side or both sides of a part based on the location of a bend line. The bend line will be the location where the bend or the centerline of a bend is bent evenly in both directions. After creating the bend, it will appear in the browser as a feature and can be edited like any other feature. To bend a part, follow these steps.

1. Create or open a part that will be bent.
2. Create a sketch where the bend line will be placed. The plane must intersect the part or lie on a planar face of the part.
3. Draw a line that will be used as the bend line.
4. Exit the sketch.
5. Click the Bend Part tool from the panel bar, as shown in the following image on the left. The Bend Part dialog box will appear, as shown in the following image on the right.

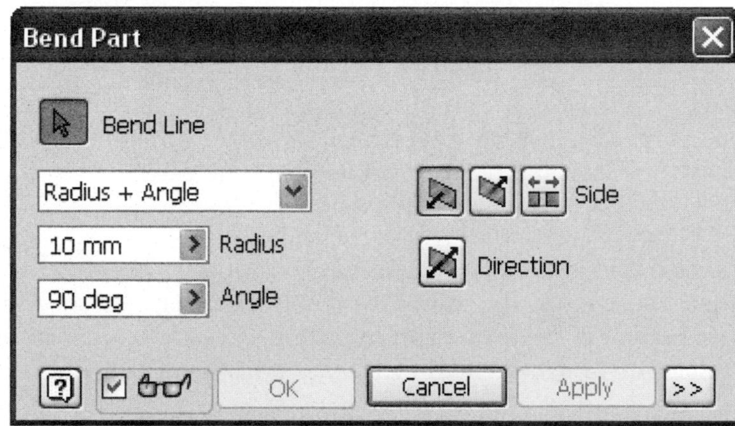

Figure 7.87

6. In the dialog box, specify how you want to calculate the bend: Radius + Angle, Radius + Arc Length, or Arc Length = Angle.
7. Enter the data to create the bend.
8. If needed, change the side and direction that the bend occurs.

The following image on the left shows the part with a bend line in the middle of the inside face, and the image on the right shows the bent part.

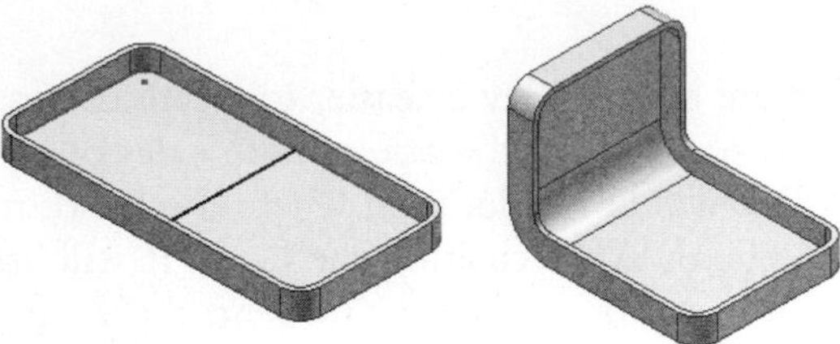

Figure 7.88

COPYING FEATURES

To increase your productivity, you can copy sketch-based features from one face of a part to another face on the same part or to a different part. If you copy a feature from one part to another, both files must either be open in Autodesk Inventor or be instances within the same assembly file. When making a copy of a feature, you can designate whether you want the feature, its dependent feature(s), and its parameters to be dependent on or independent of the parent feature(s).

DEPENDENT

The dimensions of the copied feature(s) will be equal to the dimensions of the parent feature(s). If the parent feature was defined with a value of d4=8 inches, for example, the copied feature would have a parameter similar to d14=d4. The Dependent option is only available when copying features within the same part.

INDEPENDENT

The dimensions of the copied feature(s) will contain their own values. Initially, the value will equal the value of the parent feature. If the parent feature was defined with a value of d4=8 inches, for example, the copied feature will have a parameter similar to d14=8 inches. The value of d14 can be modified independently of the value of d4.

The following rules apply when creating an independent copy of a feature:

- You can only copy selected features by default, not by children features.
- Upon pasting the feature, you can choose to copy dependent features.
- When pasting a copied feature, you can specify the plane and orientation that the new feature will have.
- The newly pasted feature is completely independent and contains its own sketches and feature definitions within the browser.

Follow these steps to copy and paste a feature within Autodesk Inventor:

1. Right-click on a feature's name in the browser, or change the Select option on the Standard toolbar to Feature Priority, as shown in the following image. Move the cursor over the feature to copy.

Figure 7.89

2. Select Copy from the menu to copy the feature to the clipboard.
3. Issue the Paste command by doing one of the following:

- Right-click, and select Paste from the menu.
- Click Paste on the Edit menu.
- Press the CTRL and V keys on the keyboard at the same time.

4. The Paste Features dialog box will appear, as shown in the following image.

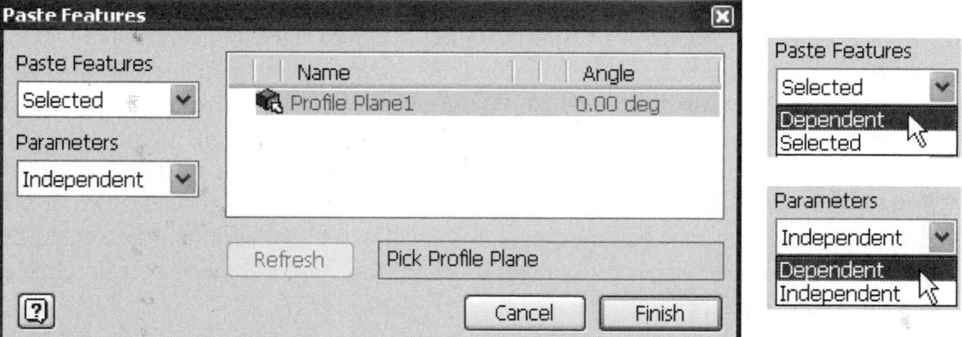

Figure 7.90

5. Move the cursor over the face or plane whose feature you want to copy, and then click a plane to place the feature. The placement plane can be any planar face on the part or a work plane.
6. You can roughly move or rotate the new feature by using the compass in the preview image, as shown in the following image on the left.
7. Click and drag the four-headed arrow to move the profile, or click and drag the circular arrow to rotate the profile as shown in the following image. You can also enter the precise value for the angle of rotation of the profile in the dialog box.
8. To locate the pasted feature accurately, edit its sketch after placement, and add constraints and dimensions as needed.
9. In the Paste Features dialog box, adjust the settings as needed.

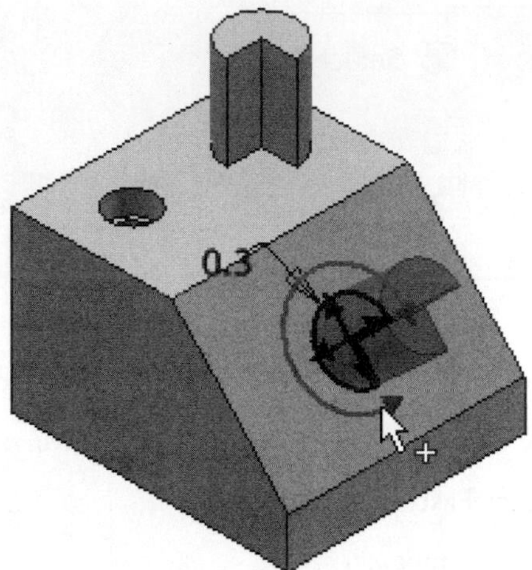

Figure 7.91

MIRRORING FEATURES

When creating a part that has features that are mirror images of one another, you can use the Mirror Feature tool to mirror a feature(s) about a planar face or work plane instead of recreating the features from scratch. Before mirroring a feature, a work plane or planar face that will be used as the mirror plane must exist. The feature(s) will be mirrored about the existing plane; it can be a planar face, a work plane on the part, or an origin plane. The mirrored feature(s) will be dependent on the parent feature—if the parent feature(s) change, the resulting mirror feature will also update to reflect the change. To mirror a feature or features, use the Mirror tool on the Part Features panel bar, as shown in the following image on the left. The Mirror dialog box will appear, as shown on the right. The following sections explain the options in the Mirror dialog box.

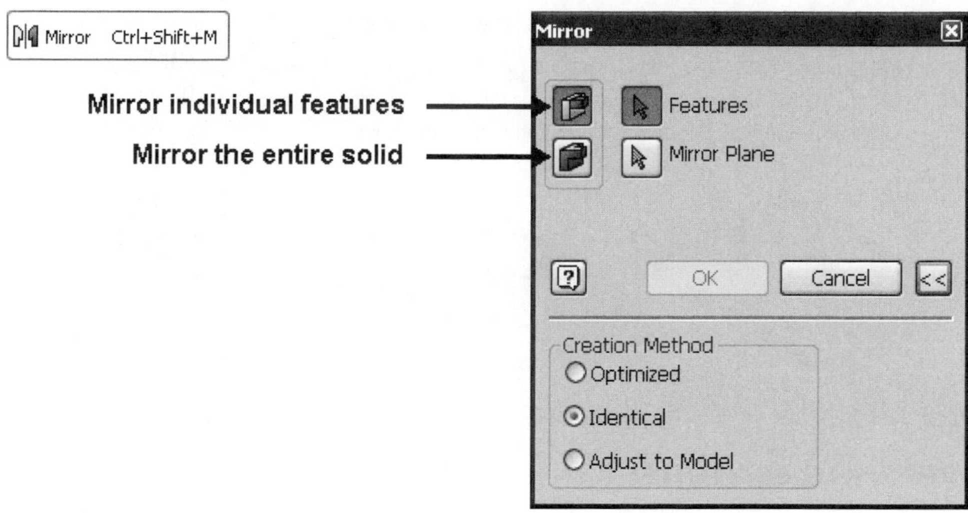

Figure 7.92

Mirror Individual Features Click this option to mirror a feature or features.

Mirror the Entire Solid Click this option to mirror the entire solid body (part).

Mirror Plane Click and choose a planar face or work plane on which to mirror the feature(s).

Creation Method Access the Creation Method tools by clicking the More (<<) button. *Optimized:* Create mirrored occurrences that are direct copies of the original features. This option will compute the fastest but will not adjust to the model.

- *Identical:* Create mirrored features that remain identical regardless of where they intersect other features on the part. Use this option when you want to mirror a large number of features that share a termination plane.
- *Adjust to Model:* Use this option when you want to calculate an independent termination for each feature that is being mirrored. Use this option if the feature(s) being mirrored uses a model face as the termination option.

To mirror a feature or features, follow these steps:

1. Click the Mirror Feature tool on the Part Features panel bar. The Mirror dialog box will appear.
2. Click the Mirror individual features option or the Mirror the entire solid option.
3. Select the feature(s) or solid body to mirror.
4. Click the Mirror Plane button, and then select the plane on which the feature will be mirrored.
5. In the More (<<) area, determine if the mirrored feature(s) termination will be optimized or identical or will adjust to the model.

6. Click the OK button to complete the operation.

The following image on the left shows a part with the features that will be re-created, the middle image shows the part with a work plane that the features will be mirrored about, and the image on the right shows the features mirrored.

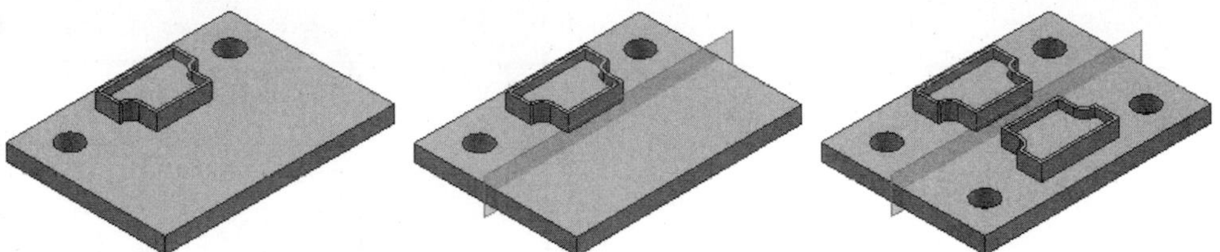

Figure 7.93

SUPPRESSING FEATURES

You can suppress a model's feature or features to temporarily turn off their display, as shown in the following image on the left. Feature suppression can be used to simplify parts, which may increase system performance. This capability also shows the part in different states throughout the manufacturing process, and it can be used to access faces and edges that you would not otherwise be able to access. If you need to dimension to a theoretical intersection of an edge that was filleted, for example, you could suppress the fillet and add the dimension and then unsuppress the fillet feature. If the feature you suppress is a parent feature for other dependent features, the child features will also be suppressed. Features that are suppressed appear gray in the browser and have a line drawn through them, as shown in the following image on the right. A suppressed feature will remain suppressed until it is unsuppressed, which will also return the features browser display to its normal state. The following image on the right also shows the suppressed child features that are dependent on the suppressed extrusions.

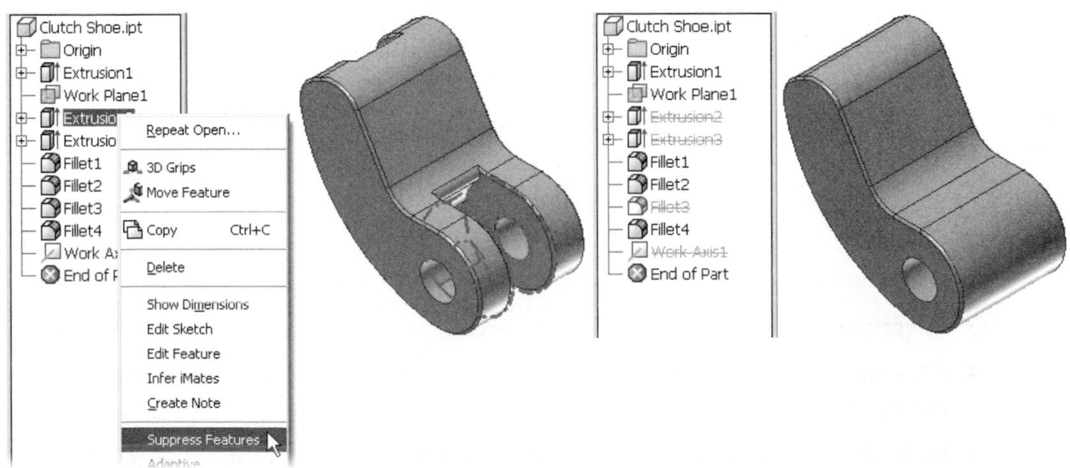

Figure 7.94

To suppress a feature on a part, use one of these methods:

1. Right-click on the feature in the browser, and select Suppress Features from the menu, as shown in the previous image on the left.

   or

2. Click the Feature Priority option from the Standard toolbar, as shown in the following image. This option allows you to select features on the parts in the graphics window.
3. Right-click on the feature in the graphics window, and click the Suppress Features option from the menu.

Figure 7.95

To unsuppress a feature, right-click on the suppressed feature's name in the browser, and select Unsuppress Features from the menu.

CONDITIONALLY SUPPRESS FEATURES

Another method to suppress features is to apply a parameter condition to a feature. For example, if a part has a feature that should be suppressed if the size of another feature is reduced, you can set the feature's properties to be suppressed when the parameter's value is changed. To conditionally suppress a feature, follow these steps:

1. Create part and rename parameters for important dimensions that will control the suppression of other features.
2. Right-click on the feature to be suppressed if a set value of a parameter is reached, and click Properties, as shown in the following image on the left.
3. In the Feature Properties dialog box from the drop-down list, select the parameter that will determine the suppression of the feature. The following image on the right shows that the Length parameter has been selected.
4. From the drop-down list, select the mathematical condition.
5. Enter a value limit for the parameter that will suppress the feature. The following image on the right shows the value of 20 mm.
6. Click OK.

To see the results of the suppression, change the value of the Length parameter to reach or exceed the limit, and update the part.

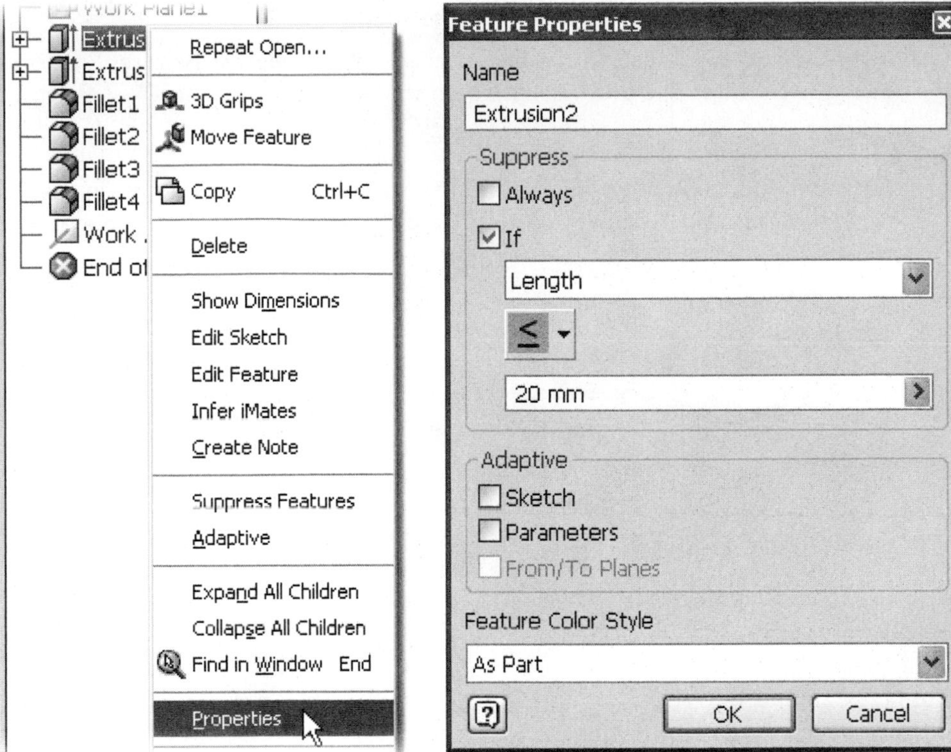

Figure 7.96

REORDERING FEATURES

You can reorder features in the browser. If you created a fillet feature using the All Fillets or All Rounds option and then created an additional extruded feature, such as a boss, you can move the fillet feature below the boss and include the edges of the new feature in the selected edges of the fillet. To reorder features, follow these steps:

1. Click the feature's name or icon in the browser. Holding down the left mouse button to click and drag, move the feature to the desired location in the browser. A horizontal line will appear in the browser to show you the feature's relative location while reordering the feature. The following image on the left shows a hole feature being reordered in the browser.
2. Release the mouse button, and the model features will be recalculated in their new sequence. The following image on the right shows the browser and the reordered hole features.

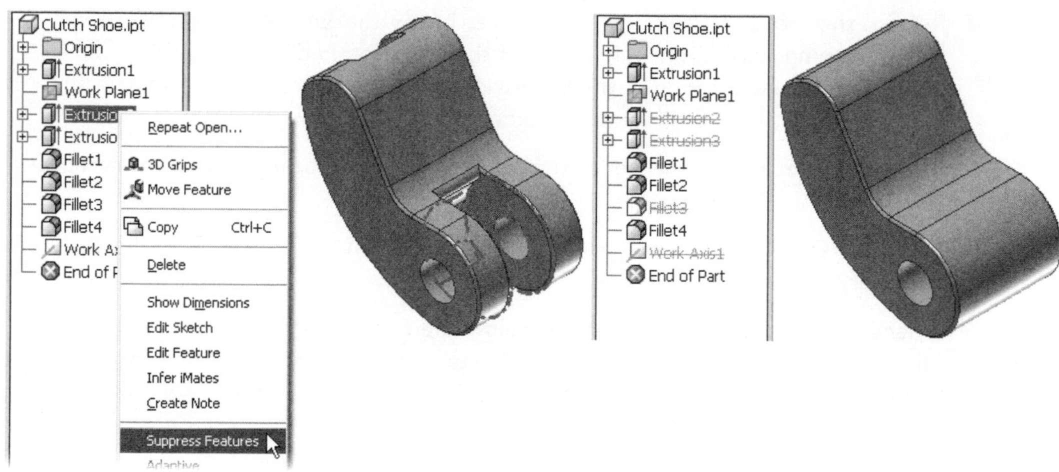

Figure 7.97

If you cannot move the feature due to parent-child relationships with other features, Autodesk Inventor will not allow you to drag the feature to the new position. In the browser, the cursor will change to a No symbol instead of a horizontal line, as shown in the following image.

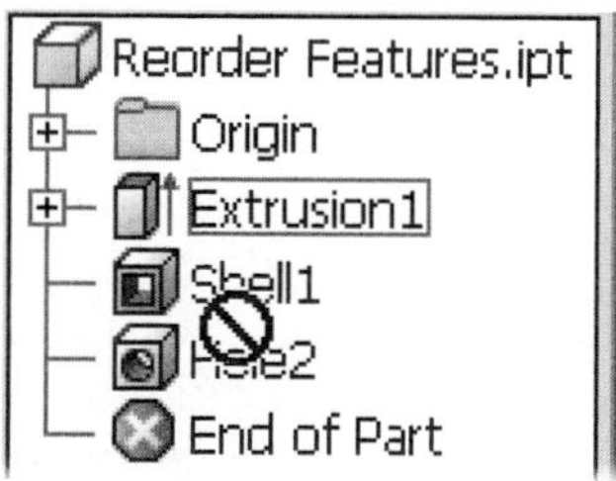

Figure 7.98

FEATURE ROLLBACK

While designing, you may not always place features in the order that your design later needs. Earlier in this chapter, you learned how to reorder features, but reordering features will not always allow you to create the desired results. To solve this problem, Autodesk Inventor allows you to roll back the design to an earlier state and then place the additional new features. To roll back a design, drag the End of Part marker in the browser to the location where the new feature will be placed. To move the End of Part marker, follow these steps:

1. Move the cursor over the End of Part marker in the browser.
2. With the left mouse button depressed, drag the End of Part marker to the new location in the browser. While dragging the marker, a line will appear, as shown in the following image on the left.

3. Release the mouse button, and the features below the End of Part marker are removed temporarily from calculation of the part. The End of Part marker will be moved to its new location in the browser. The following image in the middle shows the End of Part marker in its new location.
4. Create new features as needed. The new features will appear in the browser above the End of Part marker.
5. To return the part to its original state, including the new features, drag the End of Part marker below the last feature in the browser.
6. If needed, you can delete all features below the End of Part marker by right-clicking on the End of Part marker. Then click Delete all features below EOP, as shown in the following image on the right.

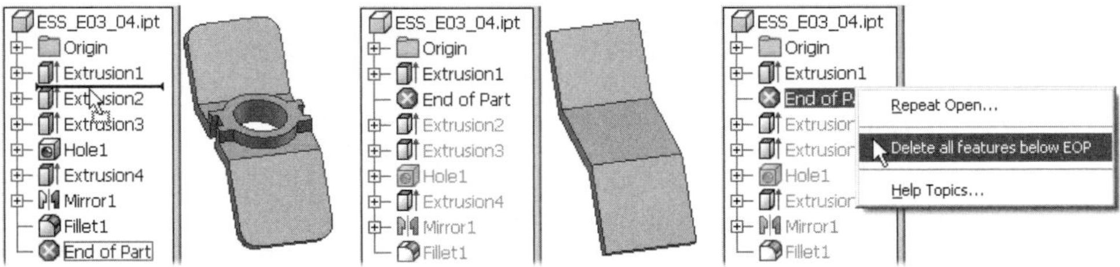

Figure 7.99

DERIVED PARTS

Derived parts are used to capture the design intent of a part, assembly, specific sketch, work geometry, surface, parameter, or iMate by either creating a new part file or by importing any of the aforementioned selections. A derived part is a new part file that uses a selected part, assembly, or sketch as its base feature. Additional features can then be added to the derived part. If the original part changes, the derived part will update to include any changes that are made to the original file. This associativity can be broken if you do not want changes made to the original file to be included in the new part file. You can derive from other files at any time in the design process, but only one solid body can be derive into a part either from another part or from an assembly.

Some possible uses for a derived part or assembly are as follows:

- Create many different parts that are machined from a single casting blank.
- Scale or mirror a derived part upon creation: note that this does not apply to for derived assemblies. You can derive sketches to be used within an assembly as a layout.
- Derive a part as a work surface that can be used to simplify the representation of a part

You create a derived part using the Derived Component tool located on the Part Features panel bar, as shown in the following image on the left.

Navigate to and select the Part or Assembly file that you want to derive. If you select a part file, the Derived Part dialog box opens, as shown in the following image in the middle. If you select an assembly file, the Derived Assembly dialog box opens to provide you with different options, as shown in the following image on the right.

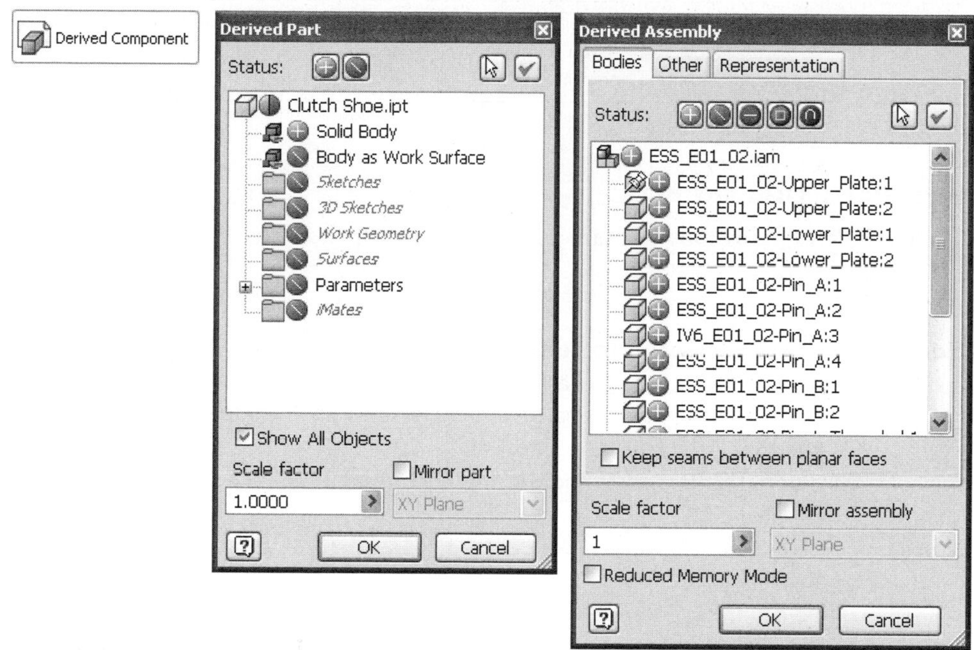

Figure 7.100

DERIVED PART SYMBOLS

The symbols show you the current status of geometry that will be contained in the derived part and whether or not it will be included or excluded from the derived part. The Derived Part Symbols table shows the a

Derived Part Symbols

Symbol	Function	Meaning
	Mixed	Indicates that both included and excluded geometry types are contained in the derived part.
	Plus	Indicates that the selected geometry will be included in the derived part.
	Exclude	Indicates that the selected geometry will not be included in the derived part. If a change is made to this type of geometry in the parent part, it will not be incorporated or updated in the derived part.

Derived Part Symbols

Symbol	Function	Meaning
	Select	Switches focus to the base part window so that you can take advantage of the selection tools and keep the Derived Part dialog open.
	Accept	Causes the Derived Part dialog to absorb the selections you made in the base part window, returns you to the Part environment, and highlights your selections in the dialog tree control.

Figure 7.101

DERIVED PART DIALOG BOX

The Derived Part dialog box contains the following options, as shown in the following chart.

Solid Body	Select this option to derive the part as a base solid.
Body as Work Surface	Select this option to derive the part body as a work surface. The body is brought in, behaves, and appears as an Autodesk Inventor surface.
Sketches	Select this option to include any unconsumed 2D sketches from the original part.
3-D Sketches	Select this option to include any unconsumed 3D sketches from the original part.
Work Geometry	Select this option to include any work features from the original file. They can then be used to create new geometry or to constrain a part in an assembly.
Surfaces	Select this option to include any surfaces that exist in the original file.
Exported Parameters	Select this option to include any parameters designated to be exported parameters in the original file.
iMates	Select this option to include any iMates that exist in the original part.

Note: If the original file consists only of surfaces, the surface option will be the only available option.

Scale Factor	Select a scale factor in percentage to scale the derived part. The default scale factor is 1.0 or the same size as the original file.
Mirror Part	Select this option to mirror the original part about the XY, XZ, or YZ origin work planes upon derived part creation. You can select the plane about which to mirror the derived part from the drop-down list.

DERIVED ASSEMBLY SYMBOLS

The symbols available when deriving an assembly are different from the symbols available when deriving a part file. The following table shows the derived assembly symbols.

Derived Assembly Symbols

Symbols	Function	Meaning
	Combination	Indicates that the selected item contains a mixture of included and excluded components.
	Plus	Indicates that the selected component will be included in the derived part.
	Exclude	Indicates that the selected component will not be included in the derived part. Changes to this type of component in the parent assembly will not be incorporated or updated in the derived part.
	Subtract	Indicates that the selected component will be subtracted from the derived part. If the subtracted component intersects another portion of an included part, the result will be a void or cavity in the derived part.
	Bounding Box	Indicates that the selected component will be represented as a bounding box. The bounding box size is determined by the extents of the component. The bounding box is used to represent a component as a placeholder and reduces memory. You can add features to a bounding box, and it will update when changes are made to the original part.
	Intersect	Intersects the selected component with the derived part. One component must have an Include status. If the component does not intersect the derived part, the result is not a solid.
	Select	Switches focus to the base assembly window so that you can take advantage of the selection tools and keep the Derived Assembly dialog open.
	Accept	Causes the Derived Assembly dialog to absorb the selections you made in the assembly environment, returns to the Part environment, and highlights your selections in the dialog tree control.

Figure 7.102

Keep seams between planar faces	Select this option to create seams between any adjacent coplanar faces of parts. If the box is clear, the faces are merged.
Scale Factor	Select a scale factor in percentage to scale the derived part. The default scale factor is 1.0 or the same size as the original file.
Mirror Part	Select this option to mirror the original part about the XY, XZ, or YZ origin work planes upon derived part creation. You can select the plane about which to mirror the derived part from the drop-down list.
Reduced Memory Mode	When checked, Autodesk Inventor uses less memory by excluding bodies of the parts that are cached in memory. When the link is broke or suppressed the memory savings is removed.

The Other tab of the Derived Assembly dialog box contains options that allow you to select sketches, work geometry, surfaces, and exported parameters from the assembly file or any of the components that are in the assembly. Using the plus and minus icons, you can select which items you want included in the file.

The Representation tab of the Derived Assembly dialog box contains options that allow you to select available design view representation, positional representation, or level of detail representations present in the assembly.

Because you are using geometry or parts based on another file, any modifications to the original parts are incorporated into the derived component. If you modify a parent file, the update icon next to the derived component in the browser displays an update symbol, as shown in the following image on the left. The Update button on the Standard toolbar also becomes active as shown in the following image on the right. To update the file, click the Update tool on the Standard toolbar.

Figure 7.103

You can break the link to the parent file by right-clicking on the derived parent component in the browser and selecting Break Link With Base Assembly from the menu, as shown in the following image on the left. Once the link is broken, the derived component icon will display a broken chain link to signify that the link to the parent file no longer exists. Once you have broken the link, you cannot reestablish it. You can temporarily suppress the changes from the base assembly from affecting the derived part by clicking Suppress Link With Base Component from the menu. To reestablish the link, click Unsuppress Link With Base Component, as shown in the following image on the right.

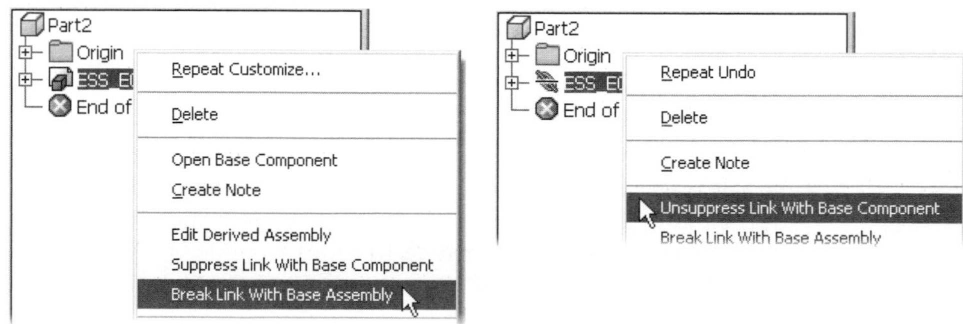

Figure 7.104

EXERCISE 7-7: CREATING A DERIVED PART

In this exercise, you create a mold cavity by deriving an assembly consisting of a part that you'll create a mold of and a mold base from which the part cavity will be removed. Assembly constraints have already been applied to the parts in the assembly to center the parts. You then edit the part and update the derived part.

1. Click the New tool, click the Metric tab, and double-click *Standard(mm).ipt*.
2. Exit the sketch environment by clicking Return on the Standard toolbar.
3. Click the Derive Component tool from the Panel bar.
4. In the Open dialog box, click the file *ESS_E0_07_Assembly.iam* from the Chapter 07 folder, and then click the Open button.
5. In the Derived Assembly dialog box, click the icon next to the entry *ESS_E07_07-Part:1* so that the subtract symbol appears, as shown in the following image on the left.
6. Click OK to create the mold cavity.
7. Change to an isometric view, and your part should resemble the following image on the right.

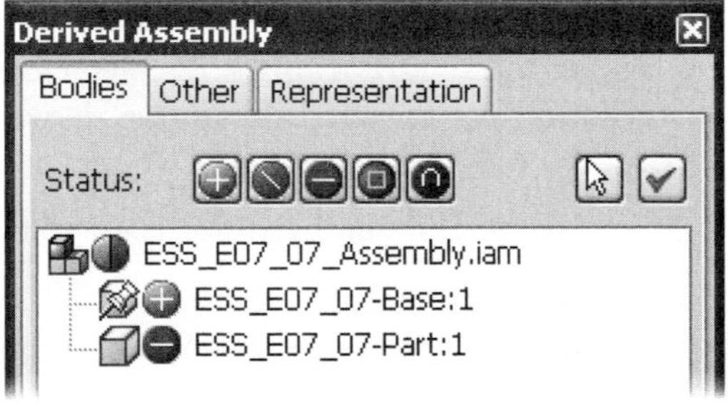

Figure 7.105

8. Next make an extrusion for the material to flow into the cavity. Create a new sketch on the front-right face of the mold base.
9. Draw a circle on the midpoint of the top-right edge.
10. Add a **2 mm** diameter dimension to the circle.
11. Press the E key to start the Extrude tool, and follow these steps:

- Select the circle as the profile.
- Click the Cut operation.
- Change the extents to the To option.

- Select the inside face of the cavity to terminate the extrusion.
- Click the More tab, and click the Minimum Solution option. Your screen should resemble the following image on the left.
- Click OK to create the extrusion. The part should resemble the following image on the right.

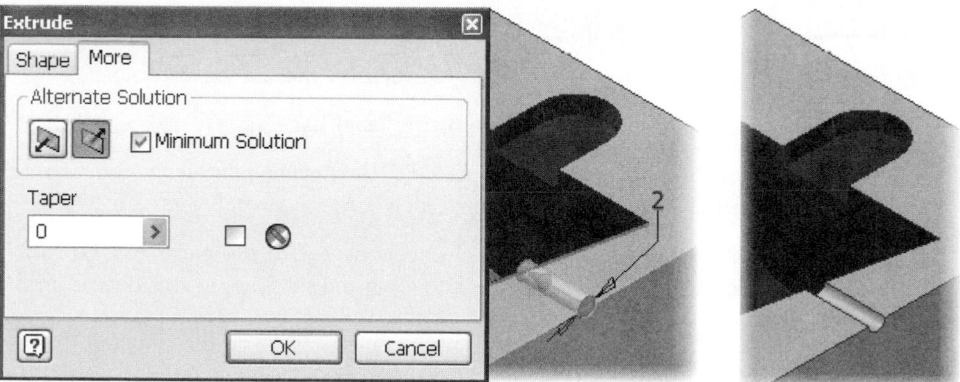

Figure 7.106

12. Save the file as *ESS_E07_07_MoldCavity.ipt* to the Chapter 07 folder.

13. Move the cursor in the browser over the entry *ESS_E07_07_Assembly.iam,* and click Open Base Component from the menu. Notice that the extrusion in the derived part does not exist in the original part.

14. In the browser. double-click on *ESS_E07_07-Part:1,* and edit Extrusion1 and Extrusion2 to the distance of **10 mm**.

15. Click the Return tool on the Standard toolbar, and save the assembly and changes to the part files.

16. Close the assembly file, and verify that the file *ESS_E07_07_MoldCavity.ipt* is the current file.

17. In the browser, notice the red lightning bolt next to *ESS_E07_07_Assembly.iam.*

18. Click the Update tool in the Standard toolbar, and the cavity should resemble the following image.

Figure 7.107

19. To change the mold cavity to the male portion, move the cursor in the browser over the entry *ESS_E07_07_Assembly.iam*. Click Edit Derived Assembly from the menu.

20. In the Derived Assembly dialog box, click the icon next to the entry *ESS_E07_07-Part:1* so that the plus symbol appears, as shown in the following image on the left.

21. Click the OK button to update the part. Your part should resemble the following image on the right.

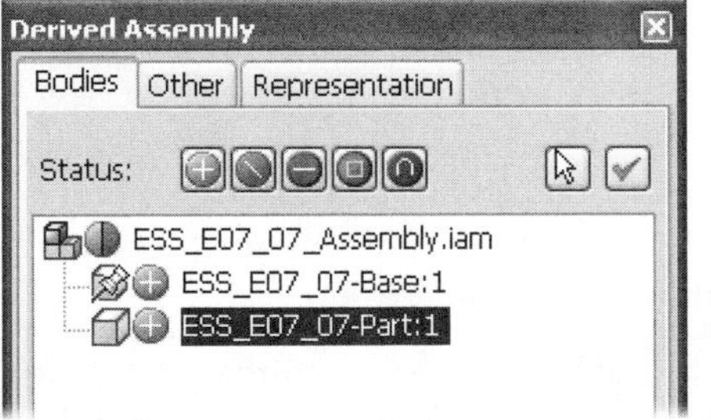

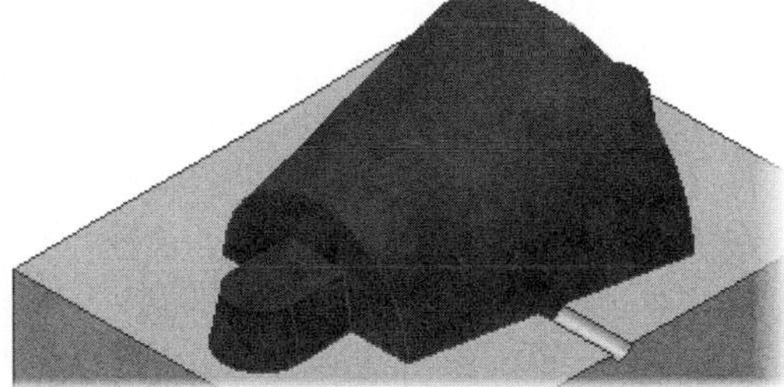

Figure 7.108

22. Close the file. Do not save changes. End of exercise.

PART MATERIALS AND COLORS

To change a part's physical properties and appearance to a specific material, click the iProperties option on the File menu, or right-click on the part's name in the browser, and select a material from the material drop-down list on the Physical tab, as shown in the following image. As you learned in the File Properties section in this chapter, the properties will reflect the attributes of this material. Click the OK button to complete the operation, and the color of the part changes to reflect the material. If the part color does not update when the material is edited, click the As Material option in the Color area of the Standard toolbar, as shown in the following image. If you selected a color from the Color drop-down list on the Standard toolbar, it will change only the color of the part, not the physical material or material properties of the part.

Figure 7.109

OVERRIDING MASS AND VOLUME PROPERTIES

While designing, you may not always draw parts that are 100% complete. For example, you may model only the bounding area and critical features of a purchased part. However, you still want the mass and volume to be represented accurately in the Properties dialog box. You can override the mass and volume values of a part or assembly by following these steps:

1. Click iProperties on the File menu, or right-click on the part's name in the browser, and select iProperties from the menu.
2. The Properties dialog box will appear. Click the Physical tab.
3. If a material has not been assigned, select a material from the Material drop-down list.
4. Click the Update button in the dialog box to update the properties if they are displaying N/A. Notice the calculator symbol next to the Mass and Volume values; the symbol shows that Autodesk Inventor has calculated these properties based on the material and size of the part, as shown in the following image on the left.
5. To override a value for Mass or Volume, click in its text area, and enter a new value. The symbol next to the overridden value will change to a hand to reflect that the value has been overridden, as shown in the following image on the right. Notice the * in the Inertial Properties and Center of Gravity areas, pointing to the comment, "Values do not reflect user-overridden mass or volume."
6. To change the overridden value back to the default value, delete the information in the text box area of Mass or Volume, and click the Update button in the dialog box.

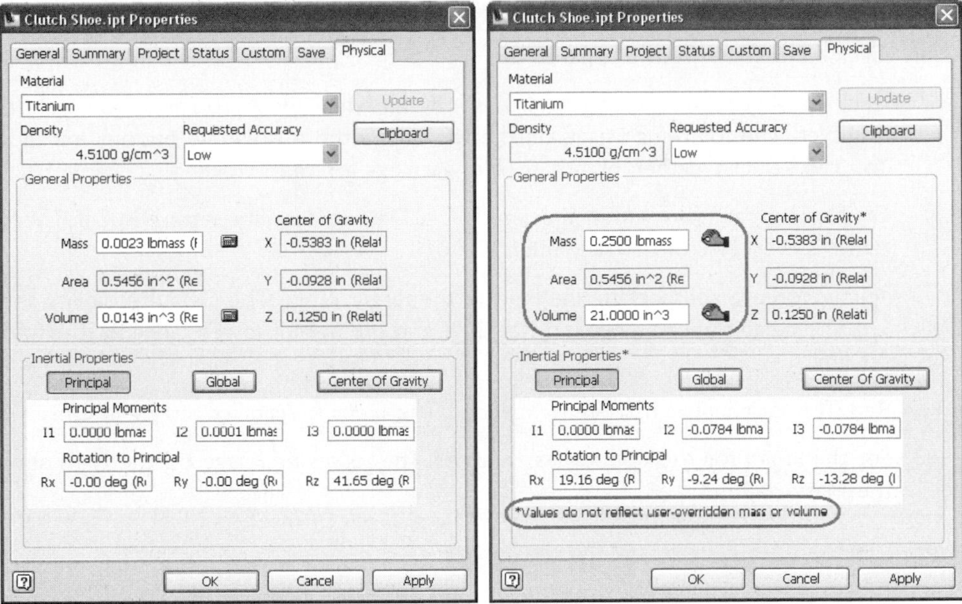

Figure 7.110

To copy the information on the Physical tab of the Properties dialog box into another application, such as Microsoft Word, click the Clipboard button on the Physical tab. Make the other application active, and paste the information into that file.

AUTOLIMITS

While working on a design, you may want to set up design limits such as dimensional or physical properties for a part or assembly and be warned if the part or assembly falls outside the selected range. An AutoLimit can be set for parts, assemblies, weldments, 2D or 3D sketches, and sheet metal parts excluding the flat pattern. To access this tool, click AutoLimits from the drop-down menu in the panel bar of a part or assembly file, as shown in the following image on the left and in the middle.

The tools in the panel bar will change, as shown in the following image on the right.

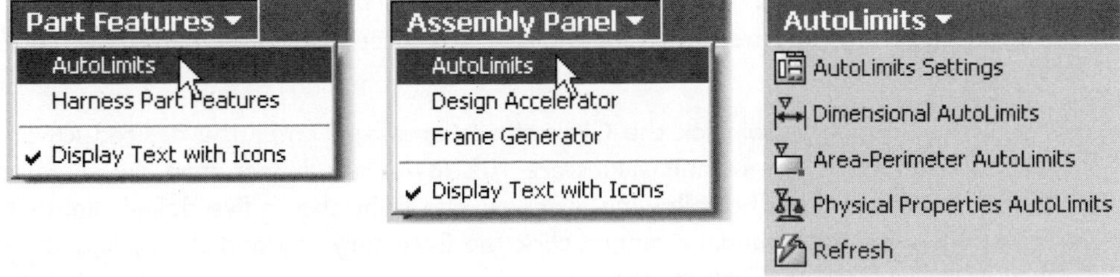

Figure 7.111

To set and use AutoLimits, follow these steps:

1. Click AutoLimits from the panel bar.
2. Click the AutoLimits Setting tool to set if and when the AutoLimits icon will be displayed. The following image on the left shows the AutoLimits options expanded. You can select whether or not to see warnings for the following.

- Green: The data falls within the limits. The Default Visibility setting is Off. The default range is 90–110% of the set limits.
- Yellow: You are approaching the limit of the target range. The Default Visibility setting is On. The default upper range is 110–130%, and the default lower range is 70–90% of the set limits.
- Red: The target limit has been exceeded. The Default Visibility setting is On.

3. Set the limits for each option as required. The following image on the right shows the Mass option.

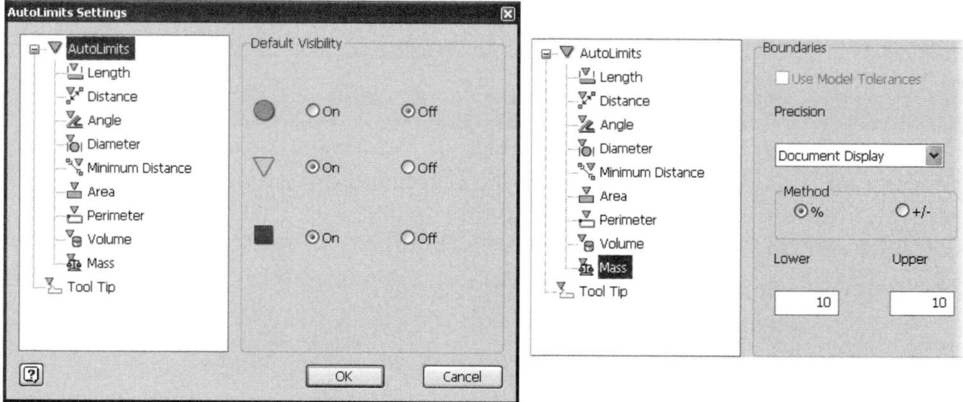

Figure 7.112

4. To set the AutoLimits for a specific property, click the Dimensional AutoLimits, Area-Perimeter AutoLimits, or the Physical Properties AutoLimits tool from the panel bar.
5. With the selection button active, select the part or subassembly to which the limit will be applied. The current value will be displayed. The following image on the left shows Physical Properties AutoLimits dialog box after selecting a part to acquire its properties.
6. Click the Boundary tab, click the Click to add area, and enter the desired lower and upper value limits. The default values are 10% lower and higher than the actual calculated property value. The following image on the right shows five default limits. Note: To add five default boundary ranges, click the Boundary tab, and then press the Alt key while clicking on Click to add.

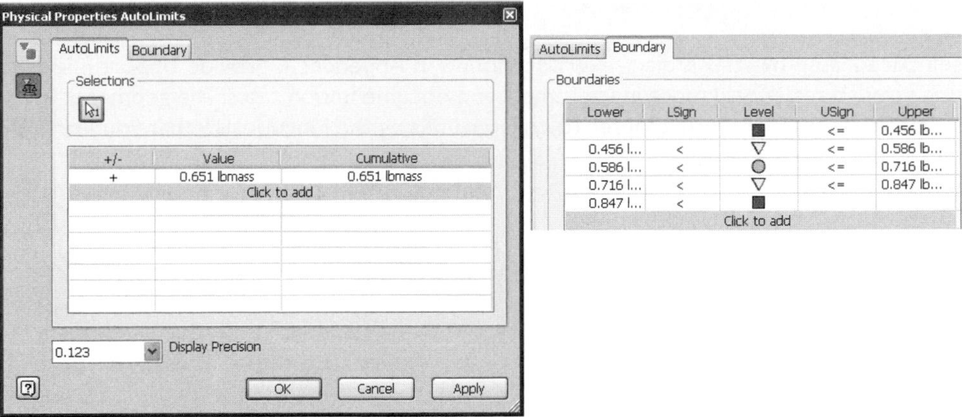

Figure 7.113

7. Click OK to create the AutoLimit. A dialog box will appear informing you if the AutoLimit has been created successfully.

8. The browser will change to display the set limits, as shown in the following image on the left.

9. To display the limit symbol in the graphics window, move the cursor over the AutoLimits option in the browser. Right-click, and click Visibility from the menu or adjust the icon for visibility in the AutoLimits setting dialog box.

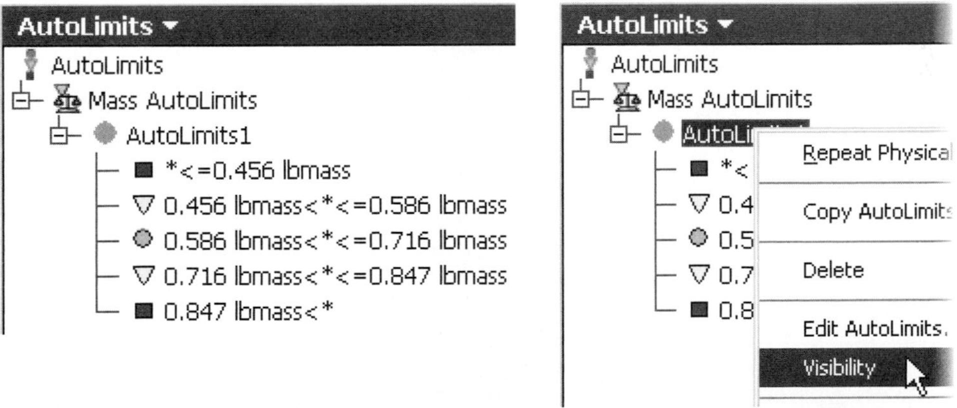

Figure 7.114

10. Change the browser and panel bar back to the model mode, and continue to design.

11. To see the status of the AutoLimit, change the browser back to AutoLimits.

Project Exercise: Chapter 7

The self-paced, step-by-step project exercises found in Appendix A provide opportunities for you to work through real-world modeling, assembly, and documentation tasks. The geometry used in the project exercises will flow from chapter to chapter, utilizing the functionality that you learned in that chapter.

Applying Your Skills

SKILL EXERCISE 7–1

In this exercise, you use complex part modeling techniques to create a housing for an electronic device. Using the knowledge you gained through this course, you will use the lofting, splitting, and embossing tools to create a joystick handle.

1. Open *ESS_E07_08.ipt* from the Chapter 07 folder.
2. Click the Loft tool.
   In the graphics window, click the three elliptical sections in order from top to bottom.

- In the Loft dialog box, select Click to add under Rails.
- In the graphics window, click the two splines. Your screen should resemble the following image.

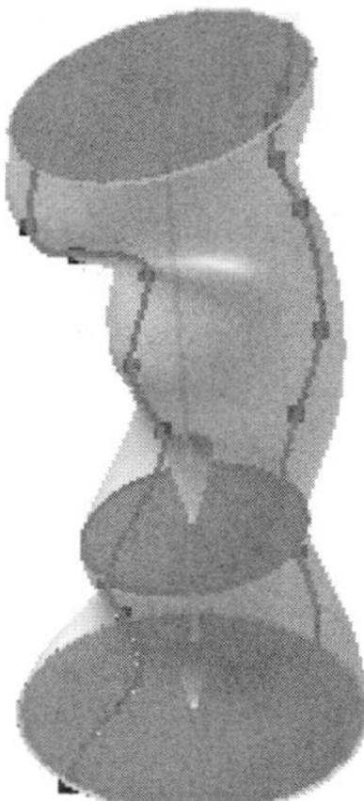

Figure 7.115

3. In the Loft dialog box, select OK.

4. Next you remove the top portion of the part using a surface, and create a fillet. In the browser, turn on the visibility of ExtrusionSrf1.

5. Click the Split tool.

- In the graphics window, click the surface as the split tool.
- In the Split dialog box, click the split part method, and click the Direction button to remove the top portion of the part, as shown in the following image on the left. Click OK.

6. From the browser, turn off the visibility of ExtrusionSrf1.

7. Use the Fillet tool to create a **3 mm** fillet around the top edge, as shown in the following image.

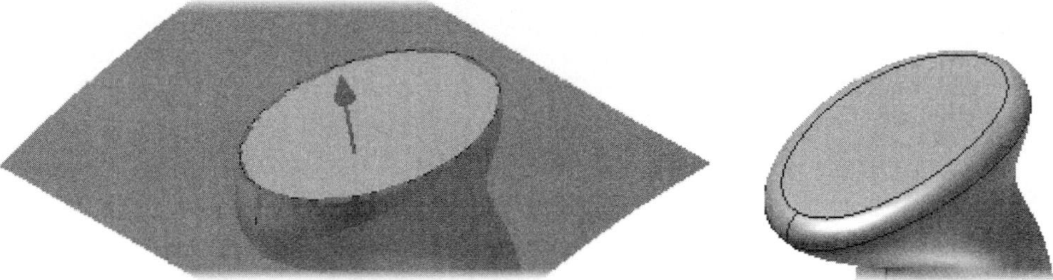

Figure 7.116

8. Next you create a sketch, and use the Emboss tool to create a button. Click the 2D Sketch tool and in the browser, click Button_Plane to define the sketch plane.

9. Click the Look At tool on the Standard toolbar, and click Button_Plane in the browser.

10. Create a circle, as shown in the following image on the left, and constrain it by doing the following:

- Project the inside ellipse.
- Add a **12 mm** diameter and a **16 mm** horizontal dimension, as shown in the following image on the left.
- Add a horizontal constraint to the center point of the circle and left quadrant of the projected ellipse.

11. Click the Return tool to finish the sketch.

12. Change to the Home View.

13. Click the Emboss tool.

- Click inside the circle you sketched in the previous step.
- Enter a Depth of **2 mm**.

- Click the Top Face Color button, and click Blue Gray.
- Verify that the Emboss from Face option is active.
- Click the Wrap to Face option. In the graphics window, click the top face of the part, as shown in the following image on the right.
- Click OK.

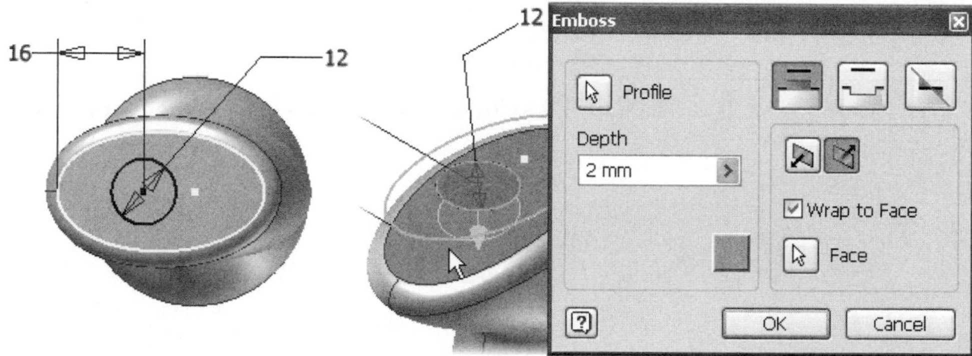

Figure 7.117

14. To finish the design, create a **1 mm** fillet around the top edge of the button. The following image shows the completed part.

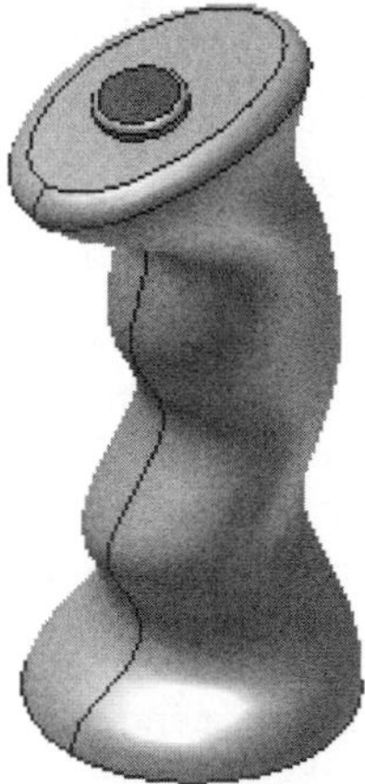

Figure 7.118

15. Close the file. Do not save changes. End of exercise.

CHECKING YOUR SKILLS

Use these questions to test your knowledge of the material covered in this chapter.

1. **True** _ **False** _ When creating a single rib or a web feature, you can only select a closed profile.
2. **True** _ **False** _ Both the Extrude and Revolve tool can use the minimum or maximum extrusion solutions.
3. **True** _ **False** _ You can only place embossed text on a planar face.
4. **True** _ **False** _ A sweep feature requires three unconsumed sketches.
5. **True** _ **False** _ You can create a 3D curve with a combination of both 2D and 3D curves.
6. Explain how to create a 3D path using geometry that intersects with a part.

___

___

7. **True** _ **False** _ The easiest way to create a helical feature is to create a 3D path and then sweep a profile along this path.
8. **True** _ **False** _ You can control the twisting of profiles in a loft by defining point sets.
9. Explain how to save both halves of a part after splitting it.

___

___

10. **True** _ **False** _ You can copy features between parts using the Copy Feature tool on the Part Features panel bar or toolbar.
11. Explain the difference between suppressing and deleting a feature.

___

___

12. **True** _ **False** _ After mirroring a feature, the mirrored feature is independent from the parent feature. If the parent feature changes, the mirrored feature will not reflect this change.

EXERCISES

13. Explain why you would want to override a part's mass and volume properties.

____

____

14. True _ False _ After changing a part's physical material properties, the part's color in the graphics window will change to match the material.

15. True _ False _ A derived part cannot be scaled.

CHAPTER 8

iComponents and Parameters

INTRODUCTION

In this chapter, you will learn how to use advanced modeling techniques. Using advanced modeling techniques, you can automate the process of assembling files by predefining assembly constraints on your parts and subassemblies as they are created. You can combine like files in a single *.ipt* or *.iam* file that stores like parts or assemblies, and you can extract features from a part to insert versions of that feature into other files. You will also learn how to display dimensions in alternate formats, set up relationships between dimensions, and create parameters. The techniques discussed will help you to become more productive and efficient when working with Autodesk Inventor.

OBJECTIVES

After completing this chapter, you will be able to perform the following:

- Create iMates
- Change the display of dimensions
- Create relationships between dimensions
- Create parameters
- Create and place iParts
- Create and place iAssemblies
- Create and place iFeatures

IMATES

Another way to apply assembly constraints is to create iMates. An iMate holds information in the component or subassembly file on how it is to be constrained when placed in an assembly. Each component or subassembly holds half of the iMate information and, when combined with a component or subassembly that contains the other half of the iMate information, they form a pair or a complete iMate. If you want the two components to assemble automatically, an iMate needs to have the same name on both components or subassemblies that are being constrained together.

When an iMate with the same name exists in two components, you can automate the assembly constraint task for those two components. iMates are useful when similar components are switched in an assembly. You may, for example, have different pins that go into an assembly. The same iMate name can be assigned to each pin or corresponding hole. When the pin is placed into the assembly, it can be automatically constrained to the corresponding hole using one of two iMate options when placing the component. An iMate can only be used once in an assembly—once used, it is consumed.

To create an iMate, follow these steps:

1. Click the Create iMate tool on either the Part Features or Assembly panel bar or from the Tools drop-down menu.
2. In the Create iMate dialog box click the type of constraint to apply: mate, angle, tangent, or insert on the Assembly tab.
3. In the graphics window, select the geometry you want to use as the primary position geometry.

The following image shows the Create iMate tool on the left, the Create iMate dialog box in the middle, and a mate-axis constraint (the iMate) being applied to the part on the right.

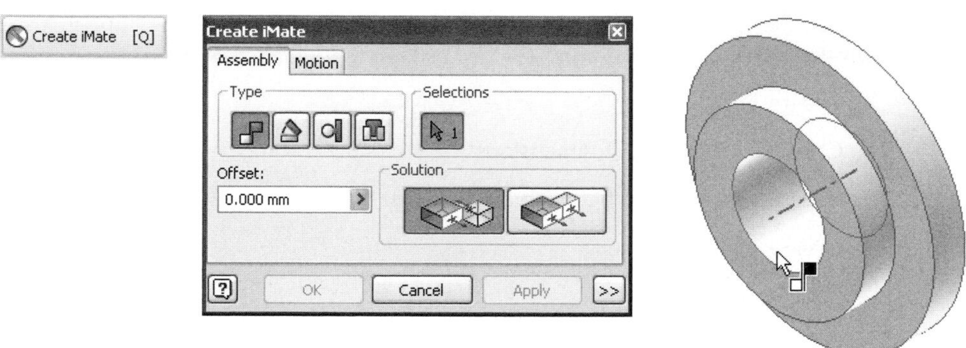

Figure 8.1

4. Click OK. An iMate glyph will appear on the component and in the browser, as shown in the following image.

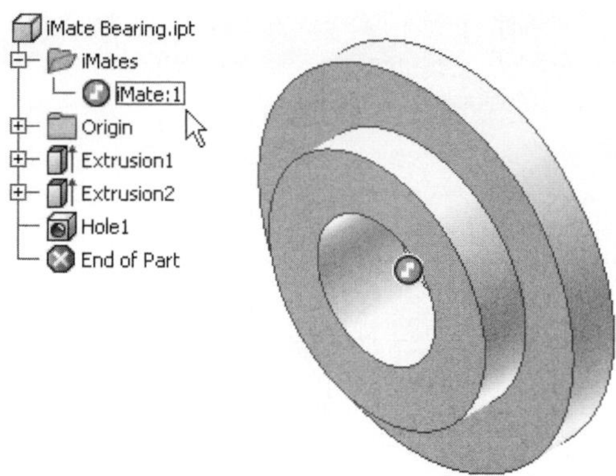

Figure 8.2

Note: Instead of clicking OK, you can click Apply and continue to create iMates.

The glyph shows the type and state of the iMate. When an iMate is created, it is given a default name according to the constraint type, such as *iInsert:1* or *iAngle:1*. The iMates can, however, be renamed to better reflect the conditions that they represent. Renaming iMates is a good idea. Providing a meaningful name that follows a common naming standard can aid when replacing components in an assembly. Matching iMate names can automatically place constraints between two components. When a component is replaced with another one that has the same iMate name, type, and offset or angle value, the constraint relationship will remain intact.

You can specify a name for an iMate using one of several methods.

- During iMate creation, expand the Create iMate dialog box, as shown in the following image, and enter a name in the field.

Figure 8.3

- Slowly double-click on the iMate name in the browser, and type in a new name.
- Right-click on the iMate in the browser, or on the iMate glyph in the graphics window, and select Properties from the menu. Type in a new name using the Name field.

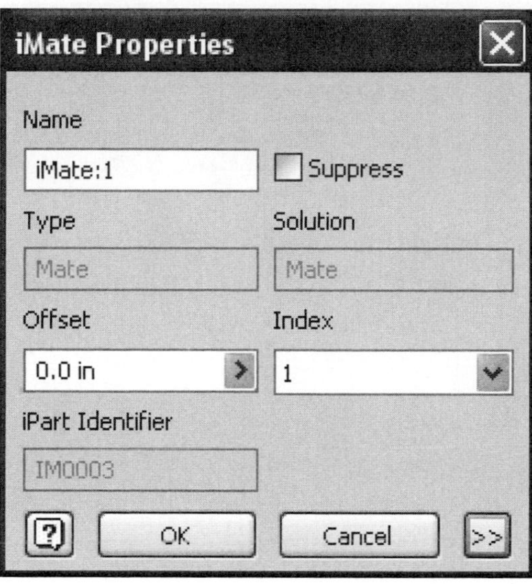

Figure 8.4

To assemble components that exist in the same assembly and have matching iMate types:

1. Use the Constraint tool on the Assembly panel bar.
2. Click an iMate on the component from the browser or graphics window, and then select a matching iMate from the browser or graphics window on another component.
3. Click the Apply button to create the constraint.

A component that has an iMate can automatically be constrained to another component that has the same iMate name and property as the component that exists in an assembly. Use the following steps to perform this operation:

1. Click the Place Component tool on the Assembly panel bar.
2. Select the component to place, and choose one of the iMates options in the lower portion of the Place Component dialog box, as shown in the following image.
3. Click the Open button.

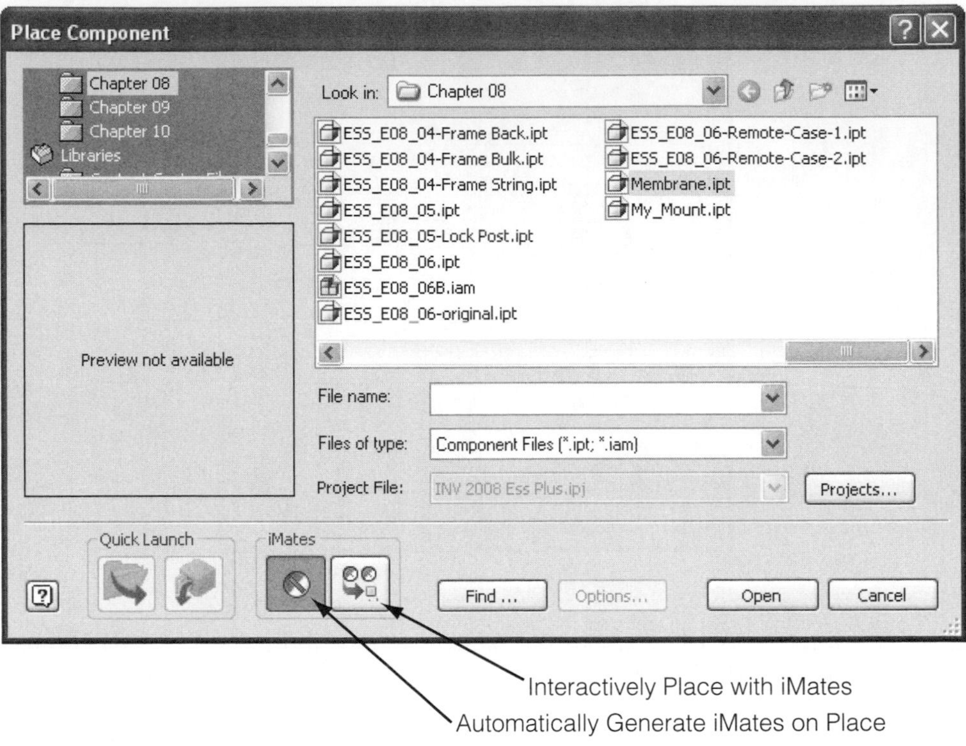

Figure 8.5

Interactively Place with iMates allows you to place one or more occurrences of a component that contains iMates. When using this mode, you can manually cycle through and apply unconsumed matching iMates or you can select to automatically place the component at all matching iMates.

The second option, Automatically Generate iMates on Place, can be used to place a single occurrence of a component that contains iMates, and have all of its iMates automatically apply to matching unconsumed iMates in other components of the assembly. If you use this option, the Place Component tool terminates once the single occurrence has been placed.

If a matching iMate on another component exists and has not yet been consumed, a preview showing how the component will be constrained will be displayed. You can left-click in the graphics window to accept the component placement. The new component will be constrained into position, the display of the assembly will be zoomed to the location of the component, and a consumed iMate icon will be shown in the browser. If multiple possible matches exist, you can continue to left-click in the graphics window to accept the additional placements, or you can right-click in the graphics window to access the iMate placement menu. The menu contains options to cycle through all valid iMate matches and component instances in the assembly. The iMate placement menu is shown in the following image.

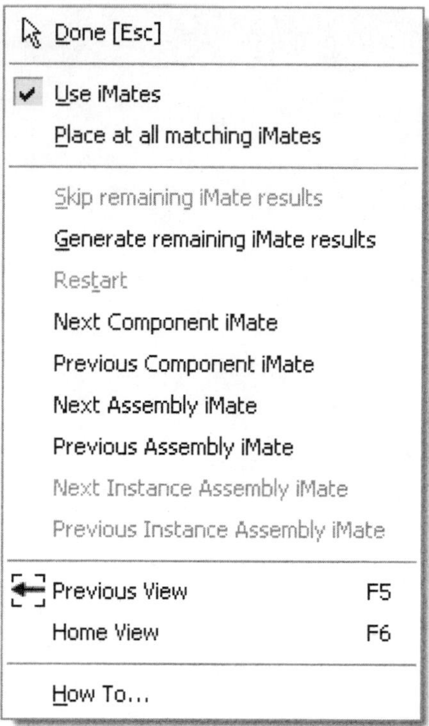

Figure 8.6

To assist and further refine how components that contain iMates are constrained to other components, you can define a Match List for the iMate. A Match List lets you define a list of names that the iMate should search for as valid matches to be constrained to. The Match List can be defined by expanding the Create iMate dialog box or by right-clicking an iMate in the browser and selecting Properties from the menu. The following image shows the iMate Properties dialog box with the Match List panel available.

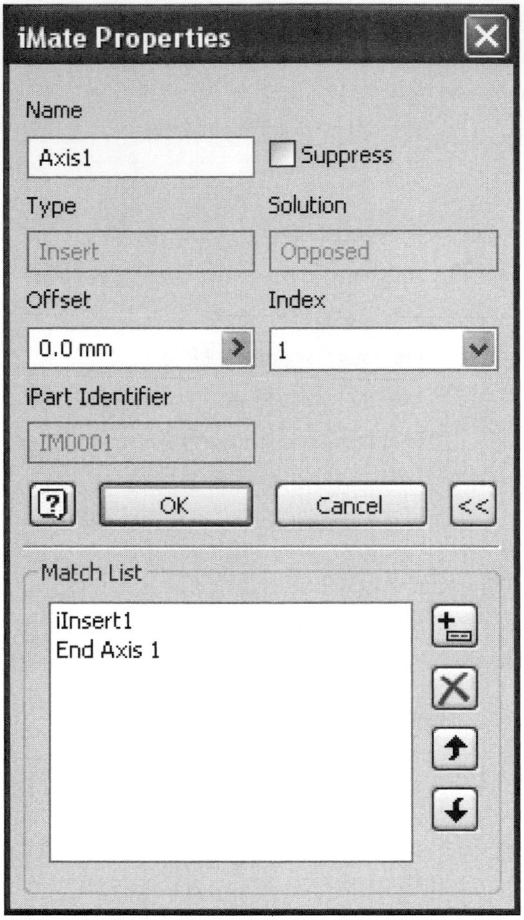

Figure 8.7

When you include names in the Match List, you can use the DELETE, move up, or move down keys to order the names in the list. The order that they appear is important when placing and automatically constraining the parts in an assembly. The match process begins at the top of the list, and if no matching name is found, proceeds to check the next name in the list, and continues this process until a match is found. If no match is found, the component is attached to the cursor and can be placed in the assembly in the same manner as a component that doesn't contain iMates.

COMPOSITE IMATES

You can group multiple iMates into a single, composite iMate. In the following image of a standard drill transmission housing, three separate iMates have been combined into a single, composite iMate. This allows you to more easily constrain components using a single constraint where multiple constraints are needed. In the following image, the three iMates that were originally created on the part are combined into a single glyph. The browser also displays the newly created composite iMate. Expanding the composite iMate will display the original iMates.

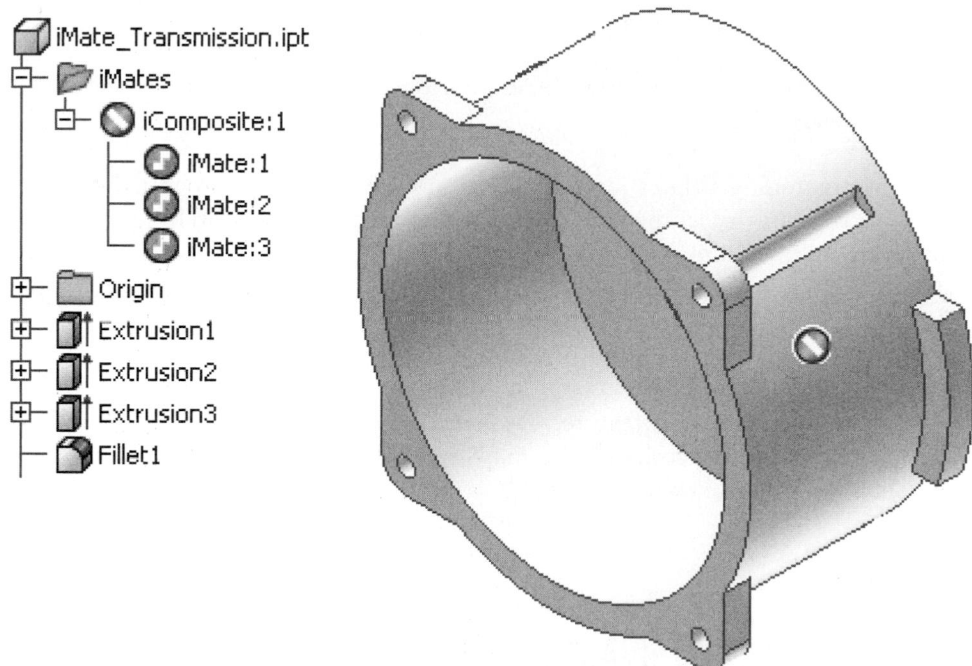

Figure 8.8

To create a composite iMate for any component, first create the individual iMates. Once created, press the CTRL key to select multiple iMates in the browser. Right-click on any iMate in the selection set, and select Create Composite from the menu.

Using Composite iMates

Similar to creating single iMates, there are two ways to orient components in your assemblies using composite iMates:

- Select one of the iMates options in the Place Component dialog box to automatically search for matching component iMates.
- Use the ALT + Drag shortcut to manually match composite iMates between two components.

To ensure a successful match between any two iMates in an assembly, check for the following criteria:

- The iMate type and values must match.

- The name at the top of the matching list is given priority when looking for a matching component in the assembly.
- The two components must have the same number and type of single iMates.
- The composites must have the same number and type of single iMates.
- The order of the single iMates that exist in the composite iMate must be identical to the other half of the composite iMate that is being matched.

Note: iMate names must match on the placed component and the unconsumed iMate in the assembly.

When the two matching iMates join in the assembly, a single consumed iMate is created. Because the relationship is specific to two components with matching iMate halves, multiple occurrences cannot be placed.

The solution that is selected for iMates (mate and flush, inside and outside, or opposed and aligned) must be the same for the matching iMates.

ALT + Drag iMate Behavior

When using the ALT + Drag feature of constraining iMates on a component part, the other component part that contains the matching iMate solution will display the iMate glyph, as shown in the following image.

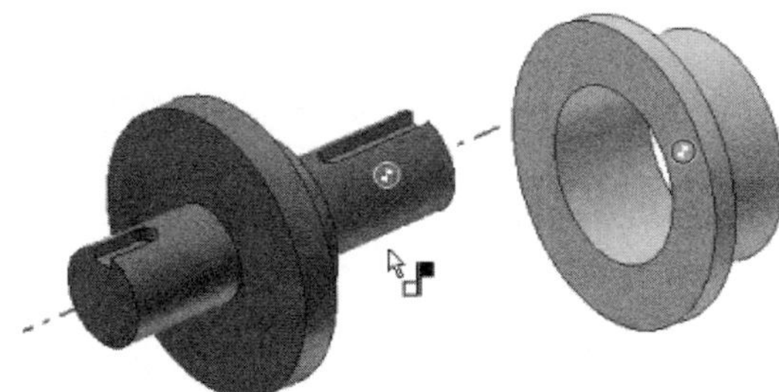

Figure 8.9

Note: iMate glyphs that are currently displayed will not display during an ALT + Drag feature if they don't match the selected iMate type.

EXERCISE 8-1: CREATING AND USING IMATES

In this exercise, you add iMates to an existing subassembly. You then use those iMates to position the subassembly. Next you open a part, combine existing iMates into a composite iMate, and automatically orient that part into the assembly using the composite iMate. You complete the exercise by replacing and automatically orienting a component in the assembly using iMates.

1. Open *ESS_E08_01.iam*.

Figure 8.10

2. Use the Zoom and Rotate tools to examine the parts, and then restore the Home View.

   In the next few steps, you create iMates on the End subassembly (the red component) to automate the process of constraining this component to another component in the assembly. These iMates can also be used to automate the process of orienting the component in other potential assemblies.

3. In the browser, expand the Connector and End components.

4. In the browser, expand iMates under the Connector component. Notice that the End component has no iMates.

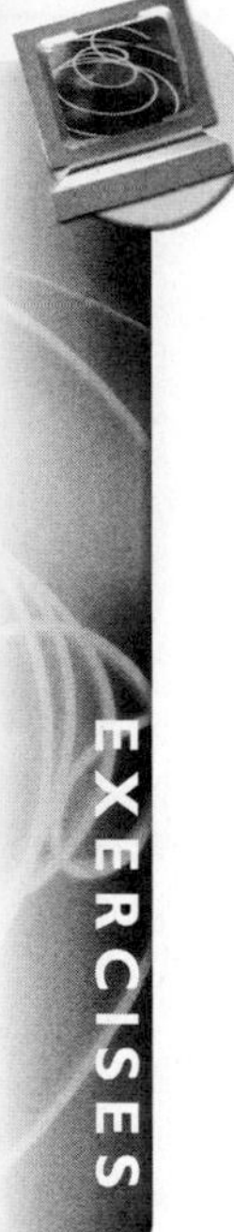

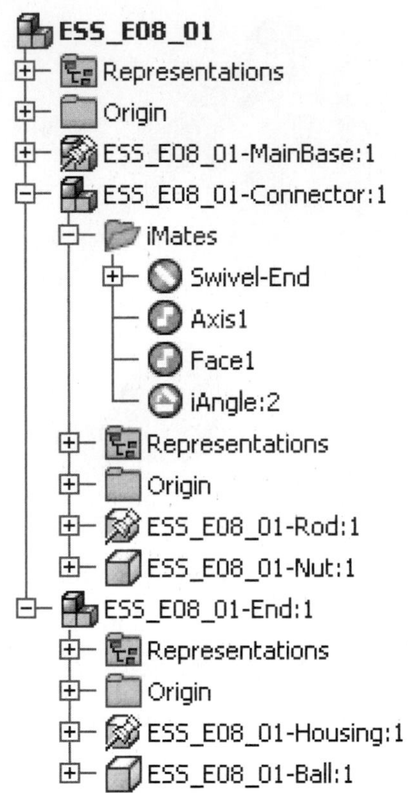

Figure 8.11

5. In the browser, double-click ESS_E08_01-End:1 to activate and edit the subassembly.

Note: iMates are created on individual components or subassemblies. You then use those iMates to position components in an assembly.

6. Click the Create iMate tool.
7. Place your cursor over the tapped hole. When the axis is highlighted, click to select it.

Figure 8.12

8. Expand the Create iMate dialog box.
9. Enter **Axis1** in the Name field, and click Apply to create the first iMate.

Note: You can rename the iMate after creation by slowly double-clicking the name in the browser or by accessing the Properties dialog box from the iMate once it is created.

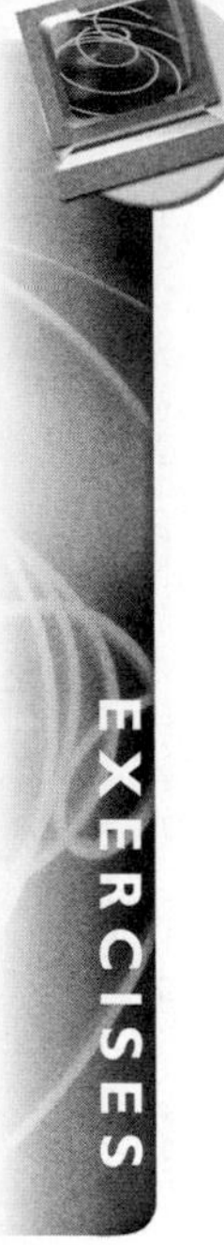

Figure 8.13

10. Place your cursor over the highlighted face, as shown in the following image. When the face is highlighted, click to select it.

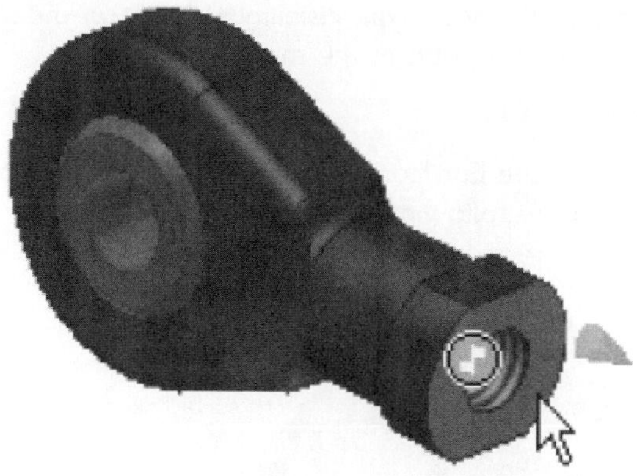

Figure 8.14

11. Enter **Face1** in the Name field of the Create iMate dialog box, and click Apply to create the second iMate.
12. In the Create iMate dialog box, select the Angle constraint button.
13. Place your cursor over the highlighted face, as shown in the following image. When the face is highlighted, click OK to create the third and final iMate.

Note: You leave the default name of iAngle for this iMate.

Figure 8.15

Next return to the assembly, and use the iMates you just created to orient the End component.

14. Click the Return tool.
15. In the browser, hold the CTRL key, and select the Connector and End components.

16. Right-click, and select iMate Glyph Visibility to turn on the iMate glyphs by placing a checkmark next to the option in the menu.
17. Click the Constraint tool.
18. Select the iMate on the End component—the one created by selecting the front face—as shown in the following image.

Figure 8.16

19. Select the iMate on the Connector component, as shown in the following image.

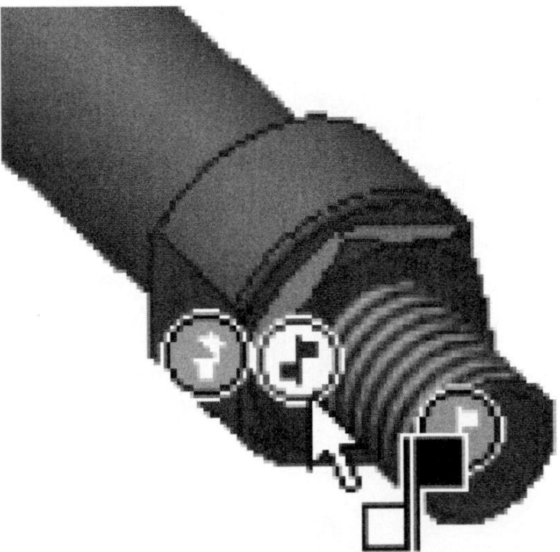

Figure 8.17

20. In the Place Constraint dialog box, click OK. You can also use the ALT + Drag shortcut rather than the Place Constraint tool to create assembly constraints.
21. In the graphics window, click and drag the End component away from the Connector.

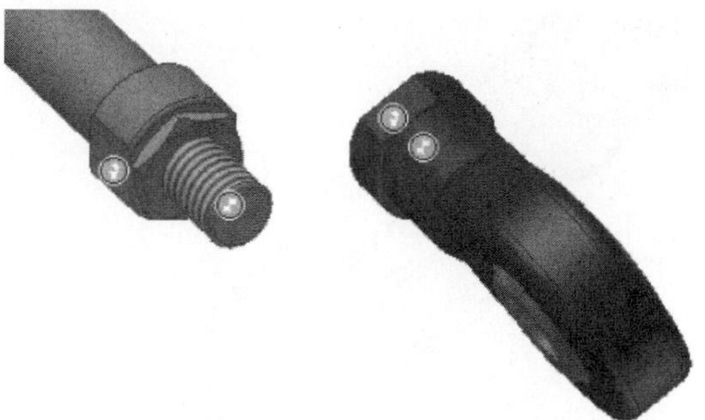

Figure 8.18

22. Press and hold the ALT key.
23. Select and drag the iMate on the End component, as shown in the following image.

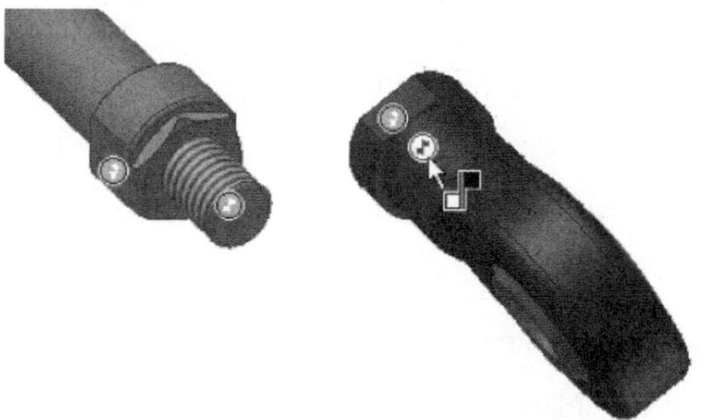

Figure 8.19

24. Drop the iMate on the Connector component iMate, as shown in the following image. Notice that after you select an iMate to drag, only iMate glyphs of the same type are visible.

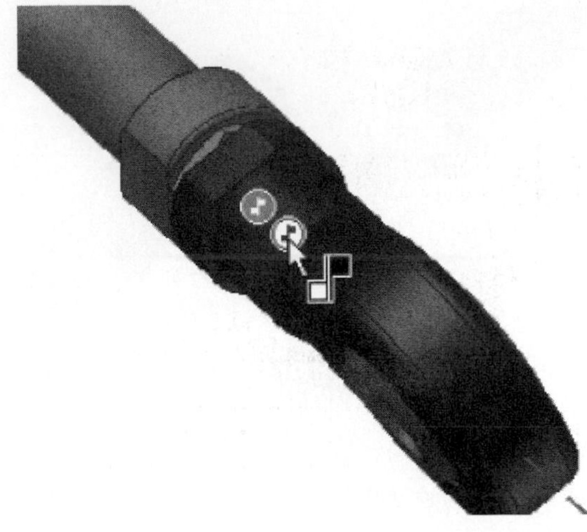

Figure 8.20

25. Select the Connector and End components.
26. Right-click, and select iMate Glyph Visibility to remove the checkmark.

    Next create the final constraint using the third iMate.

27. Select the End component. After selecting, the iMate glyph is visible.
28. Press and hold the ALT key.
29. Select and drag the iMate on the End component, and drop the iMate on the Connector component iMate, as shown in the following image.

Note: The connector may move slightly as you apply this constraint, because the connector is not constrained to a grounded component. You can temporarily ground the Connector if it makes it easier to apply the constraint. Make sure that you unground the component once you've applied the constraint.

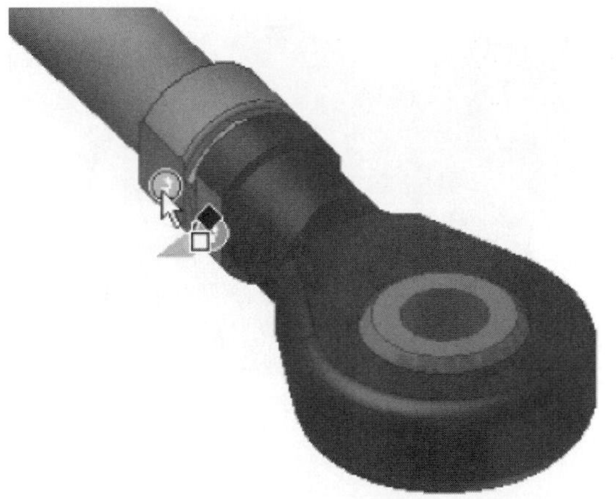

Figure 8.21

Next create a composite iMate on another part to match the composite iMate on the Connector component.

30. In the browser under ESS_E08_01-Connector:1, expand Swivel-End under the iMates folder.

Notice that this composite iMate includes two iMates, as shown in the following image.

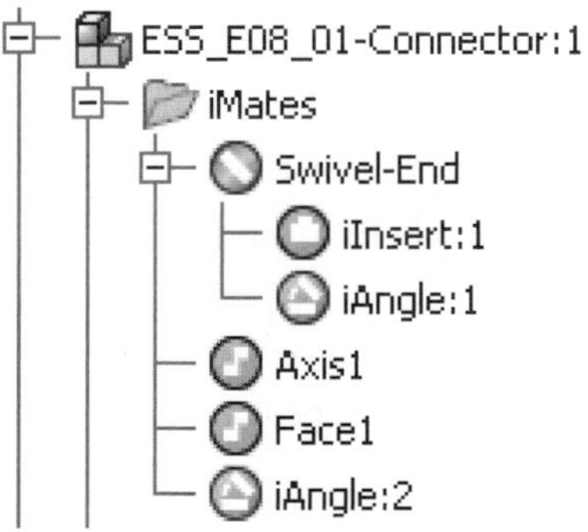

Figure 8.22

31. Open *ESS_E08_01-Mount.ipt* in a separate window.

Figure 8.23

32. In the browser, Expand iMates. Press and hold the CTRL key, and select the two iMates, as shown in the following image.

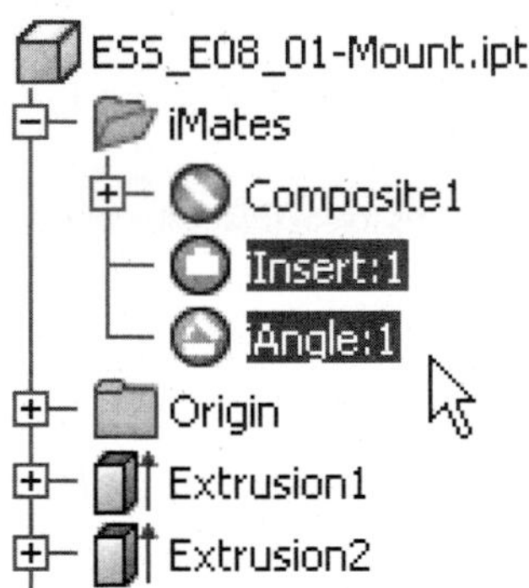

Figure 8.24

33. Right-click, and select Create Composite from the menu.

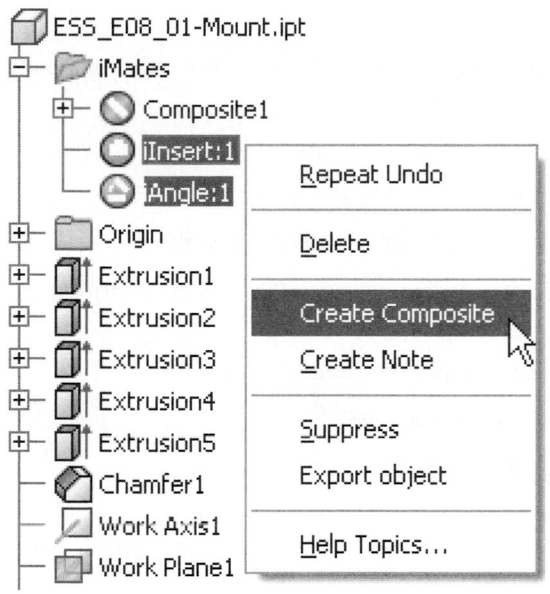

Figure 8.25

34. Right-click the new Composite iMate in the browser, and select Properties from the menu.

35. Change the Name field to **Swivel-End**. If the Suppress box is checked, unselect it.

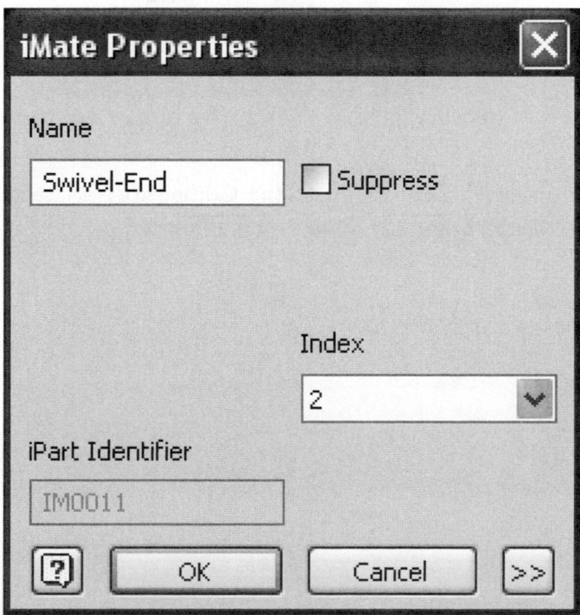

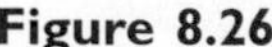

Figure 8.26

36. Click OK.

Next save the modified Mount part with the new composite iMate under a different name.

37. On the File menu, click Save Copy As.

38. Enter a file name of **My_Mount**, and then click Save.

39. Close the original *ESS_E08_01-Mount* file without saving changes.

Next insert this part into the assembly, and automatically place the part using existing iMates.

40. Click the Place Component tool.

41. In the Place Component dialog box:

    a. Select *My_Mount.ipt*. Do not double-click.

    b. Select the Interactively Place with iMates option, which is located on the left portion of the iMates section in the Place Component dialog box, as shown in the following image.

    c. Click Open.

Figure 8.27

The part previews a constraint to the Composite1 iMate that was defined in both the MainBase and My_Mount components.

Figure 8.28

42. Right-click, and select Next Component iMate from the menu. The next match is previewed. Both previews look correct. Now you will complete the placement.

Figure 8.29

43. Click in the graphics window to accept the iMate result. The preview will go to the first match that it finds. In this case, go back to the first preview of the Composite1 iMate.

Figure 8.30

44. Click in the graphics window again to accept the iMate result.
45. Right-click in the graphics window, and select Done from the menu.

Next, you replace a component with a component that has similar iMates.

46. In either the browser or the graphics window, right-click the End component, and select Component > Replace from the menu.
47. In the Place Component dialog box, select *ESS_E08_01-Clevis.ipt*. Click Open.
48. Click OK in the Possible Constraint Loss dialog box.
49. Click No when prompted to Save session edits to *ESS_E08_01-End.iam* prior to delete.

The End component is replaced with the Clevis component, and the iMates are automatically applied.

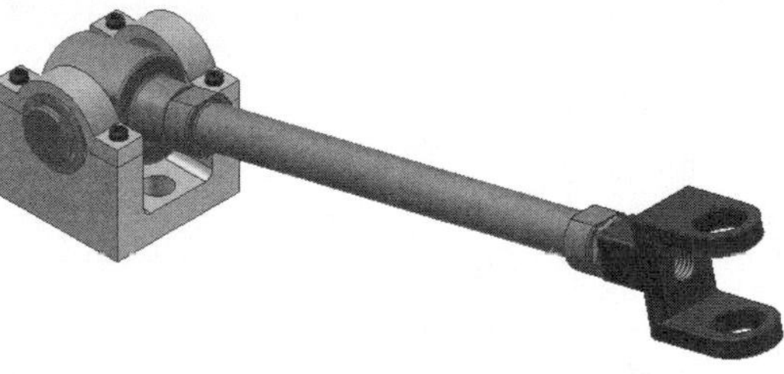

Figure 8.31

50. Close all files. Do not save changes. End of exercise.

DIMENSION DISPLAY, RELATIONSHIPS, AND EQUATIONS

When creating part features, you may want to set up relationships between features and/or sketch dimensions. The length of a part may need to be twice that of its width, for example, or a hole may always need to be in the middle of the part. In Autodesk Inventor, you can use several different methods to set up relationships between dimensions. The following sections will cover these methods.

DIMENSION DISPLAY

When you create a dimension, it is automatically tagged with a label, or parameter name, that starts with the letter "d" and a number: for example, "d0" or "d27." The first dimension created for each part is given the label "d0." Each dimension that you place for subsequent part sketches and features is sequenced incrementally, one number at a time. If you erase a dimension, the next dimension does not go back and reuse the erased value. Instead, it keeps sequencing from the last value on the last dimension created. When creating dimensional relationships, you may want to view a dimension's display style to see the underlying parameter label of the dimension. Five options for displaying a dimension's display style are available.

Display as Value Use to display the actual value of the dimensions on the screen.

Display as Name Use to display the dimensions on the screen as the parameter name: for example, d12 or Length.

Display as Expression Use to display the dimensions on the screen in the format of parameter# = value, showing each actual value: for example, d7 = 20 mm or Length = 50 mm.

Display as Tolerance Use to display the dimensions on the screen that have a tolerance style for example, 40 ± .3.

Display as Precise Value Use to display the dimensions on the screen and ignore any precision settings that are specified: for example, 40.3563123344.

To change the dimension display style, right-click in the graphics window. Click Dimension Display, as shown in the following image, or click Tools > Document Settings and, on the Units tab, change the display style. After you select a dimension display style, all visible dimensions will change to that style. As you create dimensions, they will reflect the current dimension display style.

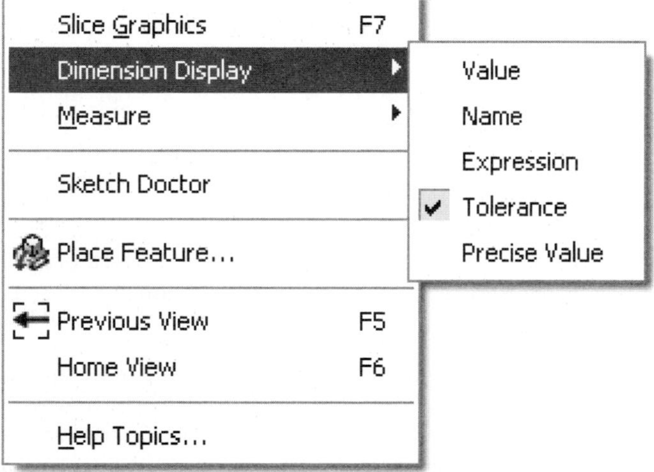

Figure 8.32

DIMENSION RELATIONSHIPS

Setting a dimensional relationship between two dimensions requires setting a relationship between the dimension you are creating and an existing dimension. When entering text in the Edit Dimension dialog box, enter the dimensions parameter name (d#) of the other dimension, or click the dimension with which you want to set the relationship in the graphics window. The following image on the left shows the Edit Dimension dialog box after selecting the 10 mm dimension to which the new dimension will be related. After establishing a relationship to another dimension, the dimension will have a prefix of fx:, as shown in the following image on the right.

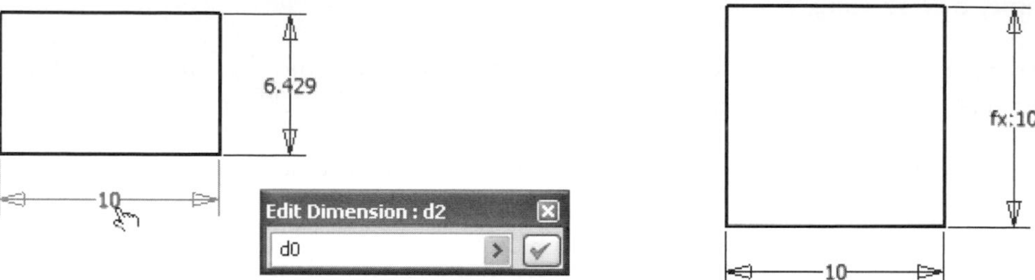

Figure 8.33

EQUATIONS

You can also use equations whenever a value is required: examples include (d9/4)*2 or 50 mm + 19 mm. When creating equations, Autodesk Inventor allows prefixes, precedence, operators, functions, syntax, and units. To see a complete listing of valid options, use the Help system, and search for Functions, Prefixes, and Algebraic Operators.

You can enter numbers with or without units; when no unit is entered, the default unit will be assumed. As you enter an equation, Autodesk Inventor evaluates it. An invalid expression will appear in red, and a valid expression will appear in black. For best results while using equations, include units for every term and factor in the equation.

To create an equation in any field, follow these steps:

1. Click in the field.
2. Enter any valid combination of numbers, parameters, operators, or built-in functions. The following image shows an example of an equation that uses both millimeters and inches for the units.
3. Press ENTER or click the green checkmark to accept the expression.

Note: Use ul (unitless) where a number does not have a unit. For example, use a unitless number when dividing, multiplying, or specifying values for a pattern count.

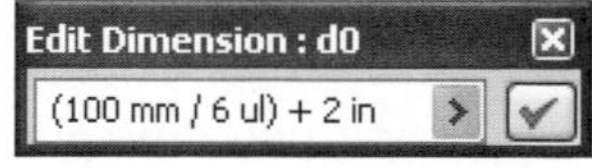

Figure 8.34

PARAMETERS

Another way to set up relationships between dimensions is to use parameters. A parameter is a user-defined name that is assigned a numeric value, either explicitly or through equations. You can use multiple parameters in an equation, and you can use parameters to define another parameter such as depth = length – width. You can use a parameter anywhere a value is required. There are four types of parameters: model, user, reference, and linked.

Note: Add-ins such as Stress Analysis and Design Accelerator can automatically add parameter groups. Depending on the Add-in, you may or may not be able to remove those groups.

Model This type is created automatically and assigned a name when you create a sketch dimension, feature parameter such as an extrusion distance, draft angle, or coil pitch, and the offset, depth, or angle value of an assembly constraint. Autodesk Inventor assigns a default name to each model parameter as you create it. The default name format is a "d" followed by an integer incremented for each new parameter. You can rename model parameters via the Parameters dialog box.

User This type is created manually from an entry in the Parameters dialog box.

Reference This type is created automatically when you create a driven dimension. Autodesk Inventor assigns a default name to each reference parameter as you create it. The default name format is a "d" followed by an integer incremented for each new parameter. You can rename reference parameters via the Parameters dialog box.

Linked This type is created via a Microsoft Excel spreadsheet and is linked into a part or assembly file.

You create and/or edit parameters by clicking the Parameters tool, as shown in the following image, on the Sketch, Part Features, or Assembly panel bars.

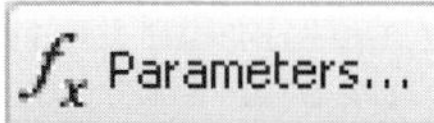

Figure 8.35

After you click the Parameters tool, the Parameters dialog box is displayed. The following image shows an example with some model, user, and reference parameters created. The Parameters dialog box is divided into two sections: Model Parameters and User Parameters. A third section called Reference Parameters is displayed if driven dimensions exist. The values from dimensions or assembly constraints used in the active document automatically fill the Model Parameters section. The User Parameters section is defined manually. You can change the names and equations of both types of parameters, and you can add comments by double-clicking in the cell and entering the new information. The column names for Model and User parameters are the same, and the following sections define them.

Parameter Name The name of the parameter will appear in this cell. To change the name of an existing parameter, click in the box, and enter a new name. When creating a new user parameter, enter a new name after clicking the Add button. When you update the model, all dependent parameters update to reflect the new name.

Unit Enter a new unit of measurement for the parameter in this cell. With Autodesk Inventor, you can build equations that include parameters of any unit type. All length parameters are stored internally in centimeters; angular parameters are stored internally in radians. This becomes important when you combine parameters of different units in equations.

Equation The equation will appear in this cell, and it will determine the value of the parameter. If the parameter is a discrete value, the value appears in rounded form to match the precision setting for the document. To change the equation, click on the existing equation, and enter the new equation.

Nominal Value The nominal tolerance result of the equation will appear in this cell, and it can only be modified by editing the equation.

Tol. (Tolerance) From the drop-down list, select a tolerance condition: upper, median, nominal, or lower.

Model Value The actual calculated model value of the equation in full precision will appear in this cell. This value reflects the current tolerance condition of the parameter.

Export Parameters Column Click to export the parameter to the Custom tab of the Properties dialog box. The parameter will also be available in the bill of materials and parts list Column Chooser dialog boxes.

Comment You may choose to enter a comment for the parameter in this cell. Click in the cell, and enter the comment.

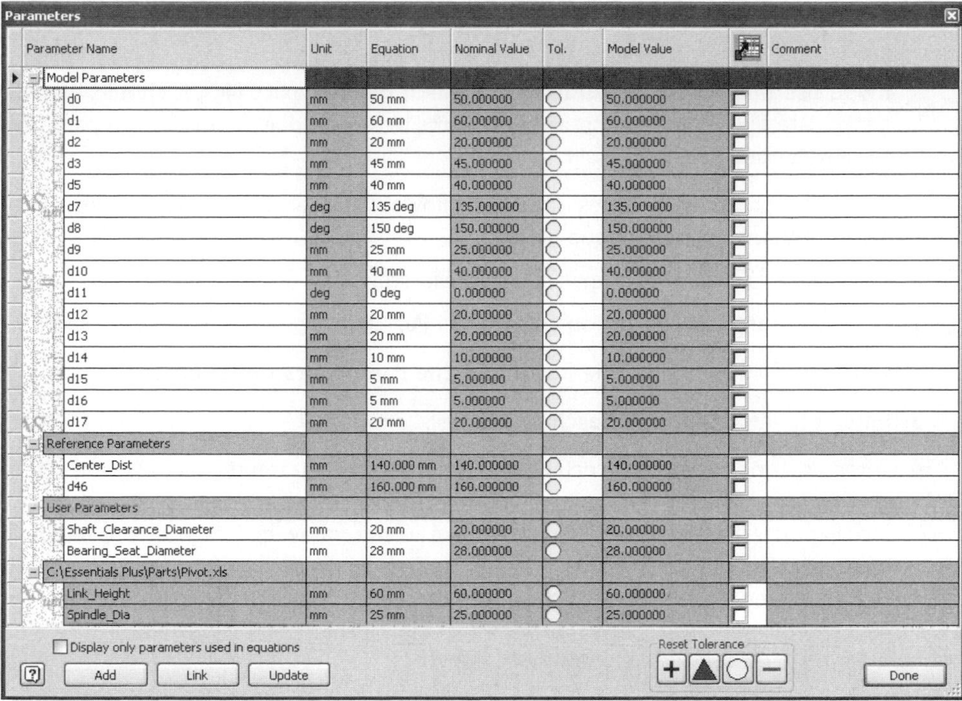

Parameters

Parameter Name	Unit	Equation	Nominal Value	Tol.	Model Value		Comment
Model Parameters							
d0	mm	50 mm	50.000000		50.000000		
d1	mm	60 mm	60.000000		60.000000		
d2	mm	20 mm	20.000000		20.000000		
d3	mm	45 mm	45.000000		45.000000		
d5	mm	40 mm	40.000000		40.000000		
d7	deg	135 deg	135.000000		135.000000		
d8	deg	150 deg	150.000000		150.000000		
d9	mm	25 mm	25.000000		25.000000		
d10	mm	40 mm	40.000000		40.000000		
d11	deg	0 deg	0.000000		0.000000		
d12	mm	20 mm	20.000000		20.000000		
d13	mm	20 mm	20.000000		20.000000		
d14	mm	10 mm	10.000000		10.000000		
d15	mm	5 mm	5.000000		5.000000		
d16	mm	5 mm	5.000000		5.000000		
d17	mm	20 mm	20.000000		20.000000		
Reference Parameters							
Center_Dist	mm	140.000 mm	140.000000		140.000000		
d46	mm	160.000 mm	160.000000		160.000000		
User Parameters							
Shaft_Clearance_Diameter	mm	20 mm	20.000000		20.000000		
Bearing_Seat_Diameter	mm	28 mm	28.000000		28.000000		
C:\Essentials Plus\Parts\Pivot.xls							
Link_Height	mm	60 mm	60.000000		60.000000		
Spindle_Dia	mm	25 mm	25.000000		25.000000		

Figure 8.36

USER PARAMETERS

User parameters are ones that you define in a part or an assembly file. Parameters defined in one environment are not directly accessible in the other environment. If parameters are to be used in both environments, use a linked parameter via a common Microsoft Excel spreadsheet. You can use a parameter any time a numeric value is required. When creating parameters, follow these guidelines:

- Assign meaningful names to parameters, as other designers may edit the part file and will need to understand your thought process. You may want to use the comments field for further clarification.
- The parameter name cannot include spaces, mathematical symbols, or special characters. You can use these to define the equation.
- The parameter name cannot consist of only numbers. It must include at least one alphabetic character, and the alphabetic character must appear first. W1 or Width1, for example, would be valid, but 123 or 1W would be invalid.
- Autodesk Inventor detects capital letters and uses them as unique characters. Length, length, and LENGTH are three different parameter names.
- When entering a parameter name where a value is requested, the upper- and lowercase letters must match the parameter name.

- When defining a parameter equation, you cannot use the parameter name to define itself; for example, Length = Length/2 would be invalid.
- If you use the same user parameter name in multiple part files, it should be defined in a template file. When you create new parts based on that template, the parameter will already be defined.
- Duplicate parameter names are not allowed. Model, User, and Spreadsheet-driven parameters must have unique names.

To create and use a User Parameter, follow these steps:

1. Click the Parameters tool on the Sketch, Part Features, or Assembly panel bars.
2. Click the Add button at the bottom of the Parameters dialog box.
3. Enter the information for each of the cells.
4. After creating the parameter(s), you can enter the parameter name(s) anywhere that a value is required. When editing a dimension, click the arrow on the right, and select List Parameters from the menu, as shown in the following image in the middle. All the available User parameters and any named parameters will appear in a list similar to that in the image on the right. Click the desired parameter from the list.

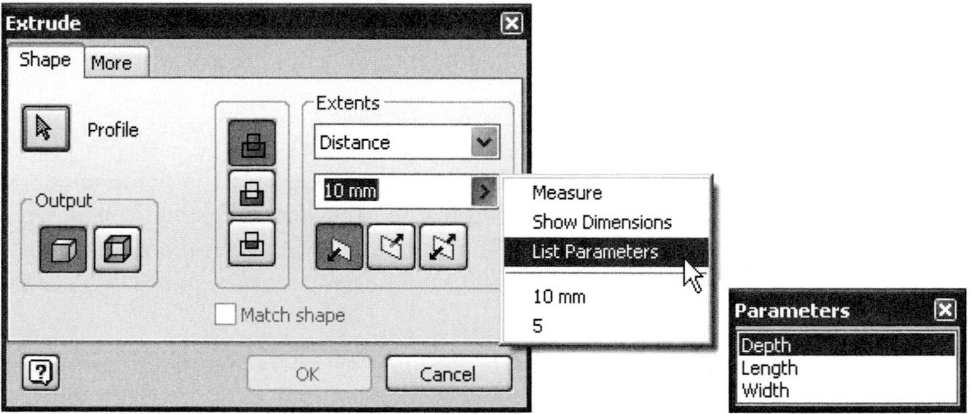

Figure 8.37

LINKED PARAMETERS

If you want to use the same parameter name and value for multiple parts, you can do so by creating a spreadsheet using Microsoft Excel. You then embed or link the spreadsheet into a part or assembly file through the Parameters dialog box. When you **embed** a Microsoft Excel spreadsheet, there is no link between the spreadsheet and the parameters in the Autodesk Inventor file, and any changes to the original spreadsheet will not be reflected in the Autodesk Inventor file. When you **link** a Microsoft Excel spreadsheet to an Autodesk Inventor part or assembly file, any changes in the original spreadsheet will update the parameters in the Autodesk Inventor file. You can link more than one spreadsheet to an Autodesk Inventor file, and you can link each spreadsheet to multiple part and assembly files. By linking a spreadsheet to an assembly file and to the part files that comprise the assembly, you can drive parameters from both environments from the same spreadsheet. There is no limit to the number of part or assembly files that can reference the same spreadsheet. Each linked spreadsheet appears in the browser under the 3rd Party folder.

When creating a Microsoft Excel spreadsheet with parameters, follow these guidelines:

- The data in the spreadsheet can start in any cell, but the cells must be specified when you link or embed the spreadsheet.
- The data can be in rows or columns, but they must be in this order: parameter name, value or equation, unit of measurement, and (if needed) a comment.
- The parameter name and value are required, but the other items are optional.
- The parameter name cannot include spaces, mathematical symbols, or special characters. You can use these to define the equation.
- Parameters in the spreadsheet must be in a continuous list. A blank row or column between parameter names eliminates all parameters after the break.
- If you do not specify a unit of measurement for a parameter, the default units for the document will be assigned when the parameter is used. To create a parameter without units, enter "ul" in the Units cell.
- Only those parameters defined on the first worksheet of the spreadsheet become linked to the Autodesk Inventor file.
- You can include column or row headings or other information in the spreadsheet, but they must be outside the block of cells that contains the parameter definitions.

The following image on the left shows three parameters that were created in rows with a name, equation, unit, and comment. The right side of the image shows the same parameters created in columns.

	A	B	C	D
1	Length	50	mm	Length of Plate
2	Width	Length/2	mm	Width of Plate
3	Depth	10	mm	Depth of Plate

	A	B	C
1	Length	Width	Depth
2	50	Length/2	10
3	mm	mm	mm
4	Length of Plate	Width of Plate	Depth of Plate

Figure 8.38

After you have created and saved the spreadsheet, you can create parameters from it by following these steps:

1. Click the Parameters tool on the Sketch, Part Features, or Assembly panel bars.
2. Click Link on the bottom of the Parameters dialog box.
3. The Open dialog box, similar to that shown in the following image, will appear.

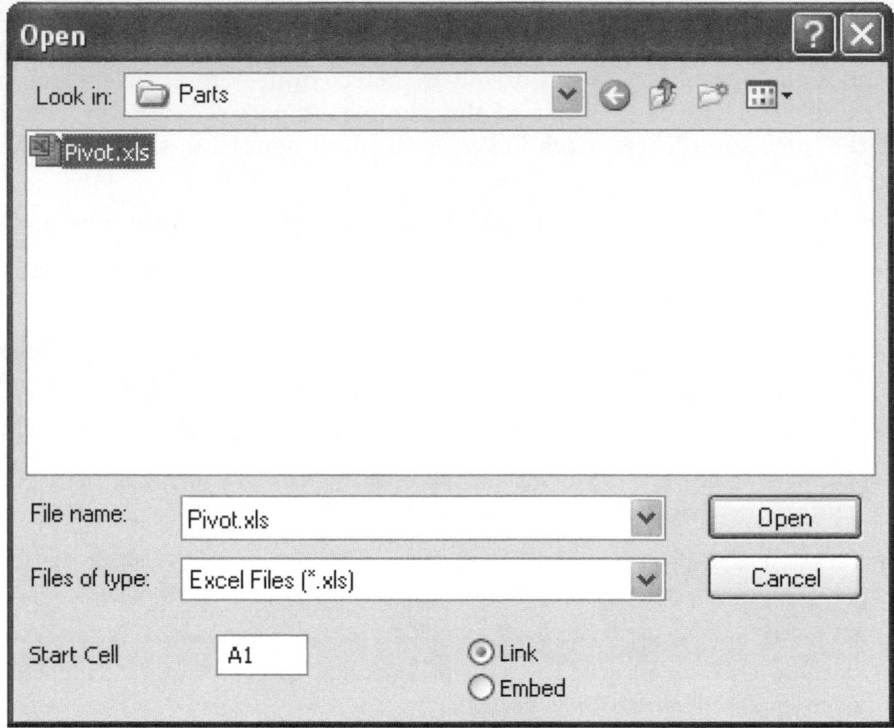

Figure 8.39

4. Navigate to and select the Microsoft Excel file to use.
5. In the lower-left corner of the Open dialog box, enter the start cell for the parameter data.
6. Select whether the spreadsheet will be linked or embedded.
7. Click the Open button: a section showing the parameters is added to the Parameters dialog box, as shown in the following image. If you embedded the spreadsheet, the new section will be titled Embedding #.
8. To complete the operation, click Done.

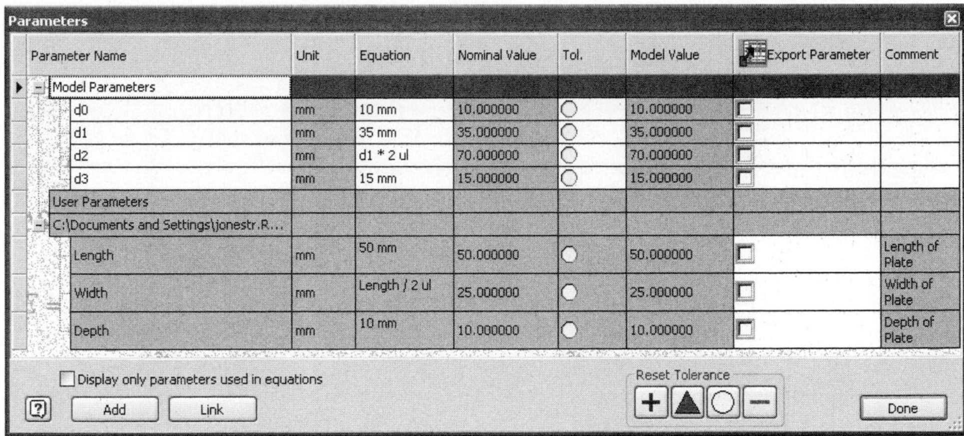

Parameters
Parameter Name
Unit
Equation
Nominal Value
Tol.
Model Value
Export Parameter
Comment
Model Parameters
d0 mm 10 mm 10.000000 10.000000
d1 mm 35 mm 35.000000 35.000000
d2 mm d1 * 2 ul 70.000000 70.000000
d3 mm 15 mm 15.000000 15.000000
User Parameters
C:\Documents and Settings\jonestr.R...
Length mm 50 mm 50.000000 50.000000 Length of Plate
Width mm Length / 2 ul 25.000000 25.000000 Width of Plate
Depth mm 10 mm 10.000000 10.000000 Depth of Plate
Display only parameters used in equations
Add
Link
Reset Tolerance
Done

Figure 8.40

To edit the parameters that are linked, follow these steps:

1. Open the Microsoft Excel file.
2. Make the required changes.
3. Save the Microsoft Excel file.
4. Open the Autodesk Inventor part or assembly file that uses the spreadsheet.

You can also follow these steps when a spreadsheet has been linked. If you chose to embed the spreadsheet, the following steps must be used to edit the parameters:

1. Open the Autodesk Inventor part or assembly file that uses the spreadsheet.
2. Expand the 3rd Party folder in the browser.
3. Either double-click the spreadsheet or right-click and select Edit from the menu, as shown in the following image.

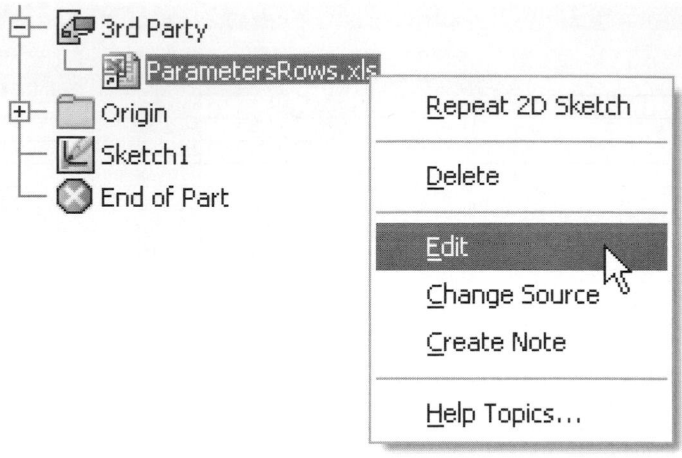

Figure 8.41

4. The Microsoft Excel spreadsheet will open in a new window for editing.
5. Make the required changes.
6. Save the Microsoft Excel file.
7. Activate the Autodesk Inventor part or assembly file that uses the spreadsheet.
8. Click the Update tool from the standard toolbar.

Note: If you embedded the spreadsheet, the changes will not be saved back to the original file but will only be saved internally to the Autodesk Inventor file.

EXERCISE 8–2: RELATIONSHIPS AND PARAMETERS

In this exercise, you create a sketch and dimension it. You then set up relationships between dimensions, and define user parameters in both the model and an external spreadsheet.

1. Unless this step has already been performed, click Tools > Application Options, and click on the Sketch tab. Check Autoproject Part Origin on Sketch Create to automatically have the origin point projected for the part.
2. Click OK to close the Application Options dialog box.
3. Start a new file based on the default *Standard (mm).ipt* file.
4. Draw the geometry, as shown in the following image. The lower-left corner should be located at the origin point, and the arc is tangent to both lines. The size should be roughly 25 mm wide by 40 mm high.

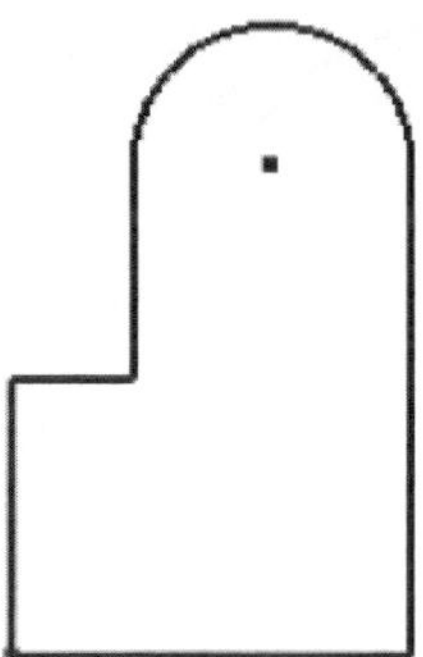

Figure 8.42

5. Add the four dimensions, as shown in the following image.

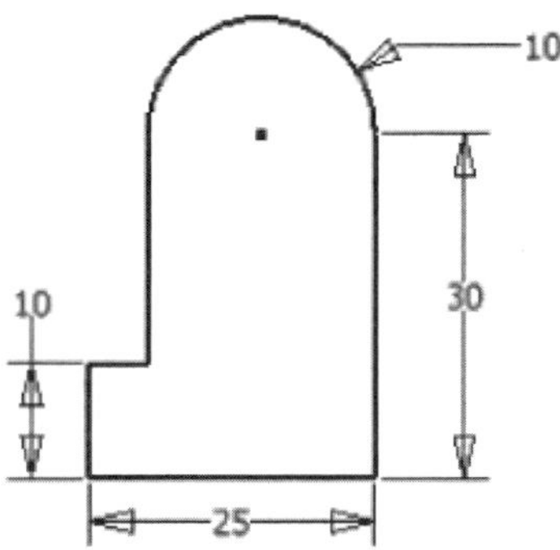

Figure 8.43

6. Double-click the 10 mm radius dimension of the arc, and for its value, select the 25 mm horizontal dimension, as shown in the following image.

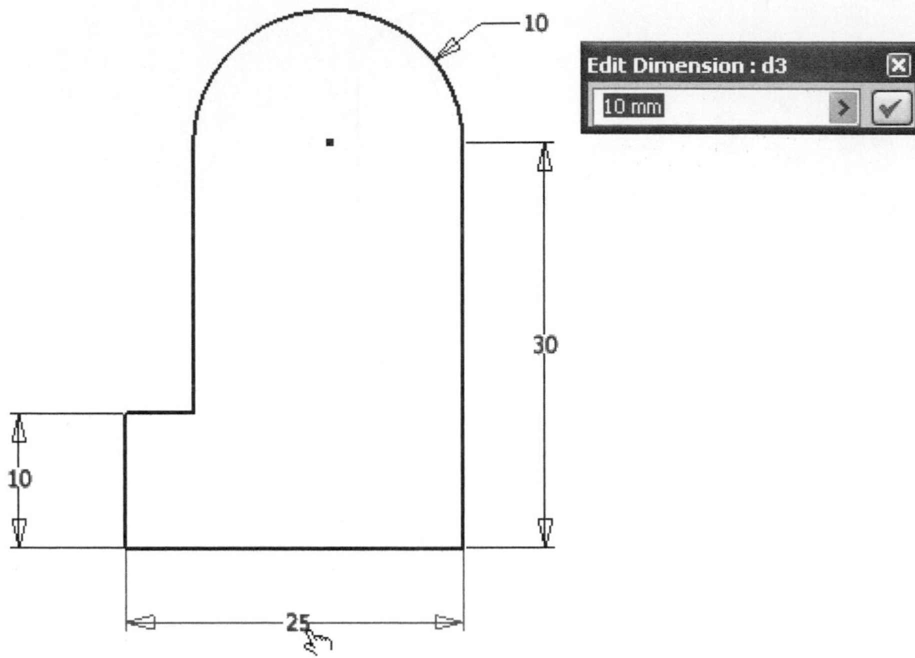

Figure 8.44

7. In the Edit Dimension dialog box, type **/4** after the d#, creating "d1/4." The 1 may be a different number depending upon the order in which your geometry was dimensioned. Click the green checkmark to accept the dimension value in the Edit Dimension dialog box.

8. Double-click the 10 mm vertical dimension, and for its value, select the 30 mm vertical dimension. Type **/2** after the d#. Click the green checkmark to accept the dimension value in the Edit Dimension dialog box.

9. Change the value of the 25 mm horizontal dimension to **50 mm** and the 30 mm vertical dimension to **40 mm**. When done, your sketch should resemble the following image. The fx: before the two dimensions denotes that they are equation-driven or have a reference to a parameter.

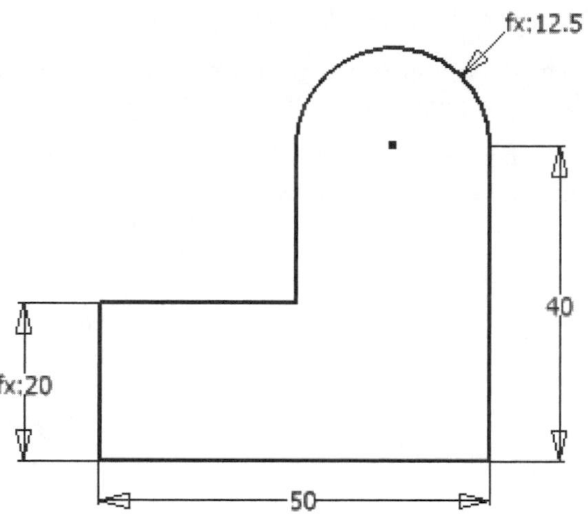

Figure 8.45

10. Next create parameters and drive the sketch from them. Click the Parameters tool from the 2D Sketch panel bar.
11. In the Parameter dialog box, create a User Parameter by clicking the Add button. Type in the following information:

- Parameter Name = **Length**
- Units = **mm**
- Equation = **75**
- Comment = **Bottom length**

12. Create another User Parameter by selecting the Add button, and type in the following information:

- Parameter Name = Height
- Units = **mm**
- Equation = **Length/2**
- Comment = **Height is half of the length**

When done, the User Parameter area should resemble the following image.

Parameters

Parameter Name	Unit	Equation	Nominal Value	Tol.	Model Value		Comment
+ Model Parameters							
− User Parameters							
Length	mm	75 mm	75.000000		75.000000		Bottom length
Height	mm	Length / 2 ul	37.500000		37.500000		Height is half of the length

Figure 8.46

13. Close the Parameter dialog box by clicking Done.
14. Change the dimension display by right-clicking the graphics window, and then click Dimension Display > Expression.
15. Double-click the bottom horizontal dimension, and change its value to the parameter Length as follows: click the arrow; from the menu, click List Parameters, click Length from the list, and then click the checkmark in the dialog box.
16. Double-click the right vertical dimension, and change its value to Height. When done, your screen should resemble the following image.

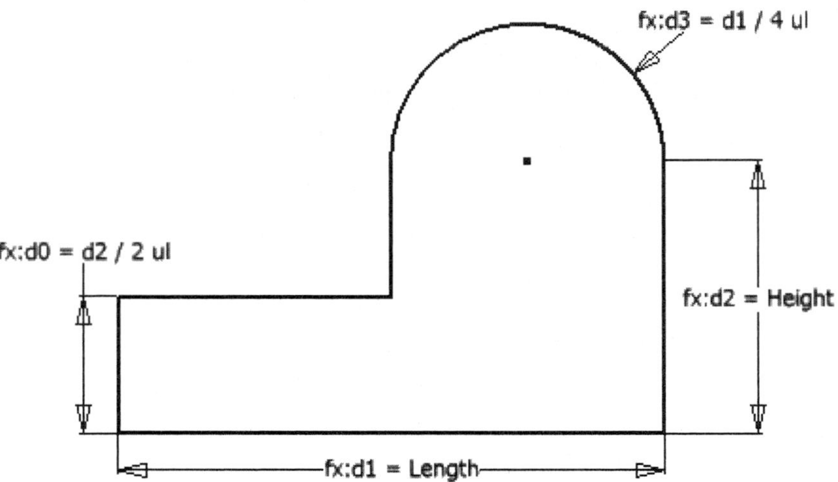

Figure 8.47

17. Click the Parameters tool in the panel bar. From the User Parameters section, click on the equation cell of the parameter name Length, and change its value to **35**.
18. Click in the equation cell of the parameter name Height; change its value to **50** and the comment field to "**Height is half of the right side**." When done making the changes, the User Parameter area should resemble the following image.

User Parameters							
Length	mm	35 mm	35.000000	○	35.000000	☐	Bottom length
Height	mm	50 mm	50.000000	○	50.000000	☐	Height is half of the right side

Figure 8.48

19. To complete the changes, click the Done button.
20. Click the Update tool, and the sketch should update to reflect the changes, as shown in the following image.

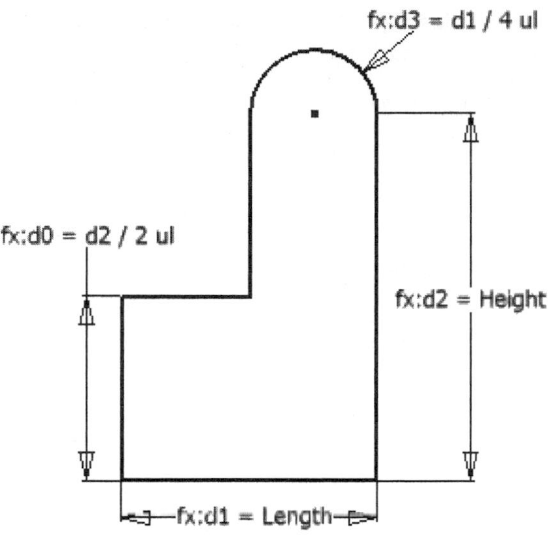

Figure 8.49

21. Save the file as *ESS_E08_02.ipt*.
22. Now create a spreadsheet that has two parameters. Create an Excel spreadsheet with the column names Parameter Name, Equation, Unit, and Comment, as shown in the following image. Enter the following data:
    - Parameter Name = BaseExtrusion
    - Equation = **30**
    - Units = **mm**
    - Comment = **Extrusion distance for the base feature**
    - Parameter Name = **Draft**
    - Equation = **5**
    - Units = **deg**
    - Comment = **Draft for all features**

	A	B	C	D
1	Parameter Name	Equation	Unit	Comment
2	BaseExtrusion	30	mm	Extrusion distance for the base feature
3	Draft	5	deg	Draft for all features

Figure 8.50

23. Save the spreadsheet as *ESS_08_Parameters.xls*.
24. Make Autodesk Inventor the current application, and then click the Parameters tool.

EXERCISES

25. In the Parameter dialog box, click the Link button and select the *ESS_08_Parameters.xls* file, but do not click Open yet.

26. For the Start Cell, enter **A2**. If you fail to do this, no parameters will be found.

27. Click the Open button, and a new spreadsheet area will appear in the Parameters dialog box, as shown in the following image. Click the Done button to complete the operation.

C:\Essentials Plus\F							
BaseExtrusion	mm	30 mm	30.000000	○	30.000000	☐	Extrusion distance for the base feature
Draft	deg	5 deg	5.000000	○	5.000000	☐	Draft for all features

Figure 8.51

28. Change to the home view.

29. Click the Extrude tool from the panel bar, and click the More tab in the Extrude dialog box, as shown in the following image. For the Taper's value, enter **Draft** or click the arrow from the menu, click List Parameters, and then click the parameter Draft.

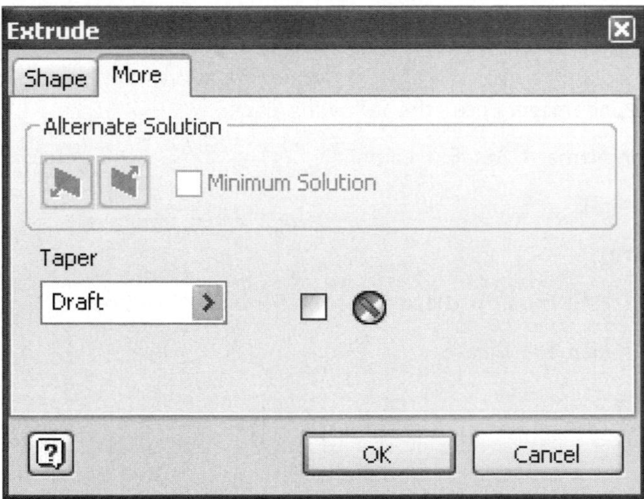

Figure 8.52

30. Click the Shape tab. For the Distance value, use the parameter BaseExtrusion, as shown in the following image.

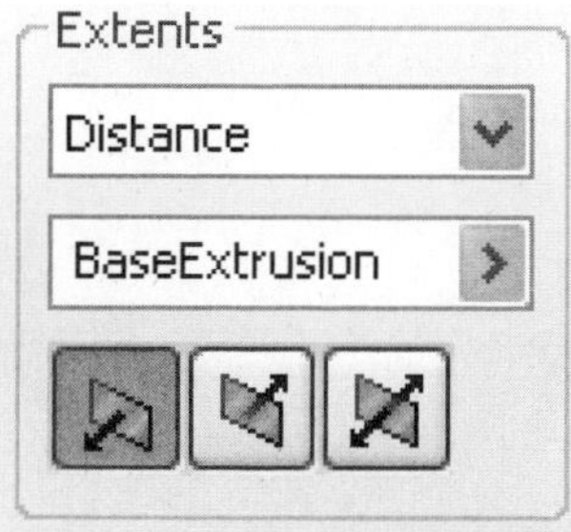

Figure 8.53

31. To complete the operation, click the OK button. Your model should resemble the following image.

Figure 8.54

32. In the browser, expand the 3rd Party icon. Either double-click or right-click on the name *ESS_08_Parameters.xls*, and select Edit from the menu.
33. In the Excel spreadsheet, make the following changes: For the Parameter Name "BaseExtrusion," change the Equation to **50 mm**. For the Parameter Name "Draft," change the Equation to **–3°**.
34. Save the spreadsheet, and close Excel.
35. Make Autodesk Inventor the current application, and click the Update tool from the Standard toolbar. When done, your model should resemble the following image.

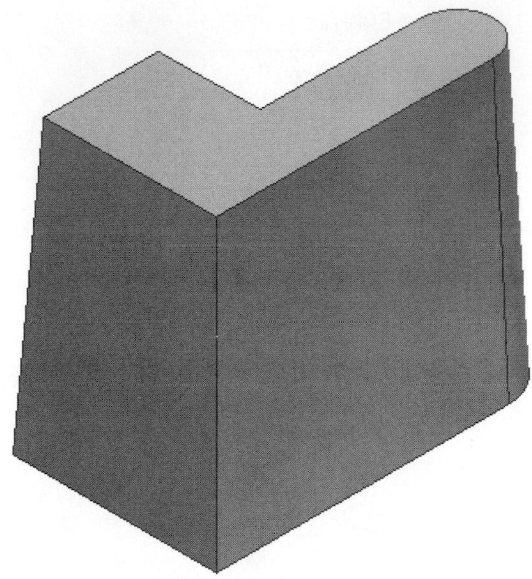

Figure 8.55

36. Close the file. Do not save changes. End of exercise.

IPARTS

iParts enable design intent and design knowledge to be shared and reused. They also enable you to store multiple part parameters and properties and then calculate unique part file versions based on certain configurations of the stored parameters and properties. Using the Parameters dialog box, you can store one value for each parameter. In the industry, configured parts have been referred to as tabulated parts, charted parts, or a family of parts. You can also use iParts to create part libraries that allow design data to be reused. An iPart is generated from a standard Autodesk Inventor part file (*.ipt*). When you activate the Create iPart tool, the standard part file is converted into an iPart. You can add individual members, or configurations, to the iPart; it is then referred to as an iPart factory.

There are two stages to the use of iParts: authoring the iPart and placing an iPart version. In the authoring or creation stage, you design the part and establish all possible versions of the design in a table. The rows of the table describe the members of the iPart factory.

In the placement stage, you select a member from the iPart factory, and an iPart version is published and inserted into your assembly.

iParts allow you to suppress specific features, control iMates and thread properties, and add or modify file properties to the different members that are contained within the iPart factory. They also allow user-defined input for specific parameters. You can further enhance inputs to include an element of control. For instance, you can apply values within a predetermined range or from a list to a parameter. After creating an iPart, you can place it into an assembly, where you can select a specific member. The following sections explain how to create and then place an iPart into an assembly.

An example of an iPart is a simple bolt. The bolt has a number of different sizes that are associated with it. An iPart allows you to define these different sizes and configurations—the material, part properties, and so on—and have them reside within a single file, or factory. When placing the bolt in an assembly, you can select which size, or version/member, of the bolt, or iPart factory, you want to use. You can then constrain the placed member in the assembly like any other part file.

CREATING IPARTS

You can create two types of iParts: standard and custom.

Standard iParts or Factories

You cannot modify iPart values; when placing a standard iPart into an assembly, you can only select the predefined members for placement. A standard iPart that you place in an assembly cannot have features added to it after placement.

Custom iParts or Factories

Custom iParts allow you to place a unique value for at least one variable. A custom iPart that you place in an assembly can have features added to it.

To create an iPart, follow these steps:

1. Create an Autodesk Inventor part or a sheet metal part.
2. Add the dimensions of the geometry of the design to be changed to the iPart Author table.
3. For easier creation of the member table, use descriptive names for the values of the parameters.
4. If you do not use parameter names, you will need to determine what each parameter (d#) represents within the parts geometry.

Tip: Any parameters with a name other than d# are added automatically to the parameter table during the creation of the iPart factory.

5. Issue the Create iPart tool on the Tools menu, as shown in the following image, to add the members or configurations to the iPart Author table.

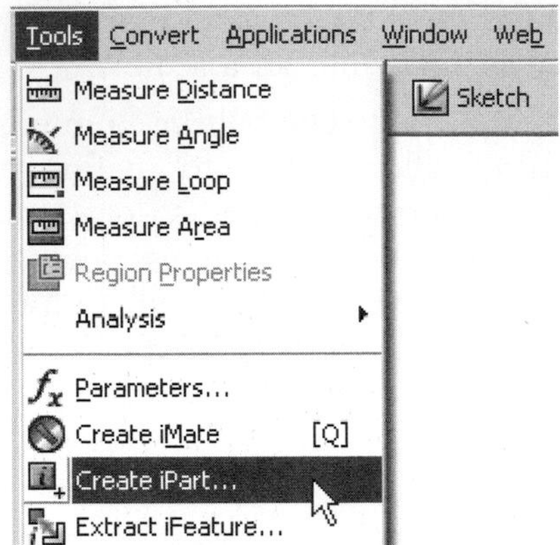

Figure 8.56

The iPart Author dialog box appears, as shown in the following image.

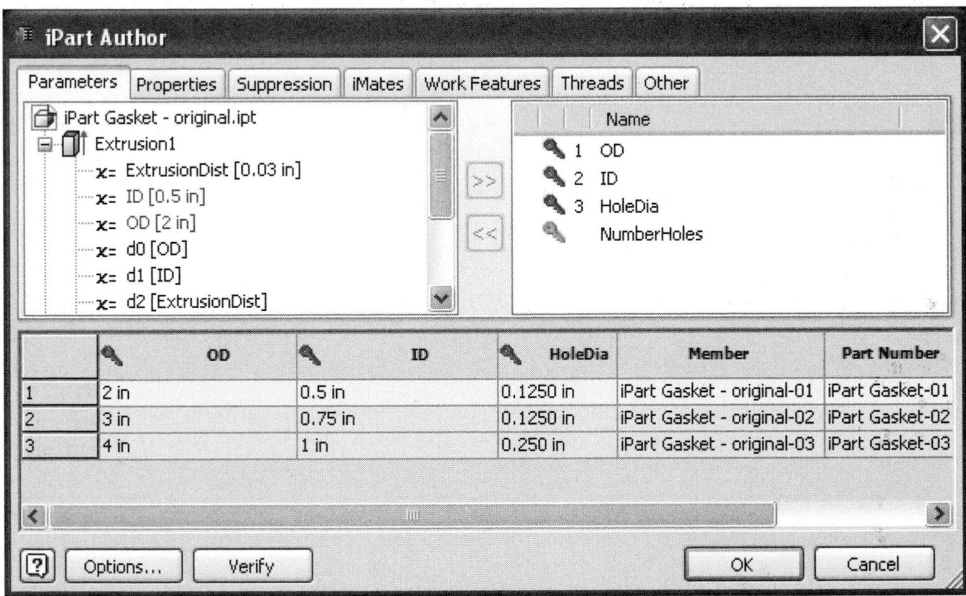

Figure 8.57

To create the iPart members, follow these steps:

6. Add to the right side of the dialog box, or the parameter list, all parameters and dimensions that will be configured. All named parameters are added automatically to the right side.
7. To add a d# dimension to the list on the right side, select it in the left column, and then click the Add parameters (>>) button. You can also double-click on the parameter to add it.
8. To remove a parameter from the right side of the dialog box, select its name, and click the Remove parameters (<<) button.
9. After you have added the parameters, you need to define the keys. Keys identify the column whose values are used to define the iPart member when the part is published or placed in an assembly. For example, setting a parameter as a primary key allows the designer placing the part to choose from all available values for that parameter in the selected items list.
10. To specify the key order, click an item in the key column of the selected parameters list to define it as a key, or right-click on the item and select the key sequence number. Selected keys are blue; items that are not selected as keys have dimmed key symbols. You can decide not to add primary keys or to add one or more additional keys as needed. You cannot specify custom table columns as keys.
11. Click on the other tabs in the dialog box to perform other specific operations. These items are not required to define the iPart factory. The following sections describe them in more detail.

Properties Tab The Properties tab lists all of the summary, project, physical, and custom properties in the file. If you use the material property on this tab to control the material of your iParts, you must use the Material Column option in the table. This option is described in the Special Table Elements section later in this chapter. You must also set the current color of the iPart to As Material prior to saving the iPart. For part properties to be used in drawings, Bill of Materials (BOMs), and other downstream purposes, you have to include the properties in the table even if their values do not vary between members.

Suppression Tab The Suppression tab allows you to suppress features of specific members of the iPart factory. If you enter Suppress in a cell for the feature, the feature will be suppressed when the version is calculated. When the cell contains Compute, the feature is not suppressed.

iMates Tab On the iMates tab, select one or more iMates to include in the iPart. iMates are included in the iPart if their status is Compute. If their status is suppressed, they are not included. Control the iMate properties for each member using the iMates, Parameters, and Suppression tabs. On the iMates tab, you can perform the following actions:

- Define different offset values for different iPart members.
- Suppress iMates or Composite iMates for different iPart members.

- Change the matching name for different assembly configurations.
- Change the sequence of the iMates.

Since iMate names are not unique, each iMate is assigned an index tag in the iPart Author. Each iMate property, except the offset value, uses this tag to identify a specific iMate property in the table. Tags are not assigned to offset values because iMate offset values are parameters, which are unique by definition.

Work Features Tab On the Work Features tab, select the user-defined work features to be included in the iPart. Origin work features are included automatically in an iPart factory and iPart members, but they are not listed. The work features are managed in the iPart if its table status is Include. If the status is Exclude, the work features are not managed in the iPart.

Threads Tab You can control thread features in the iPart factory to create table-driven items for regular or tapered thread features. Use Family to identify the thread standard, Designation to control the size and pitch, and Class to define fit based upon the standard. You can assign the thread Direction as right or left in the table. You also have the option to modify the pipe diameter.

Other Tab You can use the Other tab to create custom table items. For example, you can add a column that represents the name of the iPart member. You can create a column named Version to identify the appropriate member when it is placed in an assembly. You can also create custom values such as Color if you want to control the display style of the iPart.

Special Table Elements You can right-click on a cell of a member of the table or a column label to access additional custom table capabilities. You can define the name of the iPart file to be used when it is published using the File Name Column option. Control the color of the iPart upon publishing using the Display Style Column option. You can also specify material of the iPart using the Material Column option. These options are available only for part properties and the table elements created using the Other tab. They are not available on parameter items.

Next add a member in the iPart table by right-clicking on a row number at the bottom of the dialog box and selecting Insert Row from the menu, as shown in the following image.

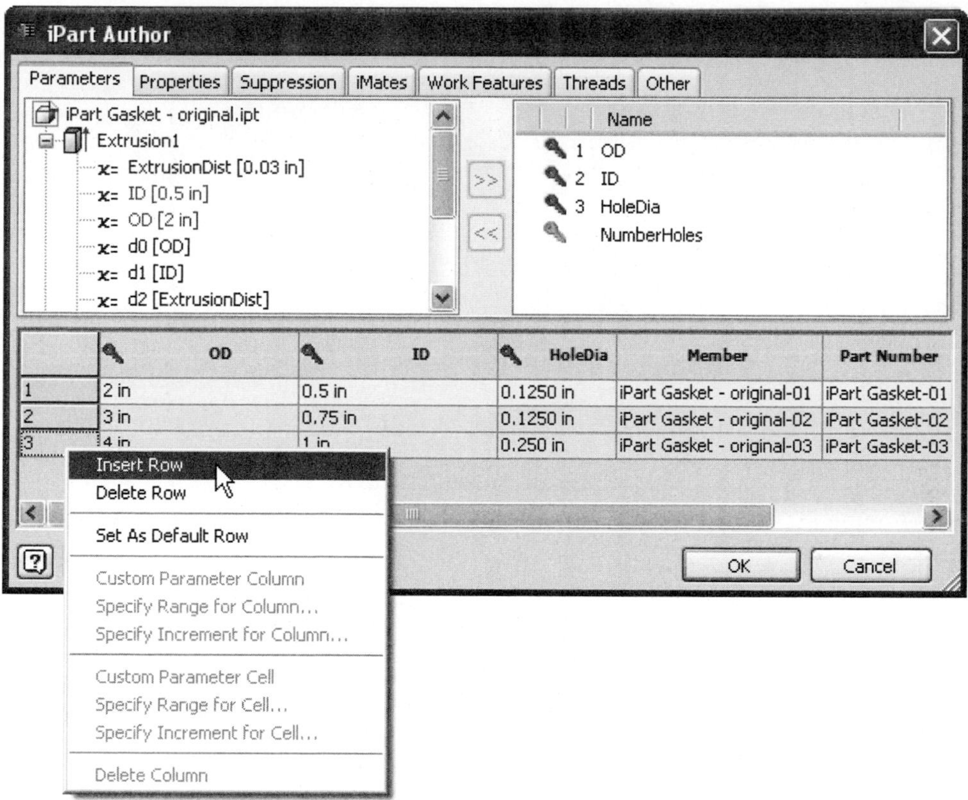

Figure 8.58

12. Edit the cell contents by clicking in the cell and typing new values or information as needed. To delete a row, right-click the row number, and select Delete Row from the menu. Each row that you add in the bottom pane of the dialog box represents an additional member within the iPart factory. The table functions similarly to a spreadsheet. In addition to adding or deleting members of the iPart factory, you can also use the table to change members by modifying cell values.

To allow the designer placing the iPart to specify a custom value for a given column, right-click on the column name, and select Custom Parameter Column from the menu, as shown in the following image. To make a specific cell custom, right-click in the individual cell, and select Custom Parameter Cell from the menu.

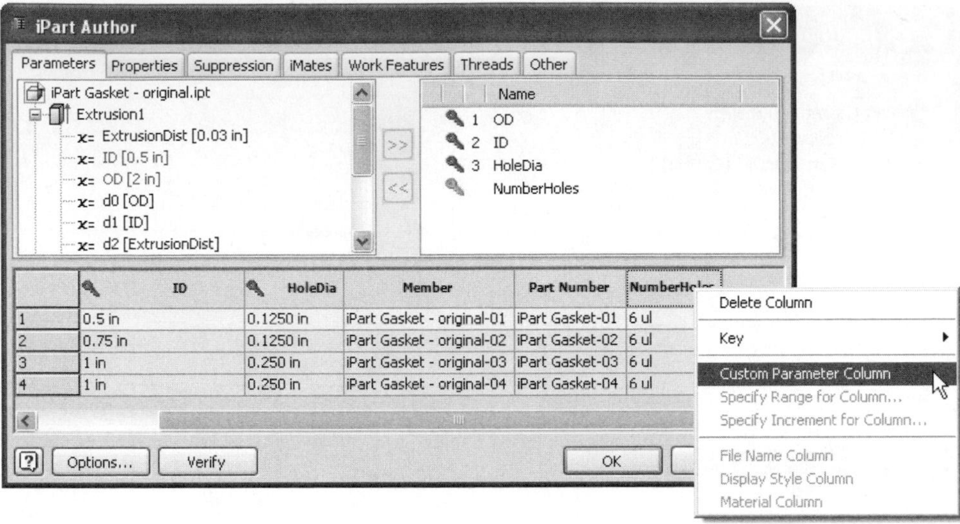

Figure 8.59

After designating a custom parameter column or cell, you can set a minimum and maximum range of values by right-clicking in the column heading or cell and selecting Specify Range for Column or Range for Cell from the menu. The Specify Range dialog box appears, as shown in the following image. Select the options as needed. Custom columns and cells are highlighted with a blue background in the table. You can also specify the increment that can be entered for a custom column or cell using the Specify Increment for Column option from the menu.

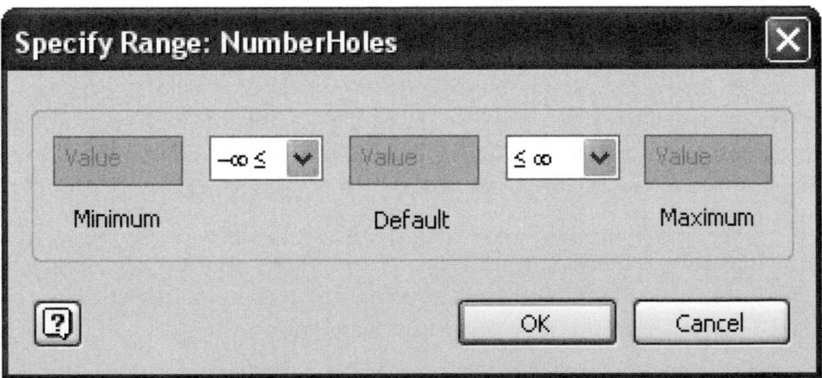

Figure 8.60

13. Set the member that will be the default by right-clicking on the row number and selecting Set As Default Row from the menu. You may select the other options as needed.

14. Click the Options button to edit the part number and member naming schemes for the iPart factory, as shown in the following image.

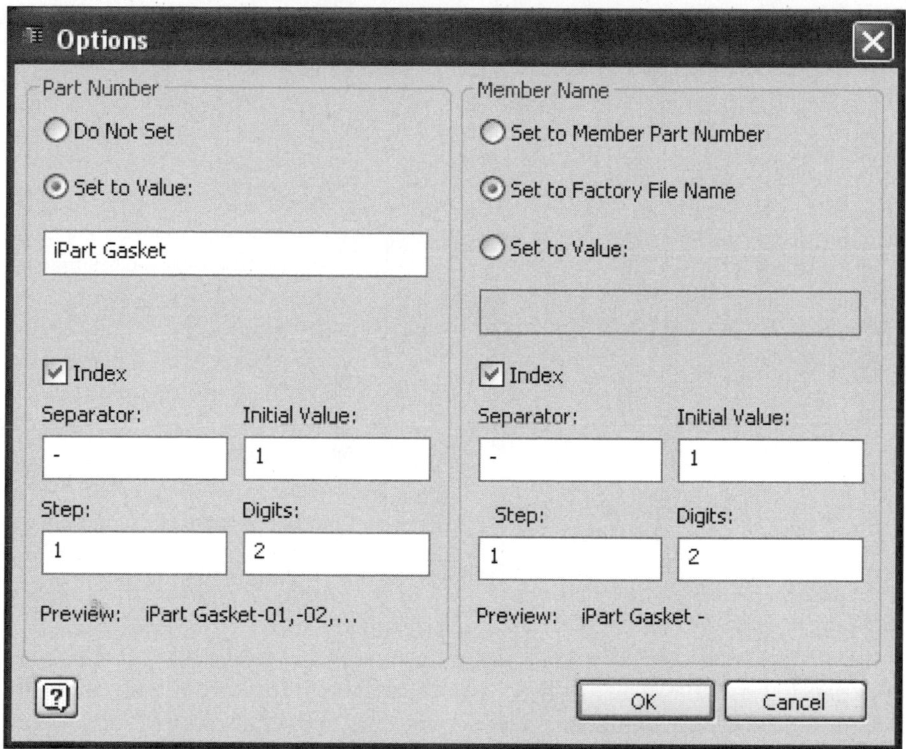

Figure 8.61

15. Click OK in the iPart Author dialog box when you have finished defining the contents of the table. The iPart Author dialog box closes, and the part is converted to an iPart factory. The table is saved, and a table icon appears in the browser, as shown in the following image.

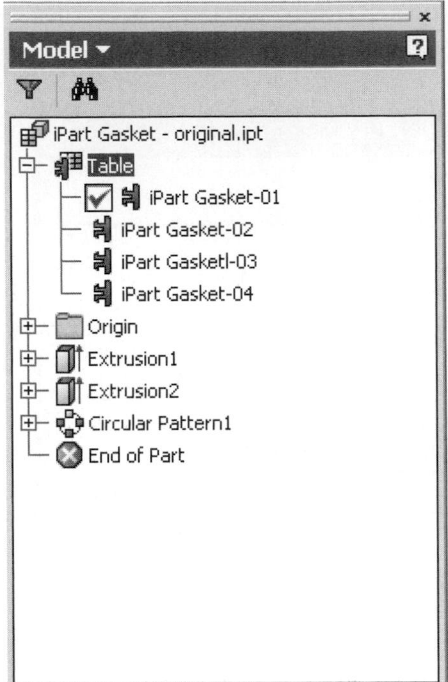

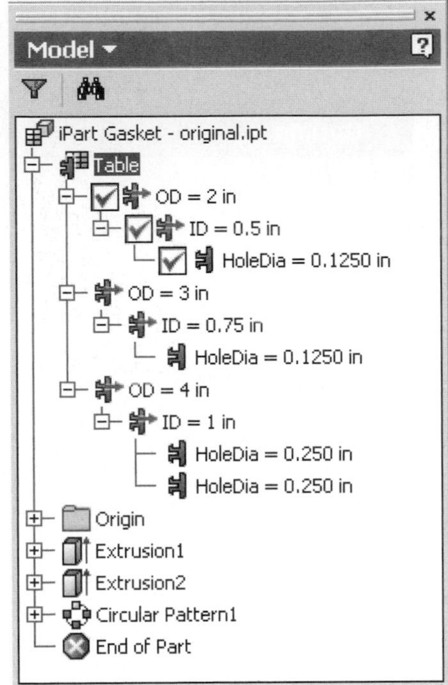

Figure 8.62

You can expand the Table icon in the browser to view the iPart members based on the member name or keys and values that you define. The active, or calculated, version appears with a checkmark, as shown in the previous image. You change the browser display to show the member name or keys by right-clicking and selecting either List by Member Name or List by Keys from the menu.

EDITING IPARTS

You can perform a number of operations on an iPart factory after you have created it. You can delete the table, modify the parameters or properties for individual members, add or delete additional members, and so on. Right-click the Table icon in the browser to delete the table and convert the iPart factory back to a part, edit the table with the iPart Author dialog box using the Edit Table option, or edit the table with Microsoft Excel using the Edit via Spread Sheet option, as shown in the following image. Make changes as needed and save the file. These options are also available from the iParts/iAssemblies toolbar.

Note: Changes made to iPart factories will not be updated automatically in members that you have previously placed in assemblies. To update the iPart members in an assembly, open the assembly, and click the Update tool on the Standard toolbar.

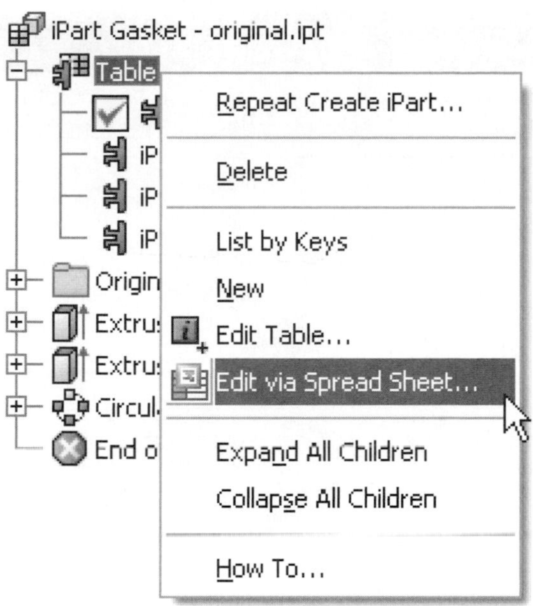

Figure 8.63

When editing the spreadsheet using Microsoft Excel, you can incorporate spreadsheet formulas, conditional statements, and multiple sheet data extractions, but you cannot modify spreadsheet formulas and conditional statements from the iPart Author dialog box. These types of cells are inactive and are highlighted in red when displayed in the iPart Author.

IPART PLACEMENT

You can place standard iParts in assemblies using the Place Component tool. When you select a standard iPart factory, an additional Place Standard iPart dialog box appears. It allows you to use the Keys tab to identify an iPart member by selecting the key values, use the Tree tab to locate a member by expanding the key values, use the Table tab to identify an iPart member by selecting a row in the table, and place multiple instances of different iPart members.

To place a standard iPart into an assembly, follow these steps:

1. Start a new assembly or open an existing assembly in which to place the iPart.
2. Click the Place Component tool, and navigate to and select the iPart to place in the assembly in the Place Component dialog box.
3. Click the Open button, and the Place Standard iPart dialog box will appear, as shown in the following image. If a custom cell(s) or column(s) exists, the Place Custom iPart dialog box appears.

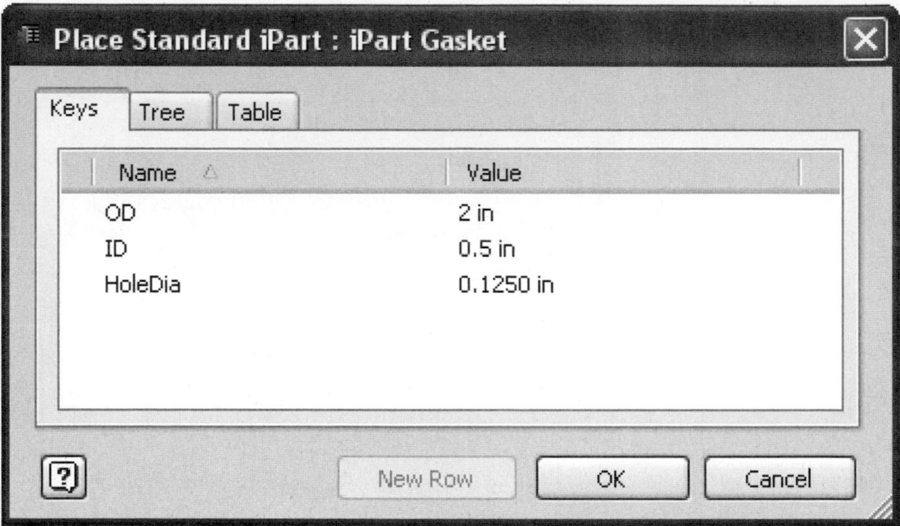

Figure 8.64

4. To select from the member list, select the Keys, Tree, or Table tab, and then select the member that defines the part you want to place.
5. If you are placing a custom part, select the Keys, Tree, or Table tab, and enter a value in the right side of the dialog box, as shown in the following image. The value must fall within the limits set in the iPart factory; otherwise, an alert appears.

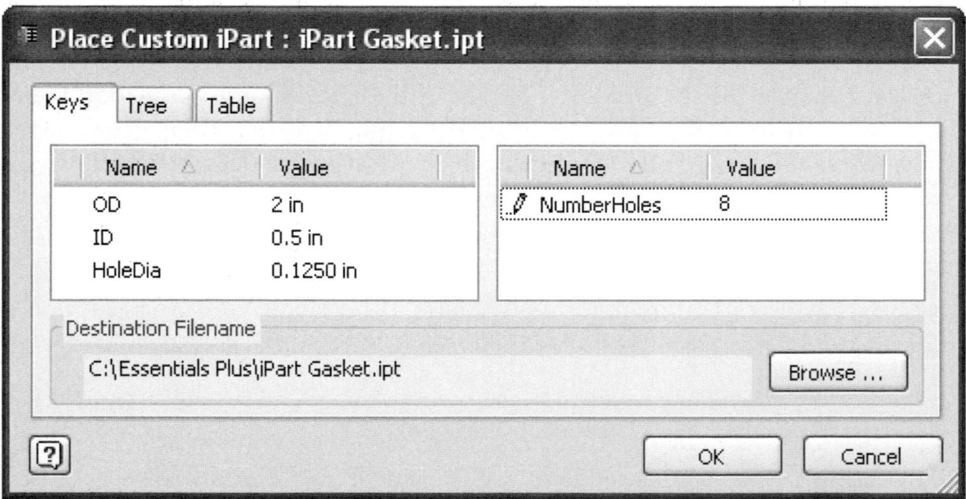

Figure 8.65

6. Place the part in the graphics window.

7. Continue placing instances of the iPart as needed.
8. To change an iPart in an assembly to a different configuration, expand the part in the browser, right-click on the table name, and select Change Component. Then select a new member from the Keys, Tree, or Table tab.

The Keys tab displays the primary and any secondary keys defined in the iPart factory. You can select the values of the keys to identify unique members. The Tree tab displays a hierarchical structure of the keys. If secondary keys exist, the values of the keys are filtered progressively as you expand the key values. The Table tab displays the entire table rather than just the keys. To identify a unique iPart member, select a row, and then click the OK button to place the iPart.

You can also dynamically create a new member for the factory when placing a standard iPart with the Table tab, as shown in the following image. In a row named "New" at the top of the table, you can enter values or select from a drop-down list if appropriate. As you enter values in the row, the table is reduced to display only the rows that match the values entered in the row. As you enter more values, the table continues to be filtered. If no member matches the entered values, a new member is created. When a unique set of values is entered, the New Row button becomes active, and you can select it to create a new row in the table. This can be done whether or not you are placing the new member in the assembly. Any columns that contain values that are not set are set to the same values as the default row.

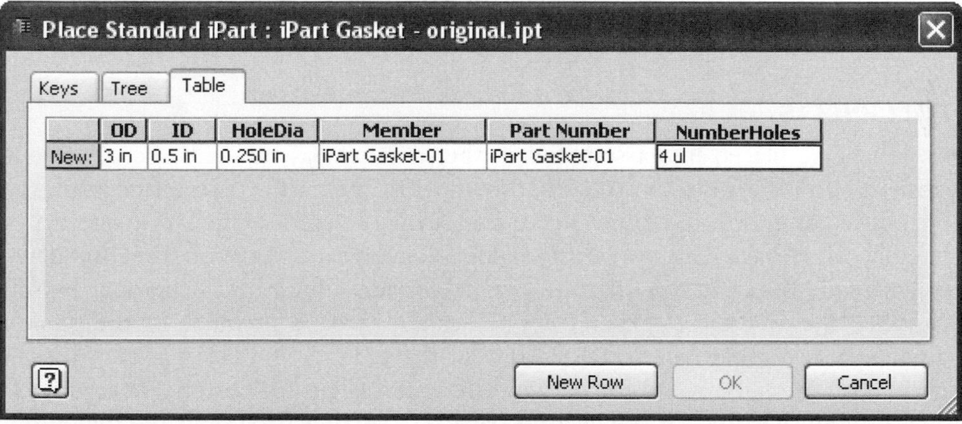

Figure 8.66

When you place the first version of a standard iPart into an assembly, a folder is created with the same name as the iPart factory. By default, this folder is created in the same folder as the iPart factory file. As you place additional standard iParts into your assemblies, this folder is checked for existing iPart files prior to creating new iPart files.

STANDARD IPART LIBRARIES

In a collaborative design environment, the best way to manage iPart factories and the iParts published from those factories is by using libraries. Define a library within your project and place your iPart factories in this library's path. You can also specify that a standard iPart factory publish standard iParts to a different, or proxy, folder. To do this, create another library entry using the same name as the iPart factory library prefixed with an underscore, as shown in the following image, or right-click on the library's name, and click Add Proxy Path.

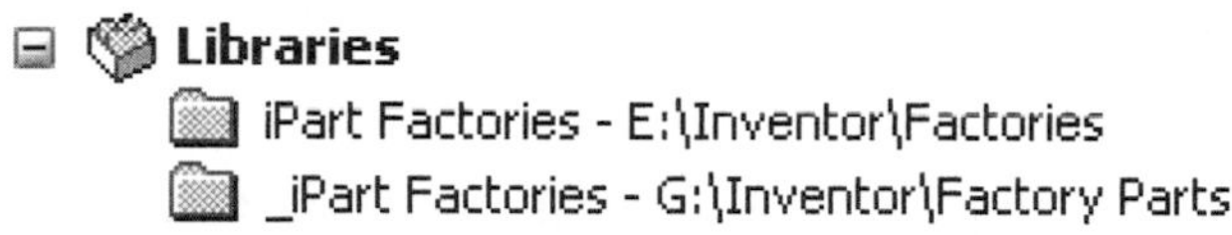

Figure 8.67

If you have an iPart factory for bolts, for example, you could store the iPart factory in a library named Bolts, and then you could define an additional library where the individual members generated from the Bolt iPart factory will be stored. The library must be named _Bolts. Autodesk Inventor will automatically store all iParts published by the factory in the library directory specified by _Bolts. If you store an iPart factory in a workspace or a workgroup search path, the published iParts are stored in a subdirectory with the same name as the factory.

Note: It is recommended that you do not store iParts in the same folders as iPart factories.

CUSTOM IPARTS

Custom iParts are placed into assemblies in the same way as standard iParts. If you select an existing custom iPart member file, that member of the part is placed into your assembly with no option to define the value of the custom parameters. When you select a custom iPart factory, the Place Custom iPart dialog box appears. As with the Standard iPart Placement dialog box, you can use the Keys tab to identify an iPart member by selecting the key value, the Tree tab to expand key values and select a member, or the Table tab to identify an iPart member by selecting a row in the table. The Place Custom iPart dialog box provides additional options so that you can enter values for custom parameters, define a destination and filename for the custom iPart by selecting Browse in the dialog box, and place multiple instances of different iPart members.

Note: Since each custom iPart is unique, you must provide different filenames as you place different members.

After you have placed a custom iPart in an assembly, you can add more features to the iPart. The following image shows the browser of a custom iPart named Gasket B that has had a fillet and extrusion feature added to it. The capability to add features to an iPart makes the custom iPart behavior similar to that of a derived part.

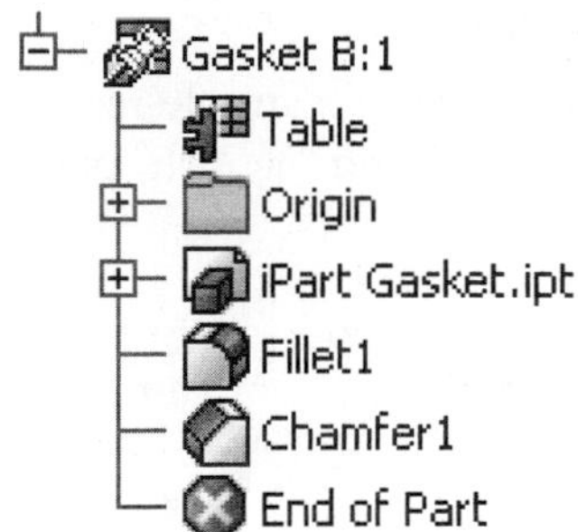

Figure 8.68

AUTO-CAPTURE

You can make changes automatically to an iPart or iAssembly, which are covered later in this chapter, using normal tools, without accessing the iPart or iAssembly Author table. You can automatically capture changes to the entire factory or to the active row. This is done using the iPart/iAssembly toolbar, as shown in the following image with the flyout menu activated to show all available tools.

Figure 8.69

The available tools on the toolbar are as follows:

Edit Member Scope	Any edit that can be configured is automatically added as one or more new columns or will edit the member's cell value if a column for the change already exists in the table. If a new column is added, the new value of the object is added for the current member's row, and the original value is used for all other members.
Edit Factory Scope	This setting works similarly to Edit Member Scope, except that only columns are modified. New values are set for the entire column. Columns are not added or removed in this mode.
iPart/iAssembly Author	Opens the iPart or iAssembly Author dialog box depending on which type of file is open.
Edit Using Spreadsheet	Opens the iPart or iAssembly table in a spreadsheet file for modification.

EDIT MEMBER SCOPE EXAMPLE

For example, say that you have an iPart of a simple rectangle made up of a single extruded rectangle whose sketch has a "Length" and "Width" parameter that define the iPart table, as shown in the following image.

	Member	Part Number	Length
1	Rectangle-01	Rectangle-01	10 mm
2	Rectangle-02	Rectangle-02	20 mm
3	Rectangle-03	Rectangle-03	30 mm

Figure 8.70

The factory is open in Autodesk Inventor, and Rectangle-02 is active. You have selected the Edit Member Scope tool, and you change the "Width" of the extrusion from 5 mm to **8 mm** by showing the dimensions and modifying them, making the change just as you would to any feature.

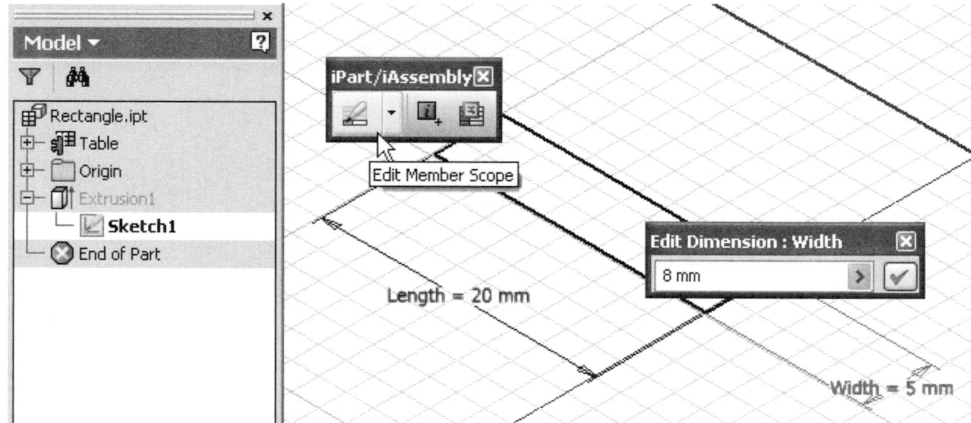

Figure 8.71

Because "Width" was not part of the original table, it is added as a new column to the table, and the original value of 5 mm is set for all other members in the factory, as shown in the following image.

	Member	Part Number	Length	Width
1	Rectangle-01	Rectangle-01	10 mm	5 mm
2	Rectangle-02	Rectangle-02	20 mm	8 mm
3	Rectangle-03	Rectangle-03	30 mm	5 mm

Figure 8.72

Next you modify "Length" from 20 mm to **25 mm**. Because the table already has length configured, only the cell for Rectangle-02 is modified. The other members remain as defined.

	Member	Part Number	Length	Width
1	Rectangle-01	Rectangle-01	10 mm	5 mm
2	Rectangle-02	Rectangle-02	25 mm	8 mm
3	Rectangle-03	Rectangle-03	30 mm	5 mm

Figure 8.73

The images above show the table in intermediate states. The workflow of changing the width to 8 mm and length to 25 mm are done in a single edit. Both changes to the table happen simultaneously.

You can open the iPart/iAssembly toolbar by right-clicking on any toolbar and selecting Customize. In the Customize dialog box, select the Toolbars tab, click iPart/iAssembly, and select Show. If an assembly file is open, the toolbar can be accessed by right-clicking the Standard toolbar and selecting iPart/iAssembly from the menu as shown in the following image.

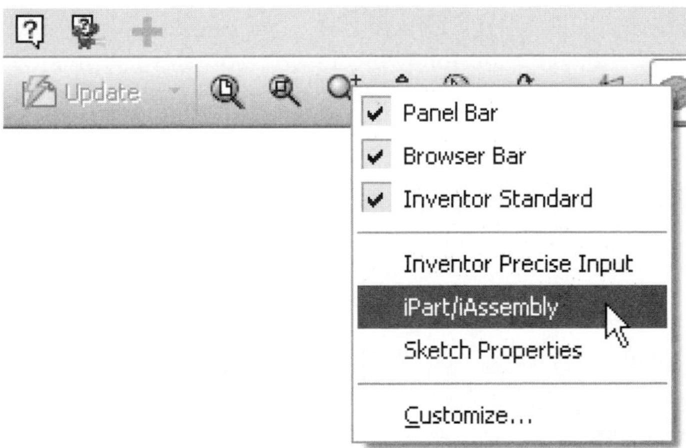

Figure 8.74

DRAWINGS

When creating a drawing view from an iPart or an iAssembly, you can select the Model State tab of the Drawing View dialog box to access a particular member for the factory, as shown in the following image.

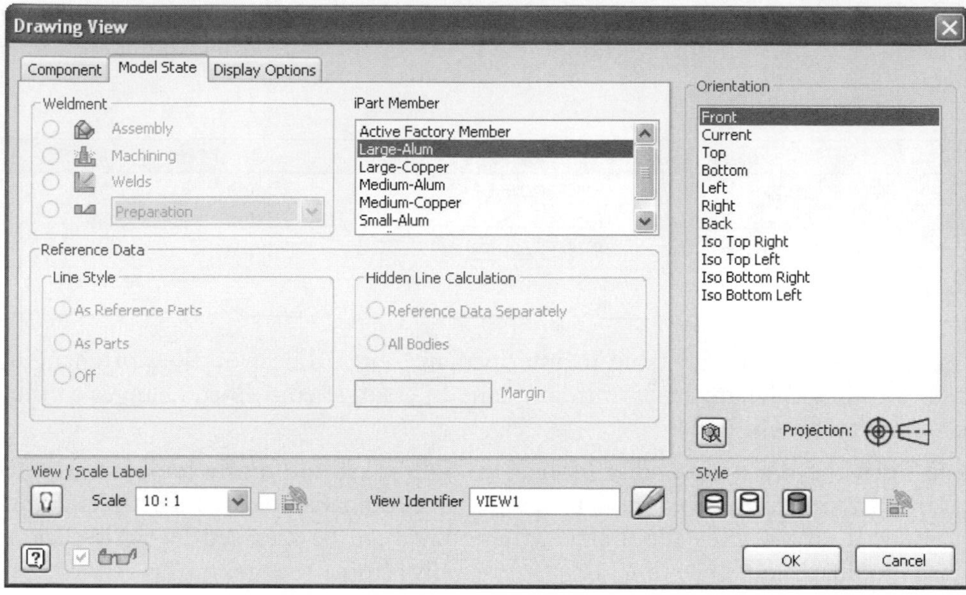

Figure 8.75

A Table command, available on the Drawing Annotation panel bar, provides access to the Table dialog box, as shown in the following image. You can use this command to create a configuration table in a drawing. The style of the table is controlled using the Style and Standard Editor.

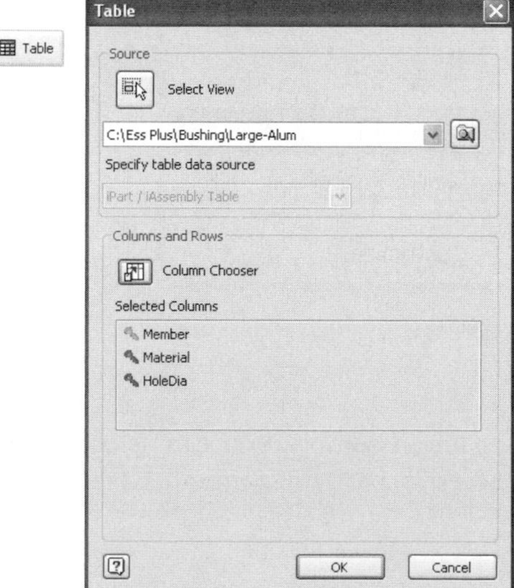

TABLE		
Member	Material	HoleDia
Large-Alum	Aluminum	8.1 mm
Medium-Alum	Aluminum	7.1 mm
Small-Alum	Aluminum	6.1 mm
Large-Copper	Copper	8.1 mm
Medium-Copper	Copper	7.1 mm
Small-Copper	Copper	6.1 mm

Figure 8.76

EXERCISE 8-3: CREATING AND PLACING IPARTS

In this exercise, you convert an existing part to a Standard iPart Factory. You then place Standard iParts from that factory into an assembly.

1. Open *ESS_E08_03.ipt*.
2. Click the Parameters tool, and review the parameter names. In this exercise, you will use the parameter HoleDia to drive the iPart factory.
3. Click Done at the bottom of the Parameters dialog box.
4. From the Tools menu, click Create iPart.
5. In the iPart Author dialog box, expand the Hole1 feature in the Parameters tab. Notice that the system parameter HoleDia has automatically been added to the list of table-driven items, and a column has been added to the table. User parameters and renamed system parameters are automatically added to the list.
6. HoleDia will be used as the primary key for this iPart Factory. Click the key next to HoleDia. It should turn blue, and the number "1" will be placed next to it.
7. You will also control the material from the iPart Factory. To add the Material property to the table, complete the following actions:

- Click the Properties tab.
- Collapse Summary.
- Expand Physical.
- Double-click Material. Notice that the material Aluminum, defined in the Properties dialog box in the Physical tab, is added to the table.

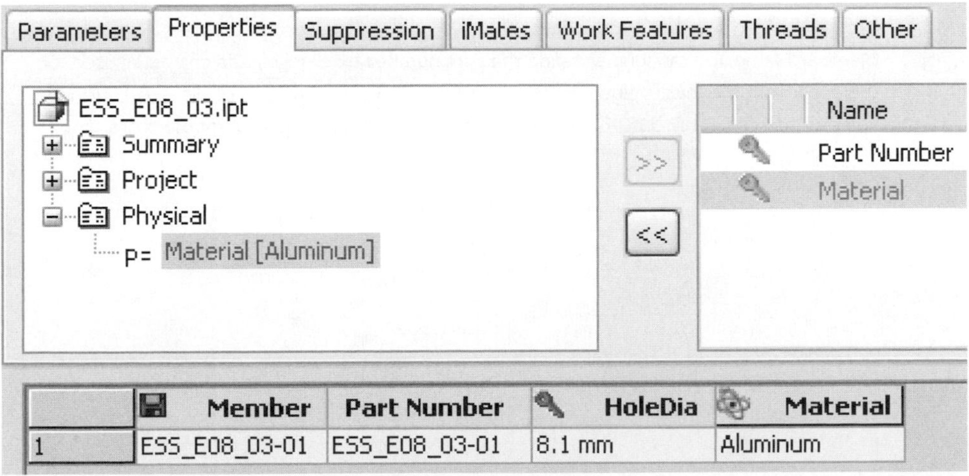

Figure 8.77

8. To identify a material when placing this Standard iPart, the Material property must also be defined as a key. To define Material as a secondary key, click the key next to Material. The number next to the key shows the key priority.
9. You can identify which column in the table is used to control the Material for each iPart version. In the table, right-click on the Material column label, and notice the checkmark next to Material Column. If no checkmark appears next to the Material Column option, select it. If one exists, do not deselect it. You'll also notice that a material icon is displayed on the column heading.

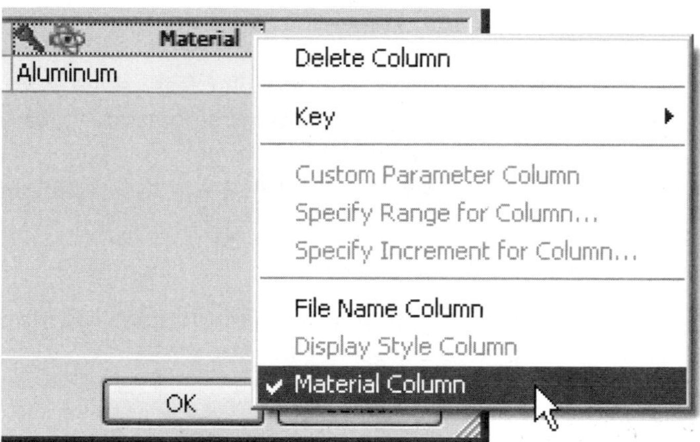

Figure 8.78

10. You will also control the Part Number of the Standard iPart when it is published. Click the Options button. Notice that the Set to Value: entry in the Part Number list is selected. Enter **Aluminum Bushing**, and then click OK.

Note: You can use the Options button to set rules and values for the Part Number and Member Name. You can further refine the part number by changing the character used for the separator, the initial value, the step increment, and the number of digits. Additionally, you can use the Verify button to ensure that the table contains no syntax errors.

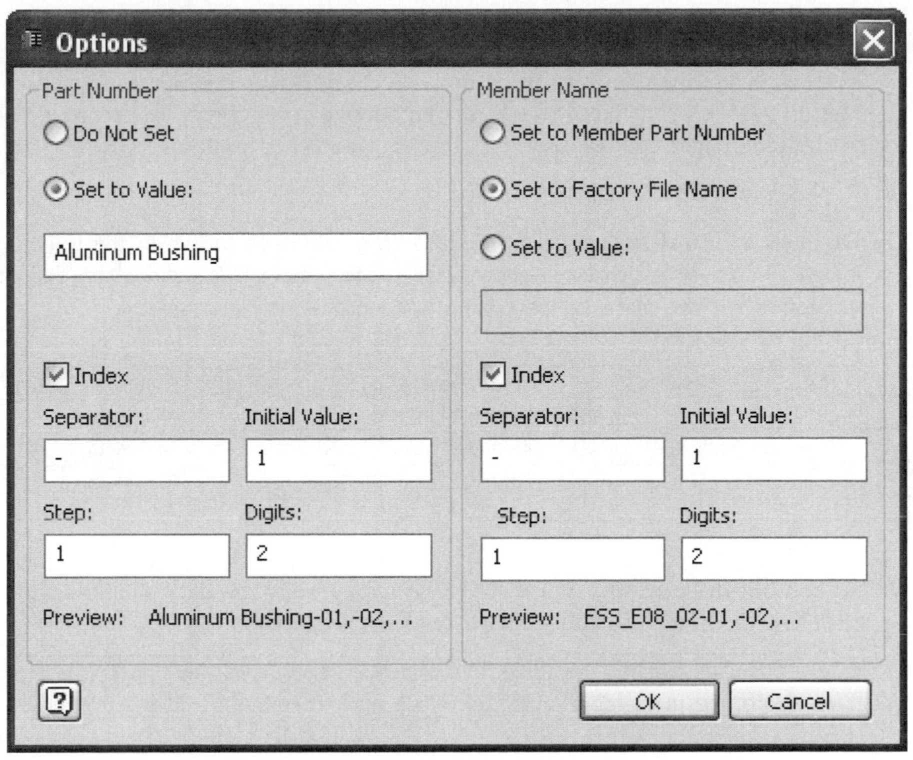

Figure 8.79

11. Click Yes when prompted to apply the part numbering scheme to all members.

12. Similar to the function of the Material Column, you can identify which column in the table is used to control the filename for each iPart version. In the table, right-click on the Member column label. There should be a checkmark next to the File Name Column option. If one exists, do not deselect it. An icon also appears in the column heading notifying you which column is used for the filename.

13. To define the filename for the first iPart version, double-click on the first cell in the Member column, enter **Large-Alum**, and then press ENTER.

	Member	Part Number	HoleDia	Material
1	Large-Alum	Aluminum Bushing-01	8.1 mm	Aluminum

Figure 8.80

14. Click OK when prompted that the filename has changed.

15. The iPart Factory will contain three different hole sizes and two different materials from which to choose. You will create the first three versions using the table at the bottom of the iPart Author dialog box. The last three versions are defined in a later step using Microsoft Excel to edit the embedded spreadsheet. To create a new row in the table, right-click on the first row label, and then click Insert Row.

16. Repeat the last step to insert a third row.

17. Double-click on individual cells, and enter the values, as shown in the following image. The row highlighted in green is the default row. This defines the default iPart version when you place an iPart from this factory into an assembly.

	Member	Part Number	HoleDia	Material
1	Large-Alum	Aluminum Bushing-01	8.1 mm	Aluminum
2	Medium-Alum	Aluminum Bushing-02	7.1 mm	Aluminum
3	Small-Alum	Aluminum Bushing-03	6.1 mm	Aluminum

Figure 8.81

18. To convert the part to a Standard iPart Factory using the data defined in the table, click OK at the bottom of the iPart Author dialog box.

19. In the browser, expand the Table. The browser represents your part as a Standard iPart Factory using a Table entry. You'll see each member listed by its Member Name.

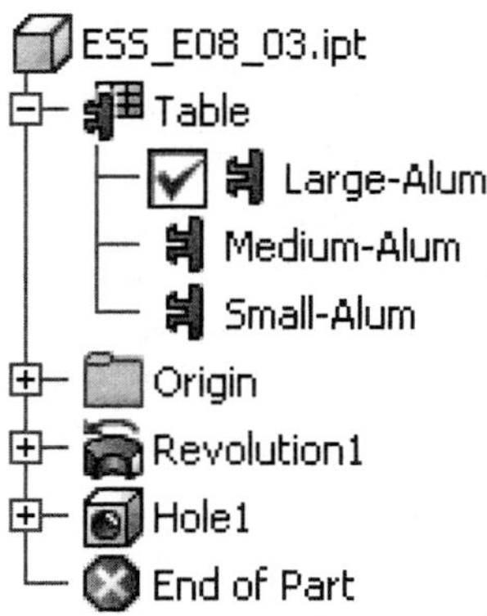

Figure 8.82

20. In the browser, right-click the table entry, and select List by Keys from the menu. Expand all entries in the browser. Each version is shown under the table sorted by the primary and secondary key names and their values, as shown in the following image. The active version has a checkmark next to it.

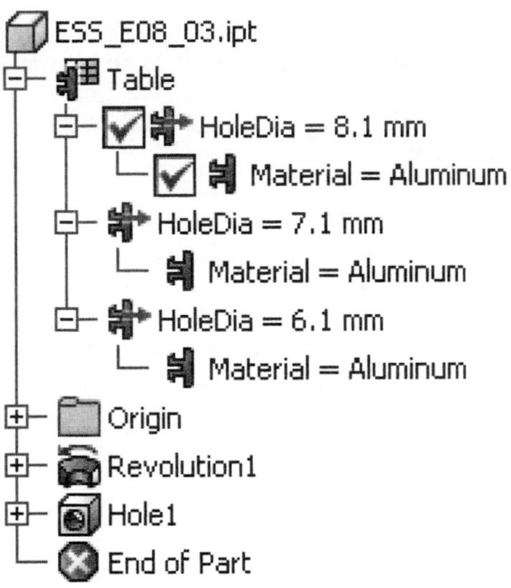

Figure 8.83

21. Notice that for each hole diameter in the table, only one material is defined, which is Aluminum. To add more versions to the table using Microsoft Excel, right-click the Table icon in the browser, and then click Edit via Spread Sheet.

Copy cells **A2** through **D4,** and paste them to cell **A5**.

- Modify cells **A5** through **D7** to be consistent with the following image.

Note: You can also use the Edit using Spreadsheet tool from the iParts iAssemblies toolbar to access Excel.

	A	B	C	D
1	Member<defaultRow>1	Part Number [Project]	HoleDia<key>0</key>	Material [Physical]<key>1</key><material></material>
2	Large-Alum	Aluminum Bushing-01	8.1 mm	Aluminum
3	Medium-Alum	Aluminum Bushing-02	7.1 mm	Aluminum
4	Small-Alum	Aluminum Bushing-03	6.1 mm	Aluminum
5	Large-Copper	Copper Bushing-01	8.1 mm	Copper
6	Medium-Copper	Copper Bushing-02	7.1 mm	Copper
7	Small-Copper	Copper Bushing-03	6.1 mm	Copper

Figure 8.84

22. When you finish modifying the contents of the spread sheet in Excel, click Save.

23. In Excel, click Close & Return to *ESS_E08_03.ipt* from the File menu.

24. You can check each version in the iPart Factory prior to publishing it for use in your assemblies by changing the active version. To test different versions, right-click on different members in the browser under the Table icon, and then click Activate.

25. Make the Large-Aluminum member the active version.

26. To save the Standard iPart Factory with a different name, click File > Save As, and enter the filename Bushing. Click to save it in the same location as the exercise files.

27. Close the file. In the next portion of this exercise, you insert iParts from this iPart factory into an assembly.

28. Open *ESS_E08_03B.iam*.

29. From the Assembly panel bar, click Place Component, select the *Bushing.ipt* file you just created, and then click Open.

30. When you select a Standard iPart Factory, the Place Standard iPart dialog box allows you to place different versions of the part. The default values are shown on the Keys tab. Place two default versions of the bushing by selecting the locations, as shown in the following image.

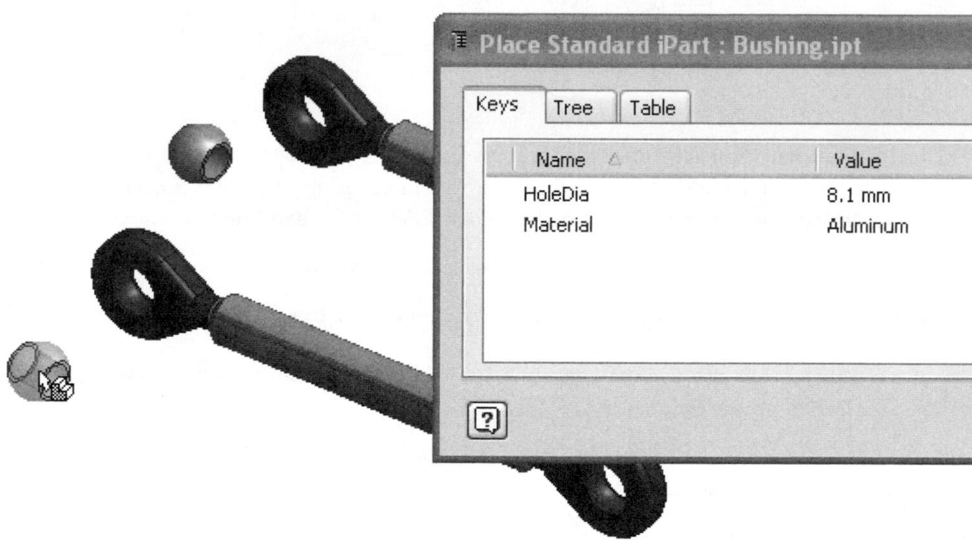

Figure 8.85

31. To place a version with a hole size of 6.1 mm and a material of copper, click the value next to HoleDia, and then click 6.1 mm from the list of available hole diameters, as shown the following image on the left.

32. Click the value next to Material, and then click Copper from the list of available Materials, as shown in the following image on the right.

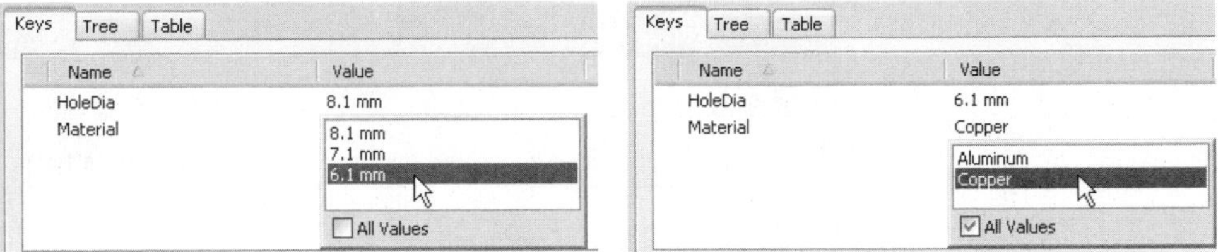

Figure 8.86

33. Click the placement location for the third iPart, as shown in the following image. Notice that the color of the iPart reflects the material you specified.

Figure 8.87

34. To place a version with a hole size of 7.1 mm and a material of aluminum, click the Table tab, and click the second row in the table. Click a location for the fourth iPart, as shown in the following image.

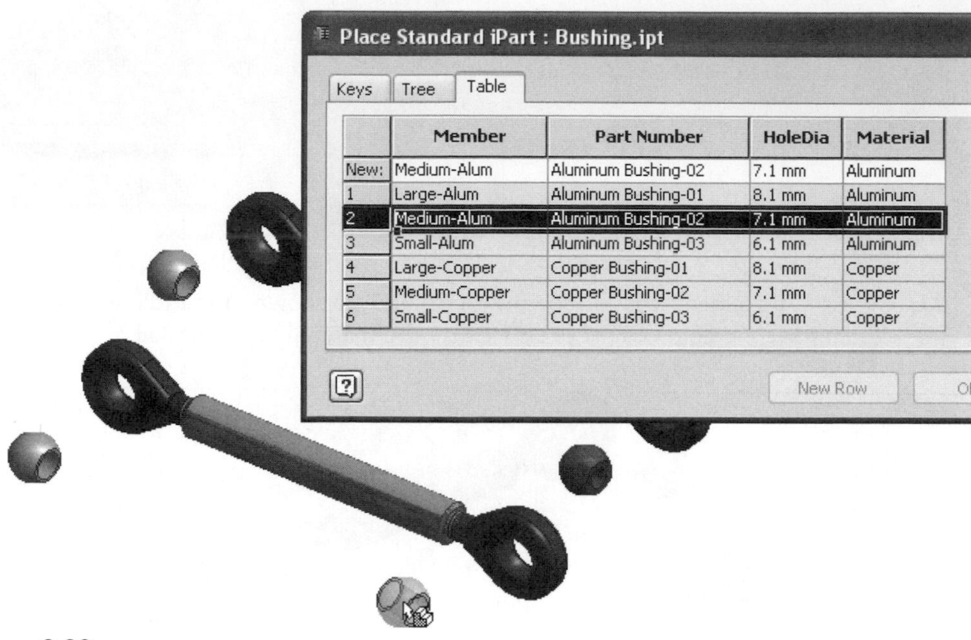

Figure 8.88

35. Click Dismiss in the Place Standard iPart dialog box.

36. To change the version of the iPart you just inserted, expand Medium-Alum:4 in the browser, right-click the Table, and click Change Component. Notice that the names of the iParts in the browser reflect the names you specified when defining the Member column in the iPart Factory.

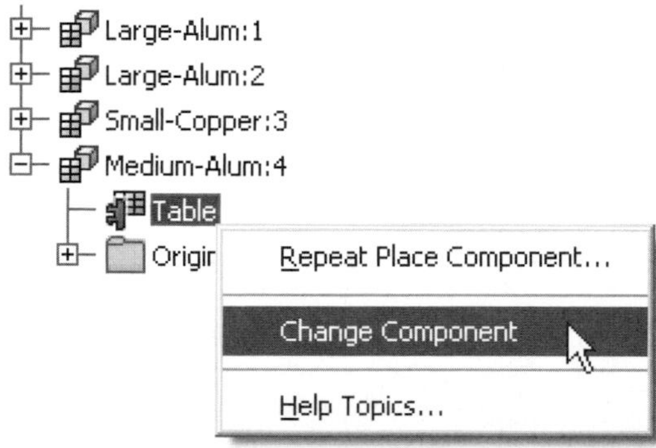

Figure 8.89

37. In the Place Standard iPart dialog box, change the HoleDia to **6.1 mm** and Material to **Copper.**

38. Click OK in the Place Standard iPart dialog box, and verify that the bushing was replaced in the graphics window.

39. Close all open files. Do not save changes. End of exercise.

IASSEMBLIES

iAssemblies are used to group a set of similar designs in a table format. These are also commonly referred to as configurations, and they function in a similar method to iParts. An iAssembly factory is the primary definition of the family. The family is defined by a table and is a set of configurations of the same design. Each row of the table defines a unique product definition, which is called a configuration or member. When you use an iAssembly factory in another assembly, you can easily change from one configuration to another.

iAssembly members are stored as separate *.iam* files, similar to iParts. They are named in the table using the File name column designation, reference designation field, or key values. When member files are generated, they are stored in a subfolder with a name identical to that for the factory—similar to iParts, but with the *.iam* file type. You can also create a proxy search path for iAssemblies to designate a set of library locations to search for iAssembly factories and a set of corresponding locations to place populated members. Refer to the iPart section above for an example of a proxy path. You can leverage iMates to assist in assembling configurations and to make sure that the assembly conditions are maintained if you change between members of the iAssembly.

CREATING IASSEMBLIES

To create an iAssembly, use the Create iAssembly tool located on the Tools menu, as shown in the following image.

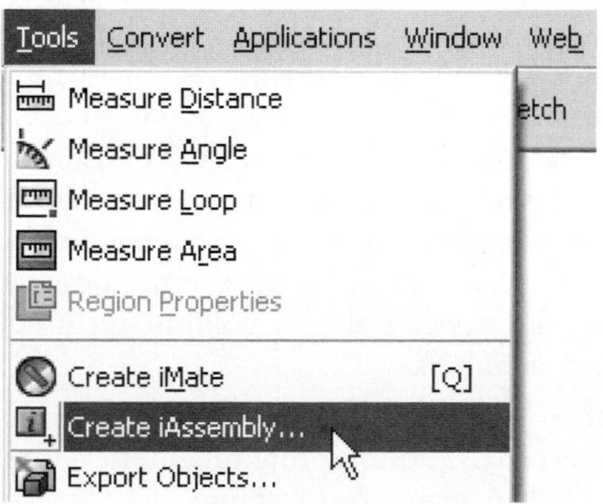

Figure 8.90

After selecting the Create iAssembly tool, the iAssembly Author dialog box is displayed, as shown in the following image. This dialog box displays the configurable items for an assembly and lists the different configurations of the assembly that have been defined—see the rows at the bottom of the dialog box. You can define what controls the iAssembly by using the seven tabs of the dialog box. Each tab has different properties that can be selected and used as configurable items.

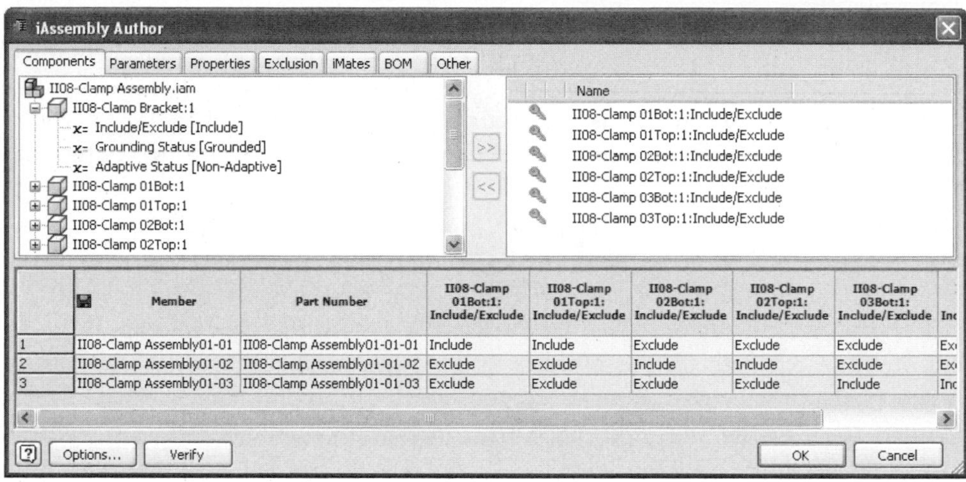

Figure 8.91

The iAssembly Author dialog box functions similar to the iPart Author dialog box. You can select a tab across the top to display objects that can be included for configuration in the iAssembly. You use the Add/Remove buttons to include objects from the top-left pane to the top-right pane of the dialog box. The list of configured objects resides in the top-right pane. The lower portion of the dialog box shows the configuration table that displays all of the configured items and defined rows, or members, of the iAssembly. You can use the Options button to set rules and values for the Part Number and Member Name. Additionally, you can use the Verify button to ensure that the table contains no syntax errors. When you have added all objects and defined all members, click OK to complete the authoring process. By right-clicking on a column in the table, you can specify which column you want to use as the filename column, designated by a disk icon in the column header. You can also specify that a column be used as a key column when establishing the iAssembly members.

In addition to creating the iAssembly factory with the iAssembly Author dialog box, you can also use Microsoft Excel or the Autocapture capabilities of the iPart/iAssembly toolbar to create configurations. When using Autocapture, you use familiar tools for editing a design, and changes made to the assembly configuration are applied either to the current configuration or to the entire factory, depending on whether Edit Factory Scope or Edit Member Scope is selected.

When using Edit Factory Scope, the value or changes you make in the assembly are not automatically written to the configuration. If you have a value or settings change that you must click to open the iAssembly Author or to activate a different configuration member, a question dialog box is displayed. This dialog box lists the values in the active row that do not match the current values, and it prompts whether or not you want to update the current member's values with the modified values. Clicking Yes updates the configuration for the active member.

The Edit Member Scope setting will automatically apply changes and edits made to the assembly to the active configuration member. If you edit a property that is not listed as a configurable item in the table, Autocapture will add it to the table definition.

When you create a configuration of an assembly, the components in that assembly can be set to adaptive or flexible. However, a given component in a configuration can only be adaptive in one member of the configuration. All other members must be set to non-adaptive. If you want a component to adjust in size, you must control the size of the component using parameters and formulas or convert the part to an iPart and define the desired sizes. The actual iAssembly can be made flexible when placed in other assembly files, but it cannot be set to adaptive.

The Components tab lists the components of the assembly with the configurable items below. Items listed can be changed in the following ways: Include/Exclude, Grounding Status, Adaptive Status, and Table Replace.

The Parameters tab lists all assembly constraints, assembly features, assembly work features, iMates, component patterns, and other parameters, such as user parameters, that can be included in the factory.

The Properties tab allows the inclusion and modification of summary, project, physical, and custom properties.

Similar to the Parameters list, the Exclusion tab shows all objects that can be excluded, including components, constraints, assembly features, assembly work features, iMates, Representations, and Component Patterns. Note that these elements can also be set on the Components tab, but the Exclusion list provides a more specific way to view and set the exclusion property.

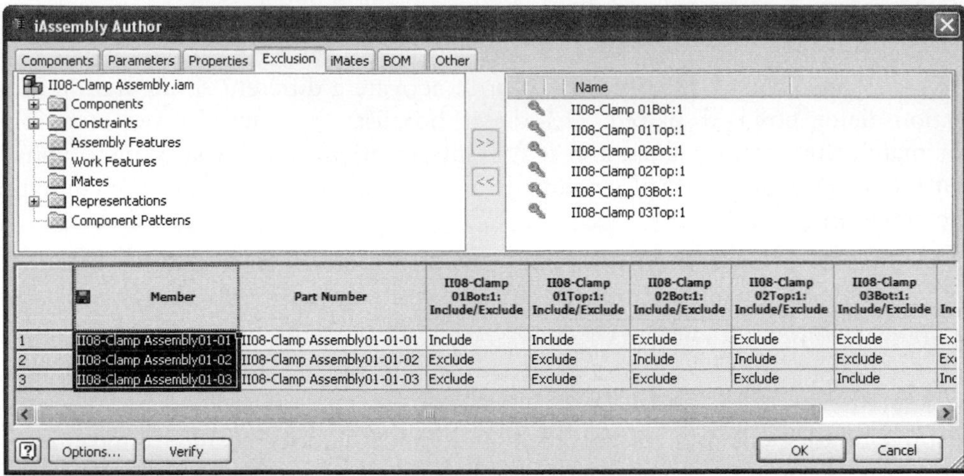

Figure 8.92

The iMates tab functions the same as it does for iParts. It lists each iMate with the offset value, include/exclude, matching name, and sequence number available for configuration.

The BOM tab can be used to work with bill of material specific properties, BOM Structure, and BOM Quantity. Two BOM viewing conditions exist for controlling what you see for an iAssembly Bill of Materials. You can view the BOM data from within an assembly that uses a specific iAssembly configuration as an occurrence of the design, or you can view BOM data from within the iAssembly file. In the first scenario, you can view all levels of its components and their details because the item number is a single configuration member.

If viewing the BOM from within the iAssembly, you can view only the top-level structure because the structure will vary based on each configuration member. As you vary the structure of the iAssembly, the item numbers will change from one configuration to another and cause issues for drawing documentation. To view the quantity information in the BOM, you have the option to display the Unit QTY values for a specific configuration member or all members, as shown in the following image.

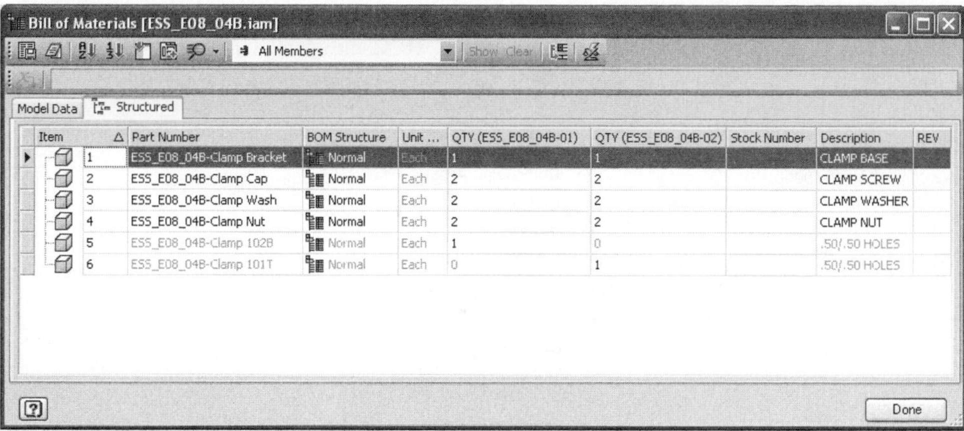

Figure 8.93

The Other tab functions the same as it does for iParts, and it can be used to specify a custom column that can contain a string value. Custom columns can be designated as keys or as a filename.

As the table is created, you will notice that different cell background colors provide information about the status of the cells as follows:

Green Background	Cells in the active table row that will be the default row when placed
Light Grey Background	Cells in the non-active row
Light Blue Background	Cells in a selected column
Dark Blue Background	Cells with a custom parameter
Mango Background	Cells that are driven by an Excel formula
Yellow Background	Cells that have an error

PLACING IASSEMBLIES

iAssemblies are placed into other assemblies using the Place Component tool, similar to placing an iPart or any other regular component. When the selected component is an iAssembly file, the Place iAssembly dialog box is displayed, as shown in the following image.

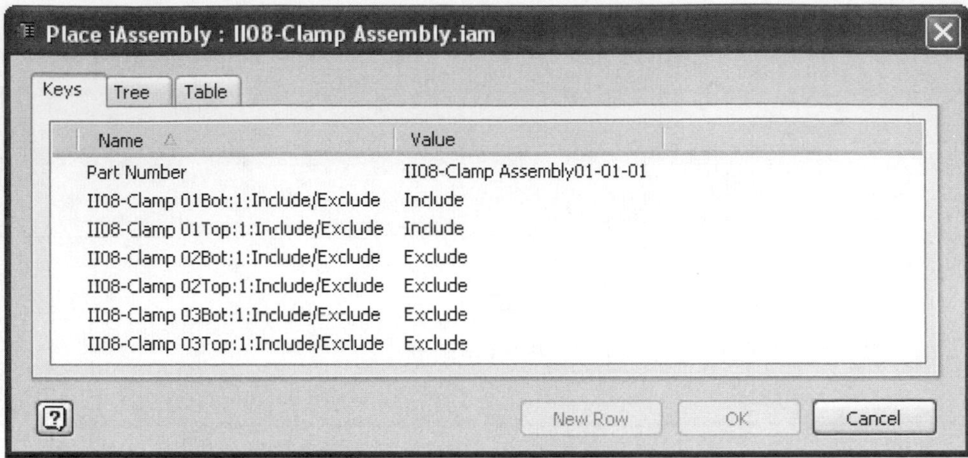

Figure 8.94

Before placing an occurrence of the iAssembly, select the configuration member from the Keys, Tree, or Table tabs. Each tab lists the same configurations, but they display the configuration members in different ways. These tabs function the same as when working with iParts. Refer to the iPart Placement section of this chapter for additional information. If iMates are used in the definition of the iAssembly and also in the iAssembly file itself, you can select one of the two iMates options in the Place Component dialog box to speed up the assembly process.

After an iAssembly configuration has been placed, you can change to another configuration by right-clicking the Table node in the browser and selecting Change Component, as shown in the following image.

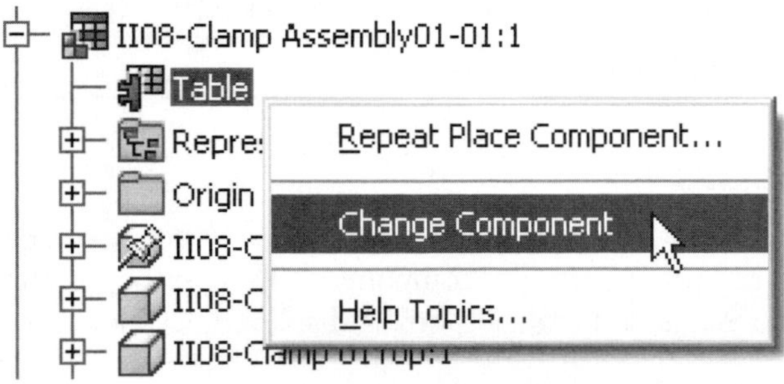

Figure 8.95

When a member of an iAssembly configuration is referenced, the member file with the specified and defined properties is generated. This file has either already been created—because it has been previously referenced—or a new member file is created and referenced. Instead of having the member files created when you select to use them in an assembly, you can pre-generate the member files of an iAssembly. This is done by opening the iAssembly and under the Table node, selecting one or more configurations, right-clicking, and selecting Generate Files from the menu, as shown in the following image. You can use this same method to update member files after changes have been made to an iAssembly, and the same method can be performed for iParts to pre-generate members of the iPart factory.

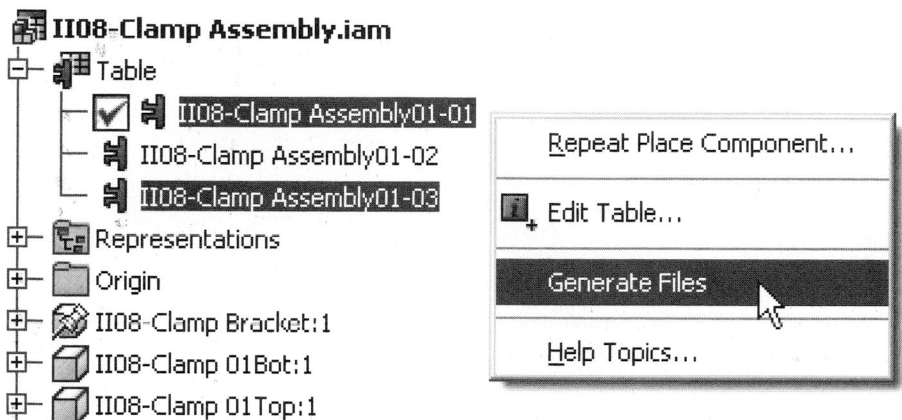

Figure 8.96

DOCUMENTING IASSEMBLIES

Drawing views of iAssemblies are created using the Base View tool. When an iAssembly file is selected, all of the iAssembly members are listed on the Model State tab of the Drawing View dialog box, as shown in the following image.

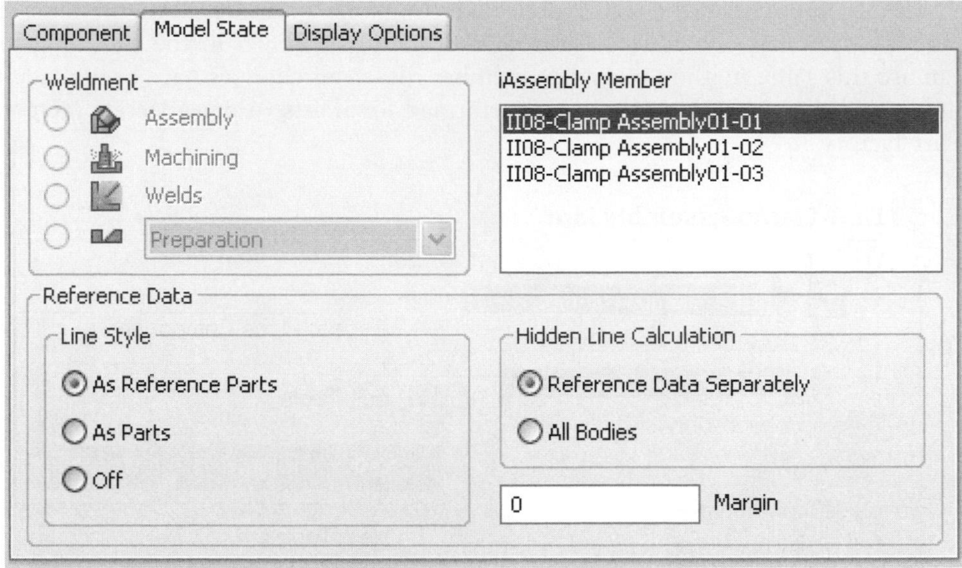

Figure 8.97

The created drawing view is based on the selected iAssembly member in that list. After the drawing view has been created, you can change it to a different iAssembly member by editing the view and selecting a different one from the list.

When adding a parts list to a drawing that is based on an iAssembly, the quantity (QTY) column is based on the method used to create it. If you select an existing drawing view, the configuration member for that view is displayed. If you browse to an iAssembly file, the active configuration member in the file is displayed. Configuration members can be added to a parts list by editing the existing parts list and selecting the Member Selection button, as shown in the following image on the left. In the Select Member dialog box, shown in the following image on the right, you can select the configuration member's checkbox to include it in the parts list.

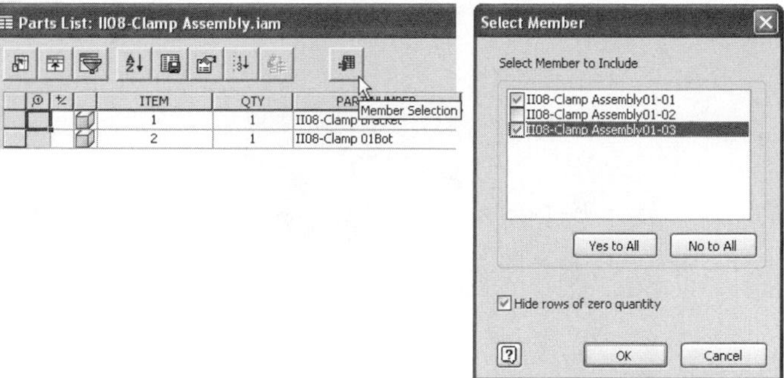

Figure 8.98

You can create tables that list iAssembly configuration values using the Table tool. After selecting the Table tool, you can choose what the table is based on by selecting an iAssembly drawing view or by choosing an iAssembly file. After selecting the source iAssembly, the Table dialog box lists the currently selected columns, as shown in the following image. By selecting Column Chooser, you can specify the attributes that you want to be included in the table.

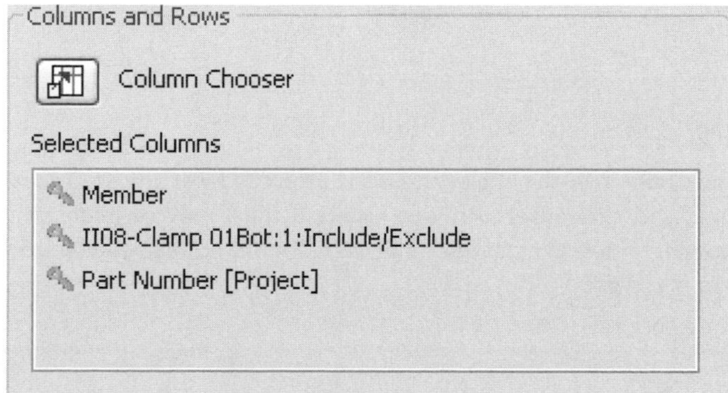

Figure 8.99

EXERCISE 8-4: WORKING WITH IASSEMBLIES

In this exercise, you will work with the functionality of iAssemblies. The exercise is meant to familiarize you with the functionality of iAssemblies and enable you to effectively create and work with them. First you will review the iAssembly Author tool.

1. Open *ESS_E08_04.iam*.

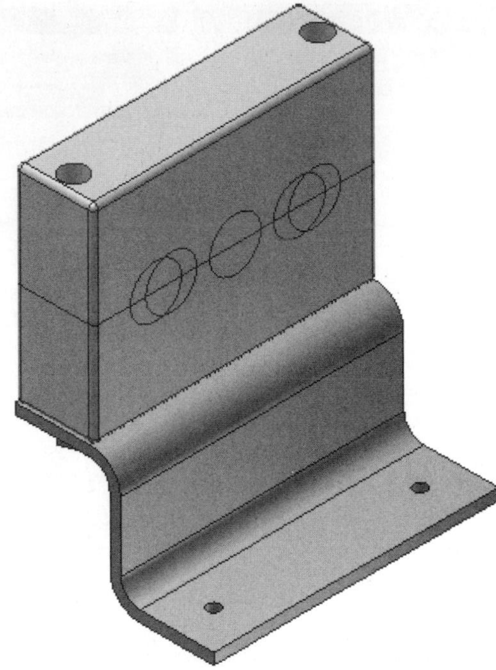

Figure 8.100

2. From the Tools menu, click Create iAssembly.
3. Right-click row 1 in the table, and click Insert Row from the menu. Right-click row number 2, and click Insert Row to create a third row of information, as shown in the following image on the left. The table should appear similar to the following image on the right.

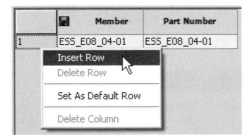

	Member	Part Number
1	ESS_E08_04-01	ESS_E08_04-01

Insert Row
Delete Row
Set As Default Row
Delete Column

	Member	Part Number
1	ESS_E08_04-01	ESS_E08_04-01
2	ESS_E08_04-02	ESS_E08_04-02
3	ESS_E08_04-03	ESS_E08_04-03

Figure 8.101

4. Expand the ESS_E08_04-Clamp01Bot:1 component, and double-click the Include/Exclude [Include] property to add it to the right pane of the dialog box, as shown in the following image.

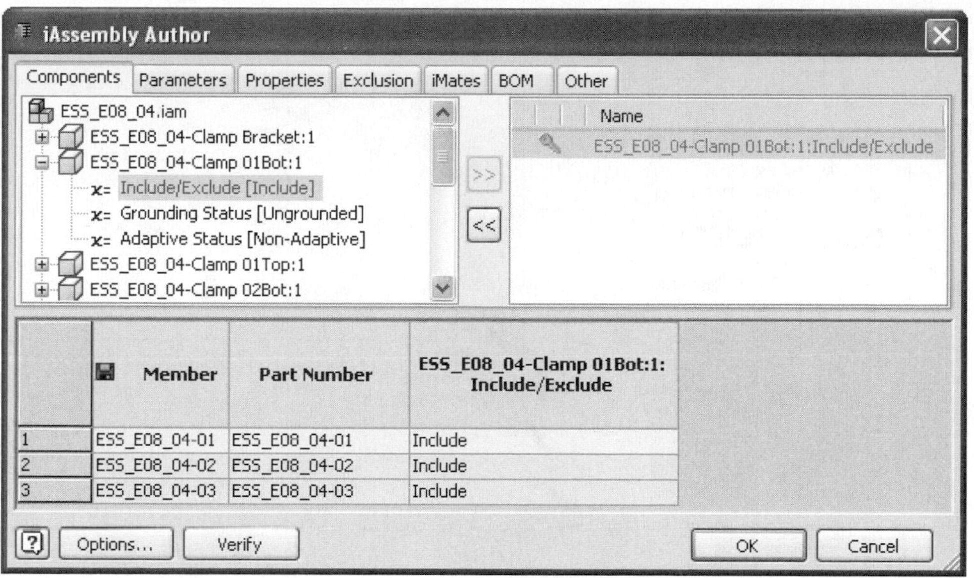

Figure 8.102

5. Continue adding the Include/Exclude property for the following parts:

- ESS_E08_04-Clamp 01Top
- ESS_E08_04-Clamp 02Bot
- ESS_E08_04-Clamp 02Top
- ESS_E08_04-Clamp 03Bot
- ESS_E08_04-Clamp 03Top

As shown in the following image, notice that all components are available in the iAssembly table area.

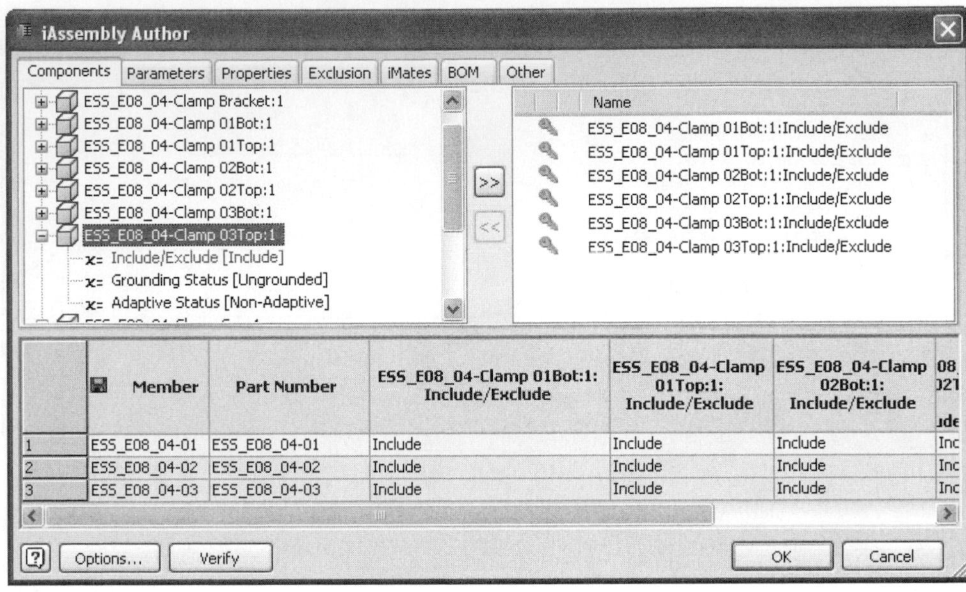

Figure 8.103

6. Now turn off or Exclude certain components, depending on the iAssembly member. For member ESS_08_04-01, set the Exclude property for the following parts, as shown in the following image:

- ESS_E08_04-Clamp 02Bot
- ESS_E08_04-Clamp 02Top
- ESS_E08_04-Clamp 03Bot
- ESS_E08_04-Clamp 03Top

	Member	Part Number	ESS_E08_04-Clamp 01Bot:1: Include/Exclude	ESS_E08_04-Clamp 01Top:1: Include/Exclude	ESS_E08_04-Clamp 02Bot:1: Include/Exclude	ESS_E08_04-Clamp 02Top:1: Include/Exclude	ESS I
1	ESS_E08_04-01	ESS_E08_04-01	Include	Include	Include	Exclude	Excl
2	ESS_E08_04-02	ESS_E08_04-02	Include	Include	Include	Include	Incl
3	ESS_E08_04-03	ESS_E08_04-03	Include	Include	Exclude	Include	Incl

Figure 8.104

7. For member ESS_08_04-02, set the Exclude property for the following parts, as shown in the following image:

- ESS_E08_04-Clamp 01Bot
- ESS_E08_04-Clamp 01Top

- ESS_E08_04-Clamp 03Bot
- ESS_E08_04-Clamp 03Top

	Member	Part Number	ESS_E08_04-Clamp 01Bot:1: Include/Exclude	ESS_E08_04-Clamp 01Top:1: Include/Exclude	ESS_E08_04-Clamp 02Bot:1: Include/Exclude	ESS_E08_04-Clamp 02Top:1: Include/Exclude	ESS Ir
1	ESS_E08_04-01	ESS_E08_04-01	Include	Include	Exclude	Exclude	Excl
2	ESS_E08_04-02	ESS_E08_04-02	Exclude	Include	Include	Include	Excl
3	ESS_E08_04-03	ESS_E08_04-03	Include	Include Exclude	Include	Include	Inclu

Figure 8.105

8. For member ESS_08_04-03, set the Exclude property for the following parts:

- ESS_E08_04-Clamp 01Bot
- ESS_E08_04-Clamp 01Top
- ESS_E08_04-Clamp 02Bot
- ESS_E08_04-Clamp 02Top

The table should appear as shown in the following image.

	Part Number	ESS_E08_04-Clamp 01Bot:1: Include/Exclude	ESS_E08_04-Clamp 01Top:1: Include/Exclude	ESS_E08_04-Clamp 02Bot:1: Include/Exclude	ESS_E08_04-Clamp 02Top:1: Include/Exclude	ESS_E08_04-Clamp 03Bot:1: Include/Exclude	ESS_E08_04-Clamp 03Top:1: Include/Exclude
1	ESS_E08_04-01	Include	Include	Exclude	Exclude	Exclude	Exclude
2	ESS_E08_04-02	Exclude	Exclude	Include	Include	Exclude	Exclude
3	ESS_E08_04-03	Exclude	Exclude	Exclude	Exclude	Include	Include

Figure 8.106

9. Click OK to save the changes and exit the iAssembly Author dialog box. When you return to the model, notice the addition of the Table node in the Assembly browser.
10. Expand the table node, and double-click each node. The assembly should update to the single hole, double hole, and triple hole configurations.

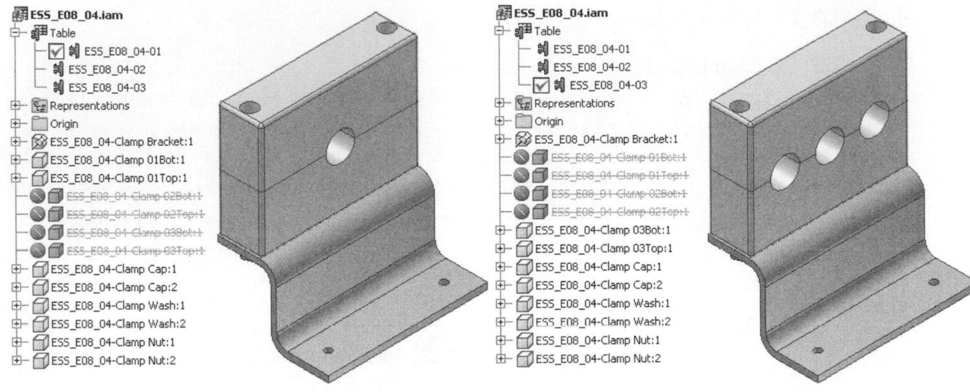

Figure 8.107

11. Save the file as *Clamp Assembly.iam.*
12. Open *ESS_E08_04-Frame Assembly.iam.*
13. Click the Place Component tool, and select the *Clamp Assembly.iam* iAssembly file that you just created and click Open.
14. In the Place iAssembly dialog box, click the Table tab, and select one of the rows in the table.
15. Click a location in the graphics window to place the component.
16. Continue to place other members of the iAssembly that represent each type of clamp, as shown in the following image.

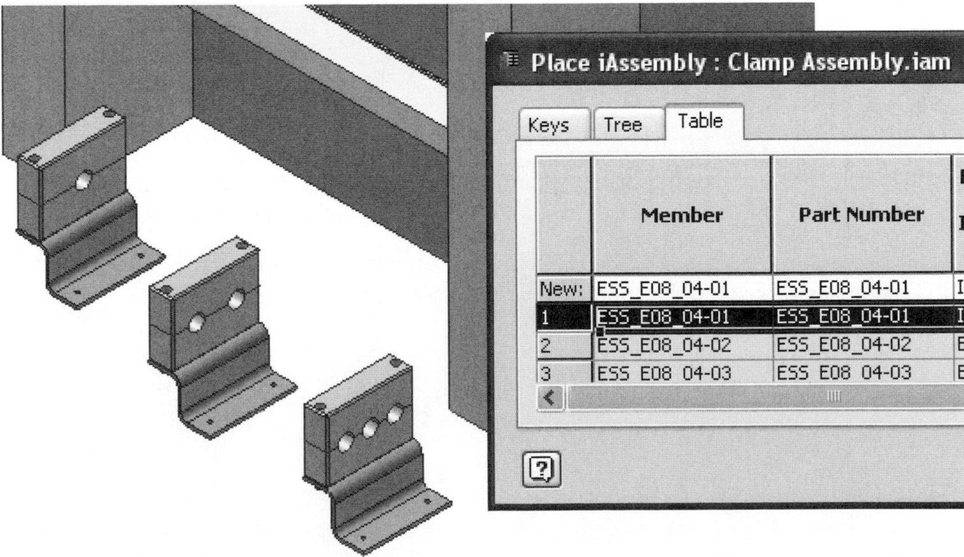

Figure 8.108

EXERCISES

17. Click Dismiss in the Place iAssembly dialog box. Close all open files. Do not save changes.

Next, you work with the creation of an iAssembly using iParts.

18. Open *ESS_E08_04B.iam*. This assembly is similar to the previous assembly with one major difference: instead of assembling all components into a single assembly and turning off certain ones, the top and bottom clamp components of this assembly were created as iParts. In the following image, the four different clamp types are shown where iParts are used to control various diameter configurations.

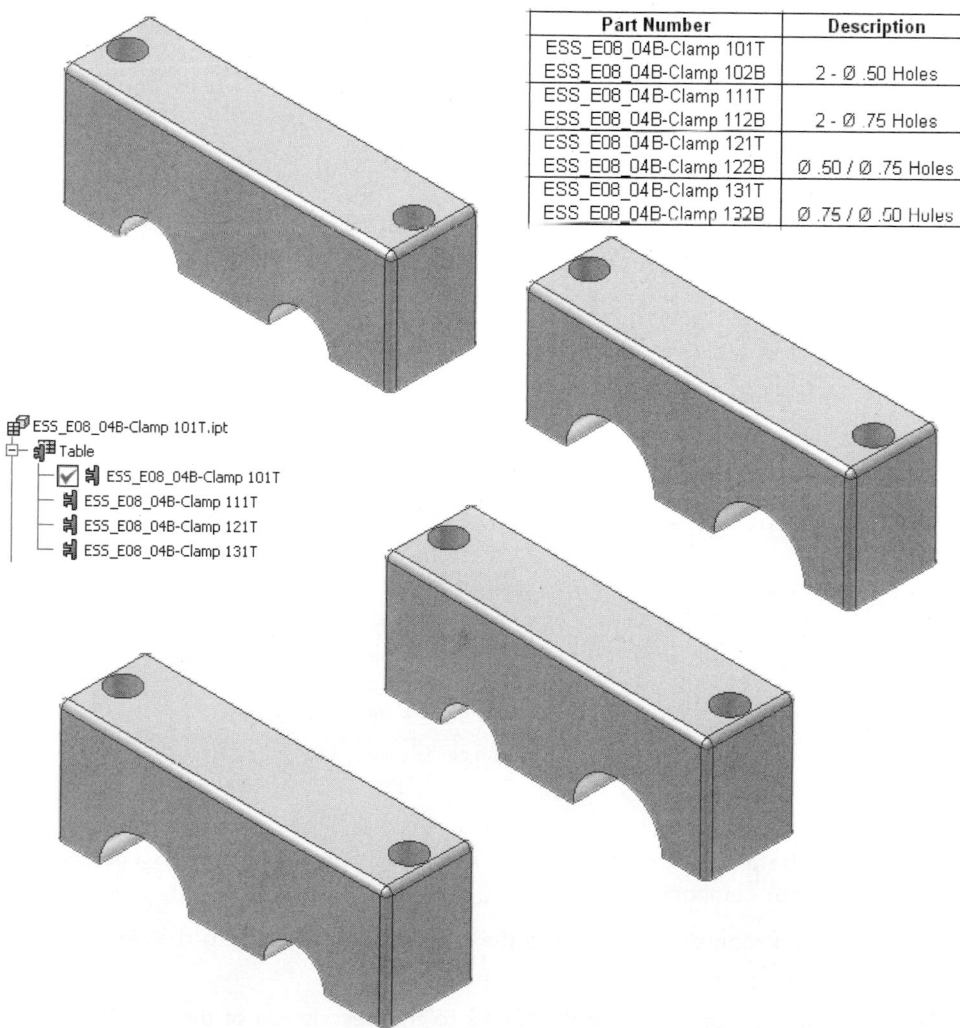

Part Number	Description
ESS_E08_04B-Clamp 101T ESS_E08_04B-Clamp 102B	2 - Ø .50 Holes
ESS_E08_04B-Clamp 111T ESS_E08_04B-Clamp 112B	2 - Ø .75 Holes
ESS_E08_04B-Clamp 121T ESS_E08_04B-Clamp 122B	Ø .50 / Ø .75 Holes
ESS_E08_04B-Clamp 131T ESS_E08_04B-Clamp 132B	Ø .75 / Ø .50 Holes

Figure 8.109

19. From the Tools menu, click on Create iAssembly.

20. In the iAssembly Author dialog box, right-click row 1 in the table, and click Insert Row from the menu.

21. Add two additional rows to the table so that you have 4 iAssembly members, as shown in the following image.

	Member	Part Number
1	ESS_E08_04B-01	ESS_E08_04B-01
2	ESS_E08_04B-02	ESS_E08_04B-02
3	ESS_E08_04B-03	ESS_E08_04B-03
4	ESS_E08_04B-04	ESS_E08_04B-04

Figure 8.110

22. On the Components tab, expand ESS_E08_04B-Clamp 102B (Bottom).

23. Double-click the Table Replace node to add the property to the right pane, as shown in the following image.

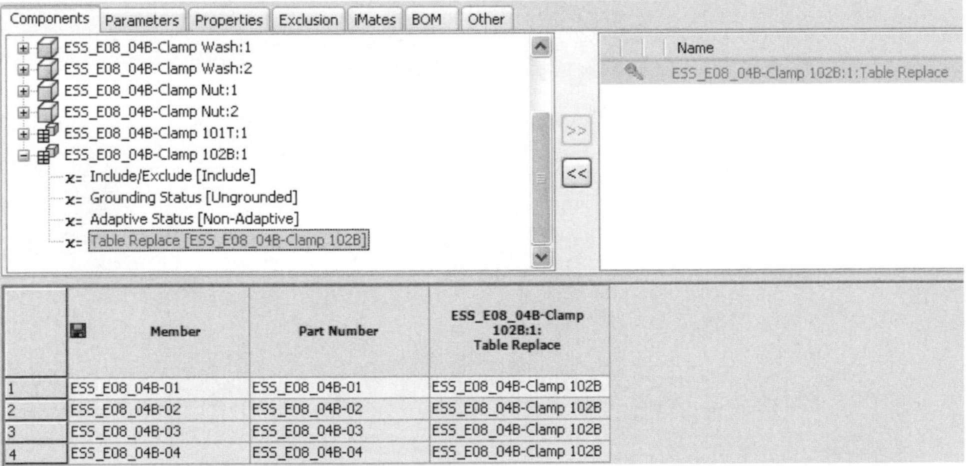

Figure 8.111

24. Repeat the process to add the Table Replace property for the ESS_E08_04B-Clamp 101T (Top) component.

25. Click the Properties tab, and add the Description property to the iAssembly definition.

26. Add the description of .50/.50 HOLES to the description of the first member, as shown in the following image.

	Member	Part Number	ESS_E08_04B-Clamp 102B:1: Table Replace	ESS_E08_04B-Clamp 101T:1: Table Replace	Description
1	ESS_E08_04B-01	ESS_E08_04B-01	ESS_E08_04B-Clamp 102B	ESS_E08_04B-Clamp 101T	.50/.50 HOLES
2	ESS_E08_04B-02	ESS_E08_04B-02	ESS_E08_04B-Clamp 102B	ESS_E08_04B-Clamp 101T	
3	ESS_E08_04B-03	ESS_E08_04B-03	ESS_E08_04B-Clamp 102B	ESS_E08_04B-Clamp 101T	
4	ESS_E08_04B-04	ESS_E08_04B-04	ESS_E08_04B-Clamp 102B	ESS_E08_04B-Clamp 101T	

Figure 8.112

EXERCISES

27. Add the remaining three descriptions to the table, as shown in the following image.

	Member	Part Number	ESS_E08_04B-Clamp 102B:1: Table Replace	ESS_E08_04B-Clamp 101T:1: Table Replace	Description
1	ESS_E08_04B-01	ESS_E08_04B-01	ESS_E08_04B-Clamp 102B	ESS_E08_04B-Clamp 101T	.50/.50 HOLES
2	ESS_E08_04B-02	ESS_E08_04B-02	ESS_E08_04B-Clamp 102B	ESS_E08_04B-Clamp 101T	.75/.75 HOLES
3	ESS_E08_04B-03	ESS_E08_04B-03	ESS_E08_04B-Clamp 102B	ESS_E08_04B-Clamp 101T	.50/.75 HOLES
4	ESS_E08_04D-04	ESS_E08_04B-04	ESS_E08_04B-Clamp 102B	ESS_E08_04B-Clamp 101T	.75/.50 HOLES

Figure 8.113

28. Select the ESS_E08_04B-02 member, and replace the following parts in the table, as shown in the following image:

- ESS_E08_04B-Clamp 102B with ESS_E08_04B-Clamp 112B
- ESS_E08_04B-Clamp 101T with ESS_E08_04B-Clamp 111T

	Member	Part Number	ESS_E08_04B-Clamp 102B:1: Table Replace	ESS_E08_04B-Clamp 101T:1: Table Replace	Description
1	ESS_E08_04B-01	ESS_E08_04B-01	ESS_E08_04B-Clamp 102B	ESS_E08_04B-Clamp 101T	.50/.50 HOLES
2	ESS_E08_04B-02	ESS_E08_04B-02	ESS_E08_04B-Clamp 112B	S_E08_04B-Clamp 101T	.75/.75 HOLES
3	ESS_E08_04B-03	ESS_E08_04B-03	ESS_E08_04B-Clamp 102B	ESS_E08_04B-Clamp 101T	.50/.75 HOLES
4	ESS_E08_04B-04	ESS_E08_04B-04	ESS_E08_04B-Clamp 102B	ESS_E08_04B-Clamp 111T	.75/.50 HOLES
				ESS_E08_04B-Clamp 121T	
				ESS_E08_04B-Clamp 131T	

Figure 8.114

29. Repeat the same replacement operations on the ESS_E08_04B-03 and ESS_E08_04B-04 members, as shown in the following image.

	Member	Part Number	ESS_E08_04B-Clamp 102B:1: Table Replace	ESS_E08_04B-Clamp 101T:1: Table Replace	Description
1	ESS_E08_04B-01	ESS_E08_04B-01	ESS_E08_04B-Clamp 102B	ESS_E08_04B-Clamp 101T	.50/.50 HOLES
2	ESS_E08_04B-02	ESS_E08_04B-02	ESS_E08_04B-Clamp 112B	ESS_E08_04B-Clamp 111T	.75/.75 HOLES
3	ESS_E08_04B-03	ESS_E08_04B-03	ESS_E08_04B-Clamp 122B	ESS_E08_04B-Clamp 121T	.50/.75 HOLES
4	ESS_E08_04B-04	ESS_E08_04B-04	ESS_E08_04B-Clamp 132B	ESS_E08_04B-Clamp 131T	.75/.50 HOLES

Figure 8.115

30. Click OK to save and exit the iAssembly Author dialog box.

31. A Table node is added to the Assembly browser. Expand the Table entry, and double-click on each assembly configuration to cycle through the different hole arrangements in the clamp, as shown in the following image.

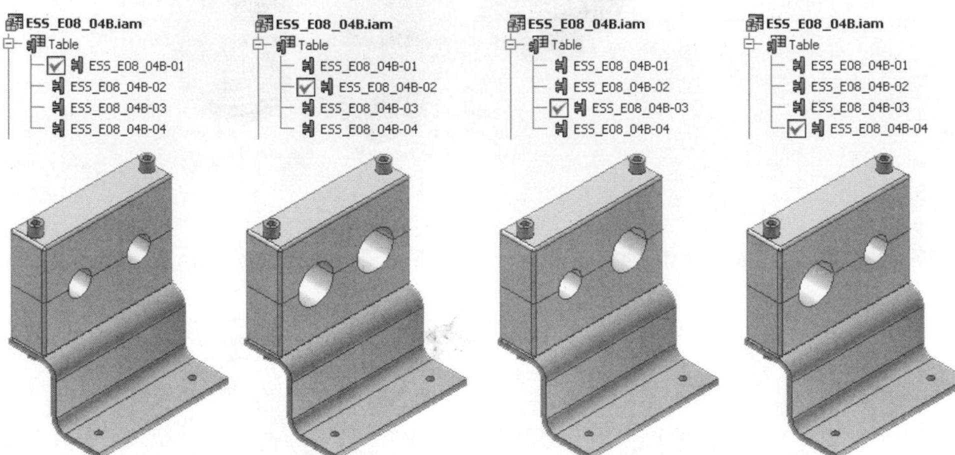

Figure 8.116

32. Close all open files. Do not save changes.

33. Open *ESS_E08_04C.iam.*

This is the same clamp that is made up of iParts with which you just finished working. The ESS_E08_04C-03 configuration is active.

When an iAssembly is created, there is an iParts/iAssemblies toolbar that can be displayed. So far, you have worked with the toolbar in Edit Factory Scope mode. Any changes made affect all members of the iAssembly. You will now work with the Edit Member Scope tool of the iParts/iAssemblies toolbar, as shown in the following image. When this mode is active and changes are made to the current iAssembly, the changes are present only in the iAssembly. The remaining iAssembly configurations are unaffected by changes.

Note: If the iParts iAssemblies toolbar is not visible, perform the following actions:

- Right-click the Standard toolbar, and click iPart/iAssembly.

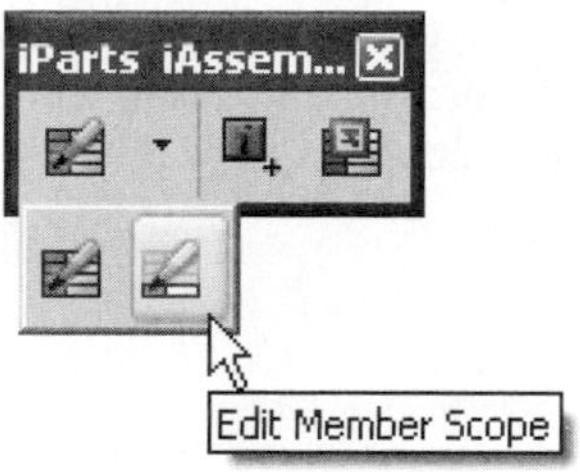

Figure 8.117

34. Set the current iAssembly to Edit Member Scope, as shown in the previous image.
35. On the Assembly panel bar, click Chamfer.
36. Chamfer both edges of both holes on both faces of the iAssembly clamp. Use a chamfer distance of **0.05** units, as shown in the following image.

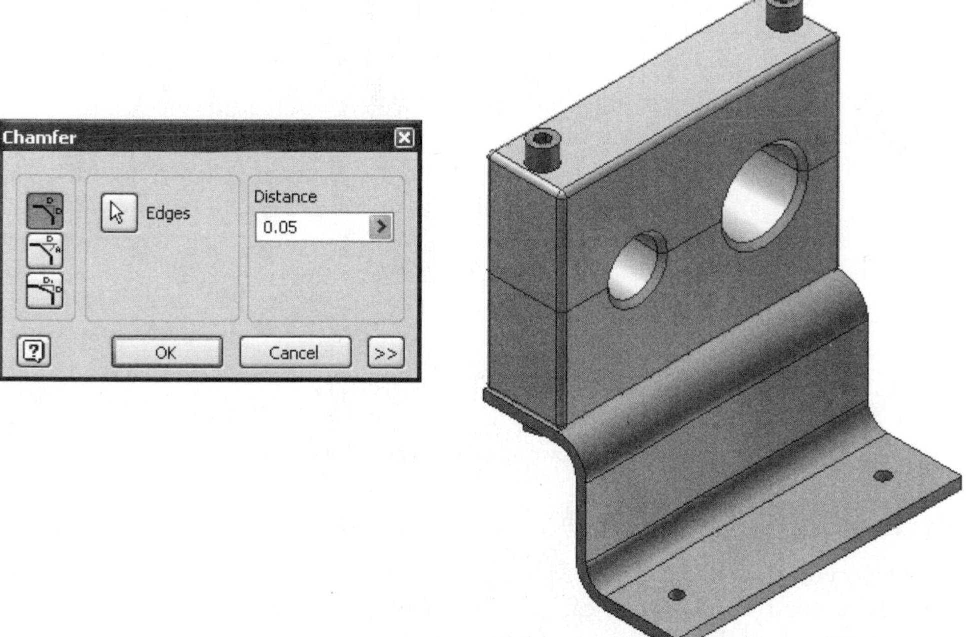

Figure 8.118

37. Double-click any of the other configurations of the iAssembly.

Since Edit Member Scope was active, the chamfers are only available in the configuration that was active when they were created. Notice also that the Chamfer in the browser is excluded from the other configurations, as shown in the following image.

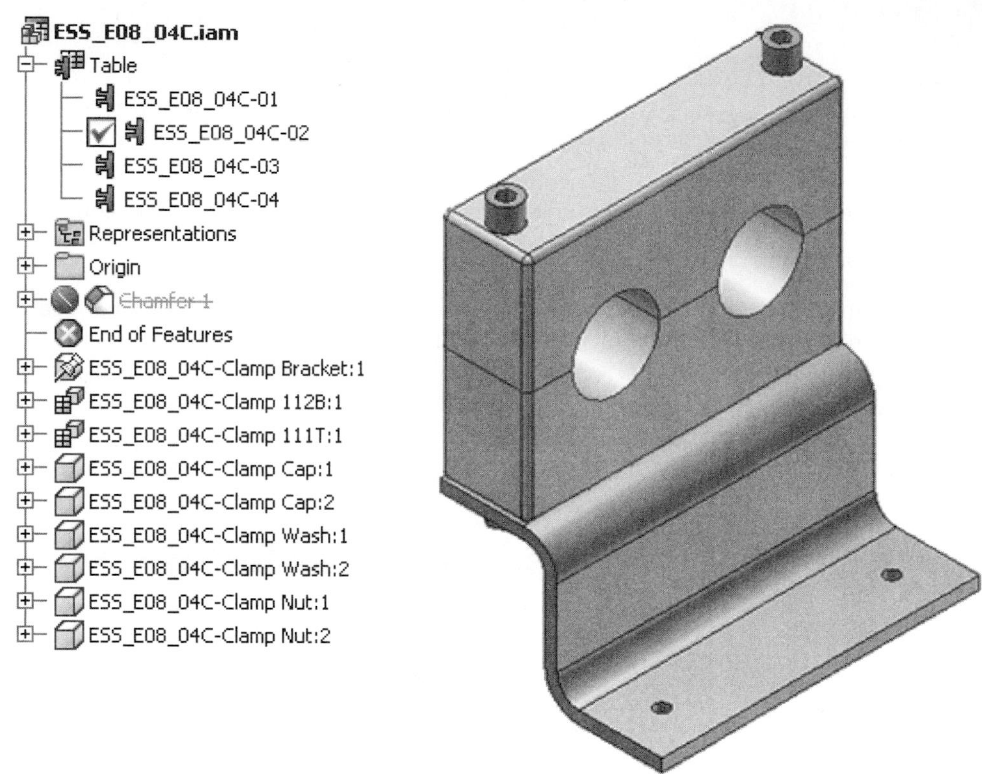

Figure 8.119

38. Close all open files. Do not save changes. End of exercise.

IFEATURES

iFeatures give you the ability to reuse single or multiple features from a part file in other Autodesk Inventor files. iFeatures capture the design intent built into a feature(s) that is going to be reused, such as the name or the size and position parameters. You can also embed or attach a file to the iFeature to be used as Placement Help when you place the iFeature into another design.

You can include reference edges as position geometry in your iFeatures. Reference edges allow you to capture additional design intent, but they require that the iFeature be positioned in the same way it was designed originally in every part in which it is placed. iFeatures are saved in their own type of file that has an *.ide* extension and a unique icon, as shown in the following image.

Figure 8.120

iFeatures are stored in a catalog. The catalog is a directory on your computer or on a server that is set to store all of the iFeatures that you create. You can browse the iFeature catalog at any time by clicking the View Catalog tool on the Part Features panel bar, as shown in the following image.

Figure 8.121

When the View Catalog tool is selected, Windows Explorer opens to the directory specified in the iFeature root setting on the iFeature tab of the Application Options dialog box. In the iFeature root folder, you can view, copy, edit, or delete iFeatures from your catalog.

CREATE IFEATURES

You create iFeatures using the Extract iFeature tool on the Tools menu, as shown in the following image.

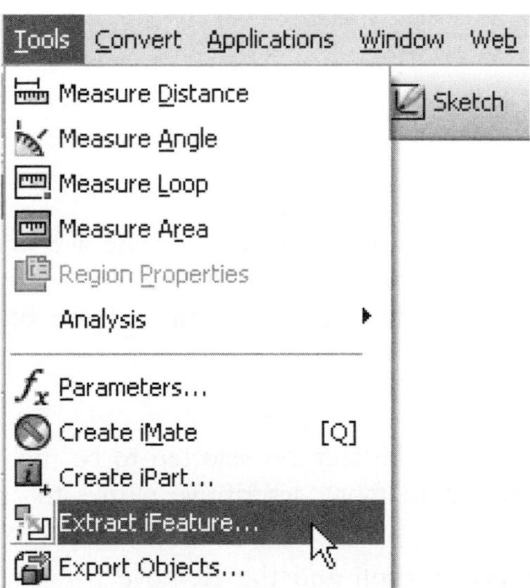

Figure 8.122

Once selected, the Extract iFeature dialog box appears, as shown in the following image.

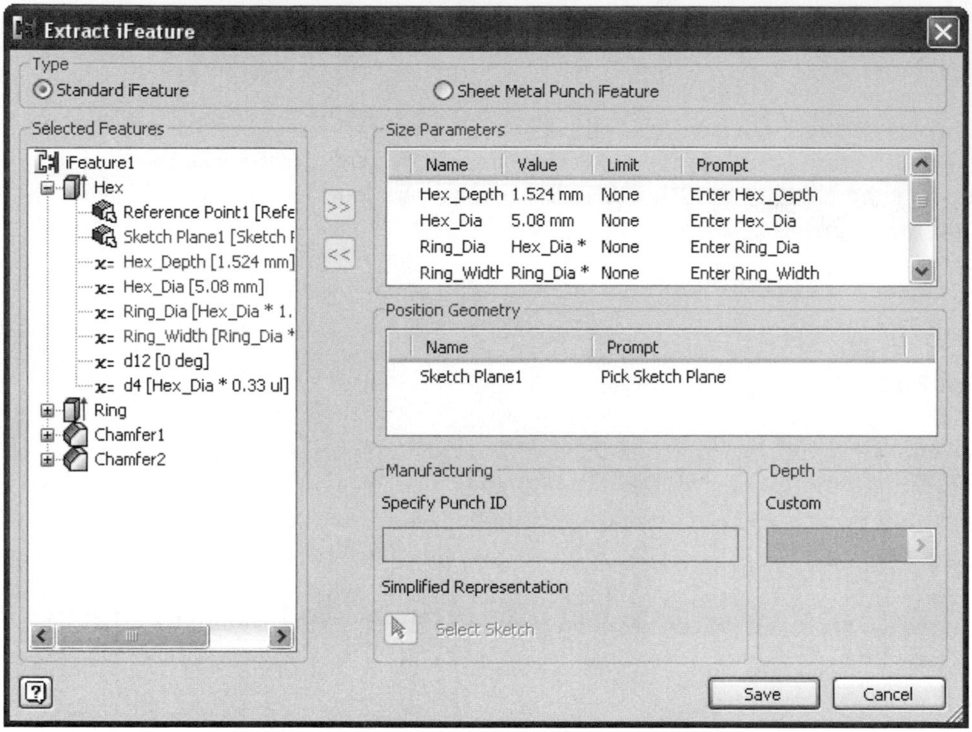

Figure 8.123

The Extract iFeature dialog box contains the following sections.

Type

In this area, you can select whether you want to create a Standard iFeature or a Sheet Metal Punch iFeature. In this chapter, we will focus on Standard iFeatures. Refer to chapter 10 for information regarding the creation and use of the Sheet Metal Punch iFeature.

Selected Features

This area of the dialog box lists the features selected to be included in the iFeature. You can rename features in the list to more descriptive names to assist in working with the iFeature at a later time.

Use the Add parameters (>>) button and the Remove parameters (<<) button to move parameters from the highlighted features in the Selected Features list to the Size Parameters table.

Size Parameters

The Size Parameters table lists all of the parameters that will be used for interface when the iFeature is placed into another file. You can select the parameters by expanding the features listed in the Selected Features list or directly in the graphics window.

Note: Any parameters that have been given a name in the Parameters dialog box appear automatically in the Size Parameters pane of the Create iFeature dialog box upon iFeature creation. Renaming parameters that you want to include in an iFeature can speed up the process of creating them.

Name Specify a descriptive name for the parameter. Use names that describe the purpose of the parameter.

Value Place a value to be the default for the parameter when inserting your iFeature into a file. The value is restricted by settings in the Limit column.

Limit Place restrictions on the values that are available for the parameter by using one of three options, None, Range, or List, as shown in the following image.

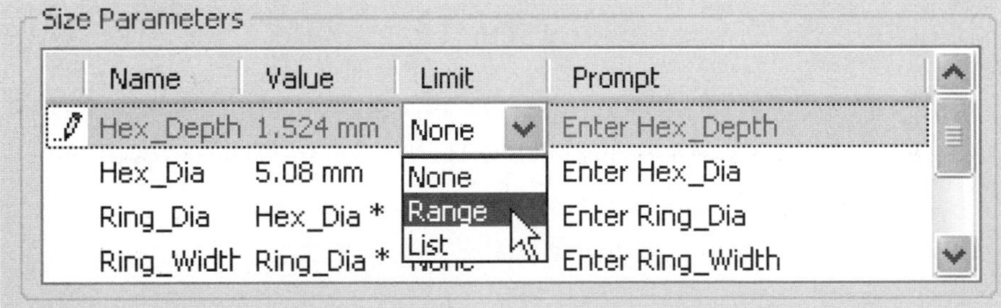

Figure 8.124

None specifies that no restrictions be placed on the Value field. Range gives you the ability to specify a minimum and maximum value, including less than, equal to, and infinity, as shown in the following image on the left. You can specify the default value to use. The List option, as shown in the following image on the right, allows you to predefine a list of values. Upon placement, you can choose from these values for the size parameters.

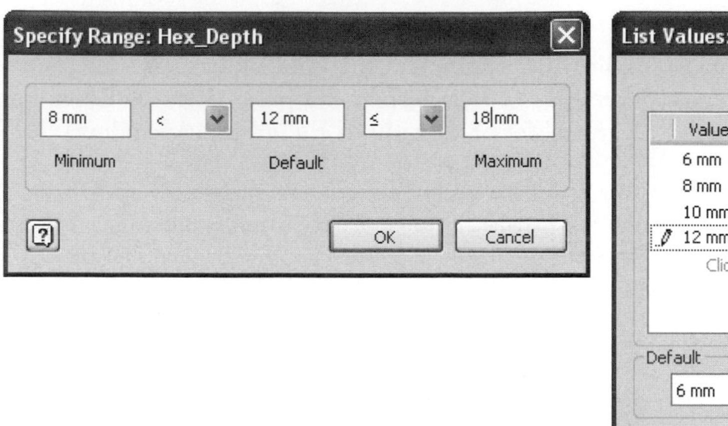

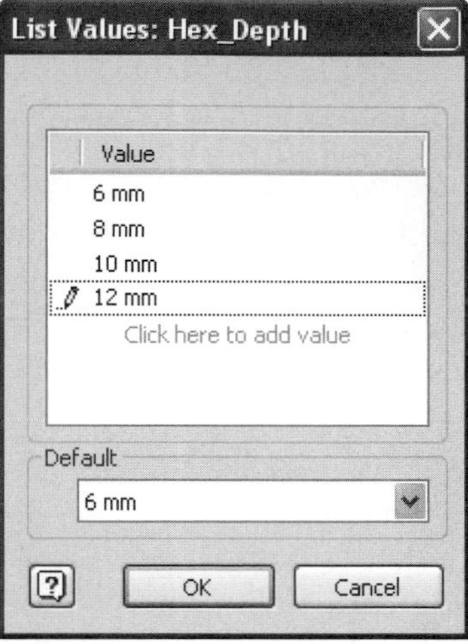

Figure 8.125

Prompt Enter descriptive instructions to explain further why the parameter is used. The text entered in the prompt field appears in a dialog box during the iFeature placement.

Position Geometry Specify the geometry of the iFeature that is necessary to position it on a part. You can add or remove geometry to or from the Position Geometry list in the Selected Features tree by right-clicking the geometry and selecting Expose Geometry. You remove geometry from the Position Geometry list by right-clicking it and selecting Remove Geometry.

Name Describes the position geometry.

Prompt Enter descriptive instructions to prompt a user for position geometry. The text entered in the prompt field appears in a dialog box during the iFeature positioning. You can customize the position geometry by right-clicking it and selecting one of the two additional options available on the menu.

Make Independent Select this option to separate entries for geometry shared by more than one feature.

Combine Geometry Select this option to combine listings of geometry shared by more than one feature in a single entry.

Manufacturing

This field is used when creating a Sheet Metal Punch iFeature. Refer to chapter 10 for information.

Depth

This field is used when creating a Sheet Metal Punch iFeature. Refer to chapter 10 for information.

INSERT IFEATURES

Insert iFeatures using the Insert iFeature tool on the Part Features panel bar, as shown in the following image.

Figure 8.126

Once selected, the Insert iFeature dialog box appears, as shown in the following image. You can select tasks from the tree structure in the left pane of the dialog box or move forward and backward using the Next and Back buttons. Click the Browse button to navigate the iFeatures folder structure and select an iFeature to place.

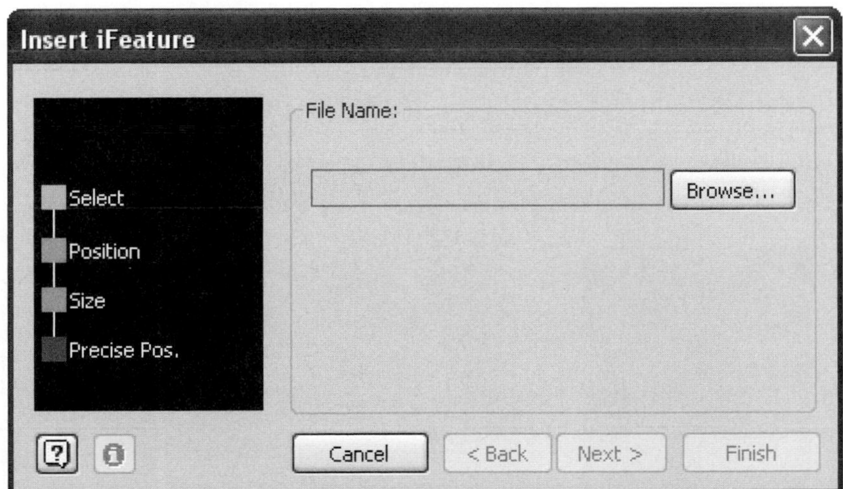

Figure 8.127

The Insert iFeature dialog contains the following sections.

Select

Choose the iFeature that you want to insert using the Browse button.

Position

This feature lists the names of the interface geometries specified during the creation process of the iFeature. After selecting the corresponding geometry in the part where you are placing the iFeature, click the arrowhead on the positioning symbol to either move or rotate the symbol, as shown in the following image.

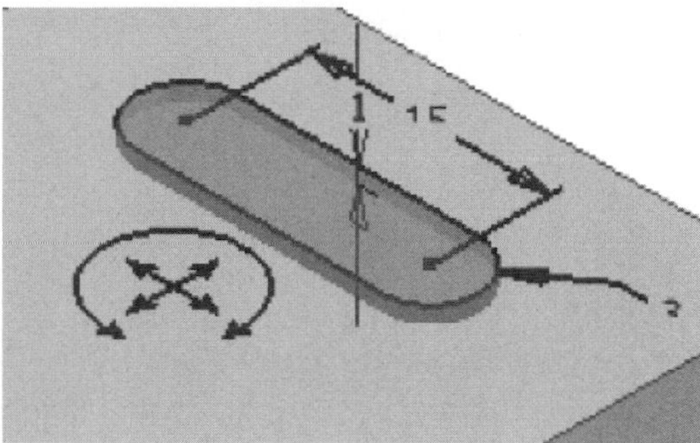

Figure 8.128

You can also specify a precise rotation value directly in the angle field of the dialog box. After the requirement has been satisfied, a checkmark will be placed in the left column, as shown in the following image.

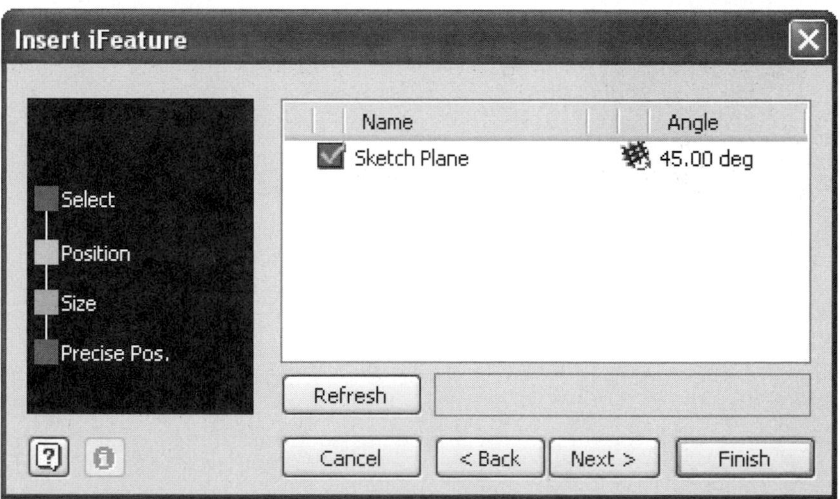

Figure 8.129

Name This section lists the named interface geometry.

Angle This section shows the default angle of the placement geometry on the iFeature.

Move Coordinate System This section defines horizontal or vertical axes when the iFeature has horizontal or vertical dimensions or constraints included.

Size

This shows the names and default values specified for the iFeature, as shown in the following image. Click in the row to edit the values, and then click Refresh to preview the changes.

Name This section lists the name of the parameter.

Value This section lists the value of the parameter.

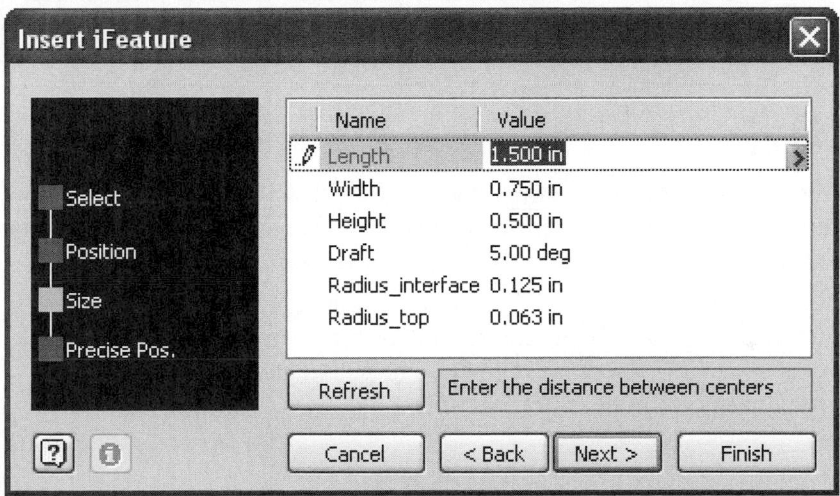

Figure 8.130

Precise Position

This feature further refines the position of the iFeature, using either dimensions or constraints after you have placed it.

Activate Sketch Edit Immediately Click this option to activate the sketch of the iFeature and the 2D Sketch panel bar. You can then apply additional dimensions and/or constraints to position the iFeature on the part.

Do not Activate Sketch Edit Click this option, as shown in the following image, to position the iFeature without applying additional constraints or dimensions.

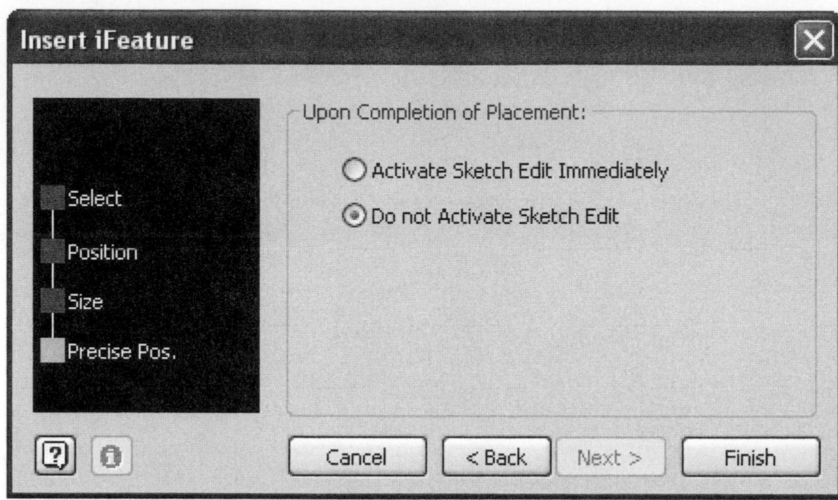

Figure 8.131

EDITING IFEATURES

You can edit an iFeature by opening the *.ide* file in Autodesk Inventor. Four tools are available in the iFeature environment when the file is opened: Edit iFeature, View Catalog, iFeature Author Table, and Edit using Spread Sheet.

The Edit iFeature tool, as shown in the following image, opens the Edit iFeature dialog box. You cannot change which parameters are used to define the iFeature after it has been created, but you can modify the size parameters and position geometry by editing the properties for the name, value, limit, and prompt.

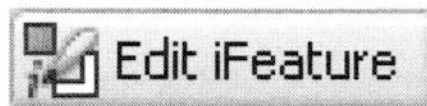

Figure 8.132

The iFeature Author Table tool, as shown in the following image, allows you to create iFeatures that are driven by a table.

Figure 8.133

Once selected, the iFeature Author dialog box appears, as shown in the following image. The table in the iFeature Author tool functions the same as it does when working with iParts. You can insert, delete, and specify the default row. You can also define custom parameter columns or cells and include a range for the value. The main difference is that there is no option to set a column as a file name, display style, or material. If you are converting an iFeature to a table-driven iFeature that contains a parameter with a list of possible values, the table is populated automatically with the different sizes from the list as rows of the table. You can also edit the table via a spreadsheet.

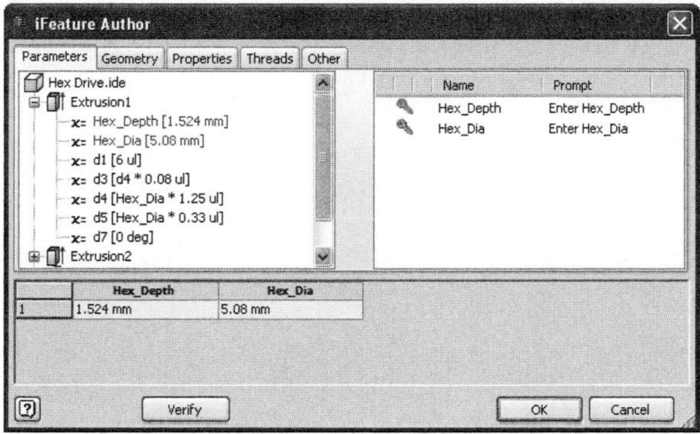

Figure 8.134

Parameters Tab

The Parameters tab is the default tab that appears in the iFeature Author dialog box, and it lists all of the parameters used to create the iFeature. You cannot change which parameters make up the iFeature after it has been created. The Size Parameters pane on the right of the dialog box displays all of the parameters used in the definition and their names and prompts. You can define keys for the parameters that function the same as when working with iParts.

Geometry Tab

The Geometry tab, as shown in the following image, allows you to add or remove interface geometry for any iFeatures created after the table-driven iFeature is saved. Previously placed iFeatures are not affected. You can also modify the labels while editing.

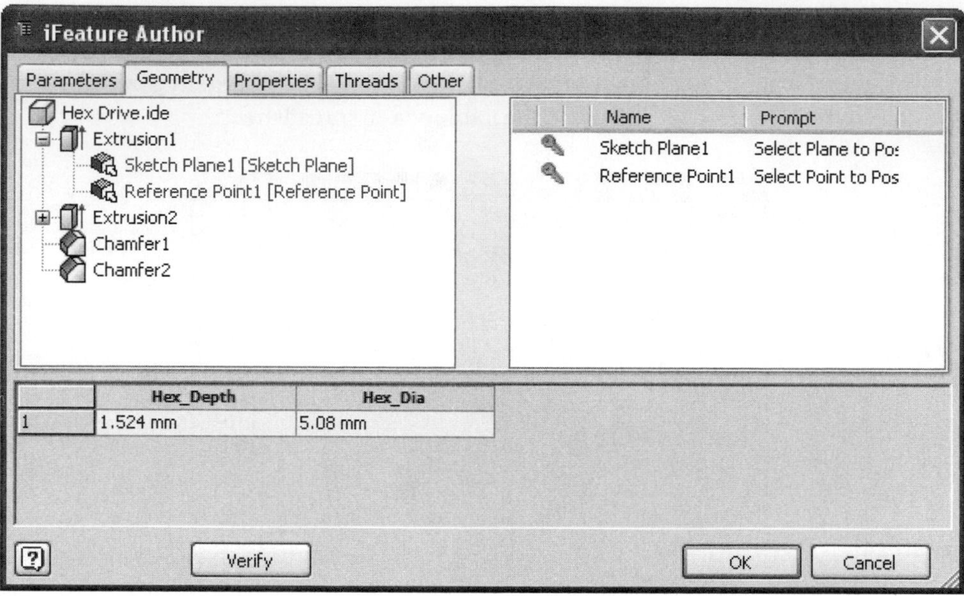

Figure 8.135

Properties Tab

The Properties tab, as shown in the following image, lists all of the Summary and Project properties for the file. This tab functions the same as it does when working with iParts, with the exception that you cannot declare a column to be a File Name, Display Style, or a Material column and that Physical Properties are not available.

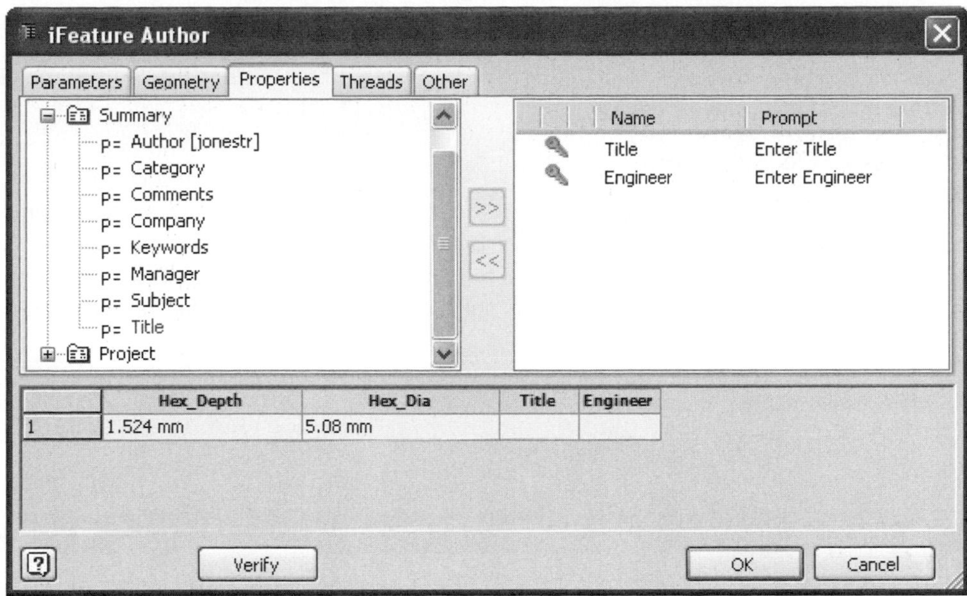

Figure 8.136

Threads Tab

You can control thread features in the iFeature similar to the way you do when working with iParts. Use Family to identify the thread standard, Designation to control the size and pitch, and Class to define fit based upon the standard. You can assign the thread Direction as right or left in the table.

Other Tab

You can use the Other tab to create custom table items. This tab functions the same as it does when working with iParts, with the exception that you cannot declare a column to be a File Name, Display Style, or Material column.

INSERTING TABLE-DRIVEN IFEATURES

You insert table-driven iFeatures the same way that you would insert a typical iFeature. The wizard displays the key parameters as a drop-down list while defining the Size parameters, and any custom parameters will have the edit field available for entry.

The browser displays table-driven iFeatures in a similar manner to that used with iParts.

When you place a table-driven iFeature in a part file, a new *.ide* file is not created. The placement is similar to that of an iFeature that is not table driven. The file in which the iFeature is placed contains an independent copy of the iFeature that is not associative to the original *.ide* file. The spreadsheet that contains the table information is not visible in the browser, and you cannot edit it once you place it in the part file.

EXERCISE 8-5: CREATING AND PLACING IFEATURES

In this exercise, you will create a hex drive iFeature from an existing component. You will then insert the iFeature into another part file.

1. Open *ESS_E08_05.ipt*.
2. Click the Parameters tool, and review the list of Model Parameters. The renamed diameter *(Hex_Dia)* and depth *(Hex_Depth)* parameters of the hex drive extrusion are the variables that will be extracted with the iFeature. The geometry of the raised ring around the hex drive is linked to the diameter of the hex drive.
3. In the Parameters dialog box, click Done.
4. From the Tools menu, click Extract iFeature.
5. Click the features named *Hex* and *Ring* in the browser.

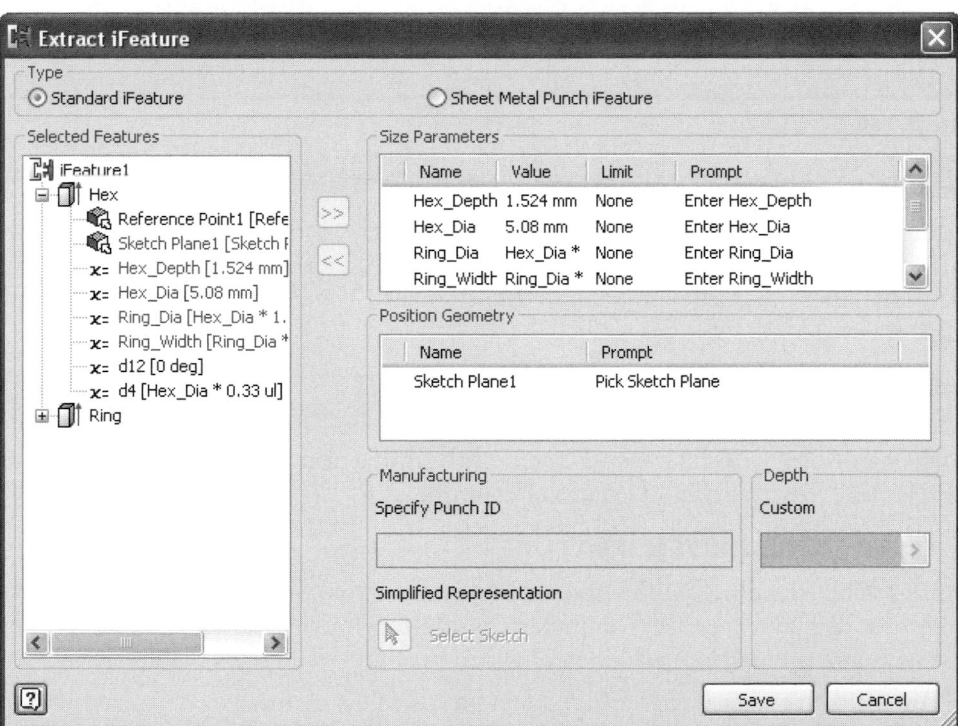

Figure 8.137

6. Click the *Pick Sketch Plane* prompt under Position Geometry, and enter **Select Plane to Position the Hex Drive** as the new prompt.
7. Expand *Hex* under Selected Features, right-click Reference Point1, and click Expose Geometry from the menu, as shown in the following image.

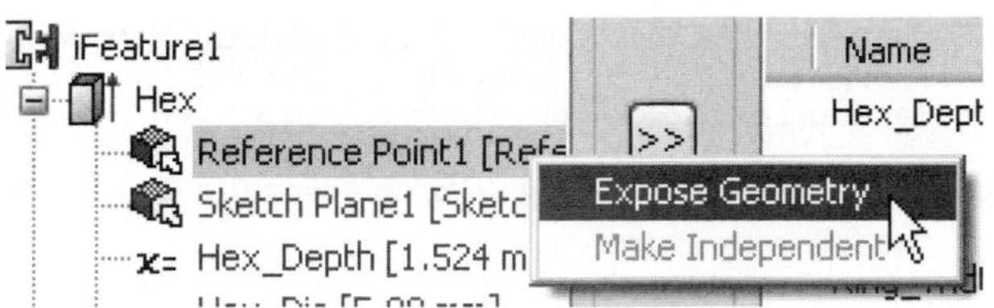

Figure 8.138

8. The point is added to the Position Geometry list. This point constrained the hex drive extrusion concentric to the circular face on the end of the cylinder. The point referenced will allow you to select model geometry to position the iFeature during placement in a new part. Click the *Pick Reference Point* prompt under Position Geometry.

- Highlight the text, and replace it with **Select Point to Position Hex Drive Centerline** as the new prompt.
- Click and drag the Reference Point1 position geometry below Sketch Plane1, as shown in the following image.

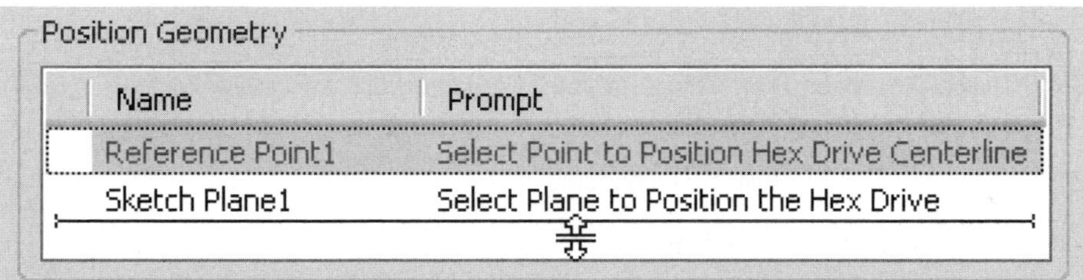

Figure 8.139

- The following image shows the completed operation.

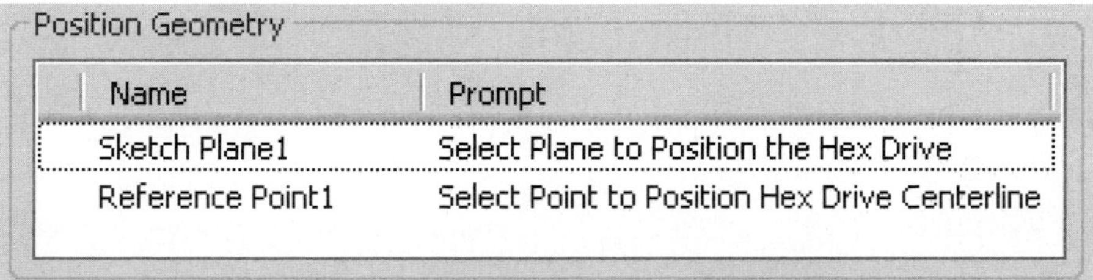

Figure 8.140

9. Three parameters, *Ring_Height, Ring_Width,* and *Ring_Dia,* are linked to the hex drive diameter and are not required in the iFeature. To remove these parameters from the list, perform the following actions:

- In the Size Parameters area, click *Ring_Height,* and click the Remove parameters button (<<).
- In the Size Parameters area, click *Ring_ Width,* and click the Remove parameters button (<<).
- In the Size Parameters area, click *Ring_ Dia,* and click the Remove parameters button (<<), as shown in the following image.

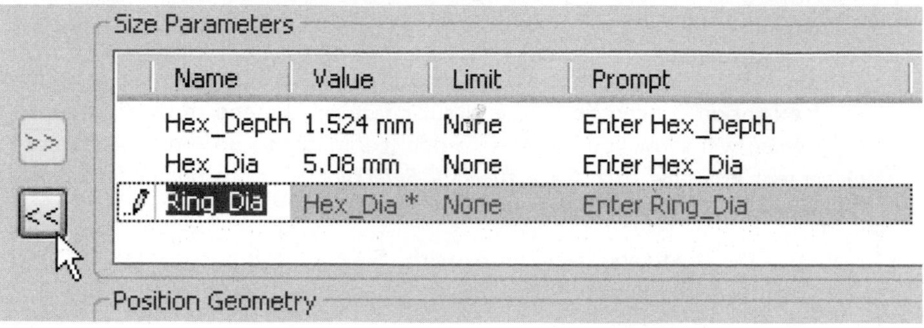

Figure 8.141

10. Only two parameters should now exist in the Size Parameters area: Hex_Depth and Hex_Dia.

11. Click Save, enter **Hex Drive** in the File name field, and place the file in the same location as the exercise files.

12. Close the file. Do not save changes.

13. Open *ESS_E08-05-Lock Post.ipt.*

14. Click the Insert iFeature tool.

- Click the Browse button in the Insert iFeature dialog box.
- Navigate to the directory where the exercises were installed.
- Select *Hex Drive.ide.*
- Click Open.

15. To locate Sketch Plane1, click the front face of the lock post, as shown in the following image.

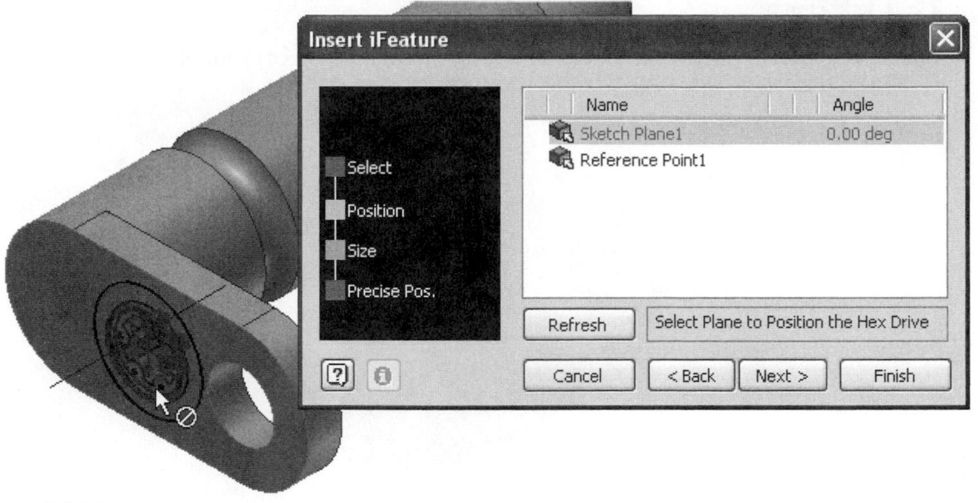

Figure 8.142

16. Click Reference Point1 in the Insert iFeature dialog box.

- Click the left, outer circular edge of the lock post.
- Click the point that is displayed on the part, as shown in the following image.

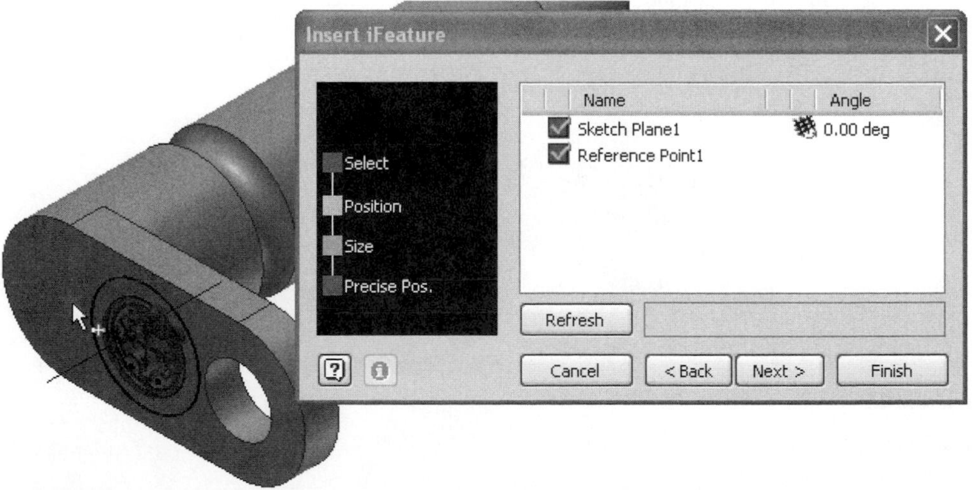

Figure 8.143

17. Click Next.

18. Leave the *Hex_Depth* and *Hex_Dia* variables as their default values.

19. Click Next.

20. Click Do not Activate Sketch Edit.

21. Click Finish.

22. The hex drive iFeature is placed with its center point coincident with the selected center point of the arc.

Figure 8.144

23. Close all open files. Do not save changes. End of exercise.

Project Exercise: Chapter 8

In this exercise, you automate an iPart factory by controlling two iMates from the table. One version of the iPart has a single button, and the other has two buttons, as shown in the following image.

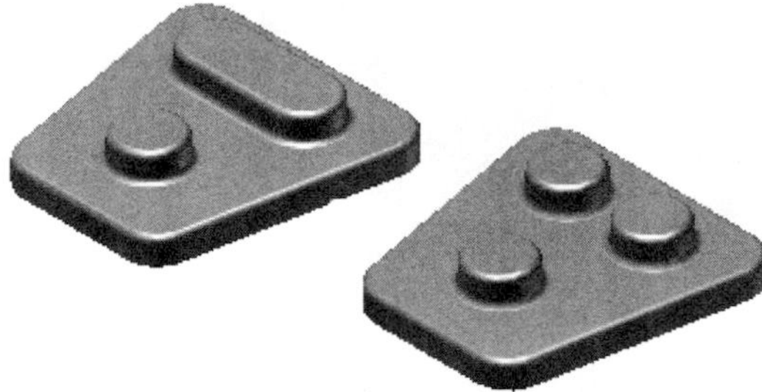

Figure 8.145

Place two different versions of the iPart in an assembly. Each version is automatically positioned to the appropriate mating part using the automated iMates, as shown in the following image.

Figure 8.146

Start with an existing iPart factory of a rubber membrane for a remote keyless entry device.

1. Open *ESS_E08_06.ipt.*

This iPart factory uses feature suppression from the table to control the number of buttons at the top of the membrane. The Single version, as shown in the following image, has one slot-shaped button. The Double version has two circular buttons.

In the following steps, you will use the button geometry to create a unique iMate for each iPart version.

Note: Ensure that the Autocapture tool on the iPart/iAssembly toolbar is set to Edit Factory Scope.

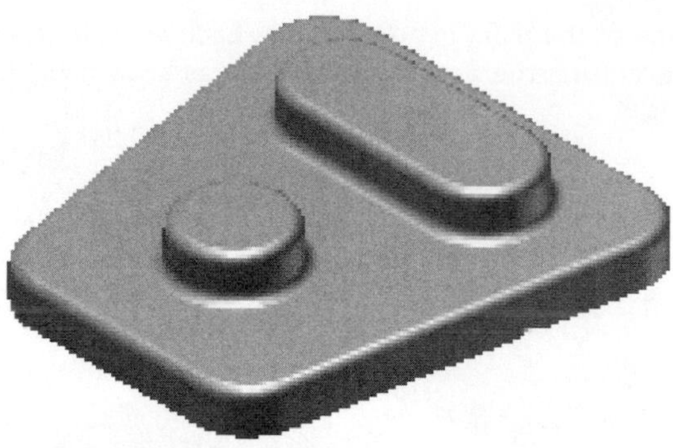

Figure 8.147

2. In the browser, expand Table.
3. Activate the Double member.

Note: In this version, Extrusion3 and Fillet3 are suppressed.

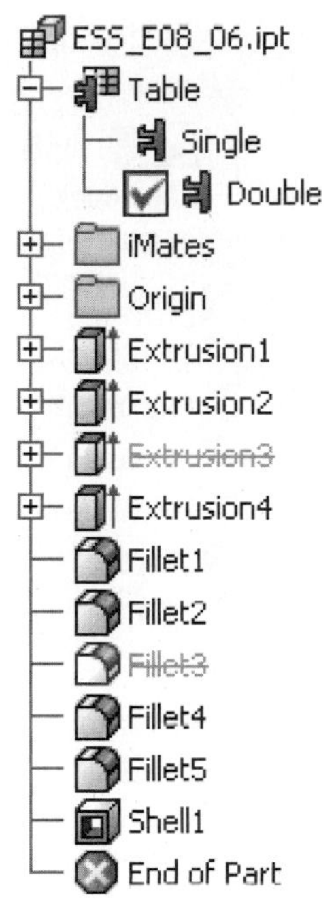

Figure 8.148

4. Activate the Single member.

Note: In this version, the Extrusion4 and Fillet4 features are suppressed.

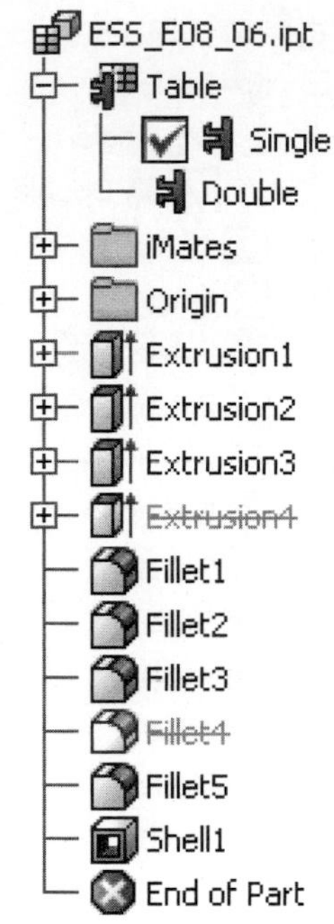

Figure 8.149

5. Click the Create iMate tool.
6. Click the cylindrical face, as shown in the following image.

Note: The axis must be highlighted.

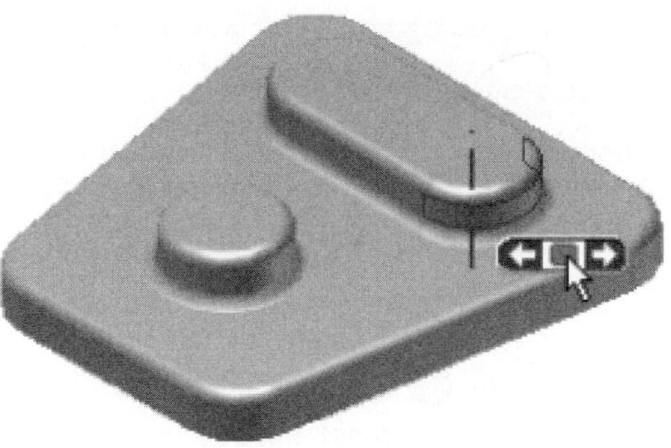

Figure 8.150

7. Expand the Create iMate dialog box, enter a name of **Axis1**, and then select OK.
8. In the browser, activate the Double member.

Note: When you make the Double member current, the Extrusion3 and Fillet3 features are suppressed. Since the Axis1 iMate is dependent on that geometry, it is also suppressed. Expand the iMates folder to view the suppression of the iMate.

9. Click the Create iMate tool.
10. Click the cylindrical face of the left button, as shown in the following image.

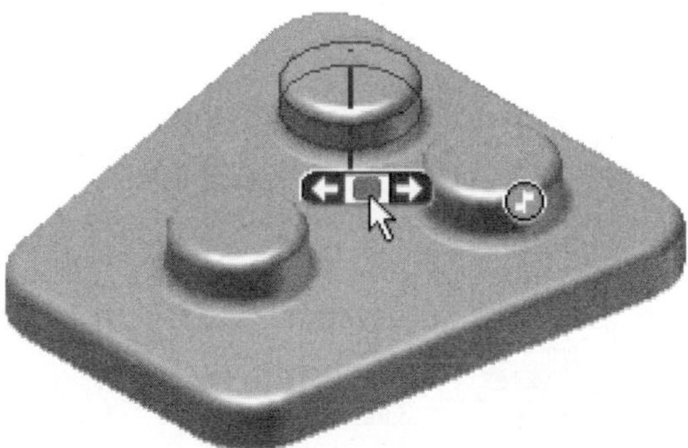

Figure 8.151

11. Expand the Create iMate dialog box, enter a name of **Axis2**, and then select OK.

12. In the browser, double-click Table.

13. In the iPart Author dialog box, click the iMates tab, and note the Axis1 and Axis2 identifying labels IM0003 and IM0004, as shown in the following image.

Figure 8.152

14. In the iPart Author dialog box, click the Suppression tab.

15. Double-click IM0003 and IM0004.

Notice that both iMates are added to the list on the right, and a column is added to the table, as shown in the following image.

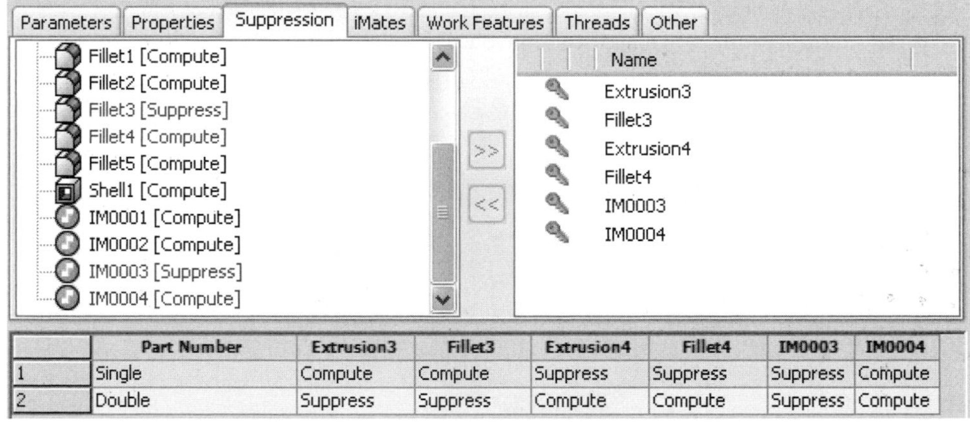

	Part Number	Extrusion3	Fillet3	Extrusion4	Fillet4	IM0003	IM0004
1	Single	Compute	Compute	Suppress	Suppress	Suppress	Compute
2	Double	Suppress	Suppress	Compute	Compute	Suppress	Compute

Figure 8.153

16. In the table, change the values for IM0003 and IM0004 to be consistent with the following image.

	Member	Part Number	Extrusion3	Fillet3	Extrusion4	Fillet4	IM0003	IM0004
1	Single	Single	Compute	Compute	Suppress	Suppress	Compute	Suppress
2	Double	Double	Suppress	Suppress	Compute	Compute	Suppress	Compute

Figure 8.154

17. Select OK in the iPart Author dialog box.

18. In the browser, double-click the Single member to make it active.

Note: In the Single version, the Axis1 iMate symbol is shown on the iPart, and the Axis2 iMate is suppressed in the browser, as shown in the following image.

19. In the browser, double-click the Double member to make it active.

Figure 8.155

Note: In the Double version, the Axis2 iMate symbol is shown on the iPart, and the Axis1 iMate is suppressed in the browser, as shown in the following image.

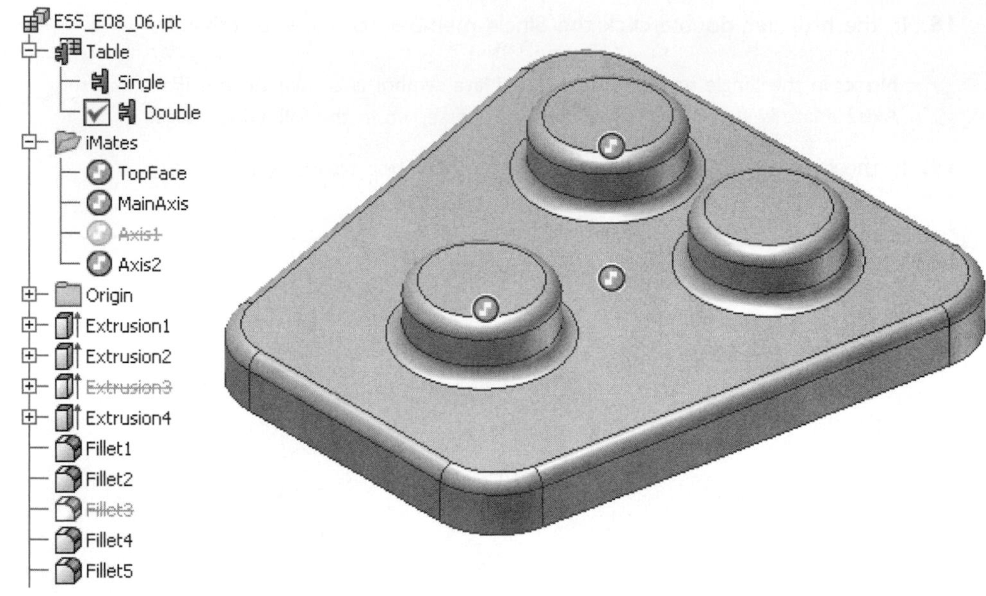

Figure 8.156

20. Double-click the Single member again to make it active.
21. From the File menu, click Save As.
22. In the Save As dialog box, save the iPart factory as Membrane.ipt.
23. Close the file.

    Next open an assembly that contains the mating cases for the Membrane iParts.

24. Open *ESS_E08_06B.iam*.

Figure 8.157

In the following steps, you place two versions of the Membrane iPart factory into this assembly. As you place the iParts, they are automatically oriented to the proper case using the iMates you created.

25. In the browser, expand both parts.

26. In the browser, expand the iMates folder for both parts, as shown in the following image.

Note: The iMate names are consistent with the membrane iPart factory.

EXERCISES

27. Click the Place Component tool.

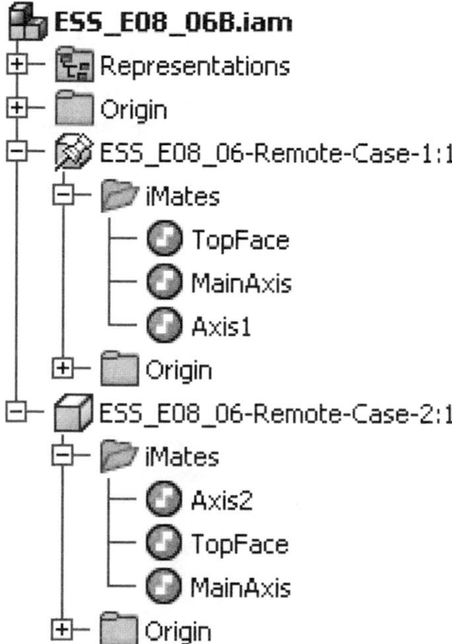

Figure 8.158

28. In the Place Component dialog box, select the Interactively Place with iMates option, click the *Membrane.ipt* file, and click Open.

29. When the Place Standard iPart dialog box is displayed, a preview of the part is displayed where the components containing the same iMate names exist, as shown in the following image. Do not click.

Figure 8.159

30. Right-click anywhere in the graphics area. Select Generate Remaining iMate Results, as shown in the following image, to place and automatically constrain the previewed member.

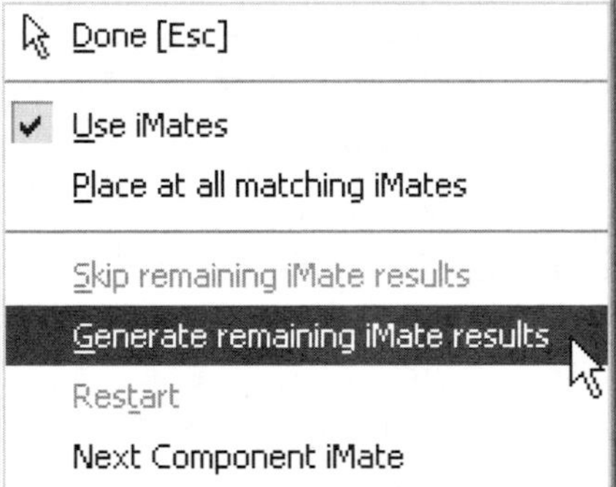

Figure 8.160

Note: After you place the iPart member, the graphics window zooms in on the placed component and previews the next match, as shown in the following image.

Figure 8.161

31. In the Place Standard iPart dialog box, change the member to the Double version.
32. Right-click the graphics windows and select Place at All Matching iMates from the menu, as shown in the following image.

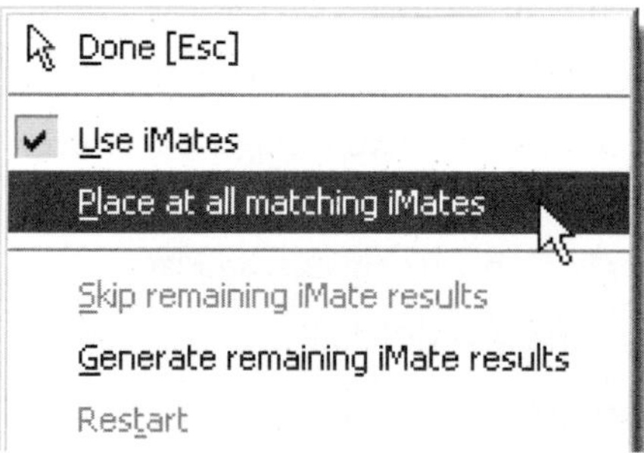

Figure 8.162

Note: This iPart version is automatically oriented with the appropriate case.

Figure 8.163

33. Close the file. Do not save changes. End of exercise.

CHECKING YOUR SKILLS

Use these questions to test your knowledge of the material covered in this chapter.

1. **True _ False _** iMates are created while inside of a part file.
2. What needs to be selected to have a single component containing iMates be placed and automatically constrained in an assembly?
   - **a.** Automatically Generate iMates on Place
   - **b.** Place Component with iMate
   - **c.** Interactively Place with iMates
   - **d.** Use Composite iMate
3. **True _ False _** When creating parameters in a spreadsheet, the data items must be in the following order: parameter name, value or equations, unit of measurement, and, if needed, a comment.
4. What is the difference between a model parameter and a user parameter?

   ____

   ____
5. **True _ False _** Multiple versions of an iPart can be placed in an assembly.
6. **True _ False _** When you make changes to an iPart factory, the changes are updated automatically in iParts that have been placed in assemblies.

7. Which tab of the Place iPart or iAssembly dialog box can be used to create a new member of the factory during placement in an assembly?

- **a.** Keys
- **b.** Table
- **c.** Tree
- **d.** None of the above

8. True _ False _ You can add features to standard iParts after they have been placed in an assembly.

9. True _ False _ Named parameters are added automatically as Size parameters during iFeature creation and cannot be removed.

10. What happens when you create a table-driven iFeature and one of the original parameters contains a list of values for the parameter?

CHAPTER 9

Advanced Assembly Modeling Techniques

INTRODUCTION

In this chapter, you will learn how to use advanced assembly modeling techniques. Using these techniques, you can work more efficiently with assemblies. The techniques discussed will help you manage different views of your assemblies, improve the performance of your system, and analyze assemblies for interference between components. They will also aid in the creation of drawings from those assemblies.

OBJECTIVES

After completing this chapter, you will be able to perform the following:

- Create design view representations
- Create flexible assemblies
- Create positional representations
- Create overlay drawing views
- Improve assembly capacity and performance
- Create an assembly substitution
- Detect contact in assemblies
- Mirror an assembly
- Copy an assembly
- Create assembly work features
- Create assembly features
- Create Skeletal Models
- Use the Frame Generator
- Locate tools related to the Content Center and Design Accelerator

DESIGN VIEW REPRESENTATIONS

While working in an assembly, you may want to save configurations that show the assembly in different states and from different viewing positions. Design view representations can store the following information:

- Component visibility: visible or not visible
- Sketch visibility: visible or not visible
- Component selection status: enabled or not enabled
- Color settings and style characteristics applied in the assembly
- Zoom magnification
- Viewing angle
- Display of origin work planes, origin work axes, origin work points, user work planes, user work axes, and user work points

You can use design view representations while working on the assembly and when creating presentation views or drawing views. You can also use them to reduce the number of items that display in an assembly, allowing large assembly files to be opened more quickly.

Once the screen orientation and part visibility is set, you can create a design view representation by clicking the Design View Representations icon at the top of the browser, clicking the arrow next to the icon, and then clicking Other, as shown in the following image, or by clicking Design View Representations on the View menu.

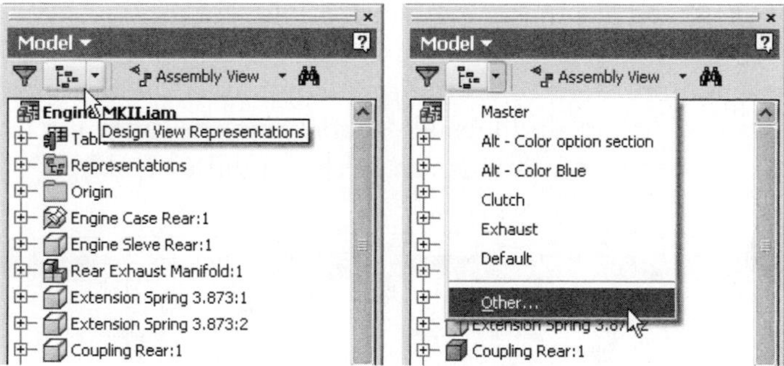

Figure 9.1

The Design View Representations dialog box will appear, as shown in the following image. In this example, four design view representations were created: Alt - Color Blue, Alt - Color option section, Clutch, and Exhaust. All names listed in the Design View Representations list box are considered public as far as the storage location is concerned. Public design view representations are saved with the assembly file and are available to all users when the assembly is accessed.

To make a design view representation current, select it from the drop-down list where you selected the Other option, or select Other to open the Design View Representations dialog box, and double-click on its name. Alternately, you can click it and select the Activate button. To delete a design view representation from the dialog box, select the design view representation's name from the list, and then click the Delete button.

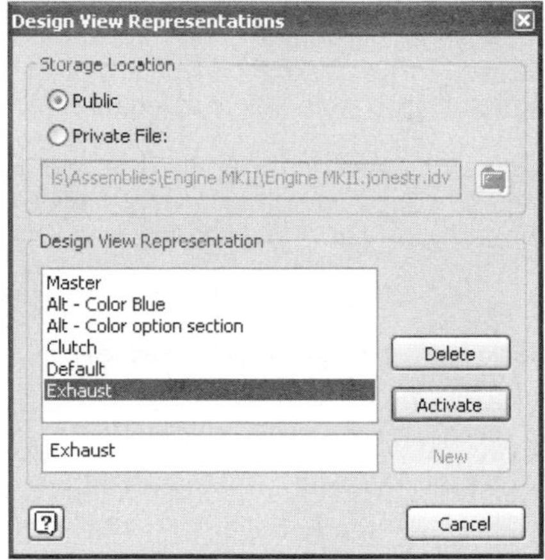

Figure 9.2

Another way to access design view representations is through the Assembly browser, as shown in the following image. Each assembly file contains a folder called Representations, as shown in the following image on the left. Expanding this folder will display all of the representations that can be used, as shown in the following image in the middle. Notice the View, Position, and Level of Detail nodes. Expanding the View node will display all design view representations defined in the assembly model. Notice also the design view representations, Master, Private, and Default, as shown in the following image on the right. These three design view representations are present in all assemblies.

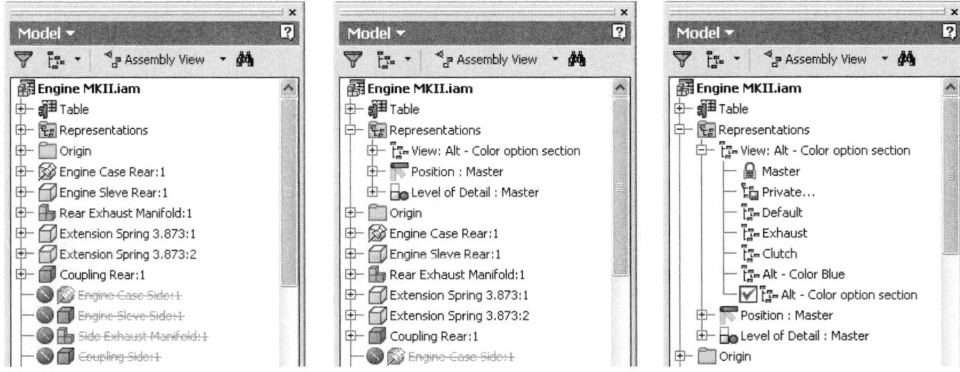

Figure 9.3

CREATING A NEW DESIGN VIEW REPRESENTATION

Begin the process of creating a new design view representation by expanding the Representations folder and the View: Default listing.

To create a design view representation, follow these steps:

1. Move your cursor over the View: Default listing under the Representations folder and right-click.
2. Select New from the menu, as shown in the following image on the left.

Completing these steps will create a new design view representation called View1 and make it current. The current design view representation is identified by the presence of a checkmark next to its name in the browser, as shown in the following image on the right. The name of the current design view representation is also shown next to the View node under the Representations folder in the Assembly browser.

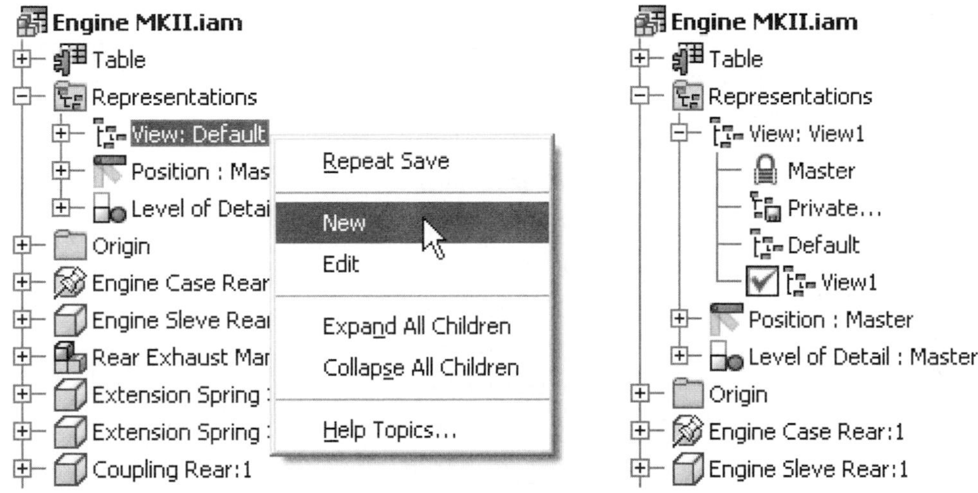

Figure 9.4

An alternative method for creation is done via the Design View Representations dialog box. Enter a name in the edit field and click New, as shown in the following image.

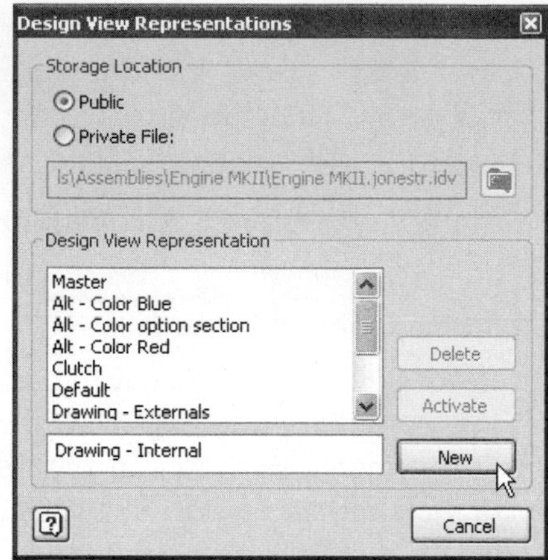

Figure 9.5

3. It is considered good practice to change the name of the design view representation to something more meaningful. As shown in the following image, View1 has been renamed Clutch.

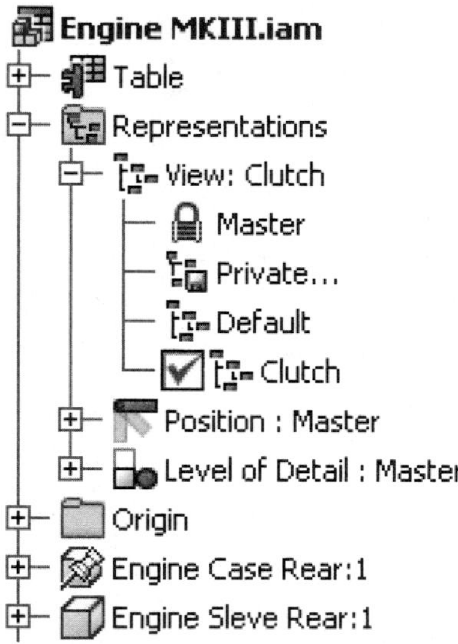

Figure 9.6

4. With a current design view representation set, various parts of the assembly can be modified. For this engine example, certain components of the engine assembly would be hidden such as the engine case, pistons, and crank shaft. These changes are saved automatically to the current design view representation. In addition to component visibility, you can also save the visibility state of sketches, work features, both user and origin, whether or not a component is enabled, color settings of components, the zoom magnification, and the viewing angle.

5. Double-clicking on the Default, or any other design view representation, name will return the engine assembly back to its original representation. In the following image, the default representation of the engine is shown on the left, and the clutch representation is shown on the right.

Figure 9.7

6. Create additional design view representations as needed. You may want to create design view representations using alternate color schemes or to assist in the creation of presentation files or drawing views where particular components are disabled or hidden.

Once you are satisfied with all of the changes made to a design view representation, you can prevent any further changes to a representation by locking it. Right-clicking on a design view representation name will display the menu shown in the following image. Selecting Lock from this menu will add a padlock icon to the design view representation, as shown next to the Clutch design view representation in the following image.

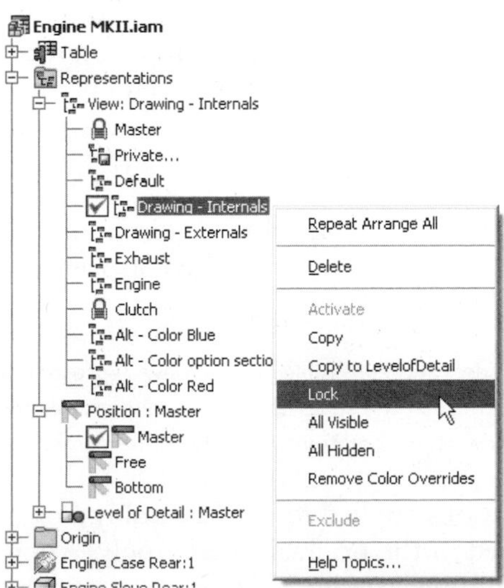

Figure 9.8

If you want to make additional changes to a locked design view representation, right-click on the name, and select Unlock from the menu.

INCREASING PERFORMANCE THROUGH DESIGN VIEW REPRESENTATIONS

Through design view representations, additional control of the visibility of each subassembly is possible, allowing you to turn off the visibility of unimportant components to increase performance. The following sections describe additional design view representation controls, as shown in the following image.

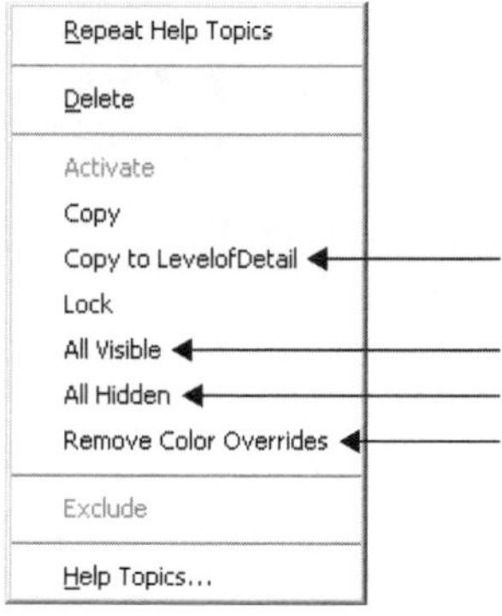

Figure 9.9

Copy to LevelofDetail Select this control to copy the settings of a design view representation to a level of detail representation. Level of Detail Representations are covered later in this chapter.

All Visible Select this control to make all of the components in the assembly visible.

All Hidden Select this control so that none of the components in the assembly will be visible. This can be beneficial for managing large assembly files. Opening a large assembly with all components hidden will allow you to manually turn on the visibility for only those components with which you need to work.

Remove Color Overrides Select this control to restore the original colors of an assembly. This works well if you applied an assembly level color override to any of the components in an assembly.

Note: The All Hidden and All Visible design view representations will override any existing visibility settings in the subassembly. You cannot alter these representations or save new design view representations with the same name.

The design view representations applied to a subassembly from within an assembly are not associated with those created in the original subassembly. After a subassembly is placed in an assembly, changes or additional design view representations in the subassembly are not automatically reflected in the placed subassembly. You must reimport the design view representation to see the new design view representation or the modifications made to a design view representation in the originating subassembly.

IMPORTING DESIGN VIEW REPRESENTATIONS FROM OTHER ASSEMBLIES

To place an assembly into another assembly with a selected design view representation, follow these steps:

1. Click the Place Component tool on the Assembly panel bar, as shown in the following image, press the hot key P, or right-click and select Place Component from the menu. The Open dialog box is displayed.
2. Navigate to and select the assembly that will be placed.
3. While still in the Open dialog box, click the Options button.
4. The File Open Options dialog box appears, showing the available design view representations, as shown in the following image.
5. Select the design view representation to apply to the subassembly.

Figure 9.10

6. Click the OK button in the File Open Options dialog box.
7. In the Open dialog box, click the Open button.
8. In the graphics window, place the subassembly. To add assembly constraints when the Nothing Visible design view representation is selected, you will need to turn on the visibility of some components or change to a design view representation that displays the necessary components. See the next section for instructions on changing from one design view representation to another.

After you have placed an assembly or created an assembly as a subassembly in a parent assembly, you can change the design view representation to a different one by following these steps:

9. In the browser, right-click on the assembly's name that you want to change, as shown in the following image on the left, or in the graphics window, place the cursor over

the assembly that you want to change and then right-click. Select Representation from the menu, as shown in the following image on the right.

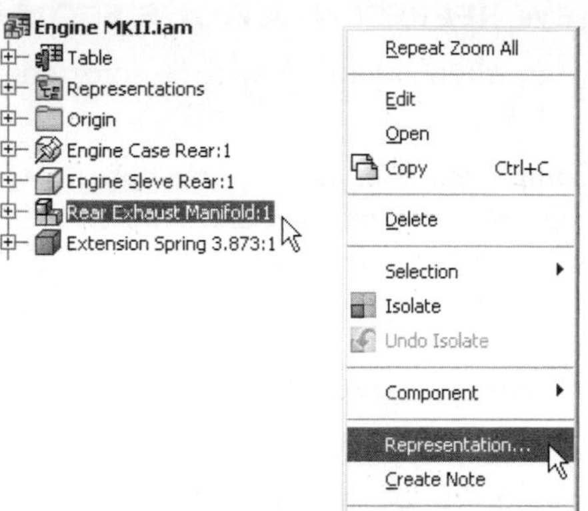

Figure 9.11

The Representation dialog box appears, as shown in the following image.

Figure 9.12

10. From the list in the dialog box, select the needed design view representation.
11. To complete the operation, click the OK button.

CREATING A PRIVATE DESIGN VIEW REPRESENTATION

In addition to creating design views internal to an assembly, you can also create a design view representation that is external to an assembly. This is called a private design view representation. When you create a private design view representation, the various assembly configurations are saved to an external file with the *.idv* extension.

To create a private design view representation, expand the Representations folder in the Assembly browser, and double-click on Private, as shown in the following image.

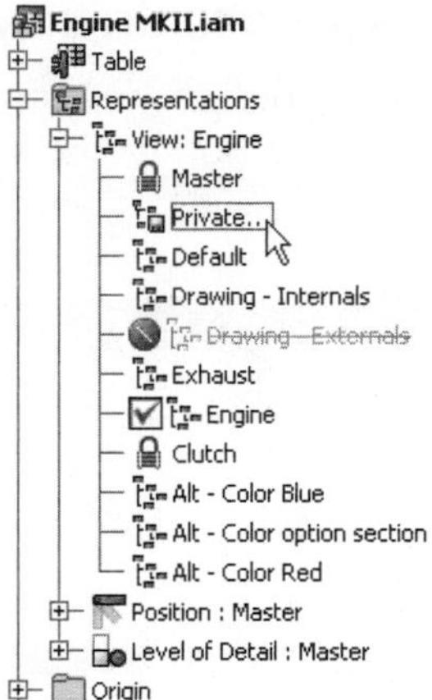

Figure 9.13

This step will launch the Design View Representations dialog box. Notice in the following image that the Storage Location option is automatically toggled to Private File. You can either accept the default location and name to which the *.idv* file will be saved or you can select a different location/name.

Figure 9.14

Note: Private design view representations are not considered associative, which means that drawing views will not reflect any changes to the settings in the private design view representation.

CREATING DRAWING VIEWS FROM DESIGN VIEW REPRESENTATIONS

To generate a drawing view based on a design view representation, activate the Drawing View dialog box, and click in the View section of the Representation box to select a valid design view, as shown in the following image.

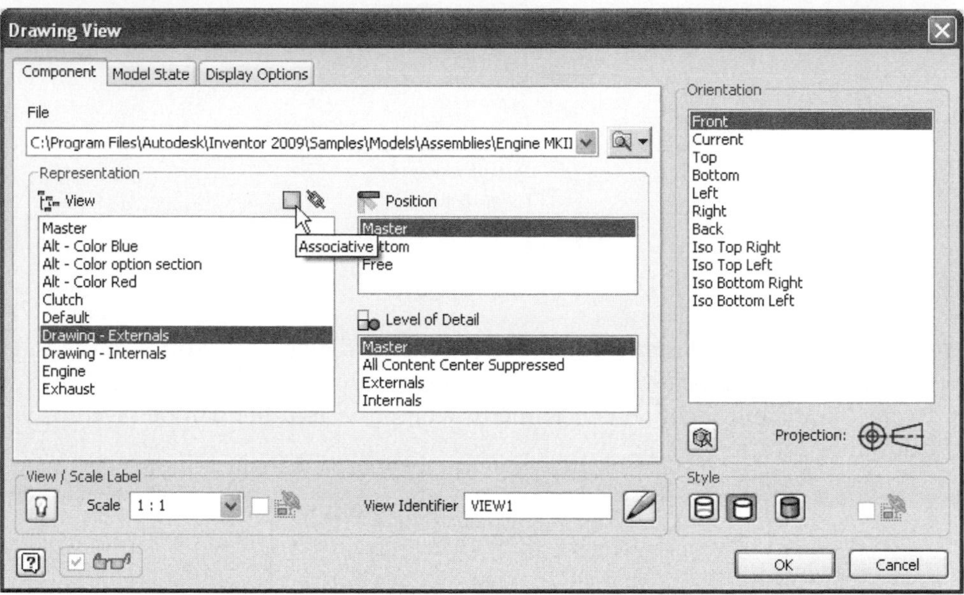

Figure 9.15

A projected view will be based on the design view representation from which its parent view was created. You can make drawing views generated from a design view representation associative by selecting the Associative checkbox, as shown in the previous image. This action will make the drawing view generated by the design view representation dependent upon the design view representation found in the assembly file. Any changes made to the design view representation in the assembly will update the drawing view.

Note: You cannot associate private design view representations identified by the *.idv* extension with drawing views.

Once a drawing view is generated based on a design view representation, you can change the drawing view or views if you select a different design view representation. To perform this operation, first select the drawing view, and then right-click. Select Apply Design View from the menu.

When the Apply Design View Representation dialog box appears, click in the list box, and select a new design view representation, as shown in the following image. If multiple views of the assembly appear on the sheet, set the Apply status for each view. Set the Associative status by changing the value from No to Yes.

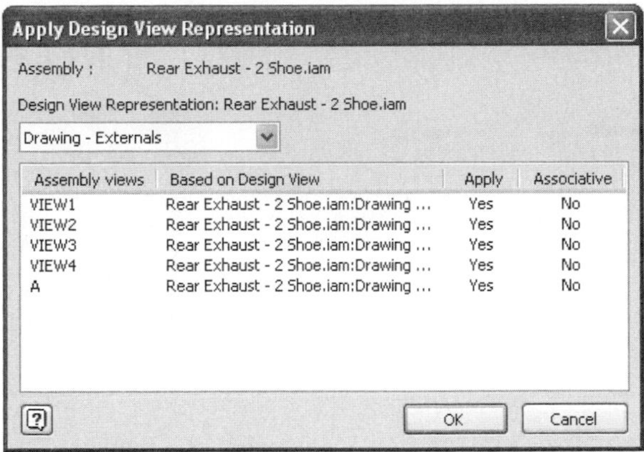

Figure 9.16

FLEXIBLE ASSEMBLIES

As already discussed, you can set the adaptivity property of a subassembly. When you place numerous instances of the same subassemby in a main assembly model, all instances of the subassemblies will act together to display the motion. In the following image, a typical industrial shovel is being controlled by three hydraulic cylinder subassemblies. When one of the cylinders is set to adaptive, all cylinders display the same adaptive positions.

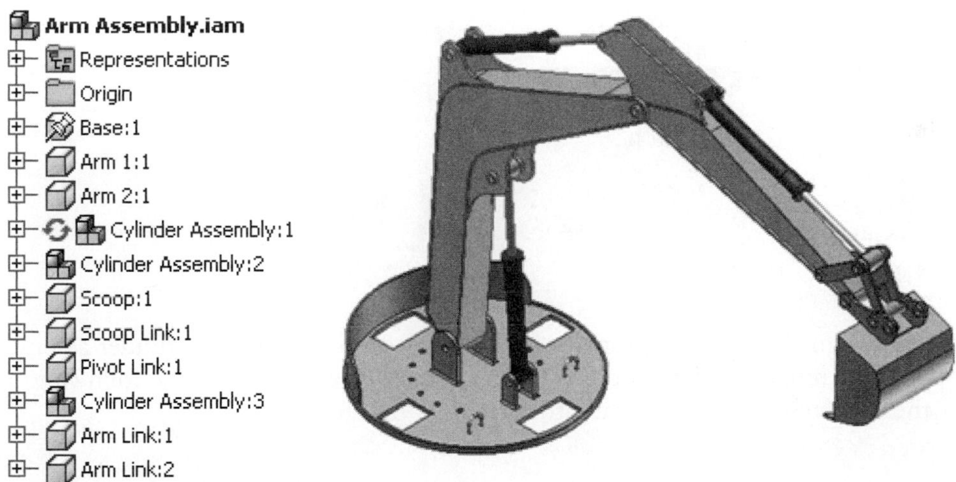

Figure 9.17

The following image illustrates the results of the adaptive cylinders. While these images may look acceptable, the problem with this solution is that when one cylinder opens or closes, the other two cylinders open or close to the same positional distance and direction. When working with assemblies of this nature, you want each cylinder to move independently.

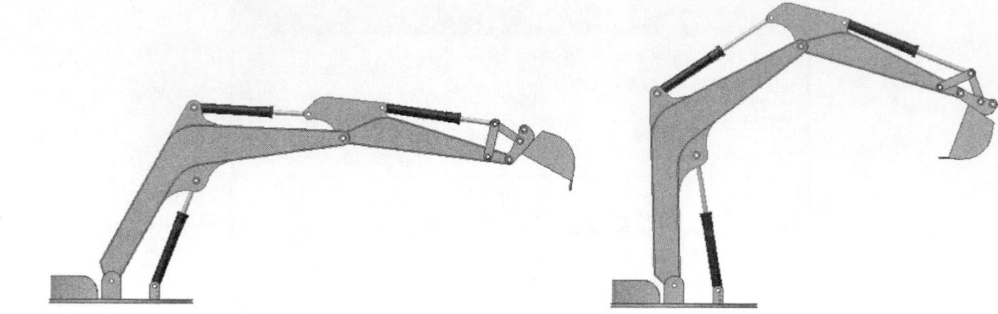

Figure 9.18

Instead of assigning the property of adaptivity that affects all cylinders, you can make each cylinder move independently by applying the Flexible property. In the following image, the Cylinder Assembly located in the Assembly browser has had the Adaptivity property removed. Right-clicking on this Cylinder Assembly will display the menu, as shown in the following image. Click on Flexible.

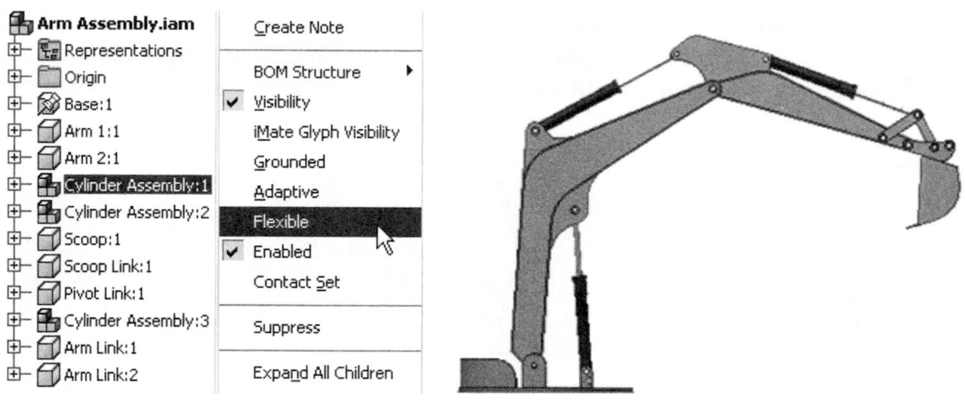

Figure 9.19

When you assign the Flexible property to the first occurrence of the cylinder assembly, an icon appears to represent flexibility for this subassembly, as shown in the following image. When the first cylinder moves, the others remain unaffected.

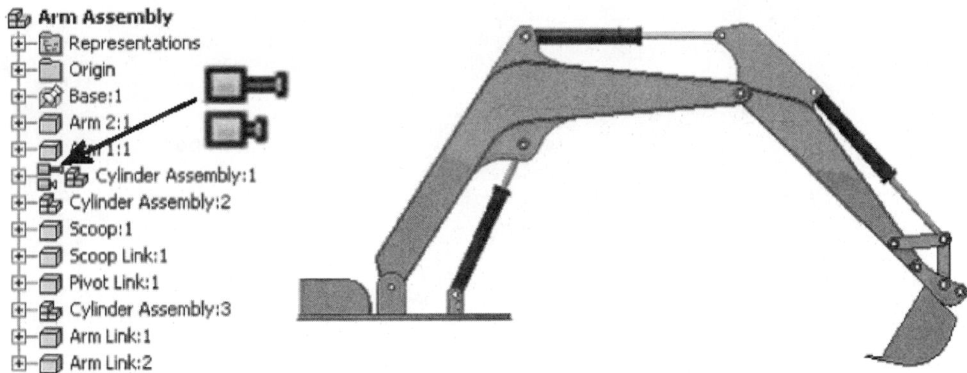

Figure 9.20

The following image shows all three cylinder assemblies set to Flexible. Now the hydraulic shovel assembly can move to any configuration, because all three cylinders can move independently.

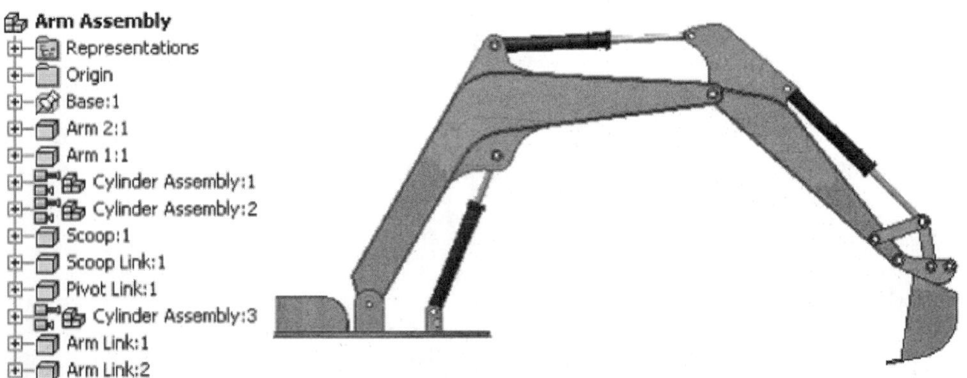

Figure 9.21

POSITIONAL REPRESENTATIONS

Positional representations can be used to override aspects of an assembly and to create different positions of an assembly. All positional representations are saved with the assembly model. The gripper mechanism shown in the following image exemplifies how positional representations are created and applied. One of the gripper fingers is assembled using an angle constraint. The opposite gripper finger is assembled using another angle constraint controlled by a parameter. When the angle constraints are driven, both grippers move in opposition to each other. In this example, positional representations are used to show the gripper mechanism in its closed, middle, and open states.

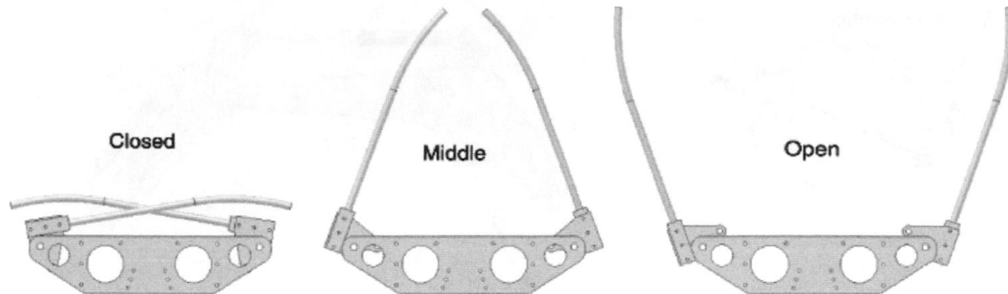

Figure 9.22

To begin creating a new positional representation, expand the Representations folder in the browser, right-click on Position, and select New from the menu, as shown in the following image.

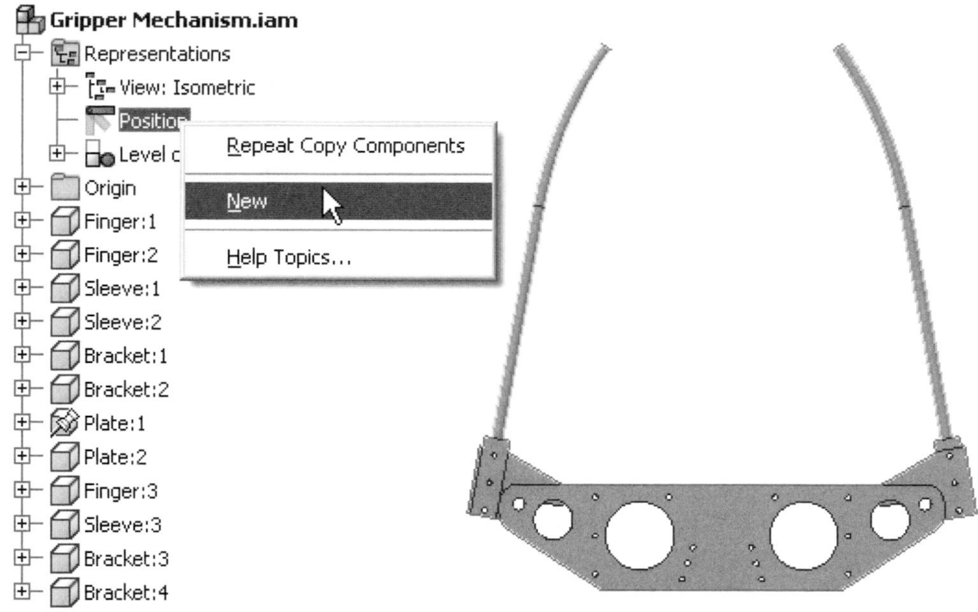

Figure 9.23

When defining a new positional representation, two representations are initially created: Master and Position1, as shown in the following image on the left. You can think of the master positional representation as the default state of the assembly. As you create additional representations, the master representation will always return you to the default state of the assembly. You cannot edit the master positional representation. Any newly created positional representations will be based on the master.

Once you have created a new positional representation, it is considered good practice to rename it with a descriptive name that better fits its intended purpose. Position1 has been renamed to Closed, as shown in the following image on the right. This name is designed to indicate the gripper assembly in its closed state.

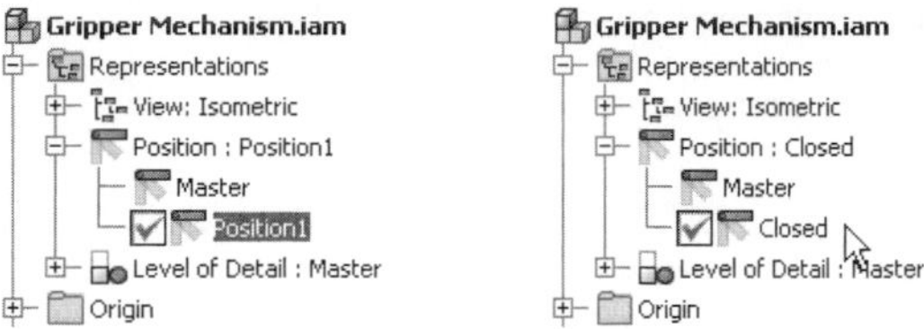

Figure 9.24

After creating a positional representation, you then set up overrides to the master state of the assembly to "move" the assembly into different positions. For example, in the following image on the right, the Angle:10 constraint appears in a suppressed state. The suppression of this constraint allows a gripper finger of the assembly to move freely. This suppressed state is required to configure the positional representation of the gripper in a closed position. Right-click on the Angle:10 constraint to display a menu, as shown in the following image on the left. Then you can select Override, as shown in the following image in the middle, to apply overrides to constraints, patterns, and some component settings.

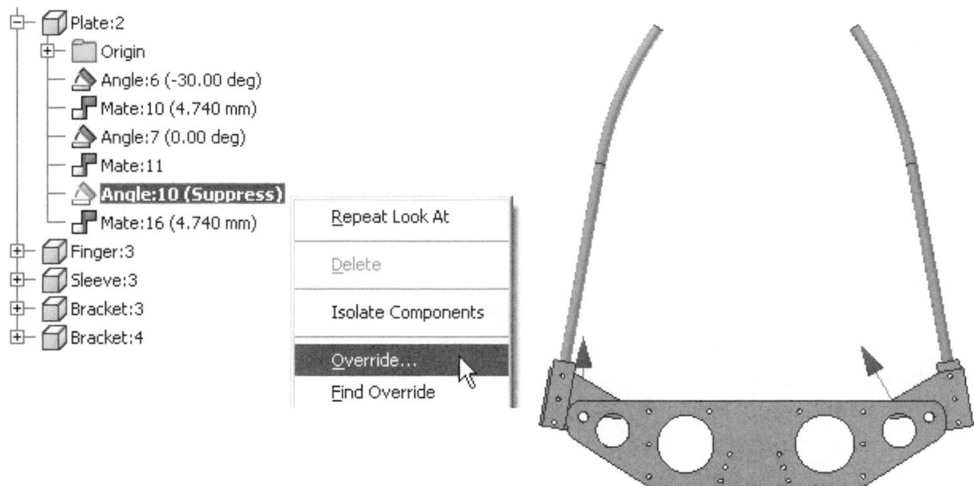

Figure 9.25

After selecting Override on an object in the assembly, the Override Object dialog box is displayed, as shown in the following image. Depending on the object, different settings will be active on the three tabs. The items you can control include the following:

- Suppression status of a constraint
- Numeric value of a constraint
- Offset values in a component pattern

- Grounded state of a component
- Position offset of a component
- The positional representation of a subassembly
- Flexible status of a component

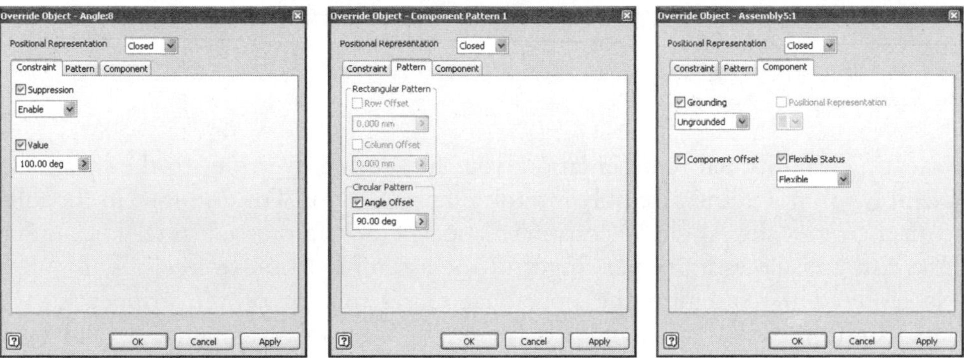

Figure 9.26

When an override is applied, the name of the associated node appears in bold in the browser, as shown in the following image on the left, to emphasize that it is overridden in a positional representation.

Your browser can reflect positional representations just as it can display components as part of an assembly or standard parts found in the Library browser. To switch the browser to positional representations, click on the arrow next to Model, and select Representations from the list, as shown in the following image in the middle. When in the Representations browser, all positional representations in the file are displayed. You can expand each positional representation to display the overrides and settings applied to that positional representation, as shown in the following image on the right. This provides a quick and easy way to review which constraints or other settings are affected by a particular positional representation override.

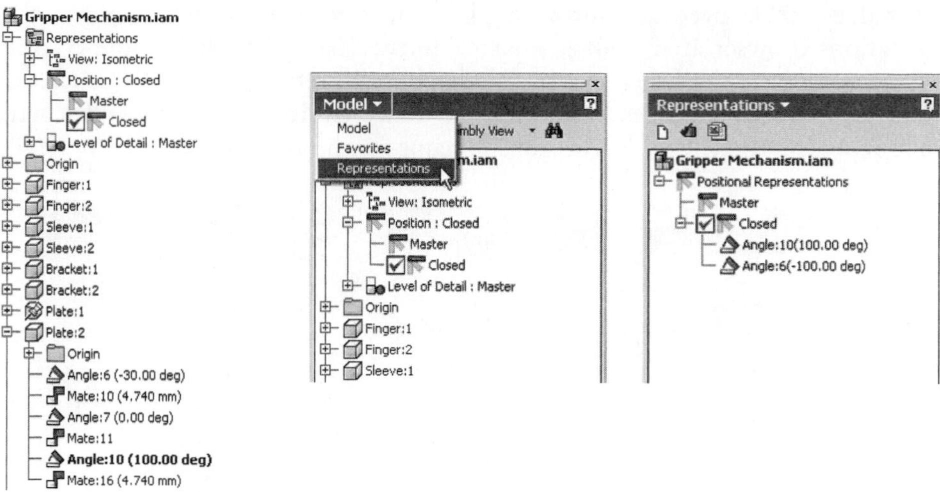

Figure 9.27

You can also copy existing representations and then modify the settings to quickly create additional positional representations showing the assembly in alternate positions. To accomplish this, right-click on a representation, and select Copy from the menu, as shown in the following image on the left. This will create a second positional representation of the same name with a number added to the end of the name, as shown in the following image in the middle. The image on the right shows an additional positional representation created and the values of the constraint overrides modified to show an assembly in three different positions.

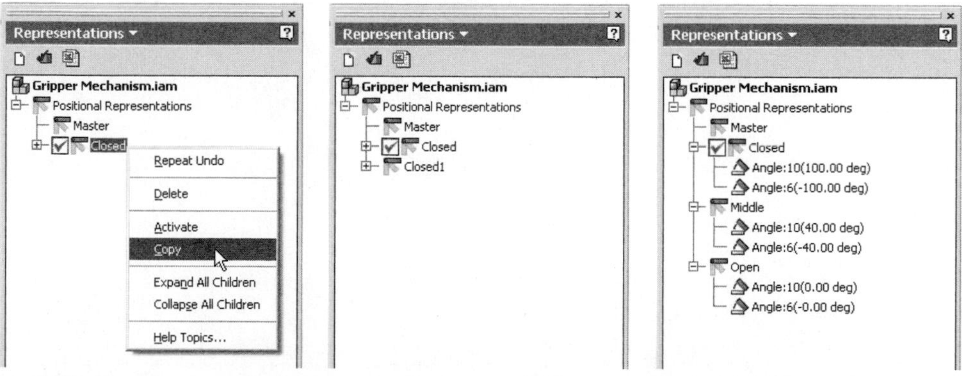

Figure 9.28

You can change override values using a special tool in the Representations browser called Edit positional representation table, as shown in the following image on the left. Clicking on this tool will launch a Microsoft Excel spreadsheet, as shown in the following image on the right. You do not have to create this spreadsheet separately; rather, it is a function of using positional representations. The spreadsheet lists all valid positional representations and the value of all overrides for each. To return to the assembly and update the Representations browser after making changes to the Excel file, click Close & Return to *Assembly Name.iam* on the File menu in Excel. This will return you to the active assembly environment, and any changes made will be reflected in the browser. To activate a positional representation, double-click on its name in the browser.

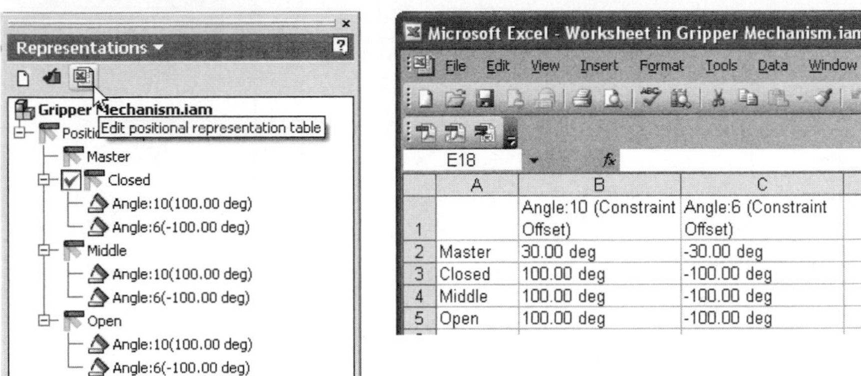

Figure 9.29

CREATING DRAWING VIEWS FROM POSITIONAL REPRESENTATIONS

When generating a base drawing view, you can specify the name of a positional representation for the view. To accomplish this, click on the Base View tool on the Drawing Views panel bar. This action will launch the Drawing View dialog box. Select the desired positional representation from the Position area in the dialog box. A drawing view will be created based upon the selected positional representation.

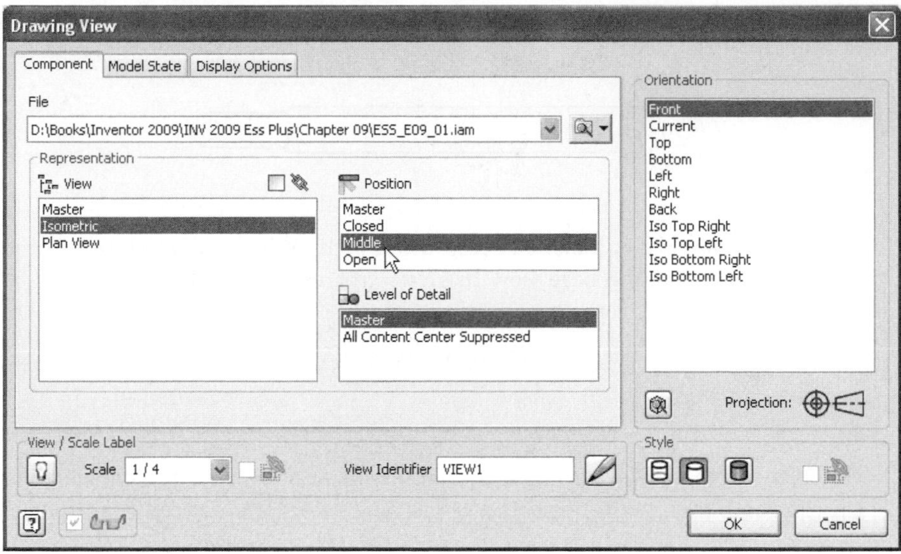

Figure 9.30

In the following image, three separate based views have been created from their corresponding positional representations: Closed, Middle, and Open.

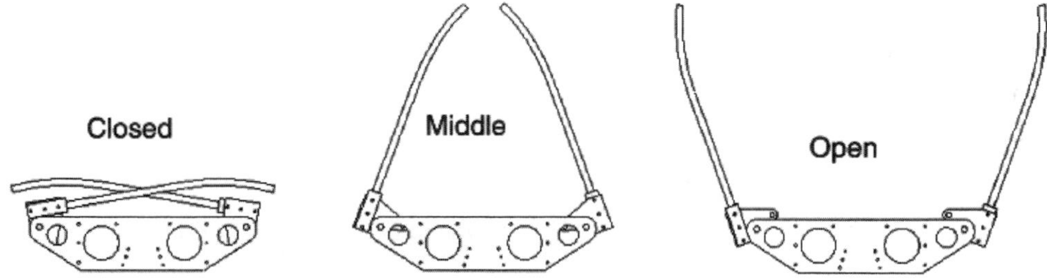

Figure 9.31

CREATING OVERLAY VIEWS

An alternate way to utilize positional representations in a drawing is to create an overlay view to show an assembly in multiple positions in a single view. Overlays are available for unbroken base, projected, and auxiliary views. Each overlay can reference a view representation independent of the parent view. Before creating an overlay view, create as many positional representations as you need to show your assembly in various positions. It is considered good practice to create a number of view representations in order to focus on certain components and to reduce potential clutter in the drawing view.

Use the following steps to create overlay drawing views:

1. On the Drawing Views panel bar, click the Base View tool, and position a view in your drawing. The following image illustrates a hydraulic cylinder set to its closed

position. It would be advantageous to also show the cylinder in its fully opened position in this base view.

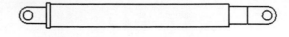

Figure 9.32

2. On the Drawing Views panel bar, click the Overlay View tool, as shown in the following image, and click the base view that was created in Step 1.

Figure 9.33

3. When the Overlay View dialog box is displayed, as shown in the following image, click in the Positional Representation box, and select the desired representation.

Note: A positional representation can be used only once per parent view.

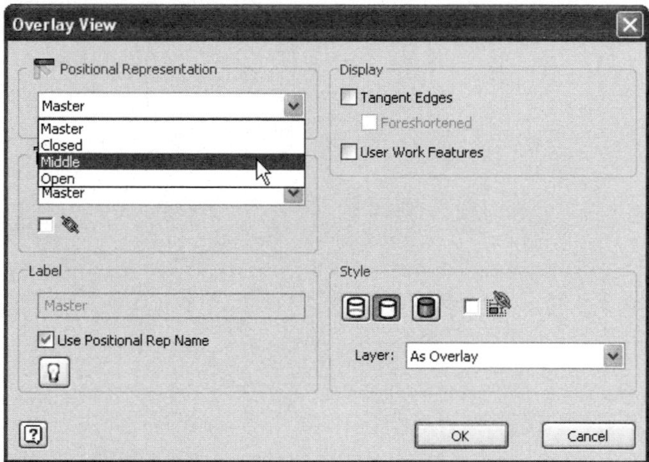

Figure 9.34

4. The new positional representation is added to the base view and is represented as a series of hidden or dashed lines, as shown in the following image. Dimensions can be added between the parent and the overlay.

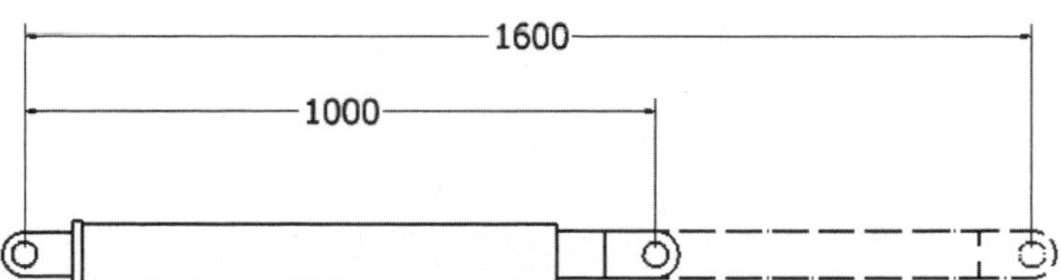

Figure 9.35

EXERCISE 9-1: POSITIONAL REPRESENTATIONS

In this exercise, you create a few positional representations of an assembly.

1. Open the existing assembly *ESS_E09_01.iam* in the Chapter 09 folder.

An angle constraint has been applied between one of the plates and the bracket that holds the gripper finger. The other side of the gripper finger is driven with a similar angle constraint that has a parameter assigned with a negative angle value, as shown in the following image on the left. In this way, the gripper fingers close and open in opposing directions, as shown in the following image on the right.

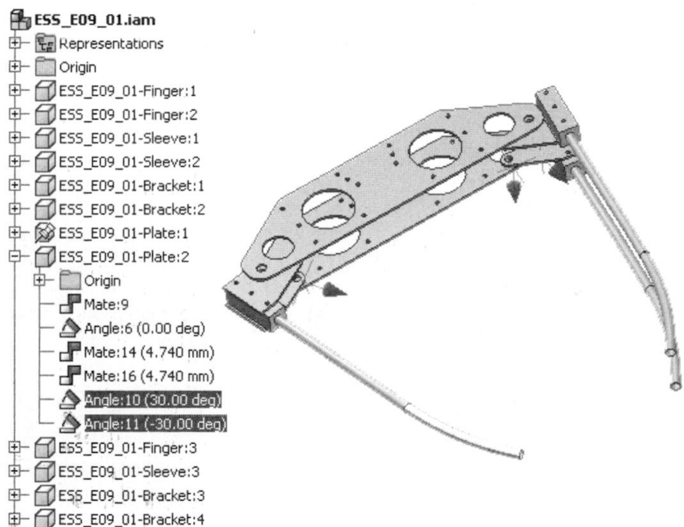

Figure 9.36

2. In the browser, expand the Representations folder. Right-click on the Position node.
3. When the menu appears, click New, as shown in the following image.

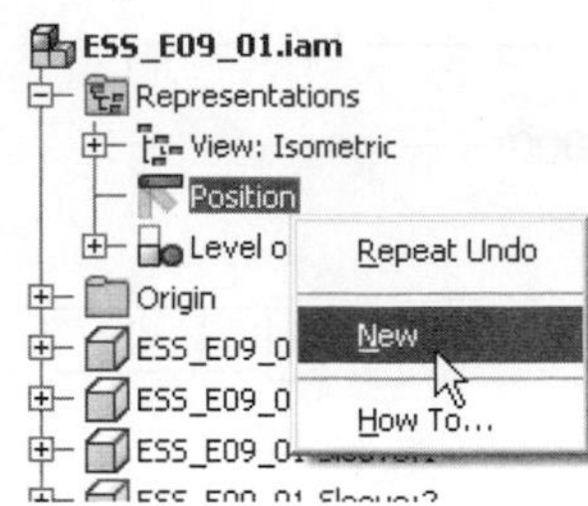

Figure 9.37

4. Expand the Position node to view the new positional representation. Creating a new positional representation for the first time creates a default representation called Master. This positional representation displays the current position of the assembly.
5. The default name of the new positional representation is Position1. Rename the default name by slowly double-clicking in entry and entering **Closed**, as shown in the following image.

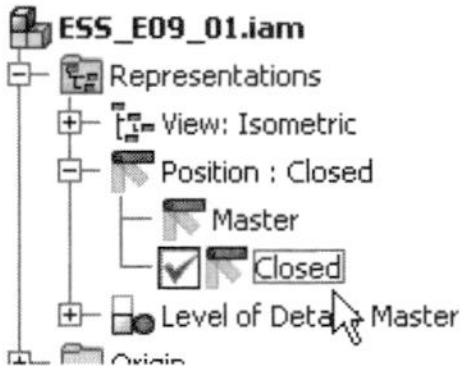

Figure 9.38

6. Next locate the constraint in the browser that will be affected by the positional representation. Expand the part file *ESS_E09_01-Plate:2*.
7. Identify the constraint Angle:10 (30 deg). This is the constraint that will be affected by the positional representation.
8. In the browser, right-click on the Angle:10 (30 deg) constraint, and select Override from the menu as shown in the following image on the left.

The Override tool is used to modify the solved state of the positional representation. You need to make two changes in order for the positional representation to function. First, the constraint needs to be enabled. Second, an offset value needs to be set.

9. While in the Override Object dialog box, click on the Constraint tab—this should be the default.
10. Check the box next to Suppression, as shown in the following image on the right.

EXERCISES

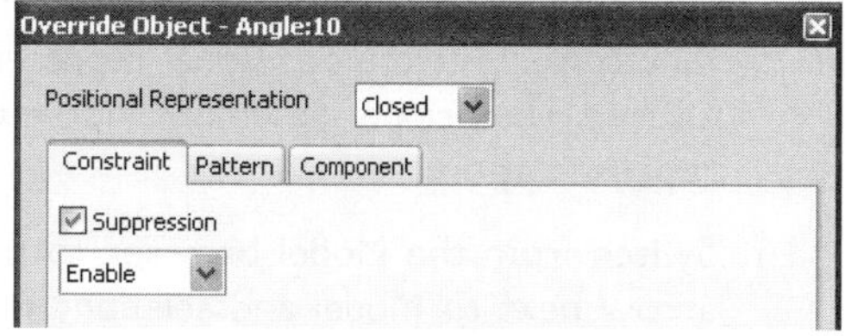

Figure 9.39

11. While still in the Override Object dialog box, check the box next to Value.

12. Change the value to **100.00 deg**, as shown in the following image.

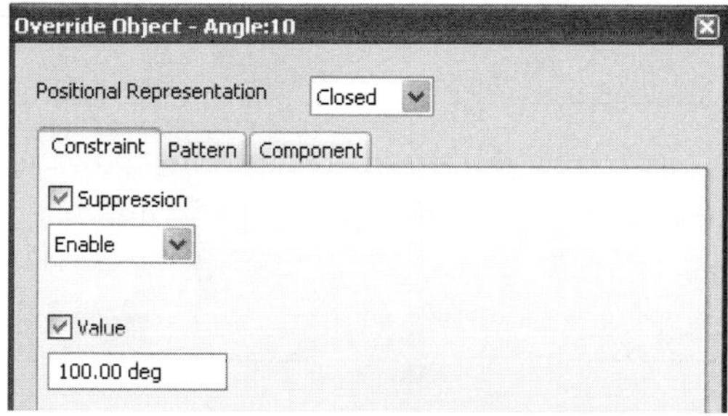

Figure 9.40

13. Click the OK button.

Note: The default master positional representation has the Angle:10 constraint set to 30 degrees. This master state displays the gripper somewhat open. Changing the angle to 100 degrees, as shown in the previous image, will close the gripper arm.

14. The Angle:10 constraint has been overridden, and this action closes the gripper, as shown in the following image. Notice in the browser that the text for the Angle:10 constraint is bold, emphasizing this change.

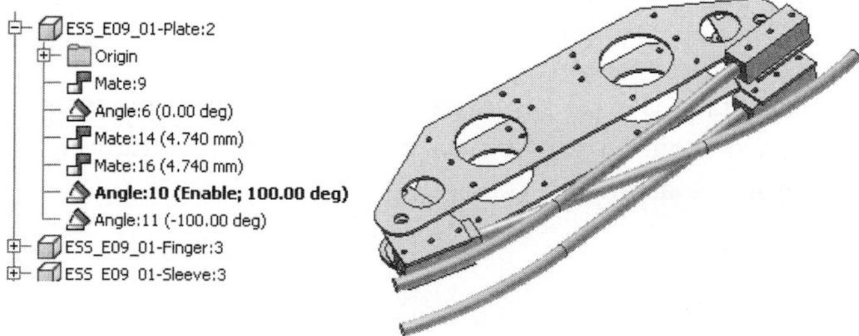

Figure 9.41

15. In the browser, expand the Representations folder and the Position node. Check the functionality of the positional representations by double-clicking on Master and Closed. The gripper assembly should update to reflect the active positional representation.

16. Switch from the Model browser to the Representations browser by clicking on the arrow next to Model and selecting Representations from the list, as shown in the following image on the left.

17. The Representations browser is displayed. Expand all items in this area of the browser, as shown in the following image on the right.

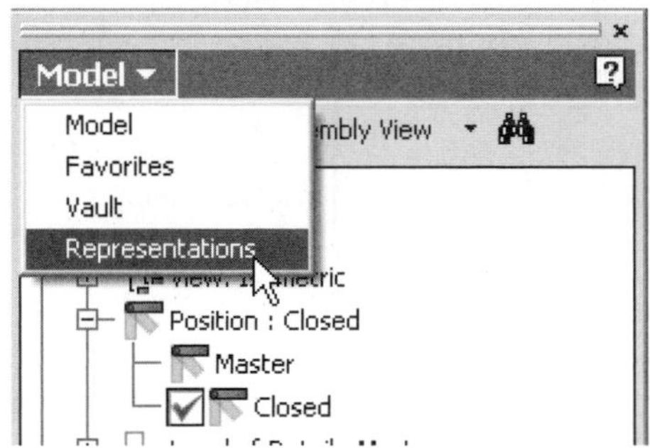

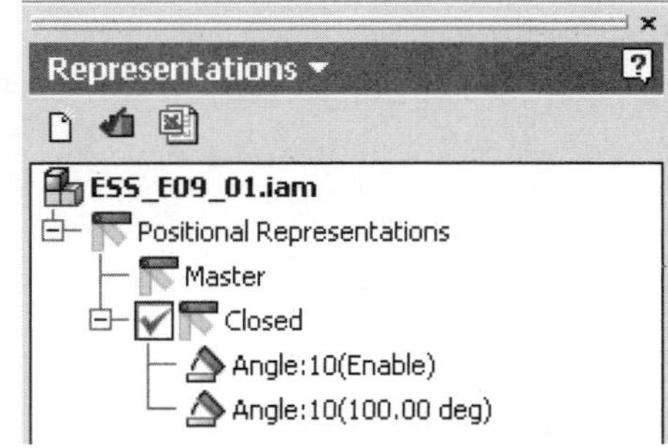

Figure 9.42

Note: The Representations browser allows you to examine all positional representations associated with an assembly. It also shows the overrides for each positional representation.

18. Two additional positional representations will now be created to show the gripper in its middle and fully opened positions. Existing positional representations can be copied and then modified in order to display other versions of an assembly. In the Representations browser, check that the Positional Representations mode is expanded.

19. Right-click on the Closed positional representation, and select Copy from the menu.

20. A new positional representation called Closed1 has been created from the existing positional representation, Closed, as shown in the following image on the left.

21. In the Representations browser, expand both positional representations. Notice that all overrides are copied into the new positional representation, Closed1, as shown in the following image as shown on the right.

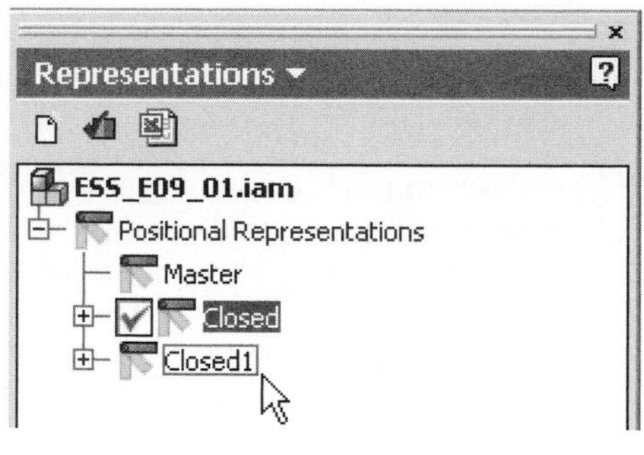

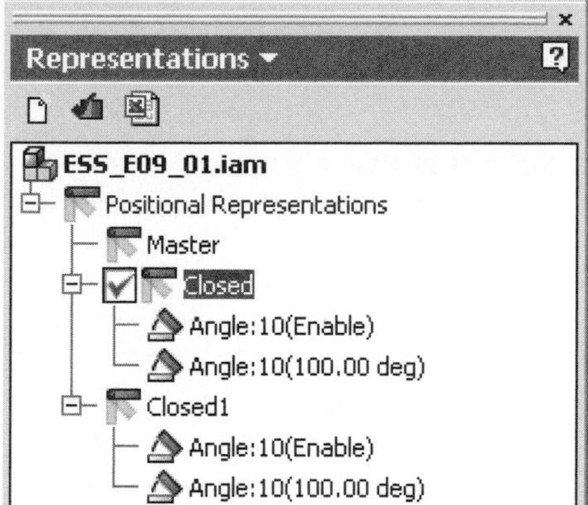

Figure 9.43

22. Create another positional reference copy. Base this copy on the existing Closed positional representation. When all copies have been made, three positional representations should be visible in the Representations browser, as shown in the following image on the left: Closed, Closed1, and Closed2.

23. The positional representation copies will now be renamed to more meaningful names. In the Representations browser, rename Closed1 to **Middle** and Closed2 to **Open**. Your Representations browser should appear similar to the following image on the right.

Note: To rename a positional representation, first single-click on its name, then single-click a second time. This will launch the rename mode.

EXERCISES

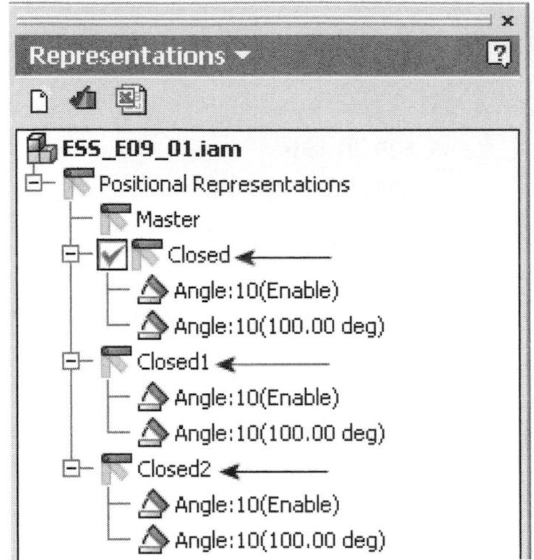

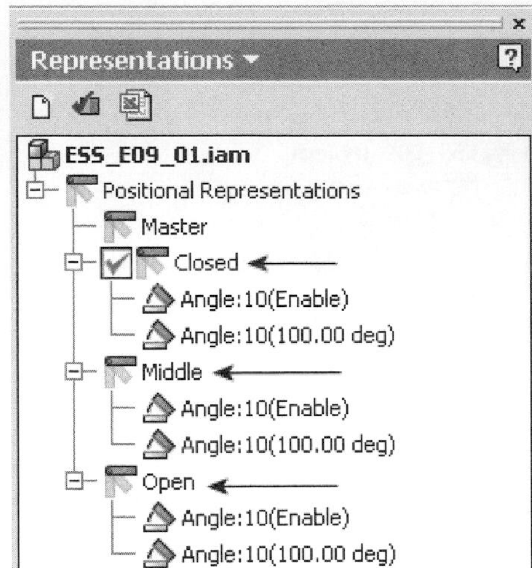

Figure 9.44

24. With the positional representations renamed, their values must now be modified in order to show different representations of the assembly. This step can be accomplished using an Excel spreadsheet. To activate the Excel spreadsheet, click on the Edit positional representation table button found at the top of the Representations browser, as shown in the following image on the left.

25. Clicking on the Edit positional representation table button will launch an Excel table similar to the following image on the right. Notice that the master and all positional representations are displayed as a series of rows and columns. Notice also the degree values present in a separate column.

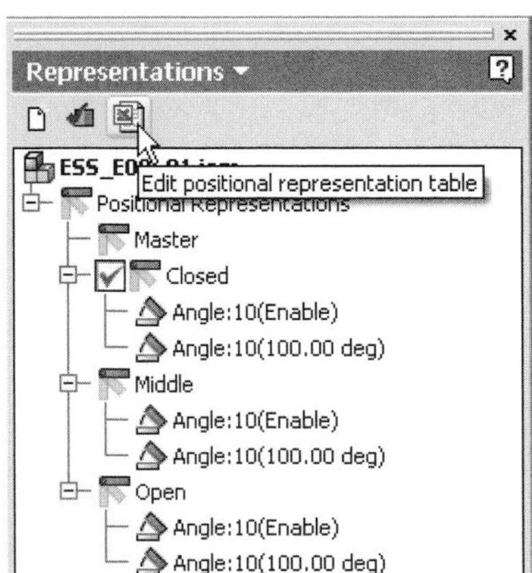

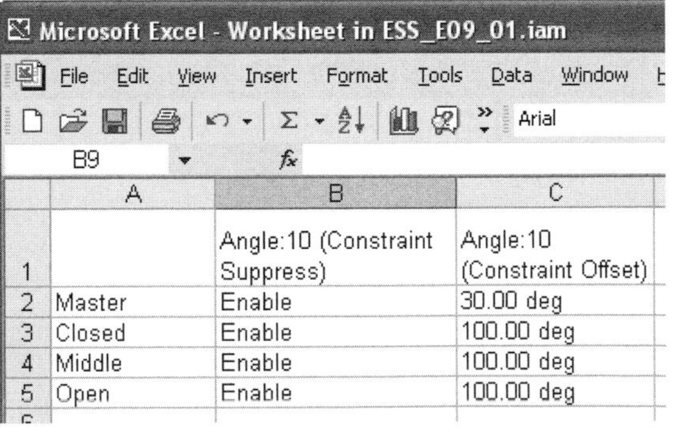

	A	B	C
1		Angle:10 (Constraint Suppress)	Angle:10 (Constraint Offset)
2	Master	Enable	30.00 deg
3	Closed	Enable	100.00 deg
4	Middle	Enable	100.00 deg
5	Open	Enable	100.00 deg

Figure 9.45

26. Make the following modifications to this table:

- For the Middle positional representation, change 100 degrees to **40** degrees.
- For the Open positional representation, change 100 degrees to **0** degrees.

Your table should appear similar to the following image on the left.

27. The changes you made in the Excel spreadsheet need to be saved before you return back to the gripper assembly. From the Excel File menu, click Close & Return to *ESS_E09_01.iam,* as shown in the following image on the right.

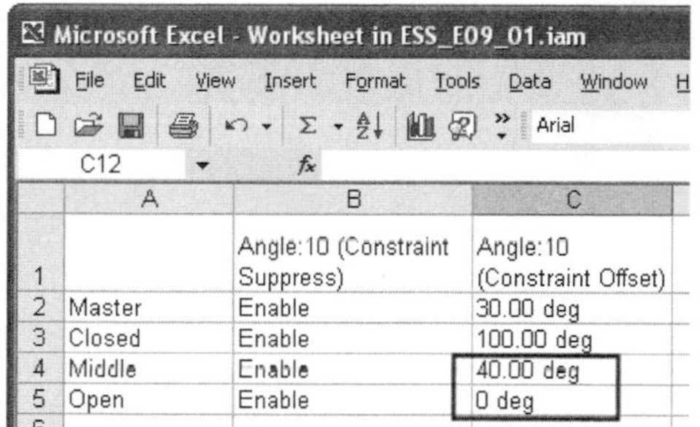

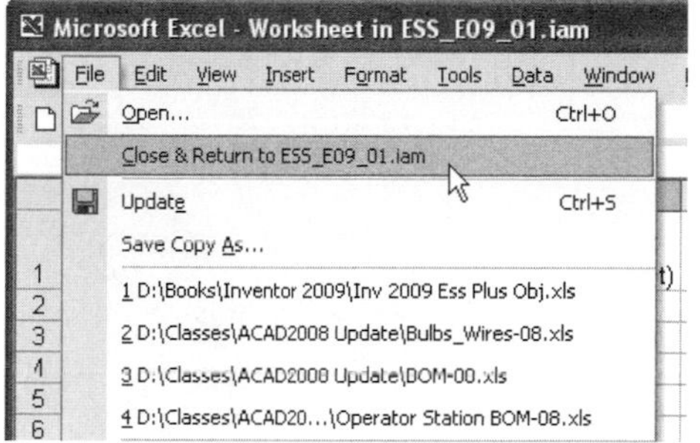

Figure 9.46

28. Saving the changes you made in the Excel spreadsheet will update each angle value for the positional representations, as shown in the following image.

29. Each positional representation can be activated by double-clicking on its name in the Representations browser. Each should show a different state of the assembly, as shown in the following image.

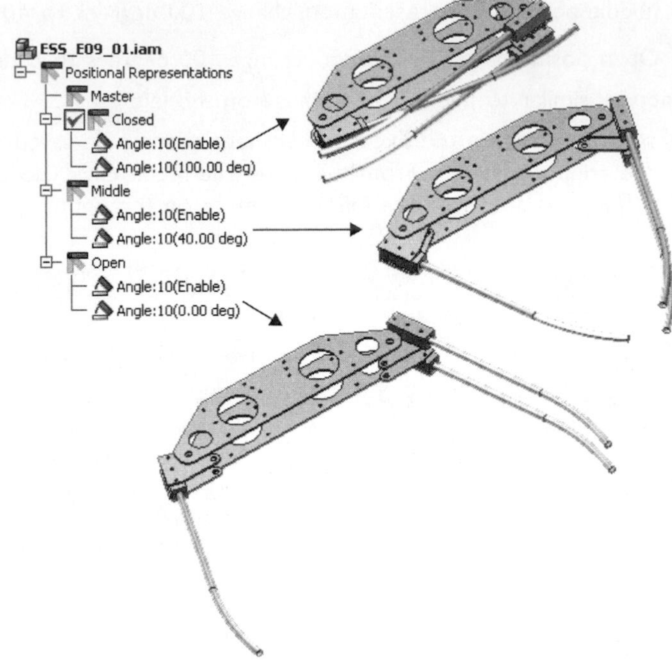

Figure 9.47

30. Double-click the Master positional representation to make it active.
31. From the File menu, click Save As.
32. Save the assembly file as Gripper Assembly.iam.
33. Create a new drawing file.
34. Click the Base View tool, select the *Gripper Assembly.iam* file, choose the Front (XY) orientation, and select the Open Positional representation.

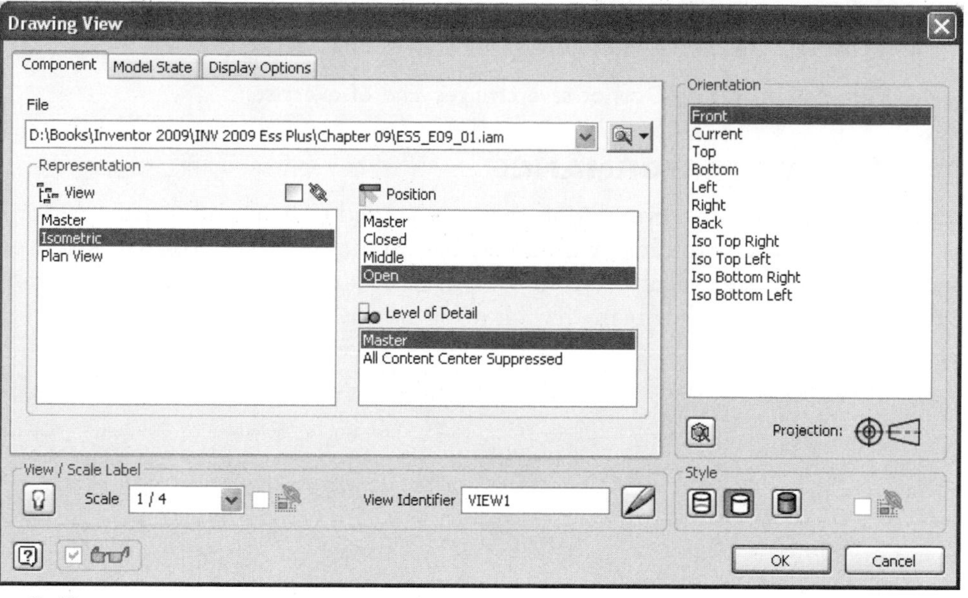

Figure 9.48

35. Click the drawing sheet to place the view on the sheet.

36. Click the Overlay View tool from the Drawing Views panel bar.

37. Select the Base view.

38. In the Overlay View dialog box, select the Middle positional representation, and click OK. The drawing view should look similar to the following image.

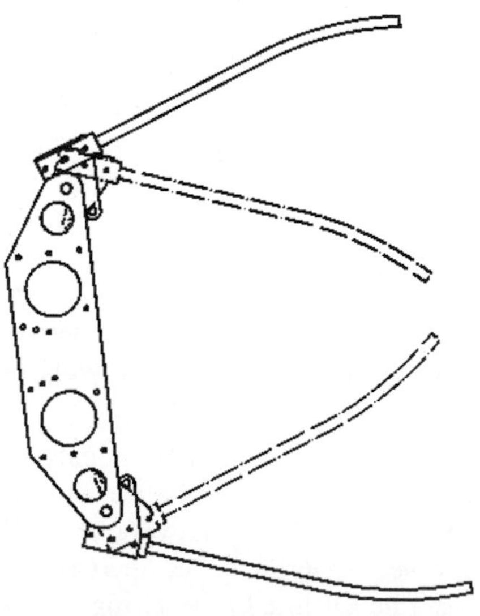

Figure 9.49

39. Add dimensions to the drawing view and experiment by placing the Closed positional representation with the Overlay View tool as desired.

40. Close all open files. Do not save changes. End of exercise.

CAPACITY AND PERFORMANCE

Autodesk Inventor contains several tools that can be used to help improve the performance of the application when working with larger files. Some of these tools will be covered in this section. Settings are available on the Drawing tab after selecting Application Options from the Tools menu, as shown in the following image.

View Preview Display

Show Preview As

All Components

Section View Preview as Uncut

Capacity

Memory Saving Mode

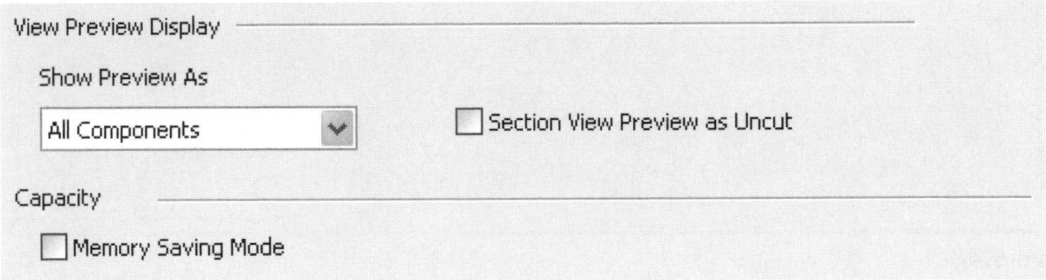

Figure 9.50

Three options are available from the Show Preview As drop-down list: All Components, Partial, and Bounding Box. These settings affect only the preview of the view and increase performance by reducing what is displayed prior to view creation. The All Components option is the default setting, and it displays a preview of all components in the assembly. The Partial option displays a bounding box of the assembly, and it shows a preview of some components in the assembly. The Bounding Box option previews the least amount of detail, but it provides the greatest capacity gains when working with drawing views. You can use the Section View Preview as Uncut checkbox to preview section views with or without cutting components. The Memory Saving Mode setting can be used to increase capacity, but it will negatively impact the amount of time Autodesk Inventor takes to compute data. When this checkbox is selected, Autodesk Inventor conserves memory by processing the way components are loaded and unloaded from memory. It should only be used when the majority of drawings that you create are of large or highly complex models and when you experience capacity limit issues. When this option is selected, view creation or modification operations cannot be undone. This setting can be overridden on a per-drawing basis, as shown in the image, from the Drawing tab in the Document Settings dialog box.

Other settings available from Tools > Document Settings > Drawing tab, as shown in the following image, can be used to support performance and capacity. The Use Bitmap for Shaded Views option has settings for either Always or Offline Only. If you choose to always use bitmaps, you may see a performance increase when using shaded views of assemblies or complex models. The Offline Only setting can be used when working offline or when reference files are missing. This setting can be used in conjunction with the Image Fidelity setting for bitmaps used to display shaded views.

Figure 9.51

You can override the component preview setting during the drawing view creation process if the Show Preview As option is set to Bounding Box or Partial. If either selection is used, the option in the Create View and Auxiliary View dialog boxes will contain an active checkbox, as shown in the following image on the left, which allows you to override the preview setting and preview all components of the assembly that is being placed on the view. If the option is set to All Components, the option becomes checked and is disabled, as shown in the following image on the right.

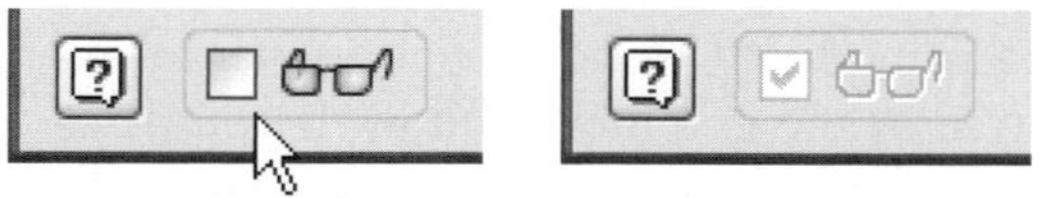

Figure 9.52

CAPACITY METER

The capacity meter, located on the lower right corner of the Autodesk Inventor application window, shows two numbers: the number on the left indicates the number of occurrences in the active file and the number on the right indicates the number of files in the session. A memory graph or meter is also included that indicates how much memory Inventor is using. The green area shows the amount of committed or used memory, and the black area represents how much free memory is available to the system, as shown in the following image on the left. When you place the cursor on top of the meter, a tooltip is displayed indicating the memory consumption in megabytes. The capacity meter color will change from green to yellow when more than 60% of memory is used and to red when more than 80% of memory is used.

The capacity meter can be turned on or off using the Capacity meter checkbox, as shown in the following image on the right. You can access the option by right-clicking a toolbar, selecting Customize, and then selecting the Toolbars tab. From this area you can also change the capacity meter to Physical Memory that shows both the memory that Inventor and other applications are using. The Inventor Only option is not available with 64 bit operating systems.

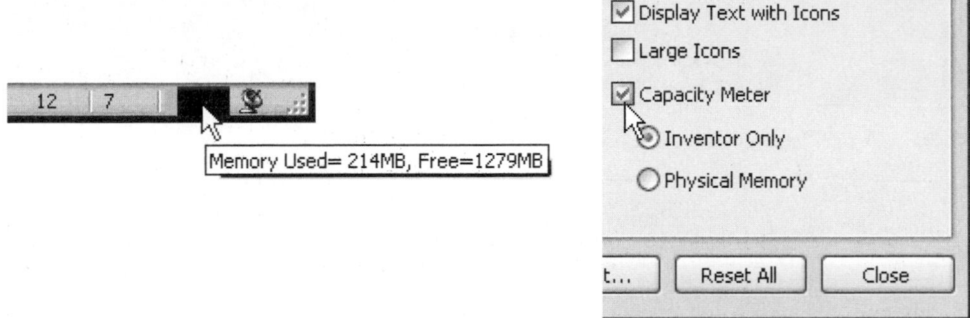

Figure 9.53

LEVELS OF DETAIL REPRESENTATIONS

The Levels of Detail Representations tool can be used to improve capacity and performance in both the modeling and drawing environments. Levels of detail use component suppression to allow you to selectively choose which models are loaded into memory. The Suppress status of a component, accessed by right-clicking a component in either the browser or graphics window, unloads the component(s) from memory, as shown in the following image on the left.

When a component is suppressed, the component's icon is modified in the browser, the top-level assembly name is modified to notify you which level of detail is active, and the component is not displayed in the graphics window. If you place your cursor over or select a suppressed component in the browser, the component bounding box is previewed in the graphics window, as shown in the following image on the right where two suppressed components have been selected.

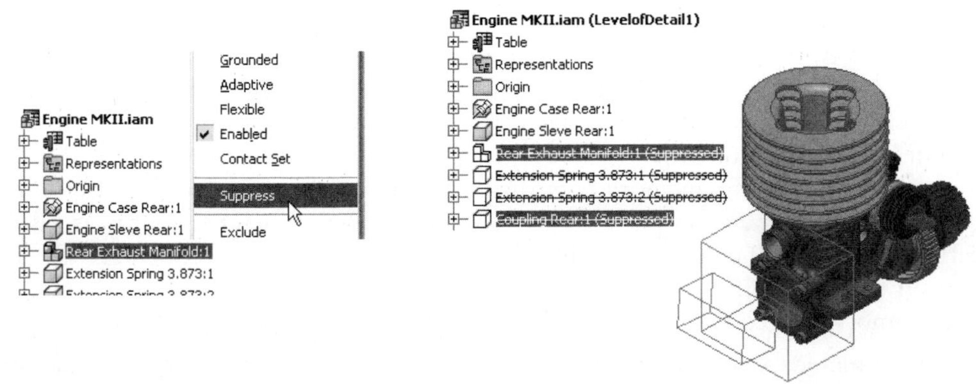

Figure 9.54

A Level of Detail node is available under the Representations folder in the browser, as shown in the following image on the left. By default, four levels of detail are created in each assembly and can be activated to increase performance, if needed. The default levels of detail are Master, which is active by default, All Components Suppressed, All Parts Suppressed, and All Content Center Suppressed. You can activate a level of detail in the same way that you activate a design view representation or a positional representation: double-click on it in the browser.

Similar to Design View Representations, the Master level of detail representation is active by default. You can make modifications to the status of children nodes of an assembly when this representation is active, but the changes are not maintained when the assembly file is saved. In order for the changes to be saved, a new or existing level of detail representation that you create must be active. Refer to the design view representation and positional representation sections of this chapter for more information about creating additional representations.

You can select a level of detail, design view, or positional representation to load when opening a file or placing a component in an assembly. This step is done by selecting the Options button in the File Open Options or Place Component dialog box and choosing the Level of Detail Representation that you want to load, as shown in the following image on the right.

Figure 9.55

In a similar manner, when creating a drawing view of an assembly, you can select a level of detail representation from the Drawing View dialog box, as shown in the following image.

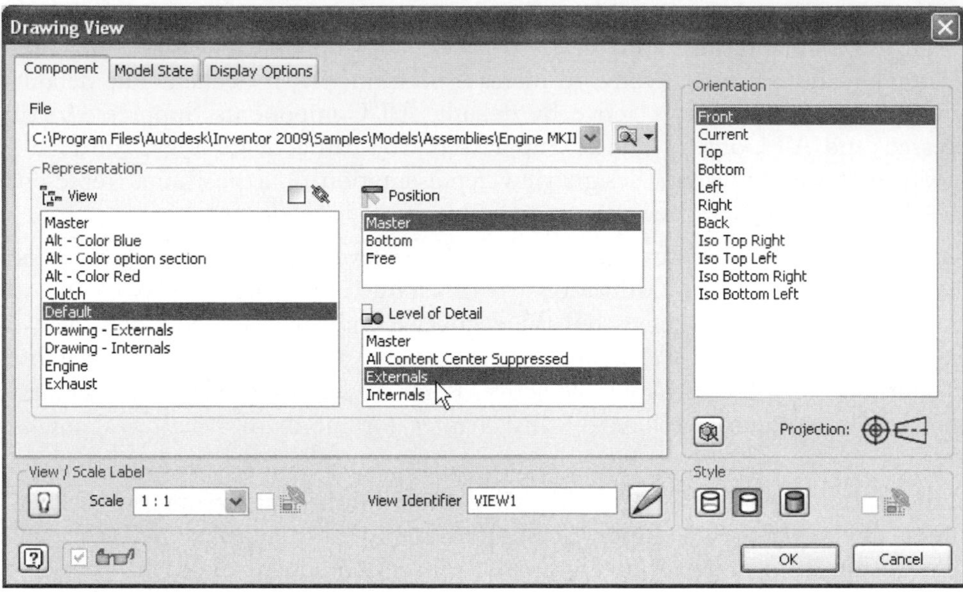

Figure 9.56

ASSEMBLY SUBSTITUTION LEVEL OF DETAIL REPRESENTATION

Another option for Level of Detail is to create an assembly substitute. While working with an assembly you can use assembly substitution to reduce the memory that the assembly uses and reduce the amount of detail. The Level of Detail needs to be created at the top level of the open assembly. You need to open an assembly to create a substitute; you cannot create a substitute Level of Detail Representation of a subassembly in the active assembly. You can create as many substitutes as possible. The iProperty data is maintained from the original assembly when a substitute is created. There are two methods for substituting assemblies; you can create a derived assembly or substitute a subassembly for an existing part file.

Create a Substitute Level of Detail Representation using Derive Assembly

To create an assembly substitute using the derived method follow these steps. The assembly constraints will be maintained.

1. Open an assembly for which you want to create a substitute.
2. In the browser expand Representation, right-click the Level of Detail, and click New Substitute > Derive Assembly to create a new derived part from the assembly. The active Level of Detail at the time of creation contains the default body representation for the derive operation.

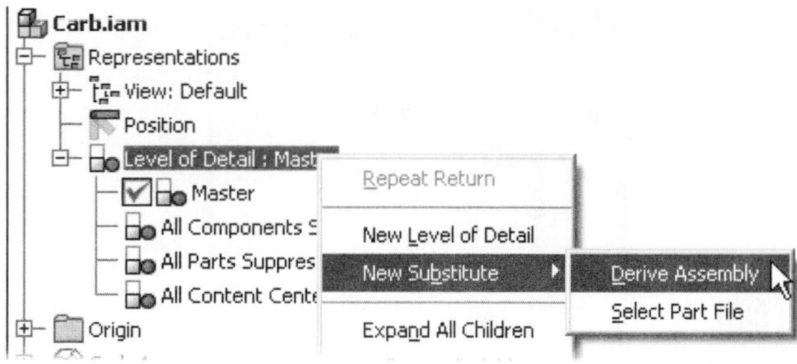

Figure 9.57

3. In the New Derived Substitute Part dialog box, enter a name for the component, select a template file to base the file on, select a location, and click OK.

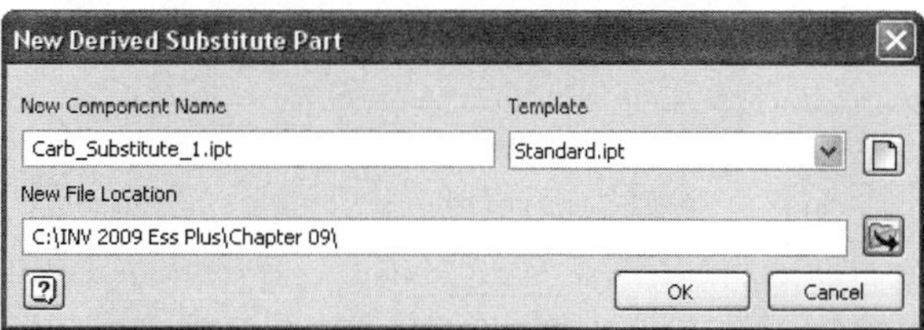

Figure 9.58

4. The Derived Assembly tool is opened.
5. In the Bodies tab select the desired option for each component or the entire assembly. See chapter 7 for more information about the derived options. The following image on the left shows the assembly in its original state. The middle image shows the Derived dialog box with the bounding box option selected for the entire assembly. The image on the right shows the resulting bounding box.
6. The Reduced Memory Mode option is on by default. This option uses less memory by not caching the source bodies. It is recommend to leave this option on.

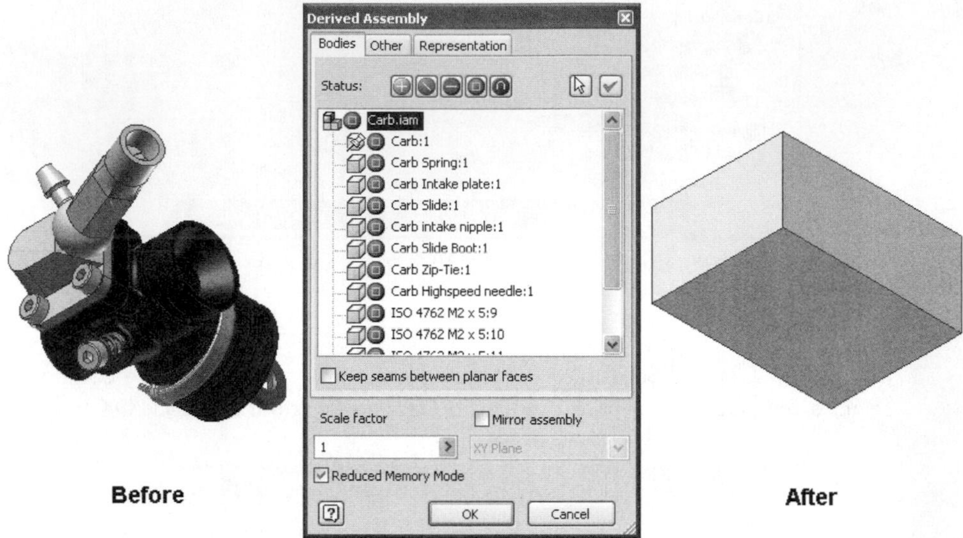

Figure 9.59

7. Click OK to create the derived part. A new entry under Level of Detail will appear in the browser as shown in the following image. If desired, you can rename the substitute. The part will appear in the graphics window as depicted in the last step.

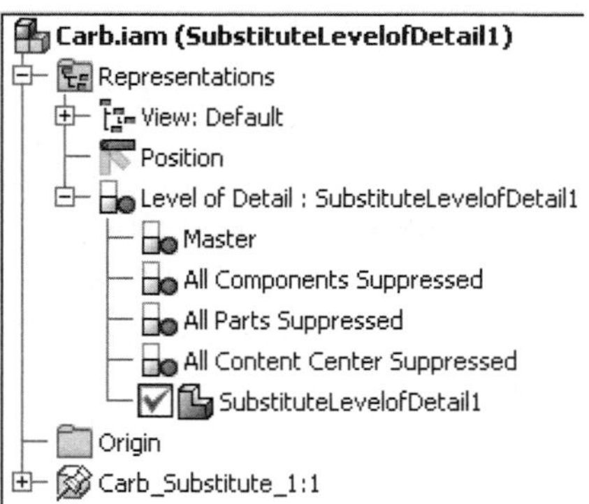

Figure 9.60

8. To edit the substitute make the derived park active by double-clicking the part in the browser, right-clicking the derived assembly, and clicking Edit Derived Assembly as shown in the following image. The Derived Assembly dialog will reappear.

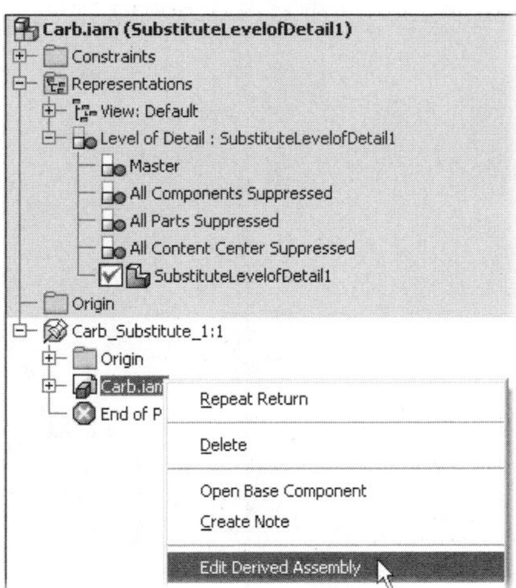

Figure 9.61

9. To use the substitute, make the assembly active for which you want to substitute. Expand the Representation folder. The following image shows the assembly with the substitute NOT active.

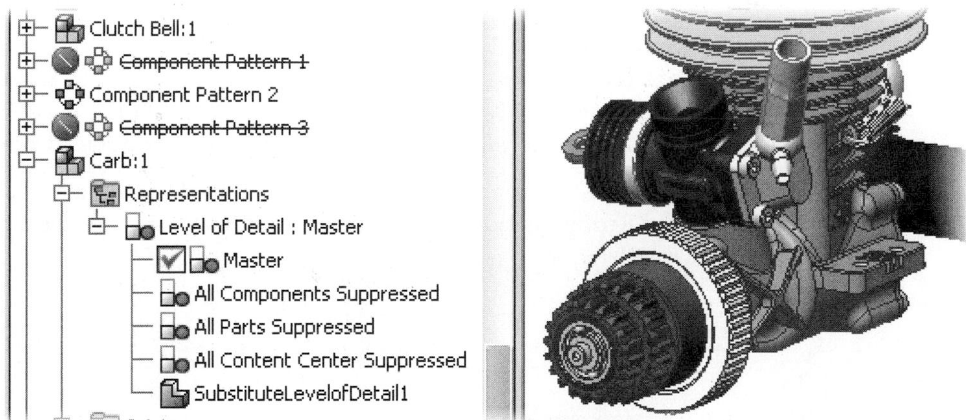

Figure 9.62

10. Make the substitution active by double-clicking the substitute's entry under the Level of Detail. The assembly constraints will be maintained.

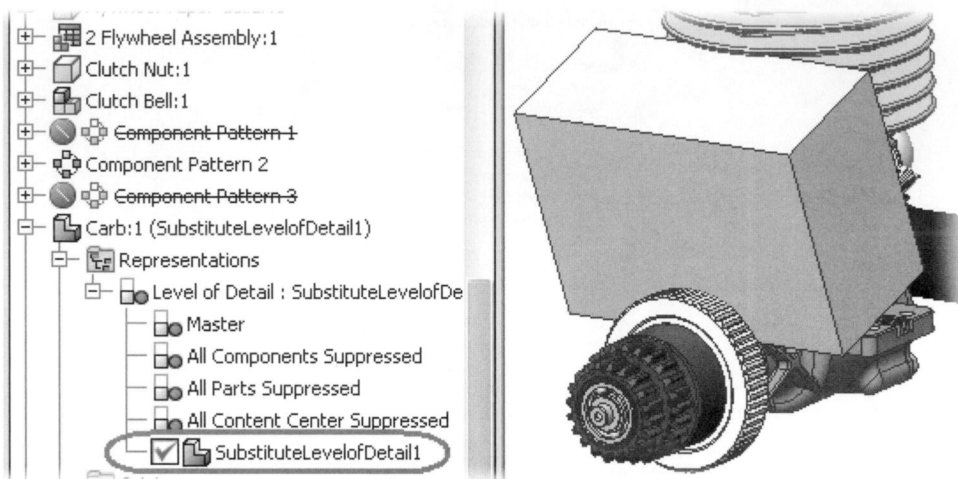

Figure 9.63

CREATE A SUBSTITUTE LEVEL OF DETAIL REPRESENTATION USING A PART FILE

With this option you manually create or derive a part and use it to substitute an assembly. The assembly constraints may not be maintained.

1. Create the part using the same XYZ location and orientation that appears in the assembly. If you do not create the part this way, you will need to use Grip Snaps or apply assembly constraints to the origin work features to re-position the part. To avoid reorienting the part, project edges from the assembly or copy bodies from the assembly.
2. In the browser expand Representation, right-click the Level of Detail, and click New Substitute > Select Part File.

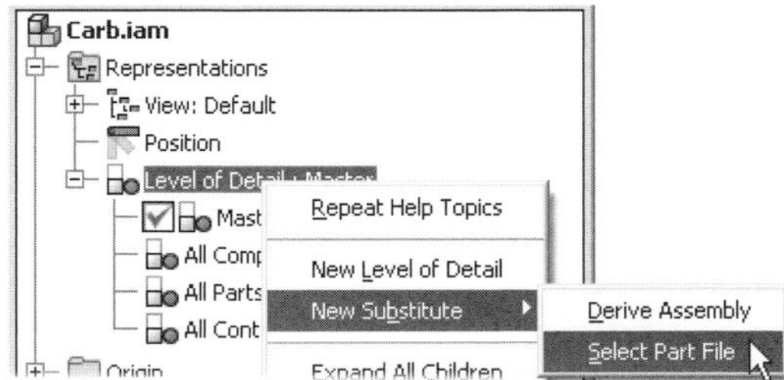

Figure 9.64

3. A dialog box will appear stating that the links to the original files will be disabled until the substitute is no longer active. Click Yes to close the dialog box. The following image shows a part being substituted. The entry in the browser was renamed.

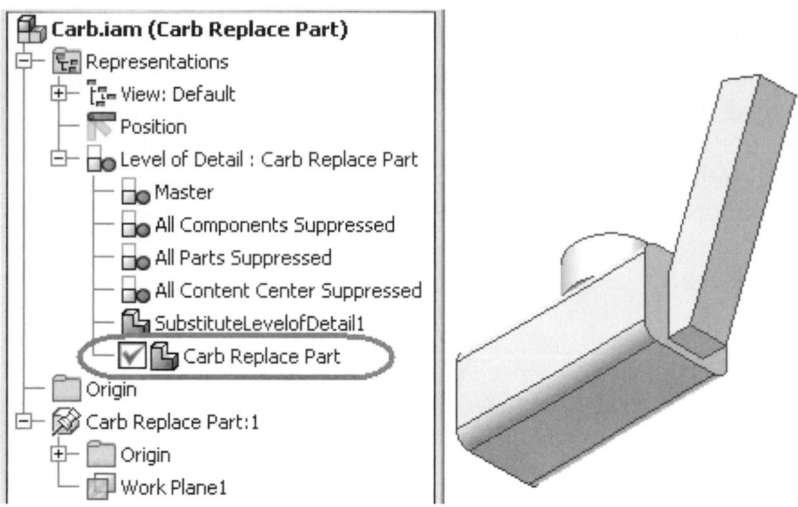

Figure 9.65

4. Save the file.
5. Open the assembly file in which the substitute will be used.
6. To enable a substitute, right-click the Level of Detail in the browser and click New Substitute > Select Part File.
7. In the Place Component dialog box select the part file to use as the substitute and click Open.
8. In the browser the Level of Detail entry now displays the Substitute icon and the name SubstituteLevelofDetail1.

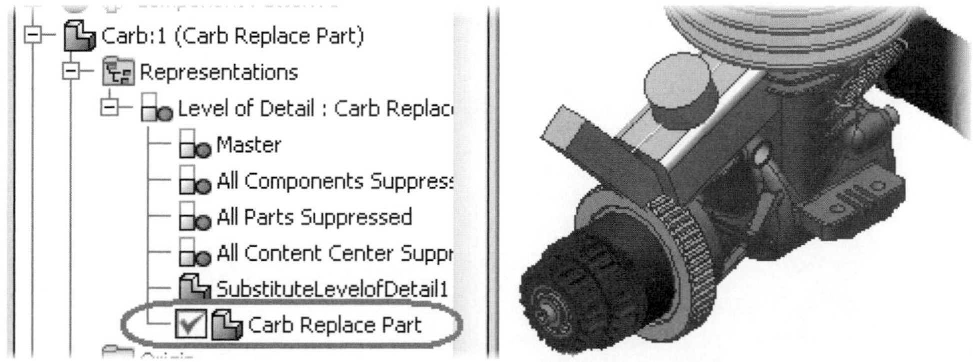

Figure 9.66

To switch between different levels of Detail representations in different subassemblies you can create a Level of Detail at the top level assembly and make the Level of Details in the different subassemblies active.

CONTACT SOLVER

The Contact Solver determines how assembled components behave when a mechanical motion is applied. To get a better idea of what happens with the Contact Solver, study the three components shown in the following image. The base component, #1, has a single pin designed to run inside of the slot found on component #2. The second component, #2, has two pins in addition to a single slot. The two pins are designed to run inside of the slots found on component #3. The third component, #3, consists of two slots cut through its wall.

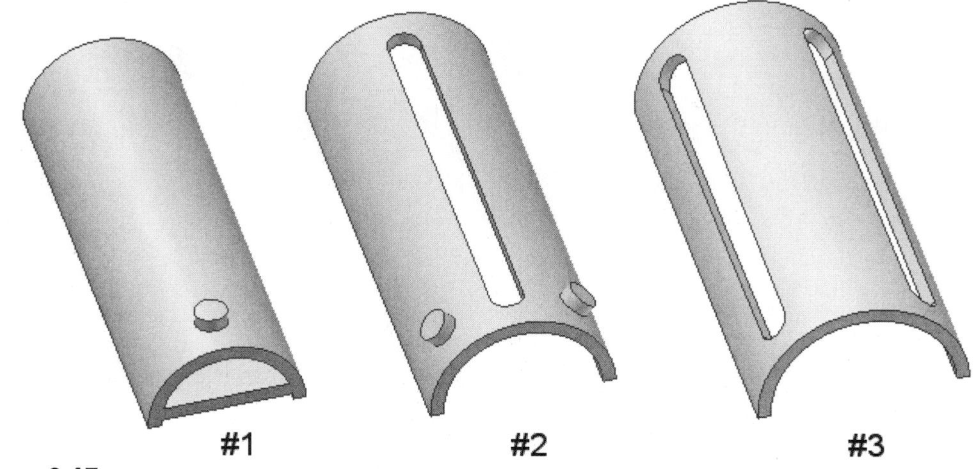

Figure 9.67

All three components are assembled with the freedom to translate along the common axis, as shown in the following image. Design intent describes that the pin features should stop at the end of a slot. However, the components continue to drag beyond the ends of the slots. The Contact Solver is based on your designation of certain components to be included in a contact set. This contact set will limit the motion of the objects, so they stop when the pin detects the end of a slot.

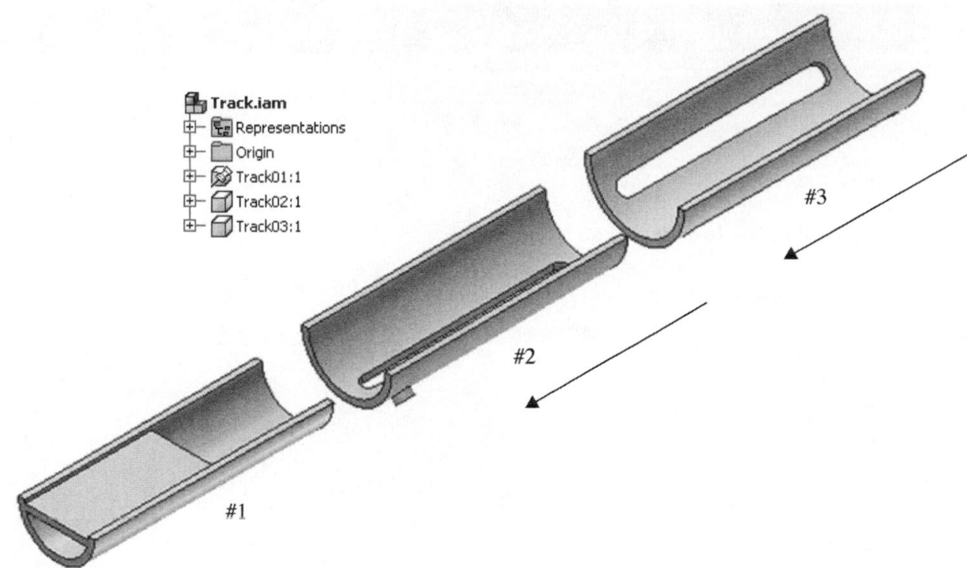

Figure 9.68

To designate components as part of a contact set and test a mechanism, follow these steps:

1. Click Tools on the Standard toolbar, and then click Document Settings to launch the Document Settings dialog box. Click the Modeling tab, as shown in the following image. By default, the Contact Solver Off option is selected. You can change the option to All Components and set the content solver to act on all components in the assembly. You can also change it to Contact Set Only and set it to act on specific, selected components.

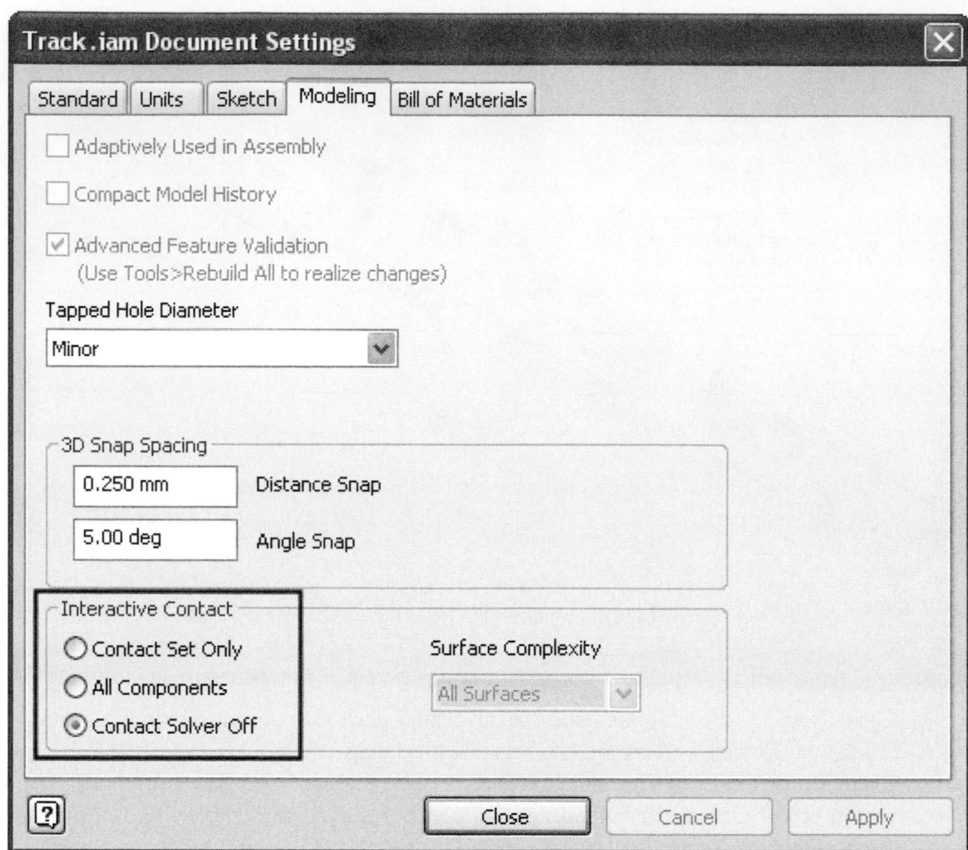

Figure 9.69

Note: Another way to activate the Contact Solver is to select the Tools menu on the Standard toolbar and then select Activate Contact Solver. This action will also turn on the Contact Solver on the Modeling tab of the Document Settings dialog box.

2. Once the Contact Solver is activated, you need to identify the components that will act upon each other, or a Contact Set. To create a Contact Set, right-click on one or more components in the browser or from the graphics window, and select Contact Set from the menu, as shown in the following image in the middle. The assembly components defined as part of the Contact Set will be identified in the browser by a Contact Set Icon, as shown in the following image on the right.

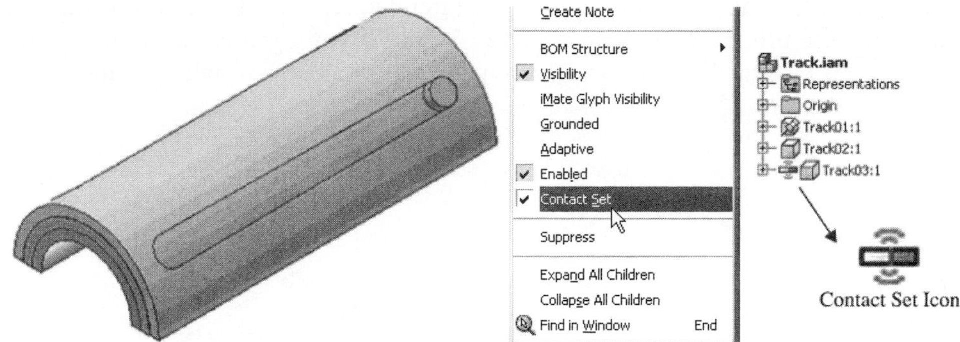

Figure 9.70

Note: Another way of creating a Contact Set is to right-click on one or more components and select Properties from the menu. When the Properties dialog box appears, click on the Occurrence tab, and select Contact Set from this dialog box, as shown in the following image.

Figure 9.71

3. You can now test the mechanism based on the Contact Solver. As shown in the following image, begin dragging components through their intended range of motion. The pins should now stop when they contact the ends of the slots. If the pins do not stop, you may have to adjust the component positions and repeat the motion.

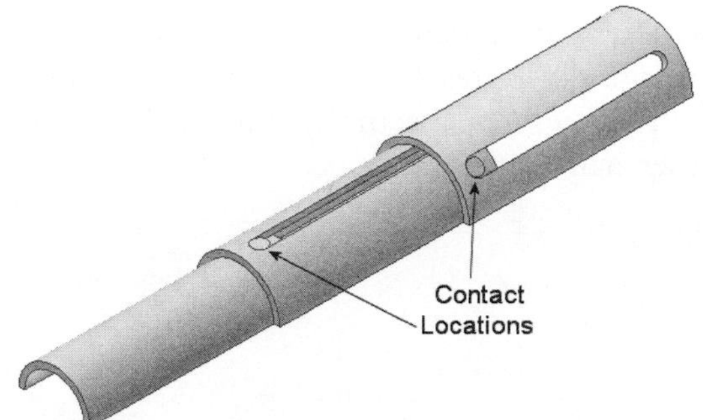

Figure 9.72

Drive constraints can also be used to produce intended motion. In the following image, a cam illustrates the motion of a Geneva drive mechanism as a pin comes into contact with a slot. Using the Contact Solver for these types of mechanisms usually produces dramatic results.

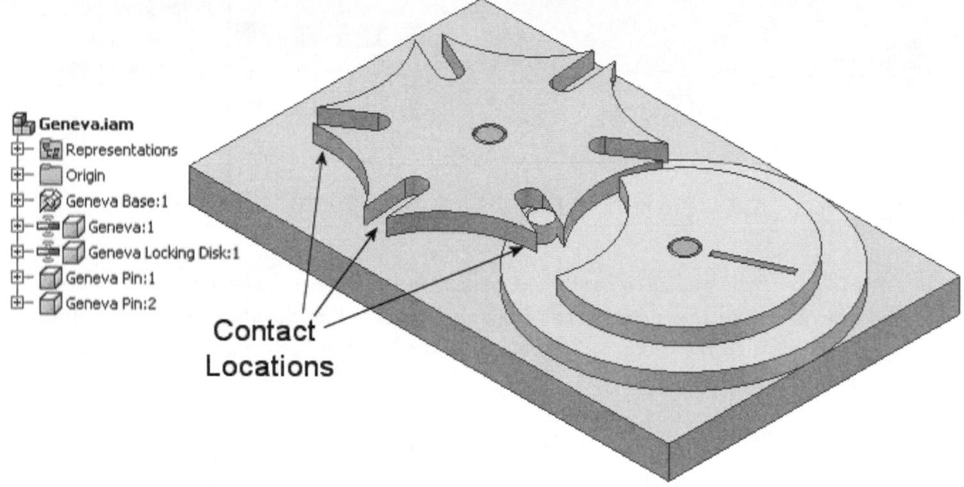

Figure 9.73

MIRRORING AN ASSEMBLY

To create right- and left-handed versions of an assembly and its components, use the Mirror Components tool. Mirrored components can be created using one of two methods:

- **Create a new assembly:** A new assembly can be created by selecting the components of a source subassembly. A mirror copy is generated from the original subassembly. The new subassembly is created relative to a mirror plane.
- **Create instances of components:** Individual components can be selected in a current assembly to create mirrored instances or copies. Each new mirrored component generates a new file. The individual components are mirrored based on a mirror plane.

To begin, open the assembly that contains the components or subassemblies you want to mirror. The following image illustrates a portion of a hold-down clamping mechanism. The image represents half of the total assembly, and the Mirror Components tool will be used to generate the other half of the clamp.

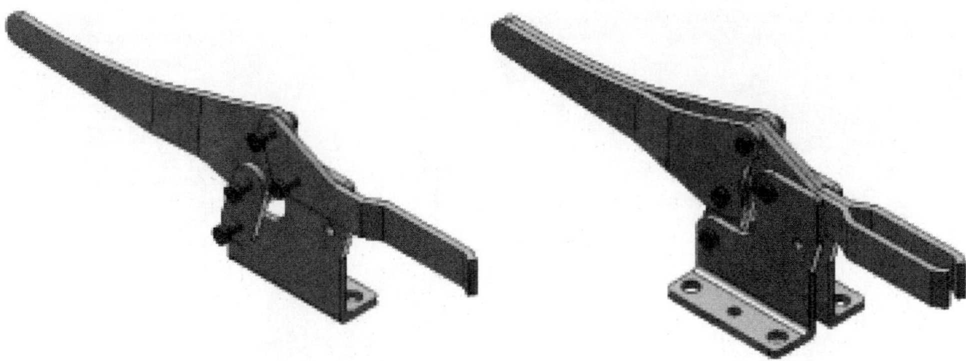

Figure 9.74

MIRRORING ASSEMBLY COMPONENTS

Follow this series of steps to create a mirrored group of components in an assembly:

1. Click the Mirror Components button, as shown in the following image, located on the Assembly panel bar.

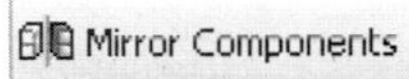

Figure 9.75

2. The Mirror Components dialog box will appear. Select all components from the graphics screen or from the browser, as shown in the following image. As each component is selected, it is added to the Mirror Components dialog box list.

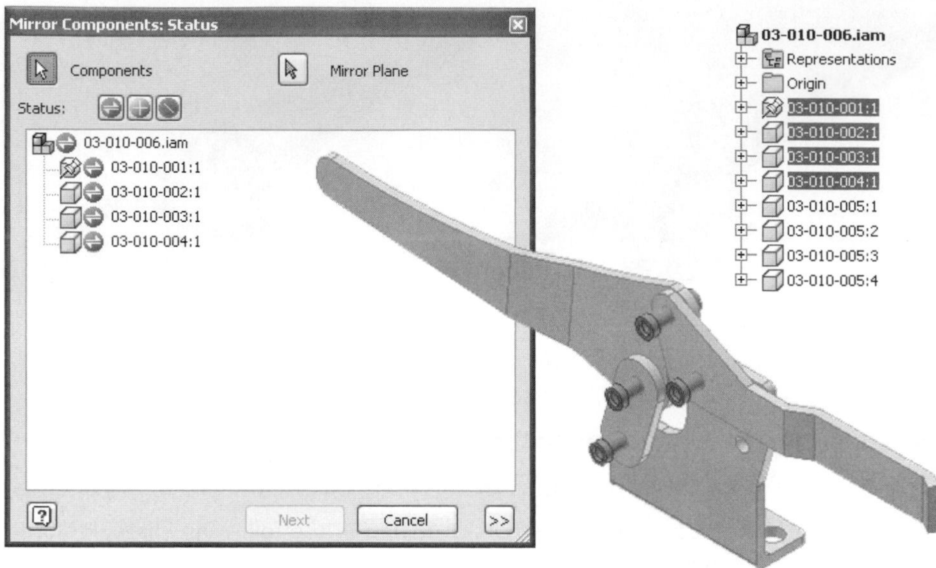

Figure 9.76

3. Next, click on the Mirror Plane button, if it is not already selected, and select a work plane or planar face from a part on the assembly or a plane in the Origin folder. Once the components and mirror plane are selected, you will see a preview of the components to be mirrored, as shown in the following image.

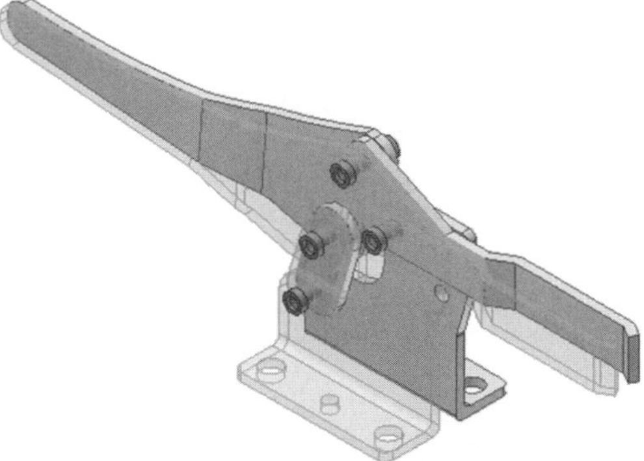

Figure 9.77

4. The following status buttons are available in the Mirror Components dialog box. Click the status button next to a component to change its status based on the following information.

Symbol	Title	Function
	Mirrored	This button will create a mirrored instance of the component. A new file is created for the mirrored component.
	Reused	This button will reuse the existing instance from the current assembly or subassembly file. A new file is not created.
	Excluded	This button will exclude a component or subassembly in the mirror operation.
	Mixed Reused/ Excluded	This button indicates that a subassembly contains components with reused and excluded status. It could also mean that the reused subassembly is not complete.

5. Click on the More (<<) button to display previewing options and controls for handling content library and factory parts.

- **Reuse Standard Content and Factory Parts:** Placing a check in this box will restrict the mirrored state for standard content and factory parts. Instances of the library parts are created in the current or new assembly file instead.
- **Preview Components:** Place a check in each box to preview the mirrored, reused, or standard content on the graphics screen. Mirrored parts are previewed in green, reused parts in yellow.

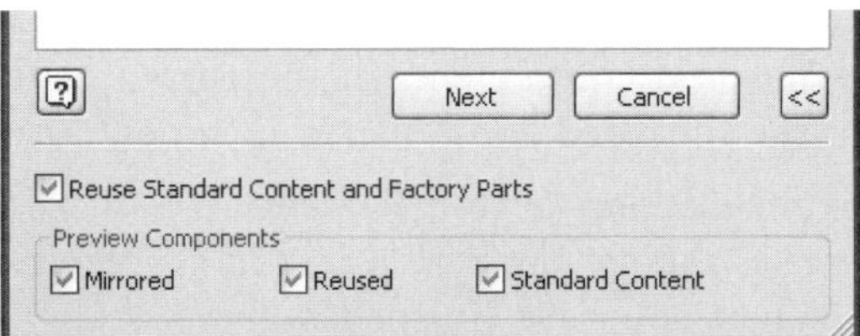

Figure 9.78

6. When you have finished making changes to settings, click the Next button to continue.
7. The Mirror Components: File Names dialog box appears. Use this dialog box to change the names of the mirrored components or to accept the default names, as shown in the following image. To change a part name, click on a current name under the New Name heading.

Note: You can also change the file location by right-clicking on Source Path under the File Location heading. However, it is considered good practice to keep the default file location in order to locate the mirrored components when reopening the assembly.

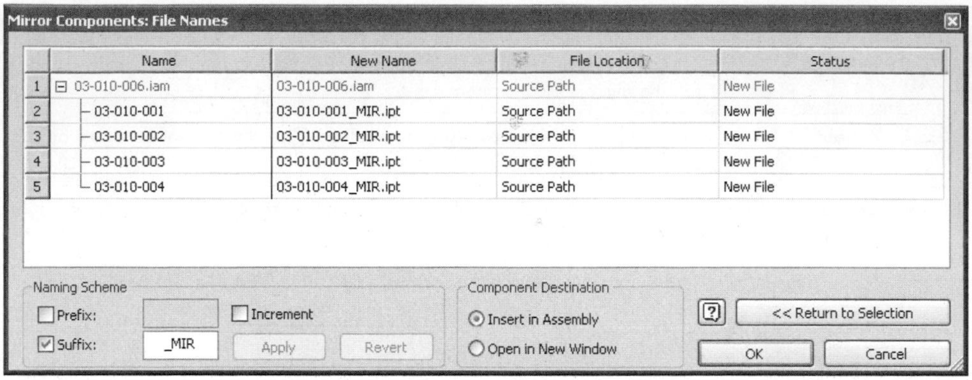

Mirror Components: File Names

	Name	New Name	File Location	Status
1	⊟ 03-010-006.iam	03-010-006.iam	Source Path	New File
2	├ 03-010-001	03-010-001_MIR.ipt	Source Path	New File
3	├ 03-010-002	03-010-002_MIR.ipt	Source Path	New File
4	├ 03-010-003	03-010-003_MIR.ipt	Source Path	New File
5	└ 03-010-004	03-010-004_MIR.ipt	Source Path	New File

Naming Scheme: Prefix: | Increment | Suffix: _MIR | Apply | Revert

Component Destination: Insert in Assembly | Open in New Window

<< Return to Selection | OK | Cancel

Figure 9.79

You can also control the listing of a prefix or suffix in the part name using the Naming Scheme area of the Mirror Components: File Names dialog box. Changing the _MIR listing to a different name activates the Apply button. Clicking on the Apply button will make the change to all component names under the New Name heading. If you want to return to the original values, this action will activate the Revert button. Placing a check in the box next to Increment will increase all filename increments if a number is used as a Suffix instead of text.

You can also make changes in the Component Destination area. Clicking Insert in Assembly will place the mirrored components in the current assembly. Clicking Open in New Window will place the mirrored components in a new assembly file.

Clicking on the Return to Selection button will return you to the Mirror Components dialog box and allow you to change the status of the mirrored components or to select additional components. When you have finished making changes, click the OK button.

The results of the mirror operation are illustrated in the following image. All selected components were mirrored about a selected plane. The browser was also updated to reflect the new part additions and any additional instances of the existing components that were reused. In this image, the default _MIR suffix was added to the end of each part name.

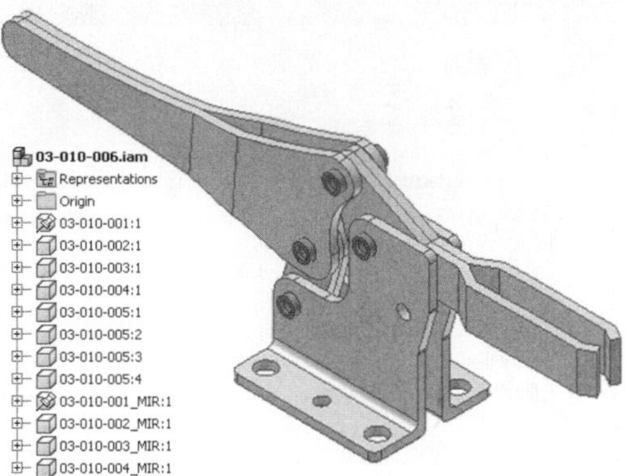

Figure 9.80

An entire assembly can be easily mirrored. To accomplish this task, follow the same process as described above but select the top-level assembly name in the assembly browser after launching the Mirror Components tool. This step will populate the Mirror Components dialog box with all assembly parts.

Note: In addition to individual components and constraints, the following items will also be included in assembly mirror operations: Assembly Features, Patterns, iMates, and Work Features.

COPYING AN ASSEMBLY

You can copy assembly components or an entire assembly in the current design file. The process is the same as that outlined for mirroring components, but you select the Copy Components tool, as shown in the following image, from the Assembly panel bar.

Figure 9.81

The Copy Components dialog box is displayed, and you can select the components or top-level assembly that you want to copy, as shown in the following image. This dialog box functions in a similar fashion to the Mirror Components dialog box previously discussed.

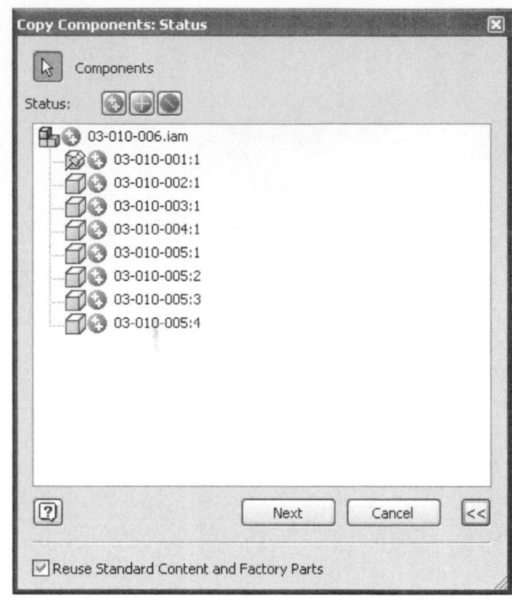

Figure 9.82

After you have finished selecting components and have made changes to the status buttons to copy, reuse, or exclude, the Copy Components: File Names dialog box can be used to change the names of the copied components or to keep the default names for the new files.

EXERCISE 9–2: MIRRORING ASSEMBLY COMPONENTS

Copy Components: File Names

	Name	New Name	File Location	Status
1	03-010-006.iam	03-010-006.iam	Source Path	New File
2	03-010-001	03-010-001_CPY.ipt	Source Path	New File
3	03-010-002	03-010-002_CPY.ipt	Source Path	New File
4	03-010-003	03-010-003_CPY.ipt	Source Path	New File
5	03-010-004	03-010-004_CPY.ipt	Source Path	New File
6	03-010-005	03-010-005_CPY.ipt	Source Path	New File

Naming Scheme
Prefix: Increment
Suffix: _CPY Apply Revert
Component Destination
Insert in Assembly
Open in New Window
<< Return to Selection
OK Cancel

Figure 9.83

In this exercise, you will create a mirrored copy of an assembly from an existing assembly.

1. To begin, open *ESS_E09_02.iam* in the Chapter 09 folder. This file consists of an assembly model in addition to a predefined work plane.
   a. Click the Mirror Components tool, which will display the Mirror Components dialog box.

b. With the Components button depressed, pick the top-level assembly from the browser (*ESS_E09_02.iam*), as shown in the following image on the left.

c. Observe the results in the Mirror Components dialog box, as shown in the following image on the right. All assembly components are added to the list.

d. Click the Mirror Plane button in the Mirror Components dialog box, as shown in the following image on the right. Then select the visible work plane as the mirror plane.

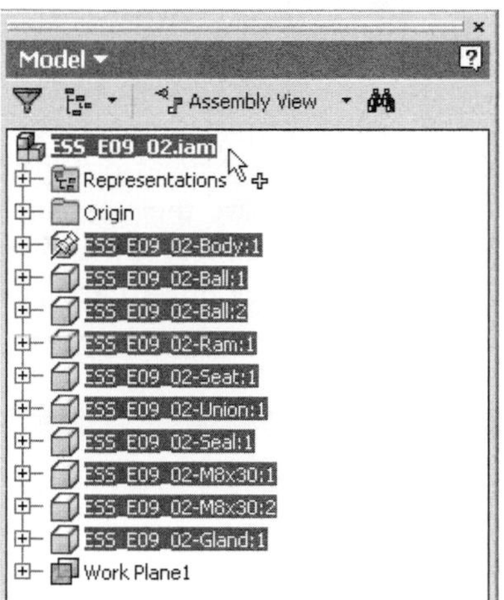

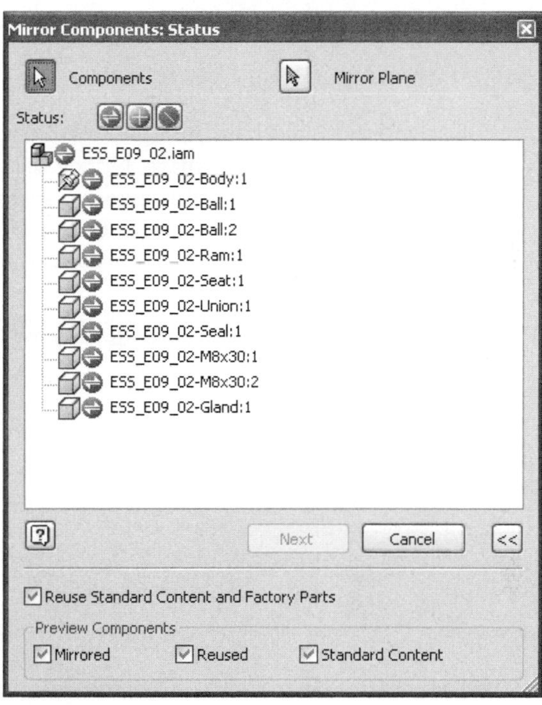

Figure 9.84

2. Clicking on the visible work plane as the mirror plane will produce a preview of the mirror operation, as shown in the following image. After reviewing this preview, click the Next button in the Mirror Components dialog box to continue with the mirror operation.

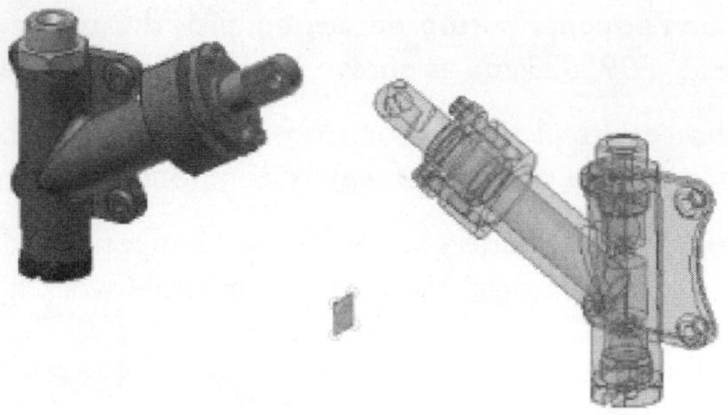

Figure 9.85

3. Clicking the Next button displays the Mirror Components: File Names dialog box, as shown in the following image. Notice that all mirrored parts have their part numbers appended with a _MIR extension, which is designed to distinguish the mirrored parts from the original parts. Click the OK button in this dialog box to accept the default settings.

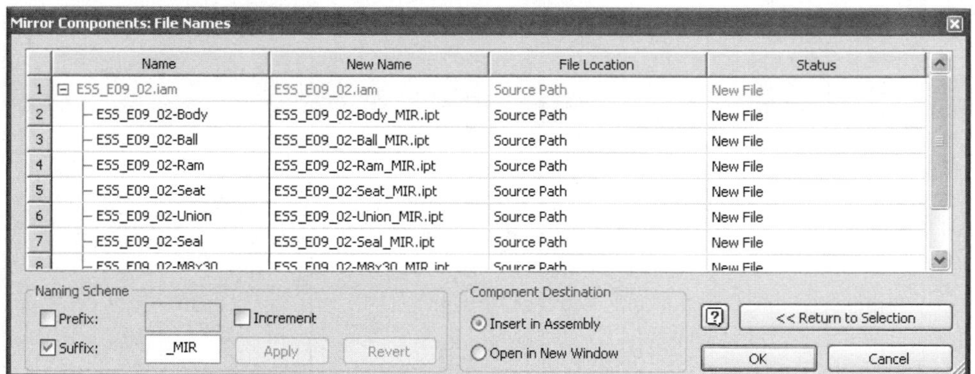

Figure 9.86

4. The results of the mirror assembly operation are shown in the following image.

Figure 9.87

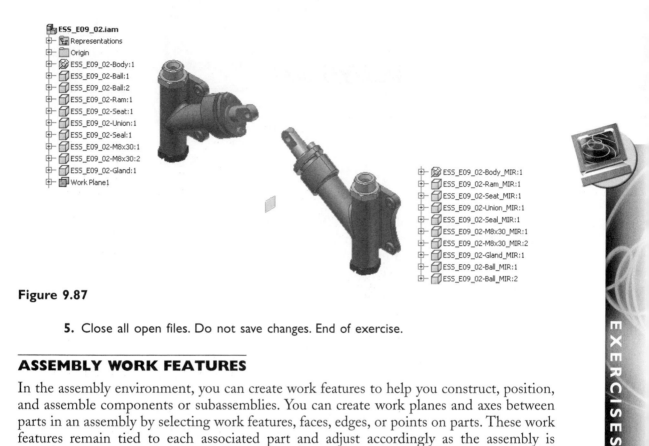

5. Close all open files. Do not save changes. End of exercise.

ASSEMBLY WORK FEATURES

In the assembly environment, you can create work features to help you construct, position, and assemble components or subassemblies. You can create work planes and axes between parts in an assembly by selecting work features, faces, edges, or points on parts. These work features remain tied to each associated part and adjust accordingly as the assembly is modified. You can use assembly work features to parametrically position new components or subassemblies, to check for clearance in an assembly, and as construction aids. You can also use work planes to help you define section views of your assemblies.

Autodesk Inventor also allows you to globally turn off the visibility of work features. This is important in the assembly environment, where the display of work features from individual parts can quickly clutter the graphics window. Options for turning off work feature visibility are available by selecting the View > Object Visibility menu, as shown in the following image.

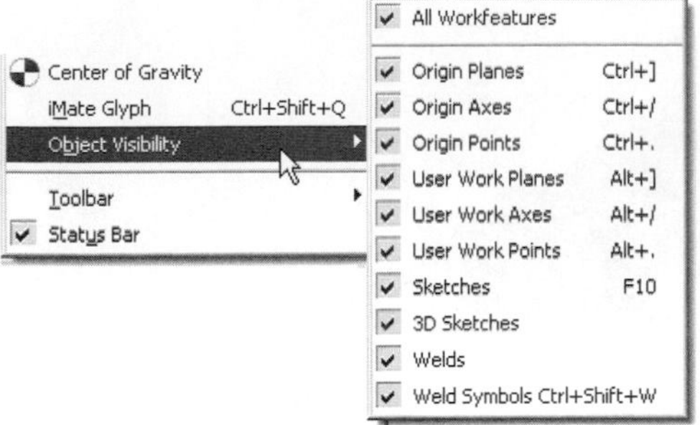

Figure 9.88

You can use these controls to turn off the visibility of work features by type. By default, all types of work geometry are initially selected for display. Thus, any work feature with its individual visibility turned on in the browser is visible in the assembly file.

To globally turn off the visibility of a particular type of work geometry, select it from the menu and clear the checkmark next to it. This action overrides the visibility setting for individual work features of that type in the assembly and in each part in the assembly. Although the components work features' visibility in the assembly are suppressed, their individual visibility control remains turned on in each of the component files. You can also control the visibility status of sketches, 3D sketches, welds, and weld symbols from the menu.

ASSEMBLY FEATURES

Assembly features are features that are defined in an assembly. They only affect a part when the part is viewed in the context of the assembly. Assembly features allow you to remove material from components after they have been assembled. Examples of material-removal processes include match-drilling operations and post-weld machining operations. Typical operations involving assembly features include cutting extrusions, drilling holes, and cutting chamfered edges.

In the following image of a post-machining operation, three blocks have been assembled through the use of mate-mate and mate-flush constraints. With all three items assembled, a feature in the form of two slots is then cut through the lower and upper faces of the assembly. A sketch is created on the front face of one of the parts, geometry is sketched and dimensioned, and the assembly feature is cut using an extrusion operation. Notice in the following image that in the completed part, a chamfer operation was applied at one end of the assembly.

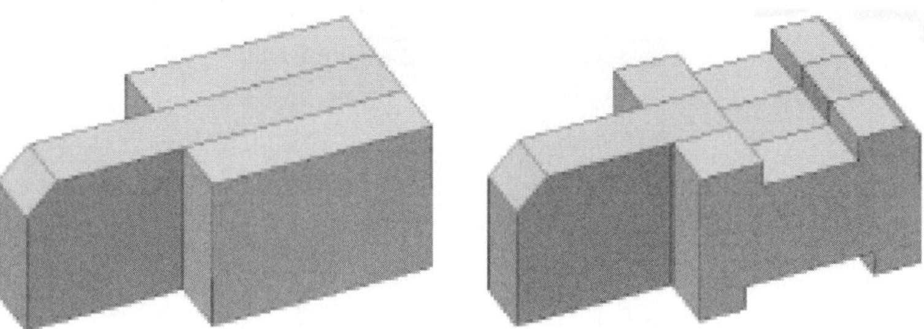

Figure 9.89

ASSEMBLY SKETCHES

Sketch geometry can be added in an assembly. You can sketch on a planar part face, a part's work plane, or an assembly work plane. Features can then be created from these sketches. Like creating features in part-modeling mode, you create a new sketch on a part or assembly plane or face—it does not matter which assembled part face or plane is selected for the sketch. This action will activate the 2D Sketch panel bar.

When you create a sketch profile, geometry can be projected from various parts to the assembly, and constraints and dimensions are added, as shown in the following image.

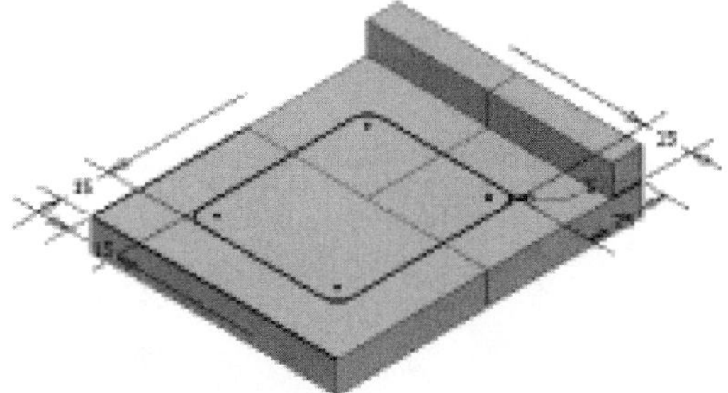

Figure 9.90

CREATING ASSEMBLY FEATURES

Once you have finished creating the desired sketch, return to the assembly mode. In the following image, a portion of the Assembly Panel bar containing tools for creating assembly features is shown. These tools include Extrude, Revolve, Hole, Sweep, Fillet, Chamfer, Move Face, Rectangular Pattern, Circular Pattern, and Mirror. The tools join the Work Plane, Work Axis, and Work Point tools that function inside of assembly models.

Note: Assembly features exist only at the assembly level. They do not affect the individual part files.

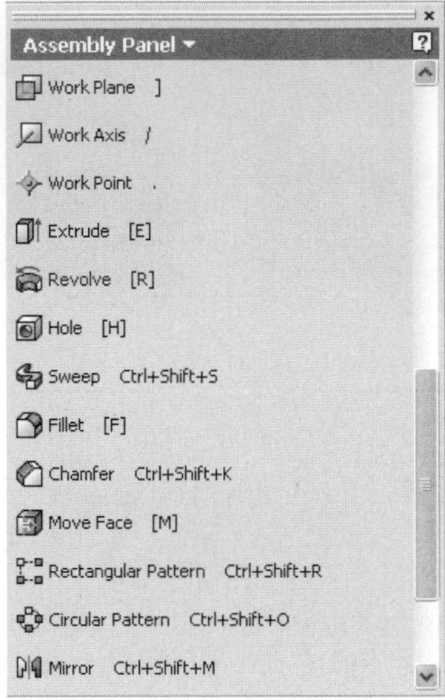

Figure 9.91

Use one of the assembly feature tools on the sketched geometry to create the desired assembly feature. In the following image, an extrusion operation is being performed by cutting the rectangular slot through the entire base of the assembly.

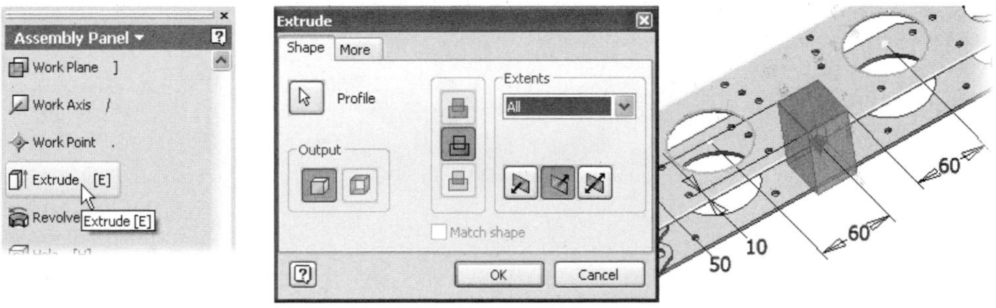

Figure 9.92

The result of the extrude cut operation on the assembly model is illustrated in the following image on the left.

Note: When extruding an assembly feature, only the cut operation is available.

An End of Features node is present in the browser to separate assembly features from assembly components. When an assembly feature is created, the feature is added above all the assembly components. In the following image on the right, an assembly feature named Extrusion 1 was added to the assembly, and the feature is placed in the browser above the

End of Features node. All parts affected by the created assembly feature are listed below that assembly feature. These affected parts are called Participants, as they are all members of the group of components being cut by the assembly feature. They are shown in the browser as *children* of the feature.

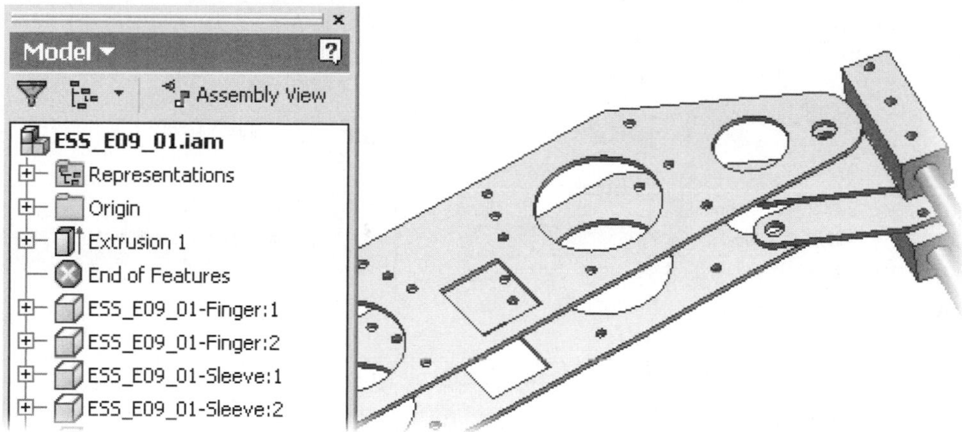

Figure 9.93

REMOVING AND ADDING PARTICIPANTS

When working with assembly features, it is possible to add and remove participants that have been affected by the assembly feature. To remove a Participant, right-click on a component listed as a Participant, and then select Remove Participant from the menu, as shown in the following image on the left. When a Participant has been removed from the assembly feature, the assembly feature no longer affects the component, as shown in the following image on the right.

To add a Participant to an assembly feature, right-click the assembly feature in the browser, and select Add Participant from the menu, as shown in the following image on the right. You can then select the part to add from either the graphic window or the browser.

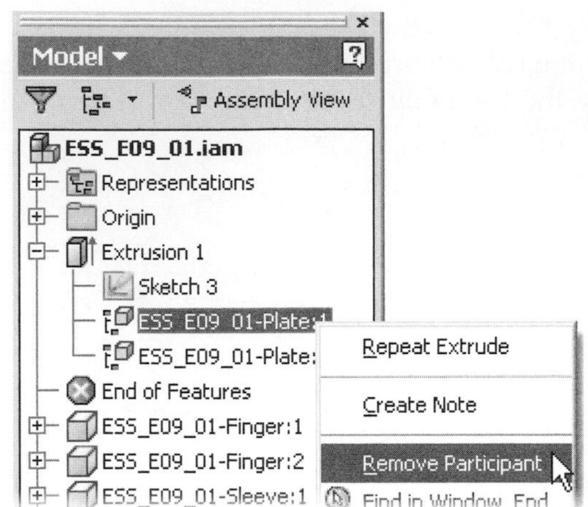

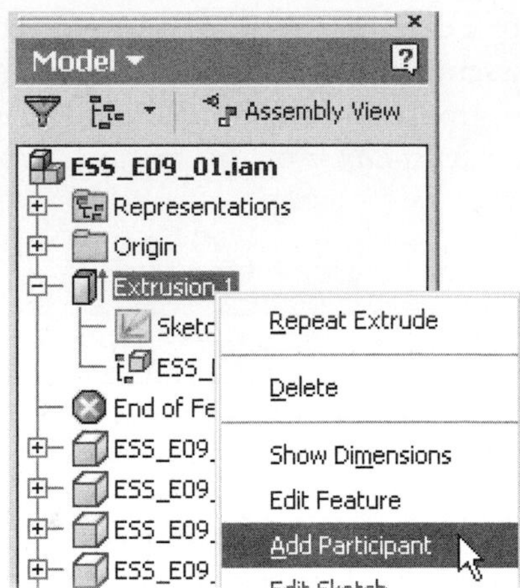

Figure 9.94

Assembly features can be suppressed by dragging the End of Features icon to a new position before the actual assembly feature, shown as Extrusion 1 in the following image on the left, or by right-clicking the feature and selecting Suppress Feature from the menu, as shown in the following image on the right.

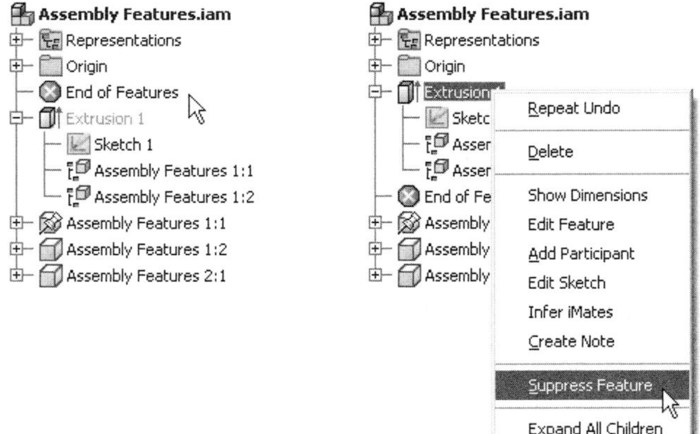

Figure 9.95

EXERCISE 9-3: CREATING ASSEMBLY FEATURES

In this exercise, you add a sketched assembly feature to an existing assembly, which is an array of index pockets spanning two assembly components. You then add an assembly hole feature and change the components that participate in the hole feature.

1. Open *ESS_E09_03.iam* in the Chapter 09 folder. The mount assembly should appear similar to the one displayed in the following image. Use the Rotate and Zoom tools to examine the assembly. Right-click in the graphics window, and select Home View.

Figure 9.96

2. Click the Sketch tool, and then select the face, as shown in the following image on the left. Next click the Project Geometry tool, and project the two edges highlighted in the following image on the right.

Figure 9.97

3. Click the Line tool. Select the Construction line style, and then sketch the two lines, as shown in the following image on the left. The upper line is horizontal, and both lines are coincident to the center point and the inner projected edge.

4. Create a third construction line, as shown in the following image on the right. The line is parallel and coincident to the endpoint of the lower line, and it is coincident to the outer projected edge.

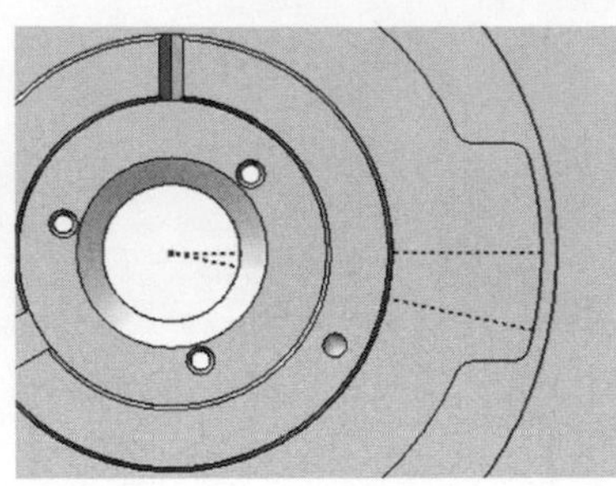

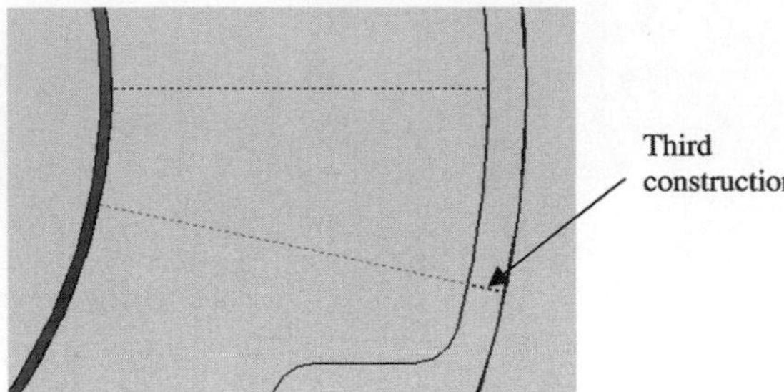

Figure 9.98

5. Click the Center Point Circle tool. Select Normal for the line style. Sketch a circle centered on the midpoint of the short construction line, as shown in the following image on the left. Click the General Dimension tool, and add the two dimensions, as shown in the following image on the right.

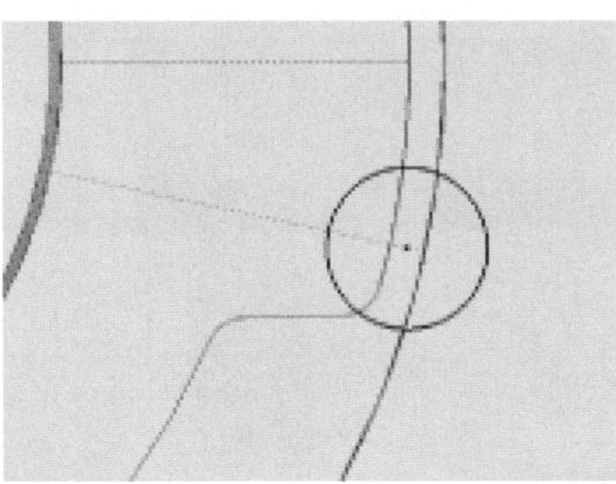

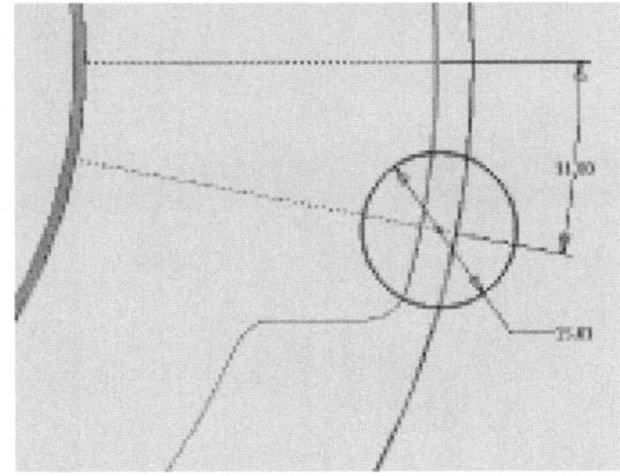

Figure 9.99

6. Click the Circular Pattern tool, and select the sketched circle. Click the Select Axis tool in the Circular Pattern dialog box, and select the projected point at the center of the assembly. Enter **3** in the Count edit box and **22 deg** in the Angle edit box. Click OK to create the pattern. Your sketch should match the following image.

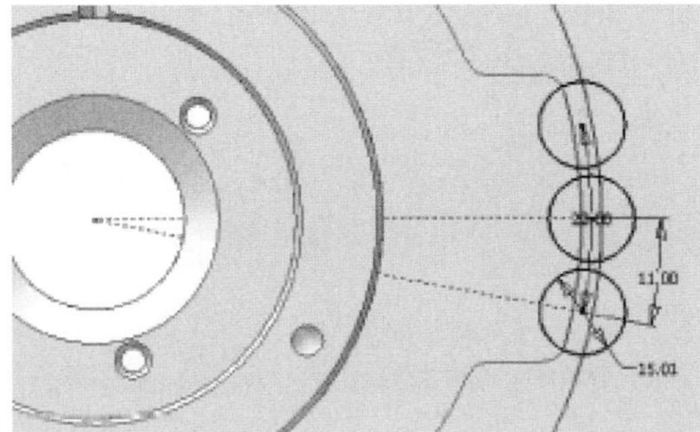

Figure 9.100

7. Right-click and select Home view from the menu; then right-click and select Finish Sketch from the menu. Click the Extrude tool from the Assembly panel bar. Click inside the three patterned circles as the areas to extrude. Enter a value of **3 mm** in the Depth edit box. By default, you will be performing a cut operation. The extrude feature previews, as shown in the following image on the left. Clicking the OK button in the Extrude dialog box will display the results shown in the following image on the right.

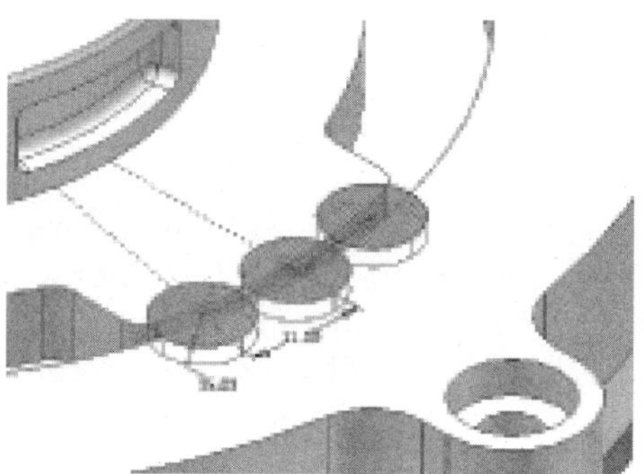

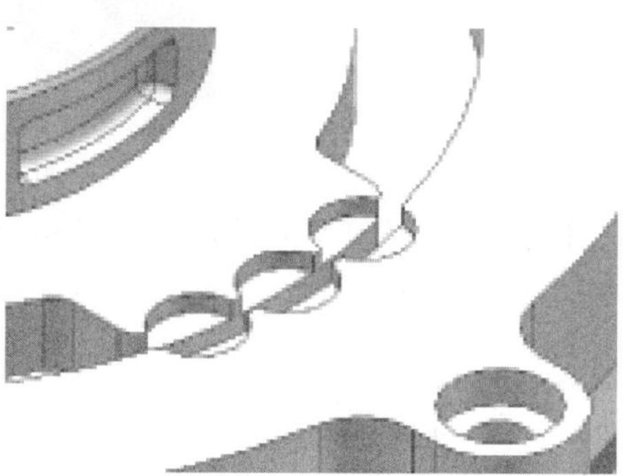

Figure 9.101

8. Create a new assembly sketch on the same face as the previous sketch. Project the edge, as shown in the following image on the left. Sketch a horizontal line from the projected point at the center of the assembly, as shown in the following image on the right. Add a Point, Center Point at the end of the line. Dimension the length of the line, and enter **78.5 mm** in the dimension edit box. Your sketch should match the one shown in the following image on the right.

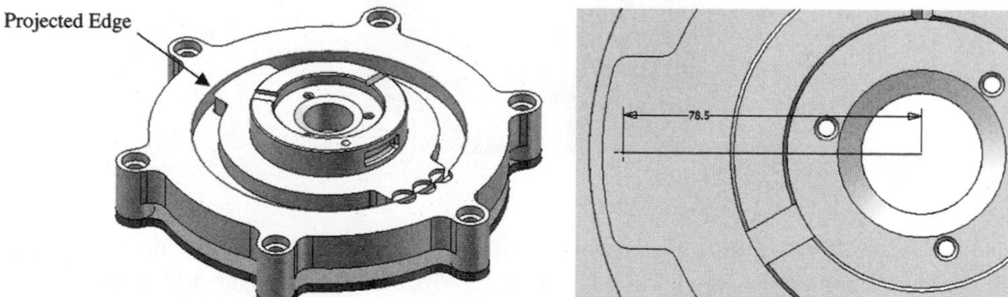

Figure 9.102

9. Right-click in the graphics window, and select Finish Sketch from the menu. Click the Hole tool and select Through All from the termination list. Enter a value of **5.01 mm** as the hole diameter. Click the OK button to cut the hole. Notice that the hole feature cuts through all components in the assembly, as shown in the following image.

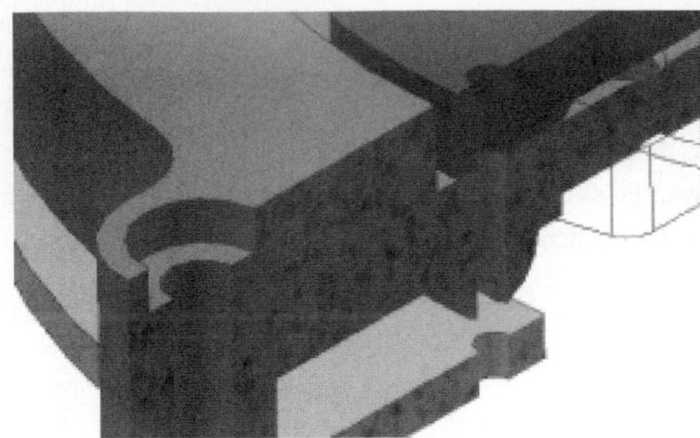

Figure 9.103

10. The hole is not required to cut through the bottom plate in the assembly. Expand Hole1 in the browser. All components affected by the feature are listed below the feature. Right-click ESS_E09_03-BasePlate:1, and select Remove Participant from the menu. The assembly feature no longer cuts through the bottom plate, as shown in the following image.

Note: To add a component as a Participant in an assembly feature, right-click the assembly feature in the browser, and select Add Participant from the menu. Select the component to include.

Figure 9.104

11. You will now change the geometry of one of the parts and examine the effect on an assembly feature. Right-click ESS_E09_03-Pivot in the browser, and select Edit from the menu. Right-click Extrusion1, select Edit Sketch, and change the diameter of the arms to **172 mm**, as shown in the following image on the left. Return to the top-level assembly. The assembly feature matches the new geometry, and the index pockets are centered between the pivot and the base, as shown in the following image on the right.

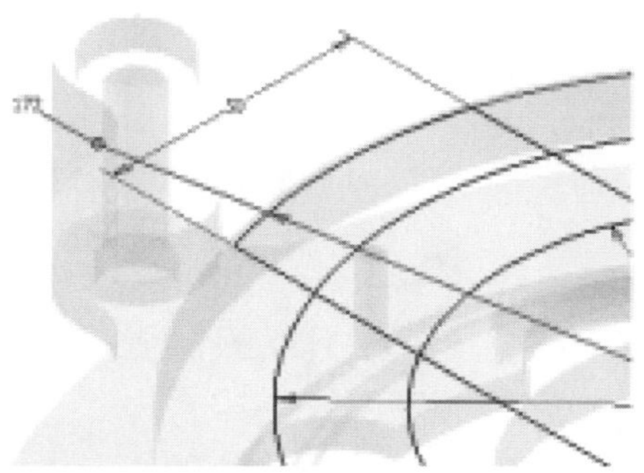
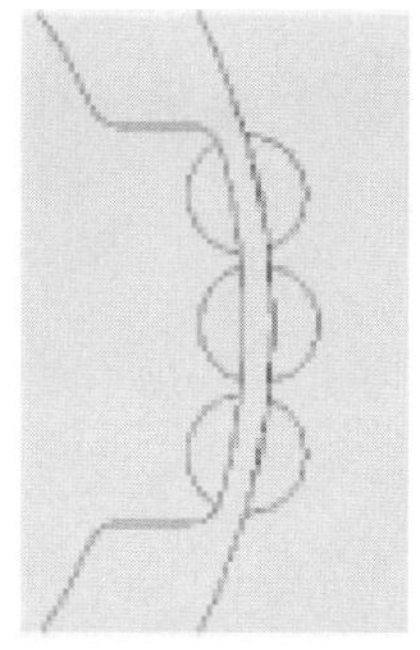
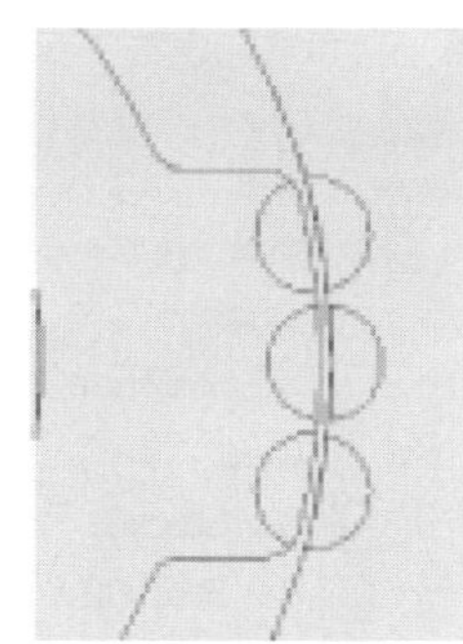

Figure 9.105

12. Close all open files. Do not save changes. End of exercise.

SKELETAL MODELING TECHNIQUES

Skeletal modeling involves creating numerous features in a single part file. In the following example, three features were created and renamed in the browser: Square Tube, Stop, and Bar Stock. You now want to capture this design intent in the form of an assembly model. You could save this part file under three different names and then remove certain features, but this action would be difficult, if not impossible, without encountering errors. Additionally, if the features could be broken down and then saved as individual parts, assembly constraints would have to be added at the assembly level. An even more important issue is what happens to the assembly when changes occur in the individual parts. This is where certain parts would need to be made adaptive, which can be tricky. All of the previously mentioned limitations point to the use of a skeletal model. When properly created, it reduces the dependency on such items as assembly constraints and making certain parts adaptive. Here is how a skeletal model works.

Start by creating a single part file that captures the design intent in the form of features. Do not worry about detailing the features at this point; in fact, it is considered good practice to keep the skeletal model simple. Once the skeletal part file is created, common practice is to rename the features to match the individual components once the assembly is made. This step is illustrated in the following image on the left of the Assembly browser.

With the skeletal model created, save this file using a name that will make it easy to identify. The file name "Skeletal_Source" will be used during this example. Exit the skeletal part file and create a new assembly. While still inside the blank assembly file, click the Create Component tool in the browser; this will launch the Create In-Place Component dialog box, as shown in the following image on the left. Add a new component name, choose the appropriate template, and click the OK button. You will need to begin this new part in relation to one of the three existing default work planes. The XY Plane will be used, as shown in the following image on the right. Once this plane is selected, it must be used throughout the creation of the individual assembly parts that reference the skeletal model. The idea is to be consistent with the use of the default work plane for all parts.

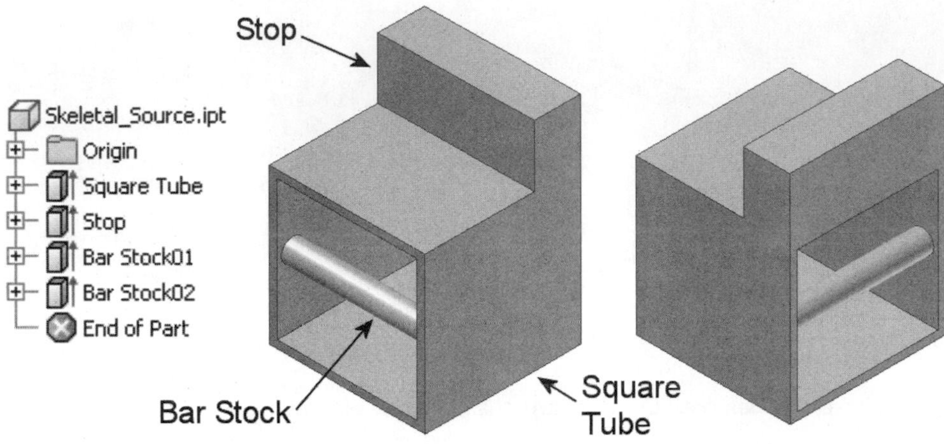

Figure 9.106

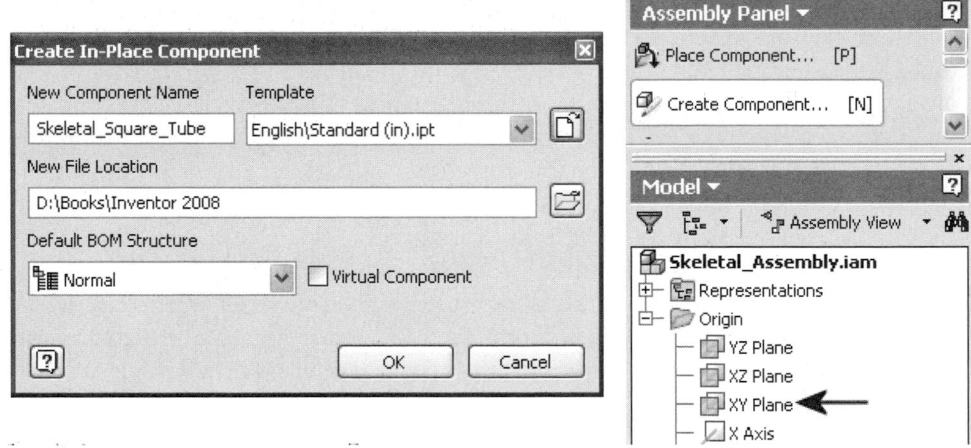

Figure 9.107

From the previous step, you would normally be designing a new component based on a component already in the assembly. However, in this case no components exist. You can use a process that creates a derived component. At this point, you should be in the Sketch environment; exit this environment by clicking the Return button. The Feature menu will now be present; click the Derived Component tool, as shown in the following image. When the Select Source File dialog box appears, select the skeletal part file that was created earlier, Skeletal_Source. When the Derived Part dialog box appears, as shown in the following image, click the status indicator to reveal the yellow plus sign next to Body as Work Surface. All of the other indicators can remain unchanged. Click the OK button to continue and place the skeletal model, as shown in the following image on the right. Notice that the skeletal model appears transparent because it consists of work surfaces and is not considered a solid object.

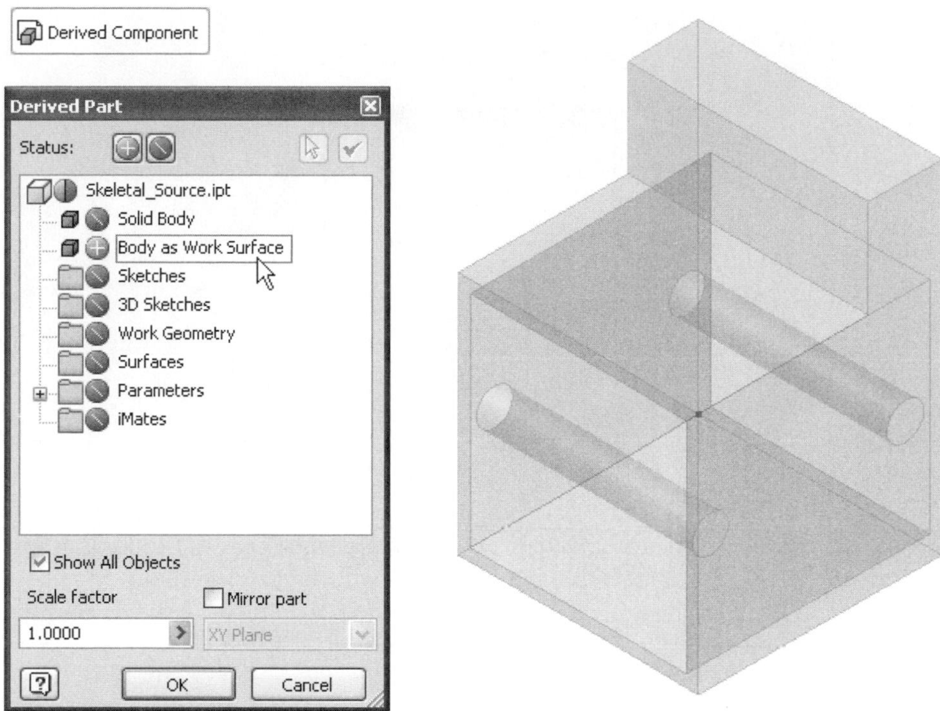

Figure 9.108

The skeletal model becomes the template for creating all individual components. First the square tube will be created from the skeletal model. Begin this process by making the front face of the square tube the current sketch plane, as shown in the following image on the left. Notice that the edges of the square tube project. These edges form the shape to be extruded. Click the Return button to exit sketch mode, and click the Extrude button. Use the To option, as shown in the following image in the middle, to extrude the square tube profile to the back side of the skeletal model, as shown in the following image on the left. The results are shown in the following image on the right.

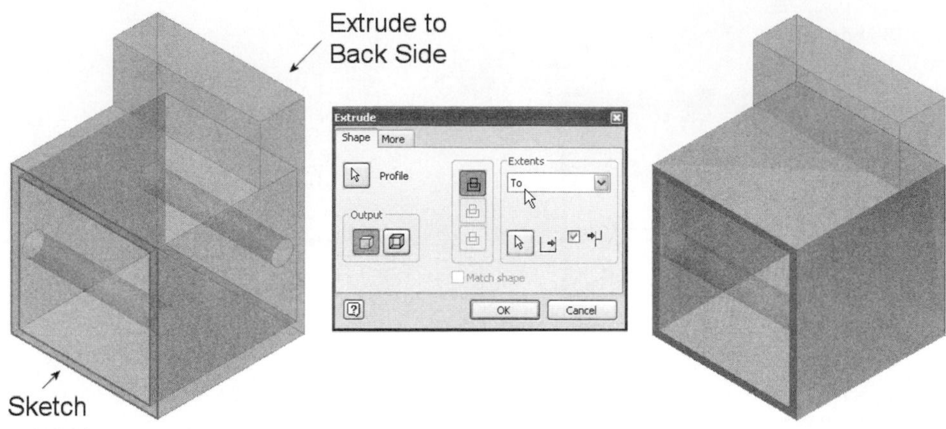

Figure 9.109

You are now finished with this first part but not finished creating the other parts. For this reason, turn off the visibility of Derived Work Body1, as shown in the following image on the left and in the middle. Right-clicking this item will display the menu to turn off the skeletal model. When finished, click the Return button to return to the assembly.

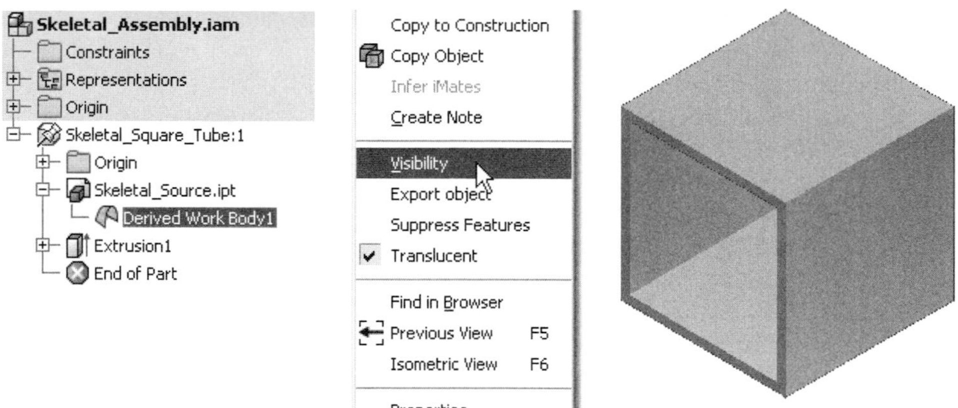

Figure 9.110

To complete the remaining stop and bar parts, repeat the process. Begin by creating a component in place, and click on the XY plane where the new component will begin. Notice the two new parts created using this derived component technique, Stop and Bar, as shown in the following image. Note that there are two bar parts. You need to derive only one of the bars from the skeletal model. The other bar was placed and constrained in the assembly using conventional methods, as discussed earlier in this chapter.

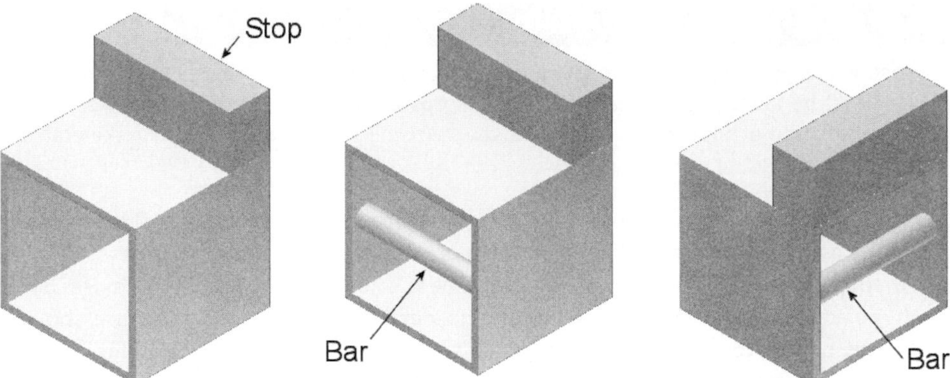

Figure 9.111

The real power of using the skeletal modeling technique comes when it is time to make changes to the original skeletal model, as shown in the following image on the left. Remember, these changes were made to the part file called Skeletal_Source. In this example, you would open the original skeletal source file. The size of the square box is changed to a rectangular shape, and the stop is reduced in size, as shown in the following image on the left. When returning to the assembly derived from the skeletal source, issuing an Update will display the associative assembly file, as shown in the following image on the right. All of these examples use a minimum of constraining and no adaptivity to achieve these results.

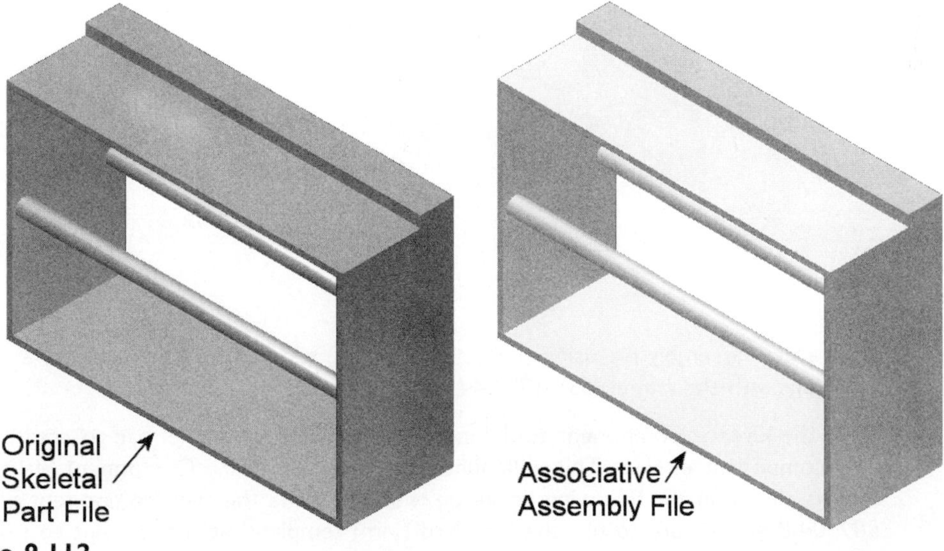

Figure 9.112

EXERCISE 9-4: CREATING A SKELETAL MODEL

This exercise walks you through the creation of a small engine lathe. Only the base, headstock, way, and tailstock will be illustrated. A majority of the steps concentrate on the first two components. You then use the same procedure to create the remaining components using skeletal modeling techniques.

1. First open the existing part *ESS_E09_04-Source.ipt* in the Chapter 09 folder. This part file actually represents four components combined into one: lathe base, headstock support, tailstock support, and lathe way, as shown in the following image. Since the objective is to capture design intent, no details of the various parts are shown at this point. Notice that the features in the browser identify the specific lathe component. When finished reviewing this assembly, close the file. Do not save changes.

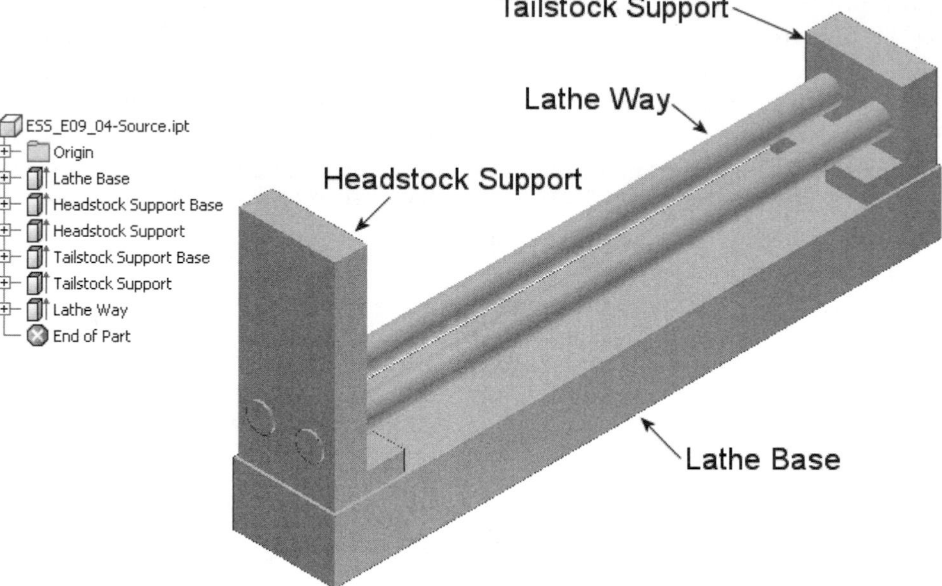

Figure 9.113

2. Start a new assembly file using the Standard (mm) template. Immediately save this assembly with the name *ESS_E09_04-Lathe*.
3. Click the Create Component tool, located in the Assembly panel bar, to create a new component in place. This will launch the Create In-Place Component dialog box, as shown in the following image on the left. Name this new component *ESS_E09_04-Base*. Be sure to use the Standard (mm) template. At the prompt to select a sketch plane for the base feature, expand the Origin folder in the browser and click in the XY Plane, as shown in the following image on the right.

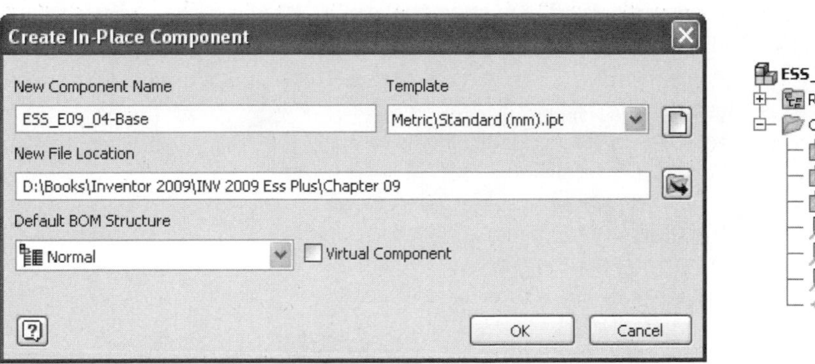

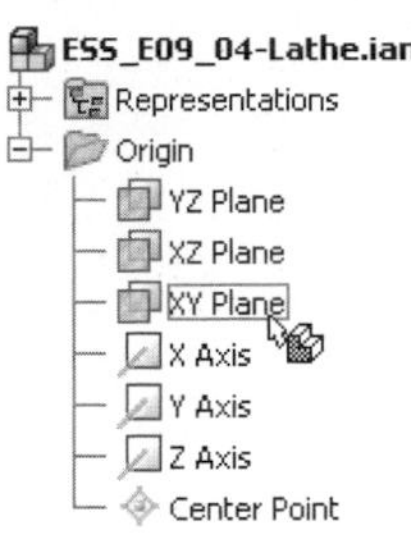

Figure 9.114

4. Switch to Home View and exit sketch mode by clicking the Return button. Click the Derived Component tool found under the Part Features panel bar. Since you will be making the base component of the lathe, you will need the *ESS_E09_04-Source.ipt* file as the skeleton. When the Derived Part dialog box appears, click Body as Work Surface to turn this mode on, as shown in the following image on the left. This action will turn off the default Solid Body mode. Notice that the derived lathe source part takes on a clear appearance since it consists of surfaces, as shown in the following image on the right. Click OK to exit the Derived Part dialog box.

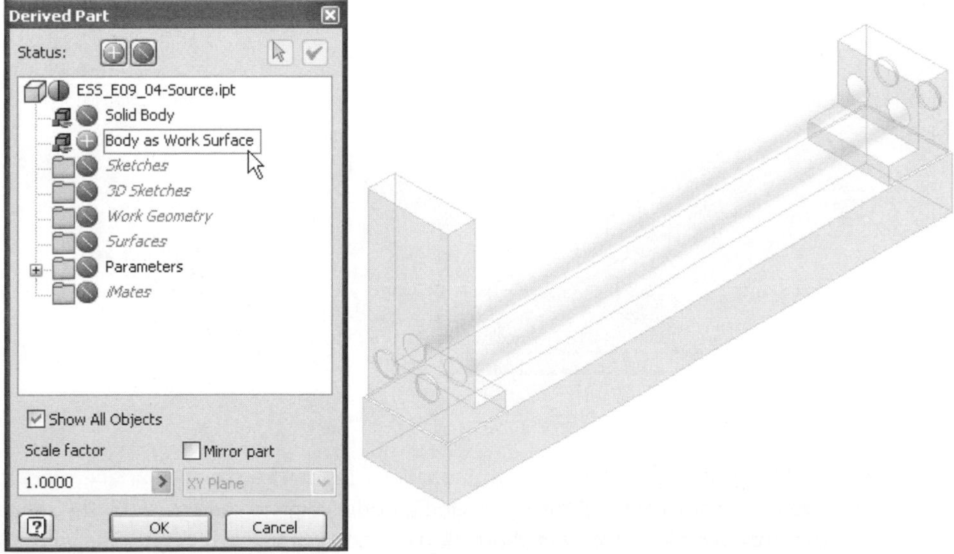

Figure 9.115

5. With the skeletal model of the lathe source present on the display screen, begin creating the lathe base by placing a sketch plane underneath the skeletal model, as shown in the following image.

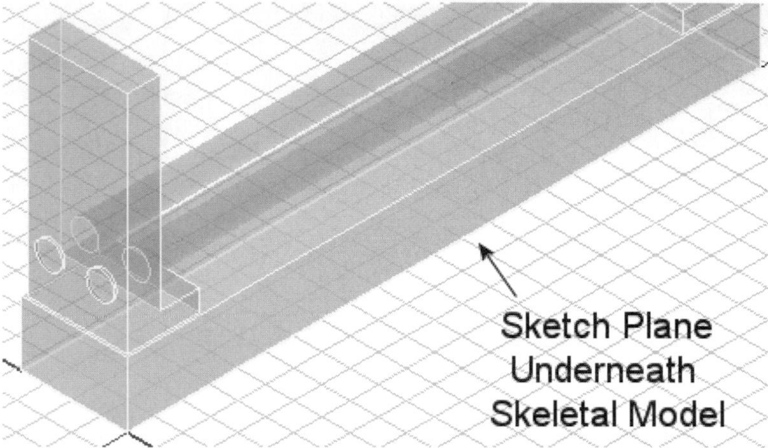

Figure 9.116

6. Click the Return button to exit sketch mode. Click the Extrude tool, and the bottom of the base should highlight as the profile to extrude. In the Extents area of the dialog box, change Distance to "To," click on the top of the base as shown in the following image, and then click OK.

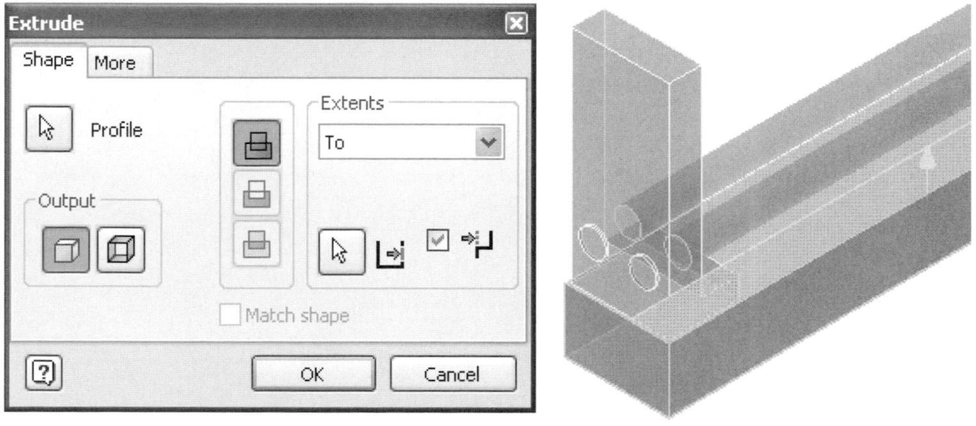

Figure 9.117

7. The part file of the lathe base is shown contained inside the skeletal surface model. Complete this part by turning off the skeletal model. Expand the *ESS_E09_04-Source.ipt* file in the browser, select Derived Work Body1, and right-click to display the menu, as shown in the following image. Click Visibility to turn off this feature.

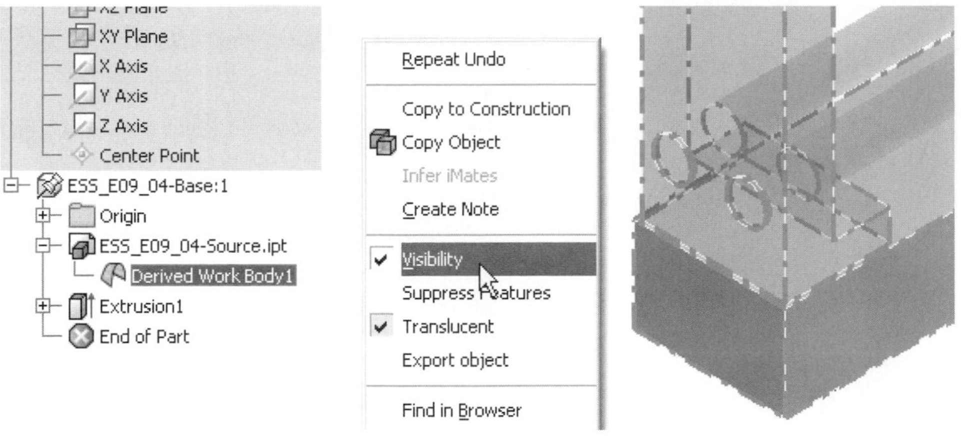

Figure 9.118

8. The results of turning off the skeletal model are displayed, as shown in the following image on the left. You can now return to the part file and add details to this part, as shown in the following image on the right. In this image, the Shell tool was used to create a wall thickness of 5 mm after the bottom and two end faces were removed. Click the Return tool to return to the top-level assembly. Notice in the Assembly browser that this first part appears grounded.

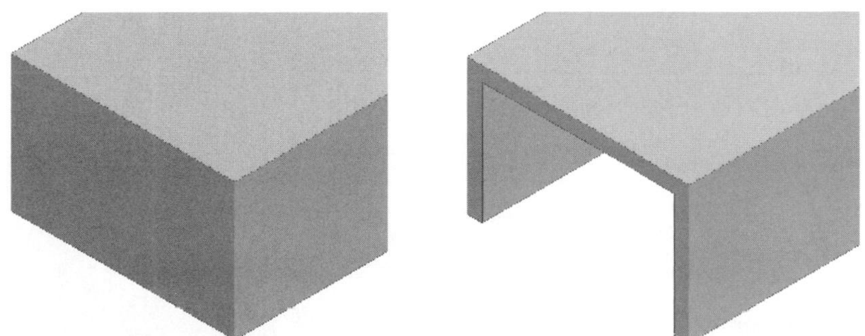

Figure 9.119

9. Begin creating the next part. Launch the Create In-Place Component dialog box, and name this new component *ESS_E09_04-Headstock*. Be sure to use the Standard (mm) template. Constrain the sketch plane by selecting the XY Plane under the Origin folder, found under the Assembly browser. Exit sketch mode, and click the Derived Component tool. Choose the file *ESS_E09_04-Source.ipt*, and choose the Body as Work Surface option of the Derived Part dialog box. The skeletal model will be overlaid on the existing part. Create a new sketch plane on the side of the headstock, as shown in the following image on the left. Click the Return button to exit sketch mode, and create an extrusion using the To option, as shown in the following image on the right.

EXERCISES

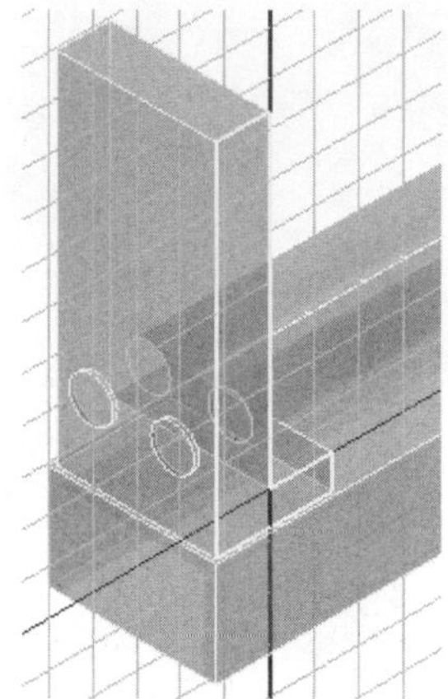

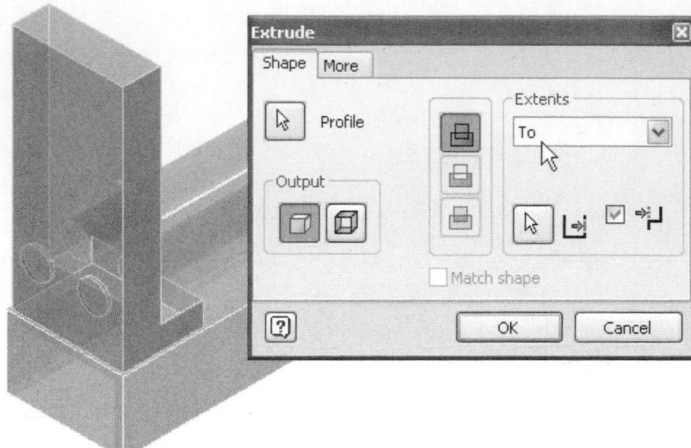

Figure 9.120

10. Complete this part by turning off the skeletal model. Expand the *ESS_E09_04-Source.ipt* file in the browser, select Derived Work Body1, and right-click to display the menu, as shown in the following image. Click Visibility to turn this feature off.

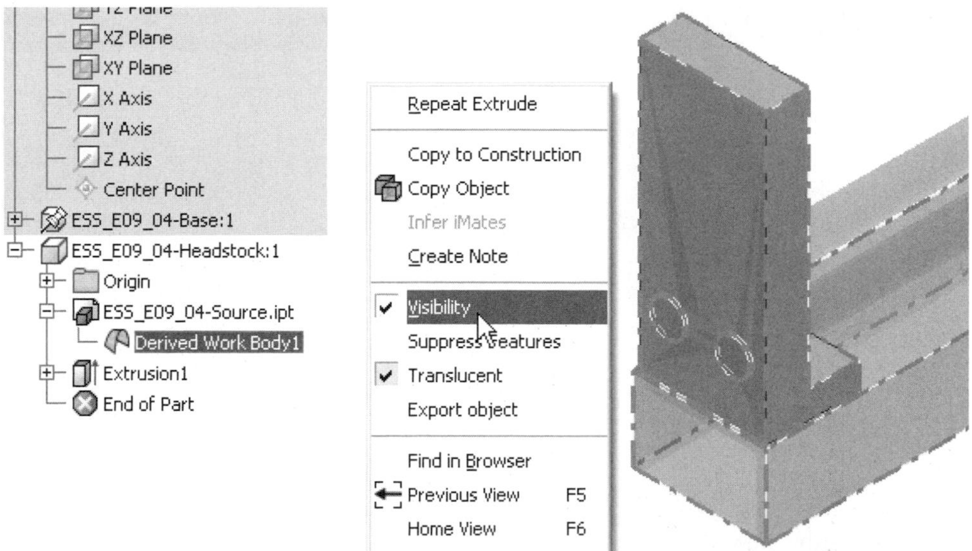

Figure 9.121

11. Clicking the Return button returns you to the main assembly, as shown in the following image on the left. Use the techniques already described to create the tailstock (*ESS_E09_04-Tailstock*) and way (*ESS_E09_04-Way*), as shown in the following image on the right.

Note: The second lathe way was placed as an individual component and assembled using an Insert constraint. One of the derived work bodies was turned back on to assist with the assembly.

12. Close all open files. Do not save changes. End of exercise.

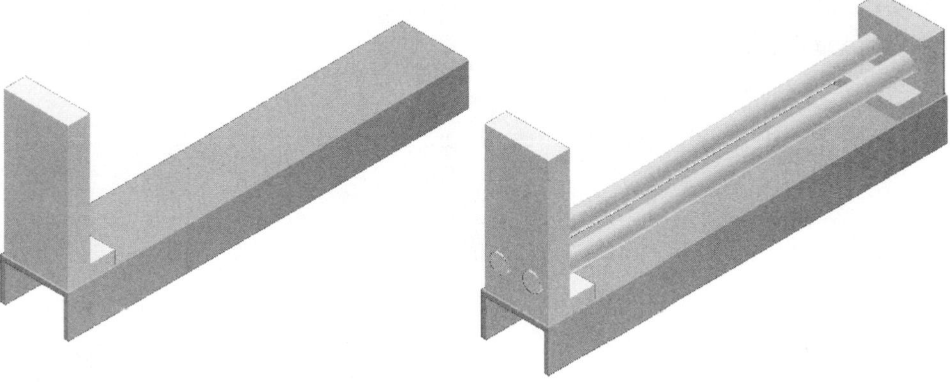

Figure 9.122

THE FRAME GENERATOR

The Frame Generator allows you to design platforms, equipment racks, and other types of structural frames. You can easily create and edit structural members. The Frame Generator has its own set of tools, which can be accessed from the Assembly or Weldment panel bar. Clicking the arrow located on the Assembly Panel will expose the Frame Generator, as shown in the following image on the left. Clicking Frame Generator displays the tools specific to this application. Before inserting structural members into an assembly, you must have a skeletal model to work from. The skeletal model acts as a guide in determining the direction in which the frame members will be placed in the assembly. An example of a structural frame assembly is shown in the following image on the right.

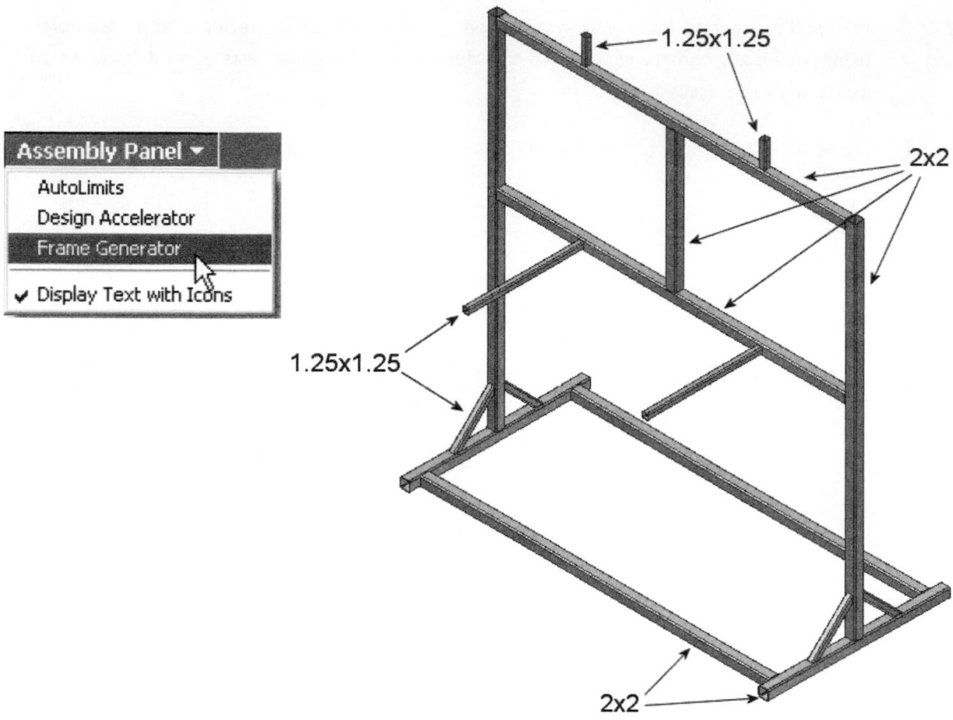

Figure 9.123

Since the Frame Generator operates in the assembly environment, any skeletal models used for generating the frames must first be created as a 2D or 3D sketch and then inserted into the assembly model. A typical example of a skeletal model is shown in the following image on the left. After the frame generator is loaded, a menu of commands is displayed, as shown in the following image on the right. To create frame members, click the Insert button.

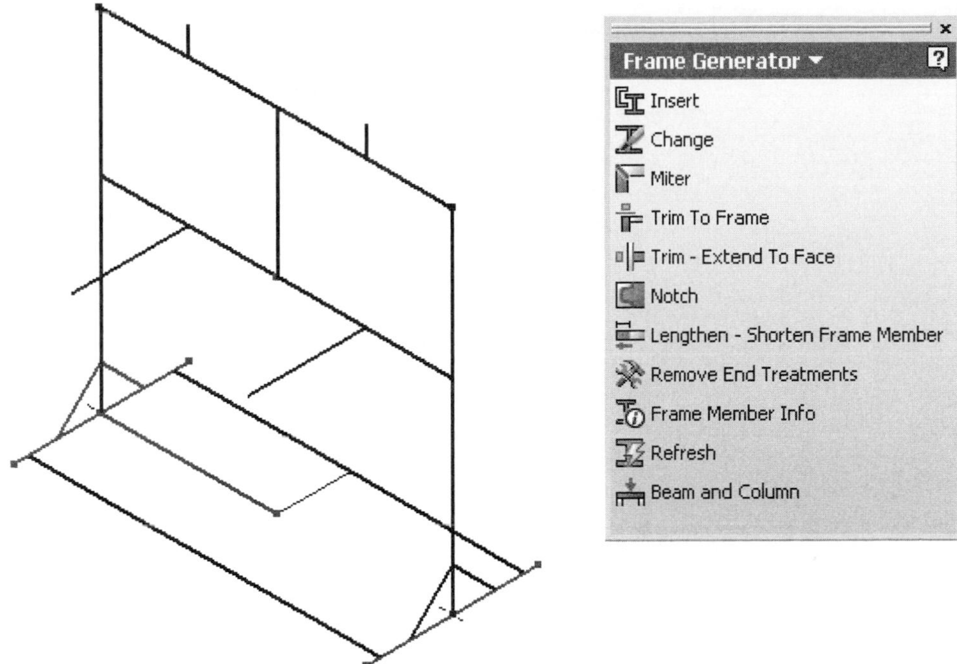

Figure 9.124

Clicking the Insert button in the panel bar will display the Insert dialog box, as shown in the following image. The following changes were made in the Frame Member Selection area in this dialog box: Standard=ANSI; Type=ANSI AISC HSS (Square); Size=2x2x1/8. The remaining default settings were left as is. Under the Orientation area, verify that the Insertion Point of the structural member is in the center of the shape. Keep all default offset and rotation values as shown. The edges of the skeletal frame were selected, as shown in the following image on the right. When finished, click OK.

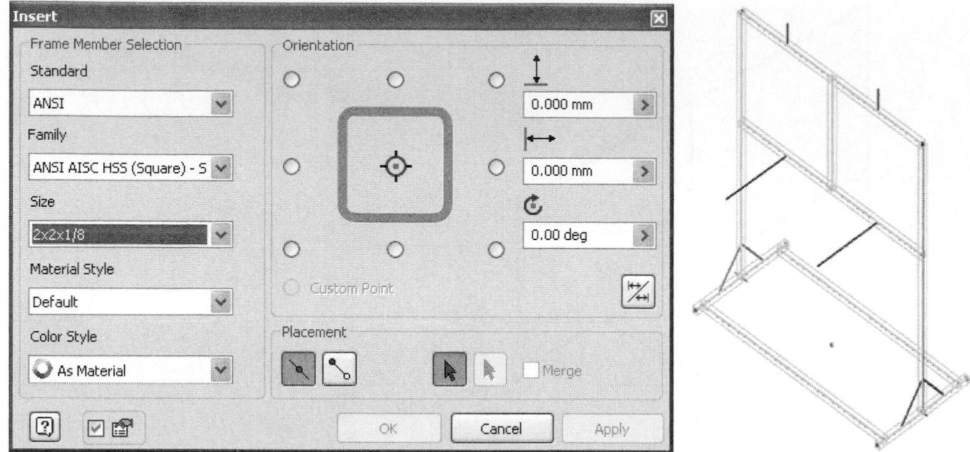

Figure 9.125

Clicking OK will launch the Create New Frame dialog box, as shown in the following image on the left. Here you define a new file location that will hold all of the frame members. Click OK in this dialog box to activate the Frame Member Naming dialog box, as shown in the following image on the right. You can keep the default display names and file names, or you can make changes to them that will be reflected in the browser.

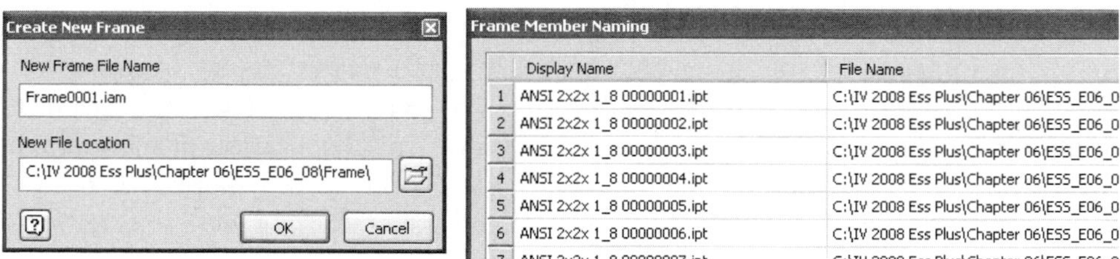

Figure 9.126

When you return to the assembly, all frame members should be properly placed. To aid in editing the frame members, turn off the skeletal frame, as shown in the following image.

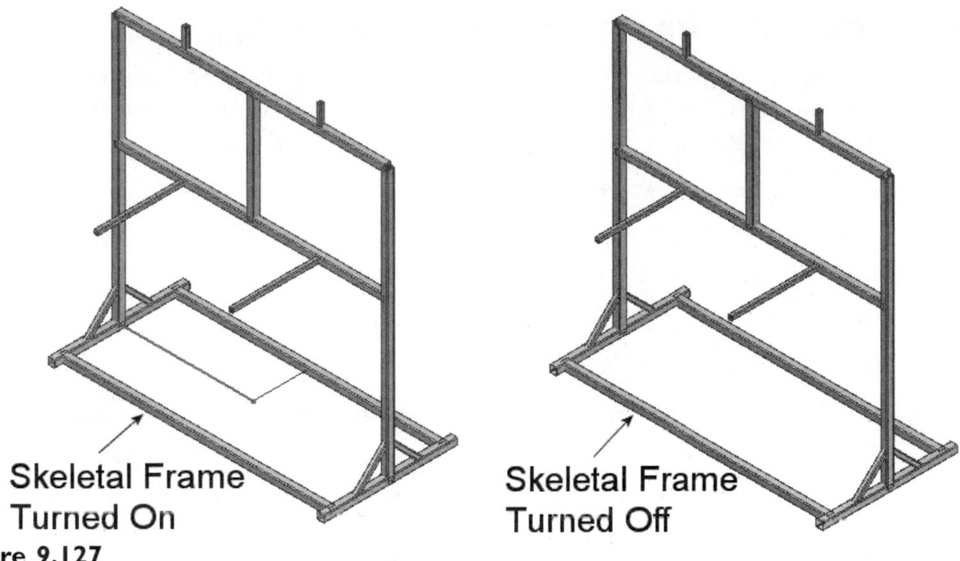

Figure 9.127

USING SOLIDS FOR FRAME GENERATION

The previous example of generating frame members relied on a wireframe skeletal model. You can also create skeletons from solids and surface bodies. In the following image of a truck bed extender, a solid model that outlines the basic shape of the extender was created, as shown in the following image on the left. Additional sketch lines were added for the vertical frame members. A new assembly file was created, and the solid skeleton was placed as the first component. As shown in the following image in the middle, frame members were generated by clicking the edges of the solid model in addition to the single line sketch objects. Once all frame members are generated, the solid skeleton is turned off, leaving the frame as shown in the following image on the right.

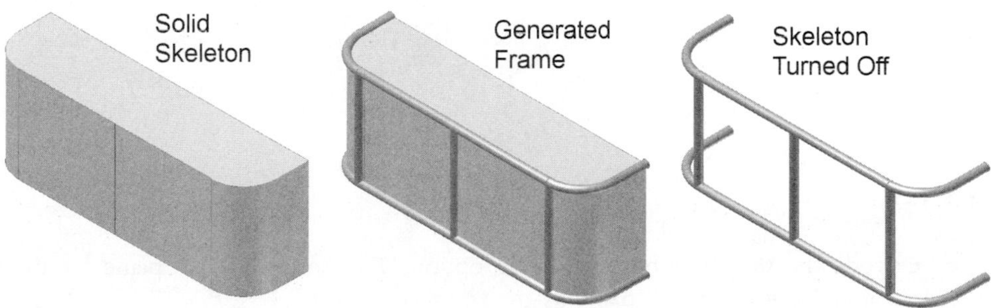

Figure 9.128

PUBLISH CUSTOM SHAPES FOR FRAME GENERATOR

If the shape (profile) that you need to use in the frame generator does not exist as a standard shape you can create your own shape(s). The shapes can be created from standard parts or iParts. If an iPart is used you can have a choice of the sizes.

PUBLISH A SHAPE FROM A STANDARD PART

To publish a shape from a standard part (a standard part has only one size option) to use in the frame generator follow these steps.

1. Create a new part file based on the required standard.
2. Sketch, constrain, and dimension a sketch as needed.
3. Create a work plane with an offset of zero based on the origin plane that the sketch was drawn on. If desired you can rename this work plane to Start.
4. Create a work plane that is offset from the work plane created in step 3. The offset distance has to affect the shape when it is used in the frame generator. The following image shows an example of a sketch with the two work planes. If desired you can rename this work plane to End.

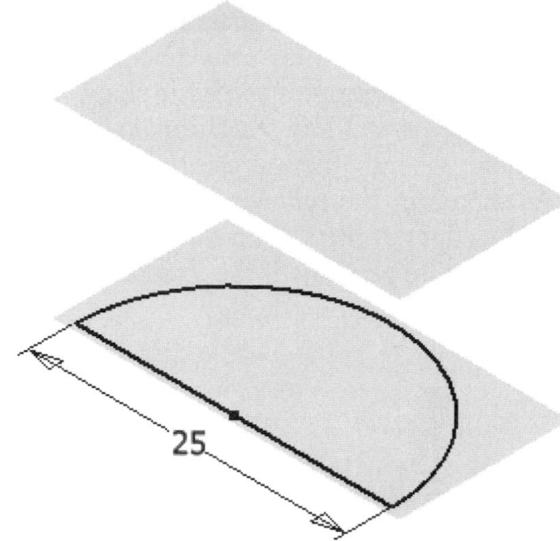

Figure 9.129

5. Extrude the sketch with the From To option. Use the Start work plane for the first plane and the End work plane as the terminating plane.

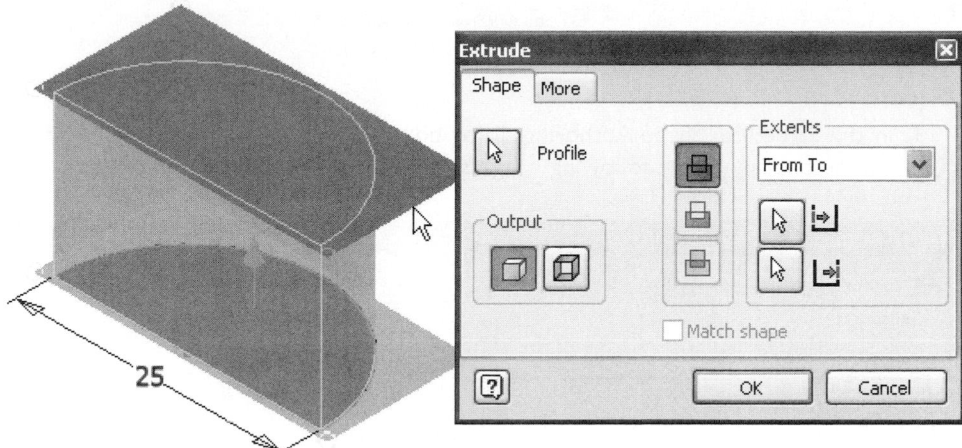

Figure 9.130

6. Click the Parameters tool from the Part Features panel bar.
7. In the Parameters dialog box rename the Model Parameter name of the length of the extrusion feature to **Base_Length.** This parameter can have a different name; however, Base_Length is consistent with the mapping when the shape is published.

Parameters

Parameter Name	Unit	Equation
Model Parameters		
d0	mm	25 mm
d1	mm	0 mm
Base_Length	mm	15 mm
d3	deg	0 deg
User Parameters		

Figure 9.131

8. Turn off the visibility of the work planes.
9. If desired change the part number via the iProperties. The part number will be displayed in a list when the shape is placed via the frame generator.
10. Save the part.
11. Before authoring the shape ensure that you have a Read/Write library.
12. Next start to publish the shape. Click the Structural Shape Authoring tool from the Part Features panel bar.

Figure 9.132

13. In the Structural Shape Authoring dialog box click the category to which the shape will be published. The following image shows the Other category selected.

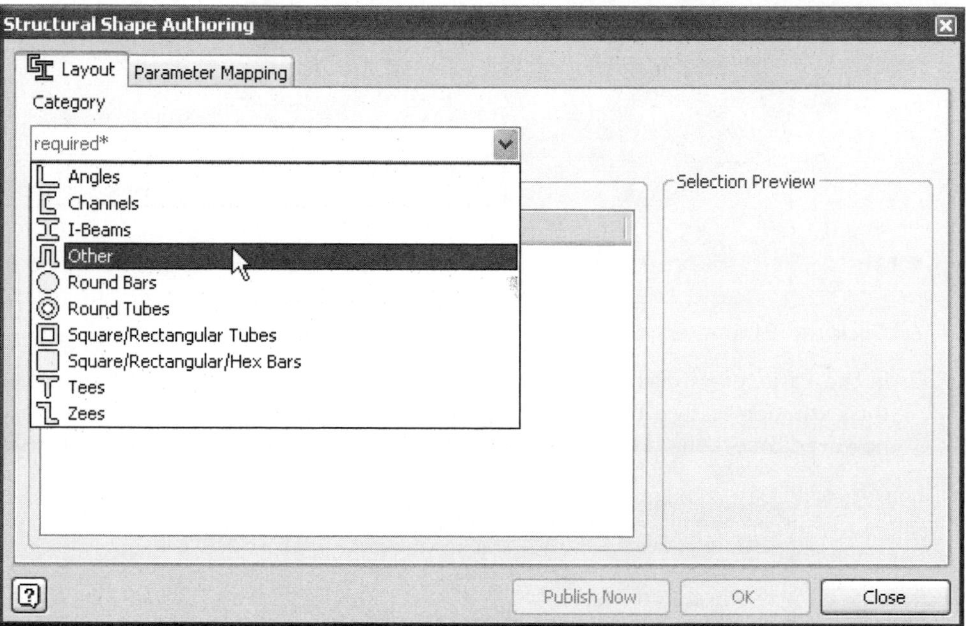

Figure 9.133

14. In the Geometry Mapping area the Base Extrusion will automatically be selected if there is a single extrusion. If there are multiple features select the feature to use as the base feature.

15. The Default Base Point that will be used as the insertion point will default to Pre-defined Point (the geometric center of the profile). You can select a point in the sketch with the Select Geometry option as shown in the following image on the left. As you move the cursor in the graphics window valid points will highlight in red like what is shown in the following image on the right.

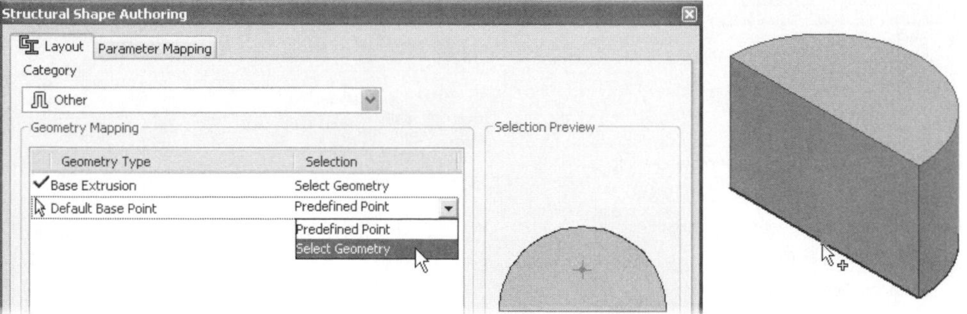

Figure 9.134

16. Next map the parameters. The Base Length parameter is the only required parameter but others can be mapped as needed. The Base Length parameter will be used in the bill of material to display the length of the frame members.

- Click the Parameter Mapping tab
- In the Base Length row click the button with the three dots
- The Part Template Parameters dialog box will be displayed
- Expand the Parameters section
- Click the Base_Length parameter as shown in the following image
- Click OK to close the Part Template Parameters dialog box

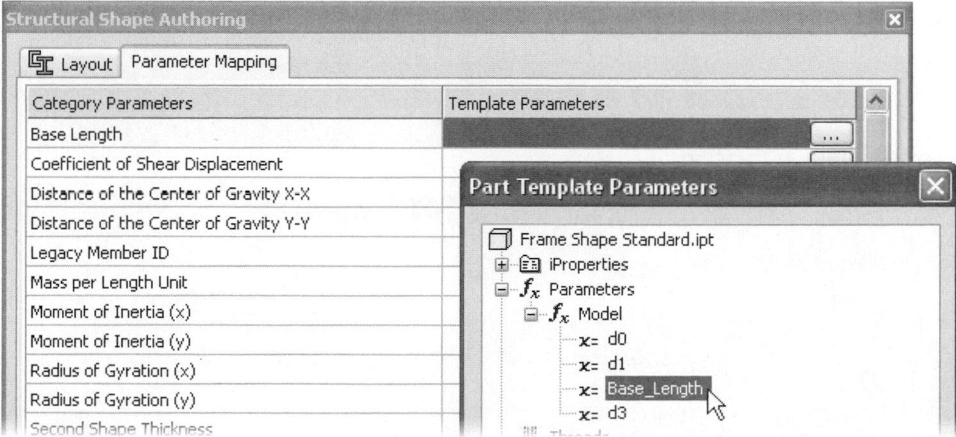

Figure 9.135

17. Next publish the shape. Click the Publish Now button on the bottom of the dialog box.

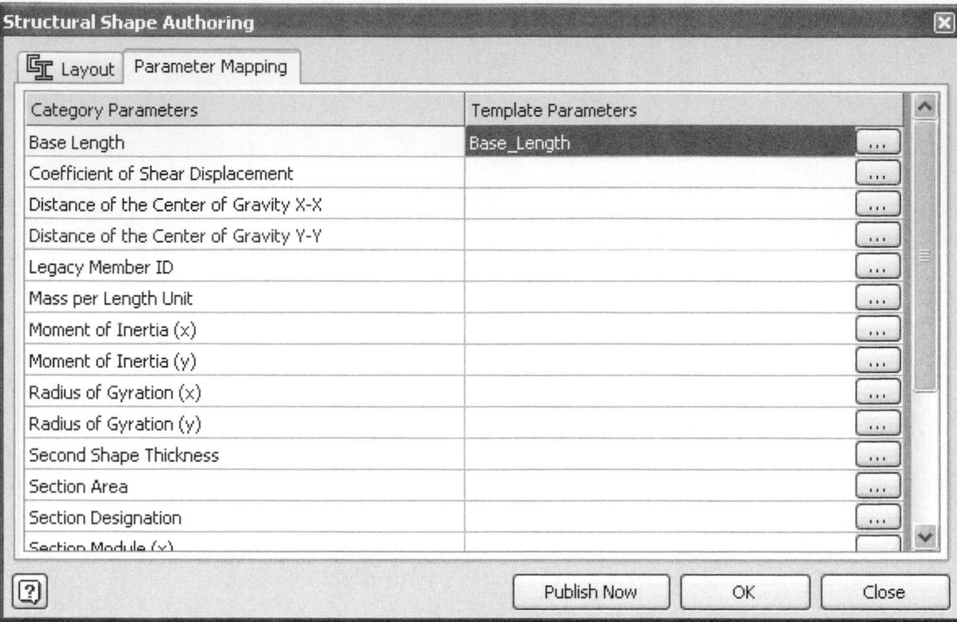

Figure 9.136

18. Click OK in the dialog box alerting you that there has been a change to the part. Inventor automatically created a few User Parameters.
19. In the Publish Guide dialog box select the Library and language to which to publish. You must have Read/Write access to the library. Click Next when done.

Figure 9.137

20. In the Publish Guide dialog box select the category to publish if different than the current category. Click Next when done.

Figure 9.138

21. In the Publish Guide dialog box verify that the needed parameters are mapped. Click Next when done.

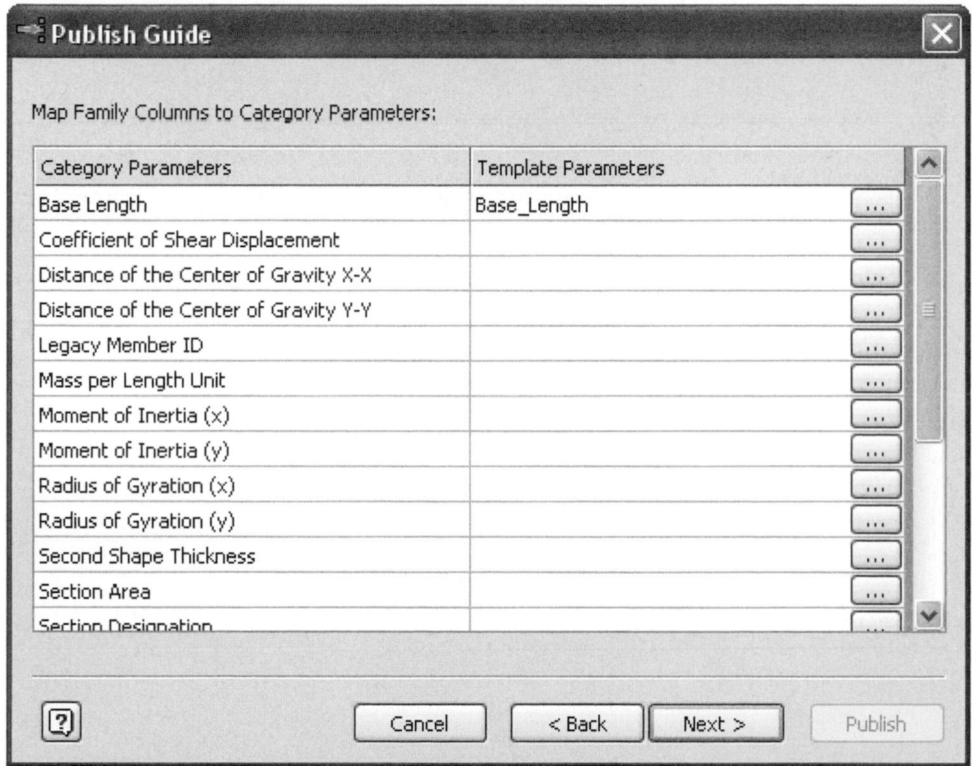

Figure 9.139

22. Next you define the key values that uniquely define a family member. At least one Key must be defined. Either double-click or click the desired key in the Table Column and click →. The following image shows the PARTNUMBER and Base Length added as key columns. Click Next when done.

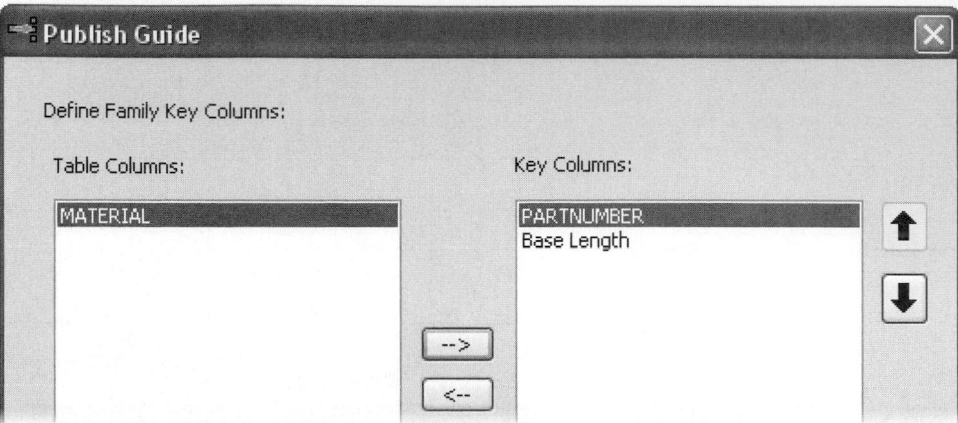

Figure 9.140

23. Enter in a name for the family. The name will appear in the Size area of the Frame Generator – Insert dialog box.
24. Enter a description; it will appear in the Family area of the Frame Generator – Insert dialog box. Enter other data as needed (they are not required). Click Next when done.
25. Select a Standard organization. If the shape does not belong to a Standard, enter custom as shown in the following image or enter another name.

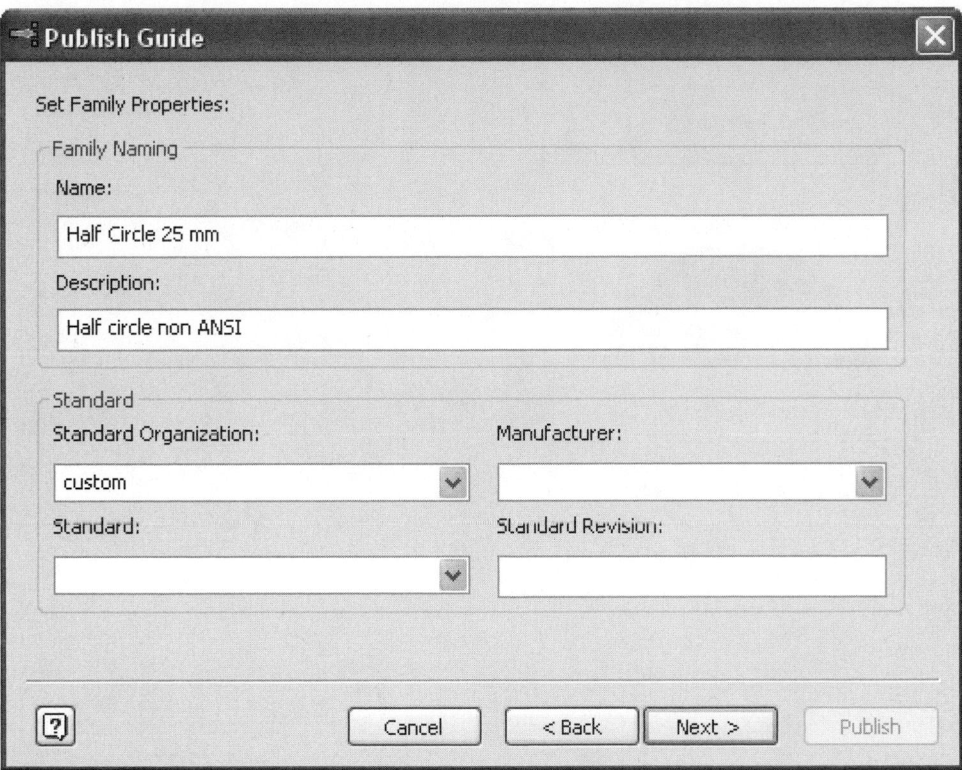

Figure 9.141

26. Lastly you can define a different thumbnail image or accept the default image.
27. To complete the publishing operation click Publish.

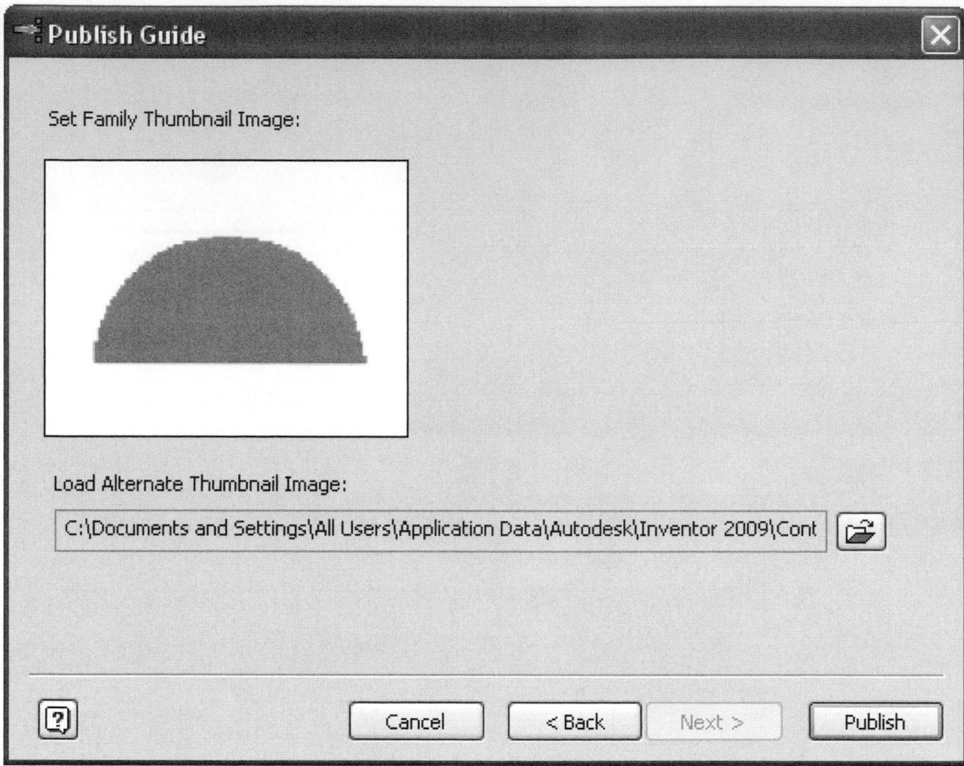

Figure 9.142

28. Use the Frame Generator - Insert tool to use the shape for a frame member. The following image shows the Half Circle shape being inserted.

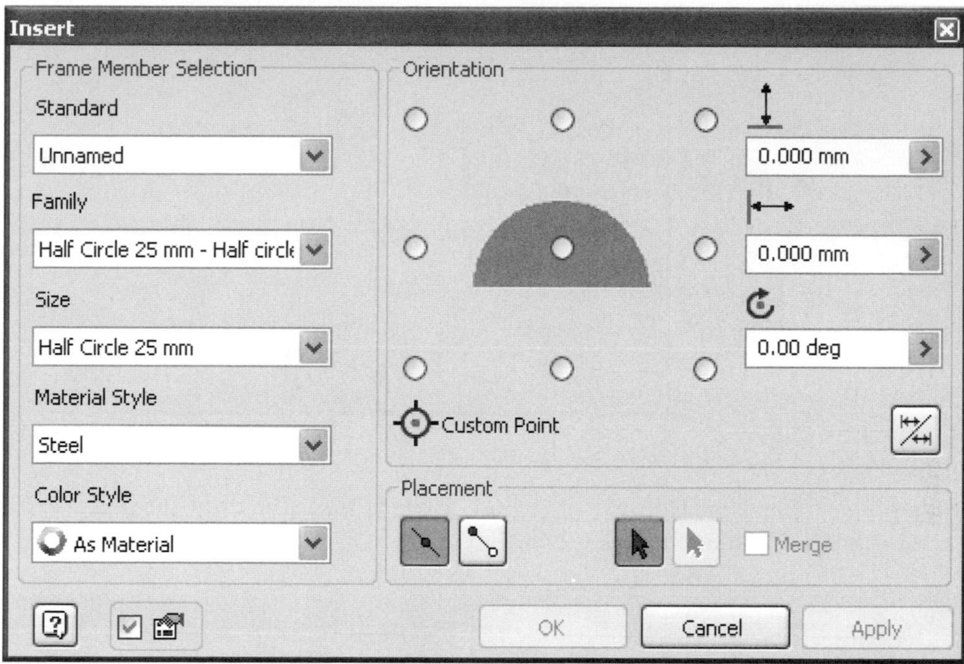

Figure 9.143

PUBLISH A SHAPE FROM AN IPART

To publish a shape from an iPart (a published iPart has multiple size options) to use in the frame generator follow the same steps as publishing a standard part with a couple of additional steps.

The biggest difference is that you must first create an iPart that has a key column defined. This column will be used in the list of sizes when a frame member is selected. The following image shows an iPart with a Diameter parameter defined as the key column and the Base_Length parameter added. You can have as many rows as required, but ensure that each cell in the key column has a unique name.

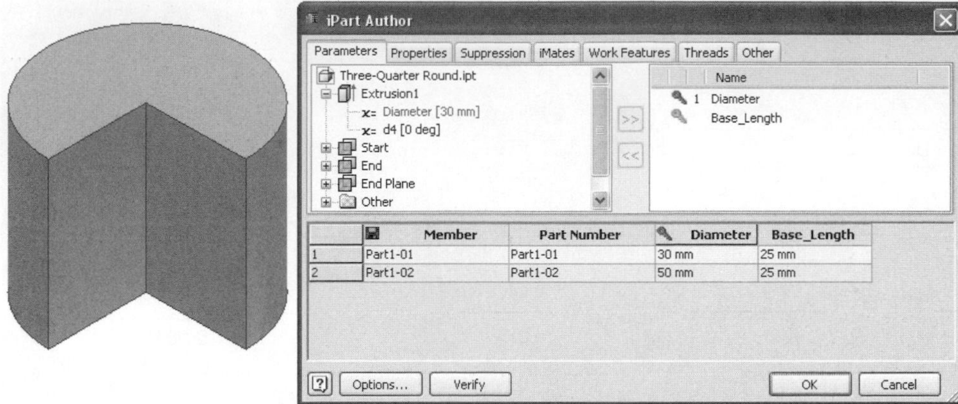

Figure 9.144

When mapping the parameters select the Base_Length from the drop-down list instead of the Part Template Parameters dialog box.

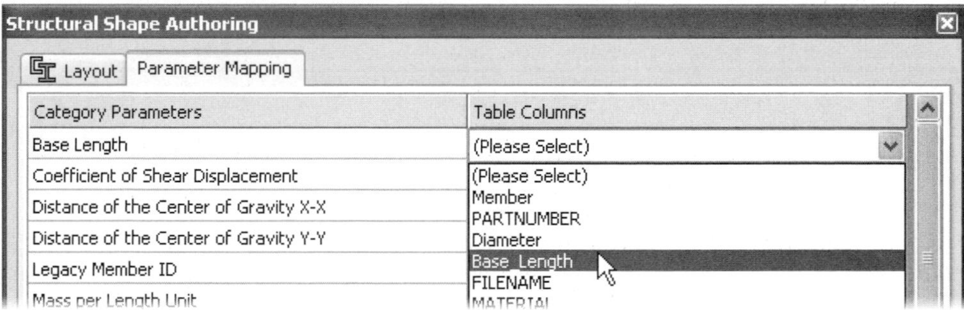

Figure 9.145

EXERCISE 9–5: USING THE FRAME GENERATOR

This exercise walks you through the creation of a frame. You will use an existing part file as a skeletal model when generating all frame members. You will then edit the frame members by trimming them to size. Finally, you will turn off the visibility of the skeleton to expose only the frame members.

1. Open the file *ESS_E09_05.iam* in the Chapter 09 folder.
2. Activate the Frame Generator by clicking the drop-down arrow in the Assembly panel bar and clicking Frame Generator as shown in the following image in the middle.
3. Click the Insert button, as shown in the following image on the right.

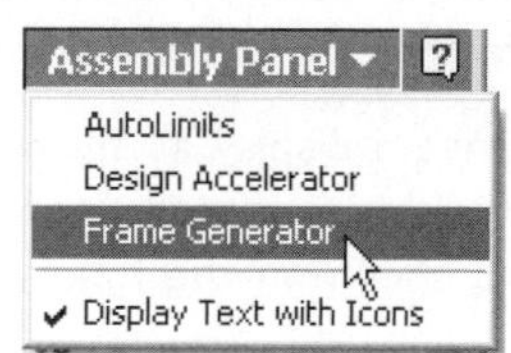

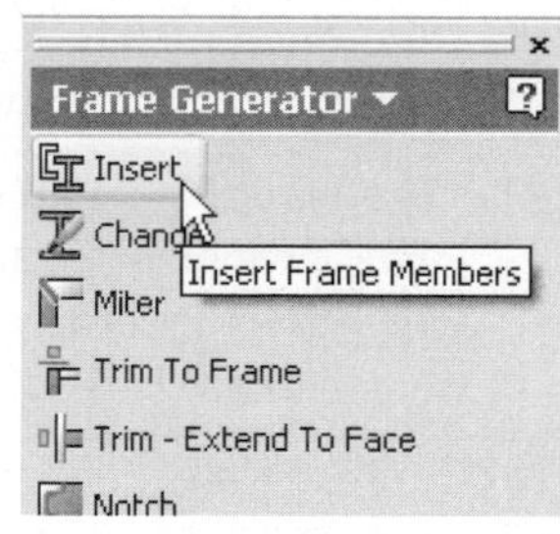

Figure 9.146

4. Make the following changes under the Frame Member Selection area of the Insert dialog box:

- Standard=ISO
- Type=ISO 657-1 hot rolled steel sections
- Size=L20x20x3.
- Keep the remaining default settings.

5. Under the Orientation area, verify that the insertion point of the structural member is in the lower-left corner of the shape. Keep all default offset values as shown in the following dialog box.

6. Click the top four edges of the skeletal frame, as shown in the following image on the right. If the frame members are not aligned, as shown in the following image on the right, click the Mirror Frame Member button to change the alignment. If mirroring does not fix the alignment, you may also have to experiment with different angle settings. When finished, click OK.

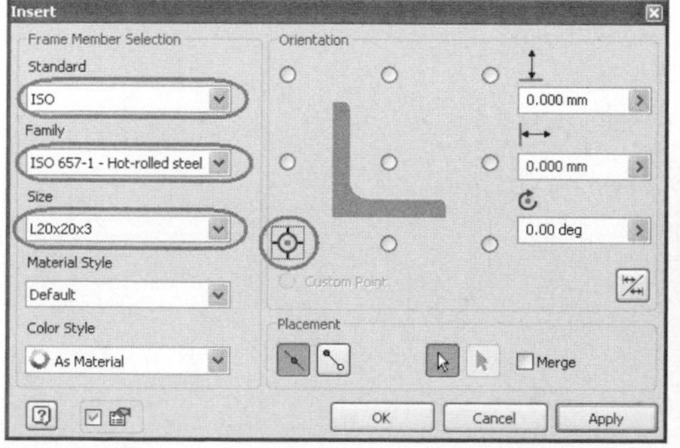

Figure 9.147

7. When the Create New Frame dialog box appears, as shown in the following image on the left, click the OK button to accept these defaults. These items dictate where the frame members will be located when created. When the Frame Member Naming dialog box appears, click the OK button to accept the display names and file names.

8. The resulting frame members display, as shown in the following image on the left, in their proper orientations. Notice also how the frame components are displayed in the Assembly browser, as shown in the following image on the right.

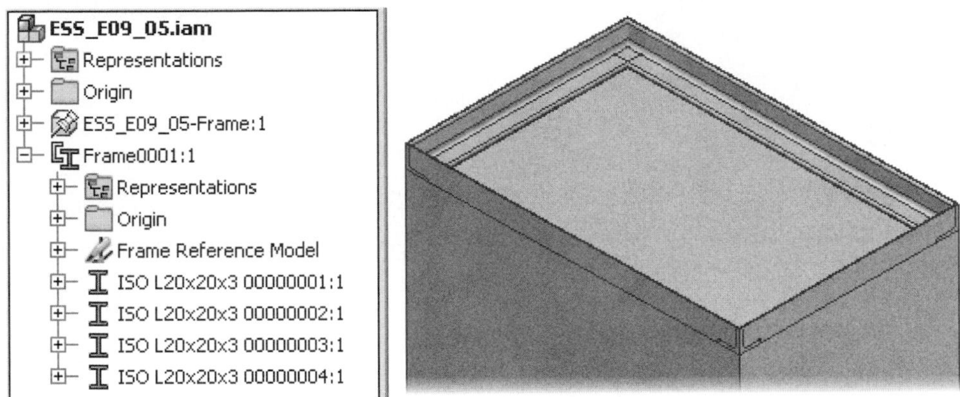

Figure 9.148

9. Use the Orbit tool to adjust the viewpoint as shown in the following image.

10. Click the Insert tool in the Frame Generator dialog box.

11. Keep all information the same under the Frame Member Selection area. Click the four bottom edges of the solid frame, as shown in the following image. Use the Mirror Frame Member button when needed to change any alignment issues. When finished, click the Apply button and accept the default names. This will create the frame members and keep you in the Insert dialog box so that you can place other members.

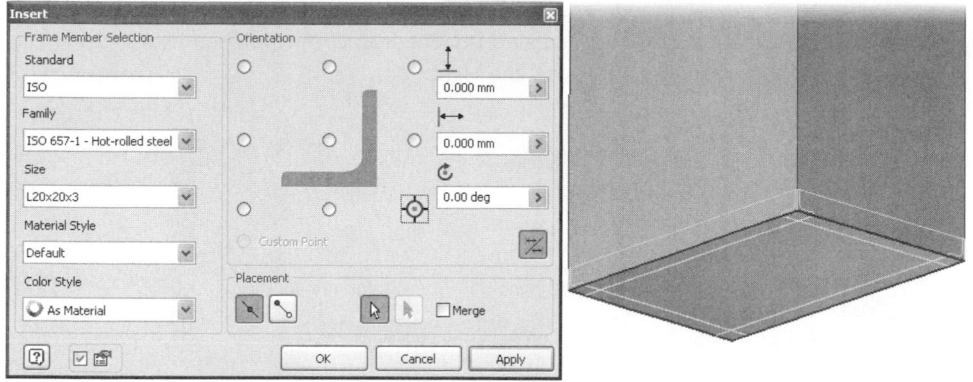

Figure 9.149

12. Press the F6 key to change the viewpoint to the Home View.

13. While still in the Insert dialog box, place the frame members using the vertical edges of the skeletal model. The frame members need to face inward. You will probably have to create each individual member separately. Use the Mirror Frame Member button and even a different angle setting to properly orient the components. When finished with each individual member, click the Apply button to remain in the Insert dialog box and accept the default names. When you have finished placing all four vertical frame members, as shown in the following image, click the OK button to exit the Insert dialog box.

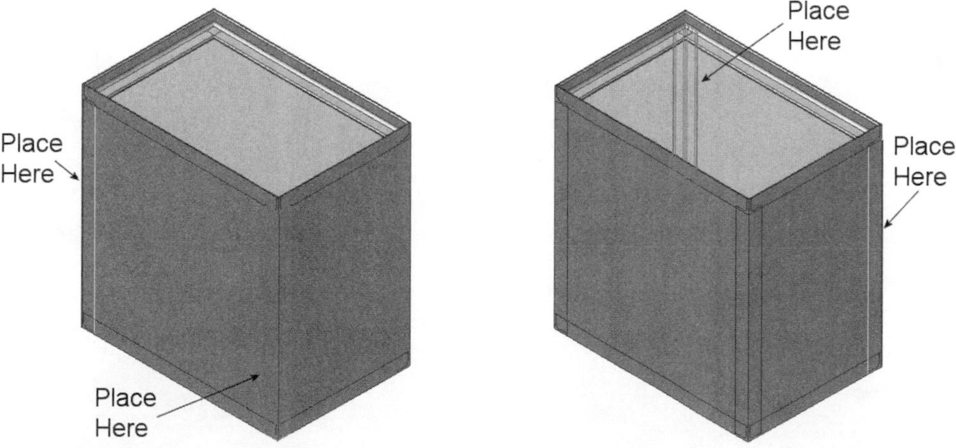

Figure 9.150

14. This completes the frame-generation process where an existing solid model was used to create the frame members, as shown in the following image on the left.

15. As it might seem difficult to interpret the appearance of the frame along with the solid model, identify *ESS_E09_05-Frame:1* in the browser, right-click, and pick Visibility

from the menu to remove the checkmark, as shown in the following image in the middle. With the solid model turned off, the frame can be seen more clearly, as shown in the following image on the right.

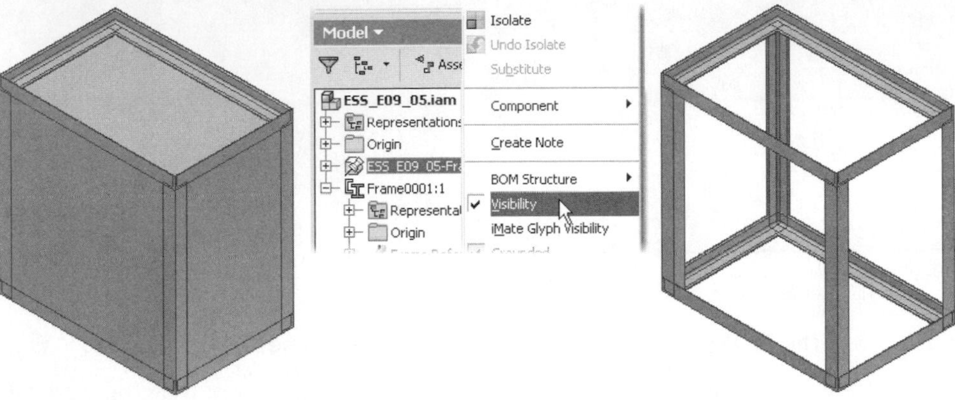

Figure 9.151

16. The existing frame has a number of members that overlap each other. Numerous frame members will need to be trimmed in order to produce the final assembly. To trim a member to a frame, click the member labeled (A); this is the member to trim to. Next click the member labeled (B); this is the member that will be trimmed, as shown in the following image.

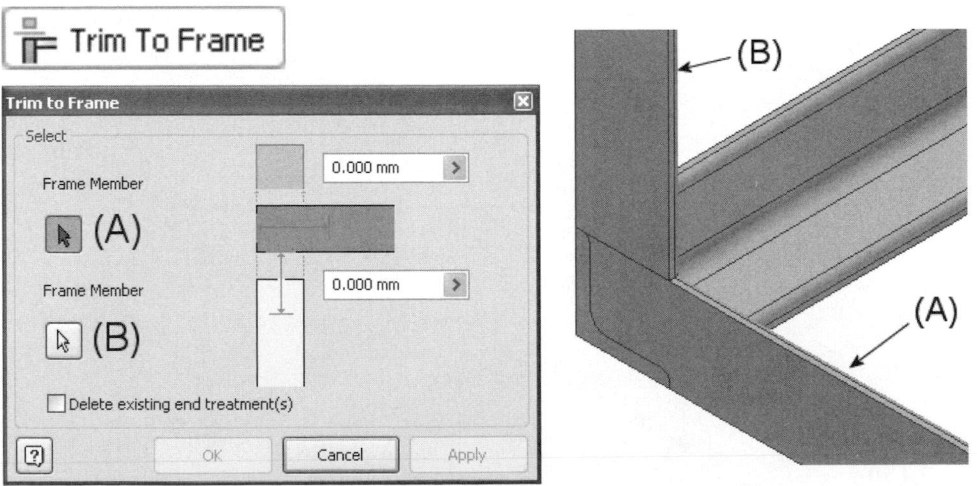

Figure 9.152

17. Continue picking individually, bottom frame members first, followed by vertical members. Your completed trim-to-frame operation should appear similar to the following image.

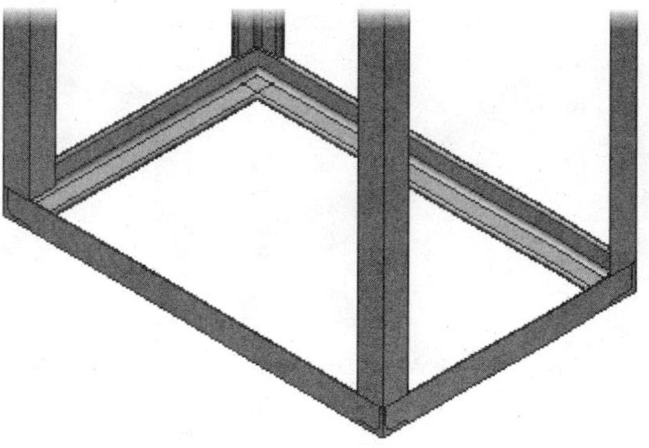

Figure 9.153

18. Next perform a mitering operation by clicking the Miter button located in the Frame Generator panel bar, as shown in the following image on the left. Click member (A) and then member (B), as shown in the following image on the right, although the order for the miter operation is not critical. Continue by mitering the remaining corners that represent the upper and lower frame.

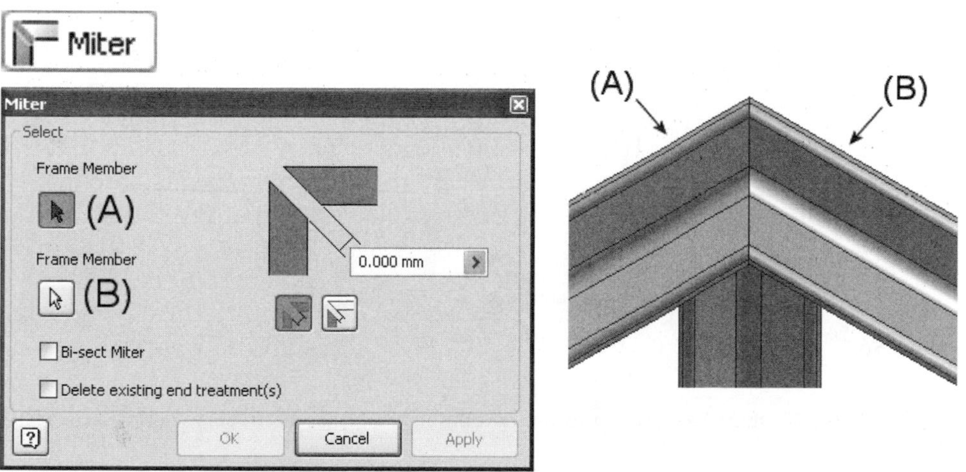

Figure 9.154

19. The finished cart frame is illustrated in the following image.

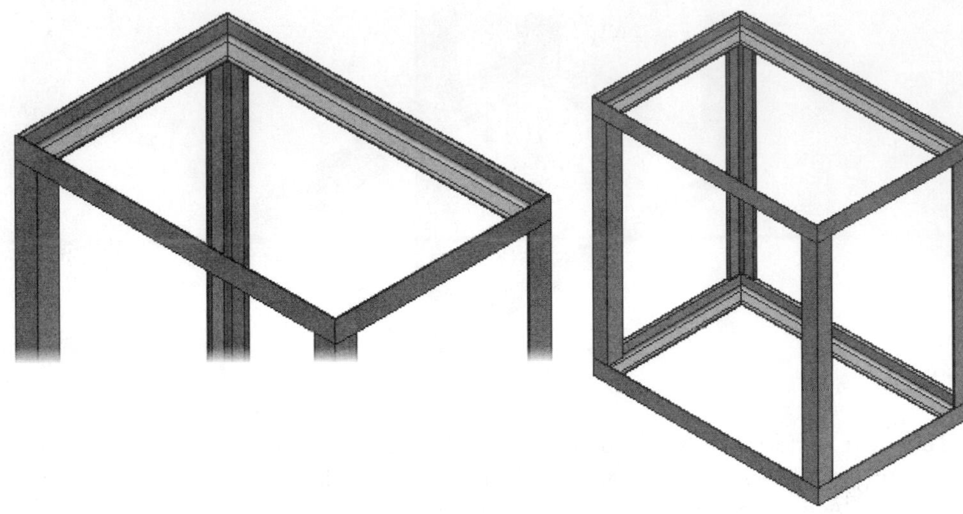

Figure 9.155

20. Close all open files. Do not save changes. End of exercise.

CONTENT CENTER

Autodesk Inventor's Content Center contains a number of standard components that are installed with the application. It contains thousands of parts such as screws, nuts, bolts, washers, pins, and so on. You can place these standard components into an existing assembly using the Place from Content Center tool, as shown in the following image.

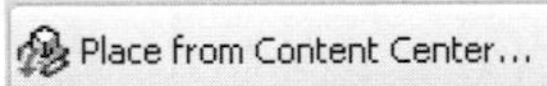

Figure 9.156

Note: You must be logged in to the Autodesk Data Management Server to access the Content Center.

After the tool is selected, the Place from Content Center dialog box opens, as shown in the following image, and you can navigate between parts and features that are either included or published to the Content Center. Select the item you want to place, click OK, and place the part as you would any other assembly component.

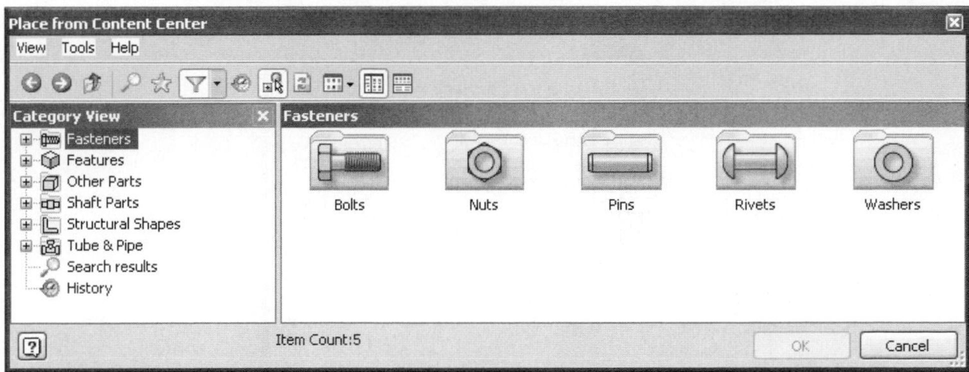

Figure 9.157

In addition to using the default items in the Content Center, you can publish your own features and parts to the Content Center. You publish parts and features using the Publish Feature, Publish Part, or Batch Publish tools, as shown in the following image. These tools are accessed from the Tools menu when a part file is open.

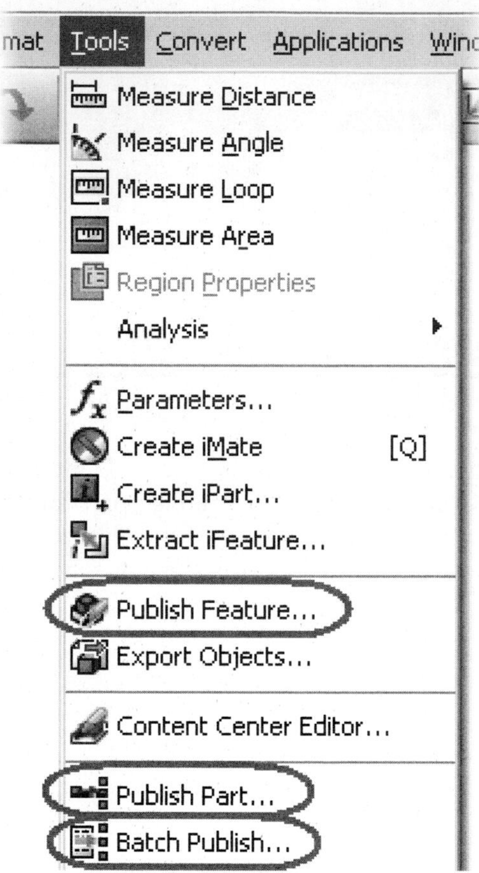

Figure 9.158

In order to publish content to the Content Center, a read/write library must exist and be added to the active project file.

Note: For more information about the Content Center, publishing features, publishing parts, or configuring libraries, refer to Autodesk Inventor's Help system.

DESIGN ACCELERATOR

Autodesk Inventor's Design Accelerator tools enable you to quickly create complex parts and features based on engineering data such as ratio, torque, power, and material properties. The Design Accelerator consists of component generators, mechanical calculators, and the Engineer's Handbook.

While working in an assembly the Design Accelerator tools are accessed from the Design Accelerator panel bar. In the Assembly panel click the down arrow at the top of the panel bar and click Design Accelerator as shown in the following image on the left. The panel bar's tools will change as shown in the following image on the right.

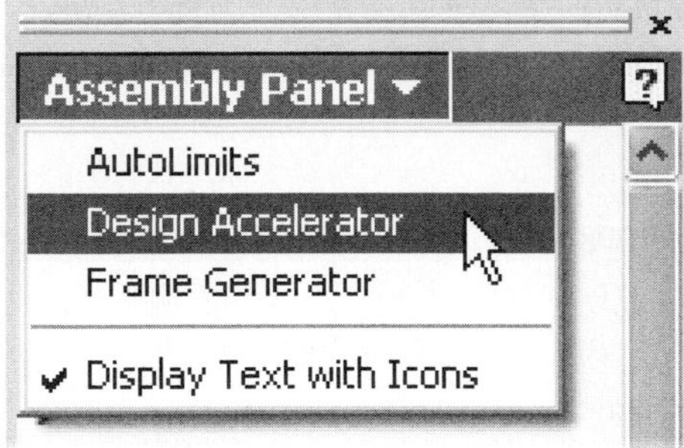

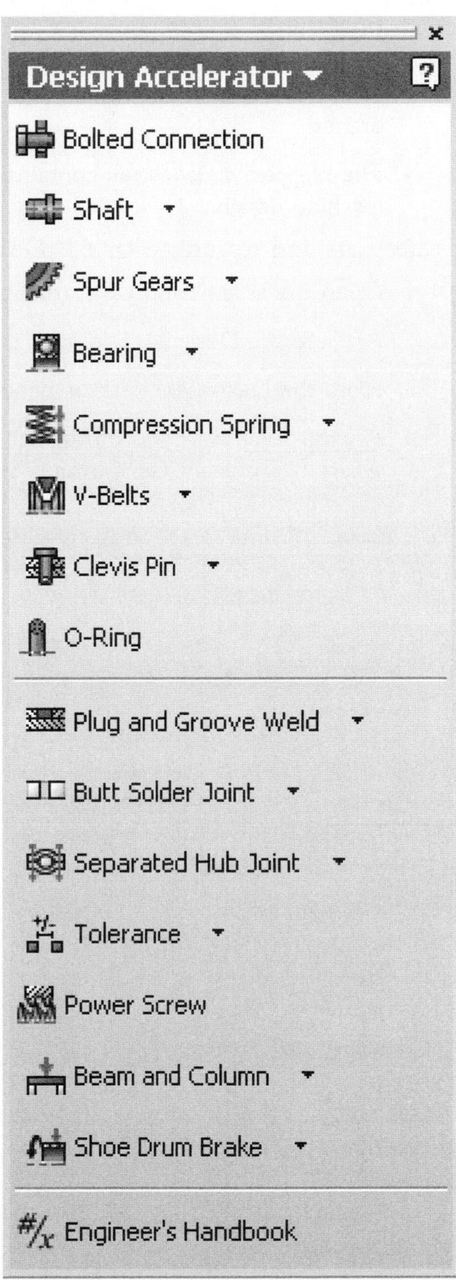

Figure 9.159

- Component and feature generators, located at the top section of the panel bar, consist of bolted connections, shafts, gears, bearings, springs, belts, and pins.

- Mechanical calculators, located in the middle section of the panel bar, use standard mathematical formulas and theories to help you design components. The calculators available include weld, solder, hub joints, fit and tolerance, power screws, beams, and brakes.
- The Engineer's Handbook contains engineering theory, formulas, and algorithms used in machine design.

To create parts and features with the Design Accelerator, follow these steps:

1. Open the assembly in which to place the part.
2. Activate the Design Accelerator panel bar.
3. From the Design Accelerator panel bar, select the operation you need to perform.
4. A wizard will walk you through the steps of the operation. An example of the Spur Gears Component Generator is shown in the following image.

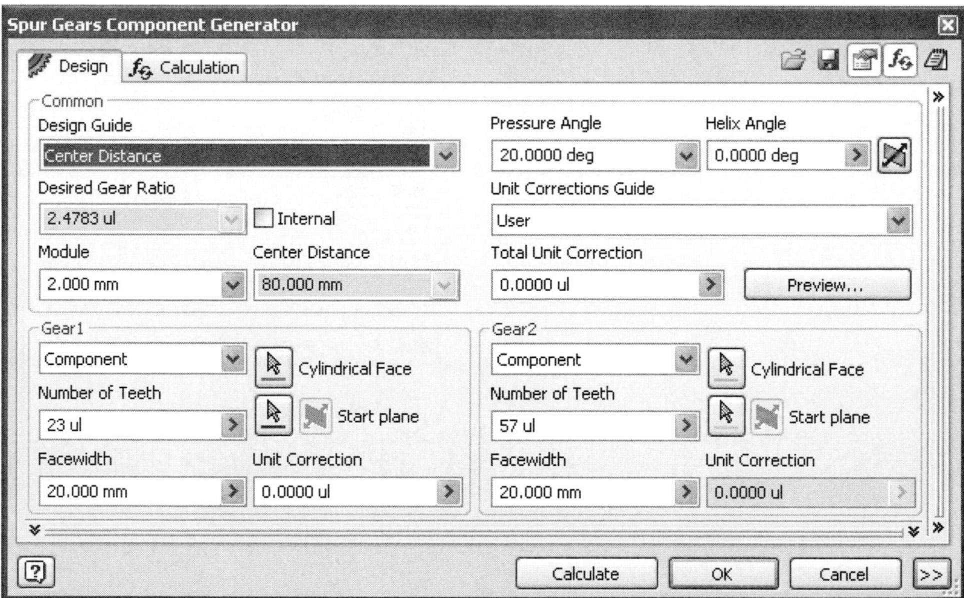

Figure 9.160

Note: For more information about the Design Accelerator, refer to Autodesk Inventor's Help system.

Applying Your Skills

SKILL EXERCISE 9–1

In this exercise, you will create a contact set, and then use the Contact Solver to simulate intermittent motion.

1. To begin, open *ESS_E09_06.iam* in the Chapter 09 folder; see the following image.

Figure 9.161

2. In the browser, right-click ESS_E09_06-0208-06P:1, and click Visibility to turn off the front plate, as shown in the following image.

Figure 9.162

3. In the browser, expand ESS_E09_06-0802-03P:1.
4. Right-click the Drive This angle constraint, and click Drive Constraint, as shown in the following image.

Figure 9.163

5. Click the Forward arrow. Notice that the roller interferes with and has no effect on other components in the assembly. This is because the components are not included in a contact set. Close the Drive Constraint dialog box.
6. If needed set the value for the Drive This angle constraint to **90.00 deg**.

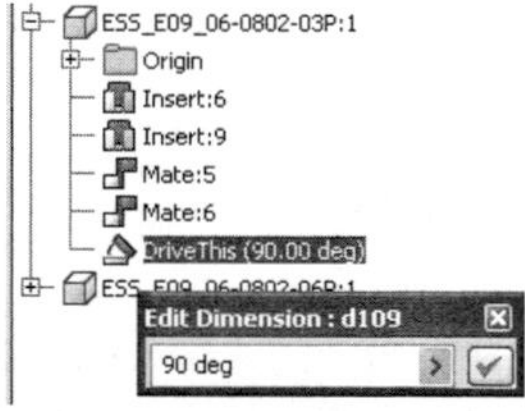

Figure 9.164

7. In the browser, select ESS_E09_06-0208-01P:1, DIN125-1 A A 8.4:1, ESS_E09_06-0802-03P:1, and ESS_E09_06-0802-06P:1.
8. Right-click one of the selected components, and select Contact Set from the menu. Notice the contact set icon that is added next to the selected assembly components.

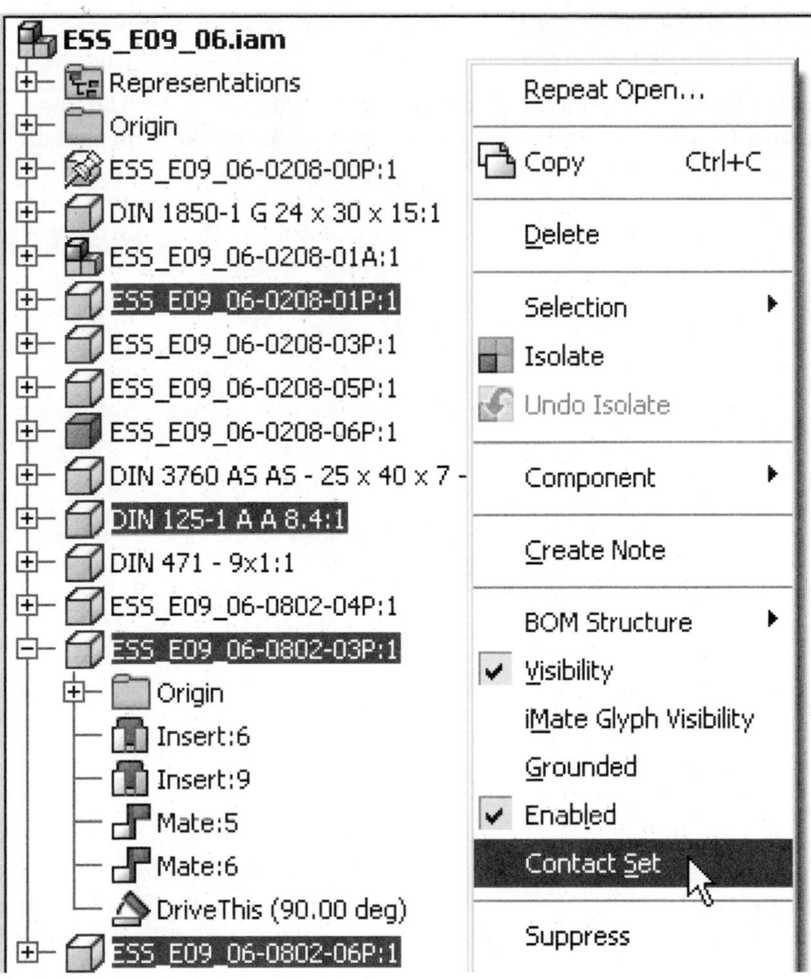

Figure 9.165

9. Click Tools > Document Settings > Modeling tab. In the Interactive contact area, click Contact Set Only. Click OK.

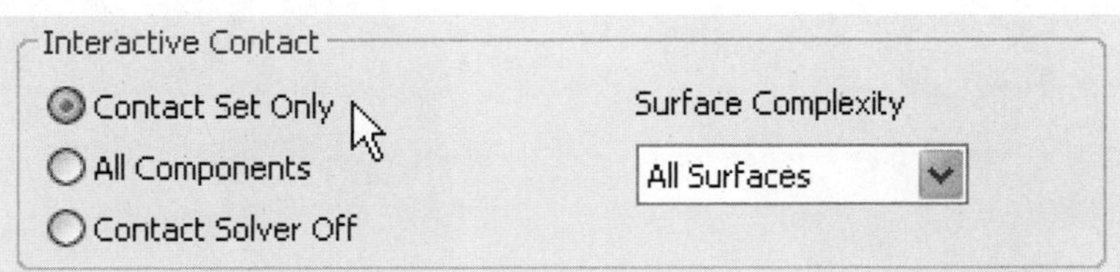

Figure 9.166

10. In the browser, expand ESS_E09_05-0802-03P:1.

11. Right-click the Drive This angle constraint, and click Drive Constraint as shown in the following image on the left.

12. In the Drive Constraint dialog box, the Start value should be **180**, the End value should be **2000**, the Increment value should be **3**, and the Repetitions value should be **2** as shown in the following image on the right.

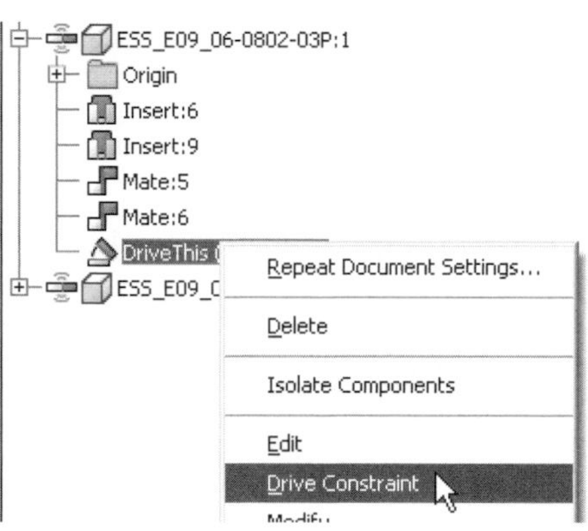

Figure 9.167

13. Click the Forward arrow. The roller moves to the 180° position and revolves until it contacts a slot in the follower component, as shown in the following image on the left. The motion continues while there is contact, as shown in the following image on the right.

Figure 9.168

14. Close the Drive Constraint dialog box.
15. In the browser, right-click the Drive This angle constraint, and select Suppress from the menu.
16. Click and drag component ESS_E09_06-0802-03P:1. In addition to driving a constraint to work with components of a contact set, you can also "constrain-drag" components to view how components will behave when contact occurs and when the components are included in a contact set.
17. Close all open files. Do not save changes. End of exercise.

CHECKING YOUR SKILLS

Use these questions to test your knowledge of the material covered in this chapter.

1. **True _ False _** A Design View Representation can control the display style, such as shaded or wireframe, of an assembly model.
2. Which of the following are types of design view representations?
   - **a.** Dynamic
   - **b.** Public
   - **c.** Private
   - **d.** Static
3. What is the purpose of a positional representation?

   ____

   ____

4. When a new positional representation is added to the browser, how many are created by default?
   - **a.** 1
   - **b.** 2
   - **c.** 3
   - **d.** 4
5. When would you want to make an assembly have the Flexible property?

   ____

   ____

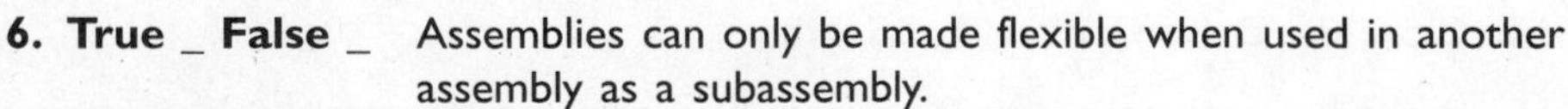

6. **True** _ **False** _ Assemblies can only be made flexible when used in another assembly as a subassembly.

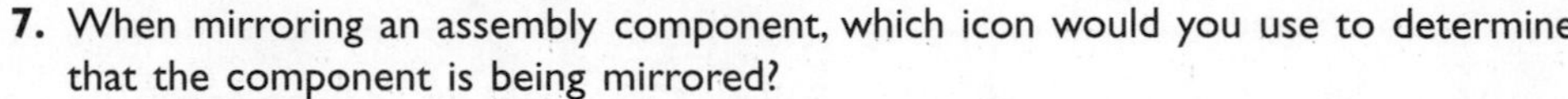

7. When mirroring an assembly component, which icon would you use to determine that the component is being mirrored?

   a. 

   b. 

   c. 

   d. 

8. What is the purpose of the Capacity Meter, and what information does it display?

   ____

   ____

9. **True** _ **False** _ Creating features in the context of an assembly will update the individual parts that make up the assembly automatically.

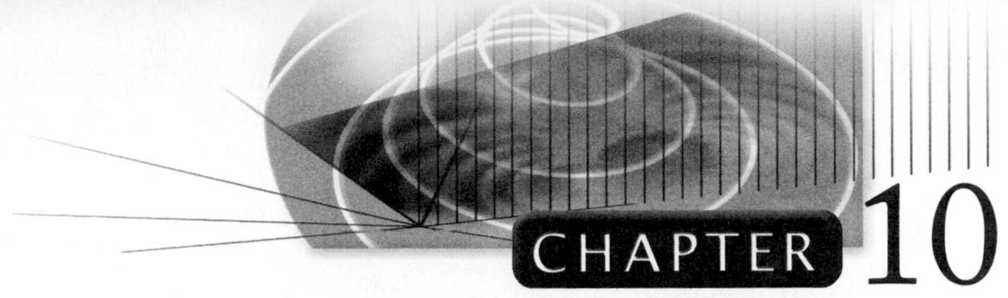

Sheet Metal Design

INTRODUCTION

In this chapter, you will learn how to create sheet metal parts. Assemblies often require components that are manufactured by bending flat metal stock to form brackets or enclosures. Cutouts, holes, and notches are cut or punched from the flat sheet, and 3D deformations such as dimples or louvers are often formed into the flat sheet. The punched sheet is then bent at specific locations using a press brake or other forming tools to create a finished part. The sheet metal environment in Autodesk Inventor presents specialized tools for creating models in both the folded and unfolded states.

OBJECTIVES

After completing this chapter, you will be able to perform the following:

- Start the Autodesk Inventor sheet metal environment
- Modify settings for sheet metal design
- Create sheet metal parts
- Modify sheet metal parts to match design requirements
- Create sheet metal flat patterns
- Create drawing views of sheet metal parts

INTRODUCTION TO SHEET METAL DESIGN

A metal blank folded into a finished shape is a common component in a mechanical assembly. Examples include enclosures, guards, simple-to-complex bracketry, and structural members. Although the term sheet metal is often associated with these components, heavy plates of 1 or more can also be formed using similar methods.

Common sheet metal components include electronic and consumer product chassis and enclosures, lighting fixtures, support brackets, frame components, and drive guards.

Sheet metal fabrication can include a number of processes:

- Drawing
- Stamping
- Punching and cutting
- Braking
- Rolling
- Other, more complex operations

The tools in the Autodesk Inventor sheet metal environment enable you to create press brake-formed models. You can also create rolled shapes, including cones and cylinders. In addition, sheet metal parts can include formed features such as nail holes, lances, and louvers. Using lofts and surfacing tools, you can create parts that are fabricated through deformation processes such as drawing or stamping.

SHEET METAL FABRICATION

Brake-formed parts must be described in a minimum of two states: the flat, blank shape prior to bending and the finished part in its folded form. Manufacturing processes are often performed on the flat sheet before folding the part into a finished shape. Holes and other openings are punched into the flat stock, and deformations such as dimples are stamped into the blank with forming tools and dies. Preparing the flat stock can be done manually or, more commonly, with CNC punching machines. High-volume sheet metal parts are often created on progressive die lines where a continuous feed of flat stock from a coil is passed through a series of forming dies that remove material or form the shape of the component. For lower-volume components, manual or CNC press brakes are the primary tools used to bend the finished blank into its bent form.

In most sheet metal designs, the folded shape of the part is known. You can create the folded model using various techniques, the most basic being a process of adding individual faces, flanges, and other sheet metal features to build the final state of the folded part. A key face is the first feature added to the part; additional "as bent" features are added and joined to open edges of the part. During feature creation, a bend is automatically created at the intersecting edge of the new feature and the existing part. You can also add bends between disjointed faces, a technique that is very helpful when designing a sheet metal component in the context of an assembly. When the folded design is complete, Autodesk Inventor can create a flattened version of the model, commonly called a flat pattern. The flat pattern locates the position of bend centerlines used to form the part in a press brake or other sheet metal forming tools. The flat pattern also contains any holes, cutouts, and other features placed on faces of the folded part. Settings in the current sheet metal rule determine the size of the flat pattern.

When a metal blank or sheet is bent in a press brake, the metal in the bend area deforms. Material on the inside of the bend compresses, and material on the outside of the bend is stretched. This deformation must be taken into account when calculating the flattened state of the folded model. The amount of deformation is dependent on the material, the radius and angle of the bend, and the process and equipment used to create the bend. In the sheet, there is a plane where the material neither compresses nor stretches. If the location of this plane is known, the length of the flat sheet can be calculated. For many materials and processes, the location of this neutral plane is known and can be expressed as a percentage of the thickness of the part, as measured from the surface on the inside of a bend. This offset value is referred to as a kFactor. The following image shows the neutral plane of a sheet metal bend. The location of the neutral plane is kFactor * Thickness, where 0 < kFactor < 1. The default sheet metal rule uses a kFactor of 0.44. This value is appropriate for many common materials and processes.

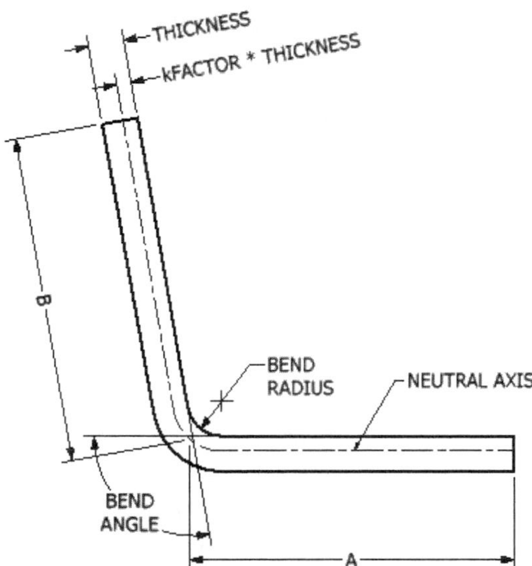

Figure 10.1

There are two methods for storing sheet metal rules. You can store sheet metal rules in a template file or you can use a style library to store sheet metal rules and "pull" or use them in your sheet metal parts as necessary based on the particular sheet metal part that you are creating. Using style libraries is the recommended method. The current sheet metal rule applies to all sheet metal features in the part. Although you cannot apply separate rules to different sheet metal features in a part, you can override default values during feature creation or editing. For example, some materials have different bending properties depending on the orientation of the bend relative to the grain of the material. As the sheet is manufactured, the material microstructure aligns with the rolling direction. This action results in a sheet with bend properties related to its orientation in the press brake. Stainless steel sheets often display this anisotropic behavior.

The use of a constant kFactor is not always appropriate. For some materials and processes, when formed parts require precise tolerances, you can use a bend table in place of the kFactor. A bend table uses empirically derived information to apply length adjustments to bends. With overrides, you can apply a separate kFactor or bend table to each bend on a sheet metal part.

BEND TABLES

In the calculation of the unfolded length of a bend, a bend table replaces the kFactor with a set of known adjustments to the unfolded length. A bend table can ensure greater precision of unfolded dimensions because the values are derived from your measurements of bend length, of a particular material, on a specific machine. When you use a bend table to calculate an unfolded length, the following formula applies:

$$L = A + B - x$$

In this equation, the variables are as follows:

L = Unfolded length

A = Length of folded face 1

B = Length of folded face 2

× = Adjustment from bend table

The value of × is dependent on the sheet thickness, the angle of the bend, and the inner radius of the bend.

Note that the measurements of A and B are to the projected intersection of the extended outer faces on either side of the bend. This intersection is used when the angle of the bend is less than or equal to 90°. The following image applies when the bend angle is greater than 90°. The measurements are parallel to the face and tangent to the outer surface of the bend. The same formula is used to determine the unfolded length of the part.

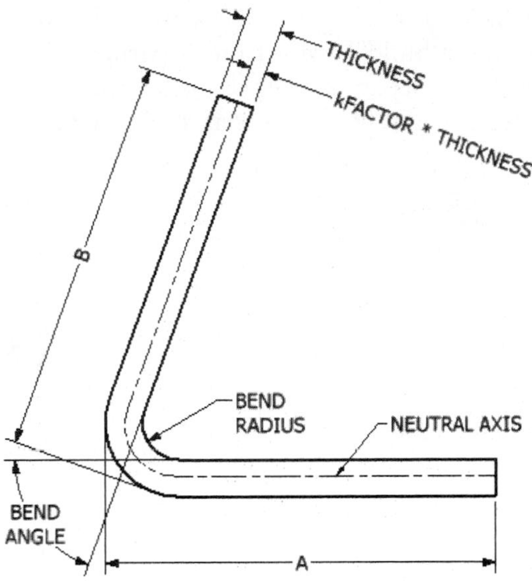

Figure 10.2

SHEET METAL PARTS

Sheet metal parts can be designed separately or in the context of an assembly. Since these parts are often used as supports or enclosures for other components, designing in the assembly environment can be advantageous.

SHEET METAL DESIGN METHODS

Sheet metal parts are most often created in the folded state, as shown in the following image on the left. The model is then unfolded into a flat sheet, as shown in the following image on the right, using the Flat Pattern tool. To create the folded model, extrude the first sketch the thickness of the sheet to create a face. Add additional faces or flanges to open edges of the part and add bends automatically between the features. Add cuts and other special sheet metal features to the part as required.

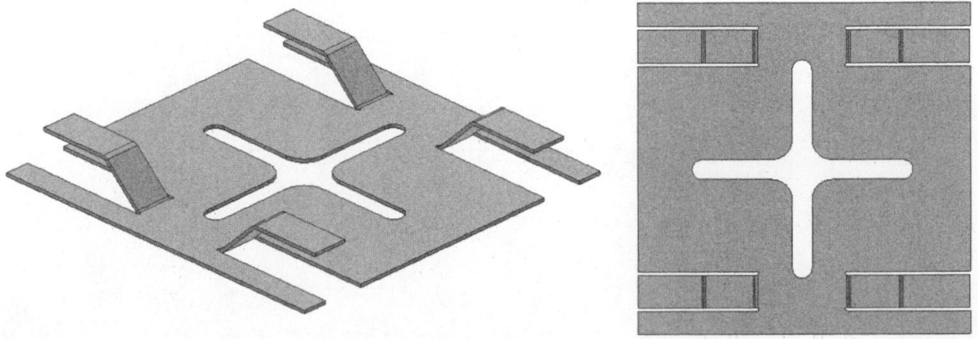

Figure 10.3

Autodesk Inventor's tools that create models with disjointed solids, or unconnected features, are very helpful when building sheet metal parts in an assembly, as shown in the following image. You can build separate faces of a sheet metal part quickly by referencing faces on other parts in the assembly. Additional sheet metal features can then join the distinct faces.

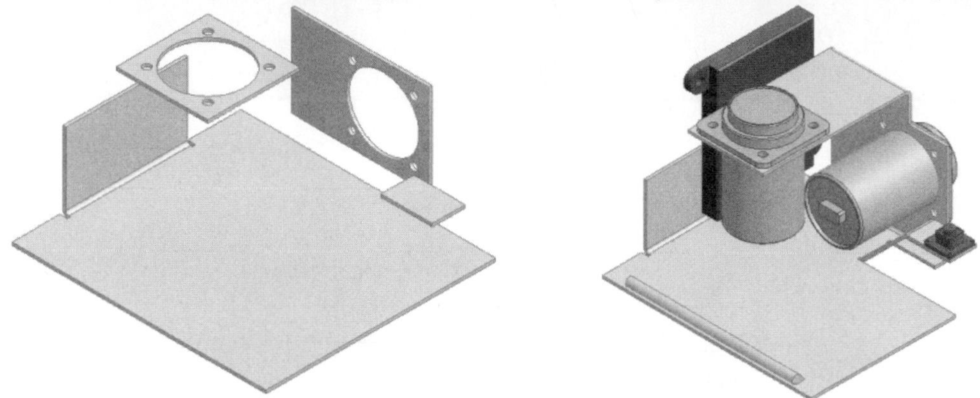

Figure 10.4

You can also create sheet metal parts from standard parts that have been shelled. All walls of the shelled part must be the same thickness and must match the thickness of the flat sheet as defined by the sheet metal rule. The solid corners of the shelled part can be ripped open to enable the model to unfold, and appropriate bends can be added along the edges between faces.

CREATING A SHEET METAL PART

The first step in creating a sheet metal part is to select a sheet metal template with which to work. Autodesk Inventor comes with a sheet metal template named *Sheet Metal.ipt*, as shown in the following image.

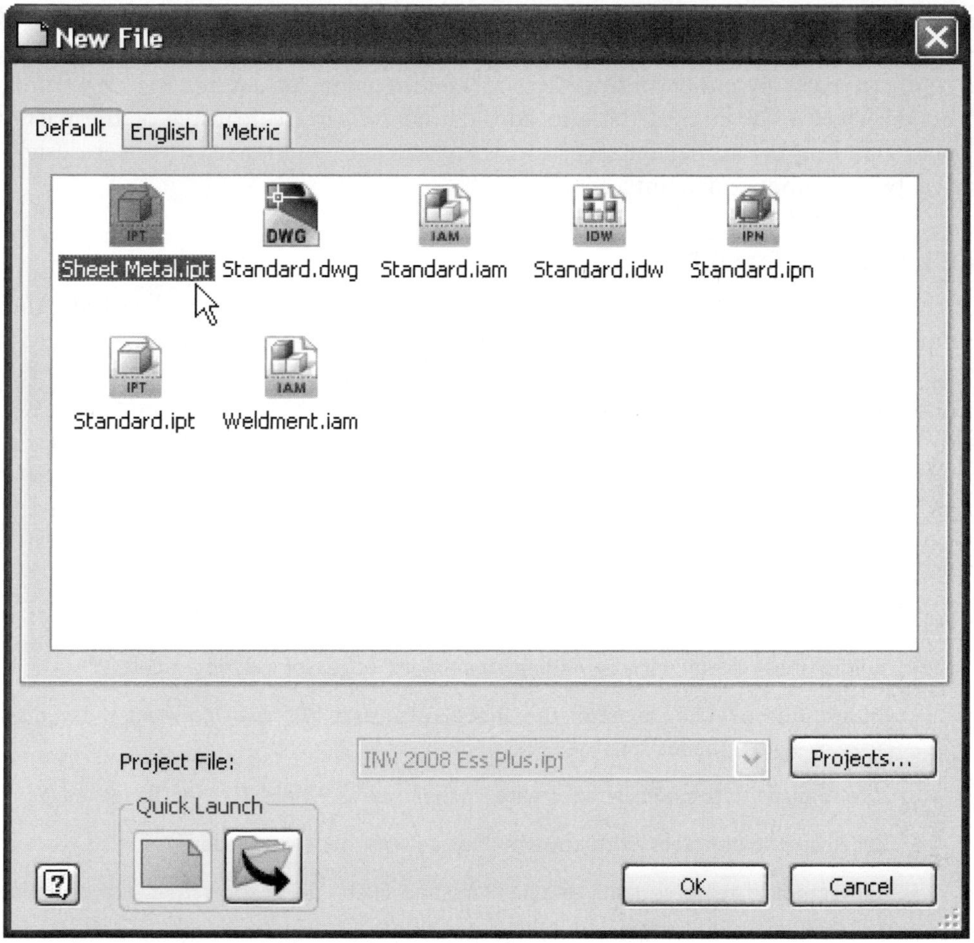

Figure 10.5

You can switch between the standard modeling environment and the sheet metal environment at any time by clicking Sheet Metal from the Convert menu, as shown in the following image. You can use Autodesk Inventor's general modeling tools to add standard part features to a sheet metal part.

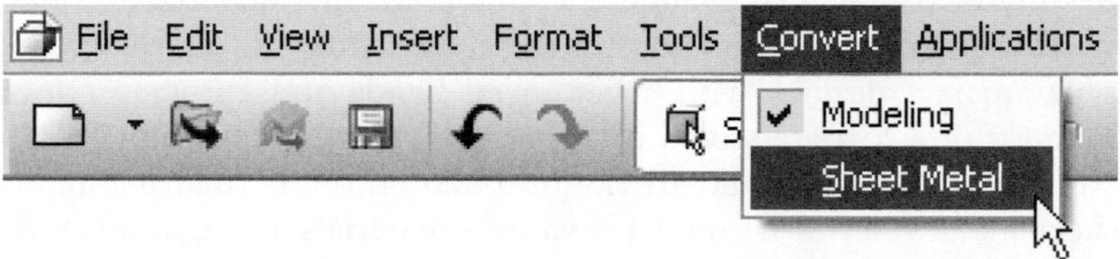

Figure 10.6

The creation of a sheet metal part usually begins by specifying the sheet metal defaults or rules such as sheet thickness, material, and bend radius. You then use sheet metal-specific tools to create parts by either adding faces at existing edges or connecting disjointed faces with bends. Use the optimized tools to add, cut, and clean up features of the part. Sheet metal faces and parts can be adaptive, and their size can be adjusted to meet design rules specified by assembly constraints.

SHEET METAL TOOLS

After creating the new sheet metal document from a sheet metal template file, Autodesk Inventor's panel bar tools change to reflect the sheet metal environment. As with standard parts, a sketch is created and becomes active. The sketch tools are common to both sheet metal and standard parts. The first sketch of a sheet metal part must be either a closed profile that is extruded the sheet thickness to create a sheet metal face or an open profile that is thickened and extruded as a contour flange. Use additional tools to add sheet metal features to the base feature. The following tools can be used to create sheet metal features. You can also add standard part features to a sheet metal part, but these features may not unfold when a flat pattern is generated from the folded model.

Use sheet metal tools to perform the following:

- Build a sheet metal part by adding faces along edges of existing faces.
- Create individual key faces of the sheet metal part and then connect these disjointed faces by adding bends between them.
- Extend faces automatically to create corner seams where face or flange edges meet.
- Cut shapes from faces with tools enhanced for sheet metal design.
- Add standard features such as chamfers and fillets, using tools that are optimized for working with thin sheets.
- Create a flat pattern model of the part with a single button click. This flat pattern model is updated automatically as features are added, removed, or edited.
- Create drawing views of the folded part and flat pattern to document your design for manufacturing.

SHEET METAL RULES

Sheet metal-specific parameters include the thickness and material of the sheet metal stock, a bend allowance factor to account for metal stretching during the creation of bends, and various parameters dealing with sheet metal bends and corners. The sheet metal-specific parameters of a part are stored in a sheet metal rule. Sheet metal rules are created and modified in the Style and Standard Editor found on the Format menu. You can create additional sheet metal rules to account for various materials and manufacturing processes or material types. If you create sheet metal rules in a template file, they are available in all sheet metal parts based on that template. Sheet metal rules can also be created and managed using style libraries, which is the recommended workflow. Use of a style library makes your template files more lightweight and makes the management of sheet metal rules and other styles more robust.

The thickness of the sheet metal stock is the key parameter in a sheet metal part. Sheet metal tools such as Face, Flange, and Cut automatically use the Thickness parameter to ensure that all features are the same wall thickness, a requirement for unfolding a model. In the default sheet metal rule, all other parameters, such as Bend Radius, are initially based on the Thickness value.

To edit or create a sheet metal rule, follow these steps:

1. Click the Style and Standard Editor from the Format menu. The Style and Standard Editor dialog box appears, as shown in the following image.
2. To create a new Sheet Metal Rule, select an existing rule and click the New button. This action creates a copy of the rule that is currently selected in the Sheet Metal Rule list.
3. Rename the rule and click the OK button.
4. Edit the values and settings on the Sheet, Bend, and Corner tabs to define the default feature properties of parts created with this sheet metal rule.
5. Click the Save button.
6. When multiple sheet metal styles exist in a part, set the active sheet metal rule in a part by selecting it in the Sheet Metal Defaults dialog box accessed from the Sheet Metal Default tools on the Sheet Metal Features panel bar, as shown in the previous image.

Changing to a different sheet metal rule or making changes to the active rule in the Style and Standard Editor updates the sheet metal part to match the new settings.

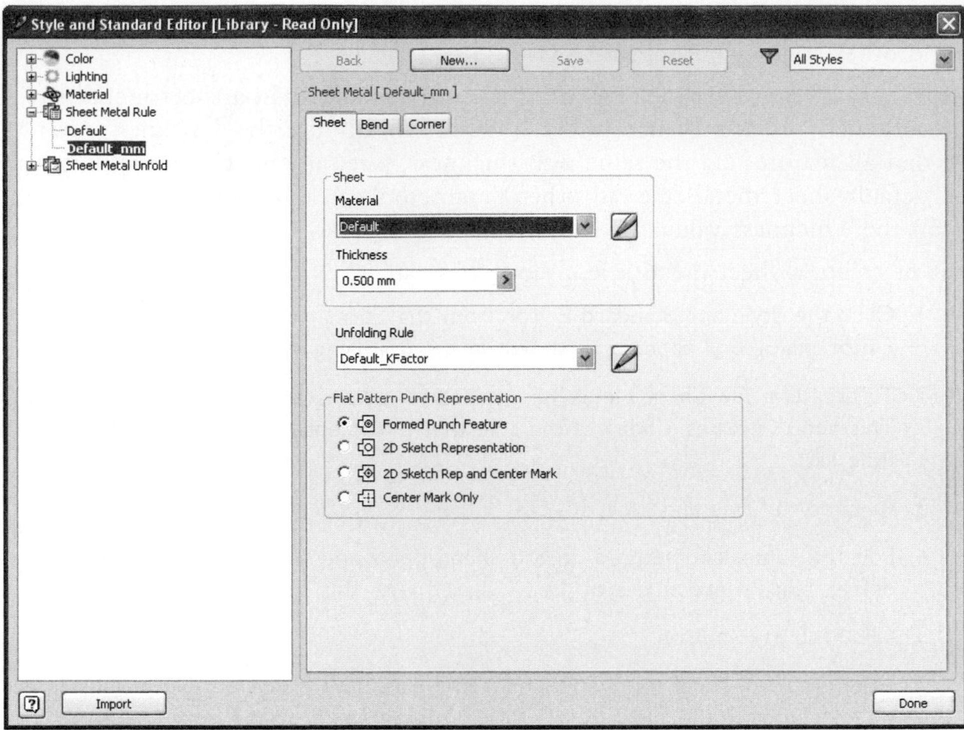

Figure 10.7

The following is a description of the settings for a sheet metal rule. The parameters with numeric values, such as Thickness and Bend Radius, are saved as Sheet Metal Parameters and are accessible in the Parameters dialog box, as shown in the following image. The parameter values are updated when you modify or activate a new sheet metal rule. Other model parameters and user-defined parameters can reference these parameters in equations.

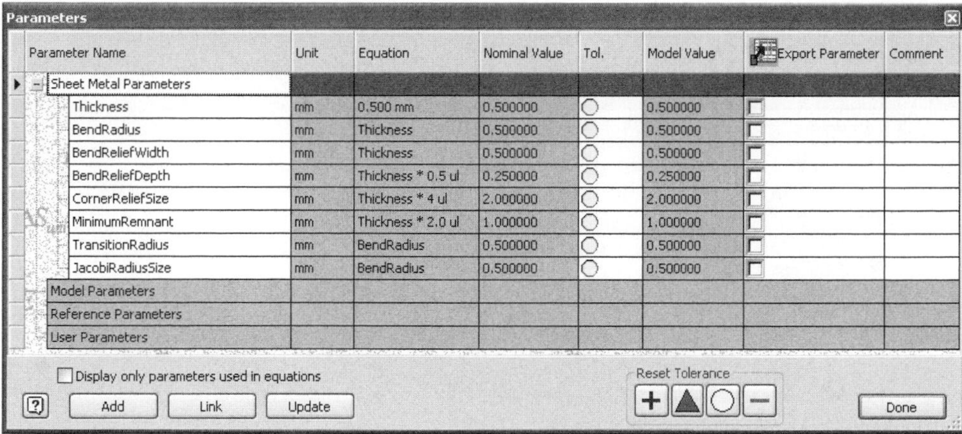

Parameters

Parameter Name	Unit	Equation	Nominal Value	Tol.	Model Value	Export Parameter	Comment
Sheet Metal Parameters							
Thickness	mm	0.500 mm	0.500000		0.500000		
BendRadius	mm	Thickness	0.500000		0.500000		
BendReliefWidth	mm	Thickness	0.500000		0.500000		
BendReliefDepth	mm	Thickness * 0.5 ul	0.250000		0.250000		
CornerReliefSize	mm	Thickness * 4 ul	2.000000		2.000000		
MinimumRemnant	mm	Thickness * 2.0 ul	1.000000		1.000000		
TransitionRadius	mm	BendRadius	0.500000		0.500000		
JacobiRadiusSize	mm	BendRadius	0.500000		0.500000		
Model Parameters							
Reference Parameters							
User Parameters							

Figure 10.8

Sheet Tab

The Sheet tab contains settings for the sheet metal material, thickness, unfolding rule, and the flat pattern punch representation. You can create multiple rules that contain different materials and thicknesses to be applied to a sheet metal model.

Material Specify a material from the list of defined materials in the library or template file, and the material color setting is applied to the part. You can define new materials using the Material node of the Style and Standard Editor.

Thickness Specify the thickness of the flat stock that will be used to create the sheet metal part.

Unfolding Rule Specify the rule or method used to calculate bend allowance, which accounts for material stretching during bending. The drop-down list provides access to the unfolding rules that are defined in the Sheet Metal Unfold node of the Style and Standard Editor.

Flat Pattern Punch Representation Choose from one of four options to specify how you want a sheet metal punch to appear when the model is displayed as a flat pattern. These options allow you to display the punch as a formed punch feature, a 2D sketch representation, a 2D sketch representation with a center mark, or as a center mark only.

Bend Tab

The Bend tab contains settings for sheet metal bends, as shown in the following image. Most bend settings are typically defined as a function of the sheet thickness. Bend Radius refers to the inside radius of the completed bend. This setting is the default for all bends, but you can override it while creating any bend.

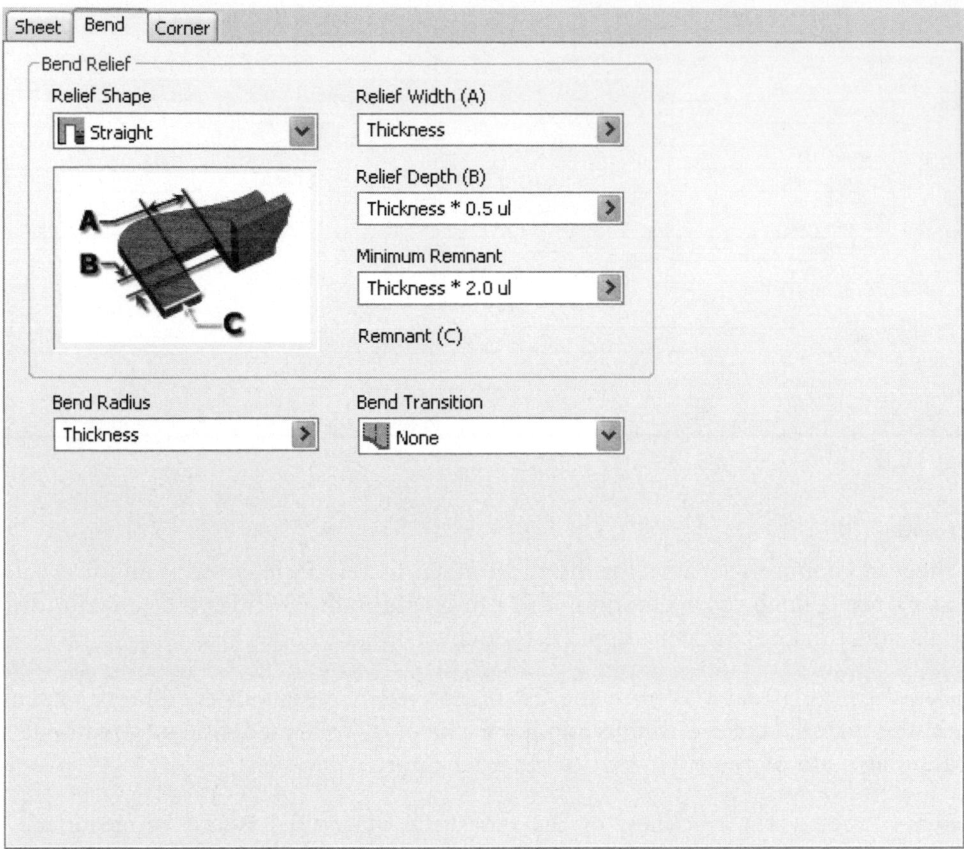

Figure 10.9

You can specify bend reliefs when a bend zone, the area deformed during a bend, does not extend completely beyond a face, as shown in the following image.

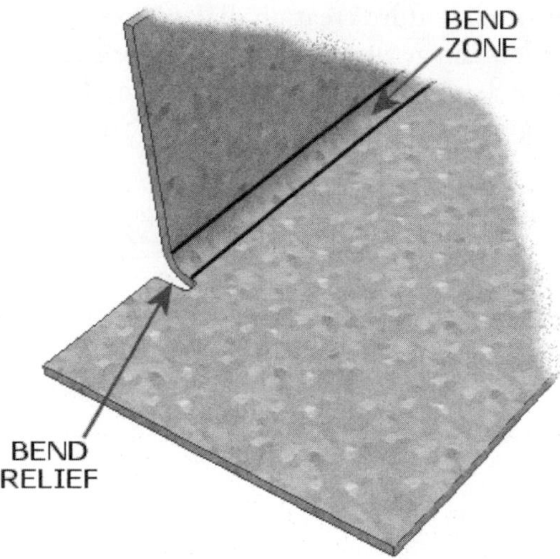

Figure 10.10

If bend reliefs are used, they are incorporated in the flat blank prior to folding. You can generate bend reliefs by punching or with laser, water-jet, or other cutting methods. Creating bend reliefs may increase production costs, and it is common practice with thin, deformable materials, such as mild steel, to add bends without bend reliefs. The material is allowed to tear or deform where the bend zone intersects the adjacent face, as shown in the following image.

Figure 10.11

Descriptions of the settings on the Bend tab follow.

Relief Shape If a bend does not extend the full width of an edge, a small notch is cut next to the end of the edge to keep the metal from tearing at the edge of the bend. Select from Straight, Round, or Tear for the shape of the relief. This setting will be displayed as the default value in the sheet metal feature creation dialog boxes. As you select from one of the three relief shapes, a preview image appears, showing the shape of the relief and how the settings for Relief Width, Relief Depth, and Minimum Remnant apply to the selected shape. The following image shows the previews of the Straight, Round, and Tear shapes.

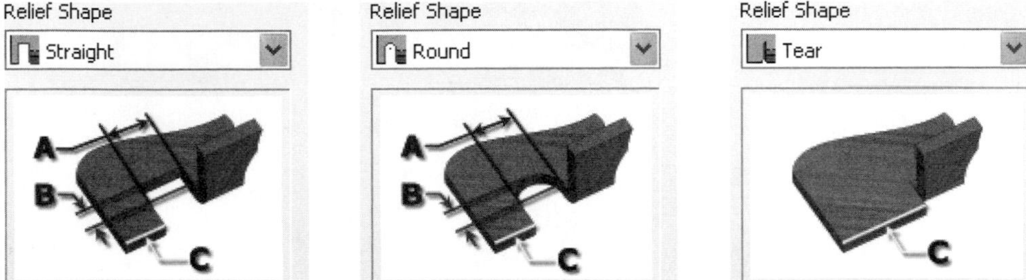

Figure 10.12

Relief Width Specifies the width of the bend relief.

Relief Depth Specifies the distance a relief is set back from an edge. Round relief shapes require this distance to be at least one-half of the Relief Width value.

Minimum Remnant Specifies the distance from the edge at which, if a bend relief cut is made, the small tab of remaining material is also removed.

Bend Radius Defines the inside bend radius value between adjacent, connected faces.

Bend Transition Controls the intersection of edges across a bend in the flattened sheet. For bends without bend relief, the unfolded shape is a complex surface. Transition settings simplify the results, creating straight lines or arcs, which can be cut in the flat sheet before bending. The following image shows the five transition types.

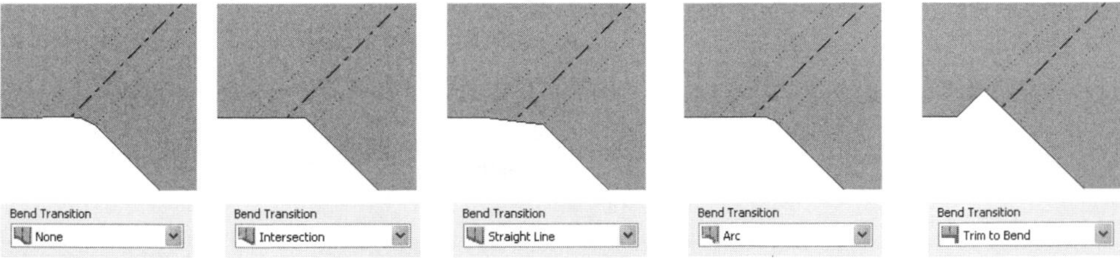

Figure 10.13

Corner Tab

A corner occurs where three faces meet. The corner seam feature controls the gap between the open faces and the relief shape at the intersection. As with bend reliefs, corner reliefs are added to the flat sheet prior to bending.

By using the Corner tab, as shown in the following image, you can set how corner reliefs will be applied to the model. You can designate the corner relief size, shape, and radius.

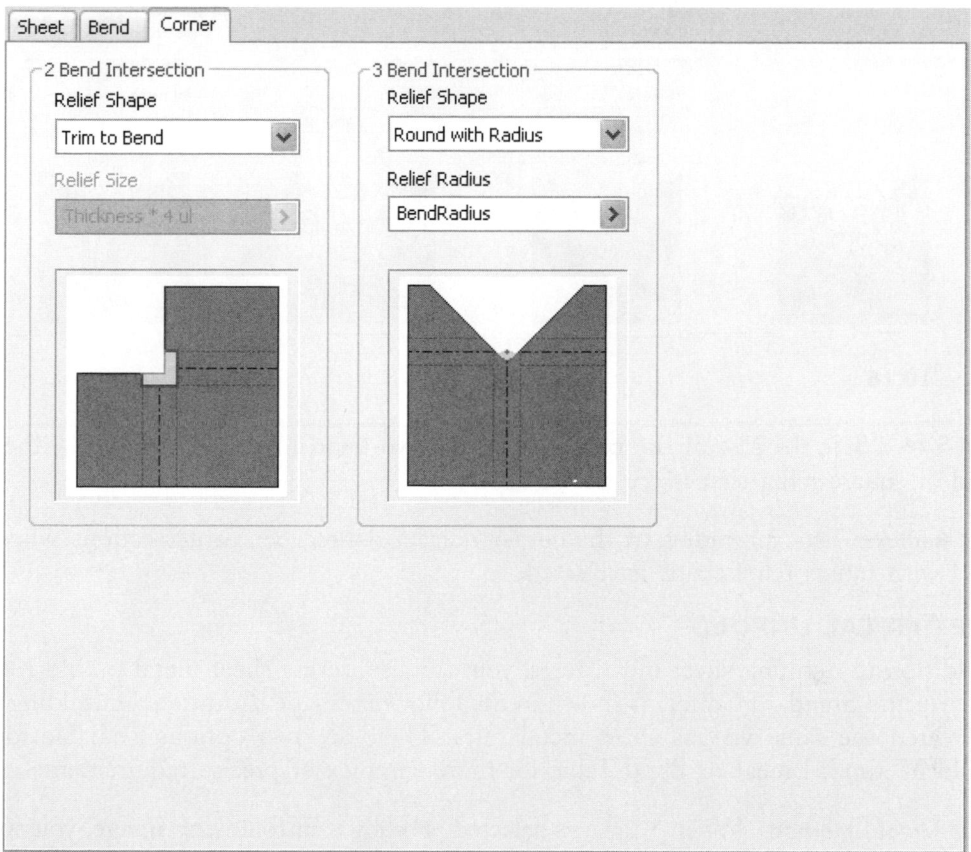

Figure 10.14

Relief Shape Specify the shape of the corner relief for either two- or three-bend intersections. When a two-bend intersection is formed in your model, you can select from one of six corner relief options, as shown in the following image.

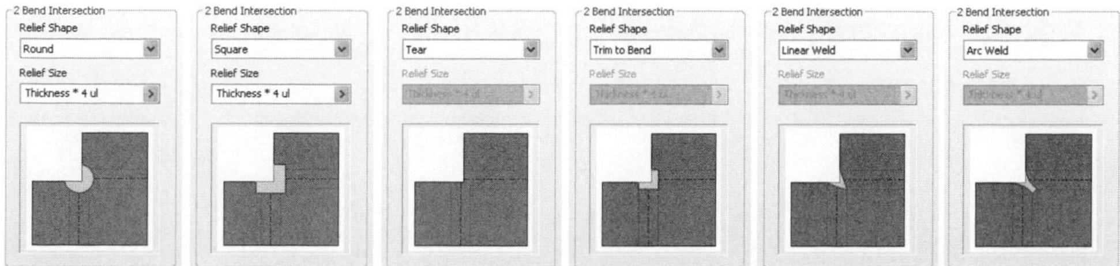

Figure 10.15

When a three-bend intersection, also referred to as a Jacobi corner, is formed in your model, four relief shapes can be used, as shown in the following image.

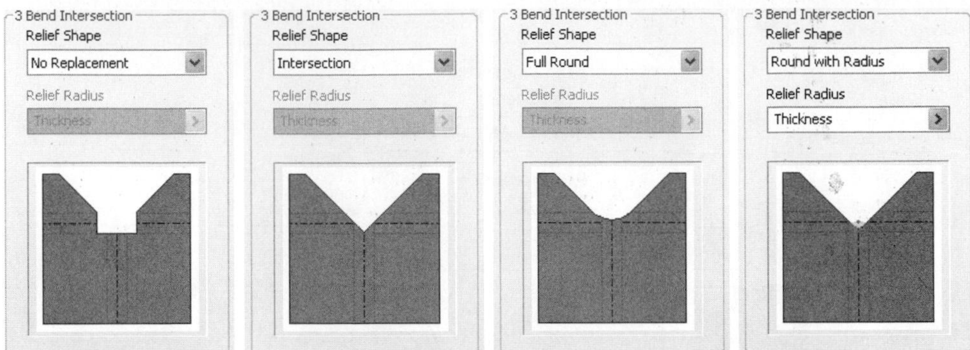

Figure 10.16

Relief Size Sets the size of the corner relief for two-bend intersections when either the round or square relief shapes are selected.

Relief Radius Sets the radius of the corner relief for three-bend intersections when the round with radius relief shape is selected.

SHEET METAL UNFOLD

In addition to defining sheet metal rules, you can also define sheet metal unfold rules in the Style and Standard Editor, as shown in the following image. Additional unfolding rules are created the same way as sheet metal rules. There are two options available for the Unfold Method: Linear or Bend Table for more complex or precise requirements.

Linear Unfold Method When Linear is selected, as shown in following image, you specify the default KFactor value used for calculating bend allowances. A KFactor is a value between 0 and 1 that indicates the relative distance from the inside of the bend to the neutral axis of the bend. A KFactor of 0.5 specifies that the neutral axis lies at the center of the material thickness. The Unfold Method is combined with the Unfolding Rule specified on the Sheet tab of the sheet metal rule.

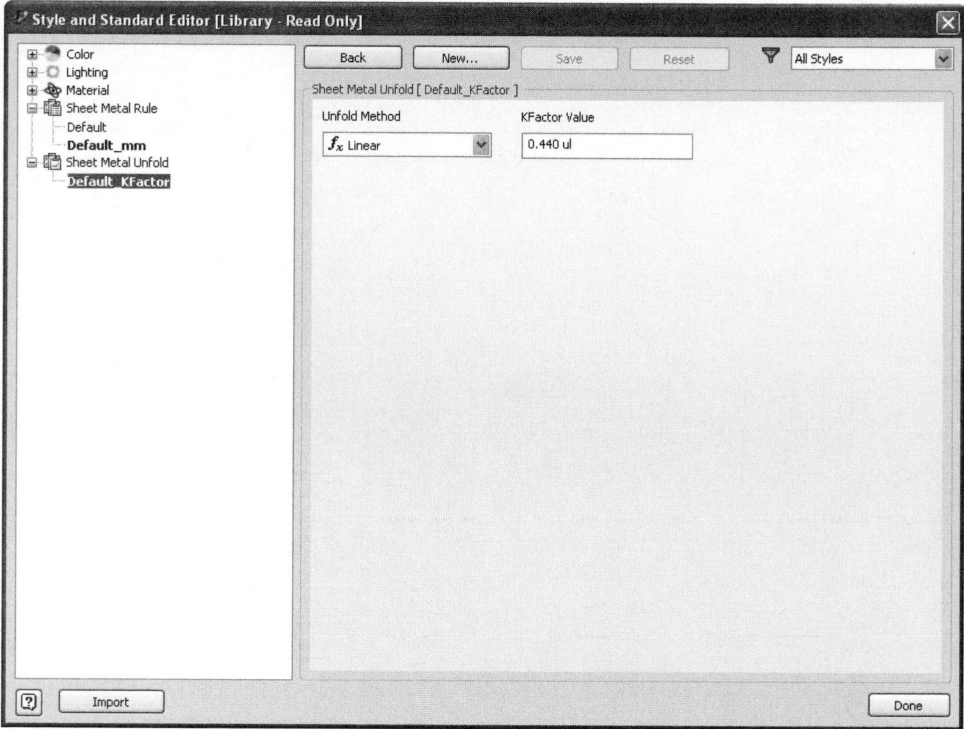

Figure 10.17

Bend Table Unfold Method When Bend Table is selected from the Unfold Method list, you have the ability to create, import, or export a bend table. Autodesk Inventor can use a bend table that is a plain text file with a *.txt* extension. You can also create or edit a spreadsheet version of the bend table by using Microsoft Excel and then saving the table in text format.

Autodesk Inventor includes both metric (mm) and imperial unit (in) bend tables that you can modify, and each bend table is supplied in both *.txt* and *.xls* formats. The text file follows a rigid format; refer to the samples that ship with the product and are located in C:\Program Files\Autodesk\Inventor <version>\Design Data\Bend Tables. The following image shows one of the sample bend tables that have been imported into the Style and Standard Editor.

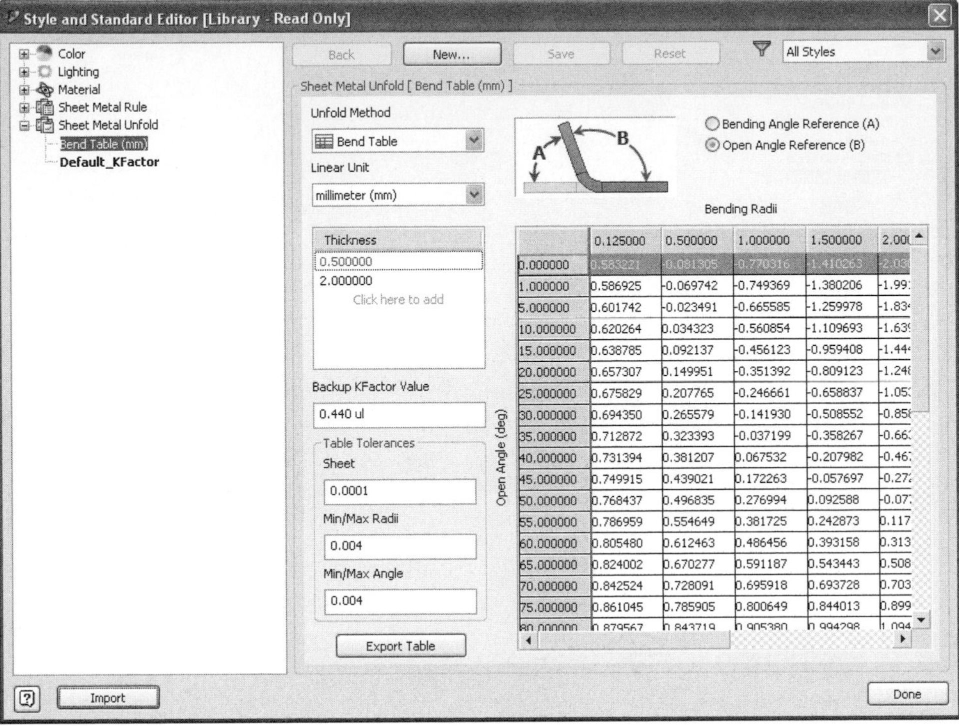

Figure 10.18

The angle information required in a bend table is shown in the following image. A flat sheet bent at an angle of 110° will have an open angle of 70°. The open angles are entered into the bend table. The sample bend tables use this method of entering the open angle, but by using the Unfold rules, you can declare how to read the angular values, whether they are open angles or bending angles, as shown in the previous image. Changing the control for the angle does not edit the bend table data, just how it is interpreted.

Unfolding rules are referenced by sheet metal rules that are created and saved in either a sheet metal template file or published to a style library for reuse in other sheet metal parts. Bend tables created in the Style and Standard Editor are stored as a Style Definition File in the *.styxml* format. It is possible to export the tables as an ascii *.txt* file. Exporting to *.txt* may be a requirement if you work in a mixed Autodesk Inventor version environment. For example, if using Inventor 2009, you would need to share *.txt* bend tables for those still using R10 to 2008.

Note: When you open an older sheet metal part, sheet metal styles are converted to sheet metal rules that have the same name as those defined in the older file. Material styles and unfold methods are also automatically converted to their Inventor 2009 counterparts and are set up in the Style and Standard Editor.

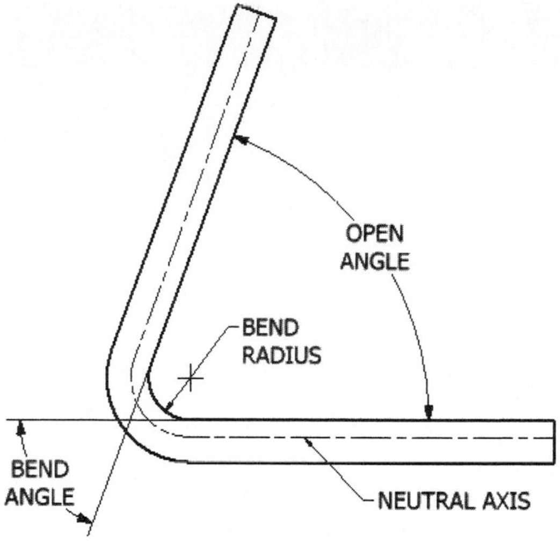

Figure 10.19

For additional information on creating or working with bend tables, refer to the Autodesk Inventor help system.

SHEET METAL DEFAULTS

After you have defined sheet metal rules, you use the Sheet Metal Defaults tool, as shown in the following image, to select the rule that is active for the sheet metal part.

Figure 10.20

Once the tool is selected, the Sheet Metal Default dialog box appears, as shown in the following image.

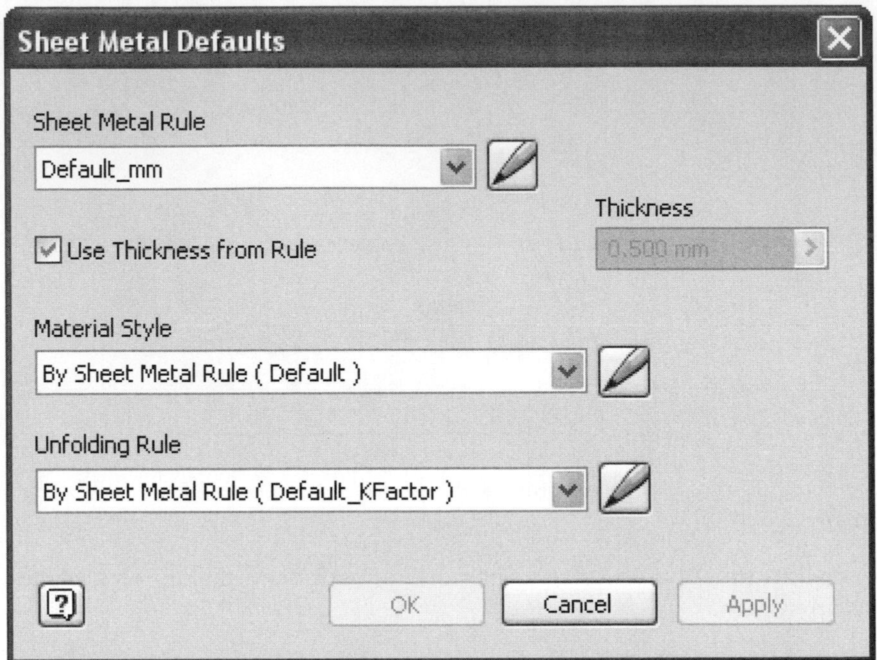

Figure 10.21

From this dialog box you can select a sheet metal rule that is defined in the Style and Standard Editor from the drop-down list. All settings defined in that rule are then applied to the sheet metal part that you are working on. The material style and unfolding rule can also be selected. By default, they are referenced from the sheet metal rule, and the text in the field tells you that the style and unfolding rule are defined By Sheet Metal Rule and then displays the specific name of the style or unfold rule in parentheses. You can change these settings so they are not the same as those defined in the sheet metal rule or you can click the Edit button (pencil icon) to be taken directly to the specific rule or style in the Style and Standard Editor to be modified. If you deselect the Use Thickness from Rule checkbox, the Thickness field becomes active. This allows you to specify a different thickness than that specified in the sheet metal rule. The ability to edit this field corresponds to the Thickness parameter in the Parameters dialog box. When the field is active in the sheet metal default, the field is also available for edit via the Parameters dialog box. When the checkbox is selected, as shown in the previous image, the Thickness parameter cannot be edited via the Parameters dialog box and is driven based on the value specified in the sheet metal rule.

Note: The items displayed when selecting one of the drop-down lists in the Sheet Metal Default dialog box are controlled by the filter specified in the top right corner of the Style and Standard Editor, as shown in the following image. When Local Styles is selected, only those rules or styles that are stored inside of the active file are displayed. If All Styles is selected, you will see all styles that are available in both the active file and the styles library.

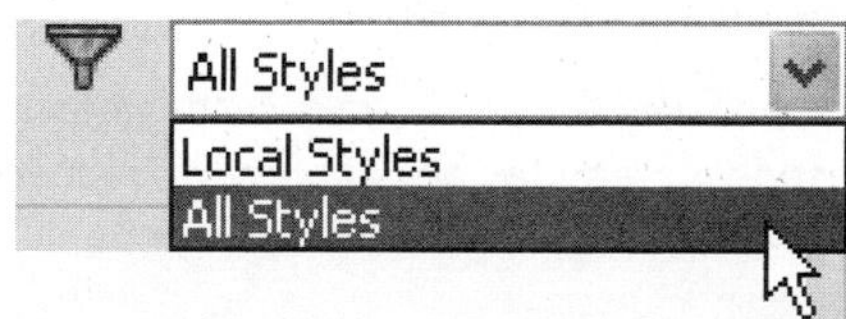

Figure 10.22

FACE

The Face tool, as shown in the following image, extrudes a closed profile for a distance equal to the sheet metal thickness. If the face is the first feature in a sheet metal part, you can flip only the direction of the extrusion. When you create a face later in the design process, you can connect an adjacent face to the new face with a bend. If the sketch shares an edge with an existing feature, the bend is added automatically. As an option, you can select a parallel edge on a disjointed face. This action will extend or trim the attached face to meet the new face, with a bend created between the two faces.

Figure 10.23

To create a sheet metal face, follow these steps:

1. Create a sketch with a single closed profile or a closed profile containing islands. The sketch is most often on a work plane created at either a specific orientation to other part features or by selecting a face on another part in an assembly.

Note: You can create a single face from multiple closed profiles.

2. Click the Face tool on the Sheet Metal Features panel bar. The Face dialog box appears, as shown in the following image.

   If a single closed profile is available, it is selected automatically. If multiple closed profiles are available, you must select which one defines the desired face area.

3. If required, flip the thickness direction to extrude the profile.
4. If the face is not the first feature and the sketch does not share an edge with an existing feature, click the Edges button and select an edge to which to connect the

face. The two faces are extended or trimmed as required, meeting at a bend. The bend is listed as a child of the new face feature in the browser. If the face attached to the selected edge is parallel to but not coplanar with the new face, the dialog box is expanded, and the double-bend options are available. An additional face is added to connect the two parallel faces. The orientation and shape of this face is determined by the selected double-bend options. The options for creating a double bend or full-radius bend can be accessed by clicking the More (<<) button. These options are discussed later in the section on bends.

The Unfold Options and Bend tabs contain settings for overriding the default values in the sheet metal style. Overrides are applied to individual features.

5. Click OK.

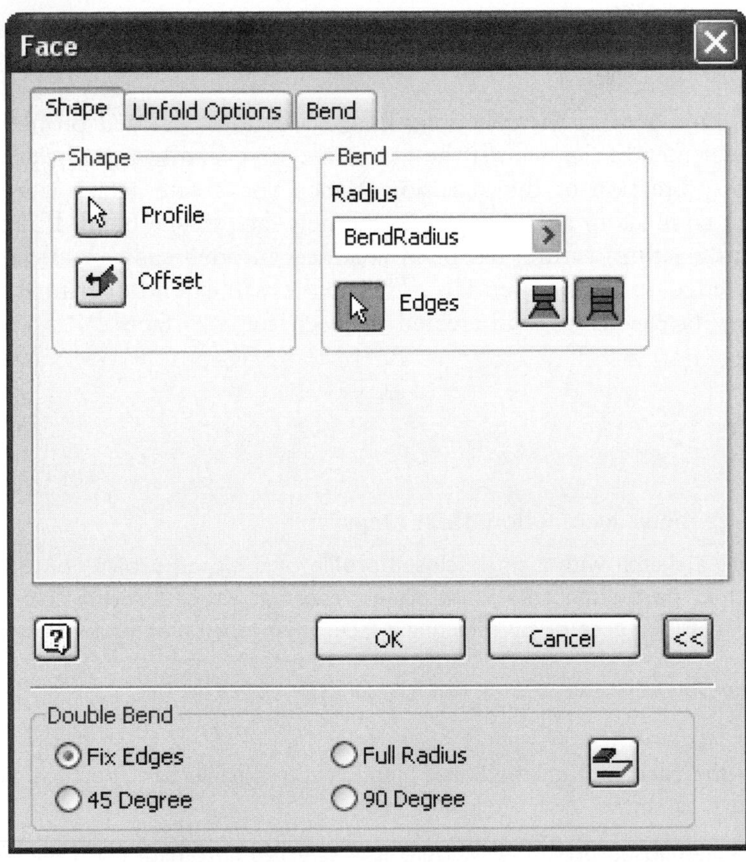

Figure 10.24

Shape Tab

By using the Shape tab, you can select the profile and direction of the face. You can also override the bend radius specified in the sheet metal rule, and you can select edges of other existing faces or edges that exist in the sheet metal part to apply a bend feature upon creation of the face. The following options are available on the Shape tab.

	Profile	Select a profile(s) to extrude a distance equal to the Thickness parameter.
	Offset	Toggle the direction for the creation of the feature.
Radius BendRadius	Radius	This tool displays the default bend radius specified in the active sheet metal style. If the face will be attached to another edge of the part, you can specify the value for the bend radius to be used. You can modify the radius value on a per-feature basis.
	Edges	Select an edge to include in the bend. When selected, the edge is extended to match the edge of the face. A bend feature is created between the two faces.
	Extend Bend Aligned to Side Faces	Click to extend material along the faces of the edges connected by the bend instead of perpendicularly to the bend axis. An image of a part before this setting is applied, as shown in the following image on the left, and an example of the part when this option is active is shown on the right.
	Extend Bend Perpendicular to Side Faces	Click to extend material along the faces of the edges to the bend axis. An example of a part demonstrating this setting is shown in the middle of the following image.

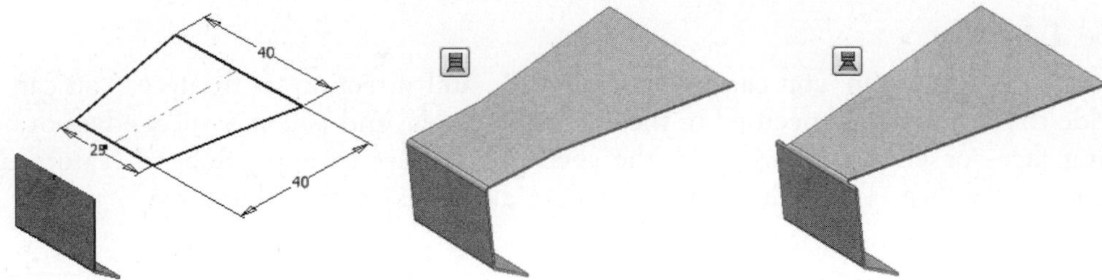

Figure 10.25

Unfold Options Tab

The Unfold Options tab, as shown in the following image, provides a tool to override the unfold style set by the sheet metal rule, if necessary. This tab is also useful when working with other sheet metal tools such as Flange, Contour Flange, Hem, and Bend. Refer to the section on Sheet Metal Unfold for a description of the unfold style. The word Default will appear next to the option specified by the active sheet metal rule. This action is consistent in other dialog boxes, and it helps you to know how the active rule is defined without having to reopen the Sheet Metal Defaults dialog box.

Figure 10.26

Bend Tab

The Bend tab, as shown in the following image, allows you to override the values set in the sheet metal rule, if necessary. As with the Unfold Options tab, the settings designated by the active sheet metal rule are displayed in parentheses with the word Default in front of them. This tab is also helpful when working with a number of the sheet metal tools such as Flange, Contour Flange, Hem, and Bend. These settings allow you to override how bends and bend reliefs are created on a per-feature basis. Refer to the section on sheet metal rules for a description of the settings.

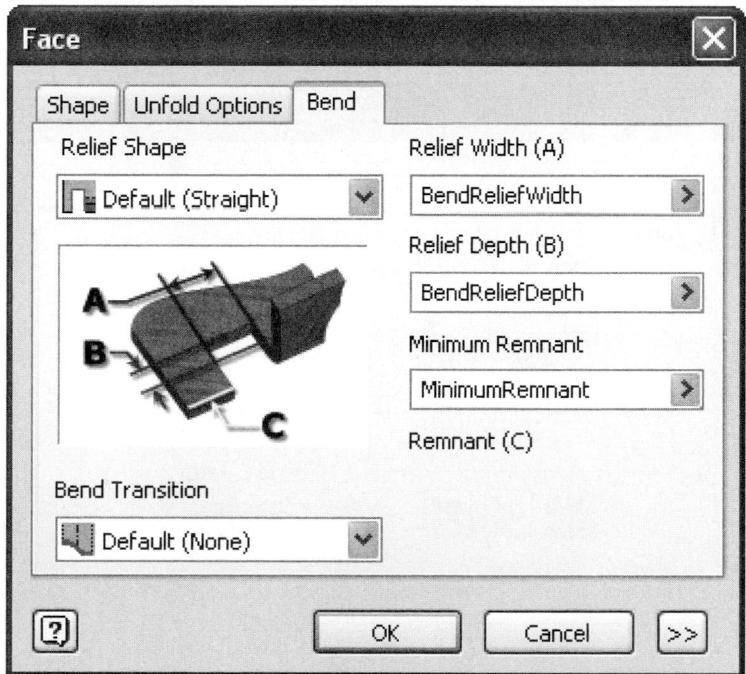

Figure 10.27

CONTOUR FLANGE

Using the Contour Flange tool shown in the following image, you create a contour flange from an open sketch profile. The sketch is extruded perpendicular to the sketch plane, and the open profile is thickened to match the sheet metal thickness. The profile does not require a sketched radius between line segments—bends are added at sharp intersections. Arc or spline segments are offset by the sheet metal thickness. A contour flange can be the first feature in a sheet metal part, or it can be added to existing features.

Figure 10.28

To create a contour flange, follow these steps:

1. Create a sketch with a single open profile. The sketch can contain line, arc, and spline segments, and it can be constrained to projected reference geometry to define a common edge between the contour flange and an existing face.
2. Click the Contour Flange tool on the Sheet Metal Features panel bar. The Contour Flange dialog box appears, as shown in the following image.

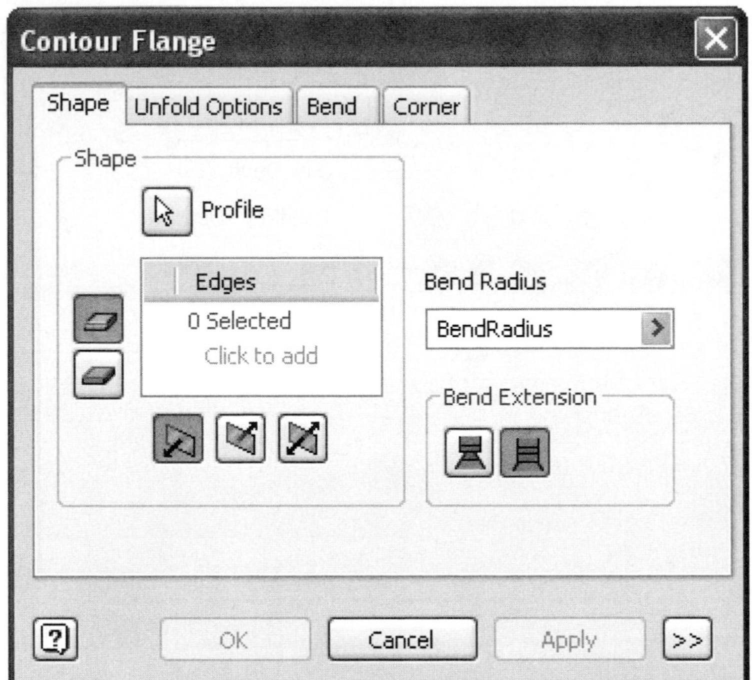

Figure 10.29

3. Click the open sketch profile to define the shape of the contour flange.
4. If required, flip the side to offset the profile or select both sides to use the selected profile as the mid-plane for the contour flange.
5. If the contour flange is the first feature in the part, enter an extrusion distance and direction. If the contour flange is not the first feature, you can select an edge perpendicular to the sketch plane to define the extrusion extents. If the sketch is

attached to an existing edge, select that edge. If the sketch is not attached to any projected geometry, the contour flange is extended or trimmed to meet the selected edge. The selected edge is typically the closest edge to the end of the sketch. You can also use Loop Select Mode to have the flange created along all edges of a selected loop.

6. Apply any unfold, bend, or corner overrides on the Unfold Options, Bend, and Corner tabs.
7. If you select an edge to define the extrusion extents, you can select from five options, available by expanding the dialog box, to further refine the length of the contour flange. The options are described below.
8. Click OK.

Shape Tab

The Shape tab allows you to select the profile, edge, and offset direction of the contour flange. You can also override the bend radius that is specified in the sheet metal rule. The following options are available on the Shape tab.

	Profile	Select an open profile to extrude for the shape of the contour flange.
Edges 0 Selected Click to add	Edge	Specify an existing edge for termination of the feature using the top button, Edge Select Mode, or Loop Select Mode.
	Flip Side/ BothSides	Toggle the direction for the creation of the feature or select mid-plane.
Bend Radius BendRadius	Bend Radius	This feature displays the default bend radius. Bends are added to sharp edges of the profile. Controls how bend extensions are calculated when a profile has edges that are not coincident to the selected edge.
Bend Extension	Bend Extension	Extend Bend Aligned to Side Faces: Aligns the bend of the shorter edge when the edges are not equal. Extend Bend Perpendicular to Side Faces: Extends the bend from the longer side when the edges are not equal.

More Button (<<)

The More button allows you to specify the type of extents for the feature, as shown in the following image.

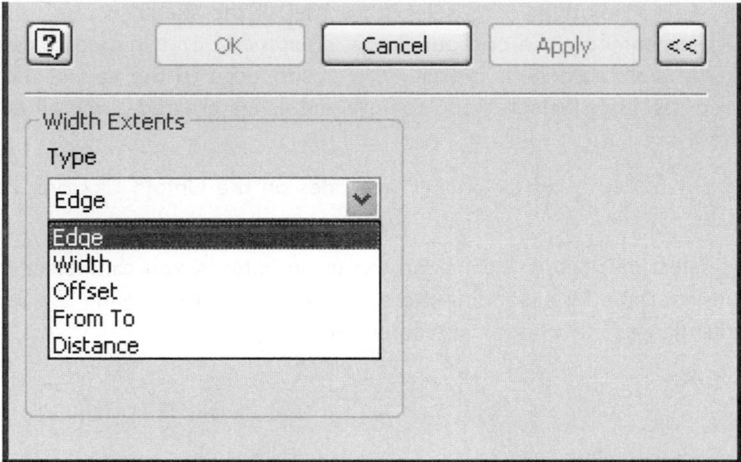

Figure 10.30

These extent types are also available for flange and hem features. However, the flange and hem features do not include the Distance option because the distance is specified in the main body of the dialog box. The five width extents options are shown in the following image, and their descriptions follow.

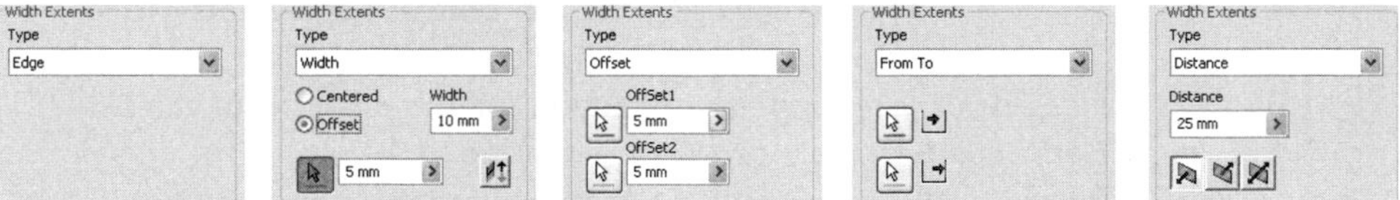

Figure 10.31

Edge Select so that the contour flange extends the full length of the selected edge.

Width Select so that a point defines one extent of the contour flange. The flange can be offset from this point and extended a fixed distance. You define the starting point by selecting an endpoint on the selected edge, a work point on a line defined by the edge, or a work plane perpendicular to the selected edge. You can also center the feature based on the selected edge and then specify the width of the created feature.

Offset Similar to the Width option, select so that two selected points define both extents of the flange. You can specify an offset distance from each point.

From To Select so that the width of the feature is defined from two selected part features. You can select work points, work planes, vertices, or planar faces that intersect the selected edge.

Distance Select to enter a distance over which you want the contour flange to be extruded and specify a direction. The following image shows samples of the Edge, Width, and Offset extent types.

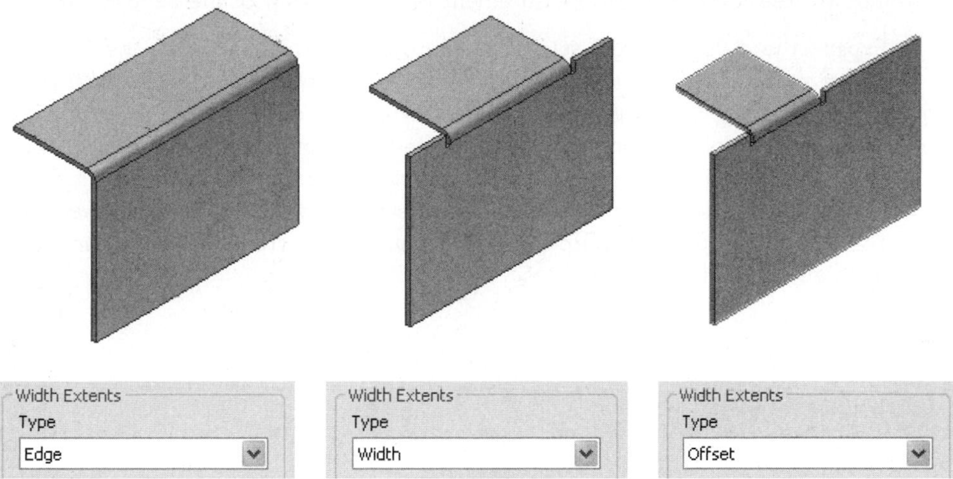

Figure 10.32

FLANGE

A sheet metal flange is a simple rectangular face created from existing face edges. A sketch is not required when creating a flange. The flange can extend the full length of the selected edge and can create automatic corner and miter features. The Flange tool, as shown in the following image, adds a new sheet metal face and bend to an existing face. If multiple edges are selected for the flange feature, corner seams and miter features are also created. You set the flange length and the angle relative to the adjacent face in the Flange dialog box. The selected edge or edges define the bend location between the selected faces.

Figure 10.33

Note: A minimum of one sheet metal face must exist before creating a flange.

To create a flange, follow these steps:

1. Click the Flange tool on the Sheet Metal Features panel bar. The Flange dialog box appears, as shown in the following image.
2. Select an edge or edges of an existing face.

3. Enter the distance and angle for the flange in the Flange dialog box. The flange preview updates to match the current values.
4. If required, flip the direction for the flange, offset the thickness, and modify the datum used for the height measurement or the position of the bend for the feature.
5. Expand the dialog box, and select the appropriate extent type. See the Contour Flange section in this chapter for additional information on extents. When creating a flange feature, a Design option is available when the dialog box is expanded, as shown in the following image. When selected, the options that were introduced specifically in Autodesk Inventor 2008 are disabled. If you are working with a file created in a previous release, the old method is selected, and the options that were used and available to create the feature in the previous release are available. Other options are disabled in the dialog box.

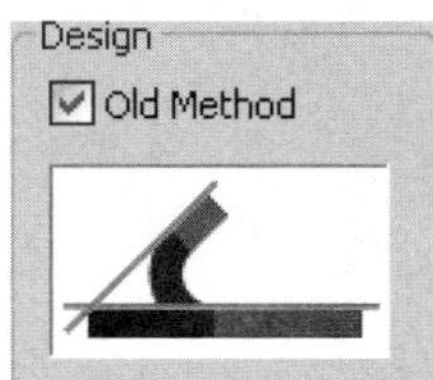

Figure 10.34

6. Click Apply to continue creating flanges, or click OK to apply the flange and exit the dialog box.

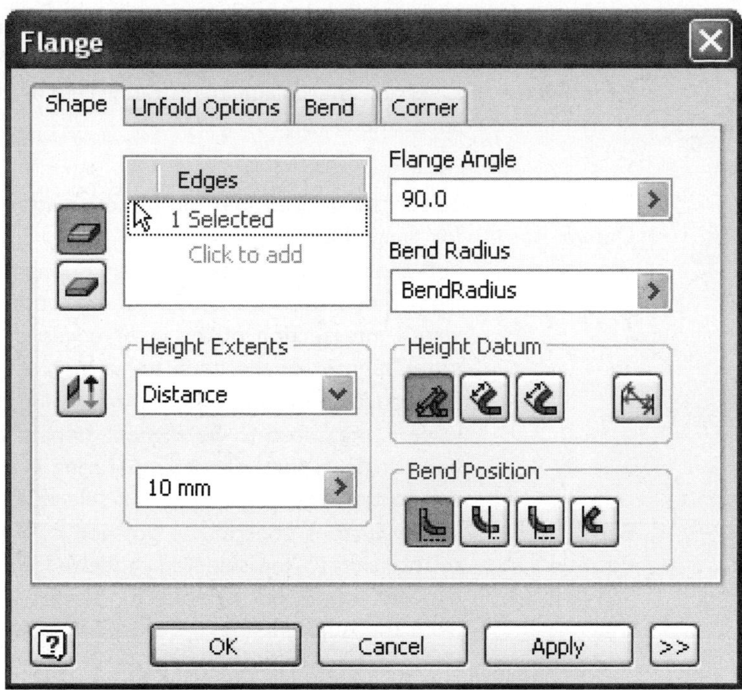

Figure 10.35

Shape Tab

The Shape tab allows you to select the edge, edge offset, flange distance, direction, and bend angle for the flange. You can also override the bend radius specified in the sheet metal style. The following options are available on the Shape tab.

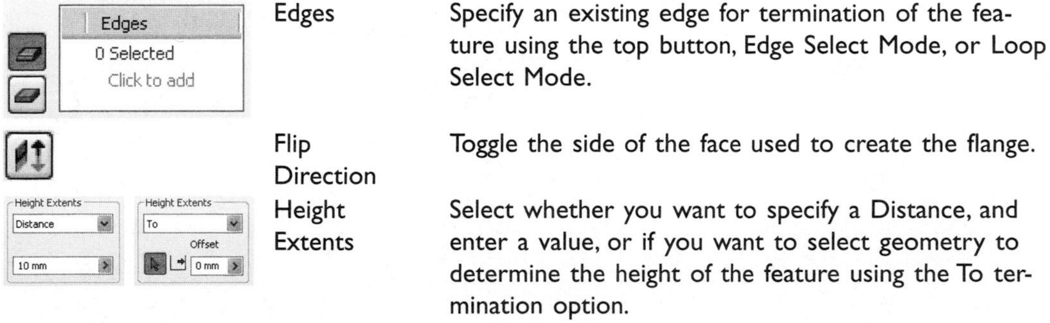

	Edges	Specify an existing edge for termination of the feature using the top button, Edge Select Mode, or Loop Select Mode.
	Flip Direction	Toggle the side of the face used to create the flange.
	Height Extents	Select whether you want to specify a Distance, and enter a value, or if you want to select geometry to determine the height of the feature using the To termination option.

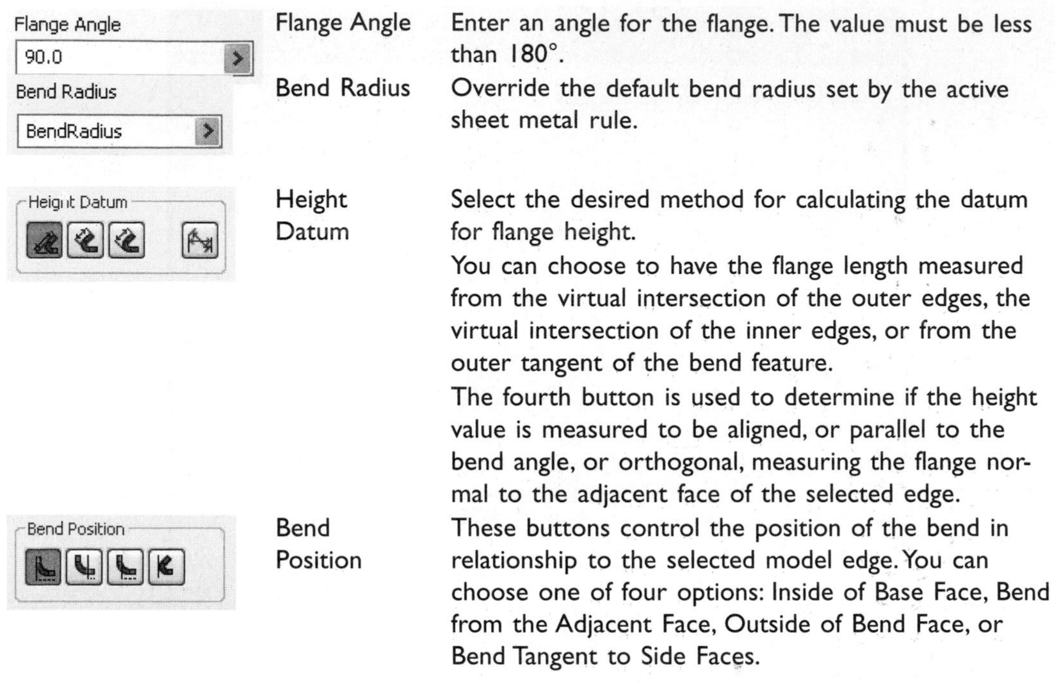

Flange Angle 90.0	Flange Angle	Enter an angle for the flange. The value must be less than 180°.
Bend Radius BendRadius	Bend Radius	Override the default bend radius set by the active sheet metal rule.
Height Datum	Height Datum	Select the desired method for calculating the datum for flange height. You can choose to have the flange length measured from the virtual intersection of the outer edges, the virtual intersection of the inner edges, or from the outer tangent of the bend feature. The fourth button is used to determine if the height value is measured to be aligned, or parallel to the bend angle, or orthogonal, measuring the flange normal to the adjacent face of the selected edge.
Bend Position	Bend Position	These buttons control the position of the bend in relationship to the selected model edge. You can choose one of four options: Inside of Base Face, Bend from the Adjacent Face, Outside of Bend Face, or Bend Tangent to Side Faces.

The options available on the Unfold Options and Bend tabs were discussed in the Face section. The Corner tab options were discussed in the Sheet Metal Rules section. The More button provides access to the Edge, Width, Offset, and From To extent types, and these are discussed in the Contour Flange section.

If you create or edit a flange or contour flange feature that has multiple edges converging in a corner, a glyph is displayed at the corner or corners. Clicking the glyph opens the Corner Edit tool as shown in the following image. The Corner Edit tool provides control over the geometry of the corner. These overrides can be set during feature creation or edit and control the type and size of relief, the overlap condition and the miter gap.

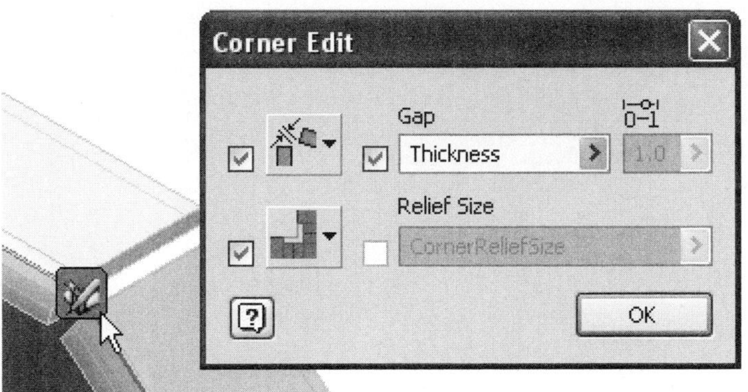

Figure 10.36

EXERCISE 10-1: CREATING SHEET METAL PARTS

In this exercise, you create sheet metal faces, flanges, and seams to build a small enclosure.

1. Open *ESS_E10_01.ipt*.
2. Click the Style and Standard Editor from the Format menu.
3. On the Sheet tab:
   a. Expand sheet metal rule.
   b. Right-click Default and click New Style from the menu.
   c. Enter Aluminum in the New Style Name dialog box and click OK.
   d. Double-click Aluminum in the Sheet Metal Rule list to make it active.
   e. Select Aluminum-6061 from the Material drop-down list.
   f. Enter a value of **1.6** in the Thickness field.
   g. Select the Bend tab.
   h. Select Round from the Relief Shape drop-down list.
   i. Select the Corner tab.
   j. Select Round from the Relief Shape drop-down list under 2 Bend Intersection.
   k. Click Save.
   l. Click Done.

The existing sheet metal face, as shown in the following image, is updated to reflect the revised sheet metal rule settings.

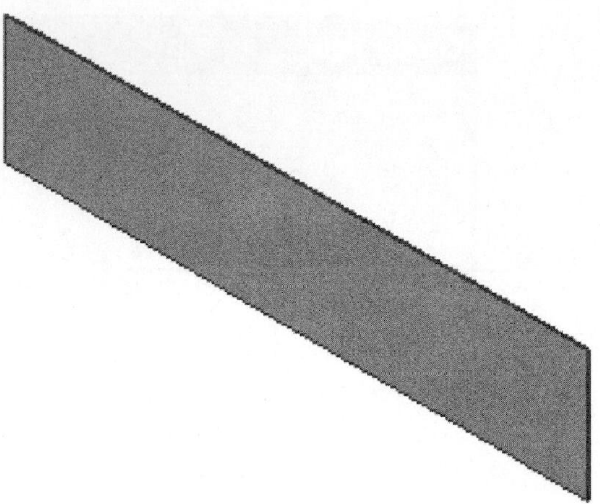

Figure 10.37

4. Create a new sketch on the long, thin face of the existing feature, as shown in the following image.

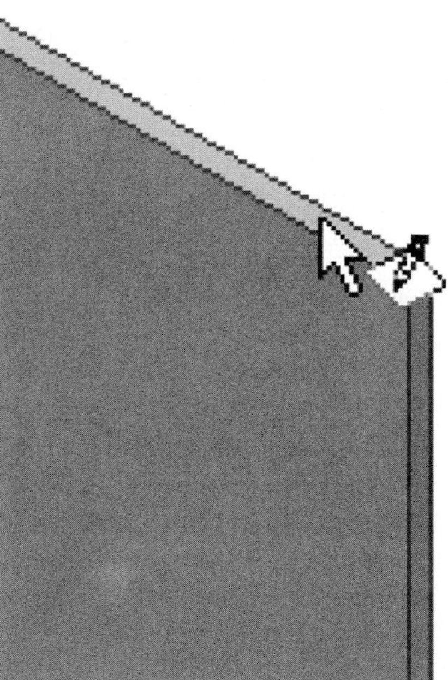

Figure 10.38

5. Click Top on the View Cube or use the Look At tool, and click Sketch2 in the browser.
6. Click the Two-Point Rectangle tool.
7. Click the upper horizontal edge when the coincident icon is displayed.
8. Create a rectangle coincident with the top edge of the profile, approximately as shown in the following image.
9. Add a **127 mm** dimension to the vertical edge of the sketch rectangle, as shown in the following image.

Figure 10.39

10. Click the Return tool to exit the sketch environment.
11. Click the Face tool.
12. Select the rectangle profile.
13. Click OK to accept the default settings.
14. Right-click the graphics window.
15. Select Home View from the menu.
16. Zoom in on a corner of the bend, and examine the bend relief notch in the original part face.
17. Use the Zoom, Rotate, and Pan tools to reorient the view, as shown in the following image.

EXERCISES

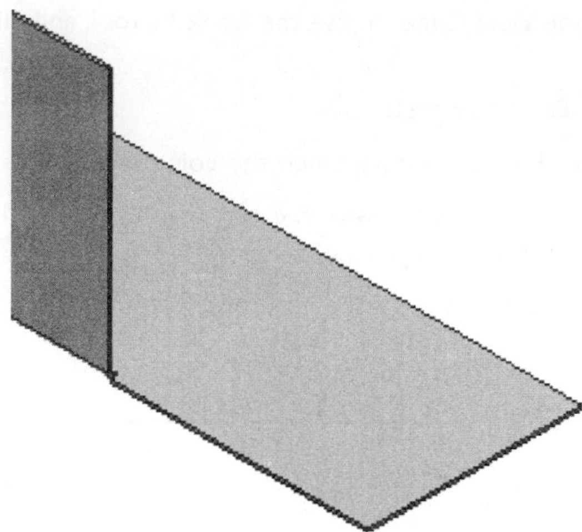

Figure 10.40

18. Click the Flange tool.

19. In the graphics window, click the edge on the horizontal surface, as shown in the following image.

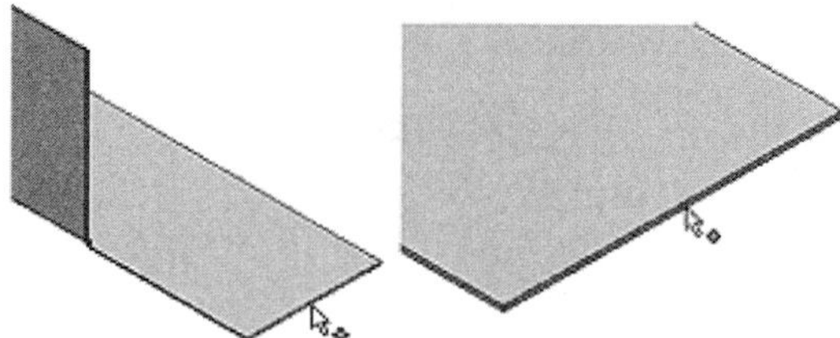

Figure 10.41

20. In the Flange dialog box:

a. Under Height Extents, verify that Distance is selected, and then click the arrow next to the Distance value field.

b. Select Show Dimensions.

c. Click the first face that you created.

d. Highlight all of the text in the Distance field.

e. Click the displayed 127 mm dimension, as shown in the following image.

f. The dimension parameter name replaces the text in the Distance field.

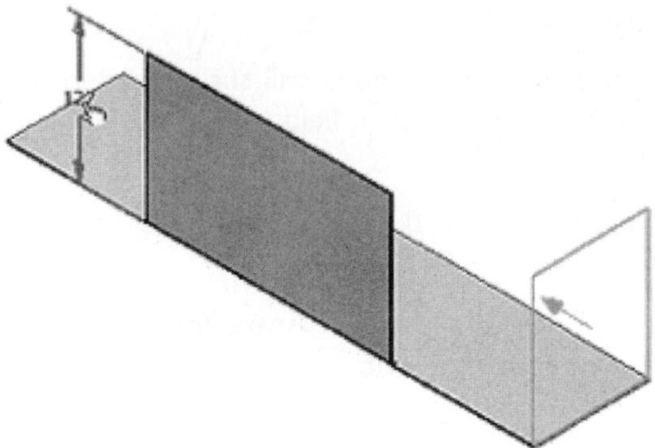

Figure 10.42

21. Click Apply to create the flange.
22. Rotate your view, as shown in the following image.

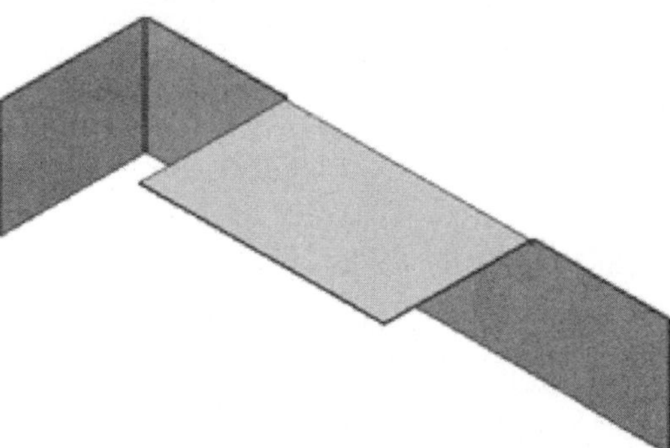

Figure 10.43

23. With the Flange dialog box still open, click the right-side (nearest) edge of the base feature, as shown in the following image.

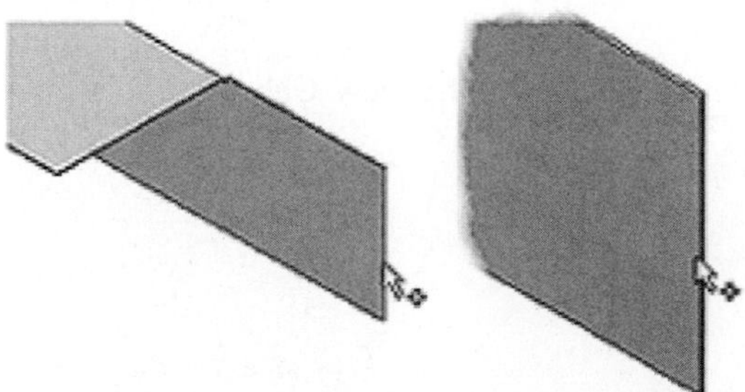

Figure 10.44

24. In the Flange dialog box:

    a. Verify that the parameter name is still shown under Height Extents. If not, click the arrow next to the Distance field, and select it from the list of recently used values.

    b. Complete the equation in the field by inserting **/sin(60)** after the dimension parameter name.

    c. Enter a value of **60** for the Flange Angle. Your dialog box should appear similar to the following image on the left.

    d. As the values are entered into the Flange dialog box, notice the sheet metal part updating to these values, as shown in the following image on the right.

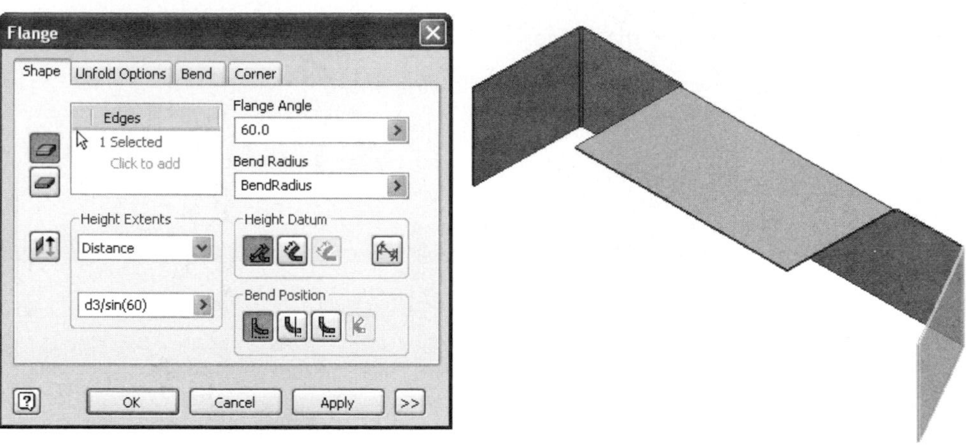

Figure 10.45

25. Click OK to complete the flange.

Note: The expressions entered for the two flange distances allow you to control the length of the face and the two flanges by editing the dimension of the first face you created.

26. Click the upper-left corner of the View Cube, as shown in the following image.

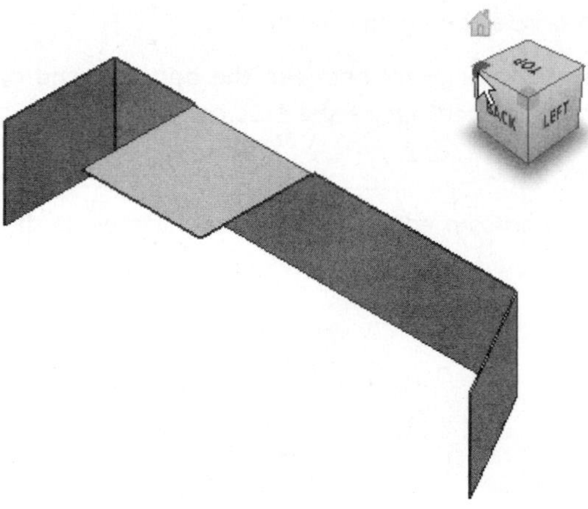

Figure 10.46

27. Click the Corner Seam tool.

a. Select the two edges, as shown in the following image.

Note: The order of selection and whether or not the top or bottom model edge is selected are not important.

28. Verify that the Thickness parameter is entered in the Gap field.

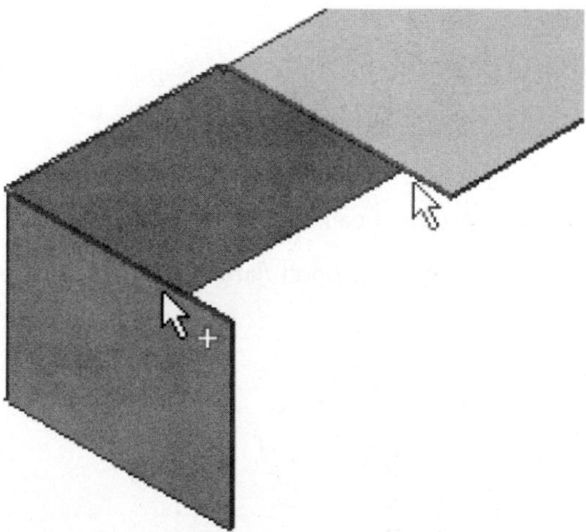

Figure 10.47

29. Click the Face/Edge Distance option in the Seam area.

30. Click each of the three Seam buttons, and examine the resulting seam preview.

31. Click the No Overlap button.

32. Click Apply to create the corner seam.
33. Create a second corner seam between the opposite end of the face and the sloped flange using the same settings as the first corner seam.
34. Click the Flange tool.
    a. Select the two bottom edges on the outer surfaces, as shown in the following image.

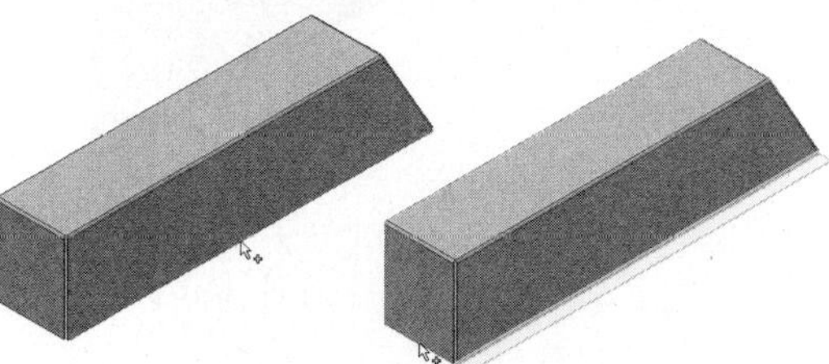

Figure 10.48

35. In the Flange dialog box:
    a. Enter a value of **20** in the Distance field.
    b. Enter a value of **90** in the Angle field.
    c. Zoom in on the lower left-hand corner of the flange where the preview is shown.
    d. Click the Height Datum and Bend Position buttons and observe the effect on the flange preview.
    e. Return the buttons to the default states: Bend from the intersection of the two outer faces and Inside of Bend Face Extents for the Bend Position.
    f. Click Apply.
36. Rotate the model so that you can see the angled flange.
37. Select the outside edge of the sloped flange, as shown in the following image.

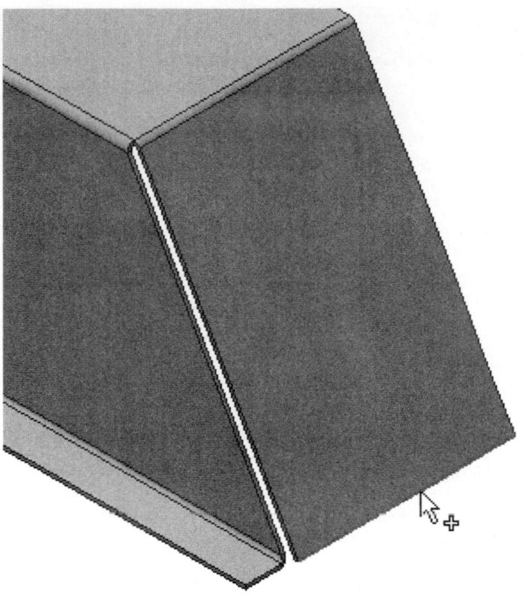

Figure 10.49

Note: This edge is selected to ensure that the bottom surfaces of the mounting flanges will be coplanar.

38. Enter a value of **20** in the Distance field.
39. Enter a value of **60** in the Angle field.
40. Click OK to complete the flange.
41. Click the Sheet Metal Default tool.
42. Select Default from the Sheet Metal Rule drop-down list. Notice that the material style is now set to "By Sheet Metal Rule" (Brass, Soft Yellow) as set by the sheet metal rule in the Style and Standard Editor. The Thickness is set to 3.00 mm and is unavailable for modification because the Use Thickness from Rule checkbox is selected, as shown in the following image.

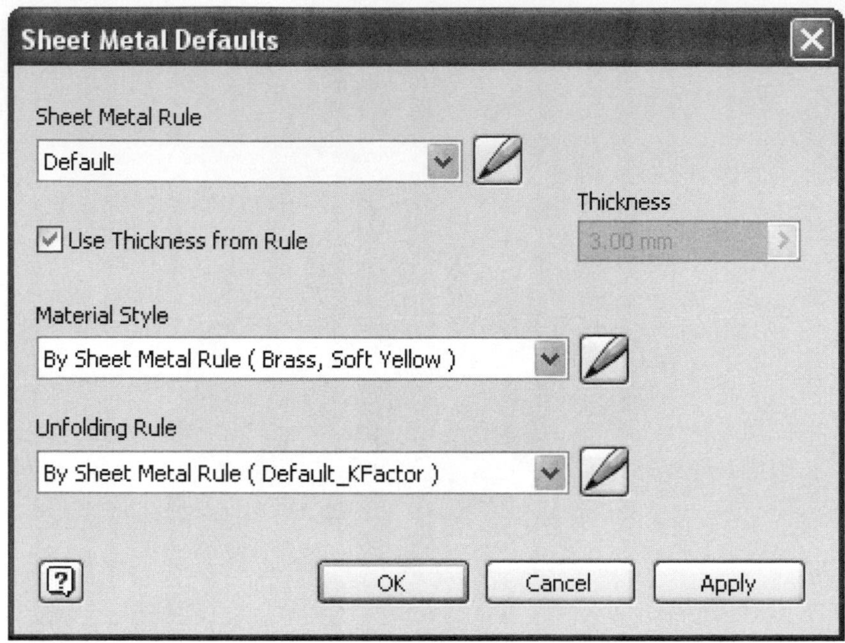

Figure 10.50

43. Click OK and the sheet metal part is updated to match the settings specified in the sheet metal rule, as shown in the following image.

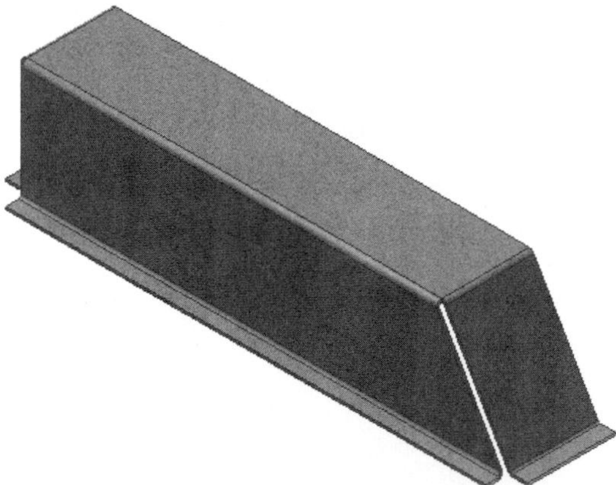

Figure 10.51

44. Create additional sheet metal rules, faces, flanges, and corner seams as desired.

45. Close the file. Do not save changes. End of exercise.

HEM

Hems eliminate sharp edges or strengthen an open edge of a face. Material is folded back over the face with a small gap between the face and the hem. A hem does not change the length of the sheet metal part; the face is trimmed so that the hem is tangent to the original length of the face. Create hems using the Hem tool, as shown in the following image.

Figure 10.52

Note: A minimum of one sheet metal face must exist before creating a hem.

To create a hem, follow these steps:

1. Click the Hem tool on the Sheet Metal Features panel bar. The Hem dialog box appears, as shown in the following image.

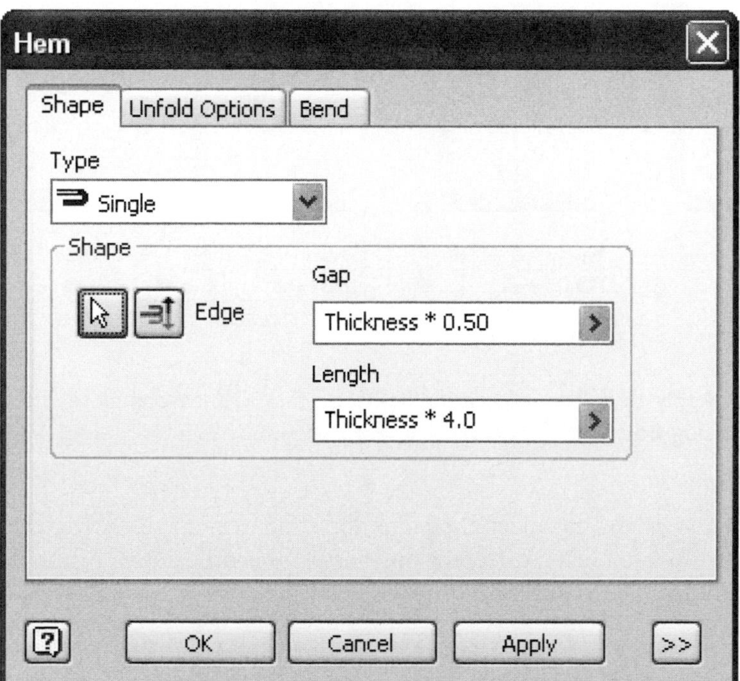

Figure 10.53

2. Select an open edge on a sheet metal face.
3. Select the hem type. Examples are shown in the following image:

- Single: A 180° flange
- Teardrop: A single hem in a teardrop shape
- Rolled: A cylindrical hem
- Double: Single hem folded 180° resulting in a double-thickness hem

4. Enter values for the hem. Teardrop and rolled hems require radius and angle values, while single and double hems require gap and length values. The hem preview changes to match the current values.
5. Expand the dialog box by clicking the More (<<) button and selecting Edge, Width, Offset, or From To for the hem extents.

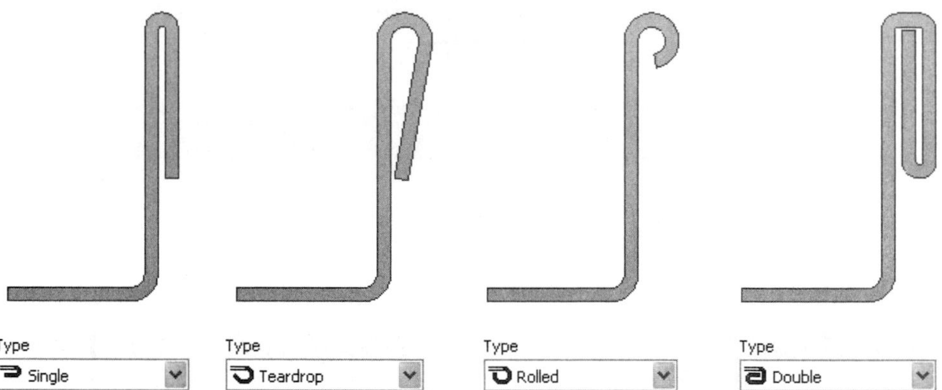

Figure 10.54

6. Click Apply to continue creating hems, or click OK to complete the hem and exit the dialog box.

Shape Tab

The Shape tab allows you to select the edge and type of hem to create. Based on the type of hem that you want to create, different options will become active in the Hem dialog box. The following options are available on the Shape tab.

	Type	Select One of Four Hem Types
	Select Edge	Select the edge along which the hem will be created.
	Flip Direction	Toggle the direction in which the hem will be created.

Gap Thickness * 0.50	Gap	Specify the distance between the inside faces of the hem. This feature is available when you select a single or double hem type.
Length Thickness * 4.0	Length	Specify the length of the hem. This feature is available when you select a single or double hem type.
Radius BendRadius	Radius	Specify the bend radius to apply at the bend. This feature is available when you select a rolled or teardrop hem type.
Angle 190.0	Angle	Specify the angle applied to the hem. This feature is available when you select a rolled or teardrop hem type.

The options available on the Unfold Options and Bend tabs were covered in the Face section. The More (<<) button provides access to the Edge, Width, Offset and From To extent types, which are discussed in the Contour Flange section.

EXERCISE 10-2: HEMS

In this exercise, you create two hem features.

1. Open *ESS_E10_02.ipt*.
2. Click the Hem tool.
   a. Click the edge, as shown in the following image.

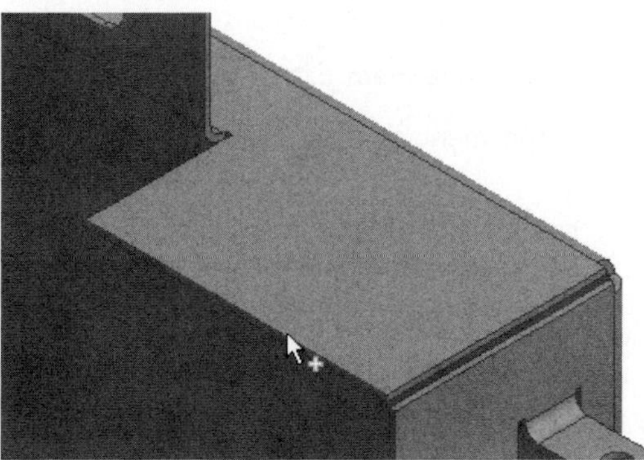

Figure 10.55

3. In the Hem dialog box:
   a. Enter **15 mm** in the Length edit box.
   b. If the hem previews to the inside of the box, click the Flip Direction button.
   c. Enter **0.5 mm** in the Gap edit box.

d. Click OK to close the Hem dialog box. Your display should appear similar to the following image.

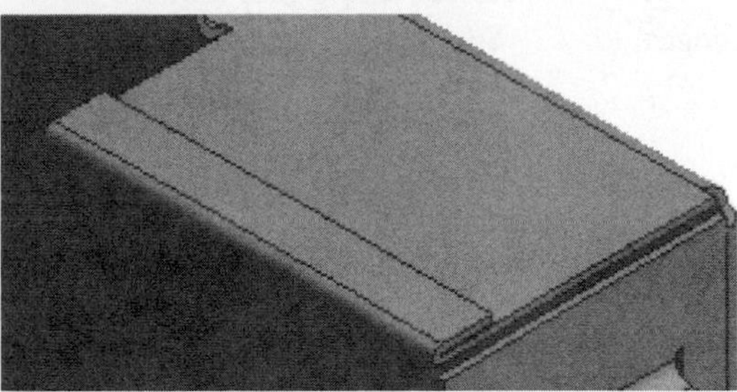

Figure 10.56

4. Use the Zoom and Rotate tools to reorient your view and match the following image.

Figure 10.57

5. In the browser, right-click the Hem Extents Work Plane.
6. Select Visibility from the menu.
7. Click the Hem tool.
8. Click the edge, as shown in the following image.

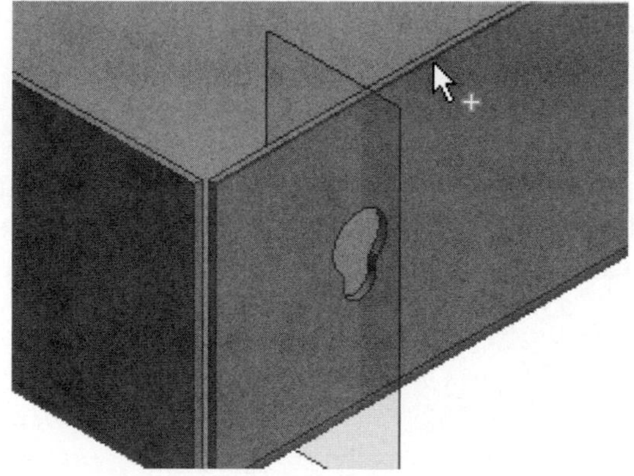

Figure 10.58

9. In the Hem dialog box:

a. Select Rolled from the Type list.

b. Enter **2 mm** in the Radius edit box.

c. Enter **265°** in the Angle edit box.

d. Click the Flip Direction button if the hem previews to the inside of the box.

e. Click the More button to expand the dialog box.

f. Select Offset from the Extents Type list.

g. Click the selection arrow next to Offset1.

h. In the graphics window, click the Hem Extents Work Plane.

i. The second offset point is automatically set to the endpoint of the selected edge.

j. Enter **0 mm** in the Offset1 edit box.

k. Enter **10 mm** in the Offset2 edit box.

l. Click OK to close the Hem dialog box. Your display should appear similar to the following image.

Figure 10.59

10. Experiment with the other Hem types and edges as desired.
11. Close the file. Do not save changes. End of exercise.

FOLD

An alternate method for creating sheet metal features is to start with a known flat pattern shape and then add folds to the sketched lines on a face. The Fold tool, as shown in the following image, can create sheet metal shapes that are difficult to create using the Face or Flange tools.

Figure 10.60

To create a fold, follow these steps:

1. Create a sketch on an existing face. Sketch a line between two open edges on the face.

Note: The sketched line endpoints must be coincident to the face edge.

2. Click the Fold tool on the Sheet Metal Features panel bar. The Fold dialog box appears, as shown in the following image.
3. Select the sketch or bend line. The fold direction and angle are previewed in the graphics window. The fold arrows extend from the face that will remain fixed. The face on the other side of the bend line will fold around the bend line.
4. If required, flip the fold direction and side.
5. Enter the angle of the fold.
6. Select the positioning of the fold with respect to the sketched line. The line can define the centerline, start, or end of the bend. The fold preview updates to match the current settings.
7. Make any needed changes in the Unfold Options or Bend tabs.

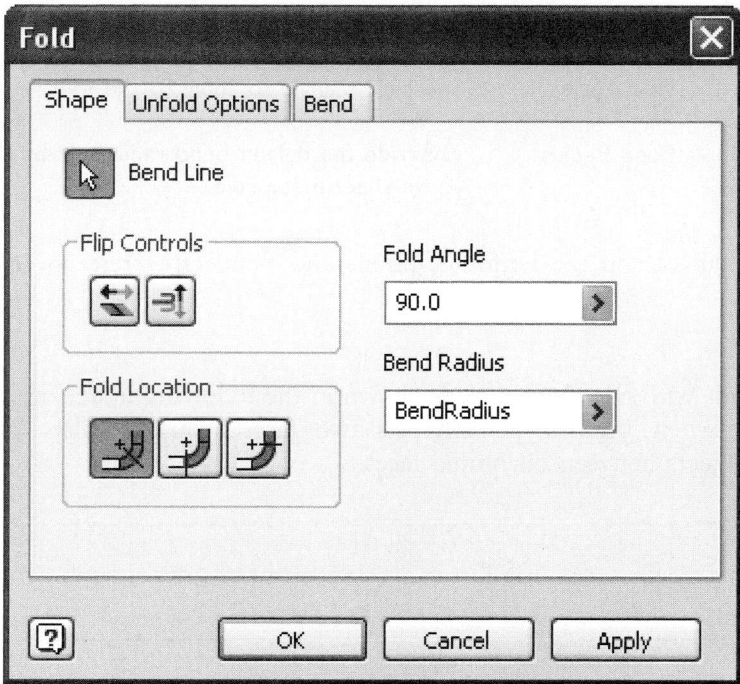

Figure 10.61

Shape Tab

The Shape tab allows you to select the bend line for the fold to be created. You can set the location of the selected bend line relative to the fold feature that determines the folded shape. The angle for the fold and direction are also specified on this tab, and you can also override the bend radius that is specified in the sheet metal rule. The following options are available on the Shape tab:

	Bend Line	Select a sketch line to use as the fold line.
	Flip Side	Flip the side used for the angle of the fold.
	Flip Direction	Toggle the direction that the fold will be created.
	Centerline of Bend	Determine the centerline of the fold from the selected sketch line.
	Start of Bend	Determine the start of the fold from the selected sketch line.

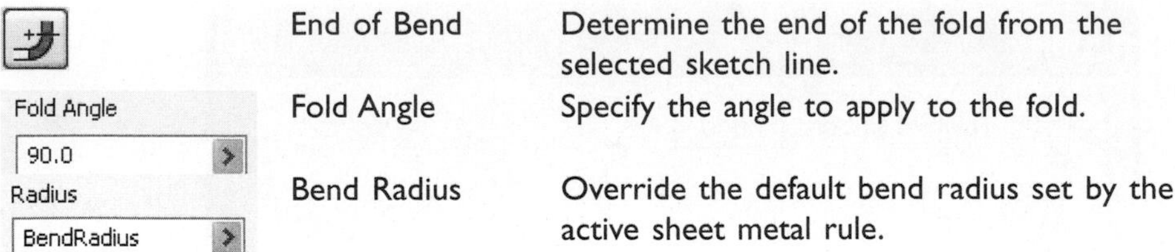

	End of Bend	Determine the end of the fold from the selected sketch line.
Fold Angle 90.0	Fold Angle	Specify the angle to apply to the fold.
Radius BendRadius	Bend Radius	Override the default bend radius set by the active sheet metal rule.

The options available on the Unfold Options and Bend tabs were covered in the Face section.

BEND

You create bends with the Bend tool, as shown in the following image, as child objects of other features when the feature connects two faces. You can also create bends as independent objects between disjointed faces.

Figure 10.62

A preview of a bend is shown in the following image, and the faces being joined are either trimmed or extended to connect the faces.

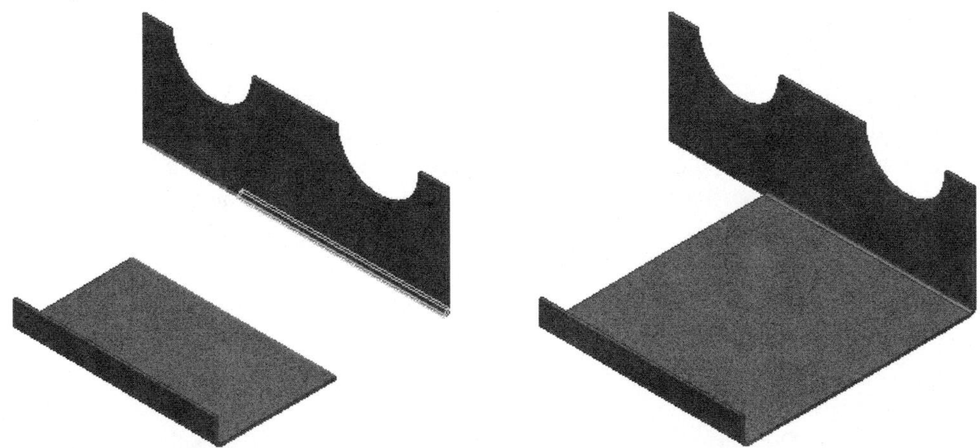

Figure 10.63

When you select parallel faces for the bend feature, a joggle or z-bend will be created, depending on the edge location. Joggles are often used to allow overlapping material. A sample of a joggle feature is shown in the following image on the left. You can also create double bends when the two faces are parallel and the selected edges face the same direction, as shown in the following image on the right.

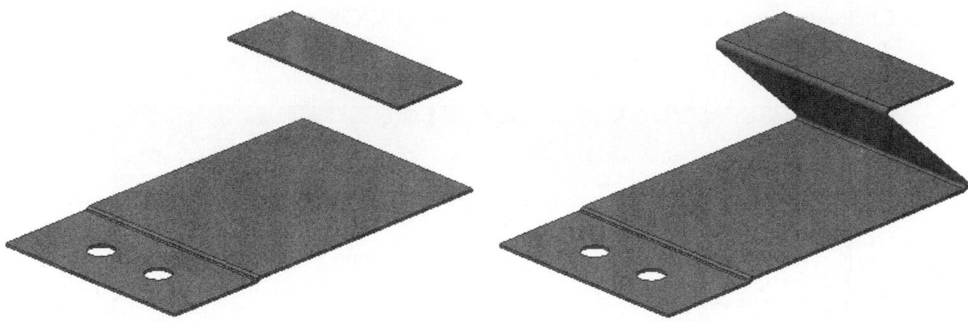

Figure 10.64

You can make the bend as two 90° bends with a tangent face between them, as shown in the following image on the left, or a single full-radius bend between the two faces, as shown in the following image on the right.

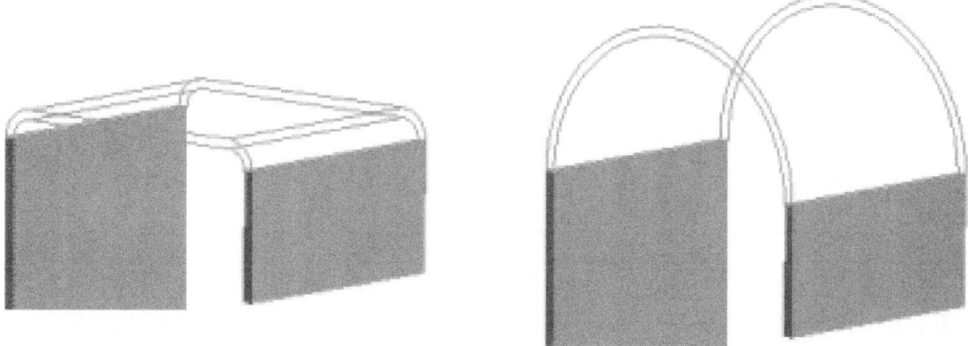

Figure 10.65

Note: To apply a bend, the sheet metal part must have two disjointed or sharp-cornered intersecting faces. An example of intersecting faces would be a shelled box that is being changed into a sheet metal part. The intersection of the box base and a wall is an edge that can be changed to a bend.

To create a bend, follow these steps:

1. Click the Bend tool on the Sheet Metal Features panel bar. The Bend dialog box appears, as shown in the following image.
2. Select the common edge of two intersecting faces, or select two parallel edges on disjointed, non-coplanar faces. If the two faces are parallel, the Double Bend options are available for selection. Depending on the position of the two faces, the allowed Double Bend options will be Fix Edges and 45 Degree or Full Radius and 90 Degree.
3. Make any changes in the Unfold Options or Bend tabs.

4. Click Apply to continue creating bends, or click OK to complete the bend and exit the dialog box.

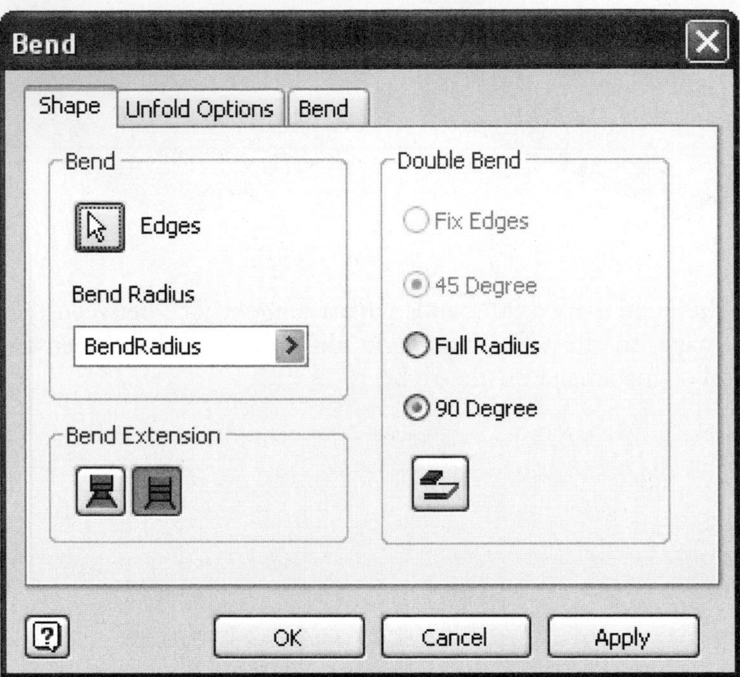

Figure 10.66

Shape Tab

The Shape tab allows you to select the edges between which to create the bend and specify the type of bend that you want to create. You can also override the bend radius specified in the sheet metal style. The following options are available on the Shape tab.

	Edges	Select the edges where the bend will be applied. The selected edges will be trimmed or extended as needed to create the bend feature.
Radius BendRadius	Bend Radius	Override the default bend radius set by the active sheet metal style.
Bend Extension	Bend Extension	Controls how bend extensions are calculated when a profile has edges that are not coincident to the selected edge.

		Extend Bend Aligned to Side Faces: Aligns the bend of the shorter edge when the edges are not equal. Extend Bend Perpendicular to Side Faces: Extends the bend from the longer side when the edges are not equal.
	Fix Edges	Click to create equal bends to the selected sheet metal edges.
	45 Degree	Click to create 45° bends on the selected edges.
	Full Radius	Click to create a single semicircular bend between the selected edges.
	90 Degree	Click to create 90° bends between the selected edges.
	Flip Fixed Edge	Click to reverse the order of the edges being fixed. Normally, the first edge selected is fixed by default, and the second edge will be trimmed or extended. This button reverses the order.

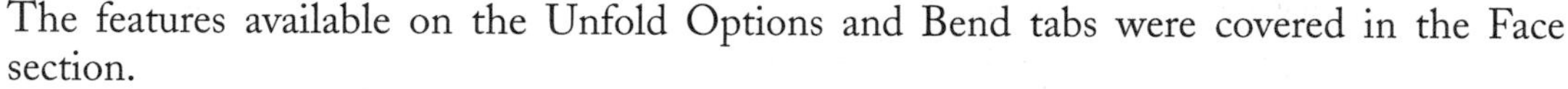

The features available on the Unfold Options and Bend tabs were covered in the Face section.

You can edit a bend that is listed under a feature in the browser at any time, allowing you to reselect the edges that define the bend. You can even edit the bend to connect two faces that were not initially joined when the bend feature was created.

EXERCISE 10-3: MODIFYING SHEET METAL PARTS

In this exercise, you complete the design of a sheet metal bracket within the context of an assembly and then modify the sheet metal part to eliminate component interference. You add bends to connect disjointed sheet metal faces and modify one bend to eliminate interference with a component in the assembly. You then add a hinge feature to the sheet metal part using general modeling tools.

1. Open *ESS_E10_03.iam*.
2. Apply the "Start" Design View Representation, as shown in the following image on the left. The assembly should appear, as shown in the following image on the right.

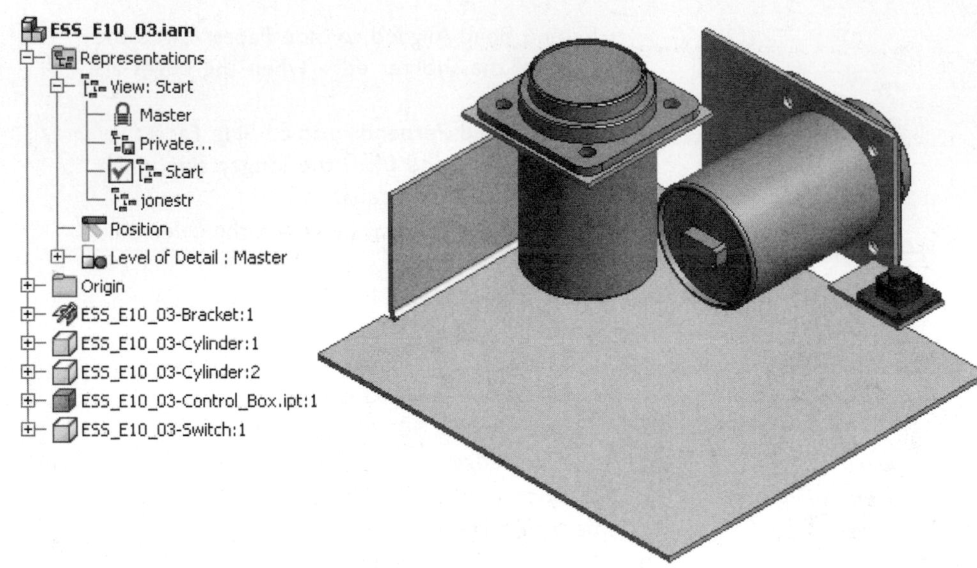

Figure 10.67

Note: The sheet metal bracket consists of several unconnected faces. Three faces act as mounting surfaces for assembly components; the two connected faces form the base of the bracket. The unique Autodesk Inventor method of creating sheet metal bends between disjointed faces is used to complete the sheet metal bracket.

3. In the browser, right-click *ESS_E10_03-Bracket:1*.
4. Select Edit from the menu.
5. Reorient the assembly, as shown in the following image.

Figure 10.68

6. Click the Bend tool.

   a. Click the two edges, as shown in the following image on the left.

7. Click Apply to place the bend. Your display should appear similar to the following image on the right.

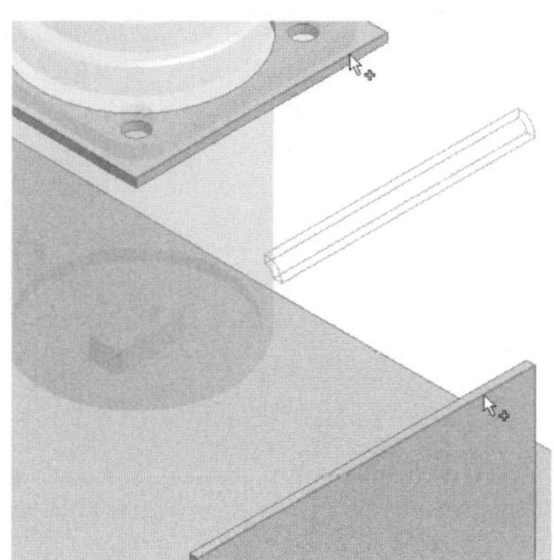

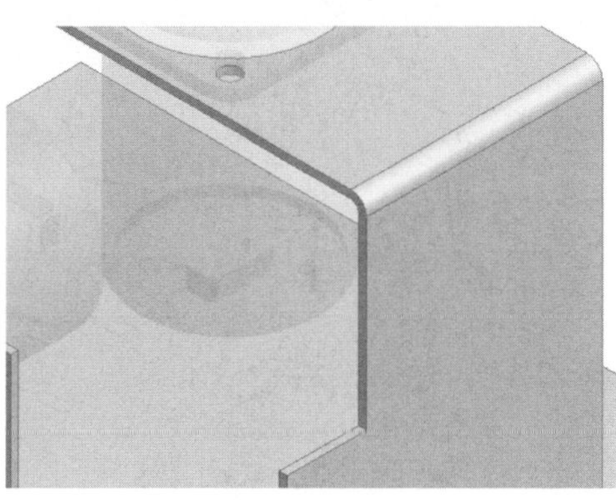

Figure 10.69

8. A second bend will now be placed. Click the two edges, as shown in the following image on the left.

9. Click Apply to place the bend. Your display should appear similar to the following image on the right.

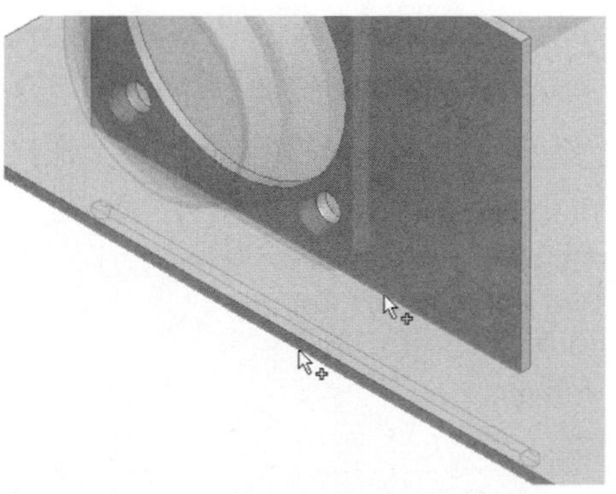

Figure 10.70

10. Reorient the view of the assembly, as shown in the following image.

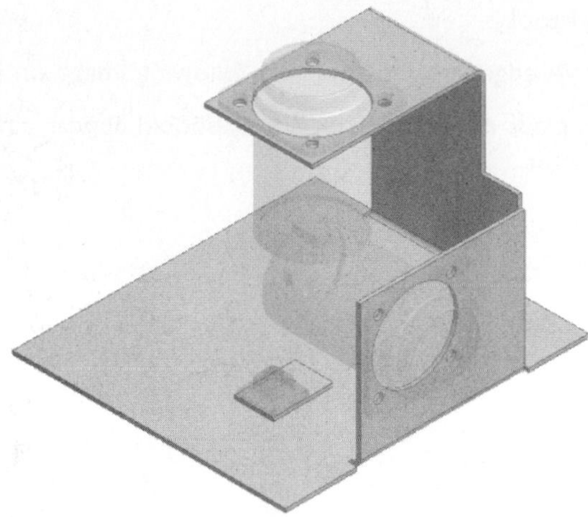

Figure 10.71

11. Click the back edge of the small face, shown as Edge 1 in the following image on the left.

   a. Click Edge 2, as shown in the following image on the left.

12. In the Bend dialog box:

   a. Click the Fix Edges option.

   b. Click OK to place the bend. Your display should appear similar to the following image on the right.

   c. The bend angles required to create the z-bend may be difficult to manufacture. You can easily edit the bend to return to the default 45° double bends.

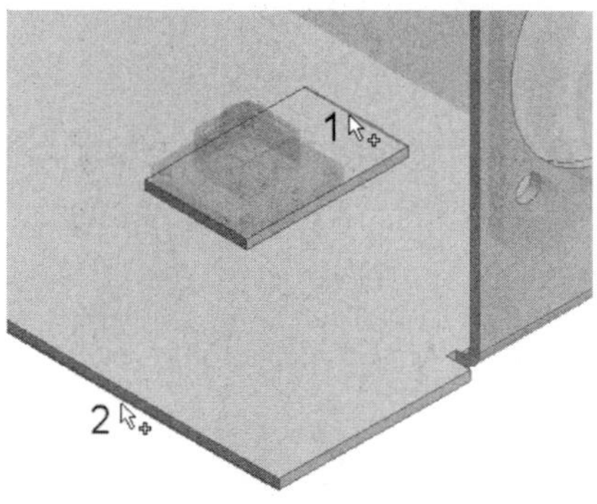

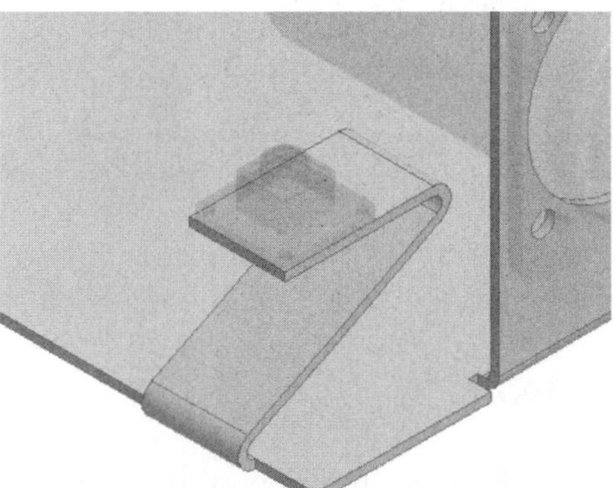

Figure 10.72

13. You will now edit the Z-shaped Bend in the browser:

   a. Expand the Folded Model node.

b. Right-click the last bend listed.

c. Select Edit Feature from the menu.

14. In the Bend dialog box:

a. Click the 45 Degree option in the Double Bend area.

b. Click OK to modify the bend. Your display should appear similar to the following image.

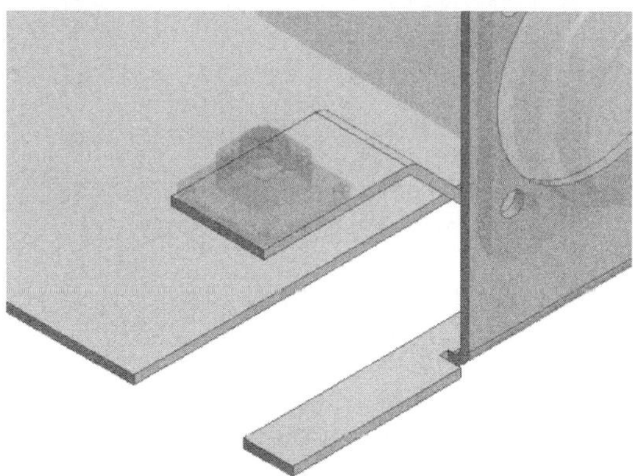

Figure 10.73

Note: The small tab next to the vertical mounting face is not required.

15. Click the large base face.

16. Press the S key to create a new sketch.

17. Use the Zoom and Pan tools to enlarge the view of the small tab.

18. Click the Line tool.

19. Draw a line connecting the corner of the bend relief and the edge of the cutout for the 45° jog bend, as shown in the following image on the left.

Note: Ensure that the line is constrained to the corner point of the bend relief and is perpendicular to the cutout edge.

20. Click the Return tool, and then click the Cut tool.

21. Select the small tab as the profile.

22. Set the Extents to Distance and a value of Thickness.

23. Click OK to complete the cut.

24. Click the Zoom All tool. Your display should appear similar to the following image on the right.

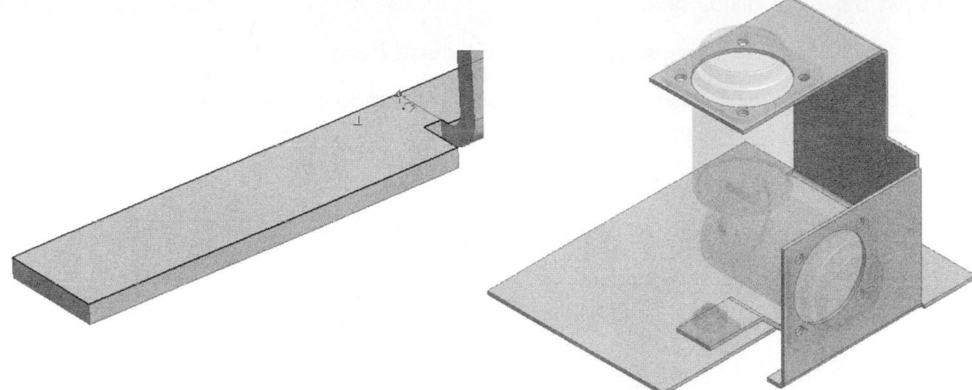

Figure 10.74

25. In the browser:

    a. Double-click *ESS_E10_03.iam* to return to the assembly environment.

    b. Right-click *ESS_E10_03-Control_Box.ipt:1*.

    c. Select Visibility from the menu.

Note: The Control Box interferes with the sheet metal part, as shown in the following image.

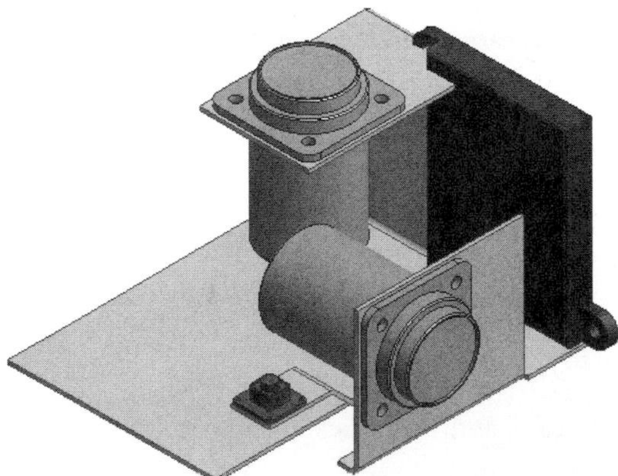

Figure 10.75

To eliminate the interference, edit the bend causing the conflict. In the graphics window:

26. Double-click *ESS_E10_03-Bracket:1* to activate it for editing.

27. Use the Rotate and Zoom tools on the Standard toolbar to orient the assembly, as shown in the following image.

Figure 10.76

28. In the browser:

a. Place the cursor over the first bend under the Folded Model node of *ESS_E10_03-Bracket.ipt.* The bend adjacent to the interference should be highlighted in the graphics window.

b. Right-click this bend, and select Edit Feature from the menu.

29. In the Bend dialog box:

a. Click the Edges button.

b. Hold down the CTRL key, and deselect the two edges that define the bend, as shown in the following image on the left.

c. Select the top edge of the face supporting the horizontal cylinder and the adjacent parallel edge on the face supporting the vertical cylinder, as shown in the following image on the right.

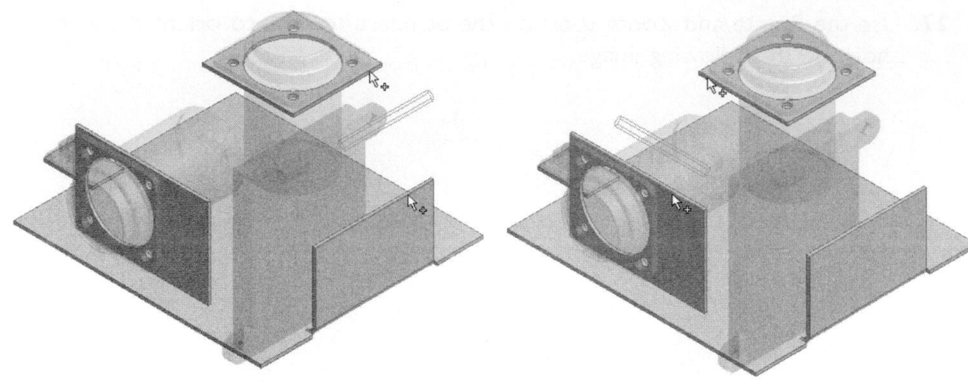

Figure 10.77

30. Click OK to create the modified bend.

The control box no longer interferes with the sheet metal bracket. The vertical face adjacent to the control box can be modified to provide support for the control box. That modification is not covered in this exercise.

31. Reorient the view of the assembly, as shown in the following image.

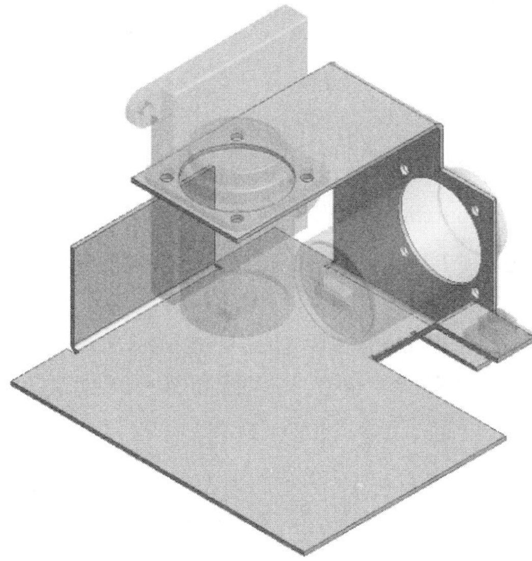

Figure 10.78

32. In the browser:

   a. Right-click Sketch_Hinge.

   b. Select Visibility from the menu to turn the sketch on, as shown in the following image on the left.

33. Activate the Part Features panel bar, and click the Extrude tool.

Verify that the selected profile matches the profile, as shown in the following image on the right. Click in the profile if it is not automatically selected.

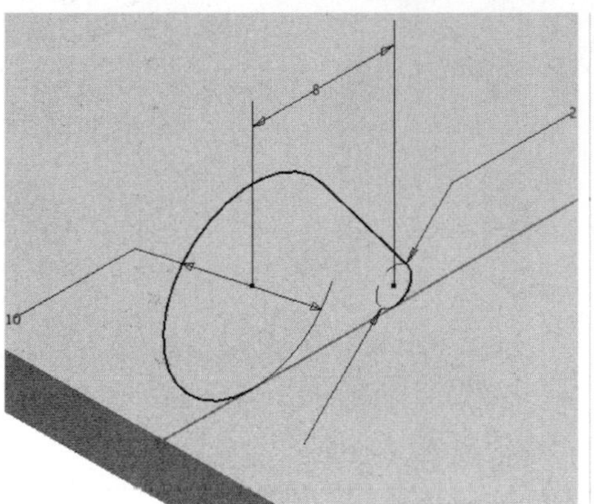

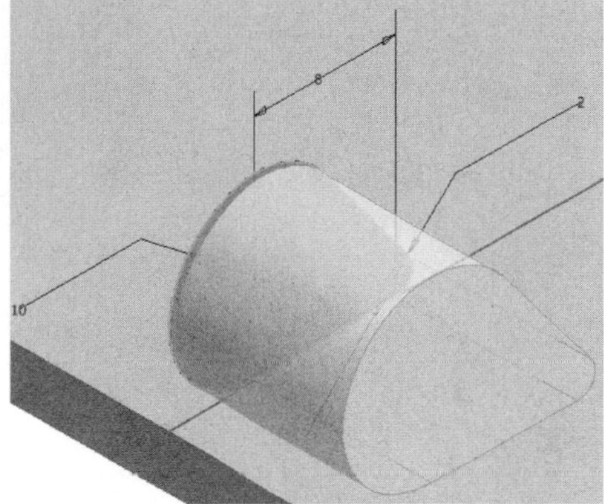

Figure 10.79

34. In the Extrude dialog box:

a. Enter a value of **200** as the Extents Distance.

b. Click the middle Extents Direction button to flip the extrusion direction.

c. Click OK to complete the weld-on hinge feature.

35. Click the Zoom All tool.

36. In the browser, double-click the top level assembly, *ESS_E10_03.iam*, to activate the assembly environment, as shown in the following image.

Figure 10.80

37. Close the file. Do not save changes. End of exercise.

CUT

The Cut tool, as shown in the following image, is a sheet metal-specific implementation of the standard Extrude tool. The Cut tool always performs an extrude cut; join and intersection cuts are not available. The default extents are a blind cut at a distance equal to the sheet metal Thickness parameter. This action ensures that the cut extends only through the face containing the sketch and not through other faces that may be folded under the sketch face. You can enter a value that is less than the thickness of the part to create a cut that does not go all of the way through the material.

Figure 10.81

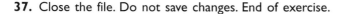

Most cuts are manufactured on the flat sheet stock before the sheet is bent to form the folded part, and cuts often cross bend lines. Because the part is modeled in the folded state, representing cuts that cross bend boundaries requires the bend be unfolded to represent the flat sheet. When the bend is refolded, the cut deforms around the bend to ensure that the extrusion remains perpendicular to the sheet metal faces on both sides of the bend and deforms throughout the bend, if required. When sketching a cut profile that will cross a bend, the Project Flat Pattern tool projects the unfolded flat pattern geometry onto the sketch face, as shown in the following image.

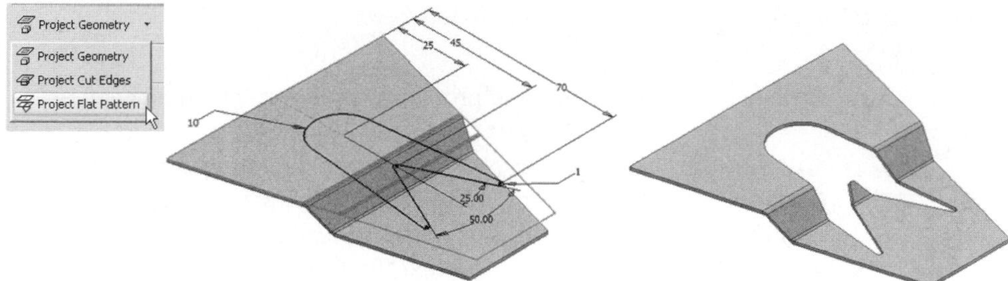

Figure 10.82

To create a cut feature, follow these steps:

1. Create a sketch on a sheet metal face that includes one or more closed profiles representing the area(s) to cut. If required, use the Project Flat Pattern tool to project the unfolded geometry of connected faces onto the sketch.
2. Click the Cut tool on the Sheet Metal Features panel bar. The Cut dialog box appears, as shown in the following image.
3. Select the profiles to cut.
4. If a profile crosses a bend, click Cut Across Bend.
5. Enter a distance for the extents of the cut.
6. Click OK to complete the cut and exit the dialog box.

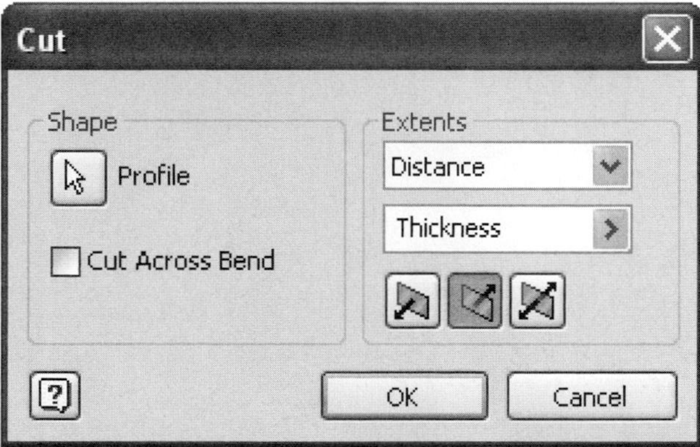

Figure 10.83

The following tools are available in the Cut dialog box:

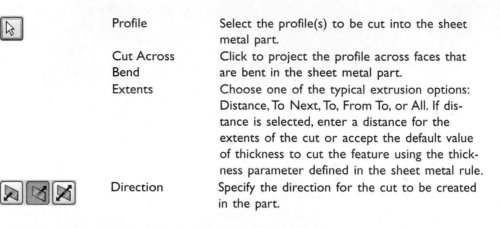

	Profile	Select the profile(s) to be cut into the sheet metal part.
	Cut Across Bend	Click to project the profile across faces that are bent in the sheet metal part.
	Extents	Choose one of the typical extrusion options: Distance, To Next, To, From To, or All. If distance is selected, enter a distance for the extents of the cut or accept the default value of thickness to cut the feature using the thickness parameter defined in the sheet metal rule.
	Direction	Specify the direction for the cut to be created in the part.

EXERCISE 10-4: CUT ACROSS BEND

In this exercise, you complete a sheet metal bracket using a fold and double bend. You then use the Project Flat Pattern tool to create a cut across bends.

1. Open *ESS_E10_04.ipt.*
2. Use the Zoom and Rotate tools to examine the part. Reorient your view to match the following image.

Figure 10.84

3. In the browser, right-click Face_Sketch, and select Visibility from the menu.

The sketch includes projected geometry from the adjacent face.

4. Click the Face tool.
5. Click OK in the Face dialog box. Your display should appear similar to the following image.

Figure 10.85

6. Reorient your view to match the following image.

Figure 10.86

You will now create a double bend between the two parallel faces.

7. Click the Bend tool.
8. Click the two edges, as shown in the following image.

Figure 10.87

9. In the Bend dialog box:
   a. Click Full Radius.
   b. Click 90 Degree.
   c. Click OK.
10. Reorient your view to resemble the following image. Create a new sketch on the face, shown highlighted in the following image.

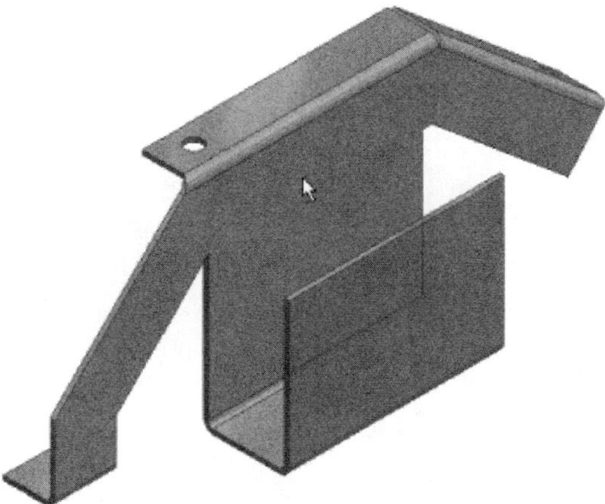

Figure 10.88

11. Click the Line tool.
12. Create a line, as shown in the following image.

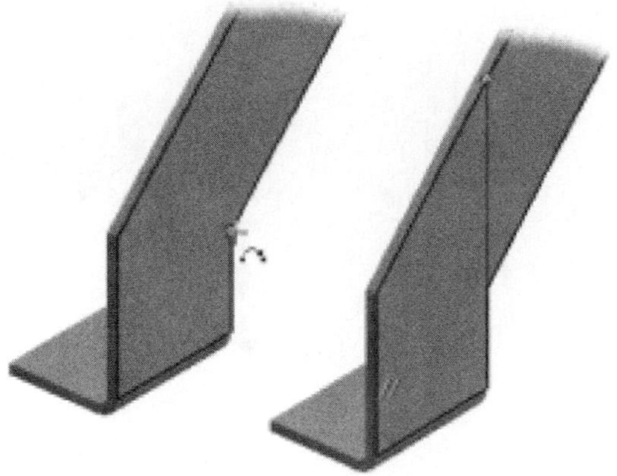

Figure 10.89

13. Right-click in the graphics window, and select Done.

14. Right-click in the graphics window, and select Finish Sketch.

15. Click the Fold tool, and select the sketched line.

16. In the Fold dialog box:

a. If required, click the Flip direction button to match the following image.

b. Click End of Bend for the Fold Location.

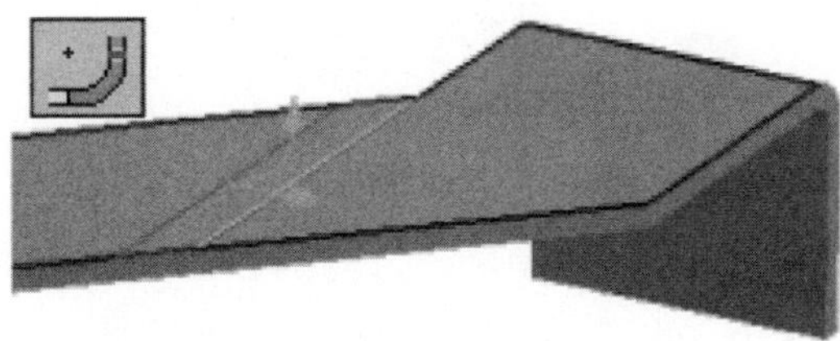

Figure 10.90

c. Click OK.

d. You will now use the Project Flat Pattern tool to create a sketch, and you use this sketch to create a cut across bends.

17. Reorient your view to resemble the following image. Create a new sketch on the face, shown highlighted in the following image.

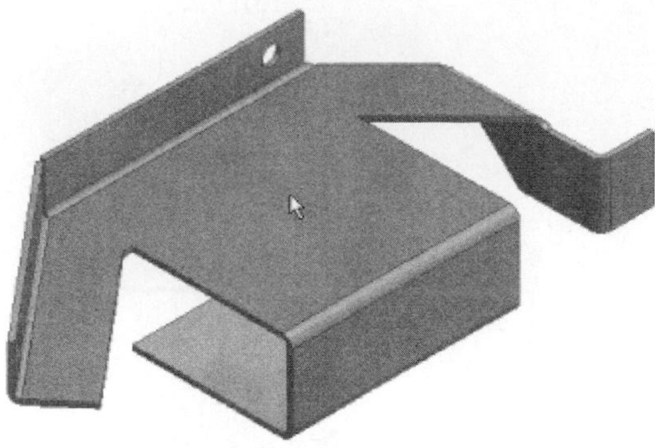

Figure 10.91

18. Click the Project Flat Pattern tool.

19. Select the face highlighted in the following image.

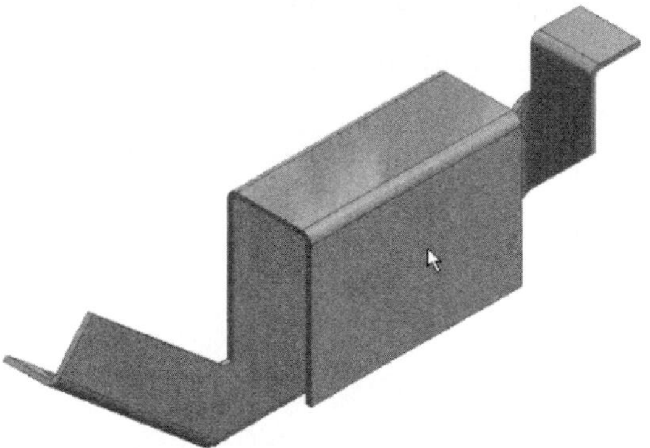

Figure 10.92

20. Click the Look At tool.

21. Click one of the projected flat pattern edges.

22. Create the sketch, as shown in the following image.

Note: The sketch is centered on the face by a horizontal constraint between the midpoint of the projected and vertical sketched line and a vertical constraint between the bottom horizontal projected edge and the centerpoint of the arc.

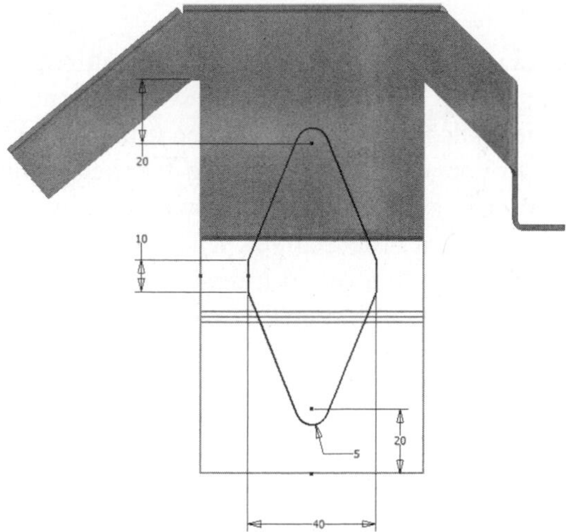

Figure 10.93

Each horizontal construction line connects the projected edge midpoint and the sketch line midpoint.

23. Right-click the graphics window, and select Finish Sketch.

24. Click the Cut tool.

25. Click inside the sketch if it is not already selected.

26. Ensure that Cut Across Bend is checked in the Cut dialog box.

27. Click OK. Your display should appear similar to the following image.

Figure 10.94

28. Close the file. Do not save changes. End of exercise.

CORNER SEAM

When three faces meet in a sheet metal part, a gap is required between two of the faces to enable unfolding. Using a box as an example, the walls of the box are connected to the floor, and gaps between the walls enable the box to be unfolded. The gap between adjacent faces is a corner seam, and you create it using the Corner Seam tool, as shown in the following image. The Corner Seam tool can work with two and three-bend intersections. When two coplanar flanges meet, the flanges will result in a three-bend intersection, referred to as a Jacobi corner. If a three-bend intersection is detected, options will become active in the Corner Seam dialog box that enable you to define how the corner feature will be treated when a flat pattern is generated. You can also create a corner seam by ripping open a corner at an intersection between faces.

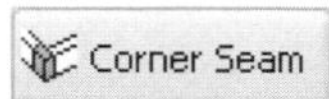

Figure 10.95

To create a corner seam, follow these steps:

1. Click the Corner Seam tool on the Sheet Metal Features panel bar. The Corner Seam dialog box appears, as shown in the following image.
2. Select face edges that meet one of the following criteria:

- Two open edges on faces that share a common connected edge: for example, two walls that share a connection to a floor.
- Two nonparallel edges on coplanar faces: for example, the edges create a mitered joint with a gap between the extended faces.
- A single edge at the intersection of two faces, such as the wall intersections of a shelled box.

3. Select the Seam option in the Corner Seam dialog box.
4. Enter a value for the corner seam Gap.
5. Make any changes to the Bend or Corner options.
6. Click Apply to continue creating corner seams, or click OK to complete the corner seam and exit the dialog box.

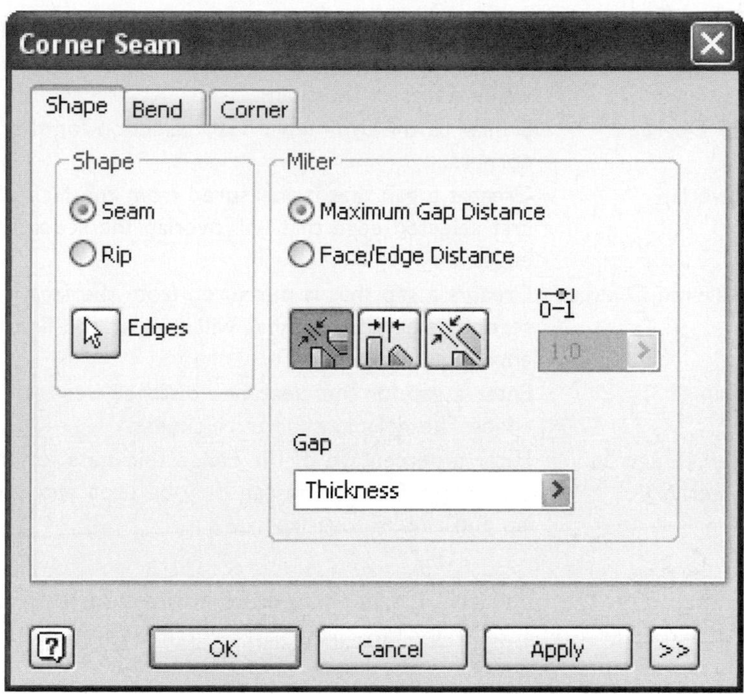

Figure 10.96

Shape Tab

The Shape tab allows you to select the edges where the corner seam will be created. You can also specify the orientation of the seam and whether or not to rip a corner. The following options are available on the Shape tab for miter corners. If you select coplanar faces, a miter corner is created and the options in the dialog box are displayed as shown in the previous image. If you select intersecting faces, a seam is created. The options available will change based on the selected geometry.

	Seam	Click to create a corner seam by extending or trimming existing coplanar or intersecting sheet metal faces.
	Rip	Click to rip open a square corner of a part.
	Edges	Select the edges of the model on which you will create a corner seam.
	Maximum Gap Distance	Select to create a corner seam gap that is measured consistent with the use of a physical inspection gauge.

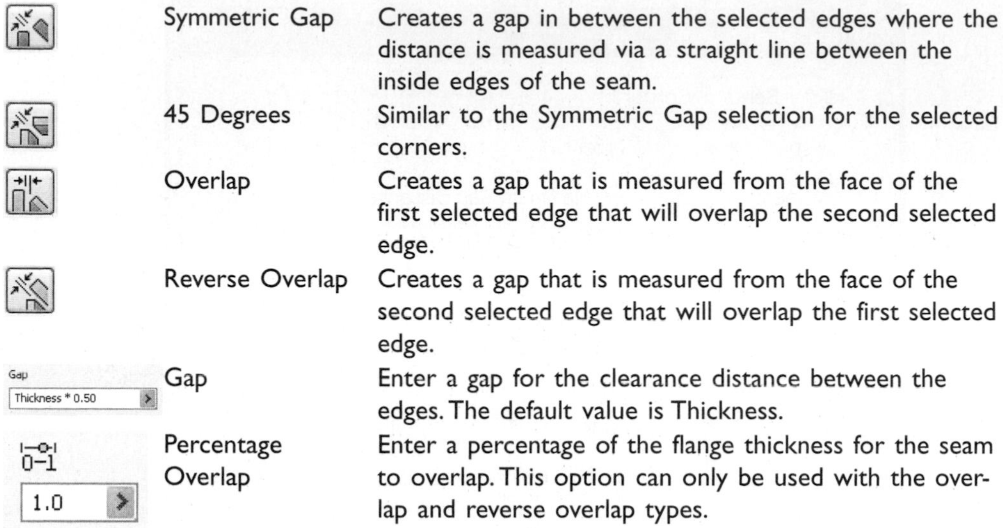

	Symmetric Gap	Creates a gap in between the selected edges where the distance is measured via a straight line between the inside edges of the seam.
	45 Degrees	Similar to the Symmetric Gap selection for the selected corners.
	Overlap	Creates a gap that is measured from the face of the first selected edge that will overlap the second selected edge.
	Reverse Overlap	Creates a gap that is measured from the face of the second selected edge that will overlap the first selected edge.
Gap Thickness * 0.50	Gap	Enter a gap for the clearance distance between the edges. The default value is Thickness.
0-1 1.0	Percentage Overlap	Enter a percentage of the flange thickness for the seam to overlap. This option can only be used with the overlap and reverse overlap types.

The options available on the Bend and Corner tabs were covered in the Face section and the Sheet Metal Rules section. On these tabs, you can override the default settings from the active sheet metal rule on a per-feature basis.

More Button

Additional options are available when you click the More (<<) button to expand the dialog box, as shown in the following image. The options are described as follows.

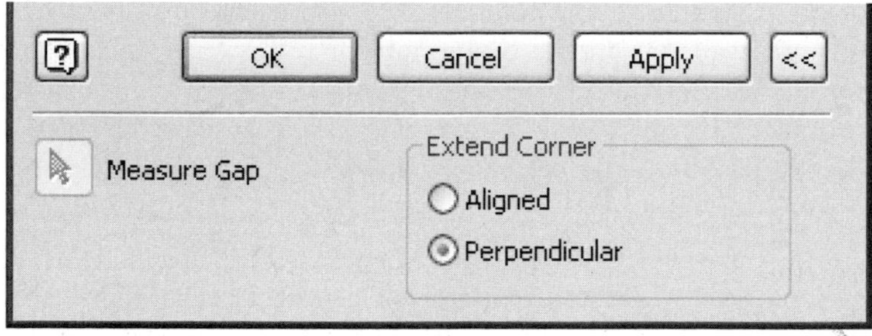

Figure 10.97

Measure Gap Select two edges to measure the distance between them.

Aligned Click to extend faces to match sloped edges adjacent to edges of the selected faces.

Perpendicular Click to extend faces at the same height as the selected edge, ignoring any adjacent sloped edges.

CORNER ROUND

The Corner Round tool, as shown in the following image, is available when no sketch is active. It is a sheet metal-specific fillet tool. All edges other than those at open corners of faces—that is, all edges having a Length = Thickness—are filtered out. This action enables you to select these small edges easily without zooming in on the part.

Figure 10.98

To create a corner round, follow these steps:

1. Click the Corner Round tool on the Sheet Metal Features panel bar. The Corner Round dialog box appears, as shown in the following image.
2. Enter a radius for the corner round.
3. Select the Corner or Feature Select Mode.
4. Select the corners or features to include.
5. Add additional corner rounds with different radii, and select corners or features for the additional corner rounds.
6. Click OK to complete the corner round and exit the dialog box.

Figure 10.99

Select Mode

Selecting edges to which you apply a corner round is similar to selecting edges for the fillet feature. Two modes enhance the feature's use in the sheet metal environment: Corner and Feature.

Corner Click to select individual thickness edges to be rounded or filleted.

Feature Click to select all thickness edges of a feature automatically.

CORNER CHAMFER

The Corner Chamfer tool, as shown in the following image, is a sheet metal-specific chamfer tool. As with the Corner Round tool, all edges other than those at open corners of faces are filtered out. Because the edges are always discontinuous, the Edge Chain and Setback settings available with the standard Chamfer tool are not available.

Figure 10.100

To create a corner chamfer, follow these steps:

1. Click the Corner Chamfer tool on the Sheet Metal Features panel bar. The Corner Chamfer dialog box appears, as shown in the following image.
2. Select the chamfer style: One Distance, Distance and Angle, or Two Distances. See the following list for explanations of these styles.
3. Select the corners, or edge and corner for Distance and Angle, that you wish to include.
4. Enter the chamfer values.
5. Click OK to complete the corner chamfer and exit the dialog box.

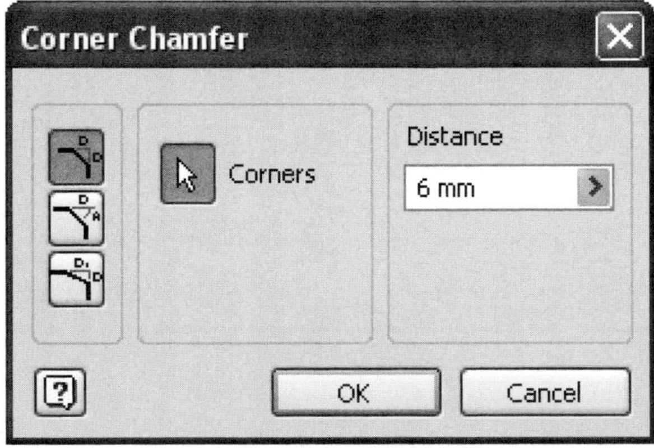

Figure 10.101

	One Distance	Creates a 45° chamfer on the selected edge. Determine the size of the chamfer by typing a distance in the dialog box. The value is offset from the two common faces.
	Distance and Angle	Creates a chamfer offset from a selected edge on a specified face at an angle from the number of degrees specified. Enter an angle and distance for the chamfer in the dialog box, click the face on which the angle is based, and specify the face edge to be chamfered.
	Two Distances	Creates a chamfer offset from two faces, each in the amount that you specify. Click a corner, and enter a value for Distance1 and Distance2. A preview image of the chamfer appears. To reverse the direction of the distances, click the Flip button.
	Corners	Click to select individual corners to chamfer.
	Edge	Click to select an edge for chamfers when using the Distance and Angle option.
	Flip	Click to flip the direction for the chamfer distance when using the Two Distances option.
	Distance	Enter a distance to be used for the chamfer feature.
	Angle	Enter an angle for the chamfer feature when using the Distance and Angle option.

PUNCHTOOL

Cuts and 3D deformation features, such as dimples and louvers, are usually added to the flat sheet metal stock in a turret punch before the sheet is bent into the folded part. Turret punch tools are positioned on the flat sheet by a center point corresponding to the tool center at an angle to a fixed coordinate system. The PunchTool tool, as shown in the following image, places specially designed iFeatures that define the center point of the tool. A streamlined interface simplifies the selection and placement of the iFeatures.

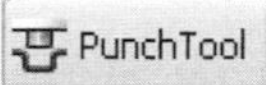

Figure 10.102

A default Punch folder is installed under the top-level Catalog folder when you install Autodesk Inventor. A selection of punch tools is included in the folder. PunchTool lists all iFeatures in the designated punch folder. You can set the default Punch folder on the iFeature tab of the Application Options dialog box, as shown in the following image.

Figure 10.103

To qualify as a PunchTool, a saved iFeature must have a single point, center point in the sketch of the first feature included in the iFeature. The point corresponds to the center of the tool, and it will be used as the point of placement when you apply a PunchTool iFeature to a part face. Asymmetrical shapes must have the center point position controlled by geometric relationships or equations to ensure that it remains centered when the iFeature parameters are changed. Create the iFeature in a sheet metal part to ensure that parameters such as Thickness are saved with the iFeature.

You create a sheet metal punch iFeature in the same way that you create standard iFeatures. Select Extract iFeature from the Tools menu, and the Extract iFeature dialog box is displayed, as shown in the following image. At the top of the dialog box, select Sheet Metal Punch iFeature from the Type area. This selection activates the Manufacturing and Depth areas.

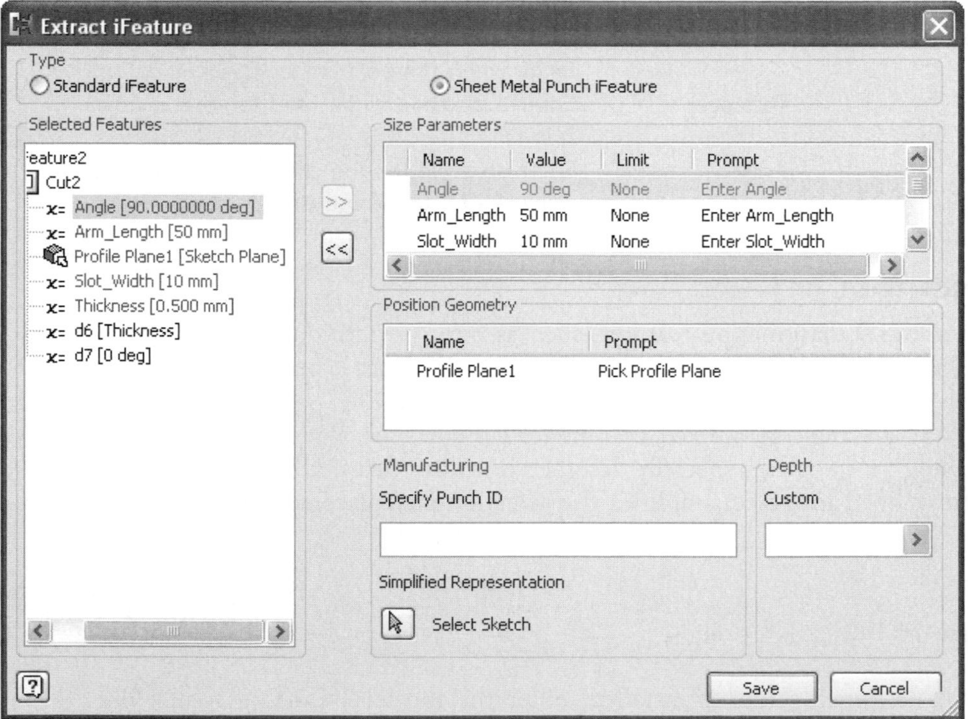

Figure 10.104

The Specify Punch ID field can be used to enter information, such as a vendor part number or designation, that will be stored with the data that defines the punch. The data can be extracted and used during drawing creation. The Custom field is used to define how far the punch penetrates past the selected surface. The Simplified Representation section allows you to select a sketch from the browser that will be used as an alternate representation for the punch in the flat pattern and when the flat pattern is placed on a drawing sheet. Similar to standard iFeatures, you can also create a table-driven iFeature to create a family of punches using the iFeature Author Table tool; this tool is located on the iFeature panel when an extracted punch or standard iFeature is opened with Autodesk Inventor.

To place a PunchTool, follow these steps:

1. Create a sketch on a sheet metal face, and place at least one sketch point in the sketch. Point, Center Points are selected automatically as punch centers during PunchTool placement. You can manually add sketch points, line or curve endpoints, and arc centers as additional punch centers.
2. Click PunchTool on the Sheet Metal Features panel bar. The PunchTool Directory dialog box appears, as shown in the following image.

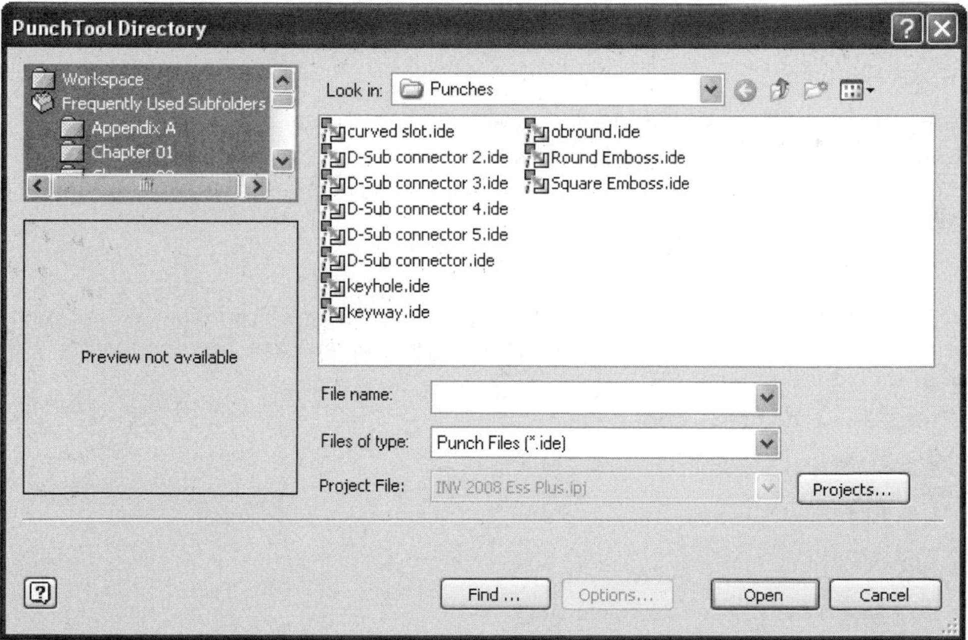

Figure 10.105

3. Select the desired PunchTool, and click Open. The PunchTool dialog box is displayed. It contains three tabs: Preview, Geometry, and Size, as shown in the following image.

4. All Point, Center Points are selected automatically. Hold down the SHIFT or CTRL key, select Point, Center Points to exclude them, and select other location geometry as required.
5. On the Geometry tab, select a rotation angle for all occurrences of the punch. At 0° rotation, the X-axis of the first feature in the saved iFeature is aligned with the X-axis of the current sketch.
6. On the Size tab, enter values for the PunchTool parameters.

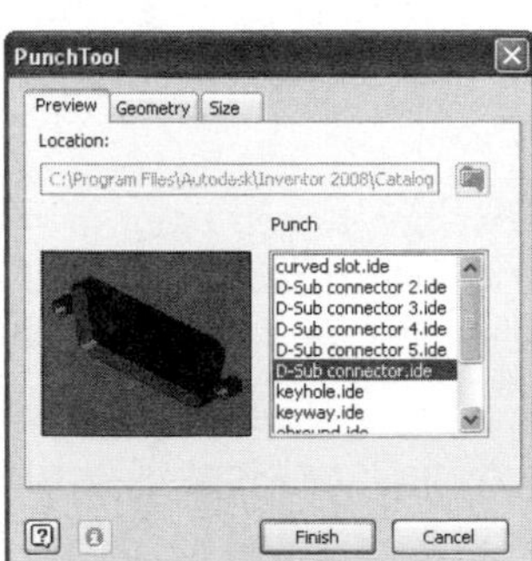

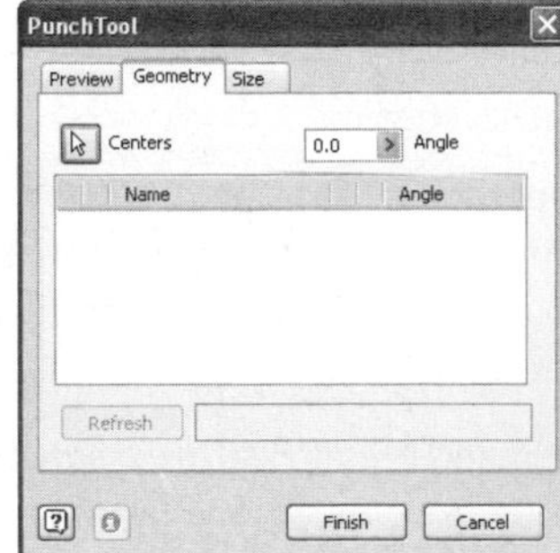

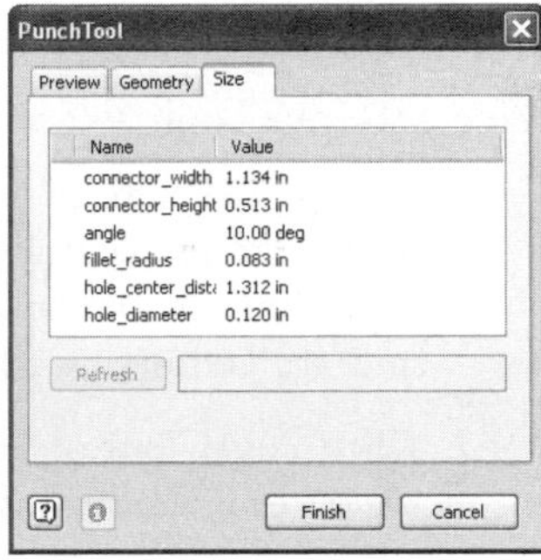

Figure 10.106

7. Click Finish to complete the PunchTool and exit the dialog box.

EXERCISE 10-5: PUNCH TOOL

In this exercise, you create an iFeature that can be used as a punch and then use the PunchTool tool in another sheet metal part.

1. Open *ESS_E10_05.ipt*.

You will complete the sketch and save the cutout as an iFeature. The visible sketch contains all geometry required for the cut. You will add construction geometry and a center point to define the center of the punch.

2. Click the Look At tool.

3. Click the face containing the sketch.

4. Right-click Sketch2 in the browser, and then select Edit Sketch.

5. Click the Line tool.

6. Select the Construction line style.

7. Sketch two lines, as shown in the following image. Endpoints for Line 1 are on the arc centers. Line 2 connects to the midpoint of Line 1.

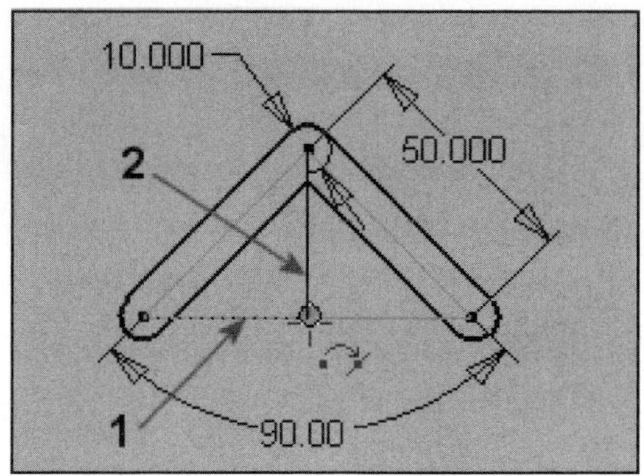

Figure 10.107

This iFeature is symmetrical in both X and Y. The midpoint of the vertical construction line is the center of the cut. For nonsymmetrical shapes, you must use equations to ensure that the center point remains at the center of the iFeature.

8. Click the Point, Center Point tool.
9. Place a point at the midpoint of the vertical construction line, as shown in the following image.

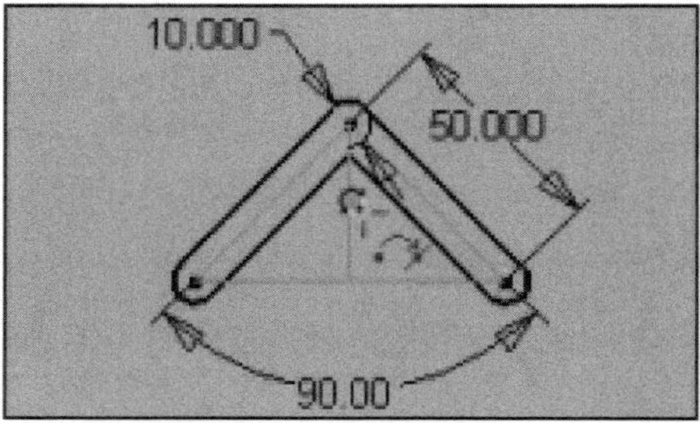

Figure 10.108

10. Press S to finish the sketch.
11. Click the Cut tool.
12. Click OK.

You rename a parameter prior to saving the iFeature. Other required parameters have previously been renamed.

13. Click the Parameters tool.
14. Click the model parameter name cell containing d1.
15. Enter **Angle** as the new parameter name.

16. Click Done.
17. Click the Extract iFeature tool, which is located under the Tools drop-down menu.
18. Click Cut1 in the browser. Your display should appear similar to the following image.

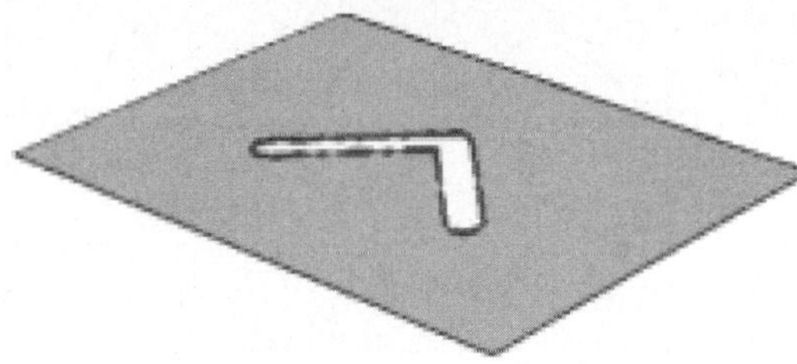

Figure 10.109

19. In the Extract iFeature dialog box:
    a. Select Sheet Metal Punch iFeature from the Type section.
    b. Under Size Parameters, click the Limit cell in the Angle row.
    c. Select Range from the in-cell list.
20. In the Specify Range:Angle dialog box:
    a. Select < (the less than operand) from the limit list between Minimum and Default.
    b. Select < (the less than operand) from the limit list between Default and Maximum.
    c. Enter **5** deg in the Minimum edit box.
    d. Leave Default set to 90.
    e. Enter **175** deg in the Maximum edit box.
    f. Click OK.
21. Click Save in the Extract iFeature dialog box.
22. Browse to the Catalog\Punches folder.
23. Enter **V Slot Punch** in the File name edit box.
24. Click Save.

Next place the punch feature into an existing part. You will create a sketch and use the V Slot punch tool.

25. Open *ESS_E10_05a.ipt.*
26. Click the top face of the part.
27. Press S to create a sketch.
28. Click the Look At tool, and select the sketch face.

29. Click the Point, Center Point tool.

30. Place a point, as shown in the following image.

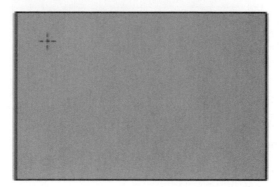

Figure 10.110

Next create a rectangular pattern of points.

31. Click the Rectangular Pattern tool.

32. Click the Point.

33. Click the selection tool under Direction 1 in the Rectangular Pattern dialog box.

34. Click the edge, as shown in the following image.

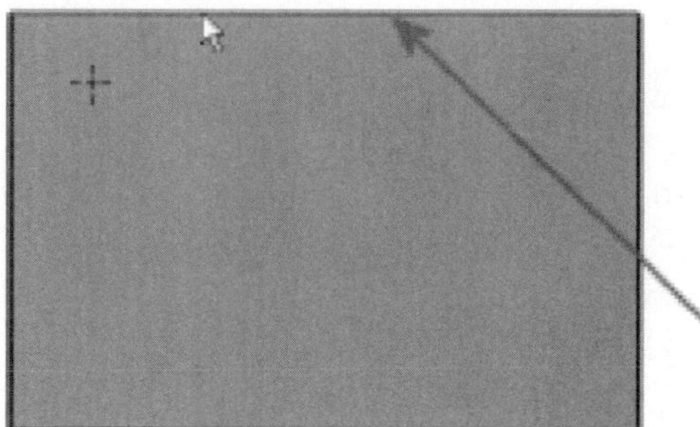

Figure 10.111

35. Enter **5** in the Direction 1 Count edit box.

36. Enter **70** in the Direction 1 Spacing edit box.

37. Click the selection tool under Direction 2.

38. Click the edge, as shown in the following image.

Figure 10.112

39. Enter **4** in the Direction 2 Count edit box.
40. Enter **50** in the Direction 2 Spacing edit box.
41. Click OK.

Sketch a line to enable manual placement of PunchTool centers.

42. Click the Line tool.
43. Sketch the line shown in the following image.

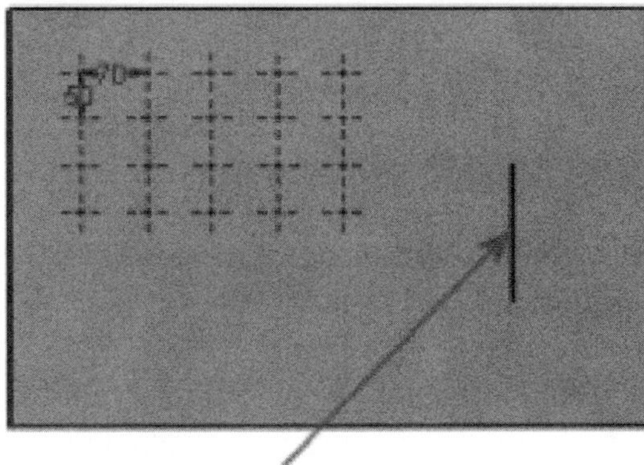

Figure 10.113

44. Press S to complete the sketch.
45. Click the PunchTool tool.
46. In the PunchTool Directory dialog box:
    a. Click V Slot *Punch.ide* in the File Name list.
    b. Click Open.
    c. The PunchTool shape is previewed at each point in the sketch.
47. Click the two line endpoints, as shown in the following image to add the line endpoints as tool centers.

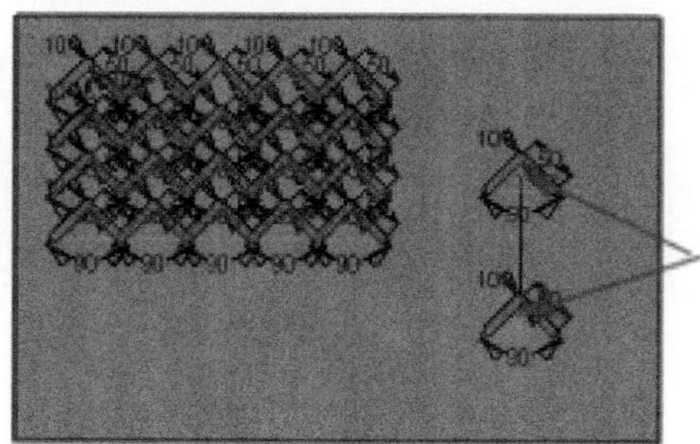

Figure 10.114

Note: The sketch plane is the only reference for this punch. Any additional geometry references must be satisfied before proceeding via the Geometry tab.

48. Click the Size tab of the PunchTool dialog box.

49. In the PunchTool dialog box:

a. Enter **60** in the Angle Value cell.

b. Enter **30** in the Arm_Length Value cell.

c. Enter **7** in the Slot_Width Value cell.

d. Click Refresh to update the preview with the new values.

e. Click Finish. Your display should appear similar to the following image.

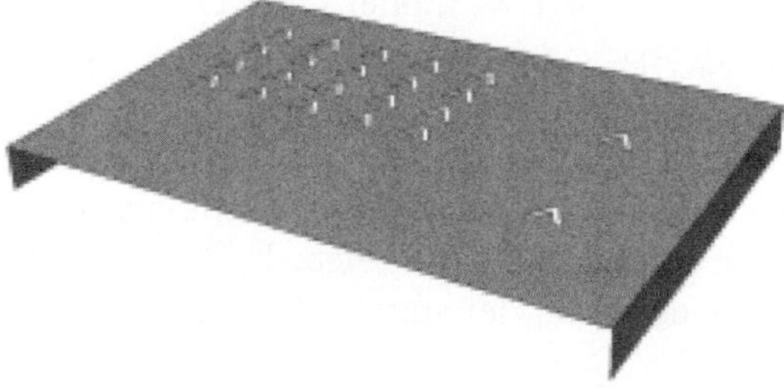

Figure 10.115

50. Close all open files. Do not save changes. End of exercise.

EXERCISES

FLAT PATTERN

You can create a sheet metal flat pattern that unfolds all of the sheet metal features. The flat pattern represents the starting point for the manufacturing of the sheet metal part. The flat pattern can appear in a 2D drawing view, complete with lines indicating bend centerlines. The Flat Pattern tool, as shown in the following image, creates a 3D model of the unfolded part.

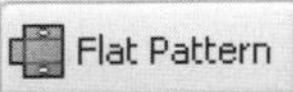

Figure 10.116

You can also export the flat pattern directly as a 3D (SAT file) or 2D (*.dxf* or *.dwg*) formats to be used to create machine tool programming for flat pattern punching or cutting. The following image on the left shows a sheet metal part in its formed state and the same part in its flat state on the right.

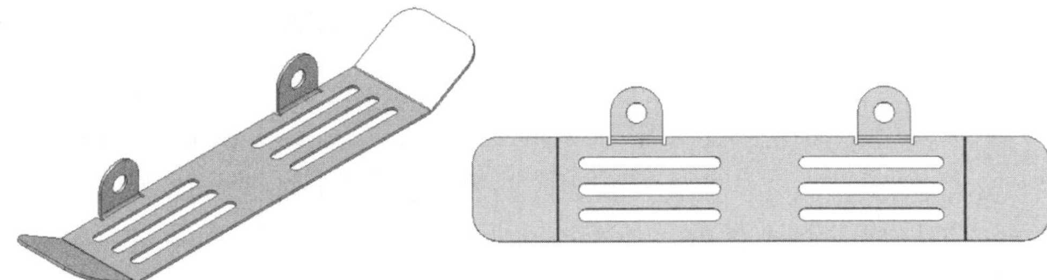

Figure 10.117

Manually modeling a flat pattern of a sheet metal part is straightforward, but with a complicated shape, the process can take considerable effort. Autodesk Inventor can create a 3D unfolded model of your sheet metal part quickly and update this model automatically as you modify the features that make up the part. You can create the flat pattern model at any time during the sheet metal modeling process.

The flat pattern model and icon are shown under the part name in the browser, as shown in the following image. The flat pattern model replaces the folded model in the graphics window, and you can toggle back and forth between the two model states by selecting either Folded Model or Flat Pattern in the browser. The flat pattern model is updated automatically as you modify sheet metal features. Double-click the flat pattern in the browser to view the current flat pattern shape after making changes to the folded model. You can also right-click on either node and select Edit Folded or Edit Flat Pattern from the menu to activate the selected model state.

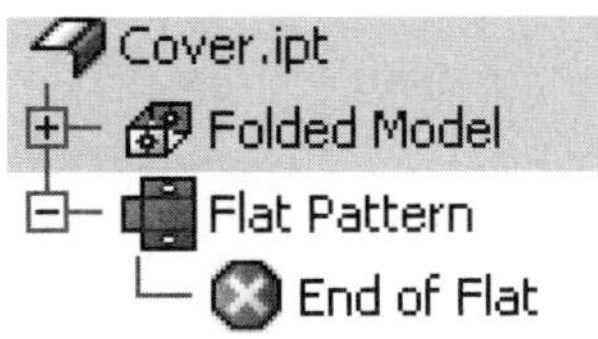

Figure 10.118

Note: If you delete the flat pattern model from the browser, you will not be able to create a drawing view of the flat pattern. If you delete the flat pattern model after the creation of a flat pattern drawing view, the drawing view will be deleted.

Right-click the flat pattern in the browser. Select Extents from the menu to display the dimensions of the smallest rectangle that encloses the flat pattern, as shown in the following image. Use this information for stock selection and for multiple flat pattern layouts on large sheets.

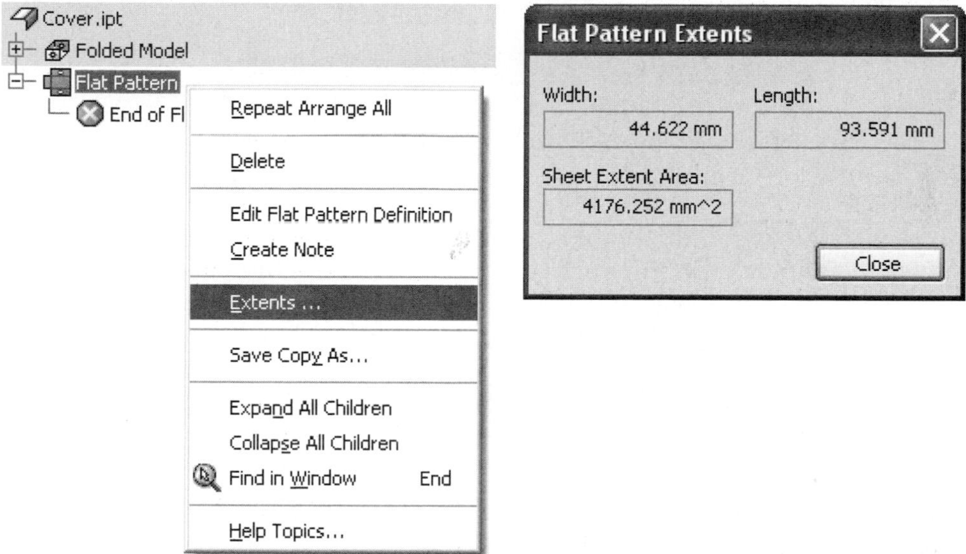

Figure 10.119

These values are also available when creating text on a drawing sheet. Select Sheet Metal Properties from the Type list and then choose flat pattern extents area, flat pattern extents length, or flat pattern extents width to include the parameters on your drawing sheet. In addition to including the flat pattern dimensions in text fields, you can also include them in a parts list. To add them to a parts list, you need to create custom iProperties that have the same name as the sheet metal properties: flat pattern area, flat pattern length, and flat pattern width. Examples of these custom iProperties are shown in the following image. Notice, from the tooltip, that the value is entered as a formula. If the extents of the flat pattern is changed, the values will update automatically.

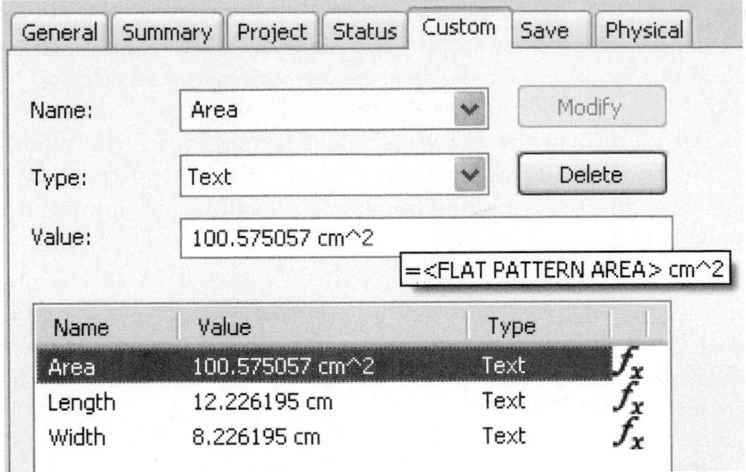

Figure 10.120

If you choose the Edit Flat Pattern Definition option from the right-click menu, a Flat Pattern dialog is displayed with three tabs: Orientation, Punch Representation, and Bend Angle as shown in the following image.

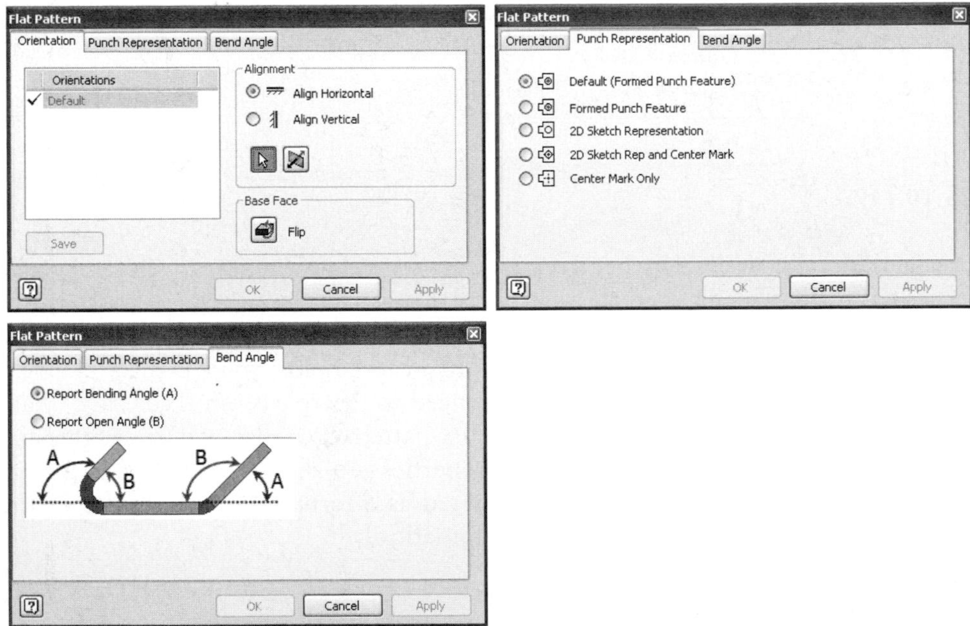

Figure 10.121

The Orientation tab lets you reorient the flat pattern. The Orientations list allows you to create, activate, delete, or rename flat pattern orientations using the right-click menu. The Alignment section allows you to select an edge and orient the flat pattern by making the edge align either horizontally or vertically. Typically, the flat pattern is created normal to the initial sketched face feature. You can flip the normal using the Flip tool in the Base Face section. On the Punch Representation tab, you can override the display of punch features as defined in the active sheet metal style. The Bend Angle tab allows you to select whether the angle reported for bends is measured from the outside (option A) or inside (option B) face of the bend. To create a flat pattern, follow these steps:

1. Select a face that you expect will not be removed in future edits. The selected face will remain fixed, and all other faces will unfold from this face.

Note: Selecting a persistent face before creating a flat pattern is good practice though not a strict requirement.

2. Click the Flat Pattern tool on the Sheet Metal Features panel bar. The flat pattern model appears in the graphic window, and a Flat Pattern node is added to the browser.
3. Right-click Flat Pattern at the top of the browser. Menu items are available for saving the flat pattern, editing the flat pattern definition, and displaying the overall dimensions or extents of the flat pattern.
4. To document a flat pattern, start a new drawing. Create a drawing view, and select Flat Pattern from the Sheet Metal View list in the Drawing View dialog box, as shown in the following image. When Flat Pattern is selected, the Recover Punch Center option can be selected to retrieve punch centers on the drawing view. The punch centers will be formatted based on settings in the active sheet metal rule or when overridden via the Edit Flat Pattern option.

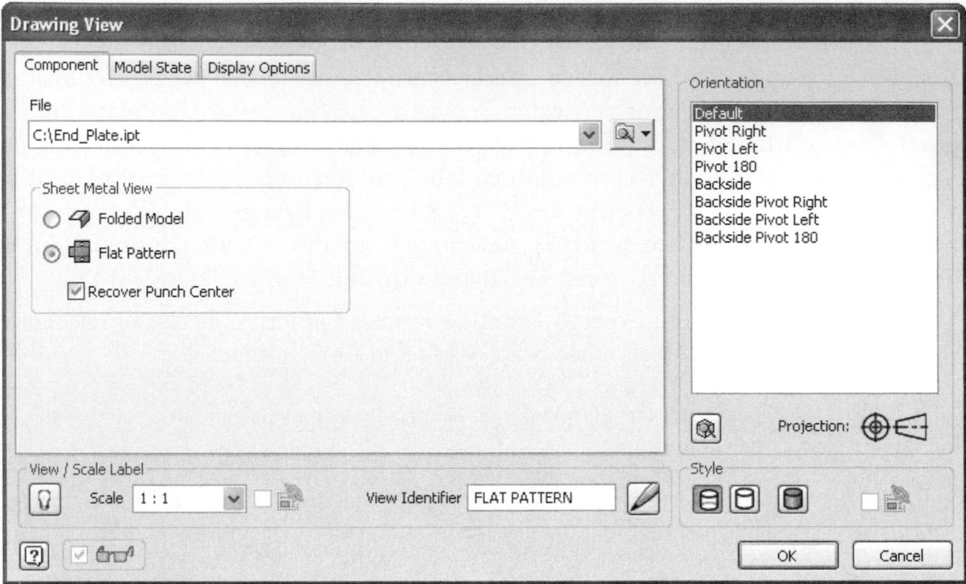

Figure 10.122

COMMON TOOLS

A number of tools on the Sheet Metal Features panel bar are common to sheet metal and standard parts. Short descriptions of how to use these tools in the sheet metal environment follow.

Work Features Work planes are often used to define sketch planes for disjointed faces.

Holes The hole feature is common to both the part modeling and the sheet metal environments. You can enter sheet metal parameters such as Thickness in numeric fields such as hole depth.

Catalog Tools Save and place iFeatures in the sheet metal environment. See the Punch-Tool section for information on special iFeatures for sheet metal.

Mirror and Feature Patterns Mirror and pattern sheet metal features like any other modeled feature. You can also pattern or mirror flanges, holes, cuts, iFeatures, and features created with the standard part feature tools.

Copy Object Import files that describe surface models, such as IGES files, in the sheet metal environment and turn them into solid bodies.

Derived Component Reference another part or assembly, regardless of whether it is a sheet metal or standard file type. You can also reference and use parameters, sketches, work features, surfaces, and iMates from other files in a sheet metal part.

Parameters Access model parameters and equations via the Parameters dialog box. You can add user parameters and link to an Excel spreadsheet to reference exported parameters from other models. Sheet metal-specific parameters are also added to the Parameters dialog box when the sheet metal environment is activated: Thickness, BendRadius, BendReliefWidth, BendReliefDepth, CornerReliefSize, MinimumRemnant, TransitionRadius, and JacobiRadiusSize.

Create iMate Apply iMates to sheet metal components prior to placement in an assembly.

DETAILING SHEET METAL DESIGNS

You can create drawing views of the 3D model and the flat pattern when detailing a sheet metal part. The flat pattern drawing view enables all bend locations, bend extents, bend and corner reliefs, and cutouts to be located and sized with dimensions. The 3D model views describe the folded shape of the part and allow the forming tool operator to validate the folded part as shown in the following image.

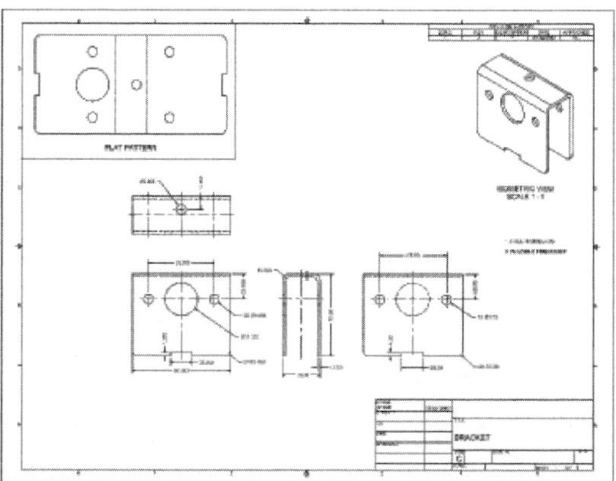

Figure 10.123

EXERCISE 10-6: DOCUMENTING SHEET METAL DESIGNS

In this exercise, you create a flat pattern model of a sheet metal part and observe the live nature of the flat pattern as you add or modify features. You obtain the flat pattern extents and create model and flat pattern drawing views of the sheet metal part.

1. Open *ESS_E10_06.ipt*.
2. Click the Flat Pattern tool. The flat pattern model of the sheet metal part is displayed in the graphics window, as shown in the following image.

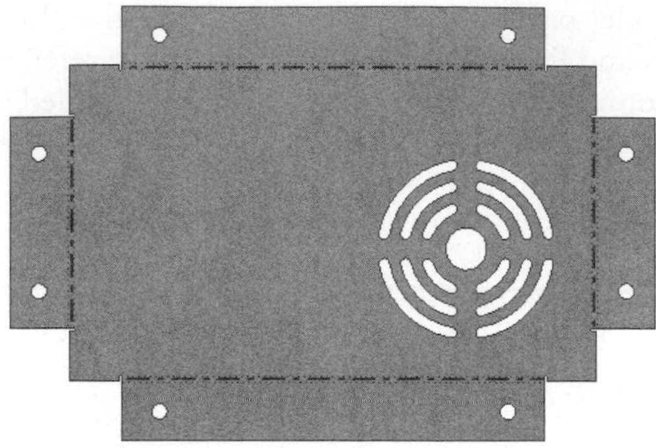

Figure 10.124

3. In the browser, double-click the Folded Model node to activate the folded model for edit.
4. Click the Corner Chamfer tool.
5. Press F4 to rotate the model and place the cursor over the top-right corner of the large face.
6. When the edge of the top-right face is highlighted, click to select it, as shown in the following image on the left.
7. Select all four corners of the large face.
8. In the Corner Chamfer dialog box, enter a value of **4 mm** in the Distance field, and verify that the One Distance creation method is selected.
9. Click OK to place the chamfers. Your display should appear similar to the following image on the right.

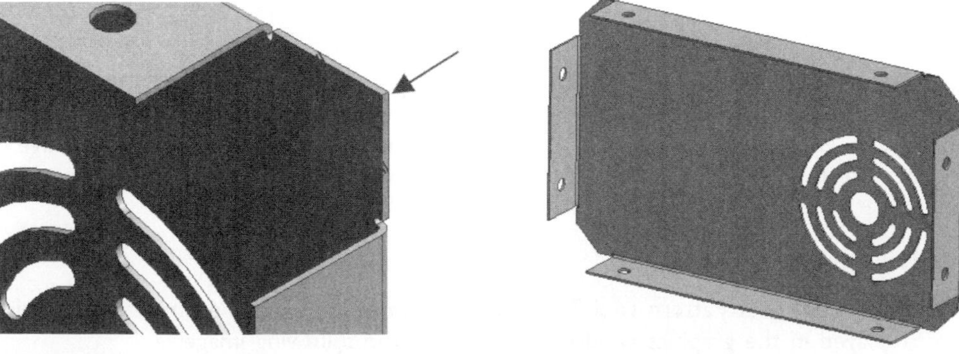

Figure 10.125

10. In the browser, right-click Flat Pattern, and select Edit Flat Pattern from the menu.
11. The flat pattern includes the chamfers, as shown in the following image.

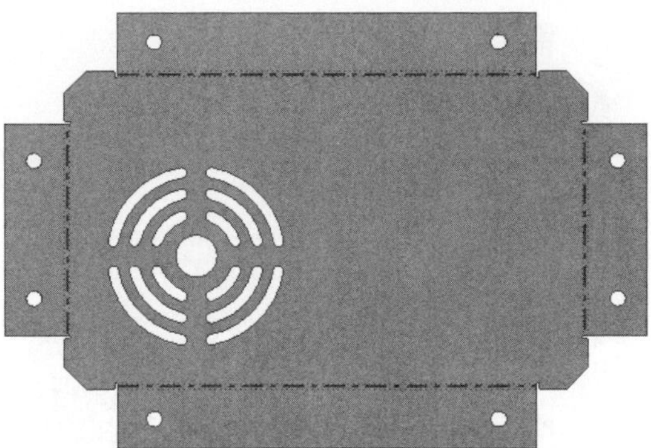

Figure 10.126

12. Activate the Folded Model by double-clicking the node in the Browser.
13. Click the Corner Round tool.
14. In the Corner Round dialog box:
    a. Enter a value of **1.5 mm** for the radius.
    b. Click the Feature button under Select Mode.
15. Place the cursor over one of the four flanges.
16. Click when the flange is highlighted.

The small corners at each end of the flange are highlighted as the flange is selected.

17. Select all four flanges.
18. Click OK to place the corner rounds.
19. Activate the Home View of the model. Your display should appear similar to the following image.

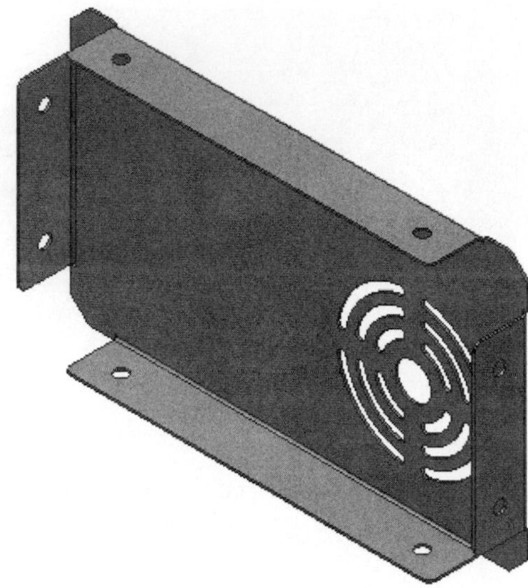

Figure 10.127

20. Click the Insert iFeature tool.
21. Browse to the exercises folder, and select *Power_Plug.ide*.
22. Click Open to place the iFeature.
23. Click the large face on the part to place the iFeature.

The placement plane is the only input for this iFeature; the plug dimensions are fixed.

24. Click the crossed arrows centered on the iFeature, and drag the shape to roughly place the iFeature, as shown in the following image on the left. Click to locate the feature.
25. Click Finish in the Insert iFeature dialog box. Your display should appear similar to the following image on the right.

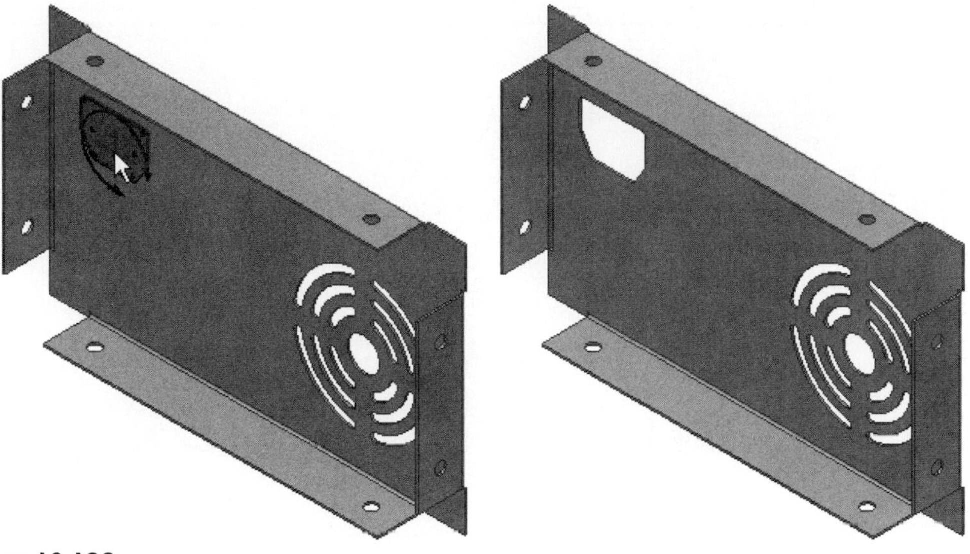

Figure 10.128

26. Double-click Flat Pattern in the browser.

27. In the browser, right-click Flat Pattern and select Extents.

Dimensions for the flat pattern sheet size are shown in the Flat Pattern Extents dialog box.

28. Click Close to close the Flat Pattern Extents dialog box.

29. Double-click Folded Model in the browser.

30. From the Menu Bar, select File > Save As.

31. Enter End_Plate.ipt as the file name.

32. Click Save.

33. Click the New tool.

34. Click the Metric tab.

35. Double-click the *DIN.idw* template.

36. Click the Base View tool.

Note: If you have more than one part or assembly file open, select *End_Plate.ipt* from the File drop-down list.

37. Select Folded Model from the Sheet Metal View list, select Bottom from the Orientation list, and click on the drawing sheet to place a view of the folded sheet metal part, as shown in the following image.

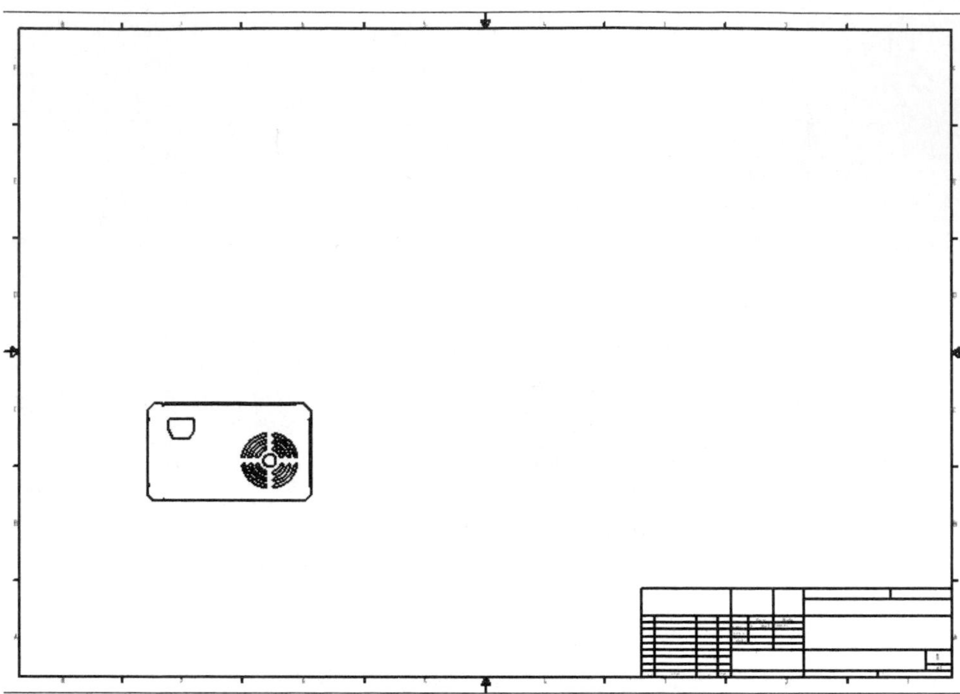

Figure 10.129

38. Click the Projected View tool.

39. Create a top and right view, as shown in the following image.

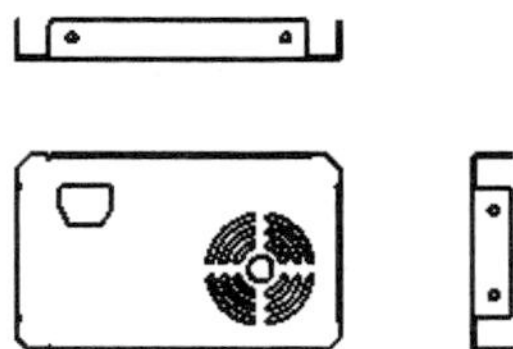

Figure 10.130

40. Click the Base View tool.

Note: If you have more than one part or assembly file open, select *End_Plate.ipt* from the File drop-down list.

41. In the Drawing View dialog box, select Flat Pattern from the Sheet Metal View list, and choose Default from the Orientation list.

42. Click on the drawing sheet to place the flat pattern view, as shown in the following image.

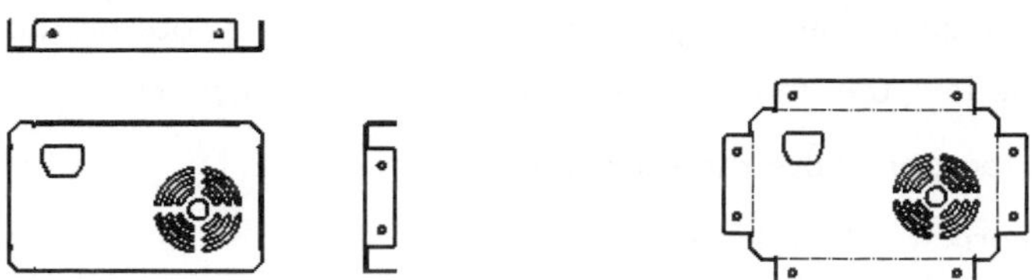

Figure 10.131

43. Click the General Dimension tool from the Drawing Annotation panel bar.
44. Dimension both tab lengths in the flat pattern view, as shown in the following image.

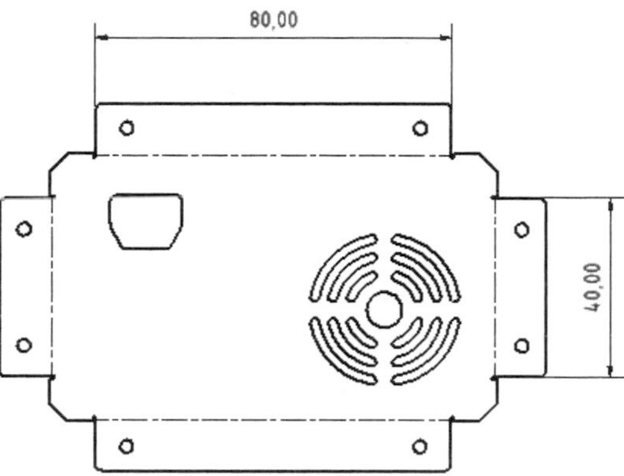

Figure 10.132

45. Click the Bend Notes tool from the Drawing Annotation panel bar.
46. Click each of the bend lines in the flat pattern view. The flat pattern should appear as shown in the following image.

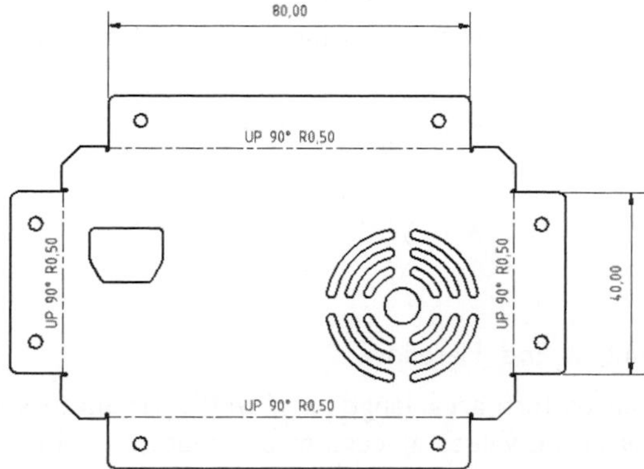

Figure 10.133

Next, you include the sheet metal length, width, and extents as a note on the drawing sheet.

47. Click the Text tool from the Drawing Annotation panel bar.
48. Click a location above the titleblock.
49. In the Format Text dialog box:
    - Type "Flat Pattern Extents:"
    - Select Sheet Metal Properties from the Type drop-down list.
    - Select Flat Pattern Extents Area from the Property drop-down list.
    - Click Add Text Parameter.
    - Press Enter to return to the next line of text.
50. Repeat the previous step to add the following line of text:
    - Type "Flat Pattern Length:"
    - Select Sheet Metal Properties from the Type drop-down list.
    - Select Flat Pattern Length from the Property drop-down list.
    - Click Add Text Parameter.
    - Press Enter to return to the next line of text.
51. Repeat the previous step to add the following line of text:
    - Type "Flat Pattern Width:"
    - Select Sheet Metal Properties from the Type drop-down list.
    - Select Flat Pattern Width from the Property drop-down list.
    - Click Add Text Parameter.
52. Click OK to close the Format Text dialog box and view the flat pattern information in the appropriate units on the drawing sheet, as shown in the following image.

Flat Pattern Extents: 10057.506 mm^2
Flat Pattern Length: 122.262 mm
Flat Pattern Width: 82.262 mm

Figure 10.134

53. Save the drawing as End_Plate.idw.

You can also include the flat pattern area, length, and width parameters in a parts list. In order to do this, you need to establish the values as custom parameters in the sheet metal part.

54. Open or make the End_Plate.ipt file active.
55. Click iProperties from the File menu.

56. Click the Custom tab and specify the following Custom parameter:

- Name: Length
- Type: Text
- Value: =<FLAT PATTERN LENGTH> cm
- Click Add

57. Repeat the previous step to add the following Custom parameter:

- Name: Width
- Type: Text
- Value: =<FLAT PATTERN WIDTH> cm
- Click Add

58. Repeat the previous step to add the following Custom parameter:

- Name: Area
- Type: Text
- Value: =<FLAT PATTERN AREA> cm^2
- Click Add

59. Click Apply in the iProperties dialog box. The value fields are updated and show the flat pattern extent dimensions.

Note: Autodesk Inventor uses cm as the native unit type for all calculations. When a unit type is set (such as inch, millimeter, etc.) a conversion is performed by the application to display the appropriate unit type as specified in the Document Settings. You entered a cm label that is included in the custom iProperty so that anyone who may look at those values will know the unit type that is being displayed. When you use these values on a drawing sheet (as done in the previous steps) or when including them in a parts list, a conversion will be made to display the values with the appropriate unit type.

60. Open or make the End_Plate.idw file active.

61. Click the Parts List tool from the Drawing Annotation panel bar.

- Select the Flat Pattern drawing view.
- Click OK in the Parts List dialog box.
- Click a location to place the parts list on the drawing sheet.

62. Right-click the parts list and select Edit Parts List from the menu.

63. Click Column Chooser in the Parts List dialog box.

64. Under Selected Properties, click DESCRIPTION, then click Remove.

65. To add the custom iProperties to the parts list:
   - Click New Property.
   - Click "Click here to add new property."
   - Enter "LENGTH" and press ENTER.
   - Click "Click here to add new property."
   - Enter "WIDTH" and press ENTER.
   - Click "Click here to add new property."
   - Enter "AREA" and press ENTER.
   - Click OK in the Define New Property dialog box.
   - Click OK in the Parts List Column Chooser dialog box.
66. The values are pulled from the iProperties of the sheet metal part and are displayed in the Parts List dialog box, as shown in the following image.

LENGTH	WIDTH	AREA
12.226195 cm	8.226195 cm	100.575057 cm^2

Figure 10.135

Next, you modify the formatting of the units in the parts list.

67. Right-click the Length column in the Parts List dialog box, and click Format Column.
68. To specify the formatting for the units:
   - Select Apply Units Formatting.
   - Select 3.123 from the Precision drop-down list.
   - Select mm from the Units drop-down list.
   - Select Period from the Decimal Marker drop-down list.
   - Click OK.
69. Repeat the previous step for the Width column.
70. Right-click the Area column and select Format Column.
71. Click Apply Units Formatting and select the Unit Type drop-down list. Because there is no area selection from the Unit Type drop-down list, the format of the units for area cannot be calculated. If you want to include the area in the parts list, it must remain shown as cm^2.
72. Click Cancel in the Format Column dialog box.
73. Click OK to close the Parts List dialog box and view the changes to the parts list as shown in the following image.

PARTS LIST					
ITEM	QTY	PART NUMBER	LENGTH	WIDTH	AREA
1	1	End_Plate	122.262 mm	82.262 mm	100.575057 cm^2

Figure 10.136

Note: You could opt to leave the area column out of the parts list and specify the area of the flat pattern, in the appropriate units, using the Text tool as explained earlier in the exercise.

74. Close all open files. Do not save changes. End of exercise.

The self-paced, step-by-step project exercises found in Appendix A provide opportunities for you to work through real-world modeling, assembly, and documentation tasks. The geometry used in the project exercises will flow from chapter to chapter, utilizing the functionality that you learned in that chapter.

Project Exercise: Chapter 10

Using the knowledge you gained in this chapter, create the sheet metal part shown in the following image. As part of the exercise, create a drawing that details both the flat and folded states of the part.

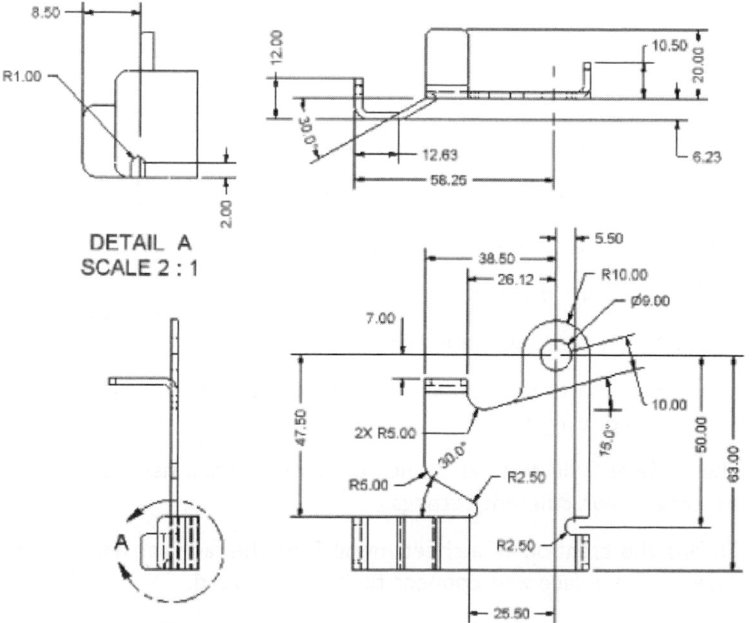

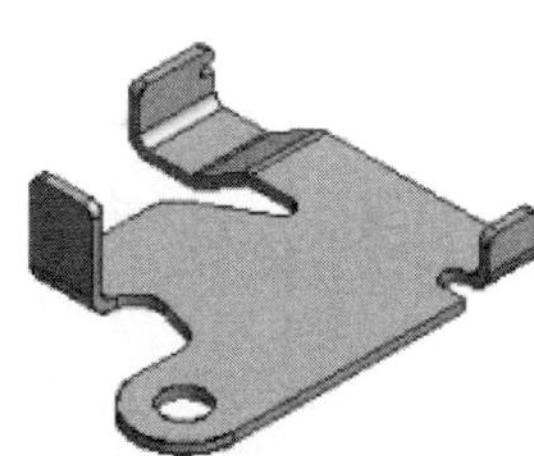

NOTES:

1. PART IS 2.00 mm. THICK.
2. ALL ROUNDS ARE 2.00 mm. UNLESS OTHERWISE SPECIFIED.
3. INSIDE BEND RADII 2.00 mm.

Figure 10.137

CHECKING YOUR SKILLS

Use these questions to test your knowledge of the material covered in this chapter.

1. The base feature of a sheet metal part is most often a:
   - a. Revolve
   - b. Face
   - c. Extrude
   - d. Flange
2. What is the correct procedure to change the edges connected by a bend feature?
   - a. Suppress the existing bend, and add a new bend.
   - b. Delete the existing bend, and add a new bend.
   - c. Edit the bend, and select the new edges.
   - d. Create a corner seam between the desired faces.
3. Which tool would you use to create a full-length rectangular face off of an existing face edge?
   - a. Flange
   - b. Extrude
   - c. Face
   - d. Bend
4. What action is required to update a flat pattern model?
   - a. Right-click on the flat pattern in the browser, and select Update.
   - b. Erase the existing flat pattern, and recreate it.
   - c. Create a new flat pattern drawing view.
   - d. None, the flat pattern is updated automatically.
5. **True _ False _** Sheet metal parts can contain features created with Autodesk Inventor modeling tools.
6. **True _ False _** Sheet Metal Style settings cannot be overridden; a new style must be created for different settings.
7. **True _ False _** During the creation of a sheet metal face, the face can extend to meet another face and connect to it with a bend.
8. **True _ False _** A sheet metal cut that used the Distance extents option must be created equal to the Thickness parameter.

EXERCISES

9. What is a common term for a three-bend intersection?

____

____

10. **True** _ **False** _ When creating a sheet metal punch, you can select an alternate 2D representation to display in flat patterns and drawing views.

APPENDIX a

Project Exercises

This appendix consists of exercises that include topics covered in the book's ten chapters. These exercises provide opportunities for you to work through real-world modeling, assembly, and documentation tasks. The geometry used in the project exercises will flow from section to section, utilizing the knowledge that you learned in the designated chapter. While going through the exercises, return to the designated chapters to review specific material as each exercise only provides an overview of the operation. The project exercises are based upon a coating system gripper station design. The specification for the design is to define a modular gripper station that can be cycled through several cleaning, coating spin, and curing stations. All stations need to work within a particulate-free, laminar airflow environment.

Project Exercise: Chapter 1

GETTING STARTED

In this exercise, you review the completed assembly via a DWF file. To review the design in detail, open the file found in the following location: C:\INV 2009 Ess Plus\Appendix A\ESS-PE-CH1-Overview.dwf. Rotate the assembly, and turn visibility of the components on and off to better see the assembly.

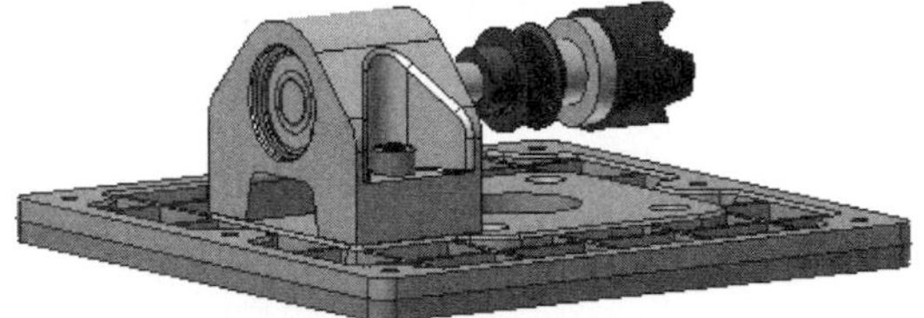

Figure A.1

Project Exercise: Chapter 2

SKETCHING, CONSTRAINING, AND DIMENSIONING

In the following exercise, you create a basic sketch that forms the profile of the gripper station assembly bearing block, as shown in the following image. The sketch captures the functional design intent of the component. This exercise uses the line, arc, geometric constraint, and dimension 2D Sketch tools.

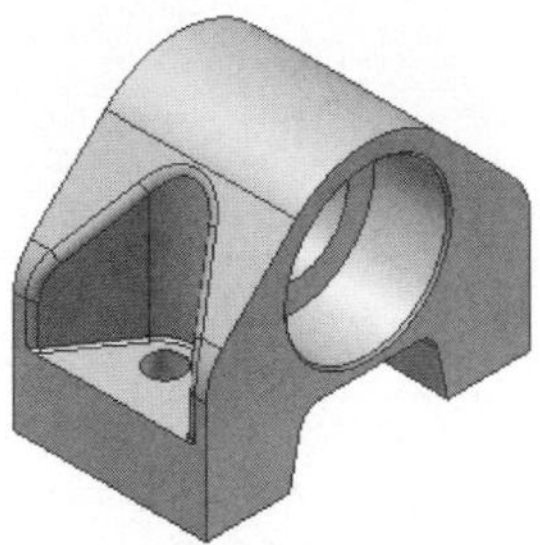

Figure A.2

1. If needed, click on the Tools drop-down menu, and click Application Options. Next click on the Sketch tab, and review the "Autoproject part origin on sketch create" setting. If the box is not checked, click to add a check. This option will automatically project the Origin folder Center Point onto the default active sketch during the creation of a new part.
2. Click the New tool, click the Metric tab, and then double-click *Standard (mm).ipt*.
3. Click to depress the Construction button on the style area of the Standard toolbar.
4. On the 2D Sketch panel, click the Center Point Circle tool to create a circle with its center point coincident to the projected origin center point and with approximately a **15 mm** radius.
5. Click the Construction button on the style area of the Standard toolbar to disable the construction style.
6. On the 2D Sketch panel, click the Two Point Rectangle tool, and create a rectangle below the circle, as shown in the following image on the left. Right-click, and click Done from the menu.
7. Right-click, select Create Constraint, and then Vertical from the menu. Click the center point of the construction circle and the midpoint of the top horizontal line of the rectangle.
8. Click the Dimension tool. Add a **60 mm** dimension to the width of the rectangle and a diameter dimension of **30 mm** to the construction circle, as shown in the following image on the right. Right-click, and click Done from the menu.

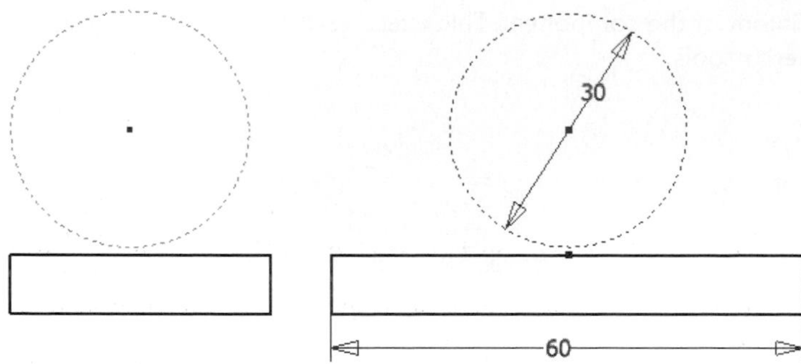

Figure A.3

9. Click the Line tool, and create the top two lines and arc, as shown in the following image on the left. Ensure that the arc is tangent to the adjacent lines. Right-click, and click Done from the menu.

10. Create a coincident constraint between the two center points by dragging the arc's center point to the center point of the construction circle. Release the mouse button when the green dot is visible.

11. Click the Dimension tool, and add a **4 mm** offset dimension between the arc and the construction circle. Next add a **14 mm** dimension between the base of the rectangle and tangent to the bottom of the construction circle, as shown in the following image on the left. Right-click, and choose Done from the menu.

12. Right-click, click Create Constraint, click Perpendicular from the menu, and select the two angled lines. When done, your sketch should resemble the following image on the right. In the lower-right corner of the screen, it should state that the sketch is fully constrained.

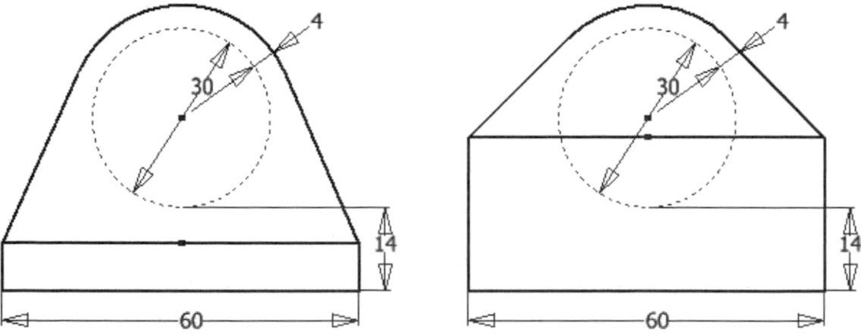

Figure A.4

13. Press the F8 key on the keyboard to display all of the sketch constraints. Review the icons to see the geometric constraint relationships in the sketch.

14. Press the F9 key to hide the sketch constraints.

15. Right-click, and click Finish Sketch from the menu to exit the active sketch.

16. Save the file as *ESS-PE-CH2-Exercise.ipt* in the Appendix A folder.

17. Close the file. End of exercise.

Project Exercise: Chapter 3

CREATING AND EDITING SKETCHED FEATURES

In the following exercise, you begin creating the gripper assembly bearing block. This exercise uses a file that contains the sketch you created in the project exercise for chapter 2 as well as additional sketches. This exercise uses the Extrude, Revolve, and 2D Sketch tools.

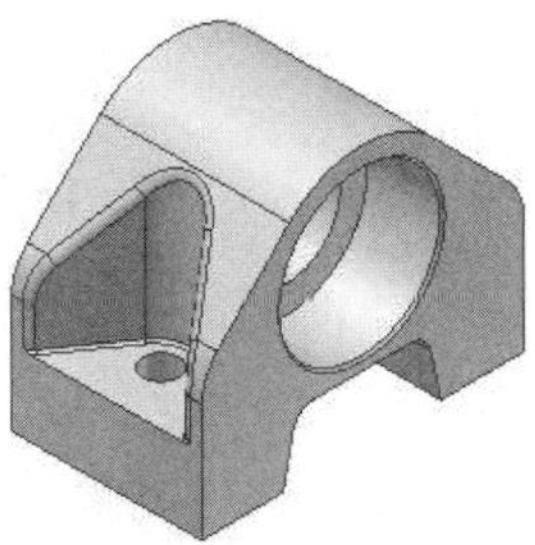

Figure A.5

1. Open the file *ESS-PE-CH3-Start.ipt* from the Appendix A folder.

2. Press the E key to start the Extrude tool. Click inside the two closed loops in Sketch1 for the profile(s), enter a distance of **35 mm**, click the midplane direction button and click OK, as shown in the following image.

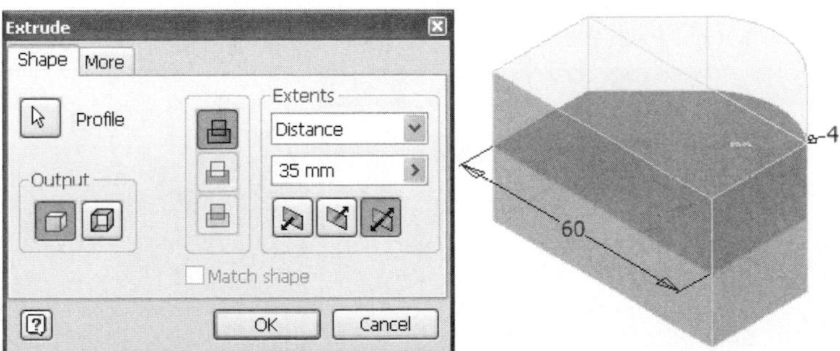

Figure A.6

3. Create a new sketch on the top face of the part, as shown in the following image on the left.

4. From the Autodesk Inventor Standard toolbar, click the Look At tool, and then click the top face of the part.

5. Create a two-point rectangle inside the sketch.
6. Create a coincident constraint from the midpoints of both lower horizontal lines, as shown in the following image on the left. This action will center the small rectangle you just drew on the existing sketch, as shown in the following image in the middle.
7. Add a **29 mm** dimension to the horizontal edge of the rectangle and a **10 mm** dimension to the vertical edge of the rectangle.
8. Change to the Home (Isometric) View, and extrude the inside of the rectangle you just created with a Cut operation and use the All option for the Extents. The following image on the right shows the completed operation.

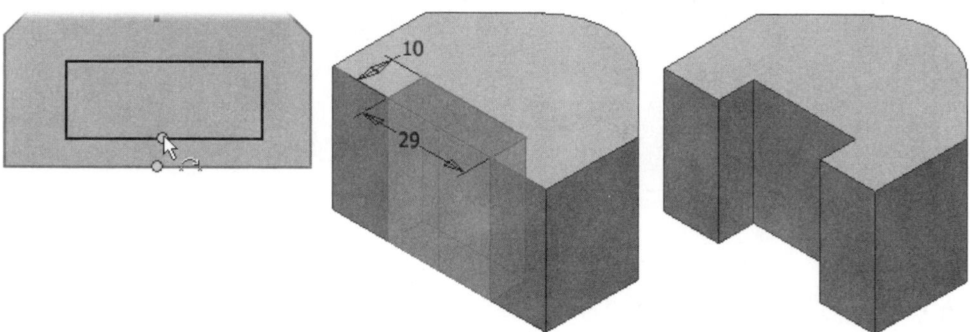

Figure A.7

9. Next create a revolve feature from an existing sketch. Right-click on Sketch2 in the browser, and click Visibility to show the sketch.
10. Press R on the keyboard to start the Revolve tool, click the Cut operation, and click OK, as shown in the following image.

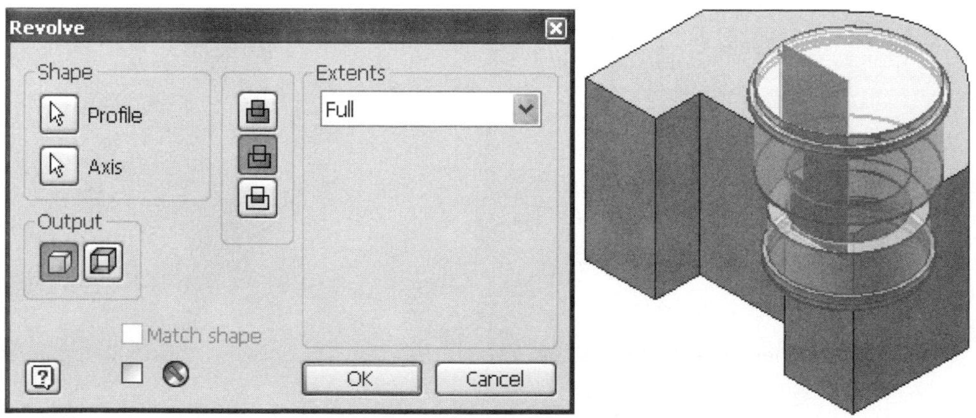

Figure A.8

11. In the browser, right-click on Work Plane1, and click New Sketch on the menu, as shown in the following image.

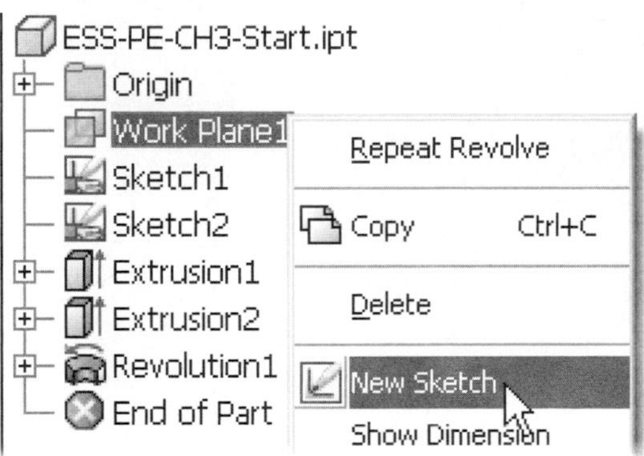

Figure A.9

12. Click the Look At tool, and click the active sketch in the browser.

13. Right-click a blank area in the graphics area, and click Slice Graphics.

14. From the 2D Sketch panel, click Project Cut Edges. This projects the part edges onto the active sketch plane.

15. Sketch the geometry as shown in the following image. Both arcs are equal in radius, and the angled lines are equal in length and at **70**° to one another. The arcs are tangent to the two lines; their center points are constrained horizontal to the midpoint of the part edge and **8 mm** from the part edge.

Note: If you are unable to add the second 70° dimension, you need to remove a sketch constraint that is preventing this dimension from being applied.

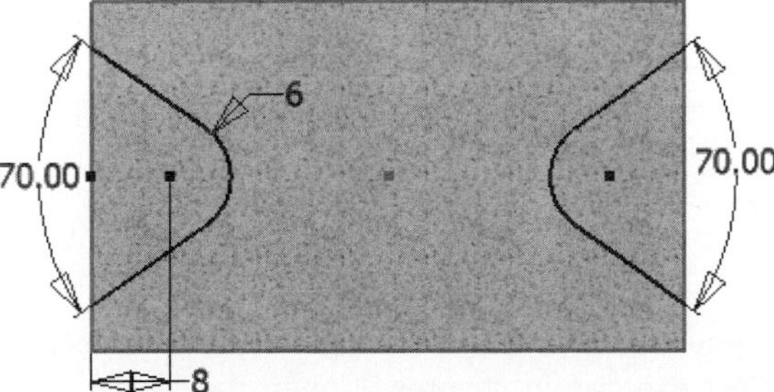

Figure A.10

16. Change to the home (Isometric) View, and then extrude both triangular sketch profiles using a Cut operation and the All Extents type. Flip the direction of the cut so that the red preview is oriented away from you, as shown in the following image on the left. The following image on the right shows the completed part.

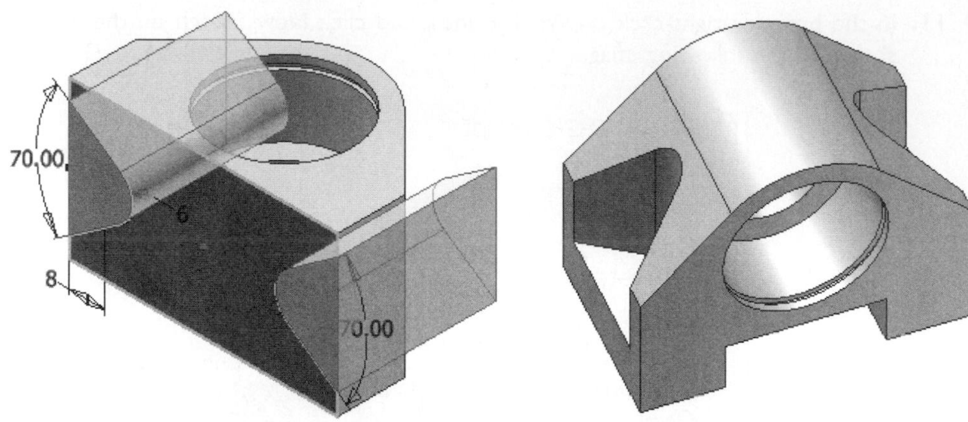

Figure A.11

17. Save this part; you will continue modeling this part in a future project exercise.

Project Exercise: Chapter 4

CREATING PLACED FEATURES

In the following exercise, you will continue creating the model of the gripper station assembly bearing block using the component model you created in the project exercise in chapter 3. This exercise uses the Hole, 2D Sketch, Rectangular Pattern, Fillet, Chamfer, and Face Draft tools.

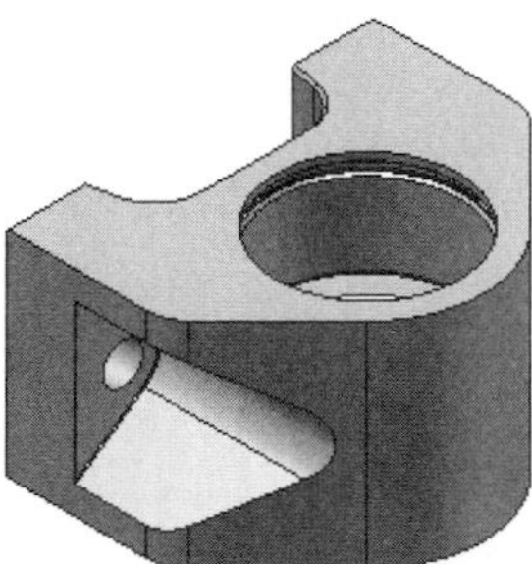

Figure A.12

1. Open the file you completed in the chapter 3 project exercise. If you didn't complete the project exercise in chapter 3, open the file *ESS-PE-CH4-Start-1.ipt* from the Appendix A folder.
2. Press the H key to start the Hole tool and place two holes with the following data.

- Placement type: Concentric
- Hole style: Drilled
- Plane: Click the face at the bottom of the triangular pocket

- Concentric reference: Click the radius edge of the packet face
- Hole type: Clearance hole
- Standard: ANSI Metric M Profile
- Fastener type: Socket Head Cap Screw
- Size: M6
- Fit: Normal
- Termination: Through All

3. Click Apply to create the hole feature and place the same hole on the opposite side of the part.

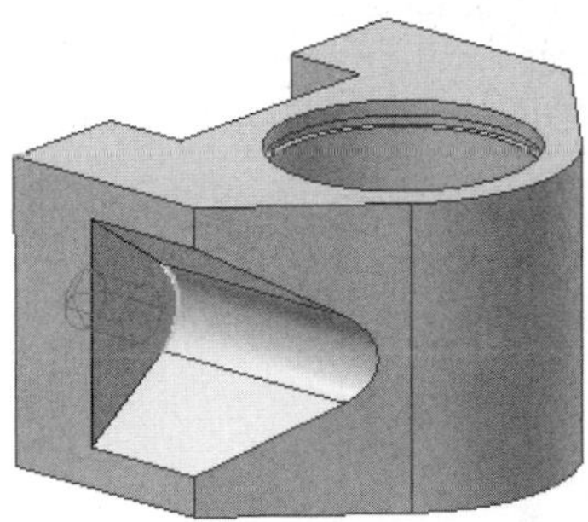

Figure A.13

4. Rotate the view to see the back of the part. Click the Look At tool, and select any bottom face on the part.
5. Create a new sketch on the face on the right side of the part.
6. Place four Point, Center Points and dimension them as shown in the following image on the left. The points are constrained horizontally and vertically to one another.
7. Press the H key on your keyboard to activate the Hole tool, and complete the dialog box using the following options.

- Placement type: From Sketch
- Hole style: Drilled
- Hole type: Simple Hole
- Termination: Distance
- Hole depth: 8 mm
- Diameter: 6 mm

8. Click OK to create the hole feature. The following image on the right shows the holes in an Isometric View.

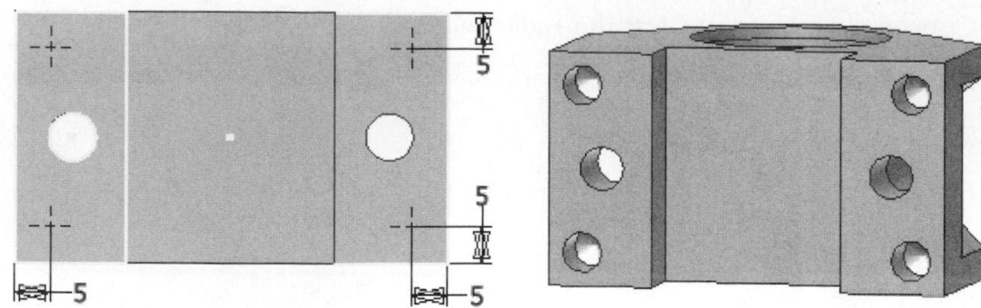

Figure A.14

9. From the Standard toolbar, click the Shaded Display tool, and click on Wireframe Display.

10. Next create fillets using the Fillet tool. From the Part Features panel, click the Fillet tool. Select the six edges, as shown in the following image, click the Radius field, and enter **0.2 mm.** Click OK to create the fillet features.

Note: You may need to use the Select Other tool to cycle through the edges to make the correct selections.

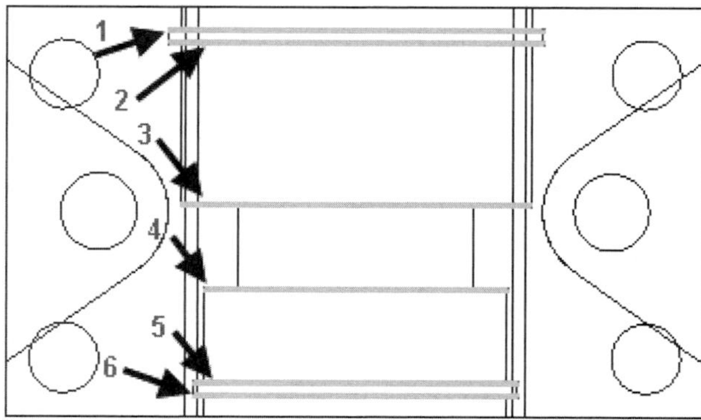

Figure A.15

11. From the Part Features panel bar, click the Chamfer tool. Select the eight horizontal outside edges, as shown in the following image, click the Distance field, and enter **0.5 mm**. Click OK to create the chamfer features.

Note: You may need to use the Select Other tool to cycle through the edges to make the correct selections.

12. Change the display back to Shaded Display.

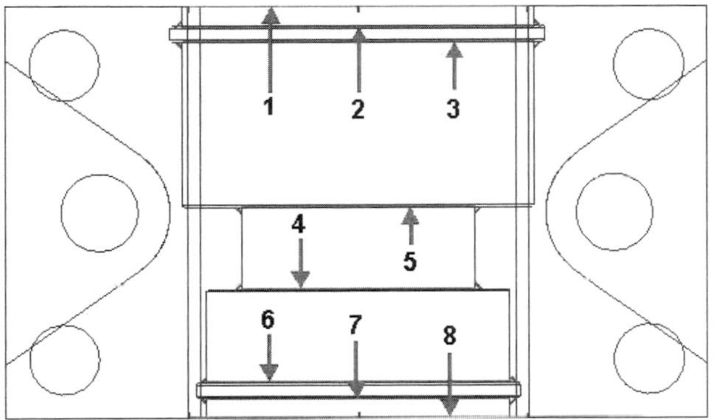

Figure A.16

13. Press and hold the F4 key. Left-click and hold to rotate the model view to the back of the part, as shown in the following image on the left.

14. Click the Face Draft tool, and then select the large middle vertical face to define the Pull Direction, as shown in the following image on the left. Click the left and right inside vertical faces near the Pull Direction face, click the Draft Angle field, and enter **12 deg** for the angle. Click OK to apply the face draft features.

Figure A.17

15. Click the Fillet tool, and place a **7 mm** fillet on the two inside edges, as shown in the following image.

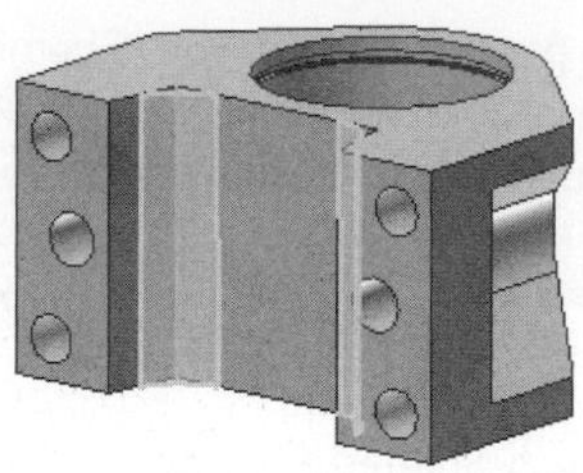

Figure A.18

16. Rotate your view so it resembles the following image.

17. Place a **7 mm** fillet on the four edges, as shown in the following image on the left. You may need to rotate your view again to select the four edges.

18. Place a **.21 mm** fillet on the two inside edges of the notch, as shown in the following image on the right. You may need to rotate your view again to select the two edges.

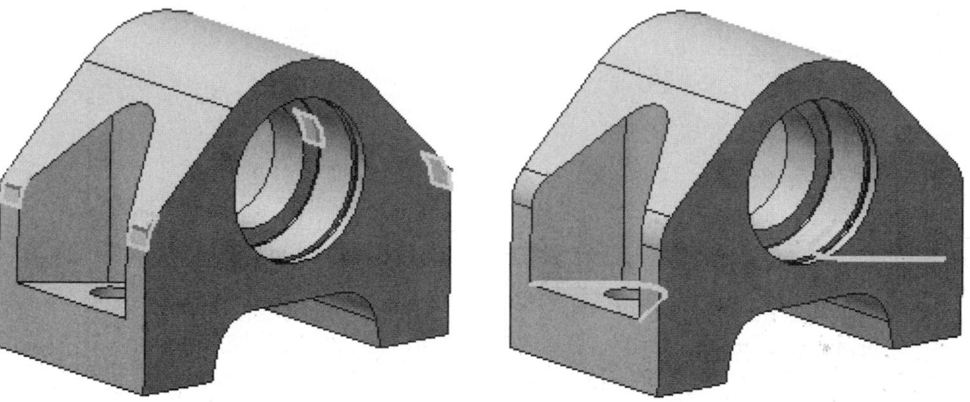

Figure A.19

19. Save the part model, as you will continue modeling the part in a future exercise. The following image shows the completed part.

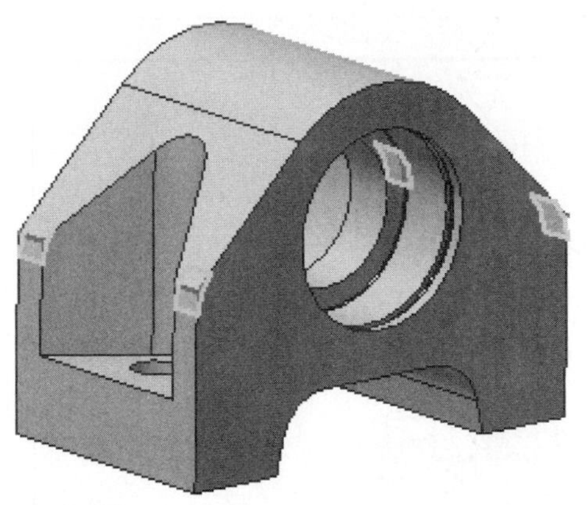

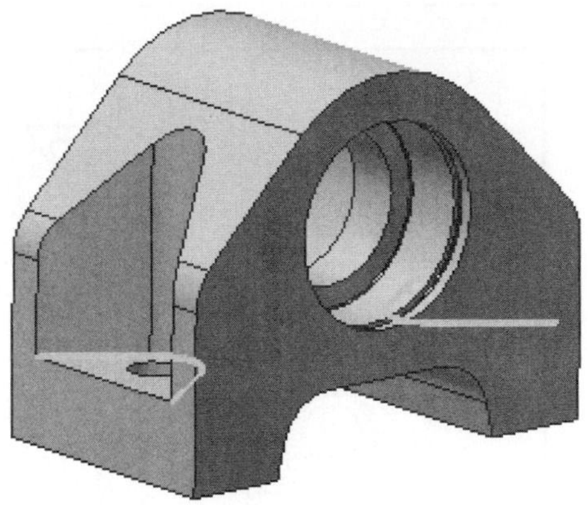

Figure A.20

20. Close the file. End of exercise.

Project Exercise: Chapter 5

CREATING AND EDITING DRAWING VIEWS

In the following exercise, you will create, detail, and annotate a drawing of the bearing block model you created in the project exercise for chapter 4. This exercise uses the Base, Projected, Section, and Detail View tools and options. You will also annotate the drawing using the Dimension, Centerline, and Hole Table tools to complete the detailed drawing, as shown in the following image.

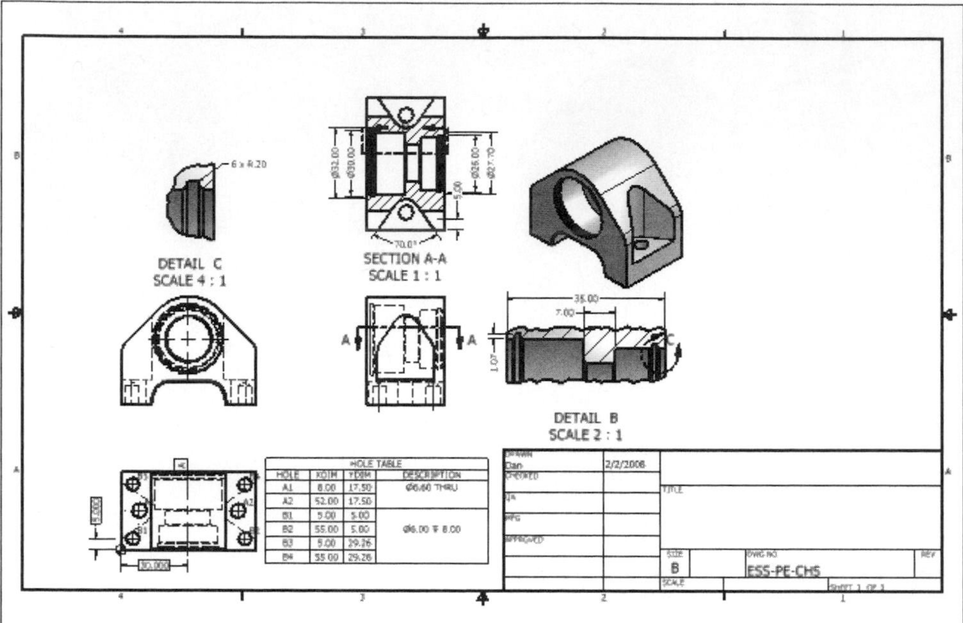

Figure A.21

1. Click on the File drop-down menu, and then click New. Click on the Metric tab, and double-click on *ANSI (mm).idw* to create a new drawing file.
2. In the browser, right-click on Sheet:1, and click Edit Sheet on the menu. In the Edit Sheet dialog box Format area, select B from the Size drop-down list, and click OK.
3. In the Drawing Views panel, click the Base View tool, and navigate to and double-click on *ESS-PE-CH5.ipt* from the Appendix A folder. For the Orientation, verify that the Front is current. Verify that the scale is set to 1:1 and that Hidden Line in the Style area is selected. Click a point in the lower-left corner of the sheet, as shown in the following image.
4. Right-click in the view boundary, click Create View, and then click Projected View from the menu. Place the three projected views, as shown in the following image.

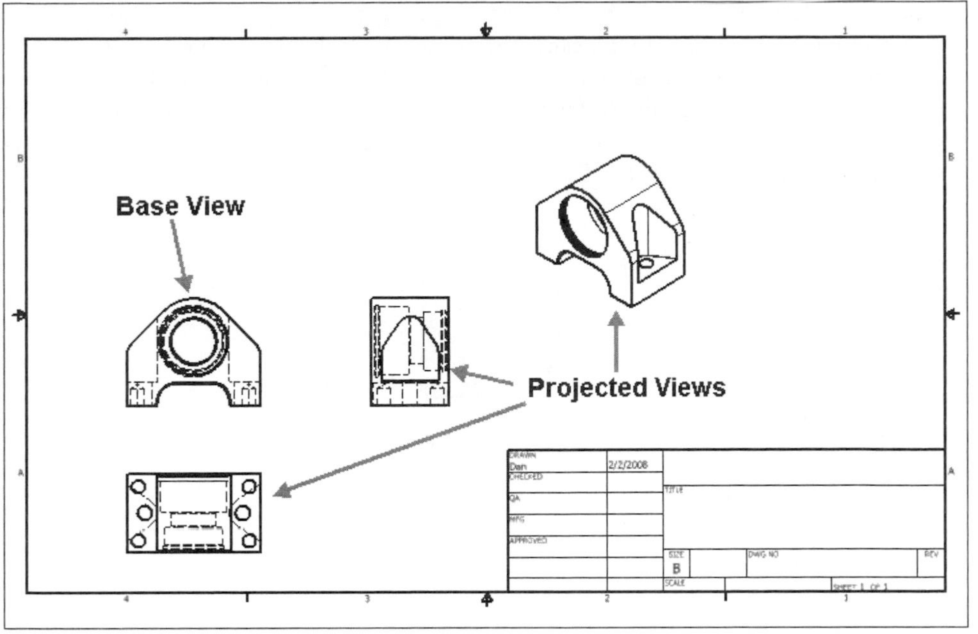

Figure A.22

5. Click the Section View tool from the panel bar, and create a section view from the right side view, as shown in the following image.

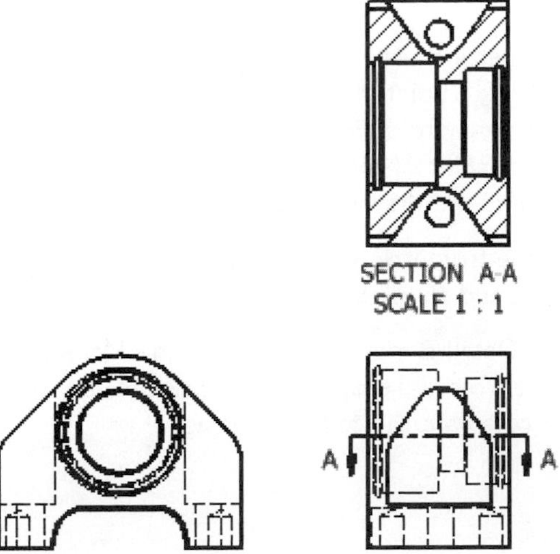

Figure A.23

6. Click Detail View tool from the panel bar, and then click inside the section view. In the Detail View dialog box, click Rectangular in the Fence Shape area. Click a point in the middle of the section view to define the center of the detail view rectangle, and pick another point to define the detail view rectangle's size, as shown in the following image on the left. Click to place the detail view to the right of the right side view, as shown in the following image on the right.

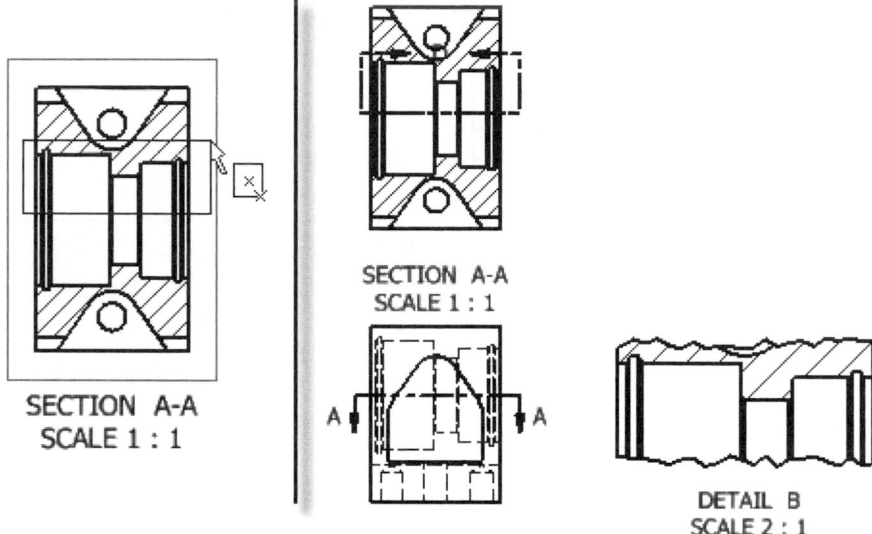

Figure A.24

7. Right-click in the Detail B view boundary, click Create View, and then click Detail View from the menu. Click a point in the detail view near the snap-ring groove in the upper-right corner of the view to define the center of the detail view circle, and pick another point to define the radius of the detail views circle. Click to place the detail view in the lower-right corner of the drawing sheet, as shown in the following image.

8. Double-click in the Isometric View boundary. In the Drawing View dialog box Style area, click Shaded, and then click OK.

9. Double-click the Detail B view boundary. In the Drawing View dialog box Style area, uncheck the Style from Base option and click Shaded, and then click OK. The drawing should now resemble the following image.

EXERCISES

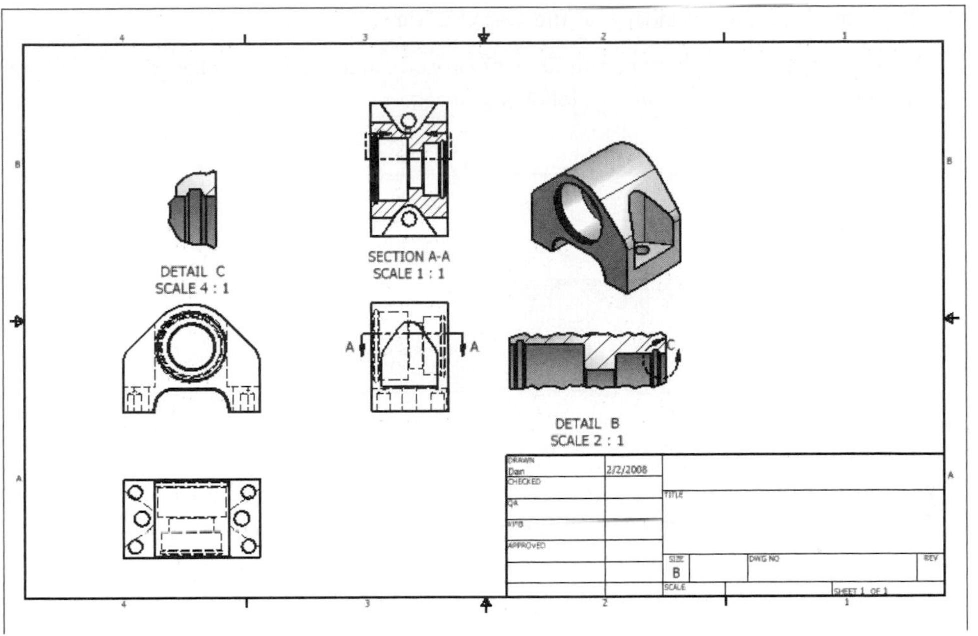

Figure A.25

10. Zoom into and right-click on the boundary of the Section A-A view. Click Retrieve Dimensions from the menu. In the Retrieve Dimensions dialog box, click Select Dimensions, select the six dimensions shown in the following image, and then click OK.
11. Click and hold down the left mouse button on the dimensions text. Drag to arrange the dimensions, as shown in the following image. Note that when moving the text and dots appear the text is centered.

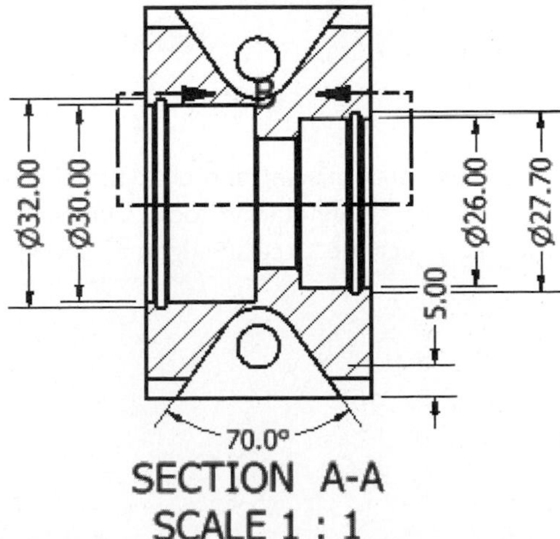

Figure A.26

12. Zoom into the boundary of the Detail B view.

13. Press the D key to start the Dimension tool, and place and align the 1.00 linear dimension, as shown in the following image.

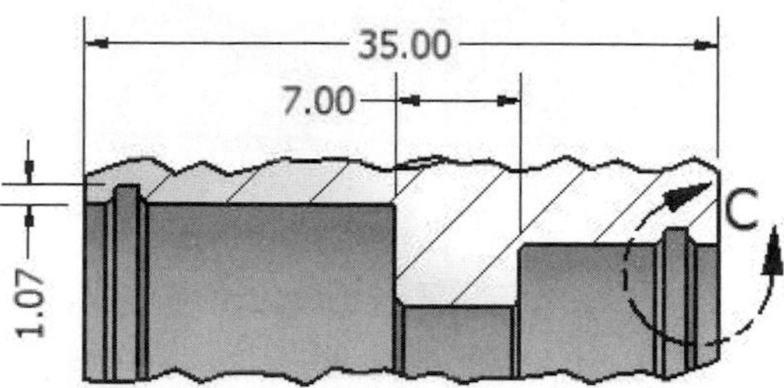

Figure A.27

14. Zoom into the Detail C view, and press the D key to start the Dimension tool. Select the arc on the right of the snap-ring groove, and click to place the R0.21 dimension.

15. Double-click on the fillet dimension. In the Edit Dimension dialog box, click in the text box window. Press the HOME key on the keyboard to place the text edit marker at the beginning of the text string. Type **6 x**, press the spacebar on the keyboard, and click OK to close the dialog box.

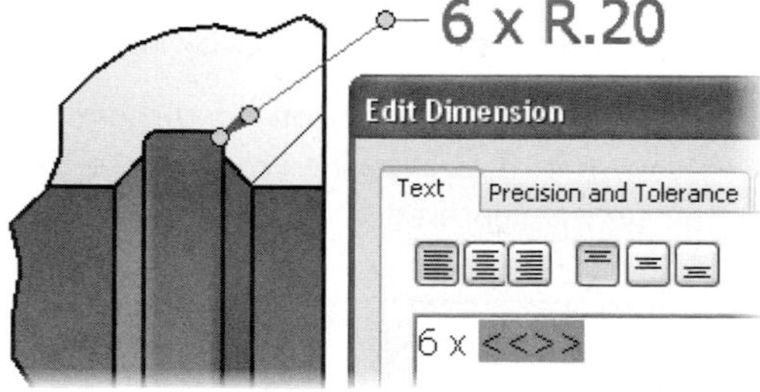

Figure A.28

16. Click the Drawing Views panel heading and click the Drawing Annotation panel from the list to expose the drawing annotation tools. On the Drawing Annotation panel, click Center Mark, and select the circular edges, as shown in the following image.

17. On the Drawing Annotation panel, click Centerline Bisector under the Center mark tool, and place a centerline between the parallel lines that represent the center of the part as shown in the bottom view and two blind holes on the view on the right, as shown in the following image.

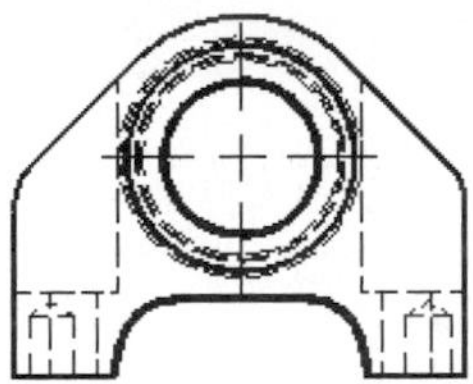

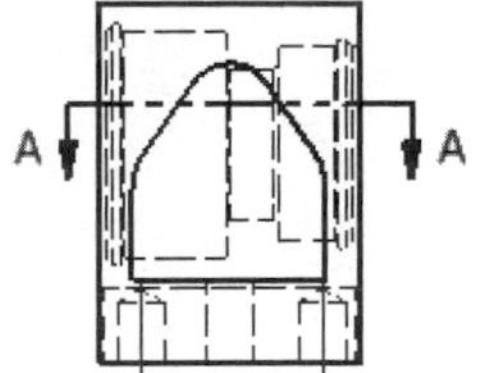

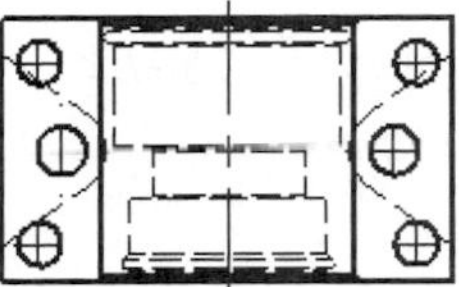

Figure A.29

18. On the Drawing Annotation panel, click the down-arrow next to the Hole Table Selection tool. Select the Hole Table View tool from the menu, click a point in the bottom view, and then place the table origin at the lower-left point of the part. Click to place the hole table to the right of the view, as shown in the following image.

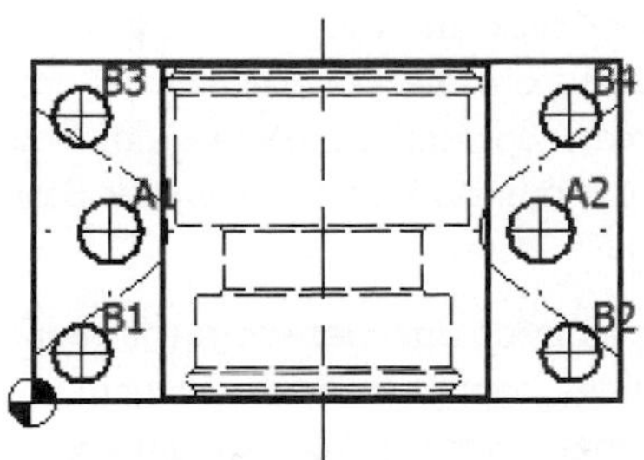

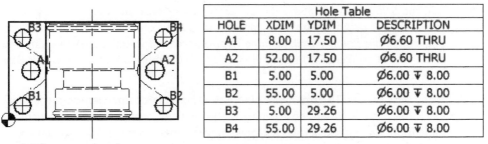

Hole Table			
HOLE	XDIM	YDIM	DESCRIPTION
A1	8.00	17.50	Ø6.60 THRU
A2	52.00	17.50	Ø6.60 THRU
B1	5.00	5.00	Ø6.00 ↧ 8.00
B2	55.00	5.00	Ø6.00 ↧ 8.00
B3	5.00	29.26	Ø6.00 ↧ 8.00
B4	55.00	29.26	Ø6.00 ↧ 8.00

Figure A.30

19. Right-click on the hole table, and click Edit Hole Table from the menu. From the Option tab, and in the Row Merge Options area, click the Combine Notes option and check Reformat Table on Custom Hole Match, as shown in the following image, and click OK.

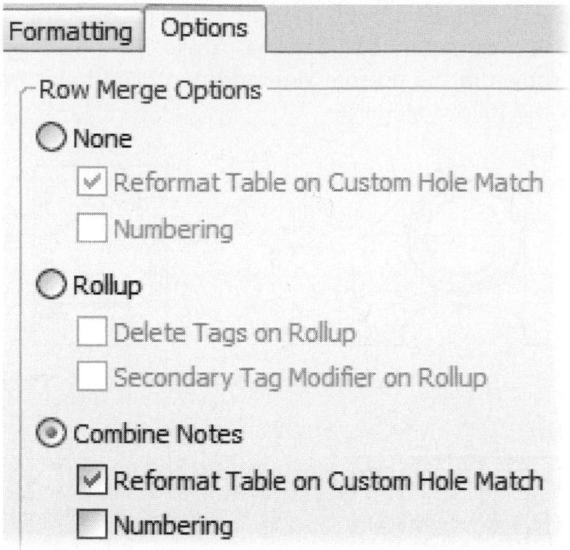

Formatting Options

Row Merge Options

None

Reformat Table on Custom Hole Match

Numbering

Rollup

Delete Tags on Rollup

Secondary Tag Modifier on Rollup

Combine Notes

Reformat Table on Custom Hole Match

Numbering

Figure A.31

20. On the Drawing Annotation panel, click the Datum Identifier Symbol tool, and select the upper end of the centerline, as shown in the following image on the left (do not click a second point). Right-click, and click Continue from the menu. In the Text Edit dialog box, click OK.

21. Press D on the keyboard to start the Dimension tool. Click to add and position the two dimensions, as shown in the following image on the left (you may need to reposition the 5.00 dimension above the top arrow). Right-click, and click Done from the menu. Double-click the **30 mm** dimension text, and from the Precision and Tolerance tab, change the Primary Unit to three decimal places, represented in the dialog box as 3.123. In the Tolerance Method area of the Edit Dimension dialog box, click Basic, and click OK.

22. Right-click on the **30 mm** dimension, and click Copy Properties from the menu. Click the **5 mm** dimension to apply the properties copied from the 30 mm dimension. Right-click, and click Done from the menu. Your display should resemble the following image on the right. If desired place, another datum on the lower edge.

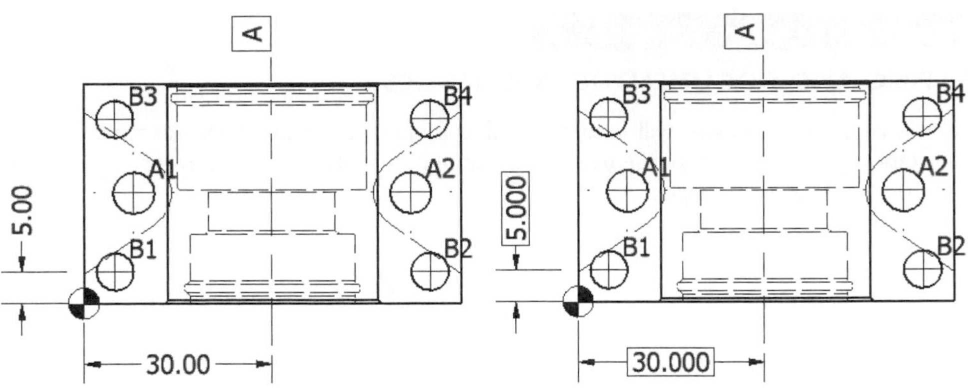

Figure A.32

23. The following image shows the completed drawing. Save the file; you will continue modeling this part in a future project exercise.

24. Close all open files. End of exercise.

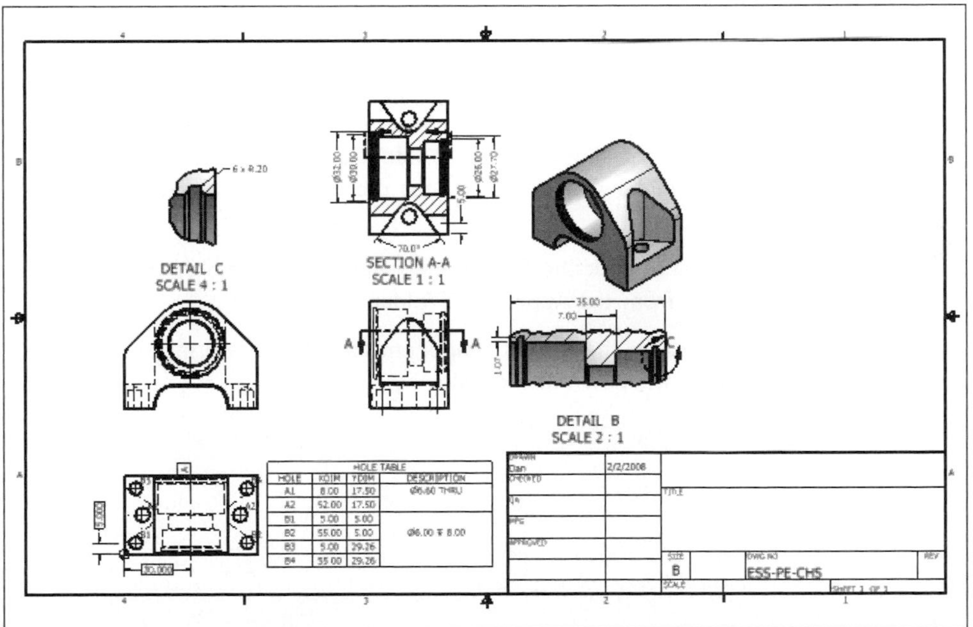

Figure A.33

Project Exercise: Chapter 6

CREATING AND DOCUMENTING ASSEMBLIES

In the following exercise, you will create and document the gripper finger assembly used in the gripper station project. This exercise uses the assembly, presentation, and drawing environments and their related tools and concepts.

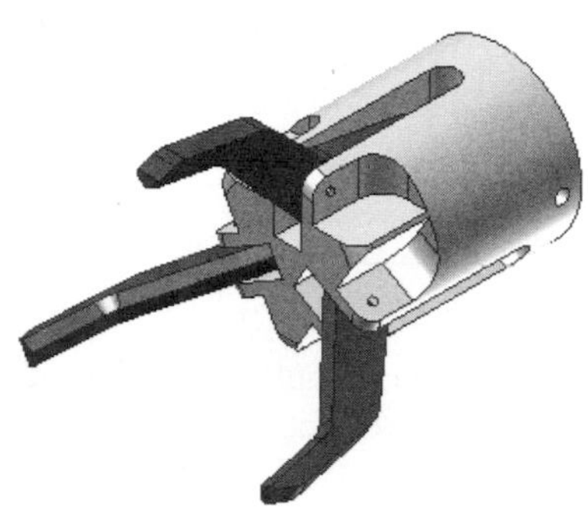

Figure A.34

1. Open the file *ESS-PE-CH6-Start.iam* from the Appendix A folder.
2. If needed, click on the Tools drop-down menu, and click Application Options. Next click on the Sketch tab, and review the "Autoproject part origin on sketch create" setting. If the box is not checked, click to add a check. This option will automatically project the Origin folder Center Point onto the default active sketch during the creation of a new part.
3. In the Assembly panel, click the Place Component tool, and place one instance of *ESS-PE-CH6-301.ipt* from the Appendix A folder.
4. Press ENTER to repeat the Place Component tool. Place two instances of *ESS-PE-CH6-300.ipt,* as shown in the following image.

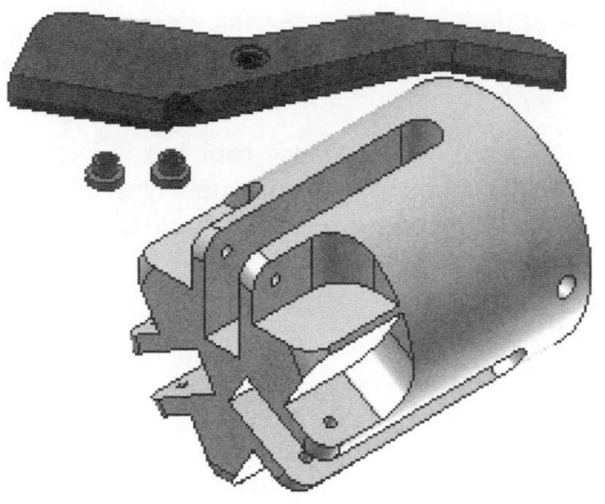

Figure A.35

5. Press the C key to start the Place Constraint tool. In the Place Constraint dialog box, click Insert as the constraint type, select the inside circular edge of the bushing and then the inside circular edge of the counterbore holes, as shown in the following image on the left, and then click Apply.
6. Repeat the same steps to constrain the other bushing to the counterbore hole on the other side of the gripper finger, and click OK.

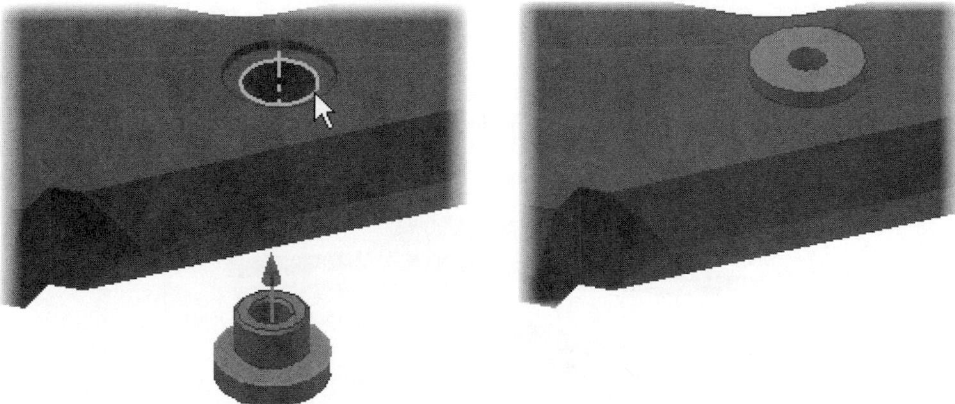

Figure A.36

7. In the Assembly panel, click the Create Component tool. In the Create In-Place Component dialog box, type **ESS-PE-CH6-FingerPin** for the New Component Name, and use the *Standard (mm).ipt* template. Change the location where the file will be saved to *C:\INV 2009 Ess Plus\Appendix A*. Uncheck the Constrain sketch plane to selected face or plane option, as shown in the following image on the left. Click OK to create the part.

8. Click a top face of the Finger to define the default sketch plane for the new part, and from the 2D Sketch panel, click the Center Point Circle tool. Draw a circle to the right of the bushing and add a **2 mm** diameter dimension.

9. Press the E key to activate the Extrude tool. In the Extrude dialog box, type **15 mm** as the extrusion distance, and the preview of the part should appear as shown in the following image on the right. Click OK.

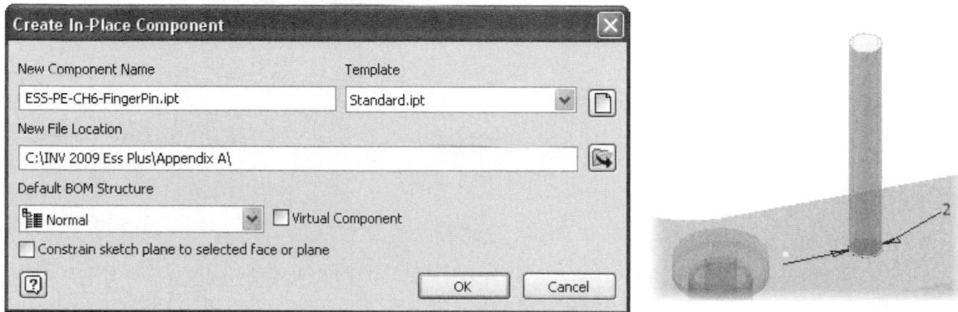

Figure A.37

10. Right-click in your display, and pick Finish Edit from the menu to return to the top-level assembly. You could also have clicked once on the Return button.

11. Rotate your view so that it resembles the following image.

12. Press C on the keyboard to start the Place Constraint tool. In the Place Constraint dialog box, click Insert as the constraint type. Select the top circular edge of the hole in the bushing and the left inside circular edge of the hole in the base component, as shown in the following image, and click OK.

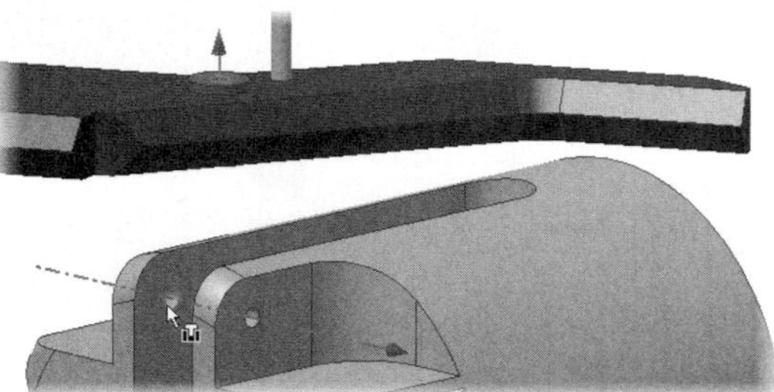

Figure A.38

13. When done, your display should resemble the following image on the left. In the graphics window, select the gripper finger and position it so it resembles the following image on the right.

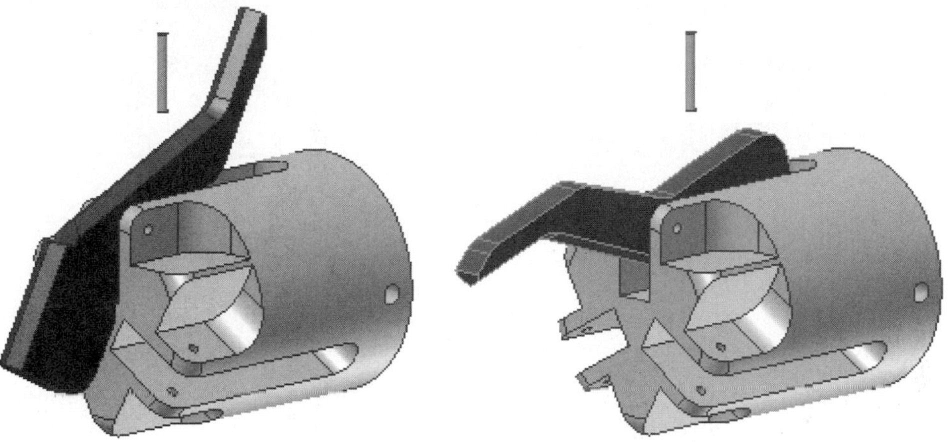

Figure A.39

14. Press C on the keyboard to activate the Place Constraint tool. In the Place Constraint dialog box, click Angle for the constraint type. Click the flat face of the finger and the flat face of the hub pocket, as shown in the following image on the left; if needed change the angle to **0** degrees and then click Apply.
15. Apply an Insert constraint between the pin and the right outside circular edge of the hole in the base component with an offset value of **-0.6 mm**. Use the Aligned Solution, as shown in the following image on the right.

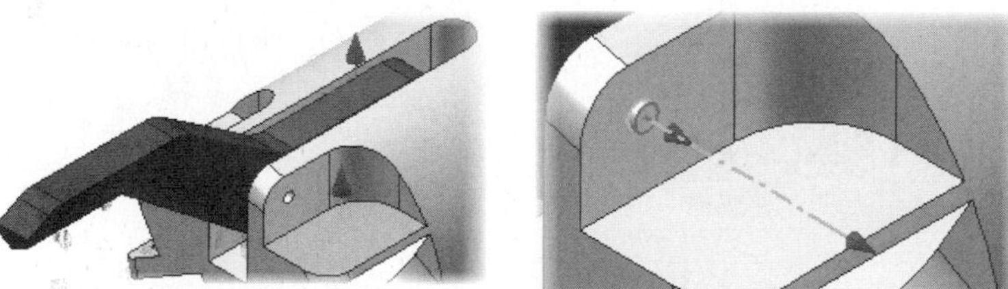

Figure A.40

16. In the Assembly panel, click the Pattern Component tool, and select the four last components in the browser as the components to pattern. In the Pattern Component dialog box, click the Circular tab, and click the Axis Direction button. Click the outside circular face of the base component, and change the number of instances to **3 ul** and the angle to **120.00 deg**, as shown in the following image. Click OK.

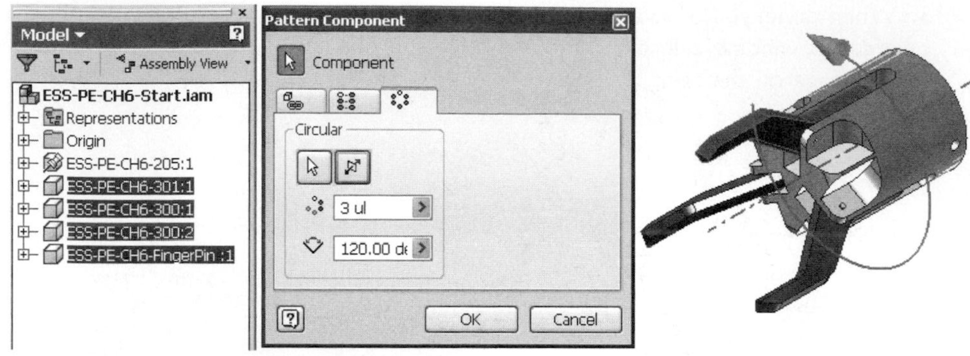

Figure A.41

17. In the browser, click the plus sign (+) next to the *ESS-PE-CH6*-205:1, the base component, to display the constraints, as shown in the following image on the left.
18. Right-click on the Angle constraint, and select Drive Constraint from the menu, as shown in the following image on the left.
19. In the Drive Constraint dialog box, type **10.00 deg** in the Start field and **-30deg** in the End field, as shown in the following image on the right.
20. Press the Forward button to drive the constraint through the defined range of motion. Press the Reverse button to drive the constraint through the defined range of motion in the opposite direction.

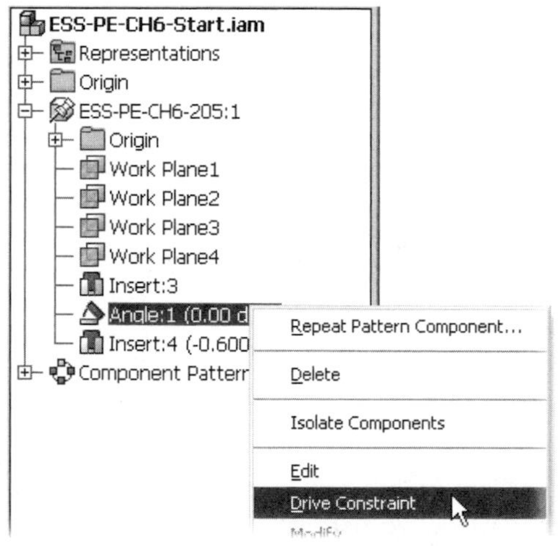

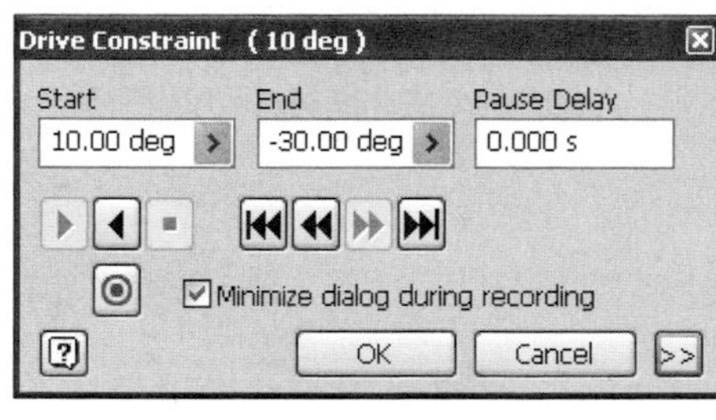

Figure A.42

21. Save the assembly. Close the file.
22. Open the file *ESS-PE-CH6-308.idw*.

This drawing contains an Isometric View based on a presentation model. You will insert a Parts List, edit the Bill of Materials, and add balloons to complete the drawing.

23. From the Drawing Annotation panel, click the Parts List tool. In the graphics window, click a point inside the drawing view, click OK and place the Parts List above the title block, as shown in the following image.

4	3	ESS-PE-CH6-FingerPin Finished	
3	6	IV11-CH6-PE-300	
2	3	IV11-CH6-PE-301	
1	1	IV11-CH6-PE-205	
ITEM	QTY	PART NUMBER	DESCRIPTION
Parts List			

DRAWN

Figure A.43

24. In the graphics window, right-click on the Parts List, and click Edit Parts List Style from the menu.

25. Next prepare the Parts List to be moved to the top right of the border. In the Heading and Table Settings section in the Direction area click Add new parts to bottom as shown in the following image on the left. Change the Heading to Top, as shown in the following image on the right, click Save, and then click Done.

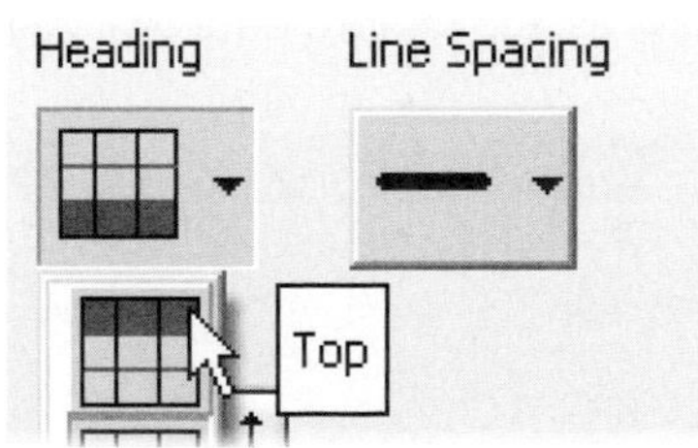

Figure A.44

26. In the graphics window, click and drag the Parts List to reposition to the top-right corner of border. The Parts List should resemble the following image.

2 1

Parts List			
ITEM	QTY	PART NUMBER	DESCRIPTION
1	1	IV11-CH6-PE-205	
2	3	IV11-CH6-PE-301	
3	6	IV11-CH6-PE-300	
4	3	ESS-PE-CH6-FingerPin Finished	

Figure A.45

EXERCISES

27. Next add a description to the parts. Place the cursor over the Parts List, right-click, and click Bill of Materials from the menu.

28. In the Bill of Materials dialog box, type the descriptions into the Description column, as shown in the following image, and click Done.

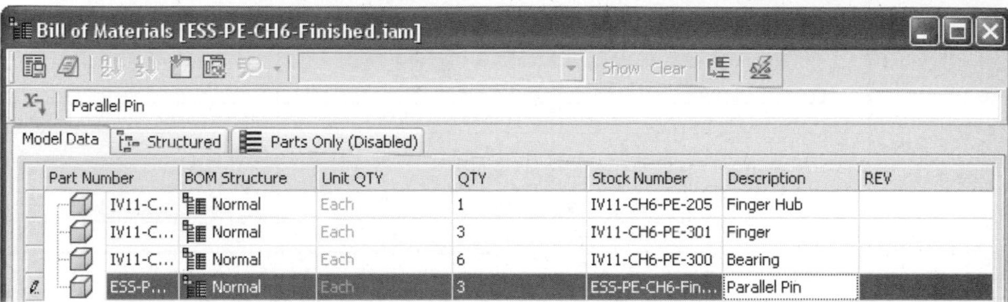

Figure A.46

29. In the Drawing Annotation panel, click the Auto Balloon tool from the Balloon drop-down list. In the graphics window, click the drawing view, and then window select all the components in the view.

30. In the Placement area of the dialog box, click Select Placement, and click the Vertical option. Type **5 mm** in the Offset Spacing field, as shown in the following image on the left, and pick to place the vertical column of balloons to the left of the drawing view, as shown in the following image on the right. Click OK.

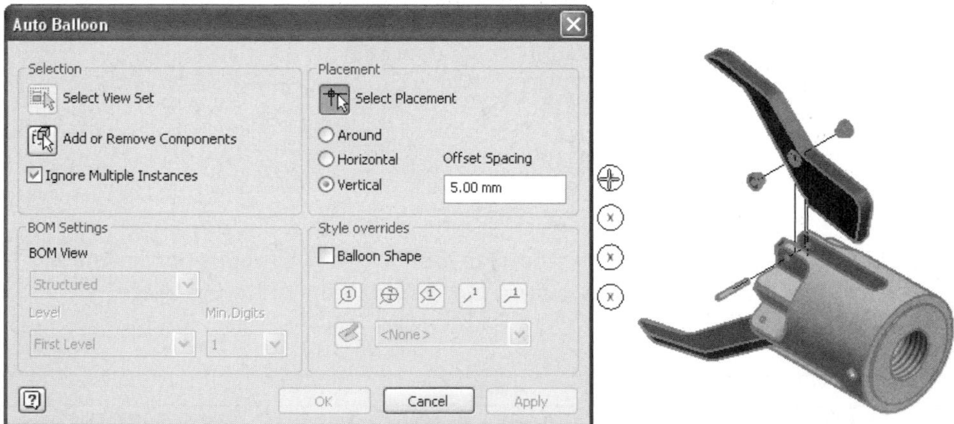

Figure A.47

31. Click and hold down and drag the green grips on a balloon and arrow to adjust the position of the balloons, as shown in the following image.

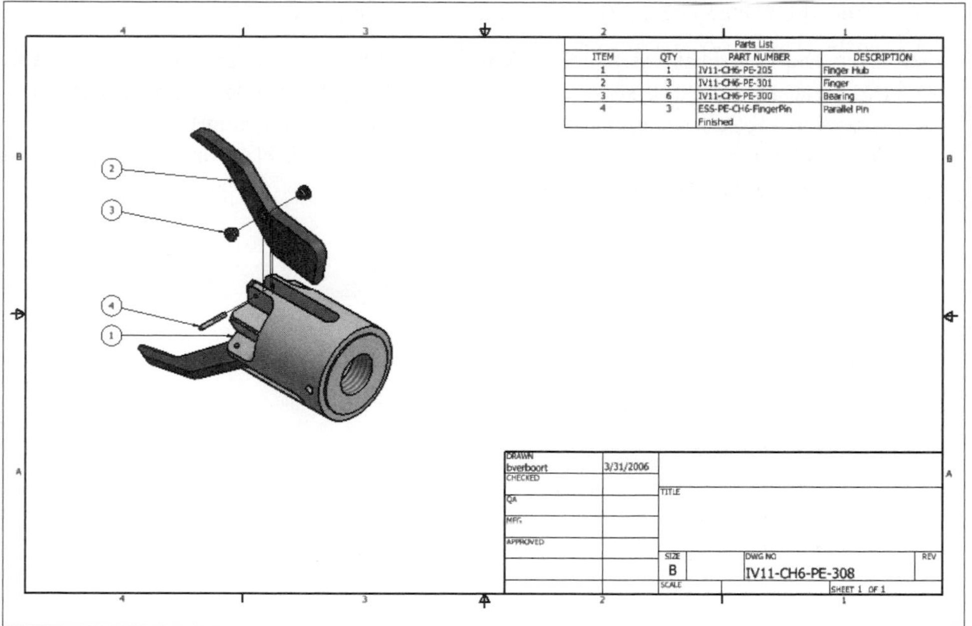

Figure A.48

32. Save the drawing. Close the file. End of exercise.

Project Exercise: Chapter 7

ADVANCED PART MODELING TECHNIQUES

In the following exercise, you edit the model of the back plate used in the gripper station assembly. This exercise uses the following advanced part modeling tools and concepts: Feature Mirror, Feature Reorder, Extrude Open Profile, Part Split, Face Draft, Rib, and Emboss.

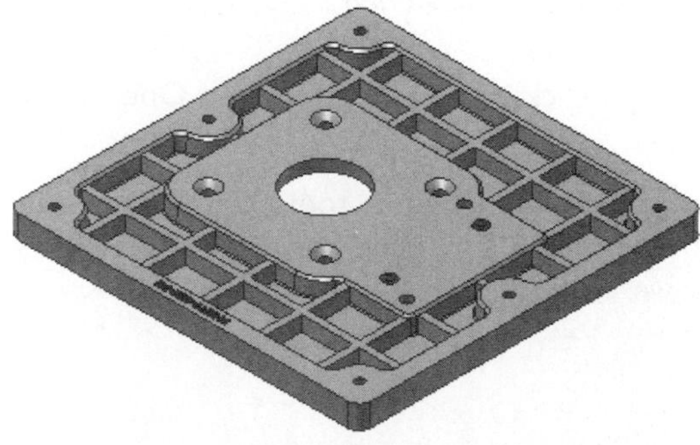

Figure A.49

EXERCISES

1. Open the file *ESS-PE-CH7-Start.ipt* from the Appendix A folder.
2. From the Part Features panel, click the mirror tool. In the graphics window, select the array of the three small threaded holes along the top left edge of the part, as shown in the following image on the left. In the Mirror dialog box, click the Mirror Plane selection arrow, and in the browser, click the YZ Plane in the Origin folder. Click OK in the Mirror dialog box.

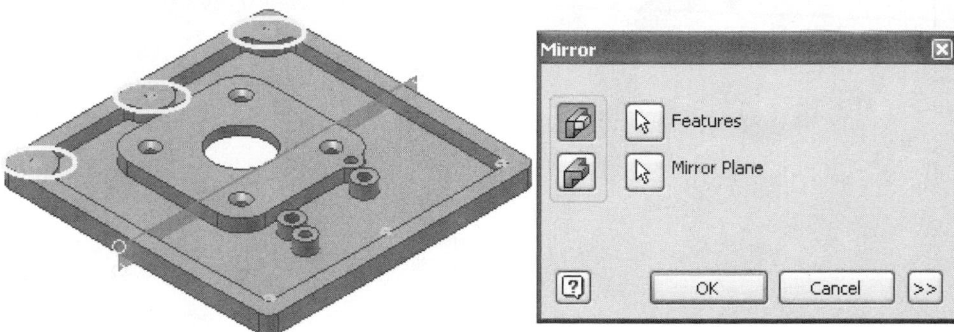

Figure A.50

3. Note that the Mirror feature appears in the browser, but there appears to be no change to the model. In the browser, click on the Mirror8 feature and hold down the left mouse button. Drag the Mirror8 feature and drop it above the Shell1feature in the list, as shown in the following image.

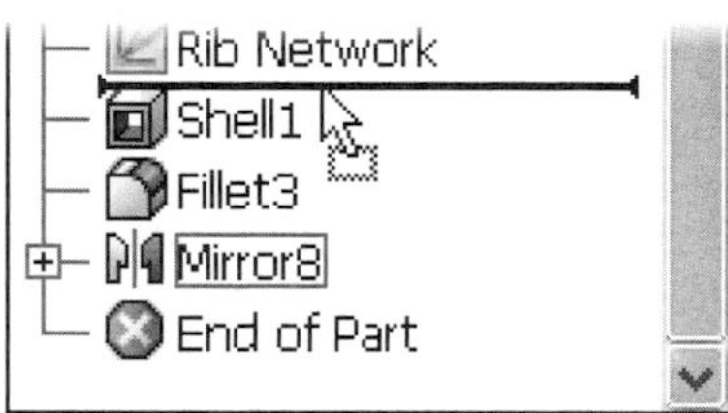

Figure A.51

4. In the browser, right-click on the sketch named Open Profile, and click Visibility from the menu.
5. Press the E key to start the Extrude tool. Select the sketch lines in the graphics window for the profile. Click in the inner rectangular boundary for the direction to extrude, as shown in the following image, and click OK.

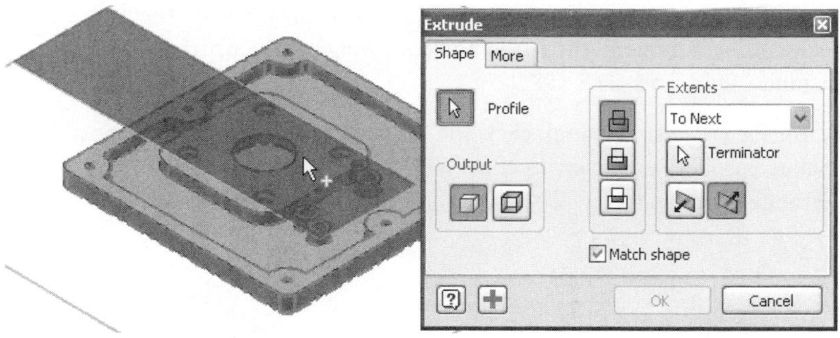

Figure A.52

6. Reorder Fillet3 so it appears below the open profile extrusion created in the previous step. In the browser click the Fillet3 feature and hold down the left mouse button. Drag the Fillet3 feature and drop it below Extrusion15.

7. In the browser, double-click on Fillet3. In the graphics window, click the additional ten inside vertical edges so that a total of 16 edges are selected, as shown in the following image. Click OK.

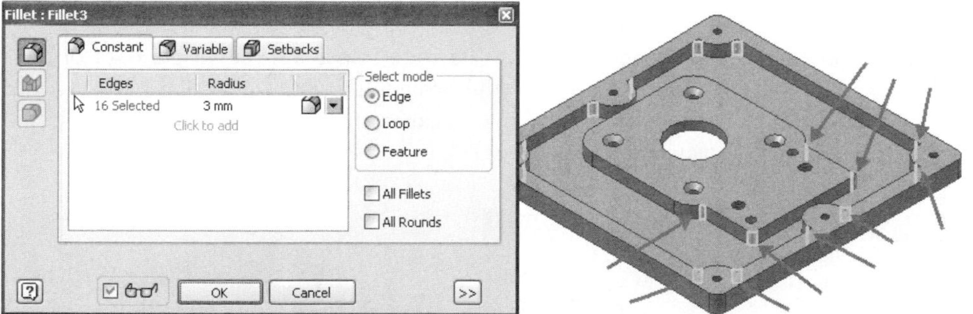

Figure A.53

8. In the Part Features panel, click the Face Draft tool, and select one of the top faces of the plate to define the Pull Direction. Select near the top of both of the inside vertical faces, change the Draft Angle to **5 deg**, as shown in the following image on the left, and click OK.

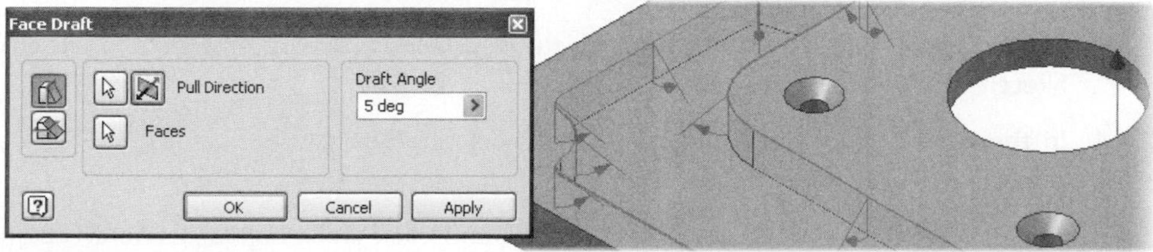

Figure A.54

EXERCISES

9. In the Part Features panel, click the Chamfer tool. Select the two top edges of the pocket to chamfer, as shown in the following image on the left. Enter **1 mm** in the Distance field, and click OK.

10. In the Part Features panel, click the Fillet tool. Select the two bottom edges of the pocket to fillet, as shown in the following image on the right. Enter **.5 mm** in the Distance field, and click OK.

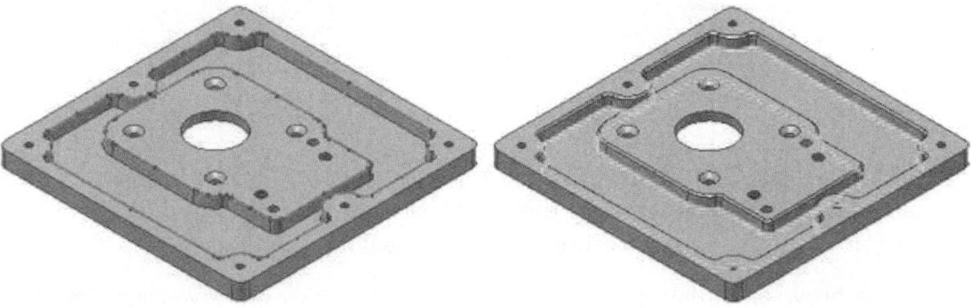

Figure A.55

11. In the browser, right-click on the sketch named Rib Network, and click Visibility from the menu.

12. In the Part Features panel, click the Rib tool. Select the ten visible lines of the sketch in the graphics window for the profile. Click the Direction button, and click below the plate. Enter **3 mm** in the Thickness field and **5** in the Taper field, as shown in the following image. Click OK.

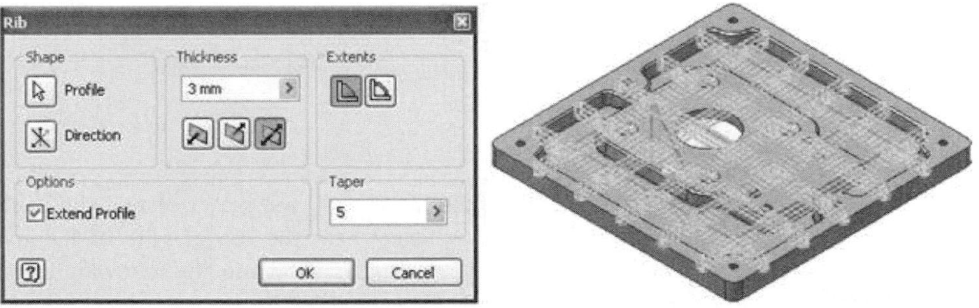

Figure A.56

13. In the graphics window, right-click on the top face of the outer wall, and click New Sketch from the menu.

14. In the view area of the Standard toolbar, click the Look At tool, and click the new sketch.

15. In the 2D Sketch Panel, click the Text tool, and click and drag to define a rectangular text area near the middle of the upper face. In the Format Text dialog box, click the Center and Top Justification buttons. Type the text **PN:INV-CH7-PE**, change the

text height to 3.50 mm, verify that Bold is selected as shown in the following image, and click OK. Right-click, and select Done from the menu.

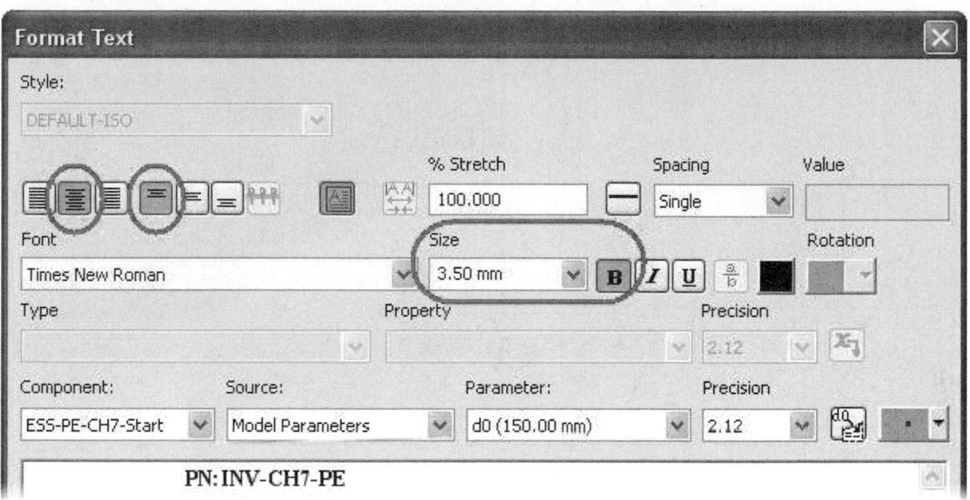

Figure A.57

16. Locate the text box by applying a **2 mm** dimension between the top horizontal line of the sketch and the top horizontal edge of the text box. Also apply a vertical sketch constraint between the midpoint of the top horizontal line of the text box and the midpoint of the top horizontal edge of the part as shown in the following image. Right-click in the graphics window, and click Done.

Figure A.58

17. Right-click in the graphics window. Click Finish Sketch from the menu. To change to an Isometric View, press the F6 key.

18. In the Part Features panel, click the Emboss tool, and then click the text in the sketch—you may need to use the Select Other tool. Change the depth to **0.5 mm**, verify the Emboss from face as the type of emboss, and change the top face color to black, as shown in the following image. Click OK.

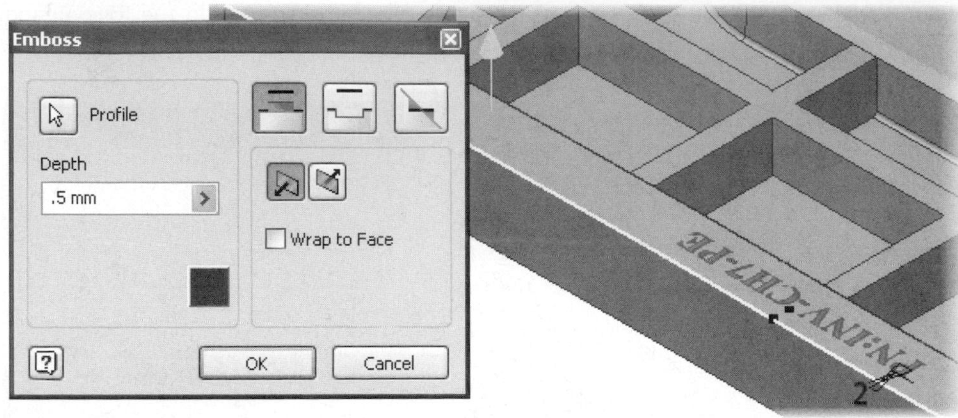

Figure A.59

19. The following image shows the completed part.
20. Save the part. Close the file. End of exercise.

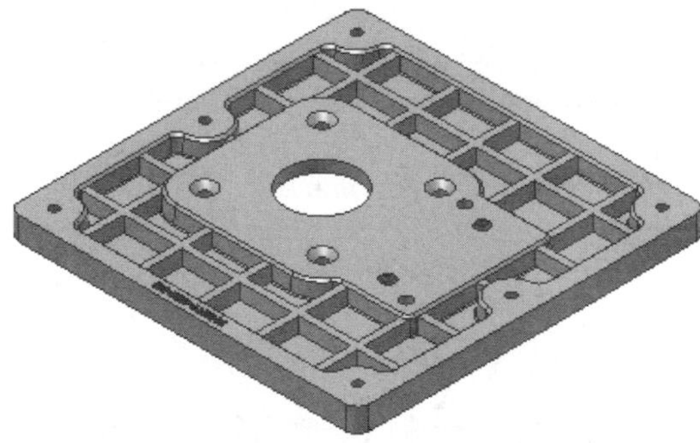

Figure A.60

Project Exercise: Chapter 8

ICOMPONENTS AND PARAMETERS

In this exercise, you edit the components used in the back plate subassembly from the gripper station. This exercise uses the following part and assembly modeling productivity tools and concepts: iMates, iParts, and iAssemblies. The following image shows the completed assembly.

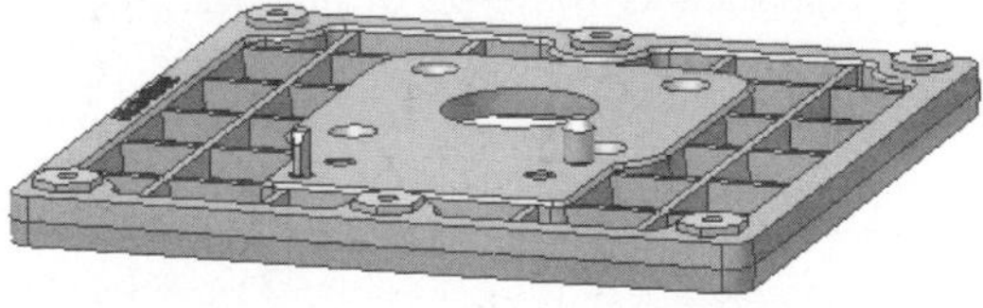

Figure A.61

IMATES

1. In this section, you will create iMates that will be used to assemble the components. Open the file *ESS-PE-CH8-Round Pin.ipt.*

2. Press Q on the keyboard to activate the Create iMate dialog box, as shown in the following image on the left. Click the Insert constraint type and then the Aligned solution. Click the More >> button, and type **Pin** in the Name field. Select the front circular edge of the pin, as shown in the following image on the right, and click OK.

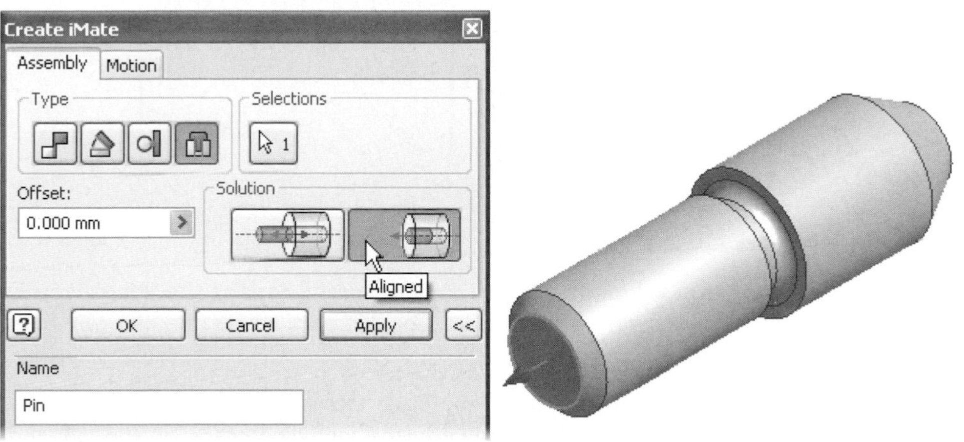

Figure A.62

3. Save the part. Close the file.

4. Open *ESS-PE-CH8-Diamond Pin.ipt.*

5. Press Q on the keyboard to activate the Create iMate dialog box. Ensure that the Mate option is selected. In the Origin folder, click the XY Plane, as shown in the following image on the left, and click Apply.

6. In the Create iMate dialog box, click the Insert constraint type, then the Aligned solution, and select the front circular edge of the pin, as shown in the following image on the right, and click OK.

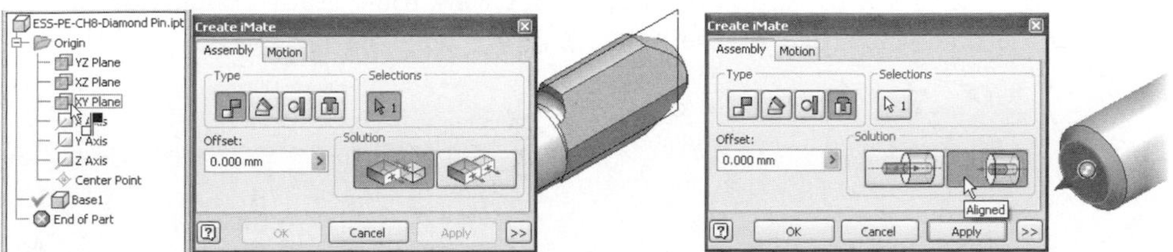

Figure A.63

7. In the browser expand the iMates folder. Hold down the CTRL key, and click the iMate:1 and iInsert:1 constraint entry. Right-click, and select Create Composite from the menu, as shown in the following image on the left.

8. Right-click on iComposite:1, and then click Properties from the menu. The iMate Properties dialog box will appear. In the Name field, type **Diamond Pin**, as shown in the following image on the right. Click OK.

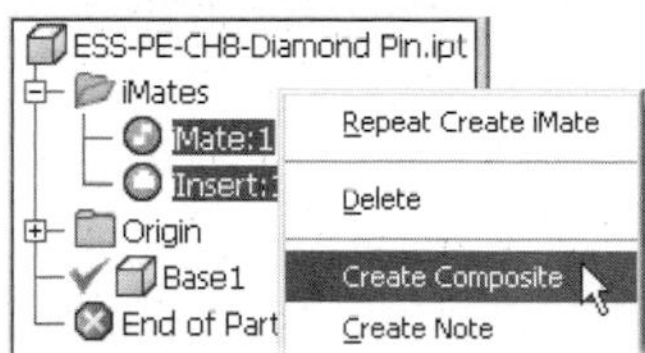

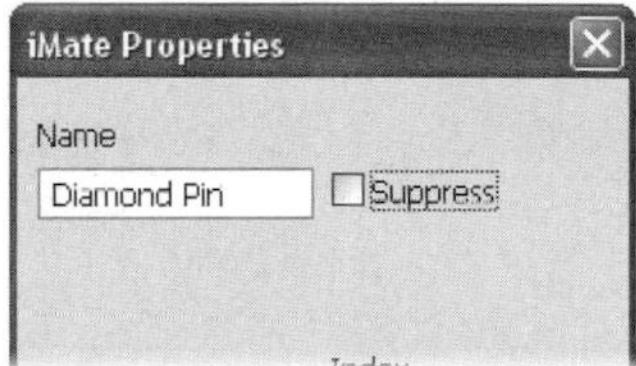

Figure A.64

9. Save the part. Close the file.

10. In this section, you will place the pins into an assembly using the iMates you created. The matching iMates were already placed on the part where the pins will be placed. Open *ESS-PE-CH8-202.iam*.

11. Press P on the keyboard to activate the Place Component tool. Click the Interactively place with iMate option, as shown in the following image on the left, and double-click the *ESS-PE-CH8-Round Pin.ipt* file from the Appendix A folder.

12. In the graphics window, the pin is previewed in its assembled location. Click anywhere in the graphics window to accept the component placement. Right-click and select Done from the menu.

13. Press the F5 key to return to the previous view.

14. Press P on the keyboard to activate the Place Component tool. Confirm that the Interactively place with iMates button is active and then double-click the *ESS-PE-CH8-Diamond Pin.ipt* file from the Appendix A folder. Click anywhere in the graphics window to accept the component placement. Right-click and select Done from the menu.

15. Press the F5 key to return to the previous view. Your display should appear similar to the following image on the right.

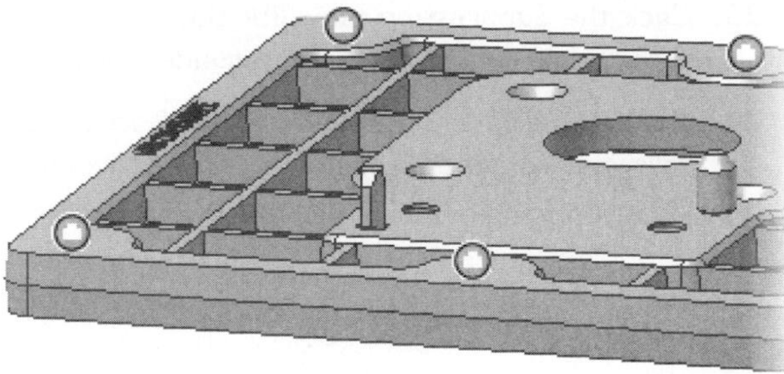

Figure A.65

16. Save the file and keep the assembly open.

IPARTS

17. Open *ESS-PE-CH8-Hub.ipt.*
18. Click on the Parameters tool in the Panel bar and verify that the following parameters exist: BoreOffset, HexDia, and RoundDia. These parameters will be used to help control the iParts size. Click Done.
19. Press the E key to start the Extrude tool. Select the three profiles inside of the hexagonal in the graphics window for the Profile, as shown in the following image on the left. In the Extents drop-down list, click To. Click in the top circular face for the termination face to extrude as shown in the following image in the middle, and click OK.
20. In the browser, right-click on the hexagonal extrusion, and select Properties from the menu. In the Name text field, rename the feature to **Hex Head**, as shown in the following image on the right.

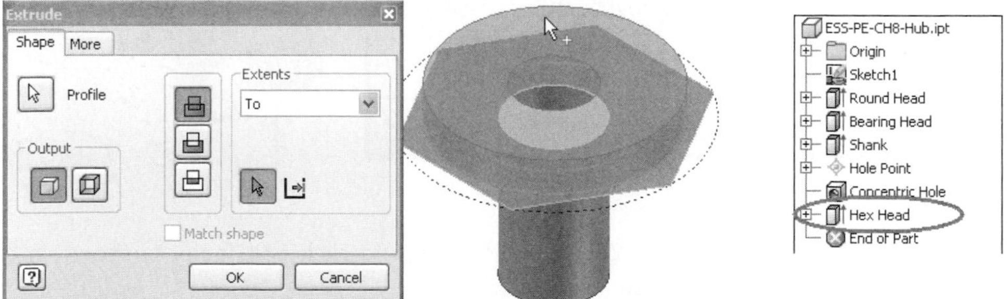

Figure A.66

21. Turn off the visibility of Sketch1. In the browser right-click Sketch1, and click Visibility from the menu.
22. Click on the Tools drop-down menu, and click Create iPart. In the iPart Author dialog box, right-click on row 1 and select Insert Row from the menu.

23. Click the Suppression tab, click the Hex Head feature from the list at the left, and click the Add (>>) button to include it in the table.

24. Click in the Member, Part Number, BoreOffset, HexDia, RoundDia, and Hex Head cells, and edit them as shown in the following image. The offset value for the Bore-Offset will offset the hexagon from the center.

	Member	Part Number	BoreOffset	HexDia	RoundDia	Hex Head
1	Hub Round	Hub-Round-12	0.0 mm	13.843 mm	12 mm	Suppress
2	Hub Hex	Hub-Hex-14	0.25 mm	14 mm	11.167 mm	Compute

Figure A.67

25. Click OK to create the iPart.

26. Next add an iMate for placing the hub into the assembly. Rotate your view to resemble the following image. Press Q on the keyboard to activate the Create iMate dialog box, as shown in the following image on the left. Click the Insert constraint type, select the middle circular edge of the hub, as shown in the following image on the right, and click OK.

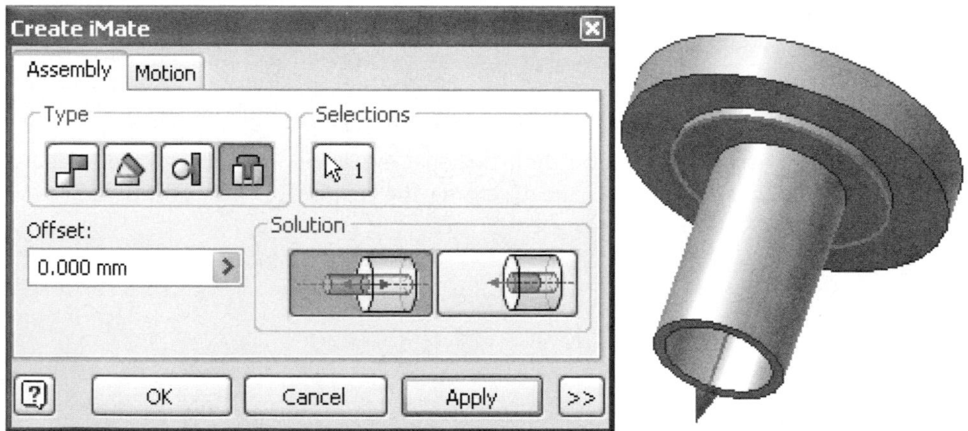

Figure A.68

27. In the browser, right-click on the Table entry, and click Edit Table, as shown in the following image on the left.

28. Right-click the Part Number column heading. Select Key and then 1 from the menu to set the field to the primary key, as shown in the following image on the right.

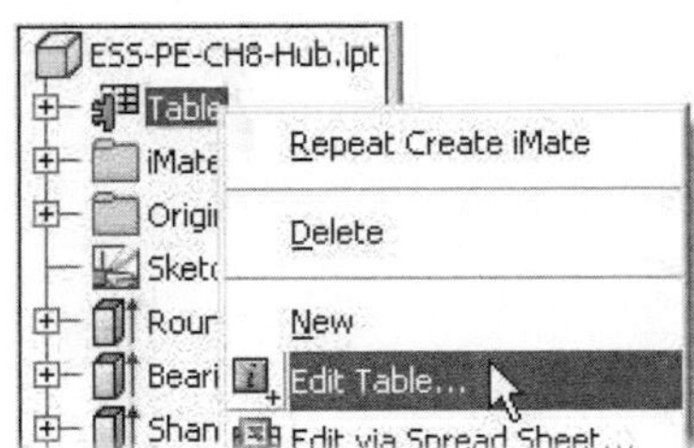

Figure A.69

29. In the iPart Author dialog box, click the iMates tab, and add the iInsert:1:IM0006 [Compute] option, as shown in the following image.

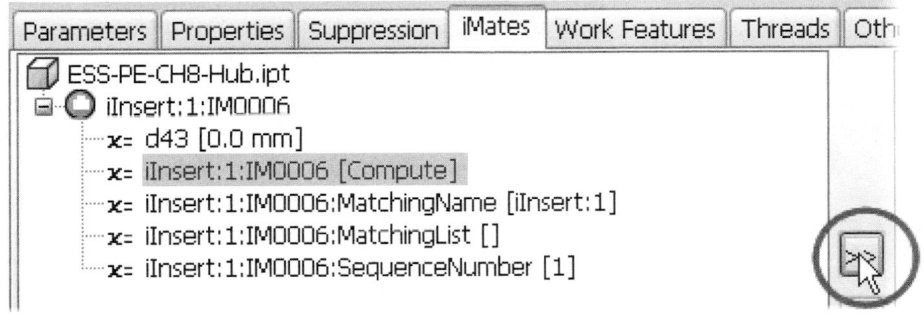

Figure A.70

30. The author area should resemble the following image. Click OK to complete the edits.

Part Number	BoreOffset	HexDia	RoundDia	Hex Head	IM0006
Hub-Round-12	0.0 mm	13.843 mm	12 mm	Suppress	Compute
Hub-Hex-14	0.25 mm	14 mm	11.167 mm	Compute	Compute

Figure A.71

31. In the browser, expand the Table entry, and double-click to activate the Hub Hex iPart member, as shown in the following image.

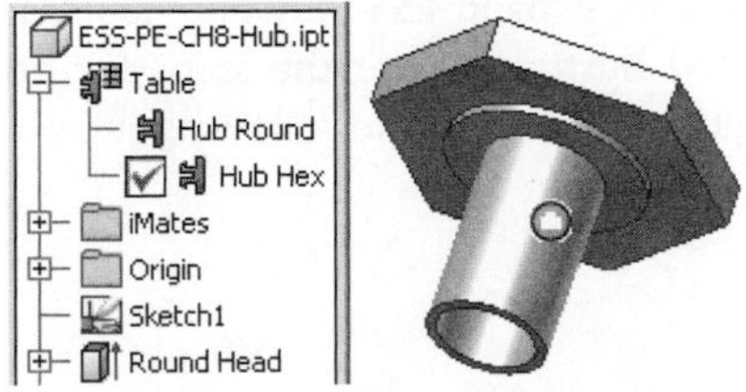

Figure A.72

32. In the table area change the iPart version back to the original iPart by double-clicking on the Hub Round iPart version.

33. Save the file. Close the file.

IASSEMBLY

34. Make the file *ESS-PE-CH8-202.iam* active or open if needed..

35. Right-click in a blank area of the graphics window, and click Place Component from the menu. Click the file *ESS-PE-CH8-Hub.ipt*.

36. Confirm that the Interactively place with iMate option is active and then Click Open and verify that the active part number for the iPart is Hub-Round-12, as shown in the following image on the left.

37. Right-click in a blank area of the graphics window and click Place at all matching iMates, as shown in the following image on the right.

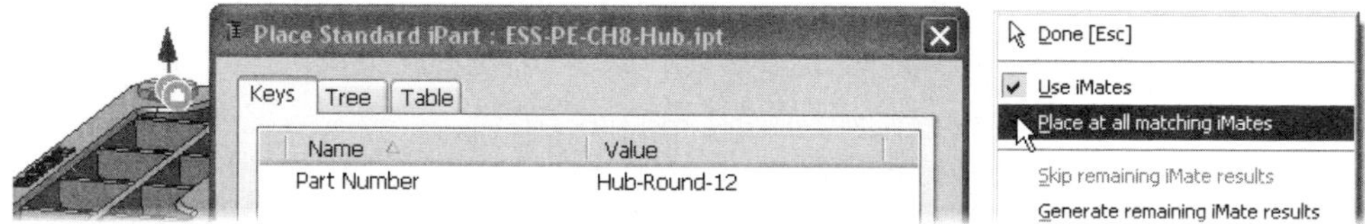

Figure A.73

38. Rotate your display, and confirm the iMate placement of the six Round Hub Concentric iParts into the assembly, as shown in the following image.

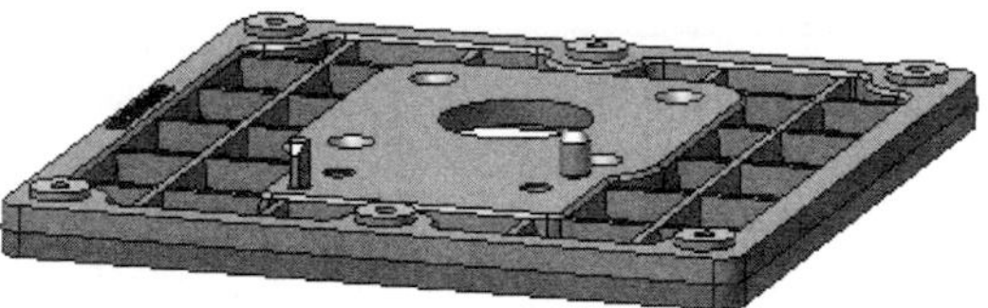

Figure A.74

39. Click on the Tools drop-down menu, and click Create iAssembly.

40. Next create an iAssembly by switching three of the iParts to the hexagon version. In the Components section, expand ESS-PE-CH8-Hub:4 and the Table Replace option, and click the Add (>>) button. Repeat the steps for both the 5th and 6th instance of the ESS-PE-CH8-Hub, as shown in the following image.

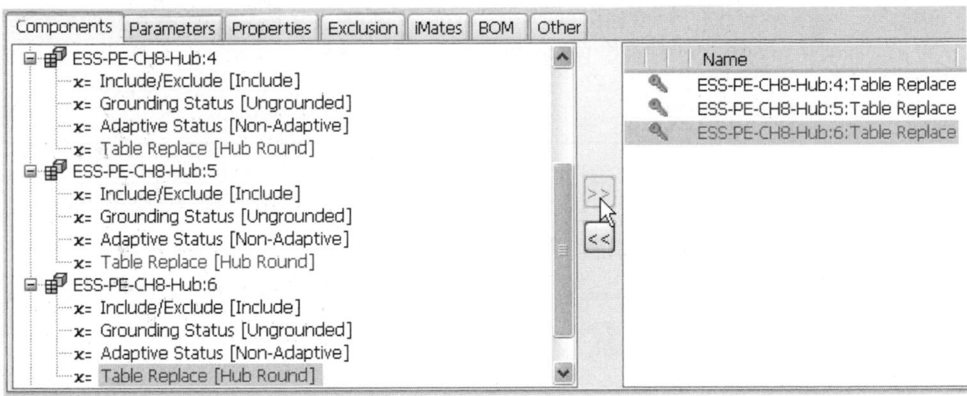

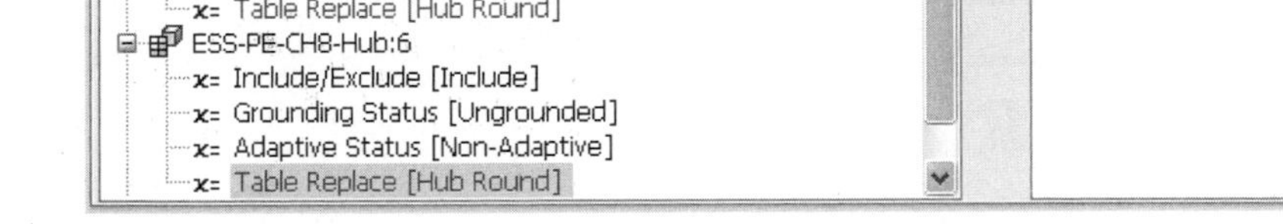

Figure A.75

41. In the iAssembly Author dialog box, right-click on row 1, and select Insert Row from the menu. Change the information for both rows, as shown in the following image.

	Member	Part Number	ESS-PE-CH8-Hub:4: Table Replace	ESS-PE-CH8-Hub:5: Table Replace	ESS-PE-CH8-Hub:6: Table Replace
1	Plate with 6 Round	6 Round Hub Sub	Hub Round	Hub Round	Hub Round
2	Plate with 3 Round 3 Hex	3 Round 3 Hex Hub Sub	Hub Hex	Hub Hex	Hub Hex

Figure A.76

42. In the iAssembly Author dialog box, click OK to create the iAssembly.

43. In the browser, expand the Table entry. Double-click to activate the Plate with 3 Round 3 Hex iAssembly version, as shown in the following image on the left.

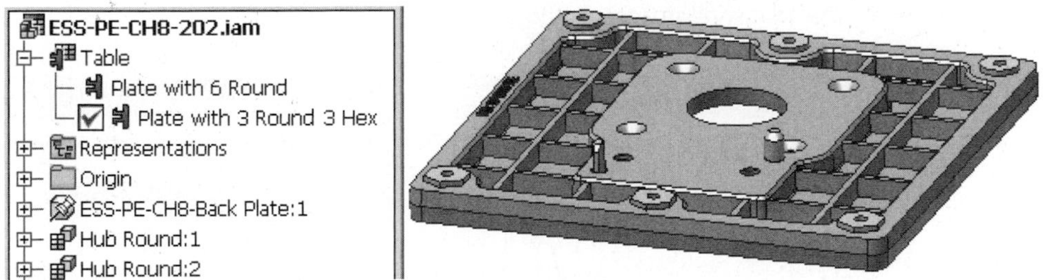

Figure A.77

44. Change the iAssembly back to Plate with 6 Round.

45. Save the assembly.

46. Create a new drawing based on the *ANSI (mm).idw* template from the Metric tab, and change the sheet to B size.

47. Place a base view of the *ESS-PE-CH8-202.iam*, based on the Plate with 6 Round iAssembly member.

48. Place a Parts List in the upper-right corner of the border, as shown in the following image. If needed, click OK to enable the BOM View.

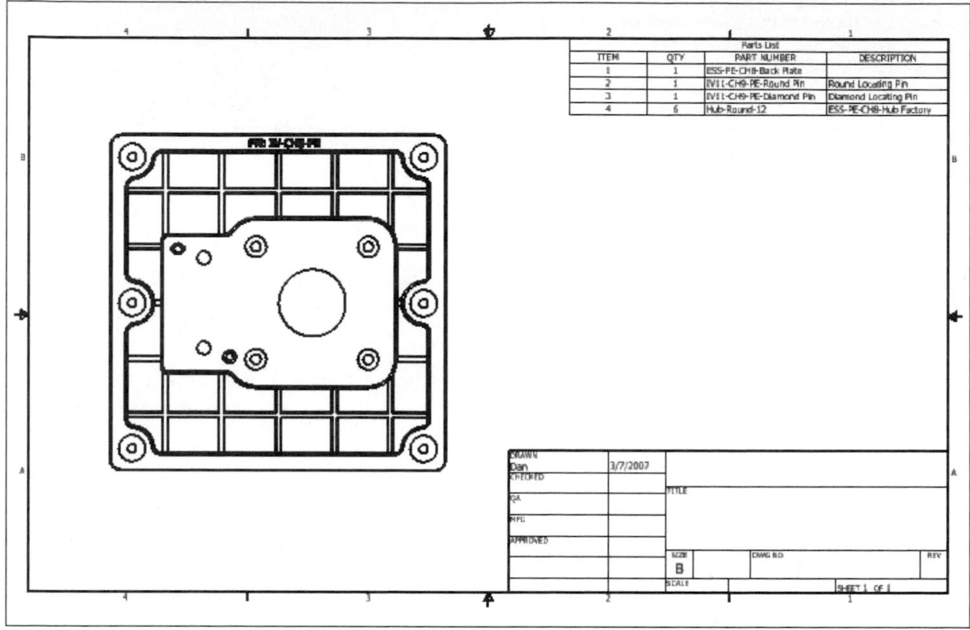

Figure A.78

49. Right-click on the Parts List, and click Edit Parts List from the menu. In the Parts List dialog box, click the Member Selection button, as shown in the following image on the left. In the Select member dialog box, click Plate with 3 Round 3 Hex, as shown in the following image on the right.

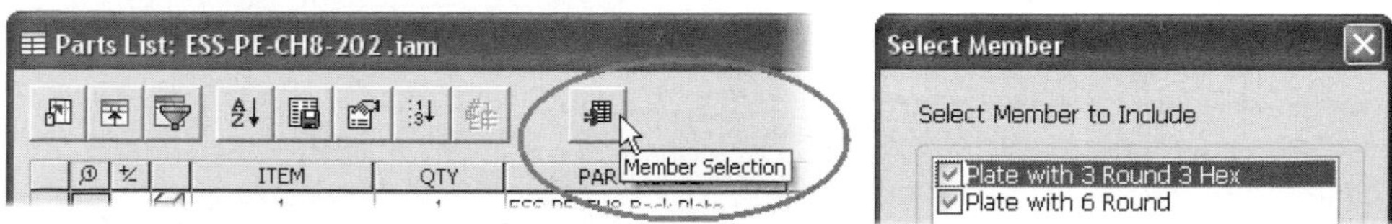

Figure A.79

50. Now edit the other fields as needed. Click OK twice to update the Parts List, as shown in the following image.

Parts List				
ITEM	QTY Plate w	QTY Plate w	PART NUMBER	DESCRIPTION
1	1	1	ESS-PE-CH8-Back Plate	
2	1	1	IV11-CH9-PE-Round Pin	Round Locating Pin
3	1	1	IV11-CH9-PE-Diamond Pin	Diamond Locating Pin
4	3	6	Hub-Round-12	ESS-PE-CH8-Hub Factory
5	3	0	Hub-Hex-14	ESS-PE-CH8-Hub Factory

Figure A.80

51. Save the assembly. Close the file.

Project Exercise: Chapter 9

DESIGN AUTOMATION TECHNIQUES

In this exercise, you edit the components used in the back plate subassembly from the gripper station. This exercise uses the following assembly modeling productivity tools and concepts: Design Views and Level of Detail Representations.

1. Open *ESS-PE-CH9-100.iam* from the Appendix A folder.

2. In this step, you create a view representation that will control the visibility of components. In the browser, click the + in front of the Representations folder and then the + in front of the View: Master heading. Right-click on View: Master, and click New, as shown in the following image on the left.

3. Slowly double-click on View1 to activate the text edit mode and rename the new view representation entry to No Magnetic Drive, as shown in the following image on the right.

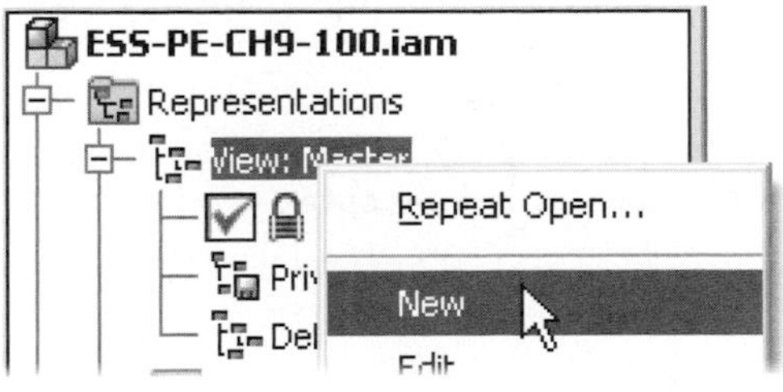

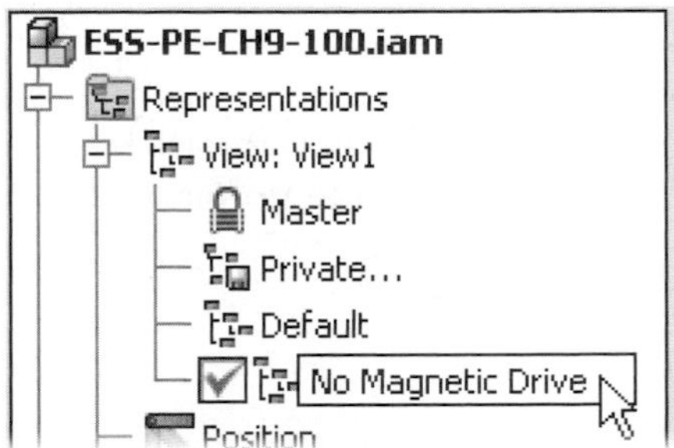

Figure A.81

4. Turn off the visibility of component. In the browser, hold down CTRL as you click the last three assembly components, right-click, and select Visibility from the menu, as shown in the following image.

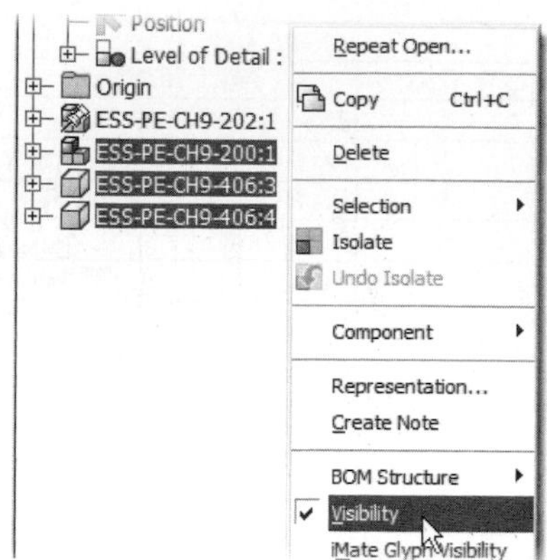

Figure A.82

5. Verify that the View Representations work. In the browser double-click the Master view representation to activate it, and double-click the No Magnetic Drive view representation to reactivate it. Note the changes to the assembly component visibility.
6. In this section, you create a Level of Detail that removes the parts from memory. In the browser, right-click the No Magnetic Drive view representation, and select Copy to LevelofDetail from the menu, as shown in the following image on the left.
7. Turn the visibility on for all the components by changing the View Representation to Master, as shown in the following image on the right. Notice that all the components are displayed in the graphics window.
8. In the browser, click to expand the Level of Detail: Master. Note the information displayed on the capacity meter, and double-click the new No Magnetic Drive level of detail. The capacity meter shows that the number of occurrences and open documents have been reduced freeing up memory, as shown in the following image on the right.

Note: The number of open documents you have may be different than shown depending upon the number of Autodesk Inventor files that are open.

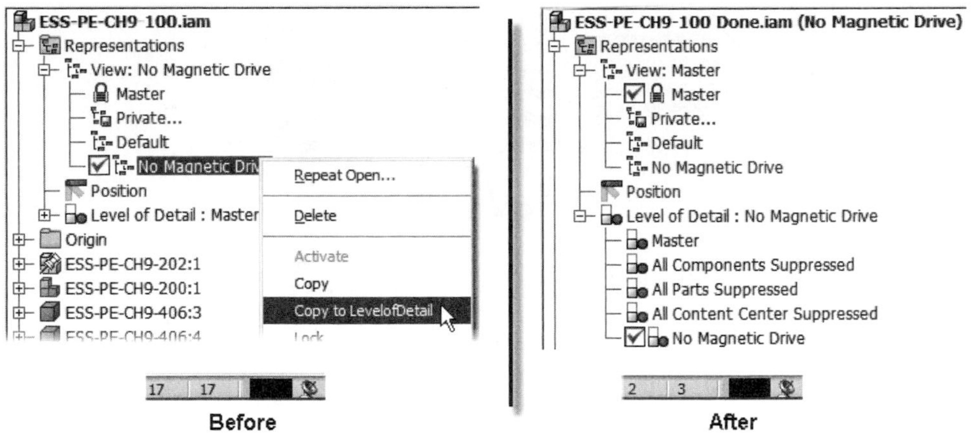

Figure A.83

9. Change the Level of Detail to Master. If prompted to save the pending changes, click OK.
10. Right-click on Level of Detail : Master, and click New Level of Detail, as shown in the following image on the left.
11. Rename the new level of detail to **No Plate**, as shown in the following image on the right.

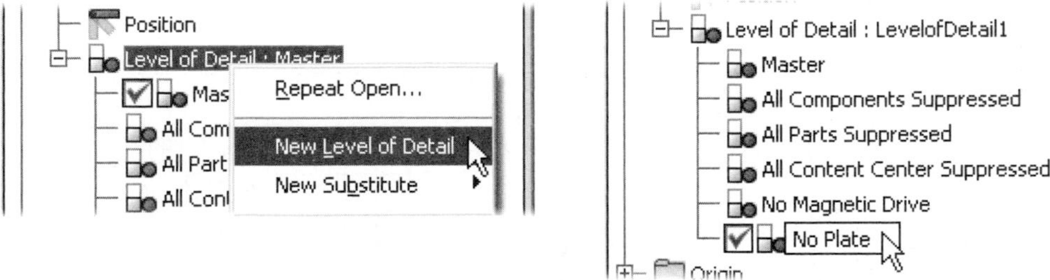

Figure A.84

12. In the browser, right-click on ESS-PE-CH9-202:1, and click Suppress from the menu, as shown in the following image. Note that the number of Occurrences and Open documents will be reduced.

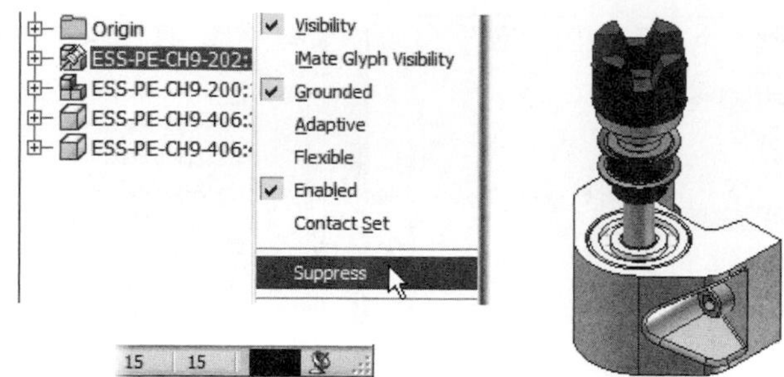

Figure A.85

13. Save the assembly.
14. Activate the different level of details. Verify the number of occurrences and open documents and the components that are visible in the graphics window changes.
15. Close the file. Do not save changes. End of exercise.

Project Exercise: Chapter 10

SHEET METAL DESIGN

In the following exercise, you will create and document a sheet metal safety cover for the magnetic drive assembly used in the gripper station project. This exercise uses the sheet metal and drawing environments and their related tools and concepts. The following shows the completed exercise.

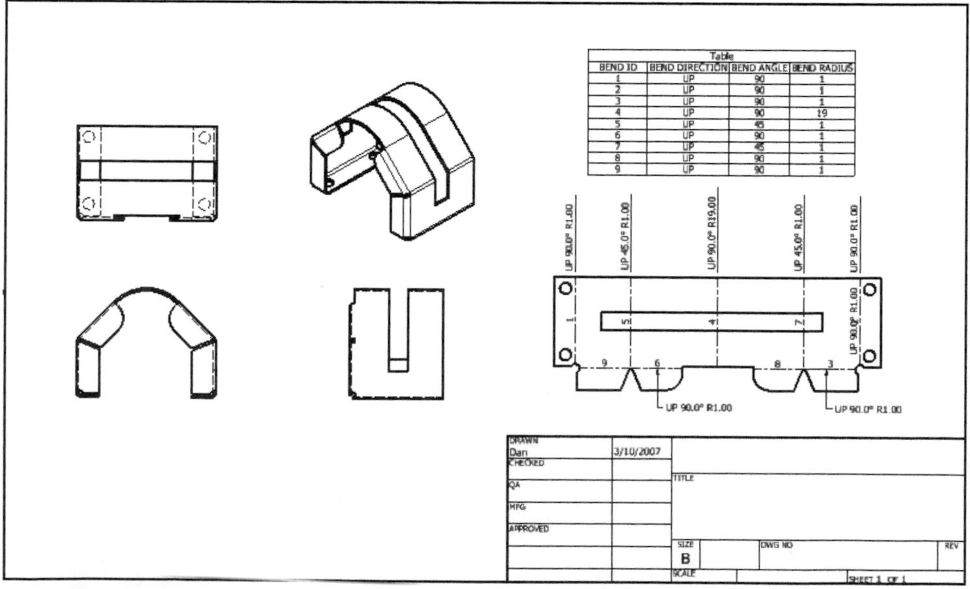

Table			
BEND ID	BEND DIRECTION	BEND ANGLE	BEND RADIUS
1	UP	90	1
2	UP	90	1
3	UP	90	1
4	UP	90	19
5	UP	45	1
6	UP	90	1
7	UP	45	1
8	UP	90	1
9	UP	90	1

Figure A.86

1. Create a new file using the *Sheet Metal (mm).ipt* template from the Metric tab.
2. Sketch and dimension the geometry, as shown in the following image on the left. The geometry is symmetrical.
3. Finish the sketch, and change to an Isometric (home) View.
4. In the Sheet Metal Features panel, click the Contour Flange tool. In the graphics window, select the sketch, and type **40 mm** in the Distance edit field, click the Both Sides button, and click OK. The following image on the right shows the part after completing the operation.

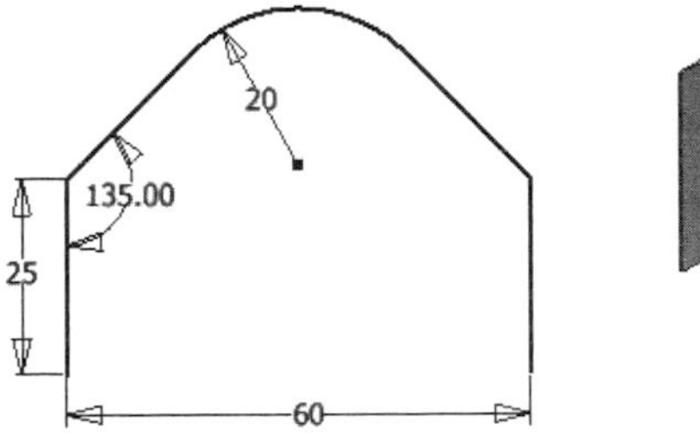

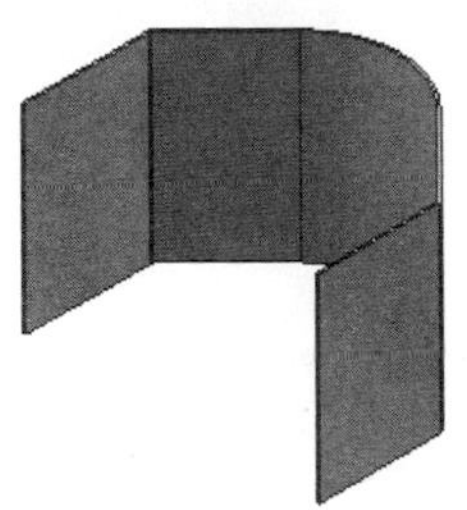

Figure A.87

5. In the Sheet Metal Features panel, click the Flange tool. Select the three left inside edges and change the Distance field to **10**, as shown in the following image. Click the Corner tab, verify that Apply Auto-Mitering is checked, and click Apply.

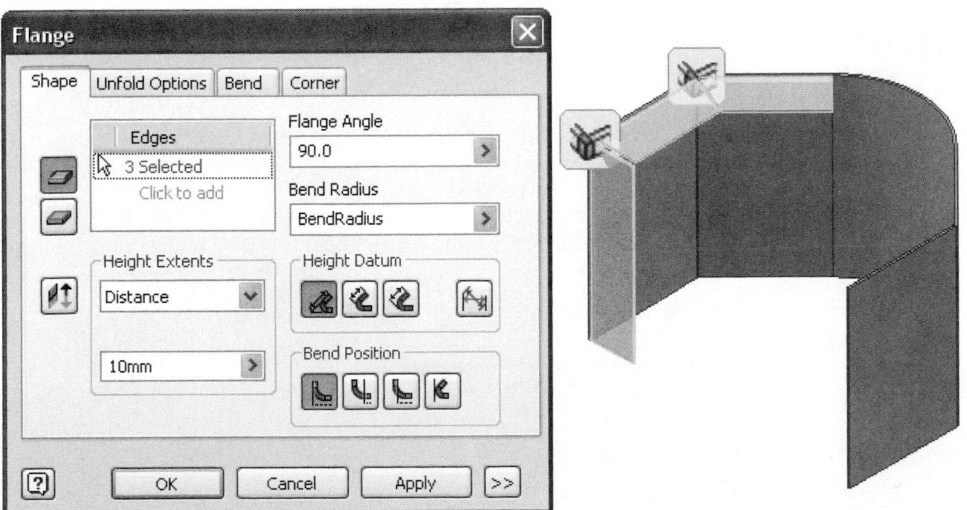

Figure A.88

6. Select the three right inside edges as shown in the following image and click OK.

EXERCISES

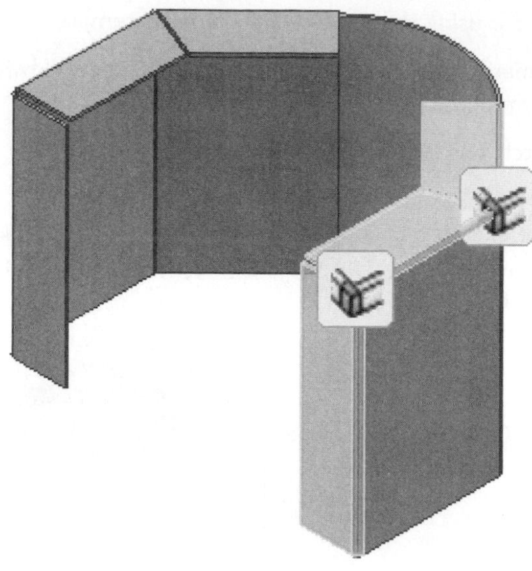

Figure A.89

7. In the Sheet Metal Features panel, click the Corner Round tool. Select the two inside edges of the last two flanges that were created, as shown in the following image on the left. Change the radius to **7 mm**, as shown in the following image on the right. Click OK to create the corner rounds.

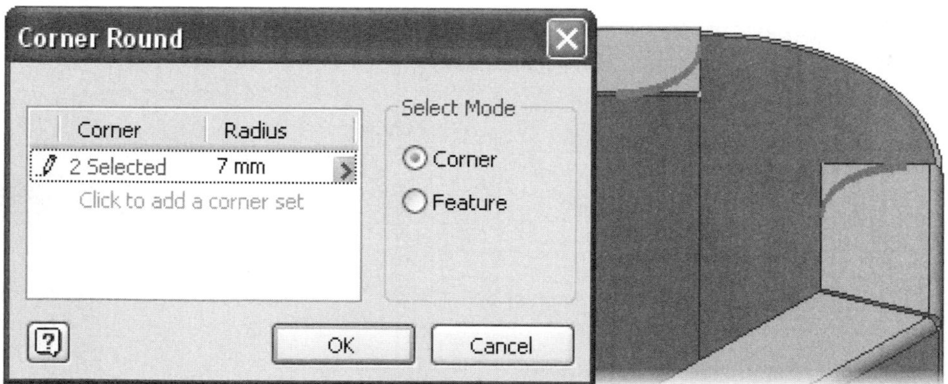

Figure A.90

8. In the Sheet Metal Features panel, click the Hole tool. Using the Linear placement option, place a **5 mm** hole that is **5 mm** from the top and inside edges of the bottom flange, as shown in the following image. Click Apply to create the hole.

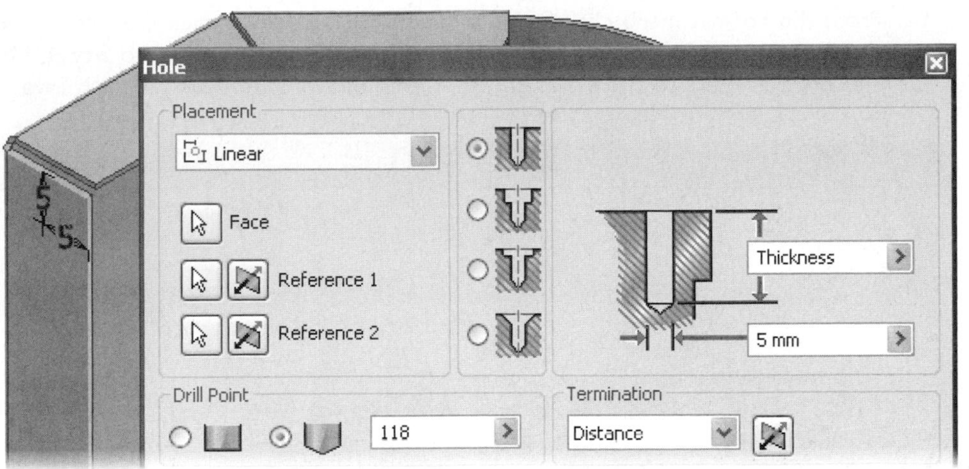

Figure A.91

9. Place **5 mm** holes with the same **5 mm** spacing on the other three corners of the two bottom flanges as shown in the following image on the left.

10. In the Sheet Metal Features panel, click the Flat Pattern tool. The flat pattern will be generated, as shown in the following image on the right. Zoom into the bends to investigate the flat pattern. In the next step, you will change how the corners are developed.

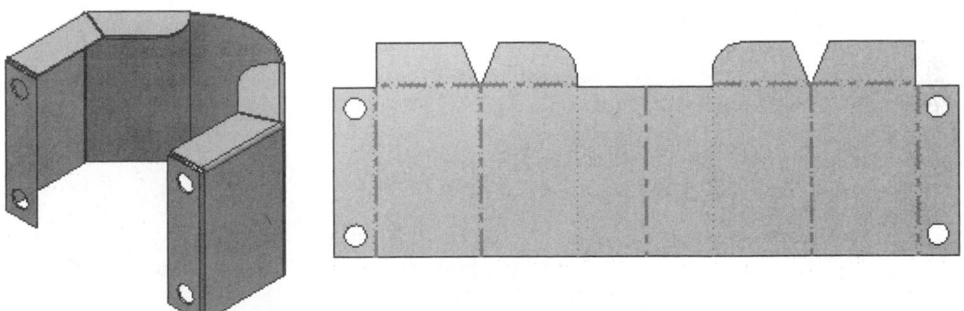

Figure A.92

11. Change the environment back to the Folded Model. In the browser, double-click on Folded Model, as shown in the following image.

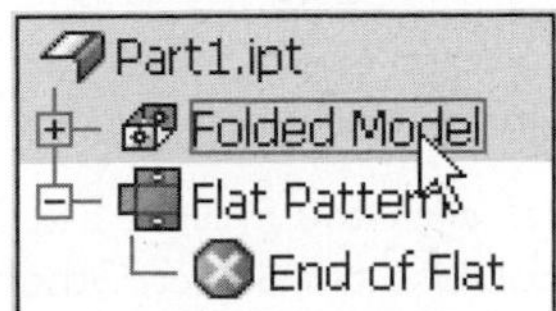

Figure A.93

12. From the Format menu click Style and Standard Editor and click Default_mm from the Sheet Metal Rule list. From the Sheet tab change the material to **Steel, Mild** and the Thickness to **1 mm** as shown in the following image. Then click Save.

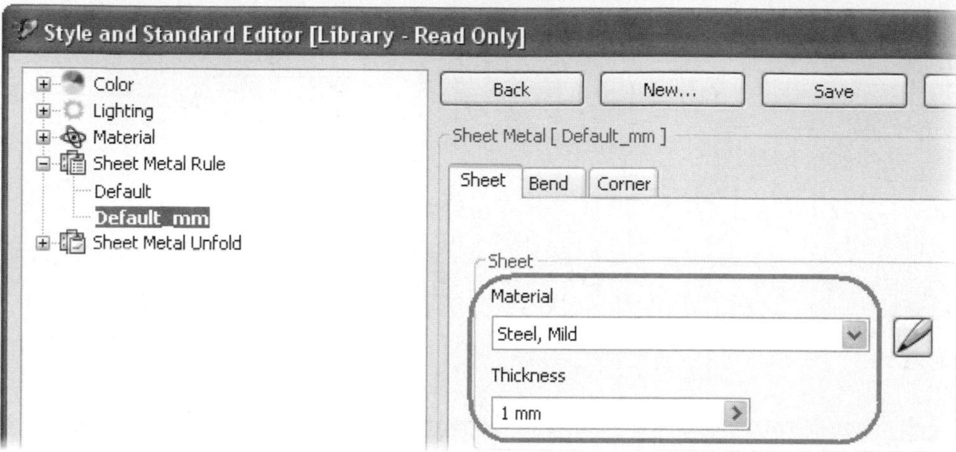

Figure A.94

13. While still in the Style and Standard Editor dialog box click the Corner tab. Change the 2 Bend Intersection Relief Shape to **Round** and the 3 Bend Intersection Relief Shape to **Round with Radius**, as shown in the following image. Click the Save button, and click Done to complete the changes.

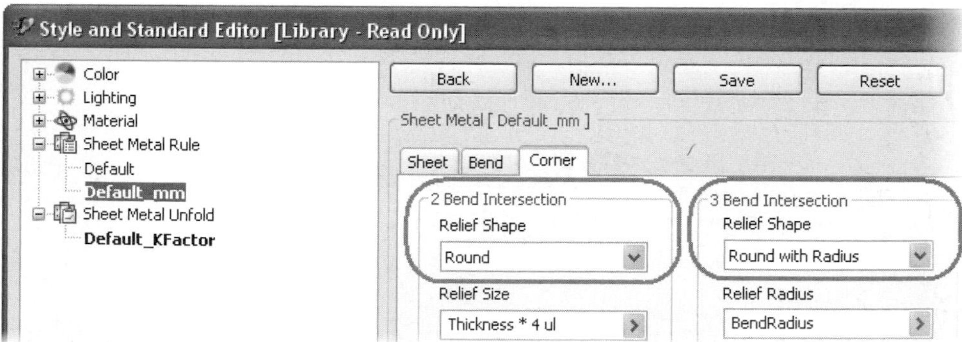

Figure A.95

14. Rotate the model and verify the changes.

15. Create a sketch on the right-vertical face, as shown in the following image on the left.

16. From the 2D Sketch Panel, click the Project Flat Pattern tool. It may be found under the Project Geometry tool. Use the Select Other tool to select the back-left vertical face, as shown in the following image on the right.

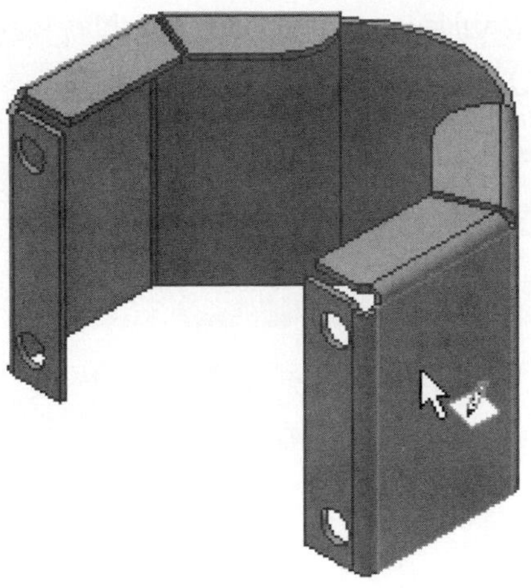
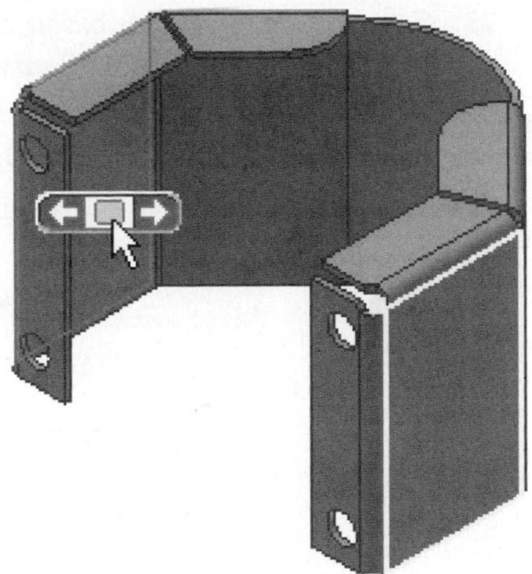

Figure A.96

17. Start the Look At tool, and then click the active sketch in the browser.
18. Place a 2-Point Rectangle, and dimension it as shown in the following image on the left. When done, right-click in the graphics window and click Finish Sketch from the menu.
19. Change to an Isometric (home) View.
20. In the Sheet Metal Features panel, click the Cut tool. Select the inside rectangle as the profile, and click the Cut Across Bend option, as shown in the following image in the middle. Note that the Extents distance is equal to the Thickness, which will cut through the part, as shown in the following image on the right. If needed, the thickness can be changed to a value that is less than the material thickness. Click OK.

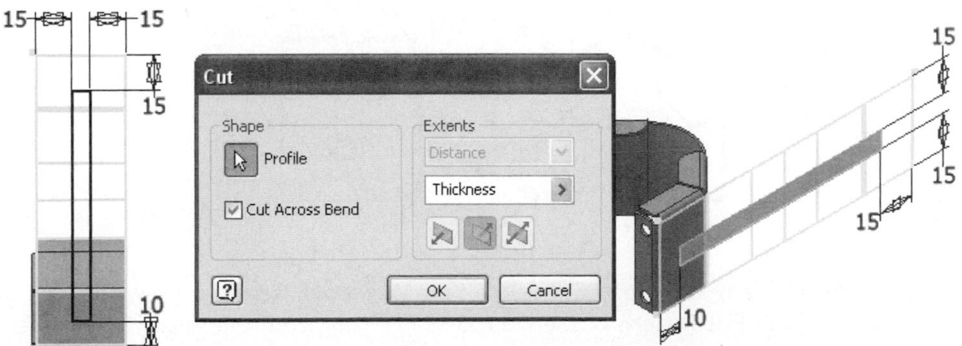

Figure A.97

21. To verify that the flat pattern is updated, double-click on the Flat Pattern entry in the browser. Return to the folded model by double-clicking on the Folded Model entry in the browser.
22. Save the file with the name *ESS-PE-CH10.ipt*, and place it in the Appendix A folder.

23. Create a new drawing based on the *ANSI (mm).idw* template from the Metric tab, and change the sheet to B size.

In the Drawing Views panel, click the Base View tool, and create a Front view of the Folded Model of the file *ESS-PE-CH10.ipt* from the Appendix A folder, as shown in the following image.

24. To place the view click a point in the lower-left corner of the drawing.

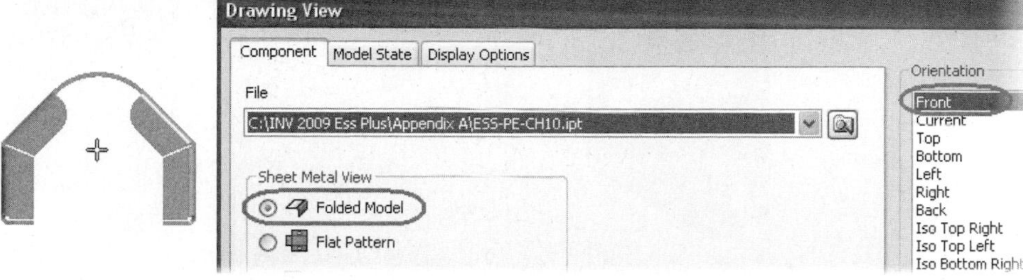

Figure A.98

25. Create three projected views (right side, top and Isometric View), as shown in the following image.

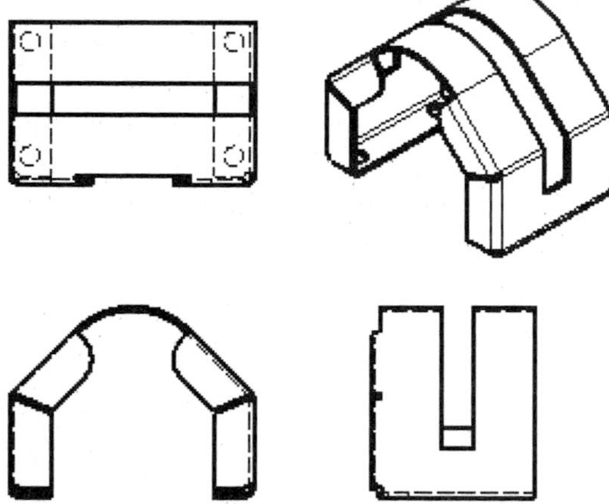

Figure A.99

26. In the Drawing Views panel, click the Base View tool. Create a view using the Default view of the Flat Pattern of the file *ESS-PE-CH10.ipt* from the Appendix A folder, as shown in the following image. Then place it in the lower-right corner of the drawing.

Note: The orientation of the flat pattern on the drawing sheet may vary from that shown.

EXERCISES

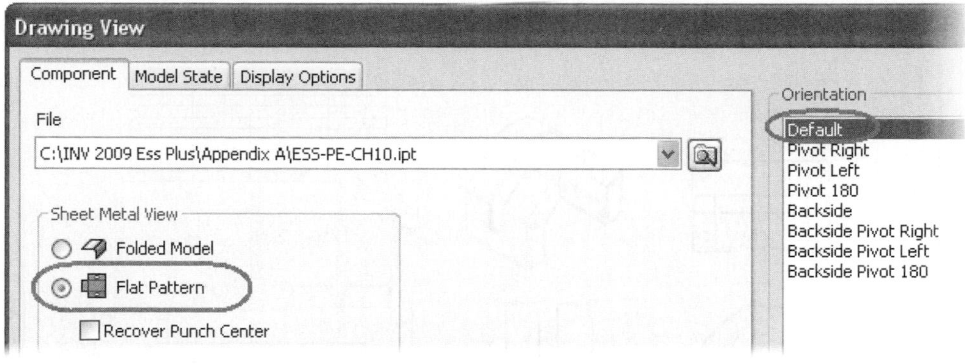

Figure A.100

27. Next annotate the bend lines. On the Drawing Annotation panel, click Bend Notes, which may be under the Punch Notes tool. Select the seven bend lines, as shown in the following image.

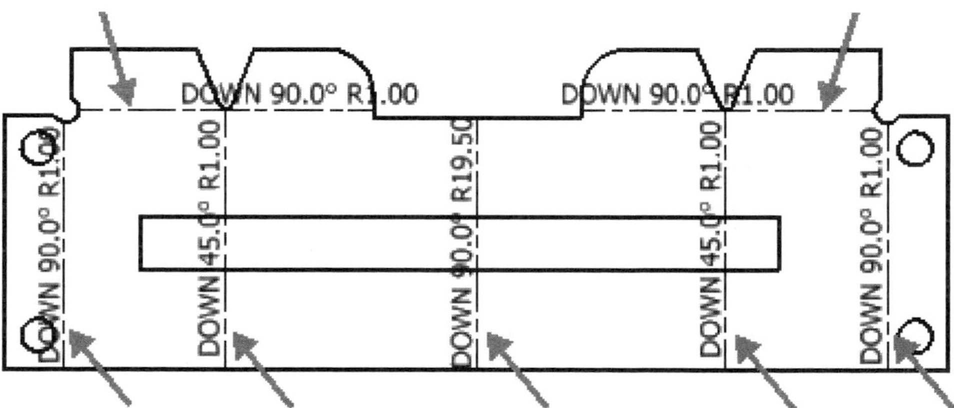

Figure A.101

28. Drag the bend notes so that they appear above and below the geometry as shown in the following image.

29. Next create a table that contains the bend information. On the Drawing Annotation panel, click the Table tool. Select the flat pattern view, and click OK in the Table dialog box. Place the table above the view of the flat pattern, as shown in the following image.

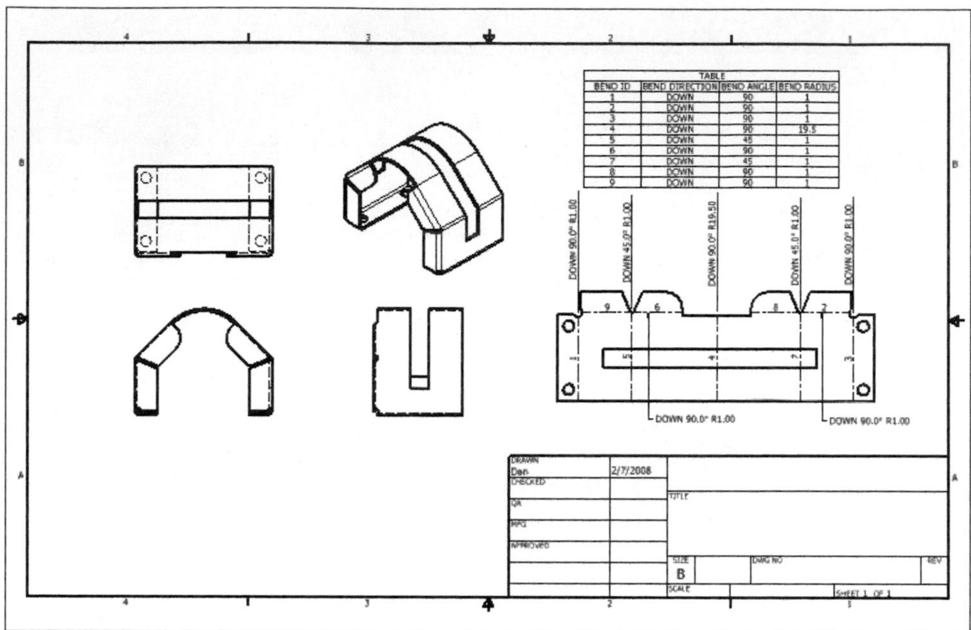

TABLE			
BEND ID	BEND DIRECTION	BEND ANGLE	BEND RADIUS
1	DOWN	90	1
2	DOWN	90	1
3	DOWN	90	1
4	DOWN	90	19.5
5	DOWN	45	1
6	DOWN	90	1
7	DOWN	45	1
8	DOWN	90	1
9	DOWN	90	1

Figure A.102

30. Save and close all files. End of exercise.

INDEX

SYMBOLS

A

D

E

F

G

H

I

J

K

L

M

N

O

P

R

S

T

U

V

W

X

Y

IMPORTANT! READ CAREFULLY: This End User License Agreement ("Agreement") sets forth the conditions by which Cengage Learning will make electronic access to the Cengage Learning-owned licensed content and associated media, software, documentation, printed materials, and electronic documentation contained in this package and/or made available to you via this product (the "Licensed Content"), available to you (the "End User"). BY CLICKING THE "I ACCEPT" BUTTON AND/OR OPENING THIS PACKAGE, YOU ACKNOWLEDGE THAT YOU HAVE READ ALL OF THE TERMS AND CONDITIONS, AND THAT YOU AGREE TO BE BOUND BY ITS TERMS, CONDITIONS, AND ALL APPLICABLE LAWS AND REGULATIONS GOVERNING THE USE OF THE LICENSED CONTENT.

1.0 SCOPE OF LICENSE

1.1 Licensed Content. The Licensed Content may contain portions of modifiable content ("Modifiable Content") and content which may not be modified or otherwise altered by the End User ("Non-Modifiable Content"). For purposes of this Agreement, Modifiable Content and Non-Modifiable Content may be collectively referred to herein as the "Licensed Content." All Licensed Content shall be considered Non-Modifiable Content, unless such Licensed Content is presented to the End User in a modifiable format and it is clearly indicated that modification of the Licensed Content is permitted.

1.2 Subject to the End User's compliance with the terms and conditions of this Agreement, Cengage Learning hereby grants the End User, a non-transferable, nonexclusive, limited right to access and view a single copy of the Licensed Content on a single personal computer system for noncommercial, internal, personal use only. The End User shall not (i) reproduce, copy, modify (except in the case of Modifiable Content), distribute, display, transfer, sublicense, prepare derivative work(s) based on, sell, exchange, barter or transfer, rent, lease, loan, resell, or in any other manner exploit the Licensed Content; (ii) remove, obscure, or alter any notice of Cengage Learning's intellectual property rights present on or in the Licensed Content, including, but not limited to, copyright, trademark, and/or patent notices; or (iii) disassemble, decompile, translate, reverse engineer, or otherwise reduce the Licensed Content.

2.0 TERMINATION

2.1 Cengage Learning may at any time (without prejudice to its other rights or remedies) immediately terminate this Agreement and/or suspend access to some or all of the Licensed Content, in the event that the End User does not comply with any of the terms and conditions of this Agreement. In the event of such termination by Cengage Learning, the End User shall immediately return any and all copies of the Licensed Content to Cengage Learning.

3.0 PROPRIETARY RIGHTS

3.1 The End User acknowledges that Cengage Learning owns all rights, title and interest, including, but not limited to all copyright rights therein, in and to the Licensed Content, and that the End User shall not take any action inconsistent with such ownership. The Licensed Content is protected by U.S., Canadian and other applicable copyright laws and by international treaties, including the Berne Convention and the Universal Copyright Convention. Nothing contained in this Agreement shall be construed as granting the End User any ownership rights in or to the Licensed Content.

3.2 Cengage Learning reserves the right at any time to withdraw from the Licensed Content any item or part of an item for which it no longer retains the right to publish, or which it has reasonable grounds to believe infringes copyright or is defamatory, unlawful, or otherwise objectionable.

4.0 PROTECTION AND SECURITY

4.1 The End User shall use its best efforts and take all reasonable steps to safeguard its copy of the Licensed Content to ensure that no unauthorized reproduction, publication, disclosure, modification, or distribution of the Licensed Content, in whole or in part, is made. To the extent that the End User becomes aware of any such unauthorized use of the Licensed Content, the End User shall immediately notify Cengage Learning. Notification of such violations may be made by sending an e-mail to delmarhelp@Cengage.com.

5.0 MISUSE OF THE LICENSED PRODUCT

5.1 In the event that the End User uses the Licensed Content in violation of this Agreement, Cengage Learning shall have the option of electing liquidated damages, which shall include all profits generated by the End User's use of the Licensed Content plus interest computed at the maximum rate permitted by law and all legal fees and other expenses incurred by Cengage Learning in enforcing its rights, plus penalties.

6.0 FEDERAL GOVERNMENT CLIENTS

6.1 Except as expressly authorized by Cengage Learning, Federal Government clients obtain only the rights specified in this Agreement and no other rights. The Government acknowledges that (i) all software and related documentation incorporated in the Licensed Content is existing commercial computer software within the meaning of FAR 27.405(b)(2); and (2) all other data delivered in whatever form, is limited rights data within the meaning of FAR 27.401. The restrictions in this section are acceptable as consistent with the Government's need for software and other data under this Agreement.

7.0 DISCLAIMER OF WARRANTIES AND LIABILITIES

7.1 Although Cengage Learning believes the Licensed Content to be reliable, Cengage Learning does not guarantee or warrant (i) any information or materials contained in or produced by the Licensed Content, (ii) the accuracy, completeness or reliability of the Licensed Content, or (iii) that the Licensed Content is free from errors or other material defects. THE LICENSED PRODUCT IS PROVIDED "AS IS," WITHOUT ANY WARRANTY OF ANY KIND AND CENGAGE LEARNING DISCLAIMS ANY AND ALL WARRANTIES, EXPRESSED OR IMPLIED, INCLUDING, WITHOUT LIMITATION, WARRANTIES OF MERCHANTABILITY OR FITNESS FOR A PARTICULAR PURPOSE. IN NO EVENT SHALL CENGAGE LEARNING BE LIABLE FOR: INDIRECT, SPECIAL, PUNITIVE OR CONSEQUENTIAL DAMAGES INCLUDING FOR LOST PROFITS, LOST DATA, OR OTHERWISE. IN NO EVENT SHALL CENGAGE LEARNING'S AGGREGATE LIABILITY HEREUNDER, WHETHER ARISING IN CONTRACT, TORT, STRICT LIABILITY OR OTHERWISE, EXCEED THE AMOUNT OF FEES PAID BY THE END USER HEREUNDER FOR THE LICENSE OF THE LICENSED CONTENT.

8.0 GENERAL

8.1 Entire Agreement. This Agreement shall constitute the entire Agreement between the Parties and supercedes all prior Agreements and understandings oral or written relating to the subject matter hereof.

8.2 Enhancements/Modifications of Licensed Content. From time to time, and in Cengage Learning's sole discretion, Cengage Learning may advise the End User of updates, upgrades, enhancements and/or improvements to the Licensed Content, and may permit the End User to access and use, subject to the terms and conditions of this Agreement, such modifications, upon payment of prices as may be established by Cengage Learning.

8.3 No Export. The End User shall use the Licensed Content solely in the United States and shall not transfer or export, directly or indirectly, the Licensed Content outside the United States.

8.4 Severability. If any provision of this Agreement is invalid, illegal, or unenforceable under any applicable statute or rule of law, the provision shall be deemed omitted to the extent that it is invalid, illegal, or unenforceable. In such a case, the remainder of the Agreement shall be construed in a manner as to give greatest effect to the original intention of the parties hereto.

8.5 Waiver. The waiver of any right or failure of either party to exercise in any respect any right provided in this Agreement in any instance shall not be deemed to be a waiver of such right in the future or a waiver of any other right under this Agreement.

8.6 Choice of Law/Venue. This Agreement shall be interpreted, construed, and governed by and in accordance with the laws of the State of New York, applicable to contracts executed and to be wholly preformed therein, without regard to its principles governing conflicts of law. Each party agrees that any proceeding arising out of or relating to this Agreement or the breach or threatened breach of this Agreement may be commenced and prosecuted in a court in the State and County of New York. Each party consents and submits to the nonexclusive personal jurisdiction of any court in the State and County of New York in respect of any such proceeding.

8.7 Acknowledgment. By opening this package and/or by accessing the Licensed Content on this Web site, THE END USER ACKNOWLEDGES THAT IT HAS READ THIS AGREEMENT, UNDERSTANDS IT, AND AGREES TO BE BOUND BY ITS TERMS AND CONDITIONS. IF YOU DO NOT ACCEPT THESE TERMS AND CONDITIONS, YOU MUST NOT ACCESS THE LICENSED CONTENT AND RETURN THE LICENSED PRODUCT TO CENGAGE LEARNING (WITHIN 30 CALENDAR DAYS OF THE END USER'S PURCHASE) WITH PROOF OF PAYMENT ACCEPTABLE TO CENGAGE LEARNING, FOR A CREDIT OR A REFUND. Should the End User have any questions/comments regarding this Agreement, please contact Cengage Learning at delmar.help@cengage.com.